What others said about this book

"Upon opening this book, I was quite delighted at both the scope and depth to which most aspects of basic quantum mechanics are covered. The most noticeable difference between this book and others on the subject is that this is a comprehensive reference text that details each topic, regardless of the conceptual complexity involved. Additionally, the authors have admirably complemented intense theoretical treatment with applications of the theory, in areas such as particle, astro-, solid state and nuclear physics."

"With the exception of field theory, everything from bound state solutions, barrier potentials, approximation methods including the JWKB method, magnetic field effects, scattering theory, bra-ket notation, through to relativistic quantum theory and more is discussed at depth. For example, I was surprised to see the double harmonic potential described theoretically with confluent hypergeometric functions and the Krönig-Penney model discussed from a QM perspective to show the formation of band structure. Moreover, nuclear decay, nuclear magnetic resonance, electron tunnelling and Dirac's prediction of the positron are but a few applications that illustrate the vast extent of the discourse, this book is indeed comprehensive and would definitely serve as a valuable reference that is worthy of addition to anyone's library."

JAMIE QUINTON (School of Chemistry, Physics and Earth Sciences, Flinders University, Australia) in *The Physicist*, Vol. 41, No. 4 July/August 2004

"There are several articles and books on Quantum Mechanics written by the founding fathers of the subject and by those who have been in the forefront of developing it for various applications. Those writings are undoubtedly the classics revealing internal beauty of the subject. However, appreciation of the value and charm of those classics requires certain amount of preparation on the part of a reader. Even teaching the subject to prepare someone for applying it to practical problems is altogether a different ball game. It is the books like the one under review which undertake and fulfill those tasks admirably."

"Since its first publication in 1975, the fact that the book in question is in to its fifth edition speaks volumes for its popularity. The students across the country have indeed found it to be inseparable part of the curricular arsenal."

"Unlike many others, this book recognizes the value of its history in understanding the quantum theory as we know it today. Hence starting with Planck's law for black body radiation advanced in 1900, the book begins by giving a brief account of the chronological development of quantum mechanics. The significance of the historical developments is brought out by the essay, 'What is Quantum Mechanics' by V.F. Weisskopf reprinted as the prologue."

"The main text is divided into two parts: I. Mathematical Preliminaries (two chapters), and II. Basic Quantum Mechanics (27 Chapters). Several mathematical formulae of interest in the book and other relevant mathematical details are contained in the Appendices A to P."

"The Mathematical preliminaries consist of a well written chapter on the concepts and the operational methods of Dirac delta function and another on Fourier transforms."

"The book deals mainly with the methods of wave-mechanics. The methods of matrix mechanics are illustrative by treating the problems of simple harmonic oscillator and the angular momentum."

. . .

"The presentation all through is lucid. It illustrates each concept and method by means of a number of solved problems and by giving problems for practice in a variety of situations. These perhaps are the unique features behind the success of the book."

R.R. PURI (Bhabha Atomic Research Center, Mumbai) in *Physics News*, Vol. 35, pp 26–27, December 2004

"The most important fact about this book is the wonderful admixture of theory and applications. This is especially important for new entrants to the field with varying backgrounds and interests. The applications cover an impressive gamut from nuclear physics, atomic and molecular spectroscopy, solid state/condensed matter physics, quantum optics and even astrophysics! The problems are treated at a level that can be understood by the novice student. The second noteworthy aspect of the book are the large number of worked problems at the end of each chapter (approximately 150). These problems range from the simple "plug-in" to ones that develop new applications and/or concepts. Therefore, this makes the book ideal for self-study. There are a number of unsolved problems also. However, an answer key is missing for these unsolved problems."

"The mathematically-minded student will learn much from the treatment of the JWKB approximation (Chapter 17), variational techniques (Chapter 21), and the development and expansion of bra and ket algebra (Chapters 11, 12, 13 and 15). The necessary vector space concepts are developed in the chapters. The book begins with a brief exposition of Fourier theory and delta functions. However, it should be pointed out that the book is aimed at practical physical examples and not specifically at heavy, mathematical formalism and rigor. The emphasis, as the authors state in their Preface, is "on making the results plausible and helping students to become familiar with methods for solving problems." This goal the book achieves easily!"

"There is plenty of material here for an instructor to pick and choose in an introductory course. Alternatively, the book can be used as an adjunct to an "applied" quantum mechanics course, given the variety of applications. The book is simple enough for students but also comprehensive enough to serve as a reference book for working quantum physicists. Its strong points include clarity of presentation, lucidity, extensive figures/diagrams, and applications. On a rather charming note, each chapter begins with an appropriate quote pertaining to the topic. There are 15 Appendices, covering special functions, complex variables, Green's functions, tunneling, hydrogen atom wavefunctions, etc. I was thrilled to discover that even modified Airy functions are described in the Appendix! Every chapter has a brief bibliography for further reading."

"A student who has mastered this book is more than well-prepared for advanced work at the level of say J.J. Sakurai, or a course on quantum field theory (at the level of Zee). This book, hitherto unknown in Europe and America is well-poised to take its rightful place alongside classic introductory texts such as Schiff, Merzbacher, Messiah and Shankar."

V. LAKSHMINARAYANAN (School of Optometry University of Missouri – St. Louis St. Louis, Missouri 63121 4499) in *Foundations of Physics* Vol. 35, No. 6, June 2005

Quantum Mechanics: Theory and Applications

I cannot tell how the truth may be
I say the tale as it was said to me
— Sir Walter Scott

Quantum Mechanics: Theory and Applications

by

Ajoy Ghatak
NASI Meghnad Saha Fellow
Formerly at IIT Delhi, India

and

S. Lokanathan
Formerly at University of Rajasthan
Jaipur, India

TRINITY
(An Imprint of Laxmi Publications Pvt. Ltd.)
An ISO 9001:2015 Company

BENGALURU • CHENNAI • GUWAHATI • HYDERABAD • JALANDHAR
KOCHI • KOLKATA • LUCKNOW • MUMBAI • RANCHI
NEW DELHI

Quantum Mechanics: Theory and Applications

by

Ajoy Ghatak
NASI Meghnad Saha Fellow
Formerly at IIT Delhi, India

and

S. Lokanathan
Formerly at University of Rajasthan,
Jaipur, India

TRINITY
PRESS
(An Imprint of Laxmi Publications Pvt. Ltd.)
An ISO 9001:2015 Company
BENGALURU • CHENNAI • GUWAHATI • HYDERABAD • JALANDHAR
KOCHI • KOLKATA • LUCKNOW • MUMBAI • RANCHI
NEW DELHI

QUANTUM MECHANICS THEORY AND APPLICATIONS

Printed and bound in India

Typeset at : Shubham Composer, Delhi

First Edition : 1975, *Second Edition* : 1977, *Reprinted* : 1979 (*twice*), 1981, 1982

Third Completely revised Edition : 1984, *Reprinted* : 1985, 1986, 1992, 1994, 1995, 1997, *Fourth Edition* : 1999

Reprinted : 2001 (*twice*), 2002, 2003 (*twice*), *Fifth Edition* : 2004, *Reprinted* : 2005 (*twice*), 2006, 2007 (*twice*) 2008, 2009 (*twice*), 2010, *Reprinted* : 2012, 2014, 2015, *Sixth Edition* : 2019, *Edition* : 2022

Seventh Edition : 2023

ISBN : 978-93-5138-296-6

Branches

Bengaluru	080-26 75 69 30
Chennai	044-24 34 47 26
Guwahati	0361-254 36 69
Hyderabad	040-27 55 53 83
Jalandhar	0181-222 12 72
Kochi	0484-405 13 03
Kolkata	033-40 04 77 79
Lucknow	0522-430 36 13
Ranchi	0651-224 24 64

Published in India by

TRINITY PRESS

(An Imprint of Laxmi Publications Pvt. Ltd.)

(An ISO 9001:2015 Company)

113, GOLDEN HOUSE, GURUDWARA ROAD, DARYAGANJ,

NEW DELHI-110002, INDIA

Telephone : 91-11-4353 2500, 4353 2501

www.laxmipublications.com

info@laxmipublications.com

C—00256/022/06

Printed at : Ajit Printing Press, Delhi.

To our teachers:

Professor H.A. Bethe
Professor D.S. Kothari
Professor R.C. Majumdar
Professor R. Serber

Preface

Quantum mechanics has long been recognized as an essential ingredient in the training of a student in Physics, Chemistry and Electrical Engineering. No matter which specialization the student chooses in his later career, it is necessary for him to develop not merely an understanding of the basic principles of quantum mechanics but an ability to apply these principles in his field.

Textbooks on quantum mechanics addressed to students at under-graduate and post-graduate levels generally have few solved problems; moreover, the problems suggested are largely of an academic nature with a greater emphasis on theory than on applications. This book makes an attempt to present the basic concepts in quantum mechanics with emphasis on applications in other areas like nuclear physics, astrophysics, solid state physics, quantum optics, etc. Each chapter is followed by a number of problems. The solutions to most of these problems have been given at the end of the chapter. This will enable the student to refer to the solutions when necessary and at the same time provide him an opportunity, if he so wishes, to solve some or all of these problems by himself before consulting the solutions. The emphasis has not been on rigour but on making results plausible and helping the students to become familiar with methods for solving problems. The book has evolved from lectures given by the authors (to students of physics and engineering) at IIT Delhi and at University of Rajasthan, Jaipur. During the past 8 years, one of the authors (AG) has been giving (every year) course of lectures on Quantum Mechanics to research scholars at IUAC (Inter University Accelerator Center) in New Delhi. Some of the analysis given here have emerged from these lectures.

As a prologue to the book we have reprinted an essay, 'What is Quantum Mechanics?', by Professor Victor F. Weisskopf. It is so well written that we felt it would motivate the reader to learn more of this fascinating subject. Following the prologue are two short chapters on mathematical preliminaries—one on the Dirac delta function and the other on Fourier transforms. The two topics are so extensively used in quantum mechanics that even if the students are familiar with them, it would be of great advantage to discuss these topics (again) in about two to three lectures right in the beginning of the course.

After the mathematical preliminaries, we have discussed (in Chapter 3) wave-particle duality and the uncertainty principle. In Chapter 4, we have introduced the Schrödinger equation and in Chapter 5, we have discussed the solutions of the Schrödinger equation corresponding to a free particle leading to a study of the time evolution of a wave packet. In Chapter 6 bound state solutions of the Schrödinger equation are discussed.

We have given a somewhat lengthy account of the Dirac notation of bras and kets; since this is used frequently in advanced work, it was felt that the student would benefit greatly if he became familiar with it right at the beginning. Immediately after solving the Schrödinger equation for the linear harmonic oscillator problem (Chapter 7), we usually introduce Dirac's bra and ket algebra (Chapter 11) and then solve again the linear harmonic oscillator problem using the bra-ket algebra (Chapter 12). It is then very straightforward to study the time evolution of the coherent state and its relationship with the classical oscillator; such an analysis brings out many salient aspects of quantum mechanics. After this, we usually discuss the angular momentum problem in detail (Chapters 9, 13 and 15), which allows us to obtain very elegantly the expressions for spherical harmonics. We then obtain solutions of the Schrödinger equation for spherically symmetric potentials—in particular, for the hydrogen atom problem and for the three-dimensional isotropic oscillator problem (Chapter 10). Chapter 14 is on the Stern-Gerlach experiment which is indeed one of the most beautiful experiments in quantum mechanics. The principle of magnetic resonance has also been discussed and, in the process, we have been able to give an exact solution of the Schrödinger equation corresponding to a time dependent Hamiltonian. Chapter 15 uses operator algebra to derive expressions for spherical harmonics. In Chapter 16 we have discussed the polarization of the photon and its intrinsic angular momentum. The chapter also discusses the entanglement of the photon and Bell's inequality.

Chapter 17 discusses the double well problem and also the Kronig-Penney model. In Chapter 18 we have also given a detailed account of the JWKB solutions of the Schrödinger equation; the JWKB methodology represents one of the very powerful approximate methods that is extensively used not only in quantum mechanics but also in many other areas.

Chapter 19 discusses addition of angular momenta and introduces the Clebsch-Gordan coefficients. Chapters 20–24 discuss time independent perturbation theory, effects of magnetic fields, the variational method and the Helium atom. In Chapter 25, we have given a detailed discussion of some select topics which range from the concept of quasi-bound states to the Thomas-Fermi model of the atom.

Chapters 26, 27, 28 and 29 discuss the elementary theory of scattering, time dependent perturbation theory, the semi-classical theory of radiation and the quantum theory of radiation. Finally in Chapter 30, the Dirac equation and some of its solutions are discussed.

In order to have a better appreciation of the theory, most of the figures correspond to actual numerical calculations; these were generated using GNUPLOT and Mathematica.

We do hope the reader enjoys going through the book; we would greatly appreciate receiving suggestions for further improvement.

We have dedicated this book to Professor H.A. Bethe who taught one of us (AG) at Cornell, to Professor R. Serber who taught the other (SL) at Columbia, and to Professor D.S. Kothari and Professor R.C. Majumdar who taught both of us at Delhi University. To them we would like to extend our gratitude and respect for introducing us to the most beautiful subject in science.

Ajoy Ghatak
S. Lokanathan

Acknowledgements

We are indebted to our colleagues at IIT Delhi and at Rajasthan University for many enlightening discussions related to many aspects of quantum mechanics.

We are indebted to various authors and publishers for their permissions to use material from their publications. We are very grateful to Professor Victor F. Weisskopf and W.H. Freeman & Co. for allowing us to reprint the essay *What is Quantum Mechanics*—in fact, Professor V. Lakshminarayanan of University of Missouri at St. Louis had brought to our attention the essay by Professor Weisskopf. We express our gratitude to the Niels Bohr Archives in Copenhagen for providing the original photograph on the Stern-Gerlach experiment. We are also grateful to Addison-Wesley Publishing Co., Reading; Mc-Graw-Hill Book Co., New York; Oliver and Boyd Ltd., Edinburgh; Oxford University Press, Oxford; Penguin Books Ltd., Middlesex and Prentice Hall Inc., Englewood Cliffs for permission to adapt figures, tables and a few analysis from the following books: G. Baym, *Lectures on Quantum Mechanics,* Addison-Wesley (1969); P.A.M. Dirac, *Principles of Quantum Mechanics,* Oxford University Press (1958); H. Enge, *Introduction to Nuclear Physics,* Addison-Wesley (1966); R.P. Feynman, R.B. Leighton and M. Sands, *The Feynman Lectures on Physics,* Vol. III, Addison-Wesley (1965); I.S. Hughes, *Elementary Particles,* Penguin (1972); V.I. Kogan and V.M. Galitsky, *Problems in Quantum Mechanics,* Prentice-Hall (1963); R.A. Newing and J. Cunningham, *Quantum Mechanics,* Oliver and Boyd (1967); L. Pauling and E.B. Wilson, *Introduction to Quantum Mechanics,* McGraw-Hill (1935); L.I. Schiff, *Quantum Mechanics,* McGraw-Hill (1955); N. Tralli and F.R. Pomilla, *Atomic Theory,* McGraw-Hill (1969).

We are very grateful to Mr Monish Das, Dr. Rajeev Jindal, Dr. Vipul Rastogi, Dr. Mukesh Pratap Singh and Mr Prem Bhawnani for their help in creating some of the figures using GNUPLOT and Mathematica. We also thank Mr. Norman Lim of National University of Singapore for carrying out numerical calculations associated with some of the figures in Chapter 8. We are also grateful to Mr Monish Das, Mr Prem Bhawnani and Mrs. Bhawna Bhardwaj, for their help in the preparation of the manuscript.

We end with the quotation (which we found in a book by G.L. Squires): *I have learnt much from my teachers, but more from my pupils.* To all our pupils, we owe a very special debt.

Ajoy Ghatak

S. Lokanathan

Number of electrons = 500

Number of electrons = 5000

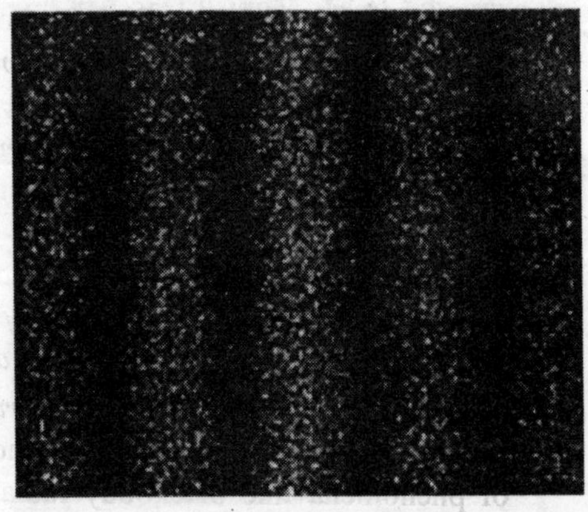

Number of electrons = 50000

Computer simulated two-slit electron interference patterns (calculations by Amrit Nandan, Mohit Malik and Ajoy Ghatak)

Chronological Development of Quantum Mechanics: A Very Brief Account

The development of quantum mechanics as a unified structure did not take place in the smooth way that is usually presented in a textbook. For a textbook to present this development in chronological fashion is, however, not really a fruitful way of presenting the subject to the student. On the other hand, we feel that every student ought to have some sense of perspective of this development. As mentioned by Ludwig [1] *'without taking account of its historical development, an existing theory often appears almost as if* it *had "fallen from heaven". However the question of the development of a theory is important not only to satisfy our curiosity but also because much can be learnt from it for the future'*. Keeping this in mind we give a very brief account of the chronological development of quantum mechanics highlighting some of the important facts associated with the development and also giving reference to books (which are usually available in most libraries) where some of the original papers have been reprinted. We hope that this introduction will induce the reader to go back to some of the articles written by the *masters* of the subject so that he is able to get a perspective of the field as it developed.

We should mention that perhaps the most comprehensive account of the chronological development of quantum mechanics is given in the book by Jammer [2] which has also given detailed references of most of the important publications during the development of the field (and where they could be found). There are numerous other books/articles written on the historical development of the subject; particular reference can be made to Born [3], Condon [4], Cropper [5] and ter Haar [6]. We give below a brief account of the chronological development:

1900: In a paper entitled *'On an improvement of Wien's equation for the spectrum'*, Planck "guessed" the blackbody radiation formula and found that it is in excellent agreement with the experimental data. In a subsequent paper entitled *'On the theory of the energy distribution law of the normal Spectrum'* he derived this formula by introducing a drastic assumption that the oscillators can only assume discrete energies. Planck wrote *"We consider, however—this is the most essential point of the whole calculation—E to be composed of a very definite number of equal parts and use thereto the constant of nature* $h = 6.65 \times 10^{-27}$ *erg sec. This constant multiplied by the common frequency ν of the resonators gives us the energy element E in erg., ..."*.The two papers of Planck are reprinted by ter Haar [6]. It may be of interest to mention that in an unpublished letter to R.W. Wood, Planck wrote (quoted from Ref. 2) that the postulate of energy quanta was 'an act of desparation' done because 'a theoretical explanation *had* to be supplied at all cost, whatever the price'. Planck was awarded the 1918 Nobel prize in Physics for his *discovery of energy quanta;* his Nobel lecture was on *'The genesis and present state of development of the quantum theory'*.

1905: In a paper entitled *'On a heuristic point of view about the creation and conversion of light'* (reprinted in Ref. 6), Einstein introduced the light quanta. In this paper he wrote that for explanation of phenomena like blackbody radiation, production of electrons by ultraviolet light (which is the photoelectric effect) it is necessary to assume that *"when a light ray starting from a point is propagated, the energy is not continuously distributed over an ever increasing volume, but it* *consists of a finite number of energy quanta, localized in space, which move without being divided and which can be absorbed or emitted only as a whole"*. Einstein received the 1921 Nobel Prize in Physics for his *discovery of the law of photoelectric effect* and *not* for his theory of relativity. It is interesting to note that Einstein did not have the symbol h in his paper; he used the quantity $R\beta\nu/N$ for the energy of the light quanta where R/N and β are actually k and h/k respectively.

1913: In a paper entitled *'On the constitution of atoms and molecules'* (reprinted in Ref. 6), Bohr gave the first successful interpretation of atomic spectra. (The experimental study of atomic spectra was started around 1860 and in 1885 Balmer gave his famous empirical formula which came to be known as Balmer series). Niels Bohr's logic of argument was somewhat as follows: For an electron rotating in a circular[1] orbit around a proton the potential energy is $-e^2/a$, the kinetic energy ($\frac{1}{2}mv^2$) is $e^2/2a$ and therefore the binding energy is given by

$$E_b = e^2/2a$$

The frequency ν' of the orbiting electron is given by

$$\nu' = \frac{v}{2\pi a} = \left(\frac{2E_b}{m}\right)^{1/2} \frac{1}{2\pi}\frac{2E_b}{e^2} = \frac{\sqrt{2}E_b^{3/2}}{\pi e^2 m^{1/2}}$$

Bohr than assumed that *"during the binding of the electron, a homogeneous radiation is emitted of a frequency* $\nu = \frac{1}{2}\nu'$*; then from Planck's theory, we might expect that the amount of energy emitted by the process considered is equal to* $\tau h \nu$*, where h is Planck's constant and* τ *an entire number"*. Thus

$$E_b = \tau h \nu = \frac{1}{2}\tau h\nu' = \frac{1}{2}\tau h\left[\left(\frac{2}{m}\right)^{1/2}\frac{E_b^{3/2}}{\pi e^2}\right]$$

which gives

$$E_b = \frac{2\pi^2 m e^4}{\tau^2 h^2}$$

and when an atom makes a transition from one stationary state ($\tau = \tau_1$) to another ($\tau = \tau_2$) a light quantum of frequency

$$\frac{2\pi^2 m e^4}{h^2}\left[\frac{1}{\tau_2^2} - \frac{1}{\tau_1^2}\right]$$

is emitted. The above formula agreed with the experimental data.

It should be mentioned that the quantization of angular momenta was put forward by Bohr in a subsequent paper in the same year. Bohr was awarded the 1922 Nobel prize in Physics for his *investigation of the structure of atoms, and of the radiation emanating from them.* His Nobel lecture on 'The structure of the atom' gives a summary of the important results in atomic theory.

1914: In a paper entitled *'On the excitation of 2536* Å *mercury resonance line by electron collisions'* (reprinted in Ref. 6), Franck and Hertz showed that by varying the kinetic energies of the electrons hitting an atom it is possible to produce controlled excitation of atoms and molecules; the experiment proved the discrete energy states of atoms, a hypothesis put forward by Bohr in 1913. Franck and

[1]Actually Bohr's consideration was for elliptical orbits; here, for the sake of simplicity we are considering circular orbits.

Hertz received the 1925 Nobel prize in Physics *for their discovery of the laws governing the impact of an electron upon an atom.*

1917: In a paper entitled *'On the quantum theory of radiation'*; (reprinted in References 6 and 7), Einstein put forward the existence of stimulated emission and introduced the famous *A* and *B* coefficients which are now known as the *Einstein coefficients.*

1922: Stern and Gerlach performed their famous experiment which demonstrated that in the presence of a strong inhomogeneous magnetic field, a beam of silver atoms split into two beamlets. This provided a confirmation of space quantization.

1923: Compton reported his studies on the scattering of X-rays by solid materials (mainly graphite) and showed that the shift of the wavelength of the scattered photon could be explained by assuming the photon having momentum equal to h/λ. Compton shared the 1927 Nobel prize in Physics *'for his discovery of the effect named after him'*. The original papers of Compton have been reprinted in a book by Shankland [8].

1924: In a paper entitled *Investigations on quantum theory* [reprinted in Ref. 1], de Broglie suggested the idea of waves associated with electrons. He recognized that the condition for the *k*th orbit

$$mv_k r_k = n\hbar$$

implies

$$2\pi r_k = h\frac{h}{mv_k} = n\,\lambda$$

i.e. the circumference of the allowed orbit contains an integral number of wavelengths. He wrote *'in order to have a stable state the length of the channel must be in resonance with the wave'*. De Broglie was awarded the 1929 Nobel prize in Physics for his *'discovery of the wave nature of electrons'*.

1925: Pauli in his paper entitled *'On the connection between the completion of" electron groups in an atom with the complex structure of spectra'* (reprinted in Ref. 6) put forward the *"fourth quantum number of the electron"* and the exclusion principle. He was awarded the 1945 Nobel prize in Physics *for the discovery of the exclusion principle.*

1925: *After* studying Pauli's paper, Uhlenbeck and Goudsmit put forward the idea of the spinning electron.

1925 to 1928: In 1925 Heisenberg put forward the *'quantum-theoretical mechanics based exclusively on relationships between quantities observable in principle'*, which was further developed by Born and Jordan in the same year. The commutation relation was first introduced by Born and Jordan. Born later wrote *'I shall never forget the thrill I experienced when I succeeded in condensing Heisenberg's ideas on quantum conditions in the mysterious equation $pq - qp = h/2\pi i$'*. Schrödinger's papers came in 1926. His three famous papers under the same title *'Quantization as an eigenvalue problem'* appeared in quick succession in 1926. In between he had also published a paper with the self-explanatory title *'On the relationship of the Heisenberg-Born-Jordan Quantum Mechanics to mine'*. The original papers of Schrödinger, Heisenberg, Born and Jordan are reprinted in Ref. 1. In 1926 Born suggested that the square of the wave function should be interpreted as the probability density of particles. In 1927 Heisenberg formulated his famous uncertainty principle[2] and in the same year Born developed the quantum mechanical theory of scattering of electrons by atoms and Dirac succeeded in developing the quantum theory of radiation which could explain satisfactorily the phenomena of interference as well as the emission and absorption of light by matter[3]. In 1928 Dirac gave his relativistic equation which led to the prediction of the *'positive electron'* which was later called the *positron.* In 1932, Anderson observed such a particle in cosmic radiation where they could be distinguished from electrons by the opposite curvature of their tracks in the presence of a magnetic field.

The 1932 Nobel prize in Physics went to Heisenberg *for the creation of quantum mechanics.* The 1933 Nobel prize in Physics was shared by Schrödinger and Dirac *for the discovery of new productive forms of atomic theory* and the 1954 Nobel prize in Physics went to Max Born *for his fundamental research in quantum mechanics, especially for his statistical interpretation of the wave function.*

1927: Davisson and Germer and independently G.P. Thomson observed electron diffraction patterns produced by metal foils which established the existence of de Broglie waves. The 1937 Nobel prize in Physics was shared by Davisson and Thomson *for their experimental discovery of the diffraction of electrons by crystals.*

1928: Gamow in Germany and Gurney and Condon in USA reported the first major application of quantum mechanics, namely, the barrier leakage interpretation of the natural radioactive decay of some elements emitting alpha particles. The theory could satisfactorily explain the fact that an alpha particle can remain inside a nucleus for about 10^9 years before it comes out. Condon in 1952 wrote: "*I think it is fair to say that the barrier-leakage idea was the opening of the modern period of the application of quantum mechanics to nuclear physics.*"

1931: In order to explain the continuous beta ray spectrum, Pauli in 1931 suggested the existence of the neutrino.

1934: Fermi developed the famous theory of beta decay.

Although the development of quantum mechanics continues even today, we felt that Fermi's beta decay theory should be an appropriate point to 'sign off' (keeping in view the scope of the present book). We conclude by mentioning the following incident (quoted from May 1982 issue of *Physics Today):*

After attending a seminar given by one of Oppenheimer's pupils on Fermi's beta ray theory, Fermi remarked to Emilio Segre, 'Emilio, I am getting rusty and old. I cannot follow the highbrow theory developed by Oppenheimer's pupils any more. I went to their seminar and was depressed by my inability to understand them; only the last sentence cheered me up. It was: "*And this is Fermi's theory of beta decay.*"

---

[2] Heisenberg's book *The Physical Principles of" Quantum Mechanics* is a very entertaining book to read on the uncertainty principle.

[3] A very nice account of Dirac's quantum theory of radiation has been given in a review paper by Fermi [9]; of course, the classic treatise by Dirac [10] should be a must for all persons trying to understand quantum mechanics.

References

1. G. Ludwig, *Wave Mechanics,* Pergamon Press, Oxford (1968).
2. M. Jammer, *The Conceptual Development of Quantum Mechanics,* Mc-Graw Hill Book Co., New York (1966).
3. M. Born, *Physics in my Generation,* Springer Verlag, New York (1969).
4. E.U. Condon, *60 years of Quantum Mechanics, Physics Today,* **15**, 45, October issue (1962).
5. W.H. Cropper, *The Quantum Physicists and an Introduction to their Physics,* Oxford University Press, New York (1970).
6. D. ter Haar, *The Old Quantum Theory,* Pergamon Press, Oxford (1967).
7. F.S. Barnes, *Laser Theory,* IEEE Press, New York (1972).
8. R.S. Shankland (Ed.), *Scientific Papers of A.H. Compton,* University of Chicago Press, Chicago (1973).
9. E. Fermi, *Quantum Theory of Radiation,* Reviews of Modern Physics, **4**, 87 (1932).
10. P.A.M. Dirac, *The Principles of Quantum Mechanics,* Oxford University Press, Oxford (1958).

Table of Contents

Prologue*

What is quantum mechanics?

Victor F. Weisskopf

There is no doubt that the most outstanding development in modern science was the conception of quantum mechanics. It showed, better than anything else, the human capability to comprehend the fundamental principles that underlie the world in which we live—even when these principles run contrary to our experience in dealing with our everyday environment. The French philosopher—mathematician Henri Poincare said, *It is hardly necessary to point out how much quantum theory deviates from everything that one has imagined until now; it is, without doubt, the greatest and deepest revolution to which natural philosophy has been subjected since Newton.*

Much happened in physics between the time of Newton and the time of quantum mechanics. The discoveries and insights over the last three centuries share a characteristic feature: seemingly unconnected phenomena turned out to be manifestations of the same fundamental principle. It was a period of unification of disparate fields of experience. Here are some of the most important steps.

Newton showed that the motion of the planets is governed by the same law as the free fall of an object on earth. Thus, he unified terrestrial and celestial mechanics. In contrast to the belief of the ancients, he showed that the world of the earth and of the heavens is governed by the same laws.

Scientists in earlier days believed that heat was some peculiar substance called caloric, which flowed from a hot object to a colder one. Physicists in the nineteenth century recognized that heat is the random motion or random vibration of the constituents of matter. Thus, thermodynamics and mechanics were unified. This feat is connected with the names of J.B. Mayer, B. Rumford, R.E. Clausius, L. Boltzmann, and J.W. Gibbs.

For a long time, the phenomena of electricity, magnetism, and light appeared to be unconnected. In the first half of the nineteenth century, one of the great unifications of physics took place. Faraday and Maxwell, together with many others, were able to show that all three phenomena are manifestations of the electromagnetic field. And so the field concept entered into physics. The simplest example is the electric field of an electric charge that exerts a force on another charge when the latter falls within its range. An electric current produces a magnetic field that exerts a force on magnetic materials. Such fields may even propagate through space independently of any charges or magnets, in the form of electromagnetic waves, of which visible light is one example. The field concept is less directly connected to our everyday experience than the concept of a particle, but it can easily be realized by our senses. For example, if one feels the attraction of a piece of iron by a magnet, one obtains the immediate impression that there is something surrounding the magnet that acts upon the iron. Finally, Einstein unified space, time, and gravity in his special and general theories of relativity.

Quantum mechanics also united two branches of science: physics and chemistry. But it did much more. In previous great developments in physics, fundamental concepts were not too different from those of our everyday experience, such as particle, position, speed, mass, force, energy, and even field. We often refer to those concepts as *classical.* The world of atoms cannot be described and understood with those concepts. For atoms and molecules, the ideas and concepts formed in dealing with the objects in our immediate environment no longer suffice. Surprising forms of behavior were observed that not only needed a different language but required new concepts to understand the properties of atoms.

Figure 1. Wolfgang Pauli, Werner Heisenberg, and Enrico Fermi at a conference in 1927, reflecting their enthusiasm for and joy of the new quantum mechanics.

A small group of people conceived of and formulated these new ideas in the middle twenties of this century. The most important among them were W. Heisenberg, a German; E. Schrödinger, an Austrian; P.A.M. Dirac, an Englishman; W. Pauli, another Austrian; and M. Born, another German. They worked at different places, but the centre of activities was in Copenhagen, where they met frequently under the leadership of the great Niels Bohr. Bohr was the ideal leader of such a group. Older than most of the others, who were then in their twenties or early thirties, he contributed enormously to the conception of the new ideas by his constant questioning, by his criticism, and by his enthusiasm. The crowd that assembled around him was a group of devoted forward-looking people who, free of the bonds of convention, attacked the deepest riddles of nature with a spirit of joy that can hardly be described. That joy of insight is a sense of involvement and awe, an elated state of mind akin to what you feel on top of a mountain after a hard climb, at the shore of the blue sea, or when you hear a great work of music. It comes not only after personal achievement, but also after finally understanding an important new insight gained by the work of others. For every real scientist, it is great compensation for the hard work and trouble he must endure (see Fig. 1).

The quantum revolution changed our old concepts of reality in many respects.

We are going to describe this drama in five parts—a prologue, plus four acts:

Prologue	*Riddles of prequantum physics*
Act I	Discovery of wave-particle duality
Act II	How wave-particle duality miraculously solves the riddles of prequantum physics
Act III	Significance of wave-particle duality and the new reality of the quantum state
Act IV	The quantum ladder: An extension of quantum mechanics to nuclear and subnuclear phenomena and to the history of the universe

PROLOGUE:

RIDDLES OF PREQUANTUM PHYSICS

Before quantum theory was conceived, physicists were unable to explain some of the most ubiquitous phenomena in our environment. Here are a few examples: A piece of iron, when heated, becomes first red, then yellow, then white, but nobody could explain why. The different colors emitted by a piece of matter come from the irregular heat motion of the electrons in the atoms of iron. Fast motion emits higher frequencies of light than slow motion. The frequency determines the color of the light. The laws of thermodynamics tell us that any form of motion should receive the same amount of energy at a given temperature, an amount that increases when the temperature rises. Thus we expect only an increase of intensity of the emitted light, not a change from red to yellow to white. This change represented an unsolved riddle, somewhat like today's ignorance of the nature of memory. We use memory constantly, but nobody knows precisely what it is.

We can consider more examples of unsolved questions of that period: Why are copper and silver metals and oxygen a gas? Why do metals have properties so different from other solids, such as rocks? Why do these properties persist, even after heating, melting, evaporating, and subsequent cooling to the original temperature? Why do oxygen atoms bind to hydrogen atoms to form water? Why does sodium gas emit yellow light when heated? Why is it that burning 1 kilogram of coal produces approximately 6000 calories? Why is the size of atoms about a hundred millionth of a finger's breadth? No one could provide any answers to these questions at the turn of the century.

More problems were generated by the discovery of the electron by J.J. Thomson in England and H.A. Lorentz in the Netherlands at the end of the last century and, in particular, by further discoveries by E. Rutherford and H.G. Mosely at the beginning of this century. They found that atoms consist of a heavy, positively charged atomic nucleus surrounded by much lighter electrons. Because the electric force between the positively charged nucleus and the negatively charged electrons is of the same form as the attraction between the sun and the planets, they concluded that atoms must be tiny planetary systems, with the nucleus as *sun* and the electrons as *planets*. Moreover, it was found that all elements seem to have this planetary structure and differ only in numbers of electrons. For example, hydrogen has one electron; helium, two; iron, twenty-six; and uranium, ninety-two.

It was difficult to understand how elements that are so different—some are gases, some are metals, some are liquids—differ in their atomic structure by only a few electrons. For example, the element neon, which has ten electrons, is a chemically inactive gas; however, the element sodium, which has eleven electrons, is one of the most chemically active metals. An electron increase of 10 percent completely changes the character of the atom! No one could explain this apparent inconsistency between quantity and quality.

Four observations defied all understanding at the turn of the century.

1. The color of objects at various stages of heating (red, yellow, white).
2. The very different specific properties of elements whose number of electrons is almost the same.
3. The fact that atoms do not change their properties in spite of the many collisions and interactions that they suffer in a gas or in an ordinary piece of matter. They quickly resume their original qualities after the perturbation. Their stability and their ability to regenerate is completely at odds with what we would expect from a planetary system. If our solar system were to collide or pass another star at a close distance its orbits and patterns would be completely changed and it would not return to its original form.
4. The energy content of the atom is *quantized.* An atom can assume a series of definite energies only and never a value in between. This most surprising fact was found at the beginning of this century and is also completely foreign to a planetary system. There is no reason why the energy of planetary motion cannot change by arbitrarily small amounts, for example, when a meteorite hits a planet. An atom, however, can accept or lose only definite amounts of energy, those that would change its energy from one of the values in the series to another.

It became clear to prequantum physicists that the analogy between an atom and a planetary system breaks down completely when atomic properties and processes are examined in detail. On the other hand, these observations left no doubt that the atom consisted of a positively charged nucleus surrounded by electrons, which ought to form a planetary system, according to the laws of mechanics known at that time. Everything, including our own bodies, consists of atoms. Obviously, the most urgent problem for physics at that time was to resolve these contradictions and to achieve a better comprehension of the structure and behavior of atoms.

ACT 1: DISCOVERY OF WAVE-PARTICLE DUALITY

Not until the first quarter of this century did physicists find the path to the solution. It all began with a series of startling discoveries that seemingly had no direct connection with atomic structure. These discoveries showed that light, believed to be a wave, exhibited particle properties and that electrons, believed to be particles, exhibited wave properties. Let us call it the wave-particle duality.

Often, when it is difficult to find an explanation for one group of strange findings, another group of unexplained observations helps us to resolve both difficulties. It's often easier to solve two riddles than one. Two different disturbing observations may lead more readily than one such observation to a solution. That was the case with the wave-particle duality, since it showed the way to understand the strange properties of the atomic world.

Let us first look at the duality itself. We have all learned in school that light is an electromagnetic wave. We perceive the different wavelengths, or frequencies of vibration of these waves, as different colors. Red has a longer wavelength and lower frequency than yellow or blue. How did we know that light is a wave? Let us look at water waves to learn about a characteristic property of waves: interference. When two wave trains merge, originating from two different points, they combine in a typical way. When the crest of one wave coincides with the crest of another, the two together create a stronger motion. When the crest of one wave coincides with the trough of another, the two wave motions cancel each other in a phenomenon called *interference.* Figure 2 shows the interference of two water waves originating from two different points. We see that they reinforce each other in certain directions and wipe each other out in other directions. The same phenomenon was observed

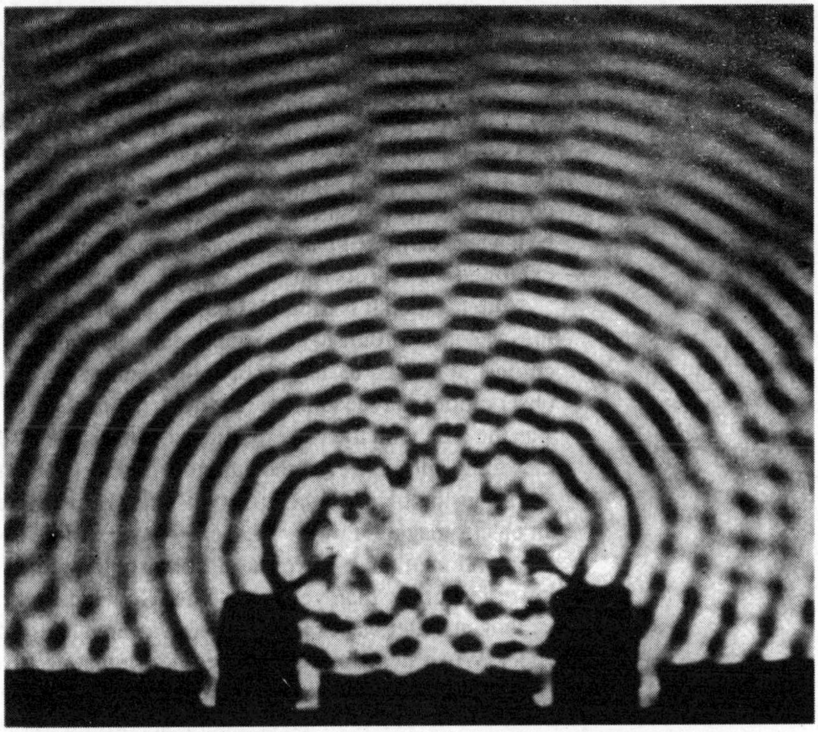

Figure 2. Interference patterns produced by two water waves originating from two neighboring points. [Physical Science Study Committee, *Physics,* D.C. Heath, Boston, 1967].

long ago by Thomas Young in light waves, as sketched in Fig. 3 and shown in Fig. 4. The two light waves are produced by illuminating a screen with two slits. Of course, there is a difference between water and light waves: In the former case, the water surface oscillates and in the latter it is the electric and magnetic field strength that oscillates in space.

Great was the surprise, therefore, when it turned out that the energy of a light beam cannot be subdivided indefinitely. An ordinary wave can be strong, weak, or still weaker, with no restrictions on its intensity or lack thereof. Not so with light waves! The observed light energy seems to occur in lumps called light quanta, or photons. The energy of a light beam is always a multiple of such lumps, never half a lump or ten-and-a-quarter lumps. This is a strange finding. How could it be that the energy of a wave cannot assume an arbitrary value, but must be a multiple of a definite quantum? The energy of a light wave is indeed quantized; this property was shown beyond any doubt. Figure 5 is an example of how it could be observed. It is a photograph of what we see when a light ray (in this case, an X-ray- light of very high frequency) penetrates through a *cloud chamber,* that is a container of saturated steam. The vapor is in a state of pressure and temperature at which it is about to form droplets. Just a bit of energy deposited here and there will form a droplet. If the bit of energy is larger, a bigger droplet is formed; if smaller, a tinier one will appear. Now, looking at the picture, we see (as we might well expect) that the light ray gets weaker penetrating the liquid. However, it gets weaker in a special way: The droplets do not become smaller, but their number does. The unchanging size of the droplets indicates that light consists of quanta: the effect of each quantum is the same all along the light beam, but there are fewer droplets farther down the road (to the right of the picture) because part of the energy has already been used up. If the ray were a

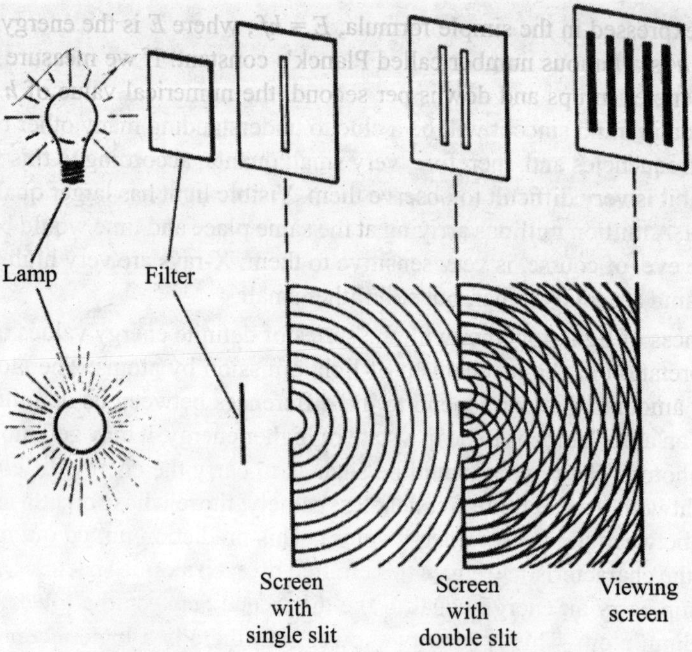

Figure 3. Arrangement for demonstrating interference of light. The first slit defines the incoming wave; the second has two openings, thus producing two waves; the third shows the interference stripes seen on the screen at right [After Atkins, *Physics,* Wiley, New York, 1965].

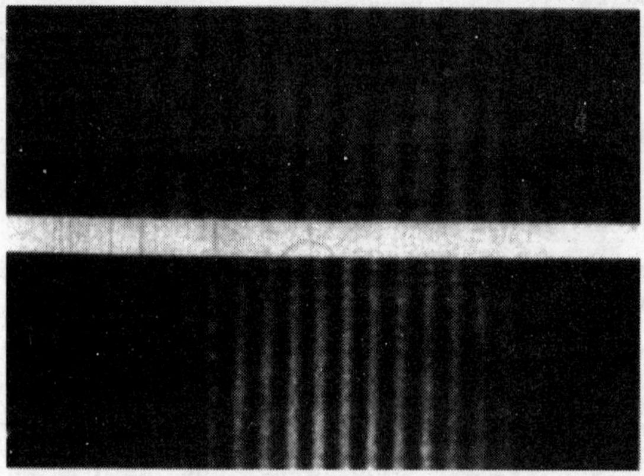

Figure 4. Patterns seen on the screen at the extreme right of the arrangement of Fig. 3. Upper pattern with red light; lower pattern with violet light. The lower pattern is narrower because of the shorter wave length. [Physical Science Study Committee, *Physics,* D.C. Heath, Boston, 1967].

true wave, its intensity would have become weaker and would have produced smaller droplets. The weakening shows itself not by smaller drops but by fewer ones. Each light quantum maintains the same strength, but there are fewer of them when the beam becomes weaker. These and other similar observations reveal the *graininess* of light.

Light *grains* (photons, or light quanta) are very small indeed. That is why they were not discovered earlier. Their energy was found to be proportional to the frequency of light. This relationship is expressed in the simple formula, $E = hf$, where E is the energy of the photon, f the frequency, and h is a famous number called Planck's constant. If we measure energies in electron volts and frequencies in ups and downs per second, the numerical value of h is 4×10^{-15}. Let us keep this formula in mind since it will be a clue to understanding many other riddles. Radio waves have very low frequencies and, therefore, very small quanta, according to this formula. The quanta are so small that it is very difficult to observe them. Visible light has larger quanta, but they are still extremely small. A million millions arriving at the same place and time would be felt as a little prick on a finger. The eye, of course, is very sensitive to them. X-rays are very high frequency radiation, and so their quanta are much larger, but still rather small.

The graininess of light, together with the series of definite energy values of an atom, forces us to a new interpretation of the mechanism of light emission by atoms. The atom may lose or gain energy only by amounts that correspond to the differences between its definite energy values. For example, when an atom happens to be in a state of higher energy, it may get into one of lower energy by emitting a photon. The photon would be such as to carry the energy difference. Thus, an atom should emit light with very definite frequencies, namely, those whose quanta are equal in energy to the *differences* between the atomic energy values. This prediction turned out indeed to be the case, and it explains the characteristic frequencies emitted by each atomic species. Quanta of yellow light of a sodium lamp carry an energy equal to the difference between the lowest and the next higher state of the sodium atom. This regularity showed that there is a logical connection between two mysterious facts: the quantized energy of atoms and the graininess of light.

Figure 5. The cloud chamber shows the absorption of photons, one by one, and the weakening of the beam as photons are removed. The photons enter from the left and are absorbed by the argon atoms of the gas. They eject photoelectrons, whose tracks produce little droplets that show white in the photograph. You can see that the beam is weakened if you divide the photograph in half; there are seventeen tracks on the left half and only nine on the right. The *light* is actually an X-ray beam of a wavelength about 0.2 Angstroms. [W. Gentner, N. Maier-Leibnitz, and W. Bothe, *Atlas Typischer Nebelkam- merilder,* Springer Verlag, Berlin, 1940].

What does this amount to? We come to the startling conclusion that light has particle properties, in spite of the fact that so many indications—interference, for one—have shown that light is a wave!

The mystery deepened considerably as a result of a second surprising discovery: particles—electrons, for example—have wave properties! The French physicist L. de Broglie predicted this duality by ingenious intuitive insight, even before it was discovered by experiment. Under certain conditions, electrons emerging from two nearby points do not add their intensities but give rise to weaker beams in some directions and to stronger beams in others, similar to the interference of two water waves or light beams, as was shown in Figs. 2 and 3. What we call a particle exhibits properties that we ordinarily ascribe only to waves. What is the electron, a particle or a wave? What is a light ray, a wave train or a beam of particles? Eddington found a telling terminology, calling photons and electrons *wavicles.*

The fundamental relation between particle energy and wave frequency, $E = hf$, holds also for particles. A similar relation exists between the particle momentum and the wave number k (the number of wave lengths in one centimeter): $p = hk$. These relations are called *de Broglie relations.* They both contain the fundamental number h, the Planck constant.

Obviously, we run into terrible difficulties and contradictions by assuming that something is a wave and particle at the same time. A school of fish swimming in a lake is completely different from a wave on its surface. Each fish in the school is localized, as is each particle in a particle beam, whereas a wave is necessarily spread out over space. These contradictions will be raised in Act III of our drama. Such everyday concepts as particle and wave are not applicable in the atomic realm without restrictions. This will lead us to a new conception of reality. For the moment, let us assume that entities such as electrons and light are indeed wavicles—exhibiting both wave and particle properties—and see what follows from this assumption.

ACT II: HOW WAVE-PARTICLE DUALITY MIRACULOUSLY SOLVES THE RIDDLES OF PREQUANTUM PHYSICS

First, we can easily see a solid material becomes first red, then yellow, then blue and white when it is heated to higher temperatures. What are heat and temperature?

In a warm material, atoms and molecules perform irregular vibrations of all kinds. The energy of these random motions, thermal energy, is proportional to the temperature: the hotter the material, the faster the particles vibrate. Atoms and molecules contain electrically charged particles; when they vibrate, they emit light. But light is quantized; it comes in lumps of energy. If the thermal energy is not sufficient to produce a light quantum of a given color, that color *cannot* show up. Thus, a piece of iron at room temperature does not emit visible light because at that temperature the thermal energy is less than the energy of the quanta of visible light. (Iron, at room temperature, does emit infrared light, whose quanta are of lower energy; we cannot see it but we can measure it, and we feel it as a heat radiation). At a higher temperature, thermal energy reaches the energy of a visible quantum. Red light has the lowest frequency and therefore the lowest quantum in the visible range; then comes yellow, then blue. Thus red is the first color to be seen when the iron is heated. As the temperature is increased, iron becomes yellow-red, then white, because the sum of red, yellow, and blue is white.

In 1900, Max Planck was the first to realize that quantum theory would lead to a correct description of the radiation of incandescent matter. His way of explaining it was different from the description given here and somewhat more involved. Nevertheless, his ideas gave impetus to the development of quantum theory.

What is much more impressive, however, is that the wave-particle duality also explains the stability and specificity of atoms and their special quantized states with definite energy values. The wave nature of electrons bestows properties on atoms that are quite different from what we expect of planetary systems.

To understand the difference, we must learn something about confined waves. When a wave is confined to a restricted region in space, it can assume only certain definite wave patterns. A violin string suspended between two end points can set up only those vibrations whose half wave length fits once or twice, or any integral number of times, into the space between the two points of attachment, as shown in Fig. 6.

Not only are the shapes of the vibrations determined, but also the frequencies (the number of ups and downs per second), once the tension of the string is kept fixed. Each of the different vibrations that can be set up has its characteristic frequency; the string can vibrate only a set of selected frequencies. The lowest of these frequencies, the easiest to set up, is the one whose half wavelength just fits the distance between the fixed ends of the string.

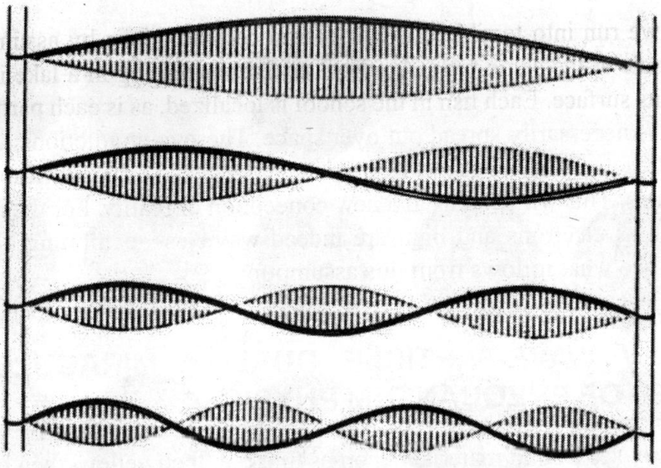

Figure 6. Waveforms of a string fixed at both ends.

The lesson learned from the string is generally true for all kinds of waves. Whenever waves are confined to a finite space, we observe special wave forms and a set of assigned frequencies that are characteristic of the system. Most musical instruments are built on this principle. Stringed instruments make use of the series of discrete frequencies characteristic of the string; a wind instrument is based on natural frequencies of air waves enclosed in the pipe of the instrument, whether it is a trumpet or an organ pipe. Another interesting example of this phenomenon is a metal plate fixed at its center. If we use a violin bow to set it in vibration, we will find again a definite set of vibrations, with characteristic frequencies and shapes. We make these shapes visible by strewing sand on the plate. The sand accumulates at places where there is no vibration, thus delineating the shapes of the vibrations in *Chladni figures,* as shown in Fig. 7. It is entirely possible to calculate the shape of these patterns and predict at which frequency of vibration they will appear. All we need to know are the shape and elastic properties of the plate.

Can electron waves be confined too? Indeed, they are confined in an atom! The positive charge of the nucleus attracts the electron and prevents it from leaving the immediate neighborhood of the nucleus. Now we come to the most astounding part of the second act of our drama. In 1926, Erwin Schrödinger calculated the shapes and frequencies of the characteristic patterns that develop when electron waves are confined by the electric attraction of the nucleus. It is a straightforward problem of the dynamics of confined waves. What did he get? As expected, he found a series of distinct vibrations, each with its own definite shape and frequency. This is a big step toward solving the riddles of the atom. It not only shows that there are indeed well-defined modes of vibration in the atom but also reveals a connection between the wave nature of the electron and the existence of discrete states in the atom. Here we are touching on the very nerve of nature. When an electron is confined to a limited region around the nucleus, wave properties of the electron permit only certain special, predetermined patterns of vibrations.

The real success of this idea came about when Schrödinger calculated the wave patterns of the hydrogen atom—the simplest of all atoms, with its single electron confined by the nucleus. The result was overwhelming. He found a series of states of vibration that correspond in every respect to the observed quantum states of the hydrogen atom. The extension of waves in space corresponded very well with the observed size of the hydrogen atom. But the most surprising and convincing result of his calculations concerned the energy of quantum states. How do we get at the energies of

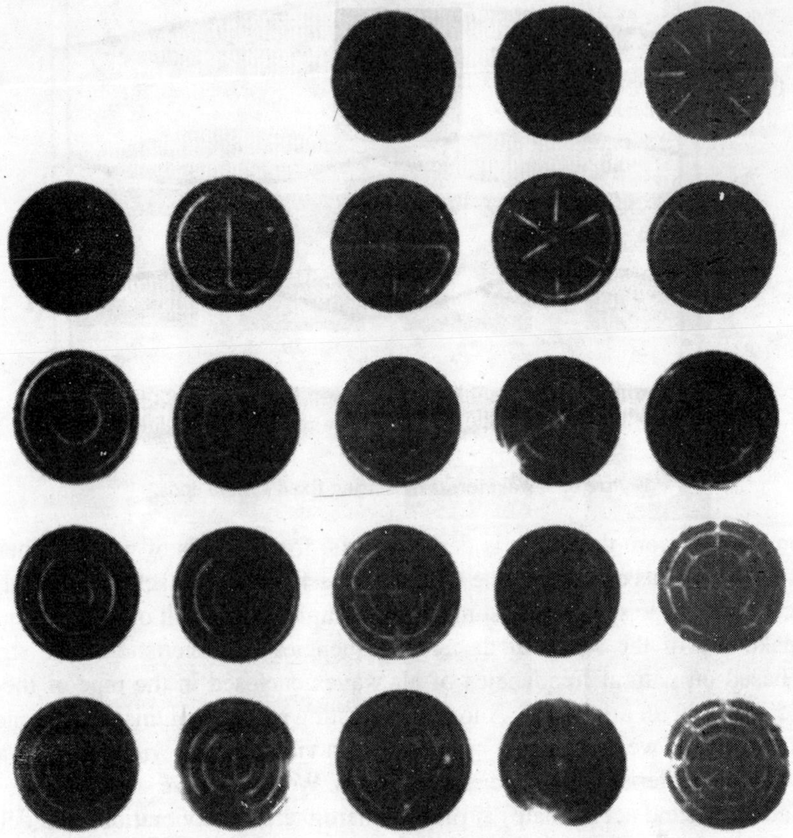

Figure 7. Vibrations of a circular metal plate fixed in the center [Chiadni figures].

the different states of vibration? We use the fundamental relation between the frequency *f* and the energy *E*: energy is frequency multiplied by Planck's constant.

When Schrödinger calculated the frequencies of these vibrations and multiplied them by Planck's constant, he got exactly the observed energies of the quantum states of hydrogen! An incredible success. Everyone who contemplates this fantastic discovery sympathizes with Italian physicist Enrico Fermi, who used to say, when presenting this calculation in his lectures, *It has no business to fit so well!*

Confined electron waves in atoms cannot be observed directly. We can measure their extension, and frequencies (to be exact, we measure the differences between frequencies, which are observed as energy differences), but we cannot see them or photograph them because they would be destroyed by the light we would have to use in order to see them. This will be the topic of the next act in our presentation.

One can make models, however, representing the results of Schrödinger's calculations. It is instructive and impressive to look at pictures of these models. Figure 8 shows the electron wave patterns, in order of increasing frequency, or energy, the successive quantum states of an electron confined by a nucleus. The lowest state, the ground state, is the simplest one; it has spherical symmetry. The next states have figure-eight forms; the higher the frequency, the more involved the patterns.

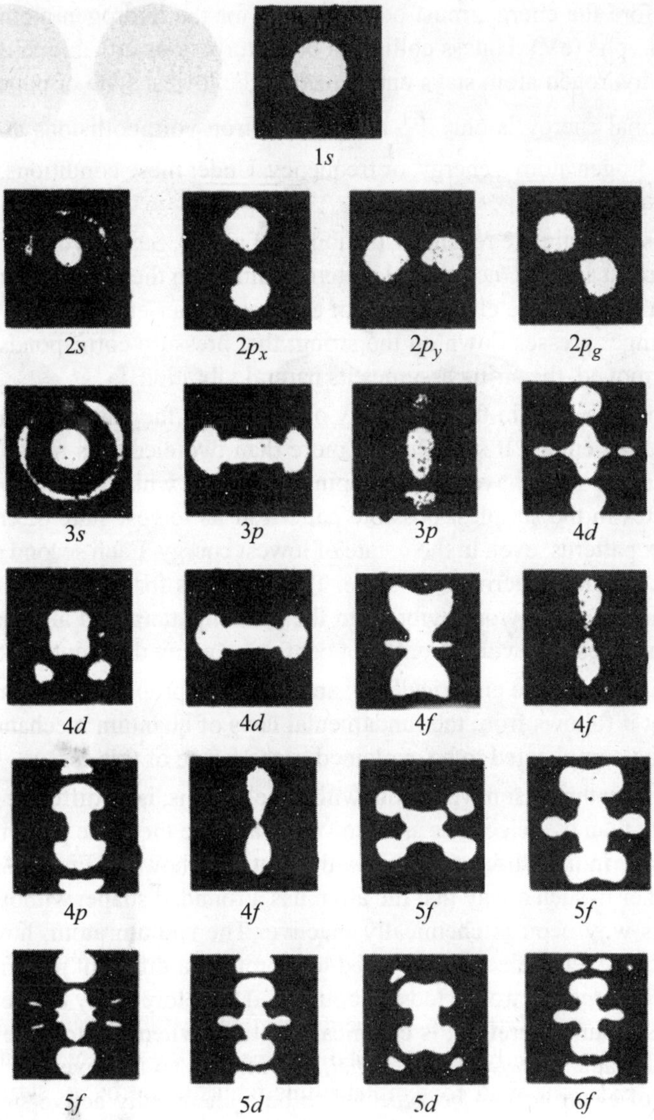

Figure 8. Mathematical models of vibrations of electron waves confined by a central charge. [Adapted from Introduction to Atomic Spectra by H.E. White, McGraw-Hill, 1934.]

These are the fundamental patterns of matter. They are the shapes, the only shapes, that the electron vibration can assume under conditions prevailing in atoms that is, under the influence of a central force (the attraction of the nucleus) that keeps the electron confined. We may call them *the primal shapes of nature.* These patterns are the original building blocks from which nature combines and forms everything we find around us.

Let us see how the other properties of atoms are explained in terms of these patterns. We begin with the stability of atoms and the regeneration of their original shapes and properties after being disturbed. Take the hydrogen atom: under ordinary conditions, when temperatures and pressures are not too high, the atom is found in the quantum state of lowest energy: it vibrates in the simplest possible pattern, the uppermost one in Fig. 8. To change from the lowest pattern to the next, the frequency, and therefore the energy, must be increased; for the hydrogen atom, the difference in energy is 10 electron volts (eV). Unless collisions or perturbations are able to deliver that amount of energy or more, a hydrogen atom stays unchanged in the lowest state or pattern. Since, at room temperature, the thermal energy is only $\left(\frac{1}{40}\right)^{\text{th}}$ of an electron volt, collisions at room temperature cannot change the hydrogen atom's energy, or frequency. Under these conditions, the atom is stable. It remains in the lowest pattern.

Wave patterns also explain the regeneration ability of atoms. Say a hydrogen atom is deformed during a collision; it must assume the original pattern again when the perturbation is over. There are no other patterns available for the electron except at much higher energies. Let us look at a violin string: the player's finger presses down on the string; that pressure corresponds to a deformation. When the finger is removed, the string assumes its natural vibration.

How do these patterns explain the specificity of atoms and the variety of elements? Here, the *Pauli exclusion principle* enters. It says that no more than two electrons may vibrate in the same pattern and that those two must have opposite spin. Therefore, while the hydrogen atom, with its single electron vibrates in the simplest possible pattern in its lowest state of energy, other atoms exhibit more complex patterns, even in their state of lowest energy. Each second additional electron must assume the next higher pattern of the scale. This accounts for the variety of nature, because otherwise electrons in all atoms would vibrate in the lowest pattern and all elements would have roughly the same properties: the world would not contain as many different materials and forms.

Pauli discovered his principle empirically by analyzing the properties of atoms. But later on he was able to show that it follows from the fundamental laws of quantum mechanics. Unfortunately, his conclusions are too complicated to be explained in the frame of this essay.

The Pauli principle is the reason why neon, with ten electrons, is so different from sodium, with eleven. Since no more than two electrons are allowed to assume the same pattern, the ten electrons in the neon atom vibrate in five different patterns, the first five shown in Fig. 8. As it happens, these five patterns fit together in such a way that the atom has a rounded shape, without any indentations or elevations. That is why neon is chemically inactive. The sodium atom, however, is irregular because the additional eleventh electron is forced to vibrate in a different pattern, according to the Pauli principle. This *isolated* electron is loosely bound and therefore ready to interact with electrons around other atoms. Sodium, therefore, is chemically active. When an atom's electron patterns fit into a nicely rounded shape, or *closed atomic shell,* the atom is an inactive gas. The atom with the next higher number of electrons can be expected to be a chemically active substance, with very different properties coming from the new occupied pattern. Atoms having one or a few electrons over and above the closed shell are those that form metals, such as sodium. The additional electrons jump easily from atom to atom and therefore give rise to the high electric and thermal conductivity of metals. That is why they feel cold when we touch them. Body heat is quickly carried away by the additional electrons.

Atomic structure provides interesting examples of how quantity determines quality. From the number of electrons in an atom, we can deduce the kind of properties a particular atom will have in its ground state by combining the various patterns of vibration. A telling instance happened when plutonium was produced in nuclear reactions for the first time. (Plutonium does not exist in nature because it is radioactive and decays into other elements in about 40,000 years.) The amounts of plutonium that were created were too small to permit any observation about its properties; however, it was known that the plutonium atom contained ninety-four electrons. Thus, we were able to determine its properties from the patterns of the electrons vibrating around the nucleus. It was found that it must be a metal; it must have specific weight of about 20 grams per cubic centimeter; it must be brown; it must have a certain electric and thermal conductivity and elasticity. All this information could be obtained from the single number ninety-four. It was a great moment when the first cubic millimeter of plutonium was finally produced and these projections were confirmed. This incident illustrates the tremendous predictive power of atomic quantum mechanics.

The connection between quantity (number of electrons) and quality (properties of the atom) points to an essential difference between quantum mechanics and classical physics. In the classical science, there is no well-defined state of lowest energy. If the electrons around the plutonium nucleus followed classical laws, the state of the plutonium atom would depend on the initial positions and velocities of the ninety-four electrons. No two plutonium atoms would be really identical. The same would be true, even for the hydrogen atom, with one electron since there are an infinite number of different ways for the one electron to circle the nucleus with a certain energy. There isn't even a lowest energy. The wave nature of electrons has changed all this. There is only *one* wave of lowest frequency (frequency determines energy), the one at the top of Fig. 8. In plutonium (as in any other atom) there is only one way to put the electrons in pairs into the wave forms such as those shown in Fig. 8—at the lowest possible energy.

We therefore conclude that atoms with a fixed number of electrons are all alike; they behave identically. Gold, with seventy-nine electrons, has the same properties in the United States as in the Soviet Union because there is only one unique way of putting seventy-nine electrons into patterns of the kind shown in Fig. 8 and obtain the lowest possible energy, the ground state. There are, of course, other ways to distribute the seventy-nine electrons, but then we would obtain higher energies: higher quantum states of gold. They represent a series of energy values—the spectrum of gold—the lowest of which is 2.3 eV above the ground state. These states occur only at very high temperatures, but there will always be the same series of states.

Quantum mechanics introduced a concept that does not exist in classical physics: the concept of exact identity between atomic systems. In the classical picture we can always imagine arbitrary, small differences between atoms. In quantum mechanics, these are either exactly alike or measurably different—an important trait of nature.

The patterns shown in Fig. 8 and their inherent symmetries determine the behaviour of atoms. They also help us to understand how atoms combine to form molecules and how atoms form regular arrays in crystals. The simple beauty of a crystal reflects the fundamental shapes of the atomic patterns on a larger scale. Ultimately, all the regularities of form and structure that we see in nature, ranging from molecular structure to the hexagonal shape of a snowflake or the intricate symmetries of living forms in flowers and animals, are based on the symmetries of these atomic patterns. The world abounds in characteristic forms and shapes, but only quantum mechanics can tell us—at least in principle—where they all come from.

Electron patterns also explain why 6000 calories of heat are produced from burning a kilogram of coal. Energy is gained when carbon dioxide is formed from a carbon atom (in coal) and an oxygen molecule (in air) because the electron patterns of carbon and oxygen fit together better in carbon dioxide and therefore have lower energy than in the oxygen molecule and the carbon crystal. The amount can be calculated, and when multiplied by the number of atoms in a kilogram of coal, the result is 6000 calories. The reason why oxygen combines so easily with hydrogen to give water is that the eight electrons of oxygen are just two electrons short of the closed shell that we found in neon. Thus, they form a round shape when the two holes are filled with the electrons of two hydrogen atoms, so we get H_2O!

Even the phenomena of life are based upon the characteristic electron patterns and their stability. The structure of DNA comes from the specific way in which electron patterns of carbon, hydrogen, oxygen, nitrogen, and other atoms fit together and form the well-known helix. They are resistant to thermal motion at room temperature; more energy is required to change those patterns and to disentangle them.

Consequently, the DNA helix has its own stability, which explains why the same flowers emerge every spring and children resemble their parents.

Schrödinger's calculations and the subsequent tremendous development of our understanding of the properties of atoms, molecules, gases, liquids, and solids have shown that the structure of matter in our environment is based on the electric force between atomic nuclei and electrons and how the wave-particle duality of the electrons respond to it.

There is an interesting historical relation. Kepler thought that planetary orbits in the solar system had very definite size ratios. He connected them with the size ratios of simple polyhedrons. In fact, it is not so, for the sizes of the planetary orbits are, to a large extent, accidental. But now Kepler's idea reappears in the atomic world, where it is clearly understood as a typical property of electrically confined waves. There is a kind of harmony among the different frequencies of the electron patterns, a *chord* that can be considered a rebirth of the Pythagorean harmony of the spheres. These harmonies may not be particularly pleasing as musical chords to our ears, but they certainly appeal to our intellectual ears when they offer so much insight into the structure of the material world.

ACT III: SIGNIFICANCE OF WAVE-PARTICLE DUALITY AND THE NEW REALITY OF THE QUANTUM STATE

Act II demonstrated that the wave-particle duality plays a decisive role in the dynamics of atoms. The wave nature of electrons permitted us to understand the puzzling properties of atoms, such as the stability, specificity, and discrete energy states. But the fundamental question remains: How can the electron be a particle and a wave at the same time? It is not a true particle, because it exhibits wave properties, and it is not a true wave, since its amplitude is fixed in the following sense: if the electron assumes one of the wave patterns of Fig. 8, it vibrates with a certain intensity. It cannot vibrate with half that intensity, since that would correspond to half an electron and there is no such thing. It is neither a wave nor a particle.

Its wave nature is exhibited not only in the patterns of Fig. 8. Beams of free electrons that are not bound to atoms also show wave and particle properties. As mentioned before, electrons emerging from two nearby points exhibit interference under certain conditions, similar to the water waves in Fig. 2. On the other hand, the intensity of the wave representing an electron beam cannot assume any value, as an ordinary wave would, but only those values corresponding to a definite number of particles. Furthermore, when two electron beams cross each other and the electrons of one beam are scattered by the other, scattered electrons have just the momenta and energies we would expect from two *particles* colliding; however, the directions in which they emerge are distributed according to the scattering laws of waves!

Why can't we find out, by exact observations, whether an electron is a wave or a particle? Couldn't we try to follow the electron motion in the atom, step by step, to see whether it is a localized particle moving along an orbit or a vibrating wave spread over a region? This question goes to the heart of quantum physics. One of the features of classical physics is the *divisibility* of physical processes, the idea that a process can be thought of as a succession of particular partial processes. According to that idea, each process can be followed, step by step, in time and space. The orbit of an electron around the nucleus would be thought of as a succession of small displacements. In contrast, a wave would be thought of as a continuous oscillatory movement spread over a region of space.

Here we meet one of the basic concepts of quantum mechanics. There are fundamental reasons why such tracing cannot be carried out. The idea of divisibility of motion is in serious trouble, and the trouble comes from the graininess of light. In order to see the details of the electron motion in the atom, we must use light waves with a very small wavelength, the size of the atom. Such light, however, has a high frequency and therefore a large energy quantum. In fact, light whose wavelength is as small as an atom has a quantum of energy that would be far more than enough to tear the electron away from the atom. When light hits an electron, it will knock it out of the atom and destroy the very object of our examination. This is so, not only when light is used to observe an object, but quite generally all measurements that could be of use to decide between the wave and the particle nature of the electron have the same result. If we attempt to perform these measurements, the object changes its state completely.

The quantum nature of light or of any other means of observation introduces a coarseness that makes it impossible to decide between wave and particle. It does not allow us to subdivide the atomic orbit into a succession of partial motions, whether they are particle displacements or wave oscillations. If we force a subdivision of the process and try to look more accurately at the wave in order to find out where the electron *really* is, we will have destroyed the subtle individuality of the atom that give rise to all its characteristic properties. The argument that an entity cannot be both a wave and a particle, because a wave is spread over space whereas a particle is concentrated at a definite position, no longer holds within atomic dimensions. The habitual concepts of *location* and *spread* are no longer applicable; any effort to observe them would destroy the object.

Here quantum mechanics enters as a new and revolutionary concept. The great new insight of quantum physics is the recognition that individual states of the atom—we call them quantum states—are each indivisible wholes that exist only as long as they are not attacked by penetrating means of observation. The quantum state is the state the atom assumes when it is left alone to adjust itself to the prevailing conditions. If we try to measure the location of an electron with those penetrating means, we not only will destroy the quantum state, but will indeed find the electron at some place, but that place is not predictable.

Electrons do not have a predictable position in the atom, because *position* is a classical concept, not directly applicable in the atomic realm. The results of our physical measurements, however, are always classical magnitudes; that is, we measure magnitudes of position, velocity, and energy. Therefore, we should not be surprised if, in some cases, all we can predict are probabilities; furthermore, the quantum state has been thoroughly changed after measurement. What we should be surprised about is that physicists were able to construct a mathematical formalism (quantum mechanics) for dealing with quantum reality that allows us to calculate those probabilities.

The restrictions to the application of classical concepts, such as position, velocity, and momentum, are codified by the famous Heisenberg uncertainty relations. When we ask what is the position or the velocity of an electron in a quantum state, we cannot always expect a definite answer. The Heisenberg uncertainty relations tell us, for example, to what degree, within a quantum state of an atom, the location and the velocity of an electron is left undetermined. They are the warning posts that advise us how far we are allowed to apply habitual classical concepts before we get into trouble with reality. *Up to here and no further may you apply your old-fashioned concepts.*

An important consequence follows. Certain statements about the atom must remain probabilistic because our concepts—for example, *electron position*—are not appropriate in the atomic quantum state. If we force the concept on the atom, we get only a probability for an answer: the electron will be found here with this probability and there with another probability. The phrase, *If you ask a silly question, you get a silly answer,* should be applied in this form: *If you ask an inappropriate question, you get a probabilistic answer.* Outside the confines of the Heisenberg relations, however, for objects much larger than atoms, we can still apply classical concepts without trouble. The orbit of an earth satellite can be calculated and predicted with practically unlimited accuracy.

The quantum state has a more subtle individuality than ordinary macroscopic states of matter. Reality exhibits different, seemingly contradictory properties, when examined in different ways. In the words of David Bohm: *The electron may be regarded as an entity that has potentialities for developing either particle properties or wave properties, depending on the type of instruments with which it interacts.*

Some philosophers like to say that, according to quantum mechanics, the world around us is not real and depends on our minds, because our observations disturb and change the objects. I disagree. True enough, typical quantum properties unfold only if atoms are not subject to methods of observation that penetrate their interior. But they are not exposed to such perturbations under ordinary conditions, and that is why characteristic quantum properties are manifest everywhere: in the color of objects; in the structure of crystals; in the mechanical, electrical, and chemical properties of different substances; and, in the last instance, in the phenomena of life. In most of our studies of material properties, we try to maintain, not to disturb, the quantum states, since they gives rise to the reality of interest to us. Sure enough, if an atomic phenomenon is observed too closely by our clumsy instruments we destroy the very object of interest, its quantum state, with all its seemingly contradictory properties. But just these properties are responsible for the behavior and appearance of matter in our environment. Hence, it is the quantum state that is *real* and exists independently from us. If close observations destroy it, and if we cannot describe it with our everyday concepts, too bad for those observations and concepts! What is real are the objects and their quantum states, since they represent nature. The inapplicability of some of our concepts does not make the objects less real.

Although an unambiguous description of atomic reality is impossible with our ordinary concepts of material processes, we have an unambiguous mathematical scheme that gives the right predictions and probabilities for any observation. A physical description, however, must make use of such seemingly contradictory concepts as the wave picture and the particle picture. Neither can give a complete description, but both are necessary to understand the phenomenon in its entirety. Each of these mutually exclusive perspectives gives only a partial, one-sided view; both are needed for a full understanding. Niels Bohr introduced the term *complementarity* to describe this novel situation. This is why we face a richer reality in the quantum world than the classical picture has ever been able to describe.

ACT IV: THE QUANTUM LADDER

An extension of quantum mechanics to nuclear and subnuclear phenomena and to the history of the universe

After its conception, quantum mechanics developed in two distinct directions. The first, broadest development was toward a better understanding of the properties of agglomeration of atoms, such as molecules, liquids, and solids. Quantum mechanics turned out to be the key to many previously unexplained properties of materials—such as their behaviour at high and low temperatures—and of chemical reactions and molecule formations. Knowing the quantum states of atoms made it possible to understand what happens when atoms are joined to molecules, liquids, or solids. Electrical and optical properties of metals, for example, turn out to be consequences of peculiar propertics of electronic quantum states in the regular array of atoms in metal. The disappearance of electric resistance in some metals in low temperature (superconductivity) could now be explained. New forms of electric conductors, semiconductors, for example, were found and understood.

These new insights led to the invention of the transistors and the microchip, which revolutionized electronic technology.

The other development of quantum mechanics is less broad but deeper: it goes into the depth of the structure of matter, into the structure of the atomic nucleus, and later into the structure of the constituents of the nucleus. The same quantum mechanics that once explained atomic behavior is now found to operate on a deeper level within atomic nuclei and even within their constituents.

Let us look at the atomic nucleus. It is a very small, massive, positively charged entity, roughly speaking, about 10,000 times smaller in diameter than the atom. It is responsible for the structure of atoms, since its positive electric charge attracts electrons and keeps them confined in the atom. That confinement produces the characteristic quantum states, such as those depicted in Fig. 8.

Rutherford, who in 1911 discovered the internal structure of atoms, observed in 1917 that atomic nuclei also have an internal structure. He found that they contained protons. The proton itself is nothing other than the nucleus of the simplest atom, the hydrogen atom. The additional constituent contained in other nuclei was recognized by Chadwick, who in 1932 discovered the neutron—a particle very similar in mass and character to the proton, except that it carries no electric charge, whereas the proton carries one unit of positive charge. Atomic nuclei were found to be composed of protons and neutrons, with the positive charge of a nucleus provided by its protons.

With this discovery, a new branch of physics began: nuclear physics. First of all, the mere existence of a conglomerate of protons and neutrons suggests the action of a new kind of force. Electric forces could not keep protons and uncharged neutrons together; indeed, protons would be driven out because charged particles of equal sign repel each other. There must be a strongly attractive force acting between neutrons and protons to keep them together in such a small entity as the nucleus. We call it the nuclear force—such a force was indeed observed and measured.

Then, in the 1930s, a repeat performance of quantum mechanics took place. Protons and neutrons are also subject to the wave-particle duality, as are all particles. The nuclear force confines them into a small area of space. Confinement produces characteristic wave patterns, not unlike those shown in Fig. 8. Thus, nuclei, like atoms, are expected to exhibit typical quantum states, with properties such as stability and regeneration, except that the relevant energies and energy differences are much larger.

Here we come to an important principle of quantum mechanics: the energy—size relation. It refers to the energies of quantum states of a confined system of particles. The smaller the size of the system, the greater are those energies and energy differences between quantum states. They also depend on the mass greater of the constituents: the higher the mass, the smaller the energies. We will concentrate on the size dependence here; it is the more important one.

The nucleus is very much smaller than the atom. Therefore, the energies of the quantum states and the energy intervals between them are much greater. We call nuclear physics the next higher rung of the quantum ladder, the lowest rung being atomic and molecular physics. Indeed, it takes millions of electron volts (a unit of one million is a mega-electron-volt or MeV) to lift a nucleus from its lowest quantum state to the next one, whereas in an atom, it takes only a few electron volts (roughly between one and ten). Nuclear reactions imply energy exchanges of millions of electron volts, whereas chemical reactions between atoms deal with exchanges of only a few. This is why nuclear explosives are so much more powerful than ordinary ones.

When a nucleus performs a transition from a higher to a lower quantum state, the energy difference is emitted in the form of a photon, just as it is in an atom, but the *nuclear light* called gamma rays is much more energetic than *atomic light.* In the nuclear regime, another kind of emission appears: sometimes, instead of a photon, a pair of particles is emitted, an electron and a neutrino, in a process known as radioactivity. It could not have occurred in the atomic regime, because it requires energies not available in atomic processes. An electron and a neutrino must be created. According to Einstein's famous equivalence of energy and mass—$E = mc^2$—the creation of an electron requires the availability of at least ½ MeV, which is the energy corresponding to the electron mass. A neutrino is massless; it does not require a minimum energy to be created. Such energies as the mass energy of the electron are not available in transitions between atomic quantum states, but they are available in transitions between nuclear quantum states.

Because of the large difference in energy, nuclear processes occur under very different conditions from atomic or molecular ones, processes that require energy exchanges of only a few electron volts. This is approximately the energy of photons coming from the sun. Therefore, the surface of the earth abounds with phenomena based on interactions of atoms and molecules. The atomic nuclei necessarily stay in their lowest quantum state under such conditions and therefore act as unchanging entities. In our terrestrial world, nuclei never get naturally excited; they are never lifted into their higher quantum states; and they appear completely stable. Nuclear physics is dormant on earth. We observe nuclear phenomena only when we construct machines—cyclotrons, synchrotrons, and other particle accelerators or nuclear reactors—that can produce the necessary energies.

The nuclear processes found on earth, apart from man-made ones, are the decays of some naturally radioactive substances, such as radium, thorium, or uranium, responsible for much of the heat produced deep in the interior of the earth. But they are not of terrestrial origin. They were produced when the matter that now constitutes the earth was ejected with tremendous energies by a supernova explosion seven billion years ago. They represent the last members of the great nuclear fire that created the elements.

The places of nuclear processes in nature are those regions where millions of electron volts are available. That happens in the interior of stars and in the explosions of supernovas. Indeed, nuclear reactions in the center of the sun are the sources of solar heat. They are the nuclear fire in which hydrogen nuclei are burnt to helium nuclei (not unlike the burning of coal to carbon dioxide in a coal fire). It will last four or five billion years before burning out.

There is an even higher rung of the quantum ladder. Certain strange occurrences in cosmic rays, unaccountable on the basis of nuclear physics, had already been observed in the 1930s. The decisive discovery was made in 1950, when E. Fermi and his collaborators found that the proton has an excited quantum state. Soon thereafter, many more quantum states of the proton and neutron were discovered, pointing toward an internal structure of nuclear particles. Like the atom and the nucleus, they seemed to be a composite system made of some constituent particles that are combined by some kind of force and form typical wave patterns that lead to discrete quantum states.

These discoveries opened up a new world of phenomena: the third rung of the quantum ladder, the subnuclear realm. As expected from the energy-size relation, the relevant energies are much larger. They are measured by the giga-electronvolt (GeV), that is, a billion electron volts. The first excited state of the proton is 0.3 GeV above the ground state. This is the reason why subnuclear phenomena are dormant, even in the nuclear realm. New types of accelerators of many GeV are needed to observe them.

It turned out to be a most surprising world, not yet fully exploited and understood. The constituents of the proton and neutron were identified. They received the ugly name *quarks*. A very strong new force was identified that holds quarks together within the proton. So far, the picture turned out to be similar to the one at the lower ranges of the quantum ladder. The different quantum states of the proton and neutron could again be interpreted as confined waves of quarks.

There are a number of unexpected new features that are to some extent still surprising and difficult to explain. For example, it seems to be impossible to extricate a quark from a proton or neutron. They cannot be liberated from confining bonds, but they can be observed to move within the proton and neutron in contrast to the situation on the lower rungs: the atom can be decomposed into nucleons and electrons; the nucleus can be decomposed into protons and neutrons.

Furthermore, a number of new short-lived entities show up: mesons and heavy electrons. Both entities need gigavolt energies to be created. They exist only for a short time. They transform themselves quickly into other known particles, such as ordinary electrons, neutrinos, and photons.

Antimatter comes to play an important role on the third rung: What is it? For every particle there is a corresponding antiparticle with exactly the same mass but opposite charge. There are antiatoms made up of positive antielectrons and negative antinuclei composed of antiprotons and antineutrons. But antimatter cannot exist together with ordinary matter, because an antiparticle and a particle annihilate each other on contrast and transform themselves into energy. This is a transformation of one kind of energy into another kind of energy. Inversely, if energy in sufficient concentration is available, it can transform itself into a pair consisting of particle and antiparticle. The energy concentrations necessary for such processes are very high and available to a sufficient degree only at the third rung of the quantum ladder, so as to make antimatter an important ingredient of the subnuclear phenomena.

The existence of antimatter is another triumph of quantum mechanics. In the early 1930s, the British physicist P.A.M. Dirac showed that the equations of quantum mechanics and relativity theory required that there be an antiparticle to every particle. It also followed that particle-antiparticle pairs can be created from other forms of energy and annihilated into energy. Dirac could not believe that result because, at that time, antimatter had not yet been found. He was deeply astonished by these strange consequences of his calculations. These equations seemed to him *more intelligent than their author.* It was an unusual feat of the human mind to predict the existence of antimatter by pure intellectual insight before it was actually observed in nature.

Let us return to the subnuclear world, which is revealed with the help of very high energy beams produced by powerful accelerators. When a particle carrying many gigavolts of energy impinges upon a target, an explosion of newly created particles takes place: protons, antiprotons, electrons, antielectrons, mesons, and other particles emerge all over. Figure 9 gives a telling picture of the variety of such events.

In the past three decades, the quantum ladder has acquired a new significance. There seems to be a connection between the infinitely small, the elementary constituents of matter, and the infinitely large, the origin and history of the universe as a whole. The connecting link is the discovery of the expansion of the universe. There are a number of indications that celestial objects move away. If this is so, we would expect that in the distant past, the concentration of mass ought to have been much greater. Indeed, about fifteen billion years ago, all space should have been filled with almost infinitely concentrated matter. Such concentration would have to be accompanied by extremely high pressure and temperature. Thus, we are led to the tentative conclusion that the universe started with a *primal bang,* where all space was filled with highly concentrated matter of extremely high temperature and energy, which decompressed and diluted itself, first quickly and then more slowly, until it reached its present state.

One of the most uncanny discoveries of the last decade was the observation of a pervading cool radiation coming from all directions in space, with all the properties of an optical reverberation of the radiation emitted during the initial hot period in the life of the universe. It was another indication of the great happening at the beginning of the universe.

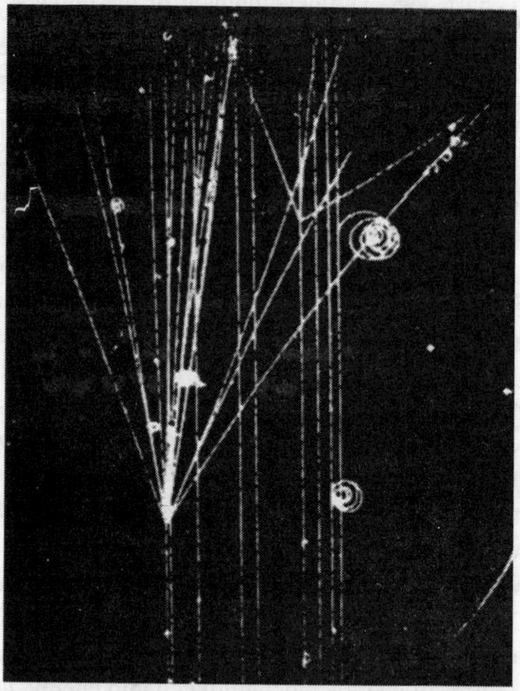

Figure 9. A bubble-chamber picture of a 25-GeV proton beam impinging on a hydrogen atom and producing new particles. [Courtesy of CERN, Geneva, Switzerland.]

Developments from the primal bang to our present state are intimately connected to the quantum ladder. In the first instants after the bang, energy concentration was indeed so high that the phenomena of the third rung of the quantum ladder was the order of the day, or more appropriately, the order of the microsecond. At that time, space was filled with quarks, antiquarks, gluons, electrons, antielectrons, and neutrinos and with an intense light radiation of extremely energetic photons. Then, as the universe expanded, energy concentration diminished rapidly. Particles and antiparticles annihilated, producing more radiation and short-lived mesons. After a few microseconds, a slight surplus of quarks over antiquarks (not quite understood today) remained, and the quarks assembled into protons and neutrons. Then the temperature dropped further because of the expansion. After about 10 seconds, some protons and neutrons assembled into helium nuclei. We now step down to the second rung of the quantum ladder: the nuclear realm. It probably took several hundred thousand years to cool down sufficiently to allow electrons to be bound by the nuclei to form atoms, arriving at the first rung. It then took billions of years to reach the present situation. Galaxies and stars were formed, and on the surface of some planets, the atomic and molecular world developed the great variety of forms and materials we now see on earth.

This view of the development of the universe is to some extent hypothetical and unproven. It is much less certain than the validity of quantum mechanics and our knowledge of the properties of molecules, atoms, and nuclei.

The quantum ladder symbolized the steady increase in depth of our knowledge about the structure of matter. It started with ordinary chunks of matter composed of atoms and molecules, whose structure was revealed on the first rung. It then proceeded to the structure of atomic nuclei and—finally today—to the quarks as constituents of the proton or neutron. There may exist yet unknown higher rungs of the ladder. The development of the universe happened in the opposite direction, down the quantum ladder, as it were. The very beginnings are shrouded in the unknown, since we know nothing about the higher rungs of the ladder that may come into play in the first, extremely short time intervals. But we have clearer ideas about the evolutionary history of the world from a microsecond after the primal bang to the present universe. It was a series of gradual steps, from the simple to the complicated, from the unordered to the organized, from the hot, formless gas of elementary particles to the cooler, structured atoms and molecules; still further to the more structured liquids and solids, and finally to sophisticated self-reproducing living organisms. Steps in this history that are still hidden in ignorance and guesswork include what happened immediately after the *primal bang,* the mechanisms forming galaxies and clusters of galaxies, and the details of the origin and development of life on our planet.

The drama of the development of quantum mechanics has not ended, but we have arrived at the present limits of our knowledge. This limit lies far beyond what could have been expected at the turn of the century, when quantum mechanics was conceived. Ingenious use of accelerators and instrumentation revealed even deeper structures of matter, still subject to the rules of quantum mechanics in more developed forms. It is certainly an encouraging thought that we were able, in our laboratories, not only to understand the deeper structure of matter, but also to recreate the processes and phenomena that, in all probability, happened in nature only in the first instants of its existence.

Part I

Mathematical Preliminaries

1 The Dirac Delta Function

... Strictly, of course, $\delta(x)$ is not a proper function of x, but can be regarded only as a limit of a certain sequence of functions. All the same one can use $\delta(x)$ as though it were a proper function for practically all the purposes of quantum mechanics without getting incorrect results. One can also use the differential coefficients of $\delta(x)$, namely, $\delta'(x)$, $\delta''(x)$,... , which are even more discontinuous and less 'proper' than $\delta(x)$ itself.

—P.A.M. DIRAC in *The Physical Interpretation of Quantum Dynamics*, Proceedings of the Royal Society of London (A) **113**, 621–641 (1926).

1.1 INTRODUCTION

The Dirac delta function is used so extensively in quantum mechanics that we felt we should discuss it right in the beginning of the book, rather than relegating it to an appendix!

The Dirac delta function is defined through the equations

$$\delta(x-a) = 0 \qquad x \neq a \tag{1}$$

$$\int_{a-\alpha}^{a+\beta} \delta(x-a)\,dx = 1 \tag{2}$$

where $\alpha, \beta > 0$. Thus the delta function has an infinite value at $x = a$ such that the area under the curve is unity. For an arbitrary function that is continuous at $x = a$, we have

$$\int_{a-\alpha}^{a+\beta} f(x)\,\delta(x-a)\,dx = f(a)\int_{a-\alpha}^{a+\beta} \delta(x-a)\,dx \qquad \text{[using Eq. (1)]}$$

$$= f(a) \tag{3}$$

It is readily seen that if x has the dimension of length, $\delta(x-a)$ would have the dimension of inverse length. Similarly, if x has the dimension of energy then $\delta(x-a)$ would have the dimension of $(\text{energy})^{-1}$.

1.2 REPRESENTATIONS OF THE DIRAC DELTA FUNCTION

There are many representations of the Dirac delta function. Perhaps the simplest representation is the limiting form of the rectangle function $R_\sigma(x)$ defined through the following equation:

$$R_\sigma(x) = \frac{1}{2\sigma} \quad \text{for} \quad -\sigma < x - a < \sigma$$
$$= 0 \quad \text{for} \quad |x - a| > \sigma \tag{4}$$

The function $R_\sigma(x)$ is plotted in Fig. 1.1 for various values of σ.

Now, $$\int_{-\infty}^{+\infty} R_\sigma(x)\,dx = \frac{1}{2\sigma}\int_{a-\sigma}^{a+\sigma} dx = 1 \qquad \text{(irrespective of the value of } \sigma)$$

For $\sigma \to 0$, the function $R_\sigma(x)$ becomes more and more sharply peaked but the area under the curve remains unity. In the limit of $\sigma \to 0$, the function $R_\sigma(x)$ has all the properties of the delta function and we may write

$$\delta(x-a) = \lim_{\sigma\to 0} R_\sigma(x) \tag{5}$$

Now,

$$\int_{-\infty}^{+\infty} f(x)\,R_\sigma(x)\,dx = \frac{1}{2\sigma}\int_{a-\sigma}^{a+\sigma} f(x)\,dx$$

We assume the function $f(x)$ to be continuous at $x = a$. Thus when $\sigma \to 0$, in the infinitesimal interval $-\sigma < x - a < \sigma$, $f(x)$ may be assumed to be a constant $[= f(a)]$ and taken out of the integral. Thus

$$\int_{-\infty}^{+\infty} f(x)\,\delta(x-a)\,dx = \lim_{\sigma\to 0}\int_{-\infty}^{+\infty} f(x)\,R_\sigma(x)\,dx$$

$$= \lim_{\sigma\to 0}\frac{1}{2\sigma} f(a)\int_{a-\sigma}^{a+\sigma} dx$$

$$= f(a)$$

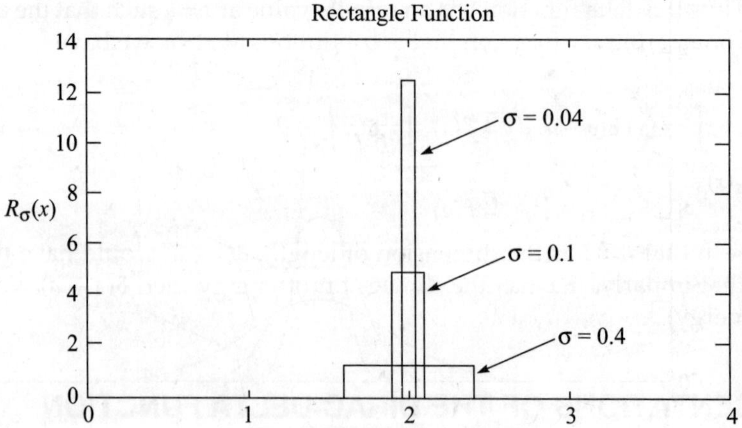

Figure 1.1: Plots of $R_\sigma(x)$ for $a = 2$ and $\sigma = 0.04$, 0.1 and 0.4. In each case the area under the curve is unity. For $\sigma \to 0$, the function $R_\sigma(x)$ has all the properties of the Dirac delta function.

Another very useful representation of the delta function is the limiting form of the Gaussian function (see Fig. 1.2)

$$G_\sigma(x) = \frac{1}{\sigma\sqrt{2\pi}} \exp\left[-\frac{(x-a)^2}{2\sigma^2}\right], \sigma > 0 \quad (6)$$

Since,

$$\int_{-\infty}^{+\infty} e^{-\alpha x^2 + \beta x}\, dx = \sqrt{\frac{\pi}{\alpha}} \exp\left[\frac{\beta^2}{4\alpha}\right]$$

[see Appendix A], we readily get

$$\int_{-\infty}^{+\infty} G_\sigma(x)\, dx = 1 \quad (7)$$

irrespective of the value of σ which is assumed to be positive. Further, the function has a width $\sim\sigma$ and at $x = a$, it has a value $1/(\sigma\sqrt{2\pi})$. Thus, in the limit of $\sigma \to 0$, $G_\sigma(x)$ has all the properties of the delta function (see Fig. 1.2) and we therefore have the following representation of the delta function.

$$\delta(x-a) = \lim_{\sigma \to 0} \frac{1}{\sigma\sqrt{2\pi}} \exp\left[-\frac{(x-a)^2}{2\sigma^2}\right], \sigma > 0 \quad (8)$$

Another important representation of the Dirac delta function is through the following integral:

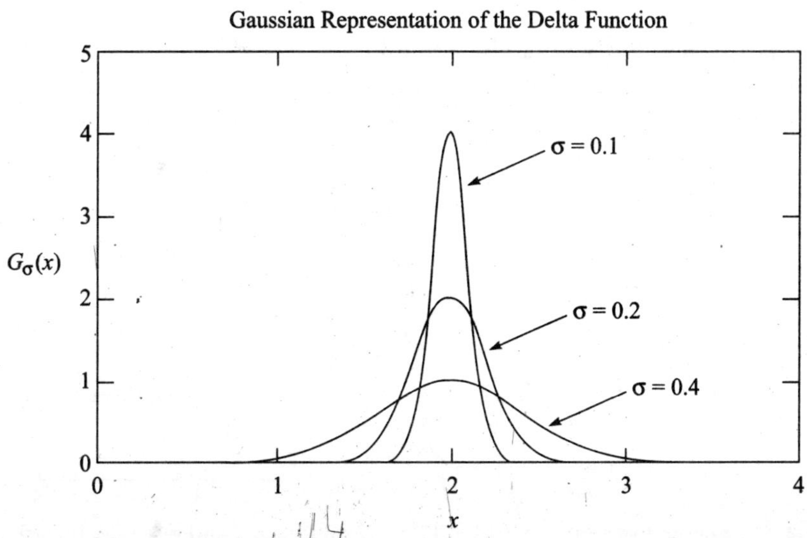

Figure 1.2: Plots of $G_\sigma(x)$ for $a = 2$ and $\sigma = 0.04$, 0.2 and 0.4. In each case the area under the curve is unity. For $\sigma \to 0$, the function $G_\sigma(x)$ has all the properties of the Dirac delta function.

$$\delta(x-a) = \frac{1}{2\pi}\int_{-\infty}^{+\infty} e^{\pm ik(x-a)}\,dk \quad \text{integral representation} \tag{9}$$

In order to prove Eq. (9) we first note that

$$\frac{1}{2\pi}\int_{-g}^{+g} e^{\pm ik(x-a)}\,dk = \frac{\sin g(x-a)}{\pi(x-a)} \tag{10}$$

Using Laplace transform or complex variable techniques one can show that (see Appendix B of Ref. 4):

$$\frac{1}{\pi}\int_{-\infty}^{+\infty} \frac{\sin g(x-a)}{(x-a)}\,dx = 1, \quad g > 0 \tag{11}$$

irrespective of the value of g which is assumed to be greater than zero.

Now, $\displaystyle\lim_{x\to 0}\frac{\sin gx}{x} = g$

Thus for a large value of g, the function

$$\frac{\sin g(x-a)}{\pi(x-a)}$$

is sharply peaked around $x = a$ (see Fig. 1.3) and has a unit area under the curve irrespective of the value of g; thus in the limit of $g \to \infty$, it has all the properties of the delta function and we may write

$$\delta(x-a) = \lim_{g\to\infty}\frac{\sin g(x-a)}{\pi(x-a)} \tag{12}$$

Equation (9) readily follows from Eqs (10) and (12).

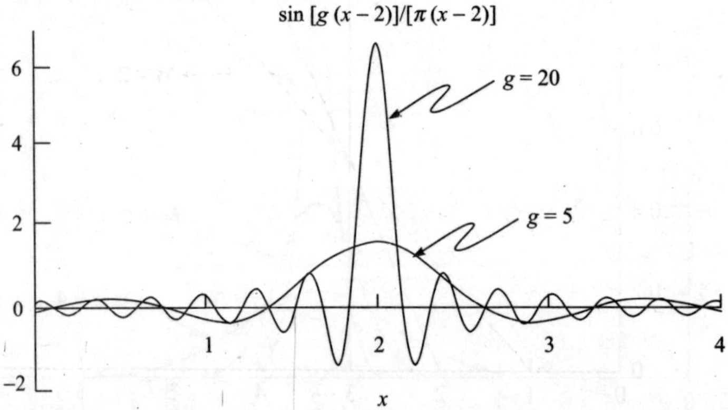

Figure 1.3: Plots of the function $[\sin g(x-a)]/[\pi(x-a)]$ for $a = 2$ and $g = 5, 20$. In each case the area under the curve is unity. For $g \to \infty$, the function is very sharply peaked at $x = a$ and has all the properties of Dirac delta function.

1.3 DERIVATIVE AT A DISCONTINUITY

We next consider a ramp function defined by the following equation (see Fig. 1.4);

$$F_\sigma(x) = \begin{cases} 0 & \text{for } x < a-\sigma \\ \frac{1}{2\sigma}(x-a+\sigma) & \text{for } |x-a| < \sigma \\ 1 & \text{for } x > a+\sigma \end{cases} \tag{13}$$

It can be readily seen that

$$\frac{dF_\sigma}{dx} = R_\sigma(x) \tag{14}$$

where $R_\sigma(x)$ is the rectangle function defined by Eq. (4).

Now,
$$\lim_{\sigma \to 0} F_\sigma(x) = H(x-a) \tag{15}$$

where $H(x-a)$ is known as the Heaviside unit step function defined by the following equation (see Fig. 1.5).

$$\left.\begin{aligned} H(x-a) &= 0 \quad x < a \\ &= 1 \quad x > a \end{aligned}\right\} \tag{16}$$

Thus,

$$\frac{d}{dx} H(x-a) = \delta(x-a) \tag{17}$$

implying that the derivative of the unit step function is the Dirac delta function which is shown in Fig. 1.5; usually $c\delta(x-a)$ is shown as a vertical arrow of height c at $x = a$. Equation (17) also follows from the fact that for a continuous function $f(x)$ (with $\alpha, \beta > 0$)

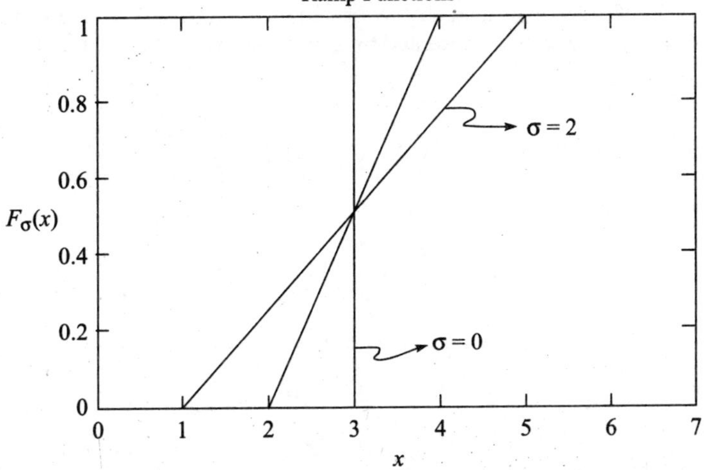

Figure 1.4: The ramp function defined by Eq. (13) for $a = 3$ and $\sigma = 0$, 1 and 2.

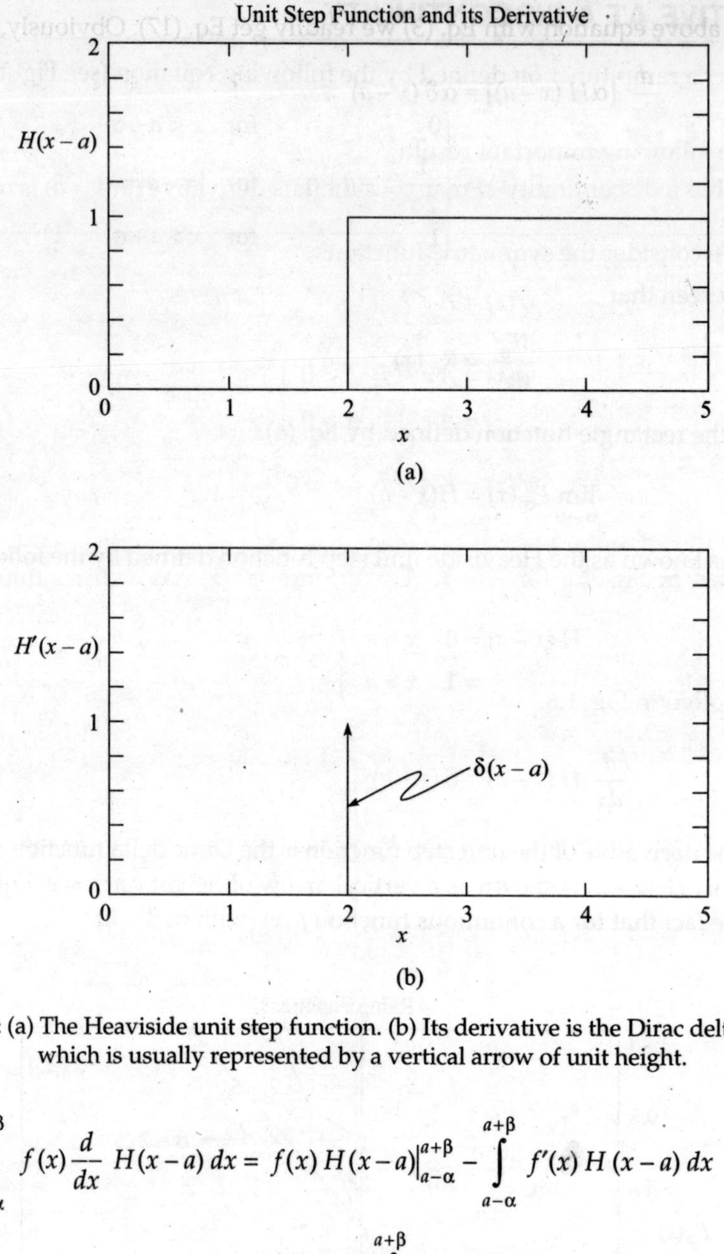

Figure 1.5: (a) The Heaviside unit step function. (b) Its derivative is the Dirac delta function which is usually represented by a vertical arrow of unit height.

$$\int_{a-\alpha}^{a+\beta} f(x)\frac{d}{dx}H(x-a)\,dx = f(x)\,H(x-a)\Big|_{a-\alpha}^{a+\beta} - \int_{a-\alpha}^{a+\beta} f'(x)\,H(x-a)\,dx$$

$$= f(a+\beta) - \int_{a}^{a+\beta} f'(x)\,dx$$

$$= f(a+\beta) - [f(a+\beta) - f(a)]$$

$$= f(a)$$

We therefore get

$$\int_{a-\alpha}^{a+\beta} f(x)\frac{d}{dx}H(x-a)\,dx = f(a);\ \ \alpha, \beta > 0 \tag{18}$$

Comparing the above equation with Eq. (3) we readily get Eq. (17). Obviously,

$$\frac{d}{dx}\left[\alpha H(x-a)\right] = \alpha\delta(x-a) \tag{19}$$

Thus we get the following important result:

If a function has a discontinuity of α at $x = a$ then its derivative (at $x = a$) is $\alpha\delta(x-a)$.

Example 1.1 We consider the symmetric function

$$\psi(x) = A\, e^{-K|x|} \tag{20}$$

Thus,

$$\left.\begin{aligned}\psi(x) &= A\, e^{Kx} \quad x < 0 \\ &= A\, e^{-Kx} \quad x > 0\end{aligned}\right\} \tag{21}$$

and

$$\left.\begin{aligned}\psi'(x) &= KAe^{Kx} \quad x < 0 \\ &= -KAe^{-Kx} \quad x > 0\end{aligned}\right\} \tag{22}$$

which are plotted in Fig. 1.6 for $K = 1$, $A = 1$. Since $\psi'(x)$ has a discontinuity of $-2KA$ at $x = 0$, we get

$$\psi''(x) = K^2\psi(x) - 2AK\delta(x) \tag{23}$$

which is also plotted in Fig. 1.6.

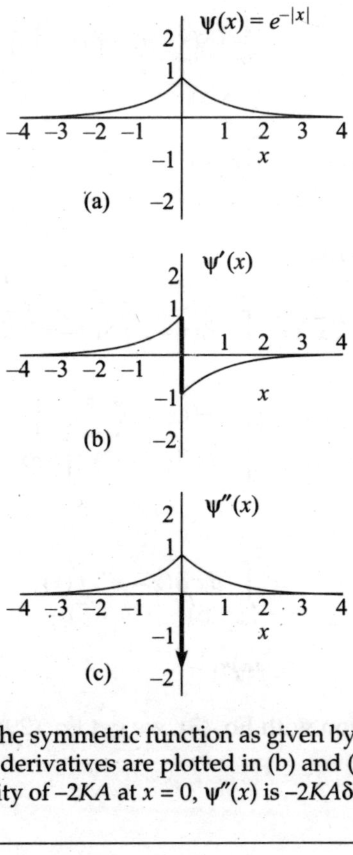

Figure 1.6: (a) A plot of the symmetric function as given by Eq. (20) for $K = 1$, $A = 1$. Its first and second derivatives are plotted in (b) and (c). Since $\psi'(x)$ has a discontinuity of $-2KA$ at $x = 0$, $\psi''(x)$ is $-2KA\delta(x)$ at $x = 0$.

1.4 IMPORTANT PROPERTIES OF THE DIRAC DELTA FUNCTION

$$\delta(x-a) = \delta(a-x) \tag{24}$$

$$f(x)\,\delta(x-a) = f(a)\,\delta(x-a) \tag{25}$$

$$\int_{-\infty}^{+\infty} f(x)\,\delta'(x-a)\,dx = -f'(a) \tag{26}$$

$$\delta[c(x-a)] = \frac{1}{|c|}\,\delta(x-a) \tag{27}$$

where c is real and not equal to zero. In Eqs. (25) and (26), $f(x)$ is assumed to be a 'well-behaved' function and continuous at $x = a$. The proof of Eq. (26) is simple:

$$\int_{-\infty}^{+\infty} f(x)\,\delta'(x-a)\,dx = f(x)\,\delta(x-a)\Big|_{-\infty}^{+\infty} - \int_{-\infty}^{+\infty} f'(x)\,\delta(x-a)\,dx$$

$$= -f'(a)$$

Further

$$\delta(x^2-a^2) = \delta[(x-a)(x+a)]$$

$$= \frac{1}{2a}[\delta(x-a) + \delta(x+a)], \quad a > 0 \tag{28}$$

We also have

$$\int_{-\infty}^{+\infty} \delta(x-x')\,\delta(x-x'')\,dx = \delta(x'-x'') \tag{29}$$

In order to prove Eq. (29), we write

$$F(x', x'') = \int_{-\infty}^{+\infty} \delta(x-x')\,\delta(x-x'')\,dx \tag{30}$$

Now,

$$\int_{-\infty}^{+\infty} f(x')\,F(x', x'')dx' = \int_{-\infty}^{+\infty} dx\,\delta(x-x'') \int_{-\infty}^{+\infty} dx'\,f(x')\,\delta(x-x')$$

$$= \int_{-\infty}^{+\infty} dx\,\delta(x-x'')\,f(x)$$

$$= f(x'')$$

If we compare the above equation with Eq. (3), we get Eq. (29).

1.5 THE COMPLETENESS CONDITION, IN TERMS OF THE DIRAC DELTA FUNCTION

Let $\psi_n(x)$ form a complete set of orthonormal functions in the domain $-\infty < x < \infty$. The orthonormality condition would require

$$\int_{-\infty}^{+\infty} \psi_m^*(x)\, \psi_n(x)\, dx = \delta_{mn} \tag{31}$$

where δ_{mn} is the Kronecker delta function defined through the following equation

$$\delta_{mn} = \begin{cases} 0 & \text{for} \quad m \neq n \\ 1 & \text{for} \quad m = n \end{cases} \tag{32}$$

Since $\psi_n(x)$ form a complete set of functions, we can expand an arbitrary 'well-behaved' function $\phi(x)$ in terms of $\psi_n(x)$:

$$\phi(x) = \sum_{n=0,1,2,\ldots}^{\infty} c_n\, \psi_n(x) \tag{33}$$

In order to determine c_n, we multiply the above equation by $\psi_m^*(x)$ and integrate to obtain

$$\int_{-\infty}^{+\infty} \psi_m^*(x)\, \phi(x)\, dx = \sum_{n=0,1,\ldots}^{\infty} c_n \int_{-\infty}^{+\infty} \psi_m^*(x)\, \psi_n(x)\, dx$$

$$= \sum_{n=0,1,2,\ldots}^{\infty} c_n\, \delta_{nm}$$

$$= c_m$$

Thus,

$$c_n = \int_{-\infty}^{+\infty} \psi_n^*(x)\, \phi(x)\, dx \tag{34}$$

Substituting in Eq. (33), we get

$$\phi(x) = \sum_{n=0,1,2,\ldots}^{\infty} \left[\int_{-\infty}^{+\infty} \psi_n^*(x')\, \phi(x')\, dx' \right] \psi_n(x)$$

where we have put a prime on the integration variable so that it does not get confused with the variable of $\psi_n(x)$. Carrying out the summation first, we get

$$\phi(x) = \int_{-\infty}^{+\infty} dx'\, \phi(x') \sum_{n=0,1,2,\ldots}^{\infty} \psi_n^*(x')\, \psi_n(x)$$

Comparing with Eq. (3), we may write

$$\sum_{n=0,1,2,\ldots}^{\infty} \psi_n^*(x')\,\psi_n(x) = \delta(x-x') \tag{35}$$

which represents the completeness condition.

As an example, the Hermite Gauss functions

$$\psi_n(x) = N_n H_n(x)\, e^{-\frac{1}{2}x^2}; \qquad n = 0, 1, 2, \ldots \tag{36}$$

with
$$N_n = \frac{1}{\sqrt{\pi^{\frac{1}{2}}\, 2^n\, n!}} \tag{37}$$

form a complete set of orthonormal functions (in the domain $-\infty < x < \infty$) satisfying Eqs. (31) and (35); in Eq. (36) $H_n(x)$ are the Hermite polynomials (see Sec. 7.3). Thus,

$$\frac{1}{\sqrt{\pi}} \sum_{n=0,1,2,\ldots}^{\infty} \frac{1}{2^n n!} H_n(x)\, H_n(x') \exp\left[-\frac{1}{2}(x^2 - x'^2)\right] = \delta(x-x') \tag{38}$$

Similarly, the functions

$$\psi_n(x) = \left(\frac{2n+1}{2}\right)^{\frac{1}{2}} P_n(x); \; n = 0, 1, \ldots \tag{39}$$

form a complete set of orthonormal functions in the domain $-1 \le x \le 1$. In Fig. 1.7, we have plotted the function

$$\sum_{n=0,1,2,\ldots}^{N} \left(\frac{2n+1}{2}\right) P_n(x)\, P_n(x')$$

for $x' = 0.6$ with $N = 12$ and 20. As $N \to \infty$, the sum will sharply peak at $x = 0.6$ and approach $\delta(x - 0.6)$.

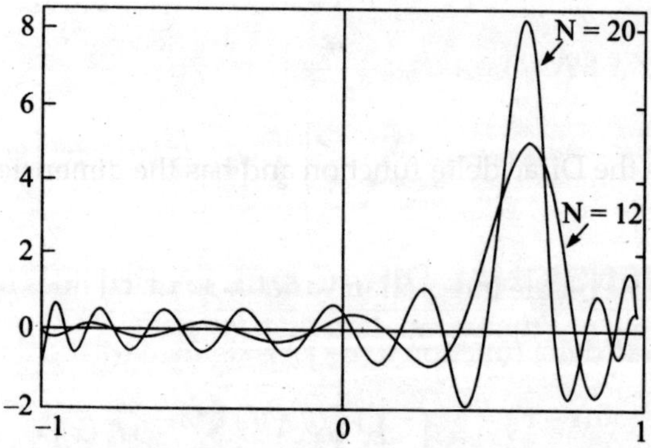

Figure 1.7: The sum $\sum_{n=0,1,\ldots}^{N} \left(\frac{2n+1}{2}\right) P_n(x)\, P_n(0.6)$ for $N = 12$ and $N = 20$. As N becomes large, the sum will sharply peak at $x = 0.6$ approaching $\delta(x - 0.6)$.

1.6 DELTA FUNCTION AS A DISTRIBUTION

We should point out that the delta function is actually a distribution. In order to understand this let us consider the Maxwellian distribution

$$N(v) = 4\pi N_0 \left(\frac{m}{2\pi kT}\right)^{\frac{2}{3}} v^2 \exp\left[-\frac{mv^2}{2kT}\right] \tag{40}$$

where k represents the Boltzmann's constant, T the absolute temperature and m the mass of each molecule. In Eq. (40), $N(v)\,dv$ represents the number of molecules whose speeds lie between v and $v + dv$. The total number of molecules is given by N_0:

$$\int_0^\infty N(v)\,dv = 4\pi N_0 \left(\frac{m}{2\pi kT}\right)^{\frac{3}{2}} \int_0^\infty \exp\left[-\frac{mv^2}{2kT}\right] dv$$

$$= 4\pi N_0 \left(\frac{m}{2\pi kT}\right)^{\frac{3}{2}} \left(\frac{2kT}{m}\right)^{\frac{3}{2}} \frac{1}{2}\int_0^\infty x^{\frac{1}{2}} e^{-x} dx$$

where $x \equiv mv^2/(2kT)$. The integral is $\Gamma\left(\frac{3}{2}\right) = \frac{1}{2}\sqrt{\pi}$.

Thus,

$$\int_0^\infty N(v)\,dv = N_0 \tag{41}$$

It may be noted that whereas N_0 is just a number, the quantity $N(v)$ has the dimensions of (velocity)$^{-1}$. Obviously, if we ask ourselves how many molecules have the precise speed v_0, the answer would be zero. This is the characteristic of a distribution. On the other hand, in addition to the distribution given by Eq. (40), if we do have N_1 molecules all of them having the same speed v_1, the corresponding distribution function would be given by

$$N(v) = 4\pi N_0 \left(\frac{m}{2\pi kT}\right)^{\frac{3}{2}} v^2 \exp\left[-\frac{mv^2}{2kT}\right] + N_1 \delta(v - v_1) \tag{42}$$

where $\delta(v - v_1)$ represents the Dirac delta function and has the dimensions of inverse velocity.

1.7 THE THREE-DIMENSIONAL DIRAC DELTA FUNCTION

The three-dimensional Dirac delta function $\delta(\mathbf{r} - \mathbf{r}')$ is defined by the following equation

$$\delta(\mathbf{r} - \mathbf{r}') = \delta(x - x')\,\delta(y - y')\,\delta(z - z') \tag{43}$$

Thus,

$$\iiint f(\mathbf{r})\,\delta(\mathbf{r} - \mathbf{r}')\,d\tau = f(\mathbf{r}') \tag{44}$$

where the domain of integration over the variable $\mathbf{r}$ should include the point $\mathbf{r}'$; $d\tau = dx\,dy\,dz$. It is obvious that we can have many representations of the three-dimensional Dirac delta function. For example, we may write [cf. Eqs (6) and (9)]

$$\delta(\mathbf{r}-\mathbf{r}') = \lim_{\sigma\to 0} \frac{1}{(2\pi)^{\frac{3}{2}}\sigma^3} \exp\left[-\frac{(x-x')^2+(y-y')^2+(z-z')^2}{2\sigma^2}\right] \quad (45)$$

and also

$$\delta(\mathbf{r}-\mathbf{r}') = \frac{1}{(2\pi)^3} \int_{-\infty}^{+\infty}\int_{-\infty}^{+\infty}\int_{-\infty}^{+\infty} e^{i\,\mathbf{k}\cdot(\mathbf{r}-\mathbf{r}')}\, dk_x\, dk_y\, dk_z \quad (46)$$

where $\mathbf{k}\cdot(\mathbf{r}-\mathbf{r}')$ is a compact form of writing

$$k_x(x-x') + k_y(y-y') + k_z(z-z')$$

Further, if (r, θ, ϕ) represent the spherical polar coordinates then,

$$\delta(\mathbf{r}-\mathbf{r}') = \frac{\delta(r-r')\,\delta(\theta-\theta')\,\delta(\phi-\phi')}{r^2\sin\theta}$$

$$= \frac{\delta(r-r')}{r^2}\,\delta(\Omega-\Omega')$$

where Ω represents the solid angle and

$$\delta(\Omega-\Omega') = \frac{\delta(\theta-\theta')\,\delta(\phi-\phi')}{\sin\theta}$$

$$= \sum_{l=0,1,\ldots} \sum_{m=-l}^{+l} Y^*_{lm}(\theta',\phi')\, Y_{lm}(\theta,\phi)$$

Further $\quad \delta(\mathbf{r}) = \delta(x)\,\delta(y)\,\delta(z) = \dfrac{\delta(r)}{4\pi r^2}$

with $\quad \displaystyle\int_0^\infty \delta(r-r')\,dr' = 1$

We also have $\quad (\nabla^2 + k^2)\,\dfrac{e^{ikr}}{r} = -4\pi\delta(\mathbf{r})$

1.8 PROBLEMS

Problem 1.1 For $f(x) = |x|$, show that $f''(x) = 2\,\delta(x)$

Problem 1.2 Consider the following

$$f_\sigma(x) = \begin{cases} 0 & x < a \\ \sigma e^{-\sigma(x-a)} & x \geq a \end{cases}$$

where $\sigma > 0$. Plot the above function for $\sigma = 1$ and 10 and show that we may write

$$\delta(x-a) = \lim_{\sigma\to\infty} f_\sigma(x)$$

Problem 1.3 Consider the function $\psi(x) = \begin{cases} -x(1-a) & 0<x<a \\ -(1-x)a & a<x<1 \end{cases}$

Show that $\frac{d^2\psi}{dx^2} = \delta(x-a)$

Problem 1.4 Prove that $x\delta'(x) = -\delta(x)$

Problem 1.5 The functions $\psi_n(x) = \sqrt{2}\sin(n\pi x)$

form a complete set of functions in the interval $0 \le x \le 1$ satisfying the boundary conditions

$$\psi(0) = 0 = \psi(1)$$

Plot

$$2\sum_{n=0,1,2,\ldots}^{N} \sin(n\pi x)\sin(n\pi x')$$

with $x' = 0.6$ (in the domain $-1 < x < 1$) for $N = 10$ and 20. As $N \to \infty$ the sum will approach $\delta(x-0.6)$.

1.9 SAMPLE QUESTIONS

Q. 1 Evaluate $\psi''(x)$,

$$\begin{aligned} \psi(x) &= 0 \quad \text{for} \quad x<0 \\ &= 4x \quad \text{for} \quad x>0 \end{aligned}$$

Q. 2 Consider a function

$$\begin{aligned} f(x) &= 20 & x<0 \\ &= (50/\sigma)x & 0<x<\sigma \\ &= 50 & x>\sigma \end{aligned}$$

Plot $f(x)$ and df/dx as a function of x.

Q. 3 Evaluate:

(a) $\underset{\sigma\to 0}{\text{Lt}} \frac{1}{\sigma}\exp\left(-\frac{(x-a)^2}{2\sigma^2}\right)$

(b) $\underset{g\to\infty}{\text{Lt}} \frac{\sin g(x-a)}{(x-a)}$

[Answer: (b) $\pi\delta(x-a)$]

Q. 4 Consider the function

$$\begin{aligned} f(x) &= \alpha x & 0<x<a \\ &= \alpha a\frac{L-x}{L-a} & a<x<L \end{aligned}$$

Evaluate and plot $f(x)$, $f'(x)$ and $f''(x)$ in the domain $0 < x < L$.

Q. 5 Consider the function $\psi(x) = A\exp[-\kappa|x|]$. Evaluate and plot $\psi(x)$, $\psi'(x)$ and $\psi''(x)$. Determine the second order differential equation satisfied by $\psi(x)$.

REFERENCES AND SUGGESTED READINGS

1. P.A.M. Dirac, *The Principles of Quantum Mechanics* (Fourth Edition), Clarendon Press, Oxford (1958).
2. J. Lighthill, *Introduction to Fourier Analysis and Generalised Functions*, Cambridge University Press (1958).
3. A.K. Ghatak, I.C. Goyal and S.J. Chua, *Mathematical Physics*, Laxmi Publications, New Delhi (2016).
4. Ajoy Ghatak, *Optics*, (6th Edition), McGraw Hill Education, New Delhi (2017).

2 Fourier Transforms

The greatest mathematics has the simplicity and inevitableness of supreme poetry and music, standing on the borderland of all that is wonderful in Science, and all that is beautiful in Art. Mathematics transfigures the fortuitous concourse of atoms into the tracery of the finger of God.

— H.W. TURNBULL
in *The Great Mathematicians*

2.1 INTRODUCTION

Fourier transforms play an important role in quantum mechanics and as such, we devote a (short) chapter giving a brief account of Fourier transforms.

2.2 FOURIER INTEGRAL THEOREM

In the previous chapter we had shown the following integral representation of the Dirac delta function

$$\delta(x-x') = \frac{1}{2\pi}\int_{-\infty}^{+\infty} e^{\pm ik(x-x')}dk \tag{1}$$

Since,

$$f(x) = \int_{-\infty}^{+\infty} \delta(x-x') f(x')\,dx' \tag{2}$$

we may write

$$f(x) = \frac{1}{2\pi}\int_{-\infty}^{+\infty}\int_{-\infty}^{+\infty} e^{\pm ik(x-x')} f(x')\,dx'\,dk \tag{3}$$

Thus if we define

$$F(k) = \frac{1}{\sqrt{2\pi}}\int_{-\infty}^{+\infty} f(x)\,e^{-ikx}\,dx \tag{4}$$

then $$f(x) = \frac{1}{\sqrt{2\pi}} \int_{-\infty}^{+\infty} F(k) e^{+ikx}\, dk \tag{5}$$

The function $F(k)$ is known as the Fourier transform of the function $f(x)$ and Eq. (5) enables us to calculate the original function from the Fourier transform. Equations (4) and (5) constitute what is known as the *Fourier Integral Theorem* which is valid when the following conditions are satisfied (see, e.g. Refs. 1 and 2):

(i) The function $f(x)$ must be a single valued function of the real variable x throughout the range $-\infty < x < \infty$. It may however have a finite number of finite discontinuities.

(ii) The integral $\int_{-\infty}^{+\infty} |f(x)|\, dx$ must exist.

From Eq. (3) it is obvious that in Eqs. (4) and (5) there is no reason why the factors e^{ikx} and e^{-ikx} cannot be interchanged, i.e., we could have defined

$$F(k) = \frac{1}{\sqrt{2\pi}} \int_{-\infty}^{+\infty} f(x)\, e^{ikx}\, dx \tag{6}$$

Then, $$f(x) = \frac{1}{\sqrt{2\pi}} \int_{-\infty}^{+\infty} F(k)\, e^{-ikx}\, dx \tag{7}$$

However, in all of what follows we will use the definitions given by Equations (4) and (5).

Example 2.1 As an example we consider a Gaussian function given by

$$f(x) = A\, e^{-x^2/2\sigma^2} \tag{8}$$

Its Fourier transform is given by

$$F(k) = \frac{A}{\sqrt{2\pi}} \int_{-\infty}^{+\infty} e^{-x^2/2\sigma^2}\, e^{-ikx}\, dx$$

or $$F(k) = A\sigma \exp\left[-\frac{1}{2}k^2\sigma^2\right] \tag{9}$$

where we have made use of the following integral (see Appendix A):

$$\int_{-\infty}^{+\infty} e^{-\alpha x^2 + \beta x}\, dx = \sqrt{\frac{\pi}{\alpha}} \exp\left[\frac{\beta^2}{4\alpha}\right];\ \text{Re}\,\alpha > 0 \tag{10}$$

As can be seen from Eq. (9), the function $F(k)$ is also a Gaussian; thus the Fourier transform of a Gaussian is a Gaussian. We may note that the Gaussian function given by Eq. (8) has a spatial width given by [see Fig. 2.1(a)]

$$\Delta x \sim \sigma \tag{11}$$

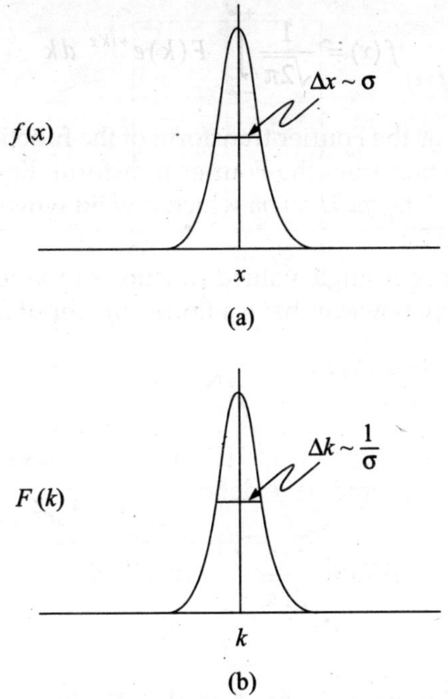

Figure 2.1: (a) The Gaussian function $f(x)$ as given by Eq. (8). (b) The Fourier transform of the Gaussian function is also a Gaussian in the *k*-space.

Its Fourier transform, $F(k)$, has a width given by [see Fig. 2.1(b)]

$$\Delta k \sim \frac{1}{\sigma} \tag{12}$$

Thus, $$\Delta x \, \Delta k \sim 1 \tag{13}$$

which is a general characteristic of the Fourier transform pair.

Example 2.2 As another example, we calculate the Fourier transform of the rectangle function

$$f(x) = \text{rect}\left(\frac{x}{a}\right) = \begin{cases} 1, & |x| < \frac{1}{2}a \\ 0, & |x| > \frac{1}{2}a \end{cases} \tag{14}$$

Its Fourier transform will be given by (see Fig. 2.2):

$$F(k) = \frac{1}{\sqrt{2\pi}} \int_{-a/2}^{+a/2} e^{-ikx}\, dx$$

or $$F(k) = \sqrt{\frac{2}{\pi}}\, \frac{\sin(ka/2)}{k} \tag{15}$$

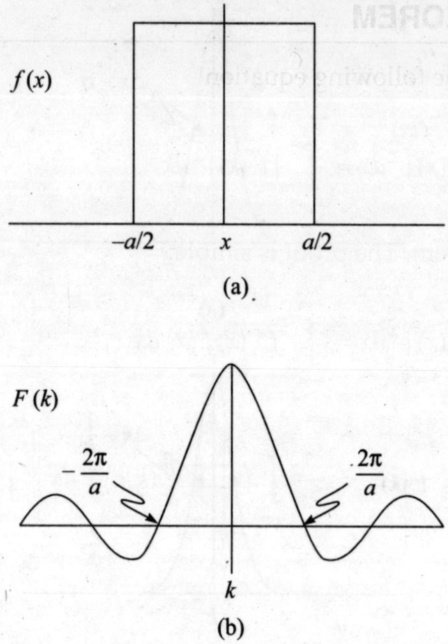

Figure 2.2: The rectangle function is shown in (a). Its Fourier transform is shown in (b).

Once again, the rectangle function has a width a and its Fourier transform has a width

$$\Delta k \sim \frac{\pi}{a}$$

In quantum mechanics, one uses the variable

$$p = \hbar k \tag{16}$$

where
$$\hbar = \frac{h}{2\pi} \tag{17}$$

h ($= 6.626 \times 10^{-34}$ J. s) represents the Planck's constant. In terms of the variable p, Eq. (3) can be written in the form

$$\psi(x) = \frac{1}{2\pi\hbar} \int_{-\infty}^{+\infty} \int_{-\infty}^{+\infty} \exp\left[\pm \frac{i}{\hbar} p\,(x - x')\right] \psi\,(x')\, dx'\, dp \tag{18}$$

Thus, if we define

$$a\,(p) = \frac{1}{\sqrt{2\pi\hbar}} \int_{-\infty}^{+\infty} \psi(x) \exp\left[-\frac{i}{\hbar} px\right] dx \tag{19}$$

then
$$\psi(x) = \frac{1}{\sqrt{2\pi\hbar}} \int_{-\infty}^{+\infty} a(p) \exp\left[\frac{i}{\hbar} px\right] dp \tag{20}$$

We will use the above equation in Chapter 5.

2.3 PARSEVAL'S THEOREM

The equality expressed by the following equation

$$\int_{-\infty}^{+\infty} |f(x)|^2\, dx = \int_{-\infty}^{+\infty} |F(k)|^2\, dk \tag{21}$$

is known as Parseval's theorem. The proof is simple:

Proof $$\int_{-\infty}^{+\infty} |f(x)|^2\, dx = \int_{-\infty}^{+\infty} f^*(x) f(x)\, dx$$

$$= \frac{1}{2\pi}\int_{-\infty}^{+\infty} dx \left[\int_{-\infty}^{+\infty} F^*(k) e^{ikx}\, dx\right]\left[\int_{-\infty}^{+\infty} F(k') e^{-ik'x} dk'\right]$$

where we have used Eq. (5) and its complex conjugate. Thus,

$$\int_{-\infty}^{+\infty} |f(x)|^2\, dx = \int_{-\infty}^{+\infty} dk \int_{-\infty}^{+\infty} dk'\, F^*(k)\, F(k') \left[\frac{1}{2\pi}\int_{-\infty}^{+\infty} e^{i(k-k')x}\, dx\right]$$

$$= \int_{-\infty}^{+\infty} dk \int_{-\infty}^{+\infty} dk'\, F^*(k)\, F(k')\, \delta(k-k')$$

$$= \int_{-\infty}^{+\infty} dk\, |F(k)|^2$$

Similarly, we can easily show from Eqs (19) and (20) that

$$\int_{-\infty}^{+\infty} |\psi(x)|^2\, dx = \int_{-\infty}^{+\infty} |a(p)|^2\, dp \tag{22}$$

The implication of the above equation will be discussed in Chapter 5.

2.4 FOURIER TRANSFORM OF A TIME DEPENDENT FUNCTION

In order to get a better physical insight to the Fourier transform, we consider a time dependent function $f(t)$. For such a function we write the Fourier transform in the following form

$$F(\omega) = \frac{1}{\sqrt{2\pi}} \int_{-\infty}^{+\infty} f(t)\, e^{i\omega t}\, dt \tag{23}$$

Thus, $$f(t) = \frac{1}{\sqrt{2\pi}} \int_{-\infty}^{+\infty} F(\omega)\, e^{-i\omega t}\, d\omega \tag{24}$$

The above equations are nothing but Eqs. (6) and (7) with x and k replaced by t and ω respectively. As a simple example, we consider an optical pulse having a Gaussian envelope:

$$f(t) = A e^{-t^2/2\tau^2} e^{-i\omega_0 t} \tag{25}$$

whose real part represents the actual electric field which is plotted in Fig. 2.3(a) for $\omega_0\tau = 15$. The Fourier transform can be readily evaluated and is given by

$$F(\omega) = A\sigma \exp\left[-\frac{(\omega-\omega_0)^2\tau^2}{2}\right] \tag{26}$$

where we have used Eq. (10). The function $F(\omega)$ is known as the frequency spectrum and is plotted in Fig. 2.3(b). One can see that the duration of the pulse is $\sim \tau$ and the corresponding spectral width is given by

$$\Delta\omega \sim \frac{1}{\tau} \tag{27}$$

implying

$$\tau\,\Delta\omega \sim 1 \tag{28}$$

Thus a pulse of duration τ has necessarily a spectral width $\sim 1/\tau$; this is indeed the result obtained in coherence theory (see, e.g. Chapters 10 and 17 of Ref. 3).

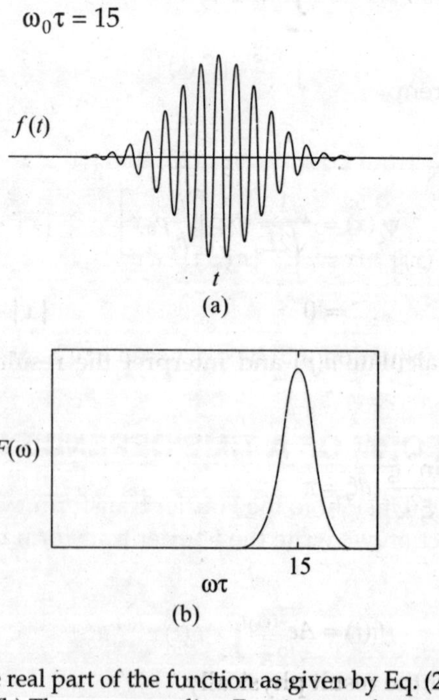

Figure 2.3: (a) The real part of the function as given by Eq. (25) for $\omega_0\tau = 15$. (b) The corresponding Fourier transform.

2.5 PROBLEMS

Problem 2.1 **(a)** Show that the Fourier transform of the function

$$f(x) = Ae^{-\alpha|x|};\ \alpha > 0 \tag{29}$$

is given by

$$F(k) = \frac{A}{\sqrt{2\pi}} \cdot \frac{2\alpha}{\alpha^2 + k^2} \tag{30}$$

which is known as the Lorentzian distribution.

(b) Hence show that

$$\int_{-\infty}^{+\infty} \frac{e^{ikx}}{\alpha^2 + k^2}\, dk = \frac{\pi}{\alpha}\, e^{-\alpha|x|} \tag{31}$$

(c) Use Parseval's theorem to prove

$$\int_{-\infty}^{+\infty} \frac{1}{(\alpha^2 + k^2)^2}\, dk = \frac{\pi}{2\alpha^3} \tag{32}$$

Problem 2.2 Using Eqs (8) and (9) show that

$$\int_{-\infty}^{+\infty} |f(x)|^2\, dx = \int_{-\infty}^{+\infty} |F(k)|^2\, dk$$

which is the Parseval's theorem.

Problem 2.3 Assume

$$\psi(x) = \frac{1}{\sqrt{2L}} \exp\left[\frac{i}{\hbar} p_0 x\right] \qquad |x| < L$$

$$= 0 \qquad |x| > L \tag{33}$$

Use Eqs. (19) and (20) to calculate $a(p)$ and interpret the result physically. Use Parseval's theorem to prove

$$\int_{-\infty}^{+\infty} \frac{\sin^2 \xi}{\xi^2}\, d\xi = \pi \tag{34}$$

Problem 2.4 Assume

$$f(t) = Ae^{-i\omega_0 t} \tag{35}$$

Calculate $F(\omega)$ and interpret the result physically.

2.6 SAMPLE QUESTIONS

Q. 1 Consider the function

$$f(t) = Ae^{-t^2/2\tau^2}\, e^{i\omega_0 t}$$

Calculate its Fourier spectrum

$$g(\omega) = \frac{1}{\sqrt{2\pi}} \int_{-\infty}^{+\infty} f(t)\, e^{-i\omega t}\, dt$$

and evaluate approximately $\Delta\omega\Delta t$. Evaluate $f(t)$ using the expression for $g(\omega)$.

Q. 2 Consider the function

$$f(x) = ae^{-x^2/2\sigma^2}\, e^{i k_0 x}$$

Calculate its Fourier spectrum

$$g(k) = \frac{1}{\sqrt{2\pi}} \int_{-\infty}^{+\infty} f(x)\, e^{-ikx}\, dx$$

and evaluate approximately $\Delta x\Delta k$. Evaluate $f(x)$ using the expression for $g(k)$.

Q. 3 Calculate the Fourier transform of the following functions

(a) $$f(x) = A\, e^{ik_0 x} \quad |x| < L/2$$
$$= 0 \quad |x| > L/2$$

(b) $$f(x) = A\, e^{-|x|/L}$$

(c) $$f(x) = A \quad -\infty < x < \infty$$

In each case make an estimate of Δx and Δk and interpret physically.

REFERENCES AND SUGGESTED READINGS

1. R.N. Bracewell, *The Fourier Transform and its Applications*, McGraw-Hill, New York (1986).
2. E.C. Titchmarsh, *Introduction to the Theory of Fourier Integrals*, Clarendon Press, Oxford (1948).
3. A. Ghatak, *Optics*, (6th Edition), McGraw Hill Education, New Delhi (2017).

Part II

Basic Quantum Mechanics

3 Particles and Waves and the Basic Quantum Mechanics

For the [illegible]

All the [illegible]
light quanta [illegible]

3.1 INTRODUCTION

As we know, the [illegible] of [illegible]
etc., are known [illegible] for [illegible]
can also be [illegible]
visualize them as [illegible] however, they also [illegible]
which can [illegible]
are experienced [illegible]
wave model [illegible]
explained if we [illegible]
like what [illegible]
particles, etc. [illegible]
abstract way [illegible]

> Newton thought that light was made up of particles, but [illegible]
> like a wave. Later, however [illegible]
> did [illegible]
> to behave like a particle [illegible]
> really [illegible]

> There is [illegible]
> along [illegible]
> 'particle-waves', or [illegible] to call them [illegible]

In this chapter, we will qualitatively discuss the wave-particle [illegible]
explained on the basis of the uncertainty principle and [illegible]
matter waves.

3 Particles and Waves and the Uncertainty Principle

For the rest of my life, I will reflect on what light is.

— ALBERT EINSTEIN, CA. 1917.

All the fifty years of conscious brooding have brought me no closer to the answer to the question, 'What are light quanta?' Of course today every rascal thinks he knows the answer, but he is deluding himself.

— ALBERT EINSTEIN, 1951.

3.1 INTRODUCTION

As we know, the values of the mass and charge of electrons, protons, neutrons, alpha-particles, etc., are known with tremendous accuracy—approximately one part in a billion. Their velocities can also be changed by application of electric and magnetic fields. Thus, we usually tend to visualize them as tiny particles. However, they also exhibit diffraction and other phenomena which can only be explained if we assume them to be *waves*. Similarly, using light beams, there are experiments on interference, diffraction, etc., which can only be explained if we assume a wave model for light. There are also phenomena, like photoelectric effect, which can only be explained if we assume a particle model of light. Thus, it is very difficult to answer questions like *what is an electron or what is light*. Indeed electrons, protons, neutrons, photons, alpha-particles, etc. are *neither particles nor waves*. Modern quantum theory describes these in a very abstract way which cannot be connected with everyday experience. To quote Feynman [Ref. 1]:

> *Newton thought that light was made up of particles, but then it was discovered that it behaves like a wave. Later, however (in the beginning of the twentieth century), it was found that light did indeed sometimes behave like a particle. Historically, the electron, for example, was thought to behave like a particle, and then it was found that in many respects it behaved like a wave. So it really behaves like neither. Now we have given up. We say: 'It is like neither'.*
>
> *There is one lucky break, however—electrons behave just like light. The quantum behaviour of atomic objects (electrons, protons, neutrons, photons, and so on) is the same for all, they are all 'particle-waves', or whatever you want to call them.*

In this chapter, we will qualitatively discuss the wave-particle duality and how it can be explained on the basis of the uncertainty principle and the probabilistic interpretation of matter waves.

3.2 THE CORPUSCULAR AND WAVE MODELS OF LIGHT

The corpuscular model is perhaps the simplest model of light. According to this model, a luminous body emits a stream of particles in all directions. The particles are assumed to be very tiny so that when two light beams overlap, a collision between two particles rarely occurs. Using the corpuscular model, one can explain the laws of reflection and refraction. Contrary to popular belief, the corpuscular model of light is due to Descartes rather than to Newton. The law of refraction was discovered experimentally in 1621 by Snell. Descartes' derivation of Snell's law was published in 1637; this derivation is equivalent to the corpuscular derivation, which is usually attributed to Newton. Newton was only about eight years old when Descartes died and the first edition of Newton's *Optiks* (in which he had discussed the corpuscular model) was published in 1704. It was probably because of the popularity of Newton's *Optiks* that the corpuscular theory is usually attributed to Newton.

Although the corpuscular model explains the propagation of light through free space and can be made to predict the correct forms of the laws of reflection and refraction, there came up a large number of experimental observations (like interference, diffraction, polarization etc.), which could not be explained on the basis of the corpuscular model of light. Indeed, in 1802, Thomas Young performed the famous two-slit interference experiment which could only be explained on assuming a wave model of light. Young showed that the wavelength of the light waves was around 6×10^{-5} cm. Because of the smallness of the wavelength, the diffraction effects are small and therefore light approximately travels in straight lines. Around 1816, Fresnel gave a satisfactory explanation of the diffraction phenomenon by means of a wave theory and calculated the diffraction patterns produced by various types of apertures and edges. In 1816, Fresnel along with Arago performed the famous experiment on the superposition of linearly polarized light waves, which was explained by Young, by assuming that light waves were transverse in character.

Around the second quarter of the nineteenth century, the wave theory seemed to be well established and since it was thought that a wave requires a medium for its propagation, the elastic ether theory was developed. Indeed, in 1832 Fresnel derived the expressions for the reflection and transmission coefficients by using models for ether vibrations. However, there were considerable difficulties in the understanding of the ether theory and since we now know that the ether does not exist, we will not go into the details of the various theories.

The nineteenth century also saw the development of electricity and magnetism. Around 1857 Maxwell generalized Ampere's law by stating that a changing electric field can also set up a changing magnetic field. He summed up all the laws of electricity and magnetism in the form of equations which are now referred to as Maxwell's equations. From these equations, he derived a wave equation and predicted the existence of electromagnetic waves (see, e.g., Chapter 23 of Ref. 2). From the derived wave equation, he showed that the velocity of the electromagnetic waves can be calculated from experiments in which a certain quantity of electric charge is measured by two different methods. These measurements were carried out in 1856 by Kohlrausch and Weber, and from their data, Maxwell found that the speed of the electromagnetic waves in air should be about 3.107×10^8 m/s. He found that this value was very close to the measured value of the speed of light which according to the measurement of Fizeau in 1849 was known to be 3.14858×10^8 m/s. Maxwell argued that these two numbers cannot be accidentally equal and with *faith in rationality of nature*, he propounded his famous electromagnetic theory of light according to which, light waves are electromagnetic in nature. Associated with a light wave, there would be changing electric and magnetic fields, the changing magnetic field produces a time and space varying electric field and the changing electric field produces a time and space varying magnetic field, this results in the propagation of the electromagnetic wave even in free space. In 1888, Heinrich Hertz carried out experiments which could produce and detect electromagnetic waves produced by discharging electrically charged plates through a spark gap. The frequency of the emitted electromagnetic waves depended on the values of the inductance and capacitance of the circuit. Hertz's experimental results provided dramatic confirmation of Maxwell's electromagnetic theory. In addition, there were so many other experimental results which were quantitatively explained by using Maxwell's theory that around the end of the nineteenth century physicists thought that one finally understood what light really was.

3.3 PARTICLE NATURE OF RADIATION: THE PHOTOELECTRIC EFFECT

In 1887, while receiving the electromagnetic waves in a coil with a spark gap, Heinrich Hertz found that the maximum spark length was reduced when the apparatus was put in a black box; this is due to what is now known as the **photoelectric effect** and the box absorbed the ultraviolet radiation which helped the electrons in jumping across the gap. Hertz reported the observations but did not pursue further and also did not make any attempt to explain them. In 1897 J J Thomson discovered the electron and in 1899 he showed that electrons are emitted when light falls on a metal surface; these are now known as photoelectrons. In 1902 Philip Lenard observed that

(a) the kinetic energy of the emitted electrons was independent of the intensity of the incident light and

(b) that the energy of the emitted electron increased when the frequency of the incident light was increased.

Later Millikan carried out very careful experiments on the photoelectric effect and the apparatus that he used was similar to the one shown in Fig. 3.1; these photoelectrons consittute a current between the plates P_1 and P_2 which can be detected by means of an ammeter A. When the voltage across the plates is varied, the current also varies; typical variations of the current with voltage are shown in Fig. 3.2. The figure corresponds to monochromatic light of a particular wavelength and different curves correspond to different intensities of the beam. From the figure one can draw the following conclusions:

(i) At zero voltage there is a finite value of the current implying that some of the emitted photoelectrons reach the metal surface P_2.

(ii) As the voltage is increased, the current increases till it reaches a saturation value; this will happen when the plate P_2 collects all the emitted photoelectrons.

(iii) If the plate P_2 is kept at a slightly negative potential, there is a weak current implying that some of the photoelectrons do manage to reach the plate P_2. However, beyond a certain voltage (which is shown as $-V_c$ in the figure) the current is zero; V_c is known as the cutoff voltage and the quantity $|q|V_c$ will represent the maximum kinetic energy of the photoelectrons (q represents the charge of the electron). For example, for sodium $V_c \approx 2.3$ volts and for copper $V_c \approx 4.7$ volts,

(iv) If we do not change the wavelength of the incident radiation but make it more intense, the magnitude of the current will become larger as shown in Fig. 3.2—implying a greater emission of photoelectrons. Notice that the value of the cutoff potential remains the same; this important result implies that the maximum kinetic energy of the emitted photoelectrons does not depend on the intensity of the incident radiation.

(v) If the frequency of the incident radiation is increased then the cutoff potential and hence the maximum kinetic energy of the electron ($= |q| V_c$) varies linearly with the frequency as shown in Fig. 3.3. Further, for frequencies less than a critical value (shown as υ_c in Fig. 3.3), there is no emission of photoelectrons no matter what the intensity of the incident radiation may be.

At first sight it appears that since electromagnetic waves carry energy, the wave model for light should be able to explain the emission of photoelectrons from a metal surface. However, there are certain peculiarities associated with photoelectric effect, which cannot be satisfactorily explained by means of a wave model:

1. The first peculiarity is the fact that the maximum kinetic energy of the electrons does not depend on the intensity of the incident radiation, it only depends on its frequency; further, a greater intensity leads to a larger number of electrons constituting a larger current. Thus, a faint violet light would eject electrons of greater kinetic energy than an intense yellow light although the latter would produce a large number of electrons. A wave model would, however, predict that a large intensity of the incident radiation would result in a greater kinetic energy of the emitted electrons.

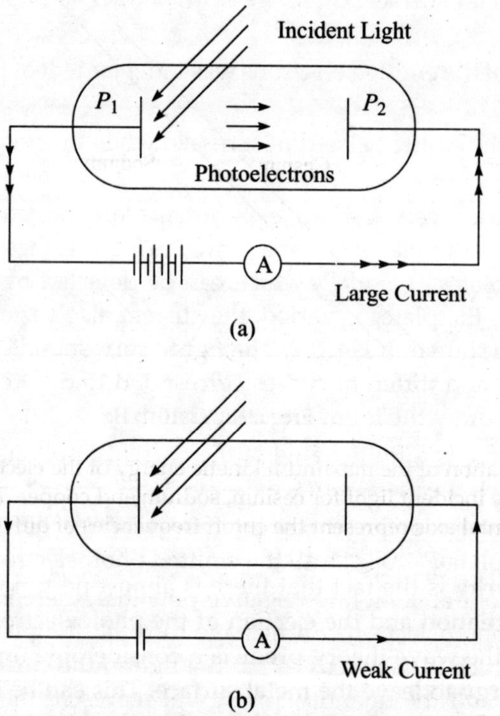

Figure 3.1: If light (of certain frequency) is allowed to fall on a metal like sodium, electrons are emitted which can be collected by the plate P_2. (a) and (b) correspond to positive and negative voltage applied to the plate P_2. Even when the plate is kept at a low negative voltage, one can detect a small current.

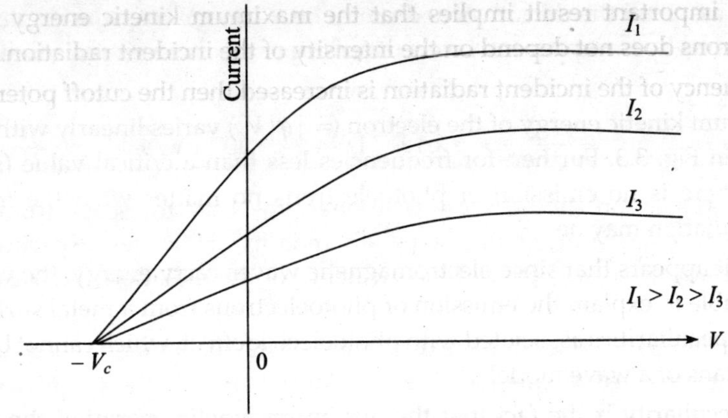

Figure 3.2: Typical variation of the photocurrent with voltage. The curves correspond to light (of the same frequency) having different intensities.

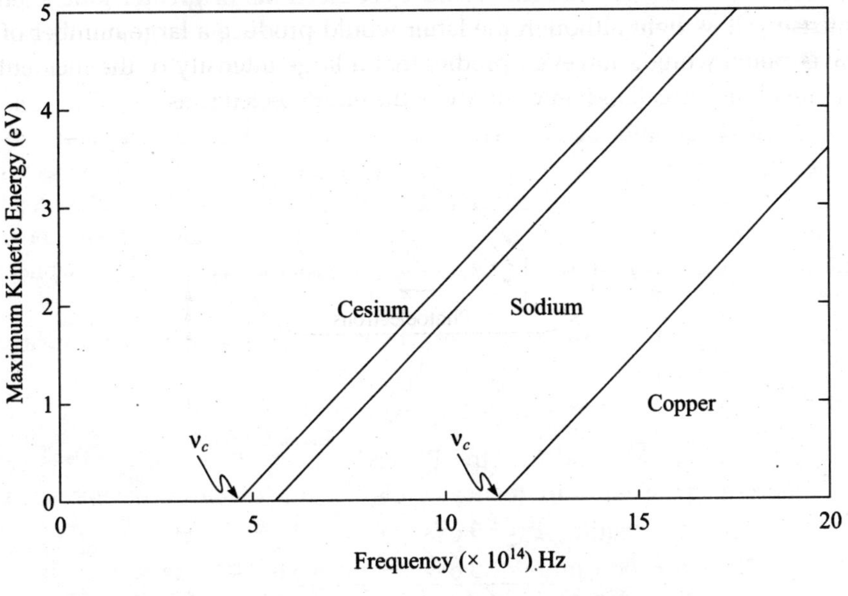

Figure 3.3: The variation of the maximum kinetic energy of the electrons as a function of frequency of the incident light for cesium, sodium and copper. The intercepts on the horizontal axis represent the cutoff frequencies of different metals.

2. The second peculiarity is the fact that there is almost no time lag between the times of incidence of the radiation and the ejection of the photoelectron. For weak intensities of the incident beam the wave theory predicts a considerable time lag for the electrons to absorb enough energy to leave the metal surface. This can be illustrated by considering a specific example. One can observe a detectable photocurrent if the surface of sodium metal is illuminated by violet light of intensity as low as 10^{-10} W/cm^2. Now, ten layers of sodium will contain

$$\frac{6\times 10^{23}\times 10\times 10^{-8}}{23} \approx 2\times 10^{15} \text{ atoms/cm}^2$$

where we have assumed the density of sodium to be ≈ 1 g/cm^3. Assuming that the energy is uniformly absorbed by the upper ten layers of sodium, each atom would receive energy at the rate of

$$\frac{10^{-10}}{2\times 10^{15}} \approx 5\times 10^{-26}\ \text{J/s} \approx 3\times 10^{-7}\ \text{eV/s}$$

Assuming that an electron should acquire an energy ~ 1 eV to escape from the metal, we should expect a time lag of order 10^7 sec (~ few months). However, experiments show that there is no detectable time lag between the incidence of the radiation and the emission of the photoelectrons.[1]

In 1905, Einstein provided a simple explanation of the above mentioned peculiarities. He argued that light consisted of quanta of energy $h\nu$ (where ν is the frequency) and that the emission of a photoelectron was the result of the interaction of a single quantum (i.e., of the photon[2]) with an electron. Einstein, in 1905, wrote in his paper,[3]

> *Monochromatic radiation behaves as if it consists of mutually independent energy·quanta of magnitude* [$h\nu$].

Einstein's *photon* theory gives a very satisfactory explanation of the photoelectric effect. According to this theory a light beam (of frequency ν) essentially consists of individual corpuscles called photons. Each photon carries an energy equal to $h\nu$. This corpuscular model can explain all the observations discussed above. Thus, for all frequencies below the cutoff ν_c, each photon will carry energy less than $h\nu_c$ which will not be sufficient to eject the electron from the metal. For $\nu > \nu_c$, a major fraction of the excess energy [$= h(\nu - \nu_c)$] appears as kinetic energy of the emitted electron. Further, the non-measurable time lag between the incidence of the radiation and the ejection of the electron follows immediately from the corpuscular nature of radiation.

Indeed, the observed maximum kinetic energy of the photoelectrons is linearly related to the frequency of the incident radiation [Fig. 3.3] and may be represented as

$$T_{\max} = -B + h\nu = h(\nu - \nu_c), \qquad (1)$$

where $B(= h\nu_c)$ is a constant and h is the Planck's constant ($= 6.627 \times 10^{-27}$ erg-sec). The frequency ν_c represents the cutoff frequency and is a characteristic of the metal. For example,

$$\begin{aligned}
&\text{for cesium} \quad B \approx 1.9\ \text{eV} \quad \Rightarrow \quad \nu_c \approx 4.6\times 10^{14}\ \text{Hz}\\
&\text{for sodium} \quad B \approx 2.3\ \text{eV} \quad \Rightarrow \quad \nu_c \approx 5.6\times 10^{14}\ \text{Hz}\\
&\text{for copper} \quad B \approx 4.7\ \text{eV} \quad \Rightarrow \quad \nu_c \approx 11.4\times 10^{14}\ \text{Hz}
\end{aligned}$$

In Fig. 3.3, ν_c is the intercept on the horizontal axis. In 1909, Einstein wrote (quoted from Ref. 3)

> *It is undeniable that there is an extensive group of data concerning radiation which shows that light has certain fundamental properties that can be understood much more readily from the standpoint of the Newton emission (particle) theory than from the standpoint of the wave theory. It is my opinion, therefore, that the next phase of the development of theoretical physics will bring us a theory of light that can be interpreted as a kind of fusion of the wave and emission theories.*

---

[1] Indeed, in 1928, Lawrence and Beams had devised an experiment to find out whether the time lag was $\leq 3\times 10^{-9}$ sec; the experiment gave a negative result.

[2] The term photon was introduced by G.N. Lewis in 1926.

[3] *Annalen der Physik*, **17**, 132, 1905.

We may note the prediction of Einstein who received the 1921 Noble prize in Physics for his *discovery of the law of photoelectric effect* and *not* for his theory of relativity. To quote Max Jammer (Ref. 4).

> *Owing to Einstein's paper of 1905, it was primarily the photoelectric effect to which physicists referred as an irrefutable demonstration of the existence of photons and which thus played an important part in the conceptual development of quantum mechanics.*

The validity of Eq. (1) was established in a series of experiments by Millikan who also made the first direct determination of Planck's constant h. In his Nobel lecture, Millikan said (see Ref. 5):

> *After ten years of testing and changing and learning and sometimes blundering, all efforts being directed from the first toward the accurate experimental measurement of the energies of emission of photoelectrons, now as a function of temperature, now of wavelength, now of material (contact e.m.f. relation), this work resulted, contrary to my own expectation, in the first direct experimental proof in 1914 of the exact validity, within narrow limits of experimental error, of the Einstein equation [Eq. (1)], and the first direct photoelectric determination of Planck's constant h.*

Millikan further wrote: '*Einstein's equation is one of exact validity (always within the present small limits of experimental error) and of very general applicability, is perhaps the most conspicuous achievement of Experimental Physics during the past decade.*'

We may mention here that a year or two after the foregoing work on photoelectric effect was completed, Duane and his associates found unambiguous proof of a relation, which is just the inverse of Einstein's. They bombarded a metal target with electrons of known and constant energy and found that the maximum frequency of the emitted X-rays was given, with much precision, by the following equation

$$\frac{1}{2}mv^2 = h\nu \tag{2}$$

In making this transition from Planck's *quantised oscillators* to *quanta of radiation*, Einstein had made a very important conceptual transition, namely, he introduced the idea of corpuscular behaviour of radiation. Although Newton had described light as a stream of particles, this view had been completely superseded by the wave picture of light, a picture that culminated in the electromagnetic theory of Maxwell. The revival of the particle picture now posed a severe conceptual problem, one of reconciling wave- and particle like behaviour of radiation. It also soon became apparent that matter also exhibited both types of behaviour. For example, an electron with an accurately measured value of mass and charge could undergo diffraction in a manner similar to that of light waves. We will now give a brief account of some of the other important experimental evidence showing wave-particle duality which led to the development of the quantum theory.

3.4 THE COMPTON EFFECT

We have seen that Einstein's explanation for the photoelectric effect implies that quanta of light (photons) carry a definite amount of energy. Compton effect provided an unambiguous example of a process in which a quantum of radiation carrying energy as well as momentum scatters off an electron (see Fig. 3.4). Now, if u represents the energy per unit volume associated with a plane electromagnetic wave, Maxwell's equations predict that the momentum per unit volume associated with the electromagnetic wave would be u/c, where c represents the speed of light in free space. Since each photon carries an energy equal to $h\nu$, it should have a momentum given by the following equation (see sections 23.9 and 32.2 of Ref. 2):

$$p = \frac{h\nu}{c} = \frac{h}{\lambda} \tag{3}$$

In 1923 Compton investigated the scattering of X-rays by a block of paraffin and found that the wavelength of the radiation scattered at an angle of 90° is greater than the wavelength of the incident radiation. In other words, the frequency ν' of the scattered wave is smaller than the frequency of the incident wave. Compton was able to explain the result[4] quantitatively as that of an elastic collision between a photon of energy $E = h\nu$ and the momentum given by Eq. (3). Compton was awarded the 1927 Nobel Prize in Physics *for his discovery of the effect named after him.*

The light quantum imparts some of its energy to the electron and emerges with lesser energy. Thus the scattered radiation has a lower frequency. The kinematics of this collision process can be worked out on elementary application of the laws of conservation of energy and momentum [see Section 3.4.1]. These calculations give the following expression for the shift in the wavelength

$$\Delta\lambda = \frac{2h}{m_0 c}\sin^2\frac{\theta}{2} \tag{4}$$

where θ is the angle of scattering of the light quantum [see Fig. 3.4] and m_0 represents the rest mass of the electron. If we substitute the values of h, m_0 and c, we obtain

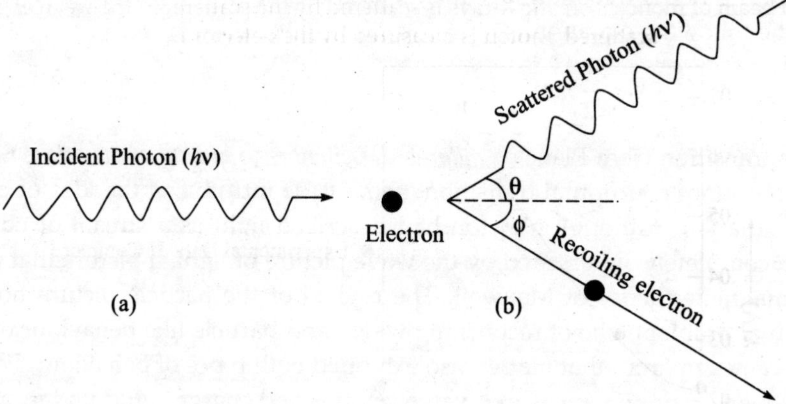

Figure 3.4: Compton scattering of a photon: (a) shows the incidence of a photon (of frequency ν) on an electron at rest, (b) shows the scattered photon (having a reduced frequency ν') propagating along the direction which makes an angle ϕ with the original direction; the electron also acquires a momentum.

---

[4] According to the classical explanation of Compton scattering, the electron undergoes oscillatory motion because of the electric field associated with the incident electromagnetic radiation. The accelerated electron emits electromagnetic waves and because of Doppler shifts due to the motion of the electron, the emitted wavelength differs from the wavelength of the incident radiation; however, the classical theory predicts that for a given angle of scattering a continuous range in the value of the scattered wavelength should be formed, which is contrary to experimental findings. The details of this analysis are given in Section 2.9 of Ref. 6.

$$\Delta\lambda = \lambda' - \lambda = 0.0485 \sin^2 \frac{\theta}{2} \tag{5}$$

where λ is measured in Angstroms. The above equation shows that the maximum change in the wavelength is about 0.05 Å, and as such for a measurable shift one must use radiation of smaller wavelength. In Fig. 3.5, we have given the schematic of the experimental arrangement for the measurement of the Compton shift. A monochromatic beam of X-rays (or γ-rays) is allowed to fall on a sample scatterer and the scattered photons were detected by means of a crystal spectrometer. The crystal spectrometer allows one to find the intensity distribution (as a function of λ) for a given value of θ. In Fig. 3.6, we have shown the wavelength of the scattered photon at different angles with respect to the primary beam as obtained by Compton in his original experiment[5] in 1923. The solid curve corresponds

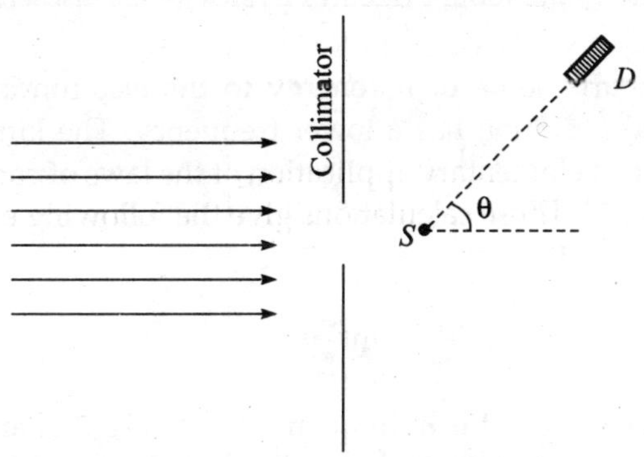

Figure 3.5: Outline of the experimental arrangement for the measurement of the Compton shift. A collimated beam of monochromatic X-rays is scattered by the scatterer S; the wavelength of the scattered photon is measured by the detector D.

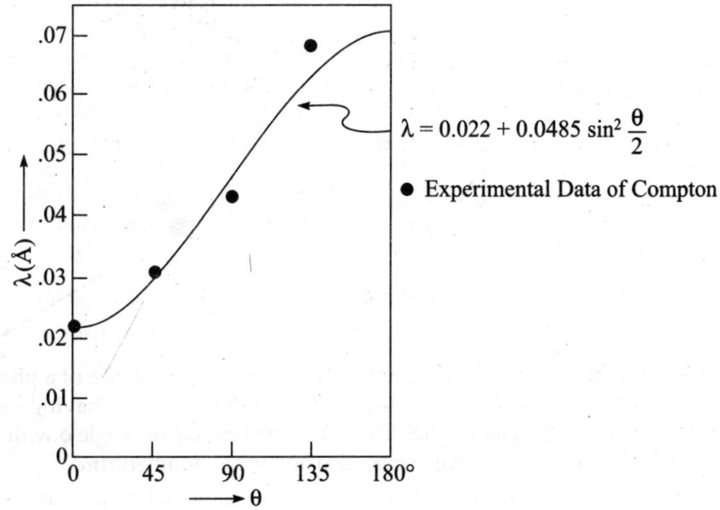

Figure 3.6: The variation of wavelength of the scattered photon with the angle of scattering. The solid curve corresponds to Eq.(5) with λ = 0.022 Å. The dots represent the experimental points obtained by Compton. The figure has been adapted from the original paper of A.H. Compton, 'A Quantum theory on the Scattering of X-rays by Light Elements', *Physical Review*, Vol. 21, 382, 1923.

[5] The original papers of Compton have been published in R.S. Shankland (ed.), *Scientific Papers of A.H. Compton: X-ray and Other Studies*, University of Chicago Press, 1975.

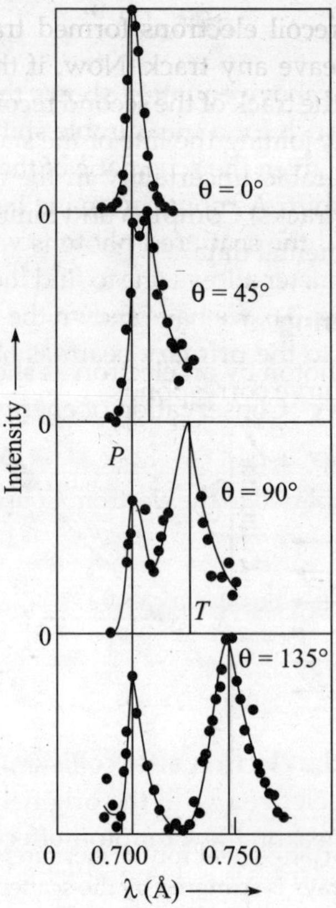

Figure 3.7: The intensity variation as a function of the wavelength of the scattered photon. The vertical line (marked P) corresponds to the unmodified wavelength $\lambda = 0.711$ Å. The second vertical line (marked T) corresponds to the wavelength as predicted by Eq. (4). The figure has been adapted from A.H. Compton's original paper "The spectrum of Scattered X-rays", *Physical Review*, **22**, 409. 1923.

to Eq. (5) with $\lambda = 0.022$ Å. Notice that the corresponding photon energy is

$$\sim \frac{6.6 \times 10^{-27} \times 3 \times 10^{10}}{2.2 \times 10^{-10}} \text{ ergs} \sim 9 \times 10^{-7} \text{ ergs} \sim 0.5 \text{ MeV}$$

which corresponds to a γ-ray. The good agreement between theory and experiment proves that radiation behaves as if it consists of corpuscles of energy $h\nu$ having a momentum $h\nu/c$.

The experimental findings of Compton are shown in Fig. 3.7; the experiment corresponds to the Molybdenum K_α line ($\lambda = 0.711$ Å). The sample used was graphite. Notice that at each value of θ, there are two peaks; the first peak appears at almost the same wavelength as the primary beam. This peak is because of the fact that the photon may be scattered by the whole atom; consequently, the quantity m_0 appearing in Eq.(4) is not the electron mass but the mass of the carbon atom (which is about 22,000 times that of the mass of the electron). Thus the wavelength shift is negligible. The second peak corresponds to the Compton shift. In each figure, the two vertical lines correspond to the unmodified wavelength and the modified wavelength as given by Eq.(5) and one can see a good agreement between the predicted and observed values.

Further evidence of the validity of the above theory was provided by the experiments carried out by Compton and Simon who studied the scattering of X-rays through supersaturated water vapor. In the scattering process, the recoil electrons formed tracks of condensed droplets; however, the light quantum did not leave any track. Now, if the light quantum undergoes another Compton scattering then from the track of the second recoil electron, one can determine the path of the light quantum by simply joining the line of the starting points of the two recoil electrons. Although there was considerable uncertainty in the analysis of the experimental data (because of the presence of many tracks) Compton and Simon could establish agreement between theoretical results and experimental data.

3.4.1 Kinematics of Compton Scattering

We, next, consider the scattering of a photon by an electron as shown in Fig. 3.4. The scattered photon is assumed to have a frequency ν'. Conservation of energy leads to

$$h\nu = h\nu' + E_k \tag{6}$$

where E_k represents the kinetic energy imparted to the electron. Conserving the x and y components of the momentum, we have

$$\frac{h\nu}{c} = \frac{h\nu'}{c}\cos\theta + p\cos\phi \tag{7}$$

and

$$0 = \frac{h\nu'}{c}\sin\theta - p\sin\phi \tag{8}$$

where p represents the momentum of the electron after collision, θ and ϕ represent the angles made by the scattered photon and the electron with the original direction of the photon (see Fig. 3.4). It will be shown that for a measurable Compton effect, the frequency ν should be in the X-ray or in the γ-ray region (for X-rays $\lambda \le 1$ Å and $h\nu \ge 10^4$ eV). For such high energy photons, the velocity imparted to the electron is comparable to the speed of light and one must use proper relativistic expressions for E_k and p.

According to the special theory of relativity (see chapter 32 of Ref. 2), the kinetic energy E_k of the scattered electron would be given by

$$E_k = E - m_0c^2 = mc^2 - m_0c^2 = \frac{m_0c^2}{\sqrt{1-\beta^2}} - m_0c^2 \tag{9}$$

where $\beta = v/c$, m_0 represents the rest mass of the electron, v the speed of the electron and c the speed of light in free-space; the quantities E and m_0c^2 are known as the total energy and the rest mass energy of the electron. Further, the relativistic momentum of the electron is given by

$$p = mv = \frac{m_0v}{\sqrt{1-\beta^2}} \tag{10}$$

Now,

$$p^2c^2 + m_0^2c^4 = \frac{m_0^2v^2c^2}{1-v^2/c^2} + m_0^2c^4$$

$$= \frac{m_0^2c^4}{1-v^2/c^2} = m^2c^4$$

or

$$p^2c^2 + m_0^2c^4 = E^2 = (E_k + m_0c^2)^2 = E_k^2 + m_0^2c^4 + 2E_km_0c^2$$

Thus,

$$E_k^2 + 2E_km_0c^2 = p^2c^2$$

Substituting for E_k from Eq. (3), we get

$$h^2(\nu - \nu')^2 + 2h(\nu - \nu')m_0c^2 = p^2c^2 \tag{11}$$

Further, Eqs. (7) and (8) can be rewritten in the form

$$p \cos \phi = \frac{h\nu}{c} - \frac{h\nu'}{c} \cos \theta \tag{12}$$

and

$$p \sin \phi = \frac{h\nu'}{c} \sin \theta \tag{13}$$

In order to eliminate ϕ, we square and add to obtain

$$p^2 = \left(\frac{h\nu}{c}\right)^2 + \left(\frac{h\nu'}{c}\right)^2 - \frac{2h^2\nu\nu'}{c^2} \cos \theta \tag{14}$$

Substituting in Eq. (11), we obtain

$$h^2(\nu^2 - 2\nu\nu' + \nu'^2) + 2h(\nu - \nu')m_0c^2$$

$$= h^2\nu^2 + h^2\nu'^2 - 2h^2\nu\nu' \cos \theta$$

or

$$\frac{2h(\nu - \nu')m_0c^2}{2\nu\nu'} = h^2 (1 - \cos \theta)$$

or

$$\Delta\lambda = \lambda' - \lambda = \frac{h}{m_0c} (1 - \cos \theta)$$

or

$$\Delta\lambda = \frac{2h}{m_0c} \sin^2 \frac{\theta}{2} \tag{15}$$

which gives us the Compton shift.[6]

3.5 WAVE NATURE OF MATTER

Wilson in his experiments with cloud chamber had clearly shown the particle-like behaviour of alpha and beta particles. These are emitted by radioactive elements and on passing through supersaturated vapour, they form tracks of condensed droplets. For alpha particles, these tracks are nearly straight lines; however, for electrons, the tracks are irregularly curved. The existence of continuous tracks suggest that the emissions from the radioactive substance can be regarded as minute particles moving with high speed. Further, the facts that electrons could be deflected by electric and magnetic fields and also that one could accurately determine the ratio of their charge to mass suggest very strongly that electrons are particles. This view remained unchallenged for a number of years. [C.T.R. Wilson was awarded the 1927 Nobel prize in Physics for his "*method of making the paths of electrically charged particles visible by condensation of vapor*"].

At this stage one could ask if matter may not show wave-like behaviour also just as light exhibited corpuscular and wave-like behaviour. In 1925, de Broglie proposed just such a hypothesis and argued that the relation given by Eq. (3), between wavelength and momentum

[6] In the derivation of the Compton shift we have assumed that the electron is free although we know that the electrons are bound to the atoms. The assumption of a free electron is justified because the binding energy ($\approx$ few eV) is usually very much smaller in comparison to the photon energy ($\gtrsim$ 1000 eV).

is applicable to electrons as well. In his 1925 paper, he wrote: *The basic idea of quantum theory is, of course, the impossibility of considering an isolated fragment of energy without assigning a certain frequency to it*; de Bröglie was awarded the 1929 Nobel prize in Physics "*for his discovery of the wave nature of electrons*". Later, de Bröglie wrote [quoted from p. 58 of Ref. 3]

> *I was convinced that the wave-particle duality discovered by Einstein in his theory of light quanta was absolutely general and extended to all of the physical world, and it seemed certain to me, therefore, that the propagation of a wave is associated with the motion of a particle of any sort—photon, electron, proton or any other.*

In 1927, Davisson and Germer studied the diffraction of electrons from single crystals of nickel and showed that the diffraction patterns could be explained if the electrons were assumed to have a wavelength given by the de Bröglie relation

$$\lambda = \frac{h}{p} \tag{16}$$

where p is the momentum of the electron. Shortly afterwards, in 1928, G.P. Thomson carried out electron diffraction experiments by passing electrons through thin polycrystalline metal targets. The diffraction pattern consisted of concentric rings similar to the Debye-Scherrer rings obtained in the X-ray diffraction pattern. By measuring the diameters of the rings and from the known structure of the crystals, Thomson calculated the wavelength associated with the electron beam which was in agreement with the de Bröglie relation. In 1937, Davisson and Thomson shared the Nobel prize for their *experimental discovery of the diffraction of electrons by crystals*. Max Jammer [Ref. 4] wrote: *Thomson, the father, was awarded the Nobel prize for having shown that the electron is a particle, and Thomson, the son, for having shown that the electron is a wave.*

3.5.1 Diffraction of X-rays by Crystals

Before we discuss the Davisson-Germer experiment, we thought it would be more appropriate to discuss briefly the diffraction of X-rays by crystals. We consider a monochromatic beam of X-rays to be incident on a crystal. In Fig. 3.8, the horizontal dotted lines represent a set of parallel crystal planes with Miller indices (hkl). W_1W_2 and W_3W_4 represent the incident and reflected wave fronts respectively. Obviously, the secondary wavelets emanating from the points A, B and C are in phase on W_3W_4 and the waves emanating from the points A_1, B_1 and C_1 will also be in phase on W_3W_4 if

$$XB_1 + B_1Y = m\lambda, \quad m = 1, 2, 3, \ldots \tag{17}$$

or when

$$2d_{hkl} \sin \theta = m\lambda \tag{18}$$

where d_{hkl} is the interplanar spacing between crystal planes of indices (hkl), $m = 1, 2, 3, \ldots$ is called the order of diffraction and θ is known as the Bragg angle.[7]

Equation (18) is known as Bragg's law and gives the angular positions of the reinforced diffracted beams in terms of the wavelength λ of the incoming X-rays and of the interplanar spacings d_{hkl} of the crystal planes. When the condition expressed by Eq. (18) is not satisfied, destructive interference occurs and no reinforced beam will be produced. Constructive interference occurs when the condition given by Eq. (18) is satisfied leading to peaks in the intensity distribution. For solids which crystallize in cubic structures, the interplanar spacing d_{hkl} between two closest parallel planes with Miller indices (hkl) is given by

[7] W.H. Bragg (the father) and W.L. Bragg (the son) were awarded the 1915 Nobel prize in Physics "*for their services in the analysis of crystal structure by means of Röntgen rays.*"

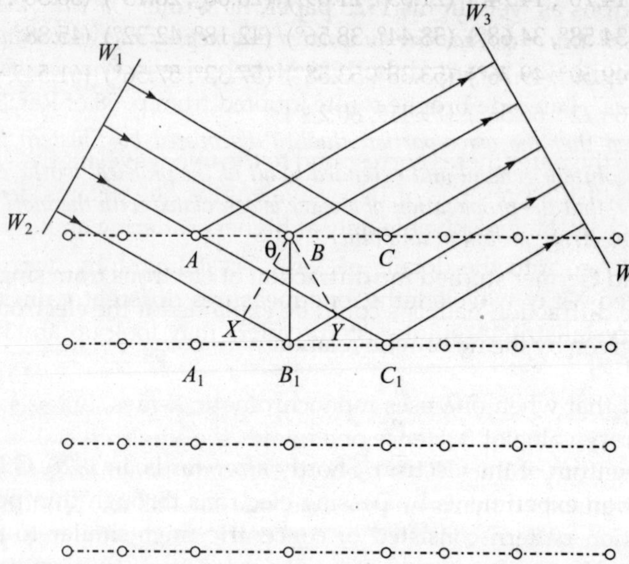

Figure 3.8: Reflection of a plane wave by a set of parallel crystal planes characterized by the Miller indices (*hkl*). When the Bragg condition $2d \sin\theta = m\lambda$ is satisfied, the waves scattered from different rows will be in phase.

$$d_{hkl} = \frac{a}{\sqrt{h^2 + k^2 + l^2}} \tag{19}$$

where a represents the lattice constant. In order to calculate the angles of diffraction we substitute Eq. (19) in the Bragg's law [Eq. (18)] to obtain

$$\frac{2a}{\sqrt{h^2 + k^2 + l^2}} \sin\theta = m\lambda \tag{20}$$

We restrict ourselves only to first order reflections ($m = 1$); higher order reflections are usually rather weak. Thus, Eq. (20) can be written in the form:

$$\sin\theta = \frac{\lambda}{2a}\sqrt{N} \tag{21}$$

where $$N \equiv h^2 + k^2 + l^2 \tag{22}$$

Now, for a simple cubic lattice, all values of (*hkl*) are possible implying the following possible values of N:

$$N = 1, 2, 3, 4, 5, 6, 7, \ldots \quad \text{(SC)} \tag{23}$$

Similarly, for a BCC lattice $h + k + l$ must be even implying

$$N = h^2 + k^2 + l^2 = 2, 4, 6, 8, 10, 12, 14, 16, 18, 20, 22, \ldots \quad \text{(BCC)} \tag{24}$$

Finally, for an FCC lattice, Miller indices are either all even or all odd implying

$$N = h^2 + k^2 + l^2 = 3, 4, 8, 11, 12, 16, 19, 20, 24, 27, \ldots \quad \text{(FCC)} \tag{25}$$

For a given structure and for given values of λ and a, one can now easily calculate the different values of θ. For example, if we consider λ = 1.540 Å and 1.544 Å (corresponding to the $CuK_{\alpha 1}$ and $CuK_{\alpha 2}$ lines) then for sodium (which is a BCC structure with a = 4.2906 Å), the various values of θ are

(14.70°, 14.74°) (21.03°, 21.09°) (26.08°, 26.15°) (30.50°, 30.59°)
(34.58°, 34.68°) (38.44°, 38.56°) (42.18°, 42.32°) (45.88°, 46.03°)
(49.59°, 49.76°) (53.38°, 53.58°) (57.33°, 57.56°) (61.54°, 61.82°)
(66.22°, 66.56°) (79.41°, 80.23°)

The two values inside the parentheses correspond to the two wavelengths 1.540 Å and 1.544 Å, respectively. Because of the presence of two wavelengths one obtains double lines for each family of planes which become resolvable only at higher scattering angles.

We do not wish to go into the details of X-ray crystallography, we would only like to mention that for given X-ray wavelengths, one measures different values of θ—an analysis of which gives the crystal structure; interested readers may look up any book on solid state physics [see, e.g., Ref. 7].

We may point out that when one uses monochromatic X-rays, Bragg's formula cannot be satisfied for an arbitrary value of θ. Hence one rotates the single crystal so that reflection can occur for a discrete set of θ values. This method can only be employed if single crystals of reasonable size are available. If this is not the case, one can still use monochromatic X-rays — provided the sample is in the powder form so that there are always enough crystallites of the right orientation available to satisfy the Bragg relation. A powder will consist of a large number of randomly oriented micro-crystals; each micro-crystal is essentially a single crystal. As the X-ray beam passes through such a polycrystalline material, the orientation of any given set of planes, with reference to the X-ray beam, changes from one microcrystal to the other. Thus, corresponding to any given set of planes there will be a large number of crystals for which Bragg's condition will be satisfied, and on the photographic plate one will obtain concentric rings [see Fig. 3.9]; each ring will correspond to a particular value of d_{hkl} and a particular value of m.

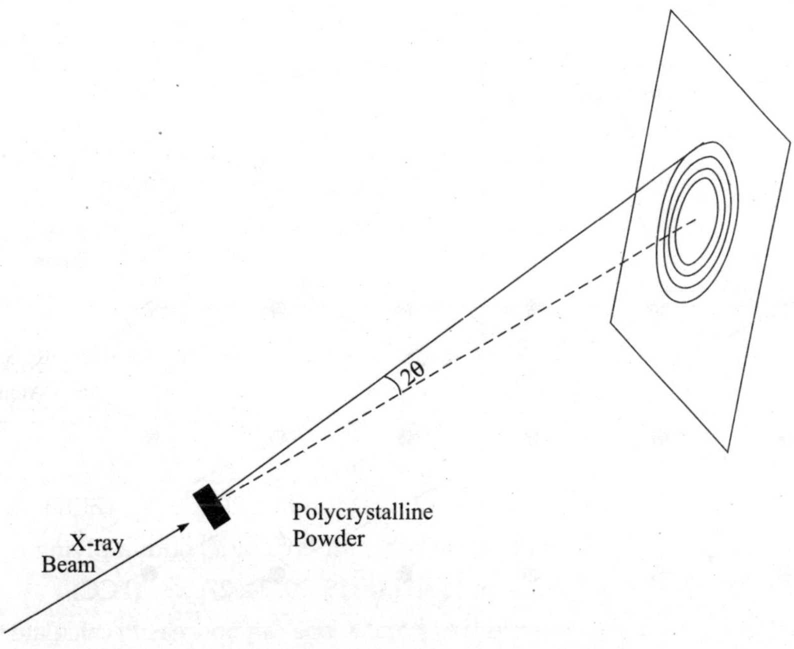

Figure 3.9: When a monochromatic X-ray beam falls on a polycrystalline sample one obtains the Debye-Scherrer rings.

3.5.2 The Davisson-Germer Experiment

Figure 3.10 gives a schematic of the original experimental set up of Davisson and Germer (some details are given in the Nobel lecture of Davisson—see Ref. 8). In the experiment, an electron beam [accelerated by a (variable) potential] was incident normally on a single crystal of nickel. A collector was designed to accept only elastically scattered electrons. Corresponding to a particular setup, Fig. 3.11 shows the intensity variation as a function of ϕ for different accelerating potentials; ϕ represents the angle between the incident and the scattered beams

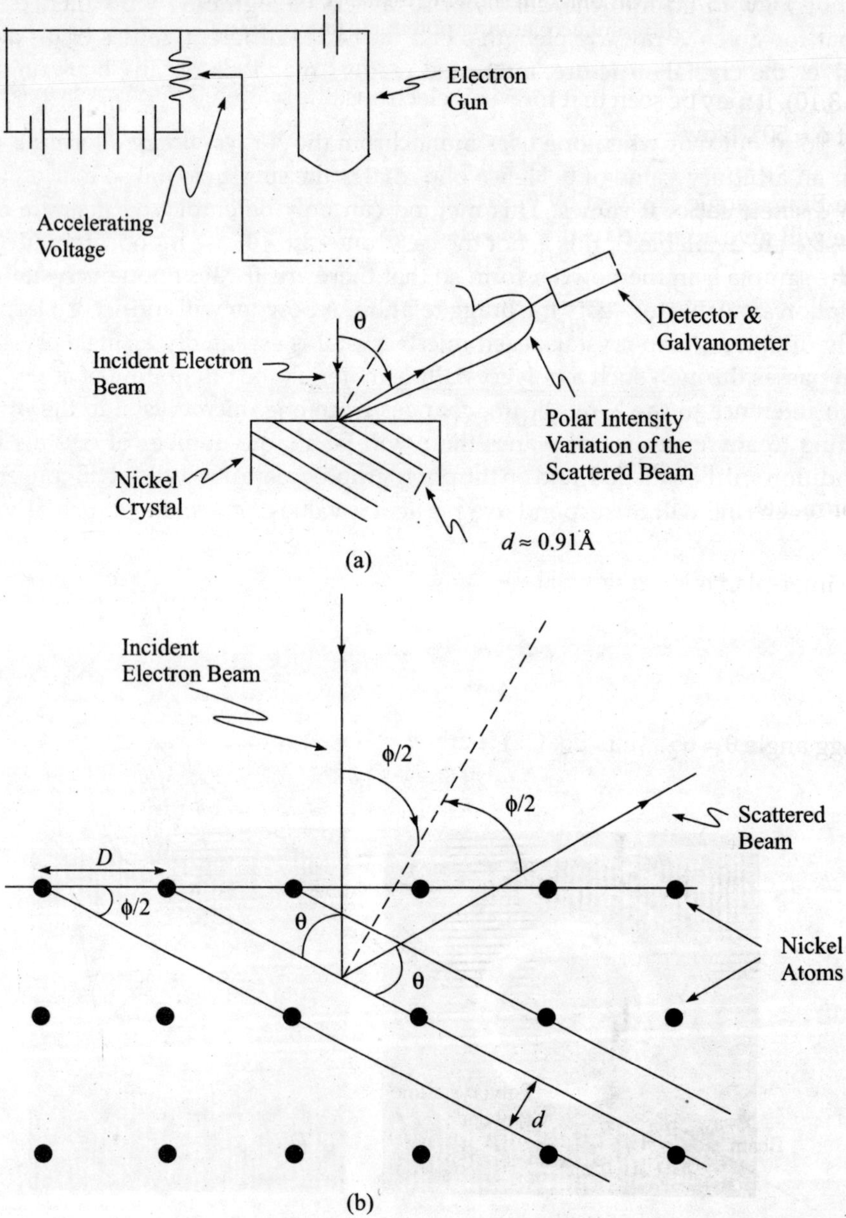

Figure 3.10: (a) Schematic of the original setup of Davisson and Germer. (b) Bragg reflection from atomic planes of nickel.

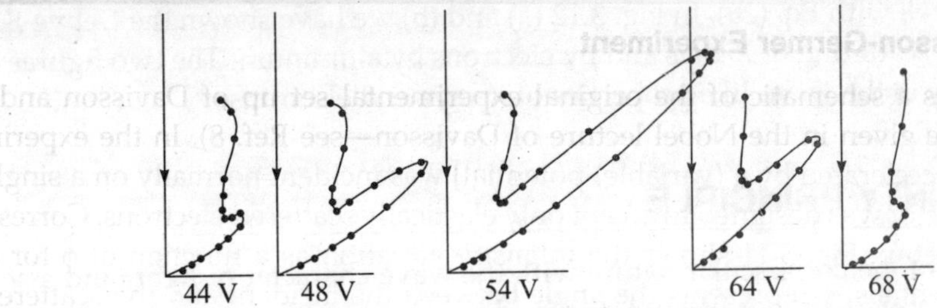

Figure 3.11: Polar diagram showing intensity variation as a function of ϕ for different accelerating potentials (adapted from Ref. 8).

(see Fig. 3.10). It may be seen that for 54 eV electrons, the scattered intensity is maximum which occurs at $\phi \approx 50°$. Now

$$2\theta = \pi - \phi$$

Thus the Bragg angle is given by $\theta \approx 65°$. Further, for 54 eV electrons use of non-relativistic formulae will give accurate results and therefore the corresponding de Bröglie wavelength is given by

$$\lambda = \frac{h}{p} = \frac{h}{\sqrt{2mE}}$$

$$= \frac{6.626\times10^{-34}}{\sqrt{2\times9.109\times10^{-34}\times(54\times1.602\times10^{-19})}} \approx 1.67\times10^{-10}\ \text{m}$$

$$\approx 1.67\ \text{Å} \tag{26}$$

Now, for nickel

$$D \approx 2.15\ \text{Å} \tag{27}$$

And the inter-planar spacing will be given by

$$d = D\sin\frac{\phi}{2}$$

$$= 2.15\sin 25° \approx 0.909\ \text{Å} \tag{28}$$

The Bragg angle $\theta = 65°$, thus Eq. (22) [with $m = 1$] will give

$$\lambda = 2d\sin\theta \approx 1.65\ \text{Å} \tag{29}$$

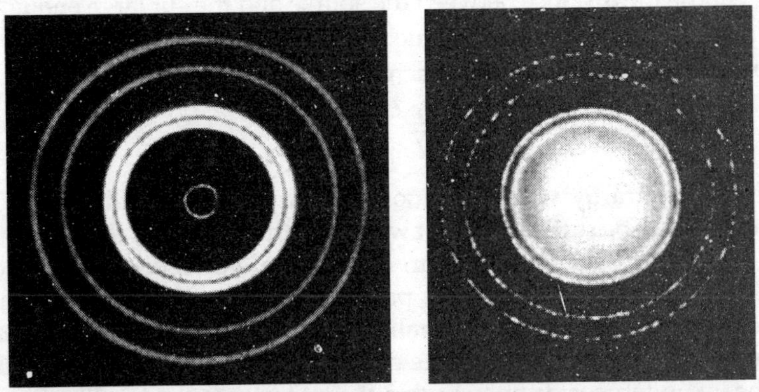

(a) (b)

Figure 3.12: The diffraction pattern of aluminum foil produced (a) by X-rays and (b) by electrons; notice the similarity in the diffraction patterns. Photographs courtesy McGraw Hill Digital Access Library.

which compares well with Eq. (26). In Fig. 3.12 (a) and (b) we have shown the Debye-Scherrer rings produced by scattering of X-rays and by electrons by aluminum. The two figures clearly show the similarity in the wave-like properties of X-rays and electrons.

3.6 UNCERTAINTY PRINCIPLE

The reconciliation of the corpuscular nature with the wave character of light (and also of the electron) has been brought about through the modern quantum theory; and perhaps the best known consequence of wave-particle duality is the uncertainty principle of Heisenberg which can be stated as follows:

> *If the x-coordinate of the position of a particle is known to an accuracy Δx, then the x-component of the momentum cannot be determined to an accuracy better than $\Delta p_x \approx h/\Delta x$, where h is the Planck's constant.*

Alternatively, one can say that if Δx and Δp_x represent the accuracies with which the x-coordinate of the position and the x-component of the momentum can be determined, then the following inequality must be satisfied

$$\Delta x \, \Delta p_x \geq h \tag{30}$$

We do not feel the effect of this inequality in our everyday experience because of the smallness of the value of Planck's constant ($\approx 6.6 \times 10^{-27}$ erg-sec). For example, for a tiny particle of mass 10^{-6} g, if the position is determined within an accuracy of about 10^{-6} cm, then according to the uncertainty principle, its velocity cannot be determined within an accuracy better than $\Delta v \approx 6 \times 10^{-16}$ cm/sec. This value is much smaller than the accuracies with which one can determine the velocity of the particle. For a particle of a greater mass, Δv will be even smaller. Indeed, had the value of Planck's constant been much larger, the world would have been totally different. In a beautifully written book, Gamow (Ref. 10) has discussed what our world would be like if the effect of the uncertainty principle were perceivable by our senses.

3.6.1 The Single-Slit Diffraction Experiment

We will now show how the diffraction of a light beam (or an electron beam) can be explained on the basis of the corpuscular nature of radiation and the uncertainty principle; see also Solution 5.7 on page 5.18. Consider a long narrow slit of width b as shown in Fig. 3.13. Now, one can always choose the distance between the source and the slit large enough so that p_x can be assumed to have an arbitrarily small value. For example, for the source at a distance d, the maximum value of p_x of the photons approaching the slit will be

$$p\frac{b}{d} = \frac{h\nu}{c}\cdot\frac{b}{d}$$

which can be made arbitrarily small by choosing a large enough value of d. Thus, we may assume the light source to be sufficiently far away from the slit so that the photons approaching the slit can be assumed to have momentum only in the y-direction. Now, according to the particle model of radiation, the number of particles reaching the point P (which lies in the geometrical shadow) will be extremely small; further, if we decrease the width of the slit, the intensity should decrease, which is quite contrary to the experimental results because we know that the beam undergoes diffraction and the intensity at a point like P would normally increase if the width of the slit is made smaller. Thus, the classical corpuscular model is quite incapable of explaining the phenomenon of diffraction. However, if we use the uncertainty

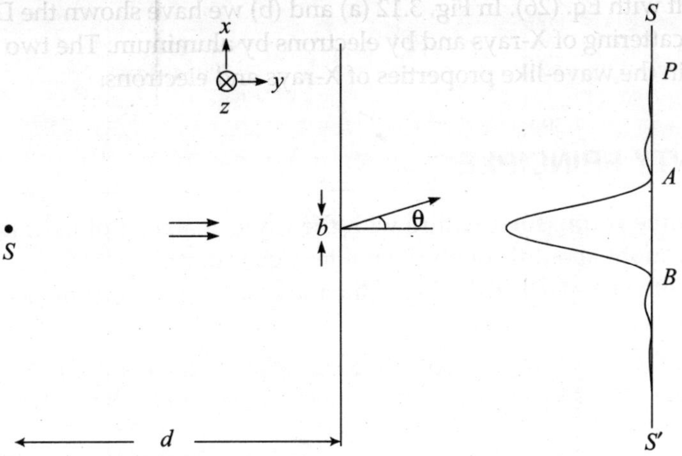

Figure 3.13: Diffraction of a photon (or an electron) beam by a long narrow slit of width b.

principle in conjunction with the corpuscular model, the diffraction phenomenon can be explained in the following manner: When a photon (or an electron) passes through the slit, one can say that

$$\Delta x \approx b$$

which implies that we can specify the position of the photon to an accuracy b. If we now use the uncertainty principle, we would have

$$\Delta p_x \approx \frac{h}{b} \tag{31}$$

that is, just by making the photon pass through a slit of width b, the slit imparts a momentum in the x-direction which is $\approx h/b$. It may be pointed out that *before* the photon entered the slit, p_x (and hence Δp_x) can be made arbitrarily small by putting the source sufficiently far away. Thus we may write $\Delta p_x \approx 0$. It would however be wrong to say that by making the photon pass through the slit, $\Delta p_x \Delta x$ is zero; this is because of the fact that $\Delta p_x \approx 0$ *before* the photon entered the slit. After the photon has entered the slit, it is confined within a distance b in the x direction and hence $\Delta p_x \approx h/b$. Further, since before the photon entered the slit $p_x \approx 0$, we will therefore have

$$|p_x| \approx \Delta p_x \approx \frac{h}{b}$$

But $p_x = p \sin \theta$, where θ is the angle that the photon coming out of the slit makes with the y-axis (see Fig. 3.13). Thus

$$p \sin \theta \approx \frac{h}{b}$$

or

$$\sin \theta \approx \frac{h}{pb} \tag{32}$$

The above equation predicts that the possibility of a photon travelling at an angle θ with the y-direction is inversely proportional to the width of the slit; i.e., smaller the value of b, greater is the value of θ and greater is the possibility of the photon to reach deep inside the geometrical shadow. This is indeed the diffraction phenomenon. Now, the momentum of a photon is given by

$$p = \frac{h}{\lambda} \tag{33}$$

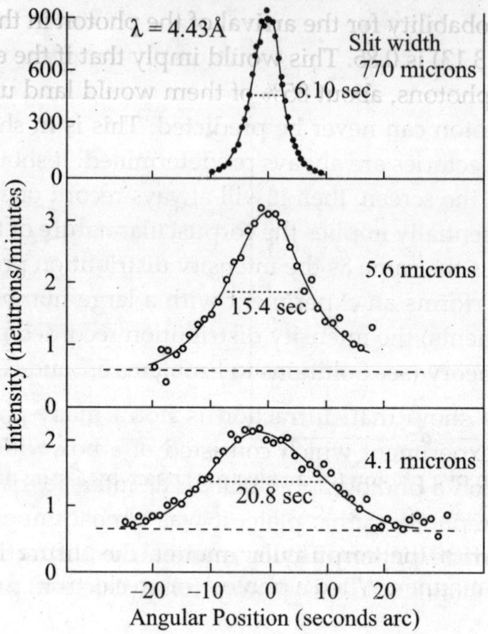

Figure 3.14: Angular broadening of a neutron beam by small slits [after Ref. 12].

Thus Eq. (32) becomes

$$\sin\theta \approx \frac{\lambda}{b} \tag{34}$$

which is the familiar diffraction theory result. We can, therefore, say that the wave-particle duality is a consequence of the uncertainty principle and the uncertainty principle is a consequence of the wave-particle duality. To quote Max Born [Ref. 11]

> *Physicists of today have learnt that not every question about the motion of an electron or a photon can be answered, but only those questions which are compatible with the uncertainty principle.*

Returning to Eq. (33), we may mention that de Bröglie had suggested that the equation $\lambda = h/p$ is not only valid for photons but is also valid for all particles like electrons, protons, neutrons, etc. Indeed, the de Bröglie relation has been verified by studying the diffraction patterns produced when electrons, neutrons, etc., pass through a single crystal; the patterns can be analysed in a manner similar to the X- ray diffractions. In Fig. 3.14, we show the experimental data of Shull (Ref. 12) who studied the Fraunhofer diffraction of neutrons by a single slit and his experimental results agree with the intensity distribution as predicted by the wave theory with λ given by Eq. (33).

3.7 PROBABILISTIC INTERPRETATION OF MATTER WAVES

In the previous section we have seen that if a photon passes through a slit of width b, then the momentum imparted in the x-direction (which is along the width of the slit) is $\approx h/b$. The question arises whether we can predict the trajectory of an individual photon. The answer is no. We cannot say where an individual photon will land up on the screen; we can only predict the probabilities of arrival of the photon in a certain region of the screen. We may, for example, say that the probability for the arrival of the photon in the region lying between the points A and B (see Fig. 3.13) is 0.85. This would imply that if the experiment was carried out with a large number of photons, about 85% of them would land up in the region AB; but the fate of an individual photon can never be predicted. This is in sharp contrast to Newtonian mechanics where the trajectories are always predetermined. It should be mentioned that if we place a light detector on the screen, then it will always record one photon or none and never half of a photon. This essentially implies the corpuscular nature of the radiation. However, the probability distribution is the same as the intensity distribution predicted by the wave theory and, therefore, if one performs an experiment with a large number of photons (as is indeed the case in most experiments) the intensity distribution recorded on the screen is the same as predicted by the wave theory (see Solutions to Problems 5.7 and 5.8).

In order to explicitly show that diffraction is not a many-photon phenomenon, Taylor in 1909 carried out an experiment which consisted of a box with a small lamp which casts the shadow of a needle on a photographic plate. The intensity of the light was so weak that between the slit and the photographic plate, it was almost impossible to find two photons. In fact to get a good fringe pattern Taylor made an exposure lasting for several months. The diffraction pattern obtained on the photographic plate was the same as predicted by the wave theory.

The corpuscular nature of radiation and the fact that one cannot predict the trajectory of an individual photon can be seen from Fig. 3.15 which consists of series of photographs obtainable from various number of photons. The photograph clearly shows that the picture is built up by the arrival of concentrated packets of energy and the point at which a particular photon will arrive is entirely a matter of chance. The figure also shows that the photograph is featureless when a small number of photons are involved and as the number of photons reaching the photographic plate increases, the intensity distribution becomes the same as would be predicted by the wave theory. To quote Feynman

> *... it would be impossible to predict what would happen. We can only predict the odds! This would mean, if it were true, that physics has given up on the problem of trying to predict exactly what will happen in a definite circumstance. Yes! physics has given up. We do not know how to predict what would happen in a given circumstance, and we believe now that it is impossible—that the only thing that can be predicted is the probability of different events. It must be recognized that this is a retrenchment in our earlier idea of understanding nature. It may be a backward step, but no one has seen a way to avoid it.*

A somewhat similar situation arises in radioactivity. Consider a radioactive nucleus having a half-life of, say, 1 hour. If we start with 1000 such nuclei, then on an average 500 of these would undergo radioactive decay in 1 hour and about 250 of them in the next 1 hour and so on. Although to start with, all nuclei are identical, some nuclei would decay in the very first minute and some nuclei can survive for hours without undergoing radioactive decay. Thus, one can never predict as to which nucleus will undergo decay in a specified period; one can only predict the probability of its undergoing decay in a certain interval of time. This is indeed a beautiful manifestation of quantum mechanics. To quote Feynman again

> *A philosopher once said it is necessary for the very existence of science that the same conditions always produced the same results. Well they don't!*

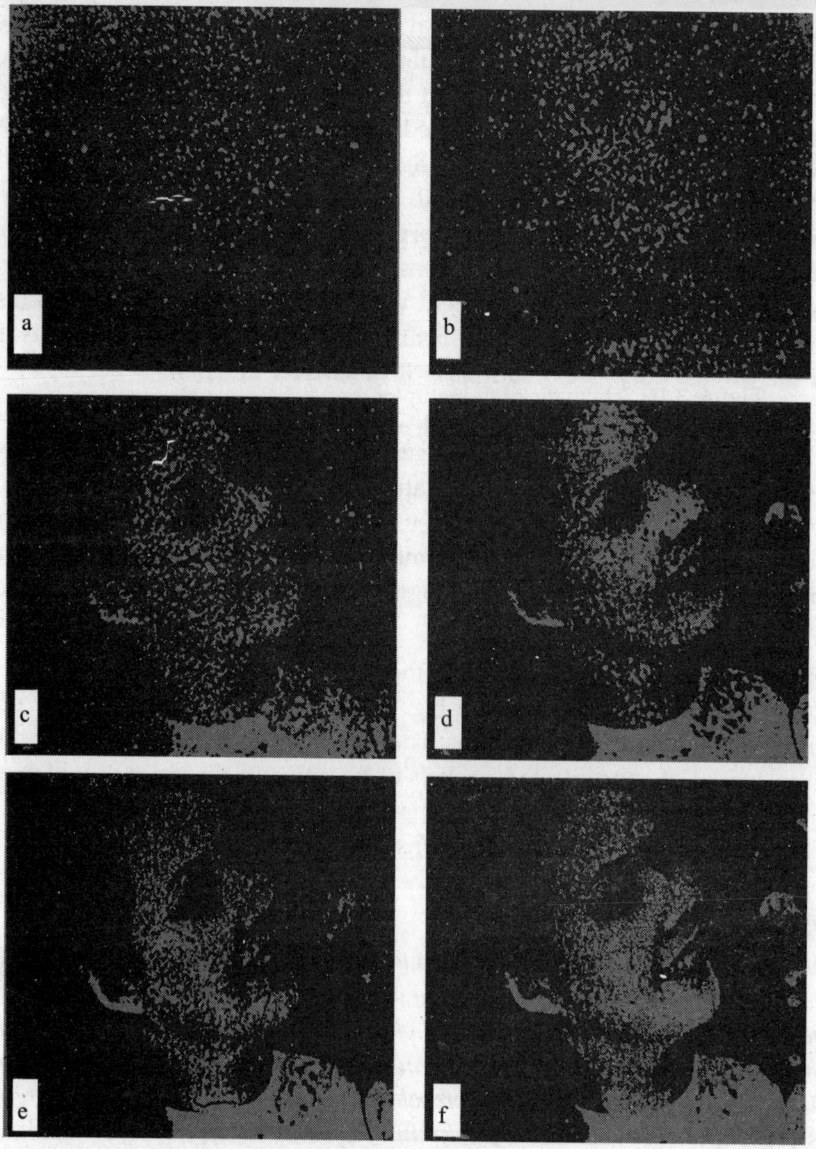

Figure 3.15: Photographs showing the quality of a picture obtainable from various numbers of photons: (a), (b), (c), (d), (e) and (f) correspond to 3×10^3 photons, 1.2×10^4 photons, 9.3×10^4 photons, 7.6×10^5 photons, 3.6×10^6 photons and 2.8×10^7 photons respectively. (From A. Rose, '*Quantum Effects in Human Vision*', *Advances in Biological and Medical Physics*, Vol. 5, Academic Press, 1957; used with permission).

3.8 AN UNDERSTANDING OF INTERFERENCE EXPERIMENTS

Let us consider the interference experiment involving the Michelson interferometer in which a light beam is partially reflected by a beam splitter and the resulting beams are made to interfere (see Fig. 3.16). According to Dirac [Ref. 13]

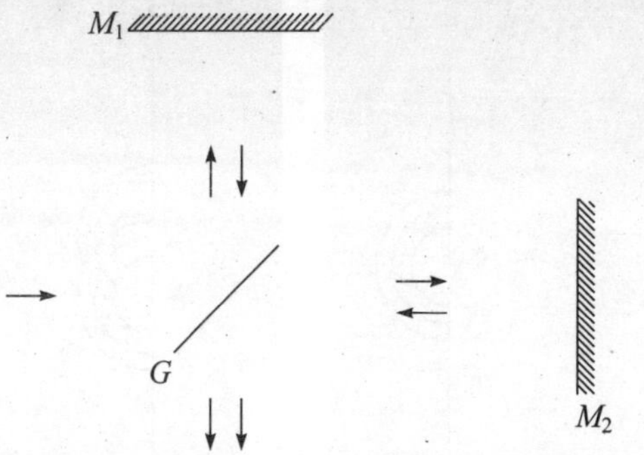

Figure 3.16: Schematic of the setup of the Michelson interferometer.
G represents a beam splitter, M_1 and M_2 represent plane mirrors.

> *... we describe the photon as going partly into each of the two components into which the incident beam is split. The photon is then, as we may say, in a translational state given by the superposition of the two translational states associated with the two components.... For a photon to be in a definite translational state it need not be associated with one single beam of light, but may be associated with two or more beams of light, which are the components into which one original beam has been split. In the accurate mathematical theory each translational state is associated with one of the wave functions of ordinary wave optics, which may describe either a single beam or two or more beams into which one original beam has been split.*

These translational states can be superposed in a manner similar to the one employed while considering the interference of two beams. Thus, each photon goes partly into each of the two components and interferes only with itself. If we try to determine the fate of a single photon by measuring the energy in one of the components then, Dirac argues:

> *The result of such a determination must be either a whole photon or nothing at all. Thus the photon must change suddenly from being partly in one beam and partly in the other to be entirely in one of the beams. This sudden change is due to the disturbance in the translational state of the photon which the observation necessarily makes. It is impossible to predict in which of the two beams the photon will be found. Only the probability of either result can be calculated... Our description of the photon allows us to infer that, after such an energy measurement, it would not be possible to bring about any interference effects between the two components. So long as the photon is partly in one beam and partly in the other, interference can occur when the two beams are superposed, but this probability disappears when the photon is forced entirely into one of the beams by an observation.*

In a similar manner, we may consider the two-hole interference experiment similar to that performed by Young. The experimental arrangement is shown in Fig. 3.17 where a weak light source S_0 illuminates the hole S and the light emerging from the holes S_1 and S_2 produces the interference pattern on the screen PP'. The intensity is assumed to be so weak that there is almost never more than one photon in the region between the planes AB and PP'. Individual photons are also counted by a detector on the screen PP' and one finds that the intensity distribution has a $\cos^2$ pattern similar to that shown in Fig. 3.17. The corpuscular nature of the radiation is evident from its detection in the form of single photons and never a fraction of a photon. The appearance of the interference pattern is because of the fact that a photon interferes with itself. The quantum theory tells us that a photon partially passes through the hole S_1 and partially through S_2. This is not the splitting of the photon into two halves but

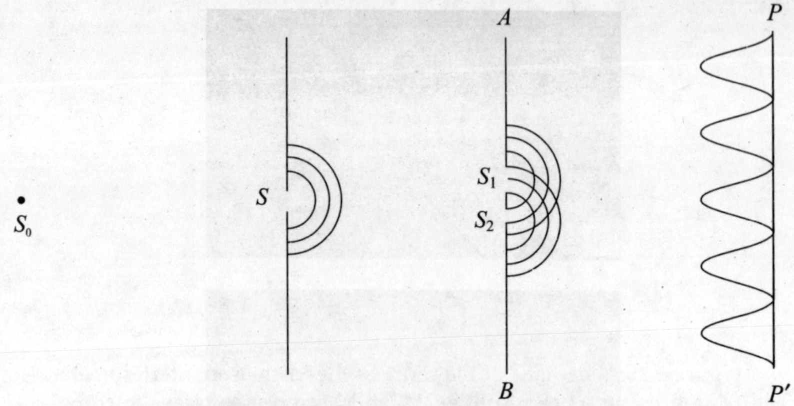

Figure 3.17: Young's double-hole experimental arrangement for obtaining the interference pattern.

only implies that if we wish to find out through which hole the photon passed, then half the time it will be found to have passed through the hole S_1 and half the time through S_2. As in the case of the Michelson interferometer, the photon is in a state which is a superposition of two states, one corresponding to the wave emanating from hole S_1 and the other to the one emanating from hole S_2. The superposed state will give rise to an intensity distribution similar to that obtained by considering the superposition of two waves. It may be noted that if we had employed a device (like a microscope) which would have determined which hole the photon had passed through, then the interference pattern on the screen would have been washed out. This is a consequence of the fact that a measurement always disturbs the system. This is beautifully discussed in Ref. 1. Thus we may say that the photons would arrive as packets of energy but the probability distribution (on the screen) will be proportional to the intensity distribution predicted by using a wave model.

In a recent paper, Tonomura and his co-workers (Ref. 14) have demonstrated the single electron build up of an interference pattern. Their results are shown in Fig. 3.18. It may be seen that when there are very few electrons they arrive randomly; however, when a large number of electrons are involved, one obtains an intensity distribution similar to the one predicted by wave theory.

Example 3.1 In this example (adapted from the Feynman Lectures), we will use the uncertainty principle to determine the size of the hydrogen atom. The example demonstrates the far-reaching consequences of the uncertainty principle.

We consider the hydrogen atom which consists of a proton and an electron. Since the proton is very much heavier than the electron, we consider only the motion of the electron. Let the electron be confined to a region of linear dimension $\approx a$. Thus, according to the uncertainty principle

$$p \approx \Delta p \approx \hbar / a \tag{35}$$

where $\hbar = h/2\pi$; the reason for using $\hbar$ rather than h will be mentioned later. The kinetic energy of the electron will be given by

$$\text{K.E.} = \frac{p^2}{2m} \approx \frac{\hbar^2}{2ma^2} \tag{36}$$

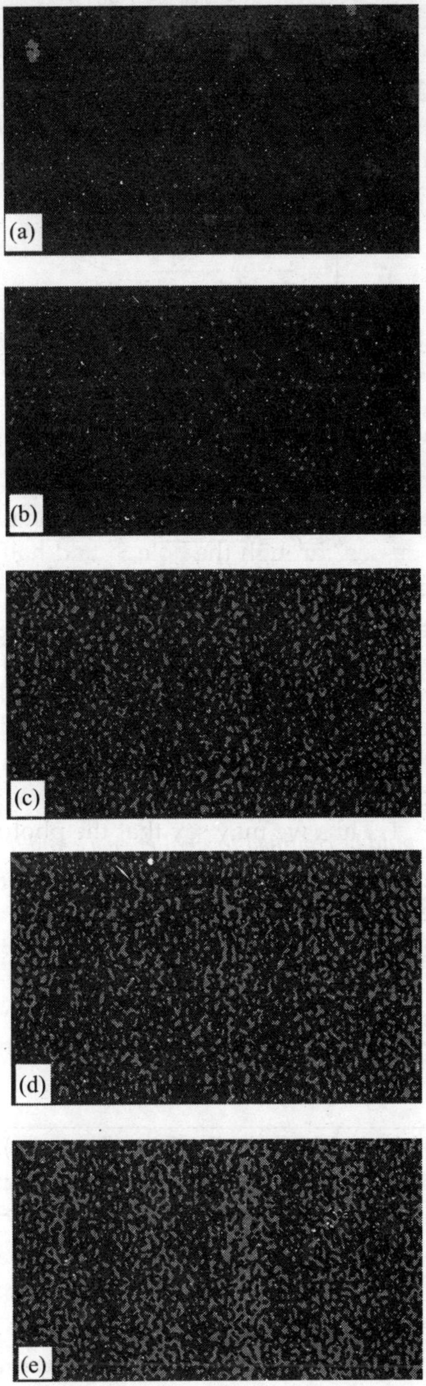

Figure 3.18: Build-up of the electron interference pattern. Number of electrons in (a), (b), (c), (d) and (e) are 10, 100, 3000, 20,000 and 70,000 respectively. [adapted from Ref. 14 used with permission].

Now, there exists an electrostatic attraction between the two particles; the corresponding potential energy being given by

$$\text{P.E.} = -\frac{q^2}{4\pi\varepsilon_0 a} \tag{37}$$

where q ($\approx 1.6 \times 10^{-19}$ C) represents the magnitude of the charge of the electron and ε_0 ($\approx 8.854 \times 10^{-12}$ CN^{-2} m^{-2}) represents the permittivity of free space. Thus, the total energy is given by

$$E = \text{K.E.} + \text{P.E.}$$

$$= \frac{\hbar^2}{2ma^2} - \frac{q^2}{4\pi\varepsilon_0 a} \tag{38}$$

The system would settle to a state of lowest energy; thus we must set dE/da equal to zero:

$$0 = \frac{dE}{da} = -\frac{\hbar^2}{ma^3} + \frac{q^2}{4\pi\varepsilon_0 a^2}$$

implying

$$a = a_0 = \frac{\hbar^2}{m\left(\frac{q^2}{4\pi\varepsilon_0}\right)} \tag{39}$$

If we substitute the values of $\hbar$ ($\approx 1.055 \times 10^{-34}$ Js), m ($\approx 9.11 \times 10^{-31}$ kg), ε_0 and q, we would obtain

$$a = a_0 \approx 0.53 \times 10^{-10} \text{ m} = 0.53 \text{ Å} \tag{40}$$

Thus, we get the remarkable result that the size of the hydrogen atom is a direct consequence of the uncertainty principle. To quote Feynman:

> *So we now understand why we do not fall through the floor .... In order to squash the atoms close together, the electrons would be confined to smaller space and by the uncertainty principle, their momenta would have to be higher on the average, and that means high energy; the resistance to atomic compression is a quantum mechanical effect...*

We next substitute the value of a from Eq. (39) in Eq. (38) to obtain

$$E = \frac{\hbar^2}{2m}\left(\frac{m}{\hbar^2}\frac{q^2}{4\pi\varepsilon_0}\right)^2 - \frac{q^2}{4\pi\varepsilon_0}\left(\frac{m}{\hbar^2}\frac{q^2}{4\pi\varepsilon_0}\right)$$

$$= -\frac{m}{2\hbar^2}\left(\frac{q^2}{4\pi\varepsilon_0}\right)^2 \tag{41}$$

Substituting the values of $\hbar$, m, q and ε_0, we get

$$E \approx -2.17 \times 10^{-19} \text{ J}$$

$$\approx -13.6 \text{ eV} \tag{42}$$

which is nothing but the ground state energy of the hydrogen atom. Thus, the fact that the ionization potential of hydrogen atom is ≈ 13.6 eV follows from the uncertainty principle. We may point out that the uncertainty principle can be used to give only an order of magnitude of the size of the hydrogen atom or its ionization potential; we had intentionally chosen the constants in such a way that the ground state energy comes out to be correct. It is for this reason that we had chosen $\hbar$ instead of h in Eq. (35).

3.9 THE FRANCK-HERTZ EXPERIMENT

We conclude this chapter with a brief discussion of the Franck-Hertz experiment. In 1914, James Franck and Gustav Hertz (nephew of Heinrich Hertz) performed a beautiful experiment demonstrating conclusively the quantized nature of the energy levels of atoms—the quantized energy states of the atom was predicted by Niels Bohr[8] in 1913. Indeed while introducing Franck and Hertz for receiving the Nobel prize, it was mentioned, "... *Bohr's hypotheses of 1913—that the atom can exist in different states, each of which is characterized by a given energy level and that these energy levels govern the spectral lines emitted by the atoms—are no longer mere hypotheses but experimentally proved facts. The methods of verifying these hypotheses are the work of James Franck and Gustav Hertz, for which they have been awarded the Physics Nobel prize for 1925*". The Nobel prize was awarded '*for their discovery of the laws governing the impact of an electron upon an atom*'.

The experimental set-up of Franck and Hertz is shown in Fig. 3.19. The electrons are produced by a heated tungsten filament. These electrons are accelerated by a wire mesh G kept at a distance of a few centimetres from the filament. The wire mesh is kept at a positive potential V which can be varied. There is a plate P behind the wire mesh which is kept at a small negative potential so that very low energy electrons (after passing through G) would not reach the plate P. A typical variation of the plate current (as obtained by Franck and Hertz) with the voltage V is shown in Fig. 3.20. As can be seen when V reaches 4.9 V (or a multiple of it) there is a sudden drop in the current. This is due to the fact that the electrons attaining 4.9 eV of energy excite the Hg atom and lose its entire energy and are no more collected by the plate P. Indeed when the Hg atom de-excites itself, it produces a photon of wavelength 2536 Å for which

$$h\nu = \frac{hc}{\lambda} = \frac{6.63\times10^{34}\times3\times10^{8}}{2.536\times10^{-7}}\,\text{J}$$
$$= 4.9 \text{ eV} \qquad (43)$$

Thus, the Franck-Hertz experiment confirms the discrete energy states of the Hg atom. Similarly, the Franck-Hertz experiment was performed using other atoms.

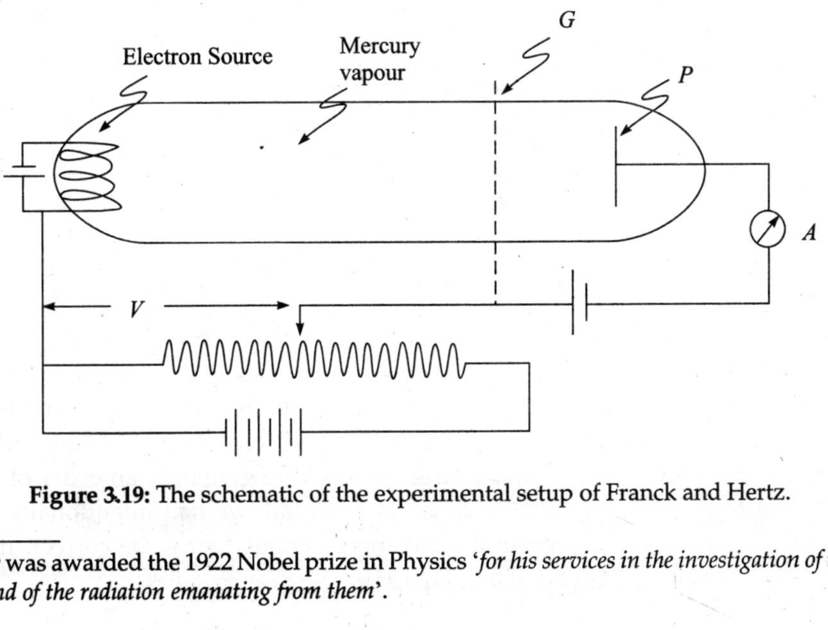

Figure 3.19: The schematic of the experimental setup of Franck and Hertz.

[8] Neils Bohr was awarded the 1922 Nobel prize in Physics '*for his services in the investigation of the structure of atoms, and of the radiation emanating from them*'.

3.10 PROBLEMS

Problem 3.1 An electron of energy 200 eV is passed through a circular hole of radius 10^{-4} cm. What is the uncertainty introduced in the angle of emergence?

Problem 3.2 In continuation of the previous problem, what would be the corresponding uncertainty for a 0.1 g lead ball thrown with a velocity 10^3 cm/sec through a hole 1 cm in radius?

Problem 3.3 Use the uncertainty principle to estimate the energy of the ground state of a two-electron atom of nuclear charge Ze.

[**Hint:** Use a method similar to that used in Example 3.1.]

Problem 3.4 Calculate the uncertainty in the momentum of a proton which is confined to a nucleus of radius equal to 10^{-13} cm. From this result, estimate the kinetic energy of the proton inside the nucleus and the strength of the nuclear interaction. What would be the kinetic energy of an electron if it had to be confined within a similar nucleus?

Problem 3.5 Let a source (with $\lambda = 5 \times 10^{-5}$ cm) of power 1 watt be used in the experimental arrangement shown in Fig. 3.17.

(a) Calculate the number of photons that are being emitted by the source per second.

(b) Assume that the radii of the holes S, S_1 and S_2 to be 0.02 cm and $S_0S = SS_1 = SS_2 = 100$ cm and the distance between the planes AB and PP' to be also 100 cm. Show that in the region between the planes AB and PP' one can almost never find two photons.

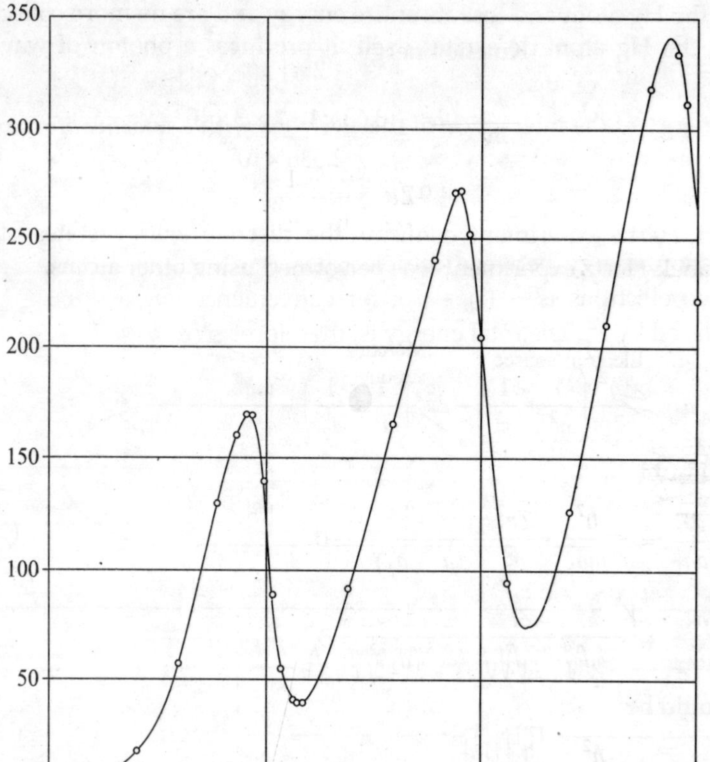

Figure 3.20: The variation of the plate current with voltage V as measured in the original experiment of Franck and Hertz (adapted from Ref. 15).

3.11 SOLUTIONS

Solution 3.1

$$p \approx \sqrt{2mE} = [2 \times 0.9 \times 10^{-27} \times 3.2 \times 10^{-10}]^{1/2} \simeq 8 \times 10^{-19} \text{ g cm/sec}$$

Now, $$\Delta p \sim \frac{\hbar}{\Delta x} \simeq \frac{10^{-27} \text{ erg sec}}{2 \times 10^{-4} \text{cm}} = 5 \times 10^{-24} \text{ g cm/sec.}$$

$$\theta \sim \frac{\Delta p_x}{p} \approx 6 \times 10^{-6} \text{ radians} \simeq 1 \text{ sec of arc}$$

Solution 3.2

$$p \approx 10^2 \text{ cm g/sec}, \Delta p \approx \frac{\hbar}{\Delta x} \approx 5 \times 10^{-28} \text{ g cm/sec.}$$

$$\theta \sim 5 \times 10^{-30} \text{ radians} \approx 10^{-34} \text{ sec of arc}$$

Solution 3.3 Let the dimensions of the region of localization for the first and second electrons be a_1 and a_2, respectively. Then the spread in momenta (and therefore the momenta themselves) would be given by

$$p_1 \sim \frac{\hbar}{a_1}, \quad p_2 \sim \frac{\hbar}{a_2}$$

Thus, $$\text{Kinetic Energy} \sim \frac{\hbar^2}{2m}\left[\frac{1}{a_1^2} + \frac{1}{a_2^2}\right] \tag{44}$$

The potential energy of the interaction of the electrons with a nucleus of charge Ze is

$$-Ze^2\left(\frac{1}{a_1} + \frac{1}{a_2}\right)$$

and the interaction energy between the electrons is $\sim e^2/(a_1 + a_2)$ because the separation between the two electrons is $\sim (a_1 + a_2)$; for convenience, we are using cgs units so that $q^2/(4\pi\varepsilon_0)$ is replaced by e^2. The total energy is, therefore, given by

$$E(a_1, a_2) = \frac{\hbar^2}{2m}\left(\frac{1}{a_1^2} + \frac{1}{a_2^2}\right) - Ze^2\left(\frac{1}{a_1} + \frac{1}{a_2}\right) + \frac{e^2}{a_1 + a_2} \tag{45}$$

For E to be minimum

$$\frac{dE}{da_1} = -\frac{\hbar^2}{ma_1^3} + \frac{Ze^2}{a_1^2} - \frac{e^2}{(a_1 + a_2)^2} = 0$$

$$\frac{dE}{da_2} = -\frac{\hbar^2}{ma_2^3} + \frac{Ze^2}{a_2^2} - \frac{e^2}{(a_1 + a_2)^2} = 0$$

The solution would be

$$a_1 = a_2 = \frac{\hbar^2}{me^2} \cdot \frac{1}{Z - \frac{1}{4}}$$

$$E \simeq -\left(Z - \frac{1}{4}\right)^2 \frac{me^4}{\hbar^2} = -\left(Z - \frac{1}{4}\right)^2 \times 27.2 \text{ eV} \tag{46}$$

The values of E calculated, using Eq. (46), compare rather well with the experimental data as shown in Table 3.1. The energy is in units of $me^4/2\hbar^2 = 13.6$ eV, the ionization potential of the hydrogen atom.

Table 3.1: Ground state energy (in units of $me^4/2\hbar^2$) of various two electron atoms. [Adapted from Ref. 16]

Ground state energy	He	Li^+	Be^{++}	B^{+++}	C^{++++}
Experimental value	–5.807	–14.56	–27.31	–44.06	–64.80
Calculated value using Eq. (50)	–6.125	–15.12	–28.12	–45.12	–66.12

Solution 3.4 The proton is confined within a sphere of radius $r_0 \approx 10^{-13}$ cm. Thus, the uncertainty in the momentum must be at least of the order of $\hbar/r_0$, or

$$p \sim \frac{\hbar}{r_0}$$

Therefore, the kinetic energy of the proton will be given by

$$E = \frac{p^2}{2m_p} \sim \frac{\hbar^2}{2m_p r_0^2}$$

where m_p is the mass of the proton. On substitution, we get

$$E \sim \frac{(1.05\times10^{-27}\,\text{erg}\cdot\text{sec})^2}{2\times1.67\times10^{-24}\,\text{g}\times(10^{-13}\,\text{cm})^2}$$

$$\simeq 3\times10^{-5}\ \text{ergs} \simeq 20\ \text{MeV}$$

Since the proton is bound inside the nucleus, the average of the potential energy $\langle V\rangle$, must be negative and greater in magnitude than the kinetic energy. Therefore,

$$-\langle V\rangle \gtrsim 20\ \text{MeV}$$

which indeed gives the correct order of the potential energy. The uncertainty in momentum for the electron is again $\hbar/r_0$; however, since the rest mass of the electron is very much smaller than that of the proton, the velocity of the electron is very close to c and we have to use the extreme relativistic formula for the energy,

$$E = cp = \frac{c\hbar}{r_0} \simeq \frac{(3\times10^{10})(1.05\times10^{-27})}{10^{-13}\times1.6\times10^{-6}}\ \text{MeV} \sim 200\ \text{MeV}$$

Although electrons do emerge from nuclei in β-decay, they seldom have energies exceeding a few million electron volts. Thus one does not expect the electron to be a basic constituent of the nucleus; the rare occasions when β-decay occurs may be attributed to the transformation of a neutron into a proton and an electron (and the neutrino) so that the electron is in fact created at the instant the decay occurs.

Solution 3.5 (a) The energy of each photon will be

$$h\nu = \frac{hc}{\lambda} \simeq \frac{6.6\times10^{-27}\times3\times10^{10}}{5\times10^{-5}} \simeq 4\times10^{-12}\ \text{ergs} = 4\times10^{-19}\ \text{J}$$

Thus, the number of photons emitted per second will be

$$\frac{1 \text{ watt}}{4\times10^{-19}\text{ J}} = 2.5\times10^{18}$$

(b) The number of photons passing through the hole S will approximately be

$$\frac{2.5\times10^{18}\times\pi\times(0.02)^2}{4\pi\times(100)^2} \approx 2.5\times10^{10} \text{ per second}$$

Similarly, the number of photons passing through either S_1 or S_2 will approximately be

$$\frac{2.5\times10^{10}\times2\times\pi\times(0.02)^2}{2\pi\times(100)^2} = 1000 \text{ per second}$$

where we have assumed that after passing through *S*, the photons are evenly distributed in the hemisphere; this is strictly not correct because the diffraction pattern is actually an Airy pattern—nevertheless, the above calculations would be qualitatively correct. The distance between the planes *AB* and *PP′* is 100 cm [see Fig. 3.17] which will be traversed by a photon in a time 3×10^{-9} sec. Thus, approximately every thousandth of a second a photon enters the region and the space is traversed much before the second photon enters. Therefore, in the region between *AB* and *PP′* one will (almost) never find two photons. This is somewhat similar to the case when, on an average, 100 persons pass through a room in one year and the time that each person takes to cross the room is ≈ 1 second, thus it will be highly improbable to have two persons simultaneously in the room.

REFERENCES AND SUGGESTED READINGS

1. R.P. Feynman, R.B. Leighton and M. Sands, *The Feynman Lectures on Physics*, Vol. III, Addison Wesley Publishing Co., Reading, Mass. (1965).
2. A. Ghatak, *Optics*, Sixth Edition, McGraw Hill Education, New Delhi (2017).
3. W.H. Cropper, *The Quantum Physicists and an Introduction to their Physics*, Oxford University Press, New York (1970).
4. M. Jammer, *The Conceptual Development of Quantum Mechanics*, McGraw-Hill, New York (1965).
5. R.A. Millikan, *The Electron and the light-quanta from the experimental point of view*, Nobel lecture delivered in May 1924, reprinted in *Nobel Lectures in Physics*, Elsevier Publishing Co., Amsterdam (1965).
6. D. Bohm, *Quantum Theory*, Prentice-Hall, Englewood Cliffs, N.J. (1951).
7. A.R. Verma and O.N. Srivastava, *Crystallography for Solid State Physics*, Wiley Eastern, New Delhi (1982).
8. C.J. Davisson, *The discovery of electron waves*, Nobel lecture delivered in December, 1937, reprinted in *Nobel Lectures in Physics*, Elsevier Publishing Co., Amsterdam (1965).
9. R. Eisberg and R. Resnick, *Quantum Physics*, John Wiley & Sons, New York (1974).
10. G. Gamow, *Mr. Tompkins in Wonderland*, Cambridge University Press, Cambridge (1940).

11. M. Born, *Atomic Physics*, Blackie & Son, London (1962).
12. C.G. Shull, *Neutron Diffraction: A General Tool in Physics* in Current Problems in Neutron Scattering, Proceedings of the Symposium held at CNEN Casaccia Center in September 1968, CNEN, Rome (1970).
13. P.A.M. Dirac, *The Principles of Quantum Mechanics*, Oxford University Press, Oxford (1958).
14. A. Tonomura, J. Endo, T. Matsuda, T. Kawasaki and H. Ezawa, *Demonstration of single-electron build up of an interference pattern*, Am. J. Phys. **57** (2). 117 (1989).
15. J. Franck, Transformations of kinetic energy of free electrons into excitation energy of atoms by impacts, Nobel lecture delivered in December, 1926, Reprinted in *Nobel Lectures in Physics*, Elsevier Publishing Co., Amsterdam (1965).
16. I.I. Goldman and V.D. Krivchenkov, *Problems in Quantum Mechanics*, Pergamon Press, London (1961).

4 Time Dependent Schrödinger Equation

The scientist builds his theories as much with a kind of inspired artistry as with the strict procedures of logic. A great theory can no more be put together entirely with straightforward reasoning processes than a great symphony can be written with textbook principles of composition. The composer must have a finely tuned ear; the scientist must have a finely tuned insight. Just as the composer 'hears' his incomplete symphony as if it were being played by a full orchestra, the scientist must 'see' his theory extending far beyond its early, primitive limitations. In the beginning the work has a strange, complex sound; its beauty is subtle and few are able to understand and appreciate it. But gradually, with patience and faith and mastery of technique, the artist-scientist manages to bring the essential elements of his art together and to weave them into a powerful and moving whole (in music this is the creation of a single artist, but in science it finally requires the efforts of many). Eventually, sometimes much later, the work plays to larger audiences and is accepted into the standard repertoire.

— W.H. Cropper in *The Quantum Physicists*, Oxford University Press, New York (1970), p. 179.

4.1 INTRODUCTION

In the previous chapter we discussed some experiments which showed that particles like electrons, protons, neutrons, atoms, etc., exhibit wavelike properties. Indeed the wavelength is related to the momentum through the de Broglie relation:

$$\lambda = \frac{h}{p} \tag{1}$$

or

$$p = \hbar k \tag{2}$$

where $k = 2\pi/\lambda$ and $\hbar = h/2\pi$, h being the Planck's constant. Thus we write

$$\mathbf{p} = \hbar \mathbf{k} \tag{3}$$

where $\mathbf{k}$ denotes the wave vector. Further, as established by Einstein's explanation of the photoelectric effect, the energy E of the particle is related to the frequency through the relation

$$E = \hbar\omega \tag{4}$$

In this chapter we would like to obtain an equation, the solution of which would describe the behaviour of particles like the electron and the proton.[1] The corresponding solutions should describe the corpuscular and wavelike behaviour of the particle and should also be consistent

[1]As discussed in the previous chapter, radiation also exhibits wave and particle like properties. The quantum theory of radiation is discussed in Chapter 29.

with the uncertainty principle. Thus we assume that the rough trajectory of a particle should be describable by a wavefunction $\Psi(\mathbf{r}, t)$ whose magnitude is large in regions where the probability of the occurrence of the particle is large; in other regions where the particle is less likely to be found, the magnitude of Ψ should be small. Indeed we will show that $|\Psi|^2\, d\tau$ can be interpreted to represent the probability of finding the particle in the volume element $d\tau$.

In Section 4.2 we will obtain an equation (known as the Schrödinger equation) satisfied by Ψ and will show that in quantum mechanics the dynamical variables like position, momentum, energy, etc., are represented by linear operators which, in general, do not commute with each other (see Section 4.3). In Section 4.4 we will give a physical interpretation of the wave function and in Section 4.5 we will give formulae which will enable us to calculate the expectation values of various observables (like position, momentum, energy, etc.); these will lead to the relation between classical and quantum motions (Ehrenfest's theorem) and also to the uncertainty principle which will be quantitatively discussed in Section 4.6. In the next chapter we will discuss the motion of a wave packet which will describe the trajectory of a free particle; the wavefunction describing the motion of the wave packet will obviously be a solution of the Schrödinger equation and will be shown to be consistent with the uncertainty principle. In Chapters 6, 7, 8 and 10 we will obtain exact solutions of the Schrödinger equation corresponding to problems of physical interest.

4.2 THE SCHRÖDINGER EQUATION

The simplest type of wave is a plane monochromatic wave described by the wavefunction

$$\Psi(\mathbf{r}, t) = A e^{i(\mathbf{k}.\mathbf{r} - \omega t)} \tag{5}$$

which represents a disturbance of amplitude A and of wavelength $\lambda = 2\pi/k$ travelling in the direction of its wave vector $\mathbf{k}$ with phase velocity ω/k. If we assume the propagation to be along the x-axis, then

$$\mathbf{k} = k\hat{\mathbf{x}} \tag{6}$$

and we have

$$\Psi(x, t) = A\, e^{i(kx - \omega t)} \tag{7}$$

which on using Eqs (2) and (4) takes the form

$$\Psi(x, t) = A \exp\left[\frac{i}{\hbar}(px - Et)\right] \tag{8}$$

Successive differentiation of Eq. (8) gives us

$$i\hbar \frac{\partial \Psi}{\partial t} = E\Psi(x, t) \tag{9}$$

$$-i\hbar \frac{\partial \Psi}{\partial x} = p\Psi(x, t) \tag{10}$$

$$-\frac{\hbar^2}{2m}\frac{\partial^2 \Psi}{\partial x^2} = \frac{p^2}{2m}\Psi(x, t) \tag{11}$$

Now for a free non-relativistic particle

$$E = \frac{p^2}{2m} \tag{12}$$

and therefore from Eqs (9) and (11), we get

$$i\hbar\frac{\partial\Psi}{\partial t} = -\frac{\hbar^2}{2m}\frac{\partial^2\Psi}{\partial x^2} \tag{13}$$

which is the one-dimensional time dependent Schrödinger equation for a free particle. In the next chapter we will discuss the most general solution of Eq. (13).

Equations (9) and (10) suggest that E and p can be represented by the operators $i\hbar\frac{\partial}{\partial t}$ and $-i\hbar\frac{\partial}{\partial x}$ respectively:

$$E \to i\hbar\frac{\partial}{\partial t}, \qquad p \to -i\hbar\frac{\partial}{\partial x} \tag{14}$$

Thus, Eq. (13) can also be written in the form

$$E\Psi = \frac{p^2}{2m}\Psi(x, t) \tag{15}$$

where it is implied that E and p are to be replaced by their operator representations.

4.2.1 The Three-Dimensional Generalization

The three-dimensional generalization is very straightforward. Instead of Eq. (8) we have

$$\begin{aligned}\Psi(\mathbf{r}, t) &= A\exp\left[\frac{i}{\hbar}(\mathbf{p}.\mathbf{r} - Et)\right] \\ &= A\exp\left[\frac{i}{\hbar}(p_x x + p_y y + p_z z - Et)\right]\end{aligned} \tag{16}$$

Thus,

$$i\hbar\frac{\partial\Psi}{\partial t} = E\Psi$$

$$-i\hbar\frac{\partial\Psi}{\partial x} = p_x\Psi$$

$$-\frac{\hbar^2}{2m}\frac{\partial^2\Psi}{\partial x^2} = \frac{p_x^2}{2m}\Psi$$

etc. Since,

$$E = \frac{p^2}{2m} = \frac{1}{2m}(p_x^2 + p_y^2 + p_z^2)$$

we immediately have

$$i\hbar\frac{\partial\Psi}{\partial t} = -\frac{\hbar^2}{2m}\left[\frac{\partial^2\Psi}{\partial x^2} + \frac{\partial^2\Psi}{\partial y^2} + \frac{\partial^2\Psi}{\partial z^2}\right] = -\frac{\hbar^2}{2m}\nabla^2\Psi \tag{17}$$

which is the three-dimensional Schrödinger equation for a free particle. Once again the operator representation of E and components of $\mathbf{p}$ would be

$$E \to i\hbar\frac{\partial}{\partial t}; \qquad \mathbf{p} \to -i\hbar\nabla \tag{18}$$

where the last step implies[2]

$$\left.\begin{aligned} p_x &\to -i\hbar\frac{\partial}{\partial x} \\ p_y &\to -i\hbar\frac{\partial}{\partial y} \\ p_z &\to -i\hbar\frac{\partial}{\partial z} \end{aligned}\right\} \tag{19}$$

Further, Eq. (17) can be written in the form

$$E\Psi = \frac{1}{2m}(p_x^2 + p_y^2 + p_z^2)\Psi = \frac{p^2}{2m}\Psi \tag{20}$$

Till now we have assumed the particle to be free. If we now assume the particle to be in a field characterised by the potential energy function $V(\mathbf{r}, t)$, then according to classical mechanics, the total energy would be given by

$$E = \frac{p^2}{2m} + V(\mathbf{r}, t) \tag{21}$$

Since the potential energy function does not depend on p and E, Eqs (18)–(21) suggest that the wavefunction should satisfy

$$i\hbar\frac{\partial\Psi}{\partial t} = \left[\frac{p^2}{2m} + V(\mathbf{r}, t)\right]\Psi = \left[-\frac{\hbar^2}{2m}\nabla^2\Psi + V\Psi\right] \tag{22}$$

which is the time dependent Schrödinger equation for a particle in a field characterised by the potential energy function $V(\mathbf{r}, t)$. Equation (22) is usually written in the form

$$i\hbar\frac{\partial\Psi}{\partial t} = H\Psi \tag{23}$$

where

$$H = \frac{p^2}{2m} + V = -\frac{\hbar^2}{2m}\nabla^2 + V \tag{24}$$

is known as the Hamiltonian operator. The corresponding one-dimensional equation is

---

[2]At this point it is necessary to mention that the entire theory could have been equally well developed if we had assumed

$$\Psi = \exp\left[-\frac{i}{\hbar}(\mathbf{p}.\mathbf{r} - Et)\right]$$

to represent a plane wave propagating in the direction of $\mathbf{p}$. The operator representations of E and p would then have been $E \to -i\hbar\frac{\partial}{\partial t}$ and $\mathbf{p} \to +i\hbar\nabla$ and the Schrödinger equation for a free particle would have been [cf. Eq. (17)]

$$-i\hbar\frac{\partial\Psi}{\partial t} = -\frac{\hbar^2}{2m}\nabla^2\Psi$$

The later analysis and the various solutions could have been obtained in a similar manner. However the sine and cosine functions of the type $\sin\left[\frac{1}{\hbar}(px - Et)\right]$, which also represent plane waves, are quite unsuitable because then the operator representation of the type given by Eqs (18) and (19) would not have been possible.

$$i\hbar\frac{\partial\Psi}{\partial t} = H\Psi = \left[-\frac{\hbar^2}{2m}\cdot\frac{\partial^2}{\partial x^2} + V(x)\right]\Psi \tag{25}$$

It should be mentioned that the method for obtaining the Schrödinger equation is far from rigorous. We have tried to put together the corpuscular and wave characteristics of the particle and have assumed that the particle can be described by a wavefunction Ψ which, in general, is complex, and satisfies Eq. (22). To quote Feynman

> *Where did we get that [equation] from? Nowhere. It is not possible to derive it from anything you know. It came out of the mind of Schrödinger.*

By solving Eq. (22) (under appropriate boundary and initial conditions) one can predict the results of various measurements which we will do in this and the following chapters. Indeed the validity of the Schrödinger equation comes from the agreement of the theoretical results (obtained by solving the Schrödinger equation) with experimental data.

4.3 THE COMMUTATOR

Because of the operator representation of p_x [see Eq. (19)], we have

$$(xp_x - p_x x)\,\Psi = -i\hbar\left[x\frac{\partial\Psi}{\partial x} - \frac{\partial}{\partial x}(x\Psi)\right]$$

or

$$(xp_x - p_x x)\,\Psi = i\hbar\Psi$$

Since Ψ is an arbitrary function, the operator $[xp_x - p_x x]/i\hbar$ is simply the unit operator. We write

$$[x, p_x] \equiv (xp_x - p_x x) = i\hbar \tag{26}$$

which is known as the commutation relation. The quantity

$$[\alpha, \beta] = \alpha\beta - \beta\alpha = -[\beta, \alpha] \tag{27}$$

is said to be the commutator of the two operators α and β. It is easy to show that

$$\begin{aligned}[x, y] &= [y, z] = [z, x] = 0\\ [x, p_y] &= [y, p_z] = [z, p_x] = 0\\ [p_x, p_y] &= [p_y, p_z] = [p_z, p_x] = 0\end{aligned}$$

But

$$[y, p_y] = [z, p_z] = i\hbar$$

Commutation relations play a very important role in quantum mechanics and we will use them in later chapters.

4.4 PHYSICAL INTERPRETATION OF Ψ AND THE PROBABILITY CURRENT DENSITY

We rewrite the Schrödinger equation [Eq. (22)]:

$$+i\hbar\frac{\partial\Psi}{\partial t} = -\frac{\hbar^2}{2m}\nabla^2\Psi + V\Psi \tag{28}$$

and its complex conjugate

$$-i\hbar\frac{\partial\Psi^*}{\partial t} = -\frac{\hbar^2}{2m}\nabla^2\Psi^* + V\Psi^* \tag{29}$$

where $V(\mathbf{r}, t)$ is assumed to be real. If we multiply Eq. (28) by Ψ^* and Eq. (29) by Ψ and subtract we would obtain

$$i\hbar\left[\Psi^*\frac{\partial \Psi}{\partial t} + \Psi\frac{\partial \Psi^*}{\partial t}\right] = -\frac{\hbar^2}{2m}[\Psi^* \nabla^2 \Psi - \Psi \nabla^2 \Psi^*] \tag{30}$$

Since,

$$\nabla^2 = \frac{\partial^2}{\partial x^2} + \frac{\partial^2}{\partial y^2} + \frac{\partial^2}{\partial z^2}$$

we can rewrite Eq. (30) in the form

$$\frac{\partial}{\partial t}(\Psi^* \Psi) + \left[\frac{\partial J_x}{\partial x} + \frac{\partial J_y}{\partial y} + \frac{\partial J_z}{\partial z}\right] = 0 \tag{31}$$

where

$$J_x = \frac{i\hbar}{2m}\left[\Psi\frac{\partial \Psi^*}{\partial x} - \Psi^*\frac{\partial \Psi}{\partial x}\right]$$

$$J_y = \frac{i\hbar}{2m}\left[\Psi\frac{\partial \Psi^*}{\partial y} - \Psi^*\frac{\partial \Psi}{\partial y}\right]$$

$$J_z = \frac{i\hbar}{2m}\left[\Psi\frac{\partial \Psi^*}{\partial z} - \Psi^*\frac{\partial \Psi}{\partial z}\right]$$

Equation (31) can be written in the form[3]

$$\frac{\partial \rho}{\partial t} + \nabla \cdot \mathbf{J} = 0 \tag{32}$$

where

$$\rho = \Psi^*\Psi \tag{33}$$

and

$$\mathbf{J} = \frac{i\hbar}{2m}[\Psi \nabla \Psi^* - \Psi^* \nabla \Psi] \tag{34}$$

Equation (32) is nothing but the equation of continuity in fluid dynamics (see Appendix B) where ρ represents the number of particles per unit volume and $\mathbf{J}$ the current density. Thus we may associate $\Psi^*\Psi$ with *position probability density* in the sense that $\Psi^*\Psi\, d\tau$ will be proportional to the probability of finding the particle in volume element $d\tau$. Now, since the Schrödinger equation is a linear equation, a multiple of Ψ would also be a solution and we may choose the multiplicative constant such that

$$\iiint |\Psi|^2\, d\tau = 1 \qquad \text{Normalization condition} \tag{35}$$

where the integration is over the entire space. Equation (35) is known as the normalization condition. If the wavefunction is normalized [i.e. if it satisfies Eq. (35)] then $|\Psi|^2\, d\tau$ can be interpreted to represent the probability of finding the particle in the volume element $d\tau$ and Eq. (35) implies the fact that the particle has to be found somewhere in space. Further,

$$\mathbf{J} = \frac{i\hbar}{2m}[\Psi \nabla \Psi^* - \Psi^* \nabla \Psi] = \text{Re}\left[\Psi^* \frac{\hbar}{im} \nabla \Psi\right] \tag{36}$$

---

[3]Equation (32) could have been directly written by using vector algebra as

$$\frac{i\hbar}{2m}\left[\Psi^* \nabla^2 \Psi + \Psi \nabla^2 \Psi^*\right] = \frac{i\hbar}{2m}\nabla \cdot [\Psi^* \nabla\Psi + \Psi\nabla \Psi^*]$$
$$= \nabla \cdot \mathbf{J}$$

can be associated with *probability current density* (see Appendix B); here Re stands for the real part of the quantity inside the bracket. It is of interest to mention that the quantity $\frac{\hbar}{i}\nabla$ represents the momentum operator **p** [(see Eq. (19)] and therefore we may write

$$\mathbf{J} = \text{Re}\,[\Psi^* \mathbf{v} \Psi] \tag{37}$$

where $\mathbf{v} = \mathbf{p}/m$ represents the velocity operator. Equations (33) and (37) are consistent with the fact that in fluid dynamics

$$\mathbf{J} = \rho \mathbf{v} \tag{38}$$

As an example we consider the wavefunction

$$\Psi = \exp\left[\frac{i}{\hbar}(\mathbf{p}\cdot\mathbf{r} - Et)\right] \tag{39}$$

which, however, cannot be normalized as the position probability density is constant everywhere.[4] But

$$\mathbf{J} = \text{Re}\left[\Psi^* \frac{\hbar}{im}\nabla\Psi\right] = \frac{\mathbf{p}}{m} = \mathbf{v} \tag{40}$$

which is consistent with Eq. (38) because $\rho = 1$. It may be noted that in Eqs (39) and (40), **p** and E are numbers.

We conclude this section by noting that

$$\begin{aligned}\frac{\partial}{\partial t}\int_\Omega \Psi^*\Psi\, d\tau &= \int_\Omega \frac{\partial\rho}{\partial t}\, d\tau = -\int_\Omega \nabla\cdot\mathbf{J}\, d\tau \\ &= -\int_S \mathbf{J}\cdot\hat{\mathbf{n}}\, dS \end{aligned} \tag{41}$$

where Ω represents an arbitrary volume bounded by the surface S with $\hat{\mathbf{n}}$ representing the outward unit normal on the surface, and in the last step we have used Gauss's theorem. Thus, if the integration is carried over entire space and if the wavefunction vanishes at infinity, then

$$\frac{\partial}{\partial t}\iiint |\Psi|^2\, d\tau = 0 \tag{42}$$

which is essentially a conservation condition.

4.5 EXPECTATION VALUES

By associating $\Psi^*\Psi$ with position probability density we can calculate the expectation or average values of the position of the particle. Thus, the expectation value of the x-coordinate would be given by

$$\langle x\rangle = \frac{\iiint x\Psi^*\Psi\, d\tau}{\iiint \Psi^*\Psi\, d\tau} \tag{43}$$

where the integration is over the entire space. If we assume the wavefunction to be normalised [see Eq. (35)], then

$$\langle x\rangle = \int \Psi^* x\Psi\, d\tau \tag{44}$$

[4]A realistic wavefunction will be a wave packet which will be discussed in the next chapter.

where for the sake of convenience we are replacing the triple integral sign by a single integral. We have also put x between Ψ^* and Ψ for reasons which will become clear later in this section. In a similar manner we can write for $\langle y\rangle$ and $\langle z\rangle$. In general, for any function $f(\mathbf{r})$, the expectation value will be given by

$$\langle f(\mathbf{r})\rangle = \int \Psi^* f(\mathbf{r})\, \Psi\, d\tau \tag{45}$$

In particular

$$\langle V(\mathbf{r}, t)\rangle = \int \Psi^* V(\mathbf{r}, t)\, \Psi\, d\tau \tag{46}$$

Now since $E = p^2/2m + V$, we should have

$$\langle E\rangle = \left\langle \frac{p^2}{2m} \right\rangle + \langle V\rangle \tag{47}$$

Further, if we multiply the Schrödinger equation [Eq. (22)] by Ψ^* and integrate over all space, we would get

$$\int \Psi^* \left(i\hbar \frac{\partial}{\partial t} \right) \Psi\, d\tau = \int \Psi^* \left(-\frac{\hbar^2}{2m} \nabla^2 \right) \Psi\, d\tau + \int \Psi^* V(\mathbf{r}, t)\, \Psi\, d\tau \tag{48}$$

The last term in this equation is simply $\langle V\rangle$ [see Eq. (46)] and therefore Eqs (46), (47) and (48) will be consistent only if we assume

$$\langle E\rangle = \int \Psi^*\, i\hbar \frac{\partial}{\partial t} \Psi\, d\tau \tag{49}$$

and

$$\left\langle \frac{p^2}{2m} \right\rangle = \int \Psi^* \left(-\frac{\hbar^2}{2m} \nabla^2 \right) \Psi\, d\tau \tag{50}$$

Since $i\hbar \frac{\partial}{\partial t}$ and $-\frac{\hbar^2}{2m}\nabla^2$ are the operator representations of E and $p^2/2m$, we can say that the general recipe of determining the expectation value of any quantity is to replace it by its operator representation, operate it on Ψ and premultiply by Ψ^* and then integrate. Thus

$$\langle p_x\rangle = \iiint \Psi^* \left(-i\hbar \frac{\partial \Psi}{\partial x} \right) d\tau \tag{51}$$

$$\langle p_x^2\rangle = \iiint \Psi^* \left(-\hbar^2 \frac{\partial^2 \Psi}{\partial x^2} \right) d\tau \tag{52}$$

$$\langle p_y\rangle = \iiint \Psi^* \left(-i\hbar \frac{\partial \Psi}{\partial y} \right) d\tau \tag{53}$$

etc. Using the above recipe for calculation of the expectation values and also using the Schrödinger equation one can show that (see Problem 4.5)

$$\left.\begin{aligned} \frac{d}{dt}\langle x\rangle &= \frac{1}{m}\langle p_x\rangle \qquad (54) \\ \frac{d}{dt}\langle p_x\rangle &= -\left\langle \frac{\partial V}{\partial x} \right\rangle \qquad (55) \end{aligned}\right\} \text{Ehrenfest's theorem}$$

The above equations show that there is agreement between quantum and classical motions in the sense that in quantum theory the expectation values satisfy Newton's laws of motion. Equations (54) and (55) were first derived by Ehrenfest and the general relationship (between classical and quantum motions) is known as Ehrenfest's theorem.

In the next section we will use various expressions for the expectation values to give an exact statement of the uncertainty principle.

4.6 EXACT STATEMENT AND PROOF OF THE UNCERTAINTY PRINCIPLE

In the previous chapter we had discussed experiments which showed that it is not possible to measure simultaneously the position of a particle along a particular (say x) direction and also the corresponding momentum p_x with unlimited accuracy. Indeed we showed that $\Delta x \, \Delta p_x \gtrsim h$, where Δx and Δp_x were uncertainties in the measurement of x and p_x respectively. In this section we will give precise definitions of Δx and Δp_x and using the results of the previous section, we will make an exact statement of the uncertainty principle. First, in analogy with the definition of standard deviation in statistics, we define the uncertainties in x and p_x as:

$$\Delta x = \left\langle (x - \langle x \rangle)^2 \right\rangle^{\frac{1}{2}} = \left[\langle x^2 \rangle - \langle x \rangle^2 \right]^{\frac{1}{2}} \tag{56}$$

$$\Delta p = \left\langle (p - \langle p \rangle)^2 \right\rangle^{\frac{1}{2}} = \left[\langle p^2 \rangle - \langle p \rangle^2 \right]^{\frac{1}{2}} \tag{57}$$

where, for the sake of convenience, we have dropped the subscript on p. We will establish that

$$\Delta x \, \Delta p \geq \frac{1}{2} \hbar \tag{58}$$

with the equality sign only for a Gaussian wavefunction. Equation (58) represents the exact statement of the uncertainty principle. In order to prove Eq. (58) we would use the following inequality

$$\int f^* f \, d\tau \int g^* g \, d\tau \geq \frac{1}{4} \left[\int (f^* g + f g^*) \, d\tau \right]^2 \tag{59}$$

where f and g are arbitrary functions of the space coordinates; however, they must be single valued and all the integrals in Eq. (59) must exist. The proof of Eq. (59) is quite simple. We define three integrals

$$a = \int f^* f \, d\tau, \quad b = \int f^* g \, d\tau, \quad c = \int g^* g \, d\tau$$

Now,

$$\int |\lambda f + g|^2 \, d\tau = a\lambda^2 + (b + b^*) \lambda + c \geq 0$$

because $|\lambda f + g|^2 \geq 0$ everywhere; we have assumed λ to be a real quantity. Obviously, for the above inequality to be satisfied we must have

$$4ac \geq (b + b^*)^2$$

which proves Eq. (59); the equality sign in Eq. (59) will occur when f is a multiple of g.

We will consider the special case[5] of $\langle x\rangle = 0$ and $\langle p\rangle = 0$ and assume

$$f = p\Psi = -i\hbar\frac{\partial\Psi}{\partial x} \tag{60}$$

$$g = ix\,\Psi \tag{61}$$

Thus,
$$\int f^* f\,d\tau = \hbar^2\iint dy\,dz\int\frac{\partial\Psi^*}{\partial x}\cdot\frac{\partial\Psi}{\partial x}\,dx$$

$$= \hbar^2\iint dy\,dz\left[\Psi^*\frac{\partial\Psi}{\partial x}\bigg|_{-\infty}^{+\infty} - \int\Psi^*\frac{\partial^2\Psi}{\partial x^2}\,dx\right]$$

$$= \iiint\Psi^*\left(-\hbar^2\frac{\partial^2\Psi}{\partial x^2}\right)dx\,dy\,dz = \langle p_x^2\rangle$$

where we have used Eq. (52) and the fact that Ψ vanishes at infinity. Further,

$$\int g^* g\,d\tau = \int\Psi^* x^2\Psi\,d\tau = \langle x^2\rangle$$

and
$$\int(f^* g + f g^*)\,d\tau = -\hbar\int\frac{\partial\Psi^*}{\partial x}\,x\Psi\,d\tau - \hbar\int\frac{\partial\Psi}{\partial x}\,x\Psi^*\,d\tau$$

$$= -\hbar\int\frac{\partial}{\partial x}(\Psi^* x\Psi)\,d\tau + \hbar\int\Psi^*\Psi\,d\tau$$

$$= \hbar$$

because Ψ is normalized and $\Psi^* x\Psi$ would vanish at $x = \pm\infty$. On substitution in Eq. (59), we immediately get the uncertainty relation given by Eq. (58). The uncertainty product is a minimum for a Gaussian wavefunction (see Problem 4.8).

4.7 EIGENVALUE EQUATION FOR THE MOMENTUM OPERATOR

For an arbitrary operator $\hat{\alpha}$, the equation represents an eigenvalue equation with λ representing

$$\hat{\alpha}\,\psi = \lambda\,\psi \tag{62}$$

the eigenvalue. In Sec. 4.2 we had shown that the x-component of the momentum can be represented by the operator given by

$$\hat{p}_x \to -i\hbar\frac{\partial}{\partial x} \tag{63}$$

where the "hat" on p_x tells us that it is an operator. We consider the eigenvalue equation of the momentum operator

$$\hat{p}_x\psi = -i\hbar\frac{\partial}{\partial x}\,\psi = p\,\psi \tag{64}$$

---

[5]For $\langle x\rangle \neq 0$ and $\langle p\rangle \neq 0$, we must choose

$$f = (p - \langle p\rangle)\,\psi$$

and
$$g = (x - \langle x\rangle)\,\psi$$

where p (on the right hand side) is a number and represents the eigenvalue. For the time being, we assume ψ to be a function of x alone so that Eq. (64) can be written in the form

$$\frac{1}{\psi(x)}\frac{d\psi}{dx} = \frac{i}{\hbar}p \tag{65}$$

The integration is simple and we get

$$\psi(x) = \psi_p(x) = \frac{1}{\sqrt{2\pi\hbar}}\exp\left[\frac{ip}{\hbar}x\right]; \quad -\infty < p < +\infty \tag{66}$$

which are known as the momentum eigenfunctions. Obviously, p has to be real, otherwise $\psi_p(x)$ will blow up either as $x \to +\infty$ or as $x \to -\infty$; p can take any real value between $-\infty$ and, $+\infty$. The factor $1/\sqrt{2\pi\hbar}$ makes the wave function normalized as we will show below. Now, in Section 1.2, we had shown:

$$\delta(x - x') = \frac{1}{2\pi}\int_{-\infty}^{+\infty} \exp\left[ik(x - x')\right] dk \tag{67}$$

If we define $p \equiv \hbar k$, we would obtain

$$\delta(x - x') = \frac{1}{2\pi\hbar}\int_{-\infty}^{+\infty} \exp\left[\frac{i}{\hbar}p(x - x')\right]dp \tag{68}$$

We can similarly write

$$\frac{1}{2\pi\hbar}\int_{-\infty}^{+\infty} \exp\left[\frac{i}{\hbar}x(p - p')\right]dx = \delta(p - p') \tag{69}$$

Thus, $$\int_{-\infty}^{+\infty} \psi_{p'}^*(x)\psi_p(x)dx = \delta(p - p') \qquad \textbf{orthonormality condition} \tag{70}$$

which represents the orthonormality condition for the momentum eigen functions. Also, using Eq. (66), Eq. (68) becomes

$$\int_{-\infty}^{+\infty} \psi_p^*(x)\psi_p(x')dp = \delta(x - x') \qquad \textbf{completeness condition} \tag{71}$$

which represents the completeness condition because an arbitrary "well behaved" function can always be written as

$$f(x) = \int_{-\infty}^{+\infty} \delta(x - x')f(x')dx' = \frac{1}{2\pi\hbar}\int_{-\infty}^{+\infty}\int_{-\infty}^{+\infty} \exp\left[\frac{i}{\hbar}p(x - x')\right]f(x')dx'dp \tag{72}$$

Thus, if we define

$$a(p) \equiv \frac{1}{\sqrt{2\pi\hbar}}\int_{-\infty}^{+\infty} f(x')\exp\left[-\frac{i}{\hbar}px'\right]dx' \tag{73}$$

then $$f(x) = \frac{1}{\sqrt{2\pi\hbar}}\int_{-\infty}^{+\infty} a(p)\exp\left[\frac{i}{\hbar}px\right]dp \tag{74}$$

or, $$f(x) = \int_{-\infty}^{+\infty} a(p)\,\psi_p(x)\,dp \tag{75}$$

The above equation tells us that an arbitrary "well behaved" function can always be written as a superposition of the eigenfunctions of the operator $\hat{p}_x$, which is the completeness condition. The Hamiltonian for a free particle (in one dimension) is given by

$$\hat{H} = \frac{\hat{p}_x^2}{2m} = -\frac{\hbar^2}{2m}\cdot\frac{\partial^2}{\partial x^2} \tag{76}$$

Thus, the eigenvalue equation

$$\hat{H}\psi = E\psi \tag{77}$$

(where E is a number) will take the form

$$-\frac{\hbar^2}{2m}\frac{d^2\psi}{dx^2} = E\psi \quad \Rightarrow \quad \frac{d^2\psi}{dx^2} + \frac{p^2}{\hbar^2}\psi = 0$$

where $p^2 \equiv 2mE$ is a number. The solution of the above equation will give us the following (normalized) eigenfunctions:

$$\psi(x) = \psi_p(x) = \frac{1}{\sqrt{2\pi\hbar}}\exp\left[\frac{ip}{\hbar}x\right]; \quad -\infty < p < +\infty$$

Thus $\Psi_p(x)$ form a simultaneous set of orthonormal eigenfunctions of the operators $\hat{p}_x$ and $\hat{H} = \frac{\hat{p}_x^2}{2m}$:

$$\hat{p}_x\,\psi_p(x) = p\psi_p(x) \tag{78}$$

$$\hat{H}\,\psi_p(x) = \frac{p^2}{2m}\,\psi_p(x) \tag{79}$$

It can easily be seen that the wavefunction.

$$\psi_p(\mathbf{r}) = \frac{1}{(2\pi\,\hbar)^{3/2}}\exp\left[\frac{i}{\hbar}(p_x x + p_y y + p_z z)\right]; \; -\infty < p_x, p_y, p_z < +\infty \tag{80}$$

represents a complete set of simultaneous eigenfunctions of the operators $\hat{p}_x$, $\hat{p}_y$, and $\hat{p}_z$; on the right hand side of the above equation, p_x, p_y and p_z are numbers. The functions $\Psi_p(\mathbf{r})$ are also eigenfunctions of the Hamiltonian operator corresponding to a free particle:

$$H = \frac{1}{2m}\hat{p}^2 = \frac{1}{2m}(\hat{p}_x^2 + \hat{p}_y^2 + \hat{p}_z^2) \tag{81}$$

Furthermore,

$$\iiint \psi^*_{\mathbf{p}'}(\mathbf{r})\,\psi_{\mathbf{p}}(\mathbf{r})\,d\mathbf{r} = \delta(\mathbf{p}-\mathbf{p}') \qquad \textbf{orthonormality condition} \tag{82}$$

and

$$\iiint \psi^*_{\mathbf{p}'}(\mathbf{r})\,\psi_{\mathbf{p}}(\mathbf{r})\,d\mathbf{p} = \delta(\mathbf{r}-\mathbf{r}') \qquad \textbf{completeness condition} \tag{83}$$

which represent the orthonormality and the completeness conditions respectively. In the above equations

$$\delta(\mathbf{p}-\mathbf{p}') \equiv \delta(p_x - p_x')\,\delta(p_y - p_y')\,\delta(p_z - p_z') \tag{84}$$

$$\delta(\mathbf{r}-\mathbf{r}') \equiv \delta(x - x')\,\delta(y - y')\,\delta(z - z') \tag{85}$$

are 3-dimensional delta functions defined in Chapter 1.

4.8 EIGENVALUE EQUATION FOR THE POSITION OPERATOR $\hat{x}$

In chapter 1, we had shown that for a well behaved function $f(x)$

$$f(x)\,\delta(x-x') = f(x')\,\delta(x-x') \tag{86}$$

Consider the position operator $\hat{x}$ and the equation

$$\hat{x}\,\delta(x-x') = x'\,\delta(x-x'); \quad -\infty < x' < \infty \tag{87}$$

where x' (on the right hand side) is just a number. The above equation tells us that $\hat{x}$ has a continuous range of eigenvalues (from $-\infty$ to $+\infty$) and the corresponding eigen functions are $\delta(x - x')$. Thus, the function $\delta(x - x')$ represents a particle localized at the point $x = x'$. The equation

$$\int_{-\infty}^{+\infty} \delta(x - x')\,\delta(x - x'')\,dx = \delta(x' - x'') \qquad \textbf{orthonormality condition} \tag{88}$$

represents the orthonormality condition. Further the equation

$$f(x) = \int_{-\infty}^{+\infty} f(x')\,\delta(x - x')\,dx' \tag{89}$$

show that any arbitrary "well behaved" function $f(x)$ can be expressed as a linear combination of the eigenfunctions. Thus, the functions $\delta(x - x')$ are the eigenfunctions of the position operator $\hat{x}$ from a complete set of (orthonormal) functions.

4.9 THE TRANSLATION OPERATOR

In the next chapter we will discuss the time evolution of a Gaussian wave packet by solving the 1-dimensional Schrödinger equation. For example, the wavefunction

$$\psi(x) = \frac{1}{\left(\pi\sigma_0^2\right)^{1/4}} \exp\left[-\frac{x^2}{2\sigma_0^2}\right] \exp\left[\frac{i}{\hbar} p_0 x\right] \tag{90}$$

will represent a particle localized around the origin $x = 0$. Notice that

$$\int_{-\infty}^{+\infty} |\psi(x)|^2\,dx = 1$$

which shows that the wavefunction is normalized. For the particle to be localized around the point $x = \varepsilon$, its wavefunctions would be given by

$$\psi(x-\varepsilon) = \frac{1}{\left(\pi\sigma_0^2\right)^{1/4}} \exp\left[-\frac{(x-\varepsilon)^2}{2\sigma_0^2}\right] \exp\left[\frac{i}{\hbar} p_0 (x-\varepsilon)\right] \tag{91}$$

We define the translation operator $\hat{T}(\varepsilon)$ which translated a state from one position to another by a distance ε; thus

$$\hat{T}\,\Psi(x) = \psi(x-\varepsilon) \tag{92}$$

We assume ε to be an infinitesimal distance so that we may write

$$\psi(x-\varepsilon) = \psi(x) - \varepsilon\frac{\partial \psi}{\partial x}$$

$$= \psi(x) - \frac{i\varepsilon}{\hbar}\hat{p}_x\,\psi(x)$$

where we have used the operator representation of $\hat{p}_x$:

$$\hat{p}_x \to -i\hbar\frac{\partial}{\partial x} \tag{93}$$

Thus, $$\hat{T}(\varepsilon)\psi(x) = \psi(x-\varepsilon) = \left(1 - \frac{i\hat{p}_x}{\hbar}\varepsilon\right)\psi(x) \tag{94}$$

Thus, the translation operator, which translates a state from one position to another by an infinitesimal distance ε along the x-axis, would be given by

$$\hat{T}(\varepsilon) = 1 - \frac{i}{\hbar}\hat{p}_x\,\varepsilon \tag{95}$$

Thus the operator $\hat{T}(\varepsilon)$ represents the generator of infinitesimal translation. Many books use the above equation to define the momentum operator (see, e.g. References 5–7). In Section 11.10 we will show that if we start with Eq. (95) then we can derive the commutation relation between $\hat{x}$ and $\hat{p}_x$.

In order to determine $\hat{T}(a)$ for finite translation (along the x-axis), we note that the effect of an infinitesimal increase in x, from a to $a + \varepsilon$, is to follow the finite translation $\hat{T}(a)$ by the infinitesimal translation $\hat{T}(\varepsilon)$ to give

$$\hat{T}(a+\varepsilon) = \hat{T}(\varepsilon)\,\hat{T}(a) = \left[1 - \frac{i}{\hbar}\hat{p}_x\,\varepsilon\right]\hat{T}(a)$$

Thus, $$\hat{T}(a+\varepsilon) - \hat{T}(a) = -\frac{i}{\hbar}\hat{p}_x\,\varepsilon\,\hat{T}(a)$$

giving $$\frac{d\hat{T}(a)}{da} = \lim_{\varepsilon\to 0}\frac{\hat{T}(a+\varepsilon)-\hat{T}(a)}{\varepsilon} = -\frac{i}{\hbar}\hat{p}_x\hat{T}(a)$$

We can integrate the above equation to obtain the generator for translation by a finite distance a:

$$\hat{T}(a) = \exp\left[-\frac{ia}{\hbar}\hat{p}_x\right] \tag{96}$$

where the exponential of the operator $\mathcal{O}$ is defined through the equation

$$e^{\mathcal{O}} \equiv 1 + \frac{1}{1!}\mathcal{O} + \frac{1}{2!}\mathcal{O}\mathcal{O} + \frac{1}{3!}\mathcal{O}\mathcal{O}\mathcal{O} + \cdots \tag{97}$$

Thus,
$$\hat{T}(a) = 1 - \frac{1}{1!}\frac{ia}{\hbar}\hat{p}_x - \frac{1}{2!}\frac{a^2}{\hbar^2}\hat{p}_x\hat{p}_x + \frac{1}{3!}\frac{ia^3}{\hbar^3}\hat{p}_x\hat{p}_x\hat{p}_x + \cdots \tag{98}$$

If we assume a to be infinitesimal (equal to ε), we can retain only the first 2 terms of the above equation to obtain Eq. (95). We can also differentiate Eq. (98) term by term to obtain

$$\frac{d\hat{T}}{da} = -\frac{i}{\hbar}\hat{p}_x \hat{T}(a) \tag{99}$$

4.10 PROBLEMS

Problem 4.1 Using the operator representations of p_x, p_y and p_z prove that

$$\left.\begin{aligned} [x, p_y] &= 0 = [y, p_z] \\ [p_x, p_y] &= 0 = [p_z, p_x] \\ [y, p_y] &= i\hbar = [z, p_z] \end{aligned}\right\} \tag{100}$$

Problem 4.2 Consider a Gaussian wave packet given by (see Example 5.1):

$$\psi(x) = \frac{1}{(\pi\sigma^2)^{\frac{1}{4}}} \exp\left[-\frac{(x - x_0)^2}{2\sigma^2}\right] \exp\left(\frac{i}{\hbar}p_0 x\right) \tag{101}$$

Show that it is normalized and that

$$\mathbf{J} = \frac{1}{\sqrt{\pi\sigma^2}} \exp\left[-\frac{(x - x_0)^2}{\sigma^2}\right]\frac{p_0}{m}\hat{\mathbf{x}} \tag{102}$$

Interpret the result physically.

Problem 4.3 Assume

$$\psi(r) = \frac{1}{r}e^{ikr} \tag{103}$$

where
$$r = \sqrt{x^2 + y^2 + z^2}$$

Calculate **J** and interpret the result physically.

Problem 4.4 Show by partial integration of Eq. (51) that $\langle p_x \rangle$ is real.

Problem 4.5 Using the general recipe that the expectation value of an observable characterised by the operator $\mathcal{O}$ is given by

$$\langle \mathcal{O} \rangle = \int \psi^* \mathcal{O}\, \psi\, d\tau \tag{104}$$

prove that
$$\frac{d}{dt}\langle x \rangle = \frac{1}{m}\langle p_x \rangle \tag{105}$$

and
$$\frac{d}{dt}\langle p_x\rangle = -\left\langle \frac{\partial V}{\partial x}\right\rangle \tag{106}$$

These equations are referred to as Ehrenfest's theorem.

Problem 4.6 Using Eq. (77) prove that:

$$\frac{d}{dt}\langle \mathcal{O}\rangle = \frac{i}{\hbar}\int \psi^*[H,\mathcal{O}]\Psi\, d\tau + \int \psi^* \frac{\partial \mathcal{O}}{\partial t}\psi\, d\tau \tag{107}$$

where $[H, \mathcal{O}] = H\mathcal{O} - \mathcal{O}H$ represents the commutator of H and $\mathcal{O}$ and $\frac{\partial \mathcal{O}}{\partial t}$ is due to the explicit time dependence of the operator $\mathcal{O}$.

[**Hint:** Differentiate Eq. (77) inside the integral sign and use Eq. (24). You may use the Hermitian nature of H (see Problem 6.1) or else substitute for H and integrate by parts as in the solution of Problem 4.5.]

Problem 4.7 For the ground state and the first excited state of the harmonic oscillator the wave functions are:

$$\exp\left[-\frac{m\omega}{2\hbar}x^2\right] \quad \text{and} \quad x\exp\left[-\frac{m\omega}{2\hbar}x^2\right]$$

respectively (see Section 7.3). Show that the values of the uncertainty product $\Delta x\, \Delta p$ are $\frac{1}{2}\hbar$ and $\frac{3}{2}\hbar$ (see also Section 12.3).

Problem 4.8 Show that the uncertainty product $\Delta x\, \Delta p$ is minimum ($= \frac{1}{2}\hbar$) for a Gaussian wave-function.

4.11 SOLUTIONS

Solution 4.3 Using Eq. (34), we get

$$\mathbf{J} = \frac{\hbar k}{m}\frac{1}{r^2}\hat{\mathbf{r}} = \frac{v}{r^2}\hat{\mathbf{r}} \tag{108}$$

which corresponds to a spherical wave whose intensity decreases according to the inverse-square law.

Solution 4.4

$$\langle p_x\rangle = -i\hbar\int \Psi^* \frac{\partial \Psi}{\partial x}\, d\tau$$

$$= -i\hbar\int\int dy\, dz\left[\Psi^*\Psi\Big|_{x=-\infty}^{x=+\infty} - \int_{-\infty}^{+\infty}\Psi\frac{\partial \Psi^*}{\partial x}dx\right] = i\hbar\int \Psi\frac{\partial \Psi^*}{\partial x}d\tau$$

$$= \langle p_x\rangle^* \tag{109}$$

Solution 4.5 Differentiating Eq. (44), we get

$$\frac{d}{dt}\langle x\rangle = \int\left[\Psi^* x\frac{\partial \Psi}{\partial t} + \frac{\partial \Psi^*}{\partial t}x\Psi\right]d\tau$$

$$= \frac{i\hbar}{2m}\int[\Psi^* x\nabla^2\Psi - (\nabla^2\Psi^*)x\Psi]\, d\tau \tag{110}$$

where use has been made of Eqs (28) and (29). Now,

$$\int (\nabla^2\Psi^*)\, x\Psi d\tau = \int \nabla\cdot(x\Psi\nabla\Psi^*)\, d\tau - \int (\nabla\Psi^*)\cdot\nabla(x\Psi)\, d\tau$$

The first integral on the right can be transformed to a surface integral by use of Gauss theorem and this integral would vanish because a wave packet vanishes at infinity. Thus,

$$\begin{aligned}\int (\nabla^2\Psi^*)\, x\Psi\, d\tau &= -\int (\nabla\Psi^*)\cdot\nabla(x\Psi)\, d\tau \\ &= -\int \nabla(\Psi^*\nabla x\Psi)\, d\tau + \int \Psi^*\nabla^2(x\Psi)\, d\tau\end{aligned}$$

The first integral on the right again vanishes. On substitution in Eq. (110), we get

$$\begin{aligned}\frac{d}{dt}\langle x\rangle &= \frac{i\hbar}{2m}\int \Psi^*[x\nabla^2\Psi - \nabla^2(x\Psi)]\, d\tau \\ &= \frac{1}{m}\int \Psi^*\left(-i\hbar\frac{\partial\Psi}{\partial x}\right) d\tau = \frac{1}{m}\langle p_x\rangle \qquad (111)\end{aligned}$$

Using a method similar to that used above one gets

$$\begin{aligned}\frac{d}{dt}\langle p_x\rangle &= -i\hbar\int\left[\frac{\partial\Psi^*}{\partial t}\frac{\partial\Psi}{\partial x} + \Psi^*\frac{\partial}{\partial x}\frac{\partial\Psi}{\partial t}\right] d\tau \\ &= \int \Psi^*\left[V\frac{\partial\Psi}{\partial x} - \frac{\partial}{\partial x}(V\Psi)\right] d\tau = \left\langle -\frac{\partial V}{\partial x}\right\rangle \qquad (112)\end{aligned}$$

Solution 4.8 The uncertainty product is minimum when f is a multiple of g (see Sec. 4.6); thus,

$$-i\hbar\frac{\partial\Psi}{\partial x} = C\, ix\Psi$$

or

$$\frac{1}{\Psi}\frac{\partial\Psi}{\partial x} = (\text{const})\, x$$

The integration of the above equation gives a Gaussian function.

4.12 SAMPLE QUESTIONS

Q. 1 (a) What are the normalized eigenfunctions of the operator

$$p_x = -i\hbar\frac{d}{dx}$$

(b) Derive the orthonormality and completeness condition for these eigenfunctions.

Q. 2 Consider the Hamiltonian corresponding to a free particle in one-dimension

$$H = -\frac{\hbar^2}{2m}\cdot\frac{\partial^2}{\partial x^2}$$

What are the corresponding eigenvalues and normalized eigenfunctions? Is there any degeneracy?

REFERENCES AND SUGGESTED READINGS

1. P.A.M. Dirac, *The Principles of Quantum Mechanics*, Oxford University Press, Oxford (1958).
2. D. Bohm, *Quantum Theory*, Prentice Hall, Englewood Cliffs, N.J. (1951).
3. J L Powell and B. Craseman, *Quantum Mechanics*, Addison-Wesley Publishing Company, Inc., (1968).
4. L.I. Schiff, *Quantum Mechanics*, 3rd Edition, McGraw-Hill, New York (1968).
5. J.S. Townsend, *A Modern Approach to Quantum Mechanics*, McGraw-Hill, New York (1992).
6. J.J. Sakurai, *Modern Quantum Mechanics*, Addison-Wesley Publishing Company, Inc. (1994).
7. M. Beck, *Quantum Mechanics: Theory and Experiment*, Oxford University Press, New York (2012).

5 Propagation of Wave Packets and Concept of Group Velocity

Quantum mechanics allows us in principle to predict nearly everything around us, within the limits set by the uncertainty principle ...

— STEPHEN HAWKING in *A Brief History of Time*,
Bantam Books, New York (1988).

5.1 INTRODUCTION

We will devote this chapter to the solution of the time dependent Schrödinger equation for a free particle. In quantum mechanics, a free particle is described by a wave packet (which is nothing but a superposition of plane waves) and the solution of the time dependent Schrödinger equation will allow us to study the time evolution of the packet and how it would disperse in free space. We will also show how the uncertainty principle is contained in the solution of the Schrödinger equation.

5.2 THE GENERAL SOLUTION OF THE ONE-DIMENSIONAL SCHRÖDINGER EQUATION FOR A FREE PARTICLE

We start with the one-dimensional time dependent Schrödinger equation [see Eq. (25) of chapter 4]

$$i\hbar\frac{\partial\Psi}{\partial t} = H\Psi \tag{1}$$

with the Hamiltonian H given by

$$H = \frac{p^2}{2m} + V(x) = -\frac{\hbar^2}{2m}\cdot\frac{\partial^2}{\partial x^2} + V(x) \tag{2}$$

For a free particle

$$V(x) = 0 \tag{3}$$

and Eq. (1) becomes

$$i\hbar\frac{\partial\Psi}{\partial t} = -\frac{\hbar^2}{2m}\cdot\frac{\partial^2\Psi(x,t)}{\partial x^2} \tag{4}$$

which represents the one-dimensional time dependent Schrödinger equation for a free particle. In order to solve the above equation we use the method of separation of variables:

$$\Psi(x, t) = \psi(x)\, T(t) \tag{5}$$

Substituting in Eq. (4) and subsequent division by $\Psi(x)\, T(t)$ gives us

$$i\hbar \frac{1}{T(t)} \cdot \frac{dT}{dt} = -\frac{\hbar^2}{2m} \cdot \frac{1}{\psi} \frac{d^2\psi}{dx^2} = E \tag{6}$$

Since the LHS is a function of t alone and the RHS is a function of x alone, we have set them equal to a constant (= E) which represents the total energy of the particle. From Eq. (6) we readily have

$$\frac{dT}{dt} = -\frac{iE}{\hbar}\, T(t) \tag{7}$$

the solution of which is given by

$$T(t) = (\text{constant})\, e^{-iEt/\hbar} \tag{8}$$

Obviously, *E has to be real;* if E is complex, the solution will blow up as $t \to +\infty$ or as $t \to -\infty$. From Eq. (6) we also have

$$\frac{d^2\psi}{dx^2} + \frac{p^2}{\hbar^2}\, \psi(x) = 0 \tag{9}$$

where

$$p = \sqrt{2mE} \tag{10}$$

represents the momentum of the particle. The solution of Eq. (9) will be of the form

$$e^{ipx/\hbar}$$

Obviously, E cannot be negative because then p would be imaginary and the solutions will blow up as $x \to +\infty$ or $-\infty$. Thus, we must have

(i) E real and positive, and

(ii) p can take any real value between $-\infty$ and $+\infty$.

We write the solution of Eq. (9) as[1]

$$\psi_p(x) = \frac{1}{\sqrt{2\pi\hbar}}\, e^{ipx/\hbar}; \quad -\infty < p < +\infty \tag{11}$$

where the factor $1/\sqrt{2\pi\hbar}$ makes it normalized:

$$\int_{-\infty}^{+\infty} \psi_p^*(x)\, \psi_{p'}(x)\, dx = \delta(p - p') \qquad \text{Orthonormality condition} \tag{12}$$

which represents the orthonormality condition; the functions $\psi_p(x)$ also satisfy the completeness condition [see Eqs. (70) and (71) of the previous chapter].

The complete solution is therefore given by [see Eq. (5)]:

$$\begin{aligned}\Psi(x, t) &= \frac{1}{\sqrt{2\pi\hbar}} \exp\left[\frac{i}{\hbar}(px - Et)\right] \\ &= \frac{1}{\sqrt{2\pi\hbar}} \exp\left[\frac{i}{\hbar}\left(px - \frac{p^2}{2m}t\right)\right]\end{aligned} \tag{13}$$

[1]The functions $\psi_p(x)$ are simultaneous eigenfunctions of the momentum operator p and the Hamiltonian H (see Section 4.7).

For positive values of p, the above equation represents a plane wave propagating along the $+x$ direction; similarly, for negative values of p, it represents a plane wave propagating along the $-x$ direction. Since p can take any value from $-\infty$ to $+\infty$, *the most general solution* of Eq. (4) (i.e., of the time dependent Schrödinger equation for a free particle) is given by

$$\Psi(x, t) = \frac{1}{\sqrt{2\pi\hbar}} \int_{-\infty}^{+\infty} a(p) \exp\left[\frac{i}{\hbar}\left(px - \frac{p^2}{2m}t\right)\right] dp \qquad \text{Most general solution} \qquad (14)$$

which is essentially a superposition of plane waves. Equation (14) describes the time evolution of a one-dimensional wave packet. In Example 5.1 we will discuss the time evolution of a Gaussian wave packet; however, we will first discuss some general properties of the solution given by Eq. (14):

If $\Psi(x, 0)$ is normalized, i.e., if

$$\int_{-\infty}^{+\infty} |\Psi(x,0)|^2\, dx = 1 \qquad (15)$$

then one can readily show that (see Problem 5.1)

$$\int_{-\infty}^{+\infty} |\Psi(x,t)|^2\, dx = 1 \qquad \text{Normalization condition} \qquad (16)$$

Physically, this is obvious because $|\Psi(x, 0)|^2\, dx$ represents the probability of finding the particle between x and $x + dx$ at time t. The time evolution of the center of the wave packet (i.e., the trajectory of the particle) is given by

$$\langle x \rangle = \int_{-\infty}^{+\infty} x\,|\Psi(x,t)|^2\, dx \qquad (17)$$

The spatial spread of the wave packet is given by

$$\Delta x = \left[\langle x^2\rangle - \langle x\rangle^2\right]^{\frac{1}{2}} \qquad (18)$$

where
$$\langle x^2 \rangle = \int_{-\infty}^{+\infty} x^2\,|\Psi(x,t)|^2\, dx \qquad (19)$$

Furthermore (see Problems 5.1 and 5.2),

$$\int_{-\infty}^{+\infty} |a(p)^2|\, dp = 1 \qquad (20)$$

$$\langle p \rangle = \int_{-\infty}^{+\infty} \Psi^*(x,t)\left[-i\hbar\frac{\partial\Psi(x,t)}{\partial x}\right] dx = \int_{-\infty}^{+\infty} p\,|a(p)|^2\, dp \qquad (21)$$

$$\langle p^2 \rangle = \int_{-\infty}^{+\infty} \Psi^*(x,t)\left[-\hbar^2\frac{\partial^2\Psi(x,t)}{\partial x^2}\right] dx = \int_{-\infty}^{+\infty} p^2\,|a(p)|^2\, dp \qquad (22)$$

Equations (20)–(22) clearly show that the quantity $|a(p)|^2\, dp$ can be interpreted as the probability of the momentum lying between p and $p + dp$. The quantity $|a(p)|^2$ is known as the momentum distribution function.

5.3 THE TIME EVOLUTION OF A WAVE PACKET

Let us suppose that at $t = 0$, a particle is described by the wave function $\Psi(x, 0)$. Knowing $\Psi(x, 0)$ we wish to determine the time evolution of the wavefunction $\Psi(x, t)$ from which we would obtain the trajectory of the particle. From Eq. (14), we have

$$\Psi(x, 0) = \frac{1}{\sqrt{2\pi\hbar}} \int_{-\infty}^{+\infty} a(p) \exp\left[\frac{i}{\hbar} px\right] dp \tag{23}$$

Thus using the inverse Fourier transform relation [see Section 2.2]:

$$a(p) = \frac{1}{\sqrt{2\pi\hbar}} \int_{-\infty}^{+\infty} \Psi(x, 0) \exp\left[-\frac{i}{\hbar} px\right] dx \tag{24}$$

Therefore, the *recipe* is as follows:

Using the known expression for $\Psi(x, 0)$, we first determine $a(p)$ by carrying out the integration in Eq. (24); we then substitute $a(p)$ in Eq. (14) and carry out the integration over p to obtain $\Psi(x, t)$ and hence the time evolution of the wave packet.

Example 5.1 *The Gaussian wave packet:* We assume that at $t = 0$, a free particle is described by the Gaussian wavefunction:

$$\Psi(x, 0) = \frac{1}{\left(\pi\sigma_0^2\right)^{\frac{1}{4}}} e^{-x^2/2\sigma_0^2} \exp\left(\frac{i}{\hbar} p_o x\right) \tag{25}$$

The wavefunction is normalized, i.e.

$$\int_{-\infty}^{+\infty} |\Psi(x, 0)|^2\, dx = 1 \tag{26}$$

Thus, $$P(x)\, dx = |\Psi(x, 0)|^2\, dx = \frac{1}{\sqrt{\pi}\,\sigma_0} e^{-x^2/\sigma_0^2}\, dx \tag{27}$$

would represent the probability of finding the particle between x and $x + dx$ (at $t = 0$). The following average values are readily obtained (see Problem 5.4):

$$\langle x \rangle = 0 \tag{28}$$

$$\Delta x = \left[\langle x^2\rangle - \langle x\rangle^2\right]^{\frac{1}{2}} = \frac{\sigma_0}{\sqrt{2}} \tag{29}$$

$$\langle p \rangle = p_0 \tag{30}$$

$$\Delta p = \left[\langle p^2\rangle - \langle p\rangle^2\right]^{\frac{1}{2}} = \frac{\hbar}{\sqrt{2}\,\sigma_0} \tag{31}$$

Thus (at $t = 0$) the wavefunction given by Eq. (25) describes a particle which is localized within a distance ~ σ_0 around $x = 0$ and moving with an average momentum of p_0 (in the $+x$ direction) with a momentum spread approximately equal to $\hbar/\sigma_0$. Thus, the uncertainty principle is contained in Eq. (25)!

In order to study the time evolution of the wave packet, we use Eq. (24) to calculate $a(p)$:

$$a(p) = \frac{1}{\sqrt{2\pi\hbar}} \cdot \frac{1}{(\pi\sigma_0^2)^{\frac{1}{4}}} \int_{-\infty}^{+\infty} e^{-x^2/2\sigma_0^2} \exp\left[-\frac{i}{\hbar}(p - p_0)x\right] dx$$

$$= \left(\frac{\sigma_0^2}{\pi\hbar^2}\right)^{\frac{1}{4}} \exp\left[-\frac{(p - p_0)^2 \sigma_0^2}{2\hbar^2}\right] \tag{32}$$

where we have used the relation (see Appendix A)

$$\int_{-\infty}^{+\infty} e^{-\alpha x^2 + \beta x} dx = \sqrt{\frac{\pi}{\alpha}} \exp\left(\frac{\beta^2}{4\alpha}\right) \tag{33}$$

The momentum probability distribution is given by

$$|a(p)|^2 = \frac{\sigma_0}{\sqrt{\pi}\hbar} \exp\left[-\frac{\sigma_0^2}{\hbar^2}(p - p_0)^2\right] \quad \text{Momentum probability distribution} \tag{34}$$

The above equation shows that the average momentum is p_0 with a spread ~ $\hbar/\sigma_0$. In Fig. 5.1, we have plotted $|\Psi(x, 0)|^2$ and $|a(p)|^2$ for

$$\sigma_0 = 5 \text{ nm} \quad \text{and} \quad 1 \text{ nm}$$

It may be seen that as we decrease the value of σ (i.e., as the wave function becomes more localized), the value of the momentum spread Δp decreases.

We next substitute for $a(p)$ from Eq. (32) in Eq. (14) and carry out the straight-forward integration to obtain

$$\Psi(X, \tau) = \frac{1}{\pi^{\frac{1}{4}} [\sigma_0 (1 + i\tau)]^{\frac{1}{2}}} \exp\left[-\frac{(X - \alpha\tau)^2}{2(1+\tau^2)} + \frac{i}{2(1+\tau^2)}[2\alpha X + \tau(X^2 - \alpha^2)]\right] \tag{35}$$

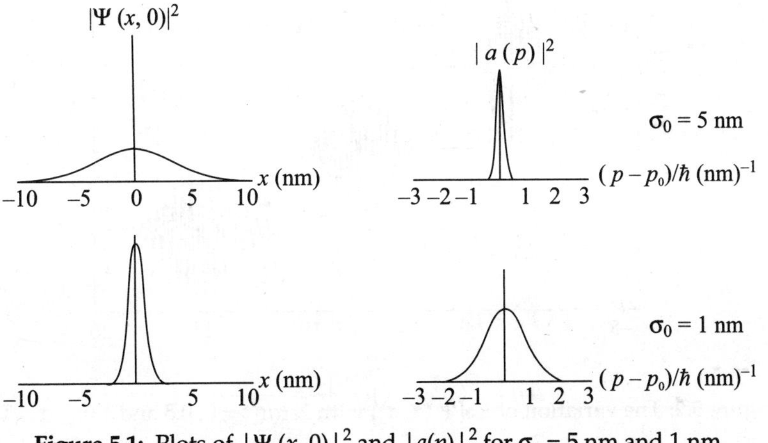

Figure 5.1: Plots of $|\Psi(x, 0)|^2$ and $|a(p)|^2$ for $\sigma_0 = 5$ nm and 1 nm.

where we have introduced the following dimensionless variables:

$$\left.\begin{aligned} X &= \frac{x}{\sigma_0} \\ \tau &= \frac{\hbar}{m\sigma_0^2}\,t \\ \alpha &= \frac{p_0\sigma_0}{\hbar} \end{aligned}\right\} \tag{36}$$

The corresponding probability distribution is given by

$$|\Psi(X,\tau)|^2 = \frac{1}{\sqrt{\pi}\,\sigma(\tau)}\exp\left[-\frac{(X-\alpha\tau)^2}{(1+\tau^2)}\right] \tag{37}$$

or

$$|\Psi(x,t)|^2 = \frac{1}{\sqrt{\pi}\,\sigma(t)}\exp\left[-\frac{\left(x-\frac{p_0}{m}t\right)^2}{\sigma^2(t)}\right] \tag{38}$$

where

$$\sigma(t) = \sigma_0\left[1+\frac{\hbar^2}{m^2\sigma_0^4}t^2\right]^{\frac{1}{2}} \tag{39}$$

which represents the spatial broadening of the wave packet with time. In Fig. 5.2, we have plotted the x dependence of the real part of the wave function [given by Eq. (35)] at

$$\tau = 0, \quad 0.5 \text{ and } 1.0$$

with $\alpha = 15$. In Fig. 5.3, we have plotted the x dependence of the probability distribution function at

$$\tau = 0, \quad 0.5, \quad 1.0 \text{ and } 1.5$$

Figures 5.2 and 5.3 show that the centre of the wave packet moves with the velocity p_0/m which is the group velocity of the wave packet (see Sec. 5.5).

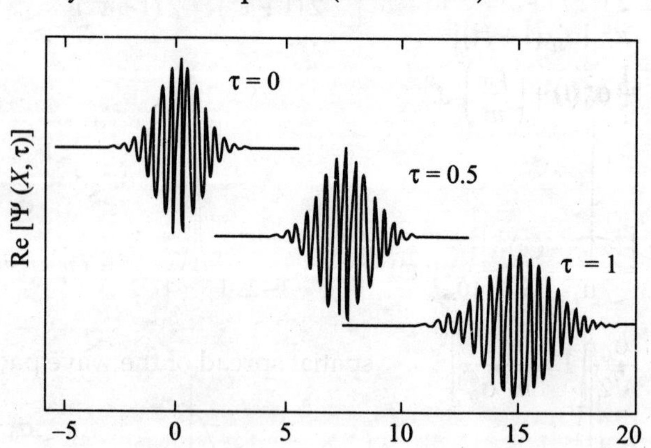

Figure 5.2: The variation of Re[Ψ (X, τ)] with X for $\tau = 0, 0.5$ and 1.0; $\alpha = 15$.

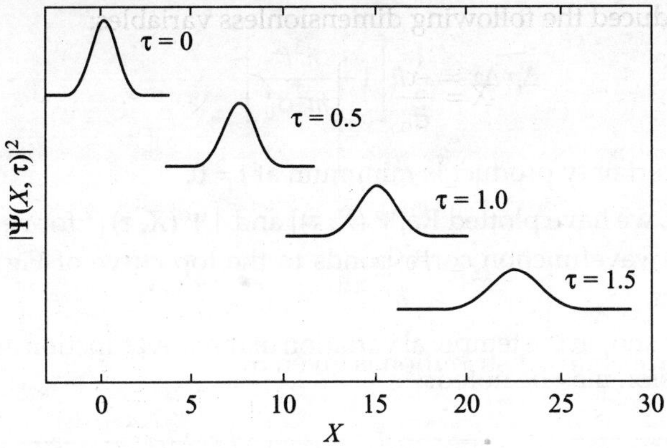

Figure 5.3: The variation of the probability distribution $|\Psi(X, \tau)|^2$ with X for $\tau = 0, 0.5, 1.0$ and 1.5; $\alpha = 15$. Notice the spreading of the wave packet.

One can readily verify that

$$\int_{-\infty}^{+\infty} |\Psi(x,t)|^2 \, dx = 1 \tag{40}$$

for all values of time as it indeed should be. It may also be noted that using Eq. (35), we readily get

$$\langle x \rangle = \int_{-\infty}^{+\infty} \Psi^*(x,t)\, x\, \Psi(x,t)\, dx$$

$$= \frac{p_0}{m} t \tag{41}$$

which represents the propagation of the center of the wavepacket with velocity p_0/m. Further,

$$\langle x^2 \rangle = \int_{-\infty}^{+\infty} x^2 |\psi(x,t)|^2 \, dx$$

$$= \frac{1}{2}\sigma_0^2(t) + \left(\frac{p_0}{m}\right)^2 t^2 \tag{42}$$

giving

$$\Delta x = \left[\langle x^2 \rangle - \langle x \rangle^2\right]^{\frac{1}{2}}$$

or,

$$\Delta x = \frac{\sigma_0}{\sqrt{2}}\left[1 + \frac{\hbar^2 t^2}{m^2 \sigma_0^4}\right]^{\frac{1}{2}} \quad \text{spatial spread of the wave packet} \tag{43}$$

The momentum distribution [as given by Eq. (34)] remains the same at all times and therefore

$$\Delta p = \frac{\hbar}{\sqrt{2}\,\sigma_0} \quad \text{for all values of } t \tag{44}$$

Thus,
$$\Delta p\,\Delta x = \frac{1}{2}\hbar\left[1+\frac{\hbar^2 t^2}{m^2\sigma_0^4}\right]^{\frac{1}{2}} \tag{45}$$

showing that the uncertainty product is minimum at $t = 0$.

In Figs 5.4 and 5.5, we have plotted Re[$\Psi(X, \tau)$] and $|\Psi(X, \tau)|^2$ for negative values of time. Thus, if at $t = 0$, the wavefunction corresponds to the top curve of Fig. 5.4, then it will first undergo compression.

In Fig. 5.6, we have shown the temporal variation of the wavefunction at different values of X. The temporal dispersion may be noted.

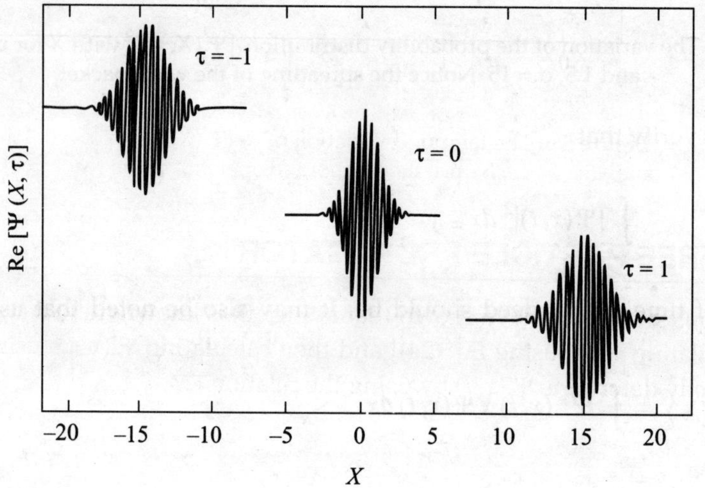

Figure 5.4: The spatial variation of Re[$\Psi(X, \tau)$] for $\tau = -1$, 0 and +1; $\alpha = 15$. Notice that if the wave packet initially corresponds to the top curve, it will first undergo spatial compression.

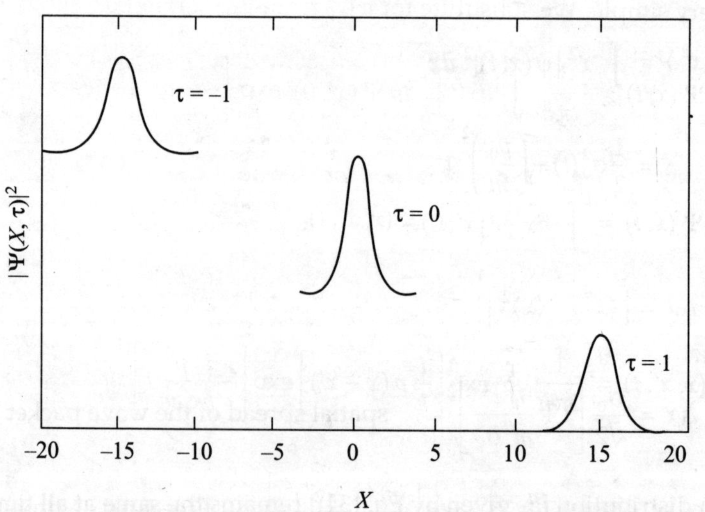

Figure 5.5: The spatial variation of $|\Psi(X, \tau)|^2$ for $\tau = -1$, 0 and +1; $\alpha = 15$. Notice the spatial compression of the wave packet.

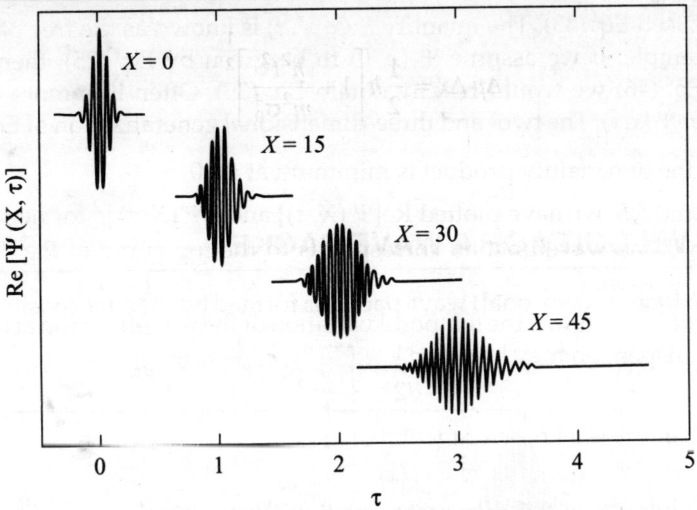

Figure 5.6: The temporal variation of Re[Ψ (X, τ)] for X = 0, 15, 30 and 45; α = 15. Notice the temporal dispersion of the wave packet.

5.4 THE FREE PARTICLE PROPAGATOR

In the previous section we have shown that if $\Psi(x, 0)$ is known then $\Psi(x, t)$ can be calculated by first calculating $a(p)$ [using Eq. (24)] and then calculating $\Psi(x, t)$ [using Eq. (14)]. Indeed, one can directly determine $\Psi(x, t)$ by using the relation

$$\Psi(x, t) = \left(\frac{m}{2\pi i\hbar t}\right)^{\frac{1}{2}} \int_{-\infty}^{+\infty} \psi(x', 0) \exp\left[\frac{im(x-x')^2}{2\hbar t}\right] dx' \tag{46}$$

without making any explicit calculation of $a(p)$. Thus, all one has to do is to substitute the form of the wavefunction at $t = 0$ and carry out the above integration to obtain $\Psi(x, t)$. The proof of Eq. (46) is very simple: We substitute for $a(p)$ from Eq. (24) in Eq. (14) to obtain

$$\Psi(x, t) = \frac{1}{2\pi\hbar} \int_{-\infty}^{+\infty} dx' \int_{-\infty}^{+\infty} dp\, \Psi(x', 0) \exp\left[-\frac{i}{\hbar} px'\right] \times \exp\left[\frac{i}{\hbar}\left(px - \frac{p^2}{2m} t\right)\right]$$

or
$$\Psi(x, t) = \int_{-\infty}^{+\infty} dx'\, \Psi(x', 0) K(x, x', t) \tag{47}$$

where

$$K(x, x', t) = \frac{1}{2\pi\hbar} \int_{-\infty}^{+\infty} \exp\left[\frac{i}{\hbar} p(x-x')\right] \exp\left[-\frac{i}{\hbar}\frac{p^2}{2m} t\right] dp$$

or
$$K(x, x', t) = \left(\frac{m}{2\pi i\hbar t}\right)^{\frac{1}{2}} \exp\left[\frac{im(x-x')^2}{2\hbar t}\right] \quad \text{Free particle propagator} \tag{48}$$

where we have used Eq. (33). The quantity $K(x, x', t)$ is known as the free particle propagator. As a simple example, if we assume $\Psi(x, 0)$ to be given by Eq. (25), then carrying out the integration in Eq. (46) we would readily obtain Eq. (35). Often it is more convenient to use Eq. (46) to obtain $\Psi(x, t)$. The two- and three-dimensional generalization of Eq. (46) is discussed in Section 5.6.

5.5 GROUP VELOCITY OF A WAVE PACKET

In wave theory a (one-dimensional) wave packet is formed by the superposition of plane waves:

$$\Psi(x, t) = \frac{1}{\sqrt{2\pi}} \int_{-\infty}^{+\infty} A(k)\, e^{i(kx - \omega t)}\, dk \tag{49}$$

where the frequency ω is a known function of k:

$$\omega = \omega(k) \tag{50}$$

Equation (50) is known as the dispersion relation. For example, for a free (non-relativistic) particle

$$E = \frac{p^2}{2m} \tag{51}$$

implying

$$\hbar\omega = \frac{\hbar^2 k^2}{2m}$$

or

$$\omega(k) = \frac{\hbar}{2m} k^2 \tag{52}$$

Now, for a purely monochromatic wave

$$A(k) = A_0 \delta(k - k_0) \tag{53}$$

If we substitute the above expression for $A(k)$ in Eq. (49), we would get

$$\Psi(x, t) = \frac{A_0}{\sqrt{2\pi}} e^{i(k_0 x - \omega_0 t)} \tag{54}$$

where

$$\omega_0 = \omega(k_0) \tag{55}$$

Equation (54) is never achieved in practice. Usually, $A(k)$ is a very sharply peaked function around $k = k_0$; see, e.g. Fig. 5.1. In all of what follows we will assume that there exists only a small range Δk (around $k = k_0$) where the amplitudes of the component waves $A(k)$ have significant values; beyond $k = k_0 \pm \Delta k$, $A(k)$ is assumed to be negligible. Under such an assumption, it is possible to assign an average velocity of the wave packet provided $\omega(k)$ is a smoothly varying function in the domain $k_0 - \Delta k \le k \le k_0 + \Delta k$. This average velocity is known as the group velocity. We write

$$A(k) = |A(k)|\, e^{i\phi(k)} \tag{56}$$

and assume that $|A(k)|$ is very sharply peaked when k lies in a small interval Δk around $k = k_0$ and negligible everywhere else so that we may write

$$\Psi(x, t) \approx \frac{1}{\sqrt{2\pi}} \int_{\Delta k} dk\, |A(k)|\, e^{i[kx - \omega(k)t + \phi(k)]} \tag{57}$$

where the integration extends over the domain Δk where $|A(k)|$ has appreciable values. Next, we make Taylor expansions of $\omega(k)$ and $\phi(k)$ around $k = k_0$:

$$\omega(k) = \omega(k_0) + \left.\frac{d\omega}{dk}\right|_{k=k_0} (k - k_0) + \frac{1}{2}\left.\frac{d^2\omega}{dk^2}\right|_{k=k_0} (k - k_0)^2 + \cdots \tag{58}$$

$$\phi(k) = \phi(k_0) + \left.\frac{d\phi}{dk}\right|_{k=k_0} (k - k_0) + \frac{1}{2}\left.\frac{d^2\phi}{dk^2}\right|_{k=k_0} (k - k_0)^2 + \cdots \tag{59}$$

We substitute the above expansion in Eq. (57) and assume that in the domain of integration Δk, $\omega(k)$ and $\phi(k)$ do not vary significantly so that the terms involving quadratic and higher powers of $(k - k_0)$ can be neglected; under such an approximation one obtains

$$\Psi(x, t) \approx f(x, t)\, e^{i(k_0 x - \omega_0 t + \phi_0)} \tag{60}$$

where $\omega_0 \equiv \omega(k_0)$, $\phi_0 \equiv \phi(k_0)$ and $f(x, t)$, which represents the envelope of the wave packet, is given by

$$f(x, t) = \frac{1}{\sqrt{2\pi}} \int_{\Delta k} dk\, |A(k)|\, e^{i(k-k_0)(x - x_0 - v_g t)} \tag{61}$$

where
$$x_0 = -\left.\frac{d\phi}{dk}\right|_{k=k_0}$$

and
$$v_g = \left.\frac{d\omega}{dk}\right|_{k=k_0} \quad \text{Group velocity} \tag{62}$$

Since the function $f(x, t)$ depends on x and t only through the combination $x - v_g t$, the wave packet propagates without any distortion with velocity v_g which is known as the *group velocity* of the packet. It should be pointed out that the packet remains undistorted as long as the neglect of second and higher order terms in Eqs (58) and (59) is justified (see Problem 5.9 and Example 5.1). However, if the ω-k relation is strictly linear, the packet will never undergo any distortion[2].

Now, from Eq. (61), we have

$$f(x, 0) = \frac{1}{\sqrt{2\pi}} \int_{\Delta k} |A(k)|\, e^{i(k - k_0)(x - x_0)}\, dk \tag{63}$$

---

[2]A medium is said to be non-dispersive when the ω-k relation is linear. For example, for electromagnetic waves in free space

$$\omega(k) = ck$$

where c represents the speed of light in free space. For such a case, Eq. (49) becomes

$$\Psi(x, t) = \frac{1}{\sqrt{2\pi}} \int_{-\infty}^{+\infty} dk\, A(k)\, e^{ik(x - ct)}$$
$$= f(x - ct)$$

implying that the packet travels undistorted with speed c. Thus, the free space is a non-dispersive medium for electromagnetic waves. On the other hand, whenever the ω-k relationship is not linear [see, e.g. Eq. (52)] the medium is said to be dispersive and a wave packet undergoes distortion (see Example 5.1).

It is immediately seen that at $x = x_0$, the integrand is everywhere positive and the value of the integral is maximum. For $|x - x_0| \geq 1/\Delta k$ the exponential function oscillates rapidly in the domain of integration and the value of the integral is very small. This immediately leads to the relation

$$\Delta k \, \Delta x \geq 1 \tag{64}$$

which when multiplied by $\hbar$ gives the uncertainty relation:

$$\Delta p \, \Delta x \geq \hbar \tag{65}$$

Thus, at $t = 0$ the wave packet is sharply peaked at $x = x_0$ and as t increases, the centre of the packet, $x_c(t)$, moves according to the equation

$$x_c(t) = x_0 + v_g t \tag{66}$$

This was explicitly seen in Example 5.1, where we discussed the spreading of a Gaussian wave packet. Now, since (see Section 4.1)

$$E = \hbar\omega \tag{67}$$

and

$$p = \hbar k \tag{68}$$

we get

$$v_g = \frac{d\omega}{dk} = \frac{dE}{dp} \tag{69}$$

Thus, for a non-relativistic particle for which $E = p^2/2m$, we have

$$v_g = \frac{dE}{dp} = \frac{p}{m} = v \tag{70}$$

Further, for a relativistic particle

$$E^2 = p^2c^2 + m^2c^4 \tag{71}$$

(where m represents the rest mass) we have

$$\frac{dE}{dp} = \frac{pc^2}{E} \tag{72}$$

But $$E = \frac{mc^2}{\sqrt{1 - v^2/c^2}} \quad \text{and} \quad p = \frac{mv}{\sqrt{1 - v^2/c^2}}$$

giving $$\frac{pc^2}{E} = v$$

Thus, $$v_g = v \tag{73}$$

Equations (70) and (73) tell us that the group velocity of the packet is to be associated with the velocity of the particle.

The phase velocity v_p is defined through the relation

$$v_p = \frac{\omega}{k} \quad \text{Phase velocity} \tag{74}$$

and represents the velocity of propagation of an infinitely long monochromatic plane wave. Only when ω is proportional to k, v_p and v_g are equal. In general, it is the group velocity of a packet which is of physical significance.

5.6 FREE PARTICLE PROPAGATOR FOR THE TWO- AND THREE-DIMENSIONAL SCHRÖDINGER EQUATION

The two- and three-dimensional generalization of Eq. (46) is obvious:

The general solution of the two-dimensional Schrödinger equation

$$i\hbar \frac{\partial \Psi}{\partial t} = -\frac{\hbar^2}{2m}\left(\frac{\partial^2}{\partial x^2} + \frac{\partial^2}{\partial y^2}\right) \Psi(x, y, t) \tag{75}$$

is given by

$$\Psi(x, y, t) = \int_{-\infty}^{+\infty}\int_{-\infty}^{+\infty} \Psi(x', y', 0)\, K_2(x, y, x' y'; t)\, dx'\, dy' \tag{76}$$

where

$$K_2(x, y, x', y'; t) = \left(\frac{m}{2\pi i \hbar t}\right) \exp\left[\frac{im}{2\hbar t}\left[(x - x')^2 + (y - y')^2\right]\right] \tag{77}$$

Similarly, the general solution of the three-dimensional Schrödinger equation

$$i\hbar \frac{\partial \Psi}{\partial t} = -\frac{\hbar^2}{2m} \nabla^2 \Psi(x, y, z, t) \tag{78}$$

is given by

$$\Psi(x, y, z, t) = \int_{-\infty}^{+\infty}\int_{-\infty}^{+\infty}\int_{-\infty}^{+\infty} \Psi(x', y', z', 0)\, K_3(x, y, z, x', y', z'; t)\, dx'\, dy'\, dz' \tag{79}$$

where

$$K_3(x, y, z, x', y', z'; t) = \left(\frac{m}{2\pi i \hbar t}\right)^{\frac{3}{2}} \exp\left[\frac{im}{2\hbar t}\left[(x - x')^2 + (y - y')^2 + (z - z')^2\right]\right] \tag{80}$$

5.7 PROBLEMS

Problem 5.1 Using Eq. (14) show that if

$$\int_{-\infty}^{+\infty} |\Psi(x, 0)|^2\, dx = 1 \tag{81}$$

then

$$\int_{-\infty}^{+\infty} |a(p)|^2\, dp = 1 \tag{82}$$

and
$$\int_{-\infty}^{+\infty} |\Psi(x,t)|^2 \, dx = 1 \tag{83}$$

Give physical interpretations of the above equations.

Problem 5.2 For an arbitrary (normalized) wavefunction $\Psi(x)$, we define

$$a(p) = \frac{1}{\sqrt{2\pi\hbar}} \int_{-\infty}^{+\infty} \Psi(x) \exp\left[-\frac{i}{\hbar} px\right] dx \tag{84}$$

Show that:

$$\langle p \rangle = \int_{-\infty}^{+\infty} a^*(p)\, p\, a(p)\, dp \tag{85}$$

and

$$\langle p^2 \rangle = \int_{-\infty}^{+\infty} a^*(p) p^2 a(p)\, dp \tag{86}$$

Problem 5.3 Using Eq. (84) show that:

$$\langle x \rangle = \int_{-\infty}^{+\infty} \Psi^*(x)\, x\, \Psi(x)\, dx$$

$$= i\hbar \int_{-\infty}^{+\infty} a^*(p) \frac{\partial a(p)}{\partial p} dp \tag{87}$$

Comparison of Eqs. (85) and (87) with Eqs. (44) and (51) of the previous chapter shows that $a(p)$ plays the same role in momentum space as x does in ordinary space. Further, Eq. (87) tells us that the operator representation of x in momentum space is $i\hbar\dfrac{\partial}{\partial p}$.

Problem 5.4 (a) For the Gaussian wave packet given by Eq. (25) evaluate $\langle x \rangle, \langle x^2 \rangle, \langle p \rangle, \langle p^2 \rangle$ and show that

$$\Delta x \, \Delta p = \frac{1}{2}\hbar \tag{88}$$

(b) Repeat the calculation for $\Psi(x, t)$ given by Eq. (35) and derive Eqs (43)–(45).

Problem 5.5 Assume $\Psi(x, 0)$ to be given by Eq. (25). Use Eq. (46) to derive Eq. (35).

Problem 5.6 Consider a two-dimensional wave packet given by

$$\Psi(x, y, 0) = \left[\left(\frac{1}{\pi\sigma_{x0}^2}\right)^{\frac{1}{4}} \exp\left[-\frac{x^2}{2\sigma_{x0}^2}\right] e^{\frac{i}{\hbar}p_0 x}\right]\left[\left(\frac{1}{\pi\sigma_{y0}^2}\right)^{\frac{1}{4}} \exp\left[-\frac{y^2}{2\sigma_{y0}^2}\right]\right] \tag{89}$$

(a) Determine $a(p_x, p_y)$ and then evaluate the probability distribution $|\Psi(x, y, t)|^2$. Interpret the results physically.

(b) Substitute Eq. (89) directly in Eq. (76) and obtain $|\Psi(x, y, t)|^2$ and show that one obtains the same result.

Problem 5.7 *Single Slit Diffraction Pattern* Consider again a two-dimensional wave packet given by

$$\Psi(x, y, 0) = \left[\left(\frac{1}{\pi\sigma_{x0}^2}\right)^{\frac{1}{4}} \exp\left[-\frac{x^2}{2\sigma_{x0}^2}\right] e^{\frac{i}{\hbar}p_0 x}\right] \psi_b(y) \tag{90}$$

where

$$\left.\begin{aligned} \psi_b(y) &= \frac{1}{\sqrt{b}}, \quad |y| < b/2 \\ &= 0, \quad |y| < b/2 \end{aligned}\right\} \tag{91}$$

(a) Determine $a(p_x, p_y)$ and interpret it physically. Show that the subsequent evaluation of $\Psi(x, t)$ is difficult to carry out.
(b) Use Eq. (76) to evaluate $|\Psi(x, y, t)|^2$ and interpret it physically.

Problem 5.8 *Double Slit Diffraction Pattern* In the previous problem assume

$$\psi_b(y) = \begin{cases} \frac{1}{\sqrt{2b}} & \text{for } (d-b)/2 < |y| < (d+b)/2 \\ 0 & \text{elsewhere} \end{cases} \tag{92}$$

which corresponds to a double slit (each of width b) separated by a distance d. Calculate $(|a(p_y)|^2 dp_y)$ and interpret the result physically.

Problem 5.9 (a) For the wave packet given by Eq. (57) show that it remains undistorted for $t \ll t_0$ where

$$t_0 = \frac{2}{(\Delta k)^2 \dfrac{d^2\omega}{dk^2}} = \frac{2\hbar}{(\Delta p)^2 \dfrac{d^2 E}{dp^2}} \simeq \frac{2m(\Delta x)^2}{\hbar} \tag{93}$$

where use has to be made of Eqs (67) and (68) and the uncertainty relation.
(b) For a particle of mass 10^{-4} g and localized within $\Delta x \sim 10^{-3}$ cm calculate t_0 and compare its value for an electron localized in a distance of 10^{-8} cm. Interpret the two results physically.
(c) Calculate t_0 for the Gaussian wave packet and show that the exact results obtained in Example 5.1 are consistent with Eq. (93).
Problem 5.10 Consider the three-dimensional wave packet describing the motion of a free particle

$$\Psi(x, y, z, 0) = N \exp\left[-\frac{x^2}{2\sigma_1^2} - \frac{y^2}{2\sigma_2^2} - \frac{z^2}{2\sigma_3^2}\right] \exp\left[\frac{i}{\hbar} p_0 x\right] \tag{94}$$

Discuss physically the motion of such a wave packet. Calculate the value of $a(p_x, p_y, p_z)$ and then $\Psi(x, y, z, t)$.

Problem 5.11 Consider the three-dimensional momentum space wavefunction

$$a(\mathbf{p}) = N \exp\left[-\frac{p}{\Delta p}\right] \tag{95}$$

Calculate $\Psi(\mathbf{r}, 0)$.

Problem 5.12 (a) From Eq. (14) show that

$$a(p) = \frac{1}{\sqrt{2\pi\hbar}} \exp\left[\frac{i}{\hbar}\frac{p^2}{2m}t\right] \int_{-\infty}^{+\infty} \psi(x,t) \exp\left[-\frac{i}{\hbar}px\right] dx \tag{96}$$

which is independent of time.

(b) Substitute the expression for $\Psi(x, t)$ from Eq. (35) and show that after evaluating the above integral one obtains Eq. (32) (obviously!).

5.8 SOLUTIONS

Solution 5.1 Equation (82) follows directly from Parseval's theorem (see Section 2.3). In order to prove Eq. (83), we take the complex conjugate of Eq. (14) and obtain

$$\int_{-\infty}^{+\infty} |\psi(x,t)|^2 \, dx = \frac{1}{2\pi\hbar} \int_{-\infty}^{+\infty} dx \int_{-\infty}^{+\infty} dp \int_{-\infty}^{+\infty} dp' a^*(p') a(p)$$

$$\times \exp\left[-\frac{i}{\hbar}\left(p'x - \frac{p'^2}{2m}t\right) + \frac{i}{\hbar}\left(px - \frac{p^2}{2m}t\right)\right]$$

Since,

$$\frac{1}{2\pi\hbar} \int_{-\infty}^{+\infty} \exp\left[\frac{i}{\hbar}(p-p')x\right] dx = \delta(p - p') \tag{97}$$

[see Eq. (9) of Chapter 1], after carrying out the integration over p' we readily get

$$\int_{-\infty}^{+\infty} |\Psi(x,t)|^2 \, dx = \int_{-\infty}^{+\infty} |a(p)|^2 \, dp = 1 \tag{98}$$

Solution 5.2

$$\Psi(x) = \frac{1}{\sqrt{2\pi\hbar}} \int_{-\infty}^{+\infty} a(p) \exp\left[\frac{i}{\hbar}px\right] dp \tag{99}$$

Thus,

$$\langle p \rangle = -i\hbar \int \Psi^* \frac{\partial \Psi}{\partial x} dx$$

$$= -i\hbar \int_{-\infty}^{+\infty} \Psi^*(x) \left[\frac{1}{\sqrt{2\pi\hbar}} \int_{-\infty}^{+\infty} \frac{i}{\hbar} p' a(p') \exp\left(\frac{i}{\hbar}p'x\right) dp'\right] dx$$

$$= \int_{-\infty}^{+\infty} dp \, a^*(p) \left[\int_{-\infty}^{+\infty} dp' \, p' a(p') \delta(p - p')\right]$$

$$= \int_{-\infty}^{+\infty} a^*(p) \, p \, a(p) \, dp \tag{100}$$

where we have used Eq. (97). Similarly, one can calculate for $\langle p^2 \rangle$.

Solution 5.3 Use Eq. (84) to evaluate $a^*(p)$ and $\frac{\partial a}{\partial p}$ and substitute in the RHS of Eq. (87). Alternatively, we can substitute the expression

$$\Psi(x) = \frac{1}{\sqrt{2\pi\hbar}} \int_{-\infty}^{+\infty} a(p) \exp\left[\frac{i}{\hbar} px\right] dp$$

and its complex conjugate in the integral

$$\int_{-\infty}^{+\infty} \Psi^*(x)\, x\, \Psi(x)\, dx$$

and carry out the integration over p by parts.

Solution 5.4

$$\langle x \rangle = \int_{-\infty}^{+\infty} \psi^*(x,0)\, x\, \psi(x,0)\, dx$$

$$= \frac{1}{\sqrt{\pi\sigma_0^2}} \int_{-\infty}^{+\infty} x \exp\left(-\frac{x^2}{\sigma_0^2}\right) dx = 0$$

because the integrand is an odd function of x

$$\langle x^2 \rangle = \frac{1}{\sqrt{\pi\sigma_0^2}} \int_{-\infty}^{+\infty} x^2 \exp\left(-\frac{x^2}{\sigma_0^2}\right) dx = \frac{2}{\sqrt{\pi\sigma_0^2}} \frac{\sigma_0^3}{2} \int_0^{\infty} \eta^{1/2} e^{-\eta} d\eta = \frac{1}{2}\sigma_0^2; \left[\eta = \frac{x^2}{\sigma_0^2}\right]$$

$$\langle p^2 \rangle = \sqrt{\pi\sigma_0^2} \int_{-\infty}^{+\infty} \exp\left[-\frac{x^2}{\sigma_0^2} - \frac{i}{\hbar} p_0 x\right]\left(-i\hbar \frac{d}{dx}\right) \exp\left[-\frac{x^2}{2\sigma_0^2} + \frac{i}{\hbar} p_0 x\right] dx$$

$$= \frac{-i\hbar}{\sqrt{\pi\sigma_0^2}} \left[-\frac{1}{\sigma_0^2} \int_{-\infty}^{+\infty} x \exp\left(-\frac{x^2}{\sigma_0^2}\right) dx + \frac{i}{\hbar} p_0 \int_{-\infty}^{+\infty} \exp\left(-\frac{x^2}{\sigma_0^2}\right) dx\right]$$

The first integral is zero because the integrand is an odd function of x and the second integral is $\sqrt{\pi\sigma_0^2}$. Thus we get $\langle p \rangle = p_0$. Similarly,

$$\langle p^2 \rangle = \int_{-\infty}^{+\infty} \Psi^*(x,0)\left[-\hbar^2 \frac{d^2}{dx^2}\right] \Psi(x,0)\, dx$$

Differentiating twice $\Psi(x, 0)$ and integrating term by term we get $\langle p^2 \rangle = p_0^2 + \frac{\hbar^2}{2\sigma_0^2}$.

Solution 5.5

(a) We define the two-dimensional Fourier transform

$$a(p_x, p_y) = \frac{1}{2\pi\hbar} \int_{-\infty}^{+\infty}\int_{-\infty}^{+\infty} \psi(x,y,0) \exp\left[\frac{i}{\hbar}(p_x x + p_y y)\right] dx\, dy \tag{101}$$

and

$$\Psi(x,y,t) = \frac{1}{2\pi\hbar} \int_{-\infty}^{+\infty}\int_{-\infty}^{+\infty} a(p_x,p_y) \exp\left[-\frac{i}{\hbar}(p_x x + p_y y)\right] dp_x\, dp_y \tag{102}$$

Since $\Psi(x, y, 0)$ is a product of a function of x and of y, the integrals over x and y in Eq. (101) separate out and we obtain

$$a(p_x, p_y) = a(p_x)\, a(p_y) \tag{103}$$

where

$$a(p_x) = \left(\frac{\sigma_{x_0}^2}{\pi \hbar^2}\right)^{\frac{1}{4}} \exp\left[-\frac{\sigma_{x_0}^2}{2\hbar^2}(p_x - p_0)^2\right] \tag{104}$$

and

$$a(p_y) = \left(\frac{\sigma_{y_0}^2}{\pi \hbar^2}\right)^{\frac{1}{4}} \exp\left[-\frac{\sigma_{y_0}^2}{2\hbar^2}{p_y}^2\right] \tag{105}$$

Obviously, the quantity

$$|a(p_y)|^2\, dp_y = \left(\frac{\sigma_{y_0}^2}{\pi \hbar^2}\right)^{\frac{1}{2}} \exp\left[-\frac{\sigma_{y_0}^2 p_y^2}{\hbar^2}\right] dp_y \tag{106}$$

represents the probability of finding p_y in the interval dp_y; this is nothing but the diffraction of a Gaussian beam.

(b) We substitute Eqs (104) and (105) in Eq. (102) and carry out the integration to obtain

$$|\Psi(x, y, t)|^2 = \left\{\frac{1}{\sqrt{\pi}\sigma_x(t)} \exp\left[-\frac{\left(x - \frac{p_0}{m}t\right)^2}{\sigma_x^2(t)}\right]\right\} \times \left\{\frac{1}{\sqrt{\pi}\,\sigma_y(t)} \exp\left[-\frac{y^2}{\sigma_y^2(t)}\right]\right\} \tag{107}$$

where

$$\left.\begin{aligned} \sigma_x(t) &= \sigma_{x0}\left[1 + \frac{\hbar^2 t^2}{m^2 \sigma_{x_0}^4}\right]^{\frac{1}{2}} \\ \sigma_y(t) &= \sigma_{y0}\left[1 + \frac{\hbar^2 t^2}{m^2 \sigma_{y_0}^4}\right]^{\frac{1}{2}} \end{aligned}\right\} \tag{108}$$

Equation (107) shows the diffraction broadening along the y-direction.

Solution 5.6 Equation (103) still remains valid with

$$a(p_y) = \sqrt{\frac{b}{2\pi\hbar}}\; \frac{\sin(p_y b)/2\hbar}{p_y b/2\hbar} \tag{109}$$

Thus, the probability that (after passing through the slit) the y-component of the momentum lies between p_y and $p_y + dp_y$ would be given by

$$|a(p_y)|^2\, dp_y = \frac{b}{2\pi\hbar}\, \frac{\sin^2\beta}{\beta^2}\, dp_y \tag{110}$$

where $$\beta = \frac{p_y b}{2\hbar} = \frac{p \sin\theta\, b}{2\hbar} = \frac{\pi b \sin\theta}{\lambda} \tag{111}$$

and use has been made of the de Bröglie relation

$$\lambda = \frac{h}{p} \tag{112}$$

Thus, far away from the slit, we will observe the intensity distribution given by

$$I = I_0 \frac{\sin^2\beta}{\beta^2} \tag{113}$$

which is nothing but the single slit Fraunhofer diffraction pattern. Using Eq. (109) one can easily verify.

$$\int_{-\infty}^{+\infty} |a(p_y)|^2 \, dp_y = 1$$

In order to get $\Psi(x, y, t)$ we must substitute for $a(p_x, p_y)$ in Eq. (102); however, the y integral would be difficult to carry out. Instead we must directly evaluate $\Psi(x, y, t)$ by using Eq. (76). The final result can be expressed in terms of Fresnel integrals (see Appendix C).

Solution 5.7

$$a(p_y) = \frac{1}{\sqrt{2\pi\hbar}} \cdot \frac{1}{\sqrt{2b}} \left[\int_{-\frac{d}{2}-\frac{b}{2}}^{-\frac{d}{2}+\frac{b}{2}} + \int_{\frac{d}{2}-\frac{b}{2}}^{\frac{d}{2}+\frac{b}{2}} \right] e^{ip_y y/\hbar}\, dy$$

Thus, $$|a(p_y)|^2 = \frac{4b}{\pi\hbar}\left[\frac{\sin^2\beta}{\beta^2}\right] \cos^2\gamma \tag{114}$$

where $$\beta = \frac{b p_y}{2\hbar} = \frac{\pi b \sin\theta}{\lambda} \tag{115}$$

and $$\gamma = \frac{d p_y}{2\hbar} = \frac{\pi d \sin\theta}{\lambda} \tag{116}$$

Equation (114) is nothing but the double slit Fraunhofer diffraction pattern.

Solution 5.8 (a) If we substitute for $\omega(k)$ from Eq. (58) in Eq. (57) we would get

$$\Psi(x, t) = \frac{e^{i(k_0 x - \omega t)}}{\sqrt{2\pi}} \int_{\Delta k} A(k)\, e^{i\phi(k)} \times \exp\left[i(x - v_g t)(k - k_0) - i\alpha (k - k_0)^2 t\right] dk$$

where $$\alpha = \frac{1}{2} \frac{d^2\omega}{dk^2}\bigg|_{k=k_0}$$

Since $|A(k)|$ is assumed to have a negligible value for $|k - k_0| \gtrsim \Delta k$, the term involving α would make a negligible contribution for $t \ll t_0$ where t_0 is given by Eq. (93).

In order to understand Eq. (93) physically, we notice that there is a spread ~ Δp in the momentum associated with the packet. Thus if we consider two portions of the packet, one corresponding to $p_0 - \frac{\Delta p}{2} < p < p_0$ and the other corresponding to $p_0 < p < p_0 + \frac{\Delta p}{2}$, then the velocity of the two packets will differ by $\frac{\Delta p}{2m}$. Thus there will be distortion in the packet when

$$\frac{t\Delta p}{2m} \approx \Delta x \qquad \text{(the initial width of the packet)}$$

The above condition is consistent with Eq. (93).

(b) For $m \approx 10^{-4}$ g, $\Delta x \approx 10^{-3}$ cm, $t_0 \sim 2 \times 10^{17}$ s ~ 10^{10} years. On the other hand, for an electron with $\Delta x \approx 10^{-8}$ cm, $t_0 \approx 10^{-16}$ sec. Thus, for a macroscopic particle the distortion is negligible and hence a classical description is adequate.

Solution 5.9 For a three-dimensional wave packet describing a free particle

$$\Psi(\mathbf{r}, t) = \frac{1}{(2\pi\hbar)^{\frac{3}{2}}} \iiint a(\mathbf{p}) \exp\left[\frac{i}{\hbar}(\mathbf{p}\cdot\mathbf{r} - Et)\right] d\mathbf{p} \tag{117}$$

and

$$a(\mathbf{p}) = \frac{1}{(2\pi\hbar)^{\frac{3}{2}}} \iiint \psi(\mathbf{r}, t) \exp\left[-\frac{i}{\hbar}(\mathbf{p}\cdot\mathbf{r} - Et)\right] d\mathbf{r} \tag{118}$$

where $\mathbf{p}\cdot\mathbf{r} = p_x x + p_y y + p_z z$, $\quad d\mathbf{r} = dx\,dy\,dz$, $\quad d\mathbf{p} = dp_x\,dp_y\,dp_z$

and

$$E = \frac{p^2}{2m} = \frac{1}{2m}(p_x^2 + p_y^2 + p_z^2)$$

On substitution of Eq. (94) in Eq. (118), the three-dimensional integral becomes a product of three one-dimensional integrals giving

$$a(\mathbf{p}) = a(p_x, p_y, p_z) = \left(\frac{\sigma_1\sigma_2\sigma_3}{\hbar^3}\right)^{\frac{1}{2}} \frac{1}{\pi^{3/4}}$$
$$\times \exp\left[-\frac{\sigma_1^2}{2\hbar^2}(p_x - p_0)^2 - \frac{\sigma_2^2}{\hbar^2}p_y^2 - \frac{\sigma_3^2}{\hbar^2}p_z^2\right]$$

where we have used the fact that the normalization constant $N = \pi^{-3/4}/(\sigma_1\sigma_2\sigma_3)^{1/2}$. If we substitute for $a(\mathbf{p})$ in Eq. (117), we would get

$$|\Psi(\mathbf{r}, t)|^2 = \frac{1}{\pi^{3/2}[\sigma_1(t)\sigma_2(t)\sigma_3(t)]} \exp\left[-\frac{\left\{x - \frac{p_0}{m}t\right\}^2}{\sigma_1^2(t)}\right]$$
$$\times \exp\left[-\frac{y^2}{\sigma_2^2(t)}\right] \exp\left[-\frac{z^2}{\sigma_3^2(t)}\right]$$

where $\quad \sigma_i(t) = \sigma_i\left[1 + \frac{\hbar^2 t^2}{m^2\sigma_i^4}\right]^{1/2}$

If we assume σ_1 to be very large then there is propagation along the x-axis and diffraction along the y- and z-directions.

Solution 5.10 In order to calculate N we use spherical polar coordinates to get

$$1 = |N|^2 \int_0^{2\pi} d\phi \int_0^{\pi} \sin\theta\, d\theta \int_0^{\infty} e^{-2p/\Delta p} p^2\, dp$$

which gives $N = (\Delta p)^{3/2}/\sqrt{\pi}$. Thus,

$$\Psi(\mathbf{r}, 0) = \frac{|N|^2}{(2\pi\hbar)^{3/2}} \int_0^{2\pi} d\phi \int_0^{\infty} p^2\, dp \exp\left[-\frac{p}{\Delta p}\right] \int_0^{\pi} \exp\left[-\frac{i\, p\, r \cos\theta}{\hbar}\right] \sin\theta\, d\theta$$

where we have assumed p_z along $\mathbf{r}$. On carrying out the integration, we get

$$\Psi(\mathbf{r}, 0) = \frac{1}{\pi}\,(2r_0)^{3/2}\,\frac{r_0}{(r^2 + r_0^2)^2}$$

where $r_0 = \hbar/\Delta p$.

5.9 SAMPLE QUESTIONS

Q. 1 (a) Write the one-dimensional time dependent Schrödinger equation for a free particle.

(b) Write the general solution of the above equation.

(c) If we know $\Psi(x, 0)$, write the recipe for determining $\Psi(x, t)$.

Q. 2 Starting from the one-dimensional Schrödinger equation for a free particle show that the solution can be written in the form

$$\Psi(x, t) = \int_{-\infty}^{+\infty} \psi(x', 0)\, K(x, x'; t)\, dx'$$

Derive the expression for $K(x, x'; t)$. What will be the form of $K(x, x'; 0)$?

Q. 3 Given

$$\Psi(x, y, 0) = e^{-x^2/2\sigma^2}\, e^{\frac{i}{\hbar}p_0 x}\, \psi(y)$$

where

$$\psi(y) = 1, \quad |y| < b/2$$
$$= 0, \quad |y| > b/2$$

(a) What is the probability of finding the particle between x and $x + dx$ and y and $y + dy$?

(b) Calculate the probability for the x-component of the momentum to lie between p_x and $p_x + dp_x$ and y-component of the momentum to lie between p_y and $p_y + dp_y$. Give a physical interpretation in terms of the single slit diffraction pattern.

REFERENCES AND SUGGESTED READINGS

1. E. Merzbacher, *Quantum Mechanics*, John Wiley (1964).
2. D. S. Saxon, *Elementary Quantum Mechanics*, Dover Edition (1968).
3. A. Ghatak, *Optics*, (6th Edition), McGraw Hill Education, New Delhi (2017).

6 Bound State Solutions of the Schrödinger Equation

Only questions about the results of experiments have a real significance and it is only such questions that theoretical physics has to consider.

— P.A.M. DIRAC
in *The Principles of Quantum Mechanics*

6.1 STATIONARY STATES

When the Hamiltonian for a system is independent of time, there is an essential simplification in that the general solution of the Schrödinger equation can be expressed as a function of spatial coordinates and a function of time. Thus, assuming the potential energy function to be independent of time, the one-dimensional time dependent Schrödinger equation [see Eq. (25) of Chapter 4] is given by

$$i\hbar \frac{\partial \Psi}{\partial t} = -\frac{\hbar^2}{2\mu}\frac{\partial^2 \Psi}{\partial x^2} + V(x)\,\Psi(x, t) \tag{1}$$

where μ represents the mass of the particle. The above equation can be solved by using the method of separation of variables

$$\Psi(x, t) = \psi(x)\, T(t) \tag{2}$$

Substituting in Eq. (1) and dividing by $\Psi(x, t)$, we obtain

$$\frac{i\hbar}{T(t)} \cdot \frac{dT}{dt} = \frac{1}{\psi(x)}\left[-\frac{\hbar^2}{2\mu}\frac{d^2\psi}{dx^2} + V(x)\,\psi(x)\right] \tag{3}$$

The variables have indeed separated out in the sense that the LHS is a function of time alone while the RHS is a function of space alone; thus each side must be equal to a constant, say E. We have therefore

$$\frac{i\hbar}{T}\frac{dT}{dt} = \frac{1}{\psi(x)}\left[-\frac{\hbar^2}{2\mu}\cdot\frac{d^2\psi}{dx^2} + V(x)\,\psi(x)\right] = E \tag{4}$$

or

$$T(t) = e^{-iEt/\hbar} \tag{5}$$

and

$$-\frac{\hbar^2}{2\mu}\frac{d^2\psi(x)}{dx^2} + V(x)\,\psi(x) = E\,\psi(x) \tag{6}$$

so that the solution can be written as

$$\Psi(x, t) = \psi(x)\, e^{-iEt/\hbar} \tag{7}$$

Now, Eq. (6) can be written as

$$H\psi(x) = E\psi(x) \tag{8}$$

where

$$H = -\frac{\hbar^2}{2\mu}\frac{d^2}{dx^2} + V(x) = \frac{p^2}{2\mu} + V(x) \tag{9}$$

represents the Hamiltonian operator and E is a *number*. Equation (8) is an eigenvalue equation and E is the energy eigenvalue. In general, the eigenvalues can have a discrete set along with a continuous range of eigenvalues and the general solution of Eq. (1) is of the form

$$\Psi(x, t) = \sum_n c_n\, \psi_n(x)\, e^{-iE_n t/\hbar} \tag{10}$$

where $\psi_n(x)$ are the eigenfunctions of H and E_n are the corresponding energy eigenvalues; c_n are constants which can be determined from the form of $\Psi(x, 0)$. In Eq. (10), Σ denotes a summation over the discrete states and integration over the continuum states[1].

Equation (6) is usually written in the form

$$\frac{d^2\psi}{dx^2} + \frac{2\mu}{\hbar^2}[E - V(x)]\,\psi(x) = 0 \tag{11}$$

which is known as the one-dimensional time independent Schrödinger equation.

In Section 6.2, we will discuss two theorems regarding the boundary and continuity conditions of the wavefunction. In Sections 6.3 and 6.4, we will discuss the degeneracy of a state and the orthogonality of the eigenfunctions and in Section 6.5, we will discuss the parity of the eigenfunctions. In Section 6.6, we will give the solution of the eigenvalue equation [Eq. (8)] for some simple problems. In Section 6.7, we will discuss the properties of the wavefunction associated with the three-dimensional Schrödinger equation and in Sec. 6.8, we will obtain solutions of the Schrödinger equation for a particle in a three-dimensional box. In Sections 6.9 and 6.10, we will discuss applications of the particle in a box problem to free electrons inside a metal and also inside a white dwarf star.

6.2 BOUNDARY AND CONTINUITY CONDITIONS

In Section 4.4, we have interpreted the wavefunction ψ as the probability amplitude, therefore ψ should be single-valued and $|\psi(x)|^2\, dx$ has to be finite for finite values of dx. Thus,

$$\lim_{dx \to 0} |\psi|^2\, dx = 0$$

In practice this is satisfied by demanding that ψ be finite everywhere. We also have the following theorems:

[1]In Chapter 5 we had solved Eq. (1) for a free particle for which $V(x) = 0$. There the eigenvalues formed a continuum $(0 < E < \infty)$ and the Σ in Eq. (10) was indeed an integral over the continuum states [see Eq. (14) of Chapter 5].

THEOREM 1. *The derivative of the wavefunction $d\psi/dx$ is always continuous as long as the potential energy $V(x)$ is finite, whether or not it is continuous*[2].

Proof We integrate the Schrödinger equation [Eq. (11)] from $x - \varepsilon$ to $x + \varepsilon$ to obtain

$$\int_{x-\varepsilon}^{x+\varepsilon} \frac{d^2\psi}{dx^2}\, dx = -\frac{2\mu}{\hbar^2} \int_{x-\varepsilon}^{x+\varepsilon} [E - V(x)]\, \psi(x)\, dx$$

or

$$\psi'(x+\varepsilon) - \psi'(x-\varepsilon) = -\frac{2\mu}{\hbar^2} \int_{x-\varepsilon}^{x+\varepsilon} [E - V(x)]\, \psi(x)\, dx \tag{12}$$

Since $V(x)$ is assumed to be finite (it could, however, be discontinuous), the RHS tends to zero as $\varepsilon \to 0$. Thus ψ' is continuous at any value of x. It is obvious that ψ has to be necessarily continuous everywhere.

Alternatively one may argue that if $d\psi/dx$ is discontinuous, then $d^2\psi/dx^2$ must be a delta function (see Section 1.3); this will be inconsistent with Eq. (11) as long as $V(x)$ does not become infinite.

THEOREM 2. *If the potential energy function $V(x)$ is infinite anywhere, the proper boundary condition is obtained by assuming $V(x)$ to be finite at that point and carrying out a limiting process making $V(x)$ tend to infinity. Such a limiting process makes the wavefunction vanish at a point where $V(x) = \infty$ (see Problem 6.2).*

6.3 DEGENERACY

If there is more than one linearly independent wavefunction belonging to the same energy eigenvalue E, the energy level is said to be degenerate. If there are g linearly independent wavefunctions ($\psi_1, \psi_2, \ldots, \psi_g$) belonging to the same energy state, then the energy level is said to be g-fold degenerate. It can be easily shown that any linear combination of the degenerate wavefunctions

$$\psi = c_1\psi_1 + c_2\psi_2 + \cdots + c_g\psi_g$$

is also an eigenfunction belonging to the same energy eigenvalue. We will prove this for $g = 2$, the generalization is straightforward. Since ψ_1 and ψ_2 are eigenfunctions belonging to same eigenvalue E_1, we have [see Eq. (6)]:

$$-\frac{\hbar^2}{2\mu}\frac{d^2\psi_1}{dx^2} + V(x)\,\psi_1 = E_1\,\psi_1$$

and

$$-\frac{\hbar^2}{2\mu}\frac{d^2\psi_2}{dx^2} + V(x)\,\psi_2 = E_1\,\psi_2$$

If we multiply the first equation by c_1 and the second equation by c_2 and add, we would get

---

[2]It may be mentioned that in many texts the continuity of ψ and $d\psi/dx$ is taken to be an axiom. This is not correct because it follows from the fact that $\psi(x)$ satisfies a second order differential equation [see Eq. (11)]. Indeed, when $V(x)$ becomes infinite, $d\psi/dx$ is *not* continuous (see Sec. 6.6.1).

$$-\frac{\hbar^2}{2\mu}\frac{d^2}{dx^2}(c_1\psi_1 + c_2\psi_2) + V(x)(c_1\psi_1 + c_2\psi_2) = E_1(c_1\psi_1 + c_2\psi_2) \tag{13}$$

which shows that the linear combination $c_1\psi_1 + c_2\psi_2$ is also an eigenfunction belonging to the *same* eigenvalue E_1.

6.4 ORTHOGONALITY OF EIGENFUNCTIONS

We shall first prove that *all energy eigenvalues E_n are real and that if $E_n \neq E_k$, the corresponding eigenfunctions are necessarily orthogonal.* We write the one-dimensional Schrödinger equation for the two states corresponding to the eigenvalues E_n and E_k:

$$-\frac{\hbar^2}{2\mu}\frac{d^2\psi_n}{dx^2} + V(x)\,\psi_n = E_n\psi_n \tag{14}$$

$$-\frac{\hbar^2}{2\mu}\frac{d^2\psi_k}{dx^2} + V(x)\,\psi_k = E_k\,\psi_k \tag{15}$$

If we multiply Eq. (14) by ψ_k^* and multiply the complex conjugate of Eq. (15) by ψ_n and subtract, we would get

$$-\frac{\hbar^2}{2\mu}\left[\psi_k^*\frac{d^2\psi_n}{dx^2} - \psi_n\frac{d^2\psi_k^*}{dx^2}\right] = (E_n - E_k^*)\,\psi_k^*\,\psi_n$$

or,

$$-\frac{\hbar^2}{2\mu}\int_{-\infty}^{\infty}\frac{d}{dx}\left[\psi_k^*\frac{d\psi_n}{dx} - \psi_n\frac{d\psi_k^*}{dx}\right]dx = (E_n - E_k^*)\int_{-\infty}^{\infty}\psi_k^*\,\psi_n\,dx \tag{16}$$

The left-hand side is simply

$$-\frac{\hbar^2}{2\mu}\left[\psi_k^*\frac{d\psi_n}{dx} - \psi_n\frac{d\psi_k^*}{dx}\right]_{-\infty}^{+\infty}$$

which would vanish if the wavefunctions are assumed to vanish at infinity. Thus we get

$$(E_n - E_k^*)\int_{-\infty}^{\infty}\psi_k^*\,\psi_n\,dx = 0 \tag{17}$$

If $k = n$, then since $\int_{-\infty}^{\infty}|\psi_n|^2\,dx$ is necessarily positive[3], we must have

$$E_n = E_n^* \tag{18}$$

proving that *all eigenvalues are real*. Further, for $E_n \neq E_k$,

$$\int_{-\infty}^{\infty}\psi_k^*\,\psi_n\,dx = 0 \qquad \text{orthogonality condition} \tag{19}$$

which represents the *orthogonality condition*.

If $E_k = E_n$, i.e., if there are two (or more) wavefunctions belonging to the same eigenvalue, then the corresponding wavefunctions are not necessarily orthogonal. Since any linear combination of degenerate state wavefunctions is also a possible eigenfunction (see Section 6.3) one can *always* construct suitable linear combinations which are mutually orthogonal (see Problems 6.3 and 6.4).

---

[3]Except in the trivial case when the wavefunction vanishes everywhere.

Since the Schrödinger equation is a linear equation, if ψ is a solution, $c\psi$ is also a solution where c is an arbitrary complex number. One can always choose the constant such that

$$\int_{-\infty}^{\infty} |\psi_n|^2 \, dx = 1 \tag{20}$$

which is known as the normalization condition. Equations (19) and (20) can be combined in the form

$$\int_{-\infty}^{\infty} \psi_k^* \psi_n \, dx = \delta_{kn} \quad \text{orthonormality condition} \tag{21}$$

where δ_{kn} is the Kronecker delta function defined through the equation

$$\delta_{kn} = \begin{cases} 0 & k \neq n \\ 1 & k = n \end{cases} \tag{22}$$

Equation (21) is known as the orthonormality condition.

6.5 PARITY

In this section we will show that if

$$V(x) = V(-x) \tag{23}$$

i.e., if the potential energy function is symmetric about $x = 0$, then the eigenfunctions of the Schrödinger equation are either symmetric functions of x [$\psi(x) = \psi(-x)$] or antisymmetric functions[4] of x [$\psi(x) = -\psi(-x)$]. In order to prove this, we first write the Schrödinger equation

$$-\frac{\hbar^2}{2\mu}\frac{d^2\psi(x)}{dx^2} + V(x)\,\psi(x) = E\psi(x) \tag{24}$$

Making the transformation $x \to -x$, we get

$$-\frac{\hbar^2}{2\mu}\frac{d^2\psi(-x)}{dx^2} + V(x)\,\psi(-x) = E\psi(-x) \tag{25}$$

where we have used the fact that $V(-x) = V(x)$. Comparing Eqs (24) and (25) we see that $\psi(x)$ and $\psi(-x)$ are eigenfunctions belonging to the same energy E. Thus, if the state is non-degenerate, then $\psi(-x)$ must be a multiple of $\psi(x)$:

$$\psi(-x) = \lambda\,\psi(x)$$

Clearly,

$$\psi(x) = \lambda\,\psi(-x) = \lambda^2\,\psi(x)$$

so that $\lambda^2 = 1$ or $\lambda = \pm 1$. Hence,

$$\psi(-x) = \pm\,\psi(x) \tag{26}$$

proving the theorem. The eigenfunctions which belong to the class $\lambda = +1$ are said to have *even parity* and those with $\lambda = -1$ to have *odd parity*.

[4]The theorem is strictly true for non-degenerate states only. For degenerate states the wavefunctions need not be symmetric or antisymmetric functions of x. However, even for degenerate states one can always construct appropriate linear combinations which are either symmetric or antisymmetric functions of x.

For a degenerate state $\psi(-x)$ need not be a multiple of $\psi(x)$. Indeed, if $\psi(-x)$ is not a multiple of $\psi(x)$ then, from Eqs (24) and (25), it immediately follows that $\psi(-x)$ must be another independent solution corresponding to the same energy level. Further any linear combination of $\psi(x)$ and $\psi(-x)$ could also be a possible eigenfunction (see Section 6.3) and we can always choose linear combinations

$$[\psi(x) + \psi(-x)] \quad \text{and} \quad [\psi(x) - \psi(-x)]$$

which are symmetric and antisymmetric functions of x respectively. Thus even for degenerate states one can always choose appropriate linear combinations which are either symmetric or antisymmetric functions of x.

6.6 SOME EXACT SOLUTIONS OF THE ONE-DIMENSIONAL SCHRÖDINGER EQUATION

In this section, we will consider two potential energy distributions for which an exact solution of the one-dimensional Schrödinger equation will be obtained. In Section 6.6.1, we will consider a particle inside an infinitely deep potential well which will be followed by considering the effects of the potential well of finite depth. In the next chapter we will consider the linear harmonic oscillator problem and in Chapter 16 we will consider the double well problem.

6.6.1 Particle in a One-dimensional Infinitely Deep Potential Well

We will determine the energy levels and the corresponding eigenfunctions of a particle of mass μ in a one-dimensional infinitely deep potential well characterised by the potential energy variation of the form

$$\begin{aligned} V(x) &= 0 \quad \text{for} \quad 0 < x < a \\ &= \infty \quad \text{for} \quad x < 0 \text{ and for } x > a \end{aligned} \tag{27}$$

(The corresponding three-dimensional problem is discussed in Section 6.8). For $0 < x < a$, the one-dimensional Schrödinger equation becomes

$$\frac{d^2\psi}{dx^2} + k^2\psi(x) = 0 \tag{28}$$

where
$$k^2 = \frac{2\mu E}{\hbar^2} \tag{29}$$

The general solution of this equation is

$$\psi = A \sin kx + B \cos kx$$

Since the boundary condition at a surface at which there is an infinite potential step is that ψ is zero (see Section 6.2, Theorem 2), we must have

$$\psi(x = 0) = \psi(x = a) = 0 \tag{30}$$

The above condition also follows from the fact that since the particle is inside an infinitely deep potential well, it is always confined in the region $0 < x < a$ and therefore ψ must vanish for $x < 0$ and $x > a$; and for ψ to be continuous, Eq. (30) must be satisfied. Using the boundary condition given by Eq. (30), we get

$$\psi(x = 0) = B = 0$$

and
$$\psi(x = a) = A \sin ka = 0$$

Thus, either

$$A = 0$$

or $\quad ka = n\pi\,; \quad n = 1, 2, \ldots \qquad (31)$

The condition $A = 0$ leads to the trivial solution of ψ vanishing everywhere, the same is the case for $n = 0$. Thus, the allowed energy levels are given by

$$E_n = \frac{\pi^2 n^2 \hbar^2}{2\mu a^2}\,; \quad n = 1, 2, 3, \ldots \quad \text{allowed energy levels} \qquad (32)$$

The corresponding eigenfunctions are

$$\psi_n = \sqrt{\frac{2}{a}} \sin\left(\frac{n\pi}{a} x\right) \quad (0 < x < a) \quad \text{normalized eigenfunctions} \qquad (33)$$

where the factor $\sqrt{2/a}$ is such that the wavefunctions form an orthonormal set [cf. Eq. (21)]:

$$\int_0^a \psi_m^*(x)\, \psi_n(x)\, dx = \delta_{mn} \quad \text{orthonormality condition} \qquad (34)$$

It may be noted that whereas $\psi_n(x)$ is continuous everywhere, $d\psi_n(x)/dx$ is discontinuous at $x = 0$ and at $x = a$. This is because of $V(x)$ becoming infinite at $x = 0$ and at $x = a$ (see Section 6.2). Figure 6.1, gives a plot of the first three eigenfunctions and one can see that the eigenfunctions are either symmetric or antisymmetric about the line $x = a/2$; this follows from the fact that $V(x)$ is symmetric about $x = a/2$ (see Section 6.5).

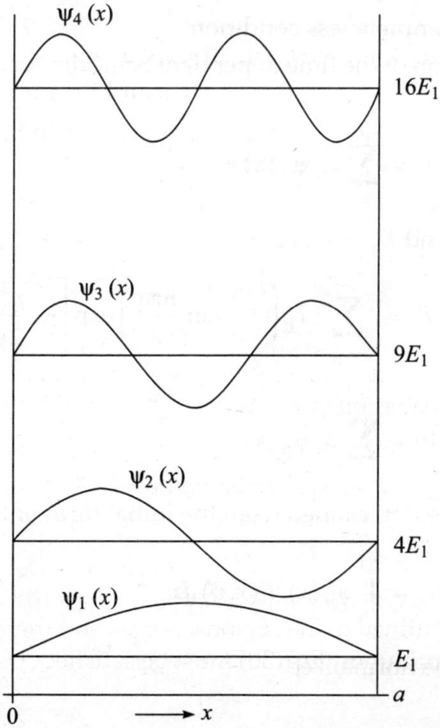

Figure 6.1: The energy eigenvalues and eigenfunctions for a particle in an infinitely deep potential well. Notice that the eigenfunctions are either symmetric or antisymmetric about $x = a/2$.

The following points are also to be noted.

(i) E cannot be negative because if we assume E to be negative then the boundary conditions at $x = 0$ and at $x = a$ cannot be simultaneously satisfied (see also Problem 7.1).

(ii) The eigenvalues form a discrete set.

(iii) The eigenfunctions given by Eq. (33) form a complete set, i.e., an arbitrary (well-behaved) function which satisfies the same boundary conditions (viz., vanishing at $x = 0$ and at $x = a$) can be expanded in terms of the eigenfunction of H:

$$f(x) = \sum_n c_n \psi_n(x) = \sqrt{\frac{2}{a}} \sum_{n=1,2,\ldots}^{\infty} c_n \sin\left(\frac{n\pi}{a} x\right) \tag{35}$$

where $f(x)$ is an arbitrary well-behaved function (with $f(0) = f(a) = 0$) and c_n are constants which can be determined by multiplying both sides of the above equation by $\psi_m^*(x)$ and integrating from 0 to a to obtain

$$\int_0^a \psi_m^*(x)\, f(x)\, dx = \sum_n c_n \int_0^a \psi_m^* \psi_n(x)\, dx = \sum_n c_n\, \delta_{mn} = c_m \tag{36}$$

where we have used the orthonormality condition given by Eq. (34). Substituting for c_m in Eq. (35) we would readily get (cf. Section 1.5):

$$\sum_n \psi_n^*(x')\, \psi_n(x) = \frac{2}{a} \sum_{n=1,2,\ldots}^{\infty} \sin\left(\frac{n\pi}{a} x\right) \sin\left(\frac{n\pi}{a} x'\right) = \delta(x - x');\ 0 < x, x' < a \tag{37}$$

which represents the completeness condition.

(iv) The most general solution of the time dependent Schrödinger equation [Eq. (1)] with $V(x)$ given by Eq. (27) would be

$$\Psi(x, t) = \sum_n^{\infty} c_n \psi_n(x)\, e^{-iE_n t/\hbar} \tag{38}$$

Substituting for $\psi_n(x)$ and E_n, we get

$$\Psi(x, t) = \sum_{n=1,2,\ldots}^{\infty} c_n \left(\sqrt{\frac{2}{a}} \sin\frac{n\pi x}{a}\right) \exp\left[-i\frac{n^2\pi^2\hbar}{2\mu a^2} t\right] \tag{39}$$

Since,

$$\Psi(x, 0) = \sum_n c_n \psi_n(x) \tag{40}$$

the coefficients c_n can be determined from the initial form of the wavefunction:

$$c_n = \int_0^a \psi_n^*(x)\, \Psi(x, 0)\, dx \tag{41}$$

(v) We assume $\psi(x, 0)$ to be normalized:

$$\int_0^a |\Psi(x, 0)|^2\, dx = 1 \tag{42}$$

This would imply

$$1 = \int_0^a \sum_n c_n^* \psi_n^*(x) \sum_m c_m \psi_m(x)\,dx$$

$$= \sum_n \sum_n c_n^* c_m \int_0^a \psi_n^*(x)\psi_m(x)\,dx$$

$$= \sum_n \sum_m c_n^* c_m \delta_{mn}$$

$$= \sum_n |c_n|^2 \tag{43}$$

where we have used the orthonormality condition given by Eq. (34). Now, using Eq. (38)

$$\int_0^a |\Psi(x,t)|^2\,dx = \sum_n \sum_m c_n^* c_m e^{i(E_n - E_m)t/\hbar} \int_0^a \psi_n^*(x)\,\psi_m(x)\,dx$$

$$= \sum_n \sum_m c_n^* c_m e^{i(E_n - E_m)t/\hbar} \delta_{mn}$$

$$= \sum_n |c_n|^2 = 1 \tag{44}$$

Thus, if the wavefunction is normalized at $t = 0$, then it will remain normalized at all times. Further, we can interpret Eq. (44) by saying that $|c_n|^2$ represents the probability of finding the system in the nth state which remains the same at all times. Thus, there are no transitions. Indeed, whenever the potential energy function is time-independent, we obtain what are known as *stationary states* and there is no transition between states.[5]

(vi) As a simple example, let us assume that the particle is described by the following normalized wavefunction (at $t = 0$):

$$\Psi(x, 0) = \sqrt{\frac{1}{6}}\,\psi_1(x) + \frac{i}{\sqrt{2}}\,\psi_2(x) + \sqrt{\frac{1}{3}}\,\psi_3(x) \tag{45}$$

Thus, if we carry out a measurement of energy, the probabilities of obtaining the values E_1, E_2 and E_3 would be $\frac{1}{6}$, $\frac{1}{2}$, and $\frac{1}{3}$ respectively. How will such a state evolve with time? Well, we just multiply each term by the appropriate time dependent factor to obtain $\Psi(x, t)$ [see Eq. (10)].

$$\Psi(x, t) = \sqrt{\frac{1}{6}}\,\psi_1(x)\,e^{-it/t_0} + \frac{i}{\sqrt{2}}\,\psi_2(x)\,e^{-4it/t_0} + \sqrt{\frac{1}{3}}\,\psi_3(x)\,e^{-9it/t_0} \tag{46}$$

where

$$t_0 = \frac{2\mu a^2}{\pi^2 \hbar}$$

Obviously,

$$\int_0^a |\Psi(x,t)|^2\,dx = 1$$

[5]A simple solution of the Schrödinger equation corresponding to a time dependent Hamiltonian is discussed in Section 14.6.

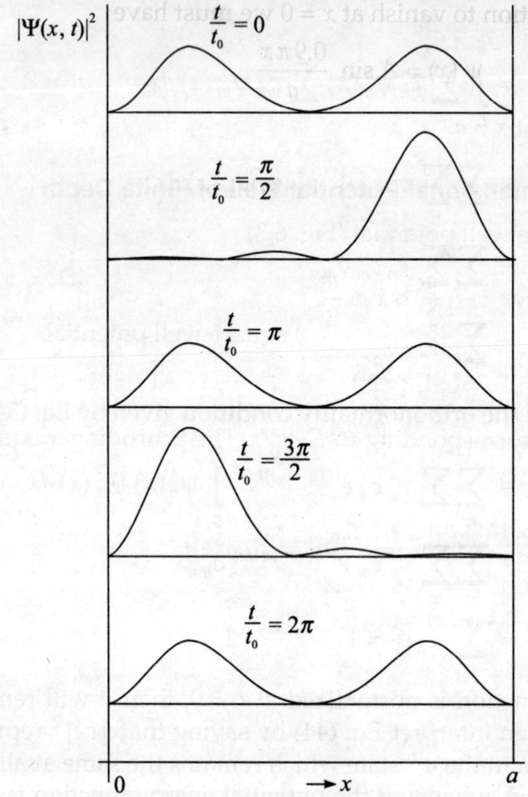

Figure 6.2: The time evolution of the probability distribution $|\Psi(x, t)|^2$ corresponding to Eq. (46) at $t/t_0 = 0, \frac{\pi}{2}, \pi, \frac{3\pi}{2}$ and 2π.

for all values of t. In Fig. 6.2, we have shown the time evolution of the probability distribution

$$P(x, t) = |\Psi(x, t)|^2$$

for $\Psi(x, t)$ given by Eq. (46) at

$$\frac{t}{t_0} = 0, \frac{\pi}{2}, \pi, \frac{3\pi}{2} \text{ and } 2\pi$$

At all values of time, the probability of finding the system in a particular state remains the same. Further, the average value of the energy is given by

$$\langle E \rangle = \frac{1}{6}E_1 + \frac{1}{2}E_2 + \frac{1}{3}E_3 = \left[\frac{1}{6} + 2 + 3\right]\frac{\pi^2\hbar^2}{2\mu a^2} = \frac{31}{6}\cdot\frac{\pi^2\hbar^2}{2\mu a^2} \tag{47}$$

Thus, if one carries out a large number of measurements (of energy) on identical systems characterized by the same wavefunction as given by Eq. (46), then the average value of the energy would be given by Eq. (47).

(vii) What happens to the wavefunction if $E \neq E_n$; i.e., if E is not one of the eigenvalues? For such a case the boundary conditions cannot be satisfied and therefore it cannot be an allowed value of energy. For example, if

$$E = \frac{0.81\pi^2\hbar^2}{2\mu a^2}$$

then for the wavefunction to vanish at $x = 0$ we must have

$$\psi(x) = A \sin \frac{0.9\pi x}{a}$$

which cannot vanish at $x = a$.

6.6.2 Particle in a One-dimensional Potential Well of Finite Depth

We next consider the square well potential (Fig. 6.3)

$$\left.\begin{aligned} V(x) &= 0 \quad \text{for} \quad -\frac{a}{2} < x < \frac{a}{2} \\ &= V_0 \quad \text{for} \quad |x| > \frac{a}{2} \end{aligned}\right\} \text{square well potential} \tag{48}$$

We first look for solutions corresponding to $E < V_0$. The Schrödinger equation can be written in the form

$$\begin{aligned} \frac{d^2\psi}{dx^2} + k^2 \psi(x) &= 0 \quad \text{for } |x| < \frac{a}{2} \\ \frac{d^2\psi}{dx^2} - \kappa^2 \psi(x) &= 0 \quad \text{for } |x| > \frac{a}{2} \end{aligned} \tag{49}$$

where

$$k^2 = \frac{2\mu E}{\hbar^2} \quad \text{and} \quad \kappa^2 = \frac{2\mu}{\hbar^2}(V_0 - E) \tag{50}$$

Since $V(-x) = V(x)$, the solutions are either symmetric or antisymmetric in x. Corresponding to the symmetric solution we may write

$$\begin{aligned} \psi(x) &= A \cos kx \qquad\qquad |x| < \frac{a}{2} \\ &= B \exp[-\kappa |x|] \quad |x| > \frac{a}{2} \end{aligned} \tag{51}$$

where in the region $|x| > a/2$, we have rejected the solution $\exp(\kappa |x|)$ because then $\psi(x)$ will go to infinity as $|x| \to \infty$. Continuity of $\psi(x)$ and $d\psi/dx$ (see Section 6.2) at $x = a/2$ leads to

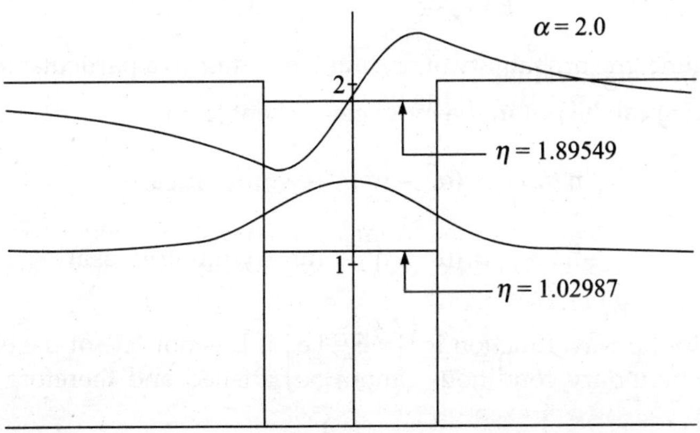

Figure 6.3: A potential well of finite depth. The eigenvalues and eigenfunctions correspond to α = 2.0.

$$A \cos \frac{ka}{2} = B \exp\left[-\frac{\kappa a}{2}\right] \tag{52}$$

and
$$-Ak \sin\frac{ka}{2} = -B\,\kappa \exp\left[-\frac{\kappa a}{2}\right] \tag{53}$$

For non-trivial solutions one gets

$$\tan\frac{ka}{2} = \frac{\kappa}{k} = \left[\frac{V_0 - E}{E}\right]^{\frac{1}{2}} \tag{54}$$

The roots of the transcendental equation (54) determine the discrete values of E (for given values of μ, a and V_0) and one has to use numerical or graphical methods to obtain the energy levels. For a particular value of E, the ratio A/B can be determined by using either Eq. (52) or Eq. (53) using which one can determine $\psi(x)$ within a multiplicative constant. The value of A (and hence of B) can be determined by using the normalization condition (see Problem 6.7). It is of interest to point out that for $V_0 \gg E$, the RHS of Eq. (54) tends to ∞ so that one obtains

$$\frac{ka}{2} = \left(p + \frac{1}{2}\right)\pi\,; \qquad p = 0, 1, 2, \ldots$$

or
$$E = \frac{(2p+1)^2\,\pi^2\,\hbar^2}{2\mu a^2}\,; \qquad p = 0, 1, 2, \ldots \tag{55}$$

which is the same as Eq. (32) with n now restricted to odd values because we have considered only the symmetric solution. Further for $V_0 \to \infty$, κ will tend to ∞ and the wavefunction would vanish in the region $|x| > a/2$.

For the antisymmetric solution (for which ψ is proportional to $\sin kx$ inside the well and to $\exp(-\kappa\,|x|)$ outside the well) the transcendental equation determining the energy eigenvalues would be [cf. Eq. (54)].

$$\cot\frac{ka}{2} = -\frac{\kappa}{k} = -\left(\frac{V_0 - E}{E}\right)^{\frac{1}{2}} \tag{56}$$

which for $V_0 \to \infty$ would give

$$E = \frac{(2p)^2\,\pi^2\,\hbar^2}{2\mu a^2}\,;\ p = 1, 2, 3, \ldots \tag{57}$$

Equations determining eigenvalues

Equations (54) and (56) can be put in the more convenient form

$$\eta \tan \eta = (\alpha^2 - \eta^2)^{\frac{1}{2}} \ \text{(symmetric case)} \tag{58}$$

$$-\eta \cot \eta = (\alpha^2 - \eta^2)^{\frac{1}{2}} \ \text{(antisymmetric case)} \tag{59}$$

where
$$\eta = \frac{ka}{2} = \left(\frac{2\mu E a^2}{4\hbar^2}\right)^{\frac{1}{2}}$$

and
$$\alpha = \left(\frac{2\mu V_0\, a^2}{4\hbar^2}\right)^{\frac{1}{2}}$$

For a given value of α, solutions of the transcendental equations (58) and (59) will determine the allowed energy eigen values of the problem.

For a given value of α, the right-hand sides of Eqs (58) and (59) are a portion of a circle and therefore the roots can easily be determined using numerical methods. In Fig. 6.4, we have plotted the functions $\eta \tan \eta$ (solid curve) and $-\eta \cot \eta$ (dotted curve) as a function of η. Their points of intersection with the portion of the circle determines the eigenvalues. It is easy to see that

for $0 < \alpha < \pi/2$, there will be only one symmetric state;

for $\pi/2 < \alpha < \pi$, there will be one symmetric and one antisymmetric state;

for $\pi < \alpha < 3\pi/2$, there will be two symmetric and one antisymmetric state; etc.

Example 6.1 As an example, we consider a proton (of mass 1.67×10^{-27} kg) inside a one-dimensional well with $V_0 \simeq 25$ MeV $\simeq 4 \times 10^{-12}$ J and $a \simeq 3.65 \times 10^{-15}$ m. Thus,

$$\alpha = \left(\frac{2\mu V_0 a^2}{4\hbar^2} \right)^{\frac{1}{2}} \simeq 2.0 \tag{60}$$

Since $\alpha < \pi$, there will be one symmetric state and one antisymmetric state; the roots of Eqs (58) and (59) for $\alpha = 2$ are given by (see Fig. 6.3)

$$\eta = 1.02987 \quad (\Rightarrow \quad E \simeq 6.63 \text{ MeV})$$

for the symmetric state and

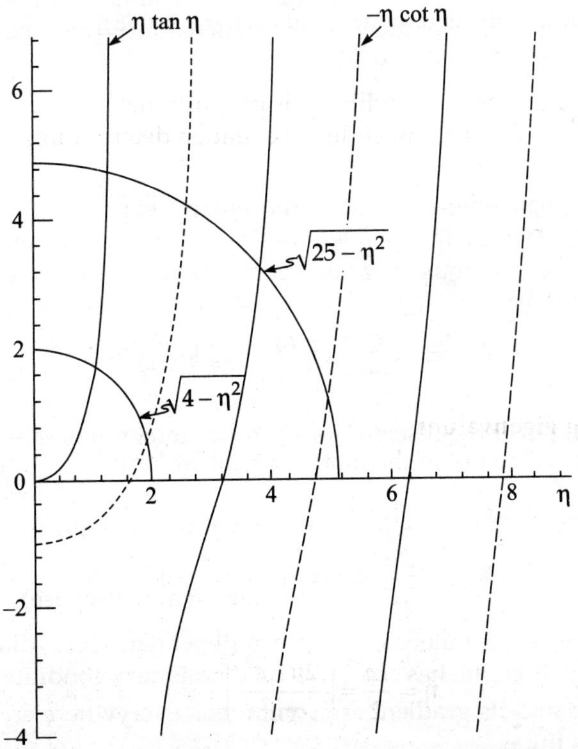

Figure 6.4: The variation of $\eta \tan \eta$ (solid curve) and $-\eta \cot \eta$ (dotted curve) as a function of η. The points of intersection of these curves with the quadrant of a circle of radius α determine the discrete eigenvalues corresponding to the potential well of finite depth.

$$\eta = 1.89549 \quad (\Rightarrow \quad E \simeq 22.45 \text{ MeV})$$
for the antisymmetric state

Till now we have assumed $E < V_0$; had we considered values of $E > V_0$, we would have found that all values of E (greater than V_0) could be possible, i.e., the energy eigenvalues would be continuously distributed in the domain $V_0 < E < \infty$. However, the wavefunction would not have vanished at large distances from the origin (see Chapter 8). Such solutions are of great importance in connection with the scattering of a particle by a force field where the energy is specified in advance and the behaviour of the wavefunction at great distances is found in terms of energy.

6.7 THREE-DIMENSIONAL SCHRÖDINGER EQUATION

In previous sections we had discussed the stationary states of the one-dimensional Schrödinger equation; the corresponding considerations for the three-dimensional equation are very straightforward. For example, by using the method of separation of variables, the solution of the time dependent Schrödinger equation

$$i\hbar \frac{\partial \Psi}{\partial t} = -\frac{\hbar^2}{2\mu} \nabla^2 \Psi + V(\mathbf{r}) \Psi(\mathbf{r}, t) \tag{61}$$

is of the form

$$\Psi(\mathbf{r}, t) = \psi(\mathbf{r}) \exp[-iE_n t/\hbar] \tag{62}$$

where $\psi(\mathbf{r})$ is the solution of the eigenvalue equation [cf. Eq. (8)]

$$H\psi(\mathbf{r}) = E\psi(\mathbf{r}) \tag{63}$$

and

$$H = -\frac{\hbar^2}{2\mu} \nabla^2 + V(\mathbf{r}) = \frac{1}{2\mu} [p_x^2 + p_y^2 + p_z^2] + V(\mathbf{r}) \tag{64}$$

represents the Hamiltonian operator and E the energy eigenvalue. The potential energy function has been assumed to be independent of time. In general, the eigenvalues can have a discrete set along with a continuous range of eigenvalues and the general solution of Eq. (61) is of the form

$$\Psi(\mathbf{r}, t) = \sum_n c_n \psi_n(\mathbf{r}) \exp[-iE_n t/\hbar] \tag{65}$$

where $\sum_n$ denotes a sum over the discrete states and an integration over the continuum.

Equation (64) is usually written in the form

$$\nabla^2 \psi + \frac{2\mu}{\hbar^2} [E - V(\mathbf{r})] \psi(\mathbf{r}) = 0 \tag{66}$$

which is known as the three-dimensional time independent Schrödinger equation. As in Section 6.2, the wavefunction satisfies continuity and boundary conditions. Thus,

(i) The wavefunction and its gradient are continuous everywhere as long as the potential energy function is finite.

(ii) The wavefunction should also be single-valued everywhere so that it corresponds to a definite physical situation.

(iii) For a degenerate state any linear combination of the degenerate eigenfunctions is also a possible eigenfunction.

The orthogonality of the wavefunctions can be proved by using a method similar to the one used in Section 6.4. Indeed, instead of Eqs (14) and (15), we will have

$$-\frac{\hbar^2}{2\mu}\nabla^2\psi_n + V(\mathbf{r})\,\psi_n = E_n\,\psi_n \tag{67}$$

$$-\frac{\hbar^2}{2\mu}\nabla^2\psi_k + V(\mathbf{r})\,\psi_k = E_k\,\psi_k \tag{68}$$

If we multiply Eq. (67) by $\psi_k{}^*$ and multiply the complex conjugate of Eq. (68) by ψ_n and subtract, we would get

$$-\frac{\hbar^2}{2\mu}\int \nabla\cdot[\psi_k^*\,\nabla\psi_n - \psi_n\,\nabla\psi_k^*]\,d\tau = (E_n - E_k{}^*)\int \psi_k^*\,\psi_n\,d\tau$$

By using the Gauss theorem, the LHS becomes a surface integral

$$-\frac{\hbar^2}{2\mu}\int [\psi_k^*\,\nabla\psi_n - \psi_n\,\nabla\psi_k^*]\cdot\hat{\mathbf{n}}\,dS$$

where $\hat{\mathbf{n}}$ represents the unit outward normal on the surface. The surface integral would vanish if the volume integral is over the entire space. Thus, as in Section 6.4, we will have

$$E_n = E_n{}^* \tag{69}$$

proving that all eigenvalues are real and the wavefunctions belonging to different energy levels are orthogonal. By appropriate normalization we can always have

$$\int \psi_k^*\,\psi_n\,d\tau = \delta_{kn} \qquad \text{orthonormality condition} \tag{70}$$

which is the orthonormality condition. For degenerate states one can always construct appropriate linear combinations so that the orthonormality condition is satisfied (see Problem 6.3).

As a simple example, we will, in the next section, solve the three-dimensional Schrödinger equation for a particle inside a cube of length L.

6.8 PARTICLE IN A THREE-DIMENSIONAL BOX

In this section, we will solve the Schrödinger equation for a potential energy variation of the form

$$\begin{aligned} V(x, y, z) &= 0 \quad \text{for} \quad 0 < x < L,\, 0 < y < L,\, 0 < z < L \\ &= \infty \text{ everywhere else} \end{aligned} \tag{71}$$

which represents a particle in an infinitely deep potential well inside a cube of side L. This problem has important applications in many areas. For example, in the free electron theory of metals, the valence electrons can be approximately assumed to be free and confined inside a box (see Section 6.9). Similarly, the free electron theory allows us to understand the basic physics behind the white drawf stars (see Section 6.10). The free electron theory is also the basis of the Thomas-Fermi model of the atom (see Section 23.6).

Now, the Schrödinger equation is given by

$$\nabla^2\psi + \frac{2\mu E}{\hbar^2}\psi = 0 \quad \begin{cases} 0 < x < L \\ 0 < y < L \\ 0 < z < L \end{cases} \tag{72}$$

with the boundary condition that ψ should vanish everywhere on the surface of the cube. We use the method of separation of variables and write $\psi = X(x)\, Y(y)\, Z(z)$ to obtain

$$\frac{1}{X}\cdot\frac{d^2X}{dx^2} + \frac{1}{Y}\cdot\frac{d^2Y}{dy^2} + \frac{1}{Z}\cdot\frac{d^2Z}{dz^2} = -\frac{2\mu E}{\hbar^2} \tag{73}$$

The first term is a function of x alone, the second term of y alone, etc., so that each term has to be set equal to a constant. We write

$$\frac{1}{X}\frac{d^2X}{dx^2} = -k_x^2 \tag{74}$$

and similar Equations for $Y(y)$ and $Z(z)$ with

$$k_x^2 + k_y^2 + k_z^2 = \frac{2\mu E}{\hbar^2} \tag{75}$$

We have set each term equal to a negative constant; otherwise the boundary conditions cannot be satisfied. The solution of Eq. (74) is

$$X(x) = A \sin k_x x + B \cos k_x x \tag{76}$$

and since ψ has to vanish on all points on the surfaces $x = 0$ and $x = L$ we must have $B = 0$ and $k_x = n_x\pi/L$ with $n_x = 1, 2, \ldots$ (cf. Section 6.6.1). Similarly, for k_y and k_z. Using Eq. (75), we get

$$E = \frac{\pi^2\hbar^2}{2\mu L^2}(n_x^2 + n_y^2 + n_z^2)\ ;\ n_x, n_y, n_z = 1, 2, 3, \ldots \qquad \text{Energy eigenvalues} \tag{77}$$

The corresponding normalized wavefunctions are:

$$\psi(x, y, z) = \left(\frac{8}{L^3}\right)^{1/2} \sin\frac{n_x\pi}{L}x \sin\frac{n_y\pi}{L}y \sin\frac{n_z\pi}{L}z \tag{78}$$

6.8.1 The Density of States

If $g(E)\,dE$ represents the number of states whose energy lies between E and $E + dE$, then $g(E)$ is known as the density of states and it represents a very important quantity in the theory of solids. In order to calculate $g(E)$ we first calculate $N(E)$ which represents the total number of states whose energies are less than E. Obviously,

$$N(E) = \int_0^E g(E)\,dE \tag{79}$$

and therefore

$$g(E) = \frac{dN(E)}{dE} \tag{80}$$

Now,

$$n_x^2 + n_y^2 + n_z^2 = \frac{2\mu L^2 E}{\pi^2\hbar^2} = R^2 \text{ (say)} \tag{81}$$

Thus, $N(E)$ will be the number of sets of integers whose sum of squares is less than R^2. In the n_x, n_y, n_z space each point corresponds to a unit volume and if we draw a sphere of radius R then the volume of the positive octant will approximately represent[6] $N(E)$; we have to take the positive octant because n_x, n_y and n_z take positive values. Thus,

$$N(E) = 2 \times \frac{1}{8} \times \frac{4\pi}{3} R^3 = \frac{(2\mu)^{3/2} L^3}{3\pi^2 \hbar^3} E^{3/2} \tag{82}$$

where an additional factor of 2 has been introduced as a state can be occupied by two electrons. Using Eq. (80), we get

$$g(E) = \frac{(2\mu)^{3/2} V}{2\pi^2 \hbar^3} E^{1/2} \qquad \text{density of states} \tag{83}$$

where V (= L^3) represents the volume of the box.

Often it is more convenient to express the density of states in momentum space. Now for a free non-relativistic particle

$$E = \frac{p^2}{2\mu} \tag{84}$$

Thus the equation

$$g(p)\, dp = g(E)\, dE \tag{85}$$

would readily give

$$g(p) = \frac{V}{\pi^2 \hbar^3} p^2 \tag{86}$$

Relativistic Considerations

Equation (84) represents the non-relativistic relation between energy and momentum. The corresponding relativistic relation is given by

$$E^2 = p^2c^2 + \mu^2c^4 \tag{87}$$

where μ represents the rest mass of the particle. Now, it so happens that if one solves Dirac's relativistic equation then the density of states in the momentum space is still given by Eq. (86) (see, e.g. Ref. 2). Since

$$E\, dE = c^2 p\, dp$$

we get

$$g(E) = \begin{cases} \dfrac{V}{\pi^2 \hbar^3 c^3} E(E^2 - \mu^2 c^4)^{1/2}, & E \geq \mu c^2 \\ 0, & E < \mu c^2 \end{cases} \tag{88}$$

In the extreme relativistic limit

$$E \gg \mu c^2$$

[6]If the reader finds it difficult to understand he may first try to make the corresponding two-dimensional calculations in which one is interested in finding the number of sets of integers such that $n_x^2 + n_y^2 < R^2$. If one takes a graph paper then each corner corresponds to a set of integers and each point can be associated with a unit area. Thus the number of sets of integers would be $\pi R^2/4$ where the factor 1/4 is because of the fact that we are interested in the positive quadrant.

we obtain

$$g(E) = \frac{V}{\pi^2 \hbar^3 c^3} E^2, \quad E \gg \mu c^2 \tag{89}$$

6.9 FREE ELECTRONS IN A METAL

The problem corresponding to a particle in a box has a very important application in the free electron theory of metals where the valence electrons can be approximately assumed to be free and confined inside the metal. Now, the probability of finding an electron with energy E is given by the Fermi-Dirac function

$$F(E) = \frac{1}{1+e^{(E-E_F)/kT}} \qquad \text{Fermi-Dirac function} \tag{90}$$

The quantity E_F is referred to as the Fermi energy. At absolute zero ($T = 0$), we have

$$F(E) = \begin{cases} 1, & E < E_{F_0} \\ 0, & E > E_{F_0} \end{cases} \tag{91}$$

where the subscript zero implies that we are calculating the Fermi energy at $T = 0$. Equation (91) tells us that all states below the Fermi level are filled up and all states above the Fermi level are empty. For such a situation (when all the lowest quantum states are occupied), the electron gas is said to be *completely degenerate*. Now the total number of electrons (inside the box) is given by

$$N = \int_0^\infty g(E)\, F(E)\, dE \tag{92}$$

which at $T = 0$ becomes

$$N = \int_0^{E_{F_0}} g(E)\, dE$$

$$= (2\mu)^{3/2} \frac{V}{2\pi^2 \hbar^3} \int_0^{E_{F_0}} E^{1/2}\, dE$$

$$= (2\mu)^{3/2} \frac{V}{3\pi^2 \hbar^3} E_{F_0}^{3/2} \tag{93}$$

Simple manipulations give us

$$E_{F_0} = \frac{\hbar^2}{2\mu} (3\pi^2 n)^{2/3} \qquad \text{Fermi energy at } T = 0 \tag{94}$$

where n (= N/V) represents the number of free electrons per unit volume. Substituting the numerical values of various constants, we get

$$E_{F_0} \approx 5.84 \times 10^{-27}\, n^{2/3} \text{ erg}$$
$$\approx 3.65 \times 10^{-15}\, n^{2/3} \text{ eV} \tag{95}$$

where n is measured in cm^{-3}. For example, for sodium with one free electron per atom and having a density of 0.97 g/cm^3, we get

$$n \approx \frac{6.023 \times 10^{23} \times 0.97}{23} \approx 2.54 \times 10^{22} \text{ electrons/cm}^3$$

Thus,

$$E_{F_0} \approx 3.2 \text{ eV}$$

The total energy of the gas at absolute zero will be given by

$$\mathcal{E} = \int_0^{E_{F_0}} E\, g(E)\, F(E)\, dE$$

$$= \frac{(2\mu)^{3/2} V}{5\pi^2 \hbar^3} E_{F_0}^{5/2} \qquad \text{Total energy at absolute zero} \tag{96}$$

If we use Eq. (93) we would get the following expression for average energy per particle

$$\langle E \rangle = \frac{\mathcal{E}}{N} = \frac{3}{5} E_{F_0} \tag{97}$$

In general, the Fermi energy E_F is dependent on temperature. However, when $kT \le E_{F_0}$, we say that the electron gas is *strongly degenerate*—or, *almost completely degenerate*. In Problem 6.13, we will discuss the variation of E with temperature for a strongly degenerate gas and we will show

$$E_F(T) = E_{F_0}\left[1 - \frac{\pi^2}{12}\left(\frac{kT}{E_{F_0}}\right)^2 + \frac{\pi^4}{720}\left(\frac{kT}{E_{F_0}}\right)^2 \cdots\right] \tag{98}$$

and the average energy per electron is

$$\langle E \rangle = \frac{3}{5} E_{F_0}\left[1 + \frac{5\pi^2}{12}\left(\frac{kT}{E_{F_0}}\right)^2 \cdots\right] \tag{99}$$

In order to get numerical appreciation of the corrections, we consider sodium for which $E_{F_0} \approx 3.15$ eV. Thus, at $T = 300$ K ($kT \approx 0.0259$ eV)

$$\frac{kT}{E_{F_0}} \approx 8.2 \times 10^{-3}$$

and the correction terms in Eqs (98) and (99) are very small. Indeed

$$E_E(300 \text{ K}) \approx E_{F_0}[1 - 5.5 \times 10^{-5} + 6 \times 10^{-10}]$$

and

$$\langle E \rangle \approx \frac{3}{5} E_{F_0}[1 - 2.8 \times 10^{-4} + \ldots]$$

Even for $T \approx 5000$ K

$$\frac{kT}{E_{F_0}} \approx 0.137$$

and

$$E_F \approx 3.10 \text{ eV}$$

Thus, even at such high temperatures, the corrections are very small. In Fig. 6.5, we have plotted the Fermi-Dirac distribution function for sodium at $T = 0$ K, 500 K and 7000 K. At $T = 0$ K,

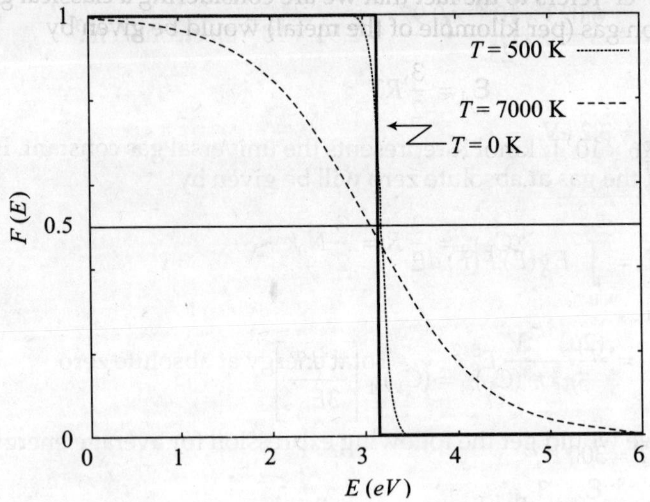

Figure 6.5: The Fermi-Dirac distribution function for sodium (E_{F_0} = 3.15 eV) at 0 K, 500 K and 7000 K.

the electron gas is in the lowest energy state and all states below E_{F_0} are occupied. At T = 500 K, the probability of occupancy remains very close to that at T = 0 K except in the close vicinity of $E = E_F$. Because of this, it is extremely unlikely that the low energy electrons will get excited to higher energy states. Thus, when the temperature is raised, only those electrons which are within kT of E_F get excited to higher energy states. It is for this reason that the specific heat for the electron gas is much below the classical value. We will explicitly show this below. On the other hand, at T = 7000 K, the occupancy probability is significantly different from that at T = 0 K and one can see that E_F (at which $F(E) = \frac{1}{2}$) is visibly smaller than E_{F_0}; at T = 500 K, this difference is extremely small.

6.9.1 The Specific Heat of the Electron Gas

Assuming 1 free electron per atom, we will have N_0 free electrons per kmol of the metal where

$$N_0 \approx 6.023 \times 10^{26} \text{ kmol}^{-1}$$

represents the Avagadro's number. Thus [using Eq. (99)] the total energy of the electron gas per kilomole of the metal would be

$$\mathcal{E} = N_0 \langle E \rangle \approx \frac{3}{5} N_0 E_{F_0} \left[1 + \frac{5\pi^2}{12} \left(\frac{kT}{E_{F_0}} \right)^2 + \ldots \right] \tag{100}$$

The specific heat (at constant volume) of the electron gas would, therefore, be given by

$$(C_V)_{\text{el}} = \left(\frac{\partial \mathcal{E}}{\partial T} \right)_V$$

$$= \frac{\pi^2 N_0 k^2 T}{2E_{F_0}} \tag{101}$$

On the other hand, if electrons are assumed to behave like a classical gas, the average energy per electron would be

$$\langle E \rangle_{\text{cl}} = \frac{3}{2} kT \tag{102}$$

where the subscript 'cl' refers to the fact that we are considering a classical gas. Thus, the total energy of the electron gas (per kilomole of the metal) would be given by

$$\mathcal{E}_{cl} = \frac{3}{2} RT \tag{103}$$

where $R = N_0 k \approx 8.316 \times 10^3$ J/kmol K represents the universal gas constant. From Eq. (103), we readily obtain

$$(C_V)_{cl} = \frac{3}{2} R = \frac{3}{2} N_0 k \tag{104}$$

Thus,

$$(C_V)_{el} = (C_V)_{cl} \left[\frac{\pi^2 kT}{3E_{F_0}} \right] \tag{105}$$

Now for sodium at $T \approx 300$ K,

$$\frac{kT}{E_{F_0}} \approx 8.2 \times 10^{-3}$$

and, therefore

$$(C_V)_{el} \approx \frac{1}{37} (C_V)_{cl}$$

In general, at normal temperatures since

$$\frac{kT}{E_{F_0}} \ll 1$$

the electronic contribution to the specific heat is much less than what would be expected if we had assumed the electrons to behave like a classical gas. This is indeed borne out by experimental data. The physical reason for this has been discussed at the end of the previous section. We may mention here that at high temperatures when $kT \gg E_{F_0}$, the Fermi energy E_F becomes negative and we may neglect the unity in the denominator of Eq. (90) to obtain

$$F(E) \approx e^{E_F/kT} e^{-E/kT} \tag{106}$$

Thus if $N(E)dE$ represents the number of electrons whose energies lie between E and $E + dE$ then

$$N(E)\, dE = g(E)\, F(E)\, dE$$
$$= \text{const. } E^{1/2} e^{-E/kT} dE$$

which corresponds to the classical Maxwellian distribution; using the above equation we would readily get Eq. (102). The total number of electrons would be given by

$$N = (2\mu)^{3/2} \frac{V}{2\pi^2 \hbar^3} e^{E_F/kT} \int_0^\infty E^{1/2} e^{-E/kT} dE$$

Simple algebra and use of Eq. (94) would give

$$E_F = -kT \ln \left[\frac{3\sqrt{\pi}}{4} \left(\frac{kT}{E_{F_0}} \right)^{3/2} \right] \tag{107}$$

showing that the Fermi energy becomes negative.

6.10 WHITE DWARF STARS

Another example of a highly degenerate electron gas is the inside of a white dwarf star where the electron densities are so high that the electron gas is in a highly degenerate state. In this section, we will discuss the basic physics associated with a white dwarf star. However, before we discuss the details, we should mention that in astrophysics almost everyone uses the cgs units and in these units, the values of some important constants (which we will be using) are given in Table 6.1.

Table 6.1: Values of some physical constants (cgs system)

Quantity	Symbol	Value
Mass of the hydrogen atom	m_H	1.673×10^{-24} g
Mass of the electron	m_e	9.109×10^{-28} g
Speed of light in vacuum	c	2.998×10^{10} cm s^{-1}
Planck's constant	$\hbar$	1.055×10^{-27} erg s
Gravitational constant	G	6.672×10^{-8} cm^3 s^{-2} g^{-1}
Mass of the sun	M_S	1.989×10^{33} g
Radius of the sun	R_S	6.960×10^{10} cm

We assume that the star to be in hydrostatic equilibrium under its own gravitation, i.e., the inward directed gravitational force exactly balances the outward directed force due to the gas pressure. We also assume the star to be of uniform density ρ (this is quite a drastic assumption; however, it allows us to obtain analytical results for important parameters associated with the star). Thus the mass of the spherical shell whose inner and outer radii are r and $r + dr$ will be given by

$$dM = \rho 4\pi r^2 dr \tag{108}$$

We next consider the gravitational force acting on this spherical shell. Obviously, the mass lying outside the shell does not contribute and the gravitational force due to the sphere of radius r would be given by

$$\frac{G\left[\rho \frac{4\pi}{3} r^3\right] dM}{r^2} \tag{109}$$

where G represents the gravitational constant. The gravitational force will be directed inwards and if dP represents the increment in pressure over the distance dr then for the star to be in hydrostatic equilibrium we must have

$$-4\pi r^2 dP(r) = \frac{G\left[\rho \frac{4\pi}{3} r^3\right] dM}{r^2} \tag{110}$$

Using Eq. (108), we get

$$\frac{dP}{dr} = -\frac{4\pi}{3} G\rho^2 r \tag{111}$$

The above equation is the basic equation in understanding stellar structures—of course, under the assumption of the constancy of density inside the star. Carrying out the straightforward integration, we get

$$P(r) = \frac{2\pi}{3} G\rho^2 (R^2 - r^2) \tag{112}$$

where R represents the radius of the star where the pressure vanishes. Thus the pressure at the center of the star is approximately given by

$$P_c = \frac{2\pi}{3} G\rho^2 R^2 \tag{113}$$

Now, for Sirius B (which is a white dwarf star)

$$\left.\begin{aligned} M &\approx 1.05\, M_s \approx 2.1 \times 10^{33}\ \text{g} \\ R &\approx 5.5 \times 10^{8}\ \text{cm} \end{aligned}\right\} \tag{114}$$

where M_s and R_s represent the mass and radius of the sun; the corresponding values are given in Table 6.1. Equation (114) gives the following value for the average density of Sirius B

$$\rho \approx 3 \times 10^{6}\ \text{g/cm}^3 \tag{115}$$

Notice that the radius of the star is about 5500 km (which is slightly smaller than that of the earth) whereas the mass is about the same as that of the sun leading to extremely high densities (about 3000 kg/cm^3). At such high densities, inspite of very high temperatures, the electrons may form a degenerate gas. Indeed, in a white dwarf star the densities are such that while the electron gas is in a degenerate state, the gas consisting of protons and other nuclei is in a non-degenerate state so that the perfect gas law is applicable to these nuclei. However, the pressure due to the gas formed by these nuclei is usually negligible in comparison to the pressure due to the degenerate electron gas and, therefore, we need only consider the pressure due to the degenerate electron gas.

Returning to Eq. (113) if we substitute the values of R and ρ (for Sirius B) we would get

$$P_c \approx 4 \times 10^{23}\ \text{dynes/cm}^2 \tag{116}$$

The above expression may be compared with the pressure at the center of the sun which is about 1.2×10^{17} dyne/cm^2. Thus the pressure at the center of the white dwarf star Sirius B is about 3 million times more than the pressure at the center of the sun! Further, because of very high temperatures, the stellar matter is in a completely ionized state; thus, if we assume the atoms are of atomic number Z and atomic weight A, the number of electrons per unit volume will be given by

$$n_e = \frac{Z}{A} \cdot \frac{\rho}{m_H} \tag{117}$$

where m_H represents the mass of the (neutral) hydrogen atom. Assuming that the stellar matter contains primarily heavy elements (which is indeed true for white dwarf stars), we may write $A/Z \approx 2$; thus, for $\rho \approx 3 \times 10^6$ g/cm^3, we get

$$n_e \approx 9 \times 10^{29}\ \text{cm}^{-3}$$

For the above value of n_e, Eq. (94) would give

$$E_{F_0} \approx 0.3\ \text{MeV}$$

We may note two points:

1. At $T \approx 10^7$ K

$$\frac{E_{F_0}}{kT} \approx 400 \gg 1$$

   Thus, even at such high temperatures, the electron gas can be assumed to be almost completely degenerate.

2. The value of E_{F_0} ($\approx$ 0.3 MeV) may be compared with the rest mass energy of the electron which is about 0.5 MeV. Thus, we should use relativistic expressions relating energy and mass. Nevertheless, we will use non-relativistic formulae and hope that their use should at least give us values which will be of the right order.

Now, in the non-relativistic theory, the pressure due to the degenerate electron gas is given by[7]

$$P = \frac{2}{5} n_e E_{F_0} = \frac{\hbar^2}{5\mu}(3\pi^2)^{2/3} n_e^{5/3} \tag{118}$$

Thus,

$$P = \frac{\hbar^2}{5\mu}(3\pi^2)^{2/3}\left[\frac{Z}{A}\frac{\rho}{m_H}\right]^{5/3} \tag{119}$$

Assuming the pressure is primarily due to the degenerate electron gas, we will have

$$\frac{2\pi}{3} G\rho^2 R^2 = \frac{\hbar^2}{5\mu}(3\pi^2)^{2/3}\left[\frac{Z}{A}\frac{\rho}{m_H}\right]^{5/3} \tag{120}$$

Since,

$$\rho = \frac{M}{\frac{4\pi}{3} R^3} \tag{121}$$

We get after simple manipulations

$$R_{wd} = \frac{3}{5}\left(\frac{3\pi^2}{2}\right)^{1/3} \frac{\hbar^2}{\mu G}\left(\frac{Z}{Am_H}\right)^{5/3} \frac{1}{M_{wd}^{1/3}} \tag{122}$$

where the subscript wd stands for white dwarf. If we substitute the numerical values of various constants (see Table 6.1), and assume $A/Z \approx 2$, we would get

$$R_{wd} = \frac{3.6\times 10^{19}}{M_{wd}^{1/3}} \text{ cgs units} \tag{123}$$

where R_{wd} is in centimeters and M_{wd} is in grams. For Sirius B, $M \approx 2.1 \times 10^{33}$ g, and Eq. (123) would give us $R \approx 2.8 \times 10^8$ cm; although it is much smaller than the actual value of R; we do get the correct order. The discrepancy is due to neglect of relativistic effects and the assumption of constancy of density of stellar matter.

Equation (123) gives the very important result that higher the mass of the white dwarf star, smaller will be the radius—leading to higher densities. Although, this is borne out by experimental data on white-dwarf stars, Eq. (122) predicts that as the mass tends to become extremely large, the radius of the star approaches zero! This is incorrect because as the densities become higher, relativistic effects become more important. Indeed, in the extreme relativistic limit, we get the following expression for the pressure [see, e.g., Ref. 2]

$$P \approx \frac{(3\pi^2)^{1/3}}{4}\hbar c\left[\frac{Z}{A}\frac{\rho}{m_H}\right]^{4/3} \tag{124}$$

Equating this to the gravitational pressure given by Eq. (113), we would get

$$M \approx \frac{3}{4}\sqrt{\frac{\pi}{2}}\left(\frac{\hbar c}{G}\right)^{3/2}\left(\frac{Z}{Am_H}\right)^2 \tag{125}$$

[7] In the elementary kinetic theory of a gas, the pressure is given by (see, e.g., Section 23.5 of Ref. 1)

$$P = \frac{2}{3} n \langle E \rangle$$

where $\langle E \rangle$ represents the average kinetic energy of the gas particles. If we now use Eqs (94) and (97), we would get Eq. (118).

The above is indeed a remarkable result—in the sense that the limiting mass depends mainly only on the values of some universal constants. If we substitute the values of various constants and assume $A/Z \approx 2$, we would get

$$M \approx 0.87 \times 10^{33} \text{ g} \approx 0.44 \, M_S \tag{126}$$

In fact, on a more careful analysis (in particular, if we take into account the density variation inside the star), we get

$$M = M_{Ch} \approx 2.86 \times 10^{33} \text{ g} \approx 1.44 \, M_S \tag{127}$$

This limiting mass is known as the *Chandrasekhar limit* [and hence the subscript 'Ch']–after the famous Indian physicist S. Chandrasekhar who had derived this result [for a detailed account of Chandrasekhar's work on white dwarf stars we refer the reader to his classic treatise entitled *Introduction to the Study of Stellar Structures* (Ref. 2)]. Equation (127) is consistent with experimental data on white dwarf stars.

6.11 NEUTRON STARS

While collapsing, if the mass of the degenerate core (of the star) is close to the Chandrasekhar limit ($\approx$ 1.44 M_S), the electron energy becomes so large that the following capture process takes place

$$p + e \rightarrow n + \nu_e \tag{128}$$

resulting in the formation of a neutron star in which the gravitational pressure is balanced by the pressure due to the degenerate neutron gas. Assuming that a neutron star consists only of neutrons, the neutron density would be given by [cf. Eq. (117)]

$$n_n = \frac{\rho}{m_n} \tag{129}$$

where m_n represents the mass of the neutron. Thus Eq. (120) takes the form

$$\frac{2\pi}{3} G\rho^2 R^2 = \frac{\hbar^2}{5m_n} (3\pi^2)^{2/3} \left[\frac{\rho}{m_n}\right]^{5/3} \tag{130}$$

If we, now, use Eq. (121) and simplify, we would get

$$R_{ns} = \frac{3}{5}\left(\frac{3\pi^2}{2}\right)^{1/3} \frac{\hbar^2}{G}\left(\frac{1}{m_n}\right)^{8/3}\left(\frac{1}{M_{ns}^{1/3}}\right) \quad \text{Radius of the neutron star} \tag{131}$$

where R_{ns} and M_{ns} represent respectively the mass and the radius of the neutron star.

For $M_{ns} \approx 1.4 \, M_S \approx 2.78 \times 10^{33}$ g, we obtain

$$R_{ns} \approx 4.4 \text{ km} \tag{132}$$

Because of the assumption of constancy of density, the above value of the radius may be an underestimate by a factor of about 2 or 3–so assuming

$$R_{ns} \approx 10 \text{ km} \tag{133}$$

we obtain the following value for the average density.

$$\rho \approx 7 \times 10^{14} \text{ g/cm}^3$$

which is indeed an enormous value—even greater than the density of nuclear matter.

6.12 CONCLUSION

In this chapter, we have obtained bound state solutions of the one- and three-dimensional Schrödinger equations for some specific potential energy distributions. In the next chapter, we will obtain solutions of the one-dimensional Schrödinger equation for the linear harmonic oscillator problem and discuss the relation with the classical oscillator.

In Chapter 10, we will show that as long as the potential is spherically symmetric (i.e., as long as the potential depends only on the magnitude of the distance from a fixed point) the angular part of the wavefunction is a spherical harmonic and the potential energy function enters only in the radial part of the equation. In the same chapter, we will also solve the hydrogen atom problem and obtain the discrete states for the Coulomb potential and will show that the results are consistent with the experimental data.

6.13 PROBLEMS

Problem 6.1 (a) Show that the eigenvectors of the Hermitian matrix

$$\begin{pmatrix} 0 & 1 & 0 \\ 1 & 0 & 0 \\ 0 & 0 & 2 \end{pmatrix}$$

are orthogonal to each other. Normalize these vectors to form a complete set of orthonormal vectors.

(b) Repeat the calculation for the matrices

$$\begin{pmatrix} 0 & -i & 0 \\ i & 0 & 0 \\ 0 & 0 & 3 \end{pmatrix}$$

and

$$\begin{pmatrix} 1 & -1 \\ -1 & 1 \end{pmatrix}$$

Problem 6.2 In order to determine the boundary condition of $V(x)$ going to infinity at a particular value of x (say $x = 0$), we consider the following limiting process: assume $V(x) = 0$ for $x < 0$ and $V(x) = V_0$ for $x > 0$; in the limit of $V_0 \to \infty$, show that ψ vanishes in the region $x > 0$.

Problem 6.3 Let $\phi_1, \phi_2, \ldots, \phi_g$ represent a set of linearly independent eigenfunctions corresponding to a degenerate state. Construct appropriate linear combinations so that the wavefunctions form an orthonormal set.

Problem 6.4 Discuss the eigenvalues and eigenfunctions of the matrix

$$\begin{pmatrix} 1 & 0 & 0 \\ 0 & 1 & 0 \\ 0 & 0 & 2 \end{pmatrix}$$

Show that the vectors

$$\begin{pmatrix} 1 \\ 0 \\ 0 \end{pmatrix} \quad \begin{pmatrix} 1 \\ 1 \\ 0 \end{pmatrix}$$

belong to the degenerate state but do not form an orthonormal set. Use the results of the previous problem to obtain an orthonormal set.

Problem 6.5 Consider the potential energy variation given by

$$V(x) = \begin{cases} \infty, & x \le 0 \\ 0, & 0 < x < b \\ V_0, & x \ge b \end{cases} \tag{134}$$

(a) Derive the transcendental equation determining the energy eigenvalues for the bound states. Show that they correspond to the antisymmetric states for the potential energy distribution given by Eq. (48) with $b = a/2$.

(b) Obtain the normalized wavefunction.

(c) Assuming

$$\frac{2\mu V_0 b^2}{\hbar^2} = 9\pi^2$$

show that there will be three bound states with the value of

$$\xi = 2.83595, 5.64146 \quad \text{and} \quad 8.33877$$

where
$$\xi = \sqrt{\frac{2\mu E b^2}{\hbar^2}}$$

Problem 6.6 For the one-dimensional potential well (see Section 6.6.2), derive the transcendental equations determining the discrete eigenvalues [i.e., Eqs (58) and (59)] without making the assumption that the wavefunctions have to be symmetric and antisymmetric functions of x.

Problem 6.7 Use the normalization condition to determine the values of A and B in Eq. (51).

Problem 6.8 For copper we may assume one free electron per atom. Its density is about 8.94 g/cm^3 and its atomic mass is 63.5. Calculate E_{F_0} and show that at room temperatures ($T \approx 300$ K), the electron gas is almost completely degenerate. **[Ans:** $E_{F_0} \approx 7.0$ eV**]**

Problem 6.9 (a) In the free electron theory if $N(E)dE$ represents the number of electrons per unit volume whose energy lie between E and $E + dE$, show that it is given by

$$N(E) = \frac{3n/2}{e^{(E-E_F)/kT}+1} E_{F_0}^{-3/2} E^{1/2} \tag{135}$$

where n represents the total number of electrons per unit volume.

(b) For sodium plot $N(E)$ as a function of E for T = 0 K and 2000 K (see Fig. 6.6). Assume $E_F(T) \approx E_{F_0} \approx 3.15$ eV. Justify this assumption.

Problem 6.10 (a) Assume the sun to consist only of free electrons and protons. Using the data given in Table 6.1, calculate E_{F_0} for proton gas and for the electron gas.

(b) Inside the sun we may assume $T \approx 10^7$ K, calculate E_{F_0}/kT in each case.

[Ans: 0.018 eV and 33 eV for the proton and electron gas**]**

Problem 6.11 Assuming the density of the neutron star to be 7×10^{14} g/cm^3, calculate, E_{F_0} and compare with rest mass energy. **[Ans:** $E_{F_0} \approx 100$ MeV**]**

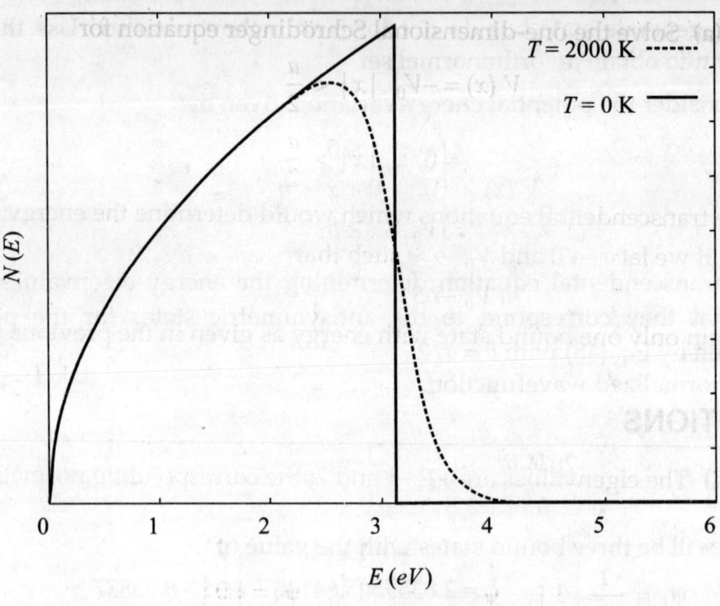

Figure 6.6: A plot of $N(E)$ for sodium at $T = 0$ K and 2000 K.

Problem 6.12 (a) Using the data given in Problem 6.8 calculate E_F for copper at 1000 K.

(b) Evaluate dF/dE and plot it for $T = 1000$ K. Show that it is a very sharply peaked function around $E = E_F$.

(c) At $T = 0$, show that dF/dE is a delta function.

Problem 6.13 Assuming $E_{F_0}/kT \gg 1$, derive Eqs. (98) and (99).
[**Hint:** Integrate Eq. (92) (and a similar equation for ε) by parts and use the previous problem].

Problem 6.14 Consider the potential energy variation given by

$$\begin{aligned} V(x) &= \infty \qquad x < 0 \\ &= \gamma x \qquad x > 0 \end{aligned} \tag{136}$$

Show that the solution of the Schrödinger equation can be written in terms of Airy functions (see Appendix D) and that the energy eigenvalues are given by

$$E_n = \left(\frac{\hbar^2\gamma^2}{2m}\right)^{1/3} \varepsilon_n \tag{137}$$

where

$$\varepsilon_1 = 2.33811; \qquad \varepsilon_2 = 4.08795; \qquad \varepsilon_3 = 5.52056; \quad \text{etc.} \tag{138}$$

Problem 6.15 Show that for

$$V(x) = -S\,\delta(x) \tag{139}$$

there is only one bound state with

$$E = -\frac{\mu S^2}{2\hbar^2} \tag{140}$$

What is the corresponding eigenfunction?

Problem 6.16 (a) Solve the one-dimensional Schrödinger equation for

$$V(x) = -V_0, \quad |x| < \frac{a}{2}$$
$$= 0, \quad |x| > \frac{a}{2} \tag{141}$$

and derive the transcendental equations which would determine the energy eigenvalues.

(b) Show that if we let $a \to 0$ and $V_0 \to \infty$ such that

$$a\, V_0 \to S$$

we would obtain only one bound state with energy as given in the previous problem.

6.14 SOLUTIONS

Solution 6.1 (a) The eigenvalues are +1, –1 and 2; the corresponding normalized eigenvectors are

$$\psi_1 = \frac{1}{\sqrt{2}}\begin{pmatrix}1\\1\\0\end{pmatrix}; \quad \psi_2 = \frac{1}{\sqrt{2}}\begin{pmatrix}1\\-1\\0\end{pmatrix}; \quad \psi_3 = \begin{pmatrix}0\\0\\1\end{pmatrix}$$

Further

$$\begin{pmatrix}a\\b\\c\end{pmatrix} = \frac{(a+b)}{\sqrt{2}}\psi_1 + \frac{(a-b)}{\sqrt{2}}\psi_2 + c\psi_3$$

showing that an arbitrary vector can be written as a linear combination of ψ_1, ψ_2 and ψ_3.

Solution 6.2 Since $V(x) = 0$ for $x < 0$ and $V = V_0$ for $x > 0$, the solution of the Schrödinger equation is

$$\psi(x) = A \sin kx + B \cos kx \qquad \text{for} \quad x < 0$$
$$= Ce^{-\kappa x} \qquad \text{for} \quad x > 0$$

where $k = (2\mu E/\hbar^2)^{1/2}$ and $\kappa = [2\mu\,(V_0 - E)/\hbar^2]^{1/2}$; we have assumed $V_0 > E$ and rejected the solution $e^{\kappa x}$ which goes to ∞ as $x \to \infty$. Continuity of ψ and $d\psi/dx$ at $x = 0$ gives

$$C = B$$

$$C = -\frac{k}{\kappa}A = -\left(\frac{1}{\frac{V_0}{E}-1}\right)^{1/2} A$$

Thus as $V_0 \to \infty$, $C \to 0$ (and hence $B \to 0$) implying that $\psi = 0$ for $x \geq 0$.

Solution 6.3 Let

$$V_1 = \phi_1, \quad \psi_1 = \frac{V_1}{l_1}, \quad l_1 = \left[\int V_1^* V_1\, dx\right]^{1/2} \tag{142}$$

l_1 is known as the length of the vector V_1. We next define

$$V_2 = \phi_2 - c_{21}\psi_1, \quad \psi_2 = \frac{V_2}{l_2}, \quad l_2 = \left[\int V_2^* V_2\, dx\right]^{1/2} \tag{143}$$

c_{21} is such that V_2 is orthogonal to ψ_1:

$$0 = \int V_2 \psi_1^* dx = \int \phi_2 \psi_1^* dx - c_{21}$$

because ψ_1 is normalized. Thus

$$c_{21} = \int \phi_2 \psi_1^* dx$$

Similarly, $$V_3 = \phi_3 - c_{31}\psi_1 - c_{32}\psi_2, \quad \psi_3 = \frac{V_3}{l_3}, \quad l_3 = \left[\int V_3^* V_3 dx\right]^{1/2} \tag{144}$$

$$c_{31} = \int \psi_1^* \phi_3 dx \quad \text{and} \quad c_{32} = \int \psi_2^* \phi_3 dx$$

are such that V_3 is orthogonal to ψ_1 and ψ_2.

In general,

$$V_{n+1} = \phi_{n+1} - \sum_{m=1}^{n} \left[\int \psi_m^* \phi_{n+1} dx\right] \psi_m$$

$$\psi_{n+1} = \frac{V_{n+1}}{l_{n+1}}, \quad l_{n+1} = \left[\int V_{n+1}^* V_{n+1} dx\right]^{1/2}$$

Thus, $\psi_1, \psi_2, \ldots$ form an orthonormal set. This is known as *Schmidt's orthogonalization procedure.*

Solution 6.4 We choose

$$\psi_1 = V_1 = \begin{pmatrix} 1 \\ 0 \\ 0 \end{pmatrix}$$

Then

$$c_{21} = (1\ 1\ 0) \begin{pmatrix} 1 \\ 0 \\ 0 \end{pmatrix} = 1$$

So that

$$V_2 = \begin{pmatrix} 1 \\ 1 \\ 0 \end{pmatrix} - \begin{pmatrix} 1 \\ 0 \\ 0 \end{pmatrix} = \begin{pmatrix} 0 \\ 1 \\ 0 \end{pmatrix} = \psi_2$$

ψ_1 and ψ_2 form an orthonormal set correspond to the degenerate eigenvalue. It may be mentioned that

$$\frac{1}{\sqrt{2}} \begin{pmatrix} 1 \\ 1 \\ 0 \end{pmatrix} \quad \text{and} \quad \frac{1}{\sqrt{2}} \begin{pmatrix} 1 \\ -1 \\ 0 \end{pmatrix}$$

also form an orthonormal set. Thus there can be infinite number of orthonormal sets.

Solution 6.5

$$\psi(x) = A \sin kx$$
$$= A \sin(kb)\, e^{-\kappa(x-b)}$$

Continuity of $d\psi/dx$ at $x = b$ will give us

$$-\xi \cot \xi = \sqrt{\alpha^2 - \xi^2} \tag{145}$$

where
$$\alpha^2 = \frac{2\mu V_0 b^2}{\hbar^2}$$

and
$$\xi^2 = \frac{2\mu E b^2}{\hbar^2}$$

The normalization condition will give us

$$A = \frac{1}{\sqrt{b}}\left[\frac{1}{2} - \frac{\sin 2\xi}{4\xi} + \frac{\sin^2 \xi}{2\sqrt{\alpha^2 - \xi^2}}\right]^{-1/2} \tag{146}$$

Solution 6.6 The wavefunction which would not blow up as $x \to \pm\infty$ would be given by

$$\psi = \begin{cases} Ae^{\kappa x} & \text{for} \quad x < -a/2 \\ B\sin kx + C\cos kx & \text{for} \quad -a/2 < x < a/2 \\ De^{-\kappa x} & \text{for} \quad x > a/2 \end{cases} \tag{147}$$

Continuity of ψ and $d\psi/dx$ at $x = \pm a/2$ will give us four equations from which we may obtain

$$2B \sin \eta = (D - A)e^{-\xi}, \quad 2kB \cos \eta = \kappa(-D + A)e^{-\xi}$$
$$2C \cos \eta = (D + A)e^{-\xi}, \quad 2kC \sin \eta = \kappa(D + A)e^{-\xi}$$

where $\eta = ka/2$ and $\xi = \kappa a/2$. From the top 2 equations we get

$$\tan \eta = -\frac{k}{\kappa}$$

And from the bottom 2 equations we get

$$\tan \eta = \frac{\kappa}{k}$$

Obviously both equations cannot be simultaneously valid and therefore *either* $B = 0$ and $D = A$ (the symmetric solution) and one obtains Eq. (58), or $C = 0$ and $D = -A$ (the antisymmetric solution) and one obtains Eq. (59).

Solution 6.7 For the symmetric mode

$$B = \left[\exp\left(\frac{\kappa a}{2}\right)\cos\left(\frac{ka}{2}\right)\right]A$$

and the normalization condition gives us

$$|A| = \left(\frac{2}{a}\right)^{1/2}\left[1 + \frac{\sin ka}{ka} + \frac{2\cos^2(ka/2)}{\kappa a}\right]^{-1/2}$$

Solution 6.8 For metals like Na, Cu, etc., $E_{F_0} \approx$ few eV, since $kT \approx 0.025$ eV (at $T \approx 300$ K), $E_{F_0}/kT \gg 1$. Now,

$$N = \int_0^\infty g(E)F(E)\,dE$$

$$= \frac{(2\mu)^{3/2}V}{2\pi^2\hbar^3}\left[\frac{2}{3}E^{3/2}F(E)\Big|_0^\infty - \frac{2}{3}\int_0^\infty E^{3/2}\frac{dF}{dE}dE\right]$$

$$= -\frac{(2\mu)^{3/2} V E_F^{3/2}}{3\pi^2\hbar^3} \int_{-E_F/kT}^{\infty} \left(1+\frac{xkT}{E_F}\right)^{3/2} \frac{dF}{dx}\, dx$$

where $x = (E - E_F)/kT$. Assuming $E_F/kT \gg 1$ and since dF/dx is very sharply peaked around $x = 0$, we replace the lower limit by $-\infty$, expand $(1 + xkT/E_F)^{3/2}$ in a binomial series and integrate term by term to obtain[8]

$$N = -\frac{(2\mu)^{3/2} V E_F^{3/2}}{3\pi^2\hbar^3} \left[[F(x)]_{-\infty}^{+\infty} + 0 - \frac{3}{8}\left(\frac{kT}{E_F}\right)^2 \int_{-\infty}^{+\infty} \frac{x^2 e^x}{(1+e^x)^2} + \ldots \right]$$

$$= \frac{(2\mu)^{3/2} V E_F^{3/2}}{3\pi^2\hbar^3} \left[1 + \frac{\pi^2}{8}\left(\frac{kT}{E_F}\right)^2\right]$$

or

$$E_F = E_{F_0}\left[1 + \frac{\pi^2}{8}\left(\frac{kT}{E_F}\right)^2\right]^{-2/3}$$

Thus,

$$E_F(T) \simeq E_{F_0}\left[1 - \frac{\pi^2}{12}\left(\frac{kT}{E_{F_0}}\right)^2 + \ldots\right]$$

Similarly,

$$\langle E \rangle = \frac{1}{N}\int_0^{\infty} E g(E) F(E)\, dE$$

$$\simeq \frac{(2m)^{3/2} V E_F^{3/2}}{3\pi^2\hbar^3 N} \int_{-\infty}^{+\infty} \left[1 + \frac{xkT}{E_F}\right]^{5/2} F(x)\, dx$$

from which we get

$$\langle E \rangle \simeq \frac{3}{5} E_{F_0}\left[1 + \frac{5\pi^2}{12}\left(\frac{kT}{E_{F_0}}\right)^2 + \cdots\right]$$

Solution 6.9 In the region $0 < x < \infty$, the Schrödinger equation is given by

$$\frac{d^2\psi}{dx^2} + \frac{2\mu}{\hbar^2}[E - \gamma x]\,\psi(x) = 0 \tag{148}$$

which can be written in the form

$$\frac{d^2\psi}{d\xi^2} - \xi\psi(\xi) = 0 \tag{149}$$

8

$$\int_{-\infty}^{\infty} \frac{x^2 e^x}{(1+e^x)^2}\, dx = 2\int_{-\infty}^{\infty} x^2 e^{-x}(1+e^{-x})^{-2}\, dx$$

$$= 2\left[\int_{-\infty}^{\infty} x^2 (e^{-x} - 2e^{-2x} + \ldots)\, dx\right]$$

$$= 4\left[1 - \frac{1}{2^2} + \frac{1}{3^2} - \ldots\right] = \frac{\pi^2}{3}$$

where

$$\xi = X - \varepsilon \tag{150}$$

with

$$X = \frac{\gamma}{\alpha}x, \quad \varepsilon = \frac{E}{\alpha} \quad \text{and} \quad \alpha = \left(\frac{\hbar^2\gamma^2}{2\mu}\right)^{1/3}$$

The solution of Eq. (149) is given by

$$\psi(\xi) = C_1 Ai(\xi) + C_2 Bi(\xi) \tag{151}$$

Since $Bi(\xi) \to \infty$, as $\xi \to \infty$ (see Fig. 2 of Appendix D), we must choose $C_2 = 0$. Further, since $V(x) = \infty$ for $x < 0$, we must have

$$\psi(x = 0) = 0 \tag{152}$$

implying

$$\psi(\xi = -\varepsilon) = 0 \tag{153}$$

or

$$Ai(-\varepsilon) = 0 \tag{154}$$

Now $Ai(\xi) = 0$ for

$$\xi = a_n = -2.33811, -4.08795, -5.52056, \ldots \tag{155}$$

using which we readily get Eq. (137), the corresponding eigenfunctions being

$$\psi_n(x) = (\text{const}) Ai(\xi) = (\text{const})\, Ai(X + a_n) \tag{156}$$

where a_n represent the zeros of the Airy function [see Eq. (155) and Table 1 of Appendix D]. The energy eigenvalues are shown as horizontal lines in Fig. 6.7 along with the corresponding eigenfunctions.

Solution 6.10 The Schrödinger equation is given by

$$\frac{d^2\psi}{dx^2} + \frac{2\mu}{\hbar^2}[E + S\delta(x)]\,\psi(x) = 0 \tag{157}$$

Thus for $x < 0$ and $x > 0$, $\psi(x)$ will satisfy the equation

$$\frac{d^2\psi}{dx^2} - \kappa^2\psi(x) = 0 \quad x < 0 \text{ and } x > 0 \tag{158}$$

where

$$\kappa^2 = -\frac{2\mu E}{\hbar^2} \tag{159}$$

For bound states E is negative and, therefore, κ^2 is a positive quantity. The solution of Eq. (158) which vanishes as $x \to +\infty$ and $x \to -\infty$ would be given by

$$\left.\begin{aligned} \psi(x) &= Ae^{-\kappa x} & x > 0 \\ &= Ae^{\kappa x} & x < 0 \end{aligned}\right\} \tag{160}$$

which takes into account the continuity of $\psi(x)$ at $x = 0$. Now, integrating Eq. (157) we obtain

$$\int_{-\varepsilon}^{+\varepsilon} \frac{d^2\psi}{dx^2}dx = -\frac{2\mu E}{\hbar^2}\int_{-\varepsilon}^{+\varepsilon}\psi(x)dx - \frac{2\mu S}{\hbar^2}\int_{-\varepsilon}^{+\varepsilon}\delta(x)\psi(x)\,dx$$

where ε is an infinitesimally small quantity. Since $\psi(x)$ is continuous at $x = 0$, the first term on the RHS of the above equation would vanish and we would get

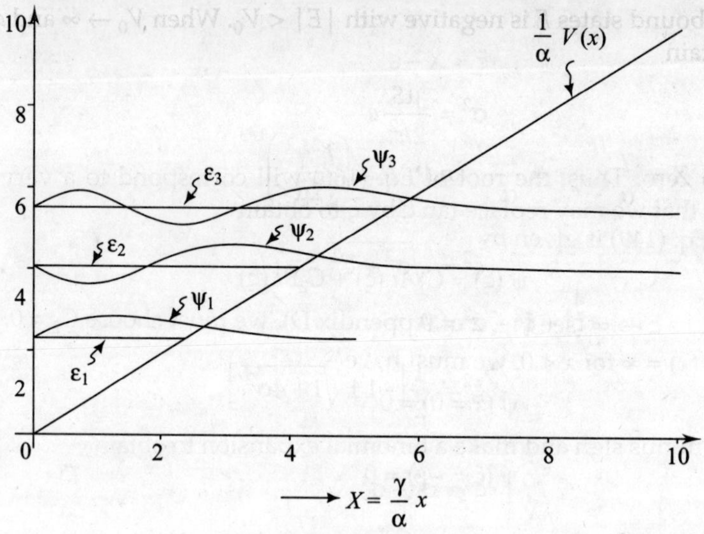

Figure 6.7: The potential energy distribution as given by Eq. (136). The first three eigenvalues and the corresponding eigenfunctions are also shown.

$$\left.\frac{d\psi}{dx}\right|_{x=\varepsilon} - \left.\frac{d\psi}{dx}\right|_{x=-\varepsilon} = -\frac{2\mu S}{\hbar^2}\psi(0) = \frac{2\mu SA}{\hbar^2} \tag{161}$$

From Eq. (160), we would get

$$\left.\frac{d\psi}{dx}\right|_{x=\varepsilon} - \left.\frac{d\psi}{dx}\right|_{x=-\varepsilon} = -2A\kappa \tag{162}$$

Thus,

$$\kappa = \frac{\mu S}{\hbar^2} \tag{163}$$

or

$$E = -\frac{\mu S^2}{2\hbar^2} \tag{164}$$

The corresponding normalized eigenfunction is given by (see also Example 1.1)

$$\psi(x) = \sqrt{\frac{\mu S}{\hbar^2}}\exp\left[-\frac{\mu S}{\hbar^2}|x|\right] \tag{165}$$

Solution 6.11 The transcendental equation determining the energy eigenvalues corresponding to symmetric states is given by

$$\xi \tan \xi = \sqrt{\sigma^2 - \xi^2} \tag{166}$$

where

$$\xi = \left[\frac{\mu}{2\hbar^2}(V_0 + E)\,a^2\right]^{1/2} \tag{167}$$

and

$$\sigma^2 = \frac{\mu}{2\hbar^2}V_0 a^2 \tag{168}$$

Notice that for bound states E is negative with $|E| < V_0$. When $V_0 \to \infty$ and $a \to 0$ such that $V_0 a \to S$ we obtain

$$\sigma^2 = \frac{\mu S}{2\hbar^2} a \tag{169}$$

which tends to zero. Thus, the root of Eq. (166) will correspond to a very small value of ξ (see Fig. 6.4) so that we may replace $\tan \xi$ by ξ to obtain

$$\xi^2 = \sqrt{\sigma^2 - \xi^2}$$

or

$$\xi^4 + \xi^2 - \sigma^2 = 0$$

or

$$\xi^2 = \frac{1}{2}\left[-1 \pm \sqrt{1 + 4\sigma^2}\right]$$

We neglect the minus sign and make a binomial expansion to obtain

$$\xi^2 \approx \sigma^2 - \sigma^4 \tag{170}$$

or

$$\frac{\mu}{2\hbar^2}[V_0 + E]a^2 \approx \frac{\mu V_0 a^2}{2\hbar^2} - \frac{\mu V_0^2 a^4}{4\hbar^4}$$

or

$$E \approx -\frac{\mu S^2}{2\hbar^2} \tag{171}$$

REFERENCES AND SUGGESTED READINGS

1. D. Halliday and R. Resnick, *Physics, Parts I & II*, John Wiley, New York (1978).
2. S. Chandrasekhar, *Introduction to the Study of Stellar Structure*, Dover Publications, New York (1957).

Linear Harmonic Oscillator I

Solution of the Schrödinger Equation and Relationship with the Classical Oscillator

In 1921 Niels Bohr wrote:[1] "As regards the general [illegible] [illegible] help of classical conceptions, are to be regarded [illegible] as laws of the quantum theory, [illegible] [illegible] the same time reproduce [illegible] the quantum theory [illegible] and, nevertheless, can be regarded as a rational generalization of classical [illegible]." [illegible] years later this consistent theory was developed. In fact there was rarely in the history of physics a comprehensive theory [illegible] Bohr's correspondence principle.

7.1 INTRODUCTION

The linear harmonic oscillator problem is one of the most fundamental problems in quantum mechanics. [illegible] and the basic features of [illegible] transition to the [illegible] studying the vibrational spectra of molecules, quantum theory of [illegible] we will first obtain solutions of the one-dimensional [illegible] the linear harmonic oscillator and then discuss the time evolution [illegible] oscillator which represent the quantum mechanical analogs of the classical oscillator.

7.2 SOLUTION OF THE TIME DEPENDENT SCHRÖDINGER EQUATION

The one-dimensional time dependent Schrödinger equation is given by

$$i\hbar\frac{\partial \Psi}{\partial t} = H\Psi(x,t)$$

where

$$H = -\frac{\hbar^2}{2\mu}\frac{d^2}{dx^2} + V(x)$$

represents the Hamiltonian and

$$V(x) = \frac{1}{2}\mu\omega^2 x^2$$

[1] N. Bohr, "On the application of the quantum theory [illegible] Philosophical Society (Supplement), Part I (Cambridge University Press, [illegible]) p. [illegible]

7 Linear Harmonic Oscillator: I

Solution of the Schrödinger Equation and Relationship with the Classical Oscillator

In 1924 Niels Bohr wrote[1]*: 'As frequently emphasized, these principles, although they are formulated by the help of classical conceptions, are to be regarded purely as laws of the quantum theory, which gives us, notwithstanding the formal nature of the quantum theory, a hope in the future of a consistent theory, which at the same time reproduces the characteristic features of the quantum theory, important for its applicability, and, nevertheless, can be regarded as a rational generalization of classical electrodynamics.' About three years later this consistent theory was accomplished.* **In fact, there was rarely in the history of physics a comprehensive theory which owed so much to one principle as quantum mechanics owed to Bohr's correspondence principle.**

— MAX JAMMER in *The Conceptual Development of Quantum Mechanics*, McGraw-Hill, New York (1966), p. 118.

7.1 INTRODUCTION

The linear harmonic oscillator problem is one of the most fascinating problems in quantum mechanics. It allows us to understand the basic features of a quantum system along with its transition to the classical domain. It has applications in many problems in physics; e.g. in studying the vibrational spectra of molecules, quantum theory of radiation, etc. In this chapter, we will first obtain solutions of the one-dimensional Schrödinger equation corresponding to the linear harmonic oscillator and then discuss the time evolution of the *coherent states* of the oscillator which represent the quantum mechanical analogs of the classical oscillator.

7.2 SOLUTION OF THE TIME DEPENDENT SCHRÖDINGER EQUATION

The one-dimensional time dependent Schrödinger equation is given by

$$i\hbar \frac{\partial \Psi}{\partial t} = H\Psi(x, t) \tag{1}$$

where

$$H = -\frac{\hbar^2}{2\mu} \frac{d^2}{dx^2} + V(x) \tag{2}$$

represents the Hamiltonian and

$$V(x) = \frac{1}{2} \mu \omega^2 x^2 \tag{3}$$

[1] N. Bohr, 'On the Application of the Quantum Theory to Atomic Structure', *Proceedings of the Cambridge Philosophical Society* (Supplement), Part I (Cambridge University Press, 1924), p. 42.

represents the potential energy of the one-dimensional harmonic oscillator[2]; in the above equation μ represents the mass of the oscillator and ω represents the *classical frequency* of the oscillator. Thus, Eq. (1) takes the form

$$i\hbar\frac{\partial \Psi}{\partial t} = \left[-\frac{\hbar^2}{2\mu}\frac{\partial^2}{\partial x^2} + \frac{1}{2}\mu\,\omega^2 x^2\right]\Psi(x,t) \quad \text{Time dependent Schrödinger equation} \tag{4}$$

which represents the time dependent Schrödinger equation for the linear harmonic oscillator problem. We try to solve the above equation by using the method of separation of variables:

$$\Psi(x,t) = \psi(x)\,T(t) \tag{5}$$

Substituting in Eq. (4) and dividing by $\psi(x)\,T(t)$, we obtain

$$\frac{i\hbar}{T(t)}\frac{dT}{dt} = \frac{1}{\psi(x)}\left[-\frac{\hbar^2}{2\mu}\frac{d^2}{dx^2} + \frac{1}{2}\mu\,\omega^2 x^2\right]\psi(x)$$

The left-hand side of the above equation is a function of t alone and the right-hand side is a function of x alone; therefore both sides must be set equal to a constant which we represent by E (the total energy of the oscillator[3])

$$\frac{i\hbar}{T(t)}\frac{dT}{dt} = \frac{1}{\psi(x)}\left[-\frac{\hbar^2}{2\mu}\frac{d^2}{dx^2} + \frac{1}{2}\mu\,\omega^2 x^2\right]\psi(x) = E \tag{6}$$

The solution of the equation

$$\frac{i\hbar}{T(t)}\frac{dT}{dt} = E \tag{7}$$

is given by

$$T(t) = (\text{constant})\exp\left[-\frac{i\,E\,t}{\hbar}\right] \tag{8}$$

Equation (6) also gives us

---

[2] The restoring force of a linear oscillator is given by

$$F = -kx$$

where $k(= \mu\omega^2)$ represents the force constant and x represents the displacement from the equilibrium position. The potential energy is the work done to bring it to the position x from the equilibrium position $x = 0$ and is given by

$$V(x) = \int_0^x F\,dx = \frac{1}{2}kx^2$$

or

$$V(x) = \frac{1}{2}\mu\omega^2 x^2$$

[3] As mentioned in Chapter 4

$$H = \frac{p^2}{2\mu} + V(x)$$

$$= -\frac{\hbar^2}{2\mu}\frac{\partial^2}{\partial x^2} + V(x)$$

represents the operator corresponding to the total energy of the system and is known as the Hamiltonian of the system. The eigenvalues E_n [see Eq. (28)] are the allowed values of the total energy.

$$H\psi = \left[-\frac{\hbar^2}{2\mu}\cdot\frac{d^2}{dx^2}+\frac{1}{2}\mu\omega^2x^2\right]\psi(x) = E\,\psi(x) \tag{9}$$

which represents an eigenvalue equation; i.e. (as we will show later) only for some specific values of E will we have 'well-behaved' solutions of Eq. (9). We rewrite Eq. (9) in the form

$$\frac{d^2\psi}{dx^2}+\frac{2\mu}{\hbar^2}\left[E-\frac{1}{2}\mu\omega^2x^2\right]\psi(x) = 0 \tag{10}$$

We may mention that the (time independent) Schrödinger equation is often written in the form

$$H\psi(x) = E\psi(x) \tag{11}$$

where H is the operator corresponding to the Hamiltonian of the system [see Eq. (2)] and the eigenvalue E is simply a number.

In order to solve Eq. (10) we introduce the variable

$$\xi = \gamma x$$

where the parameter γ will be chosen so that Eq. (10) reduces to a convenient form

Now,
$$\frac{d\psi}{dx} = \frac{d\psi}{d\xi}\frac{d\xi}{dx} = \gamma\frac{d\psi}{d\xi}$$

and
$$\frac{d^2\psi}{dx^2} = \gamma^2\frac{d^2\psi}{d\xi^2}$$

Thus, Eq. (10) becomes

$$\frac{d^2\psi}{d\xi^2}+\left[\frac{2\mu E}{\hbar^2\gamma^2}-\frac{\mu^2\omega^2}{\hbar^2}\cdot\frac{\xi^2}{\gamma^4}\right]\psi(\xi) = 0$$

We choose γ such that the coefficient of ξ^2 is unity, thus

$$\gamma = \left(\frac{\mu\omega}{\hbar}\right)^{\frac{1}{2}} \tag{12}$$

Thus, the Schrödinger equation takes the form

$$\frac{d^2\psi}{d\xi^2}+(\Lambda-\xi^2)\,\psi(\xi) = 0 \tag{13}$$

where
$$\Lambda = \frac{2\mu E}{\hbar^2\gamma^2} = \frac{2E}{\hbar\omega} \tag{14}$$

In order to solve Eq. (13), we make the transformation[4]

---

[4] The transformation given by Eq. (15) is suggested from the fact that in the region $\xi^2 \gg \Lambda$, the function

$$\psi(\xi) \sim \xi^n \exp\left(\pm\frac{1}{2}\xi^2\right)$$

indeed satisfies Eq. (13) because

$$\frac{d^2\psi}{d\xi^2} = \left[1\pm\frac{2n+1}{\xi^2}+\frac{n(n-1)}{\xi^4}\right]\xi^{n+2}\exp\left(\pm\frac{1}{2}\xi^2\right) \to \xi^2\psi(\xi)$$

for large ξ. This suggests that it might be possible to find an exact solution of Eq. (13) of the form given by Eq. (15). We reject the solution with positive exponent because then the solution would diverge as $\xi \to \pm\infty$.

$$\psi(\xi) = u(\xi)\, e^{-\frac{1}{2}\xi^2} \tag{15}$$

Thus,
$$\frac{d\psi}{d\xi} = \left[\frac{du}{d\xi} - \xi\, u(\xi)\right] e^{-\frac{1}{2}\xi^2}$$

and
$$\frac{d^2\psi}{d\xi^2} = \left[\frac{d^2u}{d\xi^2} - 2\xi \frac{du}{d\xi} + (\xi^2 - 1)\, u(\xi)\right] e^{-\frac{1}{2}\xi^2}$$

Substituting in Eq. (13) we get

$$\frac{d^2u}{d\xi^2} - 2\xi \frac{du}{d\xi} + (\Lambda - 1)\, u(\xi) = 0 \tag{16}$$

which is known as the Hermite equation. We will solve the above equation by assuming a power series solution

$$u(\xi) = \sum_{r=0,1,2,\ldots}^{\infty} c_r\, \xi^{s+r} \tag{17}$$

Substituting in Eq. (16), we get

$$\sum_{r=0,1,2,\ldots}^{\infty} c_r\,(s+r)(s+r-1)\,\xi^{s+r-2}$$

$$-2 \sum_{r=0,1,2,\ldots}^{\infty} c_r\,(s+r)\,\xi^{s+r} + (\Lambda - 1) \sum_{0,1,2,\ldots} c_r\, \xi^{s+r} = 0$$

or
$$\sum_{r=0,1,2,\ldots}^{\infty} c_r\,(s+r)(s+r-1)\xi^r$$

$$-\sum_{r=0,1,2,\ldots}^{\infty} c_r\,(2s + 2r - \Lambda + 1)\,\xi^{r+2} = 0$$

or
$$c_0\, s\,(s-1) + c_1\,(s+1)\, s\, \xi$$

$$+\sum_{r=2,3,\ldots}^{\infty} [c_r\,(s+r)(s+r-1) - c_{r-2}\,(2s + 2r - 3 - \Lambda)]\,\xi^r = 0.$$

Since, the above equation has to be valid for all values of ξ, the coefficients of various powers of ξ must be equal to zero giving

$$s\,(s-1) = 0 \tag{18}$$

$$c_1\, s\,(s+1) = 0 \tag{19}$$

$$c_r = \frac{2s + 2r - 3 - \Lambda}{(s+r)(s+r-1)}\, c_{r-2}\,;\ r \ge 2 \tag{20}$$

Equation (18) is known as the Indicial Equation and we must have

$$s = 0 \quad \text{or} \quad 1$$

The root $s = 0$ makes c_1 indeterminate and therefore the root $s = 0$ would determine both the independent solutions (see also Problem 7.6). Substituting $s = 0$ in Eq. (20), we get

$$c_r = \frac{2r - 3 - \Lambda}{r(r-1)} c_{r-2}, \quad r \geq 2 \tag{21}$$

which gives

$$c_2 = \frac{1-\Lambda}{2!} c_0, \qquad c_3 = \frac{3-\Lambda}{3!} c_1$$
$$c_4 = \frac{(1-\Lambda)(5-\Lambda)}{4!} c_0, \qquad c_5 = \frac{(3-\Lambda)(7-\Lambda)}{5!} c_1, \text{ etc.} \tag{22}$$

Because $c_2, c_4, \ldots$ are related to c_0; and $c_3, c_5, \ldots$ are related to c_1, we can split the solution in even and odd series. Thus, we may write

$$u(\xi) = [c_0 + c_2 \xi^2 + c_4 \xi^4 + \cdots] + [c_1 \xi + c_3 \xi^3 + \cdots]$$

or

$$u(\xi) = c_0 \left[1 + \frac{1-\Lambda}{2!}\xi^2 + \frac{(1-\Lambda)(5-\Lambda)}{4!}\xi^4 + \cdots\right]$$
$$+ c_1 \left[\xi + \frac{3-\Lambda}{3!}\xi^3 + \frac{(3-\Lambda)(7-\Lambda)}{5!}\xi^5 + \cdots\right] \tag{23}$$

Since the values of c_0 and c_1 can be arbitrary we have obtained two independent solutions of Eq. (16) and therefore the above equation represents the most general solution of Eq. (16). It may be readily seen that when

$$\Lambda = 1, 5, 9, \ldots$$

the even series becomes a polynomial and the odd series remains an infinite series. Similarly, for

$$\Lambda = 3, 7, 11, \ldots$$

the odd series becomes a polynomial and the even series remains an infinite series. Thus, when

$$\Lambda = 2n + 1; \quad n = 0, 1, 2, \ldots$$

one of the solutions becomes a polynomial. Returning to Eq. (21), we have

$$\frac{c_r}{c_{r-2}} = \frac{2r - 3 - \Lambda}{r(r-1)} \to \frac{2}{r} \quad \text{for larger } r \tag{24}$$

Thus,

$$\lim_{r \to \infty} \frac{c_r}{c_{r-2}} = 0$$

which shows that both the infinite series in Eq. (23) will converge for all values of ξ. We next consider the expansion of e^{ξ^2}:

$$e^{\xi^2} = \sum_{r=0,2,4,\ldots}^{\infty} b_r \xi^r; \quad b_r = \frac{1}{\left(\frac{r}{2}\right)!}$$

implying

$$\frac{b_r}{b_{r-2}} = \frac{\left(\frac{r}{2} - 1\right)!}{\frac{r}{2}!} = \frac{2}{r} \tag{25}$$

Thus the ratio c_r/c_{r-2} (in the limit of large r) is the same as that of the coefficients of ξ^r and ξ^{r+2} in the expansion of e^{ξ^2}. Consequently, if the series in Eq. (23) is not terminated, $u(\xi)$ will behave as e^{ξ^2} for large[5] ξ; therefore $\psi(\xi)$ will behave as $e^{\frac{1}{2}\xi^2}$ [see Eq. (15)]. Thus, the boundary condition

$$\psi(\xi) \to 0 \quad \text{as} \quad \xi \to \pm\infty \tag{26}$$

will not be satisfied. For the wavefunction to satisfy the boundary condition given by Eq. (26), the infinite series must be terminated to a polynomial and as mentioned earlier, this can happen only when Λ is an odd integer, i.e., when

$$\Lambda = 2n + 1; \quad n = 0, 1, 2, \ldots \tag{27}$$

Substituting in Eq. (14), we get

$$E = E_n = \left(n + \frac{1}{2}\right)\hbar\omega; \quad n = 0, 1, 2, \ldots \qquad \text{Energy eigenvalues} \tag{28}$$

which represent the (discrete) energy eigenvalues of the operator H.

7.3 THE EIGENFUNCTIONS

In order to determine the corresponding eigenfunctions we rewrite Eq. (23):

$$u(\xi) = c_0\left[1 + \frac{1-\Lambda}{2!}\xi^2 + \frac{(1-\Lambda)(5-\Lambda)}{4!}\xi^4 + \cdots\right]$$

$$+ c_1\left[\xi + \frac{3-\Lambda}{3!}\xi^3 + \frac{(3-\Lambda)(7-\Lambda)}{5!}\xi^5 + \cdots\right] \tag{29}$$

For example, for $\Lambda = 9$ ($n = 4$), the even series becomes a polynomial and the odd series remains an infinite series. We must therefore choose $c_1 = 0$; the polynomial solution is given by

$$u(\xi) = c_0\left[1 - 4\xi^2 + \frac{4}{3}\xi^4\right]$$

Similarly, for $\Lambda = 7$ ($n = 3$), the odd series becomes a polynomial and the even series remains an infinite series. We must therefore choose $c_0 = 0$; the polynomial solution is given by

$$u(\xi) = c_1\left[\xi - \frac{2}{3}\xi^3\right]$$

Now, if the multiplication constant c_0 or c_1 is chosen such that the coefficient of the highest power of ξ in the polynomial becomes 2^n, then these polynomials are known as Hermite polynomials which are denoted by $H_n(\xi)$. For example, for $\Lambda = 9 (n = 4)$, if we choose

$$c_0 = 12$$

the coefficient of ξ^4 becomes 2^4 and therefore,

$$H_4(\xi) = 16\xi^4 - 48\xi^2 + 12$$

Similarly, for $\Lambda = 7$ ($n = 3$), we choose

$$c_1 = -12$$

giving

$$H_3(\xi) = 8\xi^3 - 12\xi$$

---

[5] This is because of the fact that only for large values of ξ will the large r terms dominate in the expansion and only then will the infinite series behave as $\exp(\xi^2)$.

The first few Hermite polynomials are given by

$$\left.\begin{aligned} H_0(\xi) &= 1; \qquad H_1(\xi) = 2\xi \\ H_2(\xi) &= 4\xi^2 - 2; \qquad H_3(\xi) = 8\xi^3 - 12\xi \end{aligned}\right\} \text{Hermite polynomials} \tag{30}$$

Higher order Hermite polynomials can be determined from the following recurrence relation

$$H_{n+1}(\xi) = 2\xi\, H_n(\xi) - 2n\, H_{n-1}(\xi) \tag{31}$$

Returning to Eq. (15) we therefore have

$$\psi_n(x) = N_n H_n(\xi)\, e^{-\frac{1}{2}\xi^2}\ ; \quad n = 0, 1, 2, \ldots$$

where N_n represents a multiplicative constant. If we impose the normalization condition

$$\int_{-\infty}^{+\infty} |\psi_n(x)|^2\, dx = 1$$

we would get (see Appendix *E*)

$$N_n = \left[\frac{\gamma}{2^n\, n!\sqrt{\pi}}\right]^{\frac{1}{2}}$$

which is usually referred to as the normalization constant. The normalized eigenfunctions are given by (see also Section 7.8):

$$\psi_n(x) = \left[\frac{\gamma}{2^n\, n!\sqrt{\pi}}\right]^{\frac{1}{2}} H_n(\xi)\, e^{-\frac{1}{2}\xi^2}\ ; \quad n = 0, 1, 2, \ldots \tag{32}$$

The first four normalized eigenfunctions are given by

$$\left.\begin{aligned} \psi_0(x) &= \left(\frac{\gamma}{\sqrt{\pi}}\right)^{\frac{1}{2}} \exp\left[-\frac{1}{2}\gamma^2 x^2\right] \\ \psi_1(x) &= \left(\frac{\gamma}{2\sqrt{\pi}}\right)^{\frac{1}{2}} (2\gamma x) \exp\left[-\frac{1}{2}\gamma^2 x^2\right] \\ \psi_2(x) &= \left(\frac{\gamma}{8\sqrt{\pi}}\right)^{\frac{1}{2}} (4\gamma^2 x^2 - 2) \exp\left[-\frac{1}{2}\gamma^2 x^2\right] \\ \psi_3(x) &= \left(\frac{\gamma}{48\sqrt{\pi}}\right)^{\frac{1}{2}} (8\gamma^3 x^3 - 12\gamma x) \exp\left[-\frac{1}{2}\gamma^2 x^2\right] \end{aligned}\right\} \text{Normalized eigenfunctions} \tag{33}$$

and are plotted in Fig. 7.1. The horizontal lines correspond to the eigenvalues

$$E = \frac{1}{2}\hbar\omega, \quad \frac{3}{2}\hbar\omega, \quad \frac{5}{2}\hbar\omega, \quad \text{and} \quad \frac{7}{2}\hbar\omega \qquad \text{Eigenvalues}$$

Normalized Harmonic Oscillator Wave Functions for $n = 0, 1, 2, 3$

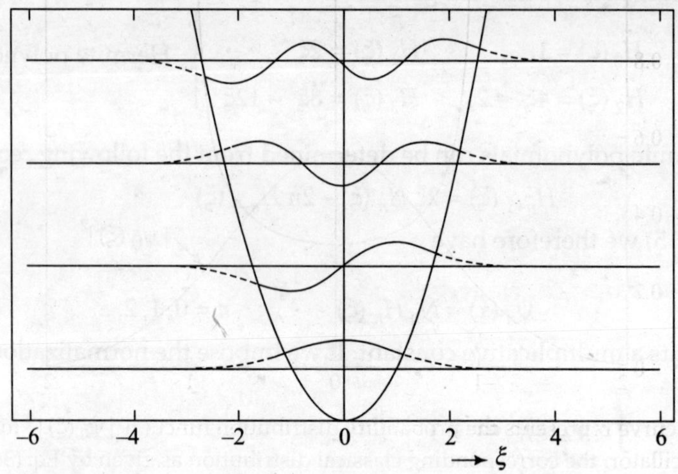

Figure 7.1: The harmonic oscillator potential and the first four eigenfunctions. The horizontal lines correspond to the eigenvalues $\frac{1}{2}\hbar\omega$, $\frac{3}{2}\hbar\omega$, $\frac{5}{2}\hbar\omega$ and $\frac{7}{2}\hbar\omega$. The dashed portions of the curves correspond to the classically forbidden regions.

For a given value of E the classically allowed region of the oscillator corresponds to

$$|x| \leq x_0$$

where
$$x_0 = \left[\frac{2E}{\mu\omega^2}\right]^{\frac{1}{2}} \tag{34}$$

represents the classical turning point of the oscillator. The corresponding value of ξ will be given by

$$\xi_0 = \gamma x_0 = \left(\frac{\mu\omega}{\hbar}\right)^{\frac{1}{2}} \left(\frac{2E}{\mu\omega^2}\right)^{\frac{1}{2}}$$

or
$$\xi_0 = \sqrt{(2n+1)}\,; \quad n = 0, 1, 2, \ldots \tag{35}$$

where we have used Eq. (28). Thus, the classically allowed region would be given by

$$|\xi| < \sqrt{2n+1}\,; \quad n = 0, 1, 2, \ldots$$

In Fig. 7.1, the dashed portions of the curves correspond to the classically forbidden regions and as can be seen, this is consistent with the above equation.

In Figs 7.2 and 7.3, we have plotted $|\psi_0(x)|^2$ and $|\psi_{12}(x)|^2$ as a function of ξ. Obviously, if the oscillator is in the nth state, $|\psi_n(x)|^2\, dx$ would represent the probability of finding the particle between x and $x + dx$. The corresponding classical distribution function is given by (see Problem 7.3)

$$P_{cl}(x) = \frac{1}{\pi\sqrt{x_0^2 - x^2}} \qquad |x| < x_0 \qquad \text{Classical distribution function} \tag{36}$$

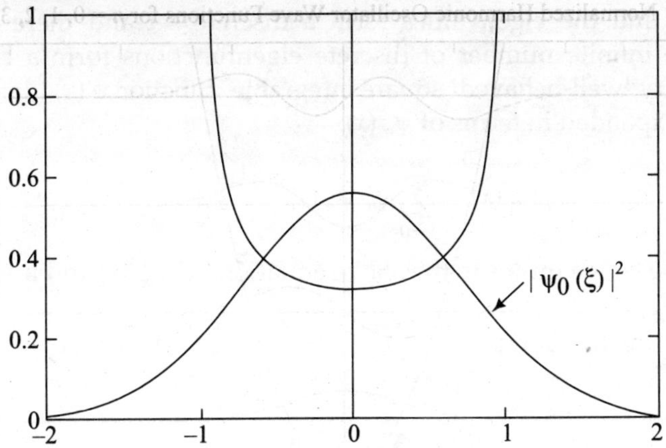

Figure 7.2: The dark curve represents the probability distribution function $|\psi_0(\xi)|^2$ for the ground state of the harmonic oscillator; the corresponding classical distribution as given by Eq. (36) is also shown.

$P_{cl}(x)\,dx$ represents the probability of finding the particle between x and $x + dx$ and represents the fractional time that it spends in the interval dx. Although $P_{cl}(x)$ blows up at $|x| = x_0$,

$$\int_{-x_0}^{+x_0} P_{cl}(x)\,dx = 1 \tag{37}$$

In Figs 7.2 and 7.3, we have shown the probability distributions as predicted by the classical and quantum mechanical theories. For large values of n, $|\psi_n(x)|^2$ undergoes very rapid oscillations and the average value is in good agreement with the corresponding classical value. Obviously, larger the value of n better will be the agreement—this is what is predicted by Böhr's Correspondence Principle.

The wavefunctions given by Eq. (32) form an orthonormal set (see Section 6.4):

$$\int_{-\infty}^{+\infty} \psi_m^*(x)\,\psi_n(x)\,dx = \delta_{mn} = \begin{cases} 0, & m \neq n \\ 1, & m = n \end{cases} \qquad \text{Orthonormality condition} \tag{38}$$

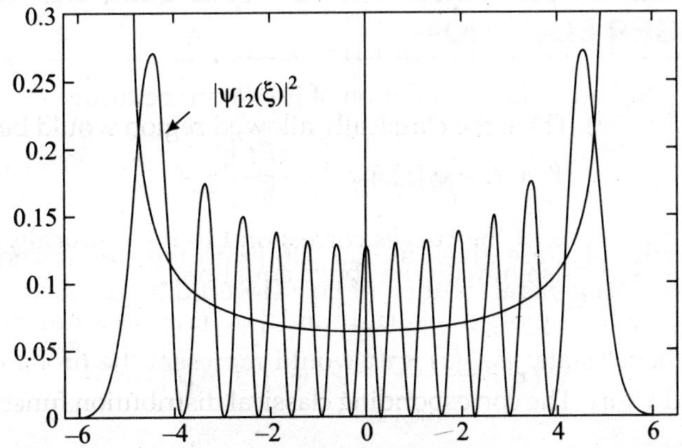

Figure 7.3: The dark curve represents the probability distribution function $|\psi_{12}(\xi)|^2$ for the $n = 12$ state of the harmonic oscillator; the average of this distribution closely resembles the classical distribution given by Eq. (36) which has also been shown.

The eigenfunctions and the eigenvalues form a discrete set and there are no continuum eigenfunctions. The infinite number of discrete eigenfunctions form a complete set in the sense that an arbitrary 'well-behaved' square integrable[6] function $\phi(x)$ (defined in the interval $-\infty < x < \infty$) can be expanded in terms of $\psi_n(x)$:

$$\phi(x) = \sum_{n=0}^{\infty} c_n \psi_n(x) \tag{39}$$

In order to determine c_n, we multiply the above equation by $\psi_m^*(x)$ and integrate to obtain

$$\int_{-\infty}^{+\infty} \psi_m^*(x)\,\phi(x)\,dx = \sum_{n=0,1,2,\ldots}^{\infty} c_n \int_{-\infty}^{+\infty} \psi_m^*(x)\,\psi_n(x)\,dx$$

$$= \sum_{n=0,1,2,\ldots}^{\infty} c_n\,\delta_{mn}$$

$$= c_m$$

Thus,

$$c_n = \int_{-\infty}^{+\infty} \psi_n^*(x)\,\phi(x)\,dx \tag{40}$$

If we substitute the above expression for c_n in Eq. (39) we would obtain

$$\phi(x) = \int_{-\infty}^{+\infty} \phi(x') \left[\sum_n \psi_n^*(x')\,\psi_n(x)\right] dx'$$

Thus (see Section 1.5):

$$\sum_{n=0}^{\infty} \psi_n^*(x')\,\psi_n(x) = \delta(x - x') \qquad \text{Completeness condition} \tag{41}$$

which represents the completeness condition of the Hermite-Gauss functions.

7.4 THE MOST GENERAL SOLUTION OF THE TIME DEPENDENT SCHRÖDINGER EQUATION

Using Eqs (5) and (8), we can write the solution of Eq. (4) in the form

$$\Psi(x, t) = \psi(x) \exp\left[-\frac{iEt}{\hbar}\right]$$

Since the allowed values of E are given by Eq. (28) [with the corresponding eigenfunctions given by Eq. (32)], the most general solution of Eq. (4) would be given by

$$\Psi(x, t) = \sum_{n=0,1,2,\ldots}^{\infty} c_n\,\psi_n(x) \exp\left[-\frac{iE_n t}{\hbar}\right] \tag{42}$$

[6] By square integrable we imply

$$\int_{-\infty}^{+\infty} |\phi(x)|^2\,dx \quad \text{should be finite}$$

or
$$\Psi(x,t) = \sum_{n=0,1,2,\ldots}^{\infty} c_n \psi_n(x)\, e^{-i\left(n+\frac{1}{2}\right)\omega t} \tag{43}$$

Determination of $\Psi(x, t)$

Thus if we know $\Psi(x, 0)$, i.e., if we know the state of the oscillator at $t = 0$, $\Psi(x, t)$ can be determined as follows:

(i) Since,
$$\Psi(x,0) = \sum_{n=0,1,2,\ldots}^{\infty} c_n \psi_n(x)$$

we get
$$c_n = \int_{-\infty}^{+\infty} \psi_n^*(x)\, \Psi(x,0)\, dx \tag{44}$$

(ii) We substitute the calculated value of c_n in Eq. (43) and sum the series to obtain[7] $\Psi(x, t)$. This will describe the time evolution of the wave packet. Obviously, if
$$\int_{-\infty}^{+\infty} |\Psi(x,0)|^2\, dx = 1 \tag{45}$$

then,
$$\sum_n |c_n|^2 = 1 \tag{46}$$

implying (see Problem 7.9)
$$\int_{-\infty}^{+\infty} |\Psi(x,t)|^2\, dx = 1 \tag{47}$$

The quantity $|c_n|^2$ can be interpreted to represent the probability of finding the oscillator in the nth state which does not change with time. As can be seen from Eq. (43), the wavefunctions superpose with different phases at different times.

As a simple example, let
$$\Psi(x,0) = \frac{1}{\sqrt{3}}\, \psi_0(x) + i\sqrt{\frac{2}{3}}\, \psi_2(x) \tag{48}$$

Obviously,
$$\int_{-\infty}^{+\infty} |\Psi(x,0)|^2\, dx = 1$$

and the time evolution of the wavefunction would be given by
$$\Psi(x,t) = \frac{1}{\sqrt{3}}\, \psi_0(x)\, e^{-i\omega t/2} + i\sqrt{\frac{2}{3}}\, \psi_2(x)\, e^{-5i\omega t/2} \tag{49}$$

---

[7] $\Psi(x, t)$ will describe the time evolution of the wave packet. The method is quite similar to the one we used for a free particle (see Section 5.3).

If we make a measurement of energy, there will be $\frac{1}{3}$rd probability of finding it in the ground state (with $E = \frac{1}{2}\hbar\omega$) and $\frac{2}{3}$rd probability of finding it in the second excited state (with $E = \frac{5}{2}\hbar\omega$). These probabilities do not change with time; however, once the measurement is made, the oscillator will jump to one of the eigenstates of the system.

For the oscillator described by Eq. (48), the average energy will be given by (which will never be measured!):

$$\langle E \rangle = \frac{1}{3}\left(\frac{1}{2}\hbar\omega\right) + \frac{2}{3}\left(\frac{5}{2}\hbar\omega\right)$$

$$= \frac{11}{6}\hbar\omega$$

7.5 THE COHERENT STATE

In this section, we will define the coherent state and in the following section we will consider its time evolution. We first define the *annihilation operator*[8] *a*:

$$a = \frac{\mu\omega x + ip}{\sqrt{2\mu\hbar\omega}} \qquad \text{Annihilation operator} \tag{50}$$

Replacing p by $-i\hbar d/dx$, we get

$$a = \sqrt{\frac{\mu\omega}{2\hbar}}\, x + \sqrt{\frac{\hbar}{2\mu\omega}} \cdot \frac{d}{dx} \tag{51}$$

We next write the eigenvalue equation for the operator a:

$$a\psi_\alpha(x) = \alpha\psi_\alpha(x) \tag{52}$$

where α represents the eigenvalue of the operator a. The eigenfunctions $\psi_\alpha(x)$ describe what are known as *coherent states*. In order to solve the eigenvalue equation Eq. (52), we substitute for a from Eq. (51) to obtain

$$\sqrt{\frac{\mu\omega}{2\hbar}}\, x\,\psi_\alpha(x) + \sqrt{\frac{\hbar}{2\mu\omega}} \cdot \frac{d\psi_\alpha}{dx} = \alpha\psi_\alpha(x)$$

or

$$\frac{1}{\psi_\alpha(x)} \cdot \frac{d\psi_\alpha(x)}{dx} = -\frac{\mu\omega}{\hbar}x + \alpha\sqrt{\frac{2\mu\omega}{\hbar}}$$

Carrying out the integration, we get

$$\ln\psi_\alpha(x) = -\frac{\mu\omega}{2\hbar}x^2 + \alpha\sqrt{\frac{2\mu\omega}{\hbar}}\,x + \text{constant}$$

$$= -\frac{1}{2}(\xi - \xi_0)^2 + \text{constant}$$

[8] The properties of creation and annihilation operators a and $\bar{a}$ are discussed in Chapter 12 while solving the harmonic oscillator problem using Dirac's bra and ket algebra.

where $$\xi = \gamma x = \sqrt{\frac{\mu\omega}{\hbar}}\, x$$

and $$\xi_0 = \sqrt{2}\,\alpha$$

Thus, $$\psi_\alpha(x) = \text{const}\, e^{-\frac{1}{2}(\xi-\xi_0)^2}$$

or $$\psi_\alpha(x) = \left(\frac{\gamma}{\sqrt{\pi}}\right)^{\frac{1}{2}} e^{-\frac{1}{2}(\xi-\xi_0)^2} \tag{53}$$

where the constant factor has been chosen such that the wavefunction is normalized. Obviously, the ground state wavefunction $\psi_0(x)$ is a coherent state corresponding to $\xi_0 = 0$.

7.6 TIME EVOLUTION OF A COHERENT STATE: RELATIONSHIP WITH THE CLASSICAL OSCILLATOR

Let us assume that at $t = 0$, the oscillator is in the coherent state, i.e.,

$$\Psi(x, 0) = \psi_\alpha(x) = \left(\frac{\gamma}{\sqrt{\pi}}\right)^{\frac{1}{2}} e^{-\frac{1}{2}(\xi-\xi_0)^2} \tag{54}$$

We wish to study its time evolution; i.e. if $\Psi(x, 0)$ is given by the above equation then what is $\Psi(x, t)$? Since for the harmonic oscillator problem

$$\Psi(x, t) = \sum_{n=0,1,2,\ldots}^{\infty} c_n \psi_n(x)\, e^{-i\left(n+\frac{1}{2}\right)\omega t} \tag{55}$$

we have $$\Psi(x, 0) = \sum_{n=0,1,2,\ldots}^{\infty} c_n \psi_n(x) \tag{56}$$

Thus [see Eq. (40)]

$$c_n = \int_{-\infty}^{+\infty} \psi_n^*(x)\, \Psi(x, 0)\, dx$$

$$= \left(\frac{\gamma^2}{2^n n!\pi}\right)^{\frac{1}{2}} \int_{-\infty}^{+\infty} \left[H_n(\xi) e^{-\frac{1}{2}\xi^2}\right] e^{-\frac{1}{2}(\xi-\xi_0)^2}\, dx$$

$$= \frac{1}{\sqrt{n!}}\left(\frac{1}{2}\xi_0^2\right)^{n/2} \exp\left(-\frac{1}{4}\xi_0^2\right) \tag{57}$$

where the explicit evaluation of the integral has been carried out in Appendix E. Substituting for c_n in Eq. (55), we get

$$\Psi(x, t) = \sum_{n=0,1,2,\ldots} \left[\frac{1}{\sqrt{n!}}\left(\frac{1}{2}\xi_0^2\right)^{n/2} \exp\left(-\frac{1}{4}\xi_0^2\right)\right] \times \left[N_n H_n(\xi) e^{-\frac{1}{2}\xi^2}\right] e^{-i\left(n+\frac{1}{2}\right)\omega t} \tag{58}$$

Indeed one can obtain an analytical expression for the sum; the final result is (the details are given in Appendix E):

$$\Psi(x,t) = \frac{\gamma^{\frac{1}{2}}}{\pi^{\frac{1}{4}}} \exp\left[-\frac{1}{2}(\xi - \xi_0 \cos\omega t)^2 - i\left(\frac{1}{2}\omega t + \xi\xi_0 \sin\omega t - \frac{1}{4}\xi_0^2 \sin 2\omega t\right)\right] \tag{59}$$

The probability distribution is, therefore, given by

$$|\Psi(x,t)|^2 = \frac{\gamma}{\sqrt{\pi}} \exp\left[-(\xi - \xi_0 \cos\omega t)^2\right] \tag{60}$$

Notice that

$$\int_{-\infty}^{+\infty} |\Psi(x,t)|^2 \, dx = 1 \tag{61}$$

for all values of t, as it indeed should be. The time evolution of the coherent state is described by Eq. (59)

In Figs 7.4 and 7.5, we have plotted $|\Psi(x,t)|^2$ as a function of ξ for different values of time and as can be seen, the centre of the wave packet executes simple harmonic motion like a classical harmonic oscillator. The average energy of the oscillator is given by

$$\langle E \rangle = \left[\langle n \rangle + \frac{1}{2}\right]\hbar\omega$$

Thus a large (average) energy of the oscillator will imply a large value of $\langle n \rangle$. Later in this section we will show that

$$\langle n \rangle = \sum_{n=1,2,\ldots} n|c_n|^2 = \frac{1}{2}\xi_0^{\,2}$$

Figures 7.4 and 7.5 correspond to

$$\langle n \rangle = 32 \;(\xi_0 = 8)$$

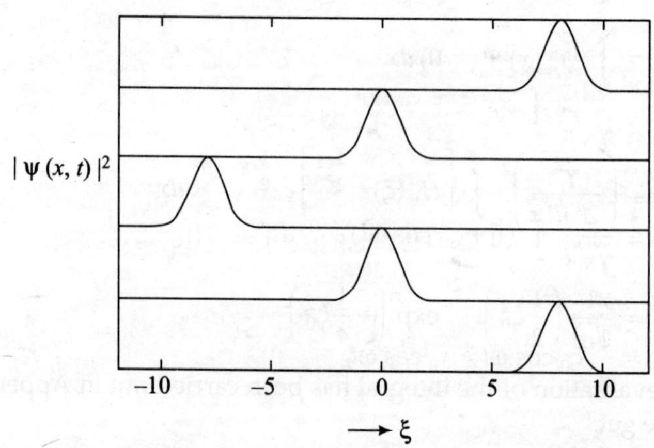

Figure 7.4: The ξ-dependence of $|\Psi(x,t)|^2$ for $\xi_0 = 8$ ($\langle n \rangle = 32$) at $\omega t = 0, \pi/2, \pi, 3\pi/2, 2\pi$ and $5\pi/2$. Notice that the centre of the wave packet executes simple harmonic motion.

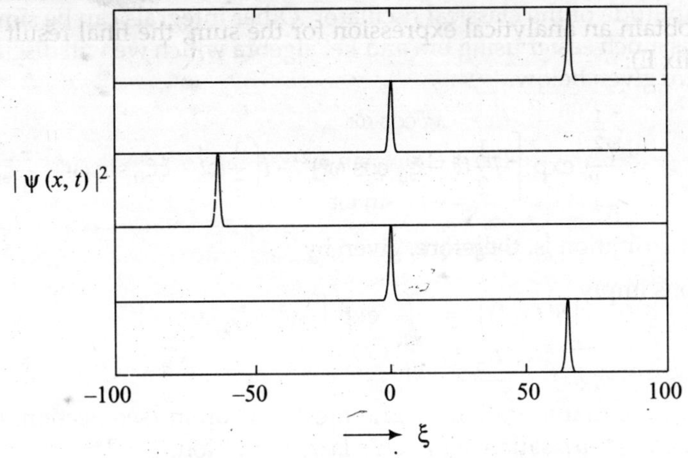

Figure 7.5: The ξ-dependence of $|\Psi(x, t)|^2$ for $\xi_0 = 64$ ($\langle n \rangle = 2048$) at $\omega t = 0, \pi/2, \pi, 3\pi/2, 2\pi$ and $5\pi/2$. Notice that the centre of the wave packet executes simple harmonic motion.

and
$$\langle n \rangle = 2048 \quad (\xi_0 = 64)$$

respectively. Obviously, as we increase the value of $\langle n \rangle$, the wave packet becomes more and more localized. We may mention that for a classical oscillator with $\mu = 0.1$ g, $\omega = 1\ \text{s}^{-1}$, $x_0 = 1$ cm

$$\xi_0 = \sqrt{\frac{\mu\omega}{\hbar}}\, x_0 \approx 10^{13}$$

so that
$$\langle n \rangle = 5 \times 10^{25}$$

which is an enormously large number. The particle would then be very localized and a classical description would be very accurate.

Using Eq. (58) one can readily calculate the various expectation values, e.g.,

$$\langle x \rangle = \int_{-\infty}^{+\infty} \Psi^*(x, t)\, x\, \Psi(x, t)\, dx$$

$$= \frac{1}{\gamma\sqrt{\pi}} \int_{-\infty}^{+\infty} \xi e^{-(\xi - \xi_0 \cos \omega t)^2}\, d\xi$$

$$= \frac{1}{\gamma\sqrt{\pi}} \int_{-\infty}^{+\infty} (\eta + \xi_0 \cos \omega t)\, e^{-\eta^2}\, d\eta \qquad [\eta \equiv \xi - \xi_0 \cos \omega t]$$

$$= \frac{1}{\gamma}\, \xi_0 \cos \omega t = x_0 \cos \omega t$$

where
$$x_0 = \frac{\xi}{\gamma_0}$$

represents the amplitude of the classical oscillator. Other integrals can be similarly evaluated, but the evalution is much easier using bra and ket algebra which we will discuss in Chapter 12. The final results are given below.

$$\langle x \rangle = x_0 \cos \omega t \tag{62}$$

$$\langle p \rangle = -\mu\omega x_0 \sin \omega t$$
$$= -p_0 \sin \omega t \tag{63}$$

where $$p_0 = \mu\omega x_0$$

The above equations imply

$$\langle p \rangle = \mu \frac{d}{dt} \langle x \rangle \tag{64}$$

This is nothing but the manifestation of Ehrenfest's theorem (see Section 4.5) according to which the expectation values satisfy Newton's laws of motion.

From Eqs (62) and (63) we find that $\langle x \rangle$ lags $\langle p \rangle$ by a phase $\frac{\pi}{2}$ just as we have for a classical oscillator. Furthermore (see Problem 12.6),

$$\langle x^2 \rangle = x_0^2 \cos^2 \omega t + \frac{1}{2\gamma^2} \tag{65}$$

$$\langle p^2 \rangle = p_0^2 \sin^2 \omega t + \frac{\gamma^2 \hbar^2}{2} \tag{66}$$

Thus,
$$\Delta x = \left[\langle x^2 \rangle - \langle x \rangle^2\right]^{\frac{1}{2}} = \frac{1}{\sqrt{2}\gamma} = \left(\frac{\hbar}{2\mu\omega}\right)^{\frac{1}{2}} \tag{67}$$

$$\Delta p = \left[\langle p^2 \rangle - \langle p \rangle^2\right]^{\frac{1}{2}} = \frac{\gamma\hbar}{\sqrt{2}} = \left(\frac{\mu\omega\hbar}{2}\right)^{\frac{1}{2}} \tag{68}$$

giving
$$\Delta x\, \Delta p = \frac{1}{2}\hbar \tag{69}$$

which represents the minimum uncertainty product[9]. Further,

$$\langle E \rangle = \frac{1}{2\mu} \langle p^2 \rangle + \frac{1}{2} \mu\omega^2 \langle x^2 \rangle$$

$$= \frac{1}{2\mu}\left(p_0^2 \sin^2 \omega t + \frac{\gamma^2\hbar^2}{2}\right) + \frac{1}{2}\mu\omega^2 \left(x_0^2 \cos^2 \omega t + \frac{1}{2\gamma^2}\right)$$

Thus,
$$\langle E \rangle = \frac{1}{2} \mu\omega^2 x_0^2 + \frac{1}{2} \hbar\omega \tag{70}$$

Notice the appearance of the zero point energy $\frac{1}{2}\hbar\omega$. For a classical oscillator the total energy is just $\frac{1}{2}\mu\omega^2 x_0^2$.

[9] The values for Δx and Δp remain the same for *all* values of time and are the same for the ground state wavefunction $\psi_0(x)$. Indeed, the ground state wavefunction $\psi_0(x)$ is also a coherent state corresponding to $\alpha = 0$ (i.e., zero eigenvalue).

In order to have a physical understanding of $\langle E \rangle$ we recall that energy eigenstates of the harmonic oscillator are given by [see Eq. (28)]

$$E = \left(n + \frac{1}{2}\right)\hbar\omega$$

Thus,

$$\langle E \rangle = \left[\langle n \rangle + \frac{1}{2}\right]\hbar\omega \tag{71}$$

Comparing the above equation with Eq. (70), we get

$$\langle n \rangle = \frac{1}{2}\gamma^2 x_0^2 = \frac{1}{2}\xi_0^2 \tag{72}$$

Now,

$$\Psi(x, t) = \sum_{n=0,1,2,\ldots}^{\infty} c_n\, \psi_n(x)\, e^{-i\left(n+\frac{1}{2}\right)\omega t}$$

Thus, the quantity

$$P_n = \left| c_n\, e^{-i\left(n+\frac{1}{2}\right)\omega t} \right|^2 = |c_n|^2 = \frac{1}{n!}\left(\frac{1}{2}\xi_0^2\right)^n e^{-\frac{1}{2}\xi_0^2} \tag{73}$$

represents the probability of finding the oscillator in the nth state[10]. This implies that if the oscillator is in a coherent state and if we make a measurement of energy then $|c_n|^2$ would represent the probability of obtaining the value $\left(n + \frac{1}{2}\right)\hbar\omega$. The average value of n is given by

$$\begin{aligned}
\langle n \rangle &= \sum_{n=0,1,2,\ldots}^{\infty} n\,|c_n|^2 \\
&= e^{-\frac{1}{2}\xi_0^2}\left(\frac{1}{2}\xi_0^2\right) \sum_{n=1,2,\ldots}^{\infty} \frac{1}{(n-1)!}\left(\frac{1}{2}\xi_0^2\right)^{n-1} \\
&= e^{-\frac{1}{2}\xi_0^2}\left(\frac{1}{2}\xi_0^2\right) e^{\frac{1}{2}\xi_0^2} \\
&= \frac{1}{2}\xi_0^2 = \frac{1}{2}\gamma^2 x_0^2
\end{aligned}$$

consistent with Eq. (72). Similarly,

$$\begin{aligned}
\langle n^2 \rangle &= \sum_{n=0,1,2,\ldots}^{\infty} n^2\,|c_n|^2 \\
&= e^{-\frac{1}{2}\xi_0^2}\left(\frac{1}{2}\xi_0^2\right) \sum_{n=1,2,\ldots}^{\infty} (n-1+1)\frac{1}{(n-1)!}\left(\frac{1}{2}\xi_0^2\, 2\right)^{n-1} \\
&= \left(\frac{1}{2}\xi_0^2\right)^2 + \frac{1}{2}\xi_0^2
\end{aligned}$$

[10] It may be seen that the probability of finding the oscillator in the nth state does not change with time. At different times, the states superpose with different phases which results in the time variation of the probability distribution $|\Psi(x, t)|^2$.

Thus, $$\Delta n = \left[\langle n^2\rangle - \langle n\rangle^2\right]^{\frac{1}{2}} = \frac{1}{\sqrt{2}}\xi_0 = \sqrt{\langle n\rangle}$$

We summarize the above results:

$$\langle n\rangle = \frac{1}{2}\xi_0^2 = \frac{1}{2}\gamma^2 x_0^2 \tag{74}$$

$$\langle n^2\rangle = \langle n\rangle^2 + \langle n\rangle \tag{75}$$

$$\Delta n = \left[\langle n^2\rangle - \langle n\rangle^2\right]^{\frac{1}{2}} = \sqrt{\langle n\rangle} \tag{76}$$

$$\frac{\Delta n}{\langle n\rangle} = \frac{1}{\sqrt{\langle n\rangle}} \tag{77}$$

the last equation implying that the fractional uncertainty in n decreases as $\langle n\rangle$ increases. The probability of finding the oscillator in the nth state is therefore given by

$$P_n = |c_n|^2 = e^{-\langle n\rangle}\frac{\langle n\rangle^n}{n!} \tag{78}$$

which is known as a Poisson distribution. In terms of $\langle n\rangle$, we express the various expectation values

$$\langle x\rangle = x_0\cos\omega t = \frac{1}{\gamma}\sqrt{2\langle n\rangle}\cos\omega t \tag{79}$$

$$\langle p\rangle = -\frac{\mu\omega}{\gamma}\sqrt{2\langle n\rangle}\sin\omega t = \mu\frac{d}{dt}\langle x\rangle \tag{80}$$

$$\langle x^2\rangle = \frac{1}{\gamma^2}\left[2\langle n\rangle\cos^2\omega t + \frac{1}{2}\right] \tag{81}$$

$$\langle p^2\rangle = \frac{\mu^2\omega^2}{\gamma^2}\left[2\langle n\rangle\sin^2\omega t + \frac{1}{2}\right] \tag{82}$$

$$\langle E\rangle = \frac{1}{2\mu}\langle p^2\rangle + \frac{1}{2}\mu\omega^2\langle x^2\rangle = \left[\langle n\rangle + \frac{1}{2}\right]\hbar\omega \tag{83}$$

$$\frac{\Delta x}{x_0} = \frac{\sqrt{\langle x^2\rangle - \langle x\rangle^2}}{x_0} = \frac{1}{\sqrt{\langle n\rangle}} \tag{84}$$

$$\frac{\Delta p}{p_0} = \frac{\sqrt{\langle p^2\rangle - \langle p\rangle^2}}{p_0} = \frac{1}{\sqrt{\langle n\rangle}} \tag{85}$$

$$\langle E^2\rangle = \left[\langle n^2\rangle + \langle n\rangle + \frac{1}{4}\right]\hbar^2\omega^2 \tag{86}$$

$$\frac{\Delta E}{\langle E\rangle} = \frac{\left[\langle E^2\rangle - \langle E\rangle^2\right]^{\frac{1}{2}}}{\langle E\rangle} = \frac{\Delta n}{\langle n\rangle + \frac{1}{2}} = \frac{\sqrt{\langle n\rangle}}{\langle n\rangle + \frac{1}{2}} \tag{87}$$

In Figs 7.6, 7.7 and 7.8, we have plotted $|c_n|^2$ as a function of n for

$$\langle n \rangle = 10 \left(\Delta n \simeq 3.1, \frac{\Delta n}{\langle n \rangle} \simeq 0.31 \right); \tag{88}$$

$$\langle n \rangle = 100 \left(\Delta n = 10, \frac{\Delta n}{\langle n \rangle} = 0.1 \right) \tag{89}$$

and

$$\langle n \rangle = 10^6 \left(\Delta n = 10^3, \frac{\Delta n}{\langle n \rangle} = 0.001 \right) \tag{90}$$

We find that as the value of $\langle n \rangle$ increases the value of Δn also increases implying that a large number of states get excited; however, the value of $\Delta n/\langle n \rangle$ (and hence of $\Delta E/\langle E \rangle$) decreases implying that the fractional spread in energy becomes very small.

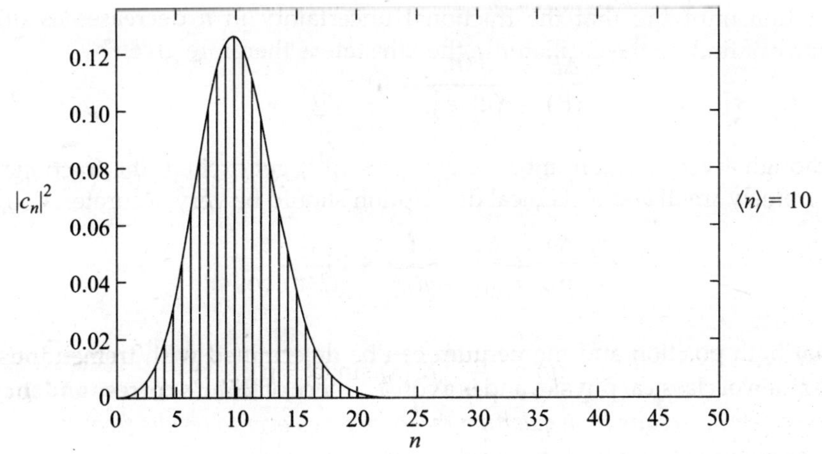

Figure 7.6: The variation of $|c_n|^2$ with n for $\langle n \rangle = 10$. Notice that although only about 20 states are excited, $\frac{\Delta x}{x_0}$, $\frac{\Delta p}{p_0}$ have an appreciable value (≈ 0.316).

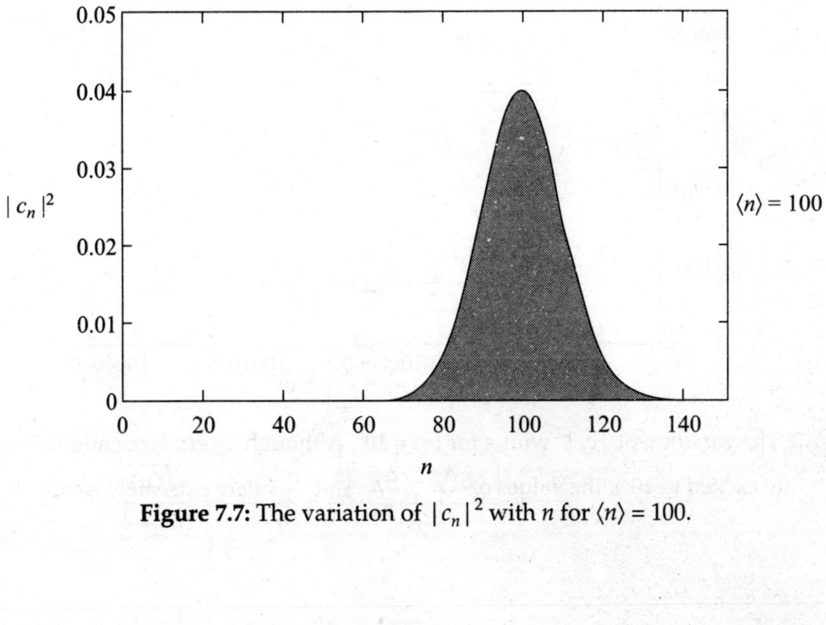

Figure 7.7: The variation of $|c_n|^2$ with n for $\langle n \rangle = 100$.

Relation with a classical oscillator:

In order to relate the above analysis to a classical oscillator, we consider a small mass ($\mu \approx 0.002$ g) executing simple harmonic motion with $\omega = 1\text{s}^{-1}$ (implying a time period of 2π seconds). If the amplitude of motion ($= x_0$) is 1 cm then,

$$\langle x \rangle = \cos t \tag{91}$$

$$\langle p \rangle = -0.002 \sin t \tag{92}$$

where $\langle x \rangle$, t and $\langle p \rangle$ are measured in cm, s and g cm/s respectively. Further

$$\langle n \rangle = \frac{1}{2}\gamma^2 x_0^2 = \frac{\mu\omega}{2\hbar} x_0^2$$

$$\approx 10^{24}$$

implying
$$\Delta n = \sqrt{\langle n \rangle} \approx 10^{12} \tag{93}$$

and
$$\frac{\Delta E}{\langle E \rangle} = \frac{\sqrt{\langle n \rangle}}{\langle n \rangle + \frac{1}{2}} \approx 10^{-12} \tag{94}$$

Thus, although a very large number of states ($\approx 10^{12}$) get excited, the fractional spread in energy is extremely small and a classical description should be very accurate. Also,

$$\frac{\Delta p}{p_0} = \frac{\Delta x}{x_0} = \frac{1}{\sqrt{\langle n \rangle}} \approx 10^{-12} \tag{95}$$

implying that both position and momentum can be determined with tremendous precision. This is the domain of classical physics and is a consequence of **Böhr's correspondence principle** according to which *the results from quantum theory must go over to results from classical mechanics when very large number of quanta are involved.*

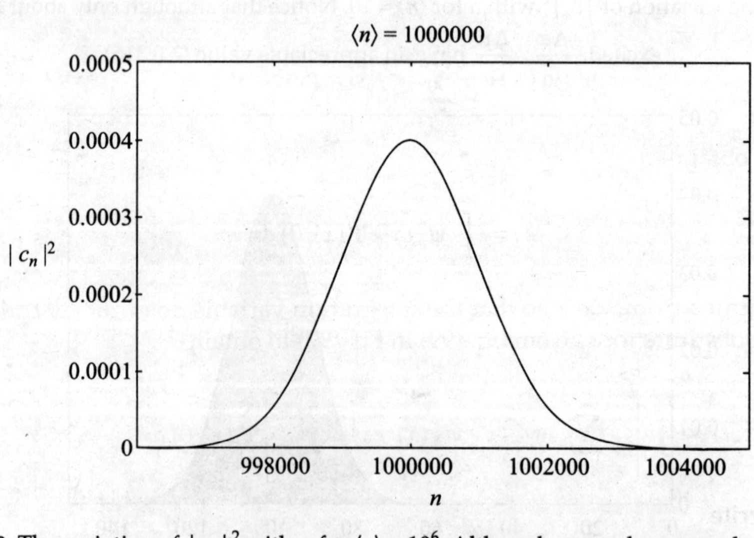

Figure 7.8: The variation of $|c_n|^2$ with n for $\langle n \rangle = 10^6$. Although a very large number of states are excited ($\approx 10^3$), the values of $\frac{\Delta x}{x_0}$, $\frac{\Delta p}{p_0}$ and $\frac{\Delta E}{E_0}$, are extremely small.

On the other hand, if the mass of the particle is so small that

$$\langle n \rangle = 10$$

then,

$$\frac{\Delta E}{E_0} \approx 0.301$$

$$\frac{\Delta p}{p_0} = \frac{\Delta x}{x_0} \approx 0.316$$

and a quantum mechanical description would be necessary.

We may mention here that when a laser is operated much beyond the threshold, it generates a coherent state excitation of the cavity mode. Thus, for such a beam if one carries out a photon counting experiment one should find that the probability of measuring n quanta in a certain interval of time is given by [see Eq. (73)]

$$P_n = e^{-\langle n \rangle} \frac{\langle n \rangle^n}{n!} \tag{96}$$

where $\langle n \rangle$ represents the average number of counts measured in that time interval. Equation (96) is known as the Poisson distribution and is indeed verified through experiments (see, e.g. Ref. 1 and 2).

7.7 THE PROPAGATOR

The general solution of the time dependent Schrödinger equation for the linear harmonic oscillator problem is given by [see Eq. (42)]:

$$\Psi(x, t) = \sum_{n=0,1,2,...}^{\infty} c_n \psi_n(x) \; e^{-i\left(n+\frac{1}{2}\right)\omega t} \tag{97}$$

Thus,

$$\Psi(x, 0) = \sum_{n=0,1,2,...}^{\infty} c_n \psi_n(x) \tag{98}$$

from which we obtain

$$c_n = \int_{-\infty}^{+\infty} \psi_n^*(x') \Psi(x', 0) \, dx' \tag{99}$$

where we have put a prime on x so that the integration variable does not get confused with x in Eq. (97). We substitute for c_n from Eq. (99) in Eq. (97) to obtain

$$\Psi(x, t) = \int_{-\infty}^{+\infty} \left[\sum_{n=0,1,2,...}^{\infty} \psi_n^*(x') \psi_n(x) \; e^{-i\left(n+\frac{1}{2}\right)\omega t} \right] \Psi(x', 0) \, dx' \tag{100}$$

Thus we may write

$$\Psi(x, t) = \int_{-\infty}^{+\infty} K(x, x', t) \Psi(x', 0) \, dx' \tag{101}$$

where

$$K(x, x', t) = \sum_{n=0,1,2,\ldots}^{\infty} \psi_n^*(x')\, \psi_n(x)\, e^{-i\left(n+\frac{1}{2}\right)\omega t} \tag{102}$$

represents the propagator. The above equation may be compared with the results obtained in Section 5.4. If we substitute for $\psi_n(x)$ from Eq. (32) then it is possible to sum the infinite series (see, e.g., Chapter 6 of Ref. 3); the final result is

$$K(x, x', t) = \frac{\gamma}{(2\pi i \sin \omega t)^{\frac{1}{2}}} \exp\left[-\frac{\gamma^2}{2i \sin \omega t}\{(x^2 + x'^2) \times \cos \omega t - 2xx'\}\right] \tag{103}$$

Thus for the linear harmonic oscillator problem, the time evolution of an arbitrary wave packet can easily be carried out by substituting for $\Psi(x, 0)$ in Eq. (101) and carrying out the integration. Indeed, if we substitute Eq. (54) for $\Psi(x, 0)$ and carry out the integration in Eq. (101) we would get Eq. (59).

7.8 SOLUTIONS IN TERMS OF THE CONFLUENT HYPERGEOMETRIC FUNCTION

By knowing the solutions to the confluent hypergeometric equation, one can very easily solve the four most important problems in the first course in quantum mechanics—namely the linear harmonic oscillator problem, the hydrogen atom problem (see Section 10.4.2) and the two- and three-dimensional isotropic oscillator problems (see Problems 10.12 and 10.13). The confluent hypergeometric equation is given by

$$x\frac{d^2y}{dx^2} + (c - x)\frac{dy}{dx} - a\,y(x) = 0 \tag{104}$$

where a and c are constants. For $c \neq 0, \pm1, \pm2, \pm3, \cdots$ the two independent solutions of the above equation are [see, Appendix F]

$$y_1(x) = {}_1F_1(a, c, x) \tag{105}$$

and

$$y_2(x) = x^{1-c}\,{}_1F_1(a - c + 1, 2 - c, x) \tag{106}$$

where

$$ {}_1F_1(a, c, x) = 1 + \frac{a}{c}\cdot\frac{x}{1!} + \frac{a(a+1)}{c(c+1)}\cdot\frac{x^2}{2!} + \cdots \tag{107}$$

is known as the confluent hypergeometric function. One can immediately see that for $c = a$

$$ {}_1F_1(a, a, x) = e^x \tag{108}$$

Indeed the infinite series given by Eq. (107) is convergent for *all* values of x and its asymptotic form is given by [see, e.g. Appendix K or Section 8.5 of Ref. 3]:

$$ {}_1F_1(a, c, x) \underset{x\to\infty}{\longrightarrow} \frac{\Gamma(c)}{\Gamma(a)} x^{a-c}\, e^x \tag{109}$$

The confluent hypergeometric series ${}_1F_1(a, c, x)$ is very easy to remember and its asymptotic forms are very easily understood. We go back to the Schrödinger equation for the linear harmonic oscillator and rewrite Eq. (16):

$$\frac{d^2u}{d\xi^2} - 2\xi\frac{du}{d\xi} + (\Lambda - 1)\,u(\xi) = 0 \tag{110}$$

If we now make the transformation $\eta = \xi^2$, we would get

$$\frac{du}{d\xi} = \frac{du}{d\eta}\,2\xi \quad\text{and}\quad \frac{d^2u}{d\xi^2} = \frac{d^2u}{d\eta^2}\,4\eta + 2\frac{du}{d\eta}$$

and Eq. (110) becomes

$$\eta\frac{d^2u}{d\eta^2} + \left(\frac{1}{2} - \eta\right)\frac{du}{d\eta} + \frac{\Lambda - 1}{4}\,u(\eta) = 0$$

which is the confluent hypergeometric equation with

$$a = \frac{1-\Lambda}{4}; \quad c = \frac{1}{2}$$

Thus the two independent solutions of Eq. (9) are

$$\psi_1(\xi) = {}_1F_1\left(\frac{1-\Lambda}{4}, \frac{1}{2}, \xi^2\right)e^{-\frac{1}{2}\xi^2} \tag{111}$$

and

$$\psi_2(\xi) = \xi\,{}_1F_1\left(\frac{3-\Lambda}{4}, \frac{3}{2}, \xi^2\right)e^{-\frac{1}{2}\xi^2} \tag{112}$$

Using the asymptotic form of the confluent hypergeometric function [Eq. (6)], one can readily see that if the series does not become a polynomial, $\psi(\xi)$ will behave as exp $\left(+\frac{1}{2}\xi^2\right)$. The even solution becomes a polynomial for $\Lambda = 1, 5, 9, \ldots$ and the odd series becomes a polynomial when $\Lambda = 3, 7, 11, \ldots$. Thus, the eigenvalues are $\left(n+\frac{1}{2}\right)\hbar\omega$ with $n = 0, 1, 2, 3, \ldots$. In addition to the quick (and easy to understand) derivation of the energy eigenvalues, one also has the following advantage.

Let us suppose we want to know the eigenfunction corresponding to $n = 7$. Then $\Lambda = 15$ and the function $\psi_2(\xi)$ becomes a polynomial. Since, the series given by Eq. (107) is easy to remember, one can readily write down

$$\psi_2(\xi) = \xi\,{}_1F_1\left(-3, \frac{3}{2}, \xi^2\right)e^{-\frac{1}{2}\xi^2} = \xi\left[1 - 2\xi^2 + \frac{4}{5}\xi^4 - \frac{8}{105}\xi^6\right]e^{-\frac{1}{2}\xi^2}$$

If we recall that the coefficient of ξ^n in $H_n(\xi)$ is 2^n, then we can immediately write down

$$H_7(\xi) = 128\xi^7 - 1344\xi^5 + 3360\xi^3 - 1680\xi$$

Indeed

$$H_n(\xi) = \begin{cases} (-1)^{n/2}\dfrac{n!}{\frac{n}{2}!}\,{}_1F_1\left(-\dfrac{n}{2}, \dfrac{1}{2}, \xi^2\right) & \text{for } n = 0, 2, 4, \ldots \\[2ex] (-1)^{(n-1)/2}\dfrac{n!}{\left(\frac{n-1}{2}\right)!}2\xi\,{}_1F_1\left(-\dfrac{n-1}{2}, \dfrac{3}{2}, \xi^2\right) & \text{for } n = 1, 3, 5, \ldots \end{cases} \tag{113}$$

The solutions expressed by Eqs. (111) and (112) are very convenient to use for a bounded parabolic potential (Problem 7.15), double oscillator (Problem 17.6), etc.

7.9 SUMMARY OF EXACT SOLUTIONS OF THE 1-DIMENSIONAL SCHRÖDINGER EQUATION

In this and earlier chapters we have given exact solutions of the Schrödinger equation for a variety of potential energy variations. We thought it would be good to give here a summary of the exact solutions of the 1- dimensional time dependent Schrödinger equation for 3 important cases that we have discussed in this and earlier chapters. This will enable the students to understand the general character of the solutions. The first will correspond to a continuous spectrum of eigenvalues and the remaining two will correspond to a discrete spectrum of eigenvalues. The 1- dimensional time dependent Schrödinger equation is given by

$$i\hbar\frac{\partial\Psi}{\partial t} = H\,\Psi\,(x,t) \text{ with } H = \frac{p^2}{2\mu} + V(x) = -\frac{\hbar^2}{2\mu}\frac{\partial^2}{\partial x^2} + V(x)$$

If we assume a separable solution

$$\Psi\,(x,t) = \psi(x)\exp\left(-\frac{iEt}{\hbar}\right)$$

we would find that $\psi(x)$ would satisfy the following eigenvalue equation

$$H\psi(x) = E\psi(x) \quad \Rightarrow \quad \frac{d^2\psi(x)}{dx^2} + \frac{2\mu}{\hbar^2}\,[E - V(x)]\,\psi(x) = 0$$

Case 1: Free Particle $V(x) = 0; \quad -\infty < x < \infty$

Normalized eigenfunctions of $H\left(= -\frac{\hbar^2}{2\mu}\frac{\partial^2}{\partial x^2}\right)$ are

$$\psi_p\,(x) = \frac{1}{\sqrt{2\pi\hbar}}\exp\left[\frac{i}{\hbar}px\right]; \qquad -\infty < p < \infty$$

Thus, $H\psi_p\,(x) = E\psi_p\,(x) = \frac{p^2}{2\mu}\,\psi_p(x); \qquad -\infty < p < \infty$

with $\int_{-\infty}^{+\infty}\psi_p^*(x)\psi_{p'}(x)dx = \delta(p-p')$ orthonormality condition

and $\int_{-\infty}^{+\infty}\psi_p^*(x')\psi_p(x)\,dp = \delta(x-x'); -\infty < x,\ x' < \infty$ completeness condition.

The functions $\psi_p\,(x)$ form a complete set of orthonormal functions. Since p can take any real value between $-\infty$ and $+\infty$, the eigenvalues $E\left(= \frac{p^2}{2\mu}\right)$ take a continuum of values: $0 < E < \infty$; and, for each value of E, there would be 2 independent solutions:

$$\psi_E(x) = \frac{1}{\sqrt{2\pi\hbar}} \exp\left[\pm\frac{i}{\hbar}\sqrt{2\mu E}\; x\right]; \quad 0 < E < \infty$$

The most general solution of time dependent Schrödinger equation is given by

$$\Psi(x, t) = \frac{1}{\sqrt{2\pi\hbar}} \int_{-\infty}^{+\infty} a(p) \exp\left[\frac{i}{\hbar}\left(px - \frac{p^2}{2\mu}t\right)\right] dp$$

which describes the propagation of a wave packet describing a free particle. If $\Psi(x, 0)$ is normalized, then $|a(p)|^2\, dp$ will represent the probability of finding the momentum between p and $p + dp$.

Case 2: Particle in a 1-dimensional Box

$$V(x) = \begin{cases} 0; & \text{for } 0 < x < a \\ \infty; & \text{for } \quad x < 0 \quad \text{and for} \quad x > a \end{cases}$$

Normalized eigenfunctions of H:

$$\psi_n(x) = \sqrt{\frac{2}{a}} \sin\left[\frac{n\pi}{a}x\right] \quad \text{for} \quad 0 < x < a; \quad n = 1, 2, 3 \ldots$$

Thus, $$H\,\psi_n(x) = E_n\,\psi_n(x) = \frac{n^2\pi^2\hbar^2}{2\mu a^2}\,\psi_n(x); \quad n = 0, 1, 2, 3, \ldots$$

with $$\int_0^a \psi_m^*(x)\psi_n(x)dx = \delta_{mn} = \begin{cases} 1 & \text{if } m = n \\ 0 & \text{if } m \neq n \end{cases}$$ orthonormality condition

and $$\sum_{n=1,2,3} \psi_n^*(x')\psi_n(x) = \delta(x - x'); \quad 0 < x,\ x' < a$$ completeness condition.

The functions $\psi_n(x)$ form a complete set of orthonormal functions in the domain $0 < x < a$. The eigenvlaues $E_n\left(= \frac{n^2\pi^2\hbar^2}{2\mu a^2}\right)$ form a discrete set and the most general solution of the time dependent Schrödinger equation is given by

$$\Psi(x,t) = \sum_{n=0,1,2,3,\ldots} c_n\,\psi_n(x)\exp\left[-\frac{iE_n t}{\hbar}\right]$$

$$= \sum_{n=0,1,2,3,\ldots} c_n\,\psi_n(x)\exp\left[-\frac{i}{\hbar}\cdot\frac{n^2\pi^2\hbar^2}{2\mu a^2}t\right]$$

If $\Psi(x, 0)$ is normalized, then $|c_n|^2$ will represent the probability of finding the system in the nth state.

Case 3: The Linear Harmonic Oscillator

$$V(x) = \frac{1}{2}\mu\omega^2 x^2; \quad -\infty < x < \infty$$

Normalized eigenfunctions of $H\left(=-\frac{\hbar^2}{2\mu}\frac{\partial^2}{\partial x^2}+\frac{1}{2}\mu\omega^2x^2\right)$ are

$$\psi_n(x) = N_n H_n(\xi)\exp\left(-\frac{1}{2}\xi^2\right);\quad n=0,1,2,3,\ldots;$$

where $\xi=\gamma x;\quad \gamma=\sqrt{\frac{\mu\omega}{\hbar}}$ and $N_n=\sqrt{\frac{\gamma}{2^n\, n!\,\sqrt{\pi}}}$

The corresponding eigenvalues are $E_n=\left(n+\frac{1}{2}\right)\hbar\omega$

Thus, $H\,\psi_n(x) = E_n\,\psi_n(x)=\left(n+\frac{1}{2}\right)\hbar\omega\;\psi_n(x);\quad n=0,1,2,3,\ldots$

with $\int\limits_{-\infty}^{+\infty}\psi_m^*(x)\psi_n(x)\,dx=\delta_{mn}=\begin{cases}1 & \text{if } m=n\\ 0 & \text{if } m\neq n\end{cases}$ orthonormality condition

and $\sum\limits_{n=0,1,2,3}\psi_n^*(x')\psi_n(x)=\delta(x-x');\ -\infty<x,x'<\infty$ completeness condition.

The functions $\psi_n(x)$ form a complete set of orthonormal functions in the domain $-\infty<x<\infty$.

The eigenvalues $E_n\left[=\left(n+\frac{1}{2}\right)\hbar\omega\right]$ form a discrete set and the most general solution of the time dependent Schrödinger equation is given by

$$\Psi(x,t)=\sum_{n=0,1,2,3} c_n\,\psi_n(x)\exp\left[-\frac{iE_n t}{\hbar}\right]$$

$$=\sum_{n=0,1,2,3} c_n\,\psi_n(x)\exp\left[-i\left(n+\frac{1}{2}\right)\omega t\right]$$

If $\Psi(x,0)$ is normalized, then $|c_n|^2$ will represent the probability of finding the oscillator in the nth state.

7.10 PROBLEMS

Problem 7.1 Show that the energy eigenvalues of the one-dimensional Schrödinger equation can never be less than the absolute minimum of $V(x)$.

Problem 7.2 For a wave packet [see Eq. (51) of Chapter 4]

$$\langle p\rangle=\int\limits_{-\infty}^{+\infty}\Psi^*(x,t)\left(-i\hbar\frac{\partial}{\partial x}\right)\Psi(x,t)\,dx$$

Use Eq. (59) for $\Psi(x,t)$ to derive Eq. (63).

Problem 7.3 (a) Derive Eq. (36) for the probability distribution function for a classical oscillator. **(b)** Using the asymptotic form of the Hermite polynomials for large values of n, show that the quantum mechanical probability distribution function agrees with the corresponding classical result. (This is a consequence of the Böhr's correspondence principle.)

Problem 7.4 For the harmonic oscillator in its ground state, show that the probability of finding it beyond the classical turning points is approximately 0.16.

Problem 7.5 (a) Substitute the function

$$\psi(x) = A\, e^{-bx^2}$$

in Eq. (10) to show that it indeed satisfies the equation only if

$$E = \frac{1}{2}\hbar\omega$$

and

$$b = \frac{1}{2}\cdot\frac{\mu\omega}{\hbar}$$

(b) Substitute the function

$$\psi(x) = A\, e^{-cx^4}$$

in Eq. (10) and show that it can never satisfy the equation.

Problem 7.6 (a) The ground state wavefunction

$$\psi_0(x) = N e^{-\frac{1}{2}\xi^2}$$

is one of the solutions of Eq. (13) for $\Lambda = 1$. Obtain the other solution in the form of an integral and show that it would behave as $\exp\left[\frac{1}{2}\xi^2\right]$ for large values of x.

[**Hint:** Write $\psi(x) = \psi_0(x)\,\phi(x)$ and obtain the differential equation satisfied by $\phi(x)$.]

(b) Repeat the above analysis for $\psi_1(x)$ and $\psi_2(x)$.

Problem 7.7 A harmonic oscillator is in a state described by the wavefunction

$$\Psi(x, 0) = \frac{1}{2}\psi_0(x) + \frac{i}{\sqrt{2}}\psi_1(x) + \frac{1}{\sqrt{2}}e^{i\pi/3}\psi_2(x)$$

where $\psi_n(x)$ are the normalized Hermite-Gauss wavefunctions [see Eq. (32)].

(a) Is the wavefunction normalized?

(b) What is $\Psi(x, t)$?

Problem 7.8 (a) Assume

$$\Psi(x, 0) = \psi_\alpha(x)$$

with $\langle n \rangle = 400$. Evaluate

$$\frac{\Delta x}{x_0}, \frac{\Delta p}{p_0}, \frac{\Delta E}{E_0} \quad \text{and} \quad P_n$$

(b) Repeat the analysis for $\langle n \rangle = 10^8$ and physically interpret the results.

Problem 7.9 Assuming the normalization condition given by Eq. (45), derive Eqs (46) and (47).

Problem 7.10 Assuming $\Psi(x, 0)$ to be given by Eq. (54), use Eq. (101) to derive Eq. (59).

Problem 7.11 Using Eqs (32) and (51) show that

$$a\psi_0(x) = 0$$
$$a\psi_1(x) = \psi_0(x)$$
$$a\psi_2(x) = \sqrt{2}\,\psi_1(x)$$

In general,

$$a\psi_n(x) = \sqrt{n}\,\psi_{n-1}(x) \tag{114}$$

Problem 7.12 Equations (50) and (51) define the operator a. The adjoint of this operator is defined by the equation

$$\bar{a} = \frac{1}{\sqrt{2\mu\hbar\omega}}(\mu\omega x - ip) \tag{115}$$

or

$$\bar{a} = \sqrt{\frac{\mu\omega}{2\hbar}}\,x - \sqrt{\frac{\hbar}{2\mu\omega}}\,\frac{d}{dx}$$

Show that

$$\bar{a}\psi_0(x) = \psi_1(x) \tag{116}$$
$$\bar{a}\psi_1(x) = \sqrt{2}\,\psi_2(x) \tag{117}$$

and

$$\bar{a}\psi_2(x) = \sqrt{3}\,\psi_3(x) \tag{118}$$

Problem 7.13 In this and the following problem replace μ by m in the definitions of a and $\bar{a}$. Solve the eigenvalue equation

$$(\mu a + \nu\bar{a})\,\psi_\beta(x) = \beta\,\psi_\beta(x)$$

to obtain

$$\psi_\beta(x) = \left(\frac{1}{\sqrt{\pi}\,\sigma_0}\right)^{\frac{1}{2}} \exp\left[-\frac{(x-x_0)^2}{2\sigma_0^{\,2}}\right] \tag{119}$$

where

$$\sigma_0 = \frac{1}{\gamma}\sqrt{\frac{\mu-\nu}{\mu+\nu}} \tag{120}$$

and

$$x_0 = \frac{1}{\gamma}\,\frac{\sqrt{2}\,\beta}{\mu+\nu} \tag{121}$$

Obviously, we must choose $\mu > \nu$; for $\mu = 1$ and $\nu = 0$ we get the coherent states. $\psi_\beta(x)$ describes what is known as a *squeezed state*.

Problem 7.14 (a) Assuming $\Psi(x, 0)$ to be $\psi_\beta(x)$ as given in the previous problem, show by using Eq. (101) that

$$|\Psi(x,t)|^2 = \frac{1}{\sqrt{\pi}\,\sigma(t)} \exp\left[-\frac{(x - x_0\cos wt)^2}{\sigma^2(t)}\right] \tag{122}$$

where

$$\sigma(t) = \frac{1}{\sqrt{2}\,\gamma^2\sigma_0}\left[(\gamma\sigma_0)^4 + 1 + \{(\gamma\sigma_0)^4 - 1\}\cos 2\omega t\right]^{\frac{1}{2}} \tag{123}$$

Discuss the time evolution of $\sigma(t)$.

(b) Show that

$$\langle x \rangle = x_0 \cos \omega t \tag{124}$$

$$\langle x^2 \rangle = \langle x \rangle^2 + \frac{1}{2}\sigma^2 (t) \tag{125}$$

and

$$\Delta x = \frac{1}{\sqrt{2}} \sigma(t) \tag{126}$$

Problem 7.15 *Bounded Harmonic Potential*[11]

(a) Consider the bounded linear harmonic potential given by

$$V(x) = \begin{cases} \frac{1}{2}\mu\omega^2 x^2, & |x| < l/2 \\ \infty, & |x| > l/2 \end{cases} \tag{127}$$

where $x = \pm l/2$ represent the two boundaries of the potential. Show that the eigenvalues Λ are obtained by solving the following transcendental equations

$$\begin{aligned} {}_1F_1\left(\frac{1-\Lambda}{4}, \frac{1}{2}, \xi_0^2\right) &= 0 \text{ for even solutions} \\ {}_1F_1\left(\frac{3-\Lambda}{4}, \frac{3}{2}, \xi_0^2\right) &= 0 \text{ for odd solutions} \end{aligned} \tag{128}$$

where $\xi_0 = \gamma l/2$.

(b) For $\xi_0^2 = 0.5$ and 2, solve the above equations numerically to obtain the first few eigenvalues Λ. [**Answer:** The eigenvalues are (5.00, 19.88, 44.57, 79.12, ...) and (1.49, 5.49, 11.73, 20.38, ...) for $\xi_0^2 = 0.5$ and 2.0 respectively, one could have guessed the eigenvalue 5.0 directly – how?].

Problem 7.16 Consider the bounded linear harmonic potential given by

$$V(x) = \begin{cases} \infty, & x = 0 \\ \frac{1}{2}\mu\omega^2 x^2, & x > 0 \end{cases} \tag{129}$$

Show that the eigenvalues are

$$\Lambda = 3, 7, 11, \ldots$$

What are the corresponding eigenfunctions?

7.11 SOLUTIONS

Solution 7.1 We write the Schrödinger equation in the form

$$\frac{d^2\psi}{dx^2} = \frac{2m}{\hbar^2} [V(x) - E]\, \psi(x)$$

[11] One of the very early works on the bounded harmonic oscillator was by D.S. Kothari and F.C. Auluck, *Science and Culture*, **6**, pp 370–1, 1940.

If $V(x) > E$ for *all* values of x, then everywhere $\psi''(x)$ has the same sign as $\psi(x)$. We assume ψ to be positive at some value of x; then $\psi''(x)$ will also be positive. Hence if ψ' is positive, then $\psi \to \infty$ as $x \to \infty$ and if ψ' is negative, ψ will tend to ∞ as $x \to -\infty$. In either case, $\psi(x)$ will blow up either at $+\infty$ or at $-\infty$. We can have a similar argument if we assume ψ to be negative at some value of x.

Solution 7.3 For the classical oscillator we may assume

$$x = x_0 \sin \omega t$$

where x_0 represents the amplitude of motion. For $0 < t < T/2$, the velocity is everywhere positive and we may write

$$v = \omega x_0 \cos \omega t = \omega \sqrt{x_0^2 - x^2}$$

If $P(x)\,dx$ represents the probability of finding the particle between x and $x + dx$, then

$$P(x)\,dx = \text{Fractional time spent between } x \text{ and } x + dx$$

$$= \frac{dx/v}{T/2}$$

Thus,
$$P(x)\,dx = \frac{1}{\pi\sqrt{x_0^2 - x^2}}\,dx \tag{130}$$

According to quantum theory (see Eq. (32))

$$P(x) = |\psi_n(x)|^2 = \frac{\gamma}{2^n\, n!\sqrt{\pi}} \exp(-\xi^2)\, H_n^2(\xi) \tag{131}$$

For large values of n, the asymptotic expression is given by (see, e.g. Reference 4)

$$H_n(\xi) \sim \frac{2^{(n+2)/2}}{\sqrt{2\cos\beta}} \sqrt{\frac{n!}{\sqrt{2n\pi}}}\, e^{n\beta^2} \cos\left[\left(2n + \frac{1}{2}\right)\beta - \frac{n\pi}{2}\right] \tag{132}$$

where β is the smallest positive angle whose sine is $(\xi/\sqrt{2n})$. For large values of n, we may take the average value of the $\cos^2$ term as $\frac{1}{2}$ Further,

$$x_0 = \left(\frac{2E}{\mu\omega}\right)^{\frac{1}{2}} = \frac{\sqrt{2n+1}}{\gamma} \sim \frac{\sqrt{2n}}{\gamma}$$

Thus,
$$\cos\beta = \sqrt{1 - \sin^2\beta} = \sqrt{1 - \frac{1}{2n}\xi^2}$$

$$\simeq \frac{1}{\xi_0}\sqrt{\xi_0^2 - \xi^2}$$

Elementary algebra gives

$$P(x) \simeq \frac{1}{\pi\sqrt{x_0^2 - x^2}}$$

Solution 7.4 The required probability is

$$2\int_{x_0}^{\infty} |\Psi_0(x)|^2\, dx = \frac{2}{\sqrt{\pi}} \int_1^{\infty} e^{-\xi^2}\, d\xi \simeq 0.16$$

Solution 7.6 $\psi_n(\xi)$ [as given by Eq. (32)] is one of the solutions of the equation

$$\frac{d^2\psi}{d\xi^2} + [(2n+1) - \xi^2]\,\psi(\xi) = 0 \tag{133}$$

In order to obtain the other solution, we write

$$\psi(\xi) = \psi_n(\xi)\,\phi(\xi)$$

Thus,
$$\psi''(\xi) = \psi_n''(\xi)\,\phi(\xi) + 2\psi_n'(\xi)\,\phi'(\xi) + \psi_n(\xi)\,\phi''(\xi)$$

Substituting in Eq. (133), we get

$$\left[\frac{d^2\psi_n}{d\xi^2} + [(2n+1)-\xi^2]\,\psi_n(\xi)\right]\phi(\xi) + 2\psi_n'(\xi)\,\phi'(\xi) + \psi_n(\xi)\,\phi''(\xi) = 0 \tag{134}$$

Since $\psi_n(\xi)$ is a solution of Eq. (133) the first term in the LHS vanishes and we obtain

$$2\,\frac{\psi_n'(\xi)}{\psi_n(\xi)} + \frac{\phi''(\xi)}{\phi'(\xi)} = 0$$

or
$$\frac{d}{d\xi}\,[2\ln\psi_n(\xi) + \ln\phi'(\xi)] = 0$$

or
$$\ln[\psi_n^2(\xi)\,\phi'(\xi)] = \text{constant}$$

Thus,
$$\phi'(\xi) = \frac{\text{constant}}{\psi_n^2(\xi)}$$

For $n = 0$, we therefore have

$$\frac{d\phi}{d\xi} = C_1\,e^{\xi^2}$$

or
$$\phi(\xi) = C_1\int^{\xi} e^{\xi^2}\,d\xi + C_2$$

Thus,
$$\psi(\xi) = C_1\,e^{-\frac{1}{2}\xi^2}\int^{\xi} e^{\xi^2}\,d\xi + C_2\,e^{-\frac{1}{2}\xi^2}$$

which represents the general solution of Eq. (133). The last term is nothing but $\psi_0(\xi)$. Since the integral behaves as $\exp(\xi^2)$, the first term would behave as $e^{\frac{1}{2}\xi^2}$. Indeed

$$e^{-\frac{1}{2}\xi^2}\int^{\xi} e^{\xi^2}\,d\xi = e^{-\frac{1}{2}\xi^2}\int\left[1+\xi^2+\frac{1}{2!}\xi^4+\frac{1}{3!}\xi^6+\cdots\right]d\xi$$

$$= e^{-\frac{1}{2}\xi^2}\left[\xi+\frac{1}{3}\xi^3+\frac{1}{10}\xi^5+\frac{1}{42}\xi^7+\cdots\right]$$

The quantity within the brackets is the term proportional to c_1 in Eq. (29) – as it indeed should be.

Solution 7.14 From Eq. (123) we may note the following:

(a) If

$$\gamma\sigma_0 = 1 \Rightarrow \sigma_0 = \frac{1}{\gamma}$$

i.e., if the initial width is the same as the ground state width, then

$$\sigma^2(t) = \sigma_0^2 = \frac{1}{\gamma} \qquad \text{(independent of time)}$$

Thus, the width does not change and we have the coherent state.

(b) In Fig. 7.9, we have plotted the time evolution of the wave packet for a squeezed state corresponding to $\gamma\sigma_0 = 1.5$ with $\xi_0 = 5$; the corresponding variation of Δx with time is shown in Fig. 7.10. As can be seen, we find that the width of the wave packet decreases below the ground state width; this is known as "squeezing" which has important applications.

$$\langle x \rangle = \int_{-\infty}^{+\infty} x|\Psi(x)|^2\, dx = \frac{1}{\sqrt{\pi}\,\sigma(t)} \int_{-\infty}^{+\infty} x e^{-(x-x_0\cos\omega t)^2/\sigma^2(t)}\, dx$$

$$= x_0 \cos\omega t$$

and $$\langle x^2 \rangle = \frac{1}{\sqrt{\pi}\,\sigma(t)} \int_{-\infty}^{+\infty} x^2 e^{-(x-x_0\cos\omega t)^2/\sigma^2(t)}\, dx$$

$$= x_0^2 \cos^2\omega t + \frac{1}{2}\sigma^2(t)$$

Thus, $$\Delta x = \frac{1}{\sqrt{2}}\sigma(t)$$

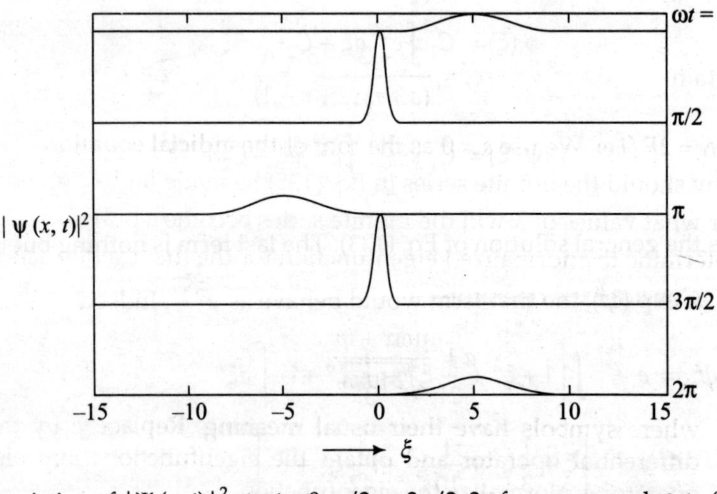

Figure 7.9: The variation of $|\Psi(x,t)|^2$ at $\omega t = 0, \pi/2, \pi, 3\pi/2, 2\pi$ for a squeezed state ($\gamma\sigma_0 = 1.5;\ \xi_0 = 5$).

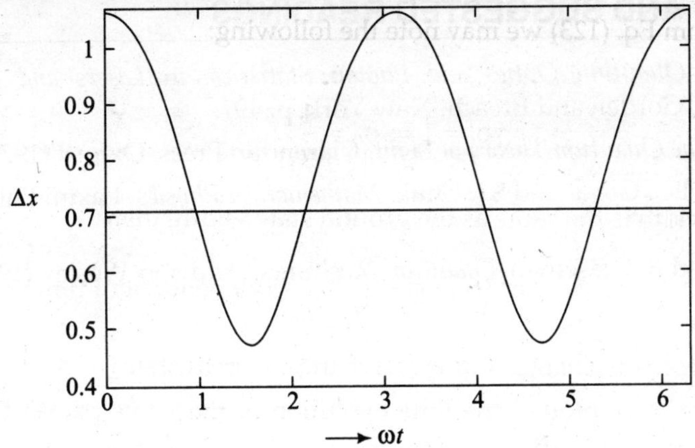

Figure 7.10: The variation of Δx with ωt for a squeezed state ($\gamma\sigma_0 = 1.5$; $\xi_0 = 5$). The horizontal line corresponds to the ground state.

showing that Δx oscillates with time (see Fig. 7.10) with its value becoming smaller than the ground state value.

7.12 SAMPLE QUESTIONS

Q. 1 (a) Write the Hamiltonian H for the linear harmonic oscillator problem.

(b) For what value of g, would the function xe^{-gx^2} be an eigenfunction of H? What is the corresponding eigenvalue?

(c) For what values of α and β would the function $(1+\alpha x^2)\,e^{-\beta x^2}$ be an eigenfunction of H? What is the corresponding eigenvalue?

Q. 2 In the linear harmonic oscillator problem we assume

$$\psi(x) = \left(\sum_{r=0}^{\infty} c_r\, \xi^{r+s}\right)\exp\left[-\frac{1}{2}\xi^2\right] \tag{135}$$

and obtain

$$c_r = \frac{2s + 2r - 3 - \Lambda}{(s+r)(s+r-1)}\, c_{r-2}, \quad r \ge 2$$

where $\Lambda = 2E/\hbar\omega$. We use $s = 0$ as the root of the indicial equation.

(a) Why should the infinite series in Eq. (135) be made into a polynomial?

(b) For what values of λ will the infinite series become a polynomial?

(c) Determine the normalized eigenfunction for the first excited state.

(a) Consider the operator

$$a = \frac{\mu\omega x + ip}{\sqrt{2\,\mu\hbar\omega}}$$

where symbols have their usual meaning. Replace p by the corresponding differential operator and obtain the eigenfunctions and eigenvalues of the operator a. Normalize the eigenfunction.

(b) Discuss the solution of the eigenvalue equation for $\bar{a}$.

REFERENCES AND SUGGESTED READINGS

1. F.T. Arechi, *Quantum Optics and Photon Statistics*, in *Lasers and Their Applications*, A. Sona (ed.), Gorden and Breach, New York, p. 497.
2. R. Loudon, *The Quantum Theory of Light*, Clarendon Press, Oxford (1973).
3. A.K. Ghatak, I.C. Goyal and S.J. Chua, *Mathematical Physics*, Laxmi Publications Limited, New Delhi (2015).
4. J.L. Powell and B. Craseman, *Quantum Mechanics*, Addison-Wesley (1961).

8 One-dimensional Barrier Transmission Problems

It is possible in quantum mechanics to sneak quickly across a region which is illegal energetically.

— RICHARD FEYNMAN[1]

8.1 INTRODUCTION

In earlier two chapters we considered the bound states corresponding to one-dimensional potentials. For example, we showed that for a particle in a one-dimensional potential well of finite depth (see Section 6.6.2) there exist a number of discrete states (for $E < V_0$), the corresponding wavefunctions vanishing at large distances from the origin. In this chapter, we will consider solutions for $E > V_0$ and will show that the corresponding wavefunctions in the region $|x| > a/2$ do not vanish at large distances from the origin and E can have any arbitrary value (greater than V_0). Thus, for such a potential well there exists a finite number of bound states (with $E < V_0$) and a continuum of states for which $E > V_0$. A good example is the neutron-proton problem; for $E < 0$ we have the deuteron nucleus (see Problem 10.4) and for $E > 0$, we have the neutron-proton scattering problem (see Section 26.6.5). The same is true for the electron-proton problem, for $E < 0$ we have discrete bound states corresponding to the hydrogen atom (see Section 10.4) and for $E > 0$ we have continuum of states corresponding to the scattering of electrons by protons.

It may be mentioned that there are situations where the continuous spectrum does not exist at all; e.g., for the infinitely deep potential well (see Section 6.6.1), the linear harmonic oscillator (see Chapter 7), etc. On the other hand, there may be situations where the discrete spectrum may not exist at all and one may have only the continuous spectrum; examples of such a situation are the potential step and the rectangular potential barrier discussed in Sections 8.2 and 8.3. Further, there may be a continuum along with an infinite number of discrete states as in the potential well problem [see Sections 6.6.2 and 8.4].

The solutions corresponding to the continuous spectrum are of great importance in connection with the scattering of a particle by a potential field. For example, in Sections 8.2 and 8.3 we will consider the incidence of a particle (of specified energy) on a potential step and on a rectangular potential barrier respectively and in each case we will calculate

[1] The author found this quotation in *The Quantum Universe* by T. Hey and P. Walters, Cambridge University Press, Cambridge (1987).

the reflection and transmission coefficients.[2] Such barrier transmission problems have many practical applications, e.g. in α-decay (see Section 18.7), field emission (see Problem 18.9), etc. However, unfortunately for most practical cases the exact solution is very difficult and one usually resorts to approximate techniques; one such technique is the JWKB approximation which will be discussed in Chapter 18 along with some practical applications.

8.2 THE POTENTIAL STEP

We first consider a potential step (see Fig. 8.1) for which

$$\begin{aligned} V(x) &= 0 \qquad x < 0 \\ &= V_0 \qquad x > 0 \end{aligned} \tag{1}$$

We consider the incidence of a particle (of energy E) from the left of the barrier. Classically, if $E > V_0$, the particle penetrates the barrier and keeps on moving (with reduced kinetic energy) in the positive x direction; on the other hand, if $E < V_0$ the particle is reflected back at $x = 0$ because the particle does not have sufficient energy to be in the region $x > 0$. Quantum mechanically, we will show that for $E > V_0$, the particle may undergo reflection and for $E < V_0$, although the reflection coefficient is unity, the wavefunction is non-zero in the region $x > 0$ implying that there is a finite probability of finding the particle in the region $x > 0$.

We will first discuss the case $E > V_0$ [see Fig. 8.1(a)]. The time-independent Schrödinger equation [see Eq. (11) of Chapter 6] takes the form:

$$\frac{d^2\psi(x)}{dx^2} + k^2\psi(x) = 0 \quad \text{for} \quad x < 0 \tag{2}$$

and

$$\frac{d^2\psi(x)}{dx^2} + k_1^2\psi(x) = 0 \quad \text{for} \quad x > 0 \tag{3}$$

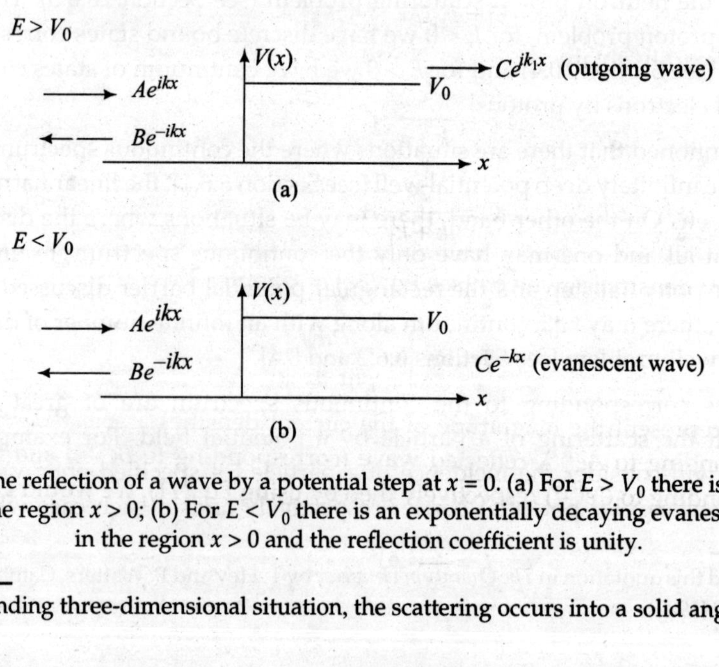

Figure 8.1: The reflection of a wave by a potential step at $x = 0$. (a) For $E > V_0$ there is an outgoing wave in the region $x > 0$; (b) For $E < V_0$ there is an exponentially decaying evanescent wave in the region $x > 0$ and the reflection coefficient is unity.

[2] In the corresponding three-dimensional situation, the scattering occurs into a solid angle $d\Omega$; see Chapter 26.

where $k = \left[\frac{2\mu E}{\hbar^2}\right]^{1/2}$ and $k_1 = \left[\frac{2\mu}{\hbar^2}(E - V_0)\right]^{1/2}$ (4)

The solutions are

$$\psi(x) = Ae^{ikx} + Be^{-ikx} \qquad x < 0 \tag{5}$$
$$= Ce^{ik_1x} + De^{-ik_1x} \qquad x > 0 \tag{6}$$

We should mention here that the complete (time-dependent) wavefunction is obtained by multiplying Eqs. (5) and (6) by

$$e^{-iEt/\hbar}$$

Thus, for $x < 0$

$$\Psi(x, t) = Ae^{i(kx - \omega t)} + Be^{-i(kx + \omega t)} \tag{7}$$

(Incident wave; propagating along + x direction) (Reflected wave; propagating along – x direction)

where $\omega \equiv \frac{E}{\hbar}$

The two terms in Eq. (7) represent the incident and reflected waves. Similarly, for $x > 0$

$$\Psi(x, t) = Ce^{i(k_1x - \omega t)} + De^{-i(k_1x + \omega t)} \tag{8}$$

(Transmitted wave propagating along + x direction) (Wave propagating along – x direction)

where the first and second terms on the RHS represent waves propagating in the $+x$ direction and $-x$ direction respectively. Since we are considering the incidence of a particle from the left of the barrier at $x = 0$, there cannot be a wave propagating in the $-x$ direction in the region $x > 0$ and hence we must put $D = 0$. We now return to Eqs (5) and (6). Continuity of ψ and $d\psi/dx$ at $x = 0$ gives us

$$A + B = C$$

and

$$ik(A - B) = ik_1C$$

from which we readily obtain

$$B = \frac{k - k_1}{k + k_1}A \tag{9}$$

$$C = \frac{2k}{k + k_1}A \tag{10}$$

Now, the current density is given by [see Eq. (36) of Chapter 4]

$$\mathbf{J} = \text{Re}\left[\psi^* \frac{\hbar}{i\mu}\frac{\partial \psi}{\partial x}\right]\hat{\mathbf{x}} \tag{11}$$

If J_i, J_r and J_t represent the magnitude of the current densities associated with the incident wave (corresponding to Ae^{ikx}), reflected wave (corresponding to Be^{-ikx}) and the transmitted wave (corresponding to Ce^{ik_1x}) respectively then by using Eq. (11), we would get

$$J_i = \frac{\hbar k}{\mu}|A|^2 \tag{12}$$

$$J_r = \frac{\hbar k}{\mu}|B|^2 \tag{13}$$

and

$$J_t = \frac{\hbar k_1}{\mu}|C|^2 \tag{14}$$

Thus, the reflection and transmission coefficients are given by

$$R = \frac{J_r}{J_i} = \frac{|B|^2}{|A|^2} = \frac{(k-k_1)^2}{(k+k_1)^2} \tag{15}$$

and

$$T = \frac{J_t}{J_i} = \frac{k_1|C|^2}{k|A|^2} = \frac{4kk_1}{(k+k_1)^2} \tag{16}$$

It is easy to see that $R + T = 1$ as it indeed should be.

We next consider the case $E < V_0$. For $E < V_0$, the Schrödinger equation becomes [cf. Eq. (3)]

$$\frac{d^2\psi}{dx^2} - \kappa^2 \psi(x) = 0 \quad x > 0 \tag{17}$$

where

$$\kappa = \left[\frac{2\mu}{\hbar^2}(V_0 - E)\right]^{1/2} \tag{18}$$

Thus, in the region $x > 0$, the solution is given by

$$\psi(x) = Ce^{-\kappa x} + De^{\kappa x} \tag{19}$$

We reject the second solution (i.e., when $D = 0$) because it increases indefinitely with x. The remaining analysis would remain the same except that k_1 has to be replaced by $i\kappa$:

$$k_1 \rightarrow i\kappa = i\left[\frac{2\mu}{\hbar^2}(V_0 - E)\right]^{1/2} \tag{20}$$

Continuity of ψ and $d\psi/dx$ leads to:

$$B = \frac{k - i\kappa}{k + i\kappa} A = e^{-2i\gamma} A \tag{21}$$

where

$$\tan\gamma = \frac{\kappa}{k} \tag{22}$$

Equation (21) tells us that there is a phase change on reflection but the amplitude of the reflected wave is the same as that of the incident wave. Thus, the reflection coefficient is given by

$$R = \frac{|B|^2}{|A|^2} = 1 \tag{23}$$

In the region $x > 0$, since $\psi(x)$ is now real, the transmitted current vanishes [see Eq. (11)] and one has

$$T = 0 \tag{24}$$

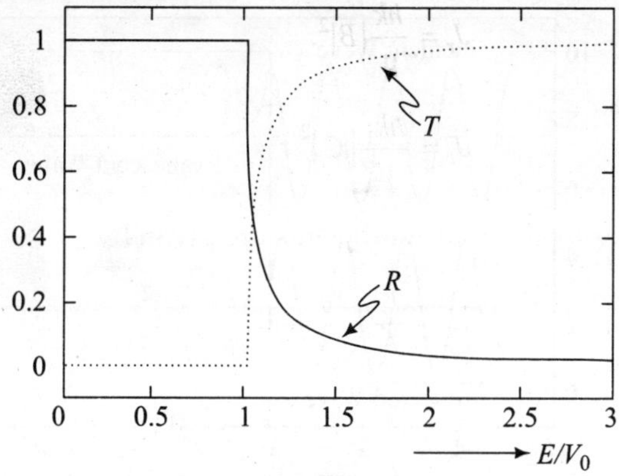

Figure 8.2: The variation of reflection and transmission coefficients with E/V_0.

Thus, $R + T = 1$, as expected. It may be noted that although the transmission coefficient vanishes, the wavefunction is not zero in the region $x > 0$, i.e., there is a finite probability of finding the particle in the *classically forbidden* region. However, in order to observe the particle in the region $x > 0$, $\Delta x \sim 1/\kappa$ (see Problem 8.4) and, therefore, the uncertainty in the momentum should be given by

$$\Delta p \gtrsim \frac{\hbar}{\Delta x} \sim \hbar\kappa \sim \sqrt{2\mu(V_0 - E)} \tag{25}$$

Thus, $$\text{uncertainty in the kinetic energy} \sim \frac{(\Delta p)^2}{2\mu} \sim (V_0 - E) \tag{26}$$

which should be sufficient to raise it to the classically allowed region. Thus, if we try to observe the particle in the region $x > 0$, we necessarily impart so much of kinetic energy to it that the total energy is greater than V_0.

One can easily see that if we plot R and T as a function of E/V_0, the plot is universal as shown in Fig. 8.2. Figures 8.3 and 8.4, give typical plots of the real part of the wavefunction for $\mu = m_p \approx 1836\, m_e$, $V_0 = 10$ MeV corresponding to $E = \frac{1}{2} V_0 = 5$ MeV and $E = 2V_0 = 20$ MeV, respectively.

In order to get a better physical insight, we should consider the reflection (and transmission) of a wave packet. This is discussed in Section 8.5.

8.3 THE RECTANGULAR POTENTIAL BARRIER

We next consider a potential barrier of the type shown in Fig. 8.5. Classically, for a particle of energy E incident on the barrier, if $E < V_0$ the particle gets reflected by the barrier and if $E > V_0$, the particle gets transmitted through the barrier. We will see that according to quantum mechanics, even for $E < V_0$ there is a finite probability of the particle *tunnelling* through the barrier. This is entirely a quantum mechanical effect and emission of α-particles from the nucleus and field emission of electrons are examples of this *tunnelling* phenomenon (see Section 18.7 and Problem 18.9). Further, for $E > V_0$, there may be a finite probability for reflection.

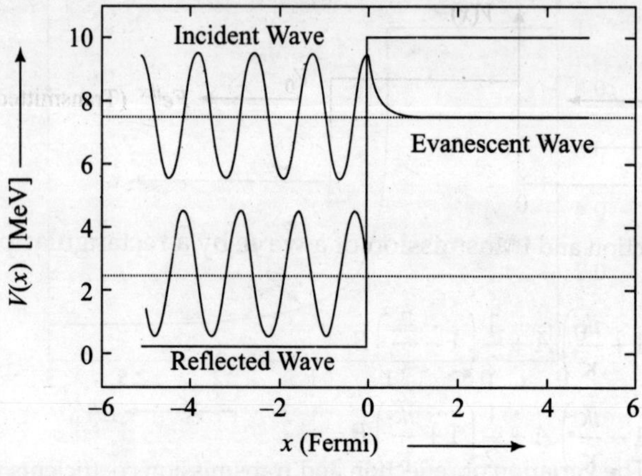

Figure 8.3: The real part of incident and reflected wavefunctions corresponding to $\mu = m_p \approx 1836\, m_e$, $V_0 = 10$ MeV and $E = \frac{1}{2} V_0 = 5$ MeV. Notice that the amplitude of the two wavefunctions are the same implying complete reflection of the wave. However, for $x > 0$, the wavefunction is exponentially decaying but nonzero.

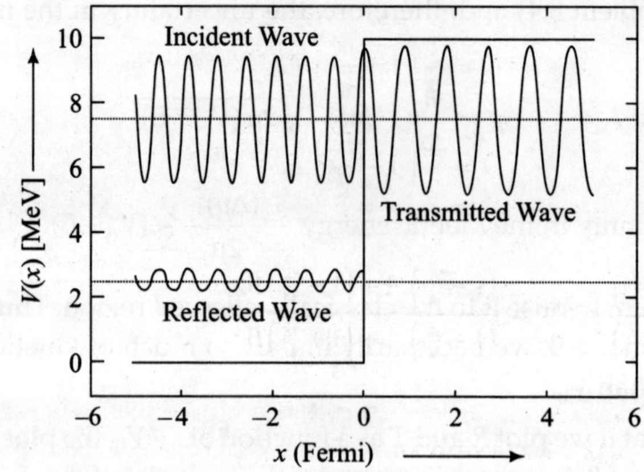

Figure 8.4: The real part of the incident, reflected and transmitted wavefunctions corresponding to $\mu = m_p \approx 1836\, m_e$, $V_0 = 10$ MeV and for $E = 2V_0 = 20$ MeV. Notice that the wavelength is increased in the region $x > 0$.

We first consider $E < V_0$. For such a case the solutions of the Schrödinger equation in the three regions are given by:

$$\left.\begin{aligned} \psi &= Ae^{ikx} + Be^{-ikx} && x < 0 \\ &= Ce^{\kappa x} + De^{-\kappa x} && 0 < x < a \\ &= Fe^{ikx} + Ge^{-ikx} && x > a \end{aligned}\right\} \tag{27}$$

where $$k = \left(\frac{2\mu E}{\hbar^2}\right)^{1/2} \quad \text{and} \quad \kappa = \left(\frac{2\mu(V_0 - E)}{\hbar^2}\right)^{1/2} \tag{28}$$

As in Section 8.2, we set $G = 0$ because there cannot be a wave propagating in the $-x$ direction in the region $x > a$. Continuity of ψ and $d\psi/dx$ at $x = 0$ gives us

$$A + B = C + D$$

$$\frac{ik}{\kappa}(A - B) = C - D$$

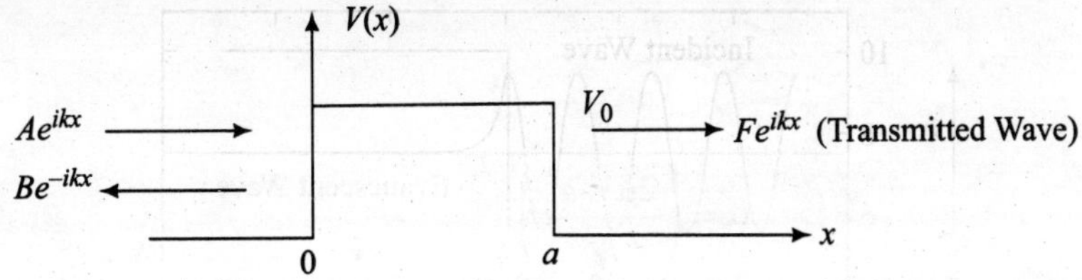

Figure 8.5: The reflection and transmission of a wave by a rectangular potential barrier.

Thus, $$C = \frac{1}{2}\left(1+\frac{ik}{\kappa}\right)A + \frac{1}{2}\left(1-\frac{ik}{\kappa}\right)B \tag{29}$$

and $$D = \frac{1}{2}\left(1-\frac{ik}{\kappa}\right)A + \frac{1}{2}\left(1+\frac{ik}{\kappa}\right)B \tag{30}$$

Continuity conditions at $x = a$, give us

$$Ce^{\kappa a} + De^{-\kappa a} = Fe^{ika}$$

and $$Ce^{\kappa a} - De^{-\kappa a} = \frac{ik}{\kappa}Fe^{ika}$$

Thus, $$C = \frac{1}{2}\left(1+\frac{ik}{\kappa}\right)Fe^{-\kappa a}e^{ika} \tag{31}$$

$$D = \frac{1}{2}\left(1-\frac{ik}{\kappa}\right)Fe^{\kappa a}e^{ika} \tag{32}$$

and, therefore,

$$\frac{C}{D} = \frac{\left(1+\frac{ik}{\kappa}\right)}{\left(1-\frac{ik}{\kappa}\right)}e^{-2\kappa a} = \frac{\left(1+\frac{ik}{\kappa}\right)A + \left(1-\frac{ik}{\kappa}\right)B}{\left(1-\frac{ik}{\kappa}\right)A + \left(1+\frac{ik}{\kappa}\right)B} \tag{33}$$

Simple manipulations give us

$$\frac{B}{A} = \frac{(k^2+\kappa^2)\sinh \kappa a}{(k^2-\kappa^2)\sinh \kappa a + 2ik\kappa\cosh \kappa a} \tag{34}$$

We also have

$$\frac{F}{A} = \frac{2ik\kappa e^{-ika}}{(k^2-\kappa^2)\sinh \kappa a + 2ik\kappa\cosh \kappa a} \tag{35}$$

The above equations will lead to the following expressions for the reflection and transmission coefficients:

$$R = \frac{J_r}{J_i} = \frac{\frac{\hbar k}{\mu}|B|^2}{\frac{\hbar k}{\mu}|A|^2} = \left[1+\frac{4\varepsilon(1-\varepsilon)}{\sinh^2(\alpha\sqrt{1-\varepsilon})}\right]^{-1} \tag{36}$$

$$T = \frac{J_t}{J_i} = \frac{\frac{\hbar k}{\mu}|C|^2}{\frac{\hbar k}{\mu}|A|^2} = \left[1+\frac{\sinh^2(\alpha\sqrt{1-\varepsilon})}{4\varepsilon(1-\varepsilon)}\right]^{-1} \tag{37}$$

where $$\varepsilon = \frac{E}{V_0} \tag{38}$$

and $$\alpha = \left(\frac{2\mu V_0 a^2}{\hbar^2}\right)^{1/2} \tag{39}$$

is a dimensionless variable characterizing the potential. In writing Eqs (36) and (37), we have used the fact that

$$\kappa a = \alpha\sqrt{1-\varepsilon} \tag{40}$$

Notice that

$$R + T = 1$$

Further, for $\kappa a \gg 1$

$$T \approx 16\varepsilon\,(1-\varepsilon)e^{-2\kappa a} \tag{41}$$

For $E > V_0$, the analysis is very similar, excepting that in the region $0 < x < a$, instead of the solutions $e^{\pm\kappa x}$ we will have solutions $e^{\pm ik_1 x}$ where

$$k_1 = \left[\frac{2\mu\,(E - V_0)}{\hbar^2}\right]^{1/2} \tag{42}$$

Thus, instead of Eq. (27) we will have

$$\left.\begin{aligned} \psi &= Ae^{ikx} + Be^{-ikx} && x < 0 \\ &= Ce^{ik_1 x} + De^{-ik_1 x} && 0 < x < a \\ &= Fe^{ikx} && x > a \end{aligned}\right\} \tag{43}$$

implying that in the entire analysis we have to replace κ by ik_1 everywhere:

$$\kappa \to ik_1 = i\left[\frac{2\mu\,(E - V_0)}{\hbar^2}\right]^{1/2} \tag{44}$$

The final results are

$$R = \left[1 + \frac{4\varepsilon\,(\varepsilon - 1)}{\sin^2(\alpha\sqrt{\varepsilon - 1})}\right]^{-1} \tag{45}$$

$$T = \left[1 + \frac{\sin^2(\alpha\sqrt{\varepsilon - 1})}{4\varepsilon\,(\varepsilon - 1)}\right]^{-1} \tag{46}$$

Once again $R + T = 1$. In writing Eqs (45) and (46), we have used the fact that

$$k_1 a = \alpha\sqrt{\varepsilon - 1}$$

As an example, we assume

$$\mu = m_p \approx 1836\, m_e, \quad a = 0.5\ \text{Å} \quad \text{and} \quad V_0 = 0.1\text{eV} \tag{47}$$

Using the above values, we would get

$$\alpha = 3.4709$$

From Eq. (46) we may readily see that the transmission coefficient is unity for

$$k_1 a \,[= \alpha\sqrt{\varepsilon - 1}] = \pi, 2\pi, 3\pi, \ldots \tag{48}$$

or

$$a = \frac{\lambda_1}{2}, 2\frac{\lambda_1}{2}, 3\frac{\lambda_1}{2}, \cdots \tag{49}$$

where

$$\lambda_1 = \frac{2\pi}{k_1} \tag{50}$$

Thus, whenever the barrier width is a multiple of $\lambda_1/2$, perfect transmission occurs. This is similar to the interference pattern observed in a Fabry-Perot etalon. Now,

$$k_1 a = \sqrt{\frac{2\mu(E - V_0)a^2}{\hbar^2}} = \alpha\sqrt{\left(\frac{E}{V_0} - 1\right)} \tag{51}$$

Thus, using Eq. (48), the transmission coefficient is unity for

$$\frac{E}{V_0} = 1 + \frac{n^2\pi^2}{\alpha^2} \tag{52}$$

For $\alpha \approx 3.4709$, the values are

$$\frac{E}{V_0} \approx 1.819, 4.277, 8.373, \ldots$$

which are consistent with the plot in Fig. 8.6.

For the values of various parameters as given by Eq. (47) the real parts of the incident, reflected and transmitted waves (for $E = 1.2V_0 = 0.12$ eV) are plotted in Fig. 8.7. The real parts of the total wavefunctions are plotted in Figs 8.8, 8.9 and 8.10 corresponding to $E = \frac{1}{2}V_0$, $1.819V_0$ and $1.2V_0$, respectively. The value $E = 1.819V_0$ corresponds to $a = \lambda_1/2$. Notice that the amplitude of the transmitted wavefunction is the same as that of the wavefunction in the region $x < 0$ implying the absence of the reflected wave. The calculations have been done using the software developed in Ref. 1.

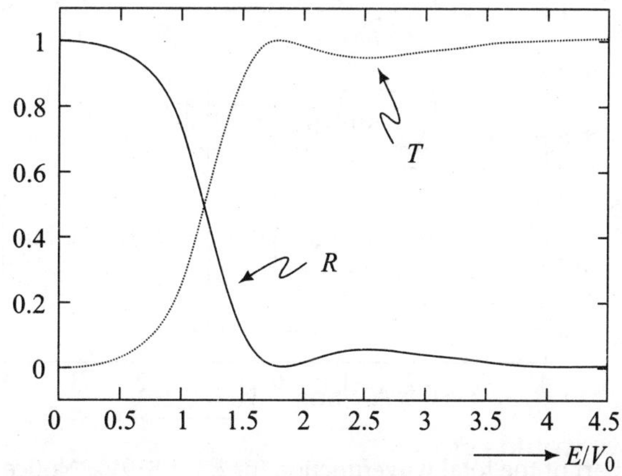

Figure 8.6: The solid and dotted curves correspond respectively to the variation of the reflection and transmission coefficients with $\varepsilon = (E/V_0)$ for a rectangular potential barrier (see Fig. 8.5) with $\alpha = 3.4709$.

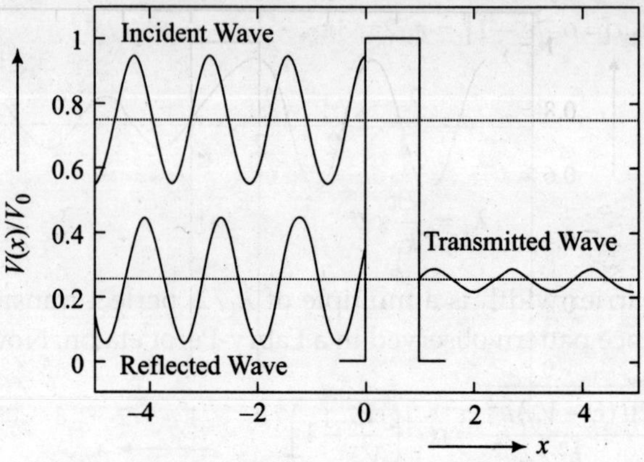

Figure 8.7: The real part of the incident, reflected and transmitted wavefunctions for $E = 1.2V_0 = 0.12$ eV.

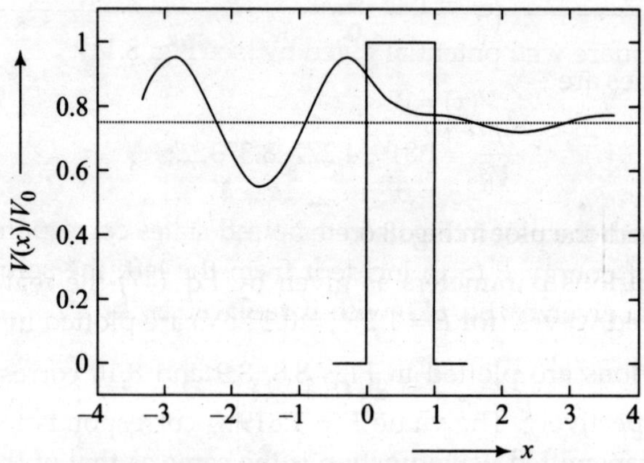

Figure 8.8: The real part of the total wavefunction for $E = \frac{1}{2}\,V_0 = 0.05$ eV.

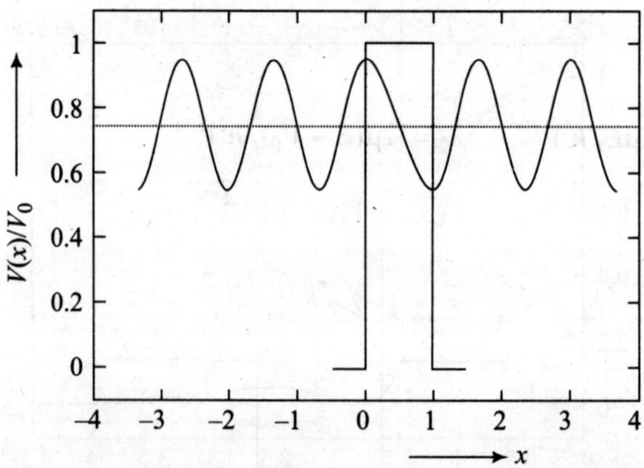

Figure 8.9: The real part of the total wavefunction for $E = 1.819V_0$. Notice that the amplitude of the transmitted wave is the same as that of the wavefunction in the region $x < 0$ implying that $T = 1$ and the absence of the reflected wave.

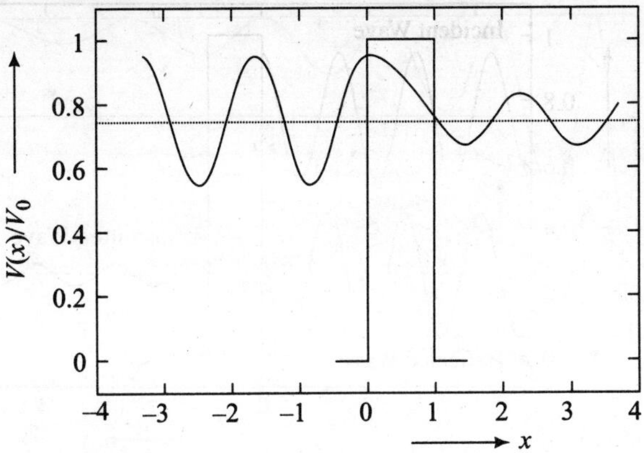

Figure 8.10: The real part of the total wavefunction for $E = 1.2V_0$.

8.4 THE SQUARE WELL POTENTIAL

We next consider a square well potential given by [see Fig. 8.11]:

$$\begin{aligned} V(x) &= 0 \qquad & x < 0 \\ &= -V_0 \qquad & 0 < x < a \\ &= 0 \qquad & x > a \end{aligned} \tag{53}$$

In Section 6.6.2, we had discussed the discrete bound states corresponding to such a potential well. For a particle of energy E (> 0) incident from the left, the solution of the Schrödinger equation will be again given by Eq. (43) with k_1 replaced by k_2:

$$k_1 \to k_2 = \left[\frac{2\mu(E + V_0)}{\hbar^2}\right]^{1/2}$$

It is left as an exercise to show that

$$T = \left[1 + \frac{1}{4}\left(\frac{k}{k_2} - \frac{k_2}{k}\right)^2 \sin^2(k_2 a)\right]^{-1} \tag{54}$$

where $\qquad k = (2\mu E/\hbar^2)^{1/2}, \quad k_2 = [2\mu(E + V_0)/\hbar^2]^{1/2}$

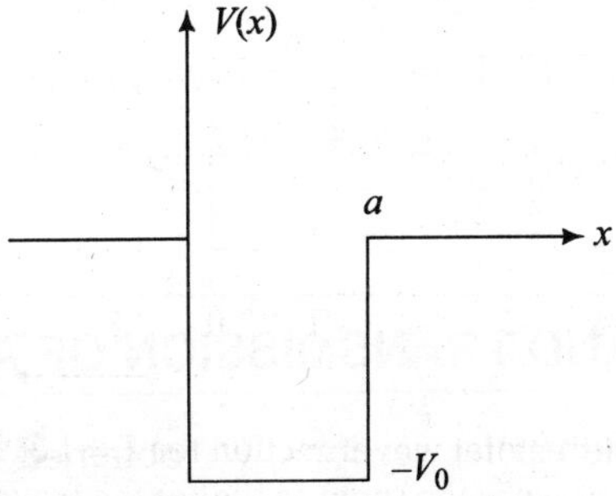

Figure 8.11: The square well potential.

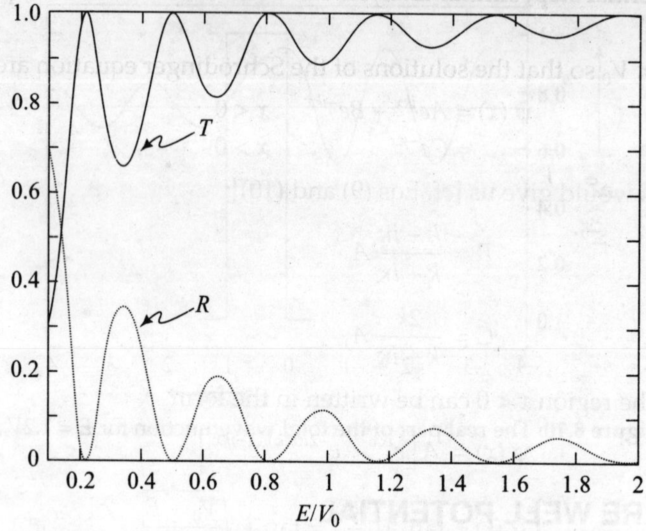

Figure 8.12: Variation of T and R as a function of E/V_0 for $\alpha = 25.598$.

Notice that transmission resonances ($T = 1$) occur when $k_2 a = n\pi$. It may be mentioned that such transmission resonances occur in the scattering of electrons from noble gases such as argon and neon. This is known as the Ramsauer-Townsend effect (see, e.g. Ref. 2, p. 133). Equation (54) can be put in the following convenient form

$$T = \left[1 + \frac{1}{4\varepsilon(\varepsilon+1)}\sin^2 \alpha\sqrt{1+\varepsilon}\right]^{-1} \tag{55}$$

where

$$\varepsilon = \frac{E}{V_0} \quad \text{and} \quad \alpha = \sqrt{\frac{2\mu V_0 a^2}{\hbar^2}}$$

It is left as an exercise to show that Eqs (46) and (55) are consistent. From Eq. (55) it readily follows that $T = 1$ for

$$\varepsilon = \frac{n^2\pi^2}{\alpha^2} - 1$$

where n has to be an integer and greater than α/π. For

$$\mu = m_e, \quad a = 10\ \text{Å}, \quad V_0 = 25\ \text{eV}$$

$$\alpha \approx 25.598$$

and, therefore, n should exceed 8.15. Thus for $T = 1$, we must have $n = 9, 10, \ldots$ and therefore

$$\varepsilon \approx 0.22, 0.51, 0.82, \ldots$$

as shown in Fig. 8.12.

8.5 THE REFLECTION AND TRANSMISSION OF A WAVE PACKET

In the analysis discussed in the earlier sections, we obtained expressions for reflection and transmission coefficients by considering the time-independent solutions of the Schrödinger equation. However, in order to get a better physical insight we consider the incidence of a wave packet on a potential barrier. For the sake of simplicity, we consider the reflection of a wave packet by a potential step; similar analysis can be carried out for other potential barriers also.

We will assume $E < V_0$ so that the solutions of the Schrödinger equation are given by

$$\begin{aligned} \psi(x) &= Ae^{ikx} + Be^{-ikx} \qquad x < 0 \\ &= Ce^{-\kappa x} \qquad x > 0 \end{aligned} \tag{56}$$

Continuity conditions would give us [cf. Eqs (9) and (10)]:

$$B = \frac{k - i\kappa}{k + i\kappa} A \tag{57}$$

$$C = \frac{2k}{k + i\kappa} A \tag{58}$$

Thus the solution in the region $x < 0$ can be written in the form

$$\psi(x) = A\left[e^{ikx} + e^{-2i\alpha(k)}\, e^{-ikx}\right] \tag{59}$$

where

$$\alpha(k) = \tan^{-1}\left(\frac{\kappa}{k}\right) = \tan^{-1}\sqrt{\frac{V_0 - E}{E}} \tag{60}$$

Since all values of E are possible, we can construct a wave packet formed by the superposition of plane waves (cf. Section 5.5):

$$\Psi(x, t) = \frac{1}{2\pi}\int_0^{\beta} A(k)\left[e^{ikx} + e^{-2i\alpha(k)}\, e^{-ikx}\right] e^{-i\omega(k)t}\, dk \tag{61}$$

where

$$\omega(k) = \frac{\hbar k^2}{2m} \tag{62}$$

and the upper limit β corresponds to $E = V_0$:

$$\beta = \sqrt{\frac{2mV_0}{\hbar^2}} \tag{63}$$

We write Eq. (61) as a superposition of incident and reflected wave packets

$$\Psi(x, t) = \Psi_i(x, t) + \Psi_r(x, t) \tag{64}$$

where

$$\Psi_i(x, t) = \frac{1}{\sqrt{2\pi}}\int_0^{\beta} |A(k)|\, e^{i[kx - \omega(k)t + \phi(k)]}\, dx \quad \text{Incident wave packet} \tag{65}$$

$$\Psi_r(x, t) = \frac{1}{\sqrt{2\pi}}\int_0^{\beta} |A(k)|\, e^{-i[kx + \omega(k)t + 2\alpha(k) - \phi(k)]}\, dk \quad \text{Reflected wave packet} \tag{66}$$

and

$$A(k) = |A(k)|\, e^{i\phi(k)} \tag{67}$$

If we assume $|A(k)|$ to be a sharply peaked function around $k = k_0$ and carry out an analysis similar to that given in Section 5.5, we would readily obtain the following expressions for the center of the wave packet associated with the incident and reflected waves

$$x_c^{\text{inc}}(t) = x_0 + v_g t \tag{68}$$

$$x_c^{\text{refl}}(t) = -x_0 - v_g t + \tau \tag{69}$$

where $$x_0 = -\left.\frac{d\phi}{dk}\right|_{k=k_0} \tag{70}$$

$$v_g = \left.\frac{d\omega}{dk}\right|_{k=k_0} = \frac{\hbar k_0}{m} \tag{71}$$

and $$\tau = -2\left.\frac{d\alpha}{dk}\right|_{k=k_0} = \frac{2}{\sqrt{\beta^2 - k_0^2}} \quad \text{Time delay} \tag{72}$$

represents the time delay[3] introduced during the reflection of the wave packet. Obviously, x_0 is negative; further, both $x_c^{\text{inc}}(t)$ and $x_c^{\text{refl}}(t)$ should be negative because the solution given by Eq. (61) corresponds to the region $x < 0$. Now at $t = 0$ we assume the packet to be far away from the barrier (see Fig. 8.13) and as t increases the packet moves towards the barrier. At a much later time the reflected packet moves towards the left (see Fig. 8.15) with the same group velocity v_g. At intermediate times the incident and reflected waves interfere near $x = 0$ and the total wavefunction is quite complicated (see Fig. 8.14). Finally, the transmitted wave packet (in the region $x > 0$) will be given by

$$\Psi_t = \frac{1}{\sqrt{2\pi}}\int_0^{\beta} |A(k)| \frac{2k}{\sqrt{k^2+\kappa^2}} \times \exp\left[-\kappa(k)\,x - i\,\{(\omega(k)\,t + \alpha(k) - \phi(k))\}\right] dk \tag{73}$$

where $\omega(k)$ is given by Eq. (62) and use has been made of the fact that the amplitude of the transmitted wave packet is given by [see Eq. (58)]

$$C = \frac{2k}{k + i\kappa} A = \frac{2k}{(k^2+\kappa^2)^{1/2}} e^{-i\alpha(k)} \tag{74}$$

As a simple example, we assume that at $t = 0$, the particle is described by a localized Gaussian wave packet given by the following equation (cf. Example 5.1)

$$\Psi(x, 0) = \frac{1}{(\pi\sigma^2)^{1/4}} e^{-(x-x_0)^2/2\sigma^2} e^{ik_0 x} \tag{75}$$

We assume $x_0 \gg \sigma$, so that the particle is localized far away from the potential step. The quantity $\hbar k_0$ represents the average momentum of the wave packet. Now,

$$A(k) \approx \frac{1}{\sqrt{2\pi}}\int_{-\infty}^{+\infty} \Psi(x, 0)\, e^{-ikx}\, dx \tag{76}$$

where we have put the limits from $-\infty$ to $+\infty$ because for $|x - x_0| \gtrsim 4\sigma$ the integrand has a negligible value. Carrying out the integration, we readily get

$$A(k) \approx \left(\frac{\sigma^2}{\pi}\right)^{1/4} \exp\left[-\frac{1}{2}\sigma^2 (k-k_0)^2 - i(k-k_0)\,x_0\right] \tag{77}$$

---

[3] This can be physically understood from the fact that since the wavefunction is finite in the region $x > 0$ (see Section 8.2) the packet can be thought of as penetrating into the region $x > 0$ for a certain distance and then coming back (see the discussion at the end of this section and Problem 8.5); the corresponding optical problem is discussed in Chapter 11 of Ref. 2.

Thus,

$$\Psi(X,\tau) = \text{const} \int_0^{\delta} \exp\left[-\frac{1}{2}\alpha^2 (K-1)^2 - i(K-1)\gamma\right] \times [e^{iKX} + r(K)\, e^{-iKX}] \exp\left[-\frac{iK^2\tau}{2}\right] dK \quad (78)$$

where we have introduced the following dimensionless variables

$$K = \frac{k}{k_0},\ X = k_0 x,\ \tau = \frac{t}{m/(\hbar k_0^2)}$$

$$\alpha = k_0\sigma,\ \gamma = k_0 x_0,\ \delta = \sqrt{\frac{V_0}{(\hbar^2 k_0^2/2m)}} \quad (79)$$

Further

$$r(K) = \frac{1 - i\Gamma(K)}{1 + i\Gamma(K)} \quad (80)$$

where

$$\Gamma(K) = \frac{\kappa}{k} = \sqrt{\frac{\delta^2}{K^2} - 1} \quad (81)$$

Figure 8.13, shows the results of the numerical integration for

$$\alpha = 10,\ \beta = -100,\ \delta = 2$$

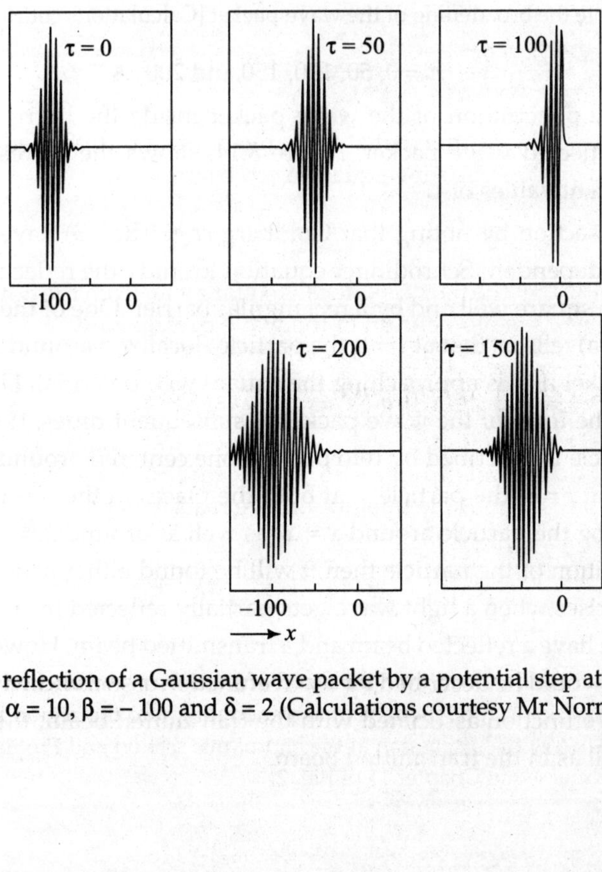

Figure 8.13: The reflection of a Gaussian wave packet by a potential step at $\tau = 0, 50, 100, 150$ and 250; $\alpha = 10$, $\beta = -100$ and $\delta = 2$ (Calculations courtesy Mr Norman Lim).

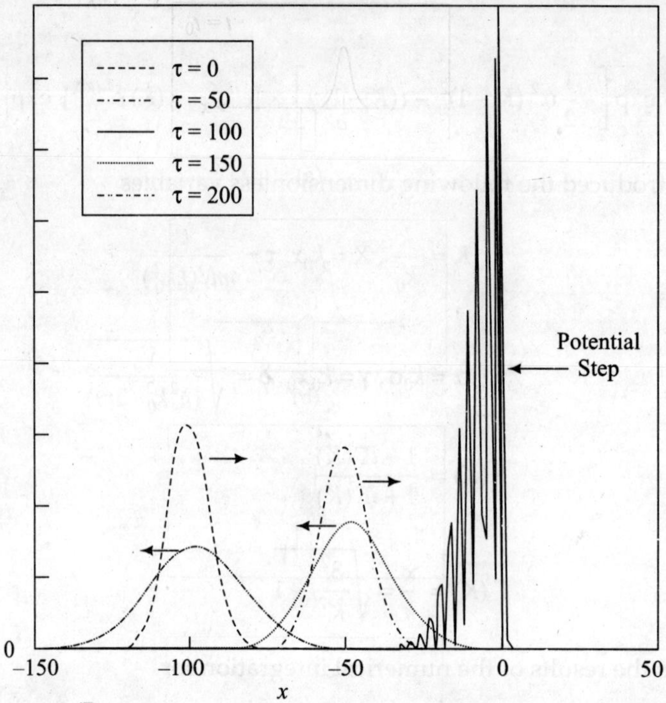

Figure 8.14: Variation of $|\Psi(X, \tau)|^2$ for the Gaussian wave packet at different values of τ; $\alpha = 10$, $\beta = -100$ and $\delta = 2$. Note the broadening of the wave packet [Calculations courtesy Mr Norman Lim].

at

$$\tau = 0, 50, 100, 150 \text{ and } 200$$

The figure shows the penetration of the wave packet inside the barrier and also the slight broadening of the reflected wave packet. Figure 8.14, shows the corresponding probability distributions at different values of τ.

We conclude this section by noting that Goldberg *et al.* (Ref. 3) have obtained numerical solutions of the time dependent Schrödinger equation to study the reflection and transmission of a wave packet by a square well and by a rectangular barrier. One of their results is shown in Fig. 8.15. Figure 8.15(a) tells us that at $t = t_0$ the particle (localized around $x = a$) is described by a Gaussian wave packet and is approaching the square well potential. Figures 8.15(b), 8.15(c) and 8.15(d) give us the form of the wave packet at subsequent times. It is interesting to note that at $t = t_1$ the particle is described by two packets one centered around $x = x_1$ and the other around $x = x_2$. Thus at $t = t_1$ the particle is at both the places in the sense that there are finite probabilities of finding the particle around $x = x_1$ as well as around $x = x_2$; however, if we try to determine the position of the particle then it will be found either near $x = x_1$ or near $x = x_2$. A similar situation arises when a light wave gets partially reflected by a dielectric (like glass). In classical optics, we have a reflected beam and a transmitted beam. However, if we consider a single photon then it would be described by a wavefunction associated with the reflected beam superposed by a wavefunction associated with the transmitted beam; thus it is present in the reflected beam as well as in the transmitted beam.

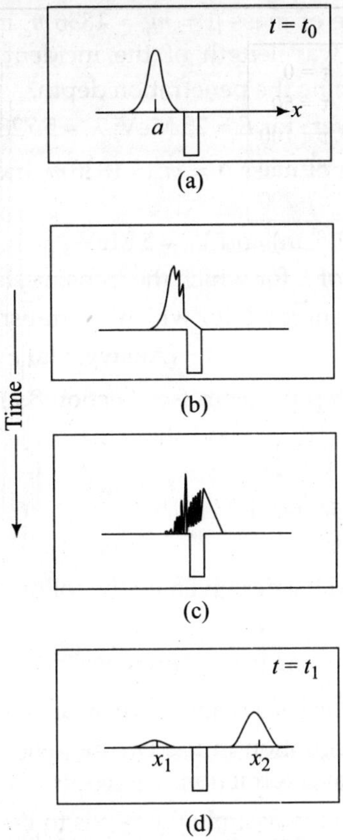

Figure 8.15: Reflection and transmission of a wave packet by a square well; the mean energy of the packet corresponds to the well depth (adapted from Ref. 3).

If we make a measurement then it will be found either in the reflected beam or in the transmitted beam.

8.6 PROBLEMS

Problem 8.1 Consider a particle of mass $\mu = m_e$ incident on a potential step (Fig. 8.1) with $V_0 = 10^3$ eV.

(a) Obtain the

(i) wavefunctions at $E = 500$ eV and 800 eV

(ii) wavelengths of the incident and the transmitted waves in each case

(iii) the penetration depth in each case

[**Answer:** For $E = 500$ eV; $\lambda_i \approx 0.549$ Å and the penetration depth $x_t \approx 0.087$ Å; for $E = 800$ eV; $\lambda_i \approx 0.434$ Å and the penetration depth $x_t \approx 0.138$ Å]

(b) Calculate the reflection and the transmission coefficients and wavefunctions at $E = 1200$ eV, 1500 eV and 2000 eV. Explain, why the amplitude of the transmitted wave is larger than that of the corresponding incident wave.

[**Answer:** $R \approx 0.1766$, 0.718 and 0.0294]

Problem 8.2 Consider a particle of mass $\mu = m_p \approx 1836 m_e$ incident on a potential step with $V_0 = 50$ MeV. Determine the wavelength of the incident and the transmitted waves at $E = 25$ MeV and 40 MeV. Determine the penetration depth.

[**Answer:** for $E = 25$ MeV; $\lambda_i \approx 5.728 \times 10^{-5}$ Å and $x_t = 9.117 \times 10^{-6}$ Å]

Problem 8.3 Consider a particle of mass $\mu = m_p \approx 1836 m_e$ incident on a rectangular potential barrier [see Fig. 8.5] with

$$a = 10 \text{ fermi } (= 10^{-14} \text{ m}) \text{ and } V_0 = 5 \text{ MeV}$$

(a) Calculate the lowest value of E for which the transmission coefficient T will be unity.

(b) Determine the next higher energy E for which the transmission coefficient will be unity.

[**Answer: (a)** $E \approx 7.05$ MeV; **(b)** $E \approx 13.203$ MeV]

Problem 8.4 For the potential step problem (see Section 8.2) the wavefunction in the region $x > 0$ is of the form $e^{-\kappa x}$. Calculate $\langle x \rangle$, $\langle x^2 \rangle$ and show that $\Delta x \sim 1/\kappa$.

[**Hint:** $\langle x \rangle = \int_0^\infty x e^{-2\kappa x}\, dx \Big/ \int_0^\infty e^{-2\kappa x}\, dx$, etc.]

Problem 8.5 Consider a potential barrier given by the following equation

$$\begin{aligned} V &= V_1 \qquad z < 0 \\ &= V_2 \qquad z > 0 \end{aligned} \tag{82}$$

with $V_2 > V_1$. Consider the incidence of a plane wave (at an angle) on the surface $z = 0$. Solve the three-dimensional Schrödinger equation in the two regions and using continuity conditions determine the laws of reflection and refraction.

[**Hint:** Without any loss of generality assume the x-axis to lie in the plane of incidence so that the y-dependence of the wavefunction can be neglected.]

Problem 8.6 Consider a delta function potential of the form

$$V(x) = S\,\delta(x) \tag{83}$$

Show that the transmission amplitude ($= F/A$) and the transmission coefficient ($= |F/A|^2$) are

$$\frac{ik}{\left(ik - \frac{mS}{\hbar^2}\right)}$$

and

$$\frac{E}{\left(E + \frac{mS^2}{2\hbar^2}\right)}$$

respectively.

[**Hint:** In the results given in Section 8.3 take the limit of $V_0 \to \infty$ and $a \to 0$ such that the product $V_0 a \to S$.]

Problem 8.7 Consider a potential barrier of the form (see Fig. 8.16)

$$\begin{aligned} V(x) &= V_0 \qquad |x| < a/2 \\ &= 0 \qquad |x| > a/2 \end{aligned} \tag{84}$$

In the region $|x| > a/2$, the solution of the Schrödinger equation can be written as

$$\begin{aligned} \psi &= Ae^{ikx} + Be^{-ikx} \qquad x < -a/2 \\ &= Fe^{ikx} + Ge^{-ikx} \qquad x > +a/2 \end{aligned} \tag{85}$$

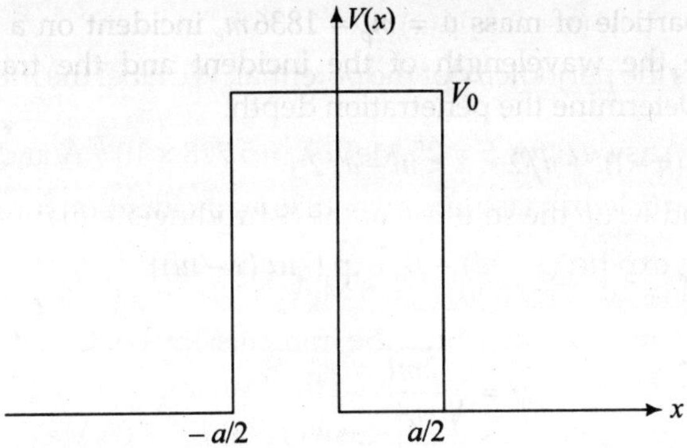

Figure 8.16: A rectangular potential barrier.

Using continuity conditions at $x = \pm a/2$ show

$$\begin{pmatrix} A \\ B \end{pmatrix} = \begin{pmatrix} R_{11} & R_{12} \\ R_{21} & R_{22} \end{pmatrix} \begin{pmatrix} F \\ G \end{pmatrix} \tag{86}$$

where

$$\left.\begin{aligned} R_{11} &= \left[\cosh \kappa a + \tfrac{i\sigma}{2} \sinh \kappa a\right] e^{ika} \\ R_{22} &= \left[\cosh \kappa a - \tfrac{i\sigma}{2} \sinh \kappa a\right] e^{-ika} \\ R_{12} &= (i\tau/2) \sinh \kappa a = -R_{21} \\ \sigma &= \frac{\kappa}{k} - \frac{k}{\kappa}, \quad \tau = \frac{\kappa}{k} + \frac{k}{\kappa} \end{aligned}\right\} \tag{87}$$

k and κ are defined through Eqs (4) and (18). For $E < V_0$, κ is real and for $E > V_0$, κ is purely imaginary.

Problem 8.8 In the previous problem if we set $G = 0$, then it will physically correspond to the situation of a wave incident from the leftside of the barrier and getting partially reflected and partially transmitted by it. Calculate the transmission coefficient for the cases $E < V_0$ and $E > V_0$ and show that the results are the same as obtained in Section 8.3.

Problem 8.9 Consider a periodic potential of the form[4] (see Fig. 8.17)

$$\begin{aligned} V(x) &= V_0 \qquad -a/2 < x < +a/2 \\ &= 0 \qquad a/2 < x < a/2 + b \end{aligned} \tag{88}$$

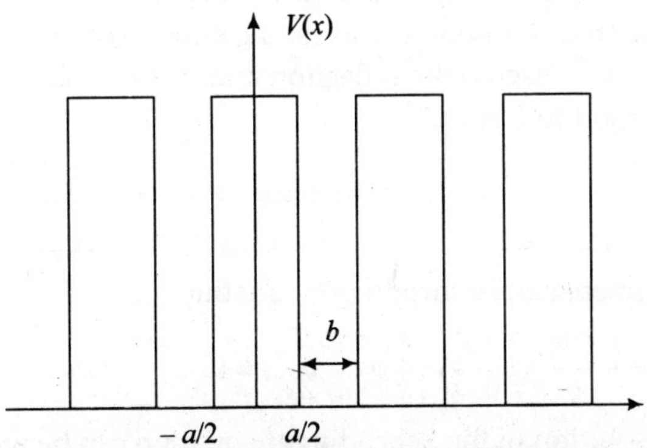

Figure 8.17: The periodic potential.

[4] Such a periodic potential is of great importance in band theory of solids (see, e.g. Ref. 4). The present method of solution for the periodic potential is adapted from Ref. 5; see also Chapter 16.

with $$V(x + l) = V(x) \tag{89}$$

where $l = a + b$ represents the periodicity of the potential. We label the nth valley (of the periodic potential) as

$$(n-1)l + a/2 < x < nl - a/2 \tag{90}$$

with $n = 0, \pm 1, \pm 2, \ldots$ and write the solution of the Schrödinger equation in the nth valley as

$$\psi(x) = A_n \exp[i\alpha(x - nl) + B_n \exp[-i\alpha(x - nl)] \tag{91}$$

where we have defined

$$\alpha = \sqrt{\frac{2mE}{\hbar^2}} \tag{92}$$

so that the notation is consistent with that in Section 16.3. We write

$$\begin{pmatrix} A_{n+1} \\ B_{n+1} \end{pmatrix} = \begin{pmatrix} R_{11} & R_{12} \\ R_{21} & R_{22} \end{pmatrix} \begin{pmatrix} A_n \\ B_n \end{pmatrix} \tag{93}$$

Use continuity conditions to determine R_{11}, R_{12}, ... and show that the matrix elements are independent of n. Thus it immediately follows that

$$\begin{pmatrix} A_n \\ B_n \end{pmatrix} = \begin{pmatrix} R_{11} & R_{12} \\ R_{21} & R_{22} \end{pmatrix}^n \begin{pmatrix} A_0 \\ B_0 \end{pmatrix} \tag{94}$$

Using the boundary condition that the wavefunction must remain finite as $x \to \pm\infty$, obtain the equation which lead to the allowed and forbidden energy bands.

Problem 8.10 Consider a potential energy variation given by

$$\left.\begin{aligned} V(x) &= 0 && \text{for} \quad x < 0 \\ &= V_0\left(1 - \frac{x}{a}\right) && \text{for} \quad 0 < x < a \\ &= 0 && \text{for} \quad x > a \end{aligned}\right\} \tag{95}$$

Obtain an exact solution of the Schrödinger equation and an expression for the transmission coefficient.

[**Hint:** In the region $0 < x < a$, the solution can be written in terms of Airy functions – see Section 17.11]

Problem 8.11 In Section 8.5, we had considered the reflection of a wave packet by a potential step and had assumed that the component waves, the superposition of which gives the wave packet, corresponds to $E < V_0$. Discuss the reflection and transmission of a wave packet whose component waves correspond to $E > V_0$.

8.7 SOLUTIONS

Solution 8.5 The three-dimensional Schrödinger equation

$$\nabla^2\psi + \frac{2m}{\hbar^2}[E - V(z)]\,\psi(x, y, z) = 0 \tag{96}$$

with
$$\left.\begin{aligned} V(z) &= V_1 \qquad z < 0 \\ &= V_2 \qquad z > 0 \end{aligned}\right\} \tag{97}$$

can easily be solved by using the method of separation of variables. The solution corresponding to the reflection and transmission of a particle is given by

$$\begin{aligned}\psi &= A \exp\,[i\,(k_{1x}\,x + k_{1z}\,z)] + B \exp\,[i\,(k_{2x}\,x - k_{2z}\,z)] \quad z < 0 \\ &= C \exp\,[i\,(k_{3x}\,x + k_{3z}\,z)] \qquad\qquad\qquad\qquad\qquad\quad z > 0\end{aligned} \tag{98}$$

where we have assumed (without any loss of generality) that $k_{1y} = 0$, i.e., the x-axis is chosen to lie in the plane of incidence (see Fig. 8.18). Further,

$$k_{1x}^2 + k_{1z}^2 = k_{2x}^2 + k_{2z}^2 = \frac{2m}{\hbar^2}\,(E - V_1) \tag{99}$$

and

$$k_{3x}^2 + k_{3z}^2 = \frac{2m}{\hbar^2}\,(E - V_2) \tag{100}$$

For the wavefunction to be continuous, at $z = 0$ (for all values of x) we must have

$$k_{1x} = k_{2x} = k_{3x} \tag{101}$$

Thus $k_{2z} = k_{1z}$ [see Eq. (99)] implying that the angles of incidence and reflection are equal. Further, if i and r represent the angles of incidence and refraction then,

$$\sin i = \frac{k_{1x}}{\left(k_{1x}^2 + k_{1z}^2\right)^{1/2}} = \left(\frac{\hbar^2}{2m(E - V_1)}\right)^{1/2} k_{1x} \tag{102}$$

and a similar relation for sin r, from which we immediately obtain

$$\frac{\sin i}{\sin r} = \left(\frac{E - V_2}{E - V_1}\right)^{1/2} \tag{103}$$

which is nothing but Snell's law. The critical angle is given by

$$\sin i_c = \left(\frac{E - V_2}{E - V_1}\right)^{1/2} \tag{104}$$

For $i > i_c$, k_{3z} is purely imaginary and there is an exponentially decaying wave in the region $z > 0$.

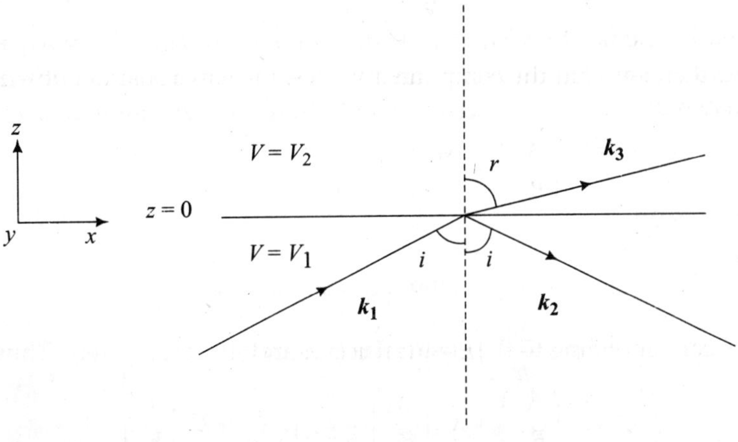

Figure 8.18: The reflection and refraction of a plane wave at the plane $z = 0$.

Now, continuity of ψ and $d\psi/dz$ at $z = 0$ gives [cf. Eqs (9) and (10)]

$$\frac{B}{A} = \frac{k_{1z} - k_{3z}}{k_{1z} + k_{3z}} = \frac{\cos i - (\sin^2 i_c - \sin^2 i)^{1/2}}{\cos i + (\sin^2 i_c - \sin^2 i)^{1/2}} \tag{105}$$

and
$$\frac{C}{A} = \frac{2k_{1z}}{k_{1z} + k_{3z}} = \frac{2\cos i}{\cos i + (\sin^2 i_c - \sin^2 i)^{1/2}} \tag{106}$$

For $i > i_c$, $|B/A| = 1$ and the reflection coefficient is unity. However, the wavefunction is given by

$$\psi = C \exp[i\,\kappa x \sin i] \exp\left[-\kappa z \frac{(\sin^2 i - \sin^2 i_c)^{1/2}}{\sin i_c}\right] \tag{107}$$

where $\kappa = [2m(E - V_2)/\hbar^2]^{1/2}$. The above wavefunction gives the following expression for the current density:

$$J_x = \frac{\hbar\kappa}{m}\frac{\sin i}{\sin i_c} |C|^2 \exp\left[-2\kappa \frac{(\sin^2 i - \sin^2 i_c)^{1/2}}{\sin i_c} z\right] \tag{108}$$

$$J_y = J_z = 0.$$

Such a wave, which is propagating in a particular direction and has an exponentially decreasing amplitude in another direction, is known as an evanescent wave and is a subject of considerable importance in optics. Because of this evanescent wave, the beam (on reflection) undergoes a lateral shift which is known as the Goos-Hanchen effect (see, e.g. Chapter 11 of Ref. 6).

Solution 8.9 The continuity conditions immediately give

$$\begin{aligned} R_{11} &= e^{i\alpha b}\left[\cosh \kappa a - \frac{i\sigma}{2}\sinh \kappa a\right] \\ R_{12} &= -\frac{i\tau}{2}\sinh \kappa a\, e^{i\alpha(a+b)} \\ R_{22} &= R_{11}^*,\ R_{21} = R_{12}^* \end{aligned} \tag{109}$$

where σ and τ are defined in Problem 8.7. Now the eigenvalues of R matrix are given by

$$\lambda_\pm = \frac{1}{2}\left[\eta \pm \sqrt{\eta^2 - 4}\right] \tag{110}$$

where we have used the fact that $R_{11}R_{22} - R_{12}R_{21} = 1$ and $\eta = R_{11} + R_{22} = R_{11} + R_{11}^*$ which is a real quantity. Further, any column vector can also be written as a sum of two eigenvectors and, therefore, we may write

$$\begin{pmatrix} A_0 \\ B_0 \end{pmatrix} = \begin{pmatrix} a_+ \\ b_+ \end{pmatrix} + \begin{pmatrix} a_- \\ b_- \end{pmatrix}$$

where
$$\begin{pmatrix} a_+ \\ b_+ \end{pmatrix} \text{ and } \begin{pmatrix} a_- \\ b_- \end{pmatrix}$$

are eigenvectors corresponding to the eigenvalues λ_+ and λ_-, respectively. Thus,

$$\begin{pmatrix} A_n \\ B_n \end{pmatrix} = R^n \begin{pmatrix} A_0 \\ B_0 \end{pmatrix} = \lambda_+^n \begin{pmatrix} a_+ \\ b_+ \end{pmatrix} + \lambda_-^n \begin{pmatrix} a_- \\ b_- \end{pmatrix} \tag{111}$$

Evidently, if $|\eta| > 2$, the two roots are real, the wavefunction will diverge either for $x \to \infty$ or $x \to -\infty$. Thus, we must have

$$\eta \le 2 \tag{112}$$

for which $|\lambda_+| = |\lambda_-| = 1$. Since η is real we can always write

$$\frac{1}{2}\eta = \cos\phi \tag{113}$$

or

$$\cos\phi = \cosh\kappa a \cos\alpha b + \frac{\sigma}{2}\sinh\kappa a \sin\alpha b \tag{114}$$

For $\alpha b = n\pi$, it immediately follows that Eqs. (112) and (114) cannot be satisfied and hence such energy values are not allowed. By plotting $\eta/2$ as a function of E, one can readily obtain regions of allowed and forbidden energies—these are the energy bands and are discussed in any book on solid state physics.

For $E > V_0$, the corresponding equation becomes

$$\cos\phi = \cos k_1 a \cos\alpha b - \frac{k_1^2 + \alpha^2}{2\alpha k_1}\sin k_1 a \sin\alpha b \tag{115}$$

which again lead to allowed and forbidden energy bands.

In order to get a numerical appreciation, we consider the limiting case of $V_0 \to \infty$ and $a \to 0$ such that

$$\frac{mb}{\hbar^2} V_0 a \to P \tag{116}$$

(This is known as the Krönig-Penney model.) For such a case, Eq. (114) becomes

$$\cos\phi = \frac{P}{\alpha b}\sin\alpha b + \cos\alpha b \tag{117}$$

Equation (117) forms the basic equation of the Krönig-Penney model, the solution of which leads to the formation of energy bands (corresponding to a periodic potential). We will again briefly discuss the formation of energy bands in Section 16.5.

Solution 8.11 For $E > V_0$, the incident, reflected and transmitted wave packets would be [cf. Eqs (65), (66) and (73)]

$$\psi_{\text{inc}}(x, t) = \frac{1}{\sqrt{2\pi}}\int_\beta^\infty |A(k)| \times \exp\left[i\left(kx - \omega(k)\,t - \phi(k)\right)\right] dk \tag{118}$$

$$\psi_{\text{refl}}(x, t) = \frac{1}{\sqrt{2\pi}}\int_\beta^\infty |A(k)| \frac{k - \sqrt{k^2 - \beta^2}}{k + \sqrt{k^2 - \beta^2}} \times \exp\left[-i\left(kx - \omega(k)\,t + \phi(k)\right)\right] dk \tag{119}$$

$$\psi_{\text{tr}}(x, t) = \frac{1}{\sqrt{2\pi}}\int_\beta^\infty |A(k)| \frac{2k}{k + \sqrt{k^2 - \beta^2}} \times \exp\left[i\left(\sqrt{k^2 - \beta^2}\, x - \omega(k)t + \phi(k)\right)\right] dx \tag{120}$$

where, we have used Eqs (9) and (10) and the fact that

$$k_1 = \left[\frac{2m(E - V_0)}{\hbar^2}\right]^{1/2} = [k^2 - \beta^2]^{1/2} \tag{121}$$

Notice that since B/A and C/A are real [see Eqs (9) and (10)], there are no additional phase factors appearing in the exponent. Thus, the center of the wave packet associated with the incident, reflected and transmitted waves would be [cf. Eqs (68) and (69)]

$$\left.\begin{aligned} x_c^{\text{inc}} &= x_0 + v_g t \\ x_c^{\text{ref}} &= -x_0 - v_g t \\ x_c^{\text{tr}} &= x_0 \frac{v'_g}{v_g} + v'_g t \end{aligned}\right\} \tag{122}$$

where $$v'_g = \frac{\hbar\sqrt{k_0^2 - \beta^2}}{m} \tag{123}$$

which represents the group velocity in the region $x > 0$. The expression for v'_g follows from the fact that

$$(k^2 - \beta^2)^{1/2} x - \omega t = \left[(k_0^2 - \beta^2)^{1/2} + \frac{k_0}{(k_0^2 - \beta^2)^{1/2}}(k - k_0) + \cdots\right] x$$

$$-\left[\omega(k_0) + (k - k_0)\frac{d\omega}{dk}\bigg|_{k=0} + \cdots\right] t$$

$$= \left[(k_0^2 - \beta^2)^{1/2} x - \omega(k_0)t\right] + (k - k_0)\frac{k_0}{\sqrt{k_0^2 - \beta^2}}$$

$$\times \left\{x - \frac{\sqrt{k_0^2 - \beta^2}\,\dfrac{d\omega}{dk}\Big|_{k=k_0}}{k_0}\, t\right\} + \ldots$$

REFERENCES AND SUGGESTED READINGS

1. A. Ghatak, *Basic Quantum Mechanics* (with CD), Macmillan India Ltd., New Delhi (2002).
2. N.F. Mott and H.S.W. Massey, *The Theory of Atomic Collisions*, Oxford University Press, London (1949).
3. A. Goldberg, H.M. Sachey and J.L. Schwartz, *Computer generated motion pictures of one-dimensional quantum mechanical transmission and reflection phenomena, Am. J. Phys.*, **35**, 177 (1967).
4. A.J. Dekker, *Solid State Physics*, Prentice-Hall, Englewood Cliffs, New Jersey (1952).
5. E. Merzbacher, *Quantum Mechanics*, John Wiley, New York (1970).
6. A.K. Ghatak and K. Thyagarajan, *Contemporary Optics*, Plenum Press, New York (1978); Reprinted by Macmillan India Limited, New Delhi.
7. P.R. Holland, *The Quantum Theory of Motion*, Cambridge University Press (1993).

9 Angular Momentum I—The Spherical Harmonics

The final criterion of truth is the agreement of a theory with experience, and it is only when all attempts to describe the facts in the frame of accepted ideas fail that new notions are formed, at first cautiously and reluctantly and then, if they are experimentally confirmed, with increasing confidence. In this way the classical philosophy of science was transformed into the modern one . . .

— From MAX BORN'S Preface in *Physics in My Generation*, Springer Verlag, New York (1969).

9.1 INTRODUCTION

In this chapter, we will discuss the basic theory of angular momentum which plays an extremely important role in the study of quantum mechanics. We will first define the angular momentum operator through the classical relation $\mathbf{L} = \mathbf{r} \times \mathbf{p}$ and replace $\mathbf{p}$ by its operator representation $-i\hbar\nabla$ [see Eq. 18 of chapter 4]. In Section 9.2, we will obtain the representations of L_x, L_y, L_z and L^2 $(= L_x^2 + L_y^2 + L_z^2)$ in spherical polar coordinates (r, θ, ϕ); for example, we will show that L_z and L^2 are given by the following differential operators:

$$L_z = -i\hbar\frac{\partial}{\partial\phi} \tag{1}$$

and

$$L^2 = -\hbar^2\left[\frac{1}{\sin\theta}\frac{\partial}{\partial\theta}\left(\sin\theta\frac{\partial}{\partial\theta}\right) + \frac{1}{\sin^2\theta}\frac{\partial^2}{\partial\phi^2}\right] \tag{2}$$

(In Section 9.5, we will describe a clever method to obtain the representations of L_x, L_y, L_z and L^2 in spherical polar coordinates). In Section 9.3, we will solve the eigenvalue equation

$$L^2 Y(\theta, \phi) = \lambda\,\hbar^2 Y(\theta, \phi) \tag{3}$$

where $\lambda\,\hbar^2$ represents the eigenvalues of the operator L^2. We will show that λ takes the value $l(l + 1)$ where $l = 0, 1, 2, 3, \ldots$ and for each value of l there is $(2l + 1)$– fold degeneracy; i.e., there are $(2l + 1)$ eigenfunctions corresponding to the same eigenvalue $l(l + 1)\hbar^2$ – these eigenfunctions are known as spherical harmonics and are denoted by $Y_{lm}(\theta, \phi)$. Thus the eigenvalue equation (3) can be written in the form

$$L^2 Y_{lm}(\theta, \phi) = l(l + 1)\,\hbar^2 Y_{lm}(\theta, \phi) \tag{4}$$

where $$l = 0, 1, 2, 3, \ldots$$

and for each value of l, we have

$$m = -l, -l+1, \dots +l$$

The spherical harmonics are actually simultaneous eigenfunctions of L^2 and L_z, i.e.,

$$L_z Y_{lm}(\theta, \phi) = m\hbar Y_{lm}(\theta, \phi) \tag{5}$$

These eigenfunctions will be used in the theory of the hydrogen atom (Chapter 10); in fact, in Section 10.2 we will show that for any problem characterized by a spherically symmetric potential $V(r)$, the angular part of the eigenfunctions are always the spherical harmonics $Y_{lm}(\theta, \phi)$. In Section 9.4, the commutation relations between L_x, L_y and L_z will be derived which will be the starting equations in Chapter 13. The relationship of angular momentum with rotations will also be discussed in Chapter 13.

9.2 THE ANGULAR MOMENTUM OPERATOR AND ITS REPRESENTATIONS IN SPHERICAL POLAR COORDINATES

In classical mechanics the angular momentum of a particle is defined through the relation

$$\mathbf{L} = \mathbf{r} \times \mathbf{p} \tag{6}$$

where **r** and **p** represent the position and linear momentum of the particle. Thus,

$$L_x = y\,p_z - z\,p_y \tag{7}$$

with similar relations for L_y and L_z. In quantum mechanics we replace p_x, p_y and p_z by the corresponding operators [see Eq. (19) of Chapter 4] to obtain the following operator representations of L_x, L_y and L_z:

$$\left.\begin{aligned} L_x &= yp_z - zp_y = -i\hbar\left(y\frac{\partial}{\partial z} - z\frac{\partial}{\partial y}\right) \\ L_y &= zp_x - xp_z = -i\hbar\left(z\frac{\partial}{\partial x} - x\frac{\partial}{\partial z}\right) \\ L_z &= xp_y - yp_x = -i\hbar\left(x\frac{\partial}{\partial y} - y\frac{\partial}{\partial x}\right) \end{aligned}\right\} \tag{8}$$

In order to determine the eigenvalues and eigenfunctions of the operator L^2 $(\equiv L_x^2 + L_y^2 + L_z^2)$ it is convenient to express L_x, L_y and L_z in spherical polar coordinates which are defined through the following equations:

$$\left.\begin{aligned} x &= r\sin\theta\cos\phi \\ y &= r\sin\theta\sin\phi \\ z &= r\cos\theta \end{aligned}\right\} \tag{9}$$

Thus,

$$r^2 = x^2 + y^2 + z^2 \tag{10}$$

$$\tan^2\theta = \frac{(x^2 + y^2)}{z^2} \tag{11}$$

$$\tan\phi = \frac{y}{x} \tag{12}$$

Now[1]

$$2r\frac{\partial r}{\partial x} = 2x \tag{13}$$

$$2\tan\theta\sec^2\theta\frac{\partial\theta}{\partial x} = \frac{2x}{z^2} = \frac{2}{r}\tan\theta\sec\theta\cos\phi \tag{14}$$

$$\sec^2\phi\frac{\partial\phi}{\partial x} = -\frac{y}{x^2} = -\frac{1}{r}\operatorname{cosec}\theta\sec\phi\tan\phi \tag{15}$$

giving

$$\frac{\partial r}{\partial x} = \sin\theta\cos\phi \tag{16}$$

$$\frac{\partial\theta}{\partial x} = \frac{1}{r}\cos\theta\cos\phi \tag{17}$$

$$\frac{\partial\phi}{\partial x} = -\frac{1}{r}\sin\phi\operatorname{cosec}\theta \tag{18}$$

Similarly, one can obtain $\frac{\partial r}{\partial y}, \frac{\partial\theta}{\partial y}$, etc.

$$\text{Now, } L_z\psi(r,\theta,\phi) = -i\hbar\left[x\frac{\partial\psi}{\partial y} - y\frac{\partial\psi}{\partial x}\right]$$

$$= -i\hbar\left[x\left(\frac{\partial\psi}{\partial r}\frac{\partial r}{\partial y} + \frac{\partial\psi}{\partial\theta}\frac{\partial\theta}{\partial y} + \frac{\partial\psi}{\partial\phi}\frac{\partial\phi}{\partial y}\right) - y\left(\frac{\partial\psi}{\partial r}\frac{\partial r}{\partial x} + \frac{\partial\psi}{\partial\theta}\frac{\partial\theta}{\partial x} + \frac{\partial\psi}{\partial\phi}\frac{\partial\phi}{\partial x}\right)\right]$$

$$= -i\hbar\left[r\sin\theta\cos\phi\left\{\frac{\partial\psi}{\partial r}(\sin\theta\sin\phi) + \frac{\partial\psi}{\partial\theta}\left(\frac{1}{r}\cos\theta\sin\phi\right)\right.\right.$$

$$\left.+\frac{\partial\psi}{\partial\phi}\left(\frac{1}{r}\cos\phi\operatorname{cosec}\theta\right)\right\} - r\sin\theta\sin\phi\left\{\frac{\partial\psi}{\partial r}\times(\sin\theta\cos\phi) + \frac{\partial\psi}{\partial\theta}\left(\frac{1}{r}\cos\theta\cos\phi\right)\right.$$

$$\left.\left.+\frac{\partial\psi}{\partial\phi}\left(-\frac{1}{r}\sin\phi\operatorname{cosec}\theta\right)\right\}\right]$$

or,

$$L_z\psi(r,\theta,\phi) = -i\hbar\frac{\partial\psi(r,\theta,\phi)}{\partial\phi} \tag{19}$$

Since the above equation holds for arbitrary $\psi(r, \theta, \phi)$, we have the following operator representation for L_z:

$$L_z = -i\hbar\frac{\partial}{\partial\phi} \tag{20}$$

---

[1] Notice, that $\frac{\partial x}{\partial r} = \sin\theta\cos\phi = \frac{x}{r}$. Thus, $\left(\frac{\partial r}{\partial x}\right) \neq 1/\left(\frac{\partial x}{\partial r}\right)$.

Similarly, (see also Sec. 9.5, Problem 9.5 and Example 13.5)

$$L_x = i\hbar\left(\sin\phi\frac{\partial}{\partial\theta} + \cot\theta\cos\phi\frac{\partial}{\partial\phi}\right) \tag{21}$$

$$L_y = i\hbar\left(-\cos\phi\frac{\partial}{\partial\theta} + \cot\theta\sin\phi\frac{\partial}{\partial\phi}\right) \tag{22}$$

The operator L^2 is defined through the following equation:

$$L^2 = L_x^2 + L_y^2 + L_z^2 \tag{23}$$

where the square of an operator is defined by

$$L_x^2 = L_x L_x$$

etc. Thus

$$L_z^2\,\psi(r, \theta, \phi) = L_z L_z \psi = (-i\hbar)^2\,\frac{\partial}{\partial\phi}\cdot\frac{\partial\psi}{\partial\phi} = -\hbar^2\frac{\partial^2\psi}{\partial\phi^2}$$

Hence, the operator representation of L_z^2 is given by

$$L_z^2 = -\hbar^2\frac{\partial^2}{\partial\phi^2} \tag{24}$$

Similarly, one can obtain the operator representations of L_x^2 and L_y^2 (although the algebra is much more involved). Adding up, one obtains (see also Problems 9.6 and 9.7)

$$L^2 = -\hbar^2\left[\frac{1}{\sin\theta}\frac{\partial}{\partial\theta}\left(\sin\theta\frac{\partial}{\partial\theta}\right) + \frac{1}{\sin^2\theta}\frac{\partial^2}{\partial\phi^2}\right] \tag{25}$$

It may be noted that (see Problem 9.8)

$$L^2 = -\hbar^2 r^2\left[\nabla^2 - \frac{1}{r^2}\frac{\partial}{\partial r}\left(r^2\frac{\partial}{\partial r}\right)\right] \tag{26}$$

9.3 EIGENVALUES AND EIGENFUNCTIONS OF L^2

In this section, we will solve the eigenvalue equation

$$L^2\,Y(\theta, \phi) = \lambda\hbar^2\,Y(\theta, \phi) \tag{27}$$

where $\lambda\hbar^2$ represents the eigenvalues of L^2 and $Y(\theta, \phi)$ the corresponding eigenfunctions. We will show that λ takes the values $l(l+1)$ where $l = 0, 1, 2, \ldots$ and the corresponding eigenfunctions are the spherical harmonics. For each value of l there will be $(2l + 1)$ fold degeneracy; i.e., there will be $(2l + 1)$ eigenfunctions belonging to the same eigenvalue $l(l + 1)\hbar^2$.

Now, substituting for L^2 from Eq. (25) in Eq. (27), we obtain

$$\frac{1}{\sin\theta}\frac{\partial}{\partial\theta}\left(\sin\theta\frac{\partial Y}{\partial\theta}\right) + \frac{1}{\sin^2\theta}\frac{\partial^2 Y}{\partial\phi^2} + \lambda Y(\theta, \phi) = 0 \tag{28}$$

In order to solve the above equation, we try the method of separation of variables and write

$$Y(\theta, \phi) = \Theta(\theta)\,\Phi(\phi) \tag{29}$$

On substitution and subsequent multiplication by $\sin^2\theta / Y(\theta, \phi)$, we obtain

$$\frac{\sin^2\theta}{\Theta}\left[\frac{1}{\sin\theta}\frac{d}{d\theta}\left(\sin\theta\frac{d\Theta}{d\theta}\right) + \lambda\Theta(\theta)\right] = -\frac{1}{\Phi}\frac{d^2\Phi}{d\phi^2} = m^2 \tag{30}$$

The variables have indeed separated out and we have set each side equal to a positive constant m^2, the reason for which will shortly become clear. Equation (30) gives us

$$\frac{d^2\Phi}{d\phi^2} + m^2\,\Phi(\phi) = 0$$

the solution of which is given by

$$\Phi(\phi) \sim e^{im\phi}$$

For the wavefunction to be a single-valued, we must have

$$\Phi(\phi + 2\pi) = \Phi(\phi) \tag{31}$$

or

$$e^{2\pi m i} = 1$$

giving[2]

$$m = 0, \pm 1, \pm 2, \ldots$$

We label the functions by the subscript m

$$\Phi_m(\phi) = \frac{1}{\sqrt{2\pi}}\, e^{im\phi}\;;\; m = 0, \pm 1, \pm 2, \ldots \tag{32}$$

where the factor $1/\sqrt{2\pi}$ ensures that

$$\int_0^{2\pi} |\Phi_m(\phi)|^2\, d\phi = 1$$

which is the normalization condition. Indeed

$$\int_0^{2\pi} \Phi^*_{m'}(\phi)\,\Phi_m(\phi)\, d\phi = \delta_{mm'} \quad \text{Orthonormality condition} \tag{33}$$

which represents the orthonormality condition for $\Phi_m(\phi)$.

The equation satisfied by $\Theta(\theta)$ is [from Eq. (30)]

$$\frac{1}{\sin\theta}\frac{d}{d\theta}\left(\sin\theta\frac{d\Theta}{d\theta}\right) + \left(\lambda - \frac{m^2}{\sin^2\theta}\right)\Theta(\theta) = 0 \tag{34}$$

Writing $\cos\theta = \mu$ and $\Theta(\theta) = F(\mu)$, we obtain

$$\frac{d}{d\mu}\left[(1-\mu^2)\frac{dF}{d\mu}\right] + \left[\lambda - \frac{m^2}{1-\mu^2}\right] F(\mu) = 0 \tag{35}$$

We first consider the case $m = 0$ and then $m \neq 0$.

9.3.1 Case I : $m = 0$

For $m = 0$, Eq. (35) reduces to

$$(1-\mu^2)\frac{d^2F}{d\mu^2} - 2\mu\frac{dF}{d\mu} + \lambda F(\mu) = 0 \tag{36}$$

---

[2] It immediately follows that we could not have set $\frac{1}{\Phi}\frac{d^2\Phi}{d\phi^2}$ to a positive (or complex) constant for then the wavefunction would not have been single-valued.

As in the case of the linear harmonic oscillator problem, we try a power series solution

$$F(\mu) = \sum_{r=0,1,\ldots}^{\infty} a_r \mu^{r+s} \tag{37}$$

Differentiating (37) term by term and substituting in (36), we obtain

$$\sum_{r=0,1,2,\ldots}^{\infty} [(r+s)(r+s-1)\, a_r \mu^{r+s-2}$$

$$- \{(r+s)(r+s-1) + 2(r+s) - \lambda\}\, a_r \mu^{r+s}] = 0$$

Since the above equation is to be valid for all values of μ, we equate the coefficients of each power of μ to zero. Thus,

$$s(s-1)\, a_0 = 0 \tag{38}$$

$$s(s+1)\, a_1 = 0 \tag{39}$$

and $$(r+s+2)(r+s+1)\, a_{r+2} = [(r+s)(r+s+1) - \lambda]\, a_r\,; \; r = 0, 1, 2, \ldots \tag{40}$$

The roots of the indicial equation [Eq. (38)] are

$$s = 0 \quad \text{or} \quad s = 1 \tag{41}$$

For $s = 0$, the recurrence relation is[3]

$$\frac{a_{r+2}}{a_r} = \frac{r(r+1) - \lambda}{(r+1)(r+2)} \tag{42}$$

Since a_{r+2} is related to a_r we may write Eq. (37) as

$$F(\mu) = a_0\left[1 + \frac{a_2}{a_0}\mu^2 + \frac{a_4}{a_0}\mu^4 + \cdots\right] + a_1\left[\mu + \frac{a_3}{a_1}\mu^3 + \frac{a_5}{a_1}\mu^5 + \cdots\right] \tag{43}$$

The quantities inside the square brackets represent the two independent solutions of Eq. (36). From Eq. (42) it is readily seen that

$$\lim_{r\to\infty} \frac{a_{r+2}}{a_r} = 1 \tag{44}$$

Thus, if the infinite series in Eq. (43) is not terminated, it will diverge at $\mu = \pm 1$ (i.e. at $\theta = 0$ and $\theta = \pi$). Since that is not admissible, the series has to be terminated which can happen only if

$$\lambda = l(l+1); \quad l = 0, 1, 2, \ldots \tag{45}$$

Thus, when $\lambda = 0, 2, 6, 12, \ldots$ one of the series will terminate at $a_l \mu^l$. For example, for $l = 4$

$$\frac{a_2}{a_0} = -10\,; \quad \frac{a_4}{a_2} = -\frac{14}{12}$$

and $$a_6 = a_8 = \ldots = 0$$

Thus the polynomial solution is given by

$$P_4(\mu) = \frac{1}{8}\,[3 - 30\,\mu^2 + 35\,\mu^4] \tag{46}$$

---

[3] Since the root $s = 0$ makes a_1 indeterminate, it provides all the necessary independent solutions, (see, e.g. Ref. 3); indeed, we do obtain two independent solutions [see Eq. (43)].

where the constant a_0 has been chosen such that

$$P_l(1) = 1 \tag{47}$$

For even values of l, the odd series will not terminate (see Problem 9.2) and we must set $a_1 = 0$. Similarly, for an odd value of l the odd series will terminate at μ^l and we must set $a_0 = 0$. The resulting polynomials are known as the Legendre polynomials $P_l(\mu)$ which are assumed to satisfy Eq. (47). The first few Legendre polynomials are

$$\left.\begin{array}{ll} P_0(\mu) = 1; & P_1(\mu) = \mu; \\ P_2(\mu) = \frac{1}{2}(3\mu^2 - 1); & P_3(\mu) = \frac{1}{2}(5\mu^3 - 3\mu) \end{array}\right\} \text{Legendre polynomials} \tag{48}$$

The Legendre polynomials satisfy the following relation (see, e.g. Reference 4: see also Problem 9.9):

$$\int_{-1}^{+1} P_l(\mu)\, P_{l'}(\mu)\, d\mu = \frac{2}{2l+1}\, \delta_{ll'} \tag{49}$$

Higher order Legendre polynomials can be obtained from the following recurrence relation

$$P_l(\mu) = \frac{2l-1}{l}\mu P_{l-1}(\mu) - \frac{(l-1)}{l} P_{l-2}(\mu);\ l \geq 2 \tag{50}$$

Thus for $m = 0$, the well-behaved orthonormal sets of solutions of Eqs (35) and (28) are

$$\Theta_l(\theta) = \left(\frac{2l+1}{2}\right)^{1/2} P_l(\cos\theta) \tag{51}$$

and

$$Y_{l,0}(\theta, \phi) = \frac{1}{\sqrt{2\pi}}\Theta_l(\theta) = \left(\frac{2l+1}{4\pi}\right)^{1/2} P_l(\cos\theta) \tag{52}$$

respectively. The subscripts of $Y(\theta, \phi)$ represent the values of l and m. The orthonormality condition is

$$\iint Y_{l',0}\, Y_{l,0} \sin\theta\, d\theta\, d\phi = \left[\frac{1}{2\pi}\int_0^{2\pi} d\phi\right]\left[\int_0^{\pi} \Theta_{l'}(\theta)\Theta_l(\theta)\sin\theta\, d\theta\right] = \delta_{ll'} \tag{53}$$

9.3.2 Case II : $m \neq 0$

The 'brute force' method for obtaining $Y_{lm}(\theta, \phi)$ is to directly solve Eq. (35); this is done in Appendix G. However, the most straightforward and elegant way of obtaining the solutions is through the use of the ladder operators:

$$L_+ = L_x + iL_y = \hbar e^{i\phi}\left[\frac{\partial}{\partial\theta} + i\cot\theta\frac{\partial}{\partial\phi}\right] \tag{54}$$

$$L_- = L_x - iL_y = \hbar e^{-i\phi}\left[-\frac{\partial}{\partial\theta} + i\cot\theta\frac{\partial}{\partial\phi}\right] \quad \text{Ladder operators} \tag{55}$$

Using operator algebra, in Chapter 13 we have shown that

$$L_+ Y_{l,m} = [(l-m)(l+m+1)]^{1/2}\hbar\, Y_{l,m+1} \tag{56}$$

and

$$L_- Y_{l,m} = [(l+m)(l-m+1)]^{1/2}\hbar\, Y_{l,m-1} \tag{57}$$

From very general considerations we have also shown that for a given value of l, m can only take values $-l, -l+1, \ldots +l$. From Eqs (56) and (57) we also see that

$$L_+ Y_{l,l} = 0 \tag{58}$$

and

$$L_- Y_{l,-l} = 0 \tag{59}$$

Now from Eqs (54) and (56) we can write

$$Y_{l,m+1} = [(l-m)(l+m+1)]^{-1/2} e^{i\phi} \left[\frac{\partial}{\partial\theta} + i\cot\theta \frac{\partial}{\partial\phi}\right] Y_{l,m} \tag{60}$$

Thus, if we start from $m = 0$, we can readily obtain $Y_{l,1}$; $Y_{l,2}$; ... $Y_{l,l}$. As an example, we consider $l = 2$. Now, from Eqs. (52) and (48)

$$Y_{2,0}(\theta,\phi) = \sqrt{\frac{5}{4\pi}}\, P_2(\cos\theta)$$

$$= \sqrt{\frac{5}{16\pi}}\, [3\cos^2\theta - 1] \tag{61}$$

Thus, $$Y_{2,1} = \sqrt{\frac{5}{96\pi}}\, e^{i\phi} \left[\frac{\partial}{\partial\theta} + i\cot\theta \frac{\partial}{\partial\phi}\right] (3\cos^2\theta - 1)$$

$$= -\sqrt{\frac{15}{8\pi}} \sin\theta\cos\theta\, e^{i\phi} \tag{62}$$

and $$Y_{2,2} = -\sqrt{\frac{15}{32\pi}} e^{i\phi} \left[\frac{\partial}{\partial\theta} + i\cot\theta \frac{\partial}{\partial\phi}\right] \left(\frac{1}{2}\sin 2\theta\, e^{i\phi}\right)$$

$$= \sqrt{\frac{15}{32\pi}} \sin^2\theta\, e^{2i\phi} \tag{63}$$

Similarly, using the relation

$$Y_{l,m-1} = [(l+m)(l-m+1)]^{-1/2} e^{-i\phi} \left[-\frac{\partial}{\partial\theta} + i\cot\theta \frac{\partial}{\partial\phi}\right] Y_{l,m} \tag{64}$$

we can obtain $Y_{l,-1}$ and $Y_{l,-2}$. The resulting wavefunctions are automatically normalized. Thus if one knows the expression of $P_l(\cos\theta)$ then by repeated application of Eqs (60) and (64) one can explicitly determine Y_{lm} for all values of m. We may mention here that one can also solve Eq. (59) to determine $Y_{l,-l}$ and then use ladder operators to determine $Y_{l,m}$ — see Chapter 15 for details. The expressions for spherical harmonics have been given in Section 15.3 and satisfy the following orthonormality condition:

$$\int_0^{2\pi} d\phi \int_0^{\pi} \sin\theta\, d\theta\, Y^*_{lm}(\theta,\phi)\, Y_{l'm'}(\theta,\phi) = \delta_{ll'}\,\delta_{mm'} \quad \text{Orthonormality condition} \tag{65}$$

Thus the eigenvalue equation for L^2 [see Eq. (24)] takes the form

$$L^2 Y_{lm}(\theta,\phi) = l(l+1)\hbar^2 Y_{lm}(\theta,\phi) \tag{66}$$

where $l(l+1)\hbar^2$ (with $l = 0, 1, 2, \ldots$) represent the eigenvalues of L^2; $Y_{lm}(\theta, \phi)$ being the corresponding eigenfunctions. The eigenvalue $l(l+1)\hbar^2$ is $(2l+1)$ fold degenerate; i.e., corresponding to a particular value of l there will be $(2l+1)$ values of $m\ (=-l, -l+1, \ldots, l-1\ l, l)$. Since the ϕ-dependence of $Y_{lm}(\theta, \phi)$ is of the form $e^{im\phi}$, one obtains

$$L_z Y_{lm}(\theta, \phi) = -i\hbar \frac{\partial}{\partial\phi}\left[\Theta_l(\theta)\frac{1}{\sqrt{2\pi}}e^{im\phi}\right]$$

or

$$L_z Y_{lm}(\theta, \phi) = m\hbar\, Y_{lm}(\theta, \phi) \tag{67}$$

where we have used Eq. (20). Thus $Y_{lm}(\theta, \phi)$ are simultaneous eigenfunctions of L^2 and L_z. It is easy to show that they are not eigenfunctions of L_x and L_y except for the case $l = 0, m = 0$. In fact, it is impossible to construct eigenfunctions of any but L^2 and one of the three components of **L**. However, by choosing appropriate linear combinations, it is possible to have simultaneous eigenfunctions of L^2 and L_x and also of L^2 and L_y (see Problems 15.5 and 15.6).

9.4 THE COMMUTATION RELATIONS

In this section, we will show that the components of the angular momentum operator do not commute. For example,

$$\begin{aligned}[L_x\ L_y] &= L_x L_y - L_y L_x\\ &= (yp_z - zp_y)(zp_x - xp_z) - (zp_x - xp_z)(yp_z - zp_y)\\ &= y\,p_z zp_x - yp_z xp_z - zp_y zp_x + zp_y xp_z - zp_x yp_z + zp_x zp_y + xp_z yp_z - xp_z zp_y\\ &= yp_x(p_z z - zp_z) + p_y x(zp_z - p_z z)\\ &= i\hbar[xp_y - yp_x] = i\hbar L_z\end{aligned}$$

where we have used the commutation relations derived in Section 4.3. Thus,

$$[L_x, L_y] = i\hbar L_z \tag{68}$$

Similarly,

$$[L_y, L_z] = i\hbar L_x \tag{69}$$

$$[L_z, L_y] = i\hbar L_y \tag{70}$$

These commutation relations will be the starting point in Chapter 13 where we will show that if we just use these commutation relations (and general axioms of quantum mechanics) then the angular momentum components can have values $0, \frac{1}{2}\hbar, \hbar, \frac{3}{2}\hbar, 2\hbar$, etc.[4] The existence of systems having components of angular momenta taking the values which are half integral multiplies of $\hbar$ lead to extremely important considerations. We will discuss these in Chapters 13 and 14.

9.5 REPRESENTATIONS OF L_x, L_y AND L_z IN SPHERICAL POLAR COORDINATES: A STRAIGHTFORWARD DERIVATION

In this section, we will give a very straightforward derivation of the representations of L_x, L_y and L_z in spherical polar coordinates*. We start with the definition of the spherical coordinate

[4] It may be noted that in this chapter we have shown that components of angular momenta, defined as $\mathbf{r}\times\mathbf{p}$, can only have eigenvalues $0, \hbar, 2\hbar, \ldots$

* Adapted from the analysis given in chapter 6 of Ref. 6.

system (r, θ, ϕ) :

$$\left.\begin{aligned} x &= r\sin\theta\cos\phi \\ y &= r\sin\theta\sin\phi \\ z &= r\cos\theta \end{aligned}\right\}$$

Now, the unit vectors $\hat{\mathbf{r}}$, $\hat{\boldsymbol{\theta}}$ and $\hat{\boldsymbol{\phi}}$ are given by

$$\hat{\mathbf{r}} = \sin\theta\cos\phi\ \hat{\mathbf{x}} + \sin\theta\sin\phi\ \hat{\mathbf{y}} + \cos\theta\ \hat{\mathbf{z}} \tag{71}$$

$$\hat{\boldsymbol{\phi}} = -\sin\phi\ \hat{\mathbf{x}} + \cos\phi\ \hat{\mathbf{y}} \tag{72}$$

$$\hat{\boldsymbol{\theta}} = -\left(\hat{\mathbf{r}}\times\hat{\boldsymbol{\phi}}\right) = \cos\theta\cos\phi\ \hat{\mathbf{x}} + \cos\theta\sin\phi\ \hat{\mathbf{y}} - \sin\theta\ \hat{\mathbf{z}} \tag{73}$$

Further, in spherical polar coordinates, the gradient of a scalar function is given by

$$\nabla T = \textbf{grad}\ T = \frac{\partial T}{\partial r}\hat{\mathbf{r}} + \frac{1}{r}\frac{\partial T}{\partial\theta}\hat{\boldsymbol{\theta}} + \frac{1}{r\sin\theta}\frac{\partial T}{\partial\phi}\hat{\boldsymbol{\phi}} \tag{74}$$

Now, $\quad \mathbf{L} = \mathbf{r}\times\mathbf{p}$

Using spherical polar co-ordinates

$$\mathbf{p}\psi = -i\hbar\,\nabla\psi = -i\hbar\left[\frac{\partial\psi}{\partial r}\hat{\mathbf{r}} + \frac{1}{r}\frac{\partial\psi}{\partial\theta}\hat{\boldsymbol{\theta}} + \frac{1}{r\sin\theta}\frac{\partial\psi}{\partial\phi}\hat{\boldsymbol{\phi}}\right]$$

Since $\mathbf{r} = \hat{\mathbf{r}}\ r$, we get

$$\mathbf{L} = -i\hbar\, r\, \hat{\mathbf{r}}\times\left[\hat{\mathbf{r}}\frac{\partial}{\partial r} + \hat{\boldsymbol{\theta}}\frac{1}{r}\frac{\partial}{\partial\theta} + \hat{\boldsymbol{\phi}}\frac{1}{r\sin\theta}\frac{\partial}{\partial\phi}\right]$$

or

$$\mathbf{L} = -i\hbar\left[\hat{\boldsymbol{\phi}}\frac{\partial}{\partial\theta} + \hat{\boldsymbol{\theta}}\,\frac{1}{\sin\theta}\frac{\partial}{\partial\phi}\right] \tag{75}$$

where we have used the relations

$$\hat{\mathbf{r}}\times\hat{\boldsymbol{\theta}} = \hat{\boldsymbol{\phi}} \quad \text{and} \quad \hat{\mathbf{r}}\times\hat{\boldsymbol{\phi}} = -\hat{\boldsymbol{\theta}}$$

Thus,

$$\mathbf{L} = -i\hbar\left[\left(-\hat{\mathbf{x}}\sin\phi + \hat{\mathbf{y}}\cos\phi\right)\frac{\partial}{\partial\theta} - \left(\cos\theta\cos\phi\,\hat{\mathbf{x}} + \cos\theta\sin\phi\,\hat{\mathbf{y}} - \sin\theta\,\hat{\mathbf{z}}\right)\frac{1}{\sin\theta}\frac{\partial}{\partial\phi}\right]$$

Thus,

$$L_x = i\hbar\left[\sin\phi\frac{\partial}{\partial\theta} + \cot\theta\cos\phi\frac{\partial}{\partial\phi}\right]$$

$$L_y = i\hbar\left[-\cos\phi\frac{\partial}{\partial\theta} + \cot\theta\sin\phi\frac{\partial}{\partial\phi}\right]$$

$$L_z = -i\hbar\frac{\partial}{\partial\phi}$$

Now, $\quad L^2 = L_x^2 + L_y^2 + L_z^2$

The evaluation of L_z^2 is very easy:

$$L_z^2\,\psi = -i\hbar\,\frac{\partial}{\partial\phi}\left(-i\hbar\frac{\partial\psi}{\partial\phi}\right) = -\hbar^2\frac{\partial^2\psi}{\partial\phi^2} \tag{76}$$

Further,

$$L_x^2\psi = -\hbar^2\left[\sin\phi\frac{\partial}{\partial\theta}+\cot\theta\cos\phi\frac{\partial}{\partial\phi}\right]\left[\sin\phi\frac{\partial\psi}{\partial\theta}+\cot\theta\cos\phi\frac{\partial\psi}{\partial\phi}\right]$$

$$= -\hbar^2\left[\sin^2\phi\frac{\partial^2\psi}{\partial\theta^2}+\sin\phi\cos\phi\left(-\text{cosec}^2\theta\frac{\partial\psi}{\partial\theta}+\cot\theta\frac{\partial^2\psi}{\partial\theta\,\partial\phi}\right)\right]$$

+ two more terms

As you can see, the evaluation of $L_x^2\,\psi$ and $L_y^2\,\psi$ is a bit cumbersome. The final result is

$$L^2\psi = -\hbar^2\left[\frac{1}{\sin\theta}\frac{\partial}{\partial\theta}\left(\sin\theta\frac{\partial\psi}{\partial\theta}\right)+\frac{1}{\sin^2\theta}\frac{\partial^2\psi}{\partial\phi^2}\right] \tag{77}$$

However, we can readily obtain the above result by evaluating **L** in spherical polar co-ordinates:

$$L^2\psi = \mathbf{L}.\,\mathbf{L}\psi = -\hbar^2\left[\left(\hat{\boldsymbol{\phi}}\frac{\partial}{\partial\theta}-\frac{\hat{\boldsymbol{\theta}}}{\sin\theta}\frac{\partial}{\partial\phi}\right).\left(\hat{\boldsymbol{\phi}}\frac{\partial\psi}{\partial\theta}-\frac{\hat{\boldsymbol{\theta}}}{\sin\theta}\frac{\partial\psi}{\partial\phi}\right)\right]$$

Thus,

$$\frac{L^2}{-\hbar^2} = \hat{\boldsymbol{\phi}}\cdot\frac{\partial\hat{\boldsymbol{\phi}}}{\partial\theta}\frac{\partial\psi}{\partial\theta}+\frac{\partial^2\psi}{\partial\theta^2}$$

$$-\hat{\boldsymbol{\phi}}\cdot\frac{\partial\hat{\boldsymbol{\theta}}}{\partial\theta}\cdot\frac{1}{\sin\theta}\frac{\partial\psi}{\partial\phi}-\frac{\hat{\boldsymbol{\theta}}}{\sin\theta}\cdot\frac{\partial\hat{\boldsymbol{\phi}}}{\partial\phi}\cdot\frac{\partial\psi}{\partial\theta}$$

$$+\frac{1}{\sin^2\theta}\frac{\partial^2\psi}{\partial\phi^2}+\hat{\boldsymbol{\theta}}\cdot\frac{\partial\hat{\boldsymbol{\theta}}}{\partial\phi}\cdot\frac{1}{\sin^2\theta}\frac{\partial\psi}{\partial\phi}$$

where we have used the fact that $\hat{\boldsymbol{\theta}}\cdot\hat{\boldsymbol{\phi}} = 0$. Now [see Eq. 72]

$$\frac{\partial\hat{\boldsymbol{\phi}}}{\partial\theta} = 0$$

and

$$\frac{\partial\hat{\boldsymbol{\phi}}}{\partial\phi} = -\cos\phi\,\hat{\mathbf{x}}-\sin\phi\,\hat{\mathbf{y}} = -\hat{\mathbf{r}}\sin\theta-\hat{\boldsymbol{\theta}}\cos\theta$$

If we multiply Eq. (71) by sin θ and Eq. (72) by cos θ and add we would get

$$\hat{\mathbf{r}}\sin\theta+\hat{\boldsymbol{\theta}}\cos\theta = \cos\phi\,\hat{\mathbf{x}}+\sin\phi\,\hat{\mathbf{y}}$$

Thus,

$$\frac{\partial\hat{\boldsymbol{\phi}}}{\partial\phi} = -\cos\phi\,\hat{\mathbf{x}}-\sin\phi\,\hat{\mathbf{y}} = -\hat{\mathbf{r}}\sin\theta-\hat{\boldsymbol{\theta}}\cos\theta$$

Further

$$\frac{\partial\hat{\boldsymbol{\theta}}}{\partial\theta} = -\sin\theta\cos\phi\,\hat{\mathbf{x}}-\sin\theta\sin\phi\,\hat{\mathbf{y}}-\cos\theta\,\hat{\mathbf{z}} = -\hat{\mathbf{r}}$$

and

$$\frac{\partial\hat{\boldsymbol{\theta}}}{\partial\phi} = -\cos\theta\sin\phi\,\hat{\mathbf{x}}+\cos\theta\cos\phi\,\hat{\mathbf{y}} = \cos\theta\,\hat{\boldsymbol{\phi}}$$

Thus,

$$\frac{L^2\psi}{-\hbar^2} = \frac{\partial^2\psi}{\partial\theta^2}+\cot\theta\frac{\partial\psi}{\partial\theta}+\frac{1}{\sin^2\theta}\frac{\partial^2\psi}{\partial\phi^2}$$

$$\Rightarrow\quad L^2\psi = -\hbar^2\left[\frac{1}{\sin\theta}\frac{\partial}{\partial\theta}\left(\sin\theta\frac{\partial\psi}{\partial\theta}\right)+\frac{1}{\sin^2\theta}\frac{\partial^2\psi}{\partial\phi^2}\right]$$

9.6 PROBLEMS

Problem 9.1

(a) In the simplest model of a rotating diatomic molecule[5], it is assumed to be a rigid rotator, i.e., the distance between the two atoms is assumed to be fixed. Show from classical considerations, that the rotational energy of such a rigid rotator is given by

$$\frac{L^2}{2\mu r_e^2}$$

where μ represents the reduced mass, r_e the distance between the two atoms (assumed to be fixed) and L represents the angular momentum of the molecule about the axis passing through the center of mass. Obtain the quantized rotational energy levels.

(b) Molecules like HCl, HF, LiH can be considered as rigid rotators and because these molecules have permanent dipole moments, transitions take place only between adjacent rotational energy levels (this is the selection rule). Show that this leads to equally spaced lines in the emission as well as in the absorption spectra.

(c) The rotational absorption spectrum of HCl appears at wave numbers 21.18, 42.36, 63.54, 84.72 and 105.91 cm^{-1}. Show that $r_e \approx 1.3$ Å; assume $M_{Cl} \simeq 35\, M_H$ and $M_H \simeq 1.68 \times 10^{-24}$ g. It may be mentioned that these lines correspond to the far-infrared region where frequencies can be measured with great accuracy; as such the accuracy in the determination of r_e is limited on account of the validity of the simple rigid rotator theory rather than the accuracy in the measurement of positions of the spectral lines.

Problem 9.2 $P_0(\mu) = 1$, $P_1(\mu) = \mu$ and $P_2(\mu) = (3\mu^2 - 1)/2$ represent one of the solutions of Eq. (36) for $l = 0, 1$ and 2 respectively. Now, the second order differential equation (36) would have two independent solutions. Obtain analytical expressions for the second solution and show that it is consistent with the solution given by Eq. (43).

[Hint: Write $F(\mu) = \phi(\mu)\, P_n(\mu)$ and solve the equation satisfied by $\phi(\mu)$.]

[Answer: The second independent solution is given by

$$Q_0(\mu) = \frac{1}{2} \ln\left(\frac{1+\mu}{1-\mu}\right) = \tanh^{-1}\mu \tag{78}$$

$$Q_1(\mu) = \frac{\mu}{2} \ln\left(\frac{1+\mu}{1-\mu}\right) - 1 = \mu \tanh^{-1}\mu - 1 \tag{79}$$

$$Q_2(\mu) = \frac{1}{4}(3\mu^2 - 1) \ln\left(\frac{1+\mu}{1-\mu}\right) - \frac{3}{2}\mu \tag{80}$$

corresponding to $l = 0, 1$ and 2 respectively. Notice that $Q_n(\mu)$ diverges as $\mu \to \pm 1$.]

Problem 9.3

(a) Using the recurrence relation Eq. (50) and Eq. (48), derive expressions for $P_5(\mu)$ and $P_6(\mu)$.

(b) Using Eqs (42), (43) and (47) derive $P_5(\mu)$ and $P_6(\mu)$ corresponding to $\lambda = 30$ and $\lambda = 42$ respectively.

---

[5] A nice elementary account of diatomic molecular spectra has been given in Ref. 5.

Problem 9.4 Derive the commutation relations

(a)

$$[L_z, x] = i\hbar y, \quad [L_z, p_x] = i\hbar p_y$$
$$[L_z, y] = -i\hbar x, \quad [L_z, p_y] = -i\hbar p_x$$
$$[L_z, z] = 0, \quad [L_z, p_z] = 0$$

(b) $$[L_x, r^2] = [L_y, r^2] = [L_z, r^2] = 0$$

Problem 9.5 Prove that

$$L^2 = L_x^2 + L_y^2 + L_z^2 = -\hbar^2 \left[\frac{1}{\sin\theta} \frac{\partial}{\partial\theta} \left(\sin\theta \frac{\partial}{\partial\theta} \right) + \frac{1}{\sin^2\theta} \frac{\partial^2}{\partial\phi^2} \right] \tag{81}$$

Problem 9.6

(a) Using Eqs (54) and (55) prove that

$$L^2 = \frac{1}{2} (L_+ L_- + L_- L_+) + L_z^2 \tag{82}$$

(b) Substitute the differential operator representations of L_+ and L_- [see Eqs (54) and (55)] to prove Eq. (25).

Problem 9.7 Using vector methods, prove that

$$L^2 \psi = (-i\hbar \mathbf{r} \times \nabla)^2 \psi = -\hbar^2 r^2 \nabla^2 \psi + \hbar^2 \frac{\partial}{\partial r} \left(r^2 \frac{\partial \psi}{\partial r} \right) \tag{83}$$

Problem 9.8 Using the following representation of p_r

$$p_r = -i\hbar \left(\frac{\partial}{\partial r} + \frac{1}{r} \right) \tag{84}$$

prove that

$$\frac{p^2}{2m} \left(= -\frac{h^2}{2m} \nabla^2 \right) = \frac{p_r^2}{2m} + \frac{L^2}{2mr^2} \tag{85}$$

Interpret the above equation physically. [We will use the operator p_r in Section 28.9]

Problem 9.9 Starting from Eq. (36) prove the orthogonality condition for Legendre polynomials

$$\int_{-1}^{+1} P_l(\mu) P_{l'}(\mu)\, d\mu = 0 \quad \text{for} \quad l \neq l'.$$

9.7 SOLUTIONS

Solution 9.1

(a) The rotational kinetic energy is $\frac{1}{2} I\omega^2$, where I (= μr_e^2) is the moment of inertia about the axis passing through the center of mass and μ being the reduced mass. Further the angular momentum $L = I\omega$. Thus

$$\text{Rotational Kinetic Energy} = \frac{L^2}{2I} = \frac{L^2}{2\mu r_e^2} \tag{86}$$

Using Eq. (66), we get the following expression for the rotational energy levels

$$E = \frac{J(J+1)\hbar^2}{2\mu r_e^2} \tag{87}$$

where $J\,(= 0, 1, 2, ...)$ is known as the rotational quantum number; we have written $J\,(J+1)$ instead of $l\,(l+1)$ so as to be consistent with the spectroscopic notation.

(b) For emission $J_f = J_i - 1$ (f and i refer to the final and initial states) and the wave number of the emitted radiation is given by

$$\nu\,(\text{cm}^{-1}) = \frac{E_i - E_f}{hc} = 2\,J_i B \tag{88}$$

where $$B = \frac{h}{8\pi^2 \mu r_e^2 c}$$

is known as the rotational constant. Thus, the lines are equally spaced. Similarly, for absorption $J_f = J_i + 1$ and

$$\nu\,(\text{cm}^{-1}) = \frac{E_f - E_i}{hc} = 2\,(J_i + 1)\,B \tag{89}$$

(c) The absorption spectra should therefore appear at $2B$, $4B$, $6B$, $8B$, ... corresponding to $J_i = 0, 1, 2, 3, ...$ respectively. From the data given one readily finds $B \approx 10.59\ \text{cm}^{-1}$. Thus,

$$B = \frac{h}{8\pi^2 \mu r_e^2 c} \approx 10.59\ \text{cm}^{-1}$$

Since, $$\mu \approx \frac{M_H\,M_{Cl}}{M_H + M_{Cl}}$$

$$\approx \frac{35}{36} M_H \approx \frac{35}{36} \times 1.68 \times 10^{-24}\ \text{g}$$

we readily get

$$r_e \approx 1.27\ \text{Å}$$

Solution 9.2 We consider the case $l = 1$ and write

$$F(\mu) = \mu\phi(\mu)$$

Substituting in Eq. (36) for $\lambda = 2$, we get

$$\frac{\phi''(\mu)}{\phi'(\mu)} = -\frac{2}{\mu} + \frac{1}{1-\mu} - \frac{1}{1+\mu}$$

Carrying out the integration, we get

$$\frac{d\phi}{d\mu} = C_1\left[\frac{1}{\mu^2} + \frac{1}{2}\left(\frac{1}{1+\mu} + \frac{1}{1-\mu}\right)\right]$$

or $$\phi(\mu) = C_1\left[-\frac{1}{\mu} + \frac{1}{2}\ln\frac{1+\mu}{1-\mu}\right] + C_2$$

where C_1 and C_2 are constants of integration. Thus,

$$F(\mu) = C_1\left[\frac{1}{2}\mu\ln\frac{1+\mu}{1-\mu} - 1\right] + C_2\mu \tag{90}$$

which represents the general solution of Legendre's equation for $l = 1$. Obviously, the second term is a multiple of $P_1(\mu)$ and the first term is $C_1Q_1(\mu)$. If we expand $Q_1(\mu)$ in a power series in μ, we would obtain the infinite series given by the first term inside the square brackets of Eq. (43).

Solution 9.6 Using Eqs. (54) and (55), we get

$$L_+ L_- \psi = (i\hbar)^2 e^{i\phi}\left[e^{-i\phi}\frac{\partial^2\psi}{\partial\theta^2} - ie^{-i\phi}\frac{\partial}{\partial\theta}\left(\cot\theta\frac{\partial\psi}{\partial\phi}\right)\right.$$

$$\left.-i\cot\theta\frac{\partial}{\partial\phi}\left(e^{-i\phi}\frac{\partial\psi}{\partial\theta}\right)+\cot^2\theta\frac{\partial}{\partial\phi}\left(e^{-i\phi}\frac{\partial\psi}{\partial\phi}\right)\right]$$

$$= -\hbar^2\left[\frac{\partial^2}{\partial\theta^2}+\frac{i}{\sin^2\theta}\frac{\partial}{\partial\phi}+\cot\theta\frac{\partial}{\partial\theta}+\cot^2\theta\frac{\partial^2}{\partial\phi^2}-i\cot^2\theta\frac{\partial^2}{\partial\phi^2}\right]$$

Therefore,

$$L_+ L_- = -\hbar^2\left[\frac{\partial^2}{\partial\theta^2}+\cot\theta\frac{\partial}{\partial\theta}+\cot^2\theta\frac{\partial^2}{\partial\phi^2}+i\frac{\partial}{\partial\phi}\right] \tag{91}$$

Similarly,

$$L_- L_+ = -\hbar^2\left[\frac{\partial^2}{\partial\theta^2}+\cot\theta\frac{\partial}{\partial\theta}+\cot^2\theta\frac{\partial^2}{\partial\phi^2}-i\frac{\partial}{\partial\phi}\right] \tag{92}$$

Since, $\quad L_z^2 = -\hbar^2\dfrac{\partial^2}{\partial\phi^2}$

one obtains

$$L^2 = \frac{1}{2}(L_+ L_- + L_- L_+) + L_z^2$$

$$= -\hbar^2\left[\frac{1}{\sin\theta}\frac{\partial}{\partial\theta}\left(\sin\theta\frac{\partial}{\partial\theta}\right)+\frac{1}{\sin^2\theta}\frac{\partial^2}{\partial\phi^2}\right]$$

Solution 9.7

$$L^2\psi = (-i\hbar\mathbf{r}\times\nabla)^2\psi = -\hbar^2(\mathbf{r}\times\nabla)\cdot(\mathbf{r}\times\nabla\psi)$$

$$= -\hbar^2\mathbf{r}\cdot[\nabla\times(\mathbf{r}\times\nabla\psi)]$$

(by interchanging the dot and cross)

$$= -\hbar^2\mathbf{r}\cdot[(\nabla\psi\cdot\nabla)\mathbf{r}-(\mathbf{r}\cdot\nabla)\nabla\psi-\nabla\psi(\nabla\cdot\mathbf{r})+\mathbf{r}(\nabla\cdot\nabla\psi)]$$

where we have used a vector identity. Thus,

$$L^2\psi = -\hbar^2\left[r\frac{\partial\psi}{\partial r}-\mathbf{r}.\left(r\frac{\partial}{\partial r}\nabla\psi\right)-r\frac{\partial\psi}{\partial r}3+r^2\nabla^2\psi\right]$$

$$= -\hbar^2\left[-2r\frac{\partial\psi}{\partial r}-r^2\frac{\partial^2\psi}{\partial r^2}+r^2\nabla^2\psi\right]$$

$$= -\hbar^2 r^2\left[\nabla^2\psi-\frac{1}{r^2}\frac{\partial}{\partial r}\left(r^2\frac{\partial\psi}{\partial r}\right)\right] = -\hbar^2\left[\frac{1}{\sin\theta}\frac{\partial}{\partial\theta}\left(\sin\theta\frac{\partial\psi}{\partial\theta}\right)+\frac{1}{\sin^2\theta}\frac{\partial^2\psi}{\partial\phi^2}\right]$$

Solution 9.8

$$\frac{p_r^2}{2m}\psi = -\frac{\hbar^2}{2m}\left(\frac{\partial}{\partial r}+\frac{1}{r}\right)\left(\frac{\partial}{\partial r}+\frac{1}{r}\right)\psi$$

$$= -\frac{\hbar^2}{2m}\left[\frac{\partial^2\psi}{\partial r^2}-\frac{\psi}{r^2}+\frac{2}{r}\frac{\partial\psi}{\partial r}+\frac{1}{r^2}\psi\right]$$

$$= -\frac{\hbar^2}{2m}\left[\frac{1}{r^2}\frac{\partial}{\partial r}\left(r^2\frac{\partial\psi}{\partial r}\right)\right] \tag{93}$$

using which Eq. (85) follows immediately. Now, in classical mechanics, the kinetic energy of a particle (in spherical polar coordinates) is $p_r^2/2m + L^2/2mr^2$ where p_r represents the radial component of the momentum and L^2 the square of the angular momentum[6]. Thus, in quantum mechanics the radial component of the angular momentum is represented by Eq. (84).

Solution 9.9 We rewrite Eq. (36) as

$$\frac{d}{d\mu}\left[(1-\mu^2)\frac{dF}{d\mu}\right]+\lambda F(\mu)=0 \tag{94}$$

which is known as the Sturm-Liouville form (see, e.g. Chapter 12 of Reference 4). Thus,

$$\frac{d}{d\mu}\left[(1-\mu^2)\frac{dP_e}{d\mu}\right]+l(l+1)P_l(\mu)=0 \tag{95}$$

and

$$\frac{d}{d\mu}\left[(1-\mu^2)\frac{dP_{l'}}{d\mu}\right]+l'(l'+1)\,P_{l'}(\mu)=0 \tag{96}$$

We multiply Eq. (95) by $P_{l'}(\mu)$ and Eq. (96) by $P_l(\mu)$ and integrate to obtain

$$[l(l+1)-l'(l'+1)]\int_{-1}^{+1}P_l(\mu)P_{l'}(\mu)\,d\mu$$

$$=\int_{-1}^{+1}\frac{d}{d\mu}\left[(1-\mu^2)\frac{dP_l}{d\mu}\right]P_{l'}(\mu)\;-\;\int_{-1}^{+1}\frac{d}{d\mu}\left[(1-\mu^2)\frac{dP_{l'}}{d\mu}\right]P_l(\mu)\,d\mu$$

$$=(1-\mu^2)\,\frac{dP_l(\mu)}{d\mu}P_{l'}(\mu)\Big|_{-1}^{+1}-\int_{-1}^{+1}(1-\mu^2)\frac{dP_l}{d\mu}\cdot\frac{dP_{l'}}{d\mu}d\mu$$

$$-(1-\mu^2)\,\frac{dP_{l'}(\mu)}{d\mu}P_l(\mu)\Big|_{-1}^{+1}+\int_{-1}^{+1}(1-\mu^2)\frac{dP_{l'}}{d\mu}\frac{dP_l}{d\mu}d\mu$$

$$=0$$

because at both limits, the Legendre polynomials have a finite value. Thus, for $l \neq l'$, $P_l(\mu)$ and $P_{l'}(\mu)$ are orthogonal in the domain $-1 \le \mu \le +1$.

---

[6] Classically,

$$\mathbf{L}=(\mathbf{r}\times\mathbf{p})=r\,p_\theta\,\hat{\phi}-rp_\phi\,\hat{\theta}$$

Therefore,

$$L^2=r^2p_\theta^2+r^2p_\phi^2$$

and

$$p^2=p_r^2+p_\theta^2+p_\phi^2=p_r^2+\frac{L^2}{r^2}$$

9.8 SAMPLE QUESTIONS

Q. 1 In the angular momentum problem, the θ dependent equation for $m = 0$ is

$$\frac{d}{d\mu}\left[(1-\mu^2)\frac{dF}{d\mu}\right] + \lambda F(\mu) = 0 \tag{97}$$

where $\mu = \cos\theta$. We assume

$$F(\mu) = \sum_{r=0}^{\infty} a_r \mu^{r+s}$$

and obtain

$$\frac{a_{r+2}}{a_r} = \frac{(r+s)(r+s+1)-\lambda}{(r+s+2)(r+s+1)} \tag{98}$$

The indicial equation gives $s = 0$.

(a) Why should the infinite series in Eq. (97) be terminated?

(b) For what values of λ will the infinite series become a polynomial?

(c) Determine the polynomial solution for $\lambda = 12$. What is the normalization condition? Normalize the polynomial solution corresponding to $\lambda = 12$.

(d) For $\lambda = 12$, obtain the second solution of Eq. (97) in terms of an integral.

(e) Substitute the polynomial solution

$$F(\mu) = a_0 + a_1\mu + a_3\mu^3 + a_5\mu^5 \tag{99}$$

in Eq. (97) and determine the values of a_0, a_1, a_3, a_5 and λ for which Eq. (99) is a solution of Eq. (97).

[**Answer:** **(e)** There will be 3 cases. If we assume $a_5 \neq 0$, then $\lambda = 30$, etc. If we assume $a_5 = 0$, then $\lambda = 12$, etc.]

Q. 2

$$L_\pm = \hbar e^{\pm i\phi}\left[\pm\frac{\partial}{\partial\theta} + i\cot\theta\frac{\partial}{\partial\phi}\right]$$

$$P_0(x) = 1;\ P_1(x) = x;\ P_2(x) = \frac{1}{2}(3x^2 - 1)$$

(a) Write $Y_{2,0}(\theta, \phi)$ (normalized).

(b) Derive expressions for $Y_{2,-1}(\theta, \phi)$ and $Y_{2,-2}(\theta, \phi)$. Are they normalized?

(c) Show that $(Y_{1,1} - Y_{1,-1})$ $(Y_{1,1} + Y_{1,-1} + \sqrt{2}\,Y_{1,0})$ and $(Y_{1,1} + Y_{1,-1} - \sqrt{2}\,Y_{1,0})$ are eigenfunctions of L_x and L^2. What are the corresponding eigenvalues?

Q. 3 Using Ladder operators write the right-hand sides in terms of spherical harmonics

(a) $L^2 Y_{4,3}(\theta, \phi) =$; (b) $L_z Y_{4,0}(\theta, \phi) =$

(c) $L_y Y_{5,1}(\theta, \phi) =$; (d) $L_- Y_{5,-5}(\theta, \phi) =$

(e) $L_z Y_{4,3}(\theta, \phi) =$; (f) $L^2 Y_{5,2}(\theta, \phi) =$

(g) $L_z Y_{5,2}(\theta, \phi) =$; (h) $L_x Y_{5,2}(\theta, \phi) =$

(i) $L_+ Y_{5,3}(\theta, \phi) =$; (j) $L_x Y_{4,3}(\theta, \phi) =$

(k) $L_x Y_{1,0}(\theta, \phi) =$; (l) $L_x Y_{2,1}(\theta, \phi) =$

(m) $L_y Y_{5,2}(\theta, \phi) =$; (n) $L_y Y_{4,3}(\theta, \phi) =$

(o) $L_y Y_{1,0}(\theta, \phi) =$; (p) $L_y Y_{2,1}(\theta, \phi) =$

[Answer: (c) $\frac{\hbar}{2i}\left[\sqrt{28}Y_{5,2}(\theta,\phi)-\sqrt{30}Y_{5,0}(\theta,\phi)\right]$**]**

REFERENCES AND SUGGESTED READINGS

1. G. Baym, *Lectures on Quantum Mechanics*, W.A. Benjamin, New York (1969).
2. M.E. Rose, *Elementary Theory of Angular Momentum*, John Wiley, New York (1957).
3. J. Irving and M. Mullineux, *Mathematics in Physics and Engineering*, Academic Press, New York (1959).
4. A.K. Ghatak, I.C. Goyal and S.J. Chua, *Mathematical Physics: I*, Macmillan India Limited, New Delhi (1995).
5. H.B. Dunford, *Elements of Diatomic Molecular Spectra*, Addison-Wesley, Reading, Massachusetts (1968).
6. B.C. Reed, *Quantum Mechanics*, Jones and Bartlett India Pvt., Ltd., New Delhi (2010).

10 Spherically Symmetric Potentials

Schrödinger's brilliant paper[1] was undoubtedly one of the most influential contributions ever made in the history of science. It deepened our understanding of atomic phenomena, served as a convenient foundation for the mathematical solution of problems in atomic physics, solid state physics, and, to some extent, also in nuclear physics, and finally opened new avenues of thought. In fact, the subsequent development of non-relativistic quantum theory was to no small extent merely an elaboration and application of Schrödinger's work.

— MAX JAMMER in *The Conceptual Development of Quantum Mechanics*, McGraw-Hill, New York (1966), p. 267.

10.1 INTRODUCTION

In Section 6.7, we had shown that for a particle of mass μ in a field described by the potential energy function $V(\mathbf{r})$, the solution of the time-dependent Schrödinger equation

$$i\hbar \frac{\partial \Psi}{\partial t} = -\frac{\hbar^2}{2\mu} \nabla^2 \Psi + V(\mathbf{r})\, \Psi(\mathbf{r}, t) \tag{1}$$

is of the form

$$\Psi(\mathbf{r}, t) = \psi(\mathbf{r})\, e^{-iEt/\hbar} \tag{2}$$

If we substitute the above solution in Eq. (1), we will see that the wavefunction $\psi(\mathbf{r})$ is the solution of the eigenvalue equation

$$H\psi(\mathbf{r}) = E\psi(\mathbf{r}) \tag{3}$$

where

$$H = -\frac{\hbar^2}{2\mu} \nabla^2 + V(\mathbf{r}) \tag{4}$$

represents the Hamiltonian operator and E the energy eigenvalue. The potential energy function has been assumed to be independent of time.

In Section 10.2, we will show that as long as the potential is spherically symmetric (i.e., as long as the potential depends only on the magnitude of the distance from a fixed point) the angular part of the wavefunction is a spherical harmonic and the potential energy function enters only in the radial part of the equation. In Section 10.3, we will show that for a system consisting of two particles, whenever the potential energy depends only on the magnitude of the distance

[1] E. Schrödinger, *Quantization as an Eigenvalue Problem*, Annalen der Physik, **79**, 361 (1926). In this paper, Schrödinger had put forward his famous equation and had given the solution corresponding to the hydrogen atom problem. The paper has been reprinted in *Wave Mechanics* by G. Ludwig, Pergamon Press, Oxford (1968).

between the two particles, the problem can be reduced to a one particle problem along with a uniform translational motion of the center of mass. In Section 10.4 ,we will solve the hydrogen atom problem and obtain the discrete states for the Coulomb potential and will show that the results are consistent with experimental data. In Section 10.5, we will discuss the vibration-rotation spectra of diatomic molecules and in Section 10.6 (and Problem 10.12) we will discuss the three-dimensional oscillator; the two-dimensional oscillator will be discussed in Problem 10.13.

10.2 SPHERICALLY SYMMETRIC POTENTIALS

One of the most important problems in quantum mechanics is that of the motion of a particle in a potential which depends only on the magnitude of the distance from a fixed point:

$$V(\mathbf{r}) = V(r) \tag{5}$$

Such a potential is referred to as a spherically symmetric potential. Now, in spherical polar coordinates

$$\begin{aligned}\nabla^2 \psi &= \frac{1}{r^2}\frac{\partial}{\partial r}\left(r^2 \frac{\partial \psi}{\partial r}\right) \\ &\quad + \frac{1}{r^2}\left[\frac{1}{\sin\theta}\frac{\partial}{\partial \theta}\left(\sin\theta \frac{\partial \psi}{\partial \theta}\right) + \frac{1}{\sin^2\theta}\frac{\partial^2 \psi}{\partial \phi^2}\right] \\ &= \frac{1}{r^2}\frac{\partial}{\partial r}\left(r^2 \frac{\partial \psi}{\partial r}\right) - \frac{L^2 \psi}{\hbar^2 r^2}\end{aligned} \tag{6}$$

where we have used the fact that the operator representation of L^2 is given by the following equation (see Section 9.2):

$$L^2 = -\hbar^2\left[\frac{1}{\sin\theta}\frac{\partial}{\partial \theta}\left(\sin\theta \frac{\partial}{\partial \theta}\right) + \frac{1}{\sin^2\theta}\frac{\partial^2}{\partial \phi^2}\right] \tag{7}$$

Thus the three-dimensional Schrödinger equation

$$\nabla^2 \psi + \frac{2\mu}{\hbar^2}\,[E - V(r)]\,\psi(r, \theta, \phi) = 0 \tag{8}$$

can be written in the form

$$\frac{1}{r^2}\frac{\partial}{\partial r}\left(r^2 \frac{\partial \psi}{\partial r}\right) + \frac{2\mu}{\hbar^2}\,[E - V(r)]\,\psi(r, \theta, \phi) = \frac{L^2 \psi}{\hbar^2 r^2} \tag{9}$$

In order to solve the above equation we use the method of separation of variables and write

$$\psi(r, \theta, \phi) = R(r)\,Y(\theta, \phi) \tag{10}$$

Substituting in Eq. (9), we get

$$\frac{Y(\theta, \phi)}{r^2}\frac{d}{dr}\left(r^2 \frac{dR}{dr}\right) + \frac{2\mu}{\hbar^2}\,[E - V(r)]\,R(r)\,Y(\theta, \phi) = \frac{R(r)}{\hbar^2 r^2}\,L^2 Y(\theta, \phi)$$

Dividing by $R(r)\,Y(\theta, \phi)/r^2$, we obtain

$$\frac{1}{R(r)}\frac{d}{dr}\left(r^2 \frac{dR}{dr}\right) + \frac{2\mu r^2}{\hbar^2}\,[E - V(r)] = \frac{1}{\hbar^2}\frac{1}{Y(\theta, \phi)}\,L^2 Y(\theta, \phi) = \lambda \tag{11}$$

where we have set the terms equal to a constant λ because the left-hand side of the above equation depends only on r while the other term depends only on θ and ϕ. The above equation gives us the eigenvalue equation

$$L^2 Y(\theta, \phi) = \lambda \hbar^2 Y(\theta, \phi) \tag{12}$$

In Section 9.3, we had shown that the eigenvalues of L^2 are $l(l+1)\hbar^2$, i.e. well-behaved solutions are obtained when

$$\lambda = l(l+1); \quad l = 0, 1, 2, \ldots \tag{13}$$

the corresponding eigenfunctions being the spherical harmonics

$$Y_{lm}(\theta, \phi); \quad m = -l, -l+1, \ldots, l-1, l \tag{14}$$

(see Sections 9.3, 9.4 and 15.3). Thus Eq. (11) can be written in the form

$$\frac{d^2R}{dr^2} + \frac{2}{r}\frac{dR(r)}{dr} + \frac{2\mu}{\hbar^2}\left[E - V(r) - \frac{l(l+1)\hbar^2}{2\mu r^2}\right]R(r) = 0 \tag{15}$$

which is known as the radial part of the Schrödinger equation. If we define a new radial function

$$u(r) = rR(r) \tag{16}$$

Equation (15) becomes

$$\frac{d^2u(r)}{dr^2} + \frac{2\mu}{\hbar^2}\left[E - V(r) - \frac{l(l+1)\hbar^2}{2\mu r^2}\right]u(r) = 0 \tag{17}$$

Comparing this with Eq. (11) of Chapter 6, we may consider the radial motion to be similar to the one-dimensional motion of a particle in a potential

$$V(r) + \frac{l(l+1)\hbar^2}{2\mu r^2} \tag{18}$$

Now, according to classical mechanics, if a particle (of mass μ) is rotating (with speed v) in a circular orbit of radius r then the angular momentum associated with such a circular motion is given by

$$L = \mu v r$$

To maintain such a circular motion there must be a centripetal force given by

$$\mathbf{F} = -\frac{\mu v^2}{r}\hat{\mathbf{r}} = -\frac{L^2}{\mu r^3}\hat{\mathbf{r}} \tag{19}$$

where the negative sign implies that the force should be directed towards the origin. The corresponding potential energy is $-L^2/2\mu r^2$ which should be supplied by $V(r)$ and hence the effective potential is $V(r) + L^2/2\mu r^2$ which is consistent with Eq. (17) if we replace L^2 by $l(l+1)\hbar^2$.

10.3 THE TWO BODY PROBLEM

We would like to solve the Schrödinger equation corresponding to the hydrogen atom which consists of two particles, viz., the electron and the proton. The interaction between the electron and the proton is described by the Coulomb potential:

$$V(r) = -\frac{q^2}{4\pi\varepsilon_0 r} \tag{20}$$

where

$$r = |\mathbf{r}_1 - \mathbf{r}_2| \tag{21}$$

represents the magnitude of the distance between the two particles. In writing Eq. (20), we have used the MKS system of units so that

$$q \simeq 1.6 \times 10^{-19}\ \text{C}$$

$$\varepsilon_0 \simeq 8.854 \times 10^{-12}\ \text{MKS units}$$

and $V(r)$ is measured in joules. In this book, we will be almost always using the MKS system of units; however, since CGS units are used in many books, we give below the corresponding expression for $V(r)$ in CGS units:

$$V(r) = -\frac{e^2}{r} \tag{22}$$

where $e \simeq 4.8 \times 10^{-10}$ esu represents the electronic charge in CGS units and $V(r)$ is measured in ergs.

We may note that the Coulomb potential described by Eq. (20) depends only on $|\mathbf{r}_1 - \mathbf{r}_2|$, i.e., on the magnitude of the distance between the two particles. Indeed, we will prove the general result that for a two body problem, whenever the potential energy depends only on the magnitude of the distance between the two particles, the problem can always be reduced to a one body problem along with a uniform translational motion of the center of mass. Now, for

$$V(\mathbf{r}_1, \mathbf{r}_2) = V(|\mathbf{r}_1 - \mathbf{r}_2|) \tag{23}$$

the Hamiltonian is given by

$$H = \frac{p_1^2}{2m_1} + \frac{p_2^2}{2m_2} + V(|\mathbf{r}_1 - \mathbf{r}_2|)$$

or

$$H = -\frac{\hbar^2}{2m_1}\nabla_1^2 - \frac{\hbar^2}{2m_2}\nabla_2^2 + V(|\mathbf{r}_1 - \mathbf{r}_2|) \tag{24}$$

where we have made use of the recipe of replacing $\mathbf{p}$ by $-i\hbar\nabla$; the subscripts 1 and 2 refer to the two particles. Further,

$$\nabla_1^2 = \frac{\partial^2}{\partial x_1^2} + \frac{\partial^2}{\partial y_1^2} + \frac{\partial^2}{\partial z_1^2}, \text{ etc.}$$

where (x_1, y_1, z_1) and (x_2, y_2, z_2) represent the coordinates of the electron and the proton respectively. Now, for the Hamiltonian given by Eq. (24), the Schrödinger equation is

$$\left[-\frac{\hbar^2}{2m_1}\nabla_1^2 - \frac{\hbar^2}{2m_2}\nabla_2^2 + V(|\mathbf{r}_1 - \mathbf{r}_2|)\right]\Psi(\mathbf{r}_1, \mathbf{r}_2) = E\Psi(\mathbf{r}_1, \mathbf{r}_2) \tag{25}$$

It can easily be shown that if we try to separate the variables by assuming

$$\Psi(\mathbf{r}_1, \mathbf{r}_2) = \Psi_1(\mathbf{r}_1)\,\Psi_2(\mathbf{r}_2)$$

the variables will not separate. However, if we introduce the center of gravity coordinates

$$\mathbf{R} = \frac{m_1\mathbf{r}_1 + m_2\mathbf{r}_2}{m_1 + m_2} \tag{26}$$

and the relative coordinate

$$\mathbf{r} = \mathbf{r}_1 - \mathbf{r}_2 \tag{27}$$

of the two particles, then the Schrödinger equation becomes (see Appendix H)

$$\left[-\frac{\hbar^2}{2(m_1+m_2)}\left(\frac{\partial^2}{\partial X^2}+\frac{\partial^2}{\partial Y^2}+\frac{\partial^2}{\partial Z^2}\right)-\frac{\hbar^2}{2\mu}\left(\frac{\partial^2}{\partial x^2}+\frac{\partial^2}{\partial y^2}+\frac{\partial^2}{\partial z^2}\right)+V(r)\right]\Psi(\mathbf{R},\mathbf{r}) = E\Psi(\mathbf{R},\mathbf{r}) \tag{28}$$

where

$$\mu = \frac{m_1 m_2}{m_1+m_2} \tag{29}$$

is known as the reduced mass. Since the potential does not depend on **R** at all, we write

$$\Psi(\mathbf{R},\mathbf{r}) = \psi(\mathbf{r})\,\Phi(\mathbf{R}) \tag{30}$$

and the Schrödinger equation is now separable giving the following equation for Φ:

$$-\frac{\hbar^2}{2M}\left(\frac{\partial^2}{\partial X^2}+\frac{\partial^2}{\partial Y^2}+\frac{\partial^2}{\partial Z^2}\right)\Phi = E_R\Phi \tag{31}$$

the solution of which is

$$\Phi(\mathbf{R}) \sim \exp[i(\mathbf{P}\cdot\mathbf{R})/\hbar];\ E_R = \frac{\hbar^2 P^2}{2M} \tag{32}$$

where $M = m_1 + m_2$

Equation (32) represents a plane wave and describes the uniform translational motion of the hydrogen atom as a whole. On the other hand, $\psi(\mathbf{r})$ describes the internal motion of the atom and satisfies the equation

$$-\frac{\hbar^2}{2\mu}\nabla^2\psi + V(r)\,\psi(\mathbf{r}) = E_r\psi(\mathbf{r}) \tag{33}$$

where $\nabla^2 = \left(\frac{\partial^2}{\partial x^2}+\frac{\partial^2}{\partial y^2}+\frac{\partial^2}{\partial z^2}\right)$ and E_r is the relative energy, i.e., it is the difference between the total energy E and the center of mass energy E_R:

$$E_r = E - E_R \tag{34}$$

Thus we have shown that whenever we have a two body problem with the potential energy function depending on $|\mathbf{r}_1 - \mathbf{r}_2|$, one need solve only one equation [Eq. (33)], which is the same as that for a single particle described by the potential energy function $V(r)$ with the difference that the mass is replaced by the reduced mass and the coordinates are the relative coordinates.

10.4 THE HYDROGEN-LIKE ATOM PROBLEM

We consider a one-electron atom with a nucleus of charge Zq. The potential energy function is given by

$$V(r) = -\frac{Zq^2}{4\pi\varepsilon_0 r} \tag{35}$$

where

$Z = 1$ for the H-atom problem,

$Z = 2$ for the singly ionized He-atom problem (He^+),

$Z = 3$ for the doubly ionized Li-atom problem (Li^{++}),

etc. In order to solve Eq. (33), we use the same method as discussed in Section 10.2 and write

$$\psi(\mathbf{r}) = R(r)\, Y_{lm}(\theta, \phi) = \frac{u(r)}{r}\, Y_{lm}(\theta, \phi) \tag{36}$$

The radial part of the wavefunction satisfies the following equation [see Eq. (15)]:

$$\frac{1}{r^2}\frac{d}{dr}\left[r^2 \frac{dR}{dr}\right] + \frac{2\mu}{\hbar^2}\left[E + \frac{Zq^2}{4\pi\varepsilon_0 r} - \frac{l(l+1)\hbar^2}{2\mu r^2}\right] R(r) = 0 \tag{37}$$

where we have removed the subscript on E since we are interested only in the energy levels associated with the relative motion and

$$\mu = \frac{m_e M_N}{m_e + M_N} \tag{38}$$

where m_e and M_N represent the mass of the electron and that of the nucleus respectively. In order to solve Eq. (37) we introduce the dimensionless variable

$$\rho = \gamma r$$

where the parameter γ is to be conveniently chosen later. Equation (37) becomes

$$\frac{1}{\rho^2}\frac{d}{d\rho}\left[\rho^2 \frac{dR}{d\rho}\right] + \left[\frac{2\mu E}{\hbar^2 \gamma^2} + \frac{2\mu}{\gamma \hbar^2}\frac{Zq^2}{4\pi\varepsilon_0 \rho} - \frac{l(l+1)}{\rho^2}\right] R(\rho) = 0 \tag{39}$$

We choose[2]

$$\gamma^2 = \frac{8\mu|E|}{\hbar^2} = -\frac{8\mu E}{\hbar^2} \tag{40}$$

and set

$$\lambda = \frac{2\mu Z}{\hbar^2 \gamma}\left(\frac{q^2}{4\pi\varepsilon_0}\right) = Zc\alpha\left(\frac{\mu}{2|E|}\right)^{1/2} \tag{41}$$

where c is the speed of light in free space and

$$\alpha \equiv \frac{q^2}{4\pi\varepsilon_0 \hbar c} \simeq \frac{1}{137} \tag{42}$$

is the fine structure coefficient. Thus, Eq. (39) becomes

$$\frac{1}{\rho^2}\frac{d}{d\rho}\left[\rho^2 \frac{dR}{d\rho}\right] + \left[-\frac{1}{4} + \frac{\lambda}{\rho} - \frac{l(l+1)}{\rho^2}\right] R(\rho) = 0 \tag{43}$$

---

[2] We are looking for energy levels corresponding to bound states; therefore $E < 0$.

In order to obtain an exact solution of Eq. (43) we introduce the function $F(\rho)$ defined through the relation[3]

$$R(\rho) = F(\rho)\, e^{-\frac{1}{2}\rho} \tag{44}$$

[A direct method for obtaining solutions of Eq. (43) in terms of confluent hypergeometric functions is discussed in Section 10.4.2]. Now, if we substitute Eq. (44) in Eq. (43), we would obtain

$$\rho^2 \frac{d^2F}{d\rho^2} + (2-\rho)\rho\,\frac{dF}{d\rho} + [(\lambda-1)\rho - l(l+1)]\,F(\rho) = 0 \tag{45}$$

We now try a series solution of the form

$$F(\rho) = \sum_{p=0,1,\ldots} C_p\, \rho^{s+p} \tag{46}$$

Substituting this into Eq. (45), we obtain

$$\sum_{p=0}^{\infty} C_p[(s+p)(s+p+1) - l(l+1)]\,\rho^{p} - \sum_{p=0}^{\infty} C_p\,[s+p-\lambda+1]\,\rho^{p+1} = 0$$

or $$C_0\,[s(s+1) - l(l+1)] + \sum_{p=0}^{\infty}\left(C_{p+1}[(s+p+1)(s+p+2)-l(l+1)] - C_p\left[s+p-\lambda+1\right]\right)\rho^{p+1} = 0$$

We, therefore, obtain the indicial equation

$$s(s+1) - l(l+1) = 0$$

Thus, $$s = l \quad \text{or} \quad -l-1 \tag{47}$$

For $F(\rho)$ to be well-behaved[4] as $\rho \to 0$ we must choose the root $s = l$. We also obtain the following recurrence relation

$$\frac{C_{p+1}}{C_p} = \frac{s+p-\lambda+1}{(s+p+1)(s+p+2) - l(l+1)}$$

$$= \frac{p+l+1-\lambda}{(p+1)(p+2l+2)} \tag{48}$$

where in the last step we have put $s = l$. For large values of p

$$\frac{C_{p+1}}{C_p} \sim \frac{1}{p}$$

---

[3] As in the harmonic oscillator problem (see Footnote 4 in Chapter 7) if we look for solutions valid for $\rho \to +\infty$, we find that $R(\rho) \sim \rho^n \exp(\pm\rho/2)$ satisfies Eq. (43) for large values of ρ. This suggests that we look for a solution of the form given by Eq. (44).

[4] For $s = -l - 1$, one of the expansion coefficients C_p becomes infinite and one has to use a special method for obtaining the second solution of Eq. (45); see, e.g. Section 2.3 (case 4) of Ref. 1. Alternatively, one can use a method to obtain directly the second solution (see Problems 10.6 and 10.7). In any case, this solution diverges at $\rho = 0$ and hence is rejected.

which is also the ratio of the terms in the expansion[5] of exp (ρ). Thus, if the series (46) is not terminated then $F(\rho)$ will behave as exp (ρ) and $R(\rho)$ will behave as exp ($\frac{1}{2}\rho$) for $\rho \to \infty$ and will diverge; it may be noted that the infinite series given by Eq. (46) is *convergent* for all values of ρ, however, it will behave as exp(ρ) for large values of ρ. We can terminate the series by requiring that

$$\lambda = n = p + l + 1 \tag{49}$$

then $C_{p+1} = 0$. Since p and l can be positive integers or zero, n can have the values 1, 2, 3, … . The number n is called the *total quantum number*. Using Eqs (41) and (49), we obtain

$$E_n = -|E_n| = -\frac{\mu Z^2}{2n^2\hbar^2}\left(\frac{q^2}{4\pi\varepsilon_0}\right)^2; \quad n = 1, 2, \ldots$$

Often it is more convenient to write the above equation as

$$E_n = -|E_n| = -\frac{\mu Z^2\alpha^2 c^2}{2n^2}; \quad n = 1, 2, \ldots \tag{50}$$

where α is the fine structure constant [see Eq. (42)]. Substituting in Eq. (40), we readily obtain

$$\gamma = \frac{2Z}{na_0} \tag{51}$$

with

$$a_0 = \frac{\hbar^2}{\mu(q^2/4\pi\varepsilon_0)} \simeq 0.52 \times 10^{-10}\,\text{m} \tag{52}$$

representing the Böhr radius. Thus,

$$\rho = \frac{2Z}{na_0}r = \frac{2}{n}\xi \tag{53}$$

where we have introduced the new variable ξ

$$\xi = \frac{Z}{a_0}r \tag{54}$$

which is independent of the total quantum number n.

Equation (50) gives the energy levels of the hydrogen atom (see Fig. 10.1). Before the development of quantum mechanics, it was known from experimental studies of the hydrogen atom spectrum, that the energy levels are given by

$$E_n = -E_H \cdot \frac{1}{n^2}, \quad n = 1, 2, \ldots \tag{55}$$

5

$$e^\rho = 1 + \rho + \frac{\rho^2}{2!} + \cdots = \sum_{p=0}^{\infty} C_p \rho^p$$

where

$$C_p = \frac{1}{p!}$$

Thus,

$$\frac{C_{p+1}}{C_p} = \frac{1}{p+1} \sim \frac{1}{p} \quad \text{for large } p.$$

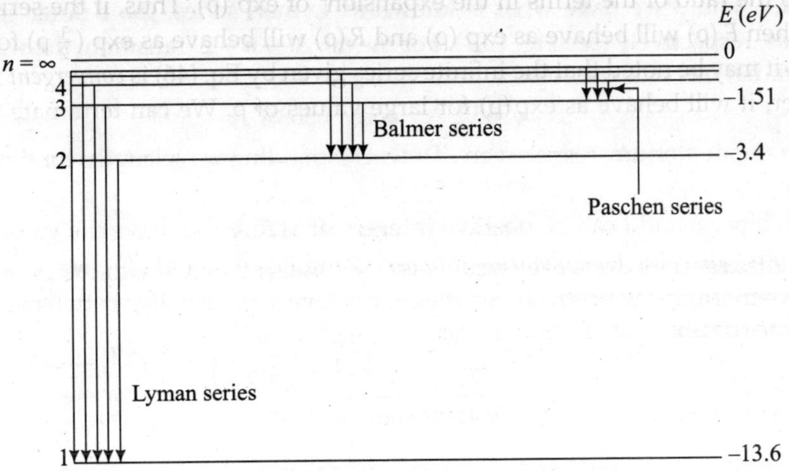

Figure 10.1: Energy levels of the hydrogen atom.

and from observations E_H was found to be 13.6 eV. Bohr then devised a model[6] which gave Eq. (55) and predicted E_H should be

$$\frac{\mu}{2\hbar^2}\left(\frac{q^2}{4\pi\varepsilon_0}\right)^2$$

It was the first (and the most dramatic) success of Schrödinger's wave mechanics that it could reproduce this result from a basic equation of motion for the system. Now,

$$\mu = \frac{m_e m_N}{m_e + m_N} \simeq m_e\left(1 - \frac{m_e}{m_N}\right)$$

For the hydrogen atom

$$m_N = m_p \simeq 1.6726 \times 10^{-27}\ \text{kg}$$

giving

$$\mu_H \simeq 9.1045 \times 10^{-31}\ \text{kg}$$

where we have taken $m_e \simeq 9.1095 \times 10^{-31}$ kg. On the other hand, for the deuterium atom

$$m_N = m_D \simeq 3.3436 \times 10^{-27}\ \text{kg}$$

giving

$$\mu_D \simeq 9.1070 \times 10^{-31}\ \text{kg}$$

Now, for the $n = n_1 \rightarrow n = n_2$ transition, the wavelength of the emitted radiation is given by

$$\lambda = \frac{hc}{E_{n_1} - E_{n_2}}$$

or

$$\lambda = \frac{2h}{\mu Z^2 \alpha^2 c^2}\left[\frac{1}{n_2^2} - \frac{1}{n_1^2}\right]^{-1} \tag{56}$$

[6] This is now known as the old quantum theory.

When n_2 = 1, 2 and 3 we have what is known as Lyman series, the Balmer series and the Paschen series, respectively (see Fig. 10.1). For the $n = 3 \rightarrow n = 2$ transition, the wavelengths of the emitted radiation come out to be

$$6565.2 \text{ Å} \quad \text{and} \quad 6563.4 \text{ Å}$$

for hydrogen and deuterium, respectively. The corresponding wavelengths for the $n = 4 \rightarrow n = 2$ transition are

$$4863.1 \text{ Å} \quad \text{and} \quad 4861.7 \text{ Å}$$

Such a small difference in the wavelength was first observed by Urey in 1932 which led to the discovery of deuterium. In spectroscopy, the energy levels are usually written in wave number units which are obtained by dividing by hc:

$$T_n = \frac{E_n}{hc} = -\frac{Z^2}{n^2} R \tag{57}$$

where
$$R = \frac{2\pi^2 \mu}{ch^3} \left(\frac{q^2}{4\pi\varepsilon_0} \right)^2 = \frac{\mu c \alpha^2}{2h} \tag{58}$$

is known as the Rydberg constant. Values of the Rydberg constant for different hydrogen-like atoms are given below.

R = 109677.58 cm^{-1} (for the hydrogen atom)

109707.56 cm^{-1} (for the deuterium atom)

109722.40 cm^{-1} (for the He^+ atom)

109728.90 cm^{-1} (for the Li^{++} atom)

The slight difference in the values is because of the difference in the values of the reduced mass μ. The eigenfunctions are usually denoted by $R_{nl}(r)$. As an example, we calculate $R_{20}(r)$ implying $n = 2$ and $l = 0$. The recurrence relation [Eq. (48)] gives

$$\frac{C_{p+1}}{C_p} = \frac{p + l + 1 - \lambda}{(p+1)(p+2l+2)} = \frac{p-1}{(p+1)(p+2)}$$

Thus,
$$C_1 = -\frac{1}{2} C_0$$

and
$$C_2 = C_3 = C_4 = \cdots = 0$$

We, therefore, have

$$F(\rho) = C_0 \left(1 - \frac{1}{2}\rho\right)$$

Since $n = 2$, we have $\rho = \xi$ [see Eq. (53)] and, therefore, if we use the normalization condition

$$\int_0^\infty |R_{nl}(r)|^2 \, r^2 \, dr = 1 \tag{59}$$

we would readily get

$$R_{20}(r) = \frac{1}{\sqrt{2}} \left(\frac{Z}{a_0} \right)^{3/2} \left(1 - \frac{1}{2}\xi \right) e^{-\xi/2} \tag{60}$$

where $$\xi = \frac{Z}{a_o} r$$

Similarly, one can calculate other wavefunctions. We give below the first few $R_{nl}(r)$

$$R_{10}(r) = 2\left(\frac{Z}{a_0}\right)^{3/2} e^{-\xi} \tag{61}$$

$$R_{21}(r) = \frac{1}{2\sqrt{6}}\left(\frac{Z}{a_0}\right)^{3/2} \xi e^{-\xi/2} \tag{62}$$

$$R_{30}(r) = \frac{2}{3\sqrt{3}}\left(\frac{Z}{a_0}\right)^{3/2} \left(1 - \frac{2}{3}\xi + \frac{2}{27}\xi^2\right) e^{-\xi/3} \tag{63}$$

$$R_{31}(r) = \frac{8}{27\sqrt{6}}\left(\frac{Z}{a_0}\right)^{3/2} \left(\xi - \frac{1}{6}\xi^2\right) e^{-\xi/3} \tag{64}$$

$$R_{32}(r) = \frac{4}{81\sqrt{30}}\left(\frac{Z}{a_0}\right)^{3/2} \xi^2 e^{-\xi/3} \tag{65}$$

All wavefunctions are normalized [see Eq. (59)]. The radial functions are usually expressed in terms of the Associated Laguerre functions (see Appendix I); however, it is more convenient to express in terms of confluent hypergeometric functions (see Section 10.4.2). The complete wavefunction is given by

$$\psi_{nlm}(r, \theta, \phi) = R_{nl}(r)\, Y_{lm}(\theta, \phi) \tag{66}$$

where $Y_{lm}(\theta, \phi)$ are the spherical harmonics tabulated in Section 15.3. Looking at Eq. (50) we see that the energy depends on the total quantum number n. Since for each n we have values of l ranging from 0 to $n-1$ [see Eq. (49)] and for each value of l, the m values range from $-l$ to $+l$ [see Section 9.3], there are

$$\sum_{l=0}^{n-1} (2l+1) = n^2 \tag{67}$$

states ψ_{nlm} belonging to a particular energy. The degeneracy with respect to m is due to spherical symmetry of the potential energy function. But the l-degeneracy is peculiar to the Coulomb field and is, in general, removed for non-Coulomb potentials (see, e.g. Problem 10.12).

10.4.1 The Probability Distribution Functions

Since $Y_{0,0} = 1/\sqrt{4\pi}$, the ground state wavefunction is spherically symmetric and is given by

$$\psi_{1,0,0} = \frac{2}{\sqrt{4\pi}}\left(\frac{Z}{a_0}\right)^{3/2} \exp(-Zr/a_0) \tag{68}$$

If $P(r)\,dr$ represents the probability of finding the electron between r and $r + dr$, then

$$P(r)\,dr = |\psi_{1,0,0}|^2\, 4\pi r^2\, dr = 4\left(\frac{Z}{a_0}\right)^3 r^2 \exp\left(-2\frac{Zr}{a_0}\right) dr \tag{69}$$

Thus, $P(r) = 0$ at $r = 0$ and at $r = \infty$ and has a maximum value at $r = a_0/Z$. This implies that the position of the electron is not certain but we can say for the hydrogen atom (for which $Z = 1$) that most probably the electron will be found around the Böhr radius. This is consistent with the uncertainty principle (see Chapter 3). It may be noted that

$$\int_0^\infty P(r)\,dr = 1 \tag{70}$$

as it indeed should be. The probability distribution functions for the first few $l = 0$ states are given below:

$$P_{1,0,0} = \left(\frac{Z}{a_0}\right) 4\xi^2 e^{-2\xi} \tag{71}$$

$$P_{2,0,0} = \left(\frac{Z}{a_0}\right)\frac{1}{2}\left(1 - \frac{1}{2}\xi\right)^2 \xi^2 e^{-\xi} \tag{72}$$

$$P_{3,0,0} = \left(\frac{Z}{a_0}\right)\frac{4}{27}\left(1 - \frac{2}{3}\xi + \frac{2}{27}\xi^2\right)^2 \xi^2 e^{-2\xi/3} \tag{73}$$

$$P_{4,0,0} = \left(\frac{Z}{a_0}\right)\frac{1}{16}\left(1 - \frac{3}{4}\xi + \frac{1}{8}\xi^2 - \frac{1}{192}\xi^3\right)^2 \xi^2 e^{-\xi/2} \tag{74}$$

For higher excited states, the probability distribution is maximum at a greater distance from the nucleus (see Fig. 10.2). In Fig. 10.3, we have given the density plots of $|\psi_{n,l,m}|^2$ on the $y = 0$ plane.

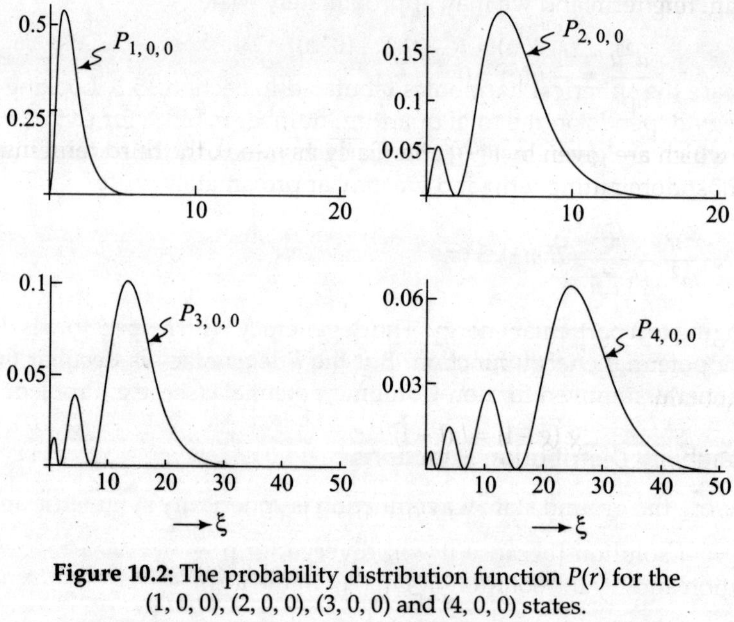

Figure 10.2: The probability distribution function $P(r)$ for the (1, 0, 0), (2, 0, 0), (3, 0, 0) and (4, 0, 0) states.

10.4.2 Solutions in Terms of the Confluent Hypergeometric Function

We rewrite Eq. (17) for the Coulomb potential [see Eq. (20)]

$$\frac{d^2u}{dr^2} + \frac{2\mu}{\hbar^2}\left[E + \frac{Zq^2}{4\pi\varepsilon_0 r} - \frac{l(l+1)\hbar^2}{2\mu r^2}\right]u(r) = 0 \tag{75}$$

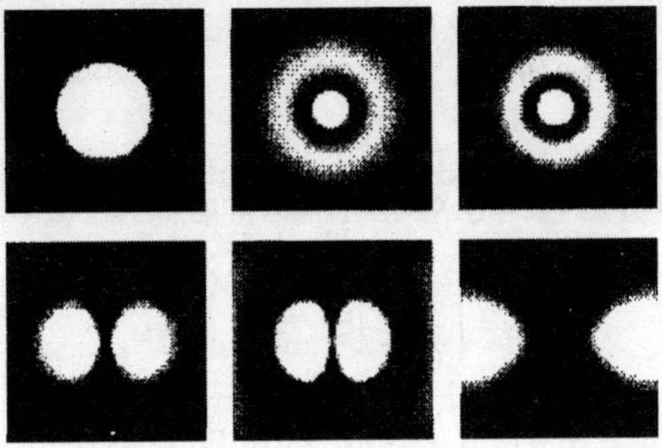

Figure 10.3: Top row: Density plots of $|\psi_{1,0,0}|^2$, $|\psi_{2,0,0}|^2$ and $|\psi_{3,0,0}|^2$. Bottom row: Density plots of $|\psi_{2,1,0}|^2$, $|\psi_{3,1,0}|^2$ and $|\psi_{3,2,0}|^2$. All density plots are on the $y = 0$ plane.

Once again, we introduce the dimensionless variable

$$\rho = \gamma r$$

[where the parameter γ is defined by Eq. (40)] to obtain

$$\frac{d^2u}{d\rho^2} + \left[-\frac{1}{4} + \frac{\lambda}{\rho} - \frac{l(l+1)}{\rho^2}\right]u(\rho) = 0 \tag{76}$$

where λ is the same as given by Eq. (41). Now as $\rho \to \infty$, the first term inside the square bracket is the most dominating term and we may approximately write

$$\frac{d^2u}{d\rho^2} - \frac{1}{4}u(\rho) = 0$$

the solutions of which are given by $e^{\pm\rho/2}$. Similarly as $\rho \to 0$, the third term inside the square brackets is the most dominating term and we may approximately write

$$\frac{d^2u}{d\rho^2} - \frac{l(l+1)}{\rho^2}u(\rho) = 0$$

If we write

$$u(\rho) \sim \rho^g$$

then we readily obtain

$$g(g-1) = l(l+1)$$

thus giving

$$g = -l \quad \text{or} \quad (l+1)$$

We reject the $g = -l$ solution [because it will diverge for $\rho \to 0$] and also the exponentially amplifying solution and try the solution of Eq. (76) in the form

$$u(\rho) = G(\rho)\, y(\rho) \tag{77}$$

where

$$G(\rho) = \rho^{l+1} e^{-\rho/2} \tag{78}$$

Simple algebra gives

$$u'' = y'' G + 2y' G' + y G'' \tag{79}$$

with

$$G'(\rho) = \left(\frac{l+1}{\rho} - \frac{1}{2}\right) G(\rho)$$

and $$G''(\rho) = \left(\frac{l(l+1)}{\rho^2} + \frac{1}{4} - \frac{l+1}{\rho}\right) G(\rho)$$

Substituting in Eq. (76), we would find that $y(\rho)$ will satisfy

$$\rho \frac{d^2 y}{d\rho^2} + (c - \rho) \frac{dy}{d\rho} - ay\,(\rho) = 0 \tag{80}$$

where $$a = l + 1 - \lambda \quad \text{and} \quad c = 2l + 2 \tag{81}$$

Equation (80) is the confluent hypergeometric equation and by using the power series method one can very easily show that one of the solutions of Eq. (80) [which is finite at the origin] is given by [see Appendix F]

$$y(\rho) = {}_1F_1\,(a, c, \rho) = 1 + \frac{a}{c}\frac{\rho}{1!} + \frac{a(a+1)}{c(c+1)}\frac{\rho^2}{2!} + \cdots \tag{82}$$

which is known as the confluent hypergeometric function. Thus, the complete solution of Eq. (76) which is well behaved at $\rho = 0$ is given by

$$u(\rho) = N\rho^{l+1} e^{-\rho/2}\, {}_1F_1\,(a, c, p) \tag{83}$$

where N represents the normalization constant so that

$$\int_0^\infty u^2(r)\, dr = \int_0^\infty R^2(r) r^2\, dr = 1 \tag{84}$$

We may mention the following:

(i) The solution given by Eq. (83) is a rigorously correct solution of Eq. (76). Writing the solution in the form of Eq. (77) does *not* involve *any* approximation.

(ii) The infinite series given by Eq. (82) is convergent for *all* values of ρ in the domain $0 < \rho < \infty$.

(iii) For $a = c$, the infinite series given in Eq. (82) is simply e^ρ; thus as $\rho \to \infty$, $u(\rho)$ will diverge as $\rho^{l+1} e^{\rho/2}$.

(iv) Indeed

$$ {}_1F_1\,(a, c, \rho) \xrightarrow[\rho\to\infty]{} \frac{\Gamma(c)}{\Gamma(a)}\, \rho^{a-c} e^\rho \tag{85}$$

(see Appendix L). Thus,

$$u\,(\rho) \xrightarrow[\rho\to\infty]{} \frac{\Gamma(c)}{\Gamma(a)}\, \rho^{l+1} \rho^{a-c} e^{\rho/2} \tag{86}$$

and will blow up as $\rho \to \infty$ [although the infinite series given by Eq. (82) is always convergent!]. In order to avoid this we must make the infinite series a polynomial which can happen only if a is a negative integer:

$$a = -n_r; \quad n_r = 0, 1, 2, \ldots$$

Thus, $$\lambda = l + 1 + n_r = n; \quad n = 1, 2, \ldots \tag{87}$$

The quantities n_r and n are usually referred to as the radial quantum number and the total quantum number, respectively. Thus, we may write

$$R_{nl}\,(\rho) = N\rho^l e^{-\rho/2}\, {}_1F_1\,(l + 1 - n, 2l + 2, \rho) \tag{88}$$

where $$N = \frac{\gamma^{3/2}}{(2l+1)!} \left\{\frac{(n+1)!}{2n(n-l-1)!}\right\}^{1/2} \tag{89}$$

We should mention that Eq. (88) is very convenient to use and Eq. (82) is quite easy to remember! For example, for $n = 3, l = 2$, we will have

$$R_{32}(\rho) = \frac{\gamma^{3/2}}{5!}\left\{\frac{5!}{6}\right\}^{1/2} e^{-\rho/2}\,\rho^2\,{}_1F_1(0, 6, \rho)$$

which can easily be seen to be identical to Eq. (65).

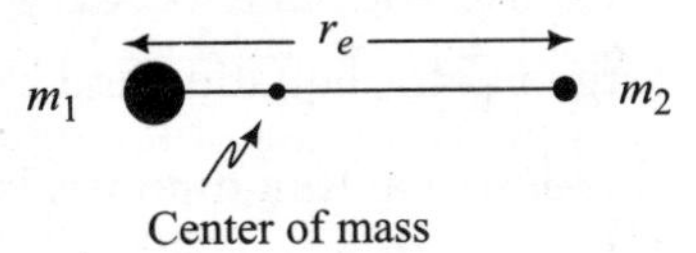

Figure 10.4: The diatomic molecule.

10.5 VIBRATION ROTATION SPECTRA OF DIATOMIC MOLECULES

In this section, we will briefly discuss the vibration rotation spectra of diatomic molecules. In the simplest model, the potential energy function (between the two atoms of the molecule) is assumed to be represented by a Hooke's law type of interaction so that

$$V(r) = \frac{1}{2}\,k(r - r_e)^2 \tag{90}$$

where k is the force constant and r_e represents the equilibrium distance between the atoms (Fig. 10.4). Since the potential is spherically symmetric, the radial part of the Schrödinger equation is given by [see Eq. (17)]

$$\frac{d^2u}{dr^2} + \frac{2u}{\hbar^2}\left[E - \frac{J(J+1)}{2\mu r^2} - \frac{1}{2}k(r - r_e)^2\right]u(r) = 0 \tag{91}$$

where $u(r) = rR(r)$ and, to be consistent with the spectroscopic notation, we have written $J(J+1)$ instead of $l(l+1)$; $J = 0, 1, 2, \ldots$. Further,

$$\mu = \frac{m_1 m_2}{m_1 + m_2}$$

represents the reduced mass with m_1 and m_2 representing the mass of the two atoms of the diatomic molecule. The boundary conditions are

$$u(r = 0) = 0 = u(r = \infty)$$

Introducing the variable

$$\rho = r - r_e$$

we get

$$\frac{d^2u}{d\rho^2} + \frac{2\mu}{\hbar^2}\left[E - \frac{J(J+1)\hbar^2}{2\mu r_e^2} - \frac{1}{2}k\rho^2\right]u(\rho) = 0 \tag{92}$$

where in the denominator we have replaced $r_e + \rho$ by r_e; the error involved is negligible for small vibrations (see, e.g. Section 10.5.1). The boundary conditions become

$$u(\rho = -r_e) = 0 = u(\rho = \infty)$$

which, for small vibrations, is approximately the same as

$$u(\rho = -\infty) = 0 = u(\rho = \infty) \tag{93}$$

Equation (92) and the above boundary conditions are equivalent to the one-dimensional harmonic oscillator problem (see Section 7.2) so that we may write

$$E - \frac{J(J+1)\hbar^2}{2\mu r_e^2} \approx \left(\nu + \frac{1}{2}\right)\hbar\omega \tag{94}$$

where $$\omega = \left(\frac{k}{\mu}\right)^{1/2}$$

and ν = 0, 1, 2, ... represents the vibration quantum number. We rewrite Eq. (94) as

$$E = E_r + E_\nu \tag{95}$$

where $$E_r = \frac{J(J+1)\hbar^2}{2\mu r_e^2}; \quad J = 0, 1, 2, \ldots$$

and $$E_\nu = \left(\nu + \frac{1}{2}\right)\hbar\omega; \quad \nu = 0, 1, 2, \ldots \tag{96}$$

represent the rotational and vibrational energies respectively. In spectroscopy, the energy levels are usually written in wave number units which are obtained by dividing by hc:

$$T\,(\text{cm}^{-1}) = \frac{E}{hc} = \left(\nu + \frac{1}{2}\right)g + BJ\,(J+1) \tag{97}$$

where $$g \equiv \frac{\omega}{2\pi c} \tag{98}$$

is known as the vibrational constant and

$$B = \frac{h}{8\pi^2 c(\mu r_e^2)} \tag{99}$$

is known as the rotational constant. Table 10.1, gives vibrational and rotational constants of some diatomic molecules.

The rotation-vibration states are schematically shown in Fig. 10.5. The upper state energy levels (ν = 1) are given by

$$T'(\text{cm}^{-1}) = BJ'\,(J'+1) + \frac{3}{2}g \tag{100}$$

and the lower state energy levels ν = 0 are given by

$$T''\,(\text{cm}^{-1}) = BJ''\,(J''+1) + \frac{1}{2}g \tag{101}$$

The transition lines for which

$$\Delta J = J'' - J' = +1 \tag{102}$$

Table 10.1. Rotational and Vibrational Constants of Some Diatomic Molecules [adapted from Ref. 4]

Molecule	r_e (Å)	$\bar{\nu} \doteq \dfrac{\omega}{2\pi c}$ (cm^{-1})	$\dfrac{\hbar^2}{2\mu r_e^2}$ (eV)
H_2	0.74	4395	7.56×10^{-3}
HD	0.74	3817	5.69×10^{-3}
N_2	1.09	2360	2.48×10^{-4}
O_2	1.21	1580	1.78×10^{-4}
LiH	1.60	1406	9.27×10^{-4}
HCl^{35}	1.27	2991	1.32×10^{-3}
$NaCl^{35}$	2.51	380	2.36×10^{-5}
KBr^{79}	2.94	231	9.1×10^{-6}

are referred to as the P branch and the lines for which

$$\Delta J = J'' - J' = -1$$

are referred to as the R branch. Now, for the P branch we have the following transitions

$$(\mathrm{v} = 1, J = J') \leftrightarrow (\mathrm{v} = 0, J = J' + 1); \quad J' = 0, 1, 2, 3, \ldots$$

and the corresponding frequencies will be given by

$$\begin{aligned}\Delta T\,(\text{cm}^{-1}) &= g + B\,[J'(J'+1) - (J'+1)(J'+2)] \\ &= g - 2B\,(J'+1) \\ &= g - 2B,\ g - 4B,\ g - 6B, \ldots \ (P \text{ branch})\end{aligned}$$

for $J' = 0, 1, 2, \ldots$ respectively. On the other hand, for the R branch we have the following transition

$$(\mathrm{v} = 1, J = J') \leftrightarrow (\mathrm{v} = 0, J = J' - 1); \ J' = 1, 2, 3, \ldots$$

for which

$$\begin{aligned}\Delta T\,(\text{cm}^{-1}) &= g + 2BJ' \\ &= g + 2B,\ g + 4B,\ g + 6B, \ldots \ (R \text{ branch})\end{aligned}$$

for $J' = 1, 2, 3, \ldots$ respectively.

Example 10.1 The experimental values of the $\mathrm{v} = 0 \rightarrow \mathrm{v} = 1$ absorption lines of HCl^{35} are given below.[7]

Branch	Line	Difference
P branch	2906.3 cm^{-1}	
		21.2 cm^{-1}
	2927.5 cm^{-1}	
		21.2 cm^{-1}
	2948.7 cm^{-1}	
		21.2 cm^{-1}
	2969.9 cm^{-1}	
		42.2 cm^{-1}
R branch	3012.2 cm^{-1}	
		21.2 cm^{-1}
	3033.4 cm^{-1}	
		21.2 cm^{-1}
	3054.6 cm^{-1}	
		21.2 cm^{-1}
	3075.8 cm^{-1}	

Thus,

$$2B \approx 21.2\ \text{cm}^{-1} \quad \Rightarrow \quad B \approx 10.6\ \text{cm}^{-1}$$

[7] Data taken from Ref. 3; this reference also gives an accurate description of infrared band analysis.

Now,
$$\mu \approx \frac{35}{36} m_H \approx \frac{35}{36} \times 1.67 \times 10^{-24} \text{ g}$$
$$\approx 1.62 \times 10^{-24} \text{ g}$$

Using Eq. (99), we get $r_e \approx 1.27$ Å. Further
$$g = \frac{\omega}{2\pi c} \approx \frac{2969.9 + 3012.2}{2} \approx 2991 \text{ cm}^{-1}$$

giving $\omega \approx 5.63 \times 10^{14} \text{ s}^{-1}$ and $k \approx 5.14 \times 10^5$ dyne/cm

The absorption lines 2948.7 cm^{-1} and 2969.9 cm^{-1} correspond to ($v = 0, J = 2$) $\rightarrow$ ($v = 1, J = 1$) and ($v = 0, J = 1$) $\rightarrow$ ($v = 1, J = 0$) transitions in the P branch, respectively. Similarly, the absorption lines 3012.2 cm^{-1} and 3033.4 cm^{-1} correspond to ($v = 0, J = 1$) $\rightarrow$ ($v = 1, J = 2$) and ($v = 0, J = 0$) $\rightarrow$ ($v = 1, J = 1$) transitions, respectively in the R branch.

Figure 10.5 shows the spectra of HCl; the slightly different frequencies at each absorption line are due to the presence of two isotopes of chlorine (HCl^{35} and HCl^{37}). Notice that the $\Delta J = 0$ transitions are forbidden.

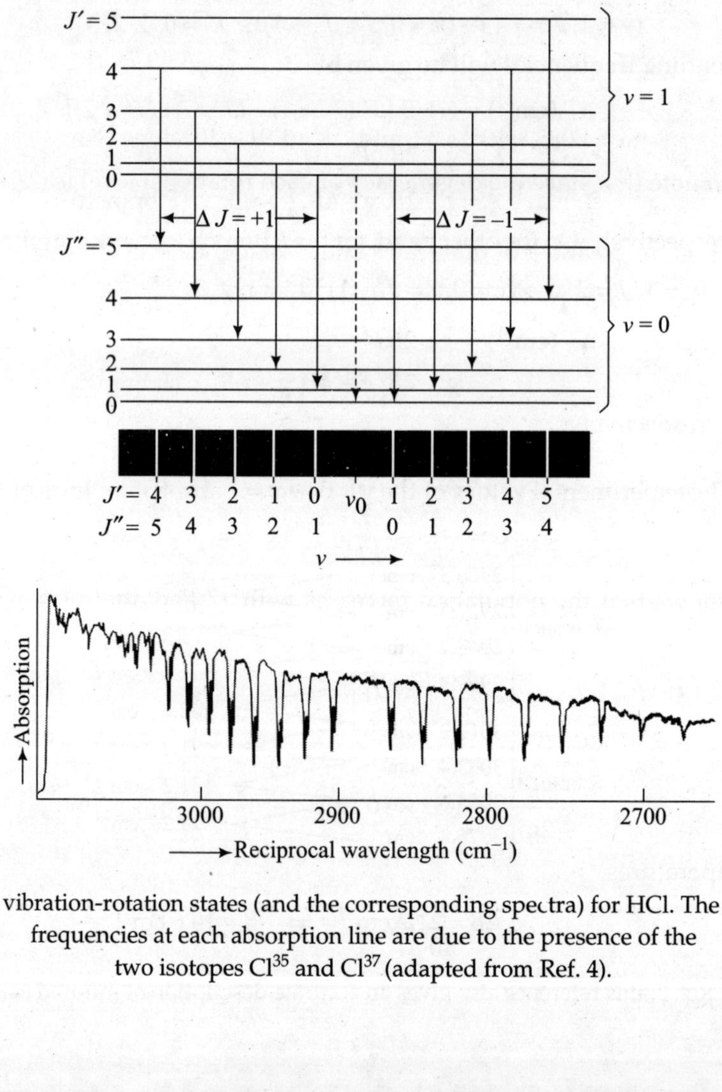

Figure 10.5: The vibration-rotation states (and the corresponding spectra) for HCl. The slightly different frequencies at each absorption line are due to the presence of the two isotopes Cl^{35} and Cl^{37} (adapted from Ref. 4).

Example 10.2 The experimental values of the $v = 0 \rightarrow v = 1$ absorption lines of $C^{12} O^{16}$ are given as follows.[8]

Line	Difference
2154.77 cm^{-1}	
	3.86 cm^{-1}
2158.63 cm^{-1}	
	3.86 cm^{-1}
2162.49 cm^{-1}	
	3.86 cm^{-1}
2166.35 cm^{-1}	
	7.72 cm^{-1}
2174.07 cm^{-1}	
	3.86 cm^{-1}
2177.93 cm^{-1}	
	3.86 cm^{-1}
2181.79 cm^{-1}	
	3.86 cm^{-1}
2185.65 cm^{-1}	

Thus,

$$B \approx 1.93 \text{ cm}^{-1}$$

Now,

$$\mu \approx \frac{12 \times 16}{28} m_H \approx 1.145 \times 10^{-23} \text{ g}$$

Using Eq. (99), we get $r_e \approx 1.13$ Å. Further

$$g = \frac{\omega}{2\pi c} \approx 2170.21 \text{ cm}^{-1}$$

giving

$$\omega \approx 4.088 \times 10^{14} \text{ s}^{-1} \quad \text{and} \quad k \approx 1.91 \times 10^{6} \text{ dyne/cm}$$

It is interesting to note that since the degeneracy of each rotational level is $(2J + 1)$, at thermal equilibrium

$$\frac{\text{Number of molecules in the rotational level } J''}{\text{Number of molecules in the rotational level } J'' = 0}$$

$$= (2J'' + 1) \exp\left[-\frac{E(J = J'', v) - E(J = 0, v)}{kT}\right]$$

$$= (2J'' + 1) e^{-\alpha J''(J'' + 1)} \tag{103}$$

where

$$\alpha = \frac{\hbar^2}{2\mu r_e^2} \cdot \frac{1}{kT} \tag{104}$$

Equation (103) shows that the population increases with J'' and attains its maximum value around

$$J'' \approx \frac{1}{2}\left[\sqrt{\frac{2}{\alpha}} - 1\right] \tag{105}$$

For HCl,

$$\frac{\hbar^2}{2\mu r_e^2} \approx 1.32 \times 10^{-3} \text{ eV}$$

and at room temperatures,

$$kT \approx \frac{1}{40} \text{ eV}$$

8 Data adapted from Ref. 3.

implying

$$\alpha \approx 0.0528 \quad \text{and} \quad J'' \approx \frac{1}{2}\left[\sqrt{\frac{2}{\alpha}} - 1\right] \approx 2.6$$

For $J'' = 1, 2, 3$ and 4, the quantity on the RHS of Eq. (103) takes the values 2.7, 3.6, 3.7 and 3.13, respectively. Thus, at room temperature, the population is maximum for the transition occurring from the level $J'' = 3$. Indeed, from the absorption spectra, it is possible to deduce the temperature of the gas.

10.5.1 A more Accurate Analysis of the Vibration Rotation Spectra

In order to obtain a more accurate expression for energy eigenvalues associated with the vibration-rotation spectra, we rewrite Eq. (91)) in the form

$$\frac{d^2u}{d\rho^2} + \frac{2\mu}{\hbar^2} F(\rho)\, u(\rho) = 0 \tag{106}$$

where

$$F(\rho) = E - \frac{J(J+1)\hbar^2}{2\mu r_e^2}\left(1 + \frac{\rho}{r_e}\right)^{-2} - \frac{1}{2}k\rho^2$$
$$\simeq E - E_r\left(1 - \frac{2\rho}{r_e} + \frac{3\rho^2}{r_e^2}\right) - \frac{1}{2}k\rho^2 \tag{107}$$

where E_r is given by Eq. (96). Thus,

$$F(\rho) \simeq E - E_r - \frac{1}{2}k'\,[(\rho - \alpha)^2 - \alpha^2] \tag{108}$$

where

$$k' = k(1 + 3\Gamma) \tag{109}$$

$$\Gamma = \frac{E_r}{\frac{1}{2}kr_e^2} \tag{110}$$

and

$$\alpha = \left(\frac{\Gamma}{1 + 3\Gamma}\right) r_e \tag{111}$$

Thus Eq. (106) can be written in the form

$$\frac{d^2u}{d\rho^2} + \frac{2\mu}{\hbar^2}\left[E' - \frac{1}{2}k'\xi^2\right] u(\xi) = 0 \tag{112}$$

where

$$\xi = \rho - \alpha \tag{113}$$

and

$$E' = E - E_r + \frac{1}{2}k'\alpha^2 \tag{114}$$

Equation (112) represents the Schrödinger equation for the linear harmonic oscillator. Assuming $u(\xi = -\infty) = 0 = u(\xi = \infty)$ we get the following expression for the energy eigenvalues

$$E' = \left(v + \frac{1}{2}\right)\hbar\sqrt{\frac{k'}{\mu}}; \quad v = 0, 1, 2, \ldots \tag{115}$$

or,

$$E = E_r\left[1 - \frac{\Gamma}{1+3\Gamma}\right] + \left(\nu + \frac{1}{2}\right)\hbar\sqrt{\frac{k}{\mu}(1+3\Gamma)} \qquad (116)$$

which represents a more accurate expression for the energy eigenvalues. However, the corrections are usually small because in most cases

$$\Gamma \ll 1$$

For example, for HCl^{35}

$$\bar{\nu} = \frac{\omega}{2\pi c} \approx 2886 \text{ cm}^{-1}$$

$$\Rightarrow \qquad \omega \approx 5.436 \times 10^{14} \text{ s}^{-1}$$

$$\Rightarrow \qquad k = \mu\omega^2$$

$$\approx \frac{35}{36} \times 1.67 \times 10^{-24} \times (5.44 \times 10^{14})^2$$

$$\approx 4.80 \times 10^5 \text{ dyne/cm}$$

where, for consistency with books on spectroscopy, we have used CGS units. Thus,

$$\frac{1}{2}kr_e^2 \approx \frac{1}{2} \times 4.8 \times 10^5 \times (1.27 \times 10^{-8})^2 \text{ erg}$$

$$\approx 24.2 \text{ eV}$$

where we have assumed $r_e \approx 1.27$ Å for the equilibrium spacing between hydrogen and chlorine atoms. Thus,

$$\Gamma = \frac{J(J+1)\dfrac{\hbar^2}{2\mu r_e^2}}{\frac{1}{2}kr_e^2} \approx 5 \times 10^{-5} J(J+1)$$

showing that the correction terms in Eq. (116) are very small. The effect of anharmonic terms is discussed in Ref. 3.

10.6 THE THREE-DIMENSIONAL HARMONIC OSCILLATOR

In this section, we will consider the three-dimensional harmonic oscillator described by the potential energy function

$$V(x, y, z) = \frac{1}{2}\mu\,(\omega_1^2 x^2 + \omega_2^2 y^2 + \omega_3^2 z^2) \qquad (117)$$

Thus in Cartesian coordinates, the three-dimensional Schrödinger equation becomes

$$\frac{\partial^2\psi}{\partial x^2} + \frac{\partial^2\psi}{\partial y^2} + \frac{\partial^2\psi}{\partial z^2} + \frac{2\mu}{\hbar^2}\left[E - \frac{1}{2}\mu\omega_1^2 x^2 - \frac{1}{2}\mu\omega_2^2 y^2 - \frac{1}{2}\mu\omega_3^2 z^2\right]\psi(x, y, z) = 0 \qquad (118)$$

We use the method of separation of variables to solve the above equation and write

$$\psi(x, y, z) = X(x)\,Y(y)\,Z(z) \qquad (119)$$

Thus,

$$YZ\frac{d^2X}{dx^2} + XZ\frac{d^2Y}{dy^2} + XY\frac{d^2Z}{dz^2} + \frac{2\mu}{\hbar^2}\left[E - \frac{1}{2}\mu\omega_1^2 x^2 - \frac{1}{2}\mu\omega_2^2 y^2 - \frac{1}{2}\mu\omega_3^2 z^2\right] XYZ = 0$$

or

$$\left[\frac{1}{X}\frac{d^2X}{dx^2}-\frac{\mu^2\omega_1^2}{\hbar^2}x^2\right]+\left[\frac{1}{Y}\frac{d^2Y}{dy^2}-\frac{\mu^2\omega_2^2}{\hbar^2}y^2\right]+\left[\frac{1}{Z}\frac{d^2Z}{dz^2}-\frac{\mu^2\omega_3^2}{\hbar^2}z^2\right]=-\frac{2\mu E}{\hbar^2} \tag{120}$$

The variables have indeed separated out. The first term inside the square brackets on the LHS of the above equation is a function of x alone; similarly the second and the third terms are functions of y and z and since their sum is equal to a constant, we must set each term equal to a constant

$$\frac{1}{X}\frac{d^2X}{dx^2}-\frac{\mu^2\omega_1^2}{\hbar^2}x^2=-g_1 \tag{121}$$

$$\frac{1}{Y}\frac{d^2Y}{dy^2}-\frac{\mu^2\omega_2^2}{\hbar^2}y^2=-g_2 \tag{122}$$

$$\frac{1}{Z}\frac{d^2Z}{dz^2}-\frac{\mu^2\omega_3^2}{\hbar^2}z^2=-g_3 \tag{123}$$

where
$$g_1+g_2+g_3=\frac{2\mu E}{\hbar^2} \tag{124}$$

Now, if we define

$$\xi=\gamma_1 x\,;\ \gamma_1=\sqrt{\frac{\mu\omega_1}{\hbar}} \tag{125}$$

Eq. (121) takes the form

$$\frac{d^2X}{d\xi^2}+(\Lambda_1-\xi^2)\,X(\xi)=0 \tag{126}$$

We had encountered the above equation in Chapter 7 while discussing the linear harmonic oscillator problem. If we apply the boundary condition that $X(\xi)\to 0$ as $\xi\to\pm\infty$ we would find that Λ_1 can only take the values 1, 3, 5, ... ; i.e.

$$\Lambda_1=(2n_1+1);\ n_1=0,1,2,\ldots \tag{127}$$

Thus,
$$g_1=(2n_1+1)\,\frac{\mu\omega_1}{\hbar} \tag{128}$$

The corresponding normalized eigenfunction would be (see Section 7.3):

$$X_{n_1}(x)=N_{n_1}H_{n_1}(\xi)\,e^{-\frac{1}{2}\xi^2} \tag{129}$$

where $H_{n_1}(\xi)$ are the Hermite polynomials and

$$N_{n_1}=\left(\frac{\gamma_1}{2^{n_1}n_1!\sqrt{\pi}}\right)^{1/2} \tag{130}$$

Similarly,
$$g_2=(2n_2+1)\,\frac{\mu\omega_2}{\hbar}$$

and $$g_3 = (2n_3 + 1)\frac{\mu\omega_3}{\hbar}$$

with

$$n_2, n_3 = 0, 1, 2, \ldots$$

The corresponding normalized eigenfunctions would be

$$Y_{n_2}(y) = N_{n_2} H_{n_2}(\eta)\, e^{-\frac{1}{2}\eta^2}$$

and

$$Z_{n_3}(z) = N_{n_3} H_{n_3}(\zeta)\, e^{-\frac{1}{2}\zeta^2}$$

where

$$\eta = \gamma_2 y \quad ; \quad \zeta = \gamma_3 z$$

$$\gamma_2 = \sqrt{\frac{\mu\omega_2}{\hbar}} \quad ; \quad \gamma_3 = \sqrt{\frac{\mu\omega_3}{\hbar}}$$

and $$N_{n_2} = \left(\frac{\gamma_2}{2^{n_2} n_2!\sqrt{\pi}}\right)^{1/2} \quad ; \quad N_{n_3} = \left(\frac{\gamma_3}{2^{n_3} n_3!\sqrt{\pi}}\right)^{1/2}$$

Thus,

$$\psi(x, y, z) = \psi_{n_1, n_2, n_3}(x, y, z)$$

$$= \left[\frac{\gamma_1\gamma_2\gamma_3}{2^n n_1! n_2! n_3! \pi^{3/2}}\right]^{1/2} [H_{n_1}(\xi)\, H_{n_2}(\eta)\, H_{n_3}(\zeta)]$$

$$\times\, e^{-\frac{1}{2}(\xi^2+\eta^2+\zeta^2)} \tag{131}$$

where $$n = n_1 + n_2 + n_3$$

The corresponding energy eigenvalues would be given by

$$E = \frac{\hbar^2}{2\mu}[g_1 + g_2 + g_3]$$

or

$$E = E_{n_1, n_2, n_3} = \left(n_1 + \frac{1}{2}\right)\hbar\omega_1 + \left(n_2 + \frac{1}{2}\right)\hbar\omega_2 + \left(n_3 + \frac{1}{2}\right)\hbar\omega_3;$$

$$n_1, n_2, n_3 = 0, 1, 2, 3, \ldots \tag{132}$$

10.6.1 The Isotropic Oscillator

For an isotropic oscillator

$$\omega_1 = \omega_2 = \omega_3 = \omega \tag{133}$$

giving

$$E = \left(n_1 + n_2 + n_3 + \frac{3}{2}\right)\hbar\omega$$

$$= \left(n + \frac{3}{2}\right)\hbar\omega\,;\; n = 0, 1, 2, \ldots \tag{134}$$

When $n = 0$, we must have

$$n_1 = n_2 = n_3 = 0$$

On the other hand, when $n = 1$, we could have

$$n_1 = 1,\ n_2 = 0,\ n_3 = 0$$

or

$$n_1 = 0,\ n_2 = 1,\ n_3 = 0$$

or

$$n_1 = 0,\ n_2 = 0,\ n_3 = 1$$

Thus, we will have a three-fold degenerate state. Similarly, the energy level corresponding to $n = 2$ will be six-fold degenerate. In general, the degree of degeneracy will be

$$1 + 2 + \cdots + n + (n + 1) = \frac{1}{2}\,(n + 1)\,(n + 2) \tag{135}$$

Now, for the isotropic oscillator, the potential energy function can be written as

$$V = V(r) = \frac{1}{2}\,\mu\omega^2 r^2 \tag{136}$$

Since the potential is spherically symmetric, we could write the solution of the Schrödinger equation in the form

$$\psi(r, \theta, \phi) = R(r)\, Y_{lm}(\theta, \phi) \tag{137}$$

where $R(r)$ satisfies the following equation

$$\frac{1}{r^2}\frac{d}{dr}\left[r^2\frac{dR}{dr}\right] + \frac{2\mu}{\hbar^2}\left[E - \frac{1}{2}\mu\omega^2 r^2 - \frac{l(l+1)\hbar^2}{2\mu r^2}\right]R(r) = 0 \tag{138}$$

One can directly solve the above equation (see Problems 10.12); however, it may be instructive to choose appropriate linear combinations[9] of the wave functions given by Eq. (131) so that they are of the form of Eq. (137). We illustrate this for the first three states:

The ground state energy is given by

$$E_0 = E_{0,0,0} = \frac{3}{2}\,\hbar\omega$$

which corresponds to

$$n_1 = n_2 = n_3 = 0$$

The corresponding eigenfunction is given by

$$\psi_0 = \psi_{0,0,0} = \left(\frac{\gamma}{\sqrt{\pi}}\right)^{3/2} e^{-\gamma^2 (x^2 + y^2 + z^2)/2} \tag{139}$$

where

$$\gamma = \gamma_1 = \gamma_2 = \gamma_3 = \sqrt{\frac{\mu\hbar}{\omega}} \tag{140}$$

We can also rewrite ψ_0 as

$$\psi_0(r, \theta, \phi) = R_{0,0}(r)\, Y_{0,0} \tag{141}$$

[9] We may recall that if $\psi_1, \psi_2, \ldots, \psi_g$ represent g linearly independent wavefunctions for a g-fold degenerate state then any linear combination of them (like $c_1\psi_1 + c_2\psi_2 + \cdots$) is also an eigenfunction corresponding to the same energy eigenvalue.

where

$$R_{0,0}(r) = \frac{2\gamma^{3/2}}{\pi^{1/4}} e^{-\gamma^2 r^2/2} \tag{142}$$

The function $R_{0,0}(r)$ is automatically normalized.
The energy eigenvalue corresponding to the first excited state is given by

$$E_1 = \frac{5}{2}\hbar\omega$$

which will occur when

$$n_1 = 1, n_2 = 0 \quad \text{and} \quad n_3 = 0$$

or, when

$$n_1 = 0, n_2 = 1 \quad \text{and} \quad n_3 = 0$$

or, when

$$n_1 = 0, n_2 = 0 \quad \text{and} \quad n_3 = 1$$

Thus, there will be three linearly independent eigenfunctions belonging to the same energy eigenvalue; hence the first excited state is said to be a three-fold degenerate state. The corresponding eigenfunctions would be

$$\psi_1 = \psi_{1,0,0} = C_1 x e^{-\gamma^2 r^2/2} = C_1 r \sin\theta \cos\phi \, e^{-\gamma^2 r^2/2} \tag{143}$$

$$\psi_2 = \psi_{0,1,0} = C_1 y e^{-\gamma^2 r^2/2} = C_1 r \sin\theta \sin\phi \, e^{-\gamma^2 r^2/2} \tag{144}$$

$$\psi_3 = \psi_{0,0,1} = C_1 z e^{-\gamma^2 r^2/2} = C_1 r \cos\theta \, e^{-\gamma^2 r^2/2} \tag{145}$$

where $$C_1 = \frac{\sqrt{2}}{\pi^{3/4}}\gamma^{5/2}$$

The eigenfunctions of the form of Eq. (137) are given by

$$\phi_1 = \frac{1}{\sqrt{2}}(\psi_1 + i\psi_2) = -R_{1,1}(r)\, Y_{1,1}(\theta, \phi) \tag{146}$$

$$\phi_2 = \frac{1}{\sqrt{2}}(\psi_1 - i\psi_2) = R_{1,1}(r)\, Y_{1,-1}(\theta, \phi) \tag{147}$$

$$\phi_3 = \psi_3 = R_{1,1}(r)\, Y_{1,0}(\theta, \phi) \tag{148}$$

where $$R_{1,1}(r) = \sqrt{\frac{8\pi}{3}} \cdot \frac{1}{\pi^{3/4}} \cdot \gamma^{5/2} r e^{-\gamma^2 r^2/2} \tag{149}$$

Similarly, the second excited state (with $E_2 = \frac{7}{2}\hbar\omega$) is six-fold degenerate with n_1, n_2 and n_3 taking the following sets of values

(2, 0, 0); (0, 2, 0); (0, 0, 2);

(1, 1, 0); (0, 1, 1); and (1, 0, 1).

Since $H_2(\xi) = (4\xi^2 - 2)$, the corresponding eigenfunctions are

$$\left.\begin{aligned}
\psi_4 &= \psi_{2,0,0} = C_2\,(4\gamma^2x^2 - 2)\,e^{-\gamma^2r^2/2}\\
\psi_5 &= \psi_{0,2,0} = C_2\,(4\gamma^2y^2 - 2)\,e^{-\gamma^2r^2/2}\\
\psi_6 &= \psi_{0,0,2} = C_2\,(4\gamma^2z^2 - 2)\,e^{-\gamma^2r^2/2}\\
\psi_7 &= \psi_{1,1,0} = \sqrt{2}C_2\,(2\gamma x)\,(2\gamma y)\,e^{-\gamma^2r^2/2}\\
\psi_8 &= \psi_{0,1,1} = \sqrt{2}C_2\,(2\gamma y)\,(2\gamma z)\,e^{-\gamma^2r^2/2}\\
\psi_9 &= \psi_{1,0,1} = \sqrt{2}C_2\,(2\gamma z)\,(2\gamma x)\,e^{-\gamma^2r^2/2}
\end{aligned}\right\} \tag{150}$$

where
$$C_2 = \frac{1}{2\sqrt{2}}\left(\frac{\gamma}{\sqrt{\pi}}\right)^{3/2} \tag{151}$$

The eigenfunctions of the form of Eq. (137) are given by

$$\phi_4 = \frac{1}{\sqrt{3}}\,(\psi_4 + \psi_5 + \psi_6) = R_{2,0}\,(r)\,Y_{0,0}\,(\theta, \phi) \tag{152}$$

$$\phi_5 = \frac{1}{\sqrt{6}}\,(-\psi_4 - \psi_5 + 2\psi_6) = R_{2,2}\,(r)\,Y_{2,0}\,(\theta, \phi) \tag{153}$$

$$\phi_6 = \frac{1}{2}\,(\psi_4 - \psi_5 + i\sqrt{2}\,\psi_7) = R_{2,2}\,(r)\,Y_{2,2}\,(\theta, \phi) \tag{154}$$

$$\phi_7 = \frac{1}{2}\,(\psi_4 - \psi_5 - i\sqrt{2}\,\psi_7) = R_{2,2}\,(r)\,Y_{2,-2}\,(\theta, \phi) \tag{155}$$

$$\phi_8 = -\frac{1}{\sqrt{2}}\,(\psi_9 + i\psi_8) = R_{2,2}\,(r)\,Y_{2,1}\,(\theta, \phi) \tag{156}$$

$$\phi_9 = \frac{1}{\sqrt{2}}\,(\psi_9 - i\psi_8) = R_{2,2}\,(r)\,Y_{2,-1}\,(\theta, \phi) \tag{157}$$

where
$$R_{2,0}\,(r) = \sqrt{\frac{\pi}{2}}\left(\frac{\gamma}{\sqrt{\pi}}\right)^{3/2}(4\gamma^2r^2 - 6)\,e^{-\gamma^2r^2/2} \tag{158}$$

$$R_{2,2}\,(r) = \sqrt{\frac{\pi}{15}}\left(\frac{\gamma}{\sqrt{\pi}}\right)^{3/2} 4\gamma^2r^2e^{-\gamma^2r^2/2} \tag{159}$$

10.7 PROBLEMS

Problem 10.1

(a) Consider a spherically symmetric potential energy function of the form

$$\left.\begin{aligned}
V(r) &= 0, \quad 0 < r < a\\
&= \infty, \quad r > a
\end{aligned}\right\} \tag{160}$$

which represents an infinitely deep potential well of radius a. Consider the $l = 0$ case and determine the energy eigenvalues and the normalized eigenfunctions.

(b) In continuation of part (a), show that for a general value of l the energy eigenvalues are given by

$$E_{n,l} = \frac{\hbar^2}{2\mu}\left(\frac{g_{l,n}}{a}\right)^2; \quad \begin{array}{l} l = 0, 1, 2, \ldots \\ n = 1, 2, \ldots \end{array} \tag{161}$$

where $g_{l,n}$ represent the zeroes of the spherical Bessel function $j_l(x)$; i.e.

$$g_{01} = \pi, \quad g_{02} = 2\pi, \quad g_{03} = 3\pi, \ldots$$

$$g_{11} = 4.4934, \quad g_{12} = 7.7253, \quad g_{13} = 10.9041, \ldots$$

$$g_{21} = 5.7635, \quad g_{22} = 9.0950, \quad g_{23} = 12.3229, \ldots$$

Show that the corresponding normalized eigenfunctions are

$$\psi(r, \theta, \phi) = N j_l\left(g_{l,n}\frac{r}{a}\right) Y_{lm}(\theta, \phi) \tag{162}$$

where
$$N = \left[\frac{2}{a^3[j_{l+1}(g_{l,n})]^2}\right]^{1/2} \tag{163}$$

represents the normalization constant.

Problem 10.2 A quantum dot can be approximately described by an electron (of a certain effective mass) inside an infinitely deep spherical well as discussed in the previous problem. Assume $a \simeq 0.4$ μm and the effective mass of the electron to be $0.065\, m_e$, calculate the frequency of radiation emitted when the electron makes a transition from the first excited state to the ground state. **[Answer:** Approximately 0.91 GHz]

Problem 10.3

(a) We next consider the electron (of effective mass μ) to be confined inside an infinitely deep cylindrical well of height b and radius a. Solve the Schrödinger equation in cylindrical coordinates and show that the energy eigenvalues are given by

$$E_{m,l,n} = \frac{\hbar^2}{2\mu}\left[\left(\frac{l\pi}{b}\right)^2 + \left(\frac{\alpha_{mn}}{a}\right)^2\right] \tag{164}$$

where $m = 0, 1, 2, \ldots; l = 1, 2, 3, \ldots$ and α_{mn} represent the zeroes of $J_m(x)$. Thus,

$$\alpha_{01} = 2.4048, \quad \alpha_{02} = 5.5201, \quad \alpha_{03} = 8.6537, \ldots$$

$$\alpha_{11} = 3.8317, \quad \alpha_{12} = 7.0156, \quad \alpha_{13} = 10.1735, \ldots$$

$$\alpha_{21} = 5.1356, \quad \alpha_{22} = 8.4172, \quad \alpha_{23} = 11.6198, \ldots$$

(b) Show that the corresponding normalized eigenfunctions are

$$\psi_{m,l,n}(r, z, \phi) = \left[\frac{\sqrt{2}}{aJ_{m+1}(\alpha_{mn})} J_m\left(\alpha_{mn}\frac{r}{a}\right)\right] \times \left[\sqrt{\frac{2}{b}}\sin\left(\frac{l\pi}{b}\right)\right]\left[\frac{1}{\sqrt{2\pi}}e^{\pm im\phi}\right] \tag{165}$$

(c) For $a = b = 0.4$ μm, $\mu = 0.065\, m_e$ calculate the frequency of emission for transition between the first excited state and the ground state.

Problem 10.4 The deuteron nucleus consists of a neutron and a proton. The neutron-proton interaction can be approximately described by the following spherically symmetric potential function[10]

$$\left.\begin{aligned} V(r) &= \infty & r &< b \\ &= -V_0 & b &< r < a+b \\ &= 0 & r &> a+b \end{aligned}\right\} \tag{166}$$

(a) Obtain the transcendental equation determining the energy eigenvalues corresponding to the $l = 0$ state and show that if we assume $V_0 = 40$ MeV and $a \approx 1.896 \times 10^{-15}$ m, we obtain $E = -2.25$ MeV which represents the binding energy of the deuteron.

(b) Also assuming $b \approx 0.4 \times 10^{-15}$ m show that for the ground state

$$\langle r^2 \rangle^{1/2} = 4.16 \times 10^{-15} \text{ m} \tag{167}$$

which compares well with the experimental value of 8.4×10^{-15} m (data quoted from Ref. 6); the value of b is suggested from the high energy experiments, see also Ref. 5.

Problem 10.5 Use the recurrence relation [Eq. (48)] and the normalization condition to derive Eqs (61)–(65) and also the following radial functions corresponding to $n = 4$:

$$R_{4,0}(r) = \frac{1}{4}\left(\frac{Z}{a_0}\right)^{3/2}\left[1 - \frac{3}{4}\xi + \frac{1}{8}\xi^2 - \frac{1}{192}\xi^3\right] e^{-\xi/4} \tag{168}$$

$$R_{4,1}(r) = \frac{1}{16}\sqrt{\frac{5}{3}}\left(\frac{Z}{a_0}\right)^{3/2}\left[\xi - \frac{1}{4}\xi^2 + \frac{1}{80}\xi^3\right] e^{-\xi/4} \tag{169}$$

$$R_{4,2}(r) = \frac{1}{64\sqrt{5}}\left(\frac{Z}{a_0}\right)^{3/2}\left[\xi^2 - \frac{1}{12}\xi^3\right] e^{-\xi/4} \tag{170}$$

$$R_{4,3}(r) = \frac{1}{768\sqrt{35}}\left(\frac{Z}{a_0}\right)^{3/2} \xi^3 e^{-\xi/4} \tag{171}$$

Show that the above equations can be written down directly using Eq. (88).

Problem 10.6 For $n = 1$ and $l = 0$, one of the solutions of Eq. (45) is a constant. Find the other solution of the equation in terms of an integral and show that it diverges for $\rho \to 0$ as well as for $\rho \to \infty$.

Problem 10.7 In continuation of the previous problem, the well-behaved solutions of Eq. (45) are

$C_0\left(1 - \frac{1}{2}\rho\right)$ for $n = 2, l = 0$

$C_0\rho$ for $n = 2, l = 1$

$C_0\left(1 - \rho + \frac{1}{6}\rho^2\right)$ for $n = 3, l = 0$

$C_0\left(\rho - \frac{1}{4}\rho^2\right)$ for $n = 3, l = 1$

$C_0\rho^2$ for $n = 3, l = 2$

Determine the other independent solution in terms of an integral.

---

[10] The potential function described by Eq. (166) is known as the *hard core* potential according to which the two particles cannot get closer than a certain distance b. This type of potential is indeed suggested by scattering experiments. It may be mentioned that the actual potential energy function describing nuclear forces is very complicated but many low energy phenomena are rather insensitive to the form of potential energy function and are reasonably well described, by a short-range attractive force of the type given by Eq. (166). For more details see Ref. 6.

Problem 10.8

(a) Consider the spherically symmetric potential well

$$V(r) = -V_0 \exp(-r/a) \tag{172}$$

For $l = 0$, solve the Schrödinger equation by making the following transformation

$$\xi = \exp(-r/2a) \tag{173}$$

which will transform Eq. (17) into a Bessel's equation. (Bessel functions are briefly discussed in Appendix J.)

(b) Obtain the energy eigenvalue assuming $\mu = 0.8369 \times 10^{-27}$ kg (corresponding to the deuteron problem), $V_0 = 40$ MeV and $a = 1.854 \times 10^{-15}$ m.

Problem 10.9 Show that $R_{0,0}(r)$, $R_{1,1}(r)$, $R_{2,0}(r)$ and $R_{2,2}(r)$ as given by Eqs (142), (149), (158) and (159) satisfy Eq. (138); determine the corresponding eigenvalues.

Problem 10.10

(a) Using the data given in the first column of Table 10.1, verify the data given in the last column.

(b) In each case calculate the value of the rotational constant B.

Problem 10.11 For HCl^{35} and HCl^{37} calculate the frequencies (in cm^{-1}) for the following transitions

$$(\nu = 1, J = J_i) \rightarrow (\nu = 0, J = J_i + 1)\ ;\ J_i = 4, 3, 2, 1, 0 \text{ (P branch)}$$

and

$$(\nu = 1, J = J_i) \rightarrow (\nu = 0, J = J_i - 1)\ ;\ J_i = 5, 4, 3, 2, 1 \text{ (R branch)}$$

and compare with the positions shown in Fig. 10.4. Assume

$$r_e = 1.27 \text{ Å and } \frac{\omega}{2\pi c} = 2885.9 \text{ cm}^{-1}$$

for both the molecules.

Problem 10.12 For the isotropic three-dimensional harmonic oscillator problem [Eq. (136)] solve the radial part of the Schrödinger equation [Eq. (138)] and show that the results are consistent with those obtained in Section 10.6.

[**Hint:** Solve Eq. (17) by determining its behavior for $r \rightarrow 0$ and $r \rightarrow \infty$ and then make the transformation $\eta = \gamma^2 r^2$ ($\gamma = \sqrt{\mu\omega/\hbar}$).]

Problem 10.13 We next consider the two-dimensional isotropic oscillator for which the potential energy variation is given by

$$V(\rho) = \frac{1}{2}\mu\omega^2\rho^2 = \frac{1}{2}\mu\omega^2(x^2 + y^2) \tag{174}$$

where

$$\rho^2 = x^2 + y^2$$

(a) Using a method similar to that used in Section 10.6, solve the two-dimensional Schrödinger equation in Cartesian coordinates to obtain the following eigenfunctions and eigenvalues

$$\psi_{n_1, n_2}(x, y) = \left[\frac{\gamma^2}{2^n\, n_1!\, n_2!\, \pi}\right]^{1/2} H_{n_1}(\gamma x)\, H_{n_2}(\gamma y)\, e^{-\frac{1}{2}\gamma^2(x^2+y^2)} \tag{175}$$

$$E = (n + 1)\,\hbar\omega \tag{176}$$

where $n = n_1 + n_2$; $n_1, n_2 = 0, 1, 2, \ldots$

(b) Solve the two-dimensional Schrödinger equation in polar coordinates and obtain solutions in terms of the confluent hypergeometric function.

[**Hint:** Remove the first derivative in the radial part of the Schrödinger equation.]

Problem 10.14 Consider the spherically symmetric potential energy function

$$V(r) = \frac{A}{r^2} - \frac{B}{r} \quad (A > 0, B > 0) \tag{177}$$

Show that the radial part of the Schrödinger equation is of the same form as for the hydrogen atom problem; the corresponding discrete energy eigenvalues being given by

$$E_p = -\frac{2B^2\mu}{\hbar^2}\left[(2p+1) + \left\{(2l+1)^2 + \frac{8\mu A}{\hbar^2}\right\}^{1/2}\right]^{-2} \tag{178}$$

where $p = 0, 1, 2, \ldots$ Determine the normalized eigenfunctions in terms of the confluent hypergeometric function.

Problem 10.15 Consider the spherically symmetric potential energy function

$$V(r) = \frac{A}{r^2} + Br^2 \quad (A > 0, B > 0) \tag{179}$$

Show that the radial part of the Schrödinger equation is of the same form as that of the isotropic oscillator [Eq. (138)]; the corresponding energy eigenvalues being given by

$$E_p = \hbar\left(\frac{B}{2\mu}\right)^{1/2}\left[4p + 2 + \left\{(2l+1)^2 + \frac{8\mu A}{\hbar^2}\right\}^{1/2}\right] \tag{180}$$

where $p = 0, 1, 2, \ldots$

10.8 SOLUTIONS

Solution 10.1 For $0 < r < a$, the radial part of the Schrödinger equation [Eq. (15)] can be written in the form

$$\frac{d^2R}{d\rho^2} + \frac{2}{\rho}\frac{dR}{d\rho} + \left[1 - \frac{l(l+1)}{\rho^2}\right]R(\rho) = 0 \tag{181}$$

where

$$\rho = kr; \quad k = \left[\frac{2\mu E}{\hbar^2}\right]^{1/2} \tag{182}$$

For $l = 0$, the transformation $u(\rho) = \rho R(\rho)$ would give

$$\frac{d^2u}{d\rho^2} + u(\rho) = 0 \tag{183}$$

the solutions of which are $\sin\rho$ and $\cos\rho$. Thus,

$$R(\rho) = A\frac{\sin\rho}{\rho} + B\frac{\cos\rho}{\rho} \quad (l = 0)$$

The second solution would blow up at $\rho = 0$ and, therefore, we must have $B = 0$.

Further,

$$R(a) = 0 \Rightarrow ka = n\pi$$

giving

$$E = E_n = \frac{n^2\pi^2\hbar^2}{2\mu a^2} \;; \quad n = 1, 2, 3, \ldots \tag{184}$$

The complete wavefunction would be

$$\psi_{n,0,0}(r, \theta, \phi) = A\frac{\sin kr}{kr}Y_{0,0} \tag{185}$$

For $l = 1$, the solution 'well-behaved' at $\rho = 0$ will be

$$R(\rho) = A\left(\frac{\sin\rho}{\rho^2} - \frac{\cos\rho}{\rho}\right) \tag{186}$$

and can be verified by direct substitution. The boundary condition $R(a) = 0$ would give the transcendental equation

$$\tan ka = ka \quad (l = 1)$$

which determines the energy eigenvalues. In general, we have the solution

$$R(\rho) = A\, j_l(\rho) \tag{187}$$

where $j_l(\rho)$ represents the spherical Bessel function (see Section 24.6.1 and Appendix J). The energy eigenvalues are determined by solving the equation

$$j_l(ka) = 0 \tag{188}$$

Solution 10.4 If we use the boundary conditions

$$u(r = b) = 0 = u(r = \infty) \tag{189}$$

the solution of Eq. (17) for $l = 0$ and for $V(r)$ given by Eq. (166) would be

$$\begin{aligned} u(r) &= A\sin k(r - b) \qquad b < r < a + b \\ &= Be^{-\kappa r} \qquad\qquad r > a + b \end{aligned} \tag{190}$$

where

$$k^2 = \frac{2\mu}{\hbar^2}(V_0 + E), \quad \kappa^2 = -\frac{2\mu E}{\hbar^2} \tag{191}$$

μ being the reduced mass; for bound states $E < 0$ and, therefore, $\kappa^2 > 0$. Obviously, $|E| < V_0$. Continuity of the wavefunction and its derivative at $r = a + b$ would readily give [cf. Eq. (59) of Chapter 6]

$$-\xi\cot\xi = \sqrt{\alpha^2 - \xi^2} \tag{192}$$

where $\qquad \xi = ka \qquad (193)$

and[11]

$$\alpha = \sqrt{\frac{2\mu V_0 a^2}{\hbar^2}}$$

---

[11] Since

$$m_n \simeq 1.67543 \times 10^{-27}\text{ kg}, \quad m_p \simeq 1.6726485 \times 10^{-27}\text{ kg}$$

we get

$$\mu = \frac{m_n m_p}{m_n + m_p} \simeq 0.8369 \times 10^{-27}\text{ kg}$$

$$\simeq \sqrt{\frac{2 \times 0.8369 \times 10^{-27} \times 40 \times 10^6 \times 1.602 \times 10^{-19} \times (1.896 \times 10^{-15})^2}{(1.055 \times 10^{-34})^2}}$$

$$\simeq 1.86 \tag{194}$$

Solving Eq. (192), we get $\xi \simeq 1.81$ giving

$$\sqrt{\frac{V_0 - |E|}{|E|}} = -\tan \xi \simeq 4.1 \tag{195}$$

$$\Rightarrow \quad E \simeq -2.25 \text{ MeV} \tag{196}$$

which compares well with the actual value of –2.225 MeV. Further

$$\langle r^2 \rangle = \int_0^\infty r^2 |u(r)|^2 \, dr = A^2 \int_b^{a+b} r^2 \sin^2 k(r-b) \, dr + B^2 \int_{a+b}^\infty r^2 \exp(-2\kappa r) \, dr \tag{197}$$

with

$$A = \left[\frac{2\kappa}{1 + \kappa a} \right]^{1/2} \quad \text{and} \quad B = A \exp[\kappa (a+b)] \sin k\alpha. \tag{198}$$

Solution 10.6 For $n = 1$ and $l = 0$, Eq. (45) can be written in the form

$$\frac{1}{dF/d\rho} \frac{d^2F}{d\rho^2} = \left(1 - \frac{2}{\rho}\right)$$

Thus,

$$\ln \left(\frac{dF}{d\rho}\right) = \rho - 2 \ln \rho + \text{constant}$$

or

$$\rho^2 \frac{dF}{d\rho} = K e^\rho$$

where K is a constant. Thus,

$$F(\rho) = K \int \frac{e^\rho}{\rho^2} \, d\rho + C_0 \tag{199}$$

which represents the general solution of Eq. (45). Integrating term by term, we get

$$-\frac{1}{\rho} + \ln \rho + \frac{1}{2!}\rho + \frac{1}{2 \cdot 3!}\rho^2 + \frac{1}{3 \cdot 4!}\rho^3 + \cdots \tag{200}$$

which represents the second independent solution of Eq. (45) for $n = 1$ and $l = 0$ which is not well behaved at $\rho = 0$.

Solution 10.7 We have to first make the transformation

$$F(\rho) = G(\rho)\, \Phi(\rho)$$

where $\Phi(\rho)$ is the known solution[12]. We illustrate the procedure by considering the $n = 2$, $l = 1$ case for which $\Phi(\rho) = \rho$; thus

$$F(\rho) = \rho G(\rho)$$

Substituting in Eq. (45), we obtain

[12] The same procedure was used in Problem 7.6. In the previous problem, $\Phi(\rho)$ was a constant and, therefore, the transformation was not necessary.

$$\frac{1}{G'}G'' = 1 - \frac{4}{\rho}$$

Carrying out the integration, we get

$$\frac{dG}{d\rho} = \frac{K}{\rho^4}e^{\rho}$$

Thus,

$$G(\rho) = K\int \frac{1}{\rho^4}e^{\rho}\,d\rho + C_0$$

or

$$F(\rho) = K\rho \int \frac{1}{\rho^4}e^{\rho}\,d\rho + C_0\rho \tag{201}$$

which represents the general solution of Eq. (45); the second term represents the known solution $\Phi(\rho)$. Integrating term by term, we obtain for the second solution

$$-\frac{1}{3\rho^2} - \frac{1}{2\rho} - \frac{1}{2!} + \frac{1}{3!}\rho \ln \rho + \frac{1}{4!}\rho^2 + \frac{1}{5!2}\rho^3 + \cdots$$

For $n = 2, l = 1$, we have

$$F(\rho) = C_0\left(\rho - \frac{1}{4}\rho^2\right)\int \frac{1}{\rho^2}e^{\rho}\,d\rho \tag{202}$$

Solution 10.8 The transformation $\xi = \exp\left(-\frac{r}{2a}\right)$ transforms Eq. (17) into the Bessel's equation, the solution of which (not diverging at $\xi = 0$ or $r = \infty$) is

$$u(\xi) = AJ_{\nu}(g\,\xi) \tag{203}$$

where

$$\nu = \left[-\frac{8\mu E a^2}{\hbar^2}\right]^{1/2} \quad \text{and} \quad g = \left[\frac{8\mu V_0 a^2}{\hbar^2}\right]^{1/2} \tag{204}$$

The condition that $u(r = 0) = 0$ gives

$$J_{\nu}(g) = 0 \tag{205}$$

which determines the energy eigenvalues. It may be mentioned that for $g < 2.405$ (the first zero of $J_0(x)$) there is no root of Eq. (205) and consequently no discrete mode exists. Now,

$$g = \left(\frac{8 \times 0.8369 \times 10^{-27} \times 40 \times 10^6 \times 1.602 \times 10^{-19} \times (1.854 \times 10^{-15})^2}{(1.055 \times 10^{-34})^2}\right)^{1/2}$$

$$\simeq 3.64$$

The solution of the transcendental equation

$$J_{\nu}(3.64) = 0$$

gives $\nu \simeq 0.858628$ which gives $E \simeq -2.223$ MeV. The corresponding wavefunction is plotted in Fig. 18.16.

Solution 10.11 Using Eqs (95) and (96), we obtain for the $(\nu = 1, J = J') \rightarrow (\nu = 0, J = J' + 1)$ transition

$$\frac{\Delta E}{hc} = \frac{\omega}{2\pi c} - 2B(J' + 1)$$

Now, assuming $r_e = 1.27$ Å, we get

$$B\,(\mathrm{HCl}^{35}) = 10.626\ \mathrm{cm}^{-1}$$

and

$$B\,(\mathrm{HCl}^{37}) = 10.610\ \mathrm{cm}^{-1}$$

Thus, if we take

$$\frac{\omega}{2\pi c} = 2885.9\ \mathrm{cm}^{-1}$$

we would get

$$\frac{\Delta E}{hc} = (2779.8,\ 2779.6)\ ;\ (2801.0,\ 2800.9)\ ;$$

$$(2822.2,\ 2822.1)\ ;\ (2843.5,\ 2843.4)\ \text{and}\ (2864.7,\ 2864.6)$$

for $J' = 4, 3, 2, 1$ and 0, respectively. The two numbers in the paranthesis correspond to HCl^{37} and HCl^{35} respectively.

Solution 10.12 For the three-dimensional harmonic oscillator,

$$V\,(r) = \frac{1}{2}\,\mu\omega^2 r^2 \tag{206}$$

and Eq. (17) can, therefore, be written in the form

$$\frac{d^2u}{d\xi^2} + \left[\Lambda - \xi^2 - \frac{l(l+1)}{\xi^2}\right]u(\xi) = 0 \tag{207}$$

where

$$\xi = \gamma r;\quad \gamma = \sqrt{\frac{\mu\omega}{\hbar}}\quad \text{and}\quad \Lambda = \frac{2E}{\hbar\omega} \tag{208}$$

Comparing Eq. (207) with Eq. (13) of Chapter 7, we find that for $\xi \to \infty$ the solution will be of the form exp $\left(\pm\frac{1}{2}\xi^2\right)$. For $\xi \to 0$, the third term inside the square brackets will dominate and the solution will be ξ^m with $m = l + 1$ or $-l$ (see the analysis given in Section 10.4.2). Thus, if we write

$$u\,(\xi) = F\,(\xi)\,y(\xi) \tag{209}$$

with

$$F\,(\xi) = \xi^{l+1}\,e^{-\frac{1}{2}\xi^2}$$

we would readily get

$$\frac{d^2y}{d\xi^2} + 2\left(\frac{l+1}{\xi} - \xi\right)\frac{dy}{d\xi} + (\Lambda - 2l - 3)\,y(\xi) = 0$$

[compare the analysis given in Section 10.4.2]. The transformation $\eta = \xi^2$ will give the confluent hypergeometric equation

$$\eta\frac{d^2y}{d\eta^2} + (c - \eta)\frac{dy}{d\eta} - a\,y(n) = 0$$

with

$$a = \frac{l}{2} + \frac{3}{4} - \frac{\Lambda}{4}\ ;\quad c = l + \frac{3}{2}$$

For the wavefunction to be bounded at infinity, the series has to terminate and we must, therefore, have $a = -n_r$ giving

$$E = \frac{1}{2}\hbar\omega\Lambda = \left(2n_r + l + \frac{3}{2}\right)\hbar\omega; \quad n_r = 0, 1, 2, \ldots \tag{210}$$

The corresponding wavefunctions are given by

$$R(r) = N\,\xi^l\, e^{-\frac{1}{2}\xi^2}\, {}_1F_1\left(-n_r, l + \frac{3}{2}, \xi^2\right) \tag{211}$$

where

$$N = \gamma^{3/2}\left[\frac{2}{\Gamma(l + \frac{3}{2})}\right]^{1/2} \quad \text{for } n_r = 0$$

$$= \gamma^{3/2}\left[\frac{2}{\Gamma(l+\frac{3}{2})\,n_r!}\left(l + \frac{3}{2}\right)\left(l + \frac{5}{2}\right)\cdots\left(l + n_r + \frac{1}{2}\right)\right]^{1/2}$$

Once again it is very straightforward to write any eigenfunction. Also, one can readily solve for a bounded three-dimensional oscillator.

Solution 10.13 For the two-dimensional isotropic oscillator

$$V(\rho) = \frac{1}{2}\mu\omega^2 r^2 = \frac{1}{2}\mu\omega^2 (x^2 + y^2) \tag{212}$$

and the corresponding Schrödinger equation is given by

$$\frac{1}{r}\frac{\partial}{\partial r}\left(r\frac{\partial\psi}{\partial r}\right) + \frac{1}{r^2}\frac{\partial^2\psi}{\partial\phi^2} + \frac{2\mu}{\hbar^2}[E - V(r)]\,\psi(r, \phi) = 0 \tag{213}$$

where (r, ϕ) now represent the polar coordinates. To remove the first derivative, we write the solution as

$$\psi(r, \phi) = \frac{u(r)}{\sqrt{r}}\left[\frac{1}{\sqrt{2\pi}}e^{im\phi}\right]; \quad m = 0, \pm 1, \pm 2, \ldots \tag{214}$$

then $u(r)$ would satisfy

$$\frac{d^2u}{d\xi^2} + \left[\Lambda - \xi^2 - \frac{m^2 - 1/4}{\xi^2}\right]u(\xi) = 0 \tag{215}$$

where $\xi = \gamma r; \quad \gamma = \sqrt{\frac{\mu\omega}{\hbar}} \quad \text{and} \quad \Lambda = \frac{2E}{\hbar\omega}$

Once again, the behaviour of the solution for $\xi \to 0$ and $\xi \to \infty$ suggests that we try a solution of the form

$$u(\xi) = F(\xi)\, y(\xi)$$

where $F(\xi) = \xi^{m+\frac{1}{2}}\, e^{-\frac{1}{2}\xi^2}$

and we have assumed $m \geq 0$. Analysis similar to that given in Solution 10.12 would give

$$u(\xi) = \xi^{|m|+\frac{1}{2}}\, e^{-\frac{1}{2}\xi^2}\, {}_1F_1(a, c, \xi^2)$$

where $$a = -\frac{\Lambda}{4} + \frac{|m|+1}{2} = -n\,; \quad c = |m| + 1$$

Thus, the eigenvalues would be given by (see also Solution 20.13)

$$E = (2n + |m| + 1)\,\hbar\omega \tag{216}$$

and the corresponding eigenfunction would be

$$\psi(r, \phi) = N\left[\xi^{|m|} e^{-\frac{1}{2}\xi^2}\, {}_1F_1(-n, |m|+1, \xi^2)\right]\left[\frac{1}{\sqrt{2\pi}} e^{im\phi}\right] \tag{217}$$

10.9 SAMPLE QUESTIONS

Q. 1 In the hydrogen atom problem the radial part of the Schrödinger equation can be written in the form

$$\frac{1}{\rho^2}\frac{d}{d\rho}\left(\rho^2 \frac{dR}{d\rho}\right) + \left(\frac{\lambda}{\rho} - \frac{1}{4} - \frac{l(l+1)}{\rho^2}\right) R(\rho) = 0 \tag{218}$$

where $$\rho = \gamma r\,;\ \gamma = \left(\frac{8\mu E}{\hbar^2}\right)^{1/2} \tag{219}$$

$$\lambda = Z\alpha\left(-\frac{\mu c^2}{2E}\right)^{1/2} \tag{220}$$

and $l = 0, 1, 2, \ldots$. If we assume a solution of the form

$$R(\rho) = e^{-\rho/2} \sum_{p=0,1,\ldots}^{\infty} C_p\, \rho^{p+s} \tag{221}$$

then we obtain

$$\frac{C_{p+1}}{C_p} = \frac{s + p - \lambda + 1}{(s+p+1)(s+p+2) - l(l+1)} \tag{222}$$

and for the solution to be well-behaved at $\rho = 0$ we must put $s = l$.

(a) For what values of ρ the infinite series in Eq. (221) be a convergent series?

(b) For what values of λ will the infinite series in Eq. (221) become a polynomial? What will be the corresponding energy eigenvalues?

(c) Determine the eigenfunctions $R_{nl}(r)$ for

$(\lambda = 1, l = 0)$, $(\lambda = 2, l = 0)$, $(\lambda = 2, l = 1)$

$(\lambda = 3, l = 0)$, $(\lambda = 3, l = 1)$ and $(\lambda = 3, l = 2)$

In each case normalize the wavefunction from first principles.

(d) For $(\lambda = 2, l = 1)$, $(\lambda = 3, l = 2)$ determine the second solution of Eq. (218) in the form of an integral and show that it is singular at $\rho = 0$.

(e) Discuss the degeneracy of the level corresponding to $\lambda = 3$.

Q. 2 Given that the 2 independent solutions of

$$\frac{d^2\phi}{dx^2} + \left\{-\frac{1}{4} + \frac{k}{x} + \frac{\frac{1}{4} - \mu^2}{x^2}\right\}\phi(x) = 0$$

are

$$y_1(x) = x^{u+\frac{1}{2}} e^{-x/2} {}_1F_1(a, c, x)$$

and

$$y_2(x) = x^{-\mu+\frac{1}{2}} e^{-x/2} {}_1F_1(a-c+1, 2-c, x)$$

where $a = \mu + \frac{1}{2} - k$, $c = 2\mu + 1$. Now, the radial part of the Schrödinger equation for the hydrogen atom problem can be transformed to the form

$$\frac{d^2u}{d\rho^2} + \left[\frac{2\mu E}{\hbar^2\gamma^2} + \frac{2\mu Ze^2}{\hbar^2\gamma\rho} - \frac{l(l+1)}{\rho^2}\right] u(\rho) = 0$$

where symbols have their usual meaning.

(a) Find the values of γ, a and c.

(b) Derive with justification the energy eigenvalues.

(c) Derive the normalized wavefunction $R_{54}(r)$ [$n = 5$, $l = 4$].

REFERENCES AND SUGGESTED READINGS

1. A.K. Ghatak, I.C. Goyal and S.J. Chua, *Mathematical Physics*, Macmillan India Limited, New Delhi (1995).
2. H.E. White, *Introduction to Atomic Spectra*, McGraw-Hill Book Co., New York (1934).
3. H.B. Dunford, *Elements of Diatomic Molecular Spectra*, Addison-Wesley Publishing Co., Reading, Massachusetts (1968).
4. R. Eisberg and R. Resnick, *Quantum Physics of Atoms, Molecules, Solids, Nuclei, and Particles*, John Wiley & Sons, New York (1974).
5. H.A. Sachs, *Nuclear Theory*, Addison-Wesley Publishing Co., Reading, Massachusetts (1953).
6. H. Enge, *Introduction to Nuclear Physics*, Addison-Wesley Publishing Co., Reading, Massachusetts (1966).

11 Dirac's Bra and Ket Algebra

11.1 INTRODUCTION

In this chapter [illegible] system [illegible] vectors [illegible] of [illegible] use the [illegible] momentum [illegible] algebra [illegible]

11.2 THE BRA AND KET VECTORS

A state of [illegible] represented [illegible] states [illegible] is described by the symbol [illegible] states [illegible]

$$[illegible] \, |A\rangle = c_1 |B\rangle [illegible]$$

is a vector in the same state [illegible] (p. 16 of Reference [illegible])

[illegible] being such that [illegible] vector [illegible]

Further, the ket [illegible] $|A\rangle$ where c is arbitrary [illegible] to the same state. In other words, the state of the system is defined by the direction [illegible] vectors. In this respect the superposition principle in classical and quantum theories differ [illegible]

---

[1] Quoted from [illegible]

[2] Portions of this chapter [illegible] University Press [illegible]

11 Dirac's Bra and Ket Algebra

Dirac to whom in my opinion we owe the most logically perfect presentation of quantum mechanics...

— ALBERT EINSTEIN[1]

11.1 INTRODUCTION

In this chapter, we will introduce Dirac's bra and ket algebra in which the states of a dynamical system will be denoted by certain vectors (which, following Dirac, will be called as bra and ket vectors) and operators representing dynamical variables (like position coordinates, components of momentum and angular momentum) by matrices.[2] In the following two chapters we will use the bra and ket algebra to solve the linear harmonic oscillator problem and the angular momentum problem. In both chapters we will show the advantage of using the operator algebra in obtaining solutions of various problems.

11.2 THE BRA AND KET NOTATION

A state of a system can be represented by a certain type of vector, which we call a ket vector and represent by the symbol $|\ \rangle$. In order to distinguish the ket vectors corresponding to different states, we insert a label; thus, the ket vector (or simply the ket) corresponding to the state A is described by the symbol $|A\rangle$. Kets form a linear vector space implying that if we have two states described by the kets $|A\rangle$ and $|B\rangle$, then the linear combination

$$c_1 |A\rangle + c_2 |B\rangle \tag{1}$$

is a vector in the same space; here c_1 and c_2 are arbitrary complex numbers. To quote Dirac (p. 16 of Reference 1):

...each state of a dynamical system at a particular time corresponds to a ket vector, the correspondence being such that if a state results from the superposition of certain other states, its corresponding ket vector is expressible linearly in terms of the corresponding ket vectors of the other states, and conversely.

Further, the ket $|A\rangle$ and $c\,|A\rangle$ (where c is an arbitrary non-zero complex number) correspond to the same state. In other words, the state of the system is defined by the 'direction' of the vectors. In this respect the superposition principle in classical and quantum theories differ.

[1] Quoted from N. Mukunda, *The World of Bohr and Dirac*, Wiley Eastern Limited, New Delhi (1993), p. 4.

[2] Portions of this chapter have been adapted from Ref. 1; this is with the kind permission of Oxford University Press.

For example, the superposition of a vibrating string on itself gives, in classical physics, a mode with twice the amplitude and four times the energy of the initial state of vibration. In contrast, in quantum mechanics, the superposition of a state on itself gives the same state.

Now, with every vector space there can be associated a dual vector space such that one can form a scalar product of the two vectors, one from each space. The vectors of the space dual to that of the ket vectors will be called bra vectors or simply 'bras' and will be denoted by $\langle \,|$. The scalar product of the ket $|A\rangle$ and bra $\langle B|$ is denoted by $\langle B | A\rangle$ and is a complex number. The logic is very similar to the one we have in theory of matrices where with every column vector there can be associated a row vector and the scalar product of the two gives a number. We will elaborate on this point in Section 11.7.

A bra is said to be a null bra if the scalar product vanishes for any ket, thus

$$\langle B | = 0 \quad \text{if} \quad \langle B | A\rangle = 0 \quad \text{for any } |A\rangle \tag{2}$$

Two bras are said to be equal if their scalar product with an arbitrary ket are equal, thus

$$\langle B_1 | = \langle B_2 | \quad \text{if} \quad \langle B_1 | A\rangle = \langle B_2 | A\rangle \quad \text{for any } |A\rangle \tag{3}$$

It is further assumed that

(i) There is a one-to-one correspondence between kets and bras in the sense that a state of a dynamical system represented by $|A\rangle$ can be equally well represented by the corresponding bra $\langle A|$. Further, if

$$|P\rangle = |A\rangle + |B\rangle \quad \text{then} \quad \langle P| = \langle A| + \langle B| \tag{4}$$

and if

$$|R\rangle = c\,|A\rangle \quad \text{then} \quad \langle R| = \bar{c}\,\langle A| \tag{5}$$

where c is a complex number and $\bar{c}$ its complex conjugate.

(ii)

$$\langle A|B\rangle = \overline{\langle B|A\rangle} \tag{6}$$

where the bar on the top implies the complex conjugate of the quantity.

Putting $|B\rangle = |A\rangle$, we get

$$\langle A|A\rangle = \overline{\langle A|A\rangle}$$

implying that the scalar product $\langle A|A\rangle$ is a real number. We further assert that

$$\langle A|A\rangle \geq 0, \tag{7}$$

the equality sign holds only when $|A\rangle = 0$, i.e., where $|A\rangle$ is a null ket.

If $\langle A|B\rangle = 0$ then the kets $|A\rangle$ and $|B\rangle$ are said to be orthogonal to each other. If

$$\langle A|A\rangle = 1$$

then the ket $|A\rangle$ is said to be normalized. Since the kets $|A\rangle$ and $c\,|A\rangle$ correspond to the same state we may always associate normalized kets to each state. It can be easily seen that a normalized ket is defined only within an arbitrary phase factor $e^{i\gamma}$ (where γ is a real number).

We may mention here the relationship between the Schrödinger wavefunctions developed in earlier chapters to the bra and kets developed in this section. If $|\psi\rangle$ and $|\phi\rangle$ represent the kets corresponding to the states described by the wavefunctions $\psi(\mathbf{r})$ and $\phi(\mathbf{r})$ respectively, then

$$\langle\phi|\psi\rangle = \int \phi^*(\mathbf{r})\psi(\mathbf{r})\,d\tau = \langle\psi|\phi\rangle \quad \text{Scalar product} \tag{8}$$

where the integration is over the entire space. The integral is usually referred to as the scalar product of the two functions.

11.3 LINEAR OPERATORS

An operator α converts a ket $|A\rangle$ into another ket $|B\rangle$

$$|B\rangle = \alpha|A\rangle \tag{9}$$

An operator is said to be *linear* if it satisfies the following equation

$$\alpha(c_1|A_1\rangle + c_2|A_2\rangle + \cdots) = c_1\alpha|A_1\rangle + c_2\alpha|A_2\rangle + \cdots \tag{10}$$

where c_1, c_2, ... are arbitrary complex numbers. Henceforth, we will consider only linear operators. An operator α is said to be a *null* operator if

$$\alpha|A\rangle = 0 \quad \text{for any } |A\rangle \quad \text{Null operator} \tag{11}$$

Thus a necessary and sufficient condition for an operator to be a null operator is

$$\langle A|\alpha|A\rangle = 0 \quad \text{for any } |A\rangle \tag{12}$$

An operator is said to be an *unit* operator if

$$\alpha|A\rangle = |A\rangle \quad \text{for any } |A\rangle \quad \text{Unit operator} \tag{13}$$

It can be easily seen that a number can be regarded as a linear operator.

Two operators α and β are said to be equal if and only if

$$\langle A|\alpha|A\rangle = \langle A|\beta|A\rangle \quad \text{for any } |A\rangle \tag{14}$$

The sum (or difference) of two operators α and β is defined through the equation

$$(\alpha \pm \beta)|A\rangle = \alpha|A\rangle \pm \beta|A\rangle \tag{15}$$

Further

$$\alpha + (\beta + \gamma) = (\alpha + \beta) + \gamma = \alpha + \beta + \gamma \quad \text{Associative Law} \tag{16}$$

and

$$(c_1\alpha)|A\rangle = c_1(\alpha|A\rangle) \tag{17}$$

where c_1 is an arbitrary complex constant. The product of two operators α and β is defined through the equation

$$(\beta\alpha)|A\rangle = \beta(\alpha|A\rangle) = \beta|B\rangle \tag{18}$$

where $|B\rangle = \alpha|A\rangle$. In general

$$\beta\alpha \neq \alpha\beta \tag{19}$$

The commutator of two operators is defined by (see Section 4.3)

$$[\alpha, \beta] = \alpha\beta - \beta\alpha = -[\beta, \alpha] \tag{20}$$

So far we have assumed a linear operator to act on kets; we can make a linear operator α operate on bras as well; the rule is that the bra has to be put on the left of the operator like $\langle P|\alpha$ and the operation is defined through the equation

$$\{\langle P|\alpha\}|A\rangle = \langle P|\{\alpha|A\rangle\} \quad \text{for any } |A\rangle$$
$$= \langle P|B\rangle$$

where

$$|B\rangle = \alpha \,|\, A\rangle$$

(The combination $\alpha\langle B \,|$ is meaningless.) Indeed because of the associative law we need not put any brackets and simply write

$$\langle P \,|\alpha\beta\gamma|\, A\rangle$$

It is interesting to note that the combination $|B\rangle\langle A|$ can be considered as a linear operator because

$$\{|B\rangle\langle A|\}|P\rangle = |B\rangle\{\langle A|P\rangle\}$$
$$= c|B\rangle$$

because $\langle A \mid P\rangle (= c)$ is just a complex number.

The adjoint of the operator α is denoted by $\bar{\alpha}$ and is *defined* through the equation

$$\langle A|\bar{\alpha}|B\rangle = \overline{\langle B|\alpha|A\rangle} \text{ Adjoint of an operator} \tag{21}$$

where $\overline{\langle B|\alpha|A\rangle}$ is the complex conjugate of the number $\langle B|\alpha|A\rangle$. Now

$$\langle A|\bar{\bar{\alpha}}|B\rangle = \langle A|\bar{\beta}|B\rangle \qquad (\beta \equiv \bar{\alpha})$$
$$= \overline{\langle B|\beta|A\rangle} = \overline{\langle B|\bar{\alpha}|A\rangle}$$
$$= \overline{\overline{\langle A|\alpha|B\rangle}} = \langle A \mid \alpha \mid B\rangle \tag{22}$$

because the complex conjugate of the complex conjugate of a number is the original number itself. Since Eq. (22) holds for arbitrary $\mid A\rangle$ and $\mid B\rangle$ we have

$$\bar{\bar{\alpha}} = \alpha \tag{23}$$

i.e., the adjoint of the adjoint of a linear operator is the original operator itself.

Further, the adjoint of the product of the two operators α and β is the product of the adjoint of the two operators in reverse order, i.e.,

$$\overline{\alpha\beta} = \bar{\beta}\bar{\alpha} \quad \text{Adjoint of a product of two operators} \tag{24}$$

The proof is simple. Let

$$\alpha|\, A\rangle = |P\rangle$$

then

$$\langle A|\bar{\alpha}|B\rangle = \overline{\langle B|\alpha|A\rangle} = \overline{\langle B|P\rangle}$$
$$= \langle P|B\rangle \qquad \text{[using Eq. (6)]}$$

Since the above equation holds for arbitrary $\mid B\rangle$, we have

$$\langle P \mid = \langle A|\,\bar{\alpha} = \text{conjugate of } \alpha|A\rangle \tag{25}$$

Thus, if

$$|Q\rangle = \alpha\beta|A\rangle$$

Then,

$$\langle Q| = \langle A|\overline{\alpha\beta} \tag{26}$$

But, if we write

$$|Q\rangle = \alpha\beta|A\rangle = \alpha|R\rangle$$

where $|R\rangle = \beta|A\rangle$ (consequently $\langle R| = \langle A|\bar{\beta}$), then

$$\langle Q| = \langle R|\bar{\alpha} = \langle A|\bar{\beta}\bar{\alpha} \tag{27}$$

Comparing Eqs (26) and (27), we get Eq. (24). In general

$$\overline{\alpha\beta\gamma\ldots} = \ldots\bar{\gamma}\bar{\beta}\bar{\alpha} \tag{28}$$

An operator is said to be self-adjoint if

$$\bar{\alpha} = \alpha \quad \text{Self-adjoint operator} \tag{29}$$

A self-adjoint operator is also called a real operator or a Hermitian operator.

11.4 THE EIGENVALUE EQUATION

For the linear operator α, consider the equation

$$\alpha|A_n\rangle = a_n|A_n\rangle \tag{30}$$

where a_n is an arbitrary complex number. Equation (30) represents an eigenvalue equation; $|A_n\rangle$ is said to be an eigenket of the operator α; a_n being the corresponding eigenvalue. It can be easily seen that $c|A_n\rangle$ (where c is an arbitrary complex number) is also an eigenket belonging to the same eigenvalue a_n. Now, if there is more than one ket (and they are not linearly dependent on each other[3]) belonging to the same eigenvalue, i.e., if

$$\left.\begin{aligned}\alpha|A_1\rangle &= a_1|A_1\rangle \\ \alpha|A_2\rangle &= a_1|A_2\rangle\end{aligned}\right\} \quad \text{Degeneracy} \tag{31}$$

then the state is said to be a degenerate state (compare discussion in Section 6.3). If there are g linearly independent kets belonging to the same eigenvalue then the state is said to be g-fold degenerate. For the sake of simplicity, let us consider a two-fold degenerate state described by Eqs (31) and (32). If we multiply Eq. (31) by c_1 and Eq. (32) by c_2 and add we would get

$$\alpha|P\rangle = a_1|P\rangle, \tag{32}$$

where $$|P\rangle = c_1\,|A_1\rangle + c_2|A_2\rangle$$

implying that the linear combination $c_1|A_1\rangle + c_2|A_2\rangle$ is also an eigenket belonging to the same eigenvalue. Similarly, one can discuss for a g-fold degenerate state.

11.4.1 Orthogonality of Eigenfunctions

When α is *real*, it can easily be shown that all the eigenvalues are real and for two different eigenvalues (i.e., for $a_n \neq a_m$) the corresponding eigenfunctions are necessarily orthogonal, i.e.,

$$\langle A_n|A_m\rangle = 0 \qquad \text{for} \qquad a_n \neq a_m \tag{33}$$

Further, one can always normalize the kets and choose an appropriate linear combination for the kets belonging to a degenerate state such that (see Problem 6.3)

$$\langle A_n|A_m\rangle = \delta_{nm} \tag{34}$$

The proof is very simple. Premultiplying Eq. (30) by $\langle A_n|$ we get

$$\langle A_n|\alpha|A_n\rangle = a_n\langle A_n|A_n\rangle$$

Now $\langle A_n|A_n\rangle$ is always real and not a null ket [otherwise Eq. (30) has no meaning]. Further, since α is real

---

[3] A ket $|P\rangle$ is said to be linearly dependent on the ket $|A_1\rangle$, $|A_2\rangle$, ... and $|A_N\rangle$ if we can write

$$|P\rangle = \sum_{n=1}^{N} c_n|A_n\rangle$$

$$\langle A_n|\alpha|A_n\rangle = \langle A_n|\bar{\alpha}|A_n\rangle = \overline{\langle A_n|\alpha|A_n\rangle}$$

implying that $\langle A_n|\alpha|A_n\rangle$ is also real and hence a_n must be real. Further, in order to prove Eq. (33) we consider

$$\alpha|A_1\rangle = a_1|A_1\rangle \tag{35}$$

$$\alpha|A_2\rangle = a_2|A_2\rangle \tag{36}$$

If we put $|P\rangle = \alpha|A_2\rangle$, then

$$\langle P| = \langle A_2|\bar{\alpha} = \langle A_2|\alpha$$

Also,

$$\langle P| = \bar{a}_2\langle A_2| = a_2\langle A_2|$$

because a_2 is real. Thus,

$$\langle A_2|\alpha = a_2\langle A_2| \tag{37}$$

Premultiplying Eq. (35) by $\langle A_2|$ and postmultiplying Eq. (37) by $|A_1\rangle$ gives

$$\langle A_2|\,\alpha|\,A_1\rangle = a_1\langle A_2|A_1\rangle = a_2\langle A_2|A_1\rangle$$

which immediately gives the orthogonality condition given by Eq. (33).

Since the formalism is symmetrical with respect to bras and kets we also have the eigenvalue equation

$$\langle B_n|\alpha = \langle B_n|b_n = b_n\langle B_n|$$

where $\langle B_n|$ are the eigenbras and b_n the corresponding eigenvalues. It can be easily seen that when α is a real operator and if $|A\rangle$ is an eigenket, then $\langle A|$ is an eigenbra belonging to the same eigenvalue. Equation (37) tells us that $\langle A_2|$ is an eigenbra of the operator α belonging to the same eigenvalue a_2.

11.5 OBSERVABLES

Any dynamical quantity (like the position coordinates, or components of the momentum or angular momentum, etc.) that can be measured is known as an observable. We assume that observables can be represented by linear operators and the operators corresponding to different observables need not commute. Further, the result of measurement of any observable must be an eigenvalue of the operator corresponding to the observable and since the measured value must be a real number we assume that **an observable is always represented by a real linear operator.** We denote the eigenkets of the observable α by $|\alpha_n\rangle$ which we assume to form an orthonormal set:

$$\langle\alpha_m|\alpha_n\rangle = \delta_{mn}$$

Also, as a result of the measurement of an observable, the system would collapse to one of the eigenstates of the observable and therefore the eigenkets $|\alpha_n\rangle$ must form a complete set in the sense that an arbitrary ket $|P\rangle$ can be expanded in terms of them

$$|P\rangle = \sum_n c_n|\alpha_n\rangle$$

If we now make a measurement of α, the probability of the outcome α_n is $|c_n|^2$; we have assumed $|P\rangle$ to be normalized. Further, as a result of the measurement, the system would *collapse* to one of the states $|\alpha_n\rangle$. For degenerate states, one can always choose an appropriate linear combination so that they form an orthonormal set (see Section 6.3 and Problem 6.3).

Let us assume that a dynamical system is in a state which is an eigenket of the observable α belonging to the eigenvalue α_1. Now, if a measurement of the observable α is made, then it is certain that the value α_1 will be obtained. On the other hand, if the system is in a state described by the normalized ket

$$|P\rangle = c_1|\alpha_1\rangle + c_2|\alpha_2\rangle$$

$[\langle P|P\rangle = 1]$, then a measurement of α would lead either to α_1 or α_2 with probabilities $|c_1|^2$ and $|c_2|^2$ respectively. Since $\langle P|P\rangle = 1$, it immediately follows that

$$|c_1|^2 + |c_2|^2 = 1 \tag{38}$$

In general, the outcome of the measurement of an observable is one of its eigenvalues and therefore, if the system is in an arbitrary state, then a measurement of α will make the dynamical system jump to one of the eigenstates of α. We further assume that the possible eigenstates of α to which the dynamical system may jump are such that the original state should be linearly expressible in terms of the eigenstates. Thus, any state can be expressed as a linear combination of the eigenkets of the observable and, therefore, **the eigenkets of an observable must form a complete set.**

A beautiful example of the above argument is the famous Stern-Gerlach experiment where an inhomogenous magnetic field (in the *z*-direction) splits the silver beam into two components, one going in the $+z$ direction and the other going in the $-z$ direction; the experiment tries to make a measurement of the z-component of the angular momentum (which we denote by J_z) and the result shows that J_z has only two eigenvalues. We will have a detailed discussion of this experiment in Section 14.4.

11.6 THE COMPLETENESS CONDITION

We have just stated that the eigenkets of an observable form a complete set. Let $|n\rangle$, $n = 0, 1, 2, \ldots$ denote these eigenkets and let $|P\rangle$ denote an arbitrary ket. Thus,

$$|P\rangle = \sum_n c_n|n\rangle$$

where Σ denotes a sum over the discrete states and an integration over the continuum states. Since the eigenkets can be assumed to form an orthonormal set ($\langle m|n\rangle = \delta_{mn}$) we have

$$\langle m|P\rangle = \sum_n c_n\langle m|n\rangle = \sum_n c_n\,\delta_{mn} = c_m$$

Thus,

$$|P\rangle = \sum_n |n\rangle\, c_n = \left\{\sum_n |n\rangle\langle n|\right\}|P\rangle \tag{39}$$

Since the above equation holds for an arbitrary ket $|P\rangle$, the quantity inside the curly brackets must be a unit operator

$$\sum_n |n\rangle\langle n| = 1 \tag{40}$$

which is usually referred to as the *completeness condition.* The above equation may be compared with the completeness condition of the Schrödinger wavefunctions [see Eq. (41) of Chapter 7].

11.7 EXAMPLES FROM MATRIX ALGEBRA

The various axiomatic rules discussed so far can be readily understood if we consider some simple examples from the theory of matrices. Let us consider a 3×3 Hermitian matrix α given by

$$\alpha = \begin{pmatrix} 0 & 1 & 0 \\ 1 & 0 & 0 \\ 0 & 0 & 2 \end{pmatrix}$$

We have assumed α to be a 3×3 matrix; we could have assumed it to be of an arbitrary order. Let

$$|A\rangle = \begin{pmatrix} a_1 \\ a_2 \\ a_3 \end{pmatrix} \quad \text{and} \quad |B\rangle = \begin{pmatrix} b_1 \\ b_2 \\ b_3 \end{pmatrix}$$

denote two arbitrary column vectors in the same field (i.e., having 3 rows). Now corresponding to every ket vector there is a bra vector which, in this case, is nothing but a row vector. Thus,

$$\langle A| = (\overline{a_1}, \overline{a_2}, \overline{a_3})$$

and

$$\langle B| = (\overline{b_1}, \overline{b_2}, \overline{b_3})$$

where $\overline{a_1}, \overline{a_2}, \ldots$ are the complex conjugates of $a_1, a_2, \ldots$. The scalar product is given by

$$\langle A|B\rangle = (\overline{a_1} \;\; \overline{a_2} \;\; \overline{a_3}) \begin{pmatrix} b_1 \\ b_2 \\ b_3 \end{pmatrix}$$

$$= \overline{a_1}\, b_1 + \overline{a_2}\, b_2 + \overline{a_3}\, b_3$$

which is a complex number. Obviously,

$$\langle A|B\rangle = \overline{\langle B|A\rangle}$$

Further,

$$\langle A|A\rangle = |a_1|^2 + |a_2|^2 + |a_3|^2 \geq 0$$

The scalar product $\langle A|A\rangle = 0$ if and only if

$$a_1 = a_2 = a_3 = 0$$

which corresponds to a 'null vector'. Since,

$$\alpha \begin{pmatrix} 0 \\ 0 \\ 0 \end{pmatrix} = \begin{pmatrix} 0 \\ 0 \\ 0 \end{pmatrix}$$

implying that a linear operator operating on a null vector will always give a null vector. Further

$$|A\rangle = |B\rangle$$

if and only if

$$a_1 = b_1, a_2 = b_2 \quad and \quad a_3 = b_3$$

Now, it can be easily shown that the eigenvalues of the operator α are +1, –1 and +2. The corresponding normalized eigenvectors are

$$|1\rangle = \frac{1}{\sqrt{2}} \begin{pmatrix} 1 \\ 1 \\ 0 \end{pmatrix}$$

$$|2\rangle = \frac{1}{\sqrt{2}}\begin{pmatrix}1\\-1\\0\end{pmatrix}$$

and $$|3\rangle = \begin{pmatrix}0\\0\\1\end{pmatrix}$$

One can easily see that

$$\langle i\,|\,j\rangle = \delta_{ij}; \quad i, j = 1, 2, 3$$

which represents the orthonormality condition. Further, they form a complete set of functions in the sense that an arbitrary vector (in the three-dimensional space) can be expressed as a linear combination of $|1\rangle$, $|2\rangle$ and $|3\rangle$. Thus,

$$|A\rangle = \frac{a_1 + a_2}{\sqrt{2}}\,|1\rangle + \frac{a_1 - a_2}{\sqrt{2}}\,|2\rangle + a_3\,|3\rangle$$

Further

$$\sum_i |i\rangle\langle i| = |1\rangle\langle 1| + |2\rangle\langle 2| + |3\rangle\langle 3|$$

$$= \begin{pmatrix}1 & 0 & 0\\0 & 1 & 0\\0 & 0 & 1\end{pmatrix} \text{ Completeness condition}$$

which represents the completeness condition [see Eq. (40)].

11.8 SIMULTANEOUS EIGENKETS OF COMMUTING OBSERVABLES

In this section, we will show that if two observables commute then one can have a complete set of simultaneous eigenkets of the two observables. Let α and β represent the operators corresponding to two observables. We assume that the two operators commute, i.e.,

$$\alpha\beta = \beta\alpha \tag{41}$$

Let $|\alpha_1\rangle$ be an eigenket of α belonging to the eigenvalue α_1:

$$\alpha\,|\alpha_1\rangle = \alpha_1\,|\,\alpha_1\rangle \tag{42}$$

then $$\alpha\,\{\beta\,|\,\alpha_1\rangle\} = \beta\alpha\,|\,\alpha\rangle$$

$$= \beta\alpha_1\,|\,\alpha_1\rangle \quad \text{[using Eq. (42)]}$$

or $$\alpha\,\{\beta\,|\,\alpha_1\rangle\} = \alpha_1\,\{\beta\,|\,\alpha_1\rangle\} \tag{43}$$

where the last step follows from the fact that α_1 is a number. Equation (43) tells us that $\beta\,|\,\alpha_1\rangle$ is also an eigenket of α belonging to the *same* eigenvalue α_1. If $|\alpha_1\rangle$ corresponds to a non-degenerate state, then Eq. (43) tells us that

$$\beta\,|\,\alpha_1\rangle \quad \text{must be a multiple of } |\,\alpha_1\rangle$$

implying

$$\beta\,|\,\alpha_1\rangle = \beta_1\,|\,\alpha_1\rangle \tag{44}$$

where β_1 is a number. Thus $|\alpha_1\rangle$ must be an eigenket of β. Hence, all eigenkets of α which correspond to nondegenerate states, must also be eigenkets of β and hence these will be simultaneous eigenkets.

We next consider the case when the eigenvalue α_1 corresponds to a degenerate state. To keep the analysis simple let the state be two-fold degenerate and let the eigenkets be $|\alpha_1\rangle$ and $|\alpha_2\rangle$; i.e.,

$$\alpha|\alpha_1\rangle = \alpha_1|\alpha_1\rangle \tag{45}$$

$$\alpha|\alpha_2\rangle = \alpha_1|\alpha_2\rangle \tag{46}$$

The kets $|\alpha_1\rangle$ and $|\alpha_2\rangle$ are assumed to be orthogonal to each other. Now, since $\beta|\alpha_1\rangle$ and $\beta|\alpha_2\rangle$ are eigenkets of α belonging to the same eigenvalue α_1 [see Eq. (43)], they must be expressible as a linear combination of $|\alpha_1\rangle$ and $|\alpha_2\rangle$:

$$\beta|\alpha_1\rangle = c_{11}|\alpha_1\rangle + c_{12}|\alpha_2\rangle \tag{47}$$

and

$$\beta|\alpha_2\rangle = c_{21}|\alpha_1\rangle + c_{22}|\alpha_2\rangle \tag{48}$$

The kets $|\alpha_1\rangle$ and $|\alpha_2\rangle$ need not be eigenkets of β; however, we may always choose appropriate linear combinations so that they are eigenkets of β (and hence simultaneous eigenkets of α and β). Let the required linear combination be

$$d_1|\alpha_1\rangle + d_2|\alpha_2\rangle$$

Then,

$$\beta\,[d_1|\alpha_1\rangle + d_2|\alpha_2\rangle] = \beta'\,[d_1|\alpha_1\rangle + d_2|\alpha_2\rangle]$$

If we now use Eqs (47) and (48), we would obtain

$$d_1\,[c_{11}|\alpha_1\rangle + c_{12}|\alpha_2\rangle] + d_2[c_{21}|\alpha_1\rangle + c_{22}|\alpha_2\rangle] = \beta'\,d_1\,|\alpha_1\rangle + \beta'\,d_2|\alpha_2\rangle$$

Since $|\alpha_1\rangle$ and $|\alpha_2\rangle$ are orthogonal to each other, we get

$$d_1\,(c_{11} - \beta') + d_2\,c_{21} = 0$$

and

$$d_1\,c_{12} + d_2\,(c_{22} - \beta') = 0$$

For non-trivial solutions, we must have

$$\begin{vmatrix} c_{11}-\beta' & c_{21} \\ c_{12} & c_{22}-\beta' \end{vmatrix} = 0 \tag{49}$$

the roots of which will give the two allowed values of β'. Once β' is known, we can obtain the ratio d_1/d_2 which will give the simultaneous eigenket. As an example we consider the two matrices

$$\sigma_x = \begin{pmatrix} 0 & 1 \\ 1 & 0 \end{pmatrix} \quad \text{and} \quad \alpha = \begin{pmatrix} 0 & 1 \\ 1 & 0 \end{pmatrix}$$

which commute with each other, i.e.,

$$\sigma_x\,\alpha = \alpha\sigma_x$$

Now, the eigenvalues of σ_x are +1 and –1 and are nondegenerate; hence both the eigenvectors are also eigenvectors of α. Thus,

$$|1\rangle = \frac{1}{\sqrt{2}}\begin{pmatrix}1\\1\end{pmatrix} \quad \text{and} \quad |2\rangle = \frac{1}{\sqrt{2}}\begin{pmatrix}1\\-1\end{pmatrix}$$

(which are eigenvectors of σ_x) represent a complete set of simultaneous eigenvectors of σ_x and α. They form a complete set in the sense that an arbitrary vector can be represented as a linear combination of $|1\rangle$ and $|2\rangle$

$$\begin{pmatrix}a\\b\end{pmatrix} = \frac{(a+b)}{\sqrt{2}}\left[\frac{1}{\sqrt{2}}\begin{pmatrix}1\\1\end{pmatrix}\right] + (a-b)\left[\frac{1}{\sqrt{2}}\begin{pmatrix}1\\-1\end{pmatrix}\right]$$

It may be seen that

$$\begin{pmatrix}1\\0\end{pmatrix} \quad \text{and} \quad \begin{pmatrix}0\\1\end{pmatrix}$$

are eigenvectors of α but *not* eigenvectors of σ_x. However, $|1\rangle$ and $|2\rangle$ can be expressed as a linear combination of the above vectors. We next consider the two matrices

$$\sigma_z = \begin{pmatrix}1 & 0\\0 & -1\end{pmatrix} \quad \text{and} \quad \alpha = \begin{pmatrix}1 & 0\\0 & 1\end{pmatrix}$$

Once again

$$\sigma_z\alpha = \alpha\sigma_z$$

and the eigenvectors

$$\begin{pmatrix}1\\0\end{pmatrix} \quad \text{and} \quad \begin{pmatrix}0\\1\end{pmatrix}$$

form a complete set of simultaneous eigenvectors of σ_z and α. Similarly,

$$\frac{1}{\sqrt{2}}\begin{pmatrix}1\\i\end{pmatrix} \quad \text{and} \quad \frac{1}{\sqrt{2}}\begin{pmatrix}1\\-i\end{pmatrix}$$

form a complete set of simultaneous eigenvectors of

$$\sigma_y = \begin{pmatrix}0 & -i\\i & 0\end{pmatrix} \quad \text{and} \quad \alpha = \begin{pmatrix}1 & 0\\0 & 1\end{pmatrix}$$

Notice that although σ_x, σ_y and σ_z do not commute with each other, they commute independently with α; and one can form simultaneous eigenkets of either σ_x and α or of σ_y and α or of σ_z and α.

11.9 EIGENVALUE EQUATION FOR POSITION OPERATOR USING BRA-KET ALGEBRA

In this and the following sections, we will use bra-ket algebra to obtain eigenvalues and eigenfunctions corresponding to the position and momentum operators and write the orthonormality and completeness conditions. We will then introduce the translation operator and discuss its significance.

In the bra-ket notation, the eigenvalue equation for the position operator $\hat{x}$ will be

$$\hat{x}\,|\,x\rangle = x\,|\,x\rangle; \quad -\infty < x < \infty \tag{50}$$

where the eigenvalues x (on the right hand side of the above equation) are real numbers and $|\,x\rangle$ represents the state in which the particle is localized at the point x. The "hat" on x (and also p_x) tell us that they are operators. The Schrödinger wavefunction $\psi(x)$ corresponding to the ket vector $|\,\Psi\rangle$ is the amplitude for finding the particle at the point x and will therefore be given by

$$\psi\,(x) = \langle x\,|\,\Psi\rangle \tag{51}$$

Similarly,

$$\phi\,(x) = \langle x\,|\,\Phi\rangle \tag{52}$$

Thus,

$$\langle\Phi\,|\,\Psi\rangle = \int_{-\infty}^{+\infty} \phi^*(x)\psi(x)\,dx = \int_{-\infty}^{+\infty} \langle\Phi|x\rangle\langle x|\Psi\rangle\,dx$$

$$= \langle\Phi|\left[\int_{-\infty}^{+\infty} |x\rangle\langle x|\,dx\right]|\Psi\rangle$$

Since $|\,\Psi\rangle$ and $|\,\Phi\rangle$ represent arbitrary states, we have

$$\int_{-\infty}^{+\infty} |x\rangle\langle x|\,dx = 1 \tag{53}$$

which represents the completeness condition and the unity on the right hand side represents the "unit operator". Operating Eq. (53) on $|\,\Psi\rangle$, we get

$$\int_{-\infty}^{+\infty} |x\rangle\langle x|\Psi\rangle\,dx = |\,\Psi\rangle \tag{54}$$

Thus,

$$|\,\Psi\rangle = \int_{-\infty}^{+\infty} \psi(x)|x\rangle\,dx \tag{55}$$

Further,

$$\psi\,(x') = \langle x'\,|\,\Psi\rangle = \int_{-\infty}^{+\infty} \psi(x)\langle x'|x\rangle\,dx \tag{56}$$

Since the above equation is valid for arbitrary $\psi\,(x)$, we must have

$$\langle x'\,|\,x\rangle = \delta\,(x - x') \tag{57}$$

which is the orthonormality condition; we may compare the above equation with the results obtained in Section 4.7.

11.10 THE TRANSLATION OPERATOR USING BRA-KET ALGEBRA

As discussed in Section 4.7, the translation operator $\hat{T}\,(a)$ would change the state $|\,x\rangle$ into another state which would correspond to the particle localized at the point $x + a$; thus,

$$\hat{T}(a)\,|x\rangle = |x+a\rangle \tag{58}$$

Following, the analysis given in Ref. 7, we define

$$|\Phi\rangle = \hat{T}(a)\,|\Psi\rangle = \hat{T}(a)\left[\int |x'\rangle\langle x'|\,dx'\right]|\Psi\rangle$$

where we have introduced the unit operator [see Eq. (53)]. Since $\langle x'|\Psi\rangle = \psi(x')$, we have

$$|\Phi\rangle = \int |x'+a\rangle\,\psi(x')\,dx' \tag{59}$$

Thus,
$$\phi(x) = \langle x|\Phi\rangle = \int \delta\big(x' - (x-a)\big)\,\psi(x')\,dx' \tag{60}$$

where we have used the relation

$$\langle x|x'+a\rangle = \delta(x'+a-x) = \delta(x'-(x-a)) \tag{61}$$

Carrying out the integration in Eq. (60), we get

$$\phi(x) = \psi(x-a) \tag{62}$$

which is consistent with our discussions in Section 4.7. The above equation implies

$$\phi(x) = \hat{T}(\varepsilon)\,\psi(x) = \psi(x-\varepsilon) \tag{63}$$

Now, again following Ref. 7, we calculate the commutator $\left[\hat{x}, \hat{T}(\varepsilon)\right]$ operating on $|x\rangle$:

$$\begin{aligned}\left[\hat{x}, \hat{T}(\varepsilon)\right]|x\rangle &= \hat{x}\hat{T}(\varepsilon)|x\rangle - \hat{T}(\varepsilon)\hat{x}|x\rangle \\ &= \hat{x}|x+\varepsilon\rangle - \hat{T}(\varepsilon)x|x\rangle\end{aligned}$$

where x is now a number. Thus,

$$\begin{aligned}\left[\hat{x}, \hat{T}(\varepsilon)\right]|x\rangle &= (x+\varepsilon)\,|x+\varepsilon\rangle - x\,|x+\varepsilon\rangle \\ &= \varepsilon\,|x+\varepsilon\rangle \approx \varepsilon\,|x\rangle\end{aligned}$$

where we have neglected terms of order ε^2. Thus, we get the commutation relation

$$\left[\hat{x}, \hat{T}(\varepsilon)\right] = \varepsilon \tag{64}$$

As mentioned earlier, many authors define the operator $\hat{p}_x$ through the translation operator

$$\hat{T}(\varepsilon) = 1 - \frac{i}{\hbar}\hat{p}_x\,\varepsilon \tag{65}$$

(see, e.g., References 5–7). If we substitute the above equation in Eq. (64) we would get

$$[\hat{x}, 1] - \frac{i}{\hbar}[\hat{x}, \hat{p}_x\,\varepsilon] = \varepsilon \quad \Rightarrow \quad -\frac{i}{\hbar}[\hat{x}, \hat{p}_x]\,\varepsilon = \varepsilon \tag{66}$$

because ε is just a (small) number. Thus, we get the commutation relation

$$[\hat{x}\;\hat{p}\;] = \hat{x}\,\hat{p}_x - \hat{p}_x\,\hat{x} = i\hbar \tag{67}$$

Now, if we substitute Eq. (65) in Eq. (63), we would get

$$\left(1 - \frac{i}{\hbar}\hat{p}_x\,\varepsilon\right)\psi(x) = \psi(x-\varepsilon) \approx \psi(x) - \varepsilon\frac{\partial\psi}{\partial x}$$

Thus,
$$\hat{p}_x\,\psi(x) = -i\hbar\frac{\partial\psi}{\partial x} \tag{68}$$

which is the operator representation of $\hat{p}_x$; this is consistent with the results obtained in Section 4.2. Now,

$$\langle x \,|\, \hat{T}(\varepsilon) \,|\, \Psi \rangle = \langle x \,|\, \Phi \rangle = \phi(x)$$

But
$$\phi(x) = \psi(x-\varepsilon) = \psi(x) - \varepsilon \frac{\partial \psi}{\partial x} \qquad \text{[see Eq. (63)].}$$

Thus,
$$\psi(x) - \varepsilon \frac{\partial \psi}{\partial x} = \phi(x) = \langle x \,|\, \hat{T}(\varepsilon) \,|\, \Psi \rangle = \left\langle x \middle| 1 - \frac{i}{\hbar}\hat{p}_x \varepsilon \middle| \Psi \right\rangle$$
$$= \langle x | \Psi \rangle - \frac{i}{\hbar}\varepsilon \langle x | \hat{p}_x | \Psi \rangle$$

Thus,
$$-i\hbar \frac{\partial}{\partial x} \langle x | \Psi \rangle = \langle x | \hat{p}_x | \Psi \rangle \tag{69}$$

Since the above equation is valid for arbitrary $|\Psi\rangle$, we may write

$$\langle x \,|\, \hat{p}_x = -i\hbar \frac{\partial}{\partial x} \langle x | \tag{70}$$

We may mention here that in 3-dimensions Eqs. (57) and (53) take the form

$$\langle \mathbf{r}' \,|\, \mathbf{r} \rangle = \delta(\mathbf{r} - \mathbf{r}') = \delta(x - x')\, \delta(y - y')\, \delta(z - z') \tag{71}$$

and
$$\int |\mathbf{r}\rangle \langle \mathbf{r}| \, d\mathbf{r} = 1 \tag{72}$$

The above equations represent the othonormality condition and the completeness condition respectively. The integration in Eq. (72) is over the entire 3-dimensional space and $|\mathbf{r}\rangle$ (which is also written as $|x, y, z\rangle$) represents the state in which the particle is localized at the point **r**. Further, in 3-dimensions, the Schrödinger wavefunction for the state represented by $|\Psi\rangle$ would be

$$\psi(\mathbf{r}) = \langle \mathbf{r} \,|\, \Psi \rangle \tag{73}$$

11.11 EIGEN VALUE EQUATION FOR MOMENTUM OPERATOR USING BRA-KET ALGEBRA

We write the eigenvalue equation for the operator $\hat{p}_x$:

$$\hat{p}_x \,|\, p \rangle = p \,|\, p \rangle; \quad -\infty < p < +\infty \tag{74}$$

where the eigenvalue p on the right hand side is a number. Since $\hat{p}_x$ is an observable, the eigenvalues are real numbers and the eigenfunctions must form a set of orthonormal functions:

$$\langle p \,|\, p' \rangle = \delta(p - p') \qquad \text{orthonormality condition.} \tag{75}$$

Further, the eigenfunctions of an observable form a complete set of functions and therefore for an arbitrary $|\Psi\rangle$ we may write

$$|\Psi\rangle = \int_{-\infty}^{+\infty} \psi(p) |p\rangle \, dp \tag{76}$$

We pre-multiply by $\langle p' |$ to obtain

$$\langle p' \,|\, \Psi \rangle = \int_{-\infty}^{+\infty} \psi(p) \langle p' | p \rangle \, dp = \int_{-\infty}^{+\infty} \psi(p)\, \delta(p - p')\, dp$$
$$= \psi(p')$$

Thus, $\psi(p) = \langle p | \Psi \rangle$ (77)

and therefore $|\Psi\rangle = \int_{-\infty}^{+\infty} \langle p|\Psi\rangle |p\rangle \, dp$

$$= \left[\int_{-\infty}^{+\infty} |p\rangle\langle p| \, dp \right] |\Psi\rangle$$

Since the above equation is valid for arbitrary $|\Psi\rangle$, we may write

$$\int_{-\infty}^{+\infty} |p\rangle\langle p| \, dp = 1 \text{ completeness condition} \tag{78}$$

which represents the completeness condition. Now, from Eq. (74), we have

$$\langle x | \hat{p}_x | p \rangle = p \langle x | p \rangle = p\psi_p(x) \tag{79}$$

where
$$\psi_p(x) = \langle x | p \rangle \tag{80}$$

is the Schrödinger wavefunction corresponding to $|p\rangle$. Using Eq. (69), we readily get

$$i\hbar \frac{\partial}{\partial x} \langle x|p\rangle = p\psi_p(x) \tag{81}$$

Thus,

$$\frac{1}{\psi_p(x)} \frac{\partial}{\partial x} \psi_p(x) = \frac{ip}{\hbar} \tag{82}$$

Simple integration would give

$$\psi(x) = \psi_p(x) = \frac{1}{\sqrt{2\pi\hbar}} \exp\left[\frac{ip}{\hbar} x\right]; \quad -\infty < p < +\infty \tag{83}$$

which represent the momentum eigenfunctions and form a complete set of orthonormal functions (see discussion in Section 17.2).

11.12 PROBLEMS

Problem 11.1 Determine the orthonormal set of the eigenvectors of the matrix

$$\begin{pmatrix} 0 & -i & 0 \\ i & 0 & 0 \\ 0 & 0 & 2 \end{pmatrix}$$

Prove the completeness condition.

Problem 11.2 Repeat the above problem for the matrix

$$\begin{pmatrix} 1 & 0 & 0 \\ 0 & -1 & 0 \\ 0 & 0 & 2 \end{pmatrix}$$

Problem 11.3 Prove that:

$$\overline{|P\rangle\langle Q|} = |Q\rangle\langle P|$$

Problem 11.4 Consider the linear operator σ defined by the equation

$$\sigma^2 = 0$$

Write the eigenvalue equation for σ and show that eigenvalues of σ are ±1. Also, show that $(1 \pm \sigma)\,|P\rangle$ (where $|P\rangle$ is an arbitrary ket) are eigenkets of σ belonging to the eigenvalues ±1. Notice that an arbitrary ket can always be written as

$$|P\rangle = \frac{1}{2}(1+\sigma)|P\rangle + \frac{1}{2}(1-\sigma)|P\rangle$$

which shows that eigenkets of σ form a complete set.

Problem 11.5 Consider the matrices

$$\alpha_1 = \frac{1}{2}\begin{pmatrix} 0 & \sqrt{2} & 0 \\ \sqrt{2} & 0 & \sqrt{2} \\ 0 & \sqrt{2} & 0 \end{pmatrix}$$

$$\alpha_2 = \frac{1}{2}\begin{pmatrix} 0 & -i\sqrt{2} & 0 \\ i\sqrt{2} & 0 & -i\sqrt{2} \\ 0 & i\sqrt{2} & 0 \end{pmatrix}$$

$$\alpha_3 = \begin{pmatrix} 1 & 0 & 0 \\ 0 & 0 & 0 \\ 0 & 0 & -1 \end{pmatrix}$$

and

$$\beta = \begin{pmatrix} 1 & 0 & 0 \\ 0 & 1 & 0 \\ 0 & 0 & 1 \end{pmatrix}$$

Show that α_1, α_2 and α_3 commute independently with β but not with each other. Construct simultaneous set of eigenvectors of α_1 and β. Repeat the analysis for α_2 and β and for α_3 and β.

11.13 SOLUTIONS

Solution 11.1 The normalized eigenvectors are

$$|1\rangle = \frac{1}{\sqrt{2}}\begin{pmatrix} 1 \\ i \\ 0 \end{pmatrix}; \; |2\rangle = \frac{1}{\sqrt{2}}\begin{pmatrix} 1 \\ -i \\ 0 \end{pmatrix}; \; |3\rangle = \begin{pmatrix} 0 \\ 0 \\ 1 \end{pmatrix}$$

Solution 11.3 Let $\alpha = |P\rangle\langle Q|$ and let

$$|\beta\rangle = \alpha|A\rangle = [\langle Q|A\rangle]\,|P\rangle$$

Thus,

$$\langle B| = \overline{\langle Q|A\rangle}\langle P|$$
$$= \langle A|Q\rangle\langle P|$$

But $\langle B| = \langle A|\bar{\alpha}\rangle$. Therefore,

$$\bar{\alpha} = |Q\rangle\langle P|$$

Solution 11.5 Simultaneous eigenvectors of α_1 and β are:

$$\frac{1}{2}\begin{pmatrix}1\\ \sqrt{2}\\ 1\end{pmatrix}, \frac{1}{\sqrt{2}}\begin{pmatrix}1\\ 0\\ -1\end{pmatrix} \text{ and } \frac{1}{\sqrt{5}}\begin{pmatrix}1\\ -\sqrt{2}\\ 1\end{pmatrix}$$

Simultaneous eigenvectors of α_2 and β are

$$\frac{1}{2}\begin{pmatrix}1\\ i\sqrt{2}\\ -1\end{pmatrix}, \frac{1}{\sqrt{2}}\begin{pmatrix}1\\ 0\\ 1\end{pmatrix} \text{ and } \frac{1}{2}\begin{pmatrix}1\\ -i\sqrt{2}\\ -1\end{pmatrix}$$

Simultaneous eigenvectors of α_3 and β are

$$\begin{pmatrix}1\\ 0\\ 0\end{pmatrix}, \begin{pmatrix}0\\ 1\\ 0\end{pmatrix} \text{ and } \begin{pmatrix}0\\ 0\\ 1\end{pmatrix}$$

REFERENCES AND SUGGESTED READINGS

1. P.A.M. Dirac, *The Principles of Quantum Mechanics*, Oxford University Press, Oxford (1958).
2. G. Baym, *Lectures on Quantum Mechanics*, W.A. Benjamin, New York (1969).
3. H.S. Green, *Matrix Methods in Quantum Mechanics*, Barnes and Noble, New York (1968).
4. H.C. Ohanian, *Principles of Quantum Mechanics*, Prentice-Hall, Englewood Cliffs, New Jersey (1990).
5. J.S. Townsend, *A Modern Approach to Quantum Mechanics*, McGraw-Hill, New York (1992).
6. J.J. Sakurai, *Modern Quantum Mechanics*, Addison-Wesley Publishing Company, Inc. (1994).
7. M. Beck, *Quantum Mechanics: Theory and Experiment*, Oxford University Press, New York (2012).

12 Linear Harmonic Oscillator II

Solutions using bra and ket algebra

HEISENBERG banished the picture of electron orbits with definite radii and periods of rotation, because these quantities are not observable; he demanded that the theory should be built up by means of quadratic arrays … . Instead of describing the motion by giving a coordinate as a function of time $x(t)$, one ought to determine an array of transition probabilities x_{mn}. By consideration of known examples discovered by guesswork he found the rule for determining the array of the square $(x^2)_{mn}$ and applied it with success to simple examples such as the harmonic and anharmonic oscillator. This was in the summer of 1925. HEISENBERG, suffering from a severe attack of hay fever, took leave of absence for a course of treatment at the seaside and handed over his paper to me for publication …

The significance of the idea was immediately clear to me, and I sent the manuscript to the Zeitschrift für Physik. HEISENBERG'S rule of multiplication left me no peace, and after a week of intensive thought and trial, I suddenly remembered an algebraic theory that I had learned from my teacher, ROSANES, in Breslau. Such quadratic arrays are quite familiar to mathematicians and are called matrices, in association with a definite rule of multiplication. I applied this rule to HEISENBERG'S quantum condition and found that it agreed for the diagonal elements. It was easy to guess what the remaining elements must be, namely, null; and immediately there stood before me the strange formula

$$pq - qp = h/2\pi i$$

This meant that coordinates q and momenta p are not to be represented by the values of numbers but by symbols whose product depends on the order of multiplication—which do not 'commute', as we say.

My excitement over this result was like that of the mariner who, after long voyaging, sees the desired land from afar, and my only regret was that HEISENBERG was not with me. I was convinced from the first that we had stumbled on the truth. Yet again a large part was only guesswork, in particular the vanishing of the nondiagonal elements in the foregoing expression. For this problem I secured the collaboration of my pupil PASCUAL JORDAN, and in a few days we succeeded in showing that I had guessed correctly. The joint paper by JORDAN and myself contains the most important principles of quantum mechanics, including its extension to electrodynamics.

— MAX BORN in *Physics in my Generation*,
Springer Verlag, New York (1969).

12.1 INTRODUCTION

In Chapter 7, we had solved the Schrödinger equation for the linear harmonic oscillator problem. In this chapter, we will use Dirac's bra and ket algebra to solve the same problem. Although the final results are the same, the analysis using bra and ket algebra is extremely elegant; it allows us to determine the various expectation values and also explicit expressions for the wavefunctions with considerable ease. The time evolution of the coherent state can also be studied in a very straightforward manner. For all this, we feel that the use of the operator algebra in solving the harmonic oscillator problem should be understood right in the beginning of the quantum mechanics course.

12.2 THE SOLUTION OF THE EIGENVALUE PROBLEM

For the linear harmonic oscillator problem, the Hamiltonian is given by

$$H = \frac{p^2}{2\mu} + \frac{1}{2}\mu\omega^2 x^2 \tag{1}$$

where the symbols have been defined in Section 7.2. Since H, p and x are observables

$$\overline{H} = H,\ \overline{p} = p,\ \overline{x} = x \tag{2}$$

Further, the operators x and p will be assumed to satisfy the commutation relation (see Section 4.3)

$$[x, p] = xp - px = i\hbar \tag{3}$$

Our objective is to solve the eigenvalue equation

$$H|H'\rangle = H'|H'\rangle \tag{4}$$

where $|H'\rangle$ is the eigenket of the operator H belonging to the eigenvalue H'. It may be noted that in the above equation H is an operator and H' is simply a number.

Definitions of a and $\overline{a}$

It is convenient to introduce the dimensionless complex operator

$$a = \frac{1}{(2\mu\hbar\omega)^{1/2}}(\mu\omega x + ip) \tag{5}$$

Since $\overline{x} = x$ and $\overline{p} = p$, the adjoint of a would be given by

$$\overline{a} = \frac{1}{(2\mu\hbar\omega)^{1/2}}(\mu\omega x - ip) \tag{6}$$

In terms of the above operators

$$\hbar\omega a\overline{a} = \frac{1}{2\mu}(\mu\omega x + ip)(\mu\omega x - ip)$$

$$= \frac{1}{2\mu}[\mu^2\omega x^2 + p^2 - i\omega\mu(xp - px)]$$

$$= H + \frac{1}{2}\hbar\omega \tag{7}$$

where, we have used Eqs (1) and (3). Similarly,

$$\hbar\omega\overline{a}a = H - \frac{1}{2}\hbar\omega \tag{8}$$

Thus,

$$H = \frac{1}{2}\hbar\omega(\overline{a}a + a\overline{a}) \tag{9}$$

and

$$a\overline{a} - \overline{a}a = [a, \overline{a}] = 1 \tag{10}$$

From Eq. (7)

$$\hbar\omega a\bar{a}a = Ha + \frac{1}{2}\hbar\omega a \tag{11}$$

and from Eq. (8)

$$\hbar\omega a\bar{a}a = aH - \frac{1}{2}\hbar\omega a \tag{12}$$

Thus,

$$aH - Ha = [a, H] = \hbar\omega a \tag{13}$$

Similarly,

$$\bar{a}H - H\bar{a} = [\bar{a}, H] = -\hbar\omega\bar{a} \tag{14}$$

The above equation can also be obtained by taking the adjoint of Eq. (13). Now, let

$$|P\rangle = a|H'\rangle$$

where $|H'\rangle$ is an eigenket of H belonging to the eigenvalue H' [see Eq. (4)]. Then

$$\langle P| = \langle H'|\bar{a}$$

and

$$\hbar\omega\langle P|P\rangle = \hbar\omega\langle H'|\bar{a}a|H'\rangle$$

$$= \langle H'|H - \frac{1}{2}\hbar\omega|H'\rangle \qquad \text{[using Eq. (8)]}$$

$$= (H' - \frac{1}{2}\hbar\omega)\langle H'|H'\rangle \qquad \text{[using Eq. (4)]}$$

where $(H' - \frac{1}{2}\hbar\omega)$ has been taken outside as it is just a number. But $\langle P|P\rangle$ and $\langle H'|H'\rangle$ are positive numbers (see Section 11.2) and therefore,

$$H' \geq \frac{1}{2}\hbar\omega \tag{15}$$

and

$$H' = \frac{1}{2}\hbar\omega$$

if and only if

$$|P\rangle = a|H'\rangle = 0 \tag{16}$$

Conversely,

$$a|H'\rangle = 0 \text{ only when } H' = \frac{1}{2}\hbar\omega \tag{17}$$

That H' should be positive follows from the fact that the expectation values of x^2 and p^2 should be positive or zero for *any* state of the system. This can be seen from the following:

$$\langle Q|H|Q\rangle = \frac{1}{2\mu}\langle Q|p^2|Q\rangle + \frac{1}{2}\mu\omega^2\langle Q|x^2|Q\rangle$$

where $|Q\rangle$ represents an arbitrary state of the system. Let

$$|R\rangle = p|Q\rangle$$

then

$$\langle R| = \langle Q|\bar{p} = \langle Q|p$$

because p is a Hermitian operator. Thus,

$$\langle R|R\rangle = \langle Q|p^2|Q\rangle \geq 0$$

the equality sign if and only if $|R\rangle$ is a null ket. Similarly,

$$\langle Q|x^2|Q\rangle \geq 0$$

Thus,

$$\langle Q|H|Q\rangle \geq 0$$

proving that the expectation value of the energy of the oscillator should be positive for *any* state of the system.

Next, let us consider the operator Ha operating on $|H'\rangle$:

$$Ha|H'\rangle = (aH - \hbar\omega a)|H'\rangle \quad \text{[using Eq. (13)]}$$

$$= (aH' - \hbar\omega a)|H'\rangle$$

$$= (H' - \hbar\omega)a|H'\rangle \tag{18}$$

Thus, if $|H'\rangle$ is an eigenket of H

then $a|H'\rangle$ is also an eigenket of H

belonging to the eigenvalue $H' - \hbar\omega$ provided

$$H' \neq \frac{1}{2}\hbar\omega$$

because if $H' = \frac{1}{2}\hbar\omega$, $a|H'\rangle$ is a null ket. Thus, if

$$H' \neq \frac{1}{2}\hbar\omega$$

then $H' - \hbar\omega$ is also an eigenvalue of H.

Similarly, if $H' - \hbar\omega \neq \frac{1}{2}\hbar\omega$

then $H' - 2\hbar\omega$ is also an eigenvalue of H.

Therefore,

$$H' - \hbar\omega,\ H' - 2\hbar\omega,\ H' - 3\hbar\omega,\ \ldots$$

are all eigenvalues of H. This, however, cannot go on indefinitely because it will then contradict Eq. (15). Further, it can only terminate at $H' = \frac{1}{2}\hbar\omega$, because then

$$a|\frac{1}{2}\hbar\omega\rangle = 0$$

Thus, the possible eigenvalues can be

$$\frac{1}{2}\hbar\omega,\ \frac{3}{2}\hbar\omega,\ \frac{5}{2}\hbar\omega,\ \ldots$$

We *cannot* have any other eigenvalue. In order to prove this let us assume that $H' = \frac{7}{3}\hbar\omega$ is an eigenvalue. Then from Eq. (18), $\frac{4}{3}\hbar\omega$, $\frac{1}{3}\hbar\omega$, $-\frac{2}{3}\hbar\omega$, ... will be eigenvalues because $a|H'\rangle$ cannot be a null ket. Obviously, this will contradict Eq. (15).

We next show that we will have an infinite number of eigenvalues. Now, using Eq. (14)

$$H\bar{a}|H'\rangle = (\bar{a}H + \hbar\omega\bar{a})|H'\rangle$$

$$= (\bar{a}H' + \hbar\omega\bar{a})\,|H'\rangle$$

$$= (H' + \hbar\omega)\bar{a}\,|H'\rangle \tag{19}$$

implying that $(H' + \hbar\omega)$ is another eigenvalue of H, with $\bar{a}\,|H'\rangle$ as the eigenket belonging to it, unless $\bar{a}\,|H'\rangle = 0$. However, $\bar{a}\,|H'\rangle$ can never be a null ket, since it would lead to

$$0 = \hbar\omega a\bar{a}\,|H'\rangle = \left(H + \tfrac{1}{2}\hbar\omega\right)|H'\rangle \qquad \text{[using Eq. (7)]}$$

$$= \left(H' + \tfrac{1}{2}\hbar\omega\right)|H'\rangle.$$

giving $H' + \frac{1}{2}\hbar\omega = 0$, which contradicts Eq. (15). Thus if H' is an eigenvalue, then $H' + \hbar\omega$ is *always* another eigenvalue of H, and so are

$$H' + 2\hbar\omega, \quad H' + 3\hbar\omega, \ldots$$

and so on. Hence the eigenvalues of the Hamiltonian for the linear harmonic oscillator problem are

$$\frac{1}{2}\hbar\omega, \quad \frac{3}{2}\hbar\omega, \quad \frac{5}{2}\hbar\omega, \quad \frac{7}{2}\hbar\omega, \ldots \quad \text{Eigenvalues} \tag{20}$$

extending to infinity, which is the same as obtained in Section 7.2.

We now relabel the eigenfunctions with the index n; thus $|n\rangle$ denotes the eigenfunction corresponding to the eigenvalue $\left(n + \frac{1}{2}\right)\hbar\omega$:

$$H\,|n\rangle = \left(n + \frac{1}{2}\right)\hbar\omega\,|n\rangle;\ n = 0, 1, 2, \ldots \tag{21}$$

Since $|n\rangle$ are eigenkets of the Hermitian operator H, they must be orthogonal to each other; we further assume that the states $|n\rangle$ are normalized, so that

$$\langle m\,|\,n\rangle = \delta_{mn} \tag{22}$$

Further, since $|0\rangle$ corresponds to $H' = \frac{1}{2}\hbar\omega$, we must have

$$a\,|0\rangle = 0 \tag{23}$$

Now, for $n = 1, 2, 3, \ldots$ $a\,|n\rangle$ is an eigenket of H belonging to the eigenvalue $\left(n - \frac{1}{2}\right)\hbar\omega$ [see Eq. (18)]; therefore $a\,|n\rangle$ must be a multiple of $|n - 1)$:

$$a\,|n\rangle = \alpha_n\,|n - 1\rangle$$

In order to determine α_n, we calculate the square of the length of $a\,|n\rangle$:

$$\langle n\,|\bar{a}a\,|\,n\rangle = |\alpha_n|^2\langle n - 1\,|\,n - 1\rangle = |\alpha_n|^2$$

But
$$\hbar\omega\,\langle n\,|\bar{a}a\,|\,n\rangle = \langle n\,|\,(H - \frac{1}{2}\hbar\omega\,|\,n\rangle$$

$$= \langle n\,|\,(n + \frac{1}{2})\,\hbar\omega - \frac{1}{2}\hbar\omega\,|\,n\rangle$$

$$= n\hbar\omega\,\langle n\,|\,n\rangle = n\hbar\omega$$

Thus,
$$|\alpha_n|^2 = n$$

and therefore we may write

$$a\,|\,n\rangle = \sqrt{n}\;|\,n-1\rangle \tag{24}$$

Similarly,

$$\bar{a}\,|\,n\rangle = \sqrt{(n+1)}\;|\,n+1\rangle \tag{25}$$

Thus, if $|\,0\rangle$ denotes the ground-state eigenket, then

$$|1\rangle = \frac{\bar{a}}{\sqrt{1}}\,|0\rangle,\; |2\rangle = \frac{\bar{a}}{\sqrt{1}}\cdot\frac{\bar{a}}{\sqrt{2}}\,|0\rangle = \frac{(\bar{a})^2}{\sqrt{2!}}\,|0\rangle,\; |3\rangle = \frac{(\bar{a})^3}{\sqrt{3!}}\,|0\rangle,\ldots$$

and, in general,

$$|\,n\rangle = \frac{(\bar{a})^n}{\sqrt{n!}}\,|\,0\rangle \tag{26}$$

We can think of an excited state $|\,n\rangle$ of the oscillator as containing n quanta of energy $\hbar\omega$, in addition to the zero-point energy of $\frac{1}{2}\hbar\omega$. The operator $\bar{a}$, according to Eq. (25), creates a quantum of energy; and, therefore, $\bar{a}$ is called the *creation* operator. Similarly, the operator a, according to Eq. (24), annihilates a quantum of energy; and, therefore, a is called the *annihilation* or *destruction* operator.

The eigenkets of H form a complete set of orthogonal kets; thus (see Section 11.6)

$$\sum_{n=0,1,2,\ldots} |\,n\rangle\langle\,n\,| = 1$$

Thus, the solution of the linear harmonic oscillator problem can be summarized through the following equations:

$$\left.\begin{aligned} H\,|\,n\rangle &= \left(n+\tfrac{1}{2}\right)\hbar\omega\,|\,n\rangle;\; n = 0, 1, 2, \ldots \\ \langle m\,|\,n\rangle &= \delta_{mn} \\ \sum_{n=0,1,2,\ldots} |\,n\rangle\langle n\,| &= 1 \end{aligned}\right\} \tag{27}$$

Further,

$$a\,|\,n\rangle = \sqrt{n}\;|\,n-1\rangle$$

$$a\,|\,0\rangle = 0$$

$$\bar{a}\,|\,n\rangle = \sqrt{n+1}\;|\,n+1\rangle$$

We may pause here for a moment and appreciate the beauty of the operator algebra which has enabled us to solve the eigenvalue equation; the only assumption has been the commutation relation along with the basic axioms of quantum mechanics.

12.3 THE UNCERTAINTY PRODUCT

In this section, we will calculate the uncertainty product $\Delta x\,\Delta p$ for the harmonic oscillator states. The uncertainties Δx and Δp are given by (see Section 4.6)

$$\Delta x = [\langle x^2\rangle - \langle x\rangle^2]^{1/2} \tag{28}$$

and $$\Delta p = [\langle p^2\rangle - \langle p\rangle^2]^{1/2} \tag{29}$$

where $$\langle x\rangle = \langle n|x|n\rangle,\quad \langle x^2\rangle = \langle n|x^2|n\rangle$$

$$\langle p\rangle = \langle n|p|n\rangle,\quad \langle p^2\rangle = \langle n|p^2|n\rangle$$

represent the expectation values of the operators in the nth eigenstate of the oscillator. Now, using Eqs (5) and (6)

$$x = \left(\frac{\hbar}{2\mu\omega}\right)^{1/2}(\bar{a}+a) \tag{30}$$

$$p = i\left(\frac{\mu\hbar\omega}{2}\right)^{1/2}(\bar{a}-a) \tag{31}$$

Thus, $$\langle x\rangle = \langle n|x|n\rangle = \left(\frac{\hbar}{2\mu\omega}\right)^{1/2}\langle n|\bar{a}+a|n\rangle$$

$$= \left(\frac{\hbar}{2\mu\omega}\right)^{1/2}\left[\sqrt{n+1}\,\langle n|n+1\rangle + \sqrt{n} + \langle n|n-1\rangle\right]$$

[using Eqs (24) and (25)]

$$= 0$$

Now, $$x^2 = \frac{\hbar}{2\mu\omega}(\bar{a}+a)(\bar{a}+a)$$

$$= \frac{\hbar}{2\mu\omega}(\bar{a}\,\bar{a} + \bar{a}\,a + a\,\bar{a} + aa)$$

We must be careful while carrying out multiplication because $\bar{a}$ and a do not commute. Thus,

$$\langle x^2\rangle = \langle n|x^2|n\rangle = \frac{\hbar}{2\mu\omega}[\langle n|\bar{a}\,\bar{a}|n\rangle + \langle n|\bar{a}a|n\rangle + \langle n|a\bar{a}|n\rangle + \langle n|aa|n\rangle]$$

$$= \frac{\hbar}{2\mu\omega}(0 + n + (n+1) + 0] = \frac{\hbar}{\mu\omega}(n + \frac{1}{2}) \tag{32}$$

Similarly, $$\langle p\rangle = \langle n|p|n\rangle = 0 \quad\text{and}\quad \langle p^2\rangle = \mu\omega\hbar(n + \frac{1}{2}) \tag{33}$$

giving $$\Delta x\Delta p = (n + \frac{1}{2})\,\hbar \tag{34}$$

the minimum value ($= \frac{1}{2}\hbar$) occurring for the ground state. Once again, we may see the ease with which we can use the operator algebra to calculate the various expectation values. We may recall that if we were to use the solutions obtained in Section 7.3, then

$$\langle x^2\rangle = \int_{-\infty}^{+\infty}\psi_n^*(x)\,x^2\psi_n(x)\,dx \tag{35}$$

and $$\langle p^2 \rangle = \int_{-\infty}^{+\infty} \psi_n^*(x) \left[-\hbar^2 \frac{d^2}{dx^2} \right] \psi_n(x)\, dx \tag{36}$$

where $\psi_n(x)$ are the Hermite-Gauss functions [see Eq. (35) of Chapter 7] representing the normalized eigenfunction of the harmonic oscillator problem. Although the final results are the same, the explicit evaluation of the integrals in Eqs (35) and (36) is fairly tedious.

12.4 HARMONIC OSCILLATOR WAVEFUNCTIONS

In section 12.2, we had shown that

$$a \,|\, 0 \rangle = 0 \tag{37}$$

where $|\,0\rangle$ represents the eigenket corresponding to the ground state. Now, if $\psi_0(x)$ represents the Schrödinger wavefunction corresponding to the ground state, the above equation would imply

$$\frac{1}{\sqrt{2\mu\hbar\omega}} (\mu\omega x + ip)\, \psi_0(x) = 0 \tag{38}$$

where we have used Eq. (5) which defines the operator a. Replacing p by $-i\hbar d/dx$, we get

$$\left(\mu\omega x + \hbar \frac{d}{dx} \right) \psi_0(x) = 0$$

or $$\frac{d\psi_0}{dx} = -\frac{\mu\omega}{\hbar} x \psi_0(x)$$

or $$\frac{1}{\psi_0} \frac{d\psi_0(x)}{dx} = -\frac{\mu\omega}{\hbar} x$$

Carrying out the straightforward integration, we get

$$\ln \psi_0(x) = -\frac{\mu\omega}{2\hbar} x^2 + \text{constant}$$

or $$\psi_0(x) = N\, e^{-\frac{1}{2}\gamma^2 x^2}$$

where $$\gamma = \sqrt{\frac{\mu\omega}{\hbar}}$$

If we use the normalization condition

$$\int_{-\infty}^{+\infty} |\, \psi_0(x)\,|^2\, dx = 1$$

we would readily get

$$\psi_0(x) = \left[\frac{\gamma}{\sqrt{\pi}} \right]^{1/2} e^{-\frac{1}{2}\gamma^2 x^2} \quad \text{Normalized ground state wavefunction} \tag{39}$$

which represents the normalized ground state wavefunction [see Eq. (33) of Chapter 7]. Since,

$$\bar{a}\,|\,n\rangle = \sqrt{n+1}\,|\,n+1\rangle$$

we may write

$$|\,n+1\rangle = \frac{1}{\sqrt{n+1}}\,\bar{a}\,|\,n\rangle$$

or

$$\psi_{n+1}(x) = \frac{1}{\sqrt{n+1}}\,\frac{\mu\omega x - ip}{\sqrt{2\mu\hbar\omega}}\,\psi_n(x) \tag{40}$$

or

$$\psi_{n+1}(x) = \frac{1}{[2(n+1)]^{1/2}}\left[\sqrt{\frac{\mu\omega}{\hbar}}\,x - \sqrt{\frac{\hbar}{\mu\omega}}\cdot\frac{d}{dx}\right]\psi_n(x)$$

$$= \frac{1}{[2(n+1)]^{1/2}}\left[\xi - \frac{d}{d\xi}\right]\psi_n(x) \tag{41}$$

where

$$\xi = \gamma x = \sqrt{\frac{\mu\omega}{\hbar}}\,x$$

Thus,

$$\psi_1(x) = \frac{1}{\sqrt{2}}\left[\xi - \frac{d}{d\xi}\right]\left[\frac{\gamma}{\sqrt{\pi}}\right]^{1/2} e^{-\frac{1}{2}\xi^2}$$

$$= \left[\frac{\gamma}{2\sqrt{\pi}}\right]^{1/2} 2\xi\, e^{-\frac{1}{2}\xi^2} \tag{42}$$

which is automatically normalized! Further,

$$\psi_2(x) = \frac{1}{\sqrt{2}\sqrt{2}}\left[\frac{\gamma}{2\sqrt{\pi}}\right]^{1/2}\left[\xi - \frac{d}{d\xi}\right]\left(2\xi\, e^{-\frac{1}{2}\xi^2}\right)$$

$$= \left[\frac{\gamma}{2^2\,2\sqrt{\pi}}\right]^{1/2}(4\xi^2 - 2)\, e^{-\frac{1}{2}\xi^2} \tag{43}$$

and

$$\psi_3(x) = \frac{1}{\sqrt{2}\sqrt{3}}\left[\frac{\gamma}{2^2\,2\sqrt{\pi}}\right]^{1/2}\left(\xi - \frac{d}{d\xi}\right)(4\xi^2 - 2)\, e^{-\frac{1}{2}\xi^2}$$

$$= \left[\frac{\gamma}{2^3\,3!\sqrt{\pi}}\right]^{1/2}(8\xi^3 - 12\xi)\, e^{-\frac{1}{2}\xi^2} \tag{44}$$

In general,

$$\psi_n(x) = \left[\frac{\gamma}{2^n\,n!\sqrt{\pi}}\right]^{1/2} H_n(\xi)\, e^{-\frac{1}{2}\xi^2} \qquad \text{Wavefunction of the } n^{\text{th}} \text{ state} \tag{45}$$

which can be proved by induction (see Problem 12.7). Equations (42), (43) and (44) are automatically normalized and identical to the expressions derived in Section 7.3.

12.5 THE COHERENT STATE

In this section, we will obtain the eigenvalues and eigenfunctions of the annihilation operator α, i.e., we will solve the eigenvalue equation

$$a\,|\,\alpha\rangle = \alpha\,|\,\alpha\rangle \tag{46}$$

The eigentkets $|\,\alpha)$ describe what are known as coherent states which represent the quantum mechanical analogs of the 'classical oscillator' (see also Section 7.4). Since, the eigenkets of an observable form a complete set, we expand $|\,\alpha\rangle$ in terms of the kets $|\,n\rangle$:

$$|\,\alpha\rangle = \sum_{n=0,1,2,\ldots}^{\infty} c_n\,|\,n\rangle \tag{47}$$

Now,

$$a\,|\,\alpha\rangle = \sum_{n=0,1,2,\ldots}^{\infty} c_n\, a\,|\,n\rangle = \sum_{n=1,2,\ldots}^{\infty} c_n \sqrt{n}\,|\,n-1\rangle$$

Also,

$$a\,|\,\alpha\rangle = \alpha\,|\,\alpha\rangle = \alpha \sum_{n=0,1,2,\ldots}^{\infty} c_n\,|\,n\rangle$$

Thus,

$$c_1\,|\,0\rangle + c_2\sqrt{2}\;|\,1\rangle + c_3\sqrt{3}\;|\,2\rangle + \cdots = \alpha(c_0\,|\,0\rangle + c_1\,|\,1\rangle + c_2\,|\,2\rangle + \cdots)$$

or

$$c_1 = \alpha c_0,$$

$$c_2 = \frac{\alpha c_1}{\sqrt{2}} = \frac{\alpha^2}{\sqrt{2}}\, c_0,$$

$$c_3 = \alpha\,\frac{c_2}{\sqrt{3}} = \frac{\alpha^3}{\sqrt{3!}}\, c_0, \ldots$$

In general,

$$c_n = \frac{\alpha^n}{\sqrt{n!}}\, c_0$$

Thus,

$$|\,\alpha\rangle = c_0 \sum_{n=0,1,2,\ldots}^{\infty} \frac{\alpha^n}{\sqrt{n!}}\,|\,n\rangle$$

If we normalize $|\,\alpha\rangle$, we would get

$$1 = \langle\,\alpha\,|\,\alpha\rangle = |\,c_0\,|^2 \sum_n \sum_m \frac{\alpha^n \alpha^{*m}}{\sqrt{n!}\sqrt{m!}}\,\delta_{nm}$$

$$= |\,c_0\,|^2 \sum_{n=0,1,2,\ldots}^{\infty} \frac{(\,|\,\alpha\,|^2)^n}{n!} = |\,c_0\,|^2\, e^{|\alpha|^2}$$

or

$$c_0 = e^{-\frac{1}{2}|\alpha|^2}$$

within an arbitrary phase factor. Thus,

$$c_n = \frac{\alpha^n}{\sqrt{n!}} e^{-\frac{1}{2}|\alpha|^2} \tag{48}$$

and
$$|\alpha\rangle = e^{-\frac{1}{2}|\alpha|^2} \sum_{n=0,1,2,\ldots}^{\infty} \frac{\alpha^n}{\sqrt{n!}} |n\rangle \tag{49}$$

which represents the (normalized) coherent state. Notice that there is no restriction on the value of α, i.e., α can take *any* complex value.

12.6 TIME EVOLUTION OF THE COHERENT STATE

The time dependent Schrödinger equation for the linear harmonic oscillator problem is given by

$$i\hbar \frac{\partial}{\partial t} |\Psi\rangle = H|\Psi\rangle = \left(\frac{p^2}{2\mu} + \frac{1}{2}\mu\omega^2 x^2 \right) |\Psi\rangle \tag{50}$$

In order to solve the above equation, we use the method of separation of variables (cf. Section 7.2)

$$|\Psi\rangle = e^{-iEt/\hbar} |\psi\rangle \tag{51}$$

where $|\psi\rangle$ are the eigenkets of the Hamiltonian H:

$$H|\psi\rangle = E|\psi\rangle \tag{52}$$

In Section 12.2, we have solved the above eigenvalue equation and have shown that E can take the following discrete values:

$$E = E_n = \left(n + \frac{1}{2} \right) \hbar\omega$$

with $|n\rangle$ representing the corresponding (normalized) eigenkets. Thus, the most general solution of Eq. (50) is given by

$$|\Psi(t)\rangle = \sum_{n=0,1,2,\ldots}^{\infty} c_n \exp\left[-\frac{iE_n t}{\hbar} \right] |n\rangle$$

$$= \sum_{n=0,1,2,\ldots}^{\infty} c_n e^{-i\left(n+\frac{1}{2}\right)\omega t} |n\rangle \tag{53}$$

where the coefficients c_n are to be determined from the following equation (cf. Section 7.3):

$$|\Psi(0)\rangle = \sum_{n=0,1,2,\ldots}^{\infty} c_n |n\rangle \tag{54}$$

If $|\Psi(0)\rangle$ is normalized, i.e., if

$$\langle \Psi(0) | \Psi(0) \rangle = 1 \tag{55}$$

then
$$1 = \sum_{m=0,1,2,\ldots}^{\infty} c_m^* \langle m | \sum_{n=0,1,2,\ldots}^{\infty} c_n | n \rangle$$

$$= \sum_{m=0,1,2,\ldots}^{\infty} \sum_{n=0,1,2,\ldots}^{\infty} c_m^* c_n \langle m | n \rangle$$

$$= \sum_{m=0,1,2,\ldots}^{\infty} \sum_{n=0,1,2,\ldots}^{\infty} c_m^* c_n \, \delta_{mn}$$

Thus,
$$\sum_{n=0,1,2,\ldots}^{\infty} | c_n |^2 = 1 \tag{56}$$

and we may interpret $| c_n |^2$ to represent the probability of finding the oscillator in the nth state (which would not change with time). Furthermore, using Eq. (53)

$$\langle \Psi(t) | \Psi(t) \rangle = \sum_{n=0,1,2,\ldots}^{\infty} \sum_{m=0,1,2,\ldots}^{\infty} c_m^* c_n e^{i(m-n)\omega t} \langle m | n \rangle$$

$$= \sum_{n=0,1,2,\ldots}^{\infty} \sum_{m=0,1,2,\ldots}^{\infty} c_m^* c_n e^{i(m-n)\omega t} \, \delta_{mn}$$

$$= \sum_{n=0,1,2,\ldots}^{\infty} | c_n |^2 = 1$$

Thus if $| \Psi(0) \rangle$ is normalized, then $| \Psi(t) \rangle$ is also normalized for all values of time.

Let us assume that at $t = 0$, the oscillator is in a coherent state, i.e.,

$$| \Psi(t = 0) \rangle = | \alpha \rangle$$

For the coherent state, c_n is given by Eq. (48) and, therefore,

$$| \Psi(t) \rangle = \sum_{n=0,1,2,\ldots}^{\infty} c_n \, e^{-iE_n t/\hbar} | n \rangle$$

$$= e^{-\frac{1}{2}|\alpha|^2} \sum_{n=0,1,2,\ldots}^{\infty} \left[\frac{\alpha^n}{\sqrt{n!}} \right] e^{-i\left(n+\frac{1}{2}\right)\omega t} | n \rangle \tag{57}$$

The above equation describes the time evolution of the coherent state. We would next like to calculate the corresponding time variations of $\langle x \rangle$ and $\langle p \rangle$ which represent the expectation values of x and p:

$$\langle x \rangle = \langle \Psi(t) | x | \Psi(t) \rangle$$
$$\langle p \rangle = \langle \Psi(t) | p | \Psi(t) \rangle$$

Since,
$$x = \sqrt{\frac{\hbar}{2\mu\omega}} \, (a + \bar{a}) \tag{58}$$

and $$p = \frac{1}{i}\sqrt{\frac{\mu\hbar\omega}{2}}\,(a - \bar{a}) \tag{59}$$

We must first calculate $\langle a \rangle$ and $\langle \bar{a} \rangle$.

Now, $$\bar{a}\,|\,\Psi(t)\rangle = e^{-\frac{1}{2}|\alpha|^2} \sum_{n=0,1,2,\ldots}^{\infty} \frac{\alpha^n}{\sqrt{n!}} e^{-i\left(n+\frac{1}{2}\right)\omega t}\, \bar{a}\,|\,n\rangle$$

$$= e^{-\frac{1}{2}|\alpha|^2} \sum_{n=0,1,2,\ldots}^{\infty} \frac{\alpha^n}{\sqrt{n!}} e^{-i\left(n+\frac{1}{2}\right)\omega t} \sqrt{n+1}\,|\,n+1\rangle \tag{60}$$

Since, $$\langle \Psi(t)\,| = e^{-\frac{1}{2}|\alpha|^2} \sum_{m=0,1,2,\ldots}^{\infty} \frac{\alpha^{*m}}{\sqrt{m!}} e^{+i\left(m+\frac{1}{2}\right)\omega t} \langle m\,| \tag{61}$$

we get $$\langle \bar{a} \rangle = \langle \Psi(t)\,|\,\bar{a}\,|\,\Psi(t)\rangle$$

$$= e^{-|\alpha|^2} \sum_m \sum_n \frac{\alpha^{*m}\alpha^n}{\sqrt{m!\,n!}} e^{i(m-n)\omega t} \sqrt{n+1}\langle m\,|\,n+1\rangle$$

$$= e^{-|\alpha|^2} \sum_{n=0,1,2,\ldots}^{\infty} \frac{\alpha^*(\alpha\alpha^*)^n}{n!} e^{i\omega t}$$

$$= e^{-|\alpha|^2}\alpha^* e^{i\omega t} e^{|\alpha|^2}$$

or $$\langle \Psi(t)\,|\,\bar{a}\,|\,\Psi(t)\rangle = \alpha^* e^{i\omega t} \tag{62}$$

Similarly[1], $$\langle \Psi(t)\,|\,a\,|\,\Psi(t)\rangle = \alpha e^{-i\omega t} \tag{63}$$

Thus,

$$\langle x \rangle = \langle \Psi(t)\,|\,x\,|\,\Psi(t)\rangle$$

$$= \sqrt{\frac{\hbar}{2\mu\omega}} \langle \Psi(t)\,|\,a + \bar{a}\,|\,\Psi(t)\rangle$$

$$= \sqrt{\frac{\hbar}{2\mu\omega}} [\alpha e^{-i\omega t} + \alpha^* e^{i\omega t}]$$

$$= \sqrt{\frac{2\hbar}{\mu\omega}}\,|\,\alpha\,|\cos(\omega t - \phi)$$

[1] Alternatively, if we take the complex conjugate of Eq. (62), we would obtain Eq. (63).

where we have written

$$\alpha = |\alpha| e^{+i\phi}$$

Thus, we may write

$$\langle x \rangle = x_0 \cos(\omega t - \phi) \quad \text{Expectation value of } \boldsymbol{x} \tag{64}$$

where
$$x_0 = \sqrt{\frac{2\hbar}{\mu\omega}}\,|\alpha| \tag{65}$$

Similarly,
$$\langle p \rangle = \sqrt{\frac{\mu\hbar\omega}{2}}\frac{1}{i}\Psi(t)|\,a - \bar{a}\,|\Psi(t)\rangle$$

$$= \sqrt{2\mu\hbar\omega}\,\frac{1}{2i}[\alpha e^{-i\omega t} - \alpha^* e^{i\omega t}]$$

$$= -\mu\omega x_0 \sin(\omega t - \phi) \quad \text{Expectation value of } \boldsymbol{p} \tag{66}$$

From Eqs (64) and (66), we may write

$$\langle p \rangle = \mu \frac{d\langle x \rangle}{dt} \tag{67}$$

which is nothing but the equation of motion for a classical particle (see Section 4.5). Once again, it is much more convenient to use the bra and ket algebra to evaluate the various expectation values.

12.7 THE NUMBER OPERATOR

Consider the operator

$$N_{op} = \bar{a}a \tag{68}$$

Using Eq. (8) we may write

$$H = \left(N_{op} + \frac{1}{2}\right)\hbar\omega \tag{69}$$

Since,
$$H|n\rangle = \left(n + \frac{1}{2}\right)\hbar\omega|n\rangle$$

we have
$$N_{op}|n\rangle = n|n\rangle \tag{70}$$

Thus $|n\rangle$ are also the eigenkets of N_{op}, the corresponding eigenvalues being n and since n takes the values 0, 1, 2, . . . , the operator N_{op} is called the number operator. For this reason, $|n\rangle$ are often referred to as number states. Obviously,

$$\langle m|N_{op}|n\rangle = n\,\delta_{mn} \tag{71}$$

12.8 THE UNITARY TRANSFORMATION

Let $|\Psi_i\rangle$ represent a set of orthonormal kets, i.e.,

$$\langle \Psi_i|\Psi_j\rangle = \delta_{ij}$$

We consider a transformation

$$|\Phi_i\rangle = \sum_j U_{ij}\,|\Psi_j\rangle \tag{72}$$

If the kets $|\Phi_i\rangle$ also form a set of orthonormal kets, i.e. if $\langle \Phi_i|\Phi_j\rangle = \delta_{ij}$ then the transformation is said to be unitary and U_{ij} is said to be a unitary matrix.

Now,

$$\delta_{ij} = \langle \Phi_i|\Phi_j\rangle = \sum_m\sum_n \bar{U}_{im}U_{jn}\langle\Psi_m|\Psi_n\rangle$$

$$= \sum_m\sum_n \bar{U}_{im}U_{jn}\delta_{mn}$$

$$= \sum_n \bar{U}_{in}U_{jn}$$

or

$$\bar{U}U = 1 \tag{73}$$

where $\bar{U}$ represents the hermitian adjoint of U. Thus, if U describes an arbitrary transformation then

$$\bar{U} = U^{-1}$$

12.9 THE SCHRÖDINGER AND HEISENBERG PICTURES

While solving the linear harmonic oscillator problem in Section 12.2, we had assumed the observables x, p and H to be real operators and independent of time. This is the so-called Schrödinger picture and the time development of the ket, describing the quantum mechanical system is obtained by solving the time dependent Schrödinger equation:

$$i\hbar\frac{\partial}{\partial t}|\Psi(t)\rangle = H|\Psi(t)\rangle \tag{74}$$

If the Hamiltonian is independent of time, we can *integrate* the above equation to obtain

$$|\Psi(t)\rangle = e^{-iHt/\hbar}\,|\Psi(0)\rangle \tag{75}$$

where the exponential of an operator is defined through the power series

$$e^{\mathcal{O}} = 1 + \mathcal{O} + \frac{1}{2!}\mathcal{O}\cdot\mathcal{O} + \frac{1}{3!}\mathcal{O}\cdot\mathcal{O}\cdot\mathcal{O}\ldots \tag{76}$$

That Eq. (75) is the solution of Eq. (74) can be immediately seen by direct substitution. Next, let $|n\rangle$ represent the eigenkets of H:

$$H|n\rangle = E_n|n\rangle$$

then by inserting the unit operator [see Eq. (27)] in Eq. (75), we obtain

$$|\Psi(t)\rangle = e^{-iHt/\hbar}\left\{\sum_n |n\rangle\langle n|\right\}|\Psi(0)\rangle$$

$$= \sum_n e^{iE_nt/\hbar}\,|n\rangle\,[\langle n|\Psi(0)\rangle] \tag{77}$$

The above equation may be used to study the time development of states. Now, the expectation value of an observable characterised by the operator $\mathcal{O}$ is given by

$$\langle \mathcal{O}\rangle = \langle \Psi(t) | \mathcal{O} | \Psi(t)\rangle \tag{78}$$

If we substitute for $|\Psi(t)$ from Eq. (75), we would get

$$\langle \mathcal{O}\rangle = \langle \Psi(0) | e^{iHt/\hbar}\, \mathcal{O} e^{-iHt/\hbar} | \Psi(0)\rangle$$

$$= \langle \Psi(0) | \mathcal{O}_H(t) | \Psi(0)\rangle \tag{79}$$

where the operator $\mathcal{O}_H(t)$ is defined by the following equation:

$$\mathcal{O}_H(t) = e^{iHt/\hbar} \mathcal{O} e^{-iHt/\hbar} \tag{80}$$

Equations (78) and (79) tell us that the expectation values remain the same if we endow the operators $\mathcal{O}_H(t)$ with the entire time dependence but assume that the kets are time independent. This is known as the Heisenberg picture (and hence the subscript H) in which operators representing the observable change with time but the ket describing the system is time independent. On the other hand, in the Schrödinger picture [see Eqs (77) and (78)], we may visualize the basis vectors (here $|n\rangle$) as a fixed set of vectors and $|\Psi(t)\rangle$ (describing the system) as moving. From Eq. (80), we have

$$\frac{d}{dt}\mathcal{O}_H(t) = e^{iHt/\hbar} \frac{\partial \mathcal{O}}{\partial t} e^{-iHt/\hbar} + \frac{i}{\hbar} e^{iHt/\hbar} [H\mathcal{O} - \mathcal{O}H] e^{-iHt/\hbar}$$

where $\frac{\partial \mathcal{O}}{\partial t}$ is introduced to allow for any explicit time dependence of the operator. If there is no such explicit time dependence, we may write

$$i\hbar \frac{d}{dt}\mathcal{O}_H(t) = [e^{iHt/\hbar} \mathcal{O} e^{-iHt/\hbar}]H - H[e^{iHt/\hbar} \mathcal{O} e^{-iHt/\hbar}]$$

$$= \mathcal{O}_H(t)H - H\mathcal{O}_H(t) \tag{81}$$

or

$$i\hbar \frac{d}{dt}\mathcal{O}_H(t) = [\mathcal{O}_H(t), H] \quad \text{Heisenberg equation of motion} \tag{82}$$

Equation (82) gives the time dependence of an operator in the Heisenberg picture and is called the *Heisenberg equation of motion*.

If the Hamiltonian is assumed to be independent of time in the Schrödinger representation, then it is also independent of time in the Heisenberg representation:

$$H_H(t) = e^{-iHt/\hbar}\, H\, e^{iHt/\hbar} = e^{-iHt/\hbar}\, e^{iHt/\hbar}\, H = H \tag{83}$$

It should be mentioned that if H had an explicit time dependence, the analysis would have been much more involved (see, e.g. Chapter 5 of Ref. 2).

We next consider the operators α and β which, in the Schrödinger representation, satisfy the commutation relation

$$[\alpha, \beta] = i\gamma \tag{84}$$

or

$$\alpha\beta - \beta\alpha = i\gamma \tag{85}$$

If we multiply on the left by $e^{iHt/\hbar}$ and on the right by $e^{-iHt/\hbar}$, we obtain

$$e^{iHt/\hbar}\, \alpha\, e^{-iHt/\hbar}\, e^{-iHt/\hbar}\, \beta e^{-iHt/\hbar} - e^{iHt/\hbar}\, \beta e^{-iHt/\hbar}\, e^{iHt/\hbar}\, \alpha e^{-iHt/\hbar} = i\, e^{iHt/\hbar}\, \gamma\, e^{-iHt/\hbar} \tag{86}$$

where we have inserted $e^{-iHt/\hbar}\, e^{iHt/\hbar}$ (=1) between α and β. Using Eq. (80), we get

$$[\alpha_H(t), \beta_H(t)] = i\gamma_H(t) \tag{87}$$

which shows the physical equivalence of Heisenberg and Schrödinger pictures.

As an illustration we consider the harmonic oscillator problem. However, before we do so, we note that

$$[x, p^n] = i\hbar n p^{n-1} = i\hbar \frac{\partial}{\partial p} p^n \tag{88}$$

and

$$[p, x^n] = -i\hbar n x^{n-1} = -i\hbar \frac{\partial}{\partial x} x^n \tag{89}$$

Thus if $P(p)$ and $X(x)$ can be expanded in a power series in p and x respectively we will have

$$[X, P(p)] = i\hbar \frac{\partial P}{\partial p} \tag{90}$$

and

$$[p, X(x)] = -i\hbar \frac{\partial X}{\partial x} \tag{91}$$

Now for the harmonic oscillator problem

$$H = \frac{p^2}{2m} + \frac{1}{2} m\omega^2 x^2 = H_H = \frac{p_H^2(t)}{2m} + \frac{1}{2} m\omega^2 x_H^2(t) \tag{92}$$

where we have used Eq. (83). Thus, using Eq. (83)

$$\frac{d}{dt} x_H(t) = \frac{1}{i\hbar}[x_H, H_H] = \frac{\partial H_H}{\partial p} = \frac{p_H(t)}{m} \tag{93}$$

and

$$\frac{d}{dt} p_H(t) = \frac{1}{i\hbar}[p_H(t), H_H] = -\frac{\partial H_H}{\partial p} = -m\omega^2 x_H(t) \tag{94}$$

These are the Hamilton's equations of motion (see, e.g. Ref. 3). The solution x of the above equations are

$$x_H(t) = x \cos \omega t + \frac{p}{m\omega} \sin \omega t \tag{95}$$

$$p_H(t) = -m\omega x \sin \omega t + p \cos \omega t \tag{96}$$

where $x = x_H(t = 0)$ and $p = p_H(t = 0)$ represent the operators in the Schrödinger representation.

Further,

$$a_H(t) = \frac{1}{(2m\hbar\omega)^{1/2}} (m\omega x_H(t) + i p_H(t))$$

$$= a\, e^{-i\omega t} \tag{97}$$

where $a = a_H(t = 0)$. Similarly,

$$\bar{a}_H(t) = \bar{a} e^{i\omega t} \tag{98}$$

These relations will be used in Chapter 27.

12.10 THE DENSITY OPERATOR

Let the kets $|0\rangle, |1\rangle, |2\rangle, \ldots |n\rangle, \ldots$ form a complete set of orthonormal kets; i.e.,

$$\sum_n |n\rangle\langle n| = 1 \tag{99}$$

and
$$\langle m|n\rangle = \delta_{mn} \tag{100}$$
An arbitrary ket $|P\rangle$ can be expanded in terms of the ket $|n\rangle$

$$|P\rangle = \sum_n c_n|n\rangle \,;\, c_n = \langle n|P\rangle \tag{101}$$

A state can be characterised by the density operator ρ defined by the following equation

$$\rho = |P\rangle\langle P| \tag{102}$$

The trace[2] of an operator is defined to be equal to the sum of the diagonal matrix elements for any complete set of states; thus

$$TrO = \sum_n \langle n|O|n\rangle \tag{103}$$

Hence,
$$Tr(|P\rangle\langle Q|) = \sum_n \langle n|P\rangle\langle Q|n\rangle$$

$$= \langle Q|\left\{\sum_n |n\rangle\langle n|\right\}|P\rangle$$

or
$$Tr(|P\rangle\langle Q|) = \langle Q|P\rangle \tag{104}$$

where in the last step we have used Eq. (27). Thus,

$$Tr\rho = Tr|P\rangle\langle P| = \langle P|P\rangle = 1 \tag{105}$$

where we have assumed $|P\rangle$ to be normalized. Also,

$$\rho_{nn} = \langle n|P\rangle\langle P|n\rangle = |c_n|^2$$

implying that the diagonal matrix elements of the density operator represent the probabilities of the system in the base states.

Now, the expectation value of the operator O is given by

$$\langle O\rangle = \langle P|O|P\rangle = \sum_n \langle P|O|n\rangle\langle n|P\rangle$$

$$= \sum_n \langle n|P\rangle\langle P|O|n\rangle$$

$$= \sum_n \langle n|\rho O|n\rangle$$

$$= Tr(\rho O) \tag{106}$$

Perhaps the most important application of the density operator is in the field of statistical mechanics where we consider an ensemble of a large number of identical systems; each system may be in a different quantum state (see, e.g. Ref. 4).

If W_Ψ represents the probability of finding the system in the state characterized by $|\Psi\rangle$, then the corresponding density operator is given by

$$\rho = \sum_\Psi W_\Psi |\Psi\rangle\langle\Psi| \tag{107}$$

[2] Abbreviated as *Tr*.

where the summation is carried over all possible states of the system; the density operator contains all the information about the ensemble. Obviously, since

$$\sum_{\Psi} W_{\Psi} = 1 \tag{108}$$

we have

$$\begin{aligned} Tr\rho &= \sum_{\Psi} W_{\Psi} Tr|\Psi\rangle\langle\Psi| \\ &= \sum_{\Psi} W_{\Psi}\langle\Psi|\Psi\rangle \\ &= \sum_{\Psi} W_{\Psi} = 1 \end{aligned} \tag{109}$$

Radiation field inside a cavity

As an example, we consider the radiation field at thermal equilibrium (at temperature T) inside a cavity. We will show in Chapter 27 that the radiation field can be visualized as consisting of an infinite number of simple harmonic oscillators. The energy of each oscillator can decrease or increase by integral multiples of $\hbar\omega_\lambda$; if we consider $\hbar\omega_\lambda$ as the energy of a photon, then each oscillator can have energy corresponding to n_λ photons. Restricting ourselves to a particular mode (i.e., to a particular frequency which we will denote by ω), the probability of finding n photons (i.e. the probability of finding the system in the nth excited state) is proportional to

$$\exp\left(-\frac{E_n}{kT}\right) = \exp\left[-\frac{\left(n+\frac{1}{2}\right)\hbar\omega}{kT}\right]$$

Thus,

$$W_n = \frac{\exp\left[-\frac{\left(n+\frac{1}{2}\right)\hbar\omega}{kT}\right]}{\sum\limits_{i=0,1,\ldots}\exp\left[-\frac{\left(n+\frac{1}{2}\right)\hbar\omega}{kT}\right]} = \frac{\exp[-n\xi]}{1+e^{-\xi}+e^{-2\xi}+\cdots}$$

$$= \frac{e^{-n\xi}}{1/(1-e^{-\xi})} = e^{-n\xi}(1-e^{-\xi}) \tag{110}$$

where $\xi = \hbar\omega/kT$. Thus,

$$\rho = \left[1-e^{-\xi}\right]\sum_n e^{-n\xi}|n\rangle\langle n|$$

$$= \left[1-e^{-\xi}\right]\sum_n e^{-\xi\bar{a}a}|n\rangle\langle n|$$

(because $\bar{a}a$ is the number operator, see Section 12.7)

$$= \left[1-e^{-\xi}\right]e^{-\xi\bar{a}a}\sum_n|n\rangle\langle n|$$

or

$$\rho = \left[1 - e^{-\xi}\right] e^{-\xi \bar{a} a} \tag{111}$$

The above equation can be put in the form

$$\rho = \frac{e^{-\beta H}}{Tr(e^{-\beta H})}, \qquad \beta = \frac{1}{kT} \tag{112}$$

where H represents the Hamiltonian of the system.

12.11 PROBLEMS

Problem 12.1 Show that:

$$\langle n \,|\, \frac{1}{2}\mu\omega^2 x^2 \,|\, n \rangle = \langle n \,|\, \frac{p^2}{2\mu} \,|\, n \rangle$$

implying that the expectation values of the potential and kinetic energies are equal.

Problem 12.2 Using Eqs (25) and (26) evaluate the matrix elements

$$\langle m \,|\, a \,|\, n \rangle \quad \text{and} \quad \langle m \,|\, \bar{a} \,|\, n \rangle$$

Write the matrices representing a and $\bar{a}$ and show that they are not Hermitian.

Problem 12.3 Using Eqs (30) and (31) and the results of the previous problem show that

$$\langle m \,|\, x \,|\, n \rangle = \begin{cases} \left(\dfrac{\hbar}{2\mu\omega}\right)^{1/2} \sqrt{n+1} & \text{for } m = n+1 \\ \left(\dfrac{\hbar}{2\mu\omega}\right)^{1/2} \sqrt{n} & \text{for } m = n-1 \\ 0 & \text{otherwise} \end{cases}$$

and

$$\langle m \,|\, p \,|\, n \rangle = \begin{cases} i\left(\dfrac{\mu\hbar\omega}{2}\right)^{1/2} \sqrt{n+1} & \text{for } m = n+1 \\ -i\left(\dfrac{\mu\hbar\omega}{2}\right)^{1/2} \sqrt{n} & \text{for } m = n-1 \\ 0 & \text{otherwise} \end{cases}$$

Write the matrices representing x and p and show that they are both Hermitian.

Problem 12.4 (a) Show that:

$$\langle m \,|\, x^2 \,|\, n \rangle = \begin{cases} \dfrac{\hbar}{2\mu\omega}\sqrt{(n+1)(n+2)} & \text{for } m = n+2 \\ \dfrac{\hbar}{2\mu\omega}(2n+1) & \text{for } m = n \\ \dfrac{\hbar}{2\mu\omega}\sqrt{n(n-1)} & \text{for } m = n-2 \\ 0 & \text{otherwise} \end{cases}$$

(b) Similarly, evaluate $\langle m | p^2 | n \rangle$ and show that

$$\langle m \left| \frac{p^2}{2\mu} + \frac{1}{2}\mu\omega^2 x^2 \right| n \rangle = \left(n + \frac{1}{2}\right)\hbar\omega\delta_{mn}$$

(The above result follows directly from the fact that $| n \rangle$ are the orthonormal eigenkets of H; thus the matrix representing H would be diagonal with diagonal elements representing the energy eigenvalues.)

Problem 12.5 Show that:

$$\langle n | x^4 | n \rangle = \frac{3}{2}\left(\frac{\hbar}{\mu\omega}\right)^2 \left(n^2 + n + \frac{1}{2}\right)$$

Problem 12.6 Show that the operator $\bar{a}$ cannot have any eigenkets and similarly the operator a cannot have any eigenbras.

[Hint: Write the eigenvalue equation and express the eigenket as a linear combination of $| n \rangle$; similarly, express the eigenbra as a linear combination of $\langle n |$.]

Problem 12.7 (a) Let

$$\langle n \rangle = \langle \alpha | N_{op} | \alpha \rangle \tag{113}$$

represent the expectation value of the number operator N_{op} $(= \bar{a}a)$ corresponding to the coherent state. Show that

$$\langle n \rangle = | \alpha |^2 \tag{114}$$

and

$$| \alpha \rangle = e^{-\langle n \rangle /2} \sum_{n=0,1,2,\ldots}^{\infty} \frac{\alpha^n}{\sqrt{n!}} | n \rangle \tag{115}$$

(b) If we write

$$\xi_0 = \sqrt{2}\alpha \tag{116}$$

show that the above equations are the same as given by Eqs (56), (57) and (113) of Chapter 7.

Problem 12.8 Using Eqs (58) and (59) and assuming α to be real, prove that

$$\langle \Psi(t) | x^2 | \Psi(t) \rangle = x_0^2 \cos^2 \omega t + \frac{1}{2\gamma^2} \tag{117}$$

$$\langle \Psi(t) | p^2 | \Psi(t) \rangle = p_0^2 \sin^2 \omega t + \frac{1}{2}\gamma^2\hbar^2 \tag{118}$$

$$\langle \Psi(t) | H | \Psi(t) \rangle = \left(\frac{1}{2}\xi_0^2 + \frac{1}{2}\right)\hbar\omega \tag{119}$$

$$\langle \Psi(t) | N_{op} | \Psi(t) \rangle = \frac{1}{2}\xi_0^2 \tag{120}$$

$$\langle \Psi(t) | N_{op}^2 | \Psi(t) \rangle = \left(\frac{1}{2}\xi_0^2\right)^2 + \frac{1}{2}\xi_0^2 \tag{121}$$

$$\Delta n = [\langle N_{op}^2 \rangle - (N_{op})^2]^{1/2}$$

$$= \frac{1}{\sqrt{2}} \xi_0 = \sqrt{\langle n \rangle} \tag{122}$$

where
$$\xi_0 \left(= \sqrt{2}\alpha\right) = \gamma x_0 \tag{123}$$

Show that the above equations are the same as derived in Section 7.5.

Problem 12.9 Prove Eq. (45) by induction.

Problem 12.10 For a two-dimensional isotropic oscillator

$$H = \frac{1}{2\mu}(p_x^2 + p_y^2) + \frac{1}{2}\mu\omega^2(x^2 + y^2) \tag{124}$$

Show that:

$$H = (a_1\bar{a}_1 + a_2\bar{a}_2 + 1)\,\hbar\omega \tag{125}$$

where
$$a_1 = \frac{1}{\sqrt{2\mu\hbar\omega}}(\mu\omega x + ip_x) \tag{126}$$

and
$$a_2 = \frac{1}{\sqrt{2\mu\hbar\omega}}(\mu\omega y + ip_y) \tag{127}$$

Show that the eigenvalues would be $\hbar\omega$, $2\hbar\omega$, . . . and will be n-fold degenerate.

Problem 12.11 If $|\alpha\rangle$ and $|\beta\rangle$ are eigenkets of the operator a belonging to the eigenvalues α and β respectively, then show that:

$$|\langle \alpha | \beta \rangle|^2 = e^{-|\alpha - \beta|^2} \tag{128}$$

The above equation shows that the eigenkets are not orthogonal.

Problem 12.12

(a) Assume α to be real and show that:

$$|\alpha\rangle = e^{-\alpha^2/2} e^{\alpha\bar{a}} |0\rangle$$

(b) If

$$|\Psi(t = 0)\rangle = |\alpha\rangle$$

then show that

$$|\Psi(t)\rangle = e^{-i\omega t/2}\, e^{-\alpha^2/2}\, e^{-i\omega t \bar{a} a}\, e^{\alpha\bar{a}} |0\rangle$$

[**Hint:** $|\Psi(t)\rangle = e^{-iHt/\hbar} |\Psi(0)\rangle$ and use Eq. (8) for H.]

12.12 SOLUTIONS

Solution 12.2

$$\langle m | a | n \rangle = \sqrt{n}\,(m | n - 1) = \sqrt{n}\,\delta_{m,\,n-1} \tag{129}$$

and

$$\langle m | \bar{a} | n \rangle = \sqrt{n+1}\,(m | n + 1) = \sqrt{n+1}\,\delta_{m,\,n+1} \tag{130}$$

Thus the matrices which represent a and $\bar{a}$ are:

$$a = \begin{pmatrix} 0 & \sqrt{1} & 0 & 0 & 0 & \cdots \\ 0 & 0 & \sqrt{2} & 0 & 0 & \cdots \\ 0 & 0 & 0 & \sqrt{3} & 0 & \cdots \\ 0 & 0 & 0 & 0 & \sqrt{4} & \cdots \\ \vdots & \vdots & \vdots & \vdots & \vdots & \cdots \end{pmatrix} \tag{131}$$

and

$$\bar{a} = \begin{pmatrix} 0 & 0 & 0 & 0 & 0 & \cdots \\ \sqrt{1} & 0 & 0 & 0 & 0 & \cdots \\ 0 & \sqrt{2} & 0 & 0 & 0 & \cdots \\ 0 & 0 & \sqrt{3} & 0 & 0 & \cdots \\ \vdots & \vdots & \vdots & \vdots & \vdots & \cdots \end{pmatrix} \tag{132}$$

which are non-Hermitian matrices.

Solution 12.8 We will evaluate $\langle \Psi(t) | p^2 | \Psi(t) \rangle$; other expectation values can be evaluated in a similar manner. Using Eq. (59)

$$\begin{aligned} \langle \Psi(t) | p^2 | \Psi(t) \rangle &= -\frac{\mu\hbar\omega}{2} \langle \Psi(t) | (a - \bar{a})\,(a - \bar{a}) | \Psi(t) \rangle \\ &= -\frac{\mu\hbar\omega}{2} [\langle \Psi(t) | aa | \Psi(t) \rangle - \langle \Psi(t) | a\bar{a} | \Psi(t) \rangle - \langle \Psi(t) | \bar{a}a | \Psi(t) \rangle \\ &\qquad + \langle \Psi(t) | \bar{a}\bar{a} | \Psi(t) \rangle] \end{aligned} \tag{133}$$

Now using Eqs (57) and (61)

$$\begin{aligned} \langle \Psi(t) | aa | \Psi(t) \rangle &= e^{-\alpha^2} \sum_{m=0,1,2,\ldots}^{\infty} \sum_{n=0,1,2,\ldots}^{\infty} \frac{\alpha^{m+n}}{\sqrt{m!\,n!}} e^{i(m-n)\omega t} \langle m | aa | n \rangle \\ &= e^{-\alpha^2} \sum_{m=0,1,\ldots}^{\infty} \sum_{n=2,3,\ldots}^{\infty} \frac{\alpha^{m+n}}{\sqrt{m!\,n!}} e^{i(m-n)\omega t} \sqrt{n(n-1)}\, \delta_{m,\,n-2} \\ &= e^{-\alpha^2} \sum_{m=0,1,2,\ldots}^{\infty} \frac{\alpha^{2m+2}}{m!} e^{-2i\omega t} \\ &= \alpha^2 e^{-2i\omega t} \end{aligned} \tag{134}$$

If we take the complex conjugate of the above equation, we would get

$$\langle \Psi(t) | \bar{a}\bar{a} | \Psi(t) \rangle = \alpha^2 e^{2i\omega t} \tag{135}$$

Similarly,

$$\langle \Psi(t) | a\bar{a} | \Psi(t) \rangle = 1 + \alpha^2 \tag{136}$$

and

$$\langle \Psi(t) | \bar{a}a | \Psi(t) \rangle = \alpha^2 \tag{137}$$

Thus,
$$\langle \Psi(t) | p^2 | \Psi(t) \rangle = -\frac{\mu\hbar\omega}{2}[2\alpha^2 \cos 2\omega t - 1 - 2\alpha^2]$$
$$= 2\mu\hbar\omega\,\alpha^2 \sin^2 \omega t + \frac{\mu\hbar\omega}{2}$$
$$= p_0^2 \sin^2\omega t + \frac{1}{2}\gamma^2 \hbar^2 \qquad (138)$$

where
$$p_0 = \sqrt{2\mu\hbar\omega}\,\alpha \qquad (139)$$

consistent with Eq. (66).

Solution 12.9 We assume Eq. (45) to be valid so that

$$\Psi_{n+1}(x) = \frac{1}{[2(n+1)]^{1/2}}\left[\frac{\gamma}{2^n n!\sqrt{\pi}}\right]^{1/2}\left(\xi - \frac{d}{d\xi}\right)[H_n(\xi)e^{-\xi^2/2}]$$
$$= \left[\frac{\gamma}{2^{n+1}(n+1)!\sqrt{\pi}}\right]^{1/2} [2\xi H_n(\xi) - H_n'(\xi)]\, e^{-\xi^2/2} \qquad (140)$$

Now, if we use the well-known recurrence relations

$$2\xi H_n(\xi) = 2n H_{n-1}(\xi) + H_{n+1}(\xi) \qquad (141)$$

and

$$H_n'(\xi) = 2n H_{n-1}(\xi) \qquad (142)$$

we readily get

$$H_{n+1}(\xi) = 2\xi H_n(\xi) - H_n'(\xi) \qquad (143)$$

Equation (140), therefore, becomes

$$\Psi_{n+1}(x) = \left[\frac{\gamma}{2^{n+1}(n+1)!\sqrt{\pi}}\right]^{1/2} H_{n+1}(\xi)\, e^{-\xi^2/2} \qquad (144)$$

Thus if Eq. (45) is valid then Eq. (144) must also be valid. Since Eq. (45) is valid for $n = 0, 1$ and 2 [see Eqs (42)–(44)], it must be valid for all integer values of n.

Thus, we have been able to derive the normalized Hermite-Gauss functions in a very straightforward manner.

12.13 SAMPLE QUESTIONS

Q. 1 In the linear harmonic oscillator problem, let $|n\rangle$ represent the set of orthonormal eigenkets of H.

(a) Evaluate

$$\langle 4 | x^2 | 6 \rangle, \qquad \langle 5 | p^2 | 6 \rangle$$

(b) Are $|0\rangle$, $|1\rangle$ and $|2\rangle$ eigenkets of a?

(c) Evaluate (i) $H[\bar{a}\bar{a} | n\rangle]$, (ii) $H[aa | n\rangle]$

(d) If

$$[\Psi(t=0)\rangle = \frac{1}{\sqrt{2}}[|1\rangle + i|3\rangle]$$

then what would be $|\Psi(t)\rangle$?

[**Answer:** **(b)** Only $|0\rangle$ is an eigenket of a.

(c) (ii) $\sqrt{n(n-1)}\left(n-\frac{5}{2}\right)\hbar\omega|n\rangle$].

Q. 2 At $t = 0$, the linear harmonic oscillator is in the coherent state given by

$$|\Psi(t=0)\rangle = |\alpha\rangle = \sum c_n|n\rangle$$

(a)

$$c_n = \frac{\alpha^n}{\sqrt{n!}}c_0$$

(b) Normalize $|\alpha\rangle$ and determine c_0.

(c) What would be $|\Psi(t)\rangle$?

(d) Assume α to be real and calculate

$$\langle\Psi(t)|p|\Psi(t)\rangle$$

$$\langle\Psi(t)|x|\Psi(t)\rangle$$

$$\langle\Psi(t)|a|\Psi(t)\rangle$$

and

$$\langle\Psi(t)|\bar{a}|\Psi(t)\rangle$$

REFERENCES AND SUGGESTED READINGS

1. P.A.M. Dirac, *The Principles of Quantum Mechanics*, Oxford University Press, Oxford (1958).
2. G. Baym, *Lectures on Quantum Mechanics*, W.A. Benjamin, New York (1969).
3. H. Goldstein, *Classical Mechanics*, Addison-Wesley Publishing Co., Reading, Massachusetts (1950).
4. W.H. Louisell, *Quantum Statistical Properties of Radiation*, John Wiley, New York (1973).

13 Angular Momentum II

Using operator algebra

In a one-page letter to the Editor of Nature ... [illegible] ... proposed the idea that each electron with ... [illegible] ... a magnetic moment ... [illegible]

13.1 INTRODUCTION

In Chapter 9, we had introduced the quantum mechanical angular momentum [illegible] had shown that its components satisfy the following commutation relations:

[illegible]

We had also [illegible]

commutes with [illegible]

In this chapter, [illegible] mechanics) to [illegible] momentum operator [illegible] will always be an integer or a half integer [illegible] which is one of the remarkable features of quantum mechanics.

we had shown that [illegible]

In the next section, we will obtain the eigenvalues of a Cartesian component of the angular momentum operator and of the square of the angular momentum operator [illegible] we will derive the angular momentum matrices for [illegible] Pauli spin matrices) [illegible] we will discuss [illegible]

---

[1] Using CGS units, the value of [illegible] 14.5 and the [illegible]

13 Angular Momentum II

Using operator algebra

In a one-page letter to the Editor of Naturwissenschaften dated 17 October 1925, Samuel A. Goudsmit and I proposed the idea that each electron rotates with an angular momentum ħ/2 and carries, besides its charge e, a magnetic moment equal to one Böhr magneton[1]*, eħ/2mc ...*

— GEORGE E. UHLENBECK in *Fifty Years of Spin: Personal Reminiscences*, Physics Today, p. 43, June 1976

13.1 INTRODUCTION

In Chapter 9, we had introduced the quantum mechanical angular momentum operator $\hat{\mathbf{L}}$ and had shown that its components satisfy the following commutation relations:

$$\left.\begin{aligned} \left[\hat{L}_x, \hat{L}_y\right] &= i\hbar\hat{L}_z \\ \left[\hat{L}_y, \hat{L}_z\right] &= i\hbar\hat{L}_x \\ \left[\hat{L}_z, \hat{L}_x\right] &= i\hbar\hat{L}_y \end{aligned}\right\} \quad (1)$$

We had also shown that the square of the angular momentum operator $\hat{L}^2\left(\equiv \hat{L}_x^2 + \hat{L}_y^2 + \hat{L}_z^2\right)$ commutes with $\hat{L}_x$, $\hat{L}_y$ and $\hat{L}_z$; i.e.,

$$\left[\hat{L}^2, \hat{L}_x\right] = \left[\hat{L}^2, \hat{L}_y\right] = \left[\hat{L}^2, \hat{L}_z\right] = 0 \quad (2)$$

In this chapter, we will use just these commutation relations (and basic axioms of quantum mechanics) to show that if one makes a measurement of Cartesian component of the angular momentum operator then it will always be an integral or a half integral multiple of $\hbar$, a result which is one of the remarkable features of quantum mechanics. We may recall that in Chapter 9 we had shown that $\hat{L}_z$ (and similarly $\hat{L}_x$ and $\hat{L}_y$) can only have eigenvalue 0, $\hbar$, $2\hbar$, ...

In the next section, we will obtain the eigenvalues of a Cartesian component of the angular momentum operator and of the square of the angular momentum operator. In Section 13.3, we will derive the angular momentum matrices for $j = \frac{1}{2}$ (these matrices are known as the Pauli spin matrices); in the next chapter we will discuss the famous Stern-Gerlach experiment

[1] Using CGS units, the value of one Bohr magneton is $e\hbar/2mc$; in MKS units, the value is $q\hbar/2m$; see also Section 14.5 and the Table of Physical Constants.

which demonstrates the quantization of angular momentum corresponding to $j = \frac{1}{2}$. In Section 13.4, we will derive the angular momentum matrices for $j = 1$; the derivation of angular momentum matrices for $j = \frac{3}{2}$ will be left as a problem! In Section 13.5, we will discuss the rotation operator and its relationship to the angular momentum operator; this will be followed by a large number of examples. In Section 13.6, we will discuss the Pauli wavefunction and the Pauli equation for spin half particles. Finally in Section 13.7, we will give a bit of history of how the spin angular momentum of the electron (and its magnetic moment) were introduced by Uhlenbeck and Goudsmit in 1925.

13.2 EIGENVALUES

In this section, we will use the commutation relations [see Eqs. (1) and (2)] to determine the eigenvalues of a Cartesian component of the angular momentum operator. For reasons to be discussed later, we will denote the angular momentum operators by $\hat{J}_x$, $\hat{J}_y$ and $\hat{J}_z$ and use a system of units in which $\hbar = 1$. Thus, Eqs. (1) and (2) become

$$\left.\begin{aligned} \left[\hat{J}_x, \hat{J}_y\right] &= i\hat{J}_z \\ \left[\hat{J}_y, \hat{J}_z\right] &= i\hat{J}_x \\ \left[\hat{J}_z, \hat{J}_x\right] &= i\hat{J}_y \end{aligned}\right\} \tag{3}$$

and $$\left[\hat{J}^2, \hat{J}_x\right] = 0 = \left[\hat{J}^2, \hat{J}_y\right] = \left[\hat{J}^2, \hat{J}_z\right] \tag{4}$$

where $$\hat{J}^2 \equiv \hat{J}_x\hat{J}_x + \hat{J}_y\hat{J}_y + \hat{J}_z\hat{J}_z = \hat{J}_x^2 + \hat{J}_y^2 + \hat{J}_z^2 \tag{5}$$

It is convenient to use the bra-ket notation introduced in Chapter 11. Since $\hat{J}^2$ and $\hat{J}_z$ commute, we can construct a complete set of orthonormal eigenkets which and simultaneous eigenkets of the operators $\hat{J}^2$ and $\hat{J}_z$ belonging to the eigenvalues λ and m respectively (see Section 11.8).

Thus, $$\hat{J}^2|\lambda, m\rangle = \lambda|\lambda, m\rangle \tag{6}$$

$$\hat{J}_z|\lambda, m\rangle = m|\lambda, m\rangle \tag{7}$$

13.2.1 Ladder Operators

We define the *ladder operators*

$$\hat{J}_+ = \hat{J}_x + i\hat{J}_y \tag{8}$$

and $$\hat{J}_- = \hat{J}_x - i\hat{J}_y \tag{9}$$

Since $\hat{J}_x$, $\hat{J}_y$ and $\hat{J}_z$ are observables, they are represented by real operators:

$$\overline{\hat{J}_x} = \hat{J}_x, \ \overline{\hat{J}_y} = \hat{J}_y, \ \overline{\hat{J}_z} = \hat{J}_z$$

Thus, $$\overline{\hat{J}_+} = \hat{J}_x - i\hat{J}_y = \hat{J}_- \quad \text{and} \quad \overline{\hat{J}_-} = \hat{J}_+ \tag{10}$$

i.e., $\hat{J}_-$ and $\hat{J}_+$ are adjoints of $\hat{J}_+$ and $\hat{J}_-$ respectively. Now,

$$\left[\hat{J}_z, \hat{J}_+\right] = \left[\hat{J}_z, \hat{J}_x\right] + i\left[\hat{J}_z, \hat{J}_y\right]$$

$$= i\hat{J}_y + \hat{J}_x = \hat{J}_+$$

Thus, $$\left[\hat{J}_z, \hat{J}_+\right] = \hat{J}_+ \tag{11}$$

Similarly, $$\left[\hat{J}_z, \hat{J}_-\right] = -\hat{J}_- \tag{12}$$

We also have $$\hat{J}_+\hat{J}_- = \left[\hat{J}_x + i\hat{J}_y\right]\left[\hat{J}_x - i\hat{J}_y\right] = \hat{J}_x^2 + \hat{J}_y^2 - i\left[\hat{J}_x, \hat{J}_y\right]$$

or $$\hat{J}_+\hat{J}_- = \hat{J}^2 - \hat{J}_z^2 + \hat{J}_z \tag{13}$$

Similarly, $$\hat{J}_-\hat{J}_+ = \hat{J}^2 - \hat{J}_z^2 - \hat{J}_z \tag{14}$$

We first show that $\lambda \geq m^2$. Now,

$$\left\langle P\left|\hat{J}^2 - \hat{J}_z^2\right|P\right\rangle = \left\langle P\left|\hat{J}_x^2 + \hat{J}_y^2\right|P\right\rangle$$

$$= \left\langle P\left|\hat{J}_x\hat{J}_x\right|P\right\rangle + \left\langle P\left|\hat{J}_y\hat{J}_y\right|P\right\rangle$$

$$= \langle Q|Q\rangle + \langle R|R\rangle \geq 0 \tag{15}$$

where $$|Q\rangle \equiv \hat{J}_x|P\rangle \quad \text{and} \quad |R\rangle \equiv \hat{J}_y|P\rangle$$

and we have used the fact that since $\hat{J}_x$ and $\hat{J}_y$ are real operators so that

$$\langle Q| = \langle P|\overline{\hat{J}_x} = \langle P|\hat{J}_x \text{ and } \langle R| = \langle P|\overline{\hat{J}_y} = \langle P|\hat{J}_y$$

If we choose $|P\rangle = |\lambda, m\rangle$, then

$$0 \leq \left\langle \lambda, m\left|\hat{J}^2 - \hat{J}_z^2\right|\lambda, m\right\rangle = \left(\lambda - m^2\right)\left\langle \lambda, m|\lambda, m\right\rangle$$

Since $\langle \lambda, m \mid \lambda, m\rangle$ is always positive, we get the important result

$$\lambda \geq m^2 \tag{16}$$

We next consider the effect of the ladder operators on the ket $|\lambda, m\rangle$. Using Eq. (11), we get

$$\hat{J}_z\hat{J}_+|\lambda, m\rangle = \left(\hat{J}_+ + \hat{J}_+\hat{J}_z\right)|\lambda, m\rangle$$

or $$\hat{J}_z\left[\hat{J}_+|\lambda, m\rangle\right] = (m+1)\left[\hat{J}_+|\lambda, m\rangle\right] \tag{17}$$

Further, since $\hat{J}^2$ commutes with $\hat{J}_x$ and $\hat{J}_y$ we will have

$$\left[\hat{J}^2, \hat{J}_\pm\right] = 0 \quad \Rightarrow \quad J^2 J_\pm|\lambda, m\rangle = \hat{J}_\pm\hat{J}^2|\lambda, m\rangle$$

Thus, $$\hat{J}^2\left[\hat{J}_\pm|\lambda, m\rangle\right] = \lambda\left[\hat{J}_\pm|\lambda, m\rangle\right] \tag{18}$$

Equations (17) and (18) tell us that $\hat{J}_+|\lambda, m\rangle$ is a simultaneous eigenket of the operators $\hat{J}^2$ and $\hat{J}_z$ belonging to the eigenvalues λ and $(m + 1)$, provided $\hat{J}_+|\lambda, m\rangle$ is not a null ket. Similarly, using Eq. (12) one can show

$$\hat{J}_z\left[\hat{J}_-|\lambda,m\rangle\right]=(m-1)\left[\hat{J}_-|\lambda,m\rangle\right] \tag{19}$$

implying that $\hat{J}_-|\lambda, m\rangle$ is a simultaneous eigenket of $\hat{J}^2$ and $\hat{J}_z$ belonging respectively to the eigenvalues λ and $(m-1)$, provided $\hat{J}_-|\lambda, m\rangle$ is not a null ket. We may, therefore, write

$$\hat{J}_+|\lambda, m\rangle = C_+|\lambda, m+1\rangle \tag{20}$$

and

$$\hat{J}_-|\lambda, m\rangle = C_-|\lambda, m-1\rangle \tag{21}$$

The above equations show why $\hat{J}_\pm$ are called *ladder* operators; $\hat{J}_+$ is an operator that *raises* the eigenket to the next higher value of *m* and $\hat{J}_-$ is an operator that *lowers* the eigenket to the next lower value of *m*. In a similar manner we may argue that $\hat{J}_+|\lambda,m+1\rangle$ and $\hat{J}_-|\lambda,m-1\rangle$ are proportional to $|\lambda, m+2\rangle$ and $|\lambda, m-2\rangle$ respectively. This, however, cannot go on indefinitely; otherwise we will violate the condition $\lambda \geq m^2$. Consequently, there must exist maximum and minimum values of *m*; we write the maximum value of *m* to be equal to *j*; thus

$$\hat{J}_+|\lambda, j\rangle = 0\,; \qquad \left[(m)_{\max} = j\right] \tag{22}$$

The above equation implies that $m = j$ is an eigenvalue, but $m = j + 1$ is not an eigenvalue. Operating on the left by $\hat{J}_-$, we obtain [using Eq. (14)]

$$\left(\hat{J}^2 - \hat{J}_z^2 - \hat{J}_z\right)|\lambda, j\rangle = 0$$

or

$$\left(\lambda - j^2 - j\right)|\lambda, j\rangle = 0 \tag{23}$$

Since $|\lambda, j\rangle$, by definition, is not a null ket, the number $\lambda - j^2 - j$ must be equal to 0; and hence,

$$\lambda = j(j+1) \tag{24}$$

The reader not yet attuned to operator algebra may pause here and reflect on Eqs. (22) and (23) to see how we treat them differently. In Eq. (22) neither the operator $\hat{J}_+$ nor the ket $|\lambda, j\rangle$ need be zero in order to produce a null ket. For example,

$$\frac{d^2(x)}{dx^2} = 0$$

does not permit us to identify either d^2/dx^2 or *x* with a null quantity. But in Eq. (23), a ket is multiplied by a number implying that either the number should be zero or the ket should be a null ket. Similarly, the inequality given by Eq. (16) demands that there has to be a minimum value of *m* (say equal to *j′*); this minimum value will correspond to

$$J_-|\lambda, j'\rangle = 0$$

implying that $m = j'$ is an eigenvalue, but $m = j' - 1$ is not an eigenvalue. Operating the above equation on the left by J_+ and using Eq. (13), we obtain

$$\left(\lambda - j'^2 + j'\right)|\lambda, j'\rangle = 0$$

Since $|\lambda, j'\rangle$ is not a null ket, we must have

$$\lambda = j'(j'-1) \tag{25}$$

Using Eqs. (24) and (25), we obtain

$$j'(j'-1) = j(j+1)$$

Thus either $j' = j + 1$ (which we reject since j' is supposed to be less than j), or

$$j' = -j \tag{26}$$

Thus, we have a family of kets which correspond to the same eigenvalue $\lambda[=j(j+1)]$ of $\hat{J}^2$ but different eigenvalues of $\hat{J}_z$. These we denote by

$$|\lambda, j\rangle, |\lambda, j-1\rangle, \ldots, |\lambda, -j\rangle \tag{27}$$

Successive application of the ladder operator $\hat{J}_-$ on $|\lambda, j\rangle$ should ultimately lead to $|\lambda, -j\rangle$. (Also, successive application of $\hat{J}_+$ on $|\lambda, -j\rangle$ should lead to $|\lambda, j\rangle$.) This will happen only if $2j$ is an integer, so that the allowed values of j are

$$j = 0, \frac{1}{2}, 1, \frac{3}{2}, \ldots \qquad \text{Allowed values of } j \tag{28}$$

and for each value of j the m value go from $-j$ and j in steps of unity. Thus, there are $(2j + 1)$ eigenkets of $\hat{J}^2$, each having the same eigenvalue $j(j + 1)$. We label the eigenkets as

$$|j,m\rangle;\ m = -j, -j+1, \ldots, j-1, j$$

with $j = 0, \frac{1}{2}, 1, \frac{3}{2}, \ldots$. Since the kets $|j, m\rangle$ are eigenkets of Hermitian operators, we may assume them to satisfy the orthonormality condition

$$\langle j', m'|j, m\rangle = \delta_{jj'}\,\delta_{mm'} \tag{29}$$

Thus, we obtain the important result that the Cartesian components of the angular momentum operator (say $\hat{J}_z$) can have half-integral values. On the other hand, components of orbital angular momenta, defined as $\mathbf{r} \times \mathbf{p}$, can have only eigenvalue $0, \hbar, 2\hbar$, etc. (see Section 9.3). The question arises whether nature provides systems which do have values $\frac{1}{2}\hbar$ for angular momenta. It is remarkable that the most elementary system known, the electron is an example of this. We cannot write this spin angular momentum as an operator operating on the coordinates of the electron. Spin has no classical analogue and is introduced in non-relativistic theory in an ad hoc manner; the relativistic Dirac theory provides a natural explanation of this (see Section 30.7). The suggestion about spin came from a study of Zeeman patterns form which it appeared that the $(2l + 1)$ multiplicity of levels often turned out to be even. We may mention here that an attempt was made to attribute spin to the rotation of the electron about its axis but this exercise proved abortive (see also Section 13.7).

13.2.2 The Constants C_+ and C_-

We now determine the normalization constants C_+ and C_-. First, we note that if we define

$$|P\rangle = \hat{J}_+|j, m\rangle = C_+|j, m+1\rangle$$

then $$\langle P| = \langle j, m|\hat{J}_- = C_+^* \langle j, m+1| \qquad (\text{since } \overline{\hat{J}_+} = \hat{J}_-)$$

Thus, $$\langle P|P\rangle = \langle j, m|\hat{J}_-\hat{J}_+|j, m\rangle = |C_+|^2 \langle j, m+1|j, m+1\rangle$$

Now, $$\langle j, m|\hat{J}_-\hat{J}_+|j, m\rangle = \langle j, m|\hat{J}^2 - \hat{J}_z^2 - \hat{J}_z|j, m\rangle$$
$$= j(j+1) - m^2 - m = (j-m)(j+m+1)$$

Hence, $$C_+ = \left[(j-m)(j+m+1)\right]^{\frac{1}{2}}$$

or $$\hat{J}_+|j, m\rangle = \left[(j-m)(j+m+1)\right]^{\frac{1}{2}}|j, m+1\rangle \tag{30}$$

Similarly, $$\hat{J}_-|j, m\rangle = \left[(j+m)(j-m+1)\right]^{\frac{1}{2}}|j, m-1\rangle \tag{31}$$

Notice that the coefficient in front of $|j, m+1\rangle$ becomes zero for $m = +j$ and the coefficient in front of $|j, m-1\rangle$ becomes zero for $m = -j$. Thus,

$$\hat{J}_+|j, j\rangle = 0 \tag{32}$$

and $$\hat{J}_-|j, -j\rangle = 0 \tag{33}$$

13.3 ANGULAR MOMENTUM MATRICES CORRESPONDING TO $j = \frac{1}{2}$ AND THE PAULI SPIN MATRICES

We will consider the matrix representation of the components of the angular momentum operator (corresponding to $j = \frac{1}{2}$) in a system in which J^2 and J_z are diagonal, i.e., we will choose as basis the simultaneous eigenfunctions of J^2 and J_z and obtain the famous Pauli spin matrices. Now, corresponding to $j = \frac{1}{2}$ we have $m = \frac{1}{2}$ and $-\frac{1}{2}$; thus we have two base states which we denote by $|1\rangle$ and $|2\rangle$:

$$|1\rangle \equiv \left|\tfrac{1}{2}, \tfrac{1}{2}\right\rangle = \left|j = \tfrac{1}{2}, m = \tfrac{1}{2}\right\rangle \tag{34}$$

and $$|2\rangle \equiv \left|\tfrac{1}{2}, -\tfrac{1}{2}\right\rangle = \left|j = \tfrac{1}{2}, m = -\tfrac{1}{2}\right\rangle \tag{35}$$

Obviously,

$$\left.\begin{aligned} \hat{J}_z|1\rangle = \tfrac{1}{2}\hbar|1\rangle, \quad \hat{J}_z|2\rangle = -\tfrac{1}{2}\hbar|2\rangle \\ \hat{J}^2|1\rangle = \tfrac{3}{4}\hbar^2|1\rangle, \; \hat{J}^2|2\rangle = \tfrac{3}{4}\hbar^2|2\rangle \end{aligned}\right\} \tag{36}$$

Thus, $|1\rangle$ and $|2\rangle$ are simultaneous eigenkets of $\hat{J}_z$ (belonging to the eigenvalues $+\frac{1}{2}\hbar$ and $-\frac{1}{2}\hbar$) and of $\hat{J}^2$ (belonging to the same eigenvalue $\frac{3}{4}\hbar^2$); however, $|1\rangle$ and $|2\rangle$ are not eigenkets of $\hat{J}_x$ and $\hat{J}_y$. Now, the $(ij)^{\text{th}}$ matrix element of an operator $\hat{\mathcal{O}}$ is defined by

$$\hat{\mathcal{O}}_{ij} = \langle i|\hat{\mathcal{O}}|j\rangle$$

Thus, $$\left(\hat{J}_z\right)_{12} = \langle 1|\hat{J}_z|2\rangle = 0 = \langle 2|\hat{J}_z|1\rangle = \left(\hat{J}_z\right)_{21}$$

Similarly, $\left(\hat{J}_z\right)_{11} = \langle 1|\hat{J}_z|1\rangle = \frac{1}{2}\hbar;\ \left(\hat{J}_z\right)_{22} = \langle 2|\hat{J}_z|2\rangle = -\frac{1}{2}\hbar$

$$(J^2)_{12} = (J^2)_{21} = 0$$

and $\left(J^2\right)_{11} = \left(J^2\right)_{22} = \frac{3}{4}\hbar^2$

Thus the matrices representing $\hat{J}_z$ and $\hat{J}^2$ are

$$\hat{J}_z = \tfrac{1}{2}\hbar\begin{pmatrix}1 & 0\\ 0 & -1\end{pmatrix} \quad \text{and} \quad \hat{J}^2 = \tfrac{3}{4}\hbar^2\begin{pmatrix}1 & 0\\ 0 & 1\end{pmatrix}$$

Now, $\langle j, m' \mid \hat{J}_x \mid j, m\rangle = \langle j, m' \mid \tfrac{1}{2}\left[\hat{J}_+ + \hat{J}_-\right] \mid j, m\rangle$

$$= \tfrac{1}{2}\left[(j-m)(j+m+1)\right]^{\frac{1}{2}}\hbar\langle j, m'|j, m+1\rangle$$

$$+ \tfrac{1}{2}\left[(j+m)(j-m+1)\right]^{\frac{1}{2}}\hbar\langle j, m'|j, m-1\rangle$$

$$= \tfrac{1}{2}\left[(j-m)(j+m+1)\right]^{\frac{1}{2}}\hbar\,\delta_{m', m+1}$$

$$+ \tfrac{1}{2}\left[(j+m)(j-m+1)\right]^{\frac{1}{2}}\hbar\,\delta_{m', m-1}$$

i.e., the matrix element is zero unless $m' = m + 1$ or $m' = m - 1$. Thus,

$$\left(\hat{J}_x\right)_{11} = \langle 1|\hat{J}_x|1\rangle = \left\langle \tfrac{1}{2}, \tfrac{1}{2}\middle|\hat{J}_x\middle|\tfrac{1}{2}, \tfrac{1}{2}\right\rangle = 0$$

and similarly $\left(\hat{J}_x\right)_{22} = 0$. However,

$$\left(\hat{J}_x\right)_{12} = \langle 1|J_x|2\rangle = \left\langle \tfrac{1}{2}, \tfrac{1}{2}\middle|J_x\middle|\tfrac{1}{2}, -\tfrac{1}{2}\right\rangle$$

$$= \tfrac{1}{2}\left[\left(\tfrac{1}{2} - \tfrac{1}{2} + 1\right)\left(\tfrac{1}{2} + \tfrac{1}{2}\right)\right]^{\frac{1}{2}}\hbar = \tfrac{1}{2}\hbar = (J_x)_{21}$$

Similarly, $\left(\hat{J}_y\right)_{11} = \left(\hat{J}_y\right)_{22} = 0$

$$\left(\hat{J}_y\right)_{12} = -\frac{i}{2}\hbar \quad \text{and} \quad \left(\hat{J}_y\right)_{21} = \frac{i}{2}\hbar$$

13.3.1 Pauli Spin Matrices

Thus, the angular momentum matrices for $j = \frac{1}{2}$ are

$$\left.\begin{aligned}\hat{J}_x &= \tfrac{1}{2}\hbar\begin{pmatrix}0 & 1\\ 1 & 0\end{pmatrix}, \quad \hat{J}_y = \tfrac{1}{2}\hbar\begin{pmatrix}0 & -i\\ i & 0\end{pmatrix}\\ \hat{J}_z &= \tfrac{1}{2}\hbar\begin{pmatrix}1 & 0\\ 0 & -1\end{pmatrix}, \hat{J}^2 = \tfrac{3}{4}\hbar^2\begin{pmatrix}1 & 0\\ 0 & 1\end{pmatrix}\end{aligned}\right\} \tag{37}$$

We can also write $\mathbf{J} = \dfrac{1}{2}\hbar\sigma$ (38)

where $\sigma_x = \begin{pmatrix}0 & 1\\ 1 & 0\end{pmatrix}$ (39)

$$\sigma_y = \begin{pmatrix} 0 & -i \\ i & 0 \end{pmatrix} \tag{40}$$

and
$$\sigma_z = \begin{pmatrix} 1 & 0 \\ 0 & -1 \end{pmatrix} \tag{41}$$

are known as Pauli spin matrices. The matrices σ_x, σ_y and σ_z are all Hermitian and it can be easily shown that their eigenvalues are +1 and –1; hence the eigenvalues of $\hat{J}_x$, $\hat{J}_y$ and $\hat{J}_z$ are $+\frac{1}{2}\hbar$ and $-\frac{1}{2}\hbar$. Thus, if we make a measurement of $\hat{J}_x$ (or of $\hat{J}_y$ or $\hat{J}_z$) we will get either $+\frac{1}{2}\hbar$ or $-\frac{1}{2}\hbar$; in the next chapter we will discuss the experimental verification of this through the analysis of Stern Gerlach experiments. On the other hand, $\hat{J}^2$ has only one eigenvalue: $\frac{3}{4}\hbar^2$. Further σ_x^2, σ_y^2, and σ_z^2 are all unit matrices; for example

$$\sigma_x^2 = \begin{pmatrix} 0 & 1 \\ 1 & 0 \end{pmatrix}\begin{pmatrix} 0 & 1 \\ 1 & 0 \end{pmatrix} = \begin{pmatrix} 1 & 0 \\ 0 & 1 \end{pmatrix} = \sigma_y^2 = \sigma_z^2$$

13.3.2 Spin up and Spin Down States

The simultaneous eigenfuctions of $\hat{J}^2$ and $\hat{J}_z$ are

$$|1\rangle = \left|\tfrac{1}{2}, \tfrac{1}{2}\right\rangle = \begin{pmatrix} 1 \\ 0 \end{pmatrix} \quad \text{and} \quad |2\rangle = \left|\tfrac{1}{2}, -\tfrac{1}{2}\right\rangle = \begin{pmatrix} 0 \\ 1 \end{pmatrix} \tag{42}$$

which are known as the spin up $|\uparrow\rangle$ and spin down $|\downarrow\rangle$ states. They are also denoted by $|\hat{\mathbf{z}}\uparrow\rangle$ and $|\hat{\mathbf{z}}\downarrow\rangle$ states. Thus, if the electron, for example, is in the $|\hat{\mathbf{z}}\downarrow\rangle$ state and if the z–component of the spin angular momentum is measured then the measurement will definitely yield $-\frac{1}{2}\hbar$.

From the matrices representing $\hat{J}_x$ and $\hat{J}_y$, one can calculate the corresponding eigenvalues and eigenvectors; indeed the normalized eigenvectors of $\hat{J}_x$ are

$$|\hat{\mathbf{x}}\uparrow\rangle = \frac{1}{\sqrt{2}}\begin{pmatrix} 1 \\ 1 \end{pmatrix} = \frac{1}{\sqrt{2}}|\hat{\mathbf{z}}\uparrow\rangle + \frac{1}{\sqrt{2}}|\hat{\mathbf{z}}\downarrow\rangle \tag{43}$$

$$|\hat{\mathbf{x}}\downarrow\rangle = \frac{1}{\sqrt{2}}\begin{pmatrix} 1 \\ -1 \end{pmatrix} = \frac{1}{\sqrt{2}}|\hat{\mathbf{z}}\uparrow\rangle - \frac{1}{\sqrt{2}}|\hat{\mathbf{z}}\downarrow\rangle \tag{44}$$

We also have
$$|\hat{\mathbf{z}}\uparrow\rangle = \begin{pmatrix} 1 \\ 0 \end{pmatrix} = \frac{1}{\sqrt{2}}|\hat{\mathbf{x}}\uparrow\rangle + \frac{1}{\sqrt{2}}|\hat{\mathbf{x}}\downarrow\rangle \tag{45}$$

and
$$|\hat{\mathbf{z}}\downarrow\rangle = \begin{pmatrix} 0 \\ 1 \end{pmatrix} = \frac{1}{\sqrt{2}}|\hat{\mathbf{x}}\uparrow\rangle - \frac{1}{\sqrt{2}}|\hat{\mathbf{x}}\downarrow\rangle \tag{46}$$

Thus, if a system is in the $|\hat{\mathbf{z}}\uparrow\rangle$ state and if one makes a measurement of $\hat{J}_x$, then the probability of obtaining the value $+\frac{1}{2}\hbar$ is $\left|\frac{1}{\sqrt{2}}\right|^2 \left(=\frac{1}{2}\right)$ and the value $-\frac{1}{2}\hbar$ is also $\left|\frac{1}{\sqrt{2}}\right|^2 \left(=\frac{1}{2}\right)$.

This will become more explicit when we discuss the Stern-Gerlach experiment in the next chapter. Similarly, one can write down the normalized eigenvectors of $\hat{J}_y$:

$$|\,\hat{\mathbf{y}}\uparrow\rangle = \frac{1}{\sqrt{2}}\begin{pmatrix}1\\ i\end{pmatrix} = \frac{1}{\sqrt{2}}|\,\hat{\mathbf{z}}\uparrow\rangle + \frac{i}{\sqrt{2}}|\,\hat{\mathbf{z}}\downarrow\rangle \tag{47}$$

and

$$|\,\hat{\mathbf{y}}\downarrow\rangle = \frac{1}{\sqrt{2}}\begin{pmatrix}1\\ -i\end{pmatrix} = \frac{1}{\sqrt{2}}|\,\hat{\mathbf{z}}\uparrow\rangle - \frac{i}{\sqrt{2}}|\,\hat{\mathbf{z}}\downarrow\rangle \tag{48}$$

Thus, if a system is in the $|\,\hat{\mathbf{y}}\downarrow\rangle$ state and if one makes a measurement of $\hat{J}_z$, then there is $\frac{1}{2}$ probability of collapsing to the $|\,\hat{\mathbf{z}}\uparrow\rangle$ state (and obtaining the value $+\frac{1}{2}\hbar$) and $\frac{1}{2}$ probability of collapsing to the $|\,\hat{\mathbf{z}}\downarrow\rangle$ and obtaining the value $-\frac{1}{2}\hbar$. To summarize:

$$|\,\hat{\mathbf{x}}\uparrow\rangle\left[=\frac{1}{\sqrt{2}}\begin{pmatrix}1\\ 1\end{pmatrix}\right] \text{ and } |\,\hat{\mathbf{x}}\downarrow\rangle\left[=\frac{1}{\sqrt{2}}\begin{pmatrix}1\\ -1\end{pmatrix}\right]$$

are simultaneous eigenkets of $\hat{J}_x$ and $\hat{J}^2$ but not eigenkets of $\hat{J}_y$ and $\hat{J}_z$. Further,

$$|\,\hat{\mathbf{y}}\uparrow\rangle\left[=\frac{1}{\sqrt{2}}\begin{pmatrix}1\\ i\end{pmatrix}\right] \text{ and } |\,\hat{\mathbf{y}}\downarrow\rangle\left[=\frac{1}{\sqrt{2}}\begin{pmatrix}1\\ -i\end{pmatrix}\right]$$

are simultaneous eigenkets of $\hat{J}_y$ and $\hat{J}^2$ but not eigenkets of $\hat{J}_x$ and $\hat{J}_z$. And,

$$|\,\hat{\mathbf{z}}\uparrow\rangle\left[=\begin{pmatrix}1\\ 0\end{pmatrix}\right] \text{ and } |\,\hat{\mathbf{z}}\downarrow\rangle\left[=\begin{pmatrix}0\\ 1\end{pmatrix}\right]$$

are simultaneous eigenkets of $\hat{J}_z$ and $\hat{J}^2$ but not eigenkets of $\hat{J}_x$ and $\hat{J}_y$.

Example 13.1 We assume a spin $\frac{1}{2}$ system to be in a state described by the (normalized) vector $|\,P\rangle = \frac{1}{\sqrt{10}}\begin{pmatrix}i\\ 3\end{pmatrix}$. Now, $|\,P\rangle = \frac{1}{\sqrt{10}}\begin{pmatrix}i\\ -3\end{pmatrix} = \frac{i}{\sqrt{10}}|\,\hat{\mathbf{z}}\uparrow\rangle - \frac{3}{\sqrt{10}}|\,\hat{\mathbf{z}}\downarrow\rangle$

Thus if a system is in the state $|\,P\rangle$ and if one makes a measurement of J_z, then the probability of collapsing to the $|\,\hat{\mathbf{z}}\uparrow\rangle$ state will be $\frac{1}{10}$ (and obtaining the value $+\frac{1}{2}\hbar$) and the probability of collapsing to the $|\,\hat{\mathbf{z}}\downarrow\rangle$ state will be $\frac{9}{10}$ (and obtaining the value $-\frac{1}{2}\hbar$). On the other hand, if one makes a measurement of J_y, we should express $|\,P\rangle$ as a linear combination of $|\,\hat{\mathbf{y}}\uparrow\rangle$ and $|\,\hat{\mathbf{y}}\downarrow\rangle$ states; simple algebra will show:

$$|\,P\rangle = \frac{1}{\sqrt{10}}\begin{pmatrix}i\\ -3\end{pmatrix} = \frac{i-3}{\sqrt{20}}|\,\hat{\mathbf{y}}\uparrow\rangle + \frac{i+3}{\sqrt{20}}|\,\hat{\mathbf{y}}\downarrow\rangle$$

Thus if a system is in the state $|P\rangle$ and if one makes a measurement of J_y, then the probability of collapsing to the $|\hat{\mathbf{y}}\uparrow\rangle$ state (and obtaining the value $+\frac{1}{2}\hbar$) is $\left|\frac{i-3}{\sqrt{20}}\right|^2 \left(=\frac{1}{2}\right)$ and probability of collapsing to the $|\hat{\mathbf{y}}\downarrow\rangle$ (and obtaining the value $-\frac{1}{2}\hbar$) is $\left|\frac{i+3}{\sqrt{20}}\right|^2 \left(=\frac{1}{2}\right)$.

13.4 ANGULAR MOMENTUM MATRICES CORRESPONDING TO $j = 1$

In this section, we will consider the matrix representation of the components of the angular momentum operator corresponding to $j = 1$, in a system in which $\hat{J}^2$ and $\hat{J}_z$ are diagonal, i.e., we will choose as basis the simultaneous eigenfunctions of $\hat{J}^2$ and $\hat{J}_z$. Now, for $j = 1$ we have $m = 1$, 0 and -1; thus we have three base states which we denote by $|1\rangle$, $|2\rangle$ and $|3\rangle$:

$$\left.\begin{aligned} |1\rangle &= |j=1, m=1\rangle = |1,1\rangle \\ |2\rangle &= |j=1, m=0\rangle = |1,0\rangle \\ |3\rangle &= |j=1, m=-1\rangle = |1,-1\rangle \end{aligned}\right\} \tag{49}$$

Obviously, $\hat{J}_z|1\rangle = \hbar|1\rangle$, $\hat{J}_z|2\rangle = 0$ and $\hat{J}_z|3\rangle = -\hbar|1\rangle$

and $\hat{J}^2|1\rangle = 2\hbar^2|1\rangle$, $\hat{J}^2|2\rangle = 2\hbar^2|2\rangle$ and $\hat{J}^2|3\rangle = 2\hbar^2|3\rangle$

Thus when the base kets are $|1\rangle$, $|2\rangle$ and $|3\rangle$, the matrices representing $\hat{J}^2$ and $\hat{J}_z$ are diagonal and are given by

$$\hat{J}_z = \hbar\begin{pmatrix} 1 & 0 & 0 \\ 0 & 0 & 0 \\ 0 & 0 & -1 \end{pmatrix} \tag{50}$$

and

$$\hat{J}^2 = 2\hbar^2\begin{pmatrix} 1 & 0 & 0 \\ 0 & 1 & 0 \\ 0 & 0 & 1 \end{pmatrix} \tag{51}$$

Further $\hat{J}_+|1\rangle = 0$, $\hat{J}_+|2\rangle = \sqrt{2}\hbar|1\rangle$ and $\hat{J}_+|3\rangle = \sqrt{2}\hbar|2\rangle$ (52)

and $\hat{J}_-|1\rangle = \sqrt{2}\hbar|2\rangle$, $\hat{J}_-|2\rangle = \sqrt{2}\hbar|3\rangle$ and $\hat{J}_-|3\rangle = 0$ (53)

Thus, for example,

$$\langle 1|\hat{J}_x|2\rangle = \frac{1}{2}\langle 1|\hat{J}_+ + \hat{J}_-|2\rangle = \frac{1}{2}\left[\sqrt{2}\hbar\langle 1|1\rangle + \sqrt{2}\hbar\langle 1|3\rangle\right]$$

giving

$$\langle 1|\hat{J}_x|2\rangle = \frac{1}{\sqrt{2}}\hbar$$

Similarly, other matrix elements can be calculated. The final result is

$$\hat{J}_x = \frac{\hbar}{\sqrt{2}}\begin{pmatrix} 0 & 1 & 0 \\ 1 & 0 & 1 \\ 0 & 1 & 0 \end{pmatrix} \quad \text{(matrix representing } \hat{J}_x\text{)} \tag{54}$$

Similarly,
$$\hat{J}_y = \frac{\hbar}{\sqrt{2}}\begin{pmatrix} 0 & -i & 0 \\ i & 0 & -i \\ 0 & i & 0 \end{pmatrix} \quad \text{(matrix representing } \hat{J}_y\text{)} \tag{55}$$

The eigenvalues of the 3 square matrices in Eqs. (50), (54) and (55) are (1, 0, –1), $\left(\sqrt{2}, 0, -\sqrt{2}\right)$ and $\left(\sqrt{2}, 0, -\sqrt{2}\right)$ respectively. Thus the eigenvalues of $\hat{J}_z$, $\hat{J}_x$ and $\hat{J}_y$ are ($\hbar$, 0, $-\hbar$) and therefore if we make a measurement of any Cartesian component of the angular momentum vector, we will obtain either $\hbar$ or 0 or $-\hbar$. Further,

$$|1\rangle = |1,1\rangle = \begin{pmatrix} 1 \\ 0 \\ 0 \end{pmatrix}, \quad |2\rangle = |1,0\rangle = \begin{pmatrix} 0 \\ 1 \\ 0 \end{pmatrix} \quad \text{and} \quad |3\rangle = |1,-1\rangle = \begin{pmatrix} 0 \\ 0 \\ 1 \end{pmatrix} \tag{56}$$

represent simultaneous eigenkets of $\hat{J}_z$ and $\hat{J}^2$ (but not eigenkets of $\hat{J}_x$ and $\hat{J}_y$); they also form a complete set of orthonormal eigenkets. It is interesting to note that

$$\hat{J}_+ = \hat{J}_x + i\hat{J}_y = \frac{\hbar}{\sqrt{2}}\begin{pmatrix} 0 & 2 & 0 \\ 0 & 0 & 2 \\ 0 & 0 & 0 \end{pmatrix} \Rightarrow \hat{J}_+|1\rangle = \frac{\hbar}{\sqrt{2}}\begin{pmatrix} 0 & 2 & 0 \\ 0 & 0 & 2 \\ 0 & 0 & 0 \end{pmatrix}\begin{pmatrix} 1 \\ 0 \\ 0 \end{pmatrix} = \begin{pmatrix} 0 \\ 0 \\ 0 \end{pmatrix} \tag{57}$$

Similarly, $\hat{J}_-|3\rangle$ is also a null ket.

13.5 THE ROTATION OPERATOR AND ANGULAR MOMENTUM

We will first introduce the operator which rotates a vector by an angle α about an axis (which we will choose as the z-axis) and then we will relate the angular momentum operator to rotations.

13.5.1 Rotation of a Vector

We first consider rotation of a vector **A** about an axis by an angle α; we choose this axis as the z-axis. The rotated vector is denoted by **A′**. Using spherical polar coordinates (r, θ, ϕ), we may write

$$\left.\begin{aligned} A_x &= A \sin\theta \cos\phi \\ A_y &= A \sin\theta \sin\phi \\ A_z &= A \cos\theta \end{aligned}\right\} \tag{58}$$

and (see Fig. 13.1)

$$\left.\begin{aligned} A_x' &= A \sin\theta \cos(\phi + \alpha) = A_x \cos\alpha - A_y \sin\alpha \\ A_y' &= A \sin\theta \sin(\phi + \alpha) = A_x \sin\alpha + A_y \cos\alpha \\ A_z' &= A \cos\theta = A_z \end{aligned}\right\} \tag{59}$$

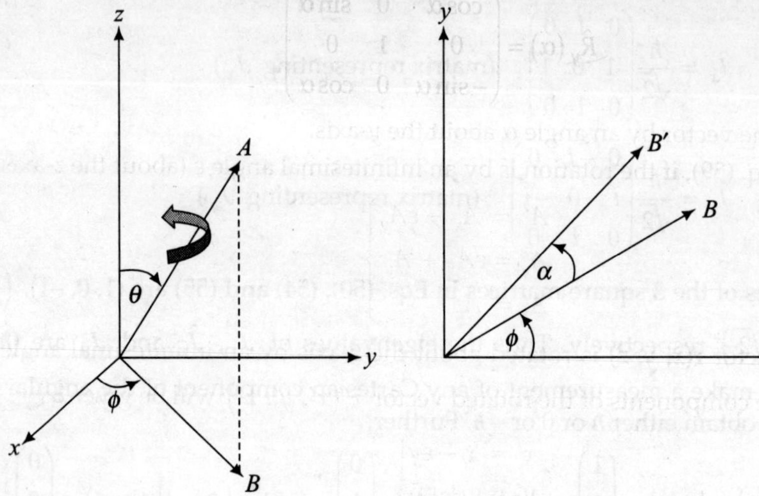

Fig. 13.1 When the vector **A** gets rotated by an angle α about the z-axis, its projection on the x-y plane (vector **B**) also gets rotated by the same angle α; however, the z-component of **A** remains the same.

Obviously, $$A_x^2 + A_y^2 + A_z^2 = A_x'^2 + A_y'^2 + A_z'^2 = A^2 \tag{60}$$

Thus we may write

$$\begin{pmatrix} A_x' \\ A_y' \\ A_z' \end{pmatrix} = \hat{R}_z(\alpha) \begin{pmatrix} A_x \\ A_y \\ A_z \end{pmatrix} \tag{61}$$

where $$\hat{R}_z(\alpha) = \begin{pmatrix} \cos\alpha & -\sin\alpha & 0 \\ \sin\alpha & \cos\alpha & 0 \\ 0 & 0 & 1 \end{pmatrix} \tag{62}$$

is the (3×3) matrix which rotates the vector by an angle α about the z-axis. In order to consider rotation of the vector **A** about the x-axis by an angle α; we will replace $z \to x$, $x \to y$ and $y \to z$ in Eq. (59) to obtain

$$\left.\begin{aligned} A_y' &= A_y \cos\alpha - A_z \sin\alpha \\ A_z' &= A_y \sin\alpha + A_z \cos\alpha \\ A_x' &= A_x \end{aligned}\right\}$$

Thus, $$\begin{pmatrix} A_x' \\ A_y' \\ A_z' \end{pmatrix} = \hat{R}_x(\alpha) \begin{pmatrix} A_x \\ A_y \\ A_z \end{pmatrix}$$

where $$\hat{R}_x(\alpha) = \begin{pmatrix} 1 & 0 & 0 \\ 0 & \cos\alpha & -\sin\alpha \\ 0 & \sin\alpha & \cos\alpha \end{pmatrix} \tag{63}$$

is the matrix which rotates the vector by an angle α about the x-axis. Similarly,

$$\hat{R}_y(\alpha) = \begin{pmatrix} \cos\alpha & 0 & \sin\alpha \\ 0 & 1 & 0 \\ -\sin\alpha & 0 & \cos\alpha \end{pmatrix} \tag{64}$$

would rotate the vector by an angle α about the y-axis.

Returning to Eq. (59), if the rotation is by an infinitesimal angle ε (about the z-axis), then

$$\left.\begin{aligned} A'_x &= A_x - \varepsilon A_y \\ A'_y &= \varepsilon A_x + A_y \\ A'_z &= A_z \end{aligned}\right\} \tag{65}$$

Thus, if the vector $\mathbf{r}(x, y, z)$ is rotated about the z-axis by an infinitesimal angle ε about the z-axis, then the components of the rotated vector $\mathbf{r}'(x', y', z')$ will be given by

$$\left.\begin{aligned} x' &= x - \varepsilon y \\ y' &= y + \varepsilon x \\ z' &= z \end{aligned}\right\} \tag{66}$$

Now, if $|x, y, z\rangle$ represents the state in which the particle is localized around the point (x, y, z) then

$$\hat{R}_z(\varepsilon)|x, y, z\rangle = |x - \varepsilon y, y + \varepsilon x, z\rangle \tag{67}$$

where the operator $\hat{R}_z(\varepsilon)$ rotates the point of localization by an infinitesimal angle ε about the z-axis. The above equation implies that the effect of an infinitesimal rotation is a translation along the x-axis by a distance $-\varepsilon y$ and a translation along the y-axis by a distance εx; thus we may write

$$\hat{R}_z(\varepsilon)|x, y, z\rangle = \hat{T}_x(-\varepsilon y)\,\hat{T}_y(\varepsilon x)\,|x, y, z\rangle \tag{68}$$

where $\hat{T}_x(-\varepsilon y)$ represents the translation operator by a distance $-\varepsilon y$ along the x-axis; similarly $\hat{T}_y(\varepsilon x)$ represents the translation operator by a distance εx along the y-axis. If we now use the results of Sections 4.7 and 11.10 we may write

$$\begin{aligned} \hat{R}_z(\varepsilon) = \hat{T}_x(-\varepsilon y)\,\hat{T}_y(\varepsilon x) &= \left[1 + \frac{i\varepsilon y}{\hbar}\hat{p}_x\right]\left[1 - \frac{i\varepsilon x}{\hbar}\hat{p}_y\right] \\ &= 1 - \frac{i\varepsilon}{\hbar}\hat{L}_z \end{aligned} \tag{69}$$

where we have neglected terms of order ε^2 and we have used the fact that (see Chapter 9)

$$\hat{L}_z \equiv x\hat{p}_y - y\hat{p}_x \tag{70}$$

Many authors *define* the angular momentum operator $\hat{J}_z$ through the rotation operator (see, e.g., References 2–4); i.e.,

$$\hat{R}_z(\varepsilon) = 1 - \frac{i\varepsilon}{\hbar}\hat{J}_z \tag{71}$$

Equation (71) represents the generator of infinitesimal rotations. In order to determine $\hat{R}_z(\alpha)$ for finite rotations (about the z-axis), we note that the effect of an infinitesimal increase in α, from α to $\alpha + \varepsilon$, is to follow the finite rotation $\hat{R}_z(\alpha)$ by the infinitesimal rotation $\hat{R}_z(\varepsilon)$ to give

$$\hat{R}_z(\alpha+\varepsilon) = \hat{R}_z(\varepsilon)\,\hat{R}_z(\alpha) = \left[1 - \frac{i\varepsilon}{\hbar}\hat{J}_z\right]\hat{R}_z(\alpha)$$

Thus,
$$\frac{d\hat{R}_z}{d\alpha} = \underset{\varepsilon\to 0}{Lt}\ \frac{\hat{R}_z(\alpha+\varepsilon) - \hat{R}_z(\alpha)}{\varepsilon} = -\frac{i}{\hbar}\hat{J}_z\,\hat{R}_z(\alpha) \tag{72}$$

We can formally integrate the above equation to obtain

$$\hat{R}_z(\alpha) = \exp\left[-\frac{i\alpha}{\hbar}\hat{J}_z\right] \tag{73}$$

where the exponential of the operator $\mathcal{O}$ is defined through the equation

$$e^{\mathcal{O}} \equiv 1 + \frac{1}{1!}\mathcal{O} + \frac{1}{2!}\mathcal{O}\mathcal{O} + \frac{1}{3!}\mathcal{O}\mathcal{O}\mathcal{O} + \cdots \tag{74}$$

13.5.2 Effect of the Rotation Operator on $|\hat{\mathbf{z}}\uparrow\rangle, |\hat{\mathbf{z}}\downarrow\rangle, |\hat{\mathbf{x}}\uparrow\rangle$ and $|\hat{\mathbf{x}}\uparrow\rangle$ States

In Section 13.3, we had considered the $j = \frac{1}{2}$ problem where we had shown that eigenvalues of $\hat{J}_z$ are $\pm\frac{1}{2}\hbar$ and the corresponding eigenkets are $|\hat{\mathbf{z}}\uparrow\rangle$ and $|\hat{\mathbf{z}}\downarrow\rangle$:

$$\hat{J}_z|\hat{\mathbf{z}}\uparrow\rangle = +\frac{1}{2}\hbar|\hat{\mathbf{z}}\uparrow\rangle \quad \text{and} \quad \hat{J}_z|z\downarrow\rangle = -\frac{1}{2}\hbar|\hat{\mathbf{z}}\downarrow\rangle \tag{75}$$

Similarly,

$$\hat{J}_x|\hat{\mathbf{x}}\uparrow\rangle = +\frac{1}{2}\hbar|\hat{\mathbf{x}}\uparrow\rangle \quad \text{and} \quad \hat{J}_x|\hat{\mathbf{x}}\downarrow\rangle = -\frac{1}{2}\hbar|\hat{\mathbf{x}}\downarrow\rangle \tag{76}$$

and
$$\hat{J}_y|\hat{\mathbf{y}}\uparrow\rangle = +\frac{1}{2}\hbar|\hat{\mathbf{y}}\uparrow\rangle \quad \text{and} \quad \hat{J}_y|\hat{\mathbf{y}}\downarrow\rangle = -\frac{1}{2}\hbar|\hat{\mathbf{y}}\downarrow\rangle \tag{77}$$

Example 13.2: In this example, we will discuss the operation of the rotation operator $\hat{R}_z(\alpha)$ on eigenkets of $\hat{J}_z$; similarly we can discuss the operation on eigenkets of $\hat{J}_x$ and $\hat{J}_y$. Let

$$|P\rangle \equiv \hat{R}_z(\alpha)|\hat{\mathbf{z}}\uparrow\rangle \tag{78}$$

Now,
$$\hat{R}_z(\alpha) = \exp\left[-\frac{i\alpha}{\hbar}\hat{J}_z\right]$$

$$= \left[1 + \frac{1}{1!}\left(-\frac{i\alpha}{\hbar}\right)\hat{J}_z + \frac{1}{2!}\left(-\frac{i\alpha}{\hbar}\right)^2\hat{J}_z\hat{J}_z + \frac{1}{3!}\left(-\frac{i\alpha}{\hbar}\right)^3\hat{J}_z\hat{J}_z\hat{J}_z + \cdots\right] \tag{79}$$

Operation of any term (inside the square brackets) on $|\hat{\mathbf{z}}\uparrow\rangle$ can easily be evaluated; for example,

$$\hat{J}_z\hat{J}_z\hat{J}_z|\hat{\mathbf{z}}\uparrow\rangle = \frac{1}{2}\hbar\,\hat{J}_z\hat{J}_z|\hat{\mathbf{z}}\rangle = \left(\frac{1}{2}\hbar\right)^2\hat{J}_z|\hat{\mathbf{z}}\uparrow\rangle = \left(\frac{1}{2}\hbar\right)^3|\hat{\mathbf{z}}\uparrow\rangle$$

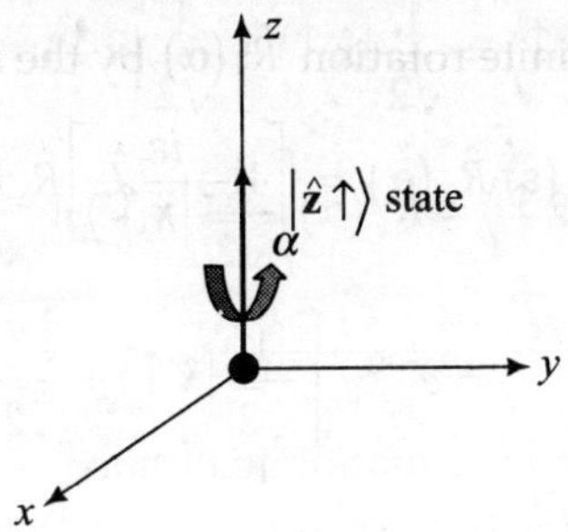

Fig. 13.2 If we apply the rotation operator $\hat{R}_z(\alpha)$ on the $|\hat{\mathbf{z}}\uparrow\rangle$ state, the state does not change except that it gets an additional phase.

Thus,
$$|P\rangle = \left[1 + \frac{1}{1!}\left(-\frac{i\alpha}{2}\right) + \frac{1}{2!}\left(-\frac{i\alpha}{2}\right)^2 + \frac{1}{3!}\left(-\frac{i\alpha}{3}\right)^3 + \cdots\right]|\hat{\mathbf{z}}\uparrow\rangle$$
$$= e^{-i\alpha/2}|\hat{\mathbf{z}}\uparrow\rangle \quad (80)$$

Thus, we get the result that the rotation operator (about the z-axis) leaves $|\hat{\mathbf{z}}\uparrow\rangle$ state unchanged except for a change of phase. However, for $\alpha = 2\pi$, we get

$$\hat{R}_z(2\pi)|\hat{\mathbf{z}}\uparrow\rangle = e^{-i\pi}|\hat{\mathbf{z}}\uparrow\rangle = -|\hat{\mathbf{z}}\uparrow\rangle \quad (81)$$

A rotation by 2π should have left the state unchanged but we find that the state picks up a phase of π. This particular consequence has been verified through an interference experiment by Werner and his collaborators (Ref. 5); a discussion of the experiment has been made in Ref. 2. Equations (78) and (80) give us

$$\hat{R}_z(\alpha)|\hat{\mathbf{z}}\uparrow\rangle = e^{-i\alpha/2}|\hat{\mathbf{z}}\uparrow\rangle \quad (82)$$

Similarly,
$$\hat{R}_z(\alpha)\,|\hat{\mathbf{z}}\downarrow\rangle = e^{+i\alpha/2}|\hat{\mathbf{z}}\downarrow\rangle \quad (83)$$

showing that $|\hat{\mathbf{z}}\uparrow\rangle$ and $|\hat{\mathbf{z}}\downarrow\rangle$ are eigenkets of the rotation operator $\hat{R}_z(\alpha)$ with eigenvalues $e^{-i\alpha/2}$ and $e^{+i\alpha/2}$ respectively.

Example 13.3: In this example, we will discuss the effect of the rotation operator $\hat{R}_x(\alpha)$ on $|\hat{\mathbf{z}}\uparrow\rangle$:

$$|Q\rangle = \hat{R}_x(\alpha)\,|\hat{\mathbf{z}}\uparrow\rangle \quad (84)$$

Now, the eigenstates of $\hat{R}_x(\alpha)$ are $|\hat{\mathbf{x}}\uparrow\rangle$ and $|\hat{\mathbf{x}}\downarrow\rangle$:

$$\hat{R}_x(\alpha)|\hat{\mathbf{x}}\uparrow\rangle = e^{-i\alpha/2}|\hat{\mathbf{x}}\uparrow\rangle \quad \text{and} \quad \hat{R}_x(\alpha)|\hat{\mathbf{x}}\downarrow\rangle = e^{+i\alpha/2}|\hat{\mathbf{x}}\downarrow\rangle \quad (85)$$

Thus, in order to evaluate $\hat{R}_x(\alpha)|\hat{\mathbf{z}}\uparrow\rangle$, we must express $|\hat{\mathbf{z}}\uparrow\rangle$ as a linear combination of $|\hat{\mathbf{x}}\uparrow\rangle$ and $|\hat{\mathbf{x}}\downarrow\rangle$ which are the eigenstates of $\hat{R}_x(\alpha)$:

$$|\hat{\mathbf{z}}\uparrow\rangle = \frac{1}{\sqrt{2}}|\hat{\mathbf{x}}\uparrow\rangle + \frac{1}{\sqrt{2}}|\hat{\mathbf{x}}\downarrow\rangle \tag{86}$$

Thus,
$$|Q\rangle = \hat{R}_x(\alpha)|\hat{\mathbf{z}}\uparrow\rangle = \hat{R}_x(\alpha)\left[\frac{1}{\sqrt{2}}|\hat{\mathbf{x}}\uparrow\rangle + \frac{1}{\sqrt{2}}|\hat{\mathbf{x}}\downarrow\rangle\right]$$
$$= e^{-i\alpha/2}\left[\frac{1}{\sqrt{2}}|\hat{\mathbf{x}}\uparrow\rangle + \frac{e^{i\alpha}}{\sqrt{2}}|\hat{\mathbf{x}}\downarrow\rangle\right]$$

If $\alpha = \pi/2$, then

$$|Q\rangle = \frac{e^{-i\pi/4}}{\sqrt{2}}\left[\frac{1}{\sqrt{2}}\begin{pmatrix}1\\1\end{pmatrix} + \frac{i}{\sqrt{2}}\begin{pmatrix}1\\-1\end{pmatrix}\right] = \frac{e^{-i\pi/4}}{\sqrt{2}}\cdot\frac{1}{\sqrt{2}}\begin{pmatrix}1+i\\1-i\end{pmatrix}$$
$$= \frac{e^{-i\pi/4}}{\sqrt{2}}\cdot\frac{1+i}{\sqrt{2}}\begin{pmatrix}1\\-i\end{pmatrix} = e^{-i\pi/4}\,.\,e^{+i\pi/4}|\hat{\mathbf{y}}\downarrow\rangle$$

where we have used the relation

$$|\hat{\mathbf{y}}\downarrow\rangle = \frac{1}{\sqrt{2}}\begin{pmatrix}1\\-i\end{pmatrix} \tag{87}$$

Thus,
$$|Q\rangle = \hat{R}_x\left(\frac{\pi}{2}\right)|\hat{\mathbf{z}}\uparrow\rangle = |\hat{\mathbf{y}}\downarrow\rangle \tag{88}$$

implying that if we rotate the $|\hat{\mathbf{z}}\uparrow\rangle$ state about the x-axis by an angle $\pi/2$ we would get the $|\hat{\mathbf{y}}\downarrow\rangle$ state – which is to be expected. Similarly, one can show that

$$\hat{R}_z(\alpha)|\hat{\mathbf{x}}\uparrow\rangle = e^{-i\alpha/2}\left[\frac{1}{\sqrt{2}}|\hat{\mathbf{z}}\uparrow\rangle + \frac{e^{i\alpha}}{\sqrt{2}}|\hat{\mathbf{z}}\downarrow\rangle\right] \tag{89}$$

Thus,
$$\hat{R}_z\left(\frac{\pi}{2}\right)|\hat{\mathbf{x}}\uparrow\rangle = e^{-i\pi/4}|\hat{\mathbf{y}}\uparrow\rangle \tag{90}$$

The recipe is: If we wish to evaluate $\hat{R}_x(\alpha)|P\rangle$, then we have to express $|P\rangle$ as a linear combination of the eigenkets of $\hat{R}_x(\alpha)$; i.e., as a linear combination of $|\hat{\mathbf{x}}\uparrow\rangle$ and $|\hat{\mathbf{x}}\downarrow\rangle$. Similarly, if we wish to evaluate $\hat{R}_y(\alpha)|P\rangle$, then we have to express $|P\rangle$ as a linear combination of the eigenkets of $\hat{R}_y(\alpha)$.

Example 13.4: In this example, we will show that the operator $\hat{R}_z(\alpha)$ is not a real operator. Let

$$|P\rangle \equiv \hat{R}_z(\alpha)|\hat{\mathbf{z}}\uparrow\rangle = e^{-i\alpha/2}|\hat{\mathbf{z}}\uparrow\rangle \tag{91}$$

then
$$\langle P| = \langle\hat{\mathbf{z}}\uparrow|\overline{\hat{R}_z}(\alpha) = e^{+i\alpha/2}\langle\hat{\mathbf{z}}\uparrow| \tag{92}$$

where $\overline{\hat{R}_z}(\alpha)$ is the adjoint of the operator $\hat{R}_z(\alpha)$. Thus,

$$\langle P|P\rangle = \langle\hat{\mathbf{z}}\uparrow|\overline{\hat{R}_z}(\alpha)\hat{R}_z(\alpha)|\hat{\mathbf{z}}\uparrow\rangle = \langle\hat{\mathbf{z}}\uparrow|e^{+i\alpha/2}e^{-i\alpha/2}|\hat{\mathbf{z}}\uparrow\rangle = 1 \tag{93}$$

Obviously if $\hat{R}_z(\alpha)$ was real, then $\overline{\hat{R}_z}(\alpha) = \hat{R}_z(\alpha)$ which would imply

$$\langle \hat{\mathbf{z}}\uparrow|\overline{\hat{R}_z}(\alpha)\hat{R}_z(\alpha)|\hat{\mathbf{z}}\uparrow\rangle = \langle \hat{\mathbf{z}}\uparrow|\hat{R}_z(\alpha)\hat{R}_z(\alpha)|\hat{\mathbf{z}}\uparrow\rangle = \langle \hat{\mathbf{z}}\uparrow|e^{-i\alpha}|\hat{\mathbf{z}}\uparrow\rangle = e^{-i\alpha}$$

which is not consistent with Eq. (93). Similarly, if

$$|P\rangle \equiv \hat{R}_z(\alpha)|\hat{\mathbf{x}}\uparrow\rangle = e^{-i\alpha/2}\left[\frac{1}{\sqrt{2}}|\hat{\mathbf{z}}\uparrow\rangle + \frac{e^{i\alpha}}{\sqrt{2}}|\hat{\mathbf{z}}\downarrow\rangle\right]$$

then
$$\langle P| = \langle \hat{\mathbf{x}}\uparrow|\overline{\hat{R}_z}(\alpha) = e^{+i\alpha/2}\left[\frac{1}{\sqrt{2}}\langle \hat{\mathbf{z}}\uparrow| + \frac{e^{-i\alpha}}{\sqrt{2}}\langle \hat{\mathbf{z}}\downarrow|\right]$$

Thus,
$$\langle P|P\rangle = \langle \hat{\mathbf{x}}\uparrow|\overline{\hat{R}_z}(\alpha)\hat{R}_z(\alpha)|\hat{\mathbf{x}}\uparrow\rangle$$
$$= \left[\frac{1}{\sqrt{2}}\langle \hat{\mathbf{z}}\uparrow| + \frac{e^{-i\alpha}}{\sqrt{2}}\langle \hat{\mathbf{z}}\downarrow|\right]\left[\frac{1}{\sqrt{2}}|\hat{\mathbf{z}}\uparrow\rangle + \frac{e^{i\alpha}}{\sqrt{2}}|\hat{\mathbf{z}}\downarrow\rangle\right] = 1$$

Thus, we may write $\overline{\hat{R}_z}(\alpha)\hat{R}_z(\alpha) = 1$.

Example 13.5: In this example, we will use the matrix representation of the rotation operators [namely, Eqs. (62), (63) and (64)] and the relationship of the rotation operator with the angular momentum operators [see Eq. (71)]:

$$\hat{R}_x(\varepsilon) = 1 - \frac{i\varepsilon}{\hbar}\hat{J}_x,\quad \hat{R}_y(\varepsilon) = 1 - \frac{i\varepsilon}{\hbar}\hat{J}_y \quad\text{and}\quad \hat{R}_z(\varepsilon) = 1 - \frac{i\varepsilon}{\hbar}\hat{J}_z \tag{94}$$

to derive the commutation relations between $\hat{J}_x, \hat{J}_y$ and $\hat{J}_z$. For rotation about the z-axis by an infinitesimal angle ε, the rotation operator is given by [see Eq. (62)]:

$$\hat{R}_z(\varepsilon) = \begin{pmatrix} 1-\frac{1}{2}\varepsilon^2 & -\varepsilon & 0 \\ \varepsilon & 1-\frac{1}{2}\varepsilon^2 & 0 \\ 0 & 0 & 1 \end{pmatrix} \tag{95}$$

where we have assumed

$$\cos\varepsilon \approx 1-\tfrac{1}{2}\varepsilon^2 \quad\text{and}\quad \sin\varepsilon \approx \varepsilon$$

and neglected terms proportional to ε^3 and the higher powers of ε. Similarly, for rotation about the x and y axes by infinitesimal angle ε, the corresponding rotation operators will be given by [see Eqs.(63) and (64)]:

$$\hat{R}_x(\varepsilon) = \begin{pmatrix} 1 & 0 & 0 \\ 0 & 1-\frac{1}{2}\varepsilon^2 & -\varepsilon \\ 0 & \varepsilon & 1-\frac{1}{2}\varepsilon^2 \end{pmatrix} \quad\text{and}\quad \hat{R}_y(\varepsilon) = \begin{pmatrix} 1-\frac{1}{2}\varepsilon^2 & 0 & \varepsilon \\ 0 & 1 & 0 \\ -\varepsilon & 0 & 1-\frac{1}{2}\varepsilon^2 \end{pmatrix} \tag{96}$$

Simple multiplication of the matrices will give us

$$\hat{R}_x(\varepsilon)\hat{R}_y(\varepsilon)-\hat{R}_y(\varepsilon)\hat{R}_x(\varepsilon) \approx \begin{pmatrix} 0 & -\varepsilon^2 & 0 \\ -\varepsilon^2 & 0 & 0 \\ 0 & 0 & 0 \end{pmatrix} \approx \hat{R}_z(\varepsilon^2)-1 \tag{97}$$

where, once again, we have neglected terms proportional to ε^3 and the higher powers of ε; and 1 on the right hand side represents a unit 3 × 3 matrix. Using Eq. (94), the right hand side of Eq. (97) will be given by

$$\hat{R}_z(\varepsilon^2)-1=-\frac{i\varepsilon^2}{\hbar}\hat{J}_z \tag{98}$$

On the other hand, the left hand side of Eq. (97) will be given by

$$\hat{R}_x(\varepsilon)\hat{R}_y(\varepsilon)-\hat{R}_y(\varepsilon)\hat{R}_x(\varepsilon) = \left(1-\frac{i\varepsilon}{\hbar}\hat{J}_x\right)\left(1-\frac{i\varepsilon}{\hbar}\hat{J}_y\right)-\left(1-\frac{i\varepsilon}{\hbar}\hat{J}_y\right)\left(1-\frac{i\varepsilon}{\hbar}\hat{J}_x\right)$$

Thus,
$$-\frac{i\varepsilon^2}{\hbar}\hat{J}_z = -\frac{\varepsilon^2}{\hbar^2}\left(\hat{J}_x\hat{J}_y-\hat{J}_y\hat{J}_x\right) \tag{99}$$

giving,
$$\left[\hat{J}_x,\hat{J}_y\right]=\left(\hat{J}_x\hat{J}_y-\hat{J}_y\hat{J}_x\right)=i\hbar\hat{J}_z \tag{100}$$

Example 13.6: If we evaluate $\hat{R}_y(\varepsilon)\hat{R}_z(\varepsilon)-\hat{R}_z(\varepsilon)\hat{R}_y(\varepsilon)$, we can derive the commutation relation $\left[\hat{J}_y,\hat{J}_z\right]=i\hbar\hat{J}_x$. Similarly, evaluation of $\hat{R}_z(\varepsilon)\hat{R}_x(\varepsilon)-\hat{R}_x(\varepsilon)\hat{R}_z(\varepsilon)$ will give us the commutation relation $\left[\hat{J}_z,\hat{J}_x\right]=i\hbar\hat{J}_y$.

Example 13.7: In Section 11.10, we had introduced the translation operator $\hat{T}(a)$ which would change the state $|x\rangle$ into another state which would correspond to the particle localized at the point $x + a$; thus

$$\hat{T}(a)|x\rangle = |x+a\rangle$$

Using the Schrödinger wavefunction, we will have (see Section 4.9):

$$\hat{T}(a)\,\psi(x) = \psi(x-a)$$

Now, as discussed in Section 13.6.1, if $|x, y, z\rangle$ represents the state in which the particle is localized around the point (x, y, z) then

$$\hat{R}_z(\varepsilon)|x,y,z\rangle = |x-\varepsilon y,\, y+\varepsilon x,\, z\rangle \tag{101}$$

where the operator $\hat{R}_z(\varepsilon)$ rotates the point of localization by an infinitesimal angle ε about the z-axis. Thus operating on the Schrödinger wavefunction,

$$\hat{R}_z(\varepsilon)\psi(x,y,z)=\psi(x+\varepsilon y,\, y-\varepsilon x,\, z) \tag{102}$$

Similarly,
$$\hat{R}_x(\varepsilon)\psi(x,y,z)=\psi(x,\, y+\varepsilon z,\, z-\varepsilon y) \tag{103}$$

and
$$\hat{R}_z(\varepsilon)\psi(x,y,z)=\psi(x-\varepsilon z,\, y,\, z+\varepsilon x) \tag{104}$$

Let us consider rotation about the x-axis:

$$\hat{R}_x(\varepsilon)\psi(x,y,z)=\psi(x,\,y+\varepsilon z,\,z-\varepsilon y)$$

Now, Change in $z=\Delta(r\cos\theta)=-\varepsilon y \;\Rightarrow -\sin\theta\,\Delta\theta=-\varepsilon\sin\theta\,\sin\phi$

$$\Rightarrow\;\Delta\theta=\varepsilon\,\sin\phi \tag{105}$$

Change in $x=\Delta(r\sin\theta\cos\phi)=0\;\Rightarrow\;\cos\theta\,\Delta\theta\,\cos\phi-\sin\theta\sin\phi\,\Delta\phi=0$

$$\Rightarrow\;\Delta\phi=\cot\theta\,\Delta\theta\,\cot\phi=\varepsilon\cot\theta\,\cos\phi \tag{106}$$

Thus, $$\hat{R}_x(\varepsilon)\psi(r,\theta,\phi)=\psi(r,\theta+\Delta\theta,\phi+\Delta\phi)$$

or $$\left(1-\frac{i\varepsilon}{\hbar}\hat{L}_x\right)\psi(r,\theta,\phi)=\psi(r,\theta,\phi)+\frac{\partial\psi}{\partial\theta}\Delta\theta+\frac{\partial\psi}{\partial\phi}\Delta\phi$$

where $\Delta\theta$ and $\Delta\phi$ are given by Eqs. (105) and (106). Thus,

$$-\frac{i\varepsilon}{\hbar}\hat{L}_x\psi=\frac{\partial\psi}{\partial\theta}\varepsilon\,\sin\phi+\frac{\partial\psi}{\partial\phi}\varepsilon\cot\theta\,\cos\phi$$

$$\Rightarrow\quad \hat{L}_x\psi=i\hbar\left(\sin\phi\frac{\partial\psi}{\partial\theta}+\cot\theta\,\cos\phi\frac{\partial\psi}{\partial\phi}\right) \tag{107}$$

which is the representation of $\hat{L}_x$ in spherical coordinates. It is left as an exercise to use a similar method to obtain

$$\hat{L}_y\psi=i\hbar\left(-\cos\phi\frac{\partial\psi}{\partial\theta}+\cot\theta\,\sin\phi\frac{\partial\psi}{\partial\phi}\right) \tag{108}$$

and $$\hat{L}_z\psi=-i\hbar\frac{\partial\psi}{\partial\phi} \tag{109}$$

The above representations of $\hat{L}_x,\hat{L}_y$ and $\hat{L}_z$ are the same as we had obtained in Section 9.2. Also, if we use the equation

$$\text{Change in } y=\Delta(r\sin\theta\sin\phi)=\varepsilon z$$

$$\Rightarrow\quad \cos\theta\,\Delta\theta\,\sin\phi+\sin\theta\,\cos\phi\,\Delta\phi=\varepsilon\cos\theta \tag{110}$$

If we substitute for $\Delta\theta$ and $\Delta\phi$ from Eqs. (105) and (106), we will get an identity.

13.6 THE PAULI WAVEFUNCTION AND THE PAULI EQUATION

The electron is endowed with two internal degrees of freedom corresponding to 2 values of the Cartesian component of its intrinsic angular momentum along a specified axis (say the z-axis); the 2 values are $\pm\frac{1}{2}\hbar$. Accordingly, we need two wavefunctions to describe the state of an electron at each point in space. Writing them in matrix form, we have

$$\psi_P=\begin{pmatrix}\psi_+(x,y,z)\\ \psi_-(x,y,z)\end{pmatrix} \tag{111}$$

Such single column matrices are known as spinors[2] and the wavefunction is known as the Pauli wavefunction (hence the subscript P on ψ). Writing

$$\psi_P = \psi_+\chi_+ + \psi_-\chi_- = \psi_+\begin{pmatrix}1\\0\end{pmatrix} + \psi_-\begin{pmatrix}0\\1\end{pmatrix} \tag{112}$$

we say that any spinor can be written as a linear combination of the basic spinors

$$\chi_+ = \begin{pmatrix}1\\0\end{pmatrix} = |\hat{z}\uparrow\rangle \quad \text{and} \quad \chi_- = \begin{pmatrix}0\\1\end{pmatrix} = |\hat{z}\downarrow\rangle \tag{113}$$

Thus,

$$\int |\psi_+|^2\, d\tau \quad \text{and} \quad \int |\psi_-|^2\, d\tau \tag{114}$$

would represent the probabilities of obtaining the $|\hat{z}\uparrow\rangle$ and $|\hat{z}\downarrow\rangle$ states respectively and we must have

$$\int |\psi_+|^2\, d\tau + \int |\psi_-|^2\, d\tau = 1 \tag{115}$$

Since the electron requires information regarding its spin as well as position coordinates, the Pauli wavefunction satisfies the equation

$$i\hbar\frac{\partial}{\partial t}\begin{pmatrix}\psi_+\\ \psi_-\end{pmatrix} = H\begin{pmatrix}\psi_+\\ \psi_-\end{pmatrix} \quad \text{Pauli equation} \tag{116}$$

where H represents the Hamiltonian. For the hydrogen atom, in the absence of any interaction other than the Coulomb interaction, H, will be of the form

$$H\psi_P = \left[-\frac{\hbar^2}{2m}\nabla^2\cdot I + V(r)\cdot I\right]\psi_P \tag{117}$$

where **I** is the 2×2 unit matrix. Thus ψ_+ and ψ_- satisfy the same equation and the spin up and spin down states have the same energy. On the other hand, if we take into account the spin-orbit interaction, H will be of the form

$$H\psi_P = \left[-\frac{\hbar^2}{2m}\nabla^2\cdot I + V(r)\cdot I + H'\right]\psi_P \tag{118}$$

where (see Section 20.5)

$$\begin{aligned}
H' &= \frac{Ze^2}{2m^2c^2}\frac{1}{r^3}\mathbf{L}\cdot\mathbf{s} \\
&= \frac{Ze^2}{2m^2c^2}\frac{1}{r^3}\frac{1}{2}\hbar\left[L_x\sigma_x + L_y\sigma_y + L_z\sigma_z\right] \\
&= \frac{Ze^2}{2m^2c^2}\frac{1}{r^3}\frac{1}{2}\hbar\begin{pmatrix}L_z & L_x - iL_y\\ L_x + iL_y & -L_z\end{pmatrix}
\end{aligned} \tag{119}$$

which is known as the Pauli operator for H'. Obviously, now the equations for ψ_+ and ψ_- will be coupled and one has to use perturbation theory to consider the effect of H'. This will be discussed in Problem 20.12.

---

[2] Spinors are characterized by special properties under rotations. We do not wish to leave the impression that they are just single column matrices.

13.7 SOME HISTORICAL REMARKS

In Section 13.2, we had shown that Cartesian components of the angular momentum operator can have values which are half integral multiples of $\hbar$. Indeed, it was in 1925 that G.E. Uhlenbeck and S. Goudsmit introduced the idea that an electron has a spin angular momentum of $\pm\frac{1}{2}\hbar$. May be it would be of interest to give a brief historical perspective of the development of the subject. To quote Uhlenbeck[3],

... He (Sam Goudsmit) had told me about Pauli's proposal to ascribe four quantum numbers to each electron. He now continued with the discussion of the famous paper of January 1925 in which Pauli formulated the exclusion principle; no two electrons could have the same **four** *quantum numbers. ... Sam himself had simplified the argument by introducing the quantum numbers, n, l, m_l and m_s instead of those used by Pauli, and he noticed that then m_s was always ±1/2.*

I was impressed, but since the whole argument was purely formal, it seemed like abracadabra to me. There was no picture that at least qualitatively connected Pauli's formalism with the old Bohr atomic model. It was then that it occurred to me that, since (as I had learned) each quantum number corresponds to a degree of freedom of the electron, the fourth quantum number must mean that the electron had an additional degree of freedom—in other words, the electron must be rotating! Sam has written that he did not know at that time what a degree of freedom was. This may be so, as Sam had not done his exam in mechanics yet; in fact, he never passed this exam, and as a result he did not have the right to teach mechanics in the Dutch high schools even after he got his Ph.D. However, this did not prevent him later from teaching the graduate course in mechanics at the University of Michigan, which he did regularly because he liked the subject so much; furthermore, it was much appreciated by the students.

In spite of this, he appreciated right away that if the angular momentum of the electron was $\frac{1}{2}\hbar$, one had a picture of the alkali doublets as the two ways the electron could rotate with respect to its orbital motion. In fact, if one assumed that the gyromagnetic ratio for the rotation was twice the classical value, so that the magnetic moment was

$$2\frac{e}{2mc}\frac{1}{2}\hbar = \text{one Böhr magneton}$$

then the properties formerly attributed to the core were now properties of the electron. ...

I remember that when this became clear to us, we had a feeling of euphoria, but we also both agreed that one could not possibly publish such stuff. Since it had not been mentioned by any of the authorities (we did not know about Kronig, of course) it must for some reason be nonsense. But, of course, we told Ehrenfest, who was immediately interested. I am not sure precisely what happened next. Sam is wrong when he writes that he was satisfied and did not think any more about how our model could be justified. I remember that he wrote me a postcard from Amsterdam very soon afterward, in which he asked whether I was sure that the gyromagnetic ratio had to be e/2mc classically—perhaps it was different for the rotation of an extended charged body. I showed this postcard to Ehrenfest, who then recalled a paper by Max Abraham about the magnetic properties of rotating electrons. I studied this paper very hard and found there to my great satisfaction that if the electron has only surface charge the gyromagnetic ratio was 2e/2mc, just as we had postulated! I think that when I showed this to Ehrenfest he thought (as he told us later) that our idea was either very important or nonsense, but that it should be published. The Abraham calculations were nonrelativistic and based on the old-fashioned rigid electron, so that they were at best

---

[3] Adapted from the article by G. H. Uhlenbeck, *Fifty Years of Spin*, Physics Today, p. 43, June 1976.

only suggestive. Anyway, Ehrenfest told us to write a short, modest Letter to Naturwissenschaften and to give it to him. **"Und dann werden wir Herrn Lorentz fragen"** *("And then we will ask Mr. Lorentz."). A letter of 16 October to H.A. Lorentz in which he mentions this (among other things) was found and shown to me by Martin Klein.*

Lorentz, who was of course the great old man of Dutch physics, was retired and lived in Haarlem but gave a lecture in Leiden every Monday at 11:00 am, in which he discussed the recent developments in physics. Everybody who could possibly make it came. So when school started in the middle of October (remember, we had long vacations) I had the opportunity to tell Lorentz about our ideas. Sam was not present because he had to resume his duties at the Zeeman laboratory. Lorentz was very kind and interested, although I also got the impression that he was rather skeptical. He said that he would think about it and that we should talk again the next Monday.

In fact, when we met that day he showed me a stack of papers full of calculations written in his beautiful handwriting, which he tried to explain to me. They were above my head but I understood enough to realize that there were serious difficulties. If the radius of the electron was

$$r_0 = \frac{e^2}{mc^2}$$

and if it rotated with an angular momentum $\frac{1}{2}\hbar$, *then the surface velocity would be about ten times the light velocity! If the electron had a magnetic moment* $e\hbar/2mc$, *its magnetic energy would have to be so big that, to keep the mass m, its radius would have to be at least ten times* r_0.

It seemed to me that if one extended the Abraham calculations properly as Lorentz had apparently done (and published in revised form), then our picture of a quantized rotation of the electron could not possibly be reconciled with classical electrodynamics. I told this to Ehrenfest, of course, and said that his second alternative had turned out to be the right one. The whole thing was nonsense, and it would be better that our Letter not be published. Then, to my surprise, Ehrenfest answered that he had already sent the Letter off quite a while ago, and that it would appear pretty soon. He added: **"Sie sind beide jung genug um sich eine Dummheit leisten zu können!"** *("You are both young enough to be able to afford a stupidity!"). This is not yet the end of the story. Our letter appeared in the middle of November, and soon afterwards (21 November) Goudsmit received a letter from Heisenberg, whom he knew quite well. In this letter Heisenberg expressed his appreciation for Sam's courageous idea and agreed that it would remove all of the difficulties of the Pauli theory.*

13.8 PROBLEMS

Problem 13.1 For an eigenstate of $\hat{J}^2$ and $\hat{J}_z$, show that

$$\langle \hat{J}_x \rangle = 0 = \langle \hat{J}_y \rangle \tag{120}$$

and

$$\langle \hat{J}_x^2 \rangle = \langle \hat{J}_y^2 \rangle = \frac{1}{2}\hbar^2\left[j(j+1) - m^2\right] \tag{121}$$

Form the above results develop a classical model for angular momentum.

Problem 13.2 In Section 13.2, we had derived the angular momentum matrices in a representation in which J^2 and J_z are diagonal. Carry out a similar analysis in a representation in which J^2 and J_z are diagonal; assume $j = \frac{1}{2}$.

Problem 13.3 Show that the Pauli spin matrices satisfy the following relation

$$\sigma_i \, \sigma_j = \delta_{ij} + i\varepsilon_{ijk}\sigma_k \tag{122}$$

where ε_{ijk}, which is known as the totally antisymmetric unit tensor, is defined by

$$\varepsilon_{123} = \varepsilon_{231} = \varepsilon_{312} = 1 \qquad \text{(cyclic indices)}$$

$$\varepsilon_{321} = \varepsilon_{213} = \varepsilon_{132} = -1 \qquad \text{(anticyclic indices)}$$

with all other components equal to zero.

Problem 13.4 Using Eq. (122) show

$$(\boldsymbol{\sigma}\cdot\mathbf{A})(\boldsymbol{\sigma}\cdot\mathbf{B}) = \mathbf{A}\cdot\mathbf{B} + i\boldsymbol{\sigma}\cdot(\mathbf{A}\times\mathbf{B}) \tag{123}$$

where **A** and **B** are arbitrary vectors. Thus,

$$(\boldsymbol{\sigma}\cdot\hat{\mathbf{n}})^2 = 1 \tag{124}$$

where $\hat{\mathbf{n}}$ is an arbitrary unit vector.

Problem 13.5 For $j = \frac{1}{2}$, obtain the eigenvectors of s_y and express them as a linear combination of $|\hat{\mathbf{z}}\uparrow\rangle$ and $|\hat{\mathbf{z}}\downarrow\rangle$ status.

Problem 13.6 We assume a spin $\frac{1}{2}$ system to be in a state described by the (normalized) vector $|P\rangle = \frac{1}{\sqrt{10}}\begin{pmatrix} i \\ 3 \end{pmatrix}$ (see Example 13.1). If we make a measurement of J_x, calculate the probability of collapsing to the $|\hat{\mathbf{x}}\uparrow\rangle$ state and to the $|\hat{\mathbf{x}}\downarrow\rangle$ state.

Problem 13.7 We assume a spin $\frac{1}{2}$ system to be in a state described by the vector

$$|P\rangle = \begin{pmatrix} 1 \\ 2 \end{pmatrix}$$

(a) If we make a measurement of J_x, calculate the probability of collapsing to the $|\hat{\mathbf{x}}\uparrow\rangle$ state and to the $|\hat{\mathbf{x}}\downarrow\rangle$ state. [**Ans:** $\frac{9}{10}$ and $\frac{1}{10}$]

(b) If we make a measurement of J_z, calculate the probability of collapsing to the $|\hat{\mathbf{z}}\uparrow\rangle$ state and to the $|\hat{\mathbf{z}}\downarrow\rangle$ state. [**Ans:** $\frac{1}{5}$ and $\frac{4}{5}$]

Problem 13.8

(a) Using Eq. (37) write the matrices for J_+ and J_- (for $j = \frac{1}{2}$) and show

$$\hat{J}_+|1\rangle = 0$$

$$\hat{J}_+|2\rangle = \hbar|1\rangle$$

$$\hat{J}_-|1\rangle = \hbar|2\rangle$$

$$\hat{J}_-|2\rangle = 0$$

where $|1\rangle$ and $|2\rangle$ are given by Eq. (42).

(b) Similarly, using Eqs. (54) and (55) write the matrices for $\hat{J}_+$ and $\hat{J}_-$ (for $j = 1$) and derive Eqs. (52) and (53).

Problem 13.9 In Sections 13.4 and 13.5, we had derived the angular momentum matrices corresponding to $j = \frac{1}{2}$ and $j = 1$ in a representation in which J^2 and J_z are diagonal. Carry out a similar analysis for $j = 3/2$.

13.9 SOLUTIONS

Solution 13.1

$$\left\langle j, m\middle|\hat{J}_x\middle|j, m\right\rangle = \tfrac{1}{2}\left\langle j, m\middle|\hat{J}_+\middle|j, m\right\rangle + \tfrac{1}{2}\left\langle j, m\middle|\hat{J}_-\middle|j, m\right\rangle = 0 \qquad \text{[using Eqs (30–31)]}$$

$$\left\langle j, m\middle|\hat{J}_x^2\middle|j, m\right\rangle = \tfrac{1}{2}\left[\left\langle j, m\middle|\hat{J}_+\hat{J}_+\middle|j, m\right\rangle + \left\langle j, m\middle|\hat{J}_+\hat{J}_-\middle|j, m\right\rangle + \left\langle j, m\middle|\hat{J}_-\hat{J}_+\middle|j, m\right\rangle + \left\langle j, m\middle|\hat{J}_-\hat{J}_-\middle|j, m\right\rangle\right]$$

$$= \tfrac{1}{2}\Bigg[0 + \{(j+m)(j-m+1)\}^{\frac{1}{2}}\left\langle j, m|J_+|j, m-1\right\rangle$$

$$+ \{(j-m)(j+m+1)\}^{\frac{1}{2}}\left\langle j, m|J_-|j, m+1\right\rangle + 0\Bigg] = \tfrac{1}{2}\hbar^2\left[j(j+1) - m^2\right] \qquad (125)$$

etc. Obviously,

$$\left\langle j, m\middle|\hat{J}_z^2\middle|j, m\right\rangle = m^2\hbar^2 \qquad (126)$$

and

$$\left\langle j, m\middle|\hat{J}^2\middle|j, m\right\rangle = j(j+1)\hbar^2 \qquad (127)$$

From the above equations we can develop a classical model for angular momentum. The total angular momentum vector **OP** (whose length is assumed to be $\sqrt{j(j+1)}\,\hbar$) is assumed to precess about the z-axis as shown in Fig. 13.2. As it precesses, the z-component of **OP** (= OQ) remains constant, which we assume to be equal to $m\hbar$. The x and y components of **OP** change with time and their average value is zero diagonal. Now,

$$PQ^2 = j(j+1)\hbar^2 - m^2\hbar^2$$

represents $\left\langle \hat{J}_x^2\right\rangle + \left\langle \hat{J}_y^2\right\rangle$ consistent with Eq. (125). Thus, for the state we may assume the total angular momentum vector **OP** precessing about the z-axis as shown in Fig. 13.3.

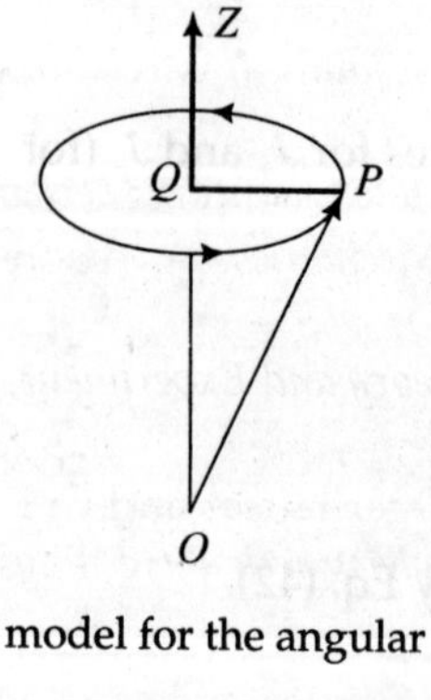

Fig. 13.3 A classical model for the angular momentum vector.

Solution 13.2

$$\hat{J}_x = \frac{1}{2}\hbar\begin{pmatrix}1 & 0\\ 0 & -1\end{pmatrix},\ \hat{J}_y = \frac{1}{2}\hbar\begin{pmatrix}0 & 1\\ 1 & 0\end{pmatrix},\ \hat{J}_z = \frac{1}{2}\hbar\begin{pmatrix}0 & -i\\ i & 0\end{pmatrix}$$

Solution 13.3 The following properties of the Pauli matrices are easily established

$$\sigma_x^2 = \sigma_y^2 = \sigma_z^2 = 1 \tag{128}$$

$$\sigma_x\sigma_y = i\sigma_z, \quad \text{etc.} \tag{129}$$

$$\sigma_x\sigma_y + \sigma_y\sigma_x = 0, \quad \text{etc.} \tag{130}$$

from which Eq. (122) immediately follows.

Solution 13.4

$$(\sigma\cdot\mathbf{A})(\sigma\cdot\mathbf{B}) = \sum_{ij}\sigma_i A_i \sigma_j B_j = (\mathbf{A}\cdot\mathbf{B}) + i\sigma(\mathbf{A}\times\mathbf{B}) \tag{131}$$

it can be verified by expanding the RHS and using Eqs. (128)–(130).

Solution 13.9

$$\hat{J}_x = \frac{1}{2}\hbar\begin{pmatrix}0 & \sqrt{3} & 0 & 0\\ \sqrt{3} & 0 & 2 & 0\\ 0 & 2 & 0 & \sqrt{3}\\ 0 & 0 & \sqrt{3} & 0\end{pmatrix}, \quad \hat{J}_y = \frac{1}{2}\hbar\begin{pmatrix}0 & -i\sqrt{3} & 0 & 0\\ i\sqrt{3} & 0 & -2i & 0\\ 0 & 2i & 0 & -i\sqrt{3}\\ 0 & 0 & i\sqrt{3} & 0\end{pmatrix}$$

$$\hat{J}_z = \hbar\begin{pmatrix}\frac{3}{2} & 0 & 0 & 0\\ 0 & \frac{1}{2} & 0 & 0\\ 0 & 0 & -\frac{1}{2} & 0\\ 0 & 0 & 0 & -\frac{3}{2}\end{pmatrix} \quad \text{and} \quad \hat{J}^2 = \frac{15}{4}\hbar^2\begin{pmatrix}1 & 0 & 0 & 0\\ 0 & 1 & 0 & 0\\ 0 & 0 & 1 & 0\\ 0 & 0 & 0 & 1\end{pmatrix}$$

The eigenvalues of $\hat{J}_x, \hat{J}_y$ and $\hat{J}_z$ are $\frac{3}{2}\hbar$, $\frac{1}{2}\hbar$, $-\frac{1}{2}\hbar$ and $-\frac{3}{2}\hbar$; the eigenvalue of $\hat{J}^2$ is $\frac{15}{4}\hbar^2$.

REFERENCES AND SUGGESTED READINGS

1. P.A.M. Dirac, *The Principles of Quantum Mechanics*, Oxford University Press, Oxford (1958).
2. J.S. Townsend, *A Modern Approach to Quantum Mechanics*, McGraw-Hill, New York (1992).
3. J.J. Sakurai, *Modern Quantum Mechanics*, Addison-Wesley Publishing Company, Inc. (1994).
4. M. Beck, *Quantum Mechanics: Theory and Experiment*, Oxford University Press, New York (2012).
5. S. A. Werner, R. Colella, A. W. Overhauser and C. F. Eagen, *Observation of the Phase Shift of a Neutron Due to Precession in a Magnetic Field*, *Phys. Rev. Lett.* 35, 1053 (1975).

14 Experiments with spin half particles

The Stern-Gerlach experiment, Larmor precession and magnetic resonance

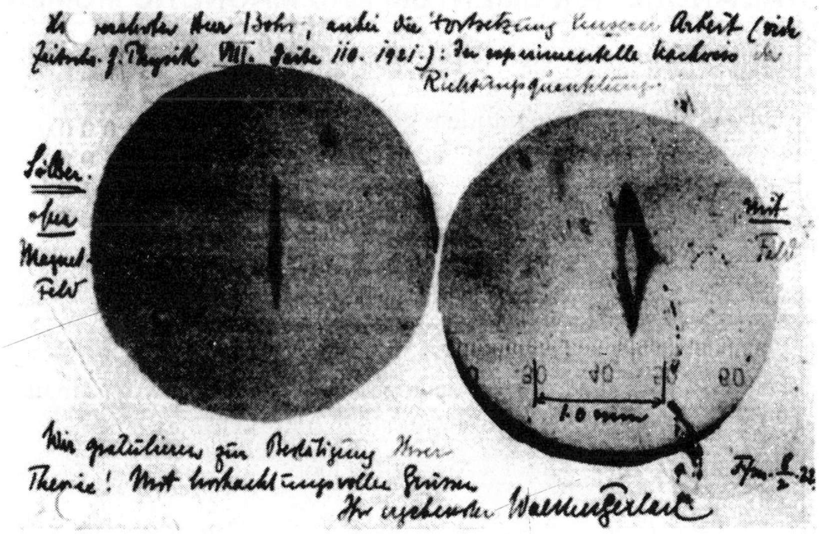

A replica of the postcard sent by Walther Gerlach to Niels Bohr on February 8, 1922. The left-hand image of the beam profile without the magnetic field shows the effect of the finite width of the collimating slit. The right-hand image shows the beam profile with the magnetic field. Only in the center of the apparatus is the magnitude of the magnetic field gradient sufficiently strong to cause splitting. The pattern is smeared because of the range of speeds of the atoms coming from the oven. Translation of the message: **'My esteemed Herr Böhr, attached is the continuation of our work [vide Zeitschr. f. Phys. 8,110 (1921)]: the experimental proof of directional quantization. We congratulate you on the confirmation of your theory! With respectful greetings. Your most humble Walther Gerlach.'**

— Adapted from J.S. Townsend, *A Modern Approach to Quantum Mechanics*, McGraw-Hill, New York (1992). The photograph was kindly provided by the Niels Böhr Archives in Copenhagen.

14.1 INTRODUCTION

The analysis of the *spin half* problem is of great significance because it is not only of tremendous practical importance but it also allows us to understand some of the fundamental postulates of quantum mechanics.

We will first discuss the magnetic moment associated with the orbital angular momentum of a charged particle, this will be followed by the magnetic moment associated with the spin angular momentum of the electron. We will then discuss the famous Stern-Gerlach experiment which had led to the concept of *space quantization*. We will then obtain solutions of the time dependent Schrödinger equation corresponding to the spin half particle in a uniform magnetic field thereby giving a quantum mechanical description of the *Larmor precession*. We will conclude this chapter with an analysis of magnetic resonance experiments and in the process we will obtain rigorous solutions of the Schrödinger equation for a time dependent Hamiltonian.

14.2 ORBITAL ANGULAR MOMENTUM AND MAGNETIC MOMENT

Consider a particle (of mass m and charge Q) rotating in a circular orbit of radius r (see Fig. 14.1). As shown in the figure, the angular momentum vector **L** is perpendicular to the plane of the circular orbit and its magnitude is given by

$$L = mvr \tag{1}$$

The rotating charge constitutes a current I which is given by

$$I = Q \times (\text{Number of rotations in a unit time})$$

$$= Q\frac{v}{2\pi r} \tag{2}$$

Now, at large distances from the loop, the magnetic field produced by the circulating current loop is the same as that by a magnet of dipole moment μ given by

$$\mu = I \times \text{Area of the loop}$$

Thus,

$$\mu = Q\frac{v}{2\pi r}\pi r^2$$

or,

$$\boldsymbol{\mu}_l = \frac{Q}{2m}\mathbf{L} \quad \text{Magnetic moment due to orbital motion} \tag{3}$$

where the subscript l implies that we are considering the magnetic moment due to the orbital motion. Equation (3) is a classical result; however, quantum mechanics also gives the same result. Equation (3) tells us that if Q is positive then μ and **L** are in the same direction and if Q is negative, μ and **L** are in opposite directions.

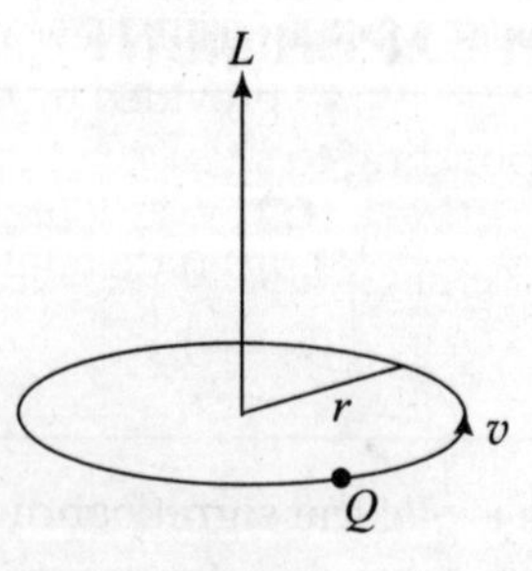

Figure 14.1: A particle (of mass m and charge Q) rotating in a circular orbit of radius r.

14.3 SPIN ANGULAR MOMENTUM AND MAGNETIC MOMENT

As mentioned in Section 13.6, Uhlenbeck and Goudsmit (in 1925) had introduced the idea that an electron has a spin angular momentum of $\pm \frac{1}{2}\hbar$; associated with this spin angular momentum, the electron has a magnetic moment which is given by

$$\mu = -g\frac{q}{2m}\mathbf{s} \tag{4}$$

where q ($\approx 1.6 \times 10^{-19}$ C) represents the magnitude of the electronic charge, m ($\approx 9.1 \times 10^{-31}$ kg) represents the electron mass and g (≈ 2.0023) is known as the *Landé g factor*; **s** represents the spin angular momentum operator. In Section 13.4, we had shown that the cartesian components of the $j = \frac{1}{2}$ angular momentum operator can be represented by 2×2 matrices; these are given by

$$s_x = \frac{1}{2}\hbar\,\sigma_x,\; s_y = \frac{1}{2}\hbar\,\sigma_y,\; s_z = \frac{1}{2}\hbar\,\sigma_z \tag{5}$$

where σ_x, σ_y and σ_z are Pauli spin matrices and are given by

$$\sigma_x = \begin{pmatrix} 0 & 1 \\ 1 & 0 \end{pmatrix},\; \sigma_y = \begin{pmatrix} 0 & -i \\ i & 0 \end{pmatrix},\; \sigma_z = \begin{pmatrix} 1 & 0 \\ 0 & -1 \end{pmatrix} \tag{6}$$

We may mention here that the magnetic moment of the proton is given by

$$\mu_p = g_p\frac{q}{2M_p}\mathbf{s} \tag{7}$$

where q($\approx 1.6 \times 10^{-19}$C) represents the proton charge, M_p($\approx 1.673 \times 10^{-27}$kg) represents the proton mass and $g_p \approx 5.56$. Even the neutron (which has no charge) has a spin magnetic moment given by

$$\mu_n = -g_n\frac{q}{2M_n}\mathbf{s} \tag{8}$$

where q ($\approx 1.6 \times 10^{-19}$C) represents the magnitude of the electron charge, M_n ($\approx 1.675 \times 10^{-27}$kg) represents the neutron mass and $g_n \approx 3.83$. The magnetic moment of the neutron is attributed to its internal charge distribution.

14.4 THE STERN-GERLACH EXPERIMENT

Let us consider the operator corresponding to s_z:

$$s_z = \frac{1}{2}\hbar\begin{pmatrix} 1 & 0 \\ 0 & -1 \end{pmatrix} \tag{9}$$

The eigenvalues of s_z are $+\frac{1}{2}\hbar$ and $-\frac{1}{2}\hbar$; the corresponding (normalized) eigenvectors are

$$|\hat{\mathbf{z}}\uparrow\rangle = \begin{pmatrix} 1 \\ 0 \end{pmatrix} \text{ and } |\hat{\mathbf{z}}\downarrow\rangle = \begin{pmatrix} 0 \\ 1 \end{pmatrix} \tag{10}$$

which are usually referred to as *spin up* and *spin down* states. They are also referred as *z up* and *z down* states. Now, if we try to make a measurement of s_z then it will be one of its eigenvalues, i.e., either $+\frac{1}{2}\hbar$ or $-\frac{1}{2}\hbar$. Indeed this is what happens in the Stern-Gerlach experiment where one passes neutral Ag atoms through an inhomogeneous magnetic field in the *z*-direction. For a silver atom in its ground state, the angular momentum is just the spin angular momentum of the electron[1] and therefore the magnetic moment of the silver atom is simply that of the electron[2]. The atoms behaving like magnets of moment μ, will be acted upon by a force given by

$$\mathbf{F} = -\nabla U = \nabla (\boldsymbol{\mu} \cdot \mathbf{B}) \tag{11}$$

where U $(= -\boldsymbol{\mu} \cdot \mathbf{B})$ represents the interaction energy. We assume the magnetic field to be predominantly in the *z*-direction and therefore[3]

$$\mathbf{F} = \mu_z \nabla B_z \approx \mu_z \frac{\partial B_z}{\partial z} \hat{\mathbf{z}} \tag{12}$$

where μ_z represents the *z*-component of the magnetic moment. Now, travelling a distance l through the region, of the magnetic field, the beam will experience a deflection Δz given by

$$\Delta z = \frac{1}{2}\left(\frac{F_z}{M}\right) t^2 = \frac{1}{2M} \mu_z \frac{\partial B}{\partial z}\left(\frac{l}{v}\right)^2 \tag{13}$$

where M is the mass and v the velocity of the atoms.

The atomic velocities, although not unique, will range around a mean value v determined by the temperature of the oven. In the absence of the magnetic field, the beam, defined by the slits, forms a narrow line. Now, according to the classical theory

$$\mu_z = \mu \cos\theta \tag{14}$$

where θ is the angle that the direction of the magnetic moment makes with the direction of the magnetic field. Thus, as the atoms come out of the oven, their magnetic moment would point in every possible direction and we should expect all values of θ (between 0 and π) to be possible. Consequently, when the field is turned on, the line ought to widen continuously in the *z*-direction, the deflection depending on the value of θ.

---

[1] The silver atom has 47 electrons; 46 electrons form a state with $L = 0$ and $S = 0$, i.e., the total angular momentum of the silver atom (in its ground state) is just that due to the last electron.

[2] The Stern-Gerlach experiment cannot be performed with electrons; this is because of the fact that since the electron has a charge, it will experience a Lorentz force. On the other hand, although the neutral Ag atom possesses the magnetic moment of the electron, it does not have any charge and therefore does not experience any Lorentz force.

[3] In Example 14.2, we will discuss the Larmor precession where we will show that the components of **s** perpendicular to **B** will vary with time and will average out to zero. On the other hand, the component parallel to **B** will remain constant. Thus, although $\left|\frac{\partial B_y}{\partial y}\right| \approx \left|\frac{\partial B_z}{\partial z}\right|$ because of the Maxwell equation $(\nabla \cdot \mathbf{B} = 0)$, since **B** is predominantly in the *z*-direction, we may neglect the terms proportional to μ_x and μ_y in Eq. (12) because they will average out to zero.

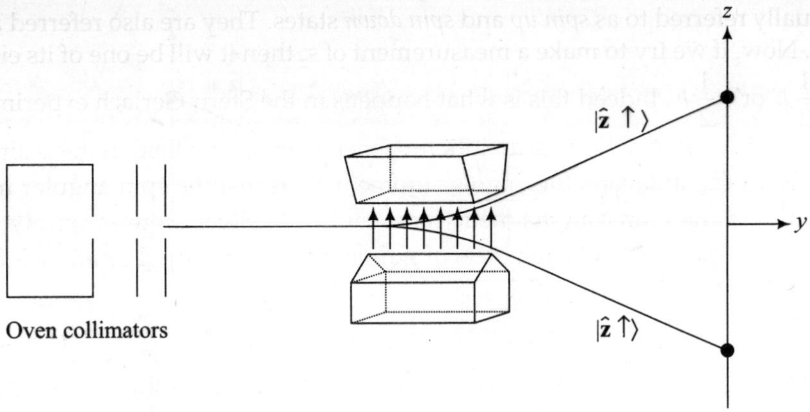

Figure 14.2: In the Stern-Gerlach experiment the beam of silver atoms (coming out of the oven) is first collimated and then passed through an inhomogenous magnetic field in the z-direction.

Thus, classically, the largest deflections will occur for atoms whose magnetic moments are aligned along or opposite the z-direction (i.e., for $\cos\theta = \pm 1$); and for atoms for which $\cos\theta = 0$, the deflection should be zero. However, when the experiment was performed (see Fig. 14.2), the detector recorded a two line pattern showing that the silver beam had split into two components on passing through the inhomogenous magnetic field. Since the displacement is proportional to μ_z, the result of the above experiment implies that *an attempt to measure* μ_z *would lead to one of the two eigenvalues* $+\mu_0$ and $-\mu_0$. The quantization of the z-component of the magnetic moment suggests that the atomic magnets would be oriented in a quantized manner in space; this is often referred to as *space quantization.* In Fig. 14.2, the two beams emerging from the magnet would correspond to the $|\hat{z}\uparrow|\rangle$ and $|\hat{z}\downarrow\rangle$ states respectively.

Now, the eigenvalues of s_x (and also of s_y) are $\pm\frac{1}{2}\hbar$. Thus, there is nothing sacred about the z-direction; if the inhomogenous magnetic field was applied in the x-direction, we would have again obtained only two beams but this time they will correspond to the $|\hat{x}\uparrow\rangle$ and $|\hat{x}\downarrow\rangle$ states.

Referring back to Fig. 14.2, let us suppose that we block the $|\hat{z}\downarrow\rangle$ beam and allow only the $|\hat{z}\uparrow\rangle$ beam to pass through. If we allow this beam to pass through an inhomogenous magnetic field in the z-direction, then since it is already in the $|\hat{z}\uparrow\rangle$ state, we will get only one spot on the screen as shown in Fig. 14.3. What would happen if we allow the $|\hat{z}\uparrow\rangle$ beam to pass through an inhomogenous magnetic field in the x-direction as shown in Fig. 14.4 ? Obviously, we are now trying to measure s_x and, therefore, we must expand the $|\hat{z}\uparrow\rangle$ state in terms of the eigenfunctions of s_x. Simple matrix algebra gives us the following (normalized) eigenfunctions of s_x:

$$|\hat{x}\uparrow\rangle = \frac{1}{\sqrt{2}}\begin{pmatrix}1\\1\end{pmatrix}$$

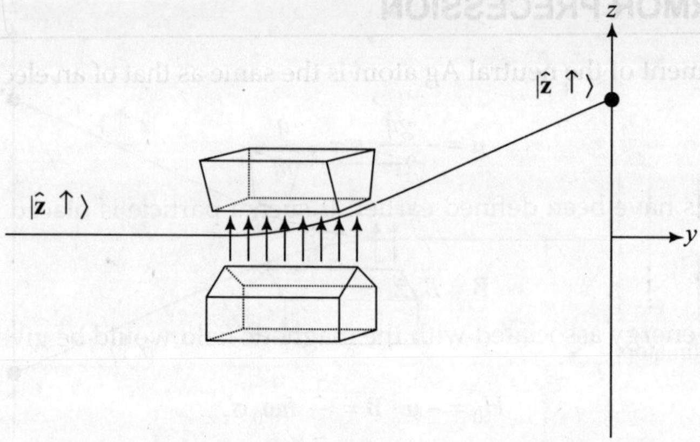

Figure 14.3: If we block the $|\hat{\mathbf{z}}\downarrow\rangle$ beam (in Fig. 14.2) and allow the $|\hat{\mathbf{z}}\uparrow\rangle$ beam to pass through an inhomogenous magnetic field in the z-direction, we would get only one spot on the screen.

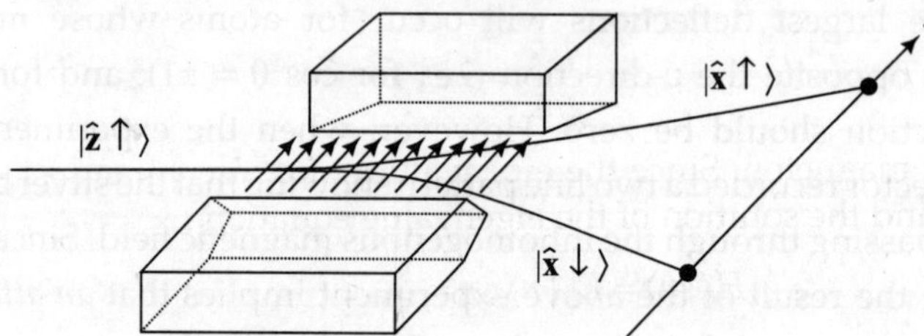

Figure 14.4: If we allow the $|\hat{\mathbf{z}}\uparrow\rangle$ beam to pass through an inhomogenous magnetic field in the x-direction, we would get two spots on the screen corresponding to $|\hat{\mathbf{x}}\uparrow\rangle$ and $|\hat{\mathbf{x}}\downarrow\rangle$ states.

and

$$|\hat{\mathbf{x}}\downarrow\rangle = \frac{1}{\sqrt{2}}\begin{pmatrix}1\\-1\end{pmatrix}$$

Now,

$$|\hat{\mathbf{z}}\uparrow\rangle = \begin{pmatrix}1\\0\end{pmatrix} = \frac{1}{\sqrt{2}}\,|\hat{\mathbf{x}}\uparrow\rangle + \frac{1}{\sqrt{2}}\,|\hat{\mathbf{x}}\downarrow\rangle$$

Thus if the system is in the $|\hat{\mathbf{z}}\uparrow\rangle$ state and if we try to measure s_x then we would get one of the two eigenvalues ($+\frac{1}{2}\hbar$ and $-\frac{1}{2}\hbar$) with equal probabilities ($=\frac{1}{2}$). In a particular measurement, one can never say for sure whether the Ag-atom will be found in the $|\hat{\mathbf{x}}\uparrow\rangle$ state or in the $|\hat{\mathbf{x}}\downarrow\rangle$ state – one can only predict the probabilities of such an event so that if the experiment is carried out with a large number of Ag atoms then (approximately) half of them will be found in the $|\hat{\mathbf{x}}\uparrow\rangle$ state and half of them in the $|\hat{\mathbf{x}}\downarrow\rangle$ state. What should we have expected classically? Well, since the magnet was initially pointing in the z-direction, the magnetic moment vector would make an angle of $\pi/2$ with the x-axis resulting in a zero force acting on the magnet. Thus, classically, we should not have expected any deflection at all!

We end this section by mentioning that in 1927 Phipps and Taylor carried out the Stern-Gerlach experiment using a beam of hydrogen atoms. This experiment conclusively showed that the observed magnetic moment is due to the electron.

14.5 THE LARMOR PRECESSION

The magnetic moment of the neutral Ag atom is the same as that of an electron and is given by

$$\boldsymbol{\mu} = -\frac{gq}{2m}\,\mathbf{s} \simeq -\frac{q}{m}\,\mathbf{s} \tag{15}$$

where the symbols have been defined earlier. If such a particle is placed in a static magnetic field given by

$$\mathbf{B} = B_0\,\hat{\mathbf{z}} \tag{16}$$

then the potential energy associated with the magnetic field would be given by

$$H_0 = -\boldsymbol{\mu}\cdot\mathbf{B} = \frac{1}{2}\,\hbar\omega_0\,\sigma_z \tag{17}$$

where
$$\omega_0 \equiv \frac{qB_0}{m} = \frac{2\mu_B B_0}{\hbar} \tag{18}$$

and
$$\mu_B = \frac{|q|\hbar}{2m} \simeq 9.274\times 10^{-24}\ \mathrm{J/T} \tag{19}$$

represents the Böhr magneton. Since the eigenvalues of σ_z are +1 and –1, this is essentially a two-state problem and the solution of the eigenvalue equation

$$H_0\,|n\rangle = E_n\,|n\rangle\,;\; n = 1,\ 2 \tag{20}$$

would be given by

$$E_1 = \frac{1}{2}\,\hbar\omega_0 \Leftrightarrow |1\rangle = |\hat{\mathbf{z}}\uparrow\rangle = \begin{pmatrix}1\\0\end{pmatrix} \tag{21}$$

$$E_2 = -\frac{1}{2}\,\hbar\omega_0 \Leftrightarrow |2\rangle = |\hat{\mathbf{z}}\downarrow\rangle = \begin{pmatrix}0\\1\end{pmatrix} \tag{22}$$

Therefore, the most general solution of the time-dependent Schrödinger equation

$$i\hbar\,\frac{\partial}{\partial t}\,|\Psi(t)\rangle = H_0\,|\Psi(t)\rangle \tag{23}$$

would be

$$|\Psi(t)\rangle = \sum_{n=1}^{2} C_n e^{-iE_n t/\hbar}\,|n\rangle$$
$$= C_1\,e^{-i\omega_0 t/2}\,|1\rangle + C_2\,e^{i\omega_0 t/2}\,|2\rangle \tag{24}$$

where the coefficients C_1 and C_2 are to be determined from the knowledge of $|\Psi(t=0)\rangle$:

$$C_1 = \langle 1\,|\,\Psi(0)\rangle$$

and
$$C_2 = \langle 2\,|\,\Psi(0)\rangle$$

Of course, if the system is initially in the $|\hat{\mathbf{z}}\uparrow\rangle$ or $|\hat{\mathbf{z}}\downarrow\rangle$ state, then it will remain in that state for all time to come; these are the *stationary states* of the problem.

Example 14.1 As an example, we assume that at $t = 0$, the atom is in the $|\hat{\mathbf{x}}\uparrow\rangle$ state. Since,

$$|\hat{\mathbf{x}}\uparrow\rangle = \frac{1}{\sqrt{2}}\begin{pmatrix}1\\1\end{pmatrix}$$

we readily get

$$C_1 = \langle\hat{\mathbf{z}}\uparrow|\hat{\mathbf{x}}\uparrow\rangle = \frac{1}{\sqrt{2}}\ (1\ 0)\begin{pmatrix}1\\1\end{pmatrix} = \frac{1}{\sqrt{2}}$$

Similarly, $C_2 = 1/\sqrt{2}$. Thus,

$$|\Psi(t)\rangle = \frac{1}{\sqrt{2}}e^{-i\theta}\begin{pmatrix}1\\0\end{pmatrix} + \frac{1}{\sqrt{2}}e^{i\theta}\begin{pmatrix}0\\1\end{pmatrix}$$

or

$$|\Psi(t)\rangle = \frac{1}{\sqrt{2}}\begin{pmatrix}e^{-i\theta}\\e^{i\theta}\end{pmatrix} \tag{25}$$

where

$$\theta \equiv \frac{1}{2}\,\omega_0 t \tag{26}$$

Equation (25) describes the time evolution of the state. For example, if we wish to determine $\langle s_x\rangle$, then

$$\langle s_x\rangle = \frac{1}{2}\hbar\,\langle\Psi(t)|\sigma_x|\Psi(t)\rangle$$

$$= \frac{1}{4}\hbar\,(e^{i\theta}\ e^{-i\theta})\begin{pmatrix}0 & 1\\1 & 0\end{pmatrix}\begin{pmatrix}e^{-i\theta}\\e^{+i\theta}\end{pmatrix}$$

or

$$\langle s_x\rangle = \frac{1}{2}\hbar\cos\omega_0 t \tag{27}$$

Similarly,

$$\langle s_y\rangle = \frac{1}{2}\hbar\sin\omega_0 t \tag{28}$$

and

$$\langle s_z\rangle = 0 \tag{29}$$

The above equations physically imply that the direction of the spin angular momentum vector rotates about the z-axis with angular velocity ω_0. We should point out that the above equations represent average values of s_x, s_y and s_z; however, if we carry out a measurement of s_z, then we would obtain either $+\frac{1}{2}\hbar$ or $-\frac{1}{2}\hbar$ (i.e., one of the eigenvalues) but since they will occur with equal probabilities the average value of s_z would be zero. On the other hand, the probabilities of obtaining the $|\hat{\mathbf{x}}\uparrow\rangle$ and $|\hat{\mathbf{x}}\downarrow\rangle$ states are

$$|\langle\hat{\mathbf{x}}\uparrow|\Psi(t)\rangle|^2 \text{ and } |\langle\hat{\mathbf{x}}\downarrow|\Psi(t)\rangle|^2$$

which can readily be evaluated to give

$$\cos^2\left(\frac{1}{2}\omega_0 t\right) \text{ and } \sin^2\left(\frac{1}{2}\omega_0 t\right)$$

respectivelv. Thus if one carries out a measurement of s_x then one would get either $+\frac{1}{2}\hbar$ or $-\frac{1}{2}\hbar$ with probabilities

$$\cos^2\left(\frac{1}{2}\omega_0 t\right) \text{ and } \sin^2\left(\frac{1}{2}\omega_0 t\right)$$

and therefore the average value of s_x would be

$$\langle s_x \rangle = \left(\frac{1}{2}\hbar\right)\cos^2\left(\frac{1}{2}\omega_0 t\right) + \left(-\frac{1}{2}\hbar\right)\sin^2\left(\frac{1}{2}\omega_0 t\right)$$

$$= \frac{1}{2}\hbar \cos \omega_0 t \tag{30}$$

consistent with Eq. (27). One can carry out a similar analysis for evaluating $\langle s_y \rangle$.

Example 14.2 We next consider the more general case when

$$|\Psi(0)\rangle = \cos\frac{\phi}{2}\,|\hat{\mathbf{z}}\uparrow\rangle + \sin\frac{\phi}{2}\,|\hat{\mathbf{z}}\downarrow\rangle \tag{31}$$

[when $\phi = \pi/2$, we obtain the results of the previous example]. Obviously,

$$C_1 = \cos\frac{\phi}{2} \quad \text{and} \quad C_2 = \sin\frac{\phi}{2}$$

so that

$$|\Psi(t)\rangle = \cos\frac{\phi}{2}\,e^{-i\theta}\,|\hat{\mathbf{z}}\uparrow\rangle + \sin\frac{\phi}{2}\,e^{i\theta}\,|\hat{\mathbf{z}}\downarrow\rangle$$

$$= \begin{pmatrix} \cos\frac{\phi}{2}e^{-i\theta} \\ \sin\frac{\phi}{2}e^{i\theta} \end{pmatrix}$$

Thus,

$$\langle s_x \rangle = \frac{1}{2}\hbar\,\langle \Psi(t)|\,\sigma_x\,|\Psi(t)\rangle$$

or

$$\langle s_x \rangle = \frac{1}{2}\hbar \sin\phi \cos\omega_0 t \tag{32}$$

Similarly,

$$\langle s_y \rangle = \frac{1}{2}\hbar \sin\phi \sin\omega_0 t \tag{33}$$

and

$$\langle s_z \rangle = \frac{1}{2}\hbar \cos\phi \tag{34}$$

From the above equations we can develop a classical model for the precession of the spin angular momentum vector about the z-axis (see Fig. 14.5).

From Eqs (29), (30) and (31) it readily follows that if at $t = 0$, **s** is assumed to lie in the x-z plane then **s** will lie in the y-z plane at $t = \pi/2\omega_0, 3\pi/2\omega_0, \ldots$. Thus, the spin angular momentum vector (represented by the vector **OP**) can be assumed to precess about the z-axis with angular velocity ω_0; this is known as *Larmor precession*. Further, the z-component of the spin angular momentum vector (represented by OQ) remains constant with

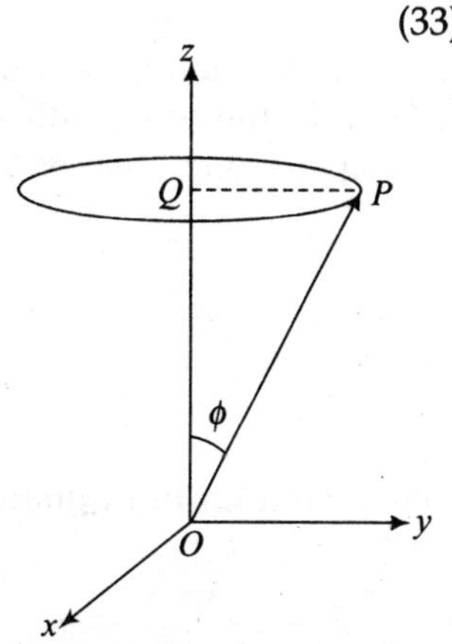

Figure 14.5: The Larmor precession.

time [see Eq. (34)]. Of course, if one measures any cartesian component of **s** one would only get $+\frac{1}{2}\hbar$ or $-\frac{1}{2}\hbar$.

Classical treatment

We may mention here that the same expression for the Larmor frequency is obtained from classical considerations. This can be seen as follows:[4] If an atomic magnet (of dipole moment μ) is suspended freely in a magnetic field **B**, it will experience a torque τ given by

$$\boldsymbol{\tau} = \boldsymbol{\mu} \times \mathbf{B} \tag{35}$$

which tries to align the magnet along the direction of **B**. Since the atomic magnet has an angular momentum (say **J**) then classical mechanics tells us that (see also Problems 14.4 and 14.7)

$$\frac{d\mathbf{J}}{dt} = \boldsymbol{\tau} = \boldsymbol{\mu} \times \mathbf{B} \tag{36}$$

or

$$\Delta J = (\mu B_0 \sin \phi)\, \Delta t \tag{37}$$

where ϕ is the angle that **μ** makes with **B** and B_0 represents the magnitude of **B**. Since μ has the same direction as **J**, Δ**J** will be at right angles to **J** and the angular momentum vector will precess about the direction of **B**.

Referring to Fig. 14.6, if we assume that at $t = 0$, **J** is along **OP** and at $t = \Delta t$, it is along **OQ**, then

$$AP = J \sin \phi$$

and

$$\Delta J = AP \times \alpha$$

$$= (J \sin \phi) \times (\omega_0\, \Delta t) \tag{38}$$

where ω_0 represents the angular velocity associated with the precession. Comparing Eqs (37) and (38), we get

$$\omega_0 = \frac{\mu B_0}{J} = g\,\frac{qB_0}{2m} \simeq \frac{qB_0}{m} \tag{39}$$

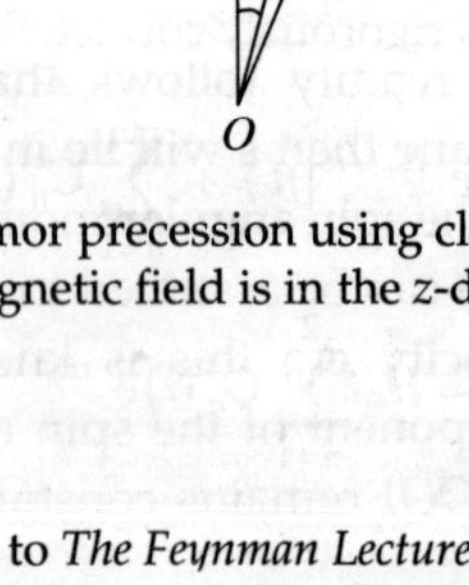

Figure 14.6: Larmor precession using classical mechanics. The magnetic field is in the z-direction.

[4] For more details, the reader is referred to *The Feynman Lectures on Physics*, Vol. II, Chapter 34.

14.6 PRINCIPLE OF THE MAGNETIC RESONANCE EXPERIMENT

In the examples discussed in the previous section, we have considered an electron in a uniform magnetic field (of magnitude B_0) in the z-direction. The energy eigenvalues were

$$+\frac{1}{2}\hbar\omega_0 \text{ and } -\frac{1}{2}\hbar\omega_0 \quad \left(\omega_0 = \frac{gqB_0}{2m} \simeq \frac{qB_0}{m}\right)$$

The corresponding eigenkets were the *spin up* and *spin down* states respectively (denoted by $|\hat{\mathbf{z}}\uparrow\rangle$ and $|\hat{\mathbf{z}}\downarrow\rangle$). In this section, we will show that in addition to the uniform magnetic field (in the z-direction), if we apply a rotating r.f. (radio frequency) field in the x-y plane given by

$$B_x = B_1 \cos \omega t,\ B_y = B_1 \sin \omega t,\ B_z = 0 \tag{40}$$

then this r.f. field would cause transitions between the *spin up* and *spin down* states.

Now, in the presence of the r.f. field, the Hamiltonian is given by

$$H = H_0 + H' \tag{41}$$

where
$$H_0 = \frac{1}{2}\hbar\omega_0\,\sigma_z \tag{42}$$

and
$$H' = -\boldsymbol{\mu}\cdot\mathbf{B}$$

$$= \frac{gq}{2m}\cdot\frac{1}{2}\hbar\boldsymbol{\sigma}\cdot[B_1\cos\omega t\,\hat{\mathbf{x}} + B_1\sin\omega t\,\hat{\mathbf{y}}]$$

or
$$H' = \mu_B B_1[\sigma_x\cos\omega t + \sigma_y\sin\omega t] \tag{43}$$

where μ_B is defined through Eq. (19). If we substitute for σ_x and σ_y from Eq. (6), we would get

$$H' = \mu_B B_1\begin{pmatrix} 0 & e^{-i\omega t} \\ e^{i\omega t} & 0 \end{pmatrix} \tag{44}$$

Our objective is to solve the Schrödinger equation

$$i\hbar\frac{\partial}{\partial t}|\Psi(t)\rangle = (H_0 + H')|\Psi(t)\rangle \tag{45}$$

We write

$$|\Psi(t)\rangle = \sum_{n=1}^{2} C_n(t)e^{-iE_nt/\hbar}|n\rangle \tag{46}$$

The physical significance of the above equation is obvious: At each instant of time, we are expanding the wavefunction as a linear combination of the eigenfunctions of H_0 which form a complete set. Thus the expansion given by Eq. (46) is rigorously correct. Substituting Eq. (46) in Eq. (45), we get

$$i\hbar\sum_{n=1}^{2}\frac{dC_n}{dt}e^{-iE_nt/\hbar}|n\rangle + \sum_{n=1}^{2}C_n(t)E_ne^{-iE_nt/\hbar}|n\rangle$$

$$= H_0\sum_{n=1}^{2}C_n(t)e^{-iE_nt/\hbar}|n\rangle + H'\sum_{n=1}^{2}C_n(t)e^{-iE_nt/\hbar}|n\rangle$$

$$= \sum_{n=1}^{2} C_n(t) e^{-iE_n t/\hbar} E_n |n\rangle + \sum_{n=1}^{2} C_n(t) e^{-iE_n t/\hbar} H'|n\rangle$$

where we have used Eq. (20). The second term on the LHS cancels with the first term on the RHS and we obtain

$$i\hbar \frac{dC_1}{dt} e^{-iE_1 t/\hbar} |1\rangle + i\hbar \frac{dC_2}{dt} e^{-iE_2 t/\hbar} |2\rangle$$
$$= C_1(t) e^{-iE_1 t/\hbar} H'|1\rangle + C_2(t) e^{-iE_2 t/\hbar} H'|2\rangle \tag{47}$$

If we premultiply by $\langle 1|$, we would get

$$i\hbar \frac{dC_1}{dt} = C_2(t) e^{i(E_1 - E_2)t/\hbar} \left[\mu_B B_1 e^{-i\omega t}\right] \tag{48}$$

where we have used the following relations $\langle 1|1\rangle = 1$, $\langle 1 \mid 2\rangle = 0$

$$\langle 1|\, H' \,|\, 1\rangle = \mu_B B_1 (1 \;\; 0) \begin{pmatrix} 0 & e^{-i\omega t} \\ e^{i\omega t} & 0 \end{pmatrix} \begin{pmatrix} 1 \\ 0 \end{pmatrix} = 0$$

and $$\langle 1 \,|\mathrm{H}'|\, 2\rangle = \mu_B B_1 (1 \;\; 0) \begin{pmatrix} 0 & e^{-i\omega t} \\ e^{i\omega t} & 0 \end{pmatrix} \begin{pmatrix} 0 \\ 1 \end{pmatrix} = \mu_B B_1 e^{-i\omega t}$$

We define $$\omega_1 \equiv \frac{\mu_B B_1}{\hbar} = \frac{qB_1}{2m} \tag{49}$$

using which Eq. (48) can be written in the following form

$$i \frac{dC_1}{dt} = \omega_1 C_2(t) e^{i(\omega_0 - \omega)t} \tag{50}$$

where we have used Eqs (21) and (22). Similarly, if we premultiply Eq. (47) by $\langle 2|$ we would get

$$i \frac{dC_2}{dt} = \omega_1 C_1(t) e^{i(\omega - \omega_0)t} \tag{51}$$

If we differentiate the above equation and substitute for dC_1/dt and $C_1(t)$ from Eqs (50) and (51) we would readily get

$$\frac{d^2 C_2}{dt^2} - 2i\Delta \frac{dC_2}{dt} + \omega_1^2 C_2(t) = 0 \tag{52}$$

where $$\Delta \equiv \frac{\omega - \omega_0}{2} \tag{53}$$

Assuming a solution of the form

$$C_2(t) \sim e^{i\alpha t} \tag{54}$$

we readily get $$\alpha = \Delta \pm \Gamma \tag{55}$$

where $$\Gamma \equiv \sqrt{\Delta^2 + \omega_1^2} \tag{56}$$

Thus the general solution of Eq. (52) is given by

$$C_2(t) = A_1 e^{i[\Delta+\Gamma]t} + A_2 e^{i[\Delta-\Gamma]t} \tag{57}$$

We assume that at $t = 0$, the atomic magnet is in the $|\hat{\mathbf{z}}\uparrow\rangle$ state. Thus,

$$C_1(0) = 1 \text{ and } C_2(0) = 0 \tag{58}$$

and Eq. (57) can be written in the form

$$C_2(t) = A\, e^{i\Delta t} \sin \Gamma t \tag{59}$$

We now use Eq. (51) to determine $C_1(t)$:

$$C_1(t) = \frac{i}{\omega_1} \frac{dC_2}{dt} e^{-2i\Delta t}$$

$$= \frac{iA\Gamma}{\omega_1}\left[\cos \Gamma t + \frac{i\Delta}{\Gamma} \sin \Gamma t\right] e^{-i\Delta t}$$

The condition $C_1(0) = 1$ gives us $A = \dfrac{\omega_1}{i\Gamma}$

Thus, we obtain $$C_1(t) = \left[\cos \Gamma t + \frac{i\Delta}{\Gamma} \sin \Gamma t\right] e^{-i\Delta t} \tag{60}$$

and $$C_2(t) = \frac{\omega_1}{i\Gamma} \sin \Gamma t\, e^{i\Delta t} \tag{61}$$

and $|\Psi(t)\rangle$ is completely determined [see Eq. (46)]. Further,

$$|C_1(t)|^2 = \cos^2 \Gamma t + \frac{\Delta^2}{\Gamma^2} \sin^2 \Gamma t \tag{62}$$

and $$|C_2(t)|^2 = \frac{\omega_1^2}{\Gamma^2} \sin^2 \Gamma t \tag{63}$$

giving the obvious result that

$$|C_1(t)|^2 + |C_2(t)|^2 = 1 \tag{64}$$

The quantity $|C_2(t)|^2$ represents the spin-flip probability, it oscillates between 0 and ω_1^2/Γ^2, the maximum value occurring at

$$t = \frac{\pi}{2\Gamma}, \frac{3\pi}{2\Gamma}, \ldots \tag{65}$$

Equation (63) can be rewritten in the form

$$P_{\text{spin-flip}} = |C_2(t)|^2 = \frac{\omega_1^2}{\omega_1^2 + \frac{1}{4}(\omega - \omega_0)^2} \sin^2 \Gamma t \tag{66}$$

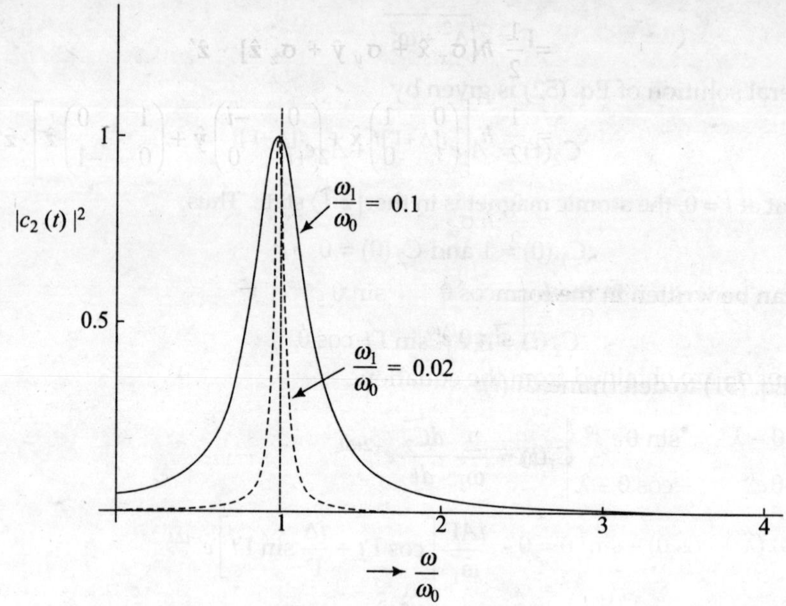

Figure 14.7: The resonance condition corresponds to $\omega = \omega_0$, where the spin flips for certainty. The solid and the dashed curves correspond to $\omega_1/\omega_0(= B_1/B_0) = 0.1$, and 0.02 respectively. The resonance becomes sharper for smaller values of B.

For $(\omega - \omega_0)/\omega_1 >> 1$, the *spin-flip* probability is small. The condition $\omega = \omega_0$ is known as the *resonance condition* and at resonance, the *spin-flips for certainty* at $t = \pi/2\Gamma, 3\pi/2\Gamma, \ldots$. The resonance condition is achieved either by varying the frequency of the r.f. field (ω) or by varying the strength of the dc magnetic field B_0 (and hence ω_0). Notice that the resonance becomes sharper as the value of B_1 is made smaller — see Fig. 14.7.

Principle of ESR and NMR experiments

Now, at a particular temperature, the number of particles in the $|\hat{\mathbf{z}}\uparrow\rangle$ state (which has higher energy) will be less in comparison to the number of particles in the $|\hat{\mathbf{z}}\downarrow\rangle$ state. Further, when a transition occurs from the $|\hat{\mathbf{z}}\downarrow\rangle$ state to the $|\hat{\mathbf{z}}\uparrow\rangle$ state, energy is supplied by the r.f. field. At resonance, the probability of spin-flip will be maximum and hence the power supplied by the r.f. field will be maximum. This is the basic principle behind the electron spin resonance (ESR) and nuclear magnetic resonance (NMR) experiments which primarily involve obtaining the resonance condition $\omega = \omega_0$. This condition can be used either to accurately measure static magnetic fields (assuming the value of the electron magnetic moment or the proton magnetic moment) or to determine the magnetic moment of the atomic system.

Example 14.3 We wish to measure $s_{z'}$ where

$$\hat{\mathbf{z}}' = \sin\theta\cos\phi\,\hat{\mathbf{x}} + \sin\theta\sin\phi\,\hat{\mathbf{y}} + \cos\theta\,\hat{\mathbf{z}}$$

which represents the unit vector along the direction defined by the polar angles (θ, ϕ). Thus,

$$s_{z'} = \mathbf{s}\cdot\hat{\mathbf{z}}'$$

$$= \frac{1}{2}\hbar[\sigma_x\,\hat{\mathbf{x}} + \sigma_y\,\hat{\mathbf{y}} + \sigma_z\,\hat{\mathbf{z}}]\cdot\hat{\mathbf{z}}'$$

$$= \frac{1}{2}\hbar\left[\begin{pmatrix}0 & 1\\ 1 & 0\end{pmatrix}\hat{\mathbf{x}} + \begin{pmatrix}0 & -i\\ +i & 0\end{pmatrix}\hat{\mathbf{y}} + \begin{pmatrix}1 & 0\\ 0 & -1\end{pmatrix}\hat{\mathbf{z}}\right]\cdot\hat{\mathbf{z}}'$$

$$= \frac{1}{2}\hbar\,\sigma_{z'}$$

where
$$\sigma_{z'} = \begin{pmatrix}\cos\theta & \sin\theta\, e^{-i\phi}\\ \sin\theta\, e^{i\phi} & -\cos\theta\end{pmatrix}$$

The eigenvalues of $\sigma_{z'}$ are obtained from the equation

$$\begin{vmatrix}\cos\theta - \lambda & \sin\theta e^{-i\phi}\\ \sin\theta e^{i\phi} & -\cos\theta - \lambda\end{vmatrix} = 0$$

or $$(\lambda - \cos\theta)(\lambda + \cos\theta) - \sin^2\theta = 0$$

or $$\lambda^2 = \cos^2\theta + \sin^2\theta = 1$$

giving $$\lambda = \pm 1$$

Thus the eigenvalues of $s_{z'}$ are $\pm\frac{1}{2}\hbar$ (obviously!). The corresponding eigenvectors are obtained by solving the equation:

$$\begin{pmatrix}\cos\theta & \sin\theta e^{-i\phi}\\ \sin\theta e^{i\phi} & -\cos\theta\end{pmatrix}\begin{pmatrix}a\\ b\end{pmatrix} = \pm 1\begin{pmatrix}a\\ b\end{pmatrix}$$

where the upper and lower signs correspond to $\lambda = +1$ and $\lambda = -1$ respectively. Elementary manipulations would give us (for $\lambda = +1$)

$$\frac{b}{a} = \frac{1-\cos\theta}{\sin\theta}\,e^{i\phi} = \tan\frac{\theta}{2}\,e^{i\phi}$$

Thus the normalized eigenkets would be

$$|\hat{\mathbf{z}}'\uparrow\rangle = \cos\frac{\theta}{2}\,|\hat{\mathbf{z}}\uparrow\rangle + \sin\frac{\theta}{2}\,e^{i\phi}\,|\hat{\mathbf{z}}\downarrow\rangle$$

Similarly, $$|\hat{\mathbf{z}}'\downarrow\rangle = \sin\frac{\theta}{2}\,|\hat{\mathbf{z}}\uparrow\rangle - \cos\frac{\theta}{2}\,e^{i\phi}\,|\hat{\mathbf{z}}\downarrow\rangle$$

Thus if the spin-half system is in the $|\hat{\mathbf{z}}'\uparrow\rangle$ state, then the probabilities of obtaining the $|\hat{\mathbf{z}}\uparrow\rangle$ and $|\hat{\mathbf{z}}\downarrow\rangle$ states would be $\cos^2\frac{\theta}{2}$ and $\sin^2\frac{\theta}{2}$, respectively.

14.7 PROBLEMS

Problem 14.1 In the Stern-Gerlach experiment using Ag atoms, assume the oven temperature to be 1000 K, with

$$l \approx 25 \text{ cm}$$

and $$\frac{\partial B_z}{\partial z} \approx 10^3 \text{ Tesla/m } (= 10^4 \text{ Gauss/cm})$$

Calculate the separation of the two components.

Problem 14.2 In Example 14.1 we had assumed the magnetic field in the z-direction. If the static magnetic field is assumed in the x-direction we will have

$$H_0 = \frac{1}{2}\hbar\,\omega_0\,\sigma_x$$

What would be the eigenvalues and eigenkets of H_0? Assume

$$|\Psi(0)\rangle = \cos\frac{\phi}{2}\,|\hat{\mathbf{z}}\uparrow\rangle + \sin\frac{\phi}{2}\,|\hat{\mathbf{z}}\downarrow\rangle$$

Evaluate $|\Psi(t)\rangle$.

Problem 14.3 For a uniform magnetic field B_0 along the y-direction the Hamiltonian is given by

$$H_0 = \frac{1}{2}\hbar\,\omega_0\sigma_y$$

(a) What are the eigenvalues and normalized eigenkets of H_0?

(b) What is the general solution of the time dependent Schrödinger equation?

(c) Assume $|\Psi(0)\rangle = |\hat{\mathbf{z}}\uparrow\rangle$. Evaluate $|\Psi(t)\rangle$ and $\langle s_z\rangle$.

Problem 14.4 Using Eqs (32), (33) and (34), prove

$$\frac{d}{dt}\langle s_x\rangle = \langle \mu_y\rangle B_z - \langle \mu_z\rangle B_y = -\frac{q}{m}[\langle s_y\rangle B_z - \langle s_z\rangle B_y] \tag{67}$$

Similarly, $$\frac{d}{dt}\langle s_y\rangle = -\frac{q}{m}[\langle s_z\rangle B_x - \langle s_x\rangle B_z] \tag{68}$$

and $$\frac{d}{dt}\langle s_z\rangle = -\frac{q}{m}[\langle s_x\rangle B_y - \langle s_y\rangle B_x] \tag{69}$$

Thus we may write

$$\frac{d}{dt}\langle \mathbf{s}\rangle = \langle \boldsymbol{\tau}\rangle = \langle \boldsymbol{\mu}\rangle \times \mathbf{B} \tag{70}$$

Problem 14.5 In the magnetic resonance experiment discussed in Section 14.6, the time evolution of the wave function can be written in the form [see Eq. (46)]:

$$|\Psi(t)\rangle = C_1(t)\,e^{-i\omega_0 t/2}\,|1\rangle + C_2(t)\,e^{i\omega_0 t/2}\,|2\rangle \tag{71}$$

Using the expressions for $C_1(t)$ and $C_2(t)$ at resonance ($\Rightarrow \Delta = 0$), show that

$$\langle s_x\rangle = \frac{1}{2}\hbar\sin(2\omega_1 t)\sin\omega_0 t \tag{72}$$

$$\langle s_y\rangle = -\frac{1}{2}\hbar\sin(2\omega_1 t)\cos\omega_0 t \tag{73}$$

and $$\langle s_z\rangle = \frac{1}{2}\hbar\cos(2\omega_1 t) \tag{74}$$

Problem 14.6 Using the results of the previous problem, verify Eqs (67), (68) and (69).

Problem 14.7 Using Eqs (46), (60) and (61), show that

$$\langle s_x \rangle = \frac{\hbar\omega_1}{2\Gamma}\left[\sin 2\Gamma t \sin \omega t - \frac{\Delta}{\Gamma}\cos \omega t + \frac{\Delta}{\Gamma}\cos 2\Gamma t \cos \omega t\right] \quad (75)$$

$$\langle s_y \rangle = \frac{\hbar\omega_1}{2\Gamma}\left[-\sin 2\Gamma t \cos \omega t + \frac{\Delta}{\Gamma}\sin \omega t - \frac{\Delta}{\Gamma}\cos 2\Gamma t \sin \omega t\right] \quad (76)$$

and

$$\langle s_z \rangle = \hbar\omega_1\left[\cos^2 \Gamma t + \frac{\Delta^2 - \omega_1^2}{\Gamma^2}\sin^2 \Gamma t\right] \quad (77)$$

Hence prove

$$\frac{d}{dt}\langle s_x \rangle = -\frac{q}{m}\left[\langle s_y \rangle B_z - \langle s_z \rangle B_y\right]$$

$$= -\omega_0 \langle s_y \rangle + 2\omega_1 \langle s_z \rangle \sin \omega t \quad (78)$$

and similar relations for

$$\frac{d}{dt}\langle s_y \rangle \text{ and } \frac{d}{dt}\langle s_z \rangle$$

14.8 SOLUTIONS

Solution 14.1 Because of collimation, the atoms entering the inhomogeneous magnetic field have their velocities in the y-direction; thus

$$\frac{3}{2}kT = \frac{1}{2}Mv^2 \approx \frac{1}{2}Mv_y^2$$

where T is the temperature of the oven. Thus,

$$v_y \approx \sqrt{\frac{3kT}{M}}$$

Thus Eq. (13) becomes

$$\Delta z \approx \frac{l^2}{6kT}\mu_z\frac{\partial B_z}{\partial z}$$

Now,

$$\mu_z = \pm\frac{q\hbar}{2m} \approx \pm 9.27 \times 10^{-24}\ \text{J/T}$$

Thus,

$$\Delta z \approx \pm\frac{(0.25)^2}{6 \times 1.38 \times 10^{-23} \times 10^3} \times 9.27 \times 10^{-24} \times 10^3\ \text{m}$$

$$\approx \pm 0.7\ \text{cm}$$

The separation is about 1.4 cm.

Solution 14.2 The eigenkets of H_0 are now given by

$$|1\rangle = |\hat{\mathbf{x}}\uparrow\rangle = \frac{1}{\sqrt{2}}\begin{pmatrix}1\\1\end{pmatrix}$$

and
$$|2\rangle = |\hat{\mathbf{x}}\downarrow\rangle = \frac{1}{\sqrt{2}}\begin{pmatrix}1\\-1\end{pmatrix}$$

Thus, the general solution of the time dependent Schrödinger equation is given by Eq. (24) with $|1\rangle$ and $|2\rangle$ given by the above equations. Now,

$$|\Psi(0)\rangle = \cos\frac{\phi}{2}\,|\hat{\mathbf{z}}\uparrow\rangle + \sin\frac{\phi}{2}\,|\hat{\mathbf{z}}\downarrow\rangle$$

$$= \cos\frac{\phi}{2}\left[\frac{1}{\sqrt{2}}|\hat{\mathbf{x}}\uparrow\rangle + \frac{1}{\sqrt{2}}|\hat{\mathbf{x}}\downarrow\rangle\right] + \sin\frac{\phi}{2}\left[\frac{1}{\sqrt{2}}|\hat{\mathbf{x}}\uparrow\rangle - \frac{1}{\sqrt{2}}|\hat{\mathbf{x}}\downarrow\rangle\right]$$

$$= \frac{1}{\sqrt{2}}\left(\cos\frac{\phi}{2} + \sin\frac{\phi}{2}\right)|\hat{\mathbf{x}}\uparrow\rangle + \frac{1}{\sqrt{2}}\left(\cos\frac{\phi}{2} - \sin\frac{\phi}{2}\right)|\hat{\mathbf{x}}\downarrow\rangle$$

Thus,
$$|\Psi(t)\rangle = \frac{1}{\sqrt{2}}\left(\cos\frac{\phi}{2} + \sin\frac{\phi}{2}\right)e^{-iE_1t/\hbar}\,|\hat{\mathbf{x}}\uparrow\rangle$$

$$+ \frac{1}{\sqrt{2}}\left(\cos\frac{\phi}{2} - \sin\frac{\phi}{2}\right)e^{-iE_2t/\hbar}\,|\hat{\mathbf{x}}\downarrow\rangle$$

$$= \frac{1}{2}\begin{pmatrix}\left(\cos\frac{\phi}{2} + \sin\frac{\phi}{2}\right)e^{-i\theta} + \left(\cos\frac{\phi}{2} - \sin\frac{\phi}{2}\right)e^{i\theta}\\ \left(\cos\frac{\phi}{2} + \sin\frac{\phi}{2}\right)e^{-i\theta} + \left(\cos\frac{\phi}{2} - \sin\frac{\phi}{2}\right)e^{i\theta}\end{pmatrix}$$

where $\theta = \omega_0 t/2$.

Solution 14.3 (b)

$$|\Psi(0)\rangle = C_1 e^{-i\theta}\,|\hat{\mathbf{y}}\uparrow\rangle + C_2 e^{i\theta}\,|\hat{\mathbf{y}}\downarrow\rangle$$

$$= \frac{1}{\sqrt{2}}\,C_1 e^{-i\theta}\begin{pmatrix}1\\i\end{pmatrix} + \frac{1}{\sqrt{2}}\,C_2 e^{i\theta}\begin{pmatrix}1\\-i\end{pmatrix}$$

where $\theta = \omega_0 t/2$.

(c)
$$C_1 = \frac{1}{\sqrt{2}};\ C_2 = -\frac{1}{\sqrt{2}}$$

Thus,
$$|\Psi(t)\rangle = \begin{pmatrix}\cos\omega_0 t/2\\ \sin\omega_0 t/2\end{pmatrix}$$

Solution 14.4 Using Eq. (32), we get

$$\frac{d}{dt}\langle s_x\rangle = -\frac{1}{2}\,\hbar\omega_0 \sin\phi\,\sin\omega_0 t$$

Since $B_x = 0 = B_y$ and $B_z = B_0$

$$\frac{q}{m}\,[\langle s_y\rangle\, B_z - \langle s_z\rangle\, B_y] = -\frac{qB_0}{m}\left[\frac{1}{2}\hbar \sin\phi \sin\omega_0 t\right]$$

$$= -\frac{1}{2}\,\hbar\,\omega_0 \sin\phi\,\sin\omega_0 t$$

Solution 14.5 At resonance

$$\omega = \omega_0\ (\Rightarrow \Delta = 0 \text{ and } \Gamma = \omega_1)$$

and Eqs (60) and (61) simplify to

$$C_1(t) = \cos\omega_1 t \tag{79}$$

and

$$C_2(t) = -i\sin\omega_1 t \tag{80}$$

Now,

$$|\Psi(t)\rangle = C_1(t)\, e^{-i\omega_0 t/2}\ |1\rangle + C_2(t)\, e^{-i\omega_0 t/2}\ |2\rangle \tag{81}$$

and

$$\langle\Psi(t)\rangle = \langle 1|\ C_1^*(t)\, e^{-i\omega_0 t/2} + \langle 2|\ C_2^*(t)\, e^{-i\omega_0 t/2} \tag{82}$$

Thus,

$$\langle s_x\rangle = \langle\Psi(t)\,|\,s_x\,|\,\Psi(t)\rangle$$

$$= \frac{1}{2}\,\hbar\,\langle\Psi(t)\,|\,\sigma_x\,|\,\Psi(t)\rangle$$

$$= C_1^*(t)\, C_2(t)\, e^{i\omega_0 t} + C_2^*(t)\, C_1(t)\, e^{-i\omega_0 t}$$

where we have used the relations

$$\langle 1|\sigma_x|1\rangle = 0 = \langle 2|\sigma_x|\,2\rangle$$

and

$$\langle 1|\sigma_x|2\rangle = 1 = \langle 2\ |\sigma_x|\ 1\rangle$$

Substituting for $C_1(t)$ and $C_2(t)$, we readily obtain

$$\langle s_x\rangle = \frac{1}{2}\,\hbar\sin(2\omega_1 t)\sin\omega_0 t \tag{83}$$

Similarly,

$$\langle s_y\rangle = -\frac{1}{2}\,\hbar\sin(2\omega_1 t)\cos\omega_0 t \tag{84}$$

and

$$\langle s_z\rangle = \frac{1}{2}\,\hbar\cos(2\omega_1 t) \tag{85}$$

The last equation shows the *spin-flip* at

$$t = \frac{\pi}{2\omega_1}, \frac{3\pi}{2\omega_1}, \frac{5\pi}{2\omega_1}, \ldots$$

Solution 14.6 From Eq. (77) we readily have

$$\frac{d}{dt}\langle s_x\rangle = -\frac{qB_0}{m}\langle s_y\rangle + \frac{qB_1 \sin\omega t}{m}\langle s_z\rangle$$

where we have used Eq. (40). Now,

$$\frac{d}{dt}\langle s_x\rangle = \frac{1}{2}\,\hbar\,\omega_0\,\sin(2\omega_1 t)\cos\omega_0 t + \hbar\omega_1\cos(2\omega_1 t)\sin\omega_0 t$$

and $$-\frac{qB_0}{m}\langle s_y\rangle+\frac{qB_1 \sin \omega t}{m}\langle s_z\rangle = \frac{1}{2}\ \hbar\omega_0 \sin(2\omega_1 t)\cos\omega_0 t + \hbar\omega_1 \cos(2\omega_1 t)\sin\omega_0 t$$

Similarly, one can prove other equations.

REFERENCES AND SUGGESTED READINGS

1. G. Baym, *Lectures on Quantum Mechanics*, W.A. Benjamin, Inc., New York (1969).
2. R.P. Feynman, R.B. Leighton and M. Sands, *The Feynman Lectures on Physics*, Vol III, Addison-Wesley, Reading, Massachusetts (1965).
3. C.P Slichter, *Principles of Magnetic Resonance*, Harper and Row, New York (1963).
4. J.S. Townsend, *A Modern Approach to Quantum Mechanics*, McGraw-Hill, New York (1992).

15 Angular momentum III

Eigenfunctions using operator algebra

It seems that if one is working from the point of view of getting beauty in one's equations, and if one has a really sound insight, one is on a sure line of progress.

— P.A.M. DIRAC in *The Evolution of the Physicist's Picture of Nature*, Scientific American, May 1963.

15.1 INTRODUCTION

In Chapter 13, we had shown that we can have a set of simultaneous eigenkets of the operators $\hat{J}^2$ and $\hat{J}_z$:

$$\hat{J}^2 \, | j, m \rangle = j\,(j+1)\hbar^2 \, | j, m \rangle \tag{1}$$

$$\hat{J}_z \, | j, m \rangle = m\hbar \, | j, m \rangle \tag{2}$$

where
$$j = 0, \frac{1}{2}, 1, \frac{3}{2}, 2, \frac{5}{2}, \ldots \tag{3}$$

and
$$m = -j, -j+1, \ldots, j \tag{4}$$

Further, we had defined the ladder operators through the equations

$$\hat{J}_+ = \hat{J}_x + i\,\hat{J}_y \tag{5}$$

and
$$\hat{J}_- = \hat{J}_+ = \hat{J}_x - i\hat{J}_y \tag{6}$$

which were shown to have the following properties:

$$\hat{J}_+ \, | j, m \rangle = [(j-m)(j+m+1)]^{\frac{1}{2}} \; \hbar \, | j, m+1 \rangle \tag{7}$$

$$\hat{J}_- \, | j, m \rangle = [(j+m)(j-m+1)]^{\frac{1}{2}} \; \hbar \, | j, m-1 \rangle \tag{8}$$

Thus,
$$\hat{J}_+ \, | j, j \rangle = 0 \tag{9}$$

and
$$\hat{J}_- \, | j, -j) = 0 \tag{10}$$

In this chapter, we will use the above equations and the operator representations of the angular momentum operators to derive explicit expressions for the eigenfunctions of $\hat{L}^2$ and $\hat{L}_z$. It will be seen that the procedure discussed in this chapter is extremely straightforward in determining the explicit expressions for the spherical harmonics.

15.2 THE SPHERICAL HARMONICS

In Section 9.2, we had shown the following operator representations of the components of the angular momentum operator

$$\hat{L}_x = i\hbar\left[\sin\phi\frac{\partial}{\partial\theta} + \cot\theta\cos\phi\frac{\partial}{\partial\phi}\right] \tag{11}$$

$$\hat{L}_y = i\hbar\left[-\cos\phi\frac{\partial}{\partial\theta} + \cot\theta\sin\phi\frac{\partial}{\partial\phi}\right] \tag{12}$$

$$\hat{L}_z = -i\hbar\frac{\partial}{\partial\phi} \tag{13}$$

Thus,

$$\hat{L}_+ = \hat{L}_x + i\hat{L}_y$$

$$= \hbar e^{i\phi}\left[\frac{\partial}{\partial\theta} + i\cot\theta\frac{\partial}{\partial\phi}\right] \tag{14}$$

and

$$\hat{L}_- = \hat{L}_x - i\hat{L}_y$$

$$= \hbar e^{-i\phi}\left[\frac{\partial}{\partial\theta} + i\cot\theta\frac{\partial}{\partial\phi}\right] \tag{15}$$

Let $Y_{lm}(\theta, \phi)$ represent the simultaneous eigenfunction of $\hat{L}^2$ and $\hat{L}_z$ [cf. Eqs (1) and (2)]:

$$\hat{L}^2 Y_{lm}(\theta, \phi) = l(l+1)\hbar^2 Y_{lm}(\theta, \phi) \tag{16}$$

$$\hat{L}_z Y_{lm}(\theta, \phi) = m\hbar Y_{lm}(\theta, \phi) \tag{17}$$

where

$$m = -l, -l+1, \ldots, l-1, l \tag{18}$$

Using Eq. (13), Eq. (17) becomes

$$-i\hbar\frac{\partial Y}{\partial\phi} = m\hbar Y(\theta, \phi) \tag{19}$$

If we use the method of separation of variables

$$Y(\theta, \phi) = F(\theta)\,\Phi(\phi) \tag{20}$$

we would obtain

$$\frac{d\Phi}{d\phi} = im\,\Phi(\phi)$$

the solution of which is given by

$$\Phi(\phi) = \Phi_m(\phi) = \frac{1}{\sqrt{2\pi}}e^{im\phi} \tag{21}$$

For $\Phi(\phi)$ to be single-valued $\Phi(\phi + 2\pi) = \Phi(\phi)$ (22)

we must have $e^{im2\pi} = 1$

implying $$m = 0, \pm 1, \pm 2, \ldots \tag{23}$$

The factor $1/\sqrt{2\pi}$ makes $\Phi_m(\phi)$ normalized:

$$\int_0^{2\pi} |\Phi_m(\phi)|^2 \, d\phi = 1 \tag{24}$$

Furthermore

$$\int_0^{2\pi} \Phi_{m'}^*(\phi)\, \Phi_m(\phi)\, d\phi = \delta_{mm'} \qquad \text{(Orthonormality condition)} \tag{25}$$

Now, using Eq. (10), we obtain

$$\hat{L}_- Y_{l,-l}(\theta, \phi) = 0 \tag{26}$$

or, using Eq. (15),
$$\hbar e^{-i\phi}\left[-\frac{\partial}{\partial\theta} + i\cot\theta \frac{\partial}{\partial\phi}\right] Y_{l,-l}(\theta, \phi) = 0$$

or
$$\left[-\frac{\partial}{\partial\theta} + i\cot\theta \frac{\partial}{\partial\phi}\right] (F(\theta)\, e^{-il\phi}) = 0$$

$$-\frac{dF}{d\theta} + l\cot\theta\, F(\theta) = 0$$

or
$$\frac{1}{F(\theta)} \cdot \frac{dF}{d\theta} = l\cot\theta$$

Integrating, we get

$$\ln F(\theta) = l \ln(\sin\theta) + \text{constant}$$

or
$$\ln\left[\frac{F(\theta)}{(\sin\theta)^l}\right] = \text{constant}$$

or
$$F(\theta) = C_- \sin^l\theta \tag{27}$$

The normalization condition
$$\int_0^{\pi} |F(\theta)|^2 \sin\theta\, d\theta = 1 \tag{28}$$

gives us
$$|C_-|^2 \int_{-1}^{+1} (1-\mu^2)^l\, d\mu = 1 \tag{29}$$

where
$$\mu = \cos\theta$$

For a particular value of l, the integral in Eq. (29) can be readily evaluated by making an expansion of the integrand and integrating term by term; the evaluation for an arbitrary value of l is a bit tedious—the final result is:

$$C_- = \frac{1}{2^l\, l!}\left[\frac{(2l+1)!}{2}\right]^{\frac{1}{2}} \tag{30}$$

within an arbitrary phase factor. Thus,

$$F(\theta) = \frac{1}{2^l\, l!}\left[\frac{(2l+1)!}{2}\right]^{\frac{1}{2}} \sin^l\theta \tag{31}$$

and using Eq. (21), we would get

$$Y_{l,-l}(\theta, \phi) = \frac{1}{2^l l!}\left[\frac{(2l+1)!}{4\pi}\right]^{\frac{1}{2}} \sin^l \theta\, e^{-il\phi} \tag{32}$$

In order to determine $Y_{l,-l+1}$; $Y_{l,-l+2}$ etc., we use Eq. (7) to write

$$\hat{L}_+ \; Y_{lm}(\theta, \phi) = [(l-m)(l+m+1)]^{\frac{1}{2}}\; \hbar\, Y_{l,\,m+1}(\theta, \phi)$$

Thus,

$$Y_{l,\,m+1} = \frac{1}{\hbar}[(l-m)(l+m+1)]^{-\frac{1}{2}}\; \hat{L}_+\; Y_{l,\,m}$$

$$= [(l-m)(l+m+1)]^{-\frac{1}{2}} e^{i\phi}\left[\frac{\partial}{\partial\theta} + i\cot\theta \frac{\partial}{\partial\phi}\right] Y_{l,m} \tag{33}$$

Starting with $m = -l$ and by repeated use of the above formula, one can readily obtain the (normalized) expressions for *all* values of *m*. We show this through two examples. We may mention here that alternatively we can start with the expression for $Y_{l,0}$ [see Eq. (52) of Chapter 9]

$$Y_{l,0}(\theta, \phi) = \sqrt{\frac{2l+1}{4\pi}}\; P_l(\cos\theta) \tag{34}$$

and use ladder operators to determine $Y_{l,\,m}$ (θ, ϕ) [see Eqs (56) and (57) of Chapter 9 and Problem 15.9].

Example 15.1 For $l = 1$, we have

$$Y_{1,-1}(\theta, \phi) = \frac{1}{2}\sqrt{\frac{6}{4\pi}}\; \sin\theta\, e^{-i\phi} = \sqrt{\frac{3}{8\pi}}\; \sin\theta\, e^{-i\phi}$$

Using Eq. (33), we get

$$Y_{1,0}(\theta, \phi) = \frac{1}{\hbar\sqrt{2}}\; L_+\; Y_{1,-1}(\theta, \phi)$$

$$= \frac{1}{\sqrt{2}} e^{i\phi}\left[\frac{\partial}{\partial\theta} + i\cot\theta\, \frac{\partial}{\partial\phi}\right]\sqrt{\frac{3}{8\pi}}\; [\sin\theta\, e^{-i\phi}]$$

$$= \sqrt{\frac{3}{16\pi}}\, e^{i\phi}\; [\cos\theta + \cos\theta]\, e^{-i\phi}$$

$$= \sqrt{\frac{3}{4\pi}}\; \cos\theta$$

Similarly,

$$Y_{1,+1}(\theta, \phi) = \frac{1}{\hbar\sqrt{2}}\hat{L}_+\; Y_{1,0}(\theta, \phi)$$

$$= \frac{1}{\sqrt{2}}\; e^{i\phi}\left[\frac{\partial}{\partial\theta} + i\cot\theta\, \frac{\partial}{\partial\phi}\right]\sqrt{\frac{3}{4\pi}}\; \cos\theta$$

$$= -\sqrt{\frac{3}{8\pi}}\; \sin\theta\, e^{i\phi}$$

Example 15.2 For $l = 2$, we have

$$Y_{2,-2}(\theta, \phi) = \left(\frac{1}{8}\sqrt{\frac{5!}{4\pi}}\right) \sin^2\theta\, e^{-2i\phi}$$

$$= \sqrt{\frac{15}{32\pi}} \sin^2\theta\, e^{-2i\phi}$$

Using Eq. (33) with $l = 2$, $m = -2$, we get

$$Y_{2,-1}(\theta, \phi) = \frac{1}{\sqrt{4}\,\hbar} \hat{L}_+ Y_{2,-2}(\theta, \phi)$$

$$= \frac{1}{2} e^{i\phi}\left[\frac{\partial}{\partial\theta} + i\cot\theta\,\frac{\partial}{\partial\phi}\right]\sqrt{\frac{15}{32\pi}}\,[\sin^2\theta\, e^{-2i\phi}]$$

$$= \sqrt{\frac{15}{8\pi}} \sin\theta\cos\theta\, e^{-i\phi}$$

Similarly,

$$Y_{2,0} = \frac{1}{\sqrt{6}\hbar} \hat{L}_+ Y_{2,-1}(\theta, \phi)$$

$$= \frac{1}{\sqrt{6}} e^{i\phi}\left[\frac{\partial}{\partial\theta} + i\cot\theta\,\frac{\partial}{\partial\phi}\right]\sqrt{\frac{15}{8\pi}}\,[\sin\theta\cos\theta\, e^{-i\phi}]$$

$$= \sqrt{\frac{5}{16\pi}}(3\cos^2\theta - 1)$$

Further,

$$Y_{2,1} = \frac{1}{\sqrt{6}\hbar} \hat{L}_+ Y_{2,0}(\theta, \phi)$$

$$= -\sqrt{\frac{15}{8\pi}} \sin\theta\cos\theta\, e^{i\phi}$$

$$Y_{2,2} = \frac{1}{\hbar\sqrt{4}} \hat{L}_+ Y_{2,1}$$

$$= \sqrt{\frac{15}{32\pi}} \sin^2\theta\, e^{2i\phi}$$

Similarly, for the $l = 3$ case, we have

$$Y_{3,-3}(\theta, \phi) = \frac{1}{2^3 3!}\left[\frac{7!}{4\pi}\right]^{\frac{1}{2}} \sin^3\theta\, e^{-3i\phi}$$

$$= \sqrt{\frac{35}{64\pi}} \sin^3\theta\, e^{-3i\phi}$$

Therefore, $$Y_{3,-2}(\theta, \phi) = \frac{1}{\hbar\sqrt{6}} \hat{L}_+ Y_{3,-3}(\theta, \phi)$$

$$= \frac{1}{\sqrt{6}} e^{i\phi} \left[\frac{\partial}{\partial\theta} + i\cot\theta \frac{\partial}{\partial\phi}\right] \sqrt{\frac{35}{64\pi}} \, [\sin^3\theta \, e^{-3i\phi}]$$

$$= \sqrt{\frac{105}{32\pi}} \sin^2\theta \cos\theta \, e^{-2i\phi}$$

We can repeat the process to obtain $Y_{3,-1}(\theta, \phi)$, $Y_{3,0}(\theta, \phi)$, etc. The final results are summarized in the next section.

We should point out that, as in the case of the harmonic oscillator problem, if the wavefunction with which we start is properly normalized, then all subsequent wavefunctions obtained by applying the ladder operator are automatically normalized.

15.3 EXPRESSIONS FOR SPHERICAL HARMONICS

$l = 0$; $2l + 1 = 1$; $m = 0$

$$Y_{0,0} = \frac{1}{\sqrt{4\pi}} \tag{35}$$

$l = 1$; $2l + 1 = 3$; $m = -1, 0, +1$

$$Y_{1,-1}(\theta, \phi) = \sqrt{\frac{3}{8\pi}} \sin\theta \, e^{-i\phi} = -Y^*_{1,1} \tag{36}$$

$$Y_{1,0}(\theta, \phi) = \sqrt{\frac{3}{4\pi}} \cos\theta \tag{37}$$

$$Y_{1,1}(\theta, \phi) = -\sqrt{\frac{3}{8\pi}} \sin\theta \, e^{i\phi} = -Y^*_{1,-1} \tag{38}$$

$l = 2$; $2l + 1 = 5$; $m = -2, -1, 0, 1, 2$

$$Y_{2,-2}(\theta, \phi) = \sqrt{\frac{15}{32\pi}} \sin^2\theta \, e^{-2i\phi} = Y^*_{2,2} \tag{39}$$

$$Y_{2,-1}(\theta, \phi) = \sqrt{\frac{15}{8\pi}} \sin\theta \cos\theta \, e^{-i\phi} = -Y^*_{2,1} \tag{40}$$

$$Y_{2,0}(\theta, \phi) = \sqrt{\frac{5}{16\pi}} (3\cos^2\theta - 1) \tag{41}$$

$$Y_{2,1}(\theta, \phi) = -\sqrt{\frac{15}{8\pi}} \sin\theta \cos\theta \, e^{i\phi} = -Y^*_{2,-1} \tag{42}$$

$$Y_{2,2}(\theta, \phi) = \sqrt{\frac{15}{32\pi}} \sin^2\theta \, e^{2i\phi} = Y^*_{2,-2} \tag{43}$$

$l = 3$; $2l + 1 = 7$; $m = -3, -2, -1, 0, 1, 2, 3$

$$Y_{3,-3}(\theta, \phi) = \sqrt{\frac{35}{64\pi}} \sin^3\theta\, e^{-3i\phi} = -Y^*_{3,3} \tag{44}$$

$$Y_{3,-2}(\theta, \phi) = \sqrt{\frac{105}{32\pi}} \sin^2\theta \cos\theta\, e^{-2i\phi} = Y^*_{3,2} \tag{45}$$

$$Y_{3,-1}(\theta, \phi) = \sqrt{\frac{21}{64\pi}} (5\cos^2\theta - 1)\sin\theta\, e^{-i\phi} = -Y^*_{3,1} \tag{46}$$

$$Y_{3,0}(\theta, \phi) = \sqrt{\frac{7}{4\pi}\cdot\frac{1}{2}}\cdot(5\cos^3\theta - 3\cos\theta) \tag{47}$$

The above equations are consistent with those derived in Section 9.3.

15.4 THE $l = \frac{1}{2}$ CASE

In the entire analysis presented in this chapter we have assumed l to be a positive integer (including the value zero) and therefore m takes positive and negative integer values. What happens if we push our luck and try to obtain the wavefunctions for $l = \frac{1}{2}, \frac{3}{2}, \frac{5}{2}, \ldots$? We will see that the procedure leads to ill-behaved and inconsistent wavefunctions.

Let us assume $l = \frac{1}{2}$; then

$$m = -\frac{1}{2} \text{ or } +\frac{1}{2}$$

Thus [see Eq. (21)]

$$\Phi_m(\phi) = \frac{1}{\sqrt{2\pi}} e^{\pm i\phi/2} \tag{48}$$

Of course, such a wavefunction is not single-valued; however, $|\Phi(\phi)|^2$ is single valued and one may argue that the wavefunction described by the above equation may be acceptable. We consider the case $m = -\frac{1}{2}$ so that Eq. (32) becomes

$$Y_{\frac{1}{2},-\frac{1}{2}}(\theta, \phi) = C\sqrt{\sin\theta}\, e^{-i\phi/2} \tag{49}$$

where C is a constant. Using Eq. (33) for $l = \frac{1}{2}$, $m = -\frac{1}{2}$, we obtain

$$Y_{\frac{1}{2},\frac{1}{2}}(\theta, \phi) = \frac{1}{\hbar\sqrt{1}} L_+ Y_{\frac{1}{2},-\frac{1}{2}}$$

$$= C e^{i\phi}\left[\frac{\partial}{\partial\theta} + i\cot\theta\frac{\partial}{\partial\phi}\right]\left[\sqrt{\sin\theta}\, e^{-i\phi/2}\right]$$

$$= C\frac{\cos\theta}{\sqrt{\sin\theta}} e^{-i\phi/2}$$

which is incorrect because we should have had (see Problem 15.8):

$$Y_{\frac{1}{2},\frac{1}{2}} = (\text{const}) \sqrt{\sin\theta}\, e^{-i\phi/2} \tag{50}$$

Thus we *must* restrict the l values to 0, 1, 2, 3, ... with m going from $-l$ to $+l$.

15.5 PROBLEMS

Problem 15.1 Show by explicit differentiation that

$$\hat{L}_+ Y_{1,1} = 0$$

$$\hat{L}_+ Y_{2,2} = 0$$

and

$$\hat{L}_+ Y_{3,3} = 0$$

where the $\hat{L}_+$ operator is given by Eq. (14) and $Y_{1,1}$, $Y_{2,2}$ and $Y_{3,3}$ are given by Eqs (38), (43) and (44) respectively.

Problem 15.2 With $l = 4$, use Eq. (26) to show that

$$Y_{4,-4} = \sqrt{\frac{315}{512\pi}} \sin^4\theta\, e^{-4i\phi} \tag{51}$$

Using Eq. (33) derive $Y_{4,-3}$, $Y_{4,-2}$, $Y_{4,-1}$, $Y_{4,0}$, $Y_{4,1}$, $Y_{4,2}$, $Y_{4,3}$ and $Y_{4,4}$.

Problem 15.3 Show that $Y_{0,0}$ is a simultaneous eigenfunction of $\hat{L}^2$, $\hat{L}_x$, $\hat{L}_y$ and $\hat{L}_z$. What are the corresponding eigenvalues?

Problem 15.4 Use

$$\hat{L}_x = \frac{1}{2}(\hat{L}_+ + \hat{L}_-)$$

to show

$$\hat{L}_x Y_{1,1} = \frac{1}{\sqrt{2}} \hbar Y_{1,0} \tag{52}$$

$$\hat{L}_x Y_{1,0} = \frac{1}{\sqrt{2}} \hbar (Y_{1,1} + Y_{1,-1}) \tag{53}$$

$$L_x Y_{1,-1} = \frac{1}{\sqrt{2}} \hbar Y_{1,0} \tag{54}$$

Verify the above equations by using the operator representation of $\hat{L}_x$ [see Eq. (11)].

Problem 15.5 The functions $Y_{1,1}$, $Y_{1,0}$ and $Y_{1,-1}$ are simultaneous eigenfunctions of $\hat{L}^2$ and $\hat{L}_z$ (and *not* of $\hat{L}_x$ and $\hat{L}_y$). Using the results of the previous problem show that

$$\hat{L}_x \psi_1 = \hbar \psi_1 \tag{55}$$

$$\hat{L}_x \psi_2 = 0 \tag{56}$$

$$\hat{L}_x \psi_3 = -\hbar \psi_3 \tag{57}$$

where
$$\psi_1 = \frac{1}{2}\,[Y_{1,1} + Y_{1,-1} + \sqrt{2}\,Y_{1,0}] \tag{58}$$

$$\psi_2 = \frac{1}{\sqrt{2}}\,[Y_{1,1} - Y_{1,-1}] \tag{59}$$

$$\psi_3 = \frac{1}{2}\,[Y_{1,1} + Y_{1,-1} - \sqrt{2}\,Y_{1,0}] \tag{60}$$

Thus ψ_1, ψ_2 and ψ_3 are simultaneous eigenfunctions of $\hat{L}^2$ and $\hat{L}_x$.

Problem 15.6 Using Eqs (7) and (8), show that

$$\hat{L}_y\,Y_{1,1} = \frac{i}{\sqrt{2}}\,\hbar\,Y_{1,0} \tag{61}$$

$$\hat{L}_y\,Y_{1,0} = -\frac{i}{\sqrt{2}}\,\hbar\,[Y_{1,1} - Y_{1,-1}] \tag{62}$$

$$\hat{L}_y\,Y_{1,-1} = -\frac{i}{\sqrt{2}}\,\hbar\,Y_{1,1} \tag{63}$$

Also verify the above equations by using the operator representation of $\hat{L}_y$ [see Eq. (12)]. Using the above equations determine the eigenfunctions of $\hat{L}_y$; these will be simultaneous eigenfunctions of $\hat{L}^2$ and $\hat{L}_y$.

Problem 15.7 The angular part of the wavefunction is described by the following equation:

$$\psi\,(\theta, \phi) = \frac{1}{2}\left[-2i\sqrt{\frac{3}{8\pi}}\,\sin\theta\,\sin\phi + \sqrt{\frac{3}{2\pi}}\,\cos\theta\right]$$

(a) If we make a measurement of $\hat{L}_z$, what values will we obtain and with what probabilities? Evaluate $\langle \hat{L}_z \rangle$.

(b) If we make a measurement of $\hat{L}_x$, what values will we obtain and with what probabilities? Evaluate $\langle \hat{L}_x \rangle$.

Problem 15.8 Using the condition

$$\hat{L}_+\;Y_{l,l} = 0 \tag{64}$$

show that
$$Y_{l,l} = (\text{const})\,\sin^l\theta\,e^{i\,l\,\phi} \tag{65}$$

Problem 15.9 For $m = 0$, we have

$$Y_{l,0}\,(\theta, \phi) = \sqrt{\frac{2l+1}{4\pi}}\;P_l\,(\cos\theta) \tag{66}$$

Thus,
$$Y_{1,0} = \sqrt{\frac{3}{4\pi}}\;\cos\theta$$

Use ladder operators to show that $Y_{1,-1}$ and $Y_{1,+1}$ are given by Eq. (36) and Eq. (38). Similarly,

$$Y_{2,0} = \sqrt{\frac{5}{4\pi}}\frac{1}{2}(3\cos^2\theta - 1)$$

Use ladder operators to determine $Y_{2,2}$, $Y_{2,1}$, $Y_{2,-1}$ and $Y_{2,-2}$.

15.6 SOLUTIONS

Solution 15.5 Let

$$\psi(\theta, \phi) = \alpha Y_{1,1} + \beta Y_{1,0} + Y_{1,-1}$$

represent the required linear combination which is an eigenfunction of L_x. Thus,

$$L_x[\alpha Y_{1,1} + \beta Y_{1,0} + \gamma Y_{1,-1}] = \lambda[\alpha Y_{1,1} + \beta Y_{1,0} + \gamma Y_{1,-1}]$$

where λ represents the eigenvalue which can be $+\hbar$, 0 or $-\hbar$. Using Eqs (51)–(53), the above equation becomes

$$\frac{\alpha+\gamma}{\sqrt{2}}\hbar Y_{1,0} + \frac{\beta}{\sqrt{2}}\hbar(Y_{1,1} + Y_{1,-1})$$

$$= \lambda[\alpha Y_{1,1} + \beta Y_{1,0} + \gamma Y_{1,-1}]$$

Thus, $$\frac{\alpha+\gamma}{\sqrt{2}}\hbar = \lambda\beta$$

$$\frac{\beta}{\sqrt{2}}\hbar = \lambda\alpha = \lambda\gamma$$

Hence, $$\lambda = \hbar \Rightarrow \alpha = \gamma = \frac{1}{\sqrt{2}}\beta$$

$$\lambda = 0 \Rightarrow \alpha = -\gamma;\ \beta = 0$$

$$\lambda = -\hbar \Rightarrow \alpha = \gamma = -\frac{1}{\sqrt{2}}\beta$$

The three eigenfunctions are therefore given by Eqs (58)–(60).

Solution 15.7

$$\psi(\theta, \phi) = \frac{1}{2}\left[-\sqrt{\frac{3}{8\pi}}\sin\theta(e^{i\phi} - e^{-i\phi}) + \sqrt{\frac{3}{2\pi}}\cos\theta\right]$$

$$= \frac{1}{2}[Y_{1,1} + Y_{1,-1} + \sqrt{2}\,Y_{1,0}]$$

If we make a measurement of $\hat{L}_z$ we would get the values $+\hbar$, 0 or $-\hbar$ with probabilities $\frac{1}{4}, \frac{1}{2}$ and $\frac{1}{4}$ respectively. Further

$$\langle \hat{L}_z \rangle = \iint \psi^* (\theta, \phi)\ \hat{L}_z\, \psi (\theta, \phi) \sin\theta\, d\theta\, d\phi$$

$$= \frac{1}{4}\hbar - \frac{1}{4}\hbar + 0 = 0.$$

On the other hand, if we make a measurement of $\hat{L}_x$, we should express $\psi(\theta, \phi)$ as a linear combination of ψ_1, ψ_2 and ψ_3 which represent the eigenfunctions of $\hat{L}_x$ [see Eqs (55) – (57)]. We find

$$\psi(\theta, \phi) = \psi_1$$

and therefore if we make a measurement $\hat{L}_x$, we will for certain obtain the value $+\hbar$. Obviously,

$$\langle \hat{L}_x \rangle = +\hbar$$

REFERENCES AND SUGGESTED READINGS

1. P.A.M. Dirac, *The Principles of Quantum Mechanics*, Oxford University Press, Oxford (1958).
2. J.S. Townsend, *A Modern Approach to Quantum Mechanics*, McGraw-Hill, New York (1992).
3. J.J. Sakurai, *Modern Quantum Mechanics*, Addison-Wesley Publishing Company, Inc. (1994).
4. M. Beck, *Quantum Mechanics: Theory and Experiment*, Oxford University Press, New York (2012).
5. S.A. Werner R. Colella, A. W. Overhauser and C. F. Eagen, *Observation of the Phase Shift of a Neutron Due to Precession in a Magnetic Field, Phys. Rev. Lett.* 35, 1053 (1975).

16 Photon Polarization and Entanglement*

16.1 INTRODUCTION

In this chapter, we will first discuss the polarization characteristics of the photon; we will do this by introducing what are known as Jones matrices (see Section 16.2). In Section 16.3, we will discuss single photon experiments using polarization beam splitters. This will be followed by a section on the angular momentum of the photon. In Section 16.5, we will discuss entanglement of two photons and the famous EPR Paradox. Finally, in Section 16.6, we will have a short discussion on the entanglement of two spin-half particles.

16.2 JONES MATRICES

In 1941, using classical electromagnetic wave theory, R C Jones introduced what is now known as Jones calculus (Ref. 1). Using Jones calculus, one can easily determine the polarization state of the light beam emerging from a polarizer or a phase retarder like a quarter wave plate (usually abbreviated as QWP) or a half wave plate (usually abbreviated as HWP). For example, an x-polarized electromagnetic wave (propagating in the + z-direction) described by

$$E_x(z,t) = E_0\, e^{i(kz - \omega t)} \quad \text{and} \quad E_y(z,t) = 0 \tag{1}$$

is represented by the vector

$$|\mathbf{E}\rangle = E_0 |x\rangle \tag{2}$$

where

$$|x\rangle = \begin{pmatrix} 1 \\ 0 \end{pmatrix} \tag{3}$$

is the normalized Jones vector representing the x-polarized wave. The actual field is the real part of the right hand sides of Eq. (1). Similarly, a y-polarized beam (propagating in the + z-direction) described by

$$E_x(z,t) = 0 \quad \text{and} \quad E_y(z,t) = E_0\, e^{i(kz - \omega t)} \tag{4}$$

would be represented by the vector

$$|\mathbf{E}\rangle = E_0 |y\rangle \tag{5}$$

where

$$|y\rangle = \begin{pmatrix} 0 \\ 1 \end{pmatrix} \tag{6}$$

*This chapter has been adapted from Ref. 19; with permission.

is the normalized Jones vector representing the y-polarized wave. In writing $|\mathbf{E}\rangle$ [see Eqs. (2) and (5)], the phase factor $e^{i(kz-\omega t)}$ is suppressed. A linearly polarized wave (polarized along the x'-axis – see Fig. 16.1) is described by the (normalized) Jones vector:

$$|x'\rangle = \begin{pmatrix} \cos\theta \\ \sin\theta \end{pmatrix} = \cos\theta|\,x\rangle + \sin\theta|\,y\rangle \tag{7}$$

where θ is the angle between the x' and the x-axes; more details about polarization and Jones vectors can be found in References 2–5.

Now, when an ordinary light beam is incident on a Polaroid, the emergent light is linearly polarized with its electric vector oscillating in a particular direction—which is known as the pass axis of the Polaroid (see Fig. 16.1); a Polaroid is a plastic-like material used for producing linearly polarized light. If an unpolarized beam propagating in the z-direction is incident on a Polaroid whose pass-axis is parallel to the x'-axis, then the electric vector associated with the emergent wave will oscillate along the x'-axis. If this x'-polarized beam is incident on a Polaroid P_2 (whose pass axis is along the x-axis which makes an angle θ with the x'-axis – see Fig. 16.1) then the intensity of the emerging beam will be given by

$$I = I_0 \cos^2\theta \tag{8}$$

where I_0 represents the intensity of the emergent beam. The above equation is referred to as the Law of Malus. Figure 16.2 shows actual photographs of two Polaroids at different relative orientations. In Fig. 16.2 (a) the two are parallel to each other and therefore almost the entire light passes through. In Fig. 16.2 (b) the two Polaroids are oriented at 45° with respect to each other and about 50% of the light passes through; because according to Malus' law $I = I_0 \cos^2 45^\circ = \frac{1}{2} I_0$. Finally, in Fig. 16.2 (c) the two Polaroids are at right angles to each other (notice the position of the blue dot) and almost no light passes through because $I = I_0 \cos^2 90^\circ = 0$.

Now, a right circularly polarized wave (usually abbreviated as an RCP wave) propagating in the z-direction is a superposition of x and y-polarized waves with a phase difference of $\pi/2$:

$$E_x(z,t) = E_0\, e^{i(kz-\omega t)} \quad \text{and} \quad E_y(z,t) = E_0\, e^{i\left(kz-\omega t+\frac{\pi}{2}\right)} = iE_0\, e^{i(kz-\omega t)} \tag{9}$$

Thus neglecting the common phase factor, the normalized Jones vector representing the RCP wave will be described by

$$|\mathrm{RCP}\rangle = \frac{1}{\sqrt{2}}\left[|x\rangle + i|y\rangle\right] = \frac{1}{\sqrt{2}}\begin{pmatrix} 1 \\ i \end{pmatrix} \tag{10}$$

where we have introduced the factor $1/\sqrt{2}$ which would normalize the vector. Similarly, the normalized Jones vector representing the left circularly polarized wave (usually abbreviated as an LCP wave) propagating in the z-direction, will be described by

$$|\mathrm{LCP}\rangle = \frac{1}{\sqrt{2}}\left[|x\rangle - i|y\rangle\right] = \frac{1}{\sqrt{2}}\begin{pmatrix} 1 \\ -i \end{pmatrix} \tag{11}$$

The right or left circularly polarized waves can be produced by passing a linearly polarized wave through an appropriately oriented QWP (Quarter Wave Plate) and a left circularly

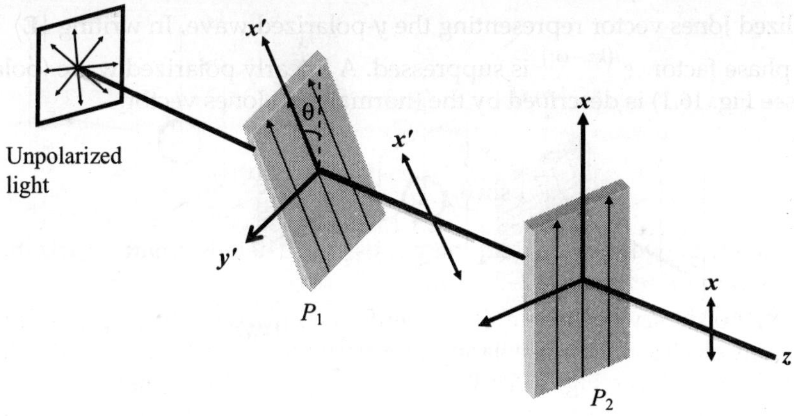

Figure 16.1 For an unpolarized wave propagating in the +z-direction, the electric vector (which lies in the x-y plane) continues to change its direction in a random manner. If an unpolarized light beam is allowed to fall on a Polaroid, then the emerging beam will be linearly polarized; i.e., the electric vector will oscillate along a particular direction. If we place another Polaroid P_2, then the intensity of the transmitted light will depend on the relative orientation of P_2 with respect to P_1.; if the pass axis of the second polaroid P_2 makes an angle θ with the x-axis, the intensity of the emerging beam will vary as $\cos^2 \theta$.

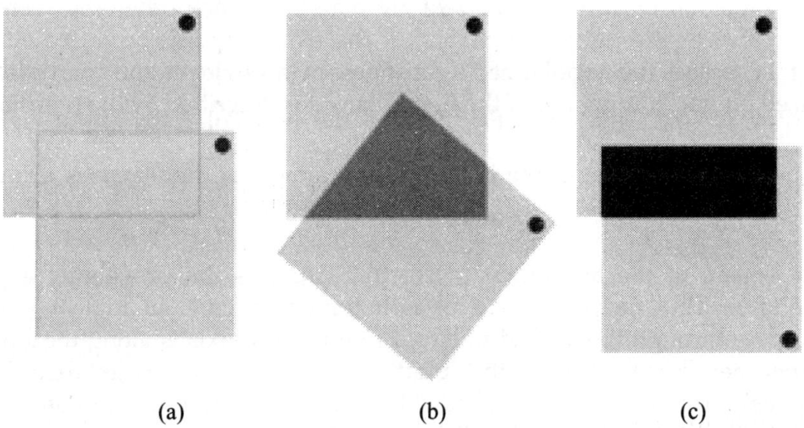

Figure 16.2 Actual photographs with two Polaroids at different angles of relative orientation. (a) If the two Polaroids are parallel to each other, almost the entire light passes through. (b) when the two Polaroids are oriented at with respect to each other about 50% of the light passes through; this is because . (c) when the two Polaroids are at right angles to each other (notice the position of the blue dot) almost no light will pass through. Photographs adapted from Alan Reed, *A- level Physics Tutor*, www.a-levelphysicstutor.com/; used with kind permission from Dr. Alan J. Reed.

polarized wave incident on a HWP (Half Wave Plate) will get converted to a right circularly polarized wave as shown in Fig. 16.3; more details can be found in any textbook on optics – see e.g., Chapter 22 of Ref. 2. Using Eqs. (10) and (11); we may write

$$|x\rangle = \begin{pmatrix} 1 \\ 0 \end{pmatrix} = \frac{1}{\sqrt{2}} \left[|\text{RCP}\rangle + |\text{LCP}\rangle \right] \tag{12}$$

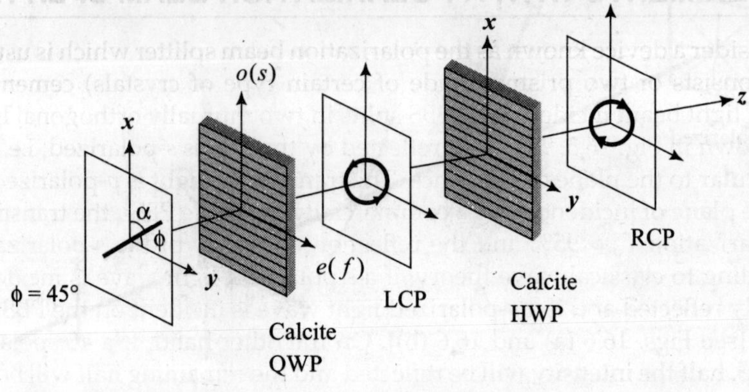

Figure 16.3 A linearly polarized beam making an angle 45° with the y-axis gets converted to a LCP after propagating through a calcite QWP; further, an LCP gets converted to a RCP after propagating through a calcite HWP. The optic axis in the QWP and HWP is along the y-direction as shown by lines parallel to the y-axis. More details can be found in Chapter 22 of Ref. 2.

which shows that an x-polarized wave can be expressed as a superposition of a right circularly polarized wave and a left circularly polarized waves. Similarly,

$$|y\rangle = \begin{pmatrix} 0 \\ 1 \end{pmatrix} = \frac{1}{i\sqrt{2}}\left[|\text{RCP}\rangle - |\text{LCP}\rangle\right] \tag{13}$$

In Fig. 16.1, if we replace the unpolarized light source by a single photon source (see Fig. 16.4), the x' polarized photon coming out of the first Polaroid is described by the (normalized) Jones vector

$$|x'\rangle = \begin{pmatrix} \cos\theta \\ \sin\theta \end{pmatrix} = \cos\theta|x\rangle + \sin\theta|y\rangle \tag{14}$$

The absolute square of the amplitude (= $\cos^2\theta$) represents the probability of detecting a x-polarized photon. Thus the probability for a photon polarized at an angle θ (with respect to the x-axis) to pass through the second polaroid (whose pass axis is along the x-axis) is $\cos^2\theta$ and if the experiment is conducted with N photons (and if N is very large) then about $N\cos^2\theta$ photons will pass through; one cannot predict the fate of an individual photon.

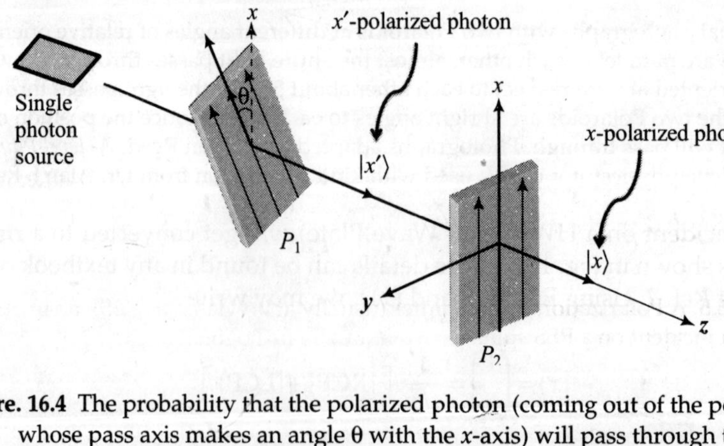

Figure. 16.4 The probability that the polarized photon (coming out of the polaroid P_1 whose pass axis makes an angle θ with the x-axis) will pass through the second polaroid P_2 (whose pass axis is along the x-axis) is $\cos^2\theta$.

16.3 EXPERIMENTS WITH A POLARIZATION BEAM SPLITTER

We next consider a device known as the polarization beam splitter which is usually abbreviated as PBS; it consists of two prisms (made of certain type of crystals) cemented together. An unpolarized light beam incident on a PBS splits in two mutually orthogonal linearly polarized beams as shown in Fig. 16.5. The light reflected by the PBS is *s*-polarized; i.e., the electric field is perpendicular to the plane of incidence. The transmitted light is *p*-polarized; i.e., the electric field is in the plane of incidence. For a commercially available PBS[1], the transmission efficiency of the *p*-polarization $T_p > 95\%$ and the reflection efficiency of the *s*-polarization $R_s > 99.9\%$. Thus, according to classical wave theory, if a *s*-polarized light wave is incident on the PBS, it will be totally reflected and if a *p*-polarized light wave is incident on the PBS, it will be totally transmitted [see Figs. 16.6 (a) and 16.6 (b)]. On the other hand, if a 45° polarized light wave falls on a PBS, half the intensity will be reflected and the remaining half will be transmitted (see Fig. 16.7). Now, what will happen if 45° polarized single photons fall on the PBS? Once again, a photon is indivisible and it cannot split in two halves. Quantum theory tells us that the photon is in both beams and in a state which is a superposition of two polarizations

$$\left|45^\circ\right\rangle = \frac{1}{\sqrt{2}}|s\rangle + \frac{1}{\sqrt{2}}|p\rangle \tag{15}$$

where $\left|45^\circ\right\rangle$ describes a 45° polarized photon, $|s\rangle$ describes a *s*-polarized photon and $|p\rangle$ describes a *p*-polarized photon. The square of the amplitude $\left|\frac{1}{\sqrt{2}}\right|^2 \left(=\frac{1}{2}\right)$ represents the probability of detecting a *s*-polarized photon, or a *p*-polarized photon. After passing through the PBS, the photon is in a superposition of the *s*-polarized state and a *p*-polarized state. There is half probability that it will be detected by the detector D_1 and half probability it will be detected by the detector D_2. No one can predict beforehand as to which detector will detect the photon. In fact, before its detection, it is in both the beams (at the same time)—but the process of detection makes it "collapse" to being either detected by the detector D_1 or by D_2. We can also use this arrangement to generate a set of random numbers (see Ref. 6; see also Chapter 2 of Ref. 2; Chapter 8 of Ref. 7).

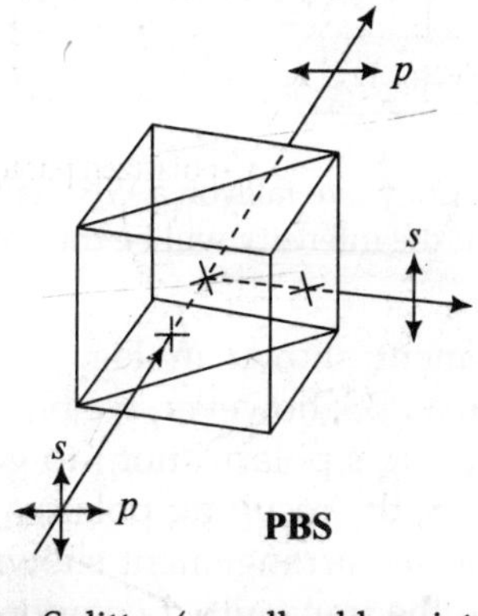

Figure. 16.5 A Polarization Beam Splitter (usually abbreviated as PBS); an unpolarized light beam incident on a PBS splits in two mutually orthogonal linearly polarized beams.

[1] See, e.g., www.chinasupply.net/optical/product/optics/beamsplitters/BBPBSC.html

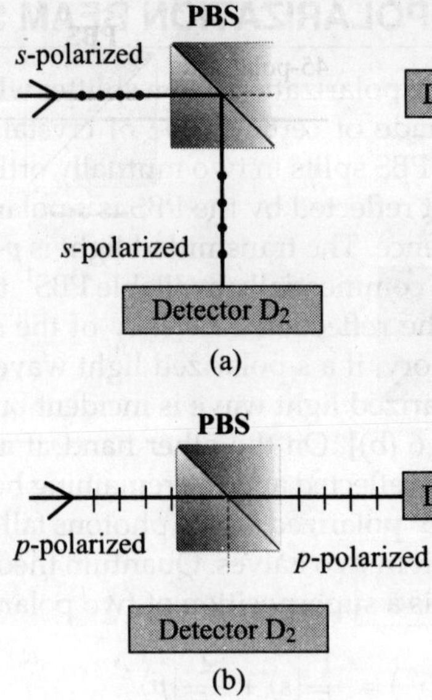

Figure. 16.6 A *s*-polarized light incident on the PBS is totally reflected and a *p*-polarized light incident on the PBS will be totally transmitted.

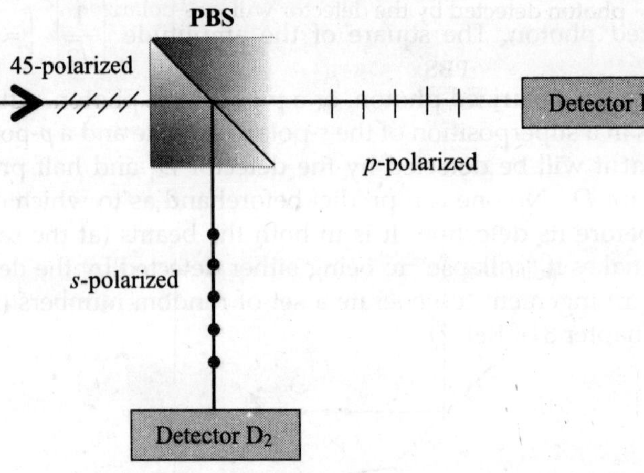

Figure. 16.7 If a 45° polarized light wave falls on a PBS, half the intensity will be reflected (as a s-polarized wave) and half the intensity will be transmitted (as a *p*-polarized wave).

We next consider the arrangement shown in Fig. 16.8 (a) where we have again 45° polarized light photons incident on a PBS; however, we block the transmitted component and allow the reflected component (having *s*-polarization) to get reflected by a mirror and then undergo another reflection by a PBS; the resulting polarization detected by the detector will be *s*-polarization. We next consider the arrangement shown in Fig. 16.8 (b) where we block the reflected component and allow the transmitted component (having *p*-polarization) to get reflected by a mirror and then undergo another transmission by a PBS; the resulting polarization detected by the detector will be *p*-polarization. Finally, we consider the arrangement shown in Fig. 16.9 where we allow both the transmitted and the reflected components to get reflected by mirrors and then incident on the second PBS; if the mirrors are properly adjusted, we would

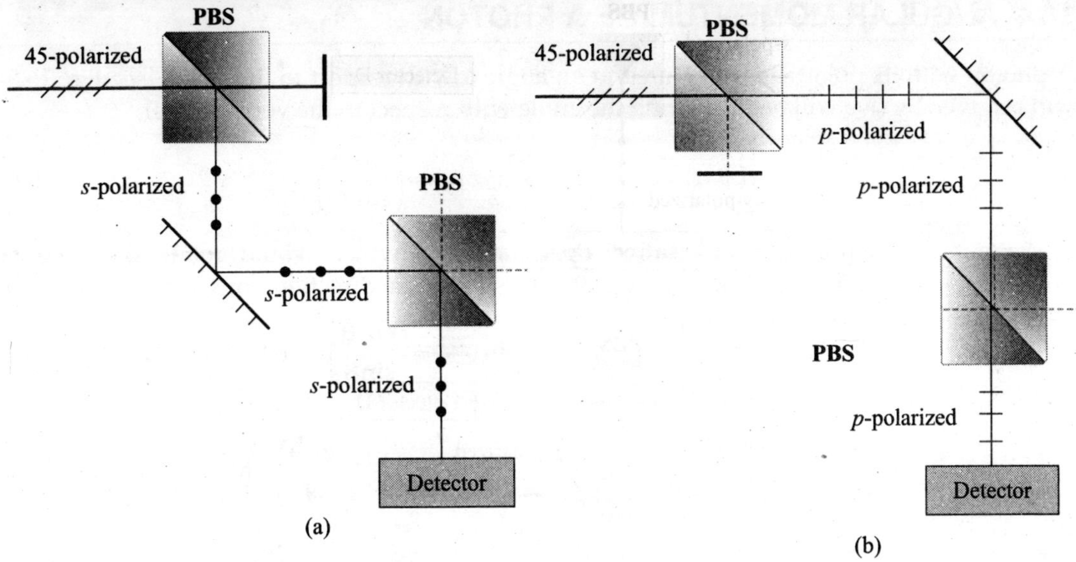

Figure. 16.8 (a) If a 45 polarized photon is incident on a PBS, and only the reflected beam is allowed to fall on the second PBS, then the photon detected by the detector will be s-polarized. (b) If a 45 polarized photon is incident on a PBS, and only the reflected beam is allowed to fall on the second PBS, then the photon detected by the detector will be s-polarized.

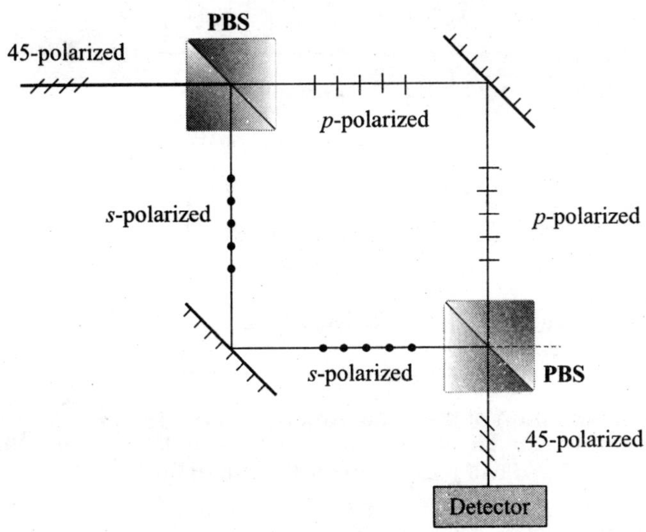

Figure. 16.9 If a 45 polarized photon is incident on a PBS, and both the reflected and transmitted beams are allowed to fall on the second PBS, then (if the distances are properly adjusted) the beam coming out of the second PBS will be 45 degree polarized; this can happen only if the photon is in both the paths; diagram adapted from Ref. 5.

obtain a photon polarized at 45°; *this can only happen if the photon is in a state of superposition of being in both paths* as described by Eq. (15). John Wheeler, the famous American physicist, said (quoted from Ref. 8):

The photon takes both paths, but it takes only one path.

16.4 ANGULAR MOMENTUM OF A PHOTON

A photon with its polarization oriented at an angle α with respect to the x-axis [see Fig. 16.3] will be given by (we will be measuring the angle with respect to the vertical axis)

$$|\text{LP } \alpha\rangle = \cos\alpha\,|x\rangle + \sin\alpha\,|y\rangle = \begin{pmatrix} \cos\alpha \\ \sin\alpha \end{pmatrix} \tag{16}$$

We define the polarization rotation operator $\hat{R}_z(\theta)$ which would rotate the state of polarization (about the z-axis) by an angle θ in the clockwise direction; thus

$$\hat{R}_z(\theta)|\,x\rangle = |\,\theta\rangle = \cos\theta|x\rangle + \sin\theta|y\rangle = \begin{pmatrix} \cos\theta \\ \sin\theta \end{pmatrix} = |\,x'\rangle \tag{17}$$

Similarly,

$$\hat{R}_z(\theta)|y\rangle = \left|\frac{\pi}{2}+\theta\right\rangle = -\sin\theta|\,x\rangle + \cos\theta\,|\,y\rangle = \begin{pmatrix} -\sin\theta \\ \cos\theta \end{pmatrix} = |y'\rangle \tag{18}$$

If we represent the polarization rotation operator $R_z(\theta)$ by a 2×2 matrix

$$\hat{R}_z(\theta) = \begin{pmatrix} a & b \\ c & d \end{pmatrix}$$

then

$$\hat{R}_z(\theta)|x\rangle = \begin{pmatrix} a & b \\ c & d \end{pmatrix}\begin{pmatrix} 1 \\ 0 \end{pmatrix} = \begin{pmatrix} a \\ c \end{pmatrix}$$

Comparing the above equation with Eq. (17), we get

$$a = \cos\theta \text{ and } c = \sin\theta$$

Similarly,

$$\hat{R}_z(\theta)|y\rangle = \begin{pmatrix} a & b \\ c & d \end{pmatrix}\begin{pmatrix} 0 \\ 1 \end{pmatrix} = \begin{pmatrix} b \\ d \end{pmatrix}$$

Comparing the above equation with Eq. (18), we get

$$b = -\sin\theta \text{ and } d = \cos\theta$$

Thus, the matrix representation of the polarization rotation operator $R_z(\theta)$ will be

$$\hat{R}_z(\theta) = \begin{pmatrix} \cos\theta & -\sin\theta \\ \sin\theta & \cos\theta \end{pmatrix} \tag{19}$$

One can easily see that

$$\hat{R}_z(\theta)|\,\phi\rangle = \begin{pmatrix} \cos\theta & -\sin\theta \\ \sin\theta & \cos\theta \end{pmatrix}\begin{pmatrix} \cos\phi \\ \sin\phi \end{pmatrix} = \begin{pmatrix} \cos(\theta+\phi) \\ \sin(\theta+\phi) \end{pmatrix} = |\,\theta+\phi\rangle \tag{20}$$

showing the rotation of the polarization state of the photon. If $|R\rangle$ and $|L\rangle$ are the unit vectors representing the right circularly polarized photon and the left circularly polarized photon, respectively, then

$$|R\rangle = \frac{1}{\sqrt{2}}\begin{pmatrix} 1 \\ i \end{pmatrix} = \frac{1}{\sqrt{2}}\left[|x\rangle + i\,|y\rangle\right] \quad \text{(RCP)} \tag{21}$$

$$|L\rangle = \frac{1}{\sqrt{2}}\begin{pmatrix}1\\-i\end{pmatrix} = \frac{1}{\sqrt{2}}\left[|x\rangle - i|y\rangle\right] \quad \text{(LCP)} \tag{22}$$

Operating $R_z(\theta)$ on $|R\rangle$, we get

$$\hat{R}_z(\theta)|R\rangle = \begin{pmatrix}\cos\theta & -\sin\theta\\ \sin\theta & \cos\theta\end{pmatrix}\frac{1}{\sqrt{2}}\begin{pmatrix}1\\i\end{pmatrix} = e^{-i\theta}\frac{1}{\sqrt{2}}\begin{pmatrix}1\\i\end{pmatrix}$$

Thus,

$$\hat{R}_z(\theta)|R\rangle = e^{-i\theta}|R\rangle \tag{23}$$

Similarly,

$$\hat{R}_z(\theta)|L\rangle = e^{+i\theta}|L\rangle \tag{24}$$

The above equations tell us that the right and left circularly polarized states are *eigenstates* of $\hat{R}_z(\theta)$; the corresponding eigenvalues are $e^{-i\theta}$ and $e^{+i\theta}$. Now, in Section 13.7 we have shown that if $R_z(\theta)$ represents the rotational operator corresponding to a rotation about the *z*-axis through an angle θ, then

$$\hat{R}_z(\theta) = \exp\left[-\frac{i}{\hbar}\theta\hat{J}_z\right] \tag{25}$$

where $\hat{J}_z$ represents the *z*-component of the angular momentum operator. Thus,

$$\exp\left[-\frac{i}{\hbar}\theta\hat{J}_z\right]|R\rangle = \hat{R}_z(\theta)|R\rangle = e^{-i\theta}|R\rangle \tag{26}$$

where, as before, the exponential of an operator $\mathcal{O}$ is defined by

$$e^{\mathcal{O}} \equiv 1 + \frac{\mathcal{O}}{1!} + \frac{\mathcal{O}\mathcal{O}}{2!} + \frac{\mathcal{O}\mathcal{O}\mathcal{O}}{3!} + \ldots \tag{27}$$

We expand the exponential on both sides of Eq. (26), and if we use the fact that Eq. (26) has to be valid for all values of θ, we must have[2]

$$\hat{J}_z|R\rangle = +\hbar|R\rangle \tag{28}$$

If we carry out a similar analysis for the left circularly polarized light, we obtain

$$\hat{J}_z|L\rangle = -\hbar|L\rangle \tag{29}$$

Thus, the right and left circularly polarized states are *eigenstates* of $\hat{J}_z$; the corresponding eigenvalues are $+\hbar$ and $-\hbar$, respectively. According to quantum mechanics, if we measure $\hat{J}_z$ of a right circularly polarized light photon, we will always obtain the value $+\hbar$; similarly, if we measure $\hat{J}_z$ of a left circularly polarized photon, we will obtain the value $-\hbar$. For an arbitrary state of polarization, if we measure $\hat{J}_z$, we will obtain one of the eigenvalues; i.e., we will obtain either the value $+\hbar$ or the value $-\hbar$. To obtain the probabilities of finding $+\hbar$ and $-\hbar$, the state should be expressed as a superposition of the eigenstates, which are the right circularly polarized state and the left circularly polarized state. For example, a *x*-polarized state can be represented as

$$|x\rangle = \begin{pmatrix}1\\0\end{pmatrix} = \frac{1}{\sqrt{2}}\left[|R\rangle + |L\rangle\right] \tag{30}$$

---

[2] Thus, for example $J_z^3|R\rangle = J_zJ_zJ_z|R\rangle = +\hbar^3|R\rangle$

Thus, if we make a measurement of $\hat{J}_z$ on a x-polarized (or on a y-polarized) state, then there will be half probability of obtaining $+\hbar$ and half probability of obtaining $-\hbar$; one can never predict the precise outcome of an experiment. As another example, let us consider a left elliptically polarized state described by the following equation

$$|LEP\rangle = \frac{1}{2}\begin{pmatrix} 1 \\ -i\sqrt{3} \end{pmatrix}$$

We must express this as a linear superposition of $|R\rangle$ and $|L\rangle$:

$$\frac{1}{2}\begin{pmatrix} 1 \\ -i\sqrt{3} \end{pmatrix} = a|R\rangle + b|L\rangle = \frac{a}{\sqrt{2}}\begin{pmatrix} 1 \\ +i \end{pmatrix} + \frac{b}{\sqrt{2}}\begin{pmatrix} 1 \\ -i \end{pmatrix}$$

Thus,
$$\frac{a+b}{\sqrt{2}} = \frac{1}{2} \quad \Rightarrow \quad a+b = \frac{1}{\sqrt{2}} \approx 0.707$$

and
$$\frac{i(a-b)}{\sqrt{2}} = \frac{-i\sqrt{3}}{2} \quad \Rightarrow \quad a-b = -\sqrt{\frac{3}{2}} \approx -1.225$$

Simple manipulations will give $a \approx -0.2588$ and $b \approx +0.9659$. Thus, if we make a measurement of $\hat{J}_z$ on such an elliptically polarized state, we obtain one of the eigenvalues—the probability of obtaining $+\hbar$ will be about 0.0670, and the probability of obtaining $-\hbar$ will be about 0.933.

16.5 EPR PARADOX AND BELL'S INEQUALITY: A SIMPLE ANALYSIS

16.5.1 Experiments with a Doubly Refracting Crystal

When an unpolarized light beam is incident normally on a calcite crystal, it usually splits up into two linearly polarized beams (see Fig. 16.10). If the calcite crystal is suitably oriented, then no matter what the incident state of polarization may be, it will always split into a x-polarized beam and a y-polarized beam (see Fig. 16.11); such an arrangement is often referred to as a (x–y) device. Thus, when a x-polarized photon (from a single photon source) is incident normally on the $(x-y)$ device, it will propagate as an x-polarized photon and when a y-polarized photon is

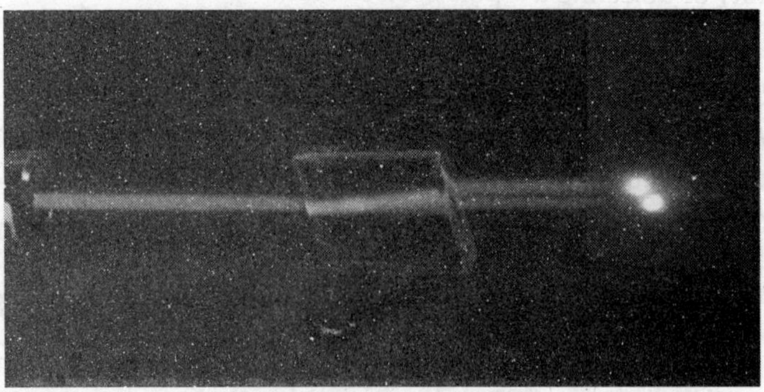

Figure. 16.10 When an unpolarized light beam is incident normally on a calcite crystal, it usually splits up into two linearly polarized beams. Photograph courtesy Professor V Lakshminarayanan and adapted from Ref. 9.

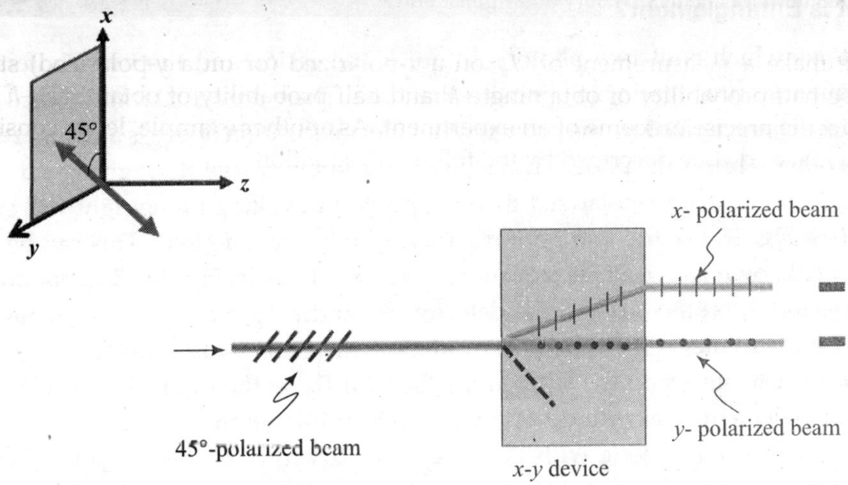

Figure. 16.11 When a linearly polarized light beam (making an angle θ with the *x*-axis) is incident on a (suitably oriented) calcite crystal, it would, in general, split up into an *x*-polarized beam and a *y*-polarized beam. In the figure we have assumed θ = 45°.

incident normally on the crystal, it will propagate as a *y*-polarized photon. What will happen if a 45° polarized photon is incident on the calcite crystal (see Fig. 16.11)? We can express a 45° polarized photon by the vector:

$$\left|\text{LP } 45^\circ\right\rangle = \begin{pmatrix} \cos 45^\circ \\ \sin 45^\circ \end{pmatrix} = \frac{1}{\sqrt{2}}|x\rangle + \frac{1}{\sqrt{2}}|y\rangle \tag{31}$$

The photon gets detected either by the detector D_1 or by D_2 (never by both); there is half probability of it getting detected by the detector D_1 and half probability of it getting detected by the detector D_2. No one can predict beforehand as to whether it will be detected by D_1 or by the detector D_2. In fact before detection, the photon is in a state of superposition of being in both paths as described by the above equation. If we rotate the calcite crystal as shown in Fig. 16.12, we will obtain an (*u*–*v*) device where the (*u*–*v*) axes are obtained by rotating the (*x*–*y*) axes as shown in Fig. 16.12. Thus, when an arbitrarily polarized photon is incident normally on the (*u*–*v*) device, there will be a certain probability of detecting an *u*-polarized photon and a certain probability of detecting a *v*-polarized photon.

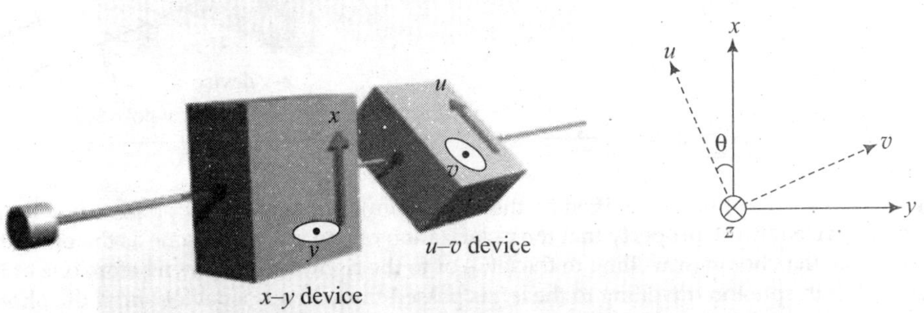

Figure. 16.12 By rotating the (*x*–*y*) device by a certain angle about the *z*-axis, we obtain an (*u*–*v*) device. Adapted from http://www.upscale.utoronto.ca/PVB/Harrison/BellsTheorem/BellsTheorem.html

16.5.2 What is Entanglement?

There are atoms which emit two photons in quick succession and the two photons are of different frequencies propagating in opposite directions. There are actual transitions where the two photons are characterized by the property that the polarization of one is always orthogonal to that of the other. Thus if the photon travelling to the left is passed through a *x-y* device and if the photon is found to be *x*-polarized then the photon travelling to the right will surely be *y*-polarized (see Fig. 16.13); the two photons are said to be "entangled". This can be seen by making what is known as "coincidence" measurements. Thus in Fig. 16.13, there are events when the detector D_1 on the left and the detector D_4 on the right click simultaneously; such simultaneous measurements are called "coincidence" measurements. However, there are no events when the detector D_1 on the left and the detector D_3 on the right click simultaneously. The word "entanglement" was introduced by Erwin Schrödinger in a paper published in 1935 (Ref. 10). In 1935, Einstein (along with Podolsky and Rosen) published a paper (Ref. 11) in which they argued that if quantum theory was correct, then two particles (which are millions of kilometers apart) can be entangled in the sense that by determining a property of one of the particles, the property of the second particle can be instantaneously changed. And special theory of relativity forbids the transmission of any signal faster than the speed of light. This came to be known as *The EPR Paradox*. About thirty years later, experiments confirmed the predictions of quantum mechanics namely Einstein's impossible proposition was in fact correct: instantaneous changes in widely separated systems did occur.

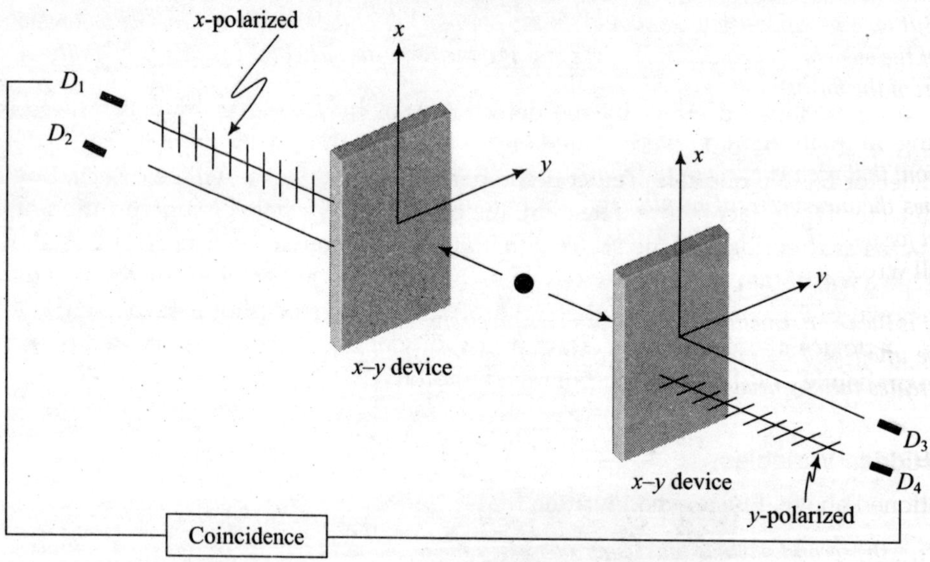

Figure. 16.13 Two photons (emitted by the same atom) propagating in opposite directions are characterized by the property that the polarization of one is always same as the other; the polarization of the photon (travelling to the left – or to the right) is not known before one of them is measured. If the photon travelling to the left is passed through a *x–y* device and if the photon is found to be *x*-polarized then the photon travelling to the right will also be *x*-polarized; we know the polarization of the photon moving to the right without measuring it.

16.5.3 Non-Locality of Quantum Theory

According to quantum theory, the polarization of the photon (traveling to the left – or to the right) is not known before one of them is measured; and, if the polarization of the photon going to the left is measured as x-polarized then the polarization of the photon going to the right collapses to state in which its polarization state is y-polarized. Similarly, if the polarization of the photon going to the left is measured as u-polarized then the polarization of the photon going to the right collapses to state in which its polarization state is for sure v-polarized.

The two photons can be hundreds of kilometers apart. This collapse of the wave function is usually referred to as the "Copenhagen interpretation" of quantum mechanics. Thus quantum theory is "non-local"—i.e., measurement on one particle can instantaneously affect the state of another particle which is very far away. Einstein could never accept this; he felt that this will contradict special theory of relativity according to which no signal can travel faster than light. Einstein called this "Spooky action at a distance". Einstein believed in "locality"; i.e., measuring the state of one particle will not affect the other. Einstein believed that if one could predict the polarization of the second photon by measuring the polarization of the first photon, then both photons had definite states of polarization all the time. Thus associated with the photons are "hidden variables" so that the polarization of the photons are known before the measurements are carried out. In their 1935 paper (Ref. 11), EPR wrote:

> *if without in any way disturbing a system, we can predict with certainty (i.e., with probability equal to unity) the value of a physical quantity, then the second particle must have possessed the measured property before the measurement was carried out.*

Einstein had also written (p. 85 in Ref. 12; see also Ref. 13)

> *But on one supposition we should, in my opinion, absolutely hold fast: the real factual situation of the system S_2 is independent of what is done with the system S_1, which is spatially separated from the former.*

According to Alan Aspect "*Einstein therefore argued for what he felt was the only reasonable description: that each particle in the pair carries a property, decided at the moment of separation, which determines the measurement results. But since entangled particles are not described separately in the quantum formalism, Einstein concluded the formalism was incomplete*". In his 1964 paper (Ref. 13), John Bell wrote

> *It is the requirement of locality, or more precisely that the result of a measurement on one system be unaffected by operations on a distant system with which it has interacted in the past, that creates the essential difficulty.*

16.5.4 Hidden Variables

As mentioned above, Einstein had written that

> "*.... the second particle must have possessed the measured property before the measurement was carried out...*".

Assuming this, we construct a theory based on "hidden variables" so that we know the number of photon pairs which have their own independent attributes when passed through a (x, y) device or a (u, v) device or a (σ, η) device; the (x, y) axes, (u, v) axes and (σ, η) axes are defined in Fig. 16.14. Thus we can assign values of all observables of a system *before* the measurement is carried out. For example, if a photon described by $\{x, u, \sigma\}$ is passed through a

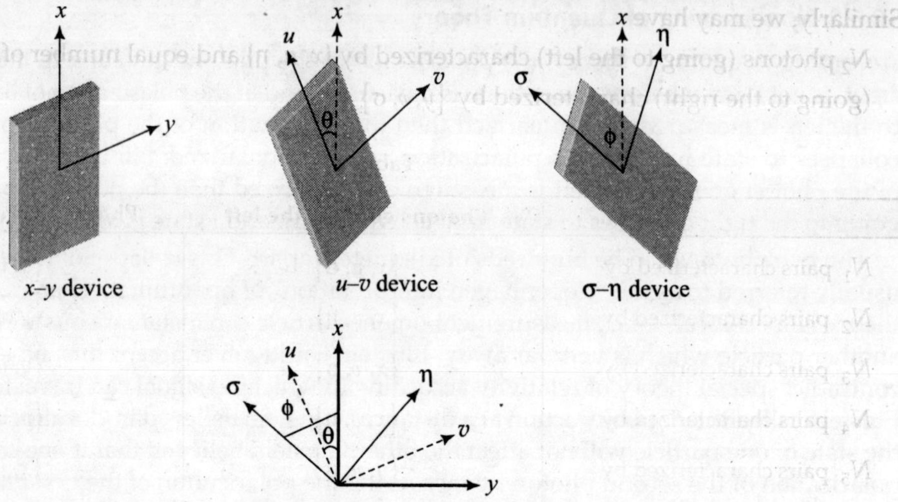

Figure. 16.14 The (u, v) axes are obtained by rotating the (x, y) axes by an angle θ; similarly the (σ, η) axes are obtained by rotating the (x, y) axes by an angle ϕ.

(x, y) device, *we will for sure* measure x-polarization; if it were to pass through a (u, v) device, *we will for sure* measure u-polarization and if it were to pass through a (σ, η) device, *we will for sure* measure σ-polarization. Further, if a photon described by $\{x, u, \sigma\}$ is propagating to the left then the photon propagating to the right will be described by $\{y, v, \eta\}$ (see Fig. 16.15) so that if the photon propagating to the right is passed through a (x, y) device, *we will for sure* measure y-polarization, if it is passed through a (u, v) device, *we will for sure* measure v-polarization and if it is passed through a (σ, η) device, *we will for sure* measure η-polarization.

Let there be

N_1 photons (going to the left) characterized by $\{x, u, \sigma\}$ and equal number of photons (going to the right) characterized by $\{y, v, \eta\}$

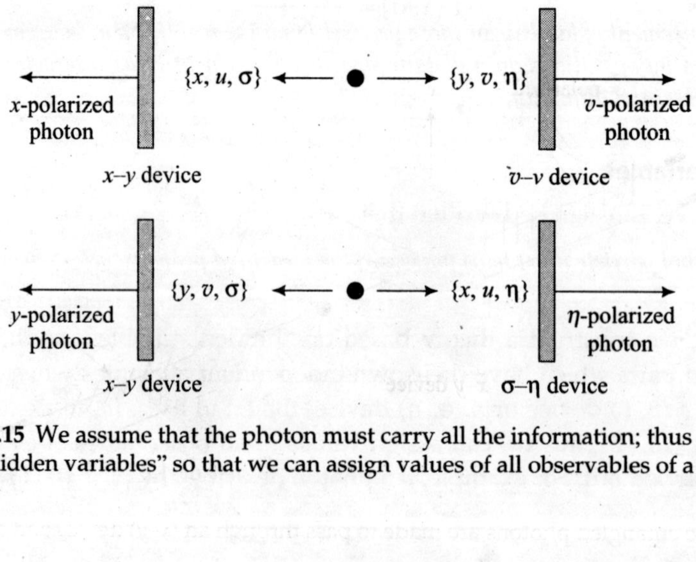

Figure. 16.15 We assume that the photon must carry all the information; thus there must exist "hidden variables" so that we can assign values of all observables of a system.

Similarly, we may have

N_2 photons (going to the left) characterized by $\{x, u, \eta\}$ and equal number of photons (going to the right) characterized by $\{y, v, \sigma\}$

Table 16.1

	Photons going to the left	Photons going to the right
N_1 pairs characterized by	$\{x, u, \sigma\}$	$\{y, v, \eta\}$
N_2 pairs characterized by	$\{x, u, \eta\}$	$\{y, v, \sigma\}$
N_3 pairs characterized by	$\{x, v, \sigma\}$	$\{y, u, \eta\}$
N_4 pairs characterized by	$\{x, v, \eta\}$	$\{y, u, \sigma\}$
N_5 pairs characterized by	$\{y, u, \sigma\}$	$\{x, v, \eta\}$
N_6 pairs characterized by	$\{y, u, \eta\}$	$\{x, v, \sigma\}$
N_7 pairs characterized by	$\{y, v, \sigma\}$	$\{x, u, \eta\}$
N_8 pairs characterized by	$\{y, v, \eta\}$	$\{x, u, \sigma\}$

Obviously, we can have 8 possible pairs as listed in Table 16.1. If $P(x, u)$ represents the probability that the photon (going to the left) is x-polarized and the photon (going to the right) is u-polarized (see Fig. 16.16) then from Table 16.1 it is obvious that

$$P(x, u) = \frac{N_3 + N_4}{N} \tag{32}$$

where

$$N = N_1 + N_2 + N_3 + N_4 + N_5 + N_6 + N_7 + N_8 \tag{33}$$

Similarly, if $P(x, \sigma)$ represents the probability that the photon (going to the left) is x-polarized and the photon (going to the right) is σ-polarized then (again from Table 16.1)

$$P(x, \sigma) = \frac{N_2 + N_4}{N} \tag{34}$$

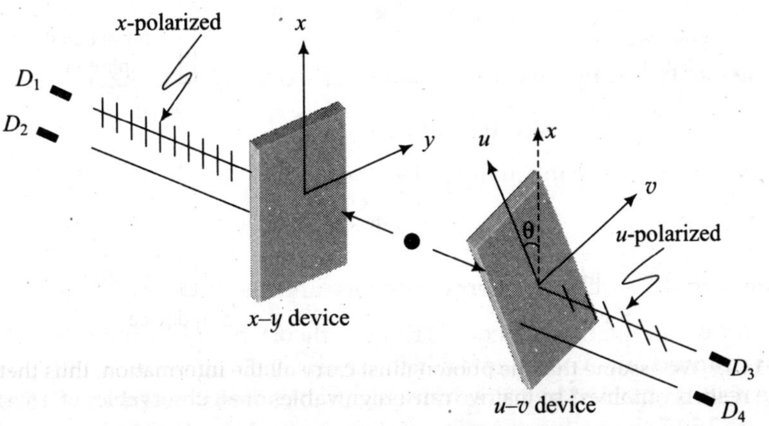

Figure. 16.16 The entangled photons are made to pass through an $(x–y)$ device and an $(u–v)$ device.

Similarly, if $P(\sigma, u)$ represents the probability that the photon (going to the left) is σ-polarized and the photon (going to the right) is u-polarized then (again from Table 16.1)

$$P(\sigma, u) = \frac{N_3 + N_7}{N} \tag{35}$$

Thus, $$P(x, \sigma) + P(\sigma, u) = \frac{N_2 + N_4 + N_3 + N_7}{N}$$

Since, $$N_3 + N_4 \leq N_2 + N_4 + N_3 + N_7$$

we will have

$$P(x, u) \leq P(x, \sigma) + P(\sigma, u) \tag{36}$$

This is a simple form of Bell's inequality. In 1964, John Bell published his famous theorem (now referred to as Bell's theorem) in a paper entitled "On the Einstein-Podolsky-Rosen Paradox" (Ref. 12). Many experiments have shown that above inequality is violated. The theorem was a revolutionary contribution to the understanding of quantum mechanics.

16.5.5 Results from Quantum Theory

We will now calculate expressions for $P(x, u)$, $P(x, \sigma)$ and $P(\sigma, u)$ using simple quantum theory. If N represents the total number of photons (going to the left) then $N/2$ photons will appear x-polarized. The corresponding photon going to the right will appear y-polarized. Thus, the probability that the photon (going to the left) is x-polarized and the photon (going to the right) is u-polarized (see Fig. 16.16) will be given by:

$$P(x, u) = \frac{1}{2} \cos^2\left(\frac{\pi}{2} + \theta\right) = \frac{1}{2} \sin^2 \theta$$

where the angle θ is defined in Fig. 16.14. Similarly,

$$P(x, \sigma) = \frac{1}{2} \sin^2 \phi$$

where the angle ϕ is defined in Fig 16.14. Similarly,

$$P(\sigma, u) = \frac{1}{2} \sin^2 (\phi - \theta)$$

Thus if Bell's inequality is consistent with quantum theory, we must have

$$\sin^2 \theta \leq \sin^2 \phi + \sin^2 (\phi - \theta) \tag{37}$$

If we assume $\theta = 2\phi$, the above inequality takes the form

$$\sin^2 2\phi \leq 2 \sin^2 \phi \tag{38}$$

which is violated for $\phi = \frac{\pi}{6}$. Thus, quantum theory and Bell's inequality are not compatible which implies that either quantum theory is right or theory based on hidden variables, but not both. Many have carried out experiments which are always in agreement with quantum theory and against the results obtained by using hidden variables (see References 14,15 and references in Ref. 16). Figure 16.17 shows the experimental set up of Aspect and his co-workers (Ref. 13)

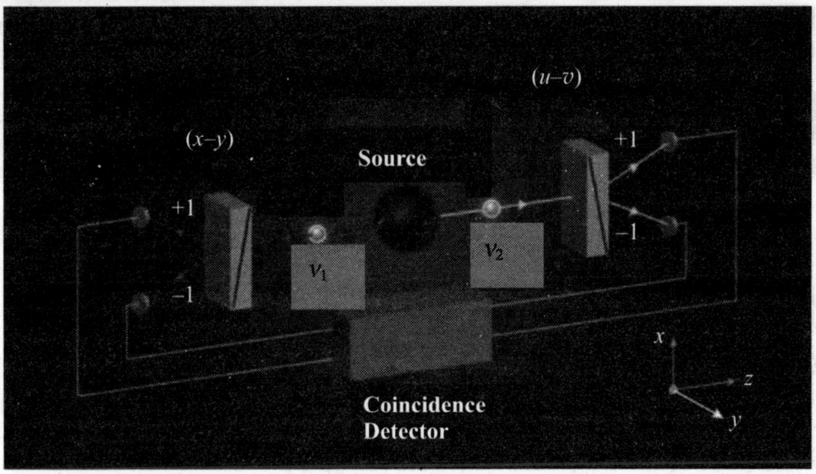

Figure. 16.17 An apparatus for performing Bell test. The photon ν_1 after passing through the $(x\text{–}y)$ device will emerge as x-polarized or, as y-polarized. Similarly, the photon ν_2 after passing through the $(u\text{–}v)$ device will emerge as u-polarized or, as v-polarized. Adapted from Ref. 13.

for performing Bell test. The photon ν_1 after passing through the $(x\text{–}y)$ device will emerge as x-polarized or, as y-polarized. Similarly, the photon ν_2 after passing through the $(u\text{–}v)$ device will emerge as u-polarized or, as v-polarized. In Ref. 17, Aspect etal write:

> *"The linear-polarization correlation of pairs of photons emitted in a radiative cascade of calcium has been measured. The new experimental scheme, using two-channel polarizers (i.e., optical analogs of Stern-Gerlach filters* [see Section 16.6]*), is a straightforward transposition of Einstein-Podolsky-Rosen-Bohm gedanken experiment. The present results, in excellent agreement with the quantum mechanical predictions, lead to the greatest violation of generalized Bell's inequalities ever achieved."*

John Bell has written:

> *It seems to me then beyond dispute that there was at least one Einstein, that of the EPR paper and the Schilpp volume [Ref. 11], who was fully committed to the view that quantum mechanics was incomplete and should be completed—which is the hidden variable program.*

Bell's theorem proved that the above view (of Einstein) is incorrect!

16.5.6 Correlation Function

We define a correlation function C

$$C = P(x, v) + P(y, u) - P(x, u) - P(y, v) \tag{39}$$

Now, according to quantum theory

$$P(x, u) = \frac{1}{2}\sin^2\theta$$

Similarly,

$$P(x, v) = \frac{1}{2}\cos^2\theta\,,\; P(y, u) = \frac{1}{2}\cos^2\theta \text{ and } P(y, v) = \frac{1}{2}\sin^2\theta$$

Thus, according to quantum theory, the correlation function C is given by

$$C = P(x, v) + P(y, u) - P(x, u) - P(y, v) = \cos^2\theta - \sin^2\theta$$

Or,
$$C = \cos 2\theta \tag{40}$$

The above variation of the correlation function C with θ was verified very accurately by experiments carried out by Freedman and Clauser in 1972 (Ref. 14).

16.6 ENTANGLEMENT OF TWO SPIN-HALF PARTICLES

Consider a molecule composed of two atoms A and B each of which has a spin $\frac{1}{2}$; i.e., angular momentum $\frac{1}{2}\hbar$. Assume that the spin of the molecule itself is 0. Suppose at time $t = 0$, the molecule dissociates and the component atoms A and B, go off in opposite directions. Soon they will be far apart so that we can think of them as isolated from each other so that we would expect that measurements on A do not disturb B and vice-versa. Now, in classical as well as in quantum mechanics, angular momenta of isolated systems stay constant. This means that each of the components of spin of the system will be conserved (i.e., will be unchanged). Let us denote the components of spin of A as $s_x(\mathrm{A})$, $s_y(\mathrm{A})$ and $s_z(\mathrm{A})$; and similarly the components of B. In Quantum Mechanics, there is then the additional feature that each of these components can only take the values $+\frac{1}{2}\hbar$ or $-\frac{1}{2}\hbar$. It will be more convenient to talk of the integer quantities $\sigma_x \equiv 2s_x/\hbar$, $\sigma_y \equiv 2s_y/\hbar$ etc., whose values can be +1 or –1. For simplicity, we shall refer to these components of σ as 'spin'.

Now the spin components of the molecule are all zero—since its total spin is zero and conservation of angular momentum demands:

$$\left.\begin{aligned}\sigma_x(\mathrm{A}) + \sigma_x(\mathrm{B}) = 0\\ \sigma_y(\mathrm{A}) + \sigma_y(\mathrm{B}) = 0\\ \sigma_z(\mathrm{A}) + \sigma_z(\mathrm{B}) = 0\end{aligned}\right\} \tag{41}$$

Thus, measurement of a component of spin of either atom determines the value of the same component of the other atom. For example, if we pass A through a Stern Gerlach apparatus and obtain it in the "*x*-up" state – i.e., obtain $\sigma_x(\mathrm{A}) = +1$, then we would have inferred that $\sigma_x(\mathrm{B}) = -1$. The spin components of A and B are correlated—a result of the conservation laws applicable to classical systems as well. But QM does impose two restrictions. The first, as we have already mentioned, is that the only possible results that we can obtain by measuring any spin component are $\pm\frac{1}{2}\hbar$ (or, in terms of the Pauli variables $\sigma, \pm 1$). The second is that only one component of spin of each atom can have definite values. For example, we cannot talk of atom A being simultaneously in the spin states specified by $\sigma_x(\mathrm{A}) = +1$ and $\sigma_y(\mathrm{A}) = -1$.

The above discussion is very similar to our discussion in Section 16.5.2 where we had 2 entangled photons (of different frequencies)—and if the photon travelling to the left is found to be *x*-polarized then the photon travelling to the right will surely be *y*-polarized. We can carry out similar entanglement experiments; however, instead of (*x*–*y*) and (*u*–*v*) devices that we used, we will now have to use the Stern Gerlach device which determines the spin state of the atom.

REFERENCES AND SUGGESTED READINGS

1. Jones, R. C., "New calculus for the treatment of optical systems I. Description and discussion of the calculus," *Journ. Opt. Soc. Am.*, **31**, 488–493 (1941).
2. Ajoy Ghatak, *Optics*, 6th Edition, McGraw Hill Education (India) Private Limited, New Delhi (2017).
3. Arun Kumar and Ajoy Ghatak, Polarization of Light With Applications in Optical Fibers, SPIE Tutorial Texts Vol. TT90, SPIE Press, USA (2011).
4. Shurcliff, W. A., "*Polarized light: Production and use*," Harward University Press, Cambridge, MA, 1962.
5. Azzam, R.M.A. and N.M. Bashara, "*Ellipsometry and polarized light*," North Holland Publishing Company, Netherlands, (1977).
6. T. Jennewein, U. Achleitner, G. Weihs, H. Weinfurter and A. Zeilinger, *A fast and compact quantum random number generator*, Rev. Sci. Instr. **71**, 1675–1680 (2000)
7. Ajoy Ghatak, Anirban Pathak and V P Sharma (Editors), *Light and Its Many Wonders*, VIVA Books, New Delhi (2015).
8. Anton Zeilenger, *Dance of the Photons: From Einstein to Quantum Teleportation*, Farrar, Straus and Giroux, New York (2010).
9. Guy Ropars, Albert Le Flocha and Vasudevan Lakshminarayanan, *The sunstone and polarised skylight: ancient Viking navigational tools?*, Contemporary Physics, 2014.
10. E. Schrödinger, *Discussion of Probability Relations between separated Systems*, Proceedings of the Cambridge Philosophical Society **31**, 555–563 (1935); **32**, 446–451 (1936). Gordon Baym, *Lectures on Quantum Mechanics*, Benjamin (1965).
11. A. Einstein, B. Podolsky, and N. Rosen, Can quantum-mechanical description of physical reality be considered complete?, Phys. Rev. **47**, 777 (1935).
12. A. Einstein, *Autobiographical Notes* (Translated from German by Paul Arthur Schilpp) in *Albert Einstein: Philosopher Scientist*, (Editor: Paul Arthur Schilpp), Second Edition, Tudor Publishing Co., New York (1951).
13. J.S. Bell, "*On the Einstein-Podolsky-Rosen Paradox*", Physics **1**, 195–200 (1964).
14. S. J. Freedman and J. F. Clauser, *Experimental Test of Local Hidden-Variable Theories*, Phys. Rev. Lett. **28**, 938–941 (1972).
15. E. S. Fry and R. C. Thompson, "Experimental Test of Local Hidden-Variable Theories, Phys. Rev. Letts. **37**, 465 (1976).
16. A. Aspect, *Closing the Door on Einstein and Bohr's Quantum Debate* in *Physics* **8**, 123, 16 Dec 2015; see https://physics.aps.org/articles/v8/123.
17. A. Aspect, P. Grangier, and G. Roger, *Experimental Realization of Einstein-Podolsky-Rosen-Bohm Gedankenexperiment: A New Violation of Bell's Inequalities*, Physical Review Letters, Vol. 49, No. **2**, pp. 91–94 (1982).
18. G. Venkataraman, *Quantum Revolution III: What Is Reality?*, Universities Press India Ltd., (1994).
19. Ajoy Ghatak, *Experiments with Light*, Viva Books, New Delhi (2015).

The Double-Well Potential and the Kronig-Penney Model

17.1 INTRODUCTION

In this chapter, [illegible]

17.2 TWO ISOLATED WELLS

Before we [illegible] Fig. 17.1. [illegible]

Figure 17.1: The first three eigenvalues and eigenfunctions of two isolated wells [illegible]

[1] Actually [illegible]

17 The Double-Well Potential and the Krönig-Penney Model

I look upon quantum mechanics with admiration and suspicion.

— ALBERT EINSTEIN in a letter[1] to his intimate friend Paul Ehrenfest, dated August 28, 1926.

17.1 INTRODUCTION

In this chapter, we will first discuss the solutions of the Schrödinger equation for the double-well problem which will be followed by a discussion of the ammonia molecule as a two-state system. We will also discuss the Krönig-Penney model leading to the formation of energy bands in a solid.

17.2 TWO ISOLATED WELLS

Before we consider a double-well structure, we consider two isolated wells as shown in Fig. 17.1. The potential energy variation corresponding to the well on the left is given by

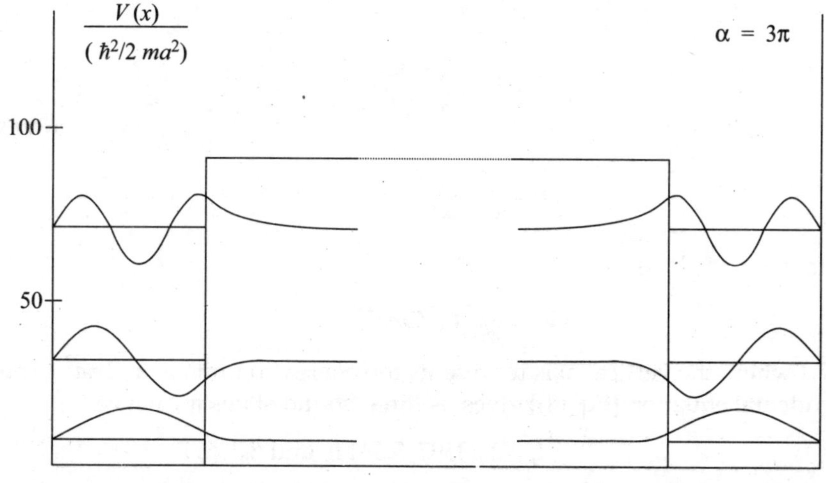

Figure 17.1: The first three eigenvalues and eigenfunctions of two isolated wells for $\alpha = 3\pi$.

[1] Actually the original letter was in German and it read *Der Quanten-Mechanik stehe ich bewundernd-misstrauisch gegenüber....* [Quoted from M. Jammer, *The Conceptual Development of Quantum Mechanics*, McGraw-Hill Book Co., New York (1966)].

$$V(x) = \begin{cases} \infty & x < 0 \\ 0 & 0 < x < a \\ V_0 & x > a \end{cases} \tag{1}$$

The Schrödinger equation in regions $0 < x < a$ and $x > a$ would be given by

$$\frac{d^2\psi}{dx^2} + k^2\psi(x) = 0 \quad \text{for} \quad 0 < x < a \tag{2}$$

and

$$\frac{d^2\psi}{dx^2} - \kappa^2\psi(x) = 0 \quad \text{for} \quad x > a \tag{3}$$

where

$$k^2 = \frac{2mE}{\hbar^2} \tag{4}$$

and

$$\kappa^2 = \frac{2m}{\hbar^2}(V_0 - E) \tag{5}$$

We are assuming $E < V_0$. As discussed in Section 6.2, the wavefunction should vanish at $x = 0$ so that

$$\psi(x) = A \sin kx \quad \text{for} \quad 0 < x < a \tag{6}$$

and

$$\psi(x) = Ce^{-\kappa(x-a)} \quad \text{for} \quad x > a \tag{7}$$

where we have rejected the exponentially amplifying solution in the region $x > a$. Continuity of $\psi(x)$ and $d\psi/dx$ at $x = a$ gives us

$$-\xi \cot \xi = \sqrt{\alpha^2 - \xi^2} \tag{8}$$

where

$$\xi = ka = \sqrt{\frac{2mEa^2}{\hbar^2}} \tag{9}$$

and

$$\alpha = \sqrt{\frac{2mV_0a^2}{\hbar^2}} \tag{10}$$

Let us assume $\alpha = 3\pi$. Thus,

$$V_0 = 9\pi^2 (\hbar^2/2ma^2)$$

(see Fig. 17.1 where the vertical axis represents the energy in units of $\hbar^2/2ma^2$.) The solution of the transcendental equation [Eq. (8)] gives us three bound states for which

$$\xi = 2.83595,\ 5.64146 \text{ and } 8.33877 \tag{11}$$

We define the normalized energy eigenvalue

$$\varepsilon = \frac{E}{\hbar^2/2ma^2} = \xi^2 \tag{12}$$

Thus, the corresponding discrete values of ε are given by

$$\varepsilon = 8.0426,\ 31.8261 \text{ and } 69.5351 \tag{13}$$

The energy eigenvalues (represented by horizontal lines) and the corresponding eigen functions are shown in Fig. 17.1. Obviously, the particle would belong to one of the wells and there will not be any tunnelling from one well to the other.

17.3 THE DOUBLE-WELL POTENTIAL

We next consider the solution of the one-dimensional Schrödinger equation (see Eq. (11) of Chapter 6) for a double-well potential characterized by the following potential energy distribution

$$V(x) = \begin{cases} V_0 & \text{for} \quad 0 < |x| < b \\ 0 & \text{for} \quad b < |x| < a + b \\ \infty & \text{for} \quad |x| > a + b \end{cases} \tag{14}$$

(see Fig. 17.2.) The analysis of such a structure is one of the most beautiful problems in quantum mechanics. It not only allows us to understand the formation of energy bands in a solid; it also allows us to understand quantitatively how an electron confined in the first well (at $t = 0$) would tunnel to the second well and then return to the first well.

Since $V(-x) = V(x)$, the solutions will be either symmetric or antisymmetric functions of x. We assume $E < V_0$. Thus, the Schrödinger equation can be written in the form

$$\frac{d^2\psi}{dx^2} - \kappa^2\psi(x) = 0 \quad \text{for} \quad |x| < b \tag{15}$$

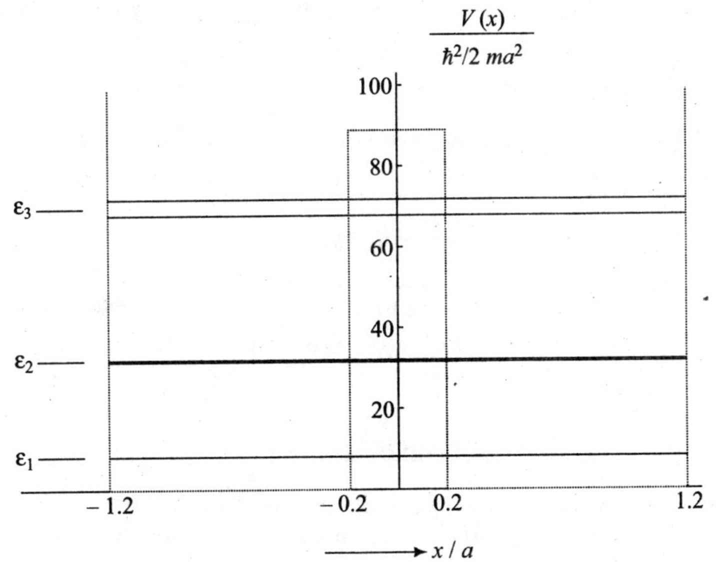

Figure 17.2: The double-well potential and the corresponding energy eigenstates for $\alpha = 3\pi$ and $b/a = 0.2$. The short horizontal lines on the extreme left represent the degenerate energy eigenvalues corresponding to two isolated wells (see Fig. 17.1); each of the levels get split because of the coupling between the two wells.

and
$$\frac{d^2\psi}{dx^2} + k^2\psi(x) = 0 \quad \text{for} \quad b < |x| < a + b \tag{16}$$

where k^2 and κ^2 have been defined earlier [see Eqs (4) and (5)]. The solution symmetric in x would be given by

$$\psi_s(x) = \begin{cases} A_s \cosh \kappa x & \text{for } |x| < b \\ A_s \dfrac{\cosh \kappa b}{\sin ka} \sin [k(a+b-|x|)] & \text{for } b < |x| < a+b \end{cases} \quad \text{Symmetric solution} \tag{17}$$

where we have assumed the continuity of $\psi(x)$ at $x = b$. Further, since $V(x)$ becomes infinite at $x = a + b$, the wave function has to vanish at $x = a + b$ which has been incorporated in the solution given by Eq. (17). Continuity of $d\psi/dx$ at $x = b$ readily gives us

$$\tan ka = -\frac{k}{\kappa} \coth (\kappa b) \tag{18}$$

which represents the transcendental equation determining the energy eigenvalues corresponding to symmetric states. Similarly, the antisymmetric solution would be given by

$$\psi_s(x) = \begin{cases} A_a \sinh \kappa x & \text{for } |x| < b \\ A_a \dfrac{\sinh \kappa b}{\sin ka} \sin[k(a+b-x)] & \text{for } b < x < a+b \\ -A_a \dfrac{\sinh \kappa b}{\sin ka} \sin[k(a+b+x)] & \text{for } -(a+b) < x < -b \end{cases} \quad \text{Antisymmetric solution} \tag{19}$$

where, once again, we have assumed $\psi(x)$ to be continuous at $x = \pm b$. Continuity of $d\psi/dx$ at $x = +b$ (or at $x = -b$) would give us

$$\tan ka = -\frac{k}{\kappa} \tanh (\kappa b) \tag{20}$$

which represents the transcendental equation determining the energy eigenvalues corresponding to antisymmetric states. Equations (18) and (20) can be rewritten in the following form

$$-\xi \cot \xi = \sqrt{\alpha^2 - \xi^2} \tanh\left[\sqrt{\alpha^2 - \xi^2}\left(\frac{b}{a}\right)\right] \quad \text{(symmetric states)} \tag{21}$$

and
$$-\xi \cot \xi = \sqrt{\alpha^2 - \xi^2} \coth\left[\sqrt{\alpha^2 - \xi^2}\left(\frac{b}{a}\right)\right] \quad \text{(antisymmetric states)} \tag{22}$$

where ξ and α have been defined through Eqs (9) and (10) respectively. For $b/a \to \infty$, Eqs (21) and (22) reduce to Eq. (8) and we have degenerate states (see Fig. 17.1).

As a numerical example[2], we assume

$$\alpha = 3\pi \quad \text{and} \quad \frac{b}{a} = 0.2$$

The ξ values [as obtained by solving Eqs (21) and (20)] are given by

$$\xi = \begin{cases} 2.82154\,(s), & 5.60021\,(s), & 8.23678\,(s) \\ 2.84990\,(a), & 5.68284\,(a), & 8.47649\,(a) \end{cases} \tag{23}$$

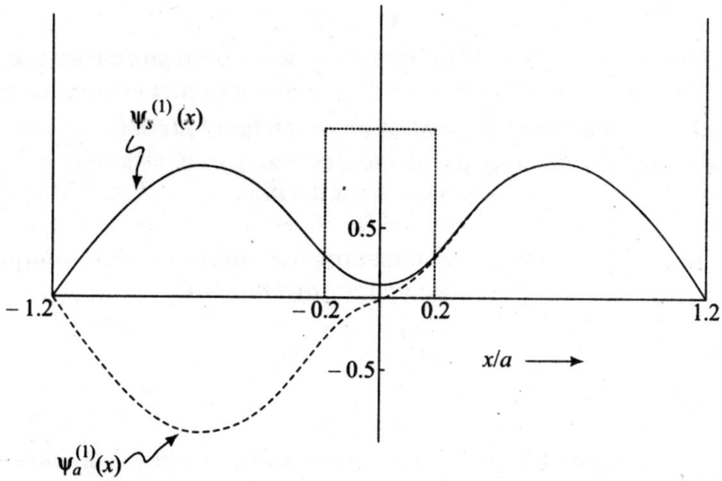

Figure 17.3: The solid and dashed curves correspond to the first energy eigenstate (ε = 7.96109) and the second eigenstate (ε = 8.12193) respectively.

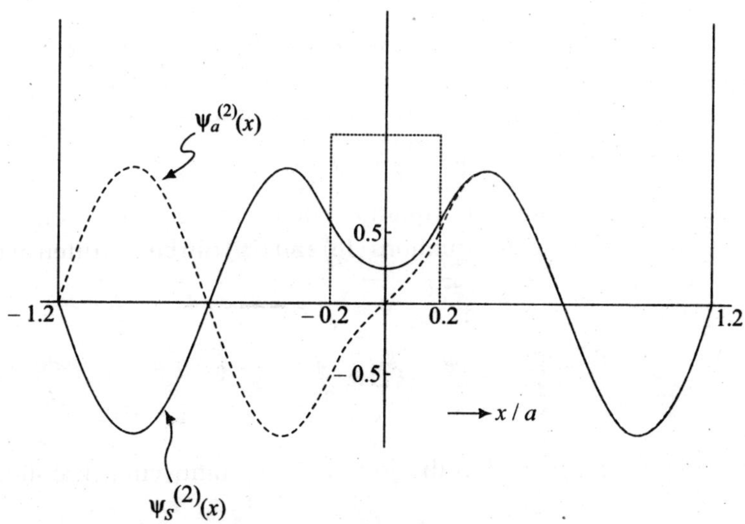

Figure 17.4: The solid and the dashed curves correspond to the third and the fourth eigenstates respectively (ε = 31.3624 and 32.2947).

[2] Most of the numerical calculations in this section have been carried out using the software given in Ref. 6.

where s and a refer to symmetric and antisymmetric states. The normalized energy eigenvalues

$$\varepsilon = \frac{E}{\hbar^2/2ma^2} = \xi^2 \tag{24}$$

would be given by

$$\varepsilon = \begin{cases} 7.96109\,(s), & 31.3624\,(s), & 67.8445\,(s) \\ 8.12193\,(a), & 32.2947\,(a), & 71.8509\,(a) \end{cases} \tag{25}$$

These are shown as horizontal lines in Fig. 17.2. From Eq. (25), we get

$$\Delta\varepsilon \simeq 0.16084,\ 0.9323,\ 4.0064 \tag{26}$$

Comparing Fig. 17.2 with Fig. 17.1, we find that there is a coupling between the wells and each level splits into two levels[3]; the splitting becoming more for higher energy states [see Eq. (26)].

Figures 17.3 and 17.4 give the spatial variation of the first four wavefunctions. In Fig. 17.3, the solid and dashed curves correspond to the first energy eigenstate (ε = 7.96109) and the second energy eigenstate (ε = 8.12193) respectively. Similarly, in Fig. 17.4, the solid and the dashed curves correspond to the third and the fourth eigenstates respectively (ε = 31.3624 and 32.2947). Notice that they are alternately symmetric and antisymmetric functions of x. Let us suppose that at $t = 0$, a particle is described by the wavefunction

$$\psi(x, t=0) = \frac{1}{\sqrt{2}}\left[\psi_s^{(1)}(x) + \psi_a^{(1)}(x)\right] \tag{27}$$

where $\psi_s^{(1)}(x)$ and $\psi_a^{(1)}(x)$ represent respectively the lowest symmetric and the lowest antisymmetric wavefunctions shown in Fig. 17.3; and the factor of $1/\sqrt{2}$ makes it normalized. How will the wave function evolve with time? Well, as in earlier chapters, we just have to multiply $\psi_s^{(1)}(x)$ and $\psi_a^{(1)}(x)$ by appropriate time dependent phase factors to obtain

$$\psi(x, t) = \frac{1}{\sqrt{2}}\left[\psi_s^{(1)}(x)\exp\left(-\frac{iE_s^{(1)}t}{\hbar}\right) + \psi_a^{(1)}(x)\exp\left(-\frac{iE_a^{(1)}t}{\hbar}\right)\right] \tag{28}$$

which represents a rigorously correct solution of the time dependent Schrödinger equation corresponding to the double-well potential. The above equation can be rewritten in the form

$$\psi(x, t) = \frac{1}{\sqrt{2}} e^{-iE_s^{(1)}t/\hbar}\left[\psi_s^{(1)}(x) + \psi_a^{(1)}(x)e^{-i\pi t/T}\right] \tag{29}$$

where

$$T = \frac{\pi\hbar}{\Delta E^{(1)}} = \left(\frac{2\pi ma^2}{\hbar}\right)\frac{1}{\Delta\varepsilon} \tag{30}$$

and

$$\Delta E^{(1)} = E_a^{(1)} - E_s^{(1)} = \frac{\hbar^2}{2ma^2}\Delta\varepsilon^{(1)} \tag{31}$$

From Eq. (29) it immediately follows that at $t = 0, 2T, 4T, \ldots$

$$|\psi(x, t|^2 = \frac{1}{2}\left[\psi_s^{(1)}(x) + \psi_a^{(1)}(x)\right]^2 \tag{32}$$

---

[3] If there are three wells then each level will split into three levels. In general, if there are N such wells (see Fig. 8.9) each level will form a band comprising of N levels; these are the energy bands in a solid; see also Section 23.3.

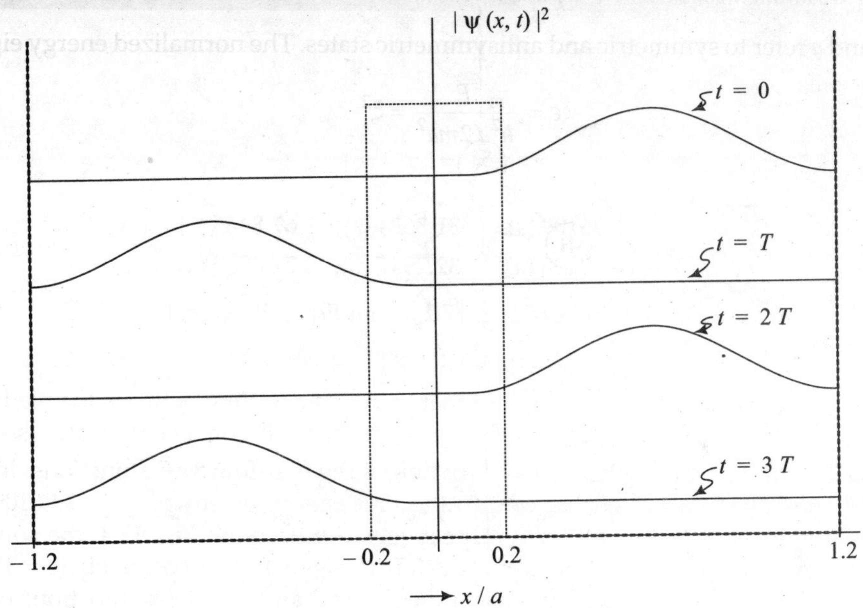

Figure 17.5: Time evolution of $|\psi(x,t)|^2$ [as given by Eq. (29)] for $t = 0, T, 2T$ and $3T$.

and the particle is most likely to be found in the second well. On the other hand, at $t = T, 3T, 5T, \ldots$

$$|\psi(x,t)|^2 = \frac{1}{2}\left[\psi_s^{(1)}(x) - \psi_a^{(1)}(x)\right]^2 \tag{33}$$

and the particle is most likely to be found in the first well. Thus the particle tunnels through the barrier in a time ~ T. Figure 17.5 gives the time evolution of $|\psi(x,t)|^2$ for $t = 0, T, 2T$ and $3T$. It may be seen that the particle which has a very high probability of being found in the second well at $t = 0$ *tunnels* to the first well at $t = T$ and then tunnels back to the initial state at $t = T$. At $t = T/2$, the probability of finding the particle in either of the wells would be approximately equal.

17.4 THE AMMONIA MOLECULE AS A TWO-STATE SYSTEM

The ammonia molecule has a nitrogen atom and three hydrogen atoms forming a pyramid (see Fig. 17.6). The potential energy function associated with the ammonia molecule can be, to a first approximation, described by a double-well potential (similar to the ones shown in Figs 17.2 and 18.9) and with x representing the distance between the nitrogen atom and the plane containing the hydrogen atoms.

We assume that only two states are needed for the dynamical description of the ammonia molecule; these two states of the molecule are:

1. the state with the nitrogen atom above the plane of the hydrogen atoms (see the top figure in Fig. 17.6), and

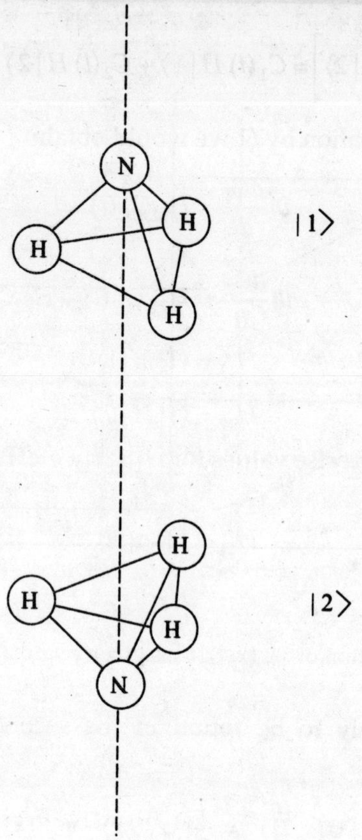

Figure 17.6: The two states of the ammonia molecule.

2. the state with the nitrogen atom below this plane (see bottom figure in Fig. 17.6).

We denote these states by $|1\rangle$ and $|2\rangle$ respectively. The state[4] of the ammonia molecule can be assumed to be described by

$$|\psi(t)\rangle = C_1(t)\,|1\rangle + C_2(t)\,|2\rangle \tag{34}$$

This is referred to as the two-state approximation. The time evolution is described by the equation

$$i\hbar\frac{\partial}{\partial t}|\Psi(t)\rangle = H\,|\Psi(t)\rangle \tag{35}$$

[4] We must point out that $|1\rangle$ and $|2\rangle$ are *not* the eigenstates of the Hamiltonian. In order to understand this point, we refer to the case of two isolated wells shown in Fig. 17.1. If we consider only the lowest state then ψ_1 and ψ_2 are the eigenfunctions of the isolated wells; the eigenfunctions of the double-well are approximately given by

$$\psi_s^{(1)}(x) \approx \frac{1}{\sqrt{2}}\,[\psi_1(x) + \psi_2(x)]$$

and

$$\psi_a^{(1)}(x) \approx \frac{1}{\sqrt{2}}\,[\psi_1(x) - \psi_2(x)]$$

(see Figs 17.1 and 17.3).

or $$i\hbar\left[\frac{dC_1}{dt}|1\rangle+\frac{dC_2}{dt}|2\rangle\right]=C_1(t)\,H|1\rangle+C_2(t)\,H|2\rangle \tag{36}$$

If we premultiply the above equation by ⟨1 we would obtain

$$i\hbar\frac{dC_1}{dt}=H_{11}C_1(t)+H_{12}C_2(t) \tag{37}$$

Similarly, $$i\hbar\frac{dC_2}{dt}=H_{21}C_1(t)+H_{22}C_2(t) \tag{38}$$

where $$H_{11}=\langle 1|H|1\rangle$$

$$H_{12}=\langle 1|H|2\rangle$$

etc. We really do not know the precise values for H_{11}, H_{12}, etc. However, from the symmetry of the problem we may write[5]

$$H_{11}=H_{22}=E_0 \tag{39}$$

and $$H_{12}=H_{21}=-A \tag{40}$$

Thus, $$i\hbar\frac{dC_1}{dt}=E_0C_1(t)-A\,C_2(t) \tag{41}$$

and

$$i\hbar\frac{dC_2}{dt}=-AC_1(t)+E_0C_2(t) \tag{42}$$

The solution of the above equations is very straightforward. We assume

$$C_1(t)=c_1e^{-iEt/\hbar}$$

and $$C_2(t)=c_2e^{-iEt/\hbar}$$

to obtain

$$(E-E_0)\,c_1+Ac_2=0$$

and $$Ac_1+(E-E_0)\,c_2=0$$

Thus, $$\begin{pmatrix}E-E_0 & A\\ A & E-E_0\end{pmatrix}=0$$

or, $$(E-E_0)^2=A^2$$

---

[5] The notation and the analysis are essentially based on the Feynman lectures (see Chapter 19 of Ref. 1); see also Ref. 2. To quote Feynman *In the discussion up to this point, we have assumed values of E_0 and A without knowing how to calculate them. According to the correct physical theory, it should be possible to calculate these constants in terms of the positions and motions of all the nuclei and electrons. But nobody has ever done it. Such a system involves ten electrons and four nuclei and that's just too complicated a problem. As a matter of fact, there is no one who knows much more about this molecule than we do.*

or,

$$E = E_0 \pm A \tag{43}$$

The levels split (similar to the splitting shown in Fig. 17.2). Further,

$$c_1 = +c_2 \quad \text{for} \quad E = E_0 - A \tag{44}$$

and

$$c_1 = -c_2 \quad \text{for} \quad E = E_0 + A \tag{45}$$

Thus the eigenstates are

$$|\psi_+\rangle = \frac{1}{\sqrt{2}}(|1\rangle + |2\rangle) \quad \text{for} \quad E = E_0 - A \tag{46}$$

and

$$|\psi_-\rangle = \frac{1}{\sqrt{2}}(|1\rangle - |2\rangle) \quad \text{for} \quad E = E_0 + A \tag{47}$$

In the two state approximation, these are the eigenfunctions of the total Hamiltonian (see, e.g. Figs 17.1 and 17.3). If

$$C_1(0) = \pm C_2(0)$$

the system will be in an eigenstate of the system. However, if

$$C_1(0) = 1$$

and

$$C_2(0) = 0$$

we would obtain

$$C_1(t) = \cos\left(\frac{At}{\hbar}\right) e^{-iE_0t/\hbar} \tag{48}$$

and

$$C_2(t) = i\sin\left(\frac{At}{\hbar}\right) e^{iE_0t/\hbar} \tag{49}$$

giving

$$|C_1(t)|^2 = \cos^2\left(\frac{At}{\hbar}\right) \tag{50}$$

and

$$|C_2(t)|^2 = \sin^2\left(\frac{At}{\hbar}\right) \tag{51}$$

We may compare this with the evolution of the wave packet discussed in Example 14.1.

In above, what we have essentially implied is that if we assume $|1\rangle$ and $|2\rangle$ as base kets, then the representation of the Hamiltonian is given by

$$H = \begin{pmatrix} H_{11} & H_{12} \\ H_{21} & H_{22} \end{pmatrix} = \begin{pmatrix} E_0 & -A \\ A & E_0 \end{pmatrix} \tag{52}$$

and the eigenvalues of the above matrix are

$$E_0 \pm A \tag{53}$$

Now, the two states $|1\rangle$ and $|2\rangle$ of the ammonia molecule (see Fig. 17.6) have equal and opposite dipole moments. Thus in the presence of an electric field ε we may write

$$\left.\begin{aligned} H_{11} &= E_0 + \mu\varepsilon \\ H_{22} &= E_0 - \mu\varepsilon \\ H_{12} &= H_{21} = -A \end{aligned}\right\} \tag{54}$$

where μ represents the electric dipole moment[6]. Thus instead of Eq. (52) the Hamiltonian matrix is

$$\begin{pmatrix} E_0 + \mu\varepsilon & -A \\ -A & E_0 - \mu\varepsilon \end{pmatrix} \tag{55}$$

The corresponding eigenvalues would be

$$E = E_0 \pm [A^2 + \mu^2\varepsilon^2]^{1/2} \simeq E_0 \pm \left[A + \frac{1}{2A}\mu^2\varepsilon^2\right] \tag{56}$$

where the last expression is valid for $\mu\varepsilon \ll A$.

Now, if a collimated beam of ammonia molecules is passed through an inhomogeneous electric field whose gradient is in a direction perpendicular to the beam then the force acting on the molecules in the lower state will be opposite in direction to the force acting on the molecules which are in the upper state. Thus the two molecular states can be separated by passing the molecules through an inhomogeneous electric field. The molecules in the upper state are then passed through a resonant cavity whose resonance frequency ω_0 equals $2A/\hbar$, then for particular lengths of the cavity all the ammonia molecules (as they pass through the cavity) will make a transition to the lower state, thereby emitting electromagnetic waves at the resonant frequency. It should be mentioned that if the two states were not separated, then the molecules in the lower state would have taken the energy from the cavity and made a transition to the upper state. This is nothing but the condition for *population inversion* which is necessary for the operation of laser or maser. For more details about the working of the maser, the reader is referred to Ref. 1 and also to the Nobel lecture of Townes (Ref. 3).

17.5 THE KRÖNIG-PENNEY MODEL

In this section, we will discuss solutions of the one-dimensional Schrödinger equation

$$\frac{d^2\psi}{dx^2} + \frac{2m}{\hbar^2}[E - V(x)]\,\psi(x) = 0 \tag{57}$$

for a periodic potential[7] given by

$$V(x + L) = V(x) \quad \text{Periodic potential} \tag{58}$$

Now, according to Bloch's theorem (see, e.g. Ref. 5), for a periodic potential given by Eq. (58), the solution of the one-dimensional Schrödinger equation is of the form

$$\psi(x) = e^{ikx}u(x) \tag{59}$$

---

[6] To quote Feynman again, *All anyone can say is that when there is an electric field, the energy of the two states is different, the difference being proportional to the electric field. We have called the coefficient of proportionality 2μ, but its value must be determined experimentally. We can also say that the molecule has the amplitude A to flip over, but this will have to be measured experimentally. Nobody can give us accurate theoretical values of μ and A, because the calculations are too complicated to do in detail.*

[7] We may recall that in Problem 8.7 we had used a matrix method to slove the schrödinger equation. We will use here a method discussed by Wolfe in Ref. 4.

where $u(x)$ has the periodicity of the lattice, i.e.,

$$u(x+L) = u(x) \tag{60}$$

Thus, $$\psi(x+L) = e^{ik(x+L)}u((x+L)) = e^{ikL}e^{ikx}u(x)$$

or $$\psi(x+L) = e^{ikL}\,\psi(x) \tag{61}$$

Now, $$\frac{d\psi}{dx} = ik\,e^{ikx}u(x) + e^{ikx}\frac{du}{dx}$$

Thus, $$\left.\frac{d\psi}{dx}\right|_{x+L} = ik\,e^{ik(x+L)}u(x) + e^{ik(x+L)}\frac{du}{dx} = e^{ikL}\left.\frac{d\psi}{dx}\right|_{x}$$

or $$\left.\frac{d\psi}{dx}\right|_{x} = e^{-ikL}\left.\frac{d\psi}{dx}\right|_{x+L}$$

or $$\left.\frac{d\psi}{dx}\right|_{x=-\varepsilon} = e^{-ikL}\left.\frac{d\psi}{dx}\right|_{L-\varepsilon} \tag{62}$$

We assume the periodic potential to be given by

$$V(x) = A\sum_{n=-\infty}^{+\infty}\delta(x-nL) \tag{63}$$

which represents a series of delta functions (see Fig. 17.7). If we refer to Fig. 8.17 and make $a \to 0$ and $V_0 \to \infty$ such that the product

$$V_0 a \to A \tag{64}$$

we will obtain a series of delta functions as shown in Fig. 17.7.

We first calculate the discontinuity of $d\psi/dx$ at a point where $V(x)$ is a delta function (compare Problems 6.15 and 6.16). Now,

$$\frac{d^2\psi}{dx^2} + \frac{2m}{\hbar^2}[E - V(x)]\,\psi(x) = 0 \tag{65}$$

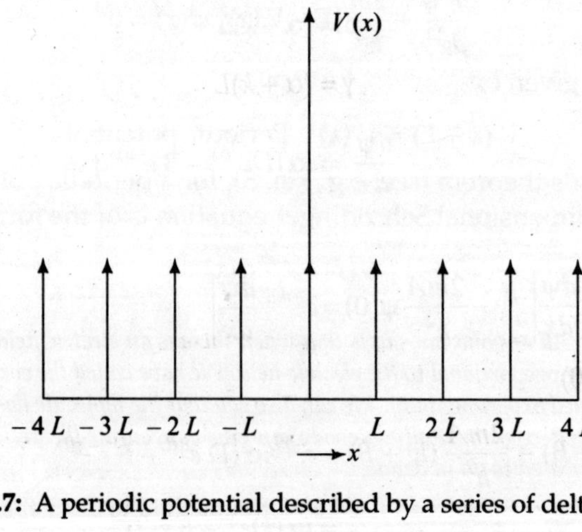

Figure 17.7: A periodic potential described by a series of delta functions.

If we integrate the above equation from $x = -\varepsilon$ to $x = +\varepsilon$ we would obtain

$$\int_{-\varepsilon}^{+\varepsilon} \frac{d^2\psi}{dx^2} dx = -\frac{2mE}{\hbar^2}\int_{-\varepsilon}^{\varepsilon} \psi(x)\, dx + \frac{2mA}{\hbar^2}\int_{-\varepsilon}^{\varepsilon} \delta(x)\, \psi(x)\, dx$$

or

$$\left.\frac{d\psi}{dx}\right|_{x=\varepsilon} - \left.\frac{d\psi}{dx}\right|_{x=-\varepsilon} = +\frac{2mA}{\hbar^2}\psi(0) \tag{66}$$

which is consistent with the results of Problem 6.15. Thus,

$$\left.\frac{d\psi}{dx}\right|_{x=-\varepsilon} = \left.\frac{d\psi}{dx}\right|_{x=\varepsilon} - \frac{2mA}{\hbar^2}\psi(0)$$

If we now use Eq. (62), we would get

$$e^{-ikL}\left.\frac{d\psi}{dx}\right|_{L-\varepsilon} = \left.\frac{d\psi}{dx}\right|_{x=\varepsilon} - \frac{2mA}{\hbar^2}\psi(0) \tag{67}$$

Let us consider the solution of the Schrödinger equation in the region (see Fig. 17.7)

$$\varepsilon < x < L - \varepsilon \quad (\varepsilon \to 0)$$

where $V(x) = 0$. Thus we may write the solution as

$$\psi(x) = De^{i\alpha x} + Be^{-i\alpha x}; \quad \varepsilon < x < L - \varepsilon \tag{68}$$

where

$$\alpha^2 = \frac{2mE}{\hbar^2} \tag{69}$$

Using the condition given by Eq. (61), we get

$$e^{-ikL}\psi(L) = \psi(0)$$

or

$$e^{-ikL}[D\, e^{i\alpha L} + B\, e^{-i\alpha L}] = D + B$$

or

$$D[1 - e^{i\beta}] + B[1 - e^{-i\gamma}] = 0 \tag{70}$$

where

$$\beta = (\alpha - k)L$$

and

$$\gamma = (\alpha + k)L$$

Now,

$$\frac{d\psi}{dx} = i\alpha[De^{i\alpha x} - B\, e^{-i\alpha x}]$$

Since [see Eq. (67)]

$$\left.\frac{d\psi}{dx}\right|_{x=\varepsilon} - \frac{2mA}{\hbar^2}\psi(0) = e^{-ikL}\left.\frac{d\psi}{dx}\right|_{L-\varepsilon}$$

we get (in the limit of $\varepsilon \to 0$)

$$i\alpha(D - B) - \frac{2mA}{\hbar^2}(D + B) = e^{-ikL} i\alpha(D\, e^{i\alpha L} - B\, e^{-i\alpha L})$$

$$= i\alpha(De^{i\beta} - B\, e^{-i\gamma})$$

or $$\left[1-e^{i\beta}+\frac{2iP}{L\alpha}\right]D+\left[1-e^{-i\gamma}-\frac{2iP}{L\alpha}\right]B=0 \tag{71}$$

where $$P=\frac{mAL}{\hbar^2} \tag{72}$$

We rewrite Eq. (70)

$$[1-e^{i\beta}]\,D+[1-e^{-i\gamma}]\,B=0 \tag{73}$$

From Eqs (71) and (73) we readily get

$$\left(1-e^{i\beta}+\frac{2iP}{L\alpha}\right)(1-e^{-i\gamma})+(1-e^{i\beta})\left(1-e^{-i\gamma}-\frac{2iP}{L\alpha}\right)=0$$

or $$2(1-e^{i\beta})\,(1-e^{-i\gamma})=\frac{2iP}{L\alpha}\,[1-e^{i\beta}-1+e^{-i\gamma}]$$

$$=-\frac{2iP}{L\alpha}\,[e^{i(\alpha-k)L}-e^{-i\,(\alpha+k)L}]=\frac{4P}{L\alpha}\,e^{-ikL}\sin\alpha L$$

On the other hand, the left-hand side is given by

$$=2\,[1-e^{i\beta}-e^{-i\gamma}+e^{i(\beta-\gamma)}]$$

$$=2\,[1-e^{-ikL}\,(e^{ikL}+e^{-i\alpha L})+e^{-2ikL}]$$

$$=2e^{-ikL}[e^{ikL}-2\cos\alpha L+e^{-ikL}]$$

$$=4e^{-ikL}\,[\cos kL-\cos\alpha L]$$

Thus, we get $$P\frac{\sin\alpha L}{\alpha L}+\cos\alpha L=\cos kL \tag{74}$$

We rewrite the above equation as

$$f(\xi)=\cos kL \tag{75}$$

where $$f(\xi)=P\frac{\sin\xi}{\xi}+\cos\xi \tag{76}$$

and $$\xi=\alpha L \tag{77}$$

From Eq. (74) it readily follows that we must have

$$-1\le f(\xi)\le +1 \tag{78}$$

In Fig. 17.8, we have plotted $f(\xi)$ as a function of ξ for $P = 10$ (top figure) and for $P = 5$ (bottom figure). In Fig. 17.9, we have made corresponding plots of $f(\xi)$ as a function of

$$\varepsilon=\xi^2=\frac{E}{(\hbar^2/2mL^2)} \tag{79}$$

The shaded regions in Fig. 17.8 correspond to $|f(\xi)|\le 1$ and therefore represent allowed values of energy. Notice the formation of allowed and forbidden energy bands. Obviously,

$$P\to 0$$

corresponds to a free particle with all values of E possible. On the other hand, for

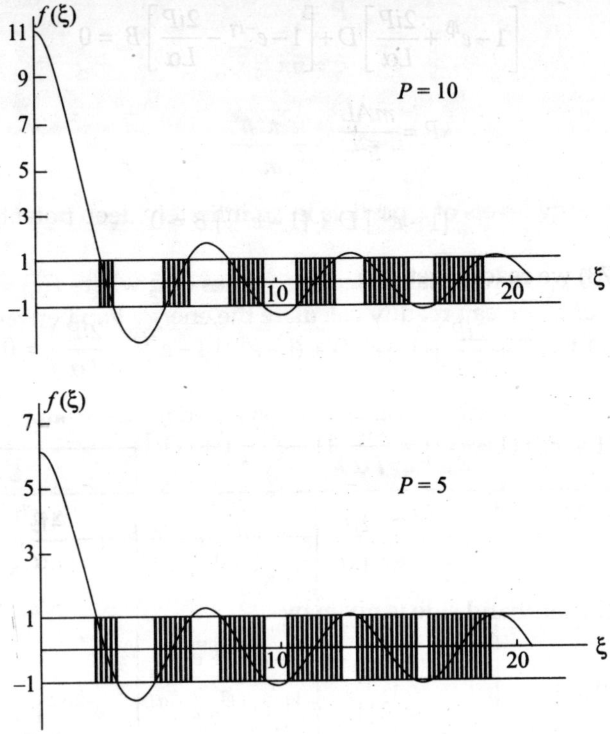

Figure 17.8: $f(\xi)$ as a function of ξ for $P = 10$ (top figure) and for $P = 5$ (bottom figure).

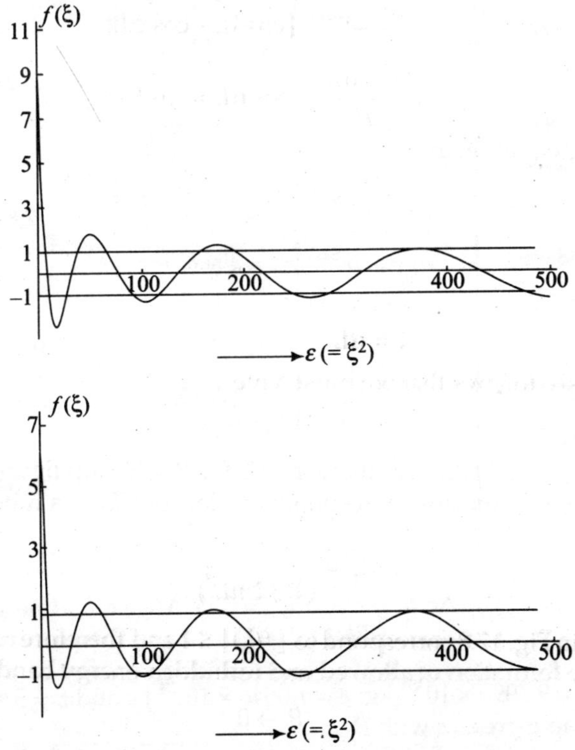

Figure 17.9: Variation of $f(\xi)$ with $\varepsilon (= \xi^2)$ for $P = 10$ (top figure) and for $P = 5$ (bottom figure).

$$P \to \infty$$

we must have
$$\xi = n\pi \tag{80}$$

implying
$$E = E_n = \frac{n^2\pi^2\hbar^2}{2mL^2} \tag{81}$$

which represent the energy levels of a particle in an infinitely deep potential well (see Section 6.6.1).

In Tables 17.1 and 17.2, we have tabulated the values of ξ where $f(\xi) = +1$ or -1. From the corresponding values of ξ^2 we can readily calculate the energy gap between successive bands from the following relation

Table 17.1: Energy gap of an electron at various wave numbers.

$P = 5$;	$A = 7.62$ eV Å;			$L = 5 \times 10^{-10}$m	
ξ	k(Å)$^{-1}$	$f(\xi)$	ξ^2	$\Delta(\xi^2)$	ΔE (eV)
0		6	0		
				5.22	0.80
2.28445	0	+1	5.21871		
π	0.628	–1	π^2		
				12.8	1.9
4.76129	0.628	–1	22.6699		
2π	1.2566	+1	$4\pi^2$		
				16.23	2.47
7.46368	1.2566	+ 1	55.7065		
3π	1.88496	–1	$9\pi^2$		
				17.81	2.71
10.3266	1.88496	–1	106.639		
4π	2.51327	+1	$16\pi^2$		
				18.61	2.84
13.2862	2.51327	+1	176.524		
5π	3.14159	–1	$25\pi^2$		
				19.05	2.90
16.3031	3.14159	–1	265.792		

$$\Delta E = \frac{\hbar}{2mL^2}\Delta(\xi^2) = 0.152\Delta(\xi^2) \text{ eV} \quad \text{Energy gap} \tag{82}$$

where we have used $m = 9.1095 \times 10^{-31}$ kg, $\hbar = 1.0546 \times 10^{-34}$ J.s and $L = 5$Å$= 5 \times 10^{-10}$ m. It may be seen that the band gap increases with P.

Table 17.2: Energy gap of an electron at various wave numbers.

$P = 10$;	$A = 15.24$ eV Å			$L = 5 \times 10^{-10}$m	
ξ	k (Å)$^{-1}$	$f(\xi)$	ξ^2	$\Delta(\xi^2)$	ΔE (eV)
0		11	0		
				6.90	1.05
2.6265	0	+1	6.90		
π	0.63	−1	π^2		
				18.30	2.79
5.30732	0.63	−1	28.17		
2π	1.26	+1	$4\pi^2$		
				25.60	3.90
8.06714	1.26	+1	65.08		
3π	1.88	−1	$9\pi^2$		
				30.17	4.60
10.9087	1.88	−1	119.0		
4π	2.51	+1	$16\pi^2$		
				33.66	5.04
13.8192	2.51	+1	191.0		
5π	3.14	−1	$25\pi^2$		
				34.92	5.32
16.7827	3.14	−1	281.7		

17.6 PROBLEMS

Problem 17.1 In the numerical example discussed in Section 17.3 calculate the value of the tunnelling time T (in seconds) for a = 2Å and a = 5Å.

Problem 17.2 Numerically solve Eqs (21) and (22) for $\alpha = 3\pi$ and $b/a = 0.5$ and show that the splitting of the energy levels become smaller.

Problem 17.3 For the Krönig-Penney model discussed in Section 17.5, calculate the band gap energies (in electron volts) for $P = 2$ and $P = 20$.

[**Answer:** For $P = 2$, the band gap energies are approximately 0.46 eV, 1.02 eV, 1.16 eV, 1.20 eV.]

Problem 17.4 For $x \gg 1$

$$\coth x \simeq 1 + 2e^{-2x}$$

and

$$\tanh x \simeq 1 - 2e^{-2x}$$

Using the above expressions, show that Eqs (21) and (22) can be written in the following approximate forms

$$\tan ka \approx -\frac{k}{\kappa} \mp \frac{2k}{\kappa} e^{-2\kappa b}$$

where the upper and lower signs correspond to the symmetric and antisymmetric states.

For $k/\kappa \ll 1$, we obtain

$$ka \approx k^{(0)}a \approx n\pi$$

giving

$$E_n^{(0)} \approx \frac{n^2\pi^2\hbar^2}{2ma^2}$$

Write $k = k^{(0)} + k^{(1)}$ and assuming $k^{(1)}a \ll 1$ show that

$$E_n \approx E_n^{(0)} - \frac{2E_n^{(0)}}{\kappa_n^{(0)}a} \mp \frac{4E_n^{(0)}}{\kappa_n^{(0)}a}\, e^{-2\kappa_n^{(0)}b}$$

which shows the splitting of the energy levels. In the above equation

$$\kappa_n^{(0)}a = \sqrt{\frac{2m}{\hbar^2}(V_0 - E_n^{(0)})\, a^2}$$

REFERENCES AND SUGGESTED READINGS

1. R.P. Feynman, R.B. Leighton and M. Sands, *The Feynman Lectures on Physics*, Vol. 3, Addison-Wesley Publishing Co., Reading, Massachusetts (1965).
2. J.S. Townsend, *A Modern Approach to Quantum Mechanics*, McGraw-Hill, New York (1992).
3. C.H. Townes, 'Production of Coherent Radiations by Atoms and Molecules', in *Nobel Lectures in Physics* (1963–1970), Elsevier Publishing Co., Amsterdam (1972). The Nobel lecture has also been reprinted in the book *LASERS: Theory and Applications* by K. Thyagarajan and A.K. Ghatak, Plenum Press, New York (1981).
4. J.C. Wolfe, *Summary of the Krönig-Penney Electron*, *Am. J. Phys.* **46** (10), pp. 1012–1014 (1978).
5. N.F. Mott and H. Jones, *The Theory of the Properties of Metals and Alloys*, Clarendon Press, Oxford (1936).
6. A. Ghatak, *Basic Quantum Mechanics* (with CD), Macmillan India Ltd., New Delhi (2002).

18 The JWKB Approximation

God is a mathematician of very high order and He used very advanced mathematics in constructing the universe.

— P.A.M. DIRAC[1]

18.1 INTRODUCTION

The exact solution of the Schrödinger equation can be obtained only for a few potential energy variations. In this chapter, we will develop a widely used approximate method which gives a direct solution of the one-dimensional Schrödinger equation

$$\frac{d^2\psi}{dx^2} + k^2(x)\,\psi(x) = 0 \tag{1}$$

where

$$k^2(x) = \frac{2\mu}{\hbar^2}[E - V(x)] \tag{2}$$

and other symbols have their usual meaning. The method, which is usually abbreviated as the JWKB (after Jeffreys, Wentzel, Kramers and Brillouin) method, is applicable to potentials which are slowly varying such that

$$\left|\frac{1}{k(x)}\frac{dk}{dx}\right| \ll k(x) \tag{3}$$

We may mention here that Eq. (1) is also encountered in many other areas of physics and engineering (see, e.g. Refs 1–3). For example, in electromagnetic theory, if a plane electromagnetic wave propagating in the *x*-direction is incident normally on an inhomogeneous medium characterized by the refractive index variation $n(x)$, then the electric field satisfies Eq. (1) with

$$k^2(x) = \frac{\omega^2}{c^2}n^2(x) = \frac{4\pi^2}{\lambda_0^2}n^2(x) \tag{4}$$

where

$$\lambda_0 = \frac{2\pi c}{\omega} \tag{5}$$

[1] Quoted from *The Quantum Physicists* by W.H. Cropper, Oxford University Press, New York (1970), p. 5.

represents the free space wavelength, c the speed of light in free space and $\omega(= 2\pi\nu)$ is the angular frequency of the wave. For such a plane wave, the wavefunction ψ is either E_y or E_z; E_x is taken to be zero (see, e.g. Ref. 1). Since,

$$k(x) = \frac{\omega}{c} n(x) \tag{6}$$

Equation (3) can be written in the form

$$\left| \frac{1}{n(x)} \frac{dn}{dx} \right| \ll \frac{2\pi}{\lambda} \tag{7}$$

where
$$\lambda(x) = \frac{2\pi}{k(x)} = \frac{\lambda_0}{n(x)} \tag{8}$$

represents the local wavelength. We rewrite Eq. (7) as

$$\left| \frac{dn}{dx} \lambda \right| \ll n \tag{9}$$

where we have omitted the factor of 2π. The quantity on the left-hand side approximately represents the change in the refractive index in a distance of one wavelength; thus Eq. (9) implies *that the refractive index should change very little over a distance* λ. This is the condition for the validity of the JWKB approximation.[2]

We may mention here that a second-order differential equation of the type

$$p(x)\frac{d^2 y}{dx^2} + q(x)\frac{dy}{dx} + r(x)\,y(x) = 0 \tag{10}$$

can always be transformed to Eq. (1) (see Problem 18.4); thus the JWKB method can be applied to obtain direct solutions of Eq. (10). In Section 18.2, we will derive the JWKB solutions and will show that they are not valid near the turning points where $k^2(x) = 0$. In Section 18.3, we will give the *connection formulae* which connect the JWKB solutions from the left of the turning point to the right of the turning point (and vice versa); a heuristic derivation of the connection formulae will be given at the end of this chapter (Section 18.8).

In Section 18.4, we will use the JWKB solutions to derive an equation which would give the energy eigenvalues of the one-dimensional Schrödinger equation. In Section 18.5, we will use the JWKB solutions to study the tunnelling problem in quantum mechanics. In Section 18.6 we will discuss the JWKB solutions of the radial part of the Schrödinger equation for spherically symmetric potentials corresponding to $l = 0$. In Section 18.7, we will discuss the tunnelling of alpha particles from the nucleus.

18.2 THE JWKB SOLUTIONS

We first note that if $k^2(x)$ is a constant, the solution of Eq. (1) would be given by

$$\psi(x) = e^{\pm ikx} \tag{11}$$

---

[2] As $\lambda \to 0$, the JWKB approximation should become more and more accurate. This is also the geometrical optics limit. In quantum mechanics this would imply the approach towards classical mechanics. It is because of this reason that the JWKB formalism is often referred to as the *semi-classical theory*.

where we have omitted the arbitrary multiplicative constant. This suggests that we may try out a solution of the form

$$\psi(x) = e^{iu(x)} \tag{12}$$

where $u(x)$ is to be determined. Substituting Eq. (12) in Eq. (1), we get

$$i\frac{d^2u}{dx^2} - \left(\frac{du}{dx}\right)^2 + k^2(x) = 0 \tag{13}$$

The solution given by Eq. (12) is exact, provided $u(x)$ is a solution of Eq. (13). Now, if $k^2(x)$ was independent of x then $d^2u/dx^2 = 0$; thus when $k^2(x)$ is slowly-varying, we may neglect the term proportional to d^2u/dx^2 to obtain

$$\left(\frac{du}{dx}\right)^2 = k^2(x) \tag{14}$$

implying

$$u(x) = \pm\int^{x} k(x)\,dx \tag{15}$$

Thus, in this approximation, we obtain

$$\psi(x) \approx \psi_0(x) = \exp\left[\pm i\int^{x} k(x)\,dx\right] \qquad \text{Zeroeth-order JWKB solution} \tag{16}$$

which is usually referred to as the zeroeth-order JWKB solution. The physical interpretation of Eq. (16) is as follows: For a slowly-varying refractive index, there is a change in phase only which is given by the integral

$$\Delta\phi = \int^{x} k(x)\,dx$$

It is for this reason that the JWKB method is often referred to as the phase-integral method (see, e.g. Ref. 4). In order to obtain the condition for the validity of the solution given by Eq. (16), we differentiate $\psi_0(x)$ twice to obtain

$$\frac{d^2\psi_0}{dx^2} + \left[k^2(x) \mp i\frac{dk}{dx}\right]\psi_0(x) = 0 \tag{17}$$

Thus $\psi_0(x)$ [as defined by Eq. (16)] is a rigorously correct solution of Eq. (17). If we compare Eq. (17) with Eq. (1) we find that the two equations are the same provided we can neglect the term proportional to dk/dx in comparison with $k^2(x)$, i.e., provided

$$\left|\frac{dk}{dx}\right| \ll k^2(x) \tag{18}$$

or

$$\left|\frac{1}{k}\cdot\frac{dk}{dx}\right| \ll k(x) \tag{19}$$

which represents the condition for the validity of the JWKB solution. Points for which $k^2(x) = 0$

are known as *turning points*[3] and obviously the JWKB solutions are not valid near the turning points.

In order to obtain a more accurate solution, we write

$$\psi(x) = F(x)\,\psi_0(x) = F(x)\exp\left[\pm i\int^x k(x)\,dx\right] \tag{20}$$

where $F(x)$ is yet to be determined. We substitute Eq. (20) in Eq. (1) to obtain

$$\frac{d^2F}{dx^2} \pm 2ik\frac{dF}{dx} \pm i\frac{dk}{dx}F(x) = 0 \tag{21}$$

Once again Eq. (20) represents a rigorously correct solution of Eq. (1) provided $F(x)$ is the solution of Eq. (21). Since $F(x)$ is expected to be slowly-varying, we neglect d^2F/dx^2 to obtain

$$\frac{1}{F}\frac{dF}{dx} + \frac{1}{2k(x)}\frac{dk}{dx} = 0$$

or

$$\frac{d}{dx}\left[\ln F(x)\sqrt{k(x)}\right] = 0$$

or

$$F(x) = \frac{\text{const}}{\sqrt{k(x)}} \tag{22}$$

The solutions

$$\psi(x) = \frac{\text{const}}{\sqrt{k(x)}}\,e^{\pm i\int^x k(x)\,dx} \tag{23}$$

are known as the JWKB solutions. The condition for the validity of the above solution is given in Problem 18.3.

18.2.1 The General JWKB Solutions

The general JWKB solution is the linear combination

$$\psi(x) = \frac{C_1}{\sqrt{k(x)}}\exp\left[i\int^x k(x)\,dx\right] + \frac{C_2}{\sqrt{k(x)}}\exp\left[-i\int^x k(x)\,dx\right] \tag{24}$$

We could also have used the sine and cosine solutions:

$$\psi(x) = \frac{C_1}{\sqrt{k(x)}}\sin\left(\int^x k(x)\,dx\right) + \frac{C_2}{\sqrt{k(x)}}\cos\left(\int^x k(x)\,dx\right) \tag{25}$$

Till now we have assumed $k^2(x)$ to be positive. For $k^2(x) < 0$, we write Eq. (1) as

---

[3] Why are they called turning points? For example, for the harmonic oscillator problem for which $V(x) = \frac{1}{2}\mu\omega^2x^2$ [see Example 18.1], classical mechanics tells us that the particle will turn back at $x = \pm\sqrt{2E/\mu\omega^2}$ which are the *turning points* [see Eq. (46)]. Similarly, if a particle is thrown up in earth's gravitational field [see Example 18.2] it will *turn back* at $x = E/\gamma$; $\gamma = \mu g$.

$$\frac{d^2\psi}{dx^2} - \kappa^2(x)\,\psi(x) = 0 \tag{26}$$

where
$$\kappa^2(x) = -k^2(x) \tag{27}$$

Using a method similar to that used in the preceding section, we obtain the following general JWKB solution

$$\psi(x) = \frac{D_1}{\sqrt{\kappa(x)}} \exp\left[\int^x \kappa(x)\,dx\right] + \frac{D_2}{\sqrt{\kappa(x)}} \exp\left[-\int^x \kappa(x)\,dx\right] \tag{28}$$

The solutions given by Eqs (24), (25) and (28) are valid when $k^2(x)$ and $\kappa^2(x)$ are slowly-varying, i.e., when

$$\left|\frac{1}{k(x)}\frac{dk}{dx}\right| \ll k(x) \quad \text{and} \quad \left|\frac{1}{\kappa(x)}\frac{d\kappa}{dx}\right| \ll \kappa(x) \tag{29}$$

A more precise condition is given in Problem 18.3.

18.3 THE CONNECTION FORMULAE

We next consider a typical variation of $k^2(x)$ (see Fig. 18.1). The point $x = a$ represents the turning point where $k^2(x) = 0$. Obviously, the JWKB solutions (as given in the preceding section) are valid far to the left of the turning point and also far to the right of the turning point. The question now arises that if we know the JWKB solutions to the left of the turning point, then how can we determine the JWKB solutions to the right of the turning point and conversely. This is done through what are known as the *connection formulae* which are given by (see Section 18.8)

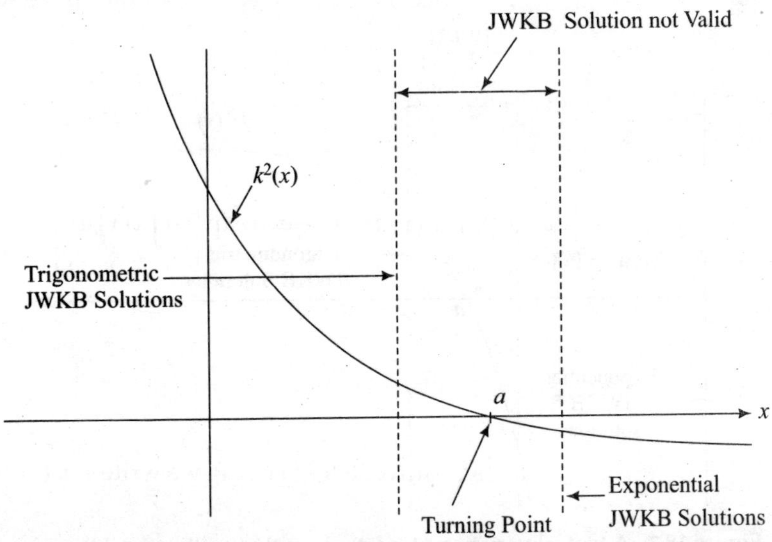

Figure 18.1: A typical variation of $k^2(x)$ where the oscillatory trigonometric solutions are to the left of the turning point at $x = a$ and exponential solutions are to the right of the turning point.

$$\frac{2}{\sqrt{k(x)}} \sin\left[\int_x^a k(x)\,dx + \frac{\pi}{4}\right] \leftrightarrow \frac{1}{\sqrt{\kappa(x)}} \exp\left[-\int_a^x \kappa(x)\,dx\right] \tag{30}$$

and

$$\frac{1}{\sqrt{k(x)}} \cos\left[\int_x^a k(x)\,dx + \frac{\pi}{4}\right] \leftrightarrow \frac{1}{\sqrt{\kappa(x)}} \exp\left[+\int_a^x \kappa(x)\,dx\right] \tag{31}$$

Thus, the JWKB solution given by

$$\frac{2}{\sqrt{k(x)}} \sin\left[\int_x^a k(x)\,dx + \frac{\pi}{4}\right]$$

in the region $x < a$ goes over to an exponentially decaying solution in the region $x > a$, and the solution

$$\frac{1}{\sqrt{k(x)}} \cos\left[\int_x^a k(x)\,dx + \frac{\pi}{4}\right]$$

in the region $x < a$ goes over to an exponentially amplifying solution in the region $x > a$.

Similarly, if the variation of $k^2(x)$ is similar to that shown in Fig. 18.2, the connection formulae would be given by

$$\frac{1}{\sqrt{\kappa(x)}} \exp\left[-\int_x^a \kappa(x)\,dx\right] \leftrightarrow \frac{2}{\sqrt{k(x)}} \sin\left[\int_a^x k(x)\,dx + \frac{\pi}{4}\right] \tag{32}$$

$$\frac{1}{\sqrt{\kappa(x)}} \exp\left[+\int_x^a \kappa(x)\,dx\right] \leftrightarrow \frac{1}{\sqrt{k(x)}} \cos\left[\int_a^x k(x)\,dx + \frac{\pi}{4}\right] \tag{33}$$

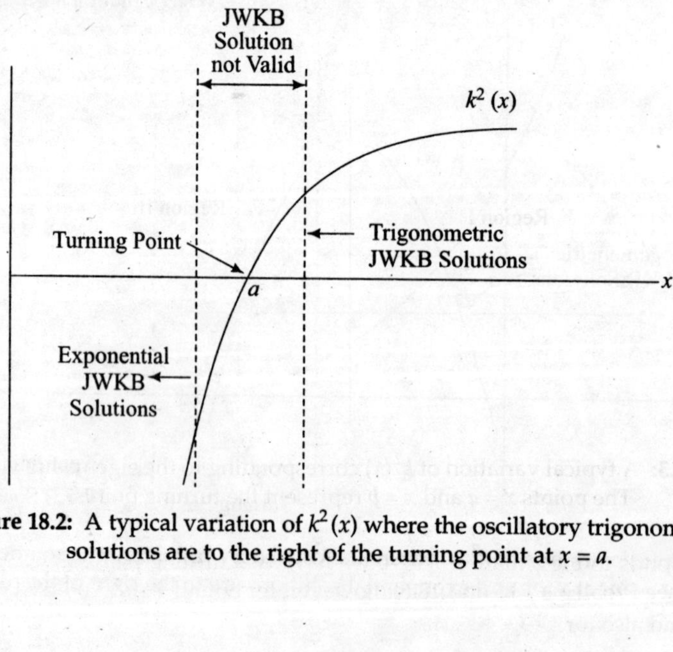

Figure 18.2: A typical variation of $k^2(x)$ where the oscillatory trigonometric solutions are to the right of the turning point at $x = a$.

18.4 APPLICATION OF THE JWKB SOLUTIONS TO EIGENVALUE PROBLEMS

One of the very important applications of the JWKB method is in the solution of the eigenvalue equations encountered in many areas like quantum mechanics, waveguide theory, etc. One usually comes across the equation

$$\frac{d^2\psi}{dx^2} + k^2(x)\,\psi(x) = 0 \tag{34}$$

with $k^2(x)$ becoming negative for $x \to +\infty$ and also for $x \to -\infty$ (see, e.g. Fig. 18.3). The general approach is as follows: We start with an exponentially decaying solution in the region where x is large and positive, and use connection formulae to obtain the solution in the region where x is negative and large. We then derive the condition so that there is only an exponentially decaying solution in the region $x \to -\infty$. This condition leads to eigenvalues of the problem which are contained in $k^2(x)$.

In most problems we have two turning points[4] which we denote by

$$x = a \quad \text{and} \quad x = b \tag{35}$$

We first look for the JWKB solution for $x > b$ (Region III) where $k^2(x)$ is negative. If we use the boundary condition that $\psi(x)$ must tend to 0 as $x \to \infty$, we must choose the exponentially decaying solution in Region III; thus

$$\psi_{III}(x) = \frac{A}{\sqrt{\kappa(x)}} \exp\left[-\int_b^x \kappa(x)\,dx\right] \tag{36}$$

where

$$\kappa^2(x) = -k^2(x) \tag{37}$$

Using the connection formula given by Eq. (30) we obtain the following JWKB solution in Region II ($a < x < b$):

$$\psi_{II}(x) = \frac{2A}{\sqrt{k(x)}} \sin\left[\int_x^b k(x)\,dx + \frac{\pi}{4}\right] \tag{38}$$

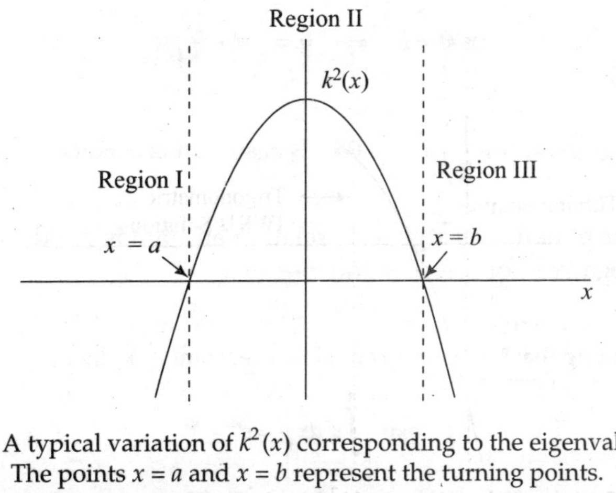

Figure 18.3: A typical variation of $k^2(x)$ corresponding to the eigenvalue problem. The points $x = a$ and $x = b$ represent the turning points.

---

[4] Figure 18.3 corresponds to the situation where we have two turning points. We can have more than two turning points (see Problems 18.5 and 18.6); however, for bound state problems, $k^2(x)$ must become negative for $x \to \infty$ and also for $x \to -\infty$.

Now in order to use the connection formulae given by Eqs (32) and (33) (to obtain the JWKB solution in Region I) the argument of the sine or cosine function should be

$$\left(\int_a^x k(x)\,dx + \frac{\pi}{4}\right).$$

Thus, we write

$$\int_x^b k(x)\,dx + \frac{\pi}{4} = \left(\int_a^b k(x)\,dx + \frac{\pi}{2}\right) - \left(\int_a^x k(x)\,dx + \frac{\pi}{4}\right) \tag{39}$$

using which, we obtain

$$\psi_{II}(x) = \frac{2A}{\sqrt{k(x)}}\cos\left[\theta - \left(\int_a^x k(x)\,dx + \frac{\pi}{4}\right)\right] \tag{40}$$

where

$$\theta = \int_a^b k(x)\,dx \tag{41}$$

Thus,

$$\psi_{II}(x) = (2A\cos\theta)\frac{1}{\sqrt{k(x)}}\cos\left[\int_a^x k(x)\,dx + \frac{\pi}{4}\right] + (2A\sin\theta)\frac{1}{\sqrt{k(x)}}\sin\left[\int_a^x k(x)\,dx + \frac{\pi}{4}\right] \tag{42}$$

Using the connection formulae given by Eqs (32) and (33), we obtain (in Region I)

$$\psi_I(x) = 2A\cos\theta\frac{1}{\sqrt{\kappa(x)}}\exp\left[+\int_x^a \kappa(x)\,dx\right] + A\sin\theta\frac{1}{\sqrt{\kappa(x)}}\exp\left[-\int_x^a \kappa(x)\,dx\right] \tag{43}$$

The first term on the term right-hand side of this equation represents an exponentially growing solution[5] that should not be present; we therefore require

$$\cos\theta = 0 \quad\Rightarrow\quad \theta = \left(n + \frac{1}{2}\right)\pi$$

or

$$\theta = \int_a^b k(x)\,dx = \left(n + \frac{1}{2}\right)\pi \qquad \text{JWKB quantization condition} \tag{44}$$

The above eigenvalue equation is a general relation and holds good for an arbitrary $k^2(x)$ as long as it becomes negative[6] for $x \to +\infty$ and $x \to -\infty$.

---

[5] This can be seen by noting that if $\kappa(x)$ were equal to a constant ($= \kappa$) then

$$\exp\left[\int_x^a \kappa\,dx\right] = e^{\kappa a}\,e^{-\kappa x}$$

For $x \to -\infty$, the above solution would blow up exponentially.

[6] If $k^2(x)$ becomes positive when x tends either to $+\infty$ or $-\infty$, we have what is known as a scattering problem; the quantization condition does not exist and a continuum of energy values is possible (see also Section 18.5).

Harmonic oscillator problem

Example 18.1 For the linear harmonic oscillator problem

$$V(x) = \frac{1}{2}\mu\omega^2 x^2$$

Thus,
$$k^2(x) = \frac{2\mu}{\hbar^2}\left[E - \frac{1}{2}\mu\omega^2 x^2\right] \tag{45}$$

and the turning points are given by

$$a = -\sqrt{\frac{2E}{\mu\omega^2}} \quad \text{and} \quad b = +\sqrt{\frac{2E}{\mu\omega^2}} \tag{46}$$

Thus the JWKB quantization condition becomes

$$\left(n+\frac{1}{2}\right)\pi = \int_a^b k(x)\,dx = \frac{\mu\omega}{\hbar}\int_a^b \left[\frac{2E}{\mu\omega^2} - x^2\right]^{1/2} dx \tag{47}$$

$$= \int_{-\sqrt{\lambda}}^{+\sqrt{\lambda}} \left[\lambda - \xi^2\right]^{1/2} d\xi$$

where
$$\xi = \gamma x; \quad \gamma = \sqrt{\frac{\mu\omega}{\hbar}}$$

and
$$\lambda = \frac{2E}{\hbar\omega} \tag{48}$$

If we make the substitution

$$\xi = \sqrt{\lambda}\sin\theta$$

we will readily get
$$\lambda\left[\frac{\theta}{2} + \frac{1}{4}\sin 2\theta\right]_{-\pi/2}^{+\pi/2} = \left(n+\frac{1}{2}\right)\pi$$

or
$$\lambda = (2n+1);\ n = 0, 1, 2, \ldots \tag{49}$$

giving
$$E = \left(n+\frac{1}{2}\right)\hbar\omega \quad \text{Energy eigenvalues} \tag{50}$$

Comparing with Eq. (28) of Chapter 7 we find that, for the linear harmonic oscillator problem, the JWKB quantization condition happens to give the exact result! Having determined the allowed values of λ, the JWKB solutions are completely known and are given by

$$\psi(x) = \frac{(-1)^n A}{\sqrt{\kappa(x)}} \exp\left[-\int_x^a \kappa(x)\,dx\right],\ x < a \tag{51a}$$

$$= \frac{(-1)^n 2A}{\sqrt{k(x)}} \sin\left[\int_a^x k(x)\,dx + \frac{\pi}{4}\right],\ a < x < b \tag{51b}$$

$$= \frac{2A}{\sqrt{k(x)}} \sin\left[\int_x^b k(x)\,dx + \frac{\pi}{4}\right],\ a < x < b \tag{51c}$$

$$= \frac{A}{\sqrt{\kappa(x)}} \exp\left[-\int_b^x \kappa(x)\,dx\right],\ x > b \tag{51d}$$

JWKB solutions

Indeed the above solution is valid for an arbitrary variation of $k^2(x)$ as long as there are two turning points and $k^2(x)$ remains negative for $x < a$ and $x > b$. For the linear harmonic oscillator problem the integrals are easy to evaluate:

$$\int_x^b k(x)\,dx = \int_\xi^{\sqrt{\lambda}} \sqrt{\lambda - \xi^2}\,d\xi$$

$$= \lambda\left[\frac{1}{2}\sin^{-1}\frac{\xi}{\sqrt{\lambda}} + \frac{\xi}{2\lambda}\sqrt{\lambda - \xi^2}\right]_\xi^{\sqrt{\lambda}}$$

$$= \lambda\left[\frac{\pi}{4} - \frac{1}{2}\sin^{-1}\frac{\xi}{\sqrt{\lambda}} - \frac{1}{2}\xi\sqrt{\lambda - \xi^2}\right] \tag{52}$$

Also,
$$\frac{1}{\sqrt{k(x)}} = \frac{1}{\sqrt{\gamma}} \cdot \frac{1}{(\lambda - \xi^2)^{1/4}}$$

For example, for $n = 1$ ($\lambda = 3$), the JWKB wavefunction (in the region $-\sqrt{\lambda} < x < \sqrt{\lambda}$ would be given by [using Eq. (51c)]:

$$\psi(\xi) = \frac{2A}{(3-\xi^2)^{1/4}} \sin\left[\pi - \frac{3}{2}\sin^{-1}\frac{\xi}{\sqrt{3}} - \frac{\xi}{2}\sqrt{3-\xi^2}\right]$$

$$= \frac{2A}{(3-\xi^2)^{1/4}} \sin\left[\frac{3}{2}\sin^{-1}\frac{\xi}{\sqrt{3}} + \frac{\xi}{2}\sqrt{3-\xi^2}\right]$$

The same result would be obtained by using Eq. (51b):

$$\psi(\xi) = -\frac{2A}{(3-\xi^2)^{1/4}} \sin\left[\pi + \frac{3}{2}\sin^{-1}\frac{\xi}{\sqrt{3}} + \frac{\xi}{2}\sqrt{3-\xi^2}\right]$$

$$= \frac{2A}{(3-\xi^2)^{1/4}} \sin\left[\frac{3}{2}\sin^{-1}\frac{\xi}{\sqrt{3}} + \frac{\xi}{2}\sqrt{3-\xi^2}\right]$$

which is an antisymmetric function of ξ. Similarly,

$$\int_{\lambda}^{x} \kappa(x)\,dx = \int_{\sqrt{\lambda}}^{\xi} \sqrt{\xi^2-\lambda}\,d\xi$$

$$= \frac{1}{2}\xi\sqrt{\xi^2-\lambda} - \frac{1}{2}\lambda \ln\left|\frac{\xi+\sqrt{\xi^2-\lambda}}{\sqrt{\lambda}}\right| \quad (53)$$

and $$\int_{\xi}^{-\sqrt{\lambda}} \sqrt{\xi^2-\lambda}\,d\xi = -\frac{1}{2}\xi\sqrt{\xi^2-\lambda} + \frac{1}{2}\lambda \ln\left|\frac{\xi+\sqrt{\xi^2-\lambda}}{\sqrt{\lambda}}\right| \quad (54)$$

Using Eqs (51a) and (51d) we obtain (for $\lambda = 3$):

$$\psi(\xi) = -\frac{A}{(\xi^2-3)^{1/4}} \exp\left[\frac{1}{2}\xi\sqrt{\xi^2-3} - \frac{3}{2}\ln\left|\frac{\xi+\sqrt{\xi^2-3}}{\sqrt{3}}\right|\right] \quad \xi < -\sqrt{3}$$

$$= \frac{A}{(\xi^2-3)^{1/4}} \exp\left[-\frac{1}{2}\xi\sqrt{\xi^2-3} + \frac{3}{2}\ln\left|\frac{\xi+\sqrt{\xi^2-3}}{\sqrt{3}}\right|\right] \quad \xi > +\sqrt{3}$$

which represents an antisymmetric function of ξ. In an exactly similar manner we can show that for $\lambda = 7, 11, 15, \ldots$, $\psi(\xi)$ will be antisymmetric and for $\lambda = 1, 5, 9, 13, \ldots$, $\psi(\xi)$ will be a symmetric function of ξ.

The JWKB wavefunctions are now completely determined except for an arbitrary multiplicative constant. We consider antisymmetric wavefunctions ($n = 1, 3, 5, \ldots$) and assume that $\psi(\xi)$ has a unit slope at $\xi = 0$. This gives

$$A = \frac{(-1)^{(n-1)/2}}{\sqrt{2n+1}};\ n = 1, 3, 5, \ldots \quad (55)$$

The corresponding exact wavefunction would be (see Section 7.3)

$$\psi_{\text{Exact}}(\xi) = \frac{1}{2}(-1)^{(n-1)/2}\frac{\left(\frac{n-1}{2}\right)!}{n!}H_n(\xi)\exp\left(-\frac{1}{2}\xi^2\right) \quad \text{Exact wavefunction} \quad (56)$$

where we have chosen the multiplicative constant so that $\psi'(0) = 1$. In Figs 18.4 and 18.5, we have plotted the exact (viz., the Hermite-Gauss functions) and the JWKB wavefunctions for $n = 3$ and 5. Except near the turning points, the JWKB wavefunctions agree well with the exact wavefunctions.

The JWKB solutions allow us to understand the general solution of the Schrödinger equation. Let us assume that E is not one of the eigenvalues, i.e. $E \neq \frac{1}{2}\hbar\omega, \frac{3}{2}\hbar\omega, \ldots$. Then, if we start with an exponentially decaying solution in the region $x < a$, we will end up with an exponentially amplifying solution in the region $x > b$. Similarly, if we start with an exponentially decaying solution in the region $x > b$, we will end up with an exponentially amplifying solution in the region $x < a$. These are the two independent solutions of the Schrödinger equation when E is not one of the eigenvalues.

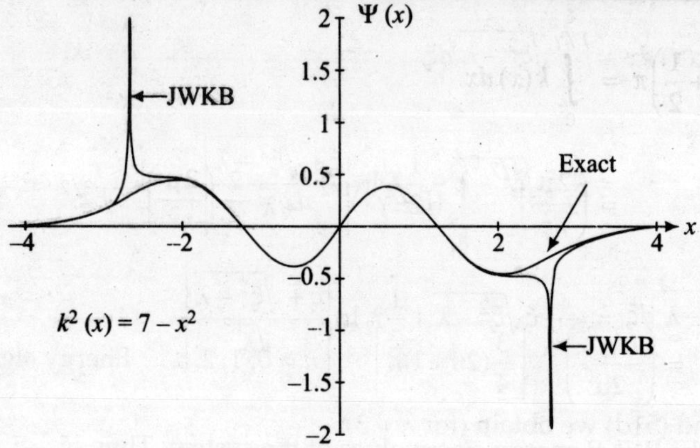

Figure 18.4: Exact and JWKB wavefunctions corresponding to $n = 3$ (adapted from Ref. 3).

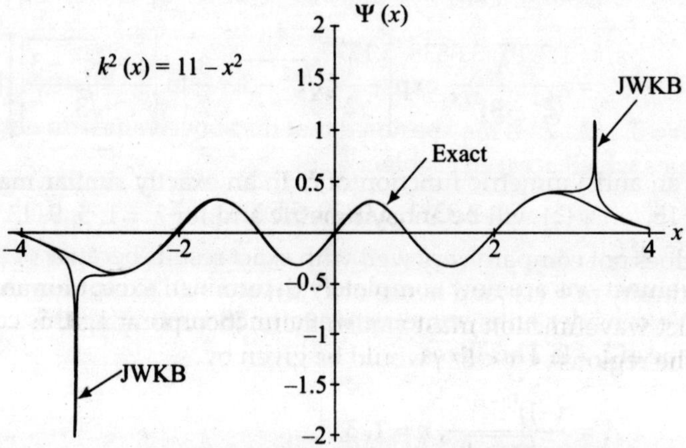

Figure 18.5: Exact and JWKB wavefunctions corresponding to $n = 5$ (adapted from Ref. 3).

Potential corresponding to free fall under gravity

Example 18.2 Consider the potential energy variation given by (see Fig. 6.5)

$$\left.\begin{aligned} V(x) &= \infty \qquad x < 0 \\ &= \gamma x \qquad x > 0 \end{aligned}\right\} \tag{57}$$

The above potential energy variation corresponds to the free fall of a particle in earth's gravitational field; obviously $\gamma = \mu g$. Since,

$$k^2(x) = \frac{2\mu}{\hbar^2}\,[E - V(x)] \tag{58}$$

the turning points are $x = 0$ and $x = E/\gamma$. Thus the eigenvalue equation [Eq. (44)] becomes

$$\left(n+\frac{1}{2}\right)\pi = \int_0^{E/\gamma} k(x)\,dx$$

$$= \left(\frac{2\mu}{\hbar^2}\right)^{1/2} \int_0^{E/\gamma} (E-\gamma x)^{1/2}\,dx = \frac{2}{3\gamma}\left(\frac{2\mu}{\hbar^2}\right)^{1/2} E^{3/2} \tag{59}$$

implying

$$E = E_n = \left(\frac{\hbar^2\gamma^2}{2\mu}\right)^{1/3}\left[\frac{3}{4}(2n+1)\pi\right]^{2/3} \quad ; n = 0, 1, 2, \ldots \quad \text{Energy eigenvalues} \tag{60}$$

which represents the JWKB energy eigenvalues of the system. Thus,

$$\varepsilon_n = \frac{E_n}{(\hbar^2\gamma^2/2\mu)^{1/3}} = \left[\frac{3}{4}(2n+1)\pi\right]^{2/3} \tag{61}$$

or

$$\varepsilon_n = 1.7707,\ 3.6838,\ 5.1775,\ \ldots \tag{62}$$

corresponding to $n = 0, 1, 2, \ldots$. It may be noted that the above values do not compare very well with the exact values which are (see Problem 6.14)

$$2.3381,\ 4.0879,\ 5.5206,\ \ldots \tag{63}$$

The JWKB result does not compare very well with exact results because of the fact that there is an infinite discontinuity of $V(x)$ [and hence of $k^2(x)$] at $x = 0$. Now, because of $V(x)$ becoming ∞ at $x = 0$, the exact wavefunction must vanish there. Incorporating this condition, the JWKB wavefunction in the region $0 < x < E/\gamma$ would be given by

$$\psi_{\text{JWKB}}(x) = \frac{A}{\sqrt{k(x)}} \sin\left[\int_0^x k(x)\,dx\right], \quad 0 < x < E/\gamma \tag{64}$$

$$= \frac{A}{\sqrt{k(x)}} \sin\left[\theta - \left(\int_x^b k(x)\,dx + \frac{\pi}{4}\right)\right] \tag{65}$$

where

$$\theta = \int_0^b k(x)\,dx + \frac{\pi}{4} \tag{66}$$

and

$$b = \frac{E}{\gamma} \tag{67}$$

Thus,

$$\psi_{\text{JWKB}}(x) = \frac{A}{\sqrt{k(x)}} \sin\theta \cos\left[\int_x^b k(x)\,dx + \frac{\pi}{4}\right] - \frac{A}{\sqrt{k(x)}} \cos\theta \sin\left[\int_x^b k(x)\,dx + \frac{\pi}{4}\right] \tag{68}$$

The first term on the right-hand side will *go over* to an exponentially amplifying solution in the region $x > b$ and therefore we must have

$$\sin\theta = 0, \text{ or } \theta = n\pi \tag{69}$$

Thus, $$\int_0^b k(x)\,dx = \left(n - \frac{1}{4}\right)\pi;\ n = 1, 2, \ldots \quad \text{Modified quantization condition} \tag{70}$$

The above equation represents the modified quantization condition and substituting for $k(x)$ we will readily obtain

$$\varepsilon_n = \frac{E_n}{(\hbar^2\gamma^2/2\mu)^{1/3}} = \left[\frac{3}{4}\left(2n - \frac{1}{2}\right)\pi\right]^{2/3} \tag{71}$$

or $$\varepsilon_n = 2.3203, 4.0818, 5.5172, \ldots \tag{72}$$

for $n = 1, 2, 3, \ldots$. The above eigenvalues compare very well with the exact values [see Eq. (63)]. The corresponding JWKB wavefunctions would be given by Eq. (64) with

$$k(x) = \left(\frac{2\mu}{\hbar^2}\right)^{1/2} (E - \gamma x)^{1/2}$$

$$= \left(\frac{2\mu\alpha}{\hbar^2}\right)^{1/2} \xi^{1/2} \tag{73}$$

where $$\xi \equiv \frac{E - \gamma x}{\alpha} = \varepsilon - X \tag{74}$$

$$X \equiv \frac{\gamma}{\alpha} x$$

and α is to be defined later. Thus,

$$\int_0^x k(x)\,dx = \left(\frac{2\mu\alpha^3}{\gamma^2\hbar^2}\right)^{1/2} \int_\xi^\varepsilon \xi^{1/2}\,d\xi \tag{75}$$

We choose α such that the factor in front of the integral is unity, i.e.,

$$\alpha \equiv \left(\frac{\hbar^2\gamma^2}{2\mu}\right)^{1/3} \tag{76}$$

[see Solution 6.14]. Thus,

$$\int_0^x k(x)\,dx = \frac{2}{3}\left[\varepsilon^{3/2} - \xi^{3/2}\right]$$

$$= \frac{2}{3}\left[\varepsilon^{3/2} - (\varepsilon - X)^{3/2}\right] \tag{77}$$

The JWKB solution [Eq. (64)] is therefore given by

$$\psi_{\text{JWKB}}(X) = \frac{1}{\varepsilon^{1/4}(\varepsilon - X)^{1/4}} \sin\left[\frac{2}{3}\left\{\varepsilon^{3/2} - (\varepsilon - X)^{3/2}\right\}\right], 0 < X < \varepsilon \tag{78}$$

where we have chosen the constant such that $d\psi/dX$ is unity at $X = 0$. For $X > \varepsilon$, the JWKB solution will be given by

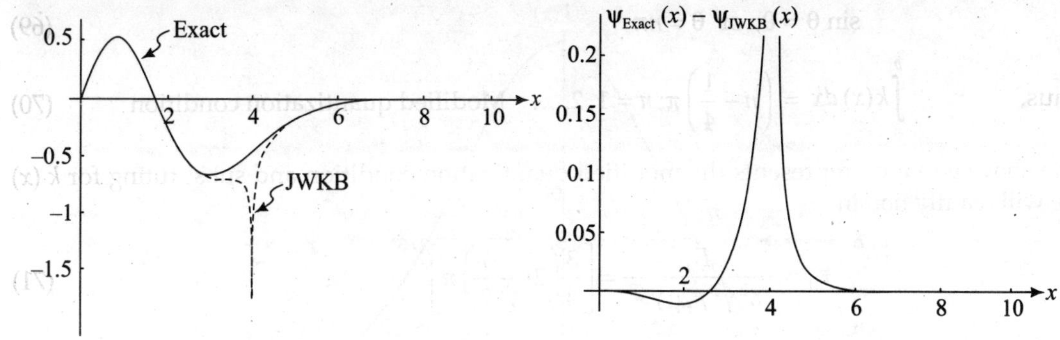

Figure 18.6: (a) Comparison of the JWKB wavefunction as given by Eqs (78) and (79) (shown as dotted curve) with the exact wavefunction as given by Eq. (81) (shown as solid curve) corresponding to the second energy state ($n = 2$) for the potential energy variation given by Eq. (57). (b) A plot of $\psi_{\text{EXACT}}(x) - \psi_{\text{JWKB}}(x)$ as a function of x (adapted from Ref. 3).

$$\psi_{\text{JWKB}}(X) = -\frac{\cos n\pi}{2\varepsilon^{1/4}(X-\varepsilon)^{1/4}} \exp\left[-\int_b^x \kappa(x)\,dx\right]$$

$$= \frac{(-1)^{n+1}}{2\varepsilon^{1/4}(X-\varepsilon)^{1/4}} \exp\left[-\frac{2}{3}(X-\varepsilon)^{3/2}\right], X > \varepsilon \tag{79}$$

Figure 18.6 gives a plot of the JWKB wavefunction for $n = 2$ corresponding to which

$$\varepsilon = \varepsilon_2 = 4.0818 \tag{80}$$

Comparison has been made with the exact wavefunction which is given by (see Solution 6.14)

$$\psi_{\text{Exact}}(X) = \frac{1}{Ai'(a_2)} Ai(X + a_2) \quad \text{Exact wavefunction} \tag{81}$$

where $a_2 = -4.08795$ represents the second zero of $Ai(x)$.

18.5 APPLICATION OF THE JWKB SOLUTIONS TO THE TUNNELLING PROBABILITY CALCULATIONS

We next consider the tunnelling problem in quantum mechanics. Consider a potential energy variation of the type shown in Fig. 18.7(a). A particle of arbitrary (positive) energy E approaches the potential energy barrier from the left[7]. Our objective is to calculate the probability for the particle to tunnel through the barrier and also the probability for its reflection[8].

In Fig. 18.7 (b), we have shown the corresponding variation of $k^2(x)$; the two turning points are denoted by a and b. In the regions $x > b$ and $x < a$, E would be greater than $V(x)$ and the Schrödinger equation will be the form

[7] This is typical of a scattering problem in which the energy is specified in advance and the behaviour of the wavefunction is found in terms of the energy. This is in contrast to the eigenvalue problem where the wavefunction vanishes as $x \to \pm\infty$ leading to discrete values of the energy of the particle.

[8] The corresponding problem in electromagnetic theory would be when a light beam is incident on a thin layer of a rarer medium at an angle of incidence greater than the critical angle.

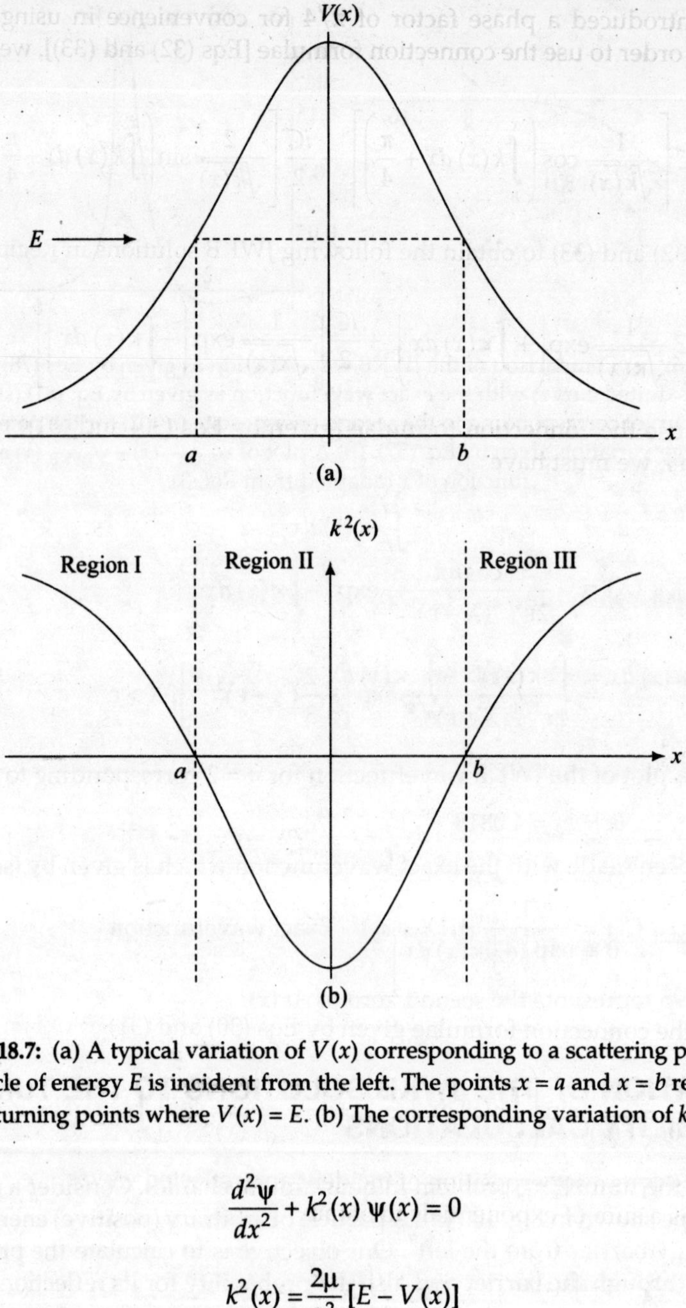

Figure 18.7: (a) A typical variation of $V(x)$ corresponding to a scattering problem. A particle of energy E is incident from the left. The points $x = a$ and $x = b$ represent the turning points where $V(x) = E$. (b) The corresponding variation of $k^2(x)$.

$$\frac{d^2\psi}{dx^2} + k^2(x)\,\psi(x) = 0 \tag{82}$$

where
$$k^2(x) = \frac{2\mu}{\hbar^2}[E - V(x)]$$

In order to describe incoming and outgoing waves, we will choose the JWKB solutions given by the exponential functions [see Eq. (24)] rather than sine and cosine solutions. Since in Region III we only have a wave propagating in the $+x$ direction, the corresponding solution will be given by

$$\psi_{\text{III}}(x) = \frac{C}{\sqrt{k(x)}} \exp\left[i\left(\int_b^x k(x)\,dx + \frac{\pi}{4}\right)\right] \tag{83}$$

where we have introduced a phase factor of $\pi/4$ for convenience in using the connection formulae. Now in order to use the connection formulae [Eqs (32) and (33)], we rewrite Eq. (83) as

$$\psi_{III}(x) = C\left[\frac{1}{\sqrt{k(x)}}\cos\left(\int_b^x k(x)\,dx + \frac{\pi}{4}\right)\right] + \frac{iC}{2}\left[\frac{2}{\sqrt{k(x)}}\sin\left(\int_b^x k(x)\,dx + \frac{\pi}{4}\right)\right] \tag{84}$$

We now use Eqs (32) and (33) to obtain the following JWKB solutions in Region II:

$$\psi_{II}(x) = C\frac{1}{\sqrt{\kappa(x)}}\exp\left(+\int_x^b \kappa(x)\,dx\right) + \frac{iC}{2}\left[\frac{1}{\sqrt{\kappa(x)}}\exp\left(-\int_x^b \kappa(x)\,dx\right)\right] \tag{85}$$

Now, in order to use the connection formulae [given by Eqs (30) and (31)] for obtaining the solution in Region I, we must have

$$\int_a^x \kappa(x)\,dx$$

in the exponent. Thus, we write

$$\int_x^b \kappa(x)\,dx = \int_a^b \kappa(x)\,dx - \int_a^x \kappa(x)\,dx \tag{86}$$

so that ψ_{II} becomes[9]

$$\psi_{II}(x) = \frac{C}{\theta}\frac{1}{\sqrt{\kappa(x)}}\exp\left(-\int_a^x \kappa(x)\,dx\right) + \frac{iC\theta}{2}\left[\frac{1}{\sqrt{\kappa(x)}}\exp\left(+\int_a^x \kappa(x)\,dx\right)\right] \tag{87}$$

where

$$\theta \equiv \exp\left[-\int_a^b \kappa(x)\,dx\right] \tag{88}$$

We can now use the connection formulae given by Eqs (30) and (31) to obtain

$$\psi_I(x) = \frac{C}{\theta}\frac{2}{\sqrt{k(x)}}\sin\left(\int_x^a k(x)\,dx + \frac{\pi}{4}\right) + \frac{iC\theta}{2}\frac{1}{\sqrt{k(x)}}\cos\left(\int_x^a k(x)\,dx + \frac{\pi}{4}\right)$$

In order to express ψ_I as a superposition of incident and reflected waves, we write the sine and cosine functions as a sum of exponentials to obtain

$$\psi_I(x) = \underset{\substack{\text{[Incident wave}\\\text{propagating to}\\\text{the right]}}}{\frac{A}{\sqrt{k(x)}}e^{-i\int_x^a k(x)\,dx}} + \underset{\substack{\text{[Wave propagating}\\\text{to the left]}}}{\frac{B}{\sqrt{k(x)}}e^{i\int_x^a k(x)\,dx}} \tag{89}$$

[9] Notice that the first term on the right-hand side of Eq. (85) is the exponentially amplifying term with respect to the point $x = b$; this term becomes an exponentially decaying term with respect to the point $x = a$ [see the first term on the right-hand side of Eq. (87)].

where $$A = iC\left(\frac{1}{\theta}+\frac{\theta}{4}\right)e^{-i\pi/4} = C\left(\frac{1}{\theta}+\frac{\theta}{4}\right)e^{i\pi/4} \tag{90}$$

and $$B = -iC\left(\frac{1}{\theta}-\frac{\theta}{4}\right)e^{i\pi/4} = C\left(\frac{1}{\theta}-\frac{\theta}{4}\right)e^{-i\pi/4} \tag{91}$$

Thus the transmission and reflection coefficients are given by

$$T = \left|\frac{C}{A}\right|^2 = \frac{1}{\left(\frac{1}{\theta}+\frac{\theta}{4}\right)^2} \quad \text{Transmission coefficient} \tag{92}$$

$$R = \left|\frac{B}{A}\right|^2 = \frac{\left(\frac{1}{\theta}-\frac{\theta}{4}\right)^2}{\left(\frac{1}{\theta}+\frac{\theta}{4}\right)^2} \quad \text{Reflection coefficient} \tag{93}$$

Notice that in spite of all the approximations made in the JWKB theory we still get

$$R + T = 1$$

In most practical cases $\theta \ll 1$ and in this approximation

$$T \approx \theta^2 = e^{-2\int_a^b \kappa(x)\,dx} \tag{94}$$

a formula that is widely used in literature.

Example 18.3 *Transmission through a parabolic barrier* as a simple application of Eq. (94), we consider a parabolic barrier of the form

$$\left.\begin{aligned} V(x) &= V_0\left[1-\frac{x^2}{d^2}\right], \quad |x| < d \\ &= 0, \quad\quad\quad\quad\quad\;\; |x| > d \end{aligned}\right\} \tag{95}$$

(see Fig. 18.8). For $E < V_0$, the turning points are the roots of the equation

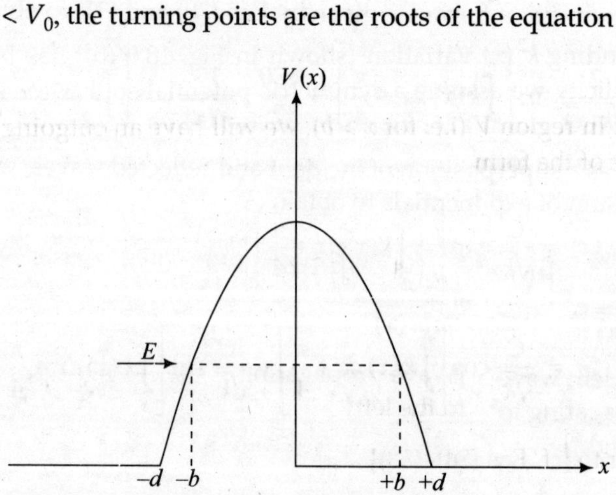

Figure 18.8: Parabolic variation of $V(x)$ as given by Eq. (95).

$$E = V_0\left[1-\frac{x^2}{d^2}\right]$$

implying

$$a = -d\left(1-\frac{E}{V_0}\right)^{1/2} \quad \text{and} \quad b = +d\left(1-\frac{E}{V_0}\right)^{1/2}$$

Thus,

$$\theta = \exp\left\{-\left(\frac{2\mu}{\hbar^2}\right)^{1/2}\int_a^b \left[V_0\left(1-\frac{x^2}{d^2}\right)-E\right]^{1/2} dx\right\}$$

$$= \exp\left\{-\frac{\pi}{2}\left(\frac{2\mu V_0 d^2}{\hbar^2}\right)^{1/2}\left(1-\frac{E}{V_0}\right)\right\}$$

$$T \approx \theta^2 = \exp\left[-\pi\left(\frac{2\mu V_0 d^2}{\hbar^2}\right)^{1/2}\left(1-\frac{E}{V_0}\right)\right] \tag{96}$$

It may be noted that $E > V_0$, there are no turning points and the JWKB formula [Eq. (94)] would predict $T = 1$ which is not correct. Rigourous analysis will show that there is a finite reflection even for $E > V_0$.

Example 18.4 In this example, we will show how to use the connection formulae when there are more than two turning points. Although the analysis does get a bit involved, we urge the students to work it out in detail so that the *tricks* in using the connection formulae are completely understood.

We consider a $V(x)$ variation of the type shown in Fig. 18.9 (a) with the incident energy E such that

$$V_1 < E < V_0$$

so that the corresponding $k^2(x)$ variation [shown in Fig. 18.9 (b)] has four turning points[10]. For the sake of simplicity we assume a symmetric potential so that the turning points are at $x = \pm a$ and at $x = \pm b$. In region V (i.e. for $x > b$), we will have an outgoing wave and hence the wavefunction will be of the form

$$\psi_V(x) = \frac{C}{\sqrt{k}}\exp\left[i\left(\int_b^x k(x)\,dx + \frac{\pi}{4}\right)\right]$$

$$= \frac{C}{\sqrt{k}}\cos\left[\int_b^x k(x)\,dx + \frac{\pi}{4}\right] + \frac{iC}{\sqrt{k}}\sin\left[\int_b^x k(x)\,dx + \frac{\pi}{4}\right] \tag{97}$$

which would go over to [cf. Eqs (84)–(88)]

[10] For $E < V_1$ we will have only two turning points and one can carry out an analysis similar to that in Section 18.5.

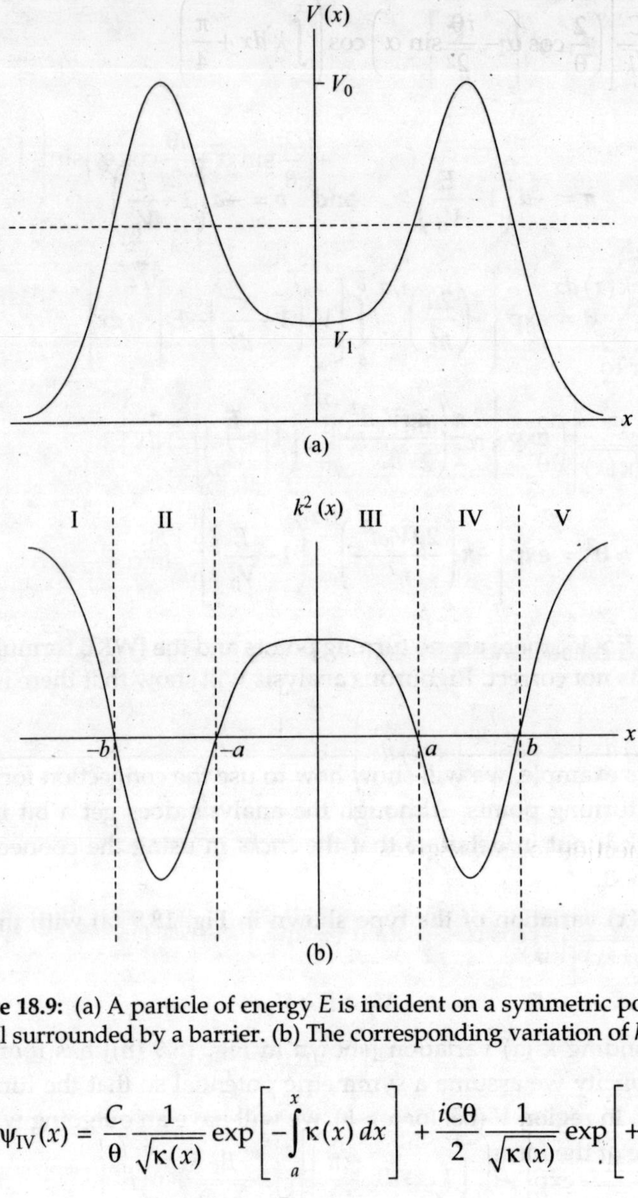

Figure 18.9: (a) A particle of energy E is incident on a symmetric potential well surrounded by a barrier. (b) The corresponding variation of $k^2(x)$.

$$\psi_{IV}(x) = \frac{C}{\theta}\frac{1}{\sqrt{\kappa(x)}}\exp\left[-\int_a^x \kappa(x)\,dx\right] + \frac{iC\theta}{2}\frac{1}{\sqrt{\kappa(x)}}\exp\left[+\int_a^x \kappa(x)\,dx\right] \tag{98}$$

where
$$\theta \equiv \exp\left[-\int_a^b \kappa(x)\,dx\right] \tag{99}$$

We again use the connection formulae [Eqs (30)–(31)] to obtain

$$\psi_{III}(x) = \frac{C}{\sqrt{k}}\left[\frac{2}{\theta}\sin\left(\int_x^a k(x)\,dx + \frac{\pi}{4}\right) + \frac{i\theta}{2}\cos\left(\int_x^a k(x)\,dx + \frac{\pi}{4}\right)\right]$$

Writing
$$\int_x^a k(x)\,dx + \frac{\pi}{4} = \left(\int_{-a}^{+a} k(x)\,dx + \frac{\pi}{2}\right) - \left(\int_{-a}^{x} k(x)\,dx + \frac{\pi}{4}\right)$$

we get

$$\psi_{III}(x) = \frac{C}{\sqrt{k}}\left[\left(\frac{2}{\theta}\cos\alpha - \frac{i\theta}{2}\sin\alpha\right)\cos\left(\int_{-a}^{x} k\,dx + \frac{\pi}{4}\right)\right.$$

$$\left. + \left(\frac{2}{\theta}\sin\alpha + \frac{i\theta}{2}\cos\alpha\right)\sin\left(\int_{-a}^{x} k\,dx + \frac{\pi}{4}\right)\right] \quad (100)$$

where $$\alpha \equiv \int_{-a}^{+a} k(x)\,dx \quad (101)$$

This would go over to

$$\psi_{II}(x) = \frac{C}{\sqrt{\kappa(x)}}\left[\left(\frac{2}{\theta}\cos\alpha - \frac{i\theta}{2}\sin\alpha\right)\frac{1}{\theta}\exp\left(-\int_{-b}^{x}\kappa(x)\,dx\right)\right.$$

$$\left. + \left(\frac{2}{\theta}\sin\alpha + \frac{i\theta}{2}\cos\alpha\right)\frac{\theta}{2}\exp\left(\int_{-b}^{x}\kappa(x)\,dx\right)\right]$$

where we have used the fact that $\kappa(x) = \kappa(-x)$ so that

$$\exp\left(-\int_{-b}^{-a}\kappa(x)\,dx\right) = \exp\left(-\int_{a}^{b}\kappa(x)\,dx\right) = \theta \quad \text{[see Eq. (99)]}$$

Further use of connection formulae gives

$$\psi_{I}(x) = \frac{C}{\sqrt{k(x)}}\left[\left(\frac{2}{\theta}\cos\alpha - \frac{i\theta}{2}\sin\alpha\right)\frac{2}{\theta}\sin\left(\int_{x}^{-b} k(x)\,dx + \frac{\pi}{4}\right)\right.$$

$$\left. + \left(\frac{2}{\theta}\sin\alpha + \frac{i\theta}{2}\cos\alpha\right)\frac{\theta}{2}\cos\left(\int_{x}^{-b} k(x)\,dx + \frac{\pi}{4}\right)\right] \quad (102)$$

$$= \frac{A}{\sqrt{k(x)}}\exp\left[-i\left(\int_{x}^{-b} k(x)\,dx + \frac{\pi}{4}\right)\right] + \frac{B}{\sqrt{k(x)}}\exp\left[i\left(\int_{x}^{-b} k(x)\,dx + \frac{\pi}{4}\right)\right] \quad (103)$$

(wave propagating to the right) (wave propagating to the left)

where $$A = \frac{1}{2}\left[2\sin\alpha + i\cos\alpha\left(\frac{4}{\theta^2} + \frac{\theta^2}{4}\right)\right]C$$

and $$B = \frac{i}{2}\cos\alpha\left(\frac{\theta^2}{4} - \frac{4}{\theta^2}\right)C \quad (104)$$

Thus the reflection and transmission coefficients are given by

$$R = \left|\frac{B}{A}\right|^2 = \frac{\left(\frac{4}{\theta^2} - \frac{\theta^2}{4}\right)^2 \cdot \cos^2\alpha}{4\sin^2\alpha + \left(\frac{4}{\theta^2} + \frac{\theta^2}{4}\right)^2 \cos^2\alpha} \quad \text{Reflection coefficient} \tag{105}$$

$$T = \left|\frac{C}{A}\right|^2 = \frac{4}{4\sin^2\alpha + \left(\frac{4}{\theta^2} + \frac{\theta^2}{4}\right)^2 \cos^2\alpha} \quad \text{Transmission coefficient} \tag{106}$$

It can be readily seen that $R + T = 1$. Further,

$$T = 1 \; (R = 0) \tag{107}$$

when

$$\alpha = \int_{-a}^{+a} k(x)\,dx = \left(n + \frac{1}{2}\right)\pi \tag{108}$$

Equation (108) represents the eigenvalue condition for the potential well and Eq. (107) tells us that the *double barrier* is completely transparent for energies corresponding to the eigenvalues of the potential well.

18.6 JWKB SOLUTIONS FOR SPHERICALLY SYMMETRIC POTENTIALS

For a spherically symmetric potential $V(r)$, the radial part of the wavefunction $R(r)$ satisfies the equation

$$\frac{1}{r^2}\frac{d}{dr}\left[r^2\frac{dR}{dr}\right] + \frac{2\mu}{\hbar^2}\left[E - V(r) - \frac{l(l+1)}{2\mu r^2}\right]R(r) = 0 \tag{109}$$

where E is the energy eigenvalue and the other symbols have their usual meanings (see Section 10.2). If we make the transformation

$$R(r) = \frac{u(r)}{r} \tag{110}$$

and consider the $l = 0$ case, we readily obtain

$$\frac{d^2u}{dr^2} + k^2(r)\,u(r) = 0, \tag{111}$$

where

$$k^2(r) = \frac{2\mu}{\hbar^2}[E - V(r)] \tag{112}$$

Since $u(0)$ must be zero, the JWKB solution in the region $0 < r < r_t$ is given by

$$u_{\mathrm{I}}(r) = \frac{A}{\sqrt{k(r)}} \sin\left[\int_0^r k(r)\,dr\right] \tag{113}$$

where r_t is the turning point where $k(r)$ is zero. In order to obtain the JWKB solution in the region $r > r_t$, we carry out the usual manipulation:

$$\int_0^r k(r)\,dr = \left(\int_0^{r_t} k\,dr + \frac{\pi}{4}\right) - \left(\int_r^{r_t} k\,dr + \frac{\pi}{4}\right)$$

Thus,
$$u_{\text{I}}(r) = \frac{A}{\sqrt{k(r)}}\left[\sin\theta\cos\left(\int_r^{r_t} k\,dr + \frac{\pi}{4}\right) - \cos\theta\sin\left(\int_r^{r_t} k\,dr + \frac{\pi}{4}\right)\right] \tag{114}$$

where
$$\theta = \int_0^{r_t} k(r)\,dr + \frac{\pi}{4}$$

The first term on the right-hand side of Eq. (114) goes over to an exponentially amplifying solution and therefore we must have

$$\theta = n\pi$$

or
$$\int_0^{r_t} k(r)\,dr = \left(n - \frac{1}{4}\right)\pi;\ n = 1, 2, 3, \ldots \tag{115}$$

The solution of the above equation would lead to the allowed energy eigenvalues [see Problem 18.8]. Further

$$u_{\text{II}}(r) = \frac{A(-1)^{n+1}}{\sqrt{k(r)}}\exp\left[-\int_{r_t}^{r} k(r)\,dr\right] \tag{116}$$

18.7 APPLICATION OF THE JWKB METHOD TO ALPHA DECAY

In this section, we will use results obtained by using the JWKB approximation to calculate the lifetime of nuclei emitting alpha particles. Now, the potential as seen by an α particle is approximately given by (see Fig. 18.10)

$$V(r) = -V_0, \quad r < R$$

$$= \frac{zZq^2}{4\pi\varepsilon_0 r}, \quad r > R \tag{117}$$

where R represents the radius of the nucleus, zq and Zq represent the charges of the α-particle and of the daughter nucleus respectively (obviously $z = 2$); assuming that a nucleus of charge $(Z + 2)q$ decays to a (daughter) nucleus of charge Zq and an α-particle of charge $2q$:

$$_{Z+2}X^{A+4} \rightarrow {}_{Z}Y^{A} + {}_{2}\text{He}^{4}$$
$$(\alpha\text{-particle})$$

The potential energy distribution given by Eq. (117) corresponds to the α-particle being *free* inside the nucleus and bound to it; outside the nucleus, the α-particle experiences the repulsive Coulomb force. The tunnelling probability is given by

$$T \approx \exp(-2G) \tag{118}$$

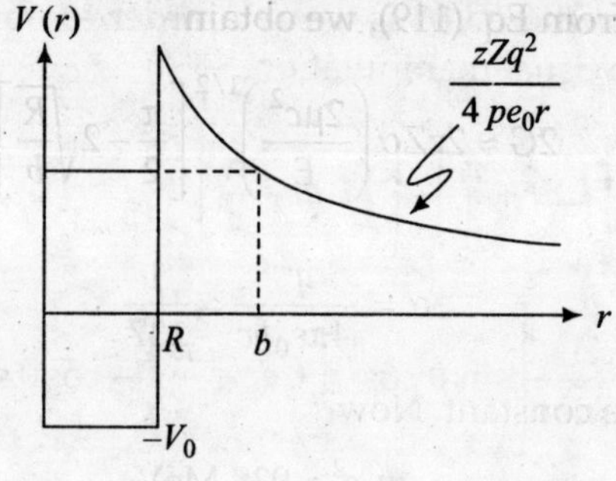

Figure 18.10 a: The potential energy variation as seen by the α-particle. Outside the nucleus (i.e., for $r > R$) the α-particle experiences a repulsive Coulomb force. The points $r = R$ and $r = b$ are the turning points.

where
$$G = \sqrt{\frac{2\mu}{\hbar^2}} \int_R^b \sqrt{\frac{zZq^2}{4\pi\varepsilon_0 r} - E}\, dr$$

$$= \left(\frac{2\mu zZq^2}{4\pi\varepsilon_0 \hbar^2}\right)^{1/2} \int_R^b \sqrt{\frac{1}{r} - \frac{1}{b}}\, dr.$$

where
$$b = \frac{zZq^2}{4\pi\varepsilon_0 E} \tag{119}$$

represents the outer turning point [see Fig. 18.10 (a)]. Making the substitution $r = b\cos^2\theta$, we readily get

$$G = \sqrt{\frac{2\mu zZq^2}{4\pi\varepsilon_0\hbar^2} b}\left[\cos^{-1}\sqrt{\frac{R}{b}} - \sqrt{\frac{R}{b} - \frac{R^2}{b^2}}\right] \tag{120}$$

Now, if the speed of the α-particles inside the nucleus is ~ v then,

$$\text{Number of times the α-particle hits the barrier} \sim \frac{v}{R} \tag{121}$$

Thus, the probability per second of escape (which is the inverse of the mean lifetime) is given by

$$\frac{1}{\tau} \approx \frac{v}{R} e^{-2G} \tag{122}$$

As we will show below, in most cases $R/b \ll 1$ so that the quantity inside the square brackets of Eq. (120) can be written as

$$\cos^{-1}\sqrt{\frac{R}{b}} - \sqrt{\frac{R}{b} - \frac{R^2}{b^2}} \approx \frac{\pi}{2} - \sqrt{\frac{R}{b}} - \sqrt{\frac{R}{b}}$$

$$\approx \frac{\pi}{2} - 2\sqrt{\frac{R}{b}}$$

Substituting the value of b from Eq. (119), we obtain

$$2G \approx 2zZ\alpha\left(\frac{2\mu c^2}{E}\right)^{1/2}\left[\frac{\pi}{2}-2\sqrt{\frac{R}{b}}\right] \tag{123}$$

where $$\alpha = \frac{q^2}{4\pi\varepsilon_0 \hbar c} \approx \frac{1}{137}$$

represents the fine structure constant. Now,

$$m_p c^2 \approx 938 \text{ MeV}$$

where m_p represents the mass of the proton and since $\mu \approx 4m_p$ we get

$$2G = \frac{2.53Z}{\sqrt{E}}\left[\frac{\pi}{2}-2\sqrt{\frac{R}{b}}\right] \tag{124}$$

where E is measured in MeV and we have assumed $z = 2$. Also,

$$b = \frac{2Zq^2}{4\pi\varepsilon_0 E}$$

$$= \frac{2Z\alpha\hbar}{m_p c}\left(\frac{m_p c^2}{E}\right)$$

$$= \frac{2Z(1.05\times10^{-34})}{(1.67\times10^{-27})(3\times10^8)}\left(\frac{1}{137}\right)\left(\frac{938}{E}\right)$$

or $$b \approx 2.87\times10^{-15}\frac{Z}{E} \tag{125}$$

where b is measured in metres and E in MeV. Now, the radius of the nucleus is approximately given by

$$R \approx (1.07\times10^{-15})A^{1/3} \text{ (m)} \tag{126}$$

where A represents the mass number and R is measured in metres (m). Thus, we obtain

$$\sqrt{\frac{R}{b}} \approx 0.611\, Z^{-1/2} A^{1/6}\sqrt{E} \tag{127}$$

Thus, $$2G \approx \frac{3.97Z}{\sqrt{E}} - 3.09Z^{1/2}A^{1/6} \tag{128}$$

Now, $$v = \sqrt{\frac{2E}{\mu}} = c\sqrt{\frac{2E}{\mu c^2}}$$

Since, $$\mu c^2 \approx 4m_p c^2 \approx 4\times938 \text{ MeV}$$

we obtain $$v \approx 6.93\times10^6\sqrt{E}\ \text{(m/s)}. \tag{129}$$

where E is measured in MeV and v is measured in metres/second. Thus,

$$\frac{v}{R} \approx 6.47\times10^{21}E^{1/2}A^{-1/3}\ (\text{s}^{-1}) \tag{130}$$

Substituting the value in Eq. (122), we get

$$\frac{1}{\tau} \approx \left[6.47 \times 10^{21} \sqrt{E} A^{-1/3}\right] \exp\left[-\frac{3.97Z}{\sqrt{E}} + 3.09\, Z^{1/2} A^{1/6}\right] \tag{131}$$

or

$$\log_{10}\tau \approx -21.8 - \frac{1}{2}\log_{10} E + \frac{1}{3}\log_{10} A + \frac{1.724\,Z}{\sqrt{E}} - 1.34 Z^{1/2} A^{1/6} \tag{132}$$

where τ is measured in seconds and E in MeV. Let us next consider some examples:
We consider the following radioactive decay

$$_{84}\text{Po}^{212} \rightarrow {}_{82}\text{Pb}^{208} + \alpha\,(8.9\text{ MeV}) \tag{133}$$

Thus, $Z = 82$, $A = 208$ and $E = 8.9$ MeV. Substituting in Eq. (27), we get

$$\log_{10}\tau \approx -21.8 - 0.5 + 0.8 + 47.4 - 29.5$$

$$\approx -3.6 \tag{134}$$

thus giving

$$\tau \approx 10^{-4}\text{ s}$$

The corresponding experimental value is ~ 3×10^{-7}s. On the other hand, for the radioactive decay

$$_{92}\text{U}^{238} \rightarrow {}_{90}\text{Th}^{234} + \alpha\,(4.2\text{ MeV}) \tag{135}$$

we get

$$\log_{10}\tau \approx -21.8 - 0.3 + 0.8 + 75.7 - 31.6$$

$$\approx 23 \tag{136}$$

or

$$\tau \approx 10^{23}\text{ s}$$

The corresponding experimental value is 10^{17} s. Although the theoretical values are not in agreement with the experimental values, one may note the enormous range in the values of τ, from ~ 10^{17} s for U^{238} to ~ 3×10^{-7} s for Po^{212} which is also predicted by our approximate theoretical analysis. One may note from Equations (134) and (135) that it is the term

$$\frac{1.724Z}{\sqrt{E}}$$

which is most important in Eq. (132). In Ref. 6 it has been shown that the empirical formula

$$\log_{10}\tau = C_1\left[\frac{Z}{\sqrt{E}} - Z^{2/3}\right] - C_2 \tag{137}$$

has been shown to agree well with the experimental data [see Fig. 18.10(b)]. In Eq. (137), τ is the half life in years, Z is the atomic number of the daughter nucleus, and E is the energy of the α-particle in MeV.

The two lines in Fig. 18.10 (b) correspond to

$$C_1 = 1.61\,(\text{MeV})^{1/2},\ C_2 = 28.9 \tag{138}$$

and

$$C_1 = 1.70\,(\text{MeV})^{1/2},\ C_2 = 30.0 \tag{139}$$

For example, for the radioactive decay given by Eq. (135)

$$E \approx 4.2\text{ MeV} \quad \text{and} \quad Z = 90$$

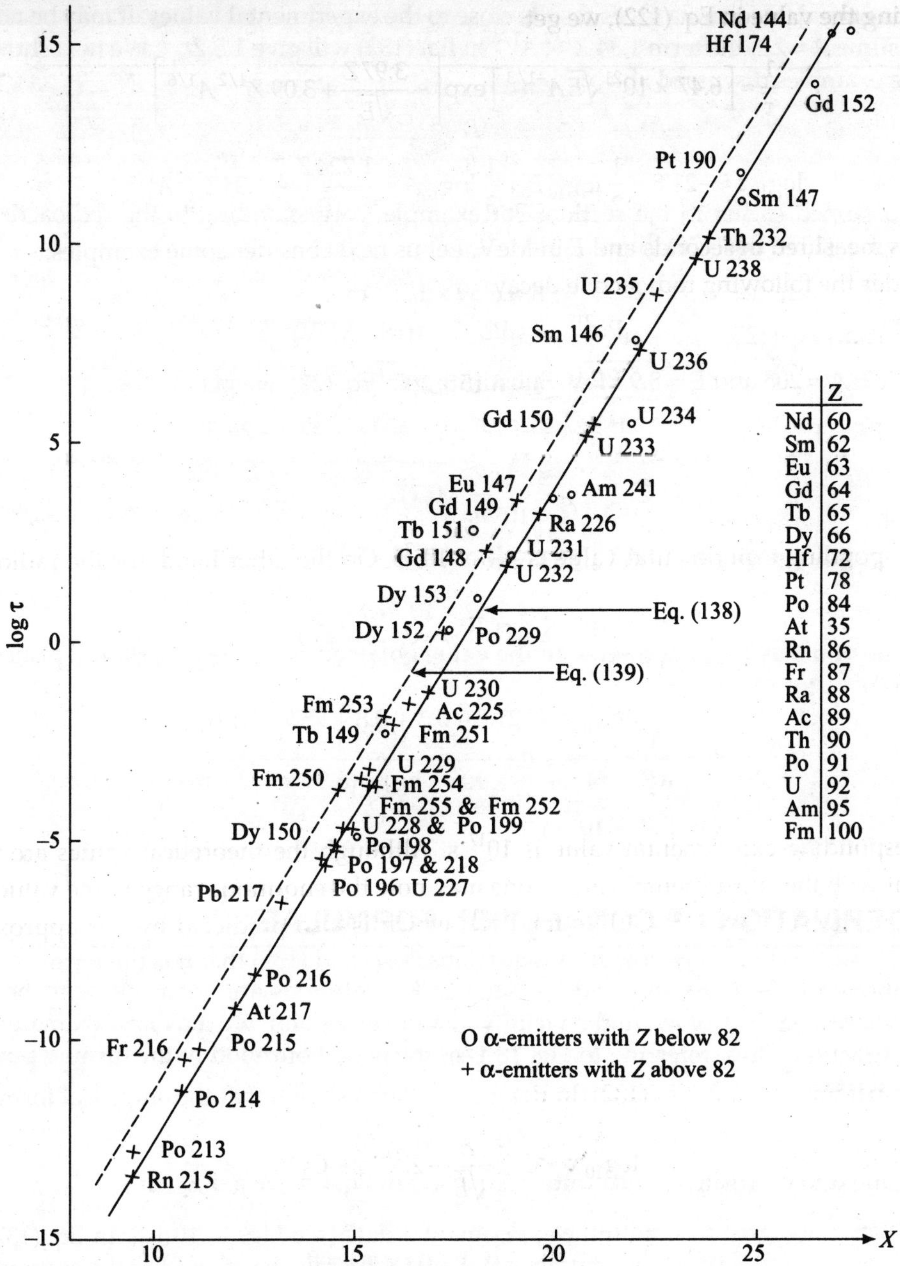

Figure 18.10 b: Variation of log τ [where τ is in years] with X [$= (Z/\sqrt{E}) - Z^{2/3}$] (adapted from Ref. 6).

Thus, using Eqs. (138) and (139), we get

$$\tau \approx 10^{9.5} \text{ years} \approx 3 \times 10^{16.5} \text{ s}$$

and

$$\tau \approx 10^{10.5} \text{ years} \approx 3 \times 10^{17.5} \text{ s}$$

Similarly, for the decay given by Eq. (133) we would get [using Eq. (138)]

$$\tau \approx 10^{-15} \text{ years} \approx 3 \times 10^{-8} \text{ s}$$

Thus, the empirical formula gives results close to the experimental values. It may be noted that if we assume $A \approx 2Z$, the term $1.34\, Z^{1/2} A^{1/6}$ in Eq. (132) will give $1.5\, Z^{2/3}$. We next show that in both the examples discussed above

$$\frac{R}{b} \ll 1$$

as was assumed earlier in the section. For example, corresponding to the radioactive decay given by Eq. (135), Eq. (126) gives

$$R \approx 6.59 \times 10^{-15} \text{ m}$$

Further, from Eq. (125)

$$b \approx 6.15 \times 10^{-14} \text{ m}$$

thus giving

$$\frac{R}{b} \approx 0.11$$

proving $R/b \ll 1$. Further, for the 4.2 MeV α-particle

$$v \approx 1.42 \times 10^7 \text{ m/s}$$

This value of v may be compared with the value obtained from the uncertainty principle (see Chapter 3)

$$v \approx \frac{h}{\mu R} \approx \frac{6.6 \times 10^{-34} \text{ J.s}}{(4 \times 1.76 \times 10^{-27} \text{ kg}) \times (6.59 \times 10^{-15} \text{ m})}$$

$$\approx 1.50 \times 10^7 \text{ m/s}$$

18.8 DERIVATION OF CONNECTION FORMULAE

As mentioned in Sections 18.1 and 18.2, the JWKB solutions are not valid near the turning points where $k^2(x) = 0$. Now, in the vicinity of a turning point, we may approximate $k^2(x)$ by a linear function. Thus, referring to Fig. 18.1, in the neighbourhood of the turning point $x = a$, we may write

$$k^2(x) \approx -\alpha(x - a) \tag{140}$$

We assume α to be positive. Substituting Eq. (140) in Eq. (1), we get

$$\frac{d^2\psi}{dx^2} - \alpha(x - \alpha)\,\psi(x) = 0 \tag{141}$$

The above equation can be written in the form

$$\frac{d^2\psi}{dz^2} - z\psi(z) = 0 \tag{142}$$

where

$$z = \alpha^{1/3}(x - \alpha) \tag{143}$$

We next write the JWKB solutions of Eq. (142).

For $z > 0$, we define $\kappa^2(z) = z$ [cf. Eq. (26)]: thus the JWKB solutions are

$$\frac{1}{z^{1/4}} \exp [\pm \zeta] \tag{144}$$

where
$$\zeta = \int_0^z \kappa(z)\, dz = \frac{2}{3} z^{3/2} \quad (z > 0) \tag{145}$$

On the other hand, for $z < 0$,

$$k^2(z) = -z = |z| \tag{146}$$

Thus the JWKB solutions are

$$\frac{1}{|z|^{1/4}} \cos \zeta \quad \text{and} \quad \frac{1}{|z|^{1/4}} \sin \zeta \tag{147}$$

where
$$\zeta = \int_z^0 (-z)^{3/2}\, dz = \frac{2}{3} (-z)^{3/2} \quad (z < 0) \tag{148}$$

Combining Eqs (145) and (148), we may write

$$\zeta = \frac{2}{3} |z|^{3/2} \tag{149}$$

Now, the exact solutions of Eq. (142) are $Ai(z)$ and $Bi(z)$ which have been discussed in Appendix D. The asymptotic forms of $Ai(z)$ and $Bi(z)$ are given by [see Eqs (20)–(23) of Appendix D]

$$Ai(z) \underset{z \to +\infty}{\longrightarrow} \frac{1}{2\sqrt{\pi}\, z^{1/4}} e^{-\zeta} \tag{150}$$

$$\underset{z \to -\infty}{\longrightarrow} \frac{1}{\sqrt{\pi}\, |z|^{1/4}} \sin\left(\zeta + \frac{\pi}{4}\right) \tag{151}$$

where $\zeta = \frac{2}{3} |z|^{3/2}$. Notice that the asymptotic forms are the JWKB solutions. Thus for $k^2(z) = -z$, the connection formula is given by

$$\frac{2}{|z|^{1/4}} \sin\left(\zeta + \frac{\pi}{4}\right) \leftrightarrow \frac{1}{z^{1/4}} e^{-\zeta} \tag{152}$$

implying that $\sin(\zeta + \frac{\pi}{4})$ goes over to the exponentially decaying solution. Similarly, the asymptotic forms of $Bi(z)$ are given by

$$Bi(z) \underset{z \to +\infty}{\longrightarrow} \frac{1}{2\sqrt{\pi}\, z^{1/4}} e^{+\zeta} \tag{153}$$

$$\underset{z \to -\infty}{\longrightarrow} \frac{1}{\sqrt{\pi}\, |z|^{1/4}} \cos\left(\zeta + \frac{\pi}{4}\right) \tag{154}$$

yielding the connection formula

$$\frac{1}{|z|^{1/4}} \cos\left(\zeta + \frac{\pi}{4}\right) \leftrightarrow \frac{1}{z^{1/4}} e^{\zeta} \tag{155}$$

Since $k^2(z) = -z$, Eqs (152) and (155) can be generalized to

$$\frac{2}{\sqrt{k(z)}} \sin\left[\int_z^a k(z)\,dz + \frac{\pi}{4}\right] \leftrightarrow \frac{1}{\sqrt{\kappa(z)}} \exp\left[-\int_a^z \kappa(z)\,dz\right] \tag{156}$$

$$\frac{1}{\sqrt{k(z)}} \cos\left[\int_z^a k(z)\,dz + \frac{\pi}{4}\right] \leftrightarrow \frac{1}{\sqrt{\kappa(z)}} \exp\left[+\int_a^z \kappa(z)\,dz\right] \tag{157}$$

Connection formulae

It is important to mention here that strictly speaking, the connection formula given by Eq. (156) is valid in the backward direction and the connection formula given by Eq. (157) is valid in the forward direction; what we imply is that in Eq. (156), if we introduce a very small phase in the sine term we would get

$$\frac{2}{\sqrt{k}} \sin\left[\int_x^a k(x)\,dx + \frac{\pi}{4} + \varepsilon\right] = \frac{2}{\sqrt{k}} \cos\varepsilon \sin\left[\int_x^a k(x)\,dx + \frac{\pi}{4}\right] + \frac{2}{\sqrt{k}} \sin\varepsilon \cos\left[\int_x^a k(x)\,dx + \frac{\pi}{4}\right]$$

$$\rightarrow \frac{1}{\sqrt{\kappa(x)}} \cos\varepsilon \exp\left[-\int_x^a \kappa(x)\,dx + \frac{\pi}{4}\right] + \frac{2}{\sqrt{\kappa(x)}} \sin\varepsilon \exp\left[\int_x^a \kappa(x)\,dx\right]$$

Thus a small change in the phase of the sine term will give rise to a cos term which would go over to an exponentially growing solution. This exponentially growing term, for sufficiently large values of x, will dominate over the exponentially decreasing term. Similarly, the presence of an exponentially decaying solution (which will be negligible for large values of x) in the right-hand side of Eq. (31) would lead to a cos term in the left-hand side. Because of this, many authors have argued that the use of connection formulae in the reverse directions is not strictly justified (see, e.g. Ref. 11). However Pauli, on the basis of conservation of current density, has argued on the validity of the connection formulae in the reverse direction also. Indeed, in the literature, the connection formulae have extensively used in the reverse direction (see, e.g. Example 18.4, Problems 18.5 and 18.6) and one finds that the JWKB results compare very well with the exact results.

18.9 SOME HISTORICAL REMARKS

The approximate solutions of Eq. (1) as given by Eqs (24)–(28) are usually referred to as the WKB solutions. However, they were first put forward by Harold Jeffreys in a 1924 paper entitled *On certain approximate solutions of linear differential equations of the second order* (Ref. 7). Even the connection formulae were given by Jeffreys. Indeed this paper along with three other papers by Jeffreys formed part of a dissertation highly commended by the adjudicators for the Adams Prize of the University of Cambridge for the year 1923. The applications of the JWKB solutions to problems in quantum mechanics came much later and were given by Wentzel (Ref. 8), Kramers (Ref. 9) and Brillouin (Ref. 10). We feel that since the solutions were first put forward by Jeffreys, it should be referred to as the JWKB method; this is indeed what has been done by Fröman and Fröman (see Ref. 11).

18.10 PROBLEMS

Problem 18.1 Consider the symmetric linear profile given by

$$V(x) = \gamma |x|, \quad \gamma > 0 \tag{158}$$

Use the JWKB quantization condition [Eq. (44)] to obtain the following

$$E_n = \left[\frac{3\left(n+\frac{1}{2}\right)\pi\hbar\gamma}{4(2m)^{1/2}}\right]^{2/3} \; ; n = 0, 1, \ldots \tag{159}$$

Problem 18.2 In continuation of Problem 18.1, consider the more general power law profile given by

$$V(x) = \gamma |x|^{\alpha}; \quad \gamma > 0, \quad \alpha > 0 \tag{160}$$

Show that the energy eigenvalues are given by

$$E_n = \left[\frac{\left(n+\frac{1}{2}\right)\pi\hbar\gamma^{1/\alpha}\,\Gamma\left(\frac{3}{2}+\frac{1}{\alpha}\right)}{2(2m)^{1/2}\,\Gamma\left(\frac{3}{2}\right)\Gamma\left(1+\frac{1}{\alpha}\right)}\right]^{2\alpha/(\alpha+2)} \tag{161}$$

[**Hint:** You may use the following relation

$$\int_0^1 t^{z-1}(1-t)^{w-1}\,dt = B(z, w) = \frac{\Gamma(z)\Gamma(w)}{\Gamma(z+w)} \tag{162}$$

Problem 18.3 Show that the JWKB solution given by Eq. (23) would represent an accurate solution of Eq. (1) provided

$$\left|\frac{1}{2k}\frac{d^2k}{dx^2} - \frac{3}{4k^2}\left(\frac{dk}{dx}\right)^2\right| \ll k^2(x) \tag{163}$$

[**Hint:** Obtain the differential equation satisfied by $\psi(x)$ as given by Eq. (23).]

Problem 18.4 Show that the equation

$$\frac{d^2y}{dx^2} + f(x)\frac{dy}{dx} + g(x)\,y(x) = 0 \tag{164}$$

can be transformed to the form given by Eq. (1) if we make the transformations

$$\psi(x) = y(x)\exp\left[\frac{1}{2}\int^{x} (x)\,dx\right] \tag{165}$$

with

$$k^2(x) = g(x) - \frac{1}{2}\frac{df}{dx} - \frac{1}{4}f^2(x) \tag{166}$$

Problem 18.5 Consider a double well symmetric potential as shown in Fig. 18.11. Using the JWKB connection formulae, show that (for $E < V_0$) the energy eigenvalues are determined by solving the following transcendental equation

$$\cot\left[\int_a^b k(x)\,dx\right] = \pm\frac{1}{2}\exp\left[-\int_{-a}^{+a}\kappa(x)\,dx\right] \tag{167}$$

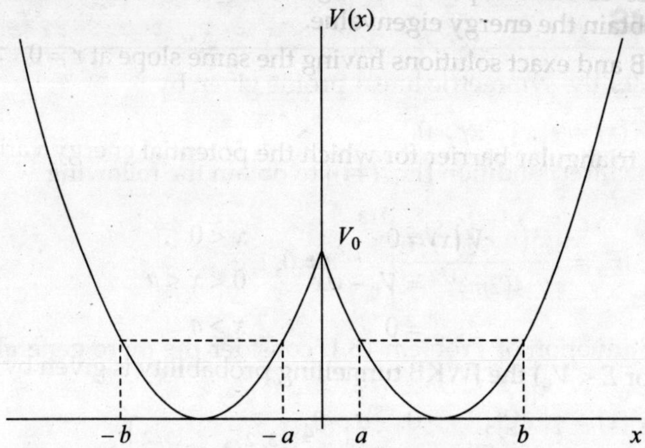

Figure 18.11: A double well potential which is symmetric about $x = 0$. The points $x = \pm a$ and $x = \pm b$ are the turning points (adapted from Ref. 3).

where the + and – signs on the right-hand side correspond to symmetric and antisymmetric solutions respectively; in the above equation $x = \pm a$ and $x = \pm b$ represent the turning points. The above equation is not expected to give accurate results for E close to V_0 (why?).

[**Hint:** Start with an exponentially decaying solution in Region V ($x > b$) and use connection formulae to obtain solutions in Region III ($-a < x < a$).]

Problem 18.6

(a) Use the results of the previous problem to determine the energy levels of a double oscillator given by

$$V(x) = \frac{1}{2} m\omega^2 [\,|x| - d]^2 \tag{168}$$

(b) Obtain the exact solution for the above potential energy distribution and compare the exact eigenvalues with the corresponding JWKB result.

Problem 18.7 Consider the three-dimensional isotropic oscillator for which (see Section 10.6.1):

$$V(r) = \frac{1}{2} \mu\omega^2 r^2$$

For $l = 0$, solve Eq. (115) to obtain

$$E = \left(2n + \frac{3}{2}\right) \hbar\omega; \quad n = 0, 1, 2, \ldots$$

which happens to be the exact result!

[**Hint:** The integration has to be carried out in a way similar to that given in Example 18.1.]

Problem 18.8

(a) The deuteron nucleus consists of a neutron and a proton and the neutron-proton interaction can be approximately described by the following potential energy variation:

$$V(r) = -V_0\, e^{-r/a} \tag{169}$$

(see also Problem 10.8). Carry out the integration in Eq. (115) and derive the transcendental equation which would give the eigenvalues.

(b) For the values of various parameters given in Problem 10.6, solve the transcendental equation to obtain the energy eigenvalue.

(c) Plot the JWKB and exact solutions having the same slope at $r = 0$.

Problem 18.9

(a) Consider the triangular barrier for which the potential energy variation is given by (see Fig. 18.12)

$$\begin{aligned} V(x) &= 0 & x < 0 \\ &= V_0 - \alpha x & 0 < x < a \\ &= 0 & x > a \end{aligned} \tag{170}$$

Show that (for $E < V_0$) the JWKB tunnelling probability is given by

$$T_{\text{JWKB}} = \exp\left[-\frac{4}{3} g^{3/2} (1-\varepsilon)^{3/2}\right] \tag{171}$$

where
$$g = \left(\frac{2m}{\hbar^2 \alpha^2}\right)^{1/3} V_0,$$

$$\varepsilon = \frac{E}{V_0}$$

(b) The above potential energy variation is applicable to the field emission of electrons from a metal surface. Then $V_0 = E_F + \phi$ and $\alpha = (-q)F$ where E_F represents the Fermi energy, ϕ the work function, $q (< 0)$ the electronic charge and F the strength of the electric field. Assume $T = 0$ K and calculate the dependence of the electric current the electric field.

Problem 18.10 The potential energy distribution given by Eq. (169) can also be (approximately) applied to a quantum-well laser and to the α-decay problem. For a quantum well laser

$$\alpha = \frac{V_0}{L_D} \tag{172}$$

where L_D is the Debye length and is typically few nanometers. Further 50 meV $\lesssim V_0 \lesssim$ 500 meV. Assuming

$$V_0 \simeq 50 \text{ meV} \quad \text{and} \quad L_D \simeq 1 \text{ nm}$$

show that $g \simeq 1.1$.

On the other hand, in the α-decay problem for $V_0 \approx 20$ MeV with $a \approx 3R$ ($R \approx 10^{-14}$ m being the radius of the nucleus) show that $g \simeq 15$.

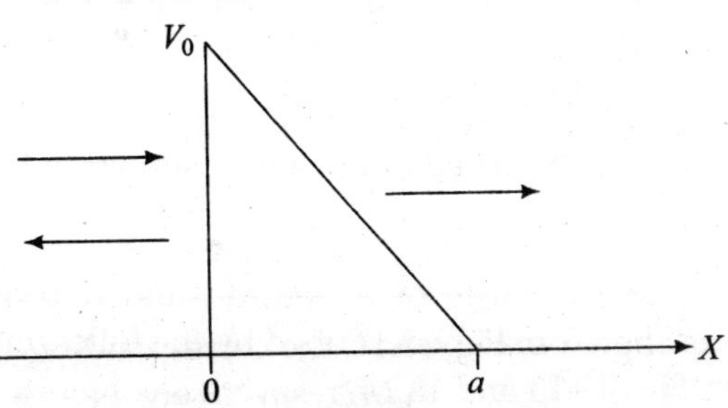

Figure 18.12: The triangular barrier.

Problem 18.11 For the potential energy function given by Eq. (169) one can obtain exact solutions of the Schrödinger equation in terms of the Airy functions. Calculate the corresponding tunnelling probability and compare with the corresponding JWKB results.

Problem 18.12 Obtain the JWKB eigenvalues for the following potential energy distribution

$$V = -V_0 \operatorname{sech}^2\left(\frac{x}{a}\right) \tag{173}$$

Problem 18.13 In the radial part of the Schrödinger equation [Eq. (109)] make the following transformations

$$R(r) = \frac{1}{r}u(r) \tag{174}$$

$$\frac{r}{a_0} = e^x \tag{175}$$

$$\chi(x) = u(x)e^{-x/2} \tag{176}$$

to obtain

$$\frac{d^2\chi}{dx^2} + k^2(x)\,\chi(x) = 0 \tag{177}$$

where

$$k^2(x) = \frac{2ma_0^2}{\hbar^2}e^{2x}\left[E - V(a_0e^x) - \frac{\left(l+\frac{1}{2}\right)^2\hbar^2}{2ma_0^2}e^{-2x}\right] \tag{178}$$

and a_0 is a conveniently chosen length parameter. Notice that as r goes from 0 to ∞, x goes from $-\infty$ to $+\infty$. Thus, show that the eigenvalue equation [Eq. (44)] leads to

$$\int_{r_1}^{r_2} k_1(r)\,dr = \left(n+\frac{1}{2}\right)\pi\,; \quad n = 0, 1, 2, \ldots. \tag{179}$$

where

$$k_1^2(r) = \frac{2m}{\hbar^2}\left[E - V(r) - \frac{\left(l+\frac{1}{2}\right)^2\hbar^2}{2mr^2}\right] \tag{180}$$

and r_1 and r_2 are the turning points of $k_1(r)$.

Problem 18.14 Using the results of the previous problem, find the eigenvalues corresponding to the Coulomb potential

$$V(r) = -\frac{Ze^2}{r} \tag{181}$$

18.11 SOLUTIONS

Solution 18.5 For $V(x)$ as shown in Fig. 18.11, the corresponding variation of $k^2(x)$ for $E < V_0$ and $E > V_0$ are shown in Figs 18.13 and 18.14, respectively. Notice that for $E < V_0$, there are 4 turning points and for $E > V_0$ there are only 2 turning points.

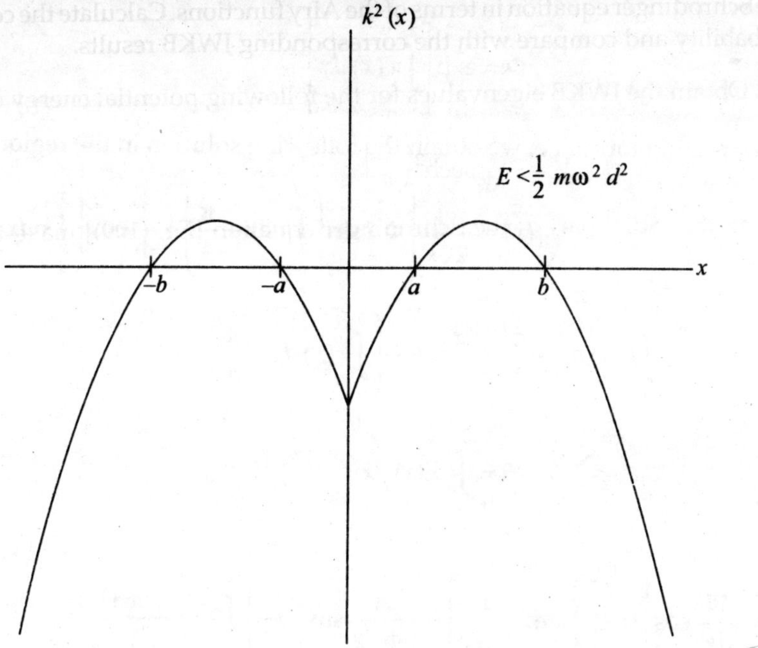

Figure 18.13: For the $V(x)$ variation as shown in Fig. 18.11, the variation of $k^2(x)$ with x for $E < V_0$.

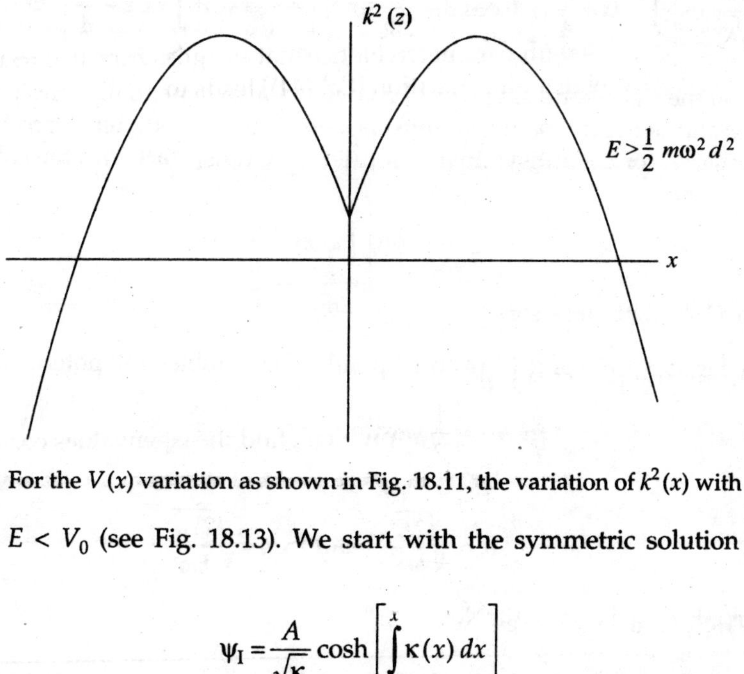

Figure 18.14: For the $V(x)$ variation as shown in Fig. 18.11, the variation of $k^2(x)$ with x for $E > V_0$.

We assume $E < V_0$ (see Fig. 18.13). We start with the symmetric solution in the region $-a < x < a$:

$$\psi_{\text{I}} = \frac{A}{\sqrt{\kappa}} \cosh\left[\int_0^x \kappa(x)\,dx\right]$$

$$= \frac{A}{2\sqrt{\kappa}}\left[\Phi \exp\left[-\int_x^a \kappa dx\right] + \frac{1}{\Phi}\exp\left[+\int_x^a \kappa dx\right]\right]$$

where $$\Phi = \exp\left[\int_0^a \kappa(x)\,dx\right]$$

Using the connection formulae, we obtain the following solution in the region $a < x < b$.

$$\psi_{II}(x) = \frac{A}{2\sqrt{k}}\left[\Phi\, 2\sin\left(\int_a^x k\,dx + \frac{\pi}{4}\right) + \frac{1}{\Phi}\cos\left(\int_a^x k\,dx + \frac{\pi}{4}\right)\right]$$

Writing $$\int_a^x k(x)\,dx + \frac{\pi}{4} = \frac{\pi}{2} + \theta - \left(\int_x^b k(x)\,dx + \frac{\pi}{4}\right)$$

where $$\theta = \int_a^b k(x)\,dx$$

we get

$$\psi_{II}(x) = \frac{A\Phi}{\sqrt{k}}\cos\left[\theta - \left(\int_x^b k\,dx + \frac{\pi}{4}\right)\right] - \frac{A}{2\Phi\sqrt{k}}\sin\left[\theta - \left(\int_x^b k\,dx + \frac{\pi}{4}\right)\right]$$

$$= \frac{A}{\sqrt{k}}\cos\left(\int_x^b k\,dx + \frac{\pi}{4}\right)\left[\Phi\cos\theta - \frac{1}{2\Phi}\sin\theta\right] + \frac{A}{\sqrt{k}}\sin\left(\int_x^b k\,dx + \frac{\pi}{4}\right)\left[\Phi\sin\theta + \frac{1}{2\Phi}\cos\theta\right]$$

The first term on the right-hand side would go over to an exponentially amplifying solution in the region $x > b$ and hence its coefficient must be zero. This immediately leads to Eq. (167) with the + sign. Similarly, for the antisymmetric solution, we must start with the solution

$$\frac{A}{\sqrt{\kappa}}\sinh\left[\int_0^x \kappa(x)\,dx\right]$$

and obtain Eq. (167) with the – sign.

Solution 18.6 Figures 18.13 and 18.14 correspond to the double-well potential with

$$V_0 = \frac{1}{2}m\omega^2 d^2 \tag{182}$$

We define

$$\alpha = \sqrt{\frac{2E}{\hbar\omega}} \quad \text{and} \quad \alpha_0 = \sqrt{\frac{2V_0}{\hbar\omega}} \tag{183}$$

We first assume $E < V_0$ (i.e. $\alpha < \alpha_0$). Now

$$\theta = \int_a^b k(x)\,dx = \sqrt{\frac{2m}{\hbar^2}}\int_a^b \left[E - \frac{1}{2}m\omega^2(x-d)^2\right]^{1/2} dx$$

$$= \int_{-\alpha}^{+\alpha}[\alpha^2 - \xi^2]^{1/2}\,d\xi = \frac{\pi}{2}\alpha^2$$

where $$\xi = \sqrt{\frac{m\omega}{\hbar}}\,(x-d)$$

Other integrations can easily be carried out. For $E > V_0$, we use Eq. (44). The final results are

(*i*) For $E < V_0 \quad (0 < \alpha < \alpha_0)$

$$\cot\left(\frac{\pi}{2}\alpha^2\right) = \pm\frac{1}{2}\exp\left[-\alpha_0(\alpha_0^2-\alpha^2)^{1/2} + \alpha^2\cosh^{-1}\frac{\alpha_0}{\alpha}\right] \tag{184}$$

and

(*ii*) For $E > V_0\ (\alpha > \alpha_0)$

$$\frac{\pi}{2}\alpha^2 + \alpha_0\,(\alpha^2-\alpha_0^2)^{1/2} + \alpha^2\sin^{-1}\frac{\alpha_0}{\alpha} = \left(m+\frac{1}{2}\right)\pi \tag{185}$$

The corresponding wavefunctions are given by (for $E < V_0$):

$$\psi_{\mathrm{V}}(\xi) = \frac{C}{(\xi^2-\alpha^2)^{1/4}}\exp\left[-\frac{1}{2}\xi(\xi^2-\alpha^2)^{1/2} + \frac{1}{2}\alpha^2\ln\left|\frac{\xi+(\xi^2-\alpha^2)^{1/2}}{\alpha}\right|\right] \tag{186}$$

Similarly,

$$\psi_{\mathrm{IV}}(\xi) = \frac{-2C}{(\alpha^2-\xi^2)^{1/4}}\sin\left[\frac{\pi}{4}+\frac{\pi}{4}\alpha^2 - \frac{1}{2}\xi(\alpha^2-\xi^2)^{1/2} - \frac{1}{2}\alpha^2\sin^{-1}\frac{\xi}{\alpha}\right] \tag{187}$$

and $$\psi_{\mathrm{III}}(\xi) = \frac{C}{(\xi^2-\alpha^2)^{1/4}}\left[\cos\left(\frac{\pi}{2}\alpha^2\right)\exp(\zeta) + \frac{1}{2}\sin\left(\frac{\pi}{2}\alpha^2\right)\exp(-\zeta)\right] \tag{188}$$

where $$\zeta(\xi) = \int_x^{x_1}\kappa(x)\,dx = \int_\xi^{-\alpha}(\xi^2-\alpha^2)^{1/2}\,d\xi$$

$$= -\frac{1}{2}\xi(\xi^2-\alpha^2)^{1/2} + \frac{1}{2}\alpha^2\ln\left|\frac{\xi+\sqrt{\xi^2-\alpha^2}}{\alpha}\right| \tag{189}$$

Figure 18.15 shows the normalized energy eigenvalues α^2 for the double oscillator as a function of α_0^2; a large value of α_0 (and hence of d) implies a large distance between the centers of the two oscillators. The solid and the dashed curves correspond to the exact and the JWKB calculations, respectively. As $d \to \infty$, $\alpha^2\,(= 2E/\hbar\omega)$ approaches $(2n + 1)$ and the levels become two-fold degenerate; thus the system becomes equivalent to two independent harmonic oscillators. Further, for $d \to 0$, we have the eigenvalue structure of a single harmonic oscillator.

The exact solution for the double oscillator problem is given in terms of the confluent hypergeometric functions. For $V(x)$ given by Eq. (167), the Schrödinger equation is

$$\frac{d^2\psi}{dx^2} + \frac{2m}{\hbar^2}\left\{E - \frac{1}{2}m\omega^2[\,|x|-d]^2\right\}\psi = 0 \tag{190}$$

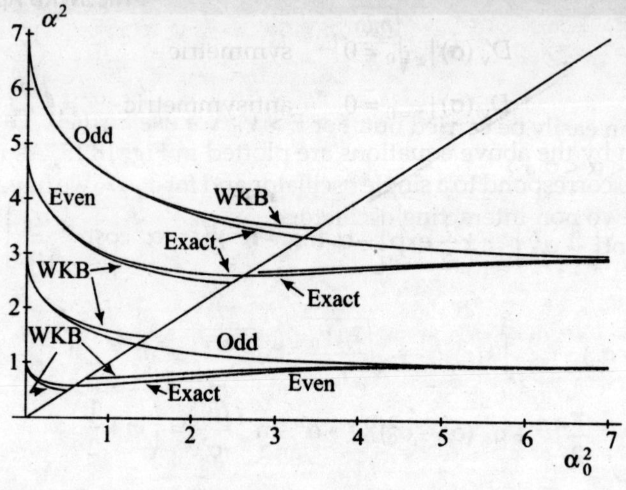

Figure 18.15: Normalized energy eigenvalues for the double oscillator as a function of α_0^2.

In terms of the variables

$$\left.\begin{array}{ll} \sigma = \left(\dfrac{2m\omega}{\hbar}\right)^{1/2} (x-d) & \text{for } x>0 \\ \tau = \left(\dfrac{2m\omega}{\hbar}\right)^{1/2} (x+d) & \text{for } x<0 \end{array}\right\} \tag{191}$$

(the second line preceded by "and")

Equation (189) becomes

$$\frac{d^2\psi}{d\sigma^2}+\left[\nu+\frac{1}{2}-\frac{1}{4}\sigma^2\right]\psi(\sigma)=0 \quad \text{for } x>0 \tag{192}$$

and

$$\frac{d^2\psi}{d\tau^2}+\left[\nu+\frac{1}{2}-\frac{1}{4}\tau^2\right]\psi(\tau)=0 \quad \text{for } x<0 \tag{193}$$

where

$$\nu=\frac{1}{2}\left(\frac{2E}{\hbar\omega}-1\right)=\frac{1}{2}(\alpha^2-1) \tag{194}$$

The boundary conditions for solving Eqs (191) and (192) are $\psi \to 0$ as $\sigma \to \infty$ and $\tau \to -\infty$. For $a = 0$, $\sigma = \tau$ and Eqs (191) and (192) become equivalent to the linear harmonic oscillator problem and ν will then be 0, 1, 2, ... The solution of Equation (191) that vanishes for $\sigma \to \infty$ is the parabolic cylinder functions $D_\nu(\sigma)$, (see Ref. 12) which are related to the confluent hypergeometric function through the relation

$$D_\nu(\sigma)=2^{\nu/2}\exp(-\sigma^2/4)\left[\frac{\Gamma\left(\frac{1}{2}\right)}{\Gamma\left(\frac{1-\nu}{2}\right)}\,{}_1F_1\left(-\frac{\nu}{2};\frac{1}{2};\frac{\sigma^2}{2}\right)+\frac{\sigma}{\sqrt{2}}\frac{\Gamma\left(-\frac{1}{2}\right)}{\Gamma\left(-\frac{\nu}{2}\right)}\,{}_1F_1\left(\frac{1-\nu}{2};\frac{3}{2};\frac{\sigma^2}{2}\right)\right] \tag{195}$$

Since $D_\nu(\sigma) \to 0$ as $\sigma \to \infty$ we must choose $D_\nu(\sigma)$ for $x > 0$ and $D_\nu(-\tau)$ for $\tau < 0$. In any case, since $V(-x) = V(x)$, the wavefunctions will be either symmetric or antisymmetric in x; thus, we need only the solution for $x > 0$. The eigenvalue equation for the symmetric and antisymmetric wavefunctions are

$$D'_v(\sigma)\,|_{x=0} = 0 \qquad \text{symmetric} \tag{196}$$

and

$$D_v(\sigma)\,|_{x=0} = 0 \qquad \text{antisymmetric} \tag{197}$$

The solution given by the above equations are plotted in Fig. 18.15. As mentioned earlier, for $a \to 0$, the solutions correspond to a single oscillator and for $a \to \infty$ we have two-fold degeneracy corresponding to two non-interacting oscillators.

Solution 18.8

$$\int_0^{r_1} k(r)\,dr = \sqrt{\frac{2\mu V_0}{\hbar^2}} \int_0^{r_1} \sqrt{e^{-r/a} - \varepsilon}\,dr$$

where

$$\varepsilon = -\frac{E}{V_0} \quad \text{and} \quad r_1 = a \ln\frac{1}{\varepsilon}$$

For a bound state $-V_0 < E < 0$ implying $0 < \varepsilon < 1$. If we make the substitution

$$\xi^2 = e^{-r/a} - \varepsilon$$

we obtain

$$\int k(r)\,dr = -g\left[\sqrt{e^{-r/a} - \varepsilon} - \sqrt{\varepsilon}\tan^{-1}\sqrt{\frac{e^{-r/a} - \varepsilon}{\varepsilon}}\right]$$

where

$$g = \sqrt{\frac{8\mu V_0 a^2}{\hbar^2}} \simeq 3.64$$

for the numerical values given in Problem 10.8. The transcendental equation determining the energy eigenvalues is given by

$$\sqrt{1-\varepsilon} - \sqrt{\varepsilon}\tan^{-1}\sqrt{\frac{1-\varepsilon}{\varepsilon}} = \left(n - \frac{1}{4}\right)\frac{\pi}{g}$$

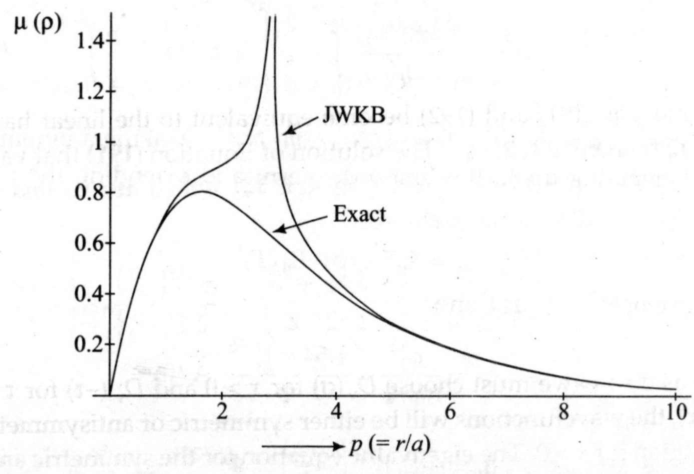

Figure 18.16: Exact and JWKB ground state wavefunctions corresponding to Eq. (111) ($l = 0$) for the potential energy distribution given by Eq. (169) (adapted from Ref. 3).

For $g = 3.64$ we obtain

$$\varepsilon \simeq 0.0593$$

Thus,

$$E \simeq -2.372 \text{ MeV}$$

which may be compared with the exact value of –2.223 MeV. In Fig. 18.16, we have plotted the JWKB and exact wavefunctions having the same slope at $r = 0$.

Solution 18.9

(b) The current density is given by

$$J = (-q_e) \int T \frac{p_x}{m} n(p_x)\, dp_x \tag{198}$$

where $n(p_x)dp_x$ represents the number of electrons per unit volume whose p_x lies between p_x and $p_x + dp_x$. Now at low temperatures (i.e., $T \ll E_F/k$), the Fermi function is approximately given by:

$$F(E) = 1 \quad \text{for } E < E_{F_0}$$

$$= 0 \quad \text{for } E > E_{F_0}$$

(see Section 6.9). Using the expression for density of states (see Section 6.8.1), we get

$$n(p_x)\, dp_x = \frac{2}{h^3} dp_x \iint dp_y dp_x$$

$$= \frac{2}{h^3} dp_x \int_0^{2\pi} d\theta \int_0^{[2m(E_F - E_x)]^{1/2}} p_1\, dp_1$$

where we have used polar coordinates and $p_1 = (p_x^2 + p_Y^2)^{1/2}$. Thus,

$$n(p_x)\, dp_x = \frac{4\pi m}{h^3} (E_F - E_x)\, dp_x \tag{199}$$

Substituting this in Eq. (198), we get

$$J = \frac{4\pi m |q|}{h^3} \int_0^{E_{F0}} \exp\left[-\beta \phi^{3/2}\left(1+\frac{\eta}{\phi}\right)^{3/2}\right] \eta\, d\eta$$

where $\eta = E_{F_0} - E_x$, $\phi = V_0 - E_{F_0}$, $\beta = (4/3\alpha)(2m/\hbar^2)^{1/2}$. Making a binomial expansion of $(1 + \eta/\phi)^{3/2}$, retaining up to the first-order term and extending the upper limit to ∞, we get

$$J \approx A_0 F^2 \exp(-B_0/F) \tag{200}$$

where we have used $\alpha = |q|F$ and

$$A_0 = \frac{|q|^2}{8\pi h \Phi} \simeq \frac{1.54 \times 10^{-6}}{\Phi} \text{ A/m}^2 \tag{201}$$

$$B_0 = \frac{4}{3}|q|^{1/2}\left(\frac{2m}{\hbar^2}\right)^{1/2} \Phi^{3/2}$$

$$\simeq 6.85 \times 10^9 \Phi^{3/2} \text{ V/m} \tag{202}$$

$\Phi = \phi/|q|$ being the workfunction measured in electron volts. Equation (200) is known as the Fowler-Nordheim field emission formula and agrees with experimental data. Notice that the field emission (sometimes referred to as cold emission) is due to the tunnelling phenomenon and hence is a purely quantum mechanical effect.

Solution 18.11 The solution of the one-dimensional Schrödinger equation is given by

$$\begin{aligned}\psi(x) &= Ae^{ikx} + Be^{-ikx}, \quad x < 0\\ &= CAi(\xi) + DBi(\xi), \quad 0 < x < a \\ &= Fe^{ik(x-a)}, \quad x > a\end{aligned} \tag{203}$$

where

$$\xi = \left(\frac{2m}{\hbar^2\alpha^2}\right)^{1/3}[V_0 - \alpha x - E]$$

and $Ai(\xi)$ and $Bi(\xi)$ are the Airy functions [see Appendix D]. Continuity of $\psi(x)$ and $d\psi/dx$ at $x = 0$ and at $x = a$ readily gives as tunnelling probability

$$T_{\text{Exact}} = \left|\frac{F}{A}\right|^2 = \left|\frac{a_0c_0^* - a_0^*c_0}{a_1c_0^* - a_0^*c_1}\right|^2 \tag{204}$$

where

$$\begin{aligned}a_0 &= Bi'(\xi_0) + i\sqrt{g\varepsilon}\, Bi(\xi_0)\\ a_1 &= Bi'(\xi_1) + i\sqrt{g\varepsilon}\, Bi(\xi_1)\\ c_0 &= Ai'(\xi_0) + i\sqrt{g\varepsilon}\, Ai(\xi_0)\\ c_1 &= Ai'(\xi_1) + i\sqrt{g\varepsilon}\, Ai(\xi_1)\\ \xi_0 &= \xi(x=0) = g(1-\varepsilon)\\ \xi_1 &= -g\varepsilon\end{aligned}$$

and use has been made of the Wronskian [see, e.g. Ref. 13]

$$W[Ai, Bi] = Ai(\xi)\, Bi'(\xi) - Ai'(\xi)\, Bi(\xi) = \frac{1}{\pi}. \tag{205}$$

In Figs 18.17 and 18.18, we have plotted the variations of T_{Exact} and T_{JWKB} with ε for $g = 0.1$ and $g = 1.0$. In Ref. 14 it has been shown that the JWKB result is accurate for $g \lesssim 0.1$.

Solution 18.12

$$k^2(x) = \frac{2mV_0}{\hbar^2}\left[\frac{E}{V_0} + \text{sech}^2\left(\frac{x}{a}\right)\right] \tag{206}$$

and the quantization condition [Eq. (44)] becomes

$$\left(\frac{2mV_0a^2}{\hbar^2}\right)^{1/2}\int_{-\alpha(\varepsilon)}^{+\alpha(\varepsilon)}[\text{sech}^2\xi - \varepsilon]^{1/2}d\xi = \left(n + \frac{1}{2}\right)\pi \tag{207}$$

where

$$\xi = \frac{x}{a}, \quad \varepsilon = -\frac{E}{V_0} \quad \text{and } \alpha = \text{sech}^{-1}\sqrt{\varepsilon} \tag{208}$$

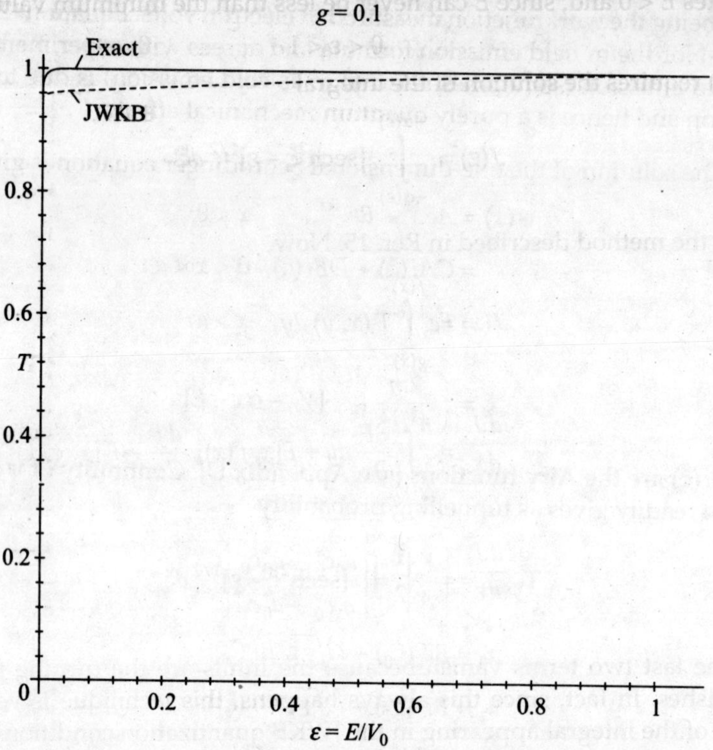

Figure 18.17: Variation of T_{EXACT} and T_{JWKB} with ε for $g = 0.1$ (after Ref. 14).

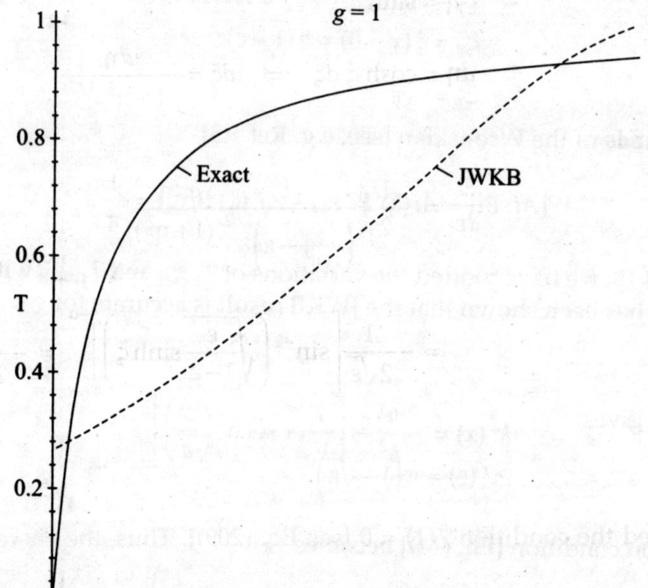

Figure 18.18: Variation of T_{Exact} and T_{JWKB} with ε for $g = 1.0$ (after Ref. 14).

For bound states $E < 0$ and, since E can never be less than the minimum value of V, we have

$$0 < \varepsilon < 1 \tag{209}$$

Equation (206) requires the solution of the integral

$$J(\varepsilon) = \int_{-\alpha(\varepsilon)}^{+\alpha(\varepsilon)} [\text{sech}^2\xi - \varepsilon]^{1/2}\, d\xi \tag{210}$$

We will adapt the method described in Ref. 15. Now,

$$J(x) = \int_{g(x)}^{f(x)} F(x, y)\, dy \tag{211}$$

then

$$\frac{dJ}{dx} = \int_{g(x)}^{f(x)} \frac{\partial F}{\partial x} dy + F[x, f(x)]\frac{df}{dx} - F[x, g(x)]\frac{dg}{dx} \tag{212}$$

Therefore,

$$\frac{dJ}{d\varepsilon} = -\frac{1}{2}\int_{-\alpha}^{+\alpha} [\text{sech}^2\xi - \varepsilon]^{1/2}\, d\xi \tag{213}$$

In our case, the last two terms vanish because the limits are the turning points where the integrand vanishes. In fact, since this always happens, this technique is very often used in the evaluation of the integral appearing in the JWKB quantization condition. If we now make the transformation

$$\eta = \sinh \xi$$

then

$$d\eta = \cosh \xi\, d\xi \quad \Rightarrow \quad d\xi = \frac{d\eta}{(1+\eta^2)^{1/2}}$$

and Eq. (212) becomes

$$\frac{dJ}{d\varepsilon} = -\frac{1}{2}\int \frac{1}{\left(\frac{1}{1+\eta^2} - \varepsilon\right)^{1/2}} \frac{d\eta}{(1+\eta^2)}$$

$$= -\frac{1}{2\sqrt{\varepsilon}}\left[\sin^{-1}\left(\sqrt{\frac{\varepsilon}{1-\varepsilon}}\sinh \xi\right)\right]_{-\alpha}^{+\alpha} = -\frac{\pi}{2\sqrt{\varepsilon}}$$

Simple integration gives

$$J(\varepsilon) = \pi\left(1 - \sqrt{\varepsilon}\right) \tag{214}$$

where we have used the condition $J(1) = 0$ [see Eq. (209)]. Thus, the energy eigenvalues are given by

$$E_n = \frac{-\hbar^2}{2ma^2}\left[\left(\frac{2mV_0a^2}{\hbar^2}\right)^{1/2} - \left(n + \frac{1}{2}\right)\right]^2 \tag{215}$$

In this case we can also obtain an exact solution if the schrödinger equation; the final result is

$$E_n = \frac{-\hbar^2}{2ma^2}\left[\frac{\nu}{2} - \left(n + \frac{1}{2}\right)\right]^2 \tag{216}$$

where $$\nu = \left(\frac{8mV_0a^2}{\hbar^2} + 1\right)^{1/2} \tag{217}$$

The JWKB eigenvalues are obtained if the first term in the brackets is large with respect to 1. The eigenfunctions are

$$\psi(\sigma) = [1 - \sigma]^{-(\nu-1)/2} W(\sigma) \tag{218}$$

where $\sigma = -\sinh^2(x/a)$ and $W(\sigma)$ is the polynomial solution of the hypergeometric equation.

Solution 18.13 The JWKB quantization condition is given by

$$\sqrt{\frac{2m}{\hbar^2}} \int_{r_1}^{r_2} \left(E + \frac{Ze^2}{r} - \frac{\left(l+\frac{1}{2}\right)^2 \hbar^2}{2mr^2}\right)^{1/2} dr = \left(n + \frac{1}{2}\right)\pi \tag{219}$$

or $$\sqrt{2}\, J(\varepsilon) = \left(n + \frac{1}{2}\right)\pi \tag{220}$$

where $$J(\varepsilon) = \int_{\xi_1}^{\xi_2} [-\varepsilon - V_{eff}(\xi)]^{1/2} d\xi \tag{221}$$

with $$V_{eff}(\xi) = -\frac{Z}{\xi} + \frac{\alpha}{\xi^2}$$

$$\varepsilon = -\frac{E}{E_0}, \quad E_0 = \frac{me^4}{\hbar^2}$$

$$\xi = \frac{r}{a_0}, \quad a_0 = \frac{\hbar^2}{me^2}$$

$$\alpha = \frac{1}{2}\left(l + \frac{1}{2}\right)^2$$

Following the method used in the previous solution, we obtain

$$\frac{dJ}{d\varepsilon} = -\frac{\pi Z}{4\varepsilon^{3/2}} \tag{222}$$

Further, the minimum value of $V_{eff}(\xi)$ is $-Z^2/4\alpha$, which occurs at $\xi = 2\alpha/Z$. Thus,

$$J\left(\varepsilon = -\frac{Z^2}{4\alpha}\right) = 0, \tag{223}$$

because ξ_1 becomes equal to ξ_2. Equation (221) can now be integrated. Using Eq. (222), we obtain

$$\varepsilon = \frac{Z^2}{2(n+l+1)^2} \tag{224}$$

which is also the *exact* result with total quantum number $(n + l + 1)$.

REFERENCES AND SUGGESTED READINGS

1. V.L. Ginzberg, *The Propagation of Electromagnetic Waves in Plasmas*, Pergamon Press, New York (1964).
2. M.J. Adams, *An Introduction to Optical Waveguides*, John Wiley, Chichester (1981).
3. A.K. Ghatak, R.L. Gallawa and I.C. Goyal, *Modified Airy Function and WKB Solutions to the Wave Equation*, National Institute of Standards and Technology Monograph #176, US Government Printing Office, Washington (1991).
4. J. Heading, *An Introduction to Phase Integral Methods*, Mathuen and Co., London (1962).
5. S. Gasiorowicz, *Quantum Physics*, John Wiley, New York (1969).
6. E.K. Hyde, I. Perlman and G.T. Seaborg, *The Nuclear Properties of the Heavy Elements*, Vol. I, Prentice-Hall, USA (1964).
7. H. Jeffreys, *On Certain Approximate Solutions of Linear Differential Equations of Second Order, Proc. London Math. Soc., Vol. 23, No.* **2**, *428–436(1923).*
8. G. Wentzel, *Eine Verallgemeinerung der quantenbedingung fur die Zweeke der Wallenmechanik, Zeits.* f. Physik, Vol. **38**, 518–529 *(1926).*
9. H.A. Kramers, *Wellenmechanik und Halbzahlige Quantisierung*, Zeits. f. Physik, Vol. **39**, 828–840 (1926).
10. L. Brillouin, *The Undulatory Mechanics of Schrödinger*, Compt. Rend. Acad. Sci., Paris, Vol. **183**, 24–26 (1926).
11. N. Fröman and P.O. Fröman, *JWKB Approximation: Contributions to the Theory*, North Holland Publishing Company, Amsterdam (1965).
12. W. Magnus, F. Oberhettinger and R.P. Soni, *Formulas and Theorems for the Special Functions of Mathematical Physics*, Springer-Verlag, Berlin (1966).
13. A.K. Ghatak, I.C. Goyal and S.J. Chua, *Mathematical Physics*, Macmillan India Limited, New Delhi (1995).
14. A.K. Ghatak, E.G. Sauter and I.C. Goyal, 'Validity of the JWKB Formula for a Triangular Potential Barrier', *European Journal of Physics*, **18**, 199–204 (1997).
15. I.I. Goldman and V.D. Krivchenkov, Problems in Quantum Mechanics, Pergamon Press (1961).

19 Addition of Angular Momenta: Clebsch-Gordan Coefficients

19.1 INTRODUCTION

We introduce [illegible]

angular momentum operators J_1 and J_2 [illegible]

and other [illegible]

units in which [illegible]

where [illegible]

Further, we assume that [illegible]

etc. Let $\psi(j_1 m_1)$ [illegible]

and

In Section 13.2, we had shown that j_1 can take the values of 0, $\frac{1}{2}$, 1, $\frac{3}{2}$, ... and for each value of

j_1, m_1 can take any of the following values $m_1 = -j_1, -j_1 + 1, \ldots$ [illegible]

[1] Abner Shimony [illegible]
Cambridge University Press, Cambridge, [illegible]

19 Addition of Angular Momenta: Clebsch-Gordan Coefficients

Quantum Mechanics is a framework theory which according to prevailing present opinion, applies to every kind of physical system.

— ABNER SHIMONY[1]

19.1 INTRODUCTION

We introduce six operators $\hat{J}_{1x}, \hat{J}_{1y}, \hat{J}_{1z}, \hat{J}_{2x}, \hat{J}_{2y}$ and $\hat{J}_{2z}$, which represent components of 2 angular momentum operators $\hat{\mathbf{J}}_1$ and $\hat{\mathbf{J}}_2$, satisfying the following commutation relations:

$$\left.\begin{array}{l}\left[\hat{J}_{1x}, \hat{J}_{1y}\right] = i\hat{J}_{1z} \\ \left[\hat{J}_{2x}, \hat{J}_{2y}\right] = i\hat{J}_{2z}\end{array}\right\} \tag{1}$$

and other equations can be written in cyclic order (see Section 9.2); we are using a system of units in which $\hbar = 1$. We also have

$$\left[\hat{J}_1^2, \hat{J}_{1x}\right] = 0 = \left[\hat{J}_2^2, \hat{J}_{2x}\right], \text{etc.} \tag{2}$$

where

$$\hat{J}_1^2 \equiv \hat{J}_{1x}^2 + \hat{J}_{1y}^2 + \hat{J}_{1z}^2 \text{ and } \hat{J}_2^2 \equiv \hat{J}_{2x}^2 + \hat{J}_{2y}^2 + \hat{J}_{2z}^2 \tag{3}$$

Further, we assume that $\hat{J}_{1x}, \hat{J}_{1y}$ and $\hat{J}_{1z}$ commute with $\hat{J}_{2x}, \hat{J}_{2y}$ and $\hat{J}_{2z}$:

$$\left[\hat{J}_{1x}, \hat{J}_{2y}\right] = 0; \ \left[\hat{J}_{2x}, \hat{J}_{1z}\right] = 0; \ \left[\hat{J}_{1z}, \hat{J}_{2x}\right] = 0; \ \left[\hat{J}_{1y}, \hat{J}_{2y}\right] = 0 \tag{4}$$

etc. Let $\psi_1(j_1, m_1)$ be the simultaneous eigenvector of $\hat{J}_1^2$ and $\hat{J}_{1z}$, i.e.,

$$\hat{J}_1^2 \psi_1(j_1, m_1) = j_1(j_1 + 1)\psi_1(j_1, m_1) \tag{5}$$

and

$$\hat{J}_{1z}\psi_1(j_1, m_1) = m_1\psi_1(j_1, m_1) \tag{6}$$

In Section 13.2, we had shown that j_1 can take the values of $0, \frac{1}{2}, 1, \frac{3}{2}, \cdots$ and for each value of j_1, m_1 can take any of the following values $m_1 = -j_1, -j_1 + 1, \cdots + j_1$. Similarly, let $\psi_2(j_2, m_2)$

[1] Abner Shimony in *Conceptual Foundations of Quantum Mechanics* in *The New Physics* (Editor: Paul Davis), Cambridge University Press, Cambridge, 1989.

represent the simultaneous eigenvectors of $\hat{J}_2^2$ and $\hat{J}_{2z}$; j_2 can also take the values of $0, \frac{1}{2}, 1, \frac{3}{2}, \ldots$ and for each value of j_2, m_2 can take any of the following values $m_2 = -j_2, -j_2 + 1, \ldots, + j_2$. Now the operators $\hat{J}_1^2, \hat{J}_2^2, \hat{J}_{1z}$ and $\hat{J}_{2z}$ commute with each other and the product functions

$$\psi(j_1, j_2, m_1, m_2) = \psi_1(j_1, m_1)\,\psi_2(j_2, m_2) \tag{7}$$

represent a set of simultaneous eigenvectors of $\hat{J}_1^2, \hat{J}_2^2, \hat{J}_{1z}$ and $\hat{J}_{2z}$; the corresponding eigenvalues are $j_1(j_1+1)\hbar^2$, $j_2(j_2+1)\hbar^2$, $m_1\hbar$ and $m_2\hbar$. Further, since for given values of j_1 and j_2 there will be $(2j_1 + 1)$ possible values of m_1 and $(2j_2 + 1)$ possible values of m_2, there will be $(2j_1 + 1)(2j_2 + 1)$ possible eigenvectors. The ket vector corresponding to $\psi(j_1, j_2, m_1, m_2)$ is written as $|\psi(j_1, j_2, m_1, m_2)\rangle$ or, just as $|j_1, j_2, m_1, m_2\rangle$ which is often written as

$$|j_1, j_2, m_1, m_2\rangle = |j_1, m_1\rangle \otimes |j_2, m_2\rangle \tag{8}$$

The symbol $\otimes$ is referred to as the "direct product" because $|j_1, m_1\rangle \otimes |j_2, m_2\rangle$ is the product of 2 vectors from two different vector spaces; we are now working in a $(2j_1 + 1) \times (2j_2 + 1)$ dimensional space. Thus, all operators will be represented by

$$(2j_1 + 1)(2j_2 + 1) \times (2j_1 + 1)(2j_2 + 1)$$

matrices. In Section 19.6, while considering two spin half particles $\left(j_1 = \frac{1}{2} = j_2\right)$, we will have 4 base vectors and will show that all operators are represented by 4×4 matrices.

The angular momentum operators $\hat{\mathbf{J}}_1$ and $\hat{\mathbf{J}}_2$ may, for example, pertain to the spin and orbital angular momentum operator of a single electron (see Section 20.5 and also Chapter 22) or, they may pertain to the spin angular momentum of two particles (like the electron and the proton - see Chapter 21). For the P state of the hydrogen atom, $\left(j_1 = l = 1 \;\&\; j_2 = s = \frac{1}{2}\right)$, we will have 6 base vectors corresponding to a 6-dimensional space.

19.2 THE TOTAL ANGULAR MOMENTUM OPERATOR

We define the 3 operators

$$\hat{J}_x \equiv \hat{J}_{1x} + \hat{J}_{2x} \tag{9}$$

$$\hat{J}_y \equiv \hat{J}_{1y} + \hat{J}_{2y} \tag{10}$$

$$\hat{J}_z \equiv \hat{J}_{1z} + \hat{J}_{2z} \tag{11}$$

We symbolically write the total angular momentum operator

$$\hat{\mathbf{J}} = \hat{\mathbf{J}}_1 + \hat{\mathbf{J}}_2 \tag{12}$$

where the components of $\hat{\mathbf{J}}$ are given by Eqs. (9), (10) and (11).

Now,

$$\left[\hat{J}_y, \hat{J}_z\right] = \hat{J}_y\hat{J}_z - \hat{J}_z\hat{J}_y = \left(\hat{J}_{1y} + \hat{J}_{2y}\right)\left(\hat{J}_{1z} + \hat{J}_{2z}\right) - \left(\hat{J}_{1z} + \hat{J}_{2z}\right)\left(\hat{J}_{1y} + \hat{J}_{2y}\right)$$
$$= \left(\hat{J}_{1y}\hat{J}_{1z} - \hat{J}_{1z}\hat{J}_{1y}\right) + \left(\hat{J}_{2y}\hat{J}_{2z} - \hat{J}_{2z}\hat{J}_{2y}\right)$$
$$= i\left(\hat{J}_{1x} + \hat{J}_{2x}\right) = i\hat{J}_x$$

Thus, $$\left[\hat{J}_y, \hat{J}_z\right] = i\hat{J}_x \tag{13}$$

Similarly, $$\left[\hat{J}_x, \hat{J}_y\right] = i\hat{J}_z \quad \text{and} \quad \left[\hat{J}_z, \hat{J}_x\right] = i\hat{J}_y \tag{14}$$

If we now define

$$\hat{J}^2 \equiv \hat{J}_x\hat{J}_x + \hat{J}_y\hat{J}_y + \hat{J}_z\hat{J}_z = \hat{J}_x^2 + \hat{J}_y^2 + \hat{J}_z^2 \tag{15}$$

then, $$\left[\hat{J}^2, \hat{J}_x\right] = \left[\hat{J}_x\hat{J}_x, \hat{J}_x\right] + \left[\hat{J}_y\hat{J}_y, \hat{J}_x\right] + \left[\hat{J}_z\hat{J}_z, \hat{J}_x\right]$$
$$= 0 + \hat{J}_y\left[\hat{J}_y, \hat{J}_x\right] + \left[\hat{J}_y, \hat{J}_x\right]\hat{J}_y + \hat{J}_y\left[\hat{J}_z, \hat{J}_x\right] + \left[\hat{J}_z, \hat{J}_x\right]\hat{J}_z$$
$$= -i\,\hat{J}_y\hat{J}_z - i\,\hat{J}_z\hat{J}_y + i\,\hat{J}_z\hat{J}_y + i\,\hat{J}_y\hat{J}_z$$

Thus, $$\left[\hat{J}^2, \hat{J}_x\right] = 0 \tag{16}$$

Similarly, $$\left[\hat{J}^2, \hat{J}_y\right] = 0 = \left[\hat{J}^2, \hat{J}_z\right] \tag{17}$$

The above equations tell us that the components of $\hat{\mathbf{J}}$ satisfy the same commutation relations as that of the angular momentum operator and therefore, if we follow the analysis of Section 13.2, we can construct simultaneous eigenvectors of $\hat{J}^2$ and $\hat{J}_z$; the corresponding eigenvalues will be $j(j+1)\hbar^2$ and $m\hbar$ respectively where j can take the values of $0, \frac{1}{2}, 1, \frac{3}{2}, \ldots$ and for each value of j, m can take any of the following values $m = -j, -j+1, \ldots +j$.

Now, $$\hat{J}^2 \equiv \hat{J}_x^2 + \hat{J}_y^2 + \hat{J}_z^2$$
$$= \left(\hat{J}_{1x} + \hat{J}_{2x}\right)^2 + \left(\hat{J}_{1y} + \hat{J}_{2y}\right)^2 + \left(\hat{J}_{1z} + \hat{J}_{2z}\right)^2$$

Thus, $$\hat{J}^2 = \hat{J}_1^2 + \hat{J}_2^2 + 2\hat{\mathbf{J}}_1.\hat{\mathbf{J}}_2 \tag{18}$$

where we have used the relation

$$\hat{\mathbf{J}}_1 \cdot \hat{\mathbf{J}}_2 \equiv \hat{J}_{1x}\hat{J}_{2x} + \hat{J}_{1y}\hat{J}_{2y} + \hat{J}_{1z}\hat{J}_{2z} = \hat{\mathbf{J}}_2 \cdot \hat{\mathbf{J}}_1$$

because components of $\hat{\mathbf{J}}_1$ commute with components of $\hat{\mathbf{J}}_2$. Further, $\hat{J}_1^2, \hat{J}_2^2, \hat{J}^2$ and $\hat{J}_z$ commute with each other (see Problem 19.1) and therefore we can construct simultaneous eigenfunctions of $\hat{J}_1^2, \hat{J}_2^2, \hat{J}^2$ and $\hat{J}_z$ which we denote by $\phi(j_1, j_2, j, m)$ and write

$$\hat{J}_1^2 \, \phi(j_1, j_2, j, m) = j_1(j_1+1) \, \phi(j_1, j_2, j, m) \tag{19}$$

$$\hat{J}_2^2 \, \phi(j_1, j_2, j, m) = j_2(j_2+1) \, \phi(j_1, j_2, j, m) \tag{20}$$

$$\hat{J}^2 \, \phi(j_1, j_2, j, m) = j(j+1) \, \phi(j_1, j_2, j, m) \tag{21}$$

$$\hat{J}_z \, \phi(j_1, j_2, j, m) = m \, \phi(j_1, j_2, j, m) \tag{22}$$

where j can take the values of $0, \frac{1}{2}, 1, \frac{3}{2}, \cdots$ and for each value of j, m can take any of the following values $m = -j, -j+1, \cdots +j$. The ket vector corresponding to $\phi(j_1, j_2, j, m)$ is written as $|\phi(j_1, j_2, j, m)\rangle$ or, just as $|j_1, j_2, j, m\rangle$. For given values of j_1 and j_2, we may just write $|\phi(j, m)\rangle$ or $|j, m\rangle$. Since the functions $\psi(j, j_2, m_1, m_2)$ are eigenfunctions of $\hat{J}_{1z}$ and $\hat{J}_{2z}$:

$$\hat{J}_{1z} \, \psi(j_1, j_2, m_1, m_2) = m_1 \, \psi(j_1, j_2, m_1, m_2) \tag{23}$$

$$\hat{J}_{2z} \, \psi(j_1, j_2, m_1, m_2) = m_2 \, \psi(j_1, j_2, m_1, m_2) \tag{24}$$

the ψ functions would also be eigenfunctions of $\hat{J}_z \left(= \hat{J}_{1z} + \hat{J}_{2z}\right)$

$$\hat{J}_z \psi(j_1, j_2, m_1, m_2) = m \, \psi(j_1, j_2, m_1, m_2) \tag{25}$$

where $$m = m_1 + m_2 \tag{26}$$

Thus, $\psi(j_1, j_2, m_1, m_2)$ are not only simultaneous eigenvectors of $\hat{J}_1^2, \hat{J}_2^2, \hat{J}_{1z}$ and $\hat{J}_{2z}$ but also of $\hat{J}_z$. Further, the maximum value of m_1 is j_1 and the maximum value of m_2 is j_2; therefore.

$$\text{maximum value of } m = j_1 + j_2 \tag{27}$$

implying that $$\text{maximum value of } j = j_1 + j_2 \tag{28}$$

and therefore $$\phi(j_1, j_2, j_1 + j_2, j_1 + j_2) = \psi(j_1, j_2, j_1, j_2)$$

In Section 19.2.1, we will show that the minimum value of j is equal to $(j_1 - j_2)$ where we will always assume $j_1 \geq j_2$.

Example 19.1 For 2 spin-half particles $j_1 = j_2 = \frac{1}{2}$, the allowed values of $m_1 = +\frac{1}{2}$ & $-\frac{1}{2}$ and the allowed values of $m_2 = +\frac{1}{2}$ and, $-\frac{1}{2}$. Thus, the possible values of $m \ (= m_1 + m_2)$ will be 1, 0, 0 and – 1. Since j is the maximum value of m, we must have $j = 1$, corresponding to which we will have $m = 1$, 0 and – 1; we will have one more $m = 0$ corresponding to which, we will have $j = 0$. Thus, for $j_1 = j_2 = \frac{1}{2}$, allowed values of j are 1 and 0. For each value of j

there will be $(2j+1)$ eigenvectors; thus there will be $(2\times1+1)+(2\times0+1)=4$ eigenvectors of the type $\phi(j_1, j_2, j, m)$. Obviously, we will have the same number of eigenvectors of the type $\psi(j_1, j_2, m_1, m_2)$ because $(2j_1+1)\times(2j_2+1)=2\times2=4$.

Example 19.2 For $j_1=1, j_2=\frac{1}{2}$, the allowed values of $m_1 = 1, 0$ and -1 and, the allowed values of $m_2 = +\frac{1}{2}$ and $-\frac{1}{2}$. Thus, the possible values of $m\ (= m_1 + m_2)$ will be $+\frac{3}{2}, +\frac{1}{2}, +\frac{1}{2}, -\frac{1}{2}, -\frac{1}{2}$ and $-\frac{3}{2}$. Once again, since j is the maximum value of m, we must have $j=\frac{3}{2}$, corresponding to which we will have $m=+\frac{3}{2}, +\frac{1}{2}, -\frac{1}{2}$ and $-\frac{3}{2}$; we will have two more $m=+\frac{1}{2}$ and $-\frac{1}{2}$ corresponding to which we will have $j=\frac{1}{2}$. Thus, for $j_1=1, j_2=\frac{1}{2}$; allowed values of j are $\frac{3}{2}, \frac{1}{2}$.

Once again, for each value of j there will be $(2j+1)$ eigenvectors; thus there will be a total of $\left(2\times\frac{3}{2}+1\right)+\left(2\times\frac{1}{2}+1\right)=6$ eigenvectors of the type $\phi(j_1, j_2, j, m)$. Since $(2j_1+1)\times(2j_2+1)=3\times2=6$, we will have the same number of eigenvectors of the type $\psi(j_1, j_2, m_1, m_2)$.

Example 19.3 It is left as an exercise to show that for $j_1 = 1, j_2 = 1$, the allowed values of j will be 2, 1 and 0 and there will be a total of $(2\times2+1)+(2\times1+1)+(2\times0+1)=9$ eigenvectors; the number is the same as $(2j_1+1)\times(2j_2+1)=3\times3=9$. Similarly, for $j_1=\frac{3}{2}, j_2=1$; allowed values of j will be $\frac{5}{2}, \frac{3}{2}$ and $\frac{1}{2}$ and there will be a total of $\left(2\times\frac{5}{2}+1\right)+\left(2\times\frac{3}{2}+1\right)+\left(2\times\frac{1}{2}+1\right)=12$ eigenvectors; the number is the same as $(2j_1+1)\times(2j_2+1)=4\times3$.

Example 19.4 For $j_1=\frac{3}{2}=j_2$ allowed values of j will be 3, 2, 1 and 0 and there will be a total of $(2\times3+1)+(2\times2+1)+(2\times1+1)+(2\times0+1)=16$ eigenvectors of the type $\phi(j_1, j_2, j, m)$. Since $(2j_1+1)\times(2j_2+1)=4\times4=16$, we will have the same number of eigenvectors of the type $\psi(j_1, j_2, m_1, m_2)$.

19.2.1 Possible Values of *j*

We have shown above that the maximum value of $m\ (= m_1 + m_2)$ and therefore the maximum value of j will be equal to $j_1 + j_2$ and

$$\phi(j_1, j_2, j_1 + j_2, j_1 + j_2) = \psi(j_1, j_2, j_1, j_2) \tag{29}$$

The next largest value of m will be $(j_1 + j_2 - 1)$ which will give rise to two ψ vectors

$$\psi(j_1, j_2, j_1, j_2 - 1) \text{ and } \psi(j_1, j_2, j_1 - 1, j_2)$$

and two ϕ vectors

$$\phi(j_1, j_2, j_1 + j_2, j_1 + j_2 - 1) \text{ and } \phi(j_1, j_2, j_1 + j_2 - 1, j_1 + j_2 - 1)$$

because for all the 4 vectors, $m\ (= m_1 + m_2)$ will be equal to $(j_1 + j_2 - 1)$. Each ϕ vector can be expressed as linear combinations of ψ vectors (and vice-versa) and the constant coefficients are known as the Clebsch-Gordan coefficients. The next largest value of m will be $(j_1 + j_2 - 2)$ which will give rise to three ψ vectors

$$\psi(j_1, j_2 - 2), \qquad \psi(j_1 - 1, j_2 - 1) \text{ and } \psi(j_1 - 2, j_2)$$

and three ϕ vectors

$$\phi(j_1 + j_2, j_1 + j_2 - 2), \phi(j_1 + j_2 - 1, j_1 + j_2 - 2) \text{ and } \phi(j_1 + j_2 - 2, j_1 + j_2 - 2)$$

because for all vectors, $m\ (= m_1 + m_2)$ will be equal to $(j_1 + j_2 -\ \ )$; we have omitted the values of j_1 and j_2 which appear in all terms. In general, for $m = j_1 + j_2 - k$, we will have the following $(k+1)$ ψ (and ϕ) vectors:

$$\psi(j_1 - k, j_2),\ \psi(j_1 - k + 1, j_2 - 1) \ldots \psi(j_1, j_2 - k)$$

$$\phi(j_1 + j_2, j_1 + j_2 - k),\ \phi(j_1 + j_2 - 1, j_1 + j_2 - k) \ldots \phi(j_1 + j_2 - k, j_1 + j_2 - k)$$

Since $j_1 \geq j_2$, the maximum value of k should be such that $j_2 - k$ equals $-j_2$ (because the minimum value of m_2 can be $-j_2$); thus $k_{max} = 2j_2$ giving

$$j_{min} = j_1 + j_2 - k_{max} = j_1 - j_2 .$$

Thus the allowed values of j will be

$$j_1 + j_2, j_1 + j_2 - 1, \ldots, j_1 - j_2 \tag{30}$$

Since for each values of j there will be $(2j + 1)$ values of m, the total number of ϕ functions (for given values of j_1 and j_2) will be

$$\sum_{j = |j_1 - j_2|}^{j = j_1 + j_2} (2j + 1) = (2j_1 + 1)(2j_2 + 1) \tag{31}$$

which is the same as the number of ψ functions. We can choose either the ψ functions or the ϕ functions as the base states and can represent each of the ψ functions as a linear combination of ϕ functions and conversely. Thus, we may write

$$\phi(j_1, j_2, j, m) = \sum_{m_1}\sum_{m_2} C(j_1, j_2, j, m, m_1, m_2)\, \psi(j_1, j_2, m_1, m_2) \tag{32}$$

which is the same as

$$|j_1, j_2, j, m\rangle = \sum_{m_1}\sum_{m_2} C(j_1, j_2; j, m, m_1, m_2)\, |j_1, j_2, m_1, m_2\rangle \tag{33}$$

The coefficients

$$C(j_1, j_2; j, m, m_1, m_2) \quad \langle j_1, j_2, m_1, m_2 | j_1, j_2, j, m\rangle \tag{34}$$

are known as the Clebsch-Gordan coefficients; the coefficients are named after Alfred Clebsch (1833 – 1872) and Paul Gordan (1837 – 1912) who did an equivalent problem in invariant theory long before the advent of quantum mechanics.

Now the ψ functions as well as the ϕ functions form orthonormal sets and therefore the transformation described by the above 2 equations are unitary. The coefficients play an extremely important role in spectroscopy and also in scattering theory. We give below the method for calculating the Clebsch-Gordan coefficients for $j_1 = \frac{1}{2} = j_2$ and for. $j_1 = 1, j_2 = \frac{1}{2}$.

The general formulae for calculating these coefficients are given in Table 19.1 for $j_2 = \frac{1}{2}$ and Table 19.2 for $j_2 = 1$.

19.3 CLEBSCH-GORDAN COEFFICIENTS FOR $j_1 = s_1 = \frac{1}{2}$ & $j_2 = s_2 = \frac{1}{2}$ (2 SPIN HALF PARTICLES)

We consider two spin half particles like an electron and a proton in a hydrogen atom. Thus,

$$j_1 = \frac{1}{2} \Rightarrow m_1 = +\frac{1}{2} \text{ and } -\frac{1}{2} \ \& \ j_2 = \frac{1}{2} \Rightarrow m_2 = +\frac{1}{2} \text{ and } -\frac{1}{2}$$

We will therefore have four $\psi(j_1, j_2, m_1, m_2)$ functions:

$$\psi\left(\frac{1}{2}, \frac{1}{2}, \frac{1}{2}, \frac{1}{2}\right),\ \psi\left(\frac{1}{2}, \frac{1}{2}, \frac{1}{2}, -\frac{1}{2}\right),\ \psi\left(\frac{1}{2}, \frac{1}{2}, -\frac{1}{2}, \frac{1}{2}\right) \text{ and } \psi\left(\frac{1}{2}, \frac{1}{2}, -\frac{1}{2}, -\frac{1}{2}\right)$$

These represent a set of simultaneous eigenvectors of $\hat{J}_1^2, \hat{J}_2^2, \hat{J}_{1z}$ and $\hat{J}_{2z}$; the corresponding eigenvalues are $j_1(j_1+1)\hbar^2, j_2(j_2+1)\hbar^2, m_1\hbar$ and $m_2\hbar$. The numbers inside the parenthesis denote the values of j_1, j_2, m_1 and m_2. Since the same values of j_1 and j_2 appear in all 4 of them, we will omit the values of j_1 and j_2 and write the ψ functions as

$$\psi\left(\frac{1}{2}, \frac{1}{2}\right),\ \psi\left(\frac{1}{2}, -\frac{1}{2}\right),\ \psi\left(-\frac{1}{2}, \frac{1}{2}\right) \text{ and } \psi\left(-\frac{1}{2}, -\frac{1}{2}\right)$$

where the numbers in the parenthesis denote the values of m_1 and m_2. The corresponding ket vectors are written as

$$\left|\psi\left(\frac{1}{2},\frac{1}{2}\right)\right\rangle \left[=\left|\hat{\mathbf{z}}\uparrow,\hat{\mathbf{z}}\uparrow\right\rangle\right], \qquad \left|\psi\left(\frac{1}{2},-\frac{1}{2}\right)\right\rangle \left[=\left|\hat{\mathbf{z}}\uparrow,\hat{\mathbf{z}}\downarrow\right\rangle\right],$$

$$\left|\psi\left(-\frac{1}{2},\frac{1}{2}\right)\right\rangle \left[=\left|\hat{\mathbf{z}}\downarrow,\hat{\mathbf{z}}\uparrow\right\rangle\right] \quad \& \quad \left|\psi\left(-\frac{1}{2},-\frac{1}{2}\right)\right\rangle \left[=\left|\hat{\mathbf{z}}\downarrow,\hat{\mathbf{z}}\downarrow\right\rangle\right]$$

Thus, for example, the ket $\left|\hat{\mathbf{z}}\downarrow,\hat{\mathbf{z}}\uparrow\right\rangle$ corresponds to the first particle being in the z-down state $\left|\hat{\mathbf{z}}\downarrow\right\rangle$, and the second particle in the z-up state $\left|\hat{\mathbf{z}}\uparrow\right\rangle$. The possible values of m $(= m_1 + m_2)$ are

$$1, 0, 0, -1$$

The maximum value of m is 1 and thus one of the values of j must be 1 which will give rise to 3 states with $m = -1, 0, +1$. The remaining one state must correspond to $j = 0$ with $m = 0$. Thus, the allowed values of j are 1 and 0 and the corresponding ϕ functions are

$$\phi\left(\frac{1}{2},\frac{1}{2},1,1\right),\ \phi\left(\frac{1}{2},\frac{1}{2},1,0\right),\ \phi\left(\frac{1}{2},\frac{1}{2},1,-1\right) \text{ and } \phi\left(\frac{1}{2},\frac{1}{2},0,0\right)$$

where the numbers in parenthesis denote the values of j_1, j_2, j and m. Once again, since the same values of j_1 and j_2 appear in all 4 of the ϕ functions, we will omit the values of j_1 and j_2 and write the ϕ functions as

$$\left.\begin{array}{l}\phi(1,1)\\ \phi(1,0)\\ \phi(1,-1)\end{array}\right\};\ j=1 \quad \text{(triplet state)} \tag{35}$$

$$\phi(0,0);\ j=0 \qquad \text{(single state)} \tag{36}$$

Now, ϕ (1, 1) and $\psi\left(\frac{1}{2},\frac{1}{2}\right)$ are the only functions which correspond to $m = 1$, we must have

$$\phi(1,1) = \psi\left(\frac{1}{2},\frac{1}{2}\right) \tag{37}$$

Further, ϕ (1, −1) and $\psi\left(-\frac{1}{2},-\frac{1}{2}\right)$ are the only state which correspond to $m = -1$, we must therefore have

$$\phi(1,-1) = \psi\left(-\frac{1}{2},-\frac{1}{2}\right) \tag{38}$$

The remaining functions ϕ (1, 0) and ϕ (0, 0) correspond to $m = 0$ and therefore must be linear combinations of $\psi\left(\frac{1}{2},-\frac{1}{2}\right)$ and $\psi\left(-\frac{1}{2},\frac{1}{2}\right)$:

$$\phi(1,0) = c_1\,\psi\left(\frac{1}{2},-\frac{1}{2}\right) + c_2\,\psi\left(-\frac{1}{2},\frac{1}{2}\right) \tag{39}$$

$$\phi(0,0) = c_3\,\psi\left(\frac{1}{2},-\frac{1}{2}\right) + c_4\,\psi\left(-\frac{1}{2},\frac{1}{2}\right) \tag{40}$$

Now, if we operate Eq. (37) by the operator $\hat{J}_-$ we would obtain:

$$\hat{J}_-\phi(1,1) = \hat{J}_-\,\psi\left(\frac{1}{2},\frac{1}{2}\right) \tag{41}$$

The LHS will be $\hat{J}_-\phi(j=1, m=1) = \sqrt{2}\,\phi(1,0)$ (42)

Further since

$$\hat{J}_- = \hat{J}_x - i\hat{J}_y = \left(\hat{J}_{1x} - i\hat{J}_{1y}\right) + \left(\hat{J}_{2x} - i\hat{J}_{2y}\right) = \left(\hat{J}_1\right)_- + \left(\hat{J}_2\right)_-$$

the RHS of Eq. (41) will be:

$$\hat{J}_-\,\psi\left(\frac{1}{2},\frac{1}{2}\right) = \left(\hat{J}_1\right)_-\,\psi_1\left(j_1=\frac{1}{2}, m_1=\frac{1}{2}\right)\psi_2\left(j_2=\frac{1}{2}, m_2=\frac{1}{2}\right)$$

$$+\left(\hat{J}_2\right)_-\,\psi_1\left(j_1=\frac{1}{2}, m_1=\frac{1}{2}\right)\psi_2\left(j_2=\frac{1}{2}, m_2=\frac{1}{2}\right)$$

$$= \sqrt{1}\,\psi\left(\frac{1}{2},\frac{1}{2},-\frac{1}{2},\frac{1}{2}\right) + \sqrt{1}\,\psi\left(\frac{1}{2},\frac{1}{2},\frac{1}{2},-\frac{1}{2}\right)$$

where we have used the fact that $\left(\hat{J}_1\right)_-$ operates only on ψ_1 and $\left(\hat{J}_2\right)_-$ operator only on ψ_2. If we now use Eq. (42), we would get

$$\phi(1,0) = \frac{1}{\sqrt{2}}\,\psi\left(\frac{1}{2},-\frac{1}{2}\right) + \frac{1}{\sqrt{2}}\,\psi\left(-\frac{1}{2},\frac{1}{2}\right) \tag{43}$$

Since $\phi(0, 0)$ must be orthogonal to $\phi(1, 0)$ we must have

$$\phi(0,0) = \frac{1}{\sqrt{2}}\,\psi\left(\frac{1}{2},-\frac{1}{2}\right) - \frac{1}{\sqrt{2}}\,\psi\left(-\frac{1}{2},\frac{1}{2}\right) \tag{44}$$

within an arbitrary phase factor. It can be easily seen that by using ladder operators on $\phi(1, 0)$ or on $\phi(1, 1)$ or $\phi(1, -1)$ one can never get $\phi(0, 0)$. Using Eqs. (37)–(44) we can write

$$\begin{pmatrix} \phi(1,1) \\ \phi(1,0) \\ \phi(1,-1) \\ \phi(0,0) \end{pmatrix} = \begin{pmatrix} 1 & 0 & 0 & 0 \\ 0 & \frac{1}{\sqrt{2}} & \frac{1}{\sqrt{2}} & 0 \\ 0 & 0 & 0 & 1 \\ 0 & \frac{1}{\sqrt{2}} & -\frac{1}{\sqrt{2}} & 0 \end{pmatrix} \begin{pmatrix} \psi\left(\frac{1}{2}, \frac{1}{2}\right) \\ \psi\left(\frac{1}{2}, -\frac{1}{2}\right) \\ \psi\left(-\frac{1}{2}, \frac{1}{2}\right) \\ \psi\left(-\frac{1}{2}, -\frac{1}{2}\right) \end{pmatrix} \tag{45}$$

In the bra-ket notation

$$\begin{pmatrix} |1,1\rangle \\ |1,0\rangle \\ |1,-1\rangle \\ |0,0\rangle \end{pmatrix} = \begin{pmatrix} 1 & 0 & 0 & 0 \\ 0 & \frac{1}{\sqrt{2}} & \frac{1}{\sqrt{2}} & 0 \\ 0 & 0 & 0 & 1 \\ 0 & \frac{1}{\sqrt{2}} & -\frac{1}{\sqrt{2}} & 0 \end{pmatrix} \begin{pmatrix} |\hat{\mathbf{z}}\uparrow, \hat{\mathbf{z}}\uparrow\rangle \\ |\hat{\mathbf{z}}\uparrow, \hat{\mathbf{z}}\downarrow\rangle \\ |\hat{\mathbf{z}}\downarrow, \hat{\mathbf{z}}\uparrow\rangle \\ |\hat{\mathbf{z}}\downarrow, \hat{\mathbf{z}}\downarrow\rangle \end{pmatrix} \tag{46}$$

The matrix elements of the square matrix are known as the Clebsch Gordan coefficients.

Example 19.5 Operate $\hat{J}_+ \left[= \left(\hat{J}_1\right)_+ + \left(\hat{J}_2\right)_+ \right]$ on Eq. (38) to obtain Eq. (43).

19.4 CLEBSCH-GORDAN COEFFICIENTS FOR $j_1 = l = 1$ & $j_2 = s = \frac{1}{2}$

In this section, we will calculate from first principles the Clebsch Gordan coefficients for

$j_1 = l = 1$ & $j_2 = s = \dfrac{1}{2}$

Thus, $\quad j_1 = 1 \Rightarrow m_1 = +1, 0$ and 1 & $j_2 = \dfrac{1}{2} \Rightarrow m_2 = +\dfrac{1}{2}$ and $-\dfrac{1}{2}$

We will therefore have six $\psi(j_1, j_2, m_1, m_2)$ functions:

$$\psi\left(1, \frac{1}{2}, 1, \frac{1}{2}\right), \quad \psi\left(1, \frac{1}{2}, 0, \frac{1}{2}\right), \quad \psi\left(1, \frac{1}{2}, -1, \frac{1}{2}\right)$$

$$\psi\left(1, \frac{1}{2}, 1, -\frac{1}{2}\right), \quad \psi\left(1, \frac{1}{2}, 0, -\frac{1}{2}\right) \text{ and } \psi\left(1, \frac{1}{2}, -1, -\frac{1}{2}\right)$$

The numbers inside the parenthesis denote the values of j_1, j_2, m_1 and m_2. Since the same values of j_1 and j_2 appear in all 4 of them, we will omit the values of j_1 and j_2 and write the ψ functions as

$$\psi\left(1, \frac{1}{2}\right), \; \psi\left(0, \frac{1}{2}\right), \; \psi\left(-1, \frac{1}{2}\right), \; \psi\left(1, -\frac{1}{2}\right), \; \psi\left(0, -\frac{1}{2}\right) \text{ and } \psi\left(-1, -\frac{1}{2}\right)$$

where the numbers in the parenthesis denote the values of m_1 and m_2. The corresponding values of m $(= m_1 + m_2)$ are

$$\frac{3}{2}, \frac{1}{2}, -\frac{1}{2}, \frac{1}{2}, -\frac{1}{2} \text{ and } -\frac{3}{2}$$

The maximum value of m is $\dfrac{3}{2}$, thus one of the values of j must be $\dfrac{3}{2}$ which will give rise to 4 values of $m = \dfrac{3}{2}, \dfrac{1}{2}, -\dfrac{1}{2}$ and $-\dfrac{3}{2}$. The remaining two states (with $m = \dfrac{1}{2}$ & $-\dfrac{1}{2}$) must correspond to $j = \dfrac{1}{2}$. Thus, the allowed values of j are $\dfrac{3}{2}$ and $\dfrac{1}{2}$ and the corresponding ϕ functions are

$$\phi\left(1,\frac{1}{2},\frac{3}{2},\frac{3}{2}\right),\phi\left(1,\frac{1}{2},\frac{3}{2},\frac{1}{2}\right),\ \phi\left(1,\frac{1}{2},\frac{3}{2},-\frac{1}{2}\right),\phi\left(1,\frac{1}{2},\frac{3}{2},-\frac{3}{2}\right)$$

(corresponding to $j=\frac{3}{2}$)

and
$$\phi\left(1,\frac{1}{2},\frac{1}{2},\frac{1}{2}\right),\ \phi\left(1,\frac{1}{2},\frac{1}{2},-\frac{1}{2}\right)$$

(corresponding to $j=\frac{1}{2}$)

where the numbers in parenthesis denote the values of j_1, j_2, j and m. Omitting the values of j_1 and j_2, we write the ϕ functions as

$$\phi\left(\frac{3}{2},\frac{3}{2}\right),\phi\left(\frac{3}{2},\frac{1}{2}\right),\ \phi\left(\frac{3}{2},-\frac{1}{2}\right),\phi\left(\frac{3}{2},-\frac{3}{2}\right)\ \text{for}\ j=\frac{3}{2} \tag{47}$$

$$\phi\left(\frac{1}{2},\frac{1}{2}\right),\ \phi\left(\frac{1}{2},-\frac{1}{2}\right)\ \text{for}\ j=\frac{1}{2} \tag{48}$$

Now, $\phi\left(\frac{3}{2},\frac{3}{2}\right)$ and $\psi\left(1,\frac{1}{2}\right)$ are the only functions which correspond to $m=\frac{3}{2}$, therefore, we must have

$$\phi\left(\frac{3}{2},\frac{3}{2}\right)=\psi\left(\frac{1}{2},\frac{1}{2}\right) \tag{49}$$

Further, $\phi\left(\frac{3}{2},-\frac{3}{2}\right)$ and $\psi\left(-1,-\frac{1}{2}\right)$ are the only state which correspond to $m=-\frac{3}{2}$, thus we must have

$$\phi\left(\frac{3}{2},-\frac{3}{2}\right)=\psi\left(-1,-\frac{1}{2}\right) \tag{50}$$

As before, if we operate Eq. (49) by the operator $\hat{J}_-$ we would obtain:

$$\hat{J}_-\phi\left(\frac{3}{2},\frac{3}{2}\right)=\hat{J}_-\ \psi\left(1,\frac{1}{2}\right)\ \Rightarrow\sqrt{3\times1}\ \phi\left(\frac{3}{2},\frac{1}{2}\right)=\sqrt{1\times1}\ \psi\left(1,-\frac{1}{2}\right)+\sqrt{2\times1}\ \psi\left(0,\frac{1}{2}\right)$$

Thus, $$\phi\left(\frac{3}{2},\frac{1}{2}\right)=\sqrt{\frac{2}{3}}\ \psi\left(0,\frac{1}{2}\right)+\sqrt{\frac{1}{3}}\ \psi\left(1,-\frac{1}{2}\right) \tag{51}$$

Example 19.6 Operate the above equation by $\hat{J}_-$ to obtain

$$\phi\left(\frac{3}{2},-\frac{1}{2}\right)=\sqrt{\frac{1}{3}}\ \psi\left(-1,\frac{1}{2}\right)+\sqrt{\frac{2}{3}}\ \psi\left(0,-\frac{1}{2}\right) \tag{52}$$

Example 19.7 Operate the above equation by $\hat{J}_-$ to obtain Eq. (50).

Example 19.8 Operate $\hat{J}_+ \left[= \left(\hat{J}_1\right)_+ + \left(\hat{J}_2\right)_+ \right]$ on Eq. (50) to get Eq. (52) and operate $\hat{J}_+$ on Eq. (52) to get Eq. (51).

The function $\phi\left(\frac{1}{2}, \frac{1}{2}\right)$ must be a linear combination of the ψ functions in Eq. (51):

$$\phi\left(\frac{1}{2}, \frac{1}{2}\right) = c_1 \psi\left(1, -\frac{1}{2}\right) + c_2 \psi\left(0, +\frac{1}{2}\right)$$

For the above function to be orthogonal to $\phi\left(\frac{3}{2}, \frac{1}{2}\right)$ we must have

$$c_1\sqrt{\frac{1}{3}} + c_2\sqrt{\frac{2}{3}} = 0 \;\Rightarrow\; c_2 = -\frac{1}{\sqrt{2}}c_1$$

Thus,
$$\phi\left(\frac{1}{2}, \frac{1}{2}\right) = c_1\left[\psi\left(1, -\frac{1}{2}\right) - \sqrt{\frac{1}{2}}\,\psi\left(0, \frac{1}{2}\right)\right]$$

Using the normalization condition, we get $c_1 = \sqrt{\frac{2}{3}}$ giving

$$\phi\left(\frac{1}{2}, \frac{1}{2}\right) = -\sqrt{\frac{1}{3}}\,\psi\left(0, \frac{1}{2}\right) + \sqrt{\frac{2}{3}}\,\psi\left(1, -\frac{1}{2}\right) \tag{53}$$

within an arbitrary phase factor. Operating the above equation by $\hat{J}_-$ we obtain by

$$\phi\left(\frac{1}{2}, -\frac{1}{2}\right) = -\sqrt{\frac{2}{3}}\,\psi\left(-1, \frac{1}{2}\right) + \sqrt{\frac{1}{3}}\,\psi\left(0, -\frac{1}{2}\right) \tag{54}$$

The Clebsch Gordan coefficients for $j_2 = \frac{1}{2}$ and an arbitrary value of $j_1 \left(\geq \frac{1}{2}\right)$ are tabulated in Table 19.1. It is left as an exercise to obtain the above equations by using Table 19.1 with $j_1 = 1$.

Table 19.1: Clebsch-Gordan coefficients

$C(j_1, j_2; j, m, m_1, m_2) \left(= \langle j_1, j_2, m_1, m_2 | j_1, j_2, j, m\rangle\right)$ **for** $j_2 = \frac{1}{2}$

$j =$	$m_2 = \frac{1}{2}$	$m_2 = -\frac{1}{2}$
$j_1 + \frac{1}{2}$	$\sqrt{\frac{j_1 + m + \frac{1}{2}}{2j_1 + 1}}$	$\sqrt{\frac{j_1 - m + \frac{1}{2}}{2j_1 + 1}}$
$j_1 - \frac{1}{2}$	$-\sqrt{\frac{j_1 - m + \frac{1}{2}}{2j_1 + 1}}$	$\sqrt{\frac{j_1 + m + \frac{1}{2}}{2j_1 + 1}}$

For $j_2 = \frac{1}{2}$, the possible values of j will $j_1 + \frac{1}{2}$ and $j_1 - \frac{1}{2}$. Thus, using the top row of Table 19.1 we may write

$$\phi\left(j, \frac{1}{2}, j_1 + \frac{1}{2}, m\right) = \sqrt{\frac{j_1 + \frac{1}{2} + m}{2j_1 + 1}}\ \psi\left(j_1, \frac{1}{2}, m_1 = m - \frac{1}{2}, m_2 = \frac{1}{2}\right)$$

$$+\sqrt{\frac{j_1 + \frac{1}{2} - m}{2j_1 + 1}}\ \psi\left(j_1, \frac{1}{2}, m_1 = m + \frac{1}{2}, m_2 = -\frac{1}{2}\right) \quad (55)$$

where one can see that $m_1 + m_2 = m$. If we now assume $j_1 = \frac{1}{2}$, then for $m = 1, 0$ and -1 we will get Eqs. (37), (43) and (38) respectively. Similarly, using the bottom row of Table 19.1, we may write

$$\phi\left(j_1, \frac{1}{2}, j_1 - \frac{1}{2}, m\right) = -\sqrt{\frac{j_1 - m + \frac{1}{2}}{2j_1 + 1}}\ \psi\left(j_1, \frac{1}{2}, m - \frac{1}{2}, \frac{1}{2}\right)$$

$$+\sqrt{\frac{j_1 + m + \frac{1}{2}}{2j_1 + 1}}\ \psi\left(j_1, \frac{1}{2}, m + \frac{1}{2}, -\frac{1}{2}\right) \quad (56)$$

Example 19.9 If we assume $j_1 = \frac{1}{2}$ (and of course $j_2 = \frac{1}{2}$), for $j = j_1 - \frac{1}{2} = 0$, m must be 0 and the above equation would give Eq. (44).

Example 19.10 Use Table 19.1 to directly obtain Eqs. (51), (52) and (53).

Example 19.11 We assume $j_1 = 1$ (and of course $j_2 = \frac{1}{2}$). For $j = \frac{1}{2}$ we use the bottom row of Table 19.1 to obtain Eqs. (53) and (54) corresponding to $m = +\frac{1}{2}$ and $m = -\frac{1}{2}$ respectively.

19.5 HYDROGEN ATOM WAVEFUNCTIONS

For an electron in the hydrogen atom $j_1 = l$ and

$$\psi\left(l, \frac{1}{2}, m_l, \frac{1}{2}\right) = R_{nl}(r)\, Y_{lm_l}(\theta, \phi)\begin{pmatrix}1\\0\end{pmatrix} = \begin{pmatrix}R_{nl}\, Y_{lm_l}\\0\end{pmatrix}$$

corresponding to the spin up state of the electron. Similarly,

$$\psi\left(l, \frac{1}{2}, m_l, -\frac{1}{2}\right) = R_{nl}(r)\, Y_{lm_l}(\theta, \phi)\begin{pmatrix}0\\1\end{pmatrix} = \begin{pmatrix}0\\R_{nl}\, Y_{lm_l}\end{pmatrix}$$

corresponding to the spin down state of the electron. We have put a subscript l on m to write it as m_l to distinguish it from m ($= m_l + m_s$) which is often written as m_j.

Example 19.12 As an example of using Table 19.1, let us consider the $n = 3$, $l = 2$ state of the hydrogen atom. Since $j_1 = l = 2$ and $j_2 = s = \frac{1}{2}$, we will have $j = \frac{5}{2}$ and $\frac{3}{2}$. For the $j = \frac{5}{2}$ state we will have to use the first row of Table 19.1 to obtain

$$\phi\left(j_1 = 2, j_2 = \frac{1}{2}, j = \frac{5}{2}, m\right) = \sqrt{\frac{2 + m + \frac{1}{2}}{5}}\, \psi\left(j_1 = 2, j_2 = \frac{1}{2}, m_l = m - \frac{1}{2}, m_s = \frac{1}{2}\right)$$

$$+ \sqrt{\frac{2 - m + \frac{1}{2}}{5}}\, \psi\left(j_1 = 2, j_2 = \frac{1}{2}, m_l = m + \frac{1}{2}, m_s = -\frac{1}{2}\right)$$

where for all wavefunctions we are assuming $n = 3$. Notice that $m_l + m_s = m$ for both ψ functions. Thus,

$$\phi\left(2, \frac{1}{2}, \frac{5}{2}, m\right) = \sqrt{\frac{\frac{5}{2} + m}{5}}\, R_{32}(r)\, Y_{2, m - \frac{1}{2}} \begin{pmatrix} 1 \\ 0 \end{pmatrix} + \sqrt{\frac{\frac{5}{2} - m}{5}}\, R_{32}(r)\, Y_{2, m + \frac{1}{2}} \begin{pmatrix} 0 \\ 1 \end{pmatrix}$$

$$= \begin{pmatrix} \sqrt{\frac{\frac{5}{2} + m}{5}}\, R_{32}\, Y_{2, m - \frac{1}{2}} \\ \sqrt{\frac{\frac{5}{2} - m}{5}}\, R_{32} Y_{2, m + \frac{1}{2}} \end{pmatrix}$$

Thus, for example, the wavefunction corresponding to the $m = -\frac{3}{2}$ state of the $3^2 D_{5/2}$ states would be

$$\phi\left(2, \frac{1}{2}, \frac{5}{2}, -\frac{3}{2}\right) = \begin{pmatrix} \sqrt{\frac{1}{5}} R_{32}(r) Y_{2,-2}(\theta, \phi) \\ \sqrt{\frac{4}{5}} R_{32}(r) Y_{2,-1}(\theta, \phi) \end{pmatrix}$$

which is usually referred to as the Pauli wavefunction for the $3^2 D_{5/2}$ state with $m = -\frac{3}{2}$.

Example 19.13 Similarly, the wavefunction corresponding to the $m = +\frac{1}{2}$ state of the $3^2 D_{3/2}$ would be (since $j = \frac{3}{2}$, we will have to use the bottom row of Table 19.1)

$$\phi\left(2, \frac{1}{2}, \frac{3}{2}, +\frac{1}{2}\right) = \begin{pmatrix} -\sqrt{\frac{2}{5}} R_{32}(r) Y_{2,0}(\theta, \phi) \\ +\sqrt{\frac{3}{5}} R_{32}(r) Y_{2,1}(\theta, \phi) \end{pmatrix}$$

The Clebsch-Gordan coefficients for arbitrary value of j_1 (≥ 1) with $j_2 = 1$ are tabulated[2] in Table 19.2. In Problem 19.4, we have outlined the procedure for calculating the Clebsch-Gordan coefficients from first principles for $j_1 = j_2 = 1$. The results obtained there can also be directly obtained by using Table 19.2.

Table 19.2: Clebsch-Gordan coefficients

$C(j_1, j_2; j, m, m_1, m_2) \left(= \langle j_1, j_2, m_1, m_2 | j_1, j_2, j, m\rangle\right)$ **for** $j_2 = 1$

$j =$	$m_2 = 1$	$m_2 = 0$	$m_2 = -1$
$j_1 + 1$	$\sqrt{\frac{(j_1+m)(j_1+m+1)}{(2j_1+1)(2j_1+2)}}$	$\sqrt{\frac{(j_1-m+1)(j_1+m+1)}{(2j_1+1)(j_1+1)}}$	$\sqrt{\frac{(j_1-m)(j_1-m+1)}{(2j_1+1)(2j_1+2)}}$
j_1	$-\sqrt{\frac{(j_1+m)(j_1-m+1)}{2j_1(j_1+1)}}$	$\frac{m}{\sqrt{j_1(j_1+1)}}$	$\sqrt{\frac{(j_1-m)(j_1+m+1)}{2j_1(j_1+1)}}$
$j_1 - 1$	$\sqrt{\frac{(j_1-m)(j_1-m+1)}{2j_1(j_1+1)}}$	$-\sqrt{\frac{(j_1-m)(j_1+m)}{j_1(2j_1+1)}}$	$\sqrt{\frac{(j_1+m+1)(j_1+m)}{2j_1(2j_1+1)}}$

19.6 REPRESENTATION OF OPERATORS BY MATRICES

We consider two spin half particles (like an electron and a proton) and replace $\hat{\mathbf{J}}_1$, $\hat{\mathbf{J}}_2$ and $\hat{\mathbf{J}}$ by $\hat{\mathbf{s}}_1$, $\hat{\mathbf{s}}_2$ and $\hat{\mathbf{S}}$; thus

$$\hat{S}_x \equiv \hat{s}_{1x} + \hat{s}_{2x}; \quad \hat{S}_y \equiv \hat{s}_{1y} + \hat{s}_{2y}; \quad \hat{S}_z \equiv \hat{s}_{1z} + \hat{s}_{2z} \tag{57}$$

and
$$\hat{S}^2 \equiv \hat{s}_x^2 + \hat{s}_y^2 + \hat{s}_z^2 = \hat{s}_1^2 + \hat{s}_2^2 + 2\hat{s}_1 \cdot \hat{s}_2 \tag{58}$$

In this section, we will obtain the matrix representations of the components of $\hat{\mathbf{s}}_1$, $\hat{\mathbf{s}}_2$ and $\hat{\mathbf{S}}$. We choose simultaneous eigenkets of $\hat{S}_1^2, \hat{S}_2^2, \hat{s}_{1z}$ and $\hat{s}_{2z}$ as our base kets:

$$|1\rangle = \left|\psi\left(\tfrac{1}{2}, \tfrac{1}{2}, \tfrac{1}{2}, \tfrac{1}{2}\right)\right\rangle = |\hat{\mathbf{z}}\uparrow, \hat{\mathbf{z}}\uparrow\rangle \tag{59}$$

$$|2\rangle = \left|\psi\left(\tfrac{1}{2}, \tfrac{1}{2}, \tfrac{1}{2}, -\tfrac{1}{2}\right)\right\rangle = |\hat{\mathbf{z}}\uparrow, \hat{\mathbf{z}}\downarrow\rangle \tag{60}$$

$$|3\rangle = \left|\psi\left(\tfrac{1}{2}, \tfrac{1}{2}, -\tfrac{1}{2}, \tfrac{1}{2}\right)\right\rangle = |\hat{\mathbf{z}}\downarrow, \hat{\mathbf{z}}\uparrow\rangle \tag{61}$$

$$|4\rangle = \left|\psi\left(\tfrac{1}{2}, \tfrac{1}{2}, -\tfrac{1}{2}, -\tfrac{1}{2}\right)\right\rangle = |\hat{\mathbf{z}}\downarrow, \hat{\mathbf{z}}\downarrow\rangle \tag{62}$$

Thus, $\hat{s}_{1z}|1\rangle = \frac{1}{2}\hbar|1\rangle$; $\hat{s}_{1z}|2\rangle = \frac{1}{2}\hbar|2\rangle$; $\hat{s}_{1z}|3\rangle = -\frac{1}{2}\hbar|3\rangle$ & $\hat{s}_{1z}|4\rangle = -\frac{1}{2}\hbar|4\rangle$

Similarly, $\hat{s}_{2z}|1\rangle = \frac{1}{2}\hbar|1\rangle$; $\hat{s}_{2z}|2\rangle = -\frac{1}{2}\hbar|2\rangle$; $\hat{s}_{2z}|3\rangle = \frac{1}{2}\hbar|3\rangle$ & $\hat{s}_{2z}|4\rangle = -\frac{1}{2}\hbar|4\rangle$

---

[2] Tables 19.1 and 19.2 have been adapted from Ref. 3.

Thus, all off-diagonal elements like $\langle 3|\hat{s}_{1z}|2\rangle$, $\langle 2|\hat{s}_{2z}|4\rangle$ will be zero and the representations of $\hat{s}_{1z}$ and $\hat{s}_{2z}$ will be

$$\hat{s}_{1z} = \frac{1}{2}\hbar\begin{pmatrix} 1 & 0 & 0 & 0 \\ 0 & 1 & 0 & 0 \\ 0 & 0 & -1 & 0 \\ 0 & 0 & 0 & -1 \end{pmatrix} \text{ and } \hat{s}_{2z} = \frac{1}{2}\hbar\begin{pmatrix} 1 & 0 & 0 & 0 \\ 0 & -1 & 0 & 0 \\ 0 & 0 & 1 & 0 \\ 0 & 0 & 0 & -1 \end{pmatrix}$$

Notice that they are represented by 4×4 matrices because we are working in a 4-dimensional space. In order to calculate the matrix representation of $\hat{S}_z$ we can just add the matrices for $\hat{s}_{1z}$ and $\hat{s}_{2z}$ or, we can use the relations

$$\hat{S}_z|1\rangle = |1\rangle,\ \hat{S}_z|2\rangle = 0 = \hat{S}_z|3\rangle \text{ and } \hat{S}_z|4\rangle = -|4\rangle$$

and calculate all elements like $\langle 3|\hat{S}_z|2\rangle$ etc. Either way, we will get the following

$$\hat{S}_z = \begin{pmatrix} 1 & 0 & 0 & 0 \\ 0 & 0 & 0 & 0 \\ 0 & 0 & 0 & 0 \\ 0 & 0 & 0 & -1 \end{pmatrix} \Rightarrow \hat{S}_z^2 = \begin{pmatrix} 1 & 0 & 0 & 0 \\ 0 & 0 & 0 & 0 \\ 0 & 0 & 0 & 0 \\ 0 & 0 & 0 & 1 \end{pmatrix}. \tag{63}$$

where we have used a system of units where $\hbar = 1$. The above equation implies that the eigenvalues of $\hat{S}_z$ are +1, 0, 0 and –1. Now,

$$\hat{S}_+|1\rangle = \hat{s}_{1+}\left|\psi\left(\frac{1}{2},\frac{1}{2},\frac{1}{2},\frac{1}{2}\right)\right\rangle + \hat{s}_{2+}\left|\psi\left(\frac{1}{2},\frac{1}{2},\frac{1}{2},\frac{1}{2}\right)\right\rangle = 0$$

where we have used the properties of the ladder operators. Similarly,

$$\hat{S}_+|2\rangle = |1\rangle;\ \hat{S}_+|3\rangle = |1\rangle;\ \hat{S}_+|4\rangle = |2\rangle + |3\rangle. \text{ Thus,}$$

$$\hat{S}_+ = \begin{pmatrix} \langle 1|\hat{S}_+|1\rangle & \langle 1|\hat{S}_+|2\rangle & \langle 1|\hat{S}_+|3\rangle & \langle 1|\hat{S}_+|4\rangle \\ \langle 2|\hat{S}_+|1\rangle & \langle 2|\hat{S}_+|2\rangle & \langle 2|\hat{S}_+|3\rangle & \langle 2|\hat{S}_+|4\rangle \\ \langle 3|\hat{S}_+|1\rangle & \langle 3|\hat{S}_+|2\rangle & \langle 3|\hat{S}_+|3\rangle & \langle 3|\hat{S}_+|4\rangle \\ \langle 4|\hat{S}_+|1\rangle & \langle 4|\hat{S}_+|2\rangle & \langle 4|\hat{S}_+|3\rangle & \langle 4|\hat{S}_+|4\rangle \end{pmatrix} = \begin{pmatrix} 0 & 1 & 1 & 0 \\ 0 & 0 & 0 & 1 \\ 0 & 0 & 0 & 1 \\ 0 & 0 & 0 & 0 \end{pmatrix}$$

Similarly,

$$\hat{S}_-|1\rangle = \hat{s}_{1-}\left|\psi\left(\frac{1}{2},\frac{1}{2},\frac{1}{2},\frac{1}{2}\right)\right\rangle + \hat{s}_{2-}\left|\psi\left(\frac{1}{2},\frac{1}{2},\frac{1}{2},\frac{1}{2}\right)\right\rangle = |3\rangle + |2\rangle;$$

$$\hat{S}_-|2\rangle = |4\rangle;\ \hat{S}_-|3\rangle = |4\rangle;\ \hat{S}_-|4\rangle = 0$$

Thus,

$$\hat{S}_- = \begin{pmatrix} 0 & 0 & 0 & 0 \\ 1 & 0 & 0 & 0 \\ 1 & 0 & 0 & 0 \\ 0 & 1 & 1 & 0 \end{pmatrix}$$

and
$$\hat{S}_x = \frac{1}{2}\left(\hat{S}_+ + \hat{S}_-\right) = \frac{1}{2}\begin{pmatrix} 0 & 1 & 1 & 0 \\ 1 & 0 & 0 & 1 \\ 1 & 0 & 0 & 1 \\ 0 & 1 & 1 & 0 \end{pmatrix} \Rightarrow \hat{S}_x^2 = \frac{1}{2}\begin{pmatrix} 1 & 0 & 0 & 1 \\ 0 & 1 & 1 & 0 \\ 0 & 1 & 1 & 0 \\ 1 & 0 & 0 & 1 \end{pmatrix} \quad (64)$$

Similarly,
$$\hat{S}_y = \frac{1}{2i}\left(\hat{S}_+ - \hat{S}_-\right) = \frac{1}{2}\begin{pmatrix} 0 & -i & -i & 0 \\ i & 0 & 0 & -i \\ i & 0 & 0 & -i \\ 0 & i & i & 0 \end{pmatrix} \Rightarrow \hat{S}_y^2 = \frac{1}{2}\begin{pmatrix} 1 & 0 & 0 & -1 \\ 0 & 1 & 1 & 0 \\ 0 & 1 & 1 & 0 \\ -1 & 0 & 0 & 1 \end{pmatrix} \quad (65)$$

Thus,
$$\hat{S}^2 = \hat{S}_x^2 + \hat{S}_y^2 + \hat{S}_z^2 = \begin{pmatrix} 2 & 0 & 0 & 0 \\ 0 & 1 & 1 & 0 \\ 0 & 1 & 1 & 0 \\ 0 & 0 & 0 & 2 \end{pmatrix} \quad (66)$$

whose eigenvalues will be 2, 2, 2 and 0.

19.7 THE CONCEPT OF ISOSPIN

In this section, we briefly introduce the concept of isospin and mention the use of Clebsch-Gordan coefficients in the addition of isospin; such considerations are of great importance in high-energy nuclear scattering studies[3].

It is well known that except for their differences in charge, neutrons and protons have similar properties. Thus neutron and proton can be thought of as eigenstates of the nucleon. Now a group of elementary particles can be thought of possessing isospin τ if there are $(2\tau + 1)$ particles possessing nearly identical properties except for their charge. Thus protons and neutrons can be thought of possessing isospin $\tau = \frac{1}{2}$ with $\tau_z = +\frac{1}{2}$ and $\tau_z = -\frac{1}{2}$ states corresponding to the proton and neutron respectively[4]. Similarly, $\pi-$ mesons have an isospin of 1 since they occur in $2\tau + 1$ (= 3) forms as positively charged $\pi^+(\tau_z = 1)$, uncharged $\pi^0(\tau_z = 0)$ and negatively charged $\pi^-(\tau_z = -1)$ mesons. The charges of the nucleons of mesons is given by

$$Q = |q|\left[\tau_z + \frac{1}{2}B\right] \quad (67)$$

where B is known as the *baryon number* and has the value + 1 for nucleons and 0 for mesons.

As an example, we consider the isospin of a system of two nucleons. Assuming that for isospin also we can add the angular momenta in a manner similar to that discussed in earlier Section 19.3, we have four possible states

[3] For a nice discussion on isospin, the reader is referred to the books by Sachs (Ref. 4), Bethe and Morrison (Ref. 5), etc. A good discussion on the analysis of high-energy nuclear scattering experiments can be found in References 2 and 5.

[4] One may see the analogy with the spin angular momentum where $s_z = +\frac{1}{2}$ and $-\frac{1}{2}$ correspond to the spin up and spin down states (see Section 14.4).

$$\phi(1, 1) = \psi\left(\frac{1}{2}, \frac{1}{2}\right) \tag{68}$$

$$\phi(1, 0) = \frac{1}{\sqrt{2}}\left[\psi\left(\frac{1}{2}, -\frac{1}{2}\right) + \psi\left(-\frac{1}{2}, \frac{1}{2}\right)\right] \tag{69}$$

$$\phi(1, -1) = \psi\left(-\frac{1}{2}, -\frac{1}{2}\right) \tag{70}$$

$$\phi(0, 0) = \frac{1}{\sqrt{2}}\left[\psi\left(\frac{1}{2}, -\frac{1}{2}\right) - \psi\left(-\frac{1}{2}, \frac{1}{2}\right)\right] \tag{71}$$

Obviously, $\psi\left(\frac{1}{2}, \frac{1}{2}\right)$ corresponds to two protons because $\tau_{1z} = \tau_{2z} = \frac{1}{2}$ and similarly ϕ (1, –1) corresponds to 2 neutrons; both ϕ (1, 0) and ϕ (0, 0) corresponds to one proton and one neutron. However, ϕ (1, 0) is a symmetric function and ϕ (0, 0) is antisymmetric. Now, the ground state of deuteron is a mixture of $^3S_1$ and $^3D_1$ states (see, example, p. 114 of Ref. 6), thus it is symmetric in coordinates. It corresponds to spin 1 and, therefore, it is also symmetric in the spins of the two nucleons. The generalized Pauli principle (see, e.g., Chapter 7 of Ref. 4), requires that the wavefunction should be antisymmetric in isospin variables and hence the isospin state must be ϕ (0, 0).

19.8 PROBLEMS

Problem 19.1 Show that the operators $\hat{J}_1^2, \hat{J}_2^2, \hat{J}^2$ and $\hat{J}_z$ commute with each other. In particular show that whereas $\hat{J}^2$ does not commute with $\hat{J}_{1z}$ and $\hat{J}_{2z}$ but it does commute with $\hat{J}_z = \hat{J}_{1z} + \hat{J}_{2z}$.

Problem 19.2 Calculate from first principles the Clebsch-Gordon coefficients for $j_1 = 1$ and $j_2 = \frac{1}{2}$ and show that the results are consistent with Table 19.1. Use the calculated coefficients to write the wavefunctions for the *P*-states of an electron $\left(j_1 = l = 1, j_2 = s = \frac{1}{2}\right)$ for

$$j = \frac{3}{2}\left(m_j = \frac{3}{2}, \frac{1}{2}, -\frac{1}{2}, -\frac{3}{2}\right) \text{ and for } j = \frac{1}{2}\left(m_j = \frac{1}{2}, -\frac{1}{2}\right).$$

Problem 19.3 Using Table 19.1 calculate the Pauli wavefunction for the following states of the hydrogen atom:

(i) $4^2 D_{5/2}$ and $4^2 D_{3/2}$ states corresponding to $m = \frac{3}{2}$.

(ii) $4^2 F_{7/2}$ and $4^2 F_{5/2}$ states corresponding to $m = \frac{5}{2}$ and $m = -\frac{1}{2}$.

Problem 19.4 Consider the case $j_1 = 1, j_2 = 1$; thus $j = 2, 1, 0$. Start with the equation

$$\phi\,(1, 1, 2, 2) = \psi\,(1, 1, 1, 1)$$

Operate the above equation by $\hat{J}_-$ to obtain

$$\phi(1,1,2,1)=\sqrt{\frac{1}{2}}\psi(1,1,0,1)+\sqrt{\frac{1}{2}}\psi(1,1,1,0)$$

Operate the above equation by $\hat{J}_-$ to obtain

$$\phi(1,1,2,0)=\sqrt{\frac{1}{6}}\psi(1,1,-1,1)+\sqrt{\frac{4}{6}}\psi(1,1,0,0)+\sqrt{\frac{1}{6}}\psi(1,1,1,-1)$$

Operate the above equation by $\hat{J}_-$ to obtain

$$\phi(1,1,2,-1)=\sqrt{\frac{1}{2}}\psi(1,1,-1,0)+\sqrt{\frac{1}{2}}\psi(1,1,0,-1)$$

Operate the above equation by $\hat{J}_-$ to obtain

$$\phi(1,1,2,-2)=\psi(1,1,-1,-1)$$

Use the normalization condition and the fact that $\phi(1,1,1,1)$ will be orthogonal to $\phi(1,1,2,1)$ to obtain

$$\phi(1,1,1,1)=-\sqrt{\frac{1}{2}}\psi(1,1,0,1)+\sqrt{\frac{1}{2}}\psi(1,1,1,0)$$

within an arbitrary phase factor. Operate the above equation by $\hat{J}_-$ to obtain

$$\phi(1,1,1,0)=-\sqrt{\frac{1}{2}}\psi(1,1,-1,1)+\sqrt{\frac{1}{2}}\psi(1,1,1,-1)$$

Operate the above equation by $\hat{J}_-$ to obtain

$$\phi(1,1,1,-1)=-\sqrt{\frac{1}{2}}\psi(1,1,-1,0)+\sqrt{\frac{1}{2}}\psi(1,1,0,-1)$$

Finally using the fact that $\phi(1,1,0,0)$ will be orthogonal to $\phi(1,1,2,0)$ and $\phi(1,1,1,0)$ and that it will be normalized, show that

$$\phi(1,1,0,0)=\sqrt{\frac{1}{3}}\psi(1,1,-1,1)-\sqrt{\frac{1}{3}}\psi(1,1,0,0)+\sqrt{\frac{1}{3}}\psi(1,1,1,-1)$$

Show that the results are consistent with the ones obtained by using Table 19.2.

Problem 19.5 Assume $j_1=\frac{3}{2}$ and $j_2=\frac{1}{2}$. Start with the equation

$$\phi\left(\frac{3}{2},\frac{1}{2},2,2\right)=\psi\left(\frac{3}{2},\frac{1}{2},\frac{3}{2},\frac{1}{2}\right)$$

Operate the above equation by $\hat{J}_-$ to obtain

$$\phi\left(\frac{3}{2},\frac{1}{2},2,1\right)=\frac{1}{2}\psi\left(\frac{3}{2},\frac{1}{2},\frac{3}{2},-\frac{1}{2}\right)+\sqrt{\frac{3}{4}}\psi\left(\frac{3}{2},\frac{1}{2},\frac{1}{2},\frac{1}{2}\right)$$

Operate the above equation by $\hat{J}_-$ to obtain

$$\phi\left(\frac{3}{2},\frac{1}{2},2,0\right)=\sqrt{\frac{1}{2}}\psi\left(\frac{3}{2},\frac{1}{2},\frac{1}{2},-\frac{1}{2}\right)+\sqrt{\frac{1}{2}}\psi\left(\frac{3}{2},\frac{1}{2},-\frac{1}{2},\frac{1}{2}\right)$$

Keep operating by $\hat{J}_-$ to obtain $\phi\left(\frac{3}{2},\frac{1}{2},2,-1\right)$ and then

$$\phi\left(\frac{3}{2},\frac{1}{2},2,-2\right)=\psi\left(\frac{3}{2},\frac{1}{2},-\frac{3}{2},-\frac{1}{2}\right)$$

Write down $\phi\left(\frac{3}{2},\frac{1}{2},1,1\right)$ which is orthogonal to $\phi\left(\frac{3}{2},\frac{1}{2},2,1\right)$ and normalized. Then operate $\phi\left(\frac{3}{2},\frac{1}{2},1,1\right)$ by $\hat{J}_-$ to obtain $\phi\left(\frac{3}{2},\frac{1}{2},1,0\right)$ and $\phi\left(\frac{3}{2},\frac{1}{2},1,-1\right)$. Show that the results are consistent with the ones obtained by using Table 19.1.

Problem 19.6 In the electron spin-orbit interaction problem $j_1 = l$, $j_2 = s = \frac{1}{2}$ and $j = l \pm \frac{1}{2}$. Assuming that the system is in a state in which $\hat{J}_1^2, \hat{J}_2^2, \hat{J}^2$ and $\hat{J}_z$ are well-defined show that

$$\langle \hat{s}_z \rangle = \pm \frac{m_j \hbar}{(2l+1)} \tag{72}$$

where the upper and lower sings correspond to $j_1 = l + \frac{1}{2}$ and $j = l - \frac{1}{2}$ respectively.

Problem 19.7 For a state in which $\hat{J}_1^2, \hat{J}_2^2, \hat{J}_{1z}$ and $\hat{J}_{2z}$ have the well-defined values $j_1(j_1+1)\hbar^2$, $j_2(j_2+1)\hbar^2$, $m_1\hbar$ and $m_2\hbar$, show that

$$\langle \hat{J}^2 \rangle = \left[j_1(j_1+1) + j_2(j_2+1) + 2m_1m_2\right]\hbar^2 \tag{73}$$

Problem 19.8 If j_1 (= 1) and $j_2\left(=\frac{1}{2}\right)$ represent the isospin of pions and nucleon, then the total isospin of a pion nucleon system j, would be either $\frac{3}{2}$ or $\frac{1}{2}$. Show that

$$|\pi^+ p\rangle = \left|j=\frac{3}{2}, m_j=\frac{3}{2}\right\rangle = \left|\frac{3}{2},\frac{3}{2}\right\rangle \tag{74}$$

$$|\pi^+ n\rangle = \sqrt{\frac{1}{3}}\left|\frac{3}{2},\frac{1}{2}\right\rangle + \sqrt{\frac{2}{3}}\left|\frac{1}{2},\frac{1}{2}\right\rangle \tag{75}$$

$$|\pi^0 p\rangle = \sqrt{\frac{2}{3}}\left|\frac{3}{2},\frac{1}{2}\right\rangle - \sqrt{\frac{1}{3}}\left|\frac{1}{2},\frac{1}{2}\right\rangle \tag{76}$$

$$|\pi^0 n\rangle = \sqrt{\frac{2}{3}}\left|\frac{3}{2},-\frac{1}{2}\right\rangle + \sqrt{\frac{1}{3}}\left|\frac{1}{2},-\frac{1}{2}\right\rangle \tag{77}$$

$$|\pi^- p\rangle = \sqrt{\frac{1}{3}}\left|\frac{3}{2},-\frac{1}{2}\right\rangle - \sqrt{\frac{2}{3}}\left|\frac{1}{2},-\frac{1}{2}\right\rangle \tag{78}$$

$$|\pi^- n\rangle = \left|\frac{3}{2},-\frac{3}{2}\right\rangle \tag{79}$$

19.9 SOLUTIONS

Solution 19.1 $\hat{J}^2 = \hat{J}_1^2 + \hat{J}_2^2 + 2\left(\hat{J}_{1x}\hat{J}_{2x} + \hat{J}_{1y}\hat{J}_{2y} + \hat{J}_{1z}\hat{J}_{2z}\right)$

Since, $\hat{J}_1^2$ and $\hat{J}_2^2$ commute with each other and also with $\hat{J}_{1x}, \hat{J}_{2x}, \hat{J}_{1y}, \ldots$, etc.

$$\left[\hat{J}^2, \hat{J}_1^2\right] = \left[\hat{J}^2, \hat{J}_2^2\right] = 0 \text{. Further, } \left[\hat{J}_1^2, \hat{J}_z\right] = \left[\hat{J}_1^2, \hat{J}_{1z}\right] + \left[\hat{J}_1^2, \hat{J}_{2z}\right] = 0$$

Similarly $\left[\hat{J}_2^2, \hat{J}_z\right] = 0$. However, $\left[\hat{J}^2, \hat{J}_{1z}\right] \neq 0$, indeed

$$\left[\hat{J}^2, \hat{J}_{1z}\right] = \left[\hat{J}_1^2, \hat{J}_{1z}\right] + \left[\hat{J}_2^2, \hat{J}_{1z}\right] + 2\left(\hat{J}_{1x}\left[\hat{J}_{2x}, \hat{J}_{1z}\right] + \left[\hat{J}_{1x}, \hat{J}_{1z}\right]\hat{J}_{2x}\right)$$
$$+2\left(\hat{J}_{1y}\left[\hat{J}_{2y}, \hat{J}_{1z}\right] + \left[\hat{J}_{1y}, \hat{J}_{1z}\right]\hat{J}_{2y}\right) + 2\left(\hat{J}_{1z}\left[\hat{J}_{2z}, \hat{J}_{1z}\right] + \left[\hat{J}_{1z}, \hat{J}_{1z}\right]\hat{J}_{2z}\right)$$
$$= \left(-i\hat{J}_{1y}\hat{J}_{2x} + i\hat{J}_{1x}\hat{J}_{2y}\right) \tag{80}$$

Similarly,
$$\left[\hat{J}^2, \hat{J}_{2z}\right] = 2\left(-i\hat{J}_{1x}\hat{J}_{2y} + i\hat{J}_{1y}\hat{J}_{2x}\right) \tag{81}$$

Thus,
$$\left[\hat{J}^2, \hat{J}_z\right] = \left[\hat{J}^2, \hat{J}_{1z} + \hat{J}_{2z}\right] = 0 \tag{82}$$

Solution 19.5 Writing l for j_1, in Eq. (55), we obtain

$$\langle \hat{s}_z \rangle_{j=l+\frac{1}{2}} = \left[\frac{l + \frac{1}{2} + m_j}{2l+1}\right]\left(\frac{1}{2}\hbar\right) + \left[\frac{l + \frac{1}{2} - m_j}{2l+1}\right]\left(-\frac{1}{2}\hbar\right) = \frac{m_j \hbar}{2l+1}$$

and similarly for $j = l - \dfrac{1}{2}$.

Solution 19.8
$$\left|j = \frac{3}{2}, m_j = \frac{3}{2}\right\rangle = \left|m_{j1} = 1, m_{j2} = \frac{1}{2}\right\rangle = \left|\pi^+ p\right\rangle \tag{83}$$

$$\left|j = \frac{3}{2}, m_j = -\frac{3}{2}\right\rangle = \left|m_{j1} = -1, m_{j2} = -\frac{1}{2}\right\rangle = \left|\pi^- n\right\rangle \tag{84}$$

Now,
$$\left|j = \frac{3}{2}, m_j = \frac{1}{2}\right\rangle = \sqrt{\frac{1}{3}}\left|m_{j1} = 1, m_{j2} = \frac{1}{2}\right\rangle + \sqrt{\frac{2}{3}}\left|m_{j1} = 0, m_{j2} = \frac{1}{2}\right\rangle \tag{85}$$
$$= \sqrt{\frac{1}{3}}\left|\pi^+ n\right\rangle + \sqrt{\frac{2}{3}}\left|\pi^0 p\right\rangle$$

and
$$\left|j = \frac{3}{2}, m_j = -\frac{1}{2}\right\rangle = \sqrt{\frac{2}{3}}\left|\pi^0 n\right\rangle + \sqrt{\frac{1}{3}}\left|\pi^- p\right\rangle, \text{ etc.} \tag{86}$$

Similarly, we may obtain other equations.

REFERENCES AND SUGGESTED READINGS

1. V.I. Korgan, and V.M. Galitsky, *Problems in Quantum Mechanics*, PrenticeHall Inc., Englewood Cliffs, New Jersey (1963).
2. R.A. Cence, *Pion Nucleon Scattering*, Princeton University Press, Princeton (1969).
3. E.U. Condon and G.H. Shortley, *The Theory of Atomic Spectra*, Cambridge University Press, Cambridge (1951).
4. H.A. Sachs, *Nuclear Theory*, Addison-Wesley Publishing Co., Reading, Massachusetts (1953).
5. H.A. Bethe and P. Morrison, *Elementary Nuclear Theory*, John Wiley, New York (1956).
6. E. Fermi, *Nuclear Physics*, University of Chicago Press, Chicago (1950).

20 Time independent perturbation theory

We can imagine that this complicated array of moving things which constitute 'the world' is something like a great chess game being played by the gods, and we are observers of the game. We do not know what the rules of the game are; all we are allowed to do is to watch the playing. Of course, if we watch long enough, we may eventually catch on to a few of the rules. ***The rules of the game*** *are what we mean by* ***fundamental physics****. Even if we know every rule, however ... what we really can explain in terms of those rules is very limited, because almost all situations are so enormously complicated that we cannot follow the plays of the game using the rules, much less tell what is going to happen next. We must, therefore, limit ourselves to the more basic question of the rules of the game. If we know the rules, we consider that we 'understand' the world.*

— RICHARD FEYNMAN[1]

20.1 INTRODUCTION

In Chapters 6, 7, 10 and 17 we had obtained exact solutions of the Schrödinger equation for specific potential energy variations. However, for most problems in quantum mechanics, it is extremely difficult to obtain exact solutions of the Schrödinger equation and one has to resort to approximate methods. The three most important approximate methods are (i) the perturbation method, (ii) the variational method, and (iii) the JWKB method; the perturbation method will be discussed in this chapter. We may mention here that the variational method (which is discussed in Chapter 21) can give good estimate of the ground state energy by choosing an appropriate trial function; however, the method becomes quite cumbersome when one has to apply to higher excited states. On the other hand, the JWKB method (discussed in Chapter 18) gives an approximate but direct solution of the Schrödinger equation. The JWKB method is applicable when the potential energy variation is smoothly varying and when the Schrödinger equation is separable to a one-dimensional equation; if these conditions are satisfied, the method gives information about all the states of the problem.

The perturbation method, which is the subject matter of this chapter, is applicable when the Hamiltonian can be written as a sum of two parts

$$H = H_0 + H' \tag{1}$$

where the eigenvalues and eigenfunctions of H_0 are known and the effect of H' can be assumed to be small; H_0 is usually referred to as the unperturbed Hamiltonian and H' is known as the perturbation. We write the eigenvalue equation for the unperturbed Hamiltonian:

---

[1] The authors found this quotation in *The Making of the Atomic Bomb* by Richard Rhodes, Simon & Schuster, New York (1986).

$$H_0\, u_n = E_n\, u_n \tag{2}$$

where E_n and u_n are *known* eigenvalues and orthonormal eigenfunctions of H_0 respectively. Our aim is to develop a perturbation method to solve the Schrödinger equation

$$H\psi_n = W_n\psi_n \tag{3}$$

and obtain approximate expressions for W_n and ψ_n. In Section 20.2 we will develop the perturbation theory for non-degenerate states which will be followed by considerations for degenerate states (Section 20.3). The theory that we will develop has been extensively applied to solve many problems in quantum mechanics; some of these applications will be discussed in this chapter. The effect of the magnetic field can also be studied by using perturbation theory which will be discussed in Chapters 21 and 22.

20.2 NON-DEGENERATE CASE

We introduce a parameter[2] g and write

$$H = H_0 + gH' \tag{4}$$

Next, we express ψ_n and W_n as power series[3] in g, i.e., we write

$$\psi_n = \psi_n^{(0)} + g\,\psi_n^{(1)} + g^2\,\psi_n^{(2)} + \ldots \tag{5}$$

$$W_n = W_n^{(0)} + g\,W_n^{(1)} + g^2\,W_n^{(2)} + \ldots \tag{6}$$

where the superscripts 0, 1, 2, ... refer to the zeroeth order, first order, second order perturbation and the parameter g is assumed to take a continuous range of values between 0 and 1. In the final result we will set $g = 1$. We substitute the expansions given by Eqs (5) and (6) in Eq. (3) to obtain

$$(H_0 + gH')\,(\psi_n^{(0)} + g\,\psi_n^{(1)} + g^2\,\psi_n^{(2)} + \ldots)$$
$$= (W_n^{(0)} + g\,W_n^{(1)} + g^2\,W_n^{(2)} + \ldots)\,(\psi_n^{(0)} + g\,\psi_n^{(1)} + g^2\psi_n^{(2)} + \ldots) \tag{7}$$

---

[2] This method is often used in mathematics. For example, if we wish to solve the equation

$$u = a + \varepsilon u^2 \tag{i}$$

where ε is a small number, we may write

$$u = u_0 + \varepsilon u_1 + \varepsilon^2 u_2 + \ldots$$

If we substitute this series in Eq. (i) above and equate powers of ε, we obtain $u_0 = a$, $u_1 = u_0^2 = a^2$; $u_2 = 2u_0\, u_1 = 2a^3$, etc. Thus,

$$u = a + \varepsilon a^2 + 2\varepsilon^2 a^3 + \ldots$$

In this particular case, the exact solution of Eq. (i) can also be found:

$$u = \frac{1}{2\varepsilon}\left[1 - (1 - 4a\varepsilon)^{\frac{1}{2}}\right] = a + \varepsilon a^2 + 2\varepsilon^2 a^3 + \ldots$$

where we have chosen that root which makes $u \to a$ as $\varepsilon \to 0$. Further, the series will be convergent only when $4a\varepsilon < 1$. For further details on perturbation techniques, see Refs 1 and 2.

[3] For example, if H' is proportional to the strength of the electric field ε then $\psi_n^{(1)}$, $\psi_n^{(2)}$ $\psi_n^{(3)}$, ... will be proportional to ε, ε^2, ε^3, respectively; $\psi_n^{(0)}$ will be independent of ε. In some cases, such a power series may not be convergent [see, e.g. Eq. (37)]; indeed if it is an asymptotic series (see Appendix K), the first few terms may give fairly good results.

Since the above equation is supposed to be valid for all values of g lying between 0 and 1, the coefficients of equal powers of g on either side of the equation must be same, i.e.,

$$H_0\psi_n^{(0)} = W_n^{(0)}\psi_n^{(0)} \tag{8}$$

$$H_0\psi_n^{(1)} + H'\psi_n^{(0)} = W_n^{0}\psi_n^{(1)} + W_n^{(1)}\psi_n^{(0)} \tag{9}$$

$$H_0\psi_n^{(2)} + H'\psi_n^{(1)} = W_n^{(0)}\psi_n^{(2)} + W_n^{(1)}\psi_n^{(1)} + W_n^{(2)}\psi_n^{(0)} \tag{10}$$

etc. Equation (8) tells us that $\psi_n^{(0)}$ and $W_n^{(0)}$ are the eigenfunctions and eigenvalues of the unperturbed Hamiltonian. Thus,

$$\psi_n^{(0)} = u_n \tag{11}$$

and

$$W_n^{(0)} = E_n \tag{12}$$

We will be considering the effect of the perturbation H' on the nth state which will be assumed to be discrete and non-degenerate[4]; however, other states need neither be discrete nor be non-degenerate.

20.2.1 First Order Perturbation

We assume that the eigenfunctions of the unperturbed Hamiltonian, H_0, form a complete set; thus we may express $\psi_n^{(1)}$ as a linear combination of the functions u_n:

$$\psi_n^{(1)} = \sum_m a_m^{(1)} u_m \tag{13}$$

where the superscript refers to the fact that we are considering first order perturbation. The summation appearing in the above equation actually represents a summation over discrete states and integration over the continuum. If we substitute the above expansion in Eq. (9), we would obtain

$$H_0 \sum_m a_m^{(1)} u_m + H'\psi_n^{(0)} = W_n^{(0)}\psi_n^{(1)} + W_n^{(1)}\psi_n^{(0)} \tag{14}$$

or

$$\sum_m a_m^{(1)} E_m u_m + H'u_n = E_n \sum_m a_m^{(1)} u_m + W_n^{(1)} u_n \tag{15}$$

where we have used Eqs (2) and (11). If we now multiply the above equation by u_k^* and integrate, we would get

$$\sum_m a_m^{(1)} E_m \delta_{km} + H'_{kn} = E_n \sum_m a_m^{(1)} \delta_{km} + W_n^{(1)} \delta_{kn}$$

or, carrying out the summations, we get

$$a_k^{(1)} (E_n - E_k) + W_n^{(1)} \delta_{kn} = \mathrm{H}'_{kn} \tag{16}$$

where

$$H'_{kn} \equiv \int u_k^* H' u_n \, d\tau = \langle k | H' | n \rangle \tag{17}$$

and we have used the orthonormality condition

$$\langle k | m \rangle = \int u_k^* u_m \, d\tau = \delta_{km};$$

[4] The corresponding theory for degenerate states is discussed in Section 20.3.

the integrations are over the entire space. For $k = n$, Eq. (16) gives

$$W_n^{(1)} = H'_{nn} = \int u_k^* H' u_n d\tau = \langle n | H' | n \rangle \quad \text{First order perturbation} \tag{18}$$

which represents the first order perturbation to the energy. For $k \neq n$, Eq. (16) gives

$$a_k^{(1)} = \frac{H'_{kn}}{(E_n - E_k)}, k \neq n \tag{19}$$

Notice that we have not been able to determine $a_n^{(1)}$. However, it can be easily seen from Eq. (9) that if we add an arbitrary multiple of u_n to $\psi_n^{(1)}$, the equation remains unaltered. Hence, without any loss of generality, we may assume

$$a_n^{(1)} = 0 \tag{20}$$

Such a choice implies

$$\int u_n^* \psi_n^{(1)} d\tau = 0 \tag{21}$$

20.2.2 Second Order Perturbation

We next expand $\psi_n^{(2)}$ as a linear combination of the eigenfunctions of H_0; thus, we write

$$\psi_n^{(2)} = \sum_m a_m^{(2)} u_m \tag{22}$$

where the superscripts refer to the fact that we are considering second order perturbation. Substituting the above expansion in Eq. (10), we get

$$H_0 \sum_m a_m^{(2)} u_m + H' \sum_m a_m^{(1)} u_m$$

$$= W_n^{(0)} \sum_m a_m^{(2)} u_m + W_n^{(1)} \sum_m a_m^{(1)} u_m + W_n^{(2)} u_n$$

Multiplying by u_k^* and integrating, we obtain

$$\sum_m a_m^{(2)} E_m \delta_{km} + \sum_m a_m^{(1)} H'_{km}$$

$$= E_n \sum_m a_m^{(2)} \delta_{km} + H'_{nn} \sum_m a_m^{(1)} \delta_{km} + W_n^{(2)} \delta_{kn}$$

or

$$a_k^{(2)} (E_n - E_k) + W_n^{(2)} \delta_{kn} = \sum_m a_m^{(1)} H'_{km} - a_k^{(1)} H'_{nn} \tag{23}$$

For $k = n$, we get

$$W_n^{(2)} = \sum_m a_m^{(1)} H'_{nm} - a_m^{(1)} H'_{nn}$$

$$= \sum_{\substack{m \\ (m \neq n)}}{}' a_m^{(1)} H'_{nm}$$

where the prime over the summation indicates that we have omitted the term $m = n$. Thus, using Eq. (19), we get

$$W_n^{(2)} = \sum_{\substack{m \\ (m \neq n)}}{}' \frac{|H'_{mn}|^2}{E_n - E_m} \quad \text{Second order perturbation} \tag{24}$$

where we have used the relation $H'_{nm} = (H'_{mn})^*$. The above equation gives the second order perturbation to the energy eigenvalue. Further, for $k \neq n$, Eq. (23) gives

$$a_k^{(2)} = \sum_m{}' \frac{H'_{km}H'_{mn}}{(E_n - E_k)(E_n - E_m)} - \frac{H'_{kn}H'_{nn}}{(E_n - E_k)^2} \tag{25}$$

Similarly, we may obtain $W_n^{(3)}$. The final result is

$$W_n = W_n^{(0)} + W_n^{(1)} + W_n^{(2)} + W_n^{(3)} + \cdots \tag{26}$$

$$= E_n + H'_{nn} + \sum_m{}' \frac{|H'_{mn}|^2}{E_n - E_m} + \left[\sum_m{}' \frac{|H'_{mn}|^2 (H'_{mm} - H'_{nn})}{(E_n - E_m)^2} \right.$$

$$\left. + \frac{1}{3} \sum_m{}' \sum_{\substack{k \\ k \neq n \\ k \neq m}} \left\{ \frac{1}{E_m - E_k} + \frac{1}{E_n - E_k} \right\} \frac{H'_{nk}H'_{km}H'_{mn} + H'_{nm}H'_{mk}H'_{kn}}{E_n - E_m} \right] \tag{27}$$

20.2.3 Two Simple Examples

In order to illustrate the use of the perturbation theory to non-degenerate states we consider a perturbation of the form $\frac{1}{2}bx^2$ to the linear harmonic oscillator problem. Thus,

$$H_0 = -\frac{\hbar^2}{2\mu} \cdot \frac{d^2}{dx^2} + \frac{1}{2}kx^2 \tag{28}$$

and

$$H' = \frac{1}{2}bx^2 \tag{29}$$

The eigenvalues and eigenfunctions of H_0 are well known (see Section 7.2)

$$H_0 u_n = E_n u_n \tag{30}$$

where

$$E_n = \left(n + \frac{1}{2}\right) \hbar\omega; \; n = 0, 1, 2, 3, \ldots \tag{31}$$

$$u_n = N_n H_n (\xi) \exp\left(-\tfrac{1}{2}\xi^2\right) \tag{32}$$

$$\omega = \sqrt{\frac{k}{\mu}}, \quad N_n = \left[\frac{\gamma}{\sqrt{\pi}2^n n!}\right]^{1/2}$$

$$\xi = \gamma x, \quad \gamma = \sqrt{\frac{\mu\omega}{\hbar}} = \left[\frac{\mu k}{\hbar^2}\right]^{1/4} \tag{33}$$

Notice that the exact solution can easily be obtained; all that one has to do is replace k by $k + b$. Thus if we write

$$H \psi_n = W_n \psi_n \tag{34}$$

where
$$H = H_0 + H' \tag{35}$$

then
$$W_n = \left(n + \frac{1}{2}\right)\hbar\sqrt{\frac{k}{\mu}}\left[1 + \frac{b}{k}\right]^{1/2} \tag{36}$$

or
$$W_n = \left(n + \frac{1}{2}\right)\hbar\sqrt{\frac{k}{\mu}}\left[1 + \frac{1}{2}\frac{b}{k} - \frac{1}{8}\frac{b^2}{k^2} + \ldots\right] \tag{37}$$

the binomial expansion being valid when $\frac{b}{k} < 1$. If we compare Eq. (37) with Eq. (6), we obtain

$$W_n^{(0)} = \left(n + \frac{1}{2}\right)\hbar\sqrt{\frac{k}{\mu}} = E_n \tag{38}$$

$$W_n^{(1)} = \frac{1}{2}\left(n + \frac{1}{2}\right)\hbar\sqrt{\frac{k}{\mu}}\,\frac{b}{k} \tag{39}$$

$$W_n^{(2)} = -\frac{1}{8}\left(n + \frac{1}{2}\right)\hbar\sqrt{\frac{k}{\mu}}\,\frac{b^2}{k^2}, \text{ etc.} \tag{40}$$

where $W_n^{(0)}$ represent the unperturbed eigenvalue, and $W_n^{(1)}$ and $W_n^{(2)}$ represent the first and the second order perturbations respectively. Notice that $W_n^{(1)}$ is linear in b, $W_n^{(2)}$ is quadratic in b, etc. Further, the power series is convergent only when $b/k < 1$.

Let us now employ perturbation theory to calculate $W_n^{(1)}$, $W_n^{(2)}$ etc., Now,

$$W_n^{(1)} = \frac{1}{2}b\int_{-\infty}^{\infty} u_n^*(x)x^2 u_n(x)\,dx = \frac{1}{2}b < n \,|\, x^2 \,|\, n >$$

$$= \frac{1}{2}\left(n + \frac{1}{2}\right)\hbar\sqrt{\frac{k}{\mu}}\,\frac{b}{k} \tag{41}$$

where we have used the results derived in section 12.3. Equation (41) agrees with Eq. (39). In order to calculate $W_n^{(2)}$ we note that

$$H'_{mn} = < m \,|\, H' \,|\, n > = \frac{1}{2}b < m \,|\, x^2 \,|\, n > \tag{42}$$

is non-zero only when $m = n - 2$, $m = n$ or $m = n + 2$ (see Problem 12.4). Thus,

$$W_n^{(2)} = \sum_m{}' \frac{|H'_{mn}|^2}{E_n - E_m}$$

$$= \left(\frac{1}{2}b\right)^2\left[\frac{|< n-2 \,|\, x^2 \,|\, n >|^2}{E_n - E_{n-2}} + \frac{|< n+2 \,|\, x^2 \,|\, n >|^2}{E_n - E_{n+2}}\right]$$

$$= -\frac{1}{8}b^2\frac{\hbar}{\mu^2\omega^3}\left(n + \frac{1}{2}\right) \tag{43}$$

which agrees with Eq. (40).

Linear harmonic oscillator perturbed by an electric field

Another simple example is to consider a linear harmonic oscillator of charge Q perturbed by an electric field ε in the + x-direction. Thus,

$$H_0 = -\frac{\hbar^2}{2\mu}\frac{d^2}{dx^2} + \frac{1}{2}\mu\omega^2 x^2 \tag{44}$$

and

$$H' = -Q\varepsilon x \tag{45}$$

Now,
$$<m|H'|n> = -Q\varepsilon <m|x|n>$$
is non-zero only when $m = n \pm 1$ (see Problem 12.3). Thus, $W_n^{(1)} = 0$ and

$$W_n^{(2)} = \sum_m{}' \frac{|H'_{mn}|^2}{E_n - E_m} = Q^2\varepsilon^2\left[\frac{|<n-1|x|n>|^2}{E_n - E_{n-1}} - \frac{|<n+1|x|n>|^2}{E_{n+1} - E_n}\right]$$

$$= -\frac{Q^2\varepsilon^2}{2\mu\omega^2}$$

In this case also one can find an exact solution of the problem by shifting the origin. This can easily be seen because

$$H = H_0 + H' = -\frac{\hbar^2}{2\mu}\frac{d^2}{dx^2} + \frac{1}{2}\mu\omega^2 x^2 - Q\varepsilon x$$

$$= -\frac{\hbar^2}{2\mu}\frac{d^2}{d\zeta^2} + \frac{1}{2}\mu\omega^2\zeta^2 - \frac{Q^2\varepsilon^2}{2\mu\omega^2}$$

where $\zeta = x - \dfrac{Q\varepsilon}{\mu\omega^2}$. Thus the exact eigenvalue spectrum is

Eigenvalue spectrum

$$W_n = \left(n + \frac{1}{2}\right)\hbar\omega - \frac{Q^2\varepsilon^2}{2\mu\omega^2} \tag{46}$$

It may be noted that in this case the second order theory happens to give the exact results.

20.3 PERTURBATION THEORY FOR DEGENERATE STATES

In the perturbation theory developed in Section 20.2, the zeroeth order wavefunction satisfies the equation

$$H_0\psi_n^{(0)} = W_n^{(0)}\psi_n^{(0)} \tag{47}$$

(see Eq. 8). The above equation implies that $\psi_n^{(0)}$ is the eigenfunction corresponding to the unperturbed Hamiltonian. If we consider the perturbation to the nth state then we put

$$W_n^{(0)} = E_n \quad \text{and} \quad \psi_n^{(0)} = u_n \tag{48}$$

where u_n and E_n represent the eigenfunction and the energy eigenvalue for the nth state respectively. Now, for a degenerate state, any linear combination of the eigenfunctions is also a possible eigenfunction (see Section 6.3); consequently, $\psi_n^{(0)}$ cannot be completely specified by its unperturbed energy. Further, we notice from Eq. (16) that corresponding to a degenerate state if $k \neq n$ but $E_k = E_n$ then H'_{kn} must be zero (otherwise $d_k^{(1)}$ will be infinite, see Eq. 19). Hence, we must choose[5] for $\psi_n^{(0)}$ such linear combinations of u_n so that $H'_{kn} = 0$ for $k \neq n$. Obviously, the choice of the linear combination will depend on the form of H′. We illustrate the procedure by considering the splitting of the first excited state of the H-atom by an electric field; the splitting of spectral lines due to an electric field is known as Stark effect. The $n = 2$ state of the H-atom is 4-fold degenerate. The eigenfunctions[6] are (see Section 10.4):

$$\left.\begin{aligned}
u_1 &= R_{20}Y_{00} = \frac{1}{\sqrt{4\pi}}R_{20}(r); \qquad (n=2, l=0, m=0) \\
u_2 &= R_{21}Y_{10} = \sqrt{\frac{3}{4\pi}}R_{21}(r)\cos\theta; \qquad (n=2, l=1, m=0) \\
u_3 &= R_{21}Y_{11} = -\sqrt{\frac{3}{8\pi}}R_{21}(r)\sin\theta\; e^{i\phi}; \; (n=2, l=1, m=1) \\
u_4 &= R_{21}Y_{1-1} = \sqrt{\frac{3}{8\pi}}R_{21}(r)\sin\theta\; e^{-i\phi}; \; (n=2, l=1, m=-1)
\end{aligned}\right\} \tag{49}$$

where $$R_{20}(r) = \frac{1}{\sqrt{2}}\frac{1}{a_0^{3/2}}\left(1-\frac{r}{2a_0}\right)e^{-r/2a_0} \tag{50}$$

and $$R_{21}(r) = \frac{1}{2\sqrt{6}}\frac{1}{a_0^{3/2}}\frac{r}{a_0}e^{-r/2a_0} \tag{51}$$

The corresponding (unperturbed) eigenvalue is [see Section 10.4]

$$W_2^{(0)} = -\frac{\mu\alpha^2c^2}{8} = -\frac{1}{4}E_H$$

where $-E_H$ (≈ -13.6 eV) represents the ground state energy of the hydrogen atom; the corresponding ground state wavefunction is denoted by u_0. Let the required linear combination be

$$\psi^{(0)} = c_1 u_1 + \ldots + c_4 u_4$$

If we substitute this linear combination in the following equation (see Eq. 9)

$$H_0\psi^{(1)} + H'\psi^{(0)} = W^{(0)}\psi^{(1)} + W^{(1)}\psi^{(0)}$$

we would obtain

$$\begin{aligned}
H_0\sum_m a_m^{(1)} u_m &+ H'[c_1u_1 + c_2u_2 + c_3u_3 + c_4u_4] \\
&= W_2^{(0)}\sum_m a_m^{(1)} u_m + W_2^{(1)}[c_1u_1 + c_2u_2 + c_3u_3 + c_4u_4]
\end{aligned} \tag{52}$$

[5] This also follows from the fact that in calculating the first order perturbation when we use the expansion

$$\psi = \psi^{(0)} + \psi^{(1)} + \psi^{(2)} + \ldots$$

we must know precisely $\psi^{(0)}$ which represents the eigenfunction to which ψ will collapse when the perturbation is made to go to zero. However, for a degenerate state, we have no *apriori* knowledge which would allow us to predict as to which linear combination will ψ collapse to. Indeed, Eq. (16) tells us that we must choose such linear combinations so that $H'_{kn} = 0$ for $k \neq n$ (among the degenerate eigenfunctions).

[6] It should be noted that we could have started with linear combinations of u_1, u_2, u_3 and u_4; however, the functions should form an orthonormal set.

where we have assumed

$$\psi_2^{(1)} = \sum_m a_m^{(1)} u_m$$

Thus,

$$\sum_m a_m^{(1)} E_m u_m + H' [c_1 u_1 + \dots + c_4 u_4] = W_2^{(0)} \sum_m a_m^{(1)} u_m + W_2^{(1)} [c_1 u_1 + \dots + c_4 u_4]$$

Multiplying by u_1^* and integrating, we obtain

$$a_1^{(1)} W_2^{(0)} + [c_1 H'_{11} + c_2 H'_{12} + c_3 H'_{13} + c_4 H'_{14}] = W_2^{(0)} a_1^{(1)} + W_2^{(1)} c_1$$

where we have used the fact that E_m $(= W_2{}^{(0)})$ represents the unperturbed energy eigenvalue. Further

$$H'_{ij} = \int u_i^* H' u_j \, d\tau$$

Rearranging, we obtain

$$c_1 (H'_{11} - W_2^{(1)}) + c_2 H'_{12} + c_3 H'_{13} + c_4 H'_{14} = 0 \tag{53}$$

Similarly, if we multiply Eq. (52) by u_2^*, u_3^* and u_4^* and integrate, we would obtain respectively the following equations

$$c_1 H'_{21} + c_2 (H'_{22} - W_2^{(1)}) + c_3 H'_{23} + c_4 H'_{24} = 0 \tag{54}$$

$$c_1 H'_{31} + c_2 H'_{32} + c_3 (H'_{33} - W_2^{(1)}) + c_4 H'_{34} = 0 \tag{55}$$

$$c_1 H'_{41} + c_2 H'_{42} + c_3 H'_{43} + c_4 (H'_{44} - W_2^{(1)}) = 0 \tag{56}$$

Obviously, for a g-fold degenerate state, we will obtain a set of g equations. Now for nontrivial solutions

$$\begin{vmatrix} H'_{11} - W_2^{(1)} & H'_{12} & H'_{13} & H'_{14} \\ H'_{21} & H'_{22} - W_2^{(1)} & H'_{23} & H'_{24} \\ H'_{31} & H'_{32} & H'_{33} - W_2^{(1)} & H'_{34} \\ H'_{41} & H'_{42} & H'_{43} & H'_{44} - W_2^{(1)} \end{vmatrix} = 0 \quad \text{Secular equation for 4-fold degeneracy} \tag{57}$$

which is known as the secular equation.

Now, for the hydrogen atom, if $\mathbf{r}_1$ and $\mathbf{r}_2$ represent the position coordinates of the electron and proton respectively, then

$$\mathbf{P} = q(\mathbf{r}_2 - \mathbf{r}_1) = -q\mathbf{r} \tag{58}$$

would represent the electric dipole moment of the atom. In the above equation $q(> 0)$ represents the charge of the proton and $\mathbf{r} = \mathbf{r}_1 - \mathbf{r}_2$ represents the relative coordinate [see Section 10.3]. Thus, in the presence of the electric field ε the interaction energy would be given by

$$H' = -\mathbf{P} \cdot \boldsymbol{\varepsilon} \tag{59}$$

Assuming a static field (of magnitude ε) in the z-direction, we get

$$H' = q\varepsilon z = q\varepsilon r \cos\theta \;(q > 0) \tag{60}$$

Now,
$$H'_{23} = \int u_2^* H' u_3 d\tau$$
$$= \iiint \left[\sqrt{\frac{3}{4\pi}} R_{21}(r)\cos\theta\right][q\varepsilon r\cos\theta]$$
$$\times\left[-\sqrt{\frac{3}{8\pi}} R_{21}(r)\sin\theta\, e^{i\phi}\right] r^2 dr \sin\theta\, d\theta\, d\phi = 0$$

because $\int_0^{2\pi} e^{i\phi}\, d\phi = 0$. Due to the same reason the matrix elements $H'_{32}, H'_{13}, H'_{31}, H'_{14}, H'_{41}$, and H'_{32} would vanish. By carrying out the integration over θ, it can easily be shown that $H'_{11}, H'_{22}, H'_{33}, H'_{44}$, and H'_{34} would vanish. The only non-vanishing[7] matrix elements would be H'_{12} and H'_{21}. Now,

$$H'^*_{21} = H'_{12} = \int u_1^* H' u_2 d\tau$$
$$= \frac{\sqrt{3}(q\varepsilon)}{4\pi}\int_0^\infty \frac{1}{\sqrt{2}}\frac{1}{2\sqrt{6}}\frac{1}{a_0^3}\frac{r^2}{a_0}\left(1-\frac{r}{2a_0}\right)e^{-r/a_0} r^2\, dr \times \int_0^\pi \cos^2\theta\sin\theta\, d\theta\int_0^{2\pi} d\phi$$
$$= -3\, q\varepsilon a_0 = -g \tag{61}$$

where $g = 3q\varepsilon a_0$. Thus the secular equation becomes

$$\begin{vmatrix} -W_2^{(1)} & -g & 0 & 0 \\ -g & -W_2^{(1)} & 0 & 0 \\ 0 & 0 & -W_2^{(1)} & 0 \\ 0 & 0 & 0 & -W_2^{(1)} \end{vmatrix} = 0 \tag{62}$$

The roots are $\qquad W_2^{(1)} = +g, -g, 0, 0$

Now, if we substitute the values of the matrix elements, Eqs (53) – (56) become

$$-c_1 W_2^{(1)} - c_2 g = 0$$
$$-c_1 g - c_2 W_2^{(1)} = 0$$
$$c_3 W_2^{(1)} = 0$$
$$c_4 W_2^{(1)} = 0$$

Clearly, for $W_2^{(1)} = g$, $c_1 = -c_2$, $c_3 = c_4 = 0$ and the required linear combination is

$$\phi_1 = \frac{1}{\sqrt{2}}(u_1 - u_2) \tag{63}$$

where the factor $1/\sqrt{2}$ is such that ϕ_1 is normalized. Similarly, for $W_2^{(1)} = -g$, $c_1 = c_2$, $c_3 = c_4 = 0$ and the required linear combination is

---

[7] This could have been directly deduced from the following reasoning. Since all the wavefunctions have definite parities and since H' has odd parity

$$\int u_1^* H' u_1 d\tau = \cdots = \int u_4^* H' u_4 d\tau = 0$$

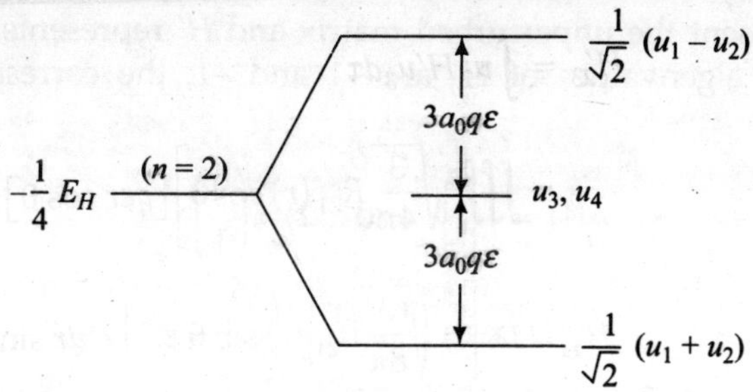

Figure 20.1: The splitting of the $n = 2$ level of the hydrogen atom by the electric field.

$$\phi_2 = \frac{1}{\sqrt{2}}(u_1 + u_2) \tag{64}$$

For the root $W_2^{(1)} = 0$, $c_1 = c_2 = 0$ and c_3 and c_4 are indeterminate. Thus,

$$\phi_3 \text{ and } \phi_4 = c_3u_3 + c_4u_4 \tag{65}$$

It can be easily seen that

$$\int \phi_1^* H'\phi_2 d\tau = 0 = \int \phi_2^* H'\phi_1 \, d\tau$$

thus if we had chosen ϕ_1 and ϕ_2 as our base kets (instead of u_1 and u_2) then H' would have been diagonal, and the diagonal elements

$$\int \phi_1^* H'\phi_1 \, d\tau = +g \text{ and } \int \phi_2^* H' \phi_2 \, d\tau = -g$$

would have represented the perturbation. The splitting of the level is shown in Fig. 20.1. As can be seen the levels characterized by the wavefunctions u_3 and u_4 remain degenerate and the degeneracy is said to be partially lifted. The above theoretical predictions agree with corresponding experimental values for electric field strengths less than about 10^6 volt/m. For higher fields, the quadratic Stark effect has to be taken into account. Eventually, for field strengths greater than 10^7 volt/m, autoionization occurs and Stark effect disappears.

20.4 EXAMPLES FROM MATRIX ALGEBRA

The use of the perturbation approach can be illustrated by considering examples from matrix algebra. We will discuss first the non-degenerate case and then the degenerate case. Obviously, while considering examples from matrix algebra one must always have Hermitian matrices.

Case 1: Non-degenerate case

Let us consider the matrix

$$H = \begin{pmatrix} 1+\varepsilon & \varepsilon \\ \varepsilon & -1+\varepsilon \end{pmatrix}$$

We write $H = H_0 + H'$, where

$$H_0 = \begin{pmatrix} 1 & 0 \\ 0 & -1 \end{pmatrix} \text{ and } H' = \begin{pmatrix} \varepsilon & \varepsilon \\ \varepsilon & \varepsilon \end{pmatrix}$$

H_0 is assumed to represent the unperturbed matrix and H' represents the perturbation and we assume $\varepsilon \ll 1$. The eigenvalues of H_0 are +1 and −1; the corresponding orthonormal eigenfunctions are

$$|1\rangle = \begin{pmatrix} 1 \\ 0 \end{pmatrix} \text{ and } |2\rangle = \begin{pmatrix} 0 \\ 1 \end{pmatrix}$$

Now,
$$H'_{11} = (1 \ 0) \begin{pmatrix} \varepsilon & \varepsilon \\ \varepsilon & \varepsilon \end{pmatrix} \begin{pmatrix} 1 \\ 0 \end{pmatrix} = \varepsilon$$

and
$$H'_{22} = (0 \ 1) \begin{pmatrix} \varepsilon & \varepsilon \\ \varepsilon & \varepsilon \end{pmatrix} \begin{pmatrix} 0 \\ 1 \end{pmatrix} = \varepsilon$$

Thus, in the first order perturbation theory, the eigenvalues are

$$1 + \varepsilon \text{ and } -1 + \varepsilon$$

The exact eigenvalues of H are given by

$$\lambda = \varepsilon \pm (1 + \varepsilon^2)^{1/2}$$

$$= 1 + \varepsilon + \frac{1}{2}\varepsilon^2 + \cdots \text{ and } -1 + \varepsilon - \frac{1}{2}\varepsilon^2 + \cdots$$

which agrees with the perturbation theory result.

Case 2: Degenerate case

As before, we write $H = H_0 + H'$, where

$$H_0 = \begin{pmatrix} 1 & 0 \\ 0 & 1 \end{pmatrix} \text{ and } H' = \begin{pmatrix} \varepsilon & \varepsilon \\ \varepsilon & \varepsilon \end{pmatrix}$$

and assume $\varepsilon \ll 1$. Clearly, the matrix H_0 has only one eigenvalue, viz., +1, which is 2-fold-degenerate. Further, any column matrix $\begin{pmatrix} a \\ b \end{pmatrix}$ is an eigenvector. We may choose

$$|u_1\rangle = \begin{pmatrix} 1 \\ 0 \end{pmatrix} \text{ and } |u_2\rangle = \begin{pmatrix} 0 \\ 1 \end{pmatrix}$$

as the base vectors. Thus,

$$H'_{11} = (1 \ 0) \begin{pmatrix} \varepsilon & \varepsilon \\ \varepsilon & \varepsilon \end{pmatrix} \begin{pmatrix} 1 \\ 0 \end{pmatrix} = \varepsilon$$

Similarly,
$$H'_{22} = H'_{12} = H'_{21} = \varepsilon$$

Therefore, the secular determinant is

$$\begin{vmatrix} \varepsilon - W & \varepsilon \\ \varepsilon & \varepsilon - W \end{vmatrix} = 0$$

which gives
$$W = 0 \text{ and } W = 2\varepsilon$$

The degeneracy is lifted and according to the first order perturbation theory the eigenvalues are

$$1 \text{ and } 1 + 2\varepsilon$$

If we denote the eigenfunction by

$$|\phi\rangle = c_1 |u_1\rangle + c_2 |u_2\rangle$$

then $\quad c_1(\varepsilon - W) + c_2\varepsilon = 0 \quad$ and $\quad c_1\varepsilon + c_2(\varepsilon - W) = 0$

For $\quad W = 0, c_1 = -c_2$

and $\quad |\phi_1\rangle = \frac{1}{\sqrt{2}}(|u_1\rangle - |u_2\rangle) = \frac{1}{\sqrt{2}}\begin{pmatrix}1\\-1\end{pmatrix}$

For $\quad W = 2\varepsilon, c_1 = c_2$

and $\quad |\phi_2\rangle = \frac{1}{\sqrt{2}}(|u_1\rangle + |u_2\rangle) = \frac{1}{\sqrt{2}}\begin{pmatrix}1\\1\end{pmatrix}$

Notice that $\quad \langle\phi_1|H'|\phi_2\rangle = \langle\phi_2|H'|\phi_1\rangle = 0,$

i.e. we have chosen such linear combinations in which H' is diagonal. Further, the diagonal elements would represent the perturbation:

$$\langle\phi_1 |H'| \phi_1\rangle = 0 \text{ and } \langle\phi_2 |H'| \phi_2\rangle = 2\varepsilon$$

It is easy to show that if we try to directly determine the eigenvalues and eigenfunctions of H, we would obtain

$$(1 + \varepsilon - \lambda)^2 = \varepsilon^2$$

or $\quad \lambda = 1 + \varepsilon \pm \varepsilon$

Thus, it so happens that the first order perturbation theory gives the exact results.

20.5 FINE STRUCTURE OF THE HYDROGEN ATOM

Important Note: In this section, in order to be consistent with the notation used in most books on spectroscopy, we will be using CGS units. In order to get the corresponding formulae in the MKS units, we must replace

$$e^2 \text{ by } \frac{q^2}{4\pi\varepsilon_0}$$

Alternatively, we can replace e^2 by $\alpha\hbar c$ where α ($\approx$ 1/137) is the fine structure constant. Then the formulae will be valid both in MKS and CGS units.

The theory of the hydrogen atom developed in Chapter 10 was based on the non-relativistic expression for the Hamiltonian given by

$$H = H_0 = \frac{p^2}{2m} + V(r) \tag{66}$$

where

$$V(r) = -\frac{Ze^2}{r} \tag{67}$$

In Section 10.4, we had solved the eigenvalue equation

$$H_0\psi = E\psi \tag{68}$$

and had shown that the discrete energy levels corresponding to the Hamiltonian H_0 are given by

$$E = E_n = -\frac{Z^2 mc^2 \alpha^2}{2n^2} = -\frac{Z^2 E_H}{n^2};\ n = 1, 2, 3,... \tag{69}$$

where $$\alpha = \frac{e^2}{\hbar c} \approx \frac{1}{137.036} \qquad (70)$$

is the fine structure constant and other symbols have been defined in Sections 10.3 and 10.4. The corresponding orthonormal wavefunctions are given by (see Section 10.4)

$$\psi = \psi_{nlm}(r, \theta, \phi) = R_{nl}(r)\, Y_{lm}(\theta, \phi) \qquad (71)$$

where $R_{nl}(r)$ are the radial part of the wavefunction (see Section 10.4 and Appendix I) and $Y_{lm}(\theta, \phi)$ are the spherical harmonics (see Chapters 9 and 15). Transitions between states (corresponding to different values of n) lead to Lyman series, Balmer series, etc. (see Fig. 10.1). However, if these spectral lines are seen through a high resolution spectroscope, usually each line is seen to consist of more than one line. This splitting is usually referred to as the fine structure of the spectral lines which is due to a combination of the effect due to relativity and also due to the interaction of the spin angular momentum of the electron with its orbital angular momentum. We will use first order perturbation theory to study the effects due to relativity and also due to spin-orbit interactions. We may mention here that when the atoms are put in a magnetic field then the spectral lines undergo further splitting—this is known as *Zeeman effect*. This will be discussed in chapters 21 and 22.

20.5.1 Relativistic Effects

Now, according to the special theory of relativity, the relation between the energy and momentum for a free particle is given by

$$E^2 = c^2 p^2 + m^2 c^4 \qquad (72)$$

where m represents the rest mass of the particle and c the speed of light in free space. In the presence of the potential $V(r)$, the above equation modifies to

$$[E - V(r)]^2 = c^2 p^2 + m^2 c^4 \qquad (73)$$

We replace $\boldsymbol{p}$ by $-i\hbar\nabla$ we obtain the relativistic Schrödinger equation

$$[-\hbar^2 c^2 \nabla^2 + m^2 c^4]\psi = [E - V(r)]^2 \psi \qquad (74)$$

If we write $E = E' + mc^2$, the above equation can be written as

$$(H_0 + H')\psi = E'\psi \qquad (75)$$

where $$H_0 = -\frac{\hbar^2}{2m}\nabla^2 + V(r) \qquad (76)$$

and $$H' = -\frac{[E' - V(r)]^2}{2mc^2} \qquad (77)$$

If we neglect H', we get the non-relativistic Schrödinger equation

$$H_0 \psi = E'\psi \qquad (78)$$

which is the same as Eq. (68). We will now use first order perturbation theory to study the effect of H'. Now, if ψ_{nlm} represent the hydrogen atom wavefunctions (see Section 10.4) then

$$\iiint \psi^*_{nl'm'}\, H' \psi_{nlm}\, d\tau = -\frac{1}{2mc^2} \iiint \psi^*_{nl'm'} \left[E_n - V(r)\right]^2 \psi_{nlm}\, d\tau$$

$$= -\frac{1}{2mc^2} \left\{ \int_0^\infty \left[E_n + \frac{Ze^2}{r} \right]^2 |R_{nl}(r)|^2 r^2\, dr \right\} \delta_{ll'}\, \delta_{mm'}$$

where in the expression for H', we have replaced E' by E_n where E_n represents the energy levels of the hydrogen atom [see Eq. (69)]. The above equation tells us that we already have a representation in which H' is diagonal. Thus, the relativity correction would be given by

$$\Delta E_{\text{rel}} = W^{(1)}_{\text{rel}} = -\frac{1}{2mc^2}\left[E_n^2 + 2E_n Ze^2\left\langle\frac{1}{r}\right\rangle + Z^2 e^4\left\langle\frac{1}{r^2}\right\rangle\right]$$

where
$$\left\langle\frac{1}{r^p}\right\rangle \equiv \int_0^\infty \frac{1}{r^p}\left[R_{nl}(r)\right]^2 r^2 dr \tag{79}$$

Substituting the expression for $\left\langle\frac{1}{r}\right\rangle$ and $\left\langle\frac{1}{r^2}\right\rangle$ from Appendix I, we obtain

$$\Delta E_{\text{rel}} = -\frac{1}{2mc^2}\left[\frac{Z^4}{n^4}E_H^2 - \frac{2Z^2}{n^2}E_H Ze^2\cdot\frac{Z}{a_0 n^2} + 4Z^4 E_H^2\frac{1}{n^3(l+\frac{1}{2})}\right]$$

$$= -\frac{Z^2 E_H}{n^2}\left(\frac{\alpha Z}{n}\right)^2\left[\frac{n}{l+\frac{1}{2}} - \frac{3}{4}\right]$$

Thus,
$$\Delta E_{\text{rel}} = -E_n\left(\frac{\alpha Z}{n}\right)^2\left[\frac{n}{l+\frac{1}{2}} - \frac{3}{4}\right] \tag{80}$$

In wave number units, we get

$$\Delta T_{\text{rel}} = \frac{\Delta E_{\text{rel}}}{hc} = -\frac{\alpha^2 Z^4}{n^4}\cdot R\left[\frac{n}{l+\frac{1}{2}} - \frac{3}{4}\right]\text{cm}^{-1} \tag{81}$$

where
$$R = \frac{mc\alpha^2}{2h} \tag{82}$$

represents the Rydberg constant whose numerical values are given in Section 10.3. For the hydrogen atom, $Z = 1$, and by substituting the values of R and α, we obtain

$$\Delta T_{\text{rel}} \approx -\frac{5.845}{n^4}\left[\frac{n}{l+\frac{1}{2}} - \frac{3}{4}\right]\text{cm}^{-1} \tag{83}$$

Thus because of relativistic effects, the eigenstates corresponding to a particular value of n will split and the degeneracy will be partially removed. For the $n = 2$ state, we have

$$\Delta T_{\text{rel}} \approx -1.1870\ \text{cm}^{-1} \quad \text{for} \quad n = 2, l = 0$$
$$\approx -0.2130\ \text{cm}^{-1} \quad \text{for} \quad n = 2, l = 1$$

Similarly, the $n = 3$ state will split into 3 levels corresponding to $l = 0$, 1 and 2. Figure 20.2 which corresponds to the $n = 2$ state of the hydrogen atom. However, the results are not in agreement

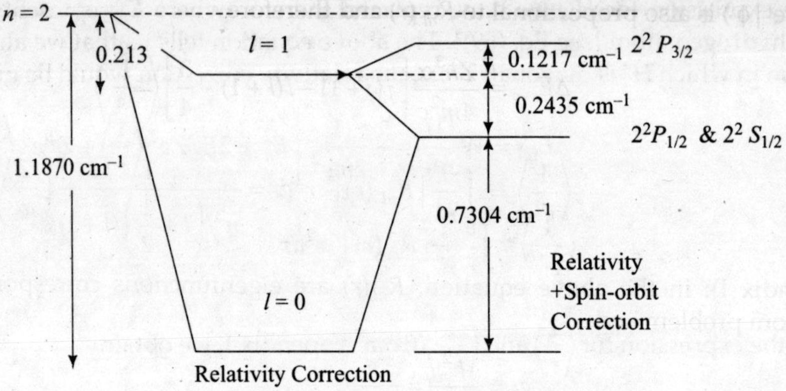

Figure 20.2: The perturbation due to relativistic effects and spin-orbit interaction for the *n*-2 state of the hydrogen atom.

with the experimental data; one has to take into account the spin-orbit interaction which will be discussed next.

20.5.2 The Spin-orbit Interaction

In a hydrogen-like atom the spin-orbit interaction H' is given by (see Sections 20.5 and 30.8)

$$H' = \frac{Ze^2}{2m^2c^2}\frac{1}{r^3}\mathbf{L}\cdot\mathbf{s} = \frac{Z\hbar\alpha}{2m^2c}\frac{1}{r^3}\mathbf{L}\cdot\mathbf{s} \tag{84}$$

Since,

$$J^2 = L^2 + s^2 + \mathbf{L}\cdot\mathbf{s}$$

Thus

$$\mathbf{L}\cdot\mathbf{s} = \frac{1}{2}\left[J^2 - L^2 - s^2\right] \tag{85}$$

$\Rightarrow$

$$H' = \frac{Z\hbar\alpha}{4m^2c}\frac{1}{r^3}\left[J^2 - L^2 - s^2\right] \tag{86}$$

Therefore we will have to choose a representation in which J^2, L^2 and s^2 are well difined. Thus we must use the ϕ functions as our eigenkets:

$$\left|\phi\left(n, j_1 = l, j_2 = \frac{1}{2}, j = l + \frac{1}{2}, m\right)\right\rangle$$

and

$$\left|\phi\left(n, j_1 = l, j_2 = \frac{1}{2}, j = l - \frac{1}{2}, m\right)\right\rangle$$

The expression for the above kets in terms of ψ functions are given in Eqs. (55) and (56) of the previous chapter. Since the ϕ functions are eigenkets of J^2, L^2 and s^2, we have

$$H'|\phi\rangle = \frac{Z\hbar\alpha}{4m^2c}\frac{1}{r^3}\left[j(j+1) - l(l+1) - \frac{3}{4}\right]\hbar^2|\phi\rangle \tag{87}$$

and we have a representation in which the off-diagonal terms are zero. Thus, the spin–orbit interaction energy will be given by

$$\Delta E_{s-o} = \langle\phi|H'|\phi\rangle = \frac{Z\hbar\alpha}{4m^2c}\left[j(j+1) - l(l+1) - \frac{3}{4}\right]\hbar^2\left\langle\phi\left|\frac{1}{r^3}\right|\phi\right\rangle \tag{88}$$

Now each ψ functions in Eqs, (55) and (56) [of the previous chapter] is proportional to $R_{nl}(r)$ and therefore $|\phi\rangle$ is also proportional to $R_{nl}(r)$ and therefore

$$\Delta E_{s\text{-}o} = \frac{Z\hbar^3\alpha}{4m^2c}\left[j(j+1) - l(l+1) - \frac{3}{4}\right]\left\langle \frac{1}{r^3} \right\rangle \tag{89}$$

where

$$\left\langle \frac{1}{r^3} \right\rangle = \int_0^\infty \frac{1}{r^3} |R_{nl}(r)|^2 r^2 dr = \frac{1}{n^3 l\left(l + \frac{1}{2}\right)(l+1)}\left(\frac{Z}{a_0}\right)^3 \tag{90}$$

(see Appendix I); in the above equation, $R_{nl}(r)$ are eigenfunctions corresponding to the hydrogen atom problem and

$$a_0 = \frac{\hbar^2}{me^2} = \frac{\hbar}{mc\alpha} \tag{91}$$

is the Böhr–radius. Thus,

$$\Delta E_{s\text{-}o} = \frac{Z\hbar^3\alpha}{4m^2c} \cdot \frac{\left[j(j+1) - l(l+1) - \frac{3}{4}\right]}{n^3\, l\left(l + \frac{1}{2}\right)(l+1)}\left(\frac{Z}{a_0}\right)^3 \tag{92}$$

Now, for $j = l + \frac{1}{2}$:

$$j(j+1) - l(l+1) - \frac{3}{4} = \left(l + \frac{1}{2}\right)\left(l + \frac{3}{2}\right) - l^2 - l - \frac{3}{4} = l$$

and, for $j = l - \frac{1}{2}$:

$$j(j+1) - l(l+1) - \frac{3}{4} = \left(l - \frac{1}{2}\right)\left(l + \frac{1}{2}\right) - l^2 - l - \frac{3}{4} = -(l+1)$$

Thus

$$\Delta E_{s\text{-}o} = \frac{Z^4\alpha^4 mc^2}{2n^3(l+1)(2l+1)} \quad \text{for } j = l + \frac{1}{2} \tag{93}$$

$$= -\frac{Z^4\alpha^4 mc^2}{2n^3 l(2l+1)} \quad \text{for } j = l - \frac{1}{2} \tag{94}$$

In wave number units $\qquad \Delta T_{s\text{-}o} = \dfrac{1}{hc}\Delta E_{s\text{-}o}$

or,

$$\Delta T_{s\text{-}o} = \begin{cases} \dfrac{Z^4\alpha^2 R}{n^3(l+1)(2l+1)} & \text{for } j = l + \dfrac{1}{2} \\ -\dfrac{Z^4\alpha^2 R}{n^3 l(2l+1)} & \text{for } j = l - \dfrac{1}{2} \end{cases} \tag{95}$$

Thus, for the hydrogen atom (Z = 1), one obtains for $n = 2$ (see Fig. 20.2).

$$\Delta T_{s\text{-}o}\left(n=2, l=1, j=\frac{3}{2}\right)=\frac{1.097\times 10^{5}}{48}\left(\frac{1}{137}\right)^{2}\ \text{cm}^{-1}$$

$$\approx 0.1217\ \text{cm}^{-1}$$

Similarly,
$$\Delta T_{s\text{-}o}\left(n=2, l=1, j=\frac{1}{2}\right)\approx -0.2435\ \text{cm}^{-1}$$

$$\Delta T_{s\text{-}o}\left(n=2, l=0, j=\frac{1}{2}\right)=+0.7304\ \text{cm}^{-1}$$

$$\Delta T_{s\text{-}o}\left(n=1, l=0, j=\frac{1}{2}\right)=+5.8435\ \text{cm}^{-1}$$

The sum of the relativity and spin-orbit correction is referred to as the fine structure splitting which we will denote by ΔT_{fs} which is given as

$$\Delta T_{fs}=\Delta T_{\text{rel}}+\Delta T_{s\text{-}o}$$

For $j = l + \frac{1}{2}$

$$\Delta T_{fs}=-\frac{\alpha^2 Z^4}{n^4}R\left[\frac{n}{l+\frac{1}{2}}-\frac{3}{4}-\frac{n}{(l+1)(2l+1)}\right]$$

where we have used Eqs (81) and (95). Simple manipulations and use of the relation $j = l + \frac{1}{2}$ would give

$$\Delta T_{fs}=-\frac{\alpha^2 Z^4}{n^4}R\left[\frac{n}{j+\frac{1}{2}}-\frac{3}{4}\right];\quad j=l\pm\frac{1}{2}\tag{96}$$

Even for $j = l - \frac{1}{2}$, the above expression is valid. We must mention here that the selection rules for transitions between doublets are

$$\Delta l = \pm 1$$

and
$$\Delta j = 0, \pm 1\tag{97}$$

The total quantum number may change by any integral amount.

20.5.3 Fine Structure of the Hydrogen Atom

For the hydrogen atom Z = 1 and by substituting the values of α and R, we get

$$\Delta T_{fs}\approx-\frac{5.845}{n^4}\left[\frac{n}{j+\frac{1}{2}}-\frac{3}{4}\right]\text{cm}^{-1}\tag{98}$$

The fine structure of various levels of the hydrogen atom are shown in Fig. 20.3. Now, the $n = 3 \rightarrow n = 2$ transition corresponds to the H_α line of the Balmer series of the hydrogen atom

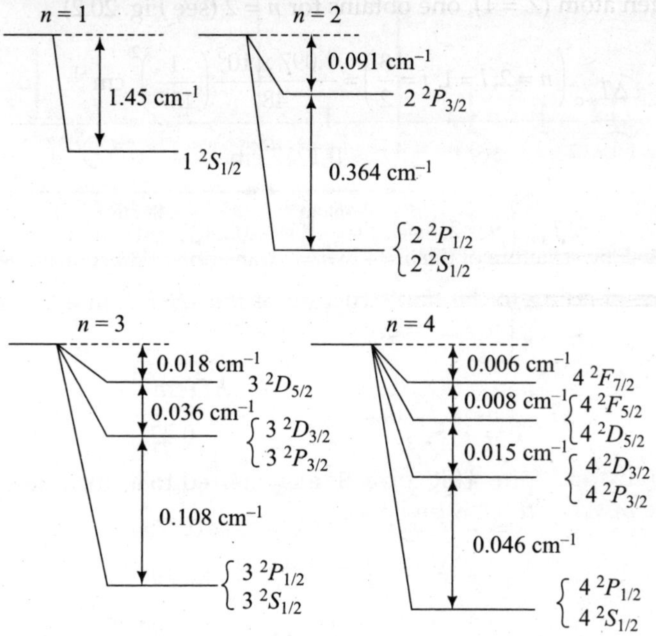

Figure 20.3: Fine structure of various levels of the hydrogen atom.

($\lambda \approx$ 6564.7 Å). The fine structure leads to 5 lines corresponding to the following transitions (see, e.g. Ref. 3):

	$\Delta T(\text{cm}^{-1})$
$3^2S_{1/2} \rightarrow 2^2P_{3/2}$	-0.071
$3^2D_{3/2} \rightarrow 2^2P_{3/2}$	$+0.037$
$3^2D_{5/2} \rightarrow 2^2P_{3/2}$	$+0.073$
$3^2P_{1/2} \rightarrow 2^2S_{1/2}$	$+0.294$
$3^2D_{3/2} \rightarrow 2^2P_{1/2}$	$+0.402$

The third and the fifth lines are the most intense–separated by 0.329 cm^{-1} consistent with experimental findings.

20.5.4 Spectra of the Ionized Helium Atom

We consider the $n = 4$ to $n = 3$ transition for the He$^+$ atom. Substituting the value of R [as given in Section 10.4], we get

$$T(n = 4) - T(n = 3) = -438889.6\left[\frac{1}{16} - \frac{1}{9}\right]\text{cm}^{-1}$$

or
$$T_{43} \approx 21335 \text{ cm}^{-1}$$

which corresponds to $\lambda \approx$ 4687.2 Å. Now, from Eq. (96)

$$\Delta T_{fs} \approx -\frac{93.49}{n^4}\left[\frac{n}{j+\frac{1}{2}} - \frac{3}{4}\right]\text{cm}^{-1}$$

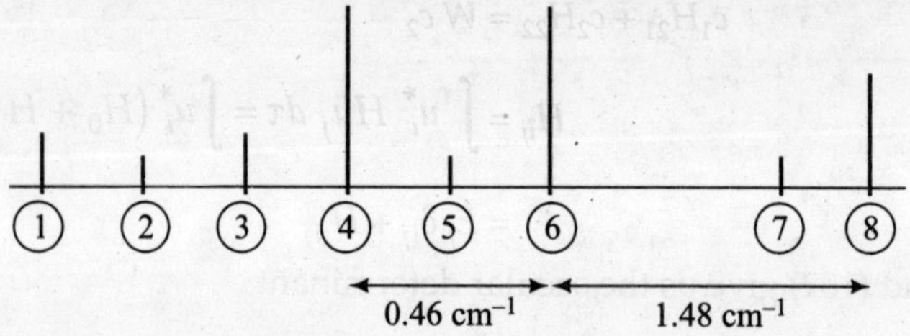

Figure 20.4: Fine structure of the $n = 4$ to $n = 3$ transition of the ionized helium atom.

Various lines, corresponding to the fine structure of the 4687 Å line, are shown in Fig. 20.4; they are

		$\Delta T(\text{cm}^{-1})$
$1 \Rightarrow 4\,^2S_{1/2}$	$\rightarrow 3\,^2P_{3/2}$	-0.321
$2 \Rightarrow 4\,^2P_{3/2}$	$\rightarrow 3\,^2D_{5/2}$	-0.168
$3 \Rightarrow 4\,^2F_{5/2}$	$\rightarrow 3\,^2D_{5/2}$	$+0.0755$
$4 \Rightarrow 4\,^2F_{7/2}$	$\rightarrow 3\,^2D_{5/2}$	$+0.197$
$5 \Rightarrow 4\,^2D_{3/2}$	$\rightarrow 3\,^2P_{3/2}$	$+0.409$
$6 \Rightarrow 4\,^2F_{5/2}$	$\rightarrow 3\,^2D_{3/2}$	$+0.653$
$7 \Rightarrow 4\,^2P_{1/2}$	$\rightarrow 3\,^2S_{1/2}$	$+1.410$
$8 \Rightarrow 4\,^2D_{3/2}$	$\rightarrow 3\,^2P_{1/2}$	$+2.140$

The wave number difference between the two most-intense lines (4 and 6) would be

$$0.653 - 0.197 \approx 0.46 \text{ cm}^{-1}$$

which agrees with experimental data; the experimental data has been given in Section 9.5 of Ref. 3.

20.6 ALMOST DEGENERATE PERTURBATION THEORY

We next consider the case when the levels are *almost degenerate*, i.e., the corresponding energy eigenvalues are almost nearly equal. For such a case the perturbation theory for non-degenerate states is strictly not valid because the corresponding expansion coefficient would become very large [see Eq. (19)]. We give here an approximate perturbation theory for a group of two levels which are almost degenerate. Let $u_1$and u_2 represent the eigenfunctions of the two levels with energy eigenvalues E_1 and E_2, i.e.,

$$H_0 u_{1,2} = E_{1,2}\, u_{1,2} \tag{99}$$

We write

$$\psi \approx c_1 u_1 + c_2 u_2 \tag{100}$$

and substitute in the eigenvalue equation

$$H\psi = W\psi$$

to obtain

$$c_1 H_{11} + c_2 H_{12} = W c_1 \tag{101}$$

and $$c_1 H_{21} + c_2 H_{22} = W c_2 \tag{102}$$

where $$H_{ij} = \int u_i^* \, H u_j \, d\tau = \int u_i^* \left(H_0 + H'\right) u_j \, d\tau$$

$$= E_j \, \delta_{ij} + H'_{ij} \tag{103}$$

Equations (101) and (102) give us the secular determinant

$$\begin{vmatrix} H_{11} - W & H_{12} \\ H_{21} & H_{22} - W \end{vmatrix} = 0 \tag{104}$$

which gives us

$$W_{1,2} = \frac{1}{2}\left[H_{11} + H_{22}\right] \pm \frac{1}{2}\left[(H_{11} - H_{22})^2 + 4|H_{12}|^2\right]^{1/2} \tag{105}$$

It is to be noted that when $E_1 = E_2$ one obtains the result for degenerate states (see Section 20.3). On the other hand, when $|H_{11} - H_{22}| \gg |H_{12}|$, one gets (by making a binomial expansion)

$$W_{1,2} \approx \frac{1}{2}\left[H_{11} + H_{22}\right] \pm \frac{1}{2}\left[(H_{11} - H_{22}) + \frac{1}{2} \cdot \frac{|H_{12}|^2}{H_{11} - H_{22}}\right] \tag{106}$$

Thus, $$W_1 = E_1 + H'_{11} + \frac{|H'_{12}|^2}{(E_1 + H'_{11}) - (E_2 + H'_{22})} \tag{107}$$

(and a similar expression for W_2) which agree with the result obtained by using non-degenerate perturbation theory to order g^2 because the term $(H'_{11} - H'_{22})$ in denominator of the third term is itself of order g; g is the parameter introduced in Eq. (4). Thus Eq. (105) represents the general expression which in the appropriate limiting forms goes over to the expressions corresponding to non-degenerate and degenerate state perturbation theory.

20.7 DALGARNO'S METHOD

If it becomes necessary to calculate the second order correction (viz., $W^{(2)}$ one has to evaluate the infinite sum given by Eq. (24) which, in general, is extremely difficult to carry out. However, in some cases it is possible to directly solve Eq. (9) and obtain an analytical expression for $\psi_n^{(1)}$ by which one can evaluate $W_n^{(2)}$, using the following expression:

$$W_n^{(2)} = \int \psi_n^{(0)*} H' \psi_n^{(1)} \, d\tau - W_n^{(1)} \int \psi_n^{(0)*} \psi_n^{(1)} \, d\tau \tag{108}$$

The above equation is obtained by multiplying Eq. (10) by $\psi_n^{(0)*}$ and integrating. This method is due to Dalgarno and is known as Dalgarno's method (see Ref. 4). We will illustrate its use by considering the Stark effect problem for the ground state of the hydrogen atom[8]. The perturbation is $H' = q\varepsilon r \cos\theta$ [see Eq. (60)]. Now $W_n^{(0)} = -E_H$ and it is easy to show that $W_1^{(1)} = 0$. We rewrite Eq. (9) as

[8] Epstein has given another method for calculating the Stark effect problem in hydrogen atom. The method has been discussed in Ref. 5, p. 195.

$$H_0\psi_1^{(1)} - W_1^{(0)}\psi_1^{(1)} = -H'\psi_1^{(0)} + W_1^{(1)}\psi_1^{(0)} \tag{109}$$

Thus,

$$\left[-\frac{\hbar^2}{2\mu}\nabla^2 - \frac{q^2}{4\pi\varepsilon_0 r} + E_H\right]\psi_1^{(1)} = -q\varepsilon r\cos\theta\,\frac{1}{\sqrt{4\pi}}\frac{2}{a_0^{3/2}}e^{-r/a_0} \tag{110}$$

where we have substituted the ground state wavefunction from Eqs (36) and (61) of Chapter 10. Since $\psi_1^{(1)}$ should approach to zero as $\varepsilon \to 0$, we need to find only the solution to the inhomogeneous part of the equation. We assume

$$\psi_1^{(1)} = R(r)\cos\theta \tag{111}$$

Thus,

$$\nabla^2\psi_1^{(1)} = \left[\frac{1}{r^2}\frac{d}{dr}\left(r^2\frac{dR}{dr}\right)\cos\theta + \frac{1}{r^2\sin\theta}\frac{d}{d\theta}\left(\sin\theta\frac{d}{d\theta}\cos\theta\right)R(r)\right]$$

$$= \left[\frac{1}{r^2}\frac{d}{dr}\left(r^2\frac{dR}{dr}\right) - \frac{2}{r^2}R(r)\right]\cos\theta \tag{112}$$

We substitute in Eq. (110) and introduce the variable [see Eq. (61) of Chapter 10] $\xi = r/a_0$, where a_0 is the Böhr radius. Simple manipulations would give

$$\frac{1}{\xi^2}\frac{d}{d\xi}\left(\xi^2\frac{dR}{d\xi}\right) - \frac{2}{\xi^2}R(\xi) + \left(\frac{2}{\xi} - 1\right)R(\xi) = C\xi e^{-\xi} \tag{113}$$

where

$$C = \frac{q\varepsilon}{E_H\sqrt{\pi a_0}} \tag{114}$$

Let us introduce a function $F(\xi)$ such that

$$R(\xi) = \xi F(\xi)e^{-\xi} \tag{115}$$

On substituting in Eq. (113), we obtain

$$\xi\frac{d^2F}{d\xi^2} + (4 - 2\xi)\frac{dF}{d\xi} - 2F(\xi) = C\xi \tag{116}$$

We are, however, interested in determining the inhomogeneous part of the solution. We find that $F(\xi) = a + b\xi$ satisfies Eq. (116), provided we choose $a = 2b = -C/2$. Thus, we obtain

$$R(\xi) = -\frac{C}{2}\xi\left(1 + \frac{1}{2}\xi\right)e^{-\xi}$$

and, therefore,

$$\psi_1^{(1)} = -\frac{C}{2}\xi\left(1 + \frac{1}{2}\xi\right)e^{-\xi}\cos\theta \tag{117}$$

Substituting in Eq. (108), we get

$$W_1^{(2)} = \int \psi_1^{(0)} H'\psi_1^{(1)}\,d\tau$$

$$= \iiint\left[\frac{1}{\sqrt{4\pi}}\frac{2}{a_0^{3/2}}e^{-r/a_0}\right][q\varepsilon r\cos\theta]\left[-\frac{C}{2}\xi\left(1 + \frac{1}{2}\xi\right)e^{-\xi}\cos\theta\right]r^2\,dr\sin\theta\,d\theta\,d\phi$$

$$= -\frac{9}{8}\frac{a_0^2}{E_H}q^2\varepsilon^2 \tag{118}$$

where E_H ($\approx$ 13.6 eV) is a positive number. The above result is in MKS units. In CGS units (which is often used in such problems), the result is

$$W_1^{(2)} = -\frac{9}{8}\frac{a_0^2}{E_H}e^2\varepsilon^2 = -\frac{9}{4}a_0^3\varepsilon^2$$

where we have used the result $E_H = e^2/2a_0$. Thus, we may write

$$W_1 = W_1^{(0)} - \frac{9}{4}a_0^3\varepsilon^2 \,(CGS \text{ units}) \tag{119}$$

Thus the polarizability[9] for the normal hydrogen atom is given by

$$\alpha = \frac{9}{2}a_0^3 = 0.677 \times 10^{-24} \text{ cm}^3 \tag{120}$$

Similarly, one can obtain expressions for $\psi_1^{(2)}(r)$ and $W_1^{(4)}$ (see, for example, Ref. 5).

20.8 PROBLEMS

Problem 20.1 Consider an electron inside an infinitely deep one-dimensional potential well:

$$V(x) = 0 \quad \text{for} \quad 0 < x < L$$

$$= \infty \quad \text{for} \quad x < 0 \text{ and } x > L$$

(see Section 6.6.1). Assuming a perturbation of the form $H' = -Q\varepsilon x$ show that in the first order perturbation theory each level gets shifted by $-\frac{1}{2}Q\varepsilon L$. Further, show that for the ground state $(n = 1)$

$$\psi^{(1)}(n = 1) = c_2\psi^{(0)}(n = 2) + c_3\psi^{(0)}(n = 3) + \ldots$$

where
$$c_2 = \frac{32}{27\pi^4}\left(\frac{\mu a^2}{\hbar^2}\right)(-Q\varepsilon L), \text{ etc.}$$

Problem 20.2 Consider the one-dimensional harmonic oscillator problem. Assuming a perturbation $H' = \beta x^4$ show that

$$W_n^{(1)} = \frac{3\beta}{2(\mu\omega/\hbar)^2}\left(n^2 + n + \frac{1}{2}\right) \tag{121}$$

[**Hint**: See Problem 12.5.]

Problem 20.3 Let the matrix

$$P = \begin{pmatrix} 1 & \varepsilon \\ \varepsilon & 3 \end{pmatrix}$$

be written in the form $P = P_0 + P'$ where

$$P_0 = \begin{pmatrix} 1 & 0 \\ 0 & 3 \end{pmatrix} \quad \text{and} \quad P' = \begin{pmatrix} 0 & \varepsilon \\ \varepsilon & 0 \end{pmatrix}$$

Consider P' as the perturbation and use second order perturbation theory to calculate the eigenvalues of P. Compare the results with the ones obtained by diagonalizing the matrix P.

---

[9] For details on poarizability one may refer to Ref. 5, Section 29e and 49f; Ref. 6, Section 5.3.

Problem 20.4 Let the matrix

$$P = \begin{pmatrix} 1 & \varepsilon & 0 \\ \varepsilon & 1 & 0 \\ 0 & 0 & 2+\varepsilon \end{pmatrix}$$

be written in the form $P = P_0 + P'$ where

$$P_0 = \begin{pmatrix} 1 & 0 & 0 \\ 0 & 1 & 0 \\ 0 & 0 & 2 \end{pmatrix} \text{ and } P' = \begin{pmatrix} 0 & \varepsilon & 0 \\ \varepsilon & 0 & 0 \\ 0 & 0 & \varepsilon \end{pmatrix}$$

Consider P' as the perturbation and use degenerate state perturbation theory to calculate the eigenvalues of P.

Problem 20.5 Consider a particle in an infinitely deep potential well inside a cube of dimension L (see Section 6.8). Consider a perturbation H' of the form

$$H' = kx^2$$

Apply first order perturbation theory to the first excited state which is 3-fold degenerate. Is the degeneracy completely lifted?

Problem 20.6 Consider a Hamiltonian of the form

$$H = H_0 + H'$$

where
$$H_0 = -\frac{\hbar^2}{2\mu}\left(\frac{\partial^2}{\partial x_1^2} + \frac{\partial^2}{\partial x_2^2}\right) + \frac{1}{2}k(x_1^2 + x_2^2) \tag{122}$$

and
$$H' = \lambda x_1 x_2 \tag{123}$$

The above Hamiltonian represents two identical oscillators (along the same axis) with an interaction between the two. (If we replace x_1 by x, and x_2 by y, we would have a two-dimensional oscillator which has been discussed in Problem 10.13.) Treat H' as a perturbation and calculate the first-order shift in the energy level of the first excited state. This problem can also be solved exactly. Obtain the exact solution and compare with the result obtained by using perturbation theory.

Problem 20.7 In continuation of the previous problem, consider again two identical oscillators with an interaction given by

$$H' = \frac{1}{2}k\varepsilon x_1 x_2 (x_1^2 + x_2^2) \tag{124}$$

Treat H' as a perturbation and show that in the first order perturbation theory, the ground state is not shifted and the twofold degenerate first excited state split whose energies are

$$\left(2 \pm \frac{3}{4}\frac{\varepsilon\hbar}{\sqrt{\mu k}}\right)\hbar\omega;$$

the corresponding eigenfunctions being

$$\frac{1}{\sqrt{2}}\left(\psi_{0,1} \pm \psi_{1,0}\right)$$

Problem 20.8 Using first order perturbation theory calculate the ground state energy of a helium-like atom; assume the interaction between the two electrons to represent the perturbation. Show that the first ionization potential for a single electron is given by

$$\left(\frac{1}{2}Z^2 - \frac{5}{8}Z\right)E_H \tag{125}$$

Assume the nucleus to be infinitely heavy. [See Section 24.3.]

Problem 20.9 While solving the Schrödinger equation for a hydrogen-like atom one assumes the nucleus to be a point charge. Using first order perturbation theory calculate the effect of the finite size of the nucleus assuming the nucleus to be a sphere of radius R_N with its charge uniformly distributed throughout its volume. (The effect of the finite size of the nucleus is usually referred to as volume effect and is important while considering *isotope shift*.)

Problem 20.10 The previous problem has an important application in muonic atoms[10] which consists of a μ-meson and a nucleus (m_μ = 207 m_e). First show that the Böhr radius for the muonic atom is comparable to the nuclear radius for Z ≈ 45; thus for $Z \gtrsim 45$ the muon orbit may be inside the nucleus. Assuming the Coulomb potential to be a perturbation, calculate the energy levels of the muonic atom.

Problem 20.11 In the n = 4 to n = 3 transition of the He^+ atom calculate the separation between the three lines corresponding to the following transitions:

$$4^2F_{7/2} \rightarrow 3^2D_{5/2}$$

$$4^2F_{5/2} \rightarrow 3^2D_{3/2}$$

and

$$4^2D_{3/2} \rightarrow 3^2P_{1/2}$$

[**Answer:** ≈ 0.45 cm^{-1}, 1.48 cm^{-1}]

Problem 20.12

(a) In the n = 3 to n = 2 transition of the hydrogen atom (the H_α line of the Balmer series) calculate the separation between the two lines corresponding to the following transitions

$$3^2D_{5/2} \rightarrow 2^2P_{3/2}$$

$$3^2D_{3/2} \rightarrow 2^2P_{1/2}$$

(b) The H_β line of the Balmer series of the hydrogen atom corresponds to the $n = 4 \rightarrow n = 2$ transition. The two most intense lines of the fine structure of the H_β line correspond to

$$4^2D_{5/2} \rightarrow 2^2P_{3/2}$$

and

$$4^2D_{3/2} \rightarrow 2^2P_{1/2}$$

transitions. Calculate the separation between the two lines.

(c) Show that there would be three lines resulting from the $^2D \rightarrow {}^2P$ transition.

Problem 20.13 A plane rotator is defined by the Hamiltonian $H = M_z^2/2I$ where I is the moment of inertia and M_z is the operator of the component of angular momentum along

---

[10] Muonic atoms may be formed when a moun shows down throght matter. Such atoms have been obtained for Z = 1 to heavy elements like uranium.

the axis of rotation z. Replace M_z *by* $-i\hbar\frac{\partial}{\partial\phi}$ and find the eigenvalues and normalized eigenfunctions of H. (You will only have to assume that the wavefunctions should be single-valued). If the rotator be perturbed by a constant field in its plane, in which its potential energy is $\varepsilon\cos\theta$, show that to the order ε^2, the displacement of the level E_m is

$$\frac{\varepsilon^2 I}{\hbar^2(4m^2-1)}.$$

Problem 20.14 A particle moves in a potential that is central except for a small non-central perturbation

$$F(r)\,(3\cos^2\theta - 1)$$

Show that in first order perturbation theory, the energies of the *s* states of the system are unaffected and that the three *p* states ($l = 1$; $m = 1, 0, -1$) are respectively perturbed by $\frac{1}{2}A$, A and $-\frac{1}{2}A$ where A is determined from the form of the central potential and of $F(r)$.

[**Hint:** The unperturbed eigenfunctions of the *p* states are of the form $R(r)Y_{1,1}(\theta,\phi)$; $R(r)\,Y_{1,0}(\theta,\phi)$ and $R(r)\,Y_{1,-1}(\theta,\phi)$.]

Problem 20.15 Consider the interaction between two hydrogen atoms in their ground state (this kind of interaction is known as Van der Waal's interaction). Let A and B denote the positions of the nuclei of the two atoms; these nuclei are assumed to be fixed in space and separated by a distance R (see Fig. 20.5). Show that the Hamiltonian can be written in the form

$$H = H_0 + H' \tag{126}$$

where

$$H_0 = -\frac{\hbar^2}{2\mu}\left(\nabla_1^2 + \nabla_2^2\right) - \frac{e^2}{r_1} - \frac{e^2}{r_2} \tag{127}$$

and

$$H' \approx \frac{e^2}{R}(x_1 x_2 + y_1 y_2 - 2 z_1 z_2) \tag{128}$$

x_1, x_2, etc., represent the x-coordinates of the electrons 1 and 2 with respect to the nuclei A and B, respectively. Use perturbation theory to calculate the effect of H'. (The variational analysis of the hydrogen molecule and the hydrogen molecule ion is discussed in Section 23.7, Problems 23.4 and 23.5.)

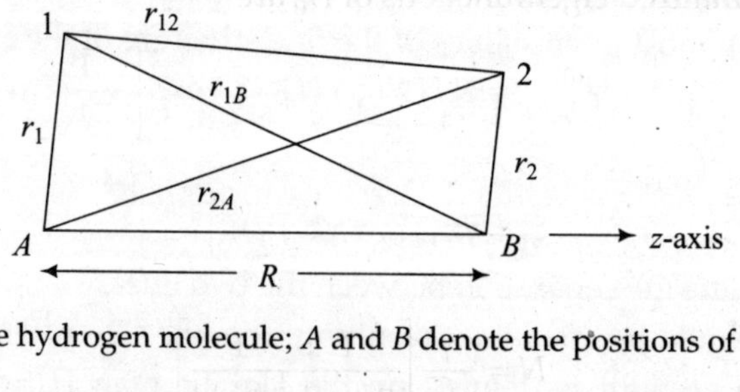

Figure 20.5: The hydrogen molecule; A and B denote the positions of the two nuclei.

20.9 SOLUTIONS

Solution 20.1 If we refer to Section 6.8, then the energy eigenvalue corresponding to the first excited state is given by

$$W_1^{(0)} = \frac{6\pi^2\hbar^2}{2\mu L^2}$$

The state is threefold degenerate; the wavefunctions are

$$u_1 = \sqrt{\frac{8}{L^3}} \sin\frac{2\pi x}{L} \sin\frac{\pi y}{L} \sin\frac{\pi z}{L}; \quad [n_x = 2, n_y = 1, n_z = 1]$$

$$u_2 = \sqrt{\frac{8}{L^3}} \sin\frac{\pi x}{L} \sin\frac{2\pi y}{L} \sin\frac{\pi z}{L}; \quad [n_x = 1, n_y = 2, n_z = 1]$$

$$u_3 = \sqrt{\frac{8}{L^3}} \sin\frac{\pi x}{L} \sin\frac{\pi y}{L} \sin\frac{2\pi z}{L}; \quad [n_x = 1, n_y = 1, n_z = 2]$$

Now,
$$H'_{ij} = \int u_i^* H' u_j \, d\tau = k \iiint u_i^* x^2 u_j \, dx\, dy\, dz$$

Simple integrations show that

$$H'_{ij} = 0 \text{ for } i \neq j$$

Thus we already have a representation in which H' is diagonal. Further

$$H'_{ij} = kL^2\left(\frac{1}{3} - \frac{1}{8\pi^2}\right) = g_1 \text{ (say)}$$

and
$$H'_{22} = H'_{33} = kL^2\left(\frac{1}{3} - \frac{1}{2\pi^2}\right) = g_2 \text{ (say)}$$

Hence the secular determinant is

$$\begin{vmatrix} g_1 - W_1^{(1)} & 0 & 0 \\ 0 & g_2 - W_1^{(1)} & 0 \\ 0 & 0 & g_2 - W_1^{(1)} \end{vmatrix} = 0$$

The roots are g_1, g_2, g_2; the degeneracy is partially lifted and the perturbed energy levels are

$$W_1^{(0)} + g_1,\ W_1^{(0)} + g_2 \text{ and } W_1^{(0)} + g_2$$

Solution 20.6 The normalized eigenfunctions of H_0 are

$$\psi^{(0)}_{n_1, n_2} = N H_{n_1}(\xi_1) H_{n_2}(\xi_2) \exp\left[-\frac{1}{2}\left(\xi_1^2 + \xi_2^2\right)\right]; \; n_1, n_2 = 0, 1, 2, \ldots$$

where
$$\xi_1 = \gamma x_1,\ \xi_2 = \gamma x_2,\ \gamma = \left[\frac{\mu k}{\hbar^2}\right]^{1/4}$$

$$N = \frac{\gamma}{\sqrt{\pi}}\left[\frac{1}{2^{n_1} n_1!}\,\frac{1}{2^{n_2} n_2!}\right]^{1/2}$$

The corresponding eigenvalues are (see also Problem 10.13)

$$\psi^{(0)}_{n_1, n_2} = (n_1 + n_2 + 1)\,\hbar\omega$$

The first excited state is twofold degenerate ($n_1 = 0, n_2 = 1$ and $n_1 = 1, n_2 = 0$) the corresponding eigenvalue being $2\hbar\omega$. If we write $u_1 = \psi^{(0)}_{0,1}$, $u_2 = \psi^{(0)}_{1,0}$ and

$$H'_{ij} = \iint u_i^* \, H' uj \, dx_1 \, dx_2, \; i, k = 1, 2$$

then $H'_{11} = H'_{22} = 0$ and

$$H'_{21} = \iint_{-\infty}^{+\infty} u_1^* \frac{(\xi_1 \xi_2)}{\gamma^2} u_2 \frac{(d\xi_1 \, d\xi_2)}{\gamma^2}$$

$$= \left(\frac{\gamma}{\sqrt{\pi}}\right)^2 \frac{1}{2} \frac{\lambda}{\gamma^4} \int_{-\infty}^{\infty} 2\xi_1^2 \, e^{-\xi_1^2} d\xi_1 \int_{-\infty}^{\infty} 2\xi_2^2 \, e^{-\xi_2^2} d\xi_2 = \frac{\lambda}{2\gamma^2} = \frac{\lambda}{2k} \hbar\omega = H'_{21}$$

On solving the secular equation, we obtain

$$W^{(1)} = \pm \frac{\lambda}{2k} \hbar\omega$$

the corresponding eigenfunctions being

$$\frac{1}{\sqrt{2}} (u_1 \pm u_2) = \frac{1}{\sqrt{2}} \left(\psi^{(0)}_{0,1} \pm \psi^{(0)}_{1,0}\right)$$

The exact eigenvalues of H can be determined if we introduce the new variables

$$\eta_1 = \frac{x_1 + x_2}{\sqrt{2}} \text{ and } \eta_2 = \frac{x_1 + x_2}{\sqrt{2}}$$

Thus,

$$H = -\frac{\hbar^2}{2\mu}\left(\frac{\partial^2}{\partial \eta_1^2} + \frac{\partial^2}{\partial \eta_2^2}\right) + \frac{1}{2}(k+\lambda)\eta_1^2 + \frac{1}{2}(k-\lambda)\eta_2^2$$

The exact energy levels of which are given by (see Problem 10.13)

$$\left(n_1 + \frac{1}{2}\right)\hbar\sqrt{\frac{k+\lambda}{\mu}} + \left(n_2 + \frac{1}{2}\right)\hbar\sqrt{\frac{k-\lambda}{\mu}} \; ; n_1, n_2 = 0, 1, 2, \ldots$$

If we make a binomial expansion of the square root in powers of λ/k, then the first-order terms will be the same as obtained from the perturbation theory.

Solution 20.9[11] The charge density inside the nucleus is

$$\frac{Ze}{\frac{4\pi}{3} R_N^3}$$

Thus the force acting on the electron will be given by

$$\left.\begin{aligned} F(r) &= -\frac{Ze^2}{\frac{4\pi}{3} R_N^3} \frac{4\pi}{3} r^3 \frac{1}{r^2}, \quad \text{for } 0 \le r \le R_N \\ &= -\frac{Ze^2}{r^2}, \quad \text{for } r \ge R_N \end{aligned}\right\} \tag{129}$$

[11] Solutions 20.9 and 20.10 have been adapted from Ref. 7.

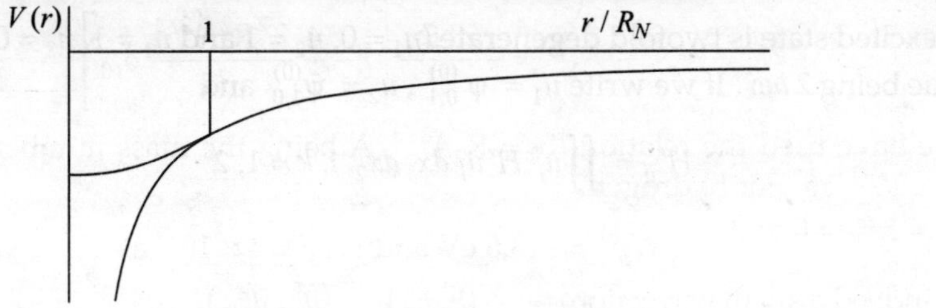

Figure 20.6: The solid curve shows the variation of $V(r)$ [as given by Eq. (130)] when the effect of the finite size of the nucleus has been taken into account. The dashed curve in the region $r < R_N$ is the extension of the $1/r$ variation.

Since $F(r) = -\partial V/\partial r$, we obtain (see Fig. 20.6)

$$\left.\begin{aligned} V(r) &= -\frac{Ze^2}{R_N}\left(\frac{3}{2}-\frac{1}{2}\frac{r^2}{R_N^2}\right), \text{ for } \quad 0 \le r \le R_N \\ &= -\frac{Ze^2}{r}, \qquad\qquad\qquad\quad \text{for } \quad r \ge R_N \end{aligned}\right\} \tag{130}$$

where the zero of the potential energy has been assumed at $r = \infty$ and use has been made of the fact that $V(r)$ is continuous at $r = R_N$. The perturbation is, therefore, given by

$$\begin{aligned} H' &= -\frac{Ze^2}{R_N}\left(\frac{3}{2}-\frac{1}{2}\frac{r^2}{R_N^2}\right)+\frac{Ze^2}{r}, \text{ for } 0 \le r \le R_N \\ &= 0, \qquad\qquad\qquad\qquad\qquad\quad \text{for } r \ge R_N \end{aligned}$$

Consequently

$$W_n^{(1)} = \int_0^{R_N} r^2\,dr \int_0^{\pi} \sin\theta\,d\theta \int_0^{2\pi} d\phi \left[\psi_n^{(0)}\right]^* H'\psi_n^{(0)}$$

Since the radius of the nucleus is usually much smaller than the Böhr radius, we may assume the wavefunction to remain constant over the region of integration; thus we may replace $\psi_n^{(0)}$ by its value at $r = 0$. Consequently,

$$W_n^{(1)} \simeq \left[R_{nl}(0)\right]^2 \int_0^{R_N} H'\,r^2\,dr \iint \left|Y_{lm}(\theta,\phi)\right|^2 \sin\theta\,d\theta\,d\phi$$

$$\simeq \frac{4}{5}\frac{Z^4}{n^3}\left(\frac{R_N}{a_0}\right)^2 E_H\,\delta_{l0}$$

where we have used the relation

$$[R_{nl}(0)]^2 = \left(\frac{Z}{a_0}\right)^3 \frac{4}{n^3}\,\delta_{l0}$$

and $E_H \simeq 13.6$ eV. A positive value of $W_n^{(1)}$ was expected because the potential energy increases due to the finite size of the nucleus (see Fig. 20.6). Thus,

$$W_n \simeq W_n^{(0)} + W_n^{(1)} = -\frac{Z^2 E_H}{n^2}\left[1 - \frac{4}{5}\frac{Z^2 R_0^2 A^{2/3}}{a_0^2 n}\delta_{l0}\right] \quad (131)$$

where we have used the relation $R_N = R_0 A^{1/3}$, A being the mass number of the nucleus and $R_0 \simeq 1.07 \times 10^{-15}$ m. For hydrogen,

$$W_n^{(0)} \simeq -13.6 \text{ eV and } W_1^{(1)} \simeq 4 \times 10^{-9} \text{ eV} \quad (132)$$

which is indeed a small correction.

Solution 20.10 For orbits inside the nucleus, we may write [see Eq. (130)]

$$H_0 = -\frac{\hbar^2}{2m_\mu}\nabla^2 - \frac{Ze^2}{R_N}\left(\frac{3}{2} - \frac{1}{2}\frac{r^2}{R_N^2}\right)$$

and

$$H' = 0 \qquad \text{for } 0 \le r \le R_N$$

$$= \frac{Ze^2}{R_N}\left(\frac{3}{2} - \frac{1}{2}\cdot\frac{r^2}{R_N^2}\right) - \frac{Ze^2}{r} \qquad \text{for } r \ge R_N$$

where m_μ represents the mass of the muon. Now, H_0 is the same as one obtains in the three-dimensional harmonic oscillator problem (see Solution 10.12). The radial part of the Schrödinger equation is of the form

$$\frac{1}{r^2}\frac{d}{dr}\left(r^2\frac{dR}{dr}\right) + \frac{2m_\mu}{\hbar^2}\left[E + \frac{3}{2}\frac{Ze^2}{R_N} - \frac{1}{2}\frac{Ze^2}{R_N^3}r^2 - \frac{l(l+1)\hbar^2}{2m_\mu r^2}\right]R(r) = 0$$

The energy levels are given by

$$E = -\frac{3}{2}\frac{Ze^2}{R_N} + \left(2p + l + \frac{3}{2}\right)\hbar\omega \quad (133)$$

where $l = 0, 1, 2, \ldots ; p = 0, 1, 2, \ldots$

and

$$\omega = \left[\frac{Ze^2}{m_\mu R_N^3}\right]^{1/2} = \left[\frac{Z\alpha\hbar c}{m_\mu R_N^3}\right]^{1/2} \quad (134)$$

Further

$$W^{(1)} = Ze^2\int_{R_N}^{\infty}|R_{pl}|^2\left[\frac{1}{R_N}\left(\frac{3}{2} - \frac{r^2}{2R_N^2}\right) - \frac{1}{r}\right]r^2\,dr \quad (135)$$

where the wavefunctions R_{pl} can be expressed in terms of confluent hypergeometric functions (see Solution 10.12 and Appendix F).

As an example[12], if we consider the $2P \to 1S$ transition in lead (Z = 82) and assume the nucleus to be a point charge, the energy liberated would be

$$\frac{m_\mu}{m_e}Z^2 E_H\left[\frac{1}{1^2} - \frac{1}{2^2}\right] = 207\, Z^2 E_H \frac{3}{4} \simeq 14 \text{ MeV}$$

If we take into account the effect due to spin and due to relativistic effects, the above value increases by 2 MeV to give 16 MeV; however, the experiments show that a photon of energy

[12] Data taken from Ref. 7. A nice discussion on muonic atoms has been given by the authors.

~ 6 MeV is emitted. If we calculate the 1S energy level from Eq. (133) and the 2P energy level from the expression

$$\frac{m_\mu Z^2 E_H}{m_e n^2}$$

then the energy difference between the 2P and 1S states comes out to be 3.6 MeV. If we now use Eq. (135) to incorporate the correction for the 1S energy state, one obtains a value of 5 MeV which is close to the experimental value.

Solution 20.15[13] The perturbation H' is given by

$$H' = \frac{e^2}{R} + \frac{e^2}{r_{12}} - \frac{e^2}{r_{1B}} - \frac{e^2}{r_{2A}} \tag{136}$$

The terms e^2/R, e^2/r_{12}, $-e^2/r_{1B}$ and $-e^2/r_{2A}$ represent the interaction between the two protons, the two electrons, the first electron with the nucleus at B and the second electron with the nucleus at A respectively. We expand H' in powers of $1/R$:

$$H' = \frac{e^2}{R}\left[1 + R\left\{\frac{1}{(x_2-x_1)^2+(y_2-y_1)^2+(z_2+R-z_1)^2}\right\}^{\frac{1}{2}} - R\left\{\frac{1}{x_1^2+y_1^2+(R-z_1)^2}\right\}^{\frac{1}{2}} - R\left\{\frac{1}{x_2^2+y_2^2+(z_2+R)^2}\right\}^{\frac{1}{2}}\right]$$

$$= \frac{e^2}{R}\left[1 + \left\{1 + \frac{2(z_2-z_1)}{R} + \frac{(x_2-x_1)^2+(y_2-y_1)^2+(z_2-z_1)^2}{R^2}\right\}^{-\frac{1}{2}} - \left\{1 - \frac{2z_1}{R} + \frac{x_1^2+y_1^2+z_1^2}{R^2}\right\}^{-\frac{1}{2}} - \left\{1 + \frac{2z_2}{R} + \frac{x_2^2+y_2^2+z_2^2}{R^2}\right\}^{-\frac{1}{2}}\right]$$

$$\approx \frac{e^2}{R^3}(x_1 x_2 + y_1 y_2 - 2z_1 z_2)$$

Let $\psi_k^{(0)}$ represent the wavefunctions of the unperturbed Hamiltonian and let $|k\rangle$ represent the corresponding eigenket; the ground state is assumed to correspond to $k = 1$. Thus,

$$\psi_1^{(0)} = u_{100}(\mathbf{r}_1)\, u_{100}(\mathbf{r}_2) \tag{137}$$

where u_{nlm} represent the hydrogen atom wavefunctions. Obviously,

$$W_1^{(1)} = \langle 1|H'|1\rangle = \iint d\tau_1\, d\tau_2 \left[\psi_1^{(0)}\right]^* H' \psi_1^{(0)} = 0 \tag{138}$$

because integrals like

$$\int d\tau_1 x_1 \left[u_{100}(\mathbf{r}_1)\right]^2$$

[13] This solution is adapted from Ref. 9 with kind permission of McGraw-Hill Book Co., New York.

would vanish. Now,

$$W_1^{(2)} = \sum_k{}' \frac{|\langle 1|H'|k\rangle|^2}{E_1 - E_k} > \frac{1}{E_1 - E_p} \sum_k{}' |\langle 1|H'|k\rangle|^2 \tag{139}$$

where the prime on Σ indicates that the term $k = 1$ has been omitted from the summation.[14] Further the pth state ($p \neq 1$) corresponds to the lowest state for which $\langle 1|H'|p\rangle$ does not vanish and the inequality follows from the fact that $E_k \geq E_p$. Now

$$\begin{aligned}\sum_k{}' |\langle 1|H'|k\rangle|^2 &= \sum_k |\langle 1|H'|k\rangle|^2 - |\langle 1|H'|1\rangle|^2 \\ &= \sum_k \langle 1|H'|k\rangle, \langle k|H'|1\rangle \\ &= \langle 1|H'^2|1\rangle\end{aligned} \tag{140}$$

where we have used Eq. (139) and the relation

$$\sum_k |k\rangle\langle k| = 1$$

[see Eq. (40) of Chapter 11]. Now,

$$\begin{aligned}\langle 1|H'^2|1\rangle &= \frac{e^4}{R^6}\langle 1|\, x_1^2x_2^2 + y_1^2y_2^2 + 4z_1^2z_2^2 + 2x_1x_2y_1y_2 \\ &\quad - 4x_1x_2z_1z_2 - 4y_1y_2z_1z_2\,|1\rangle \\ &= \frac{e^4}{R^6}\, 6\left\langle x_1^2\right\rangle^2 = \frac{6e^2a_0^4}{R^6}\end{aligned} \tag{141}$$

Further,

$$E_1 = -\frac{me^4}{2\hbar^2} - \frac{me^4}{2\hbar^2} = -\frac{e^2}{a_0};$$

and the lowest state for which $\langle 1|H'|k\rangle \neq 0$ corresponds to $n_1 = n_2 = 2$. Thus,

$$E_p = -\frac{me^4}{8\hbar^2} - \frac{me^4}{8\hbar^2} = -\frac{e^2}{4a_0}$$

and

$$E_1 - E_p = -\frac{3}{4}\cdot\frac{e^2}{a_0}$$

Hence,

$$W_1^{(2)} > -\frac{1}{\frac{3}{4}\frac{e^2}{a_0}}\,\frac{6e^4a_0^4}{R^6} = -\frac{8e^2a_0^5}{R^6}$$

or

$$W_1 \approx W_1^{(0)} + W_1^{(2)} > -\frac{e^2}{a_0}\left[1 + \frac{8a_0^6}{R^6}\right] \tag{142}$$

[14] The Σ atually represents summation over discrete states and integration over continuum.

REFERENCES AND SUGGESTED READINGS

1. R. Bellman, *Perturbation Techniques in Mathematics, Physics and Engineering*, Dover Publishing, New York (1966).
2. A.H. Nayfeh, *Introduction to Perturbation Techniques*, John Wiley & Sons, New York (1981).
3. H.W. White, *Introduction to Atomic Spectra*, McGraw-Hill Book Co., New York (1934).
4. A. Dalgarno, 'Stationary Perturbation Theory', in *Quantum Theory: 1 Elements* (Ed: D.R. Bates), Academic Press, New York (1961).
5. L. Pauling and E.B. Wilson, *Introduction to Quantum Mechanics*, McGraw-Hill Book Co., New York (1935).
6. M. Karplus and R.N. Porter, *Atoms and Molecules*, W.A. Benjamin, New York (1970).
7. A.A. Sokolov, Y.M. Loskutov and I.M. Temov, *Quantum Mechanics*, Holt, Rinehart and Winston, New York (1966).
8. F.K. Ritchmeyer, E.H. Kennard and T. Lauritsen, *Introduction to Modern Physics*, McGraw-Hill Book Co., New York (1955).
9. L.I. Schiff, *Quantum Mechanics*, McGraw-Hill Book Co., New York (1968).

21 The Hyperfine Structure of the Hydrogen Atom

21.1 INTRODUCTION

In Section 20.5, we had discussed the fine structure of the hydrogen atom which is due to effects of special theory of relativity and the spin-orbit interaction. On the other hand, the hyperfine splitting is due to the interaction between the magnetic moments of the electron and the proton. In the next section we will discuss the hyperfine splitting of the ground state of the hydrogen atom which leads to the famous "21-cm" line which is extensively used in radio astronomy. In Section 21.3, we will consider the effect of a static magnetic field which would lead to further splitting of energy levels.

21.2 THE INTERACTION TERM AND THE HYPERFINE SPLITTING

In the hydrogen atom, the magnetic field due to the proton magnetic moment interacts with the electron magnetic moment to give rise to the following interaction energy (for $l = 0$ states)

$$\hat{H}' \approx \frac{4\varepsilon}{n^3 \hbar^2} \hat{\mathbf{s}}_1 \cdot \hat{\mathbf{s}}_2 \tag{1}$$

where n represents the total quantum number and $\hat{\mathbf{s}}_1$ and $\hat{\mathbf{s}}_2$ are the spin angular momentum operators corresponding to the electron and proton respectively. Further (see Ref. 1) :

$$\varepsilon = \frac{\mu_0}{4\pi} \frac{8}{3a_0^3} \mu_p \, \mu_e \tag{2}$$

where $$\mu_0 = 4\pi \times 10^{-7} \;\; \mathrm{H\,m^{-1}}$$

is the magnetic permeability of free space;

$$a_0 = \frac{4\pi\varepsilon_0 \hbar}{m_e q^2} \approx 0.5292 \times 10^{-10} \;\; \mathrm{m} \tag{3}$$

is the Böhr radius,

$$\mu_p = \frac{g_p}{2} \cdot \frac{q\hbar}{2m_p} \approx 1.411 \times 10^{-26} \;\; \mathrm{JT^{-1}} \quad \text{and} \quad \mu_e = \frac{g_e}{2} \cdot \frac{q\hbar}{2m_e} \approx 9.285 \times 10^{-24} \;\; \mathrm{JT^{-1}} \tag{4}$$

are the magnetic moments of the proton and electron with

$$g_p \approx 5.586\,,\ \ g_e \approx 2.002 \tag{5}$$

and other symbols have their used meanings. Substituting the above values, we get

$$\varepsilon \approx 2.360 \times 10^{-25}\ \text{J} \tag{6}$$

In order to calculate the splitting of the ground state of the hydrogen atom (which corresponds to $n = 1$), we introduce the total angular momentum operator $\hat{\mathbf{S}}$

$$\hat{\mathbf{S}} = \hat{\mathbf{s}}_1 + \hat{\mathbf{s}}_2 \tag{7}$$

which actually is a short form of the equations

$$\hat{S}_x \equiv \hat{S}_{1x} + \hat{S}_{2x}\ ;\ \ \hat{S}_y \equiv \hat{S}_{1y} + \hat{S}_{2y}\quad \text{and}\quad \hat{S}_z \equiv \hat{S}_{1z} + \hat{S}_{2z} \tag{8}$$

Thus, $$\hat{S}^2 \equiv \hat{s}_x^2 + \hat{s}_y^2 + \hat{s}_z^2 = \hat{s}_1^2 + \hat{s}_2^2 + 2\hat{\mathbf{s}}_1 \cdot \hat{\mathbf{s}}_2 \tag{9}$$

Thus, $$\hat{H}' = \frac{2\varepsilon}{\hbar^2}\left[\hat{S}^2 - \hat{s}_1^2 - \hat{s}_2^2\right] \tag{10}$$

Considering the spin states of the electron and proton, the ground state ($n = 1, l = 0$) is 4 fold degenerate. We choose (see Sections 19.2 and 19.6)

$$|\,1\,\rangle = |\phi(1, 1)\rangle = |\hat{\mathbf{z}}\uparrow, \hat{\mathbf{z}}\uparrow\rangle \tag{11}$$

$$|\,2\,\rangle = |\phi(1, 0)\rangle = \frac{1}{\sqrt{2}}\left[|\hat{\mathbf{z}}\uparrow, \hat{\mathbf{z}}\downarrow\rangle + |\hat{\mathbf{z}}\downarrow, \hat{\mathbf{z}}\uparrow\rangle\right] \tag{12}$$

$$|\,3\,\rangle = |\phi(1, -1)\rangle = |\hat{\mathbf{z}}\downarrow, \hat{\mathbf{z}}\downarrow\rangle \tag{13}$$

$$|\,4\,\rangle = |\phi(0, 0)\rangle = \frac{1}{\sqrt{2}}\left[|\hat{\mathbf{z}}\uparrow, \hat{\mathbf{z}}\downarrow\rangle - |\hat{\mathbf{z}}\downarrow, \hat{\mathbf{z}}\uparrow\rangle\right] \tag{14}$$

as our base kets. On the extreme right hand side of the above equations, inside the kets the first element represents the spin state of the electron and the second element represents the spin state of the proton; thus the ket $|\hat{\mathbf{z}}\downarrow, \hat{\mathbf{z}}\uparrow\rangle$ corresponds to the electron being in the *z*-down state $|\hat{\mathbf{z}}\downarrow\rangle$, and the proton in the *z*-up state $|\hat{\mathbf{z}}\uparrow\rangle$. The kets described by the 4 equations given above are chosen as base kets because they are eigenkets of H'; indeed

$$\hat{H}'\,|1\rangle = \hat{H}'\,|\phi(1, 1)\rangle = \frac{2\varepsilon}{\hbar^2}\left[\hat{S}^2 - \hat{s}_1^2 - \hat{s}_2^2\right]\left|s_1 = \frac{1}{2}, s_2 = \frac{1}{2}, S = 1, S_z = 1\right\rangle$$

or, $$\hat{H}'\,|1\rangle = \frac{2\varepsilon}{\hbar^2}\left(2\hbar^2 - \frac{3}{4}\hbar^2 - \frac{3}{4}\hbar^2\right)|\,1\,\rangle = \varepsilon\,|\,1\,\rangle \tag{15}$$

Similarly,

$$\hat{H}'\,|2\rangle = \varepsilon\,|\,2\,\rangle \text{ and } \hat{H}'\,|3\rangle = \varepsilon\,|3\rangle \tag{16}$$

Thus, $$\hat{H}'_{11} = \langle 1\,|\,\hat{H}'\,|\,1\rangle = \varepsilon = \hat{H}'_{22} = \hat{H}'_{33} \tag{17}$$

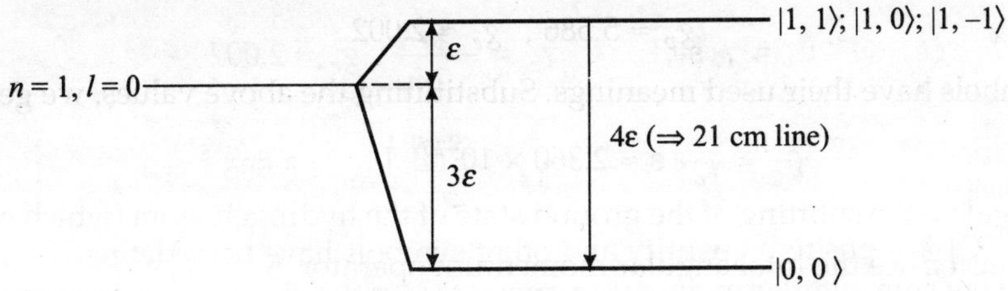

Fig. 21.1

and all off diagonal terms like $\hat{H}'_{12}$, $\hat{H}'_{23}$, $\hat{H}'_{31}, \ldots$ will be zero. On the other hand,

$$\hat{H}'|4\rangle = \frac{2\varepsilon}{\hbar^2}\left(\hat{S}^2 - \hat{s}_1^2 - \hat{s}_2^2\right)\left|s_1 = \frac{1}{2}, s_2 = \frac{1}{2}, S = 0, S_z = 0\right\rangle$$

$$= \frac{2\varepsilon}{\hbar^2}\left(0 - \frac{3}{4}\hbar^2 - \frac{3}{4}\hbar^2\right)|4\rangle = -3\varepsilon|4\rangle$$

giving

$$\hat{H}'_{44} = \langle 4|\hat{H}'|4\rangle = -3\varepsilon \tag{18}$$

Since all off diagonal terms $\hat{H}'_{ij}\ (i \neq j)$ are zero, we have a representation in which H' is diagonal. The splitting, known as the hyperfine splitting, is shown in Fig. 21.1 and leads to the emission of a photon of energy 4ε; the corresponding wavelength is given by

$$\lambda = \frac{hc}{4\varepsilon} \approx 0.2104 \text{ m} \tag{19}$$

or, $\lambda \approx 21$ cm

This is the famous 21 cm line which is extensively used by radio astronomers to measure the presence of hydrogen in the galaxy; also by measuring the Doppler shifts one can determine the velocity of the hydrogen gas. The measured value of the frequency is [Ref. 4—as quoted in Ref. 3]:

$$\nu = 1420405751.768 \pm 0.001 \text{ Hz} \tag{20}$$

which is one of the most accurately measured quantity.

21.3 HYPERFINE SPLITTING IN PRESENCE OF A STATIC MAGNETIC FIELD

In continuation of the hyperfine splitting discussed above, we consider the presence of a static magnetic field in the z-direction

$$\mathbf{B} = B_0\,\hat{\mathbf{z}} \tag{21}$$

The perturbation is now given by

$$\hat{H}_0 = \hat{H}' + \hat{H}'' \tag{22}$$

where $\hat{H}'$ is given by Eq. (1) and

$$\hat{H}'' = -\hat{\mu}_1 \cdot \mathbf{B} - \hat{\mu}_2 \cdot \mathbf{B} \tag{23}$$

$$\hat{\mu}_1 = \gamma_e \hat{\mathbf{s}}_1; \qquad \gamma_e = -\frac{g_e q}{2m_e}; \; g_e \approx 2.002 \tag{24}$$

$$\hat{\mu}_2 = \gamma_p \hat{\mathbf{s}}_2; \qquad \gamma_p = \frac{g_p q}{2m_p}; \; g_p \approx 5.586 \tag{25}$$

$q\left(\approx 1.6 \times 10^{-19}\text{ C}\right)$ is a positive quantity and other symbols have been defined earlier. Further, $\hat{\mathbf{s}}_1$ and $\hat{s}_2$ are the spin angular momentum operators for the electron and proton respectively.

Now, substituting for **B** from Eq. (21) in Eq. (23), we get

$$\hat{H}'' = \frac{2\mu_e B_0}{\hbar}\,\hat{s}_{1z} - \frac{2\mu_p B_0}{\hbar}\,\hat{s}_{2z} \tag{26}$$

where μ_e and μ_p have been defined in Eq. (4). Now,

$$\hat{s}_{1z}\,|1\rangle = \hat{s}_{1z}\,|\hat{\mathbf{z}}\uparrow, \hat{\mathbf{z}}\uparrow\rangle = \frac{1}{2}\hbar\,|1\rangle \tag{27}$$

$$\hat{s}_{1z}\,|3\rangle = \hat{s}_{1z}\,|\hat{\mathbf{z}}\downarrow, \hat{\mathbf{z}}\downarrow\rangle = -\frac{1}{2}\hbar\,|3\rangle \tag{28}$$

$$\hat{s}_{1z}\,|2\rangle = \hat{s}_{1z}\left[\frac{1}{\sqrt{2}}\left(|\hat{\mathbf{z}}\uparrow, \hat{\mathbf{z}}\downarrow\rangle + |\hat{\mathbf{z}}\downarrow, \hat{\mathbf{z}}\uparrow\rangle\right)\right]$$

$$= \frac{1}{2}\hbar.\frac{1}{\sqrt{2}}\left(|\hat{\mathbf{z}}\uparrow, \hat{\mathbf{z}}\downarrow\rangle - |\hat{\mathbf{z}}\downarrow, \hat{\mathbf{z}}\uparrow\rangle\right) = \frac{1}{2}\hbar|4\rangle \tag{29}$$

$$\hat{s}_{1z}\,|4\rangle = \frac{1}{2}\hbar\left[\frac{1}{\sqrt{2}}\left(|\hat{\mathbf{z}}\uparrow, \hat{\mathbf{z}}\downarrow\rangle + |\hat{\mathbf{z}}\downarrow, \hat{\mathbf{z}}\uparrow\rangle\right)\right] = \frac{1}{2}\hbar|2\rangle \tag{30}$$

Similarly,

$$\hat{s}_{2z}\,|1\rangle = \frac{1}{2}\hbar\,|1\rangle \tag{31}$$

$$\hat{s}_{2z}\,|3\rangle = -\frac{1}{2}\hbar\,|3\rangle \tag{32}$$

$$\hat{s}_{2z}\,|2\rangle = -\frac{1}{2}\hbar\,|4\rangle \tag{33}$$

$$\hat{s}_{2z}\,|4\rangle = -\frac{1}{2}\hbar\,|2\rangle \tag{34}$$

Since, $\hat{H}'\,|1\rangle = \varepsilon|1\rangle$; $\hat{H}'\,|2\rangle = \varepsilon|2\rangle$; $\hat{H}'\,|3\rangle = \varepsilon|3\rangle$ and $\hat{H}'\,|4\rangle = -3\varepsilon|4\rangle$

we get

$$\hat{H}_0\,|1\rangle = \hat{H}'|1\rangle + \hat{H}''|1\rangle = \left(\varepsilon + \mu_e B_0 - \mu_p B_0\right)|1\rangle = \varepsilon_1|1\rangle \tag{35}$$

$$\hat{H}_0\,|3\rangle = \left(\varepsilon - \mu_e B_0 + \mu_p B_0\right)|3\rangle = \varepsilon_2|3\rangle \tag{36}$$

$$\hat{H}_0\,|2\rangle = \varepsilon|2\rangle + \left(\mu_e B_0 + \mu_p B_0\right)|4\rangle \tag{37}$$

$$\hat{H}_0\,|4\rangle = -3\varepsilon|4\rangle + \left(\mu_e B_0 + \mu_p B_0\right)|2\rangle \tag{38}$$

where

$$\varepsilon_1 = \varepsilon + \left(\mu_e - \mu_p\right)B_0 = \varepsilon\left[1 + (\alpha - \beta)\right] \tag{39}$$

$$\varepsilon_2 = \varepsilon - (\mu_e - \mu_p)B_0 = \varepsilon[1 - (\alpha - \beta)] \tag{40}$$

$$\alpha = \frac{\mu_e B_0}{\varepsilon} \text{ and } \beta = \frac{\mu_p B_0}{\varepsilon} \tag{41}$$

Since $\mu_p \ll \mu_e$, $\beta \ll \alpha$. Thus,

$$(\hat{H}_0)_{11} = \varepsilon_1 \; ; \; (\hat{H}_0)_{33} = \varepsilon_2 \tag{42}$$

$$(\hat{H}_0)_{22} = \varepsilon \; ; \; (\hat{H}_0)_{42} = (\mu_e + \mu_p)\, B_0 = (\hat{H}_0)_{24} \tag{43}$$

$$(\hat{H}_0)_{44} = -3\varepsilon \tag{44}$$

All other matrix elements [like $(\hat{H}_0)_{12}$, $(\hat{H}_0)_{33}$, ... etc.] are zero. Thus the secular determinant is

$$\begin{vmatrix} \varepsilon_1 - W & 0 & 0 & 0 \\ 0 & \varepsilon - W & 0 & (\mu_e + \mu_p)B_0 \\ 0 & 0 & \varepsilon_2 - W & 0 \\ 0 & (\mu_e + \mu_p)B_0 & 0 & -3\varepsilon - W \end{vmatrix} = 0 \tag{45}$$

The 2 (out of the 4) roots are

$$W = W_1 = \varepsilon_1 = \varepsilon[1 + (\alpha - \beta)] \tag{46}$$

$$W = W_2 = \varepsilon_2 = \varepsilon[1 - (\alpha - \beta)] \tag{47}$$

The corresponding eigen kets are $|1\rangle$ and $|3\rangle$. Further, the equation

$$\begin{vmatrix} \varepsilon - W & (\mu_e + \mu_p)B_0 \\ (\mu_e + \mu_p)B_0 & -3\varepsilon - W \end{vmatrix} = 0 \tag{48}$$

will give us $$W^2 + 2\varepsilon W - \left[3\varepsilon^2 + (\mu_e + \mu_p)^2 B_0^2\right] = 0$$

The solutions will be

$$W = W_{3,4} = -\varepsilon \pm 2\varepsilon \sqrt{1 + \tfrac{1}{4}(\alpha + \beta)^2} \approx -\varepsilon \pm 2\varepsilon \left[1 + \tfrac{1}{8}(\alpha + \beta)^2\right] \tag{49}$$

Thus, $$W_3 \approx +\varepsilon\left[1 + \tfrac{1}{4}(\alpha + \beta)^2\right] \tag{50}$$

and $$W_4 \approx -3\varepsilon\left[1 + \tfrac{1}{12}(\alpha + \beta)^2\right] \tag{51}$$

where the binomial expansion assumes $\alpha, \beta \ll 1$ which would be true for weak magnetic fields; the corresponding eigen kets are linear combinations of $|2\rangle$ and $|4\rangle$. The splitting is shown in Fig. 21.2.

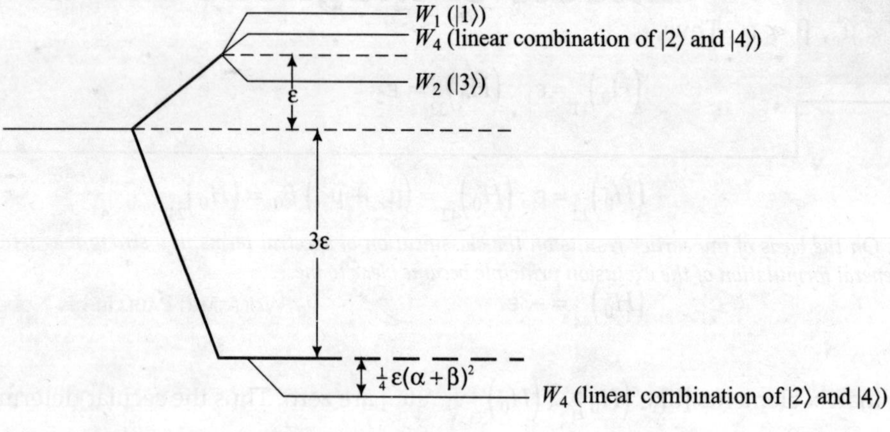

Fig 21.2

REFERENCES AND SUGGESTED READINGS

1. G. Baym, Lectures on Quantum Mechanics, W.A. Benjamin, Inc., New York (1969).
2. R.P. Feynman, R.B. Leighton and M. Sands, *The Feynman Lectures on Physics*, Vol. III, Addison Wesley Publishing Co., Reading, Mass. (1965).
3. J.S. Townsend, *A Modern Approach to Quantum Mechanics*, McGraw-Hill, New York (1992).
4. SB Crampton, D. Kleppner and NF Ramsey, *Hyperfine Separation of Ground-State Atomic Hydrogen*, Phys. Rev. Lett. 11, 338–340 (1963).

22 Effects of Magnetic Field

... On the basis of my earlier results on the classification of spectral terms in a strong magnetic field, the general formulation of the exclusion principle became clear to me.

— WOLFGANG PAULI in his Nobel lecture.

22.1 INTRODUCTION

Studies on the interaction of magnetic field with particles are of immense importance in many diverse areas. For example, in spectroscopy, the splitting of lines in a magnetic field (which leads to Zeeman and Paschen-Back effects, see Problems 22.1–22.4) is a subject of considerable importance (see, e.g. Ref. 1). In solid state physics, the calculation of diamagnetic and paramagnetic susceptibilities (see Problem 22.11) and the experiments on electron spin resonance and nuclear magnetic resonance (see Chapter 14), have immense practical importance (see, e.g. Refs 2–5).

In this chapter, we will discuss the interaction of magnetic field with particles. We will first develop the Hamiltonian of a charged particle in an electromagnetic field (Section 22.2) and will then discuss the interaction of the magnetic field with the orbital angular momentum and spin angular momentum of the electron (Sections 22.3 and 22.4). For example, we will show that the possible values of the magnetic moment component of the electron (because of its orbital motion) along the direction of the magnetic field are

$$-m_l \mu_B \tag{1}$$

where[1]

$$\mu_B \equiv \frac{q\hbar}{2m} = 9.274 \times 10^{-24} \text{ Joule/tesla}$$
$$= 9.274 \times 10^{-28} \text{ Joule/gauss} \tag{2}$$
$$(1 \text{ tesla} = 10^4 \text{ gauss})$$

is known as the Böhr magneton, m_l is the magnetic quantum number; the charge of the electron has been assumed to be $-q (q \simeq 1.602 \times 10^{-19}\text{C})$. Thus, for a d-electron ($l = 2$) the possible values of orbital magnetic moment along the direction of the magnetic field are

$$-2\mu_B, -\mu_B, 0, +\mu_B, +2\mu_B$$

corresponding to $m_l = +2, +1, 0, -1$ and -2 respectively.

[1] In Chapter 10, the magnetic quantum number is denoted by m, here it is denoted by m_l to avoid confusion with the electron mass m.

On the other hand, the magnetic moment components due to the spin angular momentum of the electron, along the direction of the magnetic field are (see Sections 14.3 and 14.4).

$$\pm\frac{1}{2}g\mu_B \simeq \pm\mu_B \tag{3}$$

where g is known as the *Lande's g factor*[2] ($\simeq$ 2.0023). The corresponding proton magnetic moment is

$$\pm\frac{1}{2}g_p\mu_p \tag{4}$$

where $g_p \simeq 5.56$, and

$$\mu_p = \frac{q\hbar}{2M} \approx 5.051 \times 10^{-31} \text{ Joule/gauss.}$$

In Section 22.5, we will discuss the spin-orbit interaction.

22.2 THE HAMILTONIAN

Scalar and Vector Potentials

We first define the vector and scalar potentials. From the Maxwell's equation

$$\nabla \cdot \mathbf{B} = 0 \tag{5}$$

it immediately follows that **B** can be expressed as the curl of a vector

$$\mathbf{B} = \nabla \times \mathbf{A} \tag{6}$$

where **A** is known as the vector potential. Further, if we substitute this in the equation

$$\nabla \times \mathbf{E} = -\frac{\partial \mathbf{B}}{\partial t} \tag{7}$$

we would get

$$\nabla \times \left(\mathbf{E} + \frac{\partial \mathbf{A}}{\partial t}\right) = 0 \tag{8}$$

where **E** represents the electric field. Thus we may set

$$\mathbf{E} = -\frac{\partial \mathbf{A}}{\partial t} - \nabla\phi \tag{9}$$

where ϕ is known as the scalar potential.

---

[2] The relativistic theory of Dirac predicts g = 2 [see Chapter 30]; however, relativistic quantum electrodynamics shows that

$$g = 2 + \frac{\alpha}{2\pi} + O(\alpha^2)$$

where $\alpha\left(\frac{q^2}{4\pi\varepsilon_0\hbar c} \approx \frac{1}{137}\right)$ is the fine structure constant. For the proton, g is very much different from 2 and hence it is not really a pure Dirac particle. It may be mentioned that neutral particles like neutrons; Λ, Σ^0 (which have spin $\frac{1}{2}$) have finite magnetic moments; e.g. for the neutron $\mu_n = -3.83\ (q/2m_n)s$, where m_n is the neutron mass. Physically this arises from internal current distributions.

Particle in an Electromagnetic Field

We next consider the motion of a charged particle (of charge Q) in an electromagnetic field. The force acting on the charged particle is given by

$$\mathbf{F} = Q[\mathbf{E} + \mathbf{v} \times \mathbf{B}] \tag{10}$$

or

$$m\frac{d\mathbf{v}}{dt} = \mathbf{F} = Q\left[-\frac{\partial \mathbf{A}}{dt} - \nabla\phi + \mathbf{v} \times (\nabla \times \mathbf{A})\right]$$

Now,

$$(\mathbf{v} \times \nabla \times \mathbf{A})_x = v_y\left(\frac{\partial A_y}{\partial x} - \frac{\partial A_x}{\partial y}\right) - v_z\left(\frac{\partial A_x}{\partial z} - \frac{\partial A_z}{\partial x}\right)$$

$$= \frac{\partial}{\partial x}(\mathbf{v}\cdot\mathbf{A}) - \frac{dA_x}{dt} + \frac{\partial A_x}{\partial t}$$

where use has been made of the following relation

$$\frac{dA_x}{dt} = \frac{\partial A_x}{\partial t} + \left(v_x\frac{\partial A_x}{\partial x} + v_y\frac{\partial A_x}{\partial y} + v_z\frac{\partial A_x}{\partial z}\right)$$

Thus,

$$\frac{d}{dt}[m\mathbf{v} + Q\mathbf{A}] = \nabla(-Q\phi + Q(\mathbf{v}\cdot\mathbf{A})) \tag{11}$$

The above equation can be written in the form

$$\frac{d}{dt}\left(\frac{\partial L}{\partial \dot{x}}\right) = \frac{\partial L}{\partial x} \tag{12}$$

(and similar equations with x replaced by y and z) where

$$L = -m\mathbf{v}\cdot\mathbf{v} - Q\phi + Q\mathbf{A}\cdot\mathbf{v} \tag{13}$$

Equation (12) is the Lagrangian equation of motion (see, e.g. Ref. 6) and L denotes the Lagrangian. The conjugate momenta are given by

$$p_x = \frac{\partial L}{\partial x} = mv_x + QA_x$$

or

$$\mathbf{p} = m\mathbf{v} + Q\mathbf{A} \tag{14}$$

The Hamiltonian is given by

$$H = \mathbf{p}\cdot\mathbf{v} - L$$

$$= \mathbf{p}\cdot\left[\frac{1}{m}(\mathbf{p} - Q\mathbf{A})\right] - \frac{1}{2}m\left[\frac{1}{m}(\mathbf{p} - Q\mathbf{A})\right]^2 + Q\phi - Q\,\mathbf{A}\cdot\frac{1}{m}(\mathbf{p} - Q\mathbf{A})$$

$$= \frac{1}{2m}[\mathbf{p} - Q\mathbf{A}]^2 + Q\phi \tag{15}$$

Equation (15) represents the Hamiltonian describing the non-relativistic motion of a charged particle in an electromagnetic field.

In order to write the Schrödinger equation, we replace the operator **p** by $-i\hbar\nabla$ to obtain

$$i\hbar \frac{\partial \Psi}{\partial t} = H\Psi \tag{16}$$

where

$$H = \frac{1}{2m}[-i\hbar\nabla - Q\mathbf{A}]^2 + Q\phi \tag{17}$$

22.3 THE SCHRÖDINGER EQUATION CORRESPONDING TO A STATIC ELECTROMAGNETIC FIELD

For a static electromagnetic field the wavefunction can be written in the form

$$\Psi(\mathbf{r}, t) = \psi(\mathbf{r})e^{-iEt/\hbar} \tag{18}$$

where $\psi(\mathbf{r})$ satisfies the equation

$$\frac{1}{2m}[-i\hbar\nabla - Q\mathbf{A}]^2\,\psi + Q\phi\psi = E\psi$$

or

$$-\frac{\hbar^2}{2m}\nabla^2\psi + \frac{iQ\hbar}{2m}[\nabla\cdot(\mathbf{A}\psi) + \mathbf{A}\cdot\nabla\psi] + \left[\frac{Q^2}{2m}A^2 + Q\phi\right]\psi = E\psi \tag{19}$$

Now, in the Coulomb gauge[3], $\nabla\cdot\mathbf{A} = 0$ and we have

$$\nabla\cdot(\mathbf{A}\psi) = (\nabla\cdot\mathbf{A})\psi + \mathbf{A}\cdot\nabla\psi = \mathbf{A}\cdot\nabla\psi$$

which gives

$$-\frac{\hbar^2}{2m}\nabla^2\psi + \frac{iQ\hbar}{m}\mathbf{A}\cdot\nabla\psi + \left[\frac{Q^2}{2m}A^2 + Q\phi\right]\psi = E\psi \tag{20}$$

Further, for a uniform magnetic field, we can always write[4]

$$\mathbf{A} = \frac{1}{2}(\mathbf{B}\times\mathbf{r}) \tag{21}$$

Thus, the second term in Eq. (20) becomes

---

[3] From Eq. (6), we see that **B** is left unchanged if we add gradient of a scalar to **A**:

$$\mathbf{A} \to \mathbf{A}' + \nabla\chi$$

We may choose χ such that $\nabla\cdot\mathbf{A} = 0$; this is known as the Coulomb gauge (for further details, see, for example, Ref. 7):

[4]

$$\nabla\times\mathbf{A} = \tfrac{1}{2}\,\nabla\times(\mathbf{B}\times\mathbf{r}) = \tfrac{1}{2}[(\nabla\cdot\mathbf{r})\,\mathbf{B} - (\mathbf{B}\cdot\nabla)\mathbf{r}]$$

$$= \tfrac{1}{2}\,[3\mathbf{B} - \mathbf{B}] = \mathbf{B}$$

Also notice that

$$\nabla\cdot\mathbf{A} = \tfrac{1}{2}\,\nabla\cdot(\mathbf{B}\times\mathbf{r}) = \tfrac{1}{2}\,[\mathbf{r}\cdot(\nabla\times\mathbf{B}) - \mathbf{B}\cdot(\nabla\times\mathbf{r})] = 0$$

because the field is uniform.

$$\frac{iQ\hbar}{m}\mathbf{A}\cdot\nabla\psi = \frac{iQ\hbar}{2m}(\mathbf{B}\times\mathbf{r})\cdot\nabla\psi$$
$$= -\frac{Q}{2m}\mathbf{B}\cdot\left[\mathbf{r}\times(-i\hbar\nabla)\right]\psi$$
$$= -\frac{Q}{2m}\mathbf{B}\cdot(\mathbf{r}\times\mathbf{p})\psi = -\frac{Q}{2m}(\mathbf{B}\cdot\mathbf{L})\psi$$
$$= -\boldsymbol{\mu}_o\cdot\mathbf{B}\psi \tag{22}$$

where $\mathbf{L} = \mathbf{r}\times\mathbf{p}$ represents the orbital angular momentum and

$$\boldsymbol{\mu}_o = \frac{Q}{2m}\mathbf{L} \tag{23}$$

can be interpreted as the magnetic moment associated with the orbital motion of the charged particle[5] (the subscript o signifies that the magnetic moment refers to the orbital motion). This follows from the fact that the interaction energy of a magnet (with magnetic moment μ_o) in a uniform magnetic field is given by[6]

$$U = -\boldsymbol{\mu}_o\cdot\mathbf{B} \tag{24}$$

Equation (20) therefore assumes the form

$$-\frac{\hbar^2}{2m}\nabla^2\psi - \left(\frac{Q}{2m}\mathbf{L}\cdot\mathbf{B}\right)\psi + \frac{Q^2}{8m}(\mathbf{B}\times\mathbf{r})^2\psi + Q\phi\psi = E\psi \tag{25}$$

The second term has just been physically interpreted as the magnetic interaction energy associated with the circulatory motion of the charged particle; the third term can also be interpreted in terms of a potential energy function (see, e.g. Ref. 10, Section 10.6), however the considerations are rather involved. Further, since the term is proportional to the square of the magnetic field, it makes a negligible contribution unless the fields are extremely high.

22.4 THE INTERACTION OF THE MAGNETIC FIELD WITH THE SPIN ANGULAR MOMENTUM OF THE ELECTRON

In Section 14.3, we had discussed that the electron is endowed with an intrinsic spin angular momentum given by

$$\mathbf{s} = \frac{1}{2}\hbar\boldsymbol{\sigma} \tag{26}$$

where σ_x, σ_y and σ_z are the Pauli matrices. Further, the magnetic moment associated with the spin angular momentum of a particle of charge Q has the form [cf. Eq. (23)]

---

[5] Equation (23) can be physically understood by noting that a charge Q rotating (in the anticlockwise direction) in a circle of radius r with speed v constitutes a current $I = Q/(2\pi r/v)$ and therefore the magnetic moment associated with the motion is given by (see, e.g. Ref. 8, Section 34.3).

$$\boldsymbol{\mu} = \frac{Qv}{2\pi r}\mathbf{A} = \frac{Qv}{2\pi r}\pi r^2\hat{\mathbf{z}} = \frac{Q}{2m}\mathbf{r}\times\mathbf{p} = \frac{Q}{2m}\mathbf{L}$$

where we have assumed the circular motion to be in the x-y plane and the area vector $\mathbf{A}$ points in the positive z-direction for anticlockwise motion—see also Section 14.2.

[6] See, e.g. Ref. 9, p.19.

$$\boldsymbol{\mu} = g\frac{Q}{2m}\mathbf{s} \tag{27}$$

As mentioned earlier, for the electron $g \approx 2$ and $Q = -q$; on the other hand, for the proton $g \simeq 5.56$, $Q = +q$ and m has to be replaced by the proton mass.

Thus, the total magnetic moment of the electron is given by

$$\boldsymbol{\mu}_{\text{total}} = -\frac{q}{2m}[\mathbf{L} + g\,\mathbf{s}] \simeq -\frac{q}{2m}[\mathbf{L} + 2\mathbf{s}] \tag{28}$$

22.5 THE SPIN-ORBIT INTERACTION

In order to physically understand the spin-orbit interaction we note that an observer who is at rest with the electron sees the nucleus moving around him; the circulating nucleus constitutes a current and produces a magnetic field at the electron given by

$$\mathbf{B} = -(\varepsilon_0\mu_0)\,\mathbf{v} \times \mathbf{E}$$

where **E** represents the electric field produced at the electron by the nucleus and **v** the velocity of the electron. For a central potential

$$-q\mathbf{E} = -\nabla V = -\frac{1}{r}\frac{dV}{dr}\mathbf{r} \tag{29}$$

Thus,
$$\mathbf{B} = \frac{1}{qc^2}\frac{1}{r}\frac{dV}{dr}\frac{1}{m}\mathbf{L} \tag{30}$$

where $\mathbf{L} = \mathbf{r} \times \mathbf{p}$ represents the angular momentum. The interaction energy with the magnetic moment of the electron is given by

$$H_{s\text{-}o} = -\frac{1}{2}\boldsymbol{\mu}\cdot\mathbf{B} = -\frac{1}{2}\left[-g\frac{q}{2m}\mathbf{s}\right]\cdot\left[\frac{1}{qc^2}\frac{1}{r}\frac{dV}{dr}\frac{1}{m}\mathbf{L}\right]$$

$$= \frac{g}{4m^2c^2}\frac{1}{r}\frac{dV}{dr}\mathbf{L}\cdot\mathbf{s} \tag{31}$$

where an additional factor of $\frac{1}{2}$ has been introduced; this is known as the Thomas procession factor and comes from a proper relativistic treatment (Ref. 11). The subscript *s-o* on H implies spin-orbit interaction. For the Coulomb potential

$$V(r) = -\frac{Zq^2}{4\pi\varepsilon_0 r}$$

and
$$H_{s\text{-}o} = \frac{gZ\hbar\alpha}{4m^2c}\frac{1}{r^3}\mathbf{L}\cdot\mathbf{s} \tag{32}$$

where
$$\alpha = \frac{q^2}{4\pi\varepsilon_0\hbar c} \simeq \frac{1}{137}$$

represents the fine structure constant. It may be mentioned that the spin-orbit interaction comes out automatically through a proper relativistic treatment (see Section 30.8).

22.6 PROBLEMS

Problem 22.1 For a hydrogen atom placed in a weak uniform magnetic field (in the z-direction) the Hamiltonian will be of the form

$$H = H_0 + H' \tag{33}$$

where
$$H_0 = \frac{p^2}{2m} + V(r) \tag{34}$$

and
$$H' = \frac{q}{2m}\mathbf{B}\cdot\mathbf{L} = \frac{\mu_B B}{\hbar}L_z \tag{35}$$

represents the interaction with the magnetic field [see Eq. (24)]; we have neglected the effects due to spin angular momentum of the electron (see Problem 22.2). Treat H' as perturbation and show that s states ($l = 0$) are not split whereas the p-states ($l = 1$) are split into three states separated by the energy interval $\mu_B B$. This is known as the *normal Zeeman effect*.
[**Note:** In this particular problem the perturbation calculation gives the exact results.]

Problem 22.2 In continuation with the previous problem, if we take into account the effects due to spin angular momentum of the electron, the Hamiltonian will be of the form

$$H = H_0 + H' + H'' \tag{36}$$

where H_0 is given by Eq. (34), H' represents the spin-orbit interaction [see Eq. (32)]

$$H' = f(r)\,\mathbf{L}\cdot\mathbf{s},\ f(r) = \frac{gZ\hbar\alpha}{4m^2c}\cdot\frac{1}{r^3} \tag{37}$$

and
$$H'' = \frac{qB}{2m}[L_z + gs_z] = \frac{\mu_B B}{\hbar}[L_z + gs_z] \tag{38}$$

represents the interaction with the magnetic field (see Eq. (28)). Assume that the effect of H'' to be small in comparison to the effect of H', use the wavefunctions corresponding to H' as the perturbation (see Section 20.5.2) to calculate the effect of H''. This is known as the weak-field *anamolous Zeeman effect*, or just *Zeeman effect*.

Problem 22.3 Consider again the hydrogen-like atom in a magnetic field **B**. Assume the magnetic field to be so strong that the effect of the spin-orbit interaction to be small compared to the effect of the magnetic field. Using the eigenfunctions of $H_0 + H''$, calculate the perturbation due to spin-orbit interaction. This is known as the *Paschen-Back effect*.

Problem 22.4 The previous two problems correspond to weak magnetic field and strong magnetic fields respectively. For the intermediate field case, $H' + H''$ has to be treated as perturbation to H_0. Use degenerate state perturbation theory to calculate the effect of $H' + H''$.

Problem 22.5 In Example 14.1, instead of solving the time dependent Schrödinger equation, solve the Heisenberg equation of motion for s_x, s_y and s_z.

For
$$|\Psi(0)\rangle = \cos\left(\frac{1}{2}\phi\right)|\hat{\mathbf{z}}\uparrow\rangle + \sin\left(\frac{1}{2}\phi\right)|\hat{\mathbf{z}}\downarrow\rangle \tag{39}$$

Calculate $\langle s_x(t)\rangle$, $\langle s_y(t)\rangle$ and $\langle s_z(t)\rangle$ and interpret the results physically.

Problem 22.6 For a uniform magnetic field in the z-direction, we may assume

$$A_x = 0, A_y = xB, A_z = 0 \tag{40}$$

(Notice that $\nabla \times \mathbf{A} = B\hat{\mathbf{z}}$ and $\nabla \cdot \mathbf{A} = 0$). Write down the corresponding Schrödinger equation for an electron and obtain an exact solution of the same[7]. How would the eigenvalues get modified if the magnetic moment associated with the spin angular momentum is also taken into account.

Problem 22.7 In the above problem, if we assume **A** to be given by Eq. (21), show that the same values for the energy eigenvalues are obtained; the wavefunctions however differ by an unimportant phase factor.

Problem 22.8 In continuation of Problem 22.7, consider an electron in a spherically symmetric potential experiencing a uniform magnetic field in the z-direction for which **A** may be assumed to be of the form

$$A_x = -\frac{1}{2} yB, A_y = \frac{1}{2} xB, A_z = 0 \tag{41}$$

Carry out a perturbation analysis and show that if $\langle H_1 \rangle$ and $\langle H_2 \rangle$ represent the perturbations proportional to B and B^2 then whenever $\langle H_1 \rangle \neq 0$,

$$\langle H_2 \rangle \ll \langle H_1 \rangle \tag{42}$$

Problem 22.9 In the above problem if we assume **A** to be given by Eq. (40) show that the final results are the same.

Problem 22.10 The potential energy of a two-dimensional harmonic oscillator can be written as

$$V = \frac{1}{2} \mu \omega_0^2 (x^2 + y^2) \tag{43}$$

Determine the energy levels and wavefunctions of the oscillator placed in a uniform magnetic field along the z-direction.

[Hint: Assuming **A** to be given by Eq. (41), solve Eq. (20) or Eq. (25)].

22.7 SOLUTIONS

Solution 22.1 The perturbation is

$$H' = \frac{\mu_B B}{\hbar} L_z = -i\mu_B B \frac{\partial}{\partial \phi} \tag{44}$$

The eigenfunctions of H_0 are (see Section 10.4)

$$u(n, l, m_l) = R_{nl}(r)\, Y_{lm_l}(\theta, \phi) \tag{45}$$

Since

$$L_z\, Y_{lm_l} = m_l \hbar\, Y_{lm_l} \tag{46}$$

we get

$$H'\, u(n, l, m_l) = m_l \hbar\, u(n, l, m_l) \tag{47}$$

and, consequently the off-diagonal elements are zero[8]. The Zeeman splitting is therefore given by

---

[7] This solution leads to the orbital diamagnetism of free electrons (see, e.g. Ref. 2, Chapter XVI).

[8] Indeed $u(n, l, m_l)$ are the exact eigenfunctions of $H (= H_0 + H')$ corresponding to the eigenvalue $E_n + \mu_B B m_l$.

$$\Delta = \int u^*(n, l, m_l) H' u(n, l, m_l)\, d\tau = \mu_B B m_l \tag{48}$$

Thus a level characterised by the orbital angular quantum number l splits up into $(2l + 1)$ levels. However, because of the selection rule

$$\Delta m_l = 0, \pm 1 \tag{49}$$

we get three lines of frequencies

$$\omega_0 - \frac{\mu_B B}{\hbar}, \omega_0, \omega_0 + \frac{\mu_B B}{\hbar} \tag{50}$$

which is known as the *normal Zeeman effect.*

Solution 22.2 Because of the spin-orbit interaction, the levels split and each level is characterised by definite values of J^2, L^2 and S^2 (see Section 19.5); the corresponding Pauli wave functions are given by

$$\phi(n, l, j_+, m_j) = \begin{pmatrix} \left[\dfrac{l + \frac{1}{2} + m_j}{2l+1}\right]^{1/2} R_{nl}(r) Y_{l, m_j - \frac{1}{2}} \\ \left[\dfrac{l + \frac{1}{2} - m_j}{2l+1}\right]^{1/2} R_{nl}(r) Y_{l, m_j + \frac{1}{2}} \end{pmatrix}_{\text{for } j = l + \frac{1}{2}} \tag{51}$$

and
$$\phi(n, l, j_-, m_j) = \begin{pmatrix} -\left[\dfrac{l + \frac{1}{2} - m_j}{2l+1}\right]^{1/2} R_{nl}(r) Y_{l, m_j - \frac{1}{2}} \\ \left[\dfrac{l + \frac{1}{2} + m_j}{2l+1}\right]^{1/2} R_{nl}(r) Y_{l, m_j + \frac{1}{2}} \end{pmatrix}_{\text{for } j = l - \frac{1}{2}} \tag{52}$$

where
$$j_\pm = l \pm \frac{1}{2}$$

The states corresponding to $j = l + \frac{1}{2}$ and $j = l - \frac{1}{2}$ are $(2l + 2)$ and $2l$-fold degenerate. Now, the Pauli operator for H'' is

$$(H'')_P = \frac{\mu_B B}{\hbar} \begin{pmatrix} L_z + \frac{1}{2} g\hbar & 0 \\ 0 & L_z - \frac{1}{2} g\hbar \end{pmatrix} \tag{53}$$

Since,
$$L_z Y_{l, m_j - \frac{1}{2}} = -i\hbar \frac{\partial}{\partial \phi} Y_{l, m_j - \frac{1}{2}} = (m_j - \tfrac{1}{2})\, \hbar Y_{l, m_j - \frac{1}{2}}$$

we immediately have (for $j = l + \frac{1}{2}$)

$$(H'')_P \phi(j_+, m_j) = \mu_B B \begin{pmatrix} \left[\dfrac{l + \frac{1}{2} + m_j}{2l+1}\right]^{1/2} (m_j - \dfrac{1}{2} + \dfrac{1}{2} g) R_{nl}(r)\, Y_{l, m_j - \frac{1}{2}} \\ \left[\dfrac{l + \frac{1}{2} - m_j}{2l+1}\right]^{1/2} (m_j + \dfrac{1}{2} - \dfrac{1}{2} g) R_{nl}(r)\, Y_{l, m_j + \frac{1}{2}} \end{pmatrix}$$

The off-diagonal elements are obviously zero and the perturbation is thus given by

$$\Delta = \int \phi^\dagger(j_+, m_j)\,(H'')_P\,\phi\,(j_+, m_j)\,d\tau$$

$$= \mu_B B \frac{(2l+g)m_j}{(2l+1)} \quad \text{for } j = l + \frac{1}{2} \tag{54}$$

and similarly,

$$\Delta = \mu_B B \frac{(2l+2-g)m_j}{(2l+1)} \quad \text{for } j = l - \frac{1}{2} \tag{55}$$

Thus the $j = l + \frac{1}{2}$ and $j = l - \frac{1}{2}$ split into $(2l + 2)$ and $2l$ levels. Further, if we assume $g = 2$, Eqs (54) and (55) can be combined to give

$$\Delta = \mu_B B \frac{j+\frac{1}{2}}{l+\frac{1}{2}} m_j\,;\; j = l \pm \frac{1}{2} \tag{56}$$

Figure 22.1 shows the splitting of the *P*-state of an alkali atom. For the *P*-state, $l = 1$ and we may have $j = 3/2$ and $1/2$. Thus,

$$\frac{\Delta}{\mu_B B} = \frac{4}{3} m_j \quad \text{for } j = \frac{3}{2}$$

$$= \frac{2}{3} m_j \quad \text{for } j = \frac{1}{2}$$

Thus the $j = 3/2$ level will split into 4 states $(m_j = -\frac{3}{2}, -\frac{1}{2}, \frac{1}{2}, \frac{3}{2})$ and the $j = \frac{1}{2}$ level will split into 2 states $(m_j = -\frac{1}{2}, \frac{1}{2})$ as shown in Fig. 22.1. In the figure the two states corresponding to $B = 0$ represent the fine structure splitting given by Eq. (96) of Chapter 20.

Solution 22.3 It is obvious that the functions

$$|\,\psi(n, l, m_l, \tfrac{1}{2})\rangle = |\,n, l, m_l, \hat{\mathbf{z}}\uparrow\rangle \tag{57}$$

and

$$|\,\psi(n, l, m_l, -\tfrac{1}{2})\rangle = |\,n, l, m_l, \hat{\mathbf{z}}\downarrow\rangle \tag{58}$$

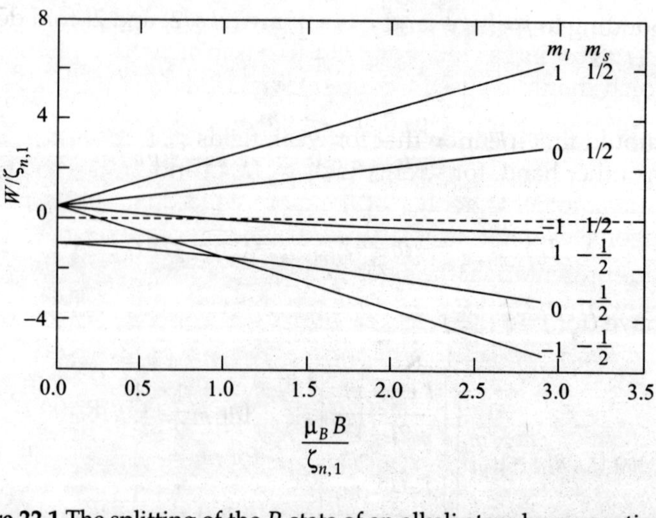

Figure 22.1 The splitting of the *P*-state of an alkali atom by magnetic field. The two states corresponding to $B = 0$ represent the fine structure splitting.

are the exact eigenfunctions of $H_0 + H''$ belonging to the eigenvalues

$$E_n + \frac{\mu_B B}{\hbar}\left[m_l\hbar + \frac{1}{2}g\hbar\right]$$

and
$$E_n + \frac{\mu_B B}{\hbar}\left[m_l\hbar - \frac{1}{2}g\hbar\right]$$

respectively. The above equations can be combined to give

$$E_n + \mu_B B(m_l + 2m_s) \tag{59}$$

where $m_s = \frac{1}{2}$ and $m_s = -\frac{1}{2}$ correspond to the spin up and spin down states and we have assumed $g = 2$. Now, for the two states given by Eqs (57) and (58), L_z and s_z are well-defined and expectation values of L_x, s_x, L_y and s_y are zero (see Problem 13.1). Thus, the perturbation is given by

$$\langle f(r)\mathbf{L}\cdot\mathbf{s}\rangle = \langle f(r)\rangle\,\langle L_z s_z\rangle$$
$$= \zeta_{n,l}\, m_l\, m_s \tag{60}$$

where
$$\zeta_{n,l} = \langle f(r)\rangle\hbar^2 = \frac{gZ\hbar\alpha\hbar^2}{4m^2c}\int\frac{1}{r^3}|R_{nl}(r)|^2\, r^2 dr$$

$$= \frac{Z^4\alpha^4 mc^2}{2n^3 l(l+\frac{1}{2})(l+1)} \tag{61}$$

where we have used Eq. (10) of Appendix I. Thus the energy levels are given by

$$E = E_n + \mu_B B(m_l + 2m_s) + \zeta_{n,l}\, m_l m_s \tag{62}$$

where
$$m_l = -l, \ldots, -l \text{ and } m_s = +\tfrac{1}{2} \text{ and } -\tfrac{1}{2} \tag{63}$$

Equation (62) represents the *Paschen-Back effect* and as can be seen the operators L^2, S^2, L_z and s_z have well-defined values for each state; this is referred to as by saying that S^2, L^2, L_z and s_z are good quantum numbers. The splitting for the *P*-state of an alkali atom is shown in Fig. 22.1. In the figure $B = \mu_B B/\zeta_{n,l}$. At $B = 0$, we have the fine structure splitting. For small values of B, Eq. (61) is to be used and for large values of B, Eq. (62) is to be used. Both equations are limiting forms of Eq. (71) which should be used for intermediate values of B.

Solution 22.4 We should first mention that for weak fields s^2, L^2, J^2 and J_z are good quantum numbers, and on the other hand, for strong fields s^2, L^2, L_z and s_z are good quantum numbers. However, for intermediate fields neither of them are good quantum numbers and one must calculate the off-diagonal elements as is done in degenerate state perturbation theory.

We choose as our representation, the eigenstates for the strong field case[9]; these are given by Eqs (57) and (58). Now, from Eq. (62)

$$\langle\psi(n, l, m_l, \pm\tfrac{1}{2})\,|H''|\,\psi(n, l, m_l, \pm\tfrac{1}{2})\rangle = (m \pm \tfrac{1}{2})\,\mu_B B \tag{64}$$

where[10]
$$m = m_l + m_s = \begin{cases} m_l + \frac{1}{2} & \text{for } m_s = \frac{1}{2} \\ m_l - \frac{1}{2} & \text{for } m_s = -\frac{1}{2} \end{cases}$$

[9] We could have equally well chosen the eigenstates for the weak field case.

[10] The quantity m is denoted by m_j in Problem 22.2. For convenience, we are dropping the subscript j.

The off diagonal elements like

$$\langle \psi(n, l, m_l, \pm \tfrac{1}{2}) | H'' | \psi(n, l, m_l, \mp \tfrac{1}{2}) \rangle$$

would vanish because $|\psi(n, l, m_l, \pm \tfrac{1}{2})\rangle$ are eigenstates of H''.

Next in order to calculate the matrix elements of

$$\mathbf{L}\cdot\mathbf{s} = \frac{J^2 - L^2 - s^2}{2} \tag{65}$$

we write

$$|\psi(n, l, m_l = m - \tfrac{1}{2}, +\tfrac{1}{2})\rangle = \left(\frac{l+m+\frac{1}{2}}{2l+1}\right)^{1/2} |\phi(n, l, j_+, m)\rangle - \left(\frac{l-m+\frac{1}{2}}{2l+1}\right)^{1/2} |\phi(n, l, j_-, m)\rangle \tag{66}$$

$$|\psi(n, l, m_l = m + \tfrac{1}{2}, -\tfrac{1}{2})\rangle = \left(\frac{l-m+\frac{1}{2}}{2l+1}\right)^{1/2} |\phi(n, l, j_+, m)\rangle + \left(\frac{l+m+\frac{1}{2}}{2l+1}\right)^{1/2} |\phi(n, l, j_-, m)\rangle \tag{67}$$

where $j_\pm = l \pm \frac{1}{2}$. Further, ϕ are eigenfunctions in the representation L^2, s^2, J^2 and J_z and we have used the Clebsch-Gordan coefficients given in Table 19.1. Now,

$$J^2 |\phi(n, l, j_+, m)\rangle = (l+\tfrac{1}{2})(l+\tfrac{3}{2})\hbar^2 |\phi(n, l, j_+, m)\rangle \tag{68}$$

$$J^2 |\phi(n, l, j_-, m)\rangle = (l-\tfrac{1}{2})(l+\tfrac{1}{2})\hbar^2 |\phi(n, l, j_-, m)\rangle \tag{69}$$

The operator L^2 has the eigenvalues $l(l+1)\hbar^2$ and s^2 has the eigenvalue $\frac{1}{2}(\frac{3}{2})\hbar^2 = \frac{3}{4}\hbar^2$. Thus, (omitting the subscripts n, l):

$$\langle \psi(m_l = m - \tfrac{1}{2}, \tfrac{1}{2}) \left| \frac{J^2 - L^2 - s^2}{2} \right| \psi(m_l = m - \tfrac{1}{2}, \tfrac{1}{2}) \rangle$$

$$= \left[\left(\frac{l+m+\frac{1}{2}}{2l+1}\right)^{1/2} \langle \phi(j_+, m)| - \left(\frac{l-m+\frac{1}{2}}{2l+1}\right)^{1/2} \langle \phi(j_-, m)|\right] \frac{J^2 - L^2 - s^2}{2}$$

$$\times \left[\left(\frac{l+m+\frac{1}{2}}{2l+1}\right)^{1/2} |\phi(j_+, m)\rangle - \left(\frac{l-m+\frac{1}{2}}{2l+1}\right)^{1/2} |\phi(j_-, m)\rangle\right]$$

$$= \frac{l+m+\frac{1}{2}}{2l+1}\frac{1}{2}\left\{j_+(j_+ + 1) - l(l+1) - \frac{3}{4}\right\}\hbar^2$$

$$+ \frac{l-m+\frac{1}{2}}{2l+1}\frac{1}{2}\left\{j_-(j_- + 1) - l(l+1) - \frac{3}{4}\right\}\hbar^2 = \frac{1}{2}\left(m - \frac{1}{2}\right)\hbar^2$$

Further, the only non-vanishing, non-diagonal matrix elements are

$$\left\langle \psi(m_l = m - \tfrac{1}{2}, \tfrac{1}{2}) \right| \mathbf{L}\cdot\mathbf{s} \left| \psi(m_l = m + \tfrac{1}{2}, -\tfrac{1}{2}) \right\rangle$$

$$= \left(\psi\left(m_l = m + \tfrac{1}{2}, -\tfrac{1}{2}\right) | \mathbf{L}\cdot\mathbf{s} | \psi\left(m_l = m - \tfrac{1}{2}, \tfrac{1}{2}\right)\right)$$

$$= \frac{1}{2}\left[\left(l + m + \frac{1}{2}\right)\left(l - m + \frac{1}{2}\right)\right]^{1/2} \hbar^2$$

Thus, considering $H' + H''$ as perturbation, secular determinant is (see Section 20.3):

$$\begin{vmatrix} \frac{1}{2}\left(m-\frac{1}{2}\right)\zeta_{n,l} + \left(m+\frac{1}{2}\right)\mu_B B - W & \frac{1}{2}\zeta_{n,l}\left[\left(l+m+\frac{1}{2}\right)\left(l-m+\frac{1}{2}\right)\right]^{1/2} \\ \frac{1}{2}\zeta_{n,l}\left[\left(l+m+\frac{1}{2}\right)\left(l-m+\frac{1}{2}\right)\right]^{1/2} & -\frac{1}{2}\left(m+\frac{1}{2}\right)\zeta_{n,l} + \left(m-\frac{1}{2}\right)\mu_B B - W \end{vmatrix} = 0 \tag{70}$$

where $\zeta_{n,l}$ is given by Eq. (61). Equation (70) gives

$$W^2 + W\left[\frac{1}{2}\zeta_{n,l} - 2m\mu_B B\right] + \left[-\left(\frac{1}{2}\zeta_{n,l}\right)^2 l(l+1) - m\mu_B B\,\zeta_{n,l} + \mu_B^2 B^2\left(m^2 - \frac{1}{4}\right)\right] = 0$$

The solutions are

$$W = -\frac{1}{4}\zeta_{n,l} + m_j\mu_B B \pm \frac{1}{4}[4\mu_B^2 B^2 + 8\zeta_{n,l} m_j \mu_B B + \zeta_{n,l}^2(2l+1)^2]^{1/2} \tag{71}$$

where we have reintroduced the subscript j on m. It may be noted that

(i) For $B = 0$

$$W = \frac{1}{4}\zeta_{n,l}[-1 \pm (2l+1)] \tag{72}$$

which is the fine structure splitting.

(ii) For small B

$$W \to -\frac{1}{4}\zeta_{n,l} + (m_j\mu_B B) \pm \frac{1}{4}\zeta_{n,l}(2l+1)\left[1 + \frac{8(m_j\mu_B B)}{\zeta_{n,l}(2l+1)^2}\right]^{1/2}$$

$$\approx \frac{1}{4}\zeta_{n,l}[-1 \pm (2l+1)] + \frac{2l+1\pm 1}{2l+1} m_j\mu_B B \tag{73}$$

which is the Zeeman effect (see Problem 22.2).

(iii) For large B

$$W \to -\frac{1}{4}\zeta_{n,l} + m_j\mu_B B \pm \frac{1}{2}\mu_B B\left[1 + \frac{m_j\zeta_{n,l}}{\mu_B B}\right]^{1/2}$$

$$\approx \left[-\frac{1}{4} \pm \frac{1}{2}m_j\right]\zeta_{n,l} + \left(m_j \pm \frac{1}{2}\right)\mu_B B \tag{74}$$

which is the same as given by Eq. (62) provided we note that $m_j = m_l + m_s$ and the upper and lower signs in Eq. (74) correspond to $m_s = +\frac{1}{2}$ and $m_s = -\frac{1}{2}$ respectively.

For the P-state of an alkali atom, $l = 1$, and we may write Eq. (71) in the form

$$\frac{W}{\zeta_{n,l}} = -\frac{1}{4} + m_j \mathcal{B} \pm \frac{1}{4}\sqrt{4\mathcal{B}^2 + 8m_j\mathcal{B} + 9} \tag{75}$$

where $\mathcal{B} = \mu_B B/\zeta_{n,l}$. In Fig. 22.1, we have plotted the splitting of the P-state of an alkali atom.

Solution 22.5 In the Heisenberg representation (see Section 12.9)

$$\left.\begin{aligned} i\hbar\frac{ds_x}{dt} &= [s_x, H] = \omega_0[s_x, s_z] = -i\hbar\omega_0 s_y(t) \\ i\hbar\frac{ds_y}{dt} &= [s_y, H] = \omega_0[s_y, s_z] = +i\hbar\omega_0 s_x(t) \\ i\hbar\frac{ds_z}{dt} &= 0 \end{aligned}\right\} \tag{76}$$

Thus, $$\frac{d^2 s_x}{dt^2} = -\omega_0\frac{ds_y}{dt} = -\omega_0^2 s_x \tag{77}$$

The solutions are therefore

$$s_x(t) = s_x(0)\cos\omega_0 t + s_y(0)\sin\omega_0 t \tag{78}$$

$$s_y(t) = -s_x(0)\sin\omega_0 t + s_y(0)\cos\omega_0 t \tag{79}$$

$$s_z(t) = s_z(0) \tag{80}$$

The expectation values are

$$\langle s_x(t)\rangle = \langle\Psi(0)\,|\,s_x(t)\,\Psi(0)\rangle = \langle s_x(0)\rangle\cos\omega_0 t + \langle s_y(0)\rangle\sin\omega_0 t \tag{81}$$

etc. For $|\Psi(0)\rangle = |\hat{\mathbf{x}}\uparrow\rangle$

$$\langle s_x(0)\rangle = \frac{1}{2}\hbar,\ \langle s_y(0)\rangle = 0 = \langle s_z(0)\rangle \tag{82}$$

and we obtain the same result as in Example 14.1. For

$$|\Psi(0)\rangle = \cos\frac{\phi}{2}\,|\hat{\mathbf{z}}\uparrow\rangle + \sin\frac{\phi}{2}\,|\hat{\mathbf{z}}\downarrow\rangle \tag{83}$$

$$s_x(0)\,|\Psi(0)\rangle = \frac{1}{2}\hbar\cos\frac{\phi}{2}\sigma_x\begin{pmatrix}1\\0\end{pmatrix} + \frac{1}{2}\hbar\sin\frac{\phi}{2}\sigma_x\begin{pmatrix}0\\1\end{pmatrix}$$

$$= \frac{1}{2}\hbar\cos\frac{\phi}{2}\,|\hat{\mathbf{z}}\downarrow\rangle + \frac{1}{2}\hbar\sin\frac{\phi}{2}\,|\hat{\mathbf{z}}\uparrow\rangle$$

Thus, $$\langle s_x(0)\rangle = \frac{1}{2}\hbar\sin\phi$$

Similarly $\langle s_y(0)\rangle = 0$ and $\langle s_z(0)\rangle = \frac{1}{2}\hbar\cos\phi$. Physically this implies that at $t = 0$, the spin is in the x-z plane at an angle with the z-axis. Using Eqs (78)–(80), we get

$$\langle s_x(t)\rangle = \frac{1}{2}\hbar\sin\phi\cos\omega_0 t \tag{84}$$

$$\langle s_y(t)\rangle = -\frac{1}{2}\hbar\sin\phi\sin\omega_0 t \tag{85}$$

$$\langle s_z(t)\rangle = \frac{1}{2}\hbar\cos\phi \tag{86}$$

which shows the precessional motion (see Example 14.2).

Solution 22.6 Equation (20) with $\phi = 0$, $Q = -q$ and **A** given by Eq. (40) becomes

$$\nabla^2\psi + \frac{2m}{\hbar^2}\left[E + \frac{iq\hbar}{m}Bx\frac{\partial\psi}{\partial y} - \frac{1}{2m}q^2B^2x^2\right]\psi = 0 \tag{87}$$

The solution of Eq. (87) is of the form

$$\psi(z) = u(x)\exp\left[\frac{i}{\hbar}(P_y y + P_z z)\right] \tag{88}$$

where $u(x)$ satisfies the following equation

$$\frac{d^2u(x)}{dx^2} + \frac{2m}{\hbar^2}\left[E - \frac{P_z^2}{2m} - \frac{q^2B^2}{2m}\left(x + \frac{P_y}{qB}\right)^2\right]u(x) = 0 \tag{89}$$

Equation (89) is very similar to the linear harmonic oscillator problem (see Section 7.2). The energy eigenvalues are

$$E = \frac{P_z^2}{2m} + \left(n + \frac{1}{2}\right)\hbar\frac{qB}{m} = \frac{P_z^2}{2m} + (2n+1)\mu_B B \tag{90}$$

where $n = 0, 1, 2, \ldots$. The corresponding $u(x)$ are the Hermite-Gauss functions. For a fixed value of n, the continuous energy bands are called Landau bands (see, e.g. Ref. 14). Notice that Eq. (90) is independent of P_y and hence there is infinite degeneracy. If spin angular momentum is also taken into account, we would get

$$E = \frac{P_z^2}{2m} + \left(2n + 1 \pm \frac{1}{2}\right)\mu_B B \tag{91}$$

Solution 22.8 Equation (20) (for $Q = -q$ and **A** given by Eq. (41)) gives the following expression for the Hamiltonian

$$H = -\frac{\hbar^2}{2m}\nabla^2 + V(r) - \frac{iq\hbar B}{2m}\left(x\frac{\partial}{\partial y} - y\frac{\partial}{\partial x}\right) + \frac{q^2B^2}{8m}(x^2 + y^2)$$

$$= H_0 + H_1 + H_2 \tag{92}$$

Thus,

$$H_0 = -\frac{\hbar^2}{2m}\nabla^2 + V(r) \tag{93}$$

$$H_1 = \frac{qB}{2m}L_z \tag{94}$$

$$H_2 = \frac{q^2B^2}{8m}(x^2 + y^2) \tag{95}$$

The term H_1 represents the interaction between the magnetic moment of the atom and the external magnetic field and gives rise to paramagnetism. The term H_2 can be interpreted as the interaction of the magnetic moment induced in the atom with the magnetic field (see, e.g. Ref. 15, Section 18.3); this gives rise to diamagnetism. Treating H_1 and H_2 as perturbations, we get

$$\langle H_1\rangle = \frac{qB}{2m}m_l\hbar = m_l\mu_B B \tag{96}$$

and
$$\langle H_2 \rangle = \frac{q^2 B^2}{8m} \langle r^2 \sin^2 \theta \rangle \tag{97}$$

If $\langle H_1 \rangle \neq 0$ then (assuming $m_l \sim 1$)

$$\frac{\langle H_2 \rangle}{\langle H_1 \rangle} = \frac{qB}{4\hbar} \langle r^2 \sin^2 \theta \rangle \sim \frac{qB}{4\hbar} a_0^2 \tag{98}$$

where $a_0 (= 0.5 \times 10^{-10}$ m) represents the Böhr radius. Thus,

$$\frac{\langle H_2 \rangle}{\langle H_1 \rangle} \simeq \frac{1.6 \times 10^{-19}}{4 \times 10^{-34}} \times 0.25 \times 10^{-20} B$$

$$\approx 10^{-6} B \text{ (} B \text{ measured in tesla)}$$

$$\approx 10^{-10} B \text{ (} B \text{ measured in Gauss)}$$

Since the laboratory magnetic fields almost never exceed 10^5 Gauss,

$$\frac{\langle H_2 \rangle}{\langle H_1 \rangle} \ll 1 \tag{99}$$

hence whenever the paramagnetic term is non-zero, it dominates over the diamagnetic term. Thus for $l > 0$, $\langle H_2 \rangle$ dominates and for $l = 0$, the diamagnetic term determines the magnetic behaviour. Indeed, for such a case $\langle \sin^2 \theta \rangle = \frac{2}{3}$ and

$$\langle H_2 \rangle = \frac{q^2 B^2}{12m} \langle r^2 \rangle \tag{100}$$

Solution 22.9 We assume **A** to be given by Eq. (41), i.e.,

$$A_x = -\frac{1}{2} yB, \quad A_y = \frac{1}{2} xB, \quad A_z = 0 \tag{101}$$

We can then use either Eq. (20) or Eq. (25). If we use Eq. (20), then the second term on the LHS would become

$$\frac{iQ\hbar}{\mu} \mathbf{A} \cdot \nabla \psi = \frac{iQ\hbar B}{2\mu} \left[-y \frac{\partial}{\partial x} + x \frac{\partial}{\partial y} \right] = -\frac{QB}{2\mu} [x p_y - y p_x]$$

$$= -\frac{QB}{2\mu} L_z = +\frac{i\hbar QB}{2\mu} \cdot \frac{\partial}{\partial \phi} \tag{102}$$

Thus Eq. (20) [or Eq. (25)] takes the following form in the 2-dimensional polar coordinates (ρ, ϕ):

$$-\frac{\hbar^2}{2\mu} \left[\frac{1}{\rho} \frac{\partial}{\partial \rho} \left(\rho \frac{\partial \psi}{\partial \rho} \right) + \frac{1}{\rho^2} \frac{\partial^2 \psi}{\partial \phi^2} \right] + \left[\frac{i\hbar QB}{2\mu} \frac{\partial \psi}{\partial \phi} + \frac{Q^2 B^2}{8\mu} \rho^2 \psi + V(\rho) \psi \right] = E \psi(\rho, \phi) \tag{103}$$

Now for ψ to be single value we must have

$$\psi(\rho, \phi) = \left[\frac{1}{\sqrt{2\pi}} e^{im\phi} \right] R(\rho); \qquad m = 0, \pm 1, \pm 2 \tag{104}$$

Thus $R(\rho)$ will satisfy the equation

$$\frac{1}{\rho}\frac{d}{d\rho}\left(\rho\frac{dR}{d\rho}\right)+\frac{2\mu}{\hbar^2}\left[\left(E+\frac{QBm\hbar}{2\mu}\right)+\frac{1}{2}\mu\omega^2\rho^2-\frac{m^2}{\rho^2}\right]R(\rho)=0 \qquad (105)$$

where

$$\omega=\left(\omega_0^2+\frac{Q^2B^2}{4\mu^2}\right)^{1/2} \qquad (106)$$

Equation (109) is of the same form as that of a 2-dimensional harmonic oscillator and we can solve either using Cartesian coordinates (see Section 10.6) or using cylindrical coordinates (see Problem 10.13). If we use Cartesian coordinates, the energy eigenvalues would be given by

$$E=(n_1+n_2+1)\,\hbar\omega-\frac{mQ\hbar B}{2\mu};\ n_1, n_2=0, 1, 2, \ldots \qquad (107)$$

The corresponding eigenfunctions would be appropriate linear combinations of the Hermite-Gauss functions which are also eigenfunctions of L_z; i.e. the ϕ dependence of the wavefunction must be of the form of $\exp(im\phi)$.

REFERENCES AND SUGGESTED READINGS

1. H.E. White, *Introduction to Atomic Spectra*, McGraw-Hill Book Co., New York (1934).
2. F. Seitz, *The Modern Theory of Solids*, McGraw-Hill Book Co., New York (1940).
3. C.P. Slichter, *Principles of Magnetic Resonance*, Harper and Row, New York (1963).
4. J.H. Van Vleck, *The Theory of Electric and Magnetic Susceptibilities*, Oxford University Press, London (1932).
5. D. Wagner, *Introduction to the Theory of Magnetism*, Pergamon Press, Oxford (1972).
6. H. Goldstein, *Classical Mechanics*, Addison-Wesley Publishing Co., Reading, Massachusetts (1950).
7. J.D. Jackson, *Classical Electrodynamics*, John Wiley, New York (1962).
8. D. Halliday, R., Resnick and J. Walker, *Fundamentals of Physics*, Fourth Edition, John Wiley & Sons, New York (1993).
9. W.K.H. Panofsky and M. Philips, *Classical Electricity and Magnetism*, Addison-Wesley Publishing Co., Reading, Massachusetts (1962).
10. J.L Powell, and B. Craseman, *Quantum Mechanics*, Addison-Wesley Publishing Co., Reading, Massachusetts (1961).
11. L.H. Thomas, 'The motion of the spinning electron', *Nature*, **117**, p. 514 (1926).
12. G. Baym, *Lectures on Quantum Mechanics*, W.A. Benjamin, Inc., New York (1969)..
13. R.P. Feynman, R.B. Leighton and M. Sands, *The Feynman Lectures on Physics*, Vol. III, Addison-Wesley Publishing Co., Reading, Massachusetts (1965).
14. H. Kroemer, *Quantum Mechanics for Engineering, Material science and Applied Physics*, Prentice-Hall, Englewood Cliffs (1994).
15. A.J. Dekker, *Solid State Physics*, Prentice-Hall, Englewood Cliffs, New Jersey (1952).

23 The variational method

All science is either physics or stamp collecting.

— ERNST RUTHERFORD.

23.1 INTRODUCTION

In many cases stationary perturbation theory cannot be applied successfully as there may not be a closely related problem which is capable of exact solution.

In this chapter, we shall describe the variational method for evaluating the energies of the ground state and the first few discrete states of a bound system (approximately). We shall then apply it to calculate the ground state energy of the helium atom and the hydrogen molecular ion.

23.2 THE BASIC PRINCIPLE

Let Ψ_n denote the exact set of eigenfunctions of the hamiltonian H

$$H\,\Psi_n = E_n\,\Psi_n \tag{1}$$

These eigenfunctions constitute a complete set of functions in the sense that an arbitrary bounded quadratically integrable function ϕ can be expanded in terms of the eigenfunctions. Thus we may write

$$\phi = \sum_n a_n\,\Psi_n \tag{2}$$

Evidently, the quantity

$$\langle H\rangle = \int \Psi_n^* H\,\Psi_n\,d\tau \tag{3}$$

yields the energy E_n, assuming that the Ψ_n are normalised. Consider, now, the quantity

$$E_\phi = \frac{\int \phi^* H\,\phi\,d\tau}{\int \phi^*\,\phi\,d\tau} = \frac{\langle\phi|H|\phi\rangle}{\langle\phi|\phi\rangle} \tag{4}$$

We can show that E_ϕ will always be larger than E_0, the lowest of the spectrum E_n. (E_0 is the ground state energy). Before proving this simple result, one can see how this comes about. The effect of H on the ket vector $|\phi\rangle$ is, in general, to rotate it, except when $|\phi\rangle$ is one of the eigenkets of H. Thus the scalar product, formed with the bra $\langle\phi|$ and $H\,|\phi\rangle$ will be an extremum for the precise case when $H\,|\phi\rangle$ and $|\phi\rangle$ point in the same direction.

We have, on substitution for ϕ from Eq. (2) in Eq. (3)

$$E_\phi = \frac{\int\left(\sum a_n^* \Psi_n^* H \sum a_n \Psi_n\right) d\tau}{\int\left(\sum_n a_n^* \Psi_n^*\right)\left(\sum_n a_n \Psi_n\right) d\tau} \tag{5}$$

Since,

$$\int \psi_n^* \psi_m \, d\tau = \delta_{nm} \tag{6}$$

we obtain

$$E_\phi = \frac{\sum_n a_n^* a_n E_n}{\sum_n a_n^* a_n} \tag{7}$$

Subtracting E_0, the energy of the ground state, from both sides of the equation, we have

$$E_\phi - E_0 = \frac{\sum_n |a_n|^2 (E_n - E_0)}{\sum_n |a_n|^2} \tag{8}$$

Now,

$$E_n \geq E_0$$

because by definition E_0 is the ground state energy, and since $|a_n|^2 \geq 0$ for all n, we must have

$$E_\phi - E_0 \geq 0$$

or

$$E_\phi \geq E_0 \tag{9}$$

This important inequality can be used to obtain an upper bound to the ground state energy E_0. We choose a suitable function ϕ known as the trial function which depends on a number of variable parameter c_i $(i = 1, \ldots, p)$. This trial function is substituted in Eq. (3) and the parameters are chosen so as to make E_ϕ a minimum by solving the p equations

$$\frac{\partial E_\phi}{\partial c_i} = 0; \quad (i = 1, 2, \ldots, p) \tag{10}$$

The minimum value of E determined in this way provides an upper bound to E_0, which will be close to the actual value if the trial function has a form closely resembling that of the actual ground state function ψ_0. Many different types of trial functions have been tried for the simple atoms and molecules, and some of these have been applied to much more complicated systems.

23.3 THE HYDROGEN ATOM AS AN EXAMPLE

In order to illustrate the use of the variational method, we apply it to the hydrogen atom problem. If we choose

$$\phi = A_1 \exp\left(-\alpha \frac{r}{a_0}\right) \quad \text{Trial wavefunction}$$

we already have the correct form of the wavefunction (see Section 10.4). We will calculate

$$E(\alpha) = \frac{\int \phi^* H \phi \, d\tau}{\int \phi^* \phi \, d\tau}$$

and set $\partial E/\partial \alpha = 0$, this will give us the value of α for which E is a minimum. We first normalize ϕ:

$$1 = \int_0^\infty 4\pi r^2 \, dr \, \phi^* \phi = 4\pi |A_1|^2 \int_0^\infty \exp\left[-\frac{2\alpha r}{a_0}\right] r^2 \, dr$$

giving
$$A_1 = \left(\frac{\alpha^3}{\pi a_0^3}\right)^{\frac{1}{2}}$$

Thus,
$$\phi = \left(\frac{\alpha^3}{\pi a_0^3}\right)^{\frac{1}{2}} \exp\left(-\alpha \frac{r}{a_0}\right)$$

Since ϕ (r) is now normalized, $E(\alpha) = \int \phi^* H\phi \, d\tau$. Further, since ϕ depends only on r, we need only keep the part of H which depends on r, i.e.,

$$H_{\text{radial}} = -\frac{\hbar^2}{2m}\frac{1}{r^2}\frac{\partial}{\partial r}\left(r^2 \frac{\partial}{\partial r}\right) - \frac{e^2}{r} \tag{11}$$

The integrals are easy to evaluate and one obtains

$$E(\alpha) = \left[\frac{\hbar^2 \alpha^2}{2ma_0^2} - \frac{e^2 \alpha}{a_0}\right]$$

Setting, $dE/d\alpha = 0$ gives $\alpha = (me^2/\hbar^2)\, a_0 = 1$. Thus

$$E(\alpha)\big|_{\min} = -\frac{e^2}{2a_0} \tag{12}$$

which is the exact result, as it indeed should be, because we had started out with the correct form of ϕ.

It is instructive to see what we would obtain if we use other forms of the trial function (see also Problem 23.1). We choose another trial wavefunction given by:

$$\phi = A \exp\left[-\alpha\left(\frac{r}{a_0}\right)^2\right]$$

Normalizing, we get
$$A = \left(\frac{2\alpha}{\pi a_0^2}\right)^{3/4}$$

The integrals are again easy to evaluate and one obtains

$$E(\alpha) = \left[\frac{3}{2}\alpha - 2^{2/3}\left(\frac{\alpha}{\pi}\right)^{1/2}\right]\frac{e^2}{a_0}$$

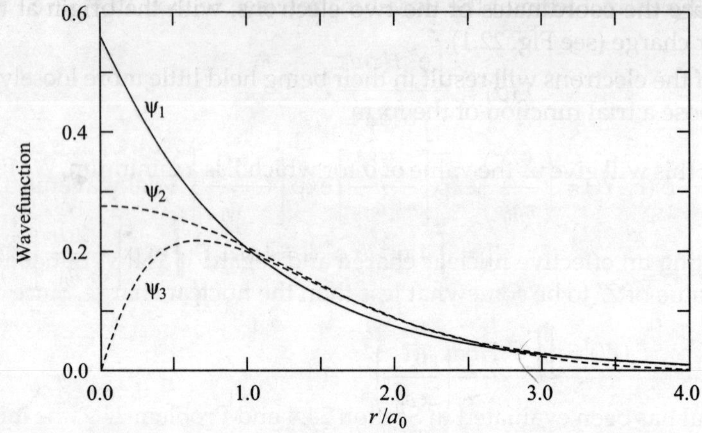

Figure 23.1: The variation of the normalized trial functions as given by Eqs (14), (15) and (16).

The minimum value of E is obtained when $\alpha = 8/9\pi$, the corresponding energy being

$$E(\alpha)\big|_{\min} = -0.85\frac{e^2}{2a_0} \tag{13}$$

In Fig. 23.1, we have plotted the following normalized functions

$$\psi_1 = \left(\frac{\alpha^3}{\pi a_0^3}\right)^{1/2} \exp\left(-\alpha\frac{r}{a_0}\right), \quad \alpha = 1 \tag{14}$$

$$\psi_2 = \left(\frac{2\alpha}{\pi a_0^2}\right)^{3/4} \exp\left(-\alpha\frac{r^2}{a_0^2}\right), \quad \alpha = \frac{8}{9\pi} \tag{15}$$

$$\psi_3 = \left(\frac{\alpha^5}{3\pi a_0^3}\right)^{1/2} \frac{r}{a_0}\exp\left(-\alpha\frac{r}{a_0}\right), \quad \alpha = \frac{3}{2} \tag{16}$$

the values of α are such that one obtains minimum energies for that particular form of function. It can be seen that although ψ_3 is closer to the actual wavefunction ψ_1 than ψ_2 for large r, the energy obtained for ψ_2 is a better estimate. The reason is that for calculating energy eigenvalues, the wavefunction for large r does not contribute as much as for smaller values of r.

This observation illustrates an important point, namely, that the 'best' form of wavefunction depends on the particular problem at hand. If, we are interested primarily in the large r behaviour of the wavefunction for calculating a physical quantity for the system, the function that yields a better energy estimate may not nevertheless be an appropriate wavefunction.

23.4 THE HELIUM ATOM

Another illustration of the variational method is the application to the ground state of the helium atom. If we assume that the two 1s electrons move independently in the field of the nucleus (i.e. neglecting their mutual repulsion), the wavefunction will be

$$\psi(\mathbf{r}_1, \mathbf{r}_2) = \left(\frac{Z^3}{\pi a_0^3}\right) e^{-Zr_1/a_0}\, e^{-Zr_2/a_0}$$

where $\mathbf{r}_1$ and $\mathbf{r}_2$ are the coordinates of the two electrons, with the origin at the nucleus and $Z = 2$, the nuclear charge (see Fig. 22.1).

The interaction of the electrons will result in their being held little more loosely by the nucleus. We therefore choose a trial function of the form

$$\phi(\mathbf{r}_1, \mathbf{r}_2) = \left(\frac{Z'^3}{\pi a_0^3}\right) \exp\left[-\frac{Z' r_1}{a_0}\right] \exp\left[-\frac{Z' r_2}{a_0}\right] \quad \text{Trial wavefunction} \tag{17}$$

with Z' representing an effective nuclear charge and regard it as a variational parameter; we expect the best value of Z' to be somewhat less than the nuclear charge. Since ϕ is normalized,

$$E(Z') = \iint \phi^* H \phi \, d\tau_1 \, d\tau_2$$

The above integral has been evaluated in Section 24.4 and Problem 24.2, the minimum value is

$$E_0 = -2\left(Z - \frac{5}{16}\right)^2 \frac{e^2}{2a_0} = -2\left(Z - \frac{5}{16}\right)^2 \times 13.6 \text{ eV} \quad \text{Minimum energy} \tag{18}$$

which occurs when

$$Z' = Z - \frac{5}{16} \tag{19}$$

This value of the energy is better than the value obtained from first order perturbation theory (see Section 24.3). Putting $Z = 2$, we get

$$E_0 = -\left(\frac{27}{16}\right)^2 \frac{e^2}{a_0} = -2.85 \frac{e^2}{a_0} \tag{20}$$

The experimental value for the minimum energy required to remove both the electrons from a helium atom is $(2.904\, e^2/a_0)$. It is worthwhile pointing out that a large number of trial functions have been used to obtain the ground state energy of the helium atom. Some of the trial functions used and the corresponding ground state energies are tabulated in Table 23.1.

It should be noted that in general, the accuracy of the calculated value of the ground state energy increases with increase in the number of terms.

23.5 APPLICATION TO EXCITED STATES

Suppose we choose a ket $|\phi\rangle$ which has no component along the *ground state axis*, i.e. $|\phi\rangle$ lies entirely in the 'plane' of the remaining eigenkets in the Hilbert space. If we now vary $|\phi\rangle$, (keeping the variation within this 'plane'), we ought to arrive at the best estimate for the energy of the first excited state. We do this by first restricting our choice ϕ to be orthogonal to ψ_0. Let

$$\phi = \chi - \psi_0 \int \psi_0^* \chi \, d\tau \tag{21}$$

Since ϕ is now orthogonal to ψ_0 (as can be easily seen by evaluating $\int \phi^* \psi_0 \, d\tau$), we have

$$\phi = \sum_{1}^{\infty} a_n \psi_n$$

the coefficient a_0 being zero.

Table 23.1: Variation Functions for the Normal Helium Atom*

(Symbols $s = \frac{r_1 + r_2}{a_0}$, $t = \frac{r_1 - r_2}{a_0}$, $u = r_{12}/a_0$

Experimental value** $W = -5.80736\ R_{He}\ hc$)

*Variation function, with best values of constants****	Energy (in units of $-R_{He}$ hc)	Deviation from experiment (in units of $-R_{He}$ hc)
1. e^{-2s}	5.50	0.31
2. $e^{-Z's}$, $Z' = \frac{27}{16} = 1.6875$	5.6953	0.1120
3. $e^{-Z's} \cosh ct$, $Z' = 1.67$, $c = 0.48$	5.7508	0.0565
4. $e^{-Z's}(1 + c_2t^2)$, $Z' = 1.69$, $c_2 = 0.142$	5.7536	0.0537
5. $e^{-Z's}\, e^{cu}$, $Z' = 1.86$, $c = 0.26$	5.7792	0.0281
6. $e^{-Z's}(1 + c_1u)$, $Z' = 1.849$, $c_1 = 0.364$	5.7824	0.0249
7. $e^{-Z's}(1 + c_1 u + c_2t^2)$, $Z' = 1.816$, $c_1 = 0.30$, $c_2 = 0.13$	5.80488	0.00245
8. $e^{-Z's}(1 + c_1u + c_2t^2 + c_3s + c_4s^2 + c_5u^2)$, $Z' = 1.816$, $c_1 = 0.353$, $c_2 = 0.128$, $c_3 = -0.101$, $c_4 = 0.033$, $c_5 = -0.032$	5.80648	0.00085

* Adapted from Section VIII–29d of Ref. 1.

** $R_{He} = \frac{2\pi^2 \mu e^4}{h^3 c} = 109722.4\ \text{cm}^{-1}$, (μ is the reduced mass of the electron and the nucleus).

*** The normalization factor is omitted of these functions, 1 is due to Unsold, 2 to Kellner, 3 to Eckart and Hylleraas and the remainder to Hylleraas.

We can now evaluate E_ϕ as before and show that

$$E_\phi - E_1 = \frac{\sum_1^\infty |a_n|^2 (E_n - E_1)}{\sum_1^\infty |a_n|^2}$$

and conclude $E_\phi > E_1$ using the same arguments as before. Thus, one chooses an appropriate form χ with a set of variable parameters, then define ϕ as in Eq. (21) and go through the variation procedure. Evidently, this procedure can be continued for higher states; we define

$$\phi = \chi - \sum_{n=0}^{N-1} \psi_n \int \psi_n^* \chi\, d\tau \tag{22}$$

ϕ is now orthogonal to the lowest $N - 1$ states and we can estimate the energy of the Nth state. However, the procedure is quite cumbersome as one goes to higher levels. It is sometimes possible to choose a trial function ϕ directly from symmetry considerations so that orthogonality to lower state eigenfunctions is ensured.

23.6 LINEAR VARIATION FUNCTIONS

A trial function can also be chosen as a linear combination of a set of complete orthonormal function u_n:

$$\psi = \sum_{n=1}^{m} c_n u_n \tag{23}$$

The variational energy is, therefore, given by

$$E = \frac{\int \psi^* H\psi \, d\tau}{\int \psi^* \psi \, d\tau} = \frac{\sum_n \sum_m c_n^* c_m H_{nm}}{\sum_n |c_n|^2}$$

or

$$E \sum_n |c_n|^2 = \sum_n \sum_m c_n^* c_m H_{nm} \tag{24}$$

where

$$H_{nm} = \int u_n^* \, H u_m \, d\tau \tag{25}$$

Differentiating Eq. (24) with respect to c_k and setting $\partial E / \partial c_k = 0$, we obtain[1]

$$\sum c_n^* H_{nk} - E c_k^* = 0 \tag{26}$$

For non-trivial solutions the determinant must vanish. Thus,

$$|H_{ij} - E\delta_{ij}| = 0; \; i, j = 1, 2, \ldots, m \quad \text{Secular equation} \tag{27}$$

Equation (27) is known as the secular equation. Eq. (27) will, in general, give m roots of E. For a particular value of E, the coefficient c_k can be found by using Eq. (26) and the normalization condition

$$\sum_n |c_n|^2 = 1 \tag{28}$$

The lowest root of E and the corresponding wavefunction represent approximations to the energy and the wavefunction of the ground state of the system. Other roots and the corresponding wavefunctions represent approximation to the excited states.

Often, it is more convenient to use a set of basis functions which are not necessarily orthogonal. In that case, the secular equation takes the form

$$|H_{ij} - S_{ij} E| = 0 \tag{29}$$

where

$$S_{ij} = \int u_i^* \, u_j \, d\tau \tag{30}$$

is known as the overlap integral.

23.7 THE H_2^+ ION

As an example of the linear combination method, we consider the H_2^+ ion. The Hamiltonian, neglecting the nuclear kinetic energy and the spin-orbit interaction, is given by

$$H = -\frac{\hbar^2}{2m}\nabla^2 - \frac{e^2}{r_A} - \frac{e^2}{r_B} + \frac{e^2}{R} \quad \text{Hamiltonian} \tag{31}$$

[1] What would be the equation if we had diferentiated eq. (24) with respect to c_k^*?

where $r_A = |\mathbf{r} - \mathbf{R}_A|$, $r_B = |\mathbf{r} - \mathbf{R}_B|$, $R = |\mathbf{R}_A - \mathbf{R}_B|$, $\mathbf{r}$ being the position of the electron, $\mathbf{R}_A$ and $\mathbf{R}_B$ being the positions of the two protons.

If R is large, then there will be a large probability for the electron to be either; near A or near B but not both. Indeed, in the former case we have approximately a 1s hydrogen atom wavefunction centred at A and in the latter case at B. This: suggests that we use a linear combination as a trial function, i.e.,

$$\psi = c_A u_{1A} + c_B u_{1B} \quad \text{Trial function} \tag{32}$$

Note that the wavefunctions u_{1A} and u_{1B} which are the 1s wavefunctions at A and B are *not* orthogonal. A single electron wavefunction as above is called a molecular orbital. This particular form of a linear combination of atomic 'orbitals' making a molecular orbital is called LCAO–MO for short. We now use Eq. (29). The normalized wavefunctions are given by

$$\begin{aligned} u_{1A} &= \left(\frac{1}{\pi a_0^3}\right)^{1/2} \exp\left[-r_A/a_0\right] \\ u_{1B} &= \left(\frac{1}{\pi a_0^3}\right)^{1/2} \exp\left[-r_B/a_0\right] \end{aligned} \tag{33}$$

The integrals in Eq. (29) should now be evaluated

$$S_{AA} = \frac{1}{\pi a_0^3}\int \exp\left[-(2r_A/a_0)\right] d\tau = 1 = S_{BB}$$

$$S_{AB} = S_{BA} = \frac{1}{\pi a_0^3}\int \exp\left[(-r_A/a_0) - (r_B/a_0)\right] d\tau = S, \text{ say}$$

$$H_{AA} = H_{BB} = \frac{1}{\pi a_0^3}\int \exp(-r_A/a_0)\left[-\frac{\hbar^2}{2m}\nabla^2 - \frac{e^2}{r_A} - \frac{e^2}{r_B} + \frac{e^2}{R}\right]\exp(-r_A/a_0)\,d\tau$$

$$= \frac{e^2}{R} - \frac{e^2}{2a_0} - \frac{e^2}{\pi a_0^3}\int \frac{1}{r_B}\exp(-2r_A/a_0)\,d\tau$$

and
$$H_{AB} = H_{BA} = \frac{1}{\pi a_0^3}\int \exp(r_A/a_0)\left[-\frac{\hbar^2}{2m}\nabla^2 - \frac{e^2}{r_A} - \frac{e^2}{r_B} + \frac{e^2}{R}\right]\exp(-r_B/a_0)\,d\tau$$

$$= \left(\frac{e^2}{R} - \frac{e^2}{2a_0}\right)S_{AB} - \frac{e^2}{\pi a_0^3}\int \frac{1}{r_A}\exp(-r_A/a_0) \times \exp(-r_B/a_0)\,d\tau$$

In order to evaluate the integrals in the expressions for S_{AB}, H_{AA} and H_{AB}, it is convenient to introduce the confocal elliptic coordinates

$$\mu = \frac{r_A + r_B}{R},\ \nu = \frac{r_A - r_B}{R} \tag{34}$$

and the azimuthal angle ϕ. Obviously,

$$1 \le \mu < \infty;\ -1 \le \nu \le +1;\ 0 \le \phi \le 2\pi$$

and
$$\int F\,d\tau = \int_1^\infty d\mu \int_{-1}^{+1} d\nu \int_0^{2\pi} d\phi\, \frac{R^3}{8}(\mu^2 - \nu^2)F(\mu, \nu, \phi) \tag{35}$$

The integrations can easily be carried out. The results are as follows:

$$S_{AB} = S_{BA} = \left[1 + \xi + \frac{1}{3}\xi^2\right] e^{-\xi} \tag{36}$$

$$H_{AA} = H_{BB} = \left[-\frac{1}{2} + \left(\frac{1}{\xi} + 1\right) e^{-2\xi}\right] \frac{e^2}{a_0}$$

$$H_{AB} = H_{BA} = \left[\left(\frac{1}{\xi} - \frac{1}{2}\right) S_{AB} - (1 + \xi) e^{-\xi}\right] \frac{e^2}{a_0}$$

where $\xi = R/a_0$. Now the secular equation [Eq. (29)] is of the form

$$\begin{vmatrix} H_{AA} - S_{AA}E & H_{AB} - S_{AB}E \\ H_{BA} - S_{BA}E & H_{BB} - S_{BB}E \end{vmatrix} = 0 \text{ Secular equation} \tag{37}$$

Thus,

$$(H_{AA} - E)^2 = (H_{AB} - S_{AB} E)^2 \tag{38}$$

or

$$E_{\pm} = \frac{H_{AA} \pm H_{AB}}{1 \pm S_{AB}} \tag{39}$$

The corresponding wavefunctions can easily be found

$$\psi_+ = \frac{1}{\sqrt{2}\,(1 + S_{AB})^{1/2}} (u_{1A} + u_{1B}) \tag{40}$$

$$\psi_- = \frac{1}{\sqrt{2}\,(1 - S_{AB})^{1/2}} (u_{1A} - u_{1B}) \tag{41}$$

Obviously, ψ_+ is symmetric with respect to the interchange of the nuclei and ψ_- is antisymmetric. Since the dissociation energy is $E_H + E_{H^+} - E_{H_2^+}$ we define

$$\varepsilon_{\pm} = E_{\pm} + \frac{e^2}{2a_0} = \frac{H_{AA} \pm H_{AB}}{1 \pm S_{AB}} + \frac{e^2}{2a_0} \tag{42}$$

We substitute the expressions for H_{AA}, H_{AB} and S_{AB} from Eq. (36) and calculate $\varepsilon_{\pm}$ as a function of ξ (see Fig. 23.2). The quantity ε_+ (which corresponds to the symmetric state) has a minimum value[2] for $\xi \approx 2.4$ giving $R \approx 1.3$ Å, the corresponding value of ε_+ being –1.76 eV. Experimentally $R \approx 1.06$ Å and the dissociation energy is 2.8 eV. A better agreement between theory and experiment is obtained by introducing a variational parameter in the trial function (see Problem 23.4).

We may point out certain shortcomings of the chosen trial function. For example, if we let $R \to 0$ we should obtain the He^+ ion the ground state energy of which is $-2e^2/a_0$, whereas Eqs (39) and(36) will give

$$E_+ \to -1.5\, e^2/a_0 \quad \text{as } R \to 0$$

The discrepancy is due to the choice of the trial function which, for $R \to 0$, becomes the 1*s* hydrogen atom wavefunction which is obviously incorrect. This is the reason why we find the binding too weak. We should, however, point out that for $R \to \infty$, ψ_+ should describe the system accurately.

[2] The wavefunction ψ_+ which leads to negative energy values is known as a 'bounding orbital'; on the other hand, since ψ_- leads to positive energy values, it is known as an 'anti-bonding orbital'.

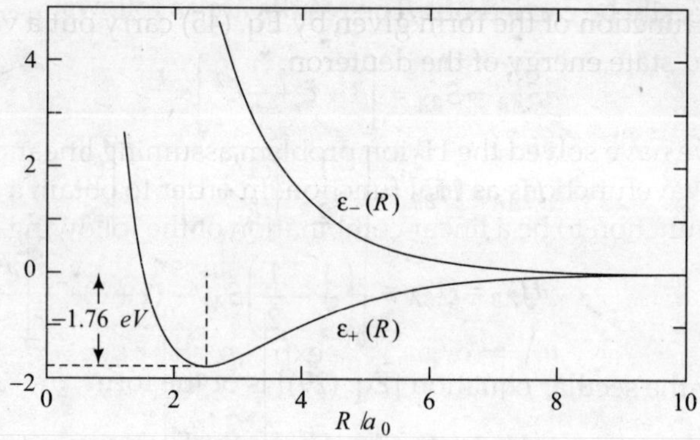

Figure 23.2: The variation of $\varepsilon_\pm$ as a function of R/a_0. The minimum of ε_+ occurs at $R \approx 2.45\, a_0 \approx 1.3$ Å at which $\varepsilon_+ \approx -1.76$ eV.

23.8 PROBLEMS

Problem 23.1 For the ground state of the hydrogen atom use the following as trial wavefunctions

$$\phi_1 \sim r \exp\left[-\alpha \frac{r}{a_0}\right], \quad \phi_2 \sim \frac{1}{\alpha^2 + (r/a_0)^2} \tag{43}$$

where α is the parameter to be varied and a_o is Böhr radius. Show that the values of α for which minimum energy is obtained are 1.5 and $\pi/4$ respectively, the corresponding energies being $-0.75\, E_H$ and $-0.81\, E_H$. Discuss with the results obtained in Section 23.3.

Problem 23.2 The neutron-proton short-range interaction can be approximately described by the square well potential

$$\left.\begin{aligned} V &= -V_0, \quad 0 < r < a \\ &= 0, \qquad\quad r > a \end{aligned}\right\} \tag{44}$$

(a) For $l = 0$ obtain the exact solution of the Schrödinger equation and show that for $V_0 = 40$ MeV and $a = 1.896 \times 10^{-13}$ cm, the ground state energy is given by $E = -2.223$ MeV which is the experimental value for the binding energy of the deuteron[3].

(b) Next assume a trial wavefunction of the form

$$\phi \sim \exp[-\beta r/a] \tag{45}$$

Using the values of V_0 and a given in part (a) carry out a variational analysis (with β as the variational parameter) to obtain the ground state energy of the deuteron and compare with the exact result.

Problem 23.3 The neutron-proton interaction can also be described by the exponential potential

$$V = -V_0\, e^{-r/a} \tag{46}$$

(a) For $l = 0$ has exact solution of the Schrödinger equation can be obtained in terms of Bessel functions (see Problem 10.8). Show that for $V_0 = 40$ MeV and $a = 1.854 \times 10^{-13}$ cm, the ground state energy is given by $E = -2.223$ MeV.

[3] In Problems 23.2(a) and 23.3(a) the values of V_0 and a are chosen such that the minimum energy eigenvalue agrees with the experimental value.

(b) Assuming a trial function of the form given by Eq. (45) carry out a variational analysis to obtain the ground state energy of the deuteron.

Problem 23.4[4]

(a) In Section 23.4, we have solved the H_2^+ ion problem assuming linear combination of the 1*s* hydrogen atom wavefunctions as trial function. In order to obtain a better estimate of $E_{\pm}$, assume the trial function to be a linear combination of the following functions (see Eqs 32 and 33)

$$u_{1A} = (\alpha/\pi a_0^3)^{1/2} \exp\left[-\alpha \frac{r_A}{a_0}\right] \tag{47}$$

$$u_{1B} = (\alpha/\pi a_0^3)^{1/2} \exp\left[-\alpha \frac{r_B}{a_0}\right] \tag{48}$$

with α as the variational parameter. Show that for $\alpha \approx 1.238$ the minimum value of E_+ ($=-15.95$ eV) is obtained; Calculate the corresponding dissociation energy and the separation between the two protons and compare with the experimental values given in Section 23.4.

(b) If the minimum value of E_+ is obtained for $R = R_0$ then

$$\left.\frac{dE_+}{dR}\right|_{R=R_0} = 0 \tag{49}$$

and we may write

$$E_+ = E_0 + \frac{1}{2}k\,(R - R_0)^2 \tag{50}$$

where $E_0 \approx -15.95$ eV. From the calculations in part (a) of the problem, obtain the value of k. Compare the above equation with Eq. (90) of Chapter 10 and obtain the zero-point energy corresponding to the vibrational motion. Add this to E_+ to obtain a better estimate for the dissociation energy. The discussion of the hydrogen ion problem (Solution 23.4) is based on the analysis given in Ref. 3.

Problem 23.5 For the hydrogen molecule problem discussed in Problem 20.15, estimate the ground state energy by assuming a trial function of the form

$$\phi = u_0\,(\mathbf{r}_1, \mathbf{r}_2)\ (1 + AH') \tag{51}$$

where H' is given by Eq. (128) of Chapter 20, A is the variational parameter[5] (assumed to be real) and

$$u_0(\mathbf{r}_1, \mathbf{r}_2) = u_{100}(\mathbf{r}_1)\, u_{100}\,(\mathbf{r}_2) \tag{52}$$

$u_{100}\,(\mathbf{r}_1)$ and $u_{100}\,(\mathbf{r}_2)$ being the wavefunctions corresponding to the ground state of the hydrogen atom, the coordinates $\mathbf{r}_1$ and $\mathbf{r}_2$ are defined in Fig. 20.5.

23.9 SOLUTIONS

Solution 23.2 (a) The solution of the radial part of the Schrödinger equation for the square well potential is discussed in Problem 10.4 (with $b = 0$). We write $E = -B$ where B is the binding energy of the deuteron. The equation that determines the energy eigenvalues for $l = 0$ is

[4] Problems 23.4 and 23.5 and their solutions have been adapted from Ref. 3.

[5] This problem is adapted from Ref. 2; the justification for such a choice of the trial function is also given there.

$$\cot k_1 a = -\kappa/k_1 = -[B/(V_0 - B)]^{1/2} \tag{53}$$

where $\kappa^2 = 2\mu B/\hbar^2$ and $k_1^2 = 2\mu\,(V_0 - B)/\hbar^2$. Now $\mu = (m_n m_p)/(m_n + m_p) \approx 0369 \times 10^{-24}$gm. For $V_0 \simeq 40$ MeV, $a \simeq 1.896 \times 10^{-13}$cm we obtain $B \simeq 2.223$ MeV.

(b) The normalized trial wavefunction is given by

$$\phi = \left[\frac{\beta^3}{\pi a^3}\right]^{1/2} \exp[-\beta r/a] \tag{54}$$

Thus,
$$E_\beta = \frac{\beta^3}{\pi a^3}\int_0^\infty e^{-\beta r/a}\left[-\frac{\hbar^2}{2\mu}\cdot\frac{1}{r^2}\frac{d}{dr}\left(r^2\frac{d}{dr}\right) - V(r)\right]e^{-\beta r/a}\, 4\pi r^2\, dr \tag{55}$$

Substituting for $V(r)$ from Eq. (44) and carrying out the integration, we get

$$E_\beta = \frac{\hbar^2}{2\mu a^2}\left[\beta^2 - 2\gamma\left\{1 - (1 + 2\beta + 2\beta^2)e^{-2\beta}\right\}\right] \tag{56}$$

where
$$\gamma = \frac{\mu V_0 a^2}{\hbar^2} \simeq 1.73 \tag{57}$$

The minimum occurs when $dE_\beta/d\beta = 0$ giving

$$\beta e^{-2\beta} = \frac{1}{4\gamma} = 0.1445 \tag{58}$$

On solving the above equation, we get $\beta \simeq 0.93$. Thus,

$$E_\beta \simeq -1.43 \text{ MeV}$$

As can be seen the error is quite large which is due to the fact that the choice of the trial function is poor.

Solution 23.3

(a) Referring to the solution of Problem 10.8, we have

$$g = \left[\frac{8\mu V_0\, a^2}{\hbar^2}\right]^{1/2} \simeq 3.64 \tag{59}$$

The solution of the equation $J_\nu\,(g) = 0$ gives

$$\nu = \left[\frac{8\mu B\, a^2}{\hbar^2}\right]^{1/2} \simeq 0.858 \tag{60}$$

giving $B = 2.223$ MeV.

(b) We substitute for $V(r)$ from Eq. (46) in Eq. (55) and carry out the integration to obtain

$$E_\beta = \frac{\hbar^2}{2\mu a^2}\left[\beta^2 - \frac{8\gamma\beta^3}{(1+2\beta)^3}\right] \tag{61}$$

where
$$\gamma = \frac{\mu V_0 a^2}{\hbar^2} \simeq 1.655$$

The minimum occurs when $dE_\beta / d\beta = 0$ giving

$$\frac{(1+2\beta)^4}{\beta} = 24\,\gamma = 39.73 \tag{62}$$

On solving the above equation, we get $\beta = 0.6088$. Thus,

$$E_\beta = -2.14 \text{ MeV}$$

Solution 23.4 The analysis is similar to the one given in Section 23.4; the result for E_+ is

$$E_+ = \left[-\frac{\alpha^2}{2} + \frac{\alpha(\alpha-1) - \frac{\alpha}{\zeta}\left\{1-(1+\zeta)e^{-2\zeta}\right\} + \alpha(\alpha-2)(1+\zeta)e^{-\zeta}}{1+\left(1+\zeta+\frac{1}{3}\zeta^2\right)e^{-\zeta}} + \frac{\alpha}{\zeta} \right] \frac{e^2}{a_0} \tag{63}$$

where $\zeta = \alpha\xi = \alpha R/a_0$. For $\alpha = 1$ we will get the results of Section 23.4. Now, we may write the above equation in the form

$$E_+ = \alpha F_1(\zeta) + \alpha^2 F_2(\zeta) \tag{64}$$

where

$$F_1(\zeta) = \left[\frac{-1-\frac{1}{\zeta}\left\{1-(1+\zeta)e^{-2\zeta}\right\} - 2(1+\zeta)e^{-\zeta}}{1+\left(1+\zeta+\frac{1}{3}\zeta^2\right)e^{-\zeta}} + \frac{1}{\zeta} \right] \frac{e^2}{a_0} \tag{65}$$

and

$$F_2(\zeta) = \left[-\frac{1}{2} + \frac{1+(1+\zeta)e^{-\zeta}}{1+\left(1+\zeta+\frac{1}{3}\zeta^2\right)e^{-\zeta}} \right] \frac{e^2}{a_0} \tag{66}$$

We take α and ζ as two independent parameters. Setting $dE_+/d\alpha = 0$, we get

$$\alpha = -F_1(\zeta)/2F_2(\zeta) \tag{67}$$

Thus,

$$E_+ = -\frac{\left[F_1(\zeta)\right]^2}{4F_2(\zeta)} \tag{68}$$

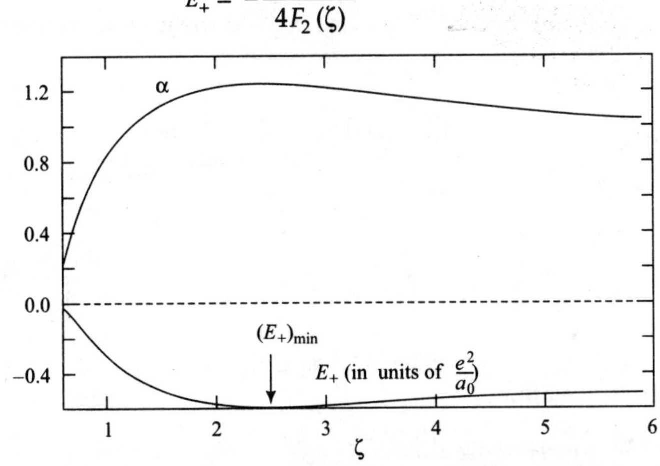

Figure 23.3: The variation of E_+ and α with ζ. The minimum value of $E_+(\simeq -0.5865\,\frac{e^2}{a_0} \simeq -15.95\text{ eV})$ occurs at $\zeta = 2.48$ at which $\alpha = 1.238$.

The above two equations enable us to plot α and E_+ as a function of ζ (see Fig. 23.3) the minimum value of $E_+ (\simeq -0.5865 \frac{e^2}{a_0} \simeq -15.95$ eV) occurs at

Thus, $$\zeta \simeq 2.48 \text{ for which } \alpha \simeq 1.238$$

Dissociation energy $= E_H - E_{H_2^+} \simeq -13.6 + 15.95 \text{ eV} = 2.35$ eV

$$R = \frac{\zeta}{\alpha} a_0 \simeq 2.003\, a_0 \approx 1.06\text{Å}$$

(b) By fitting the curve for E_+ to a parabola (around the minimum value) one can show

$$E_+ \approx [-0.5865 + 0.050\,(\zeta - 2.48)^2]\, \frac{e^2}{a_0}$$

$$\approx [-0.5865 + 2.74 \times 10^{15} (R - 1.06 \times 10^{-8})^2]\, \frac{e^2}{a_0}$$

where R is in centimeters. Thus,

$$\frac{d^2E_+}{dR^2} \approx 1.19 \times 10^5 \text{ ergs/cm}^2$$

Since $\mu = m_n m_P/(m_n + m_p) \approx 0.837 \times 10^{-24}$ gm, we have using Eq. (99) of Chapter 10, the following value for the zero-point energy

$$E_{vib} = \frac{1}{2}\hbar\sqrt{\frac{k}{\mu}} \approx \frac{1}{2} \times 10^{-27} \times \sqrt{\frac{1.19 \times 10^5}{0.837 \times 10^{-24}}}$$

$$\approx 1.89 \times 10^{-13} \text{ erg} \approx 0.12\text{eV}$$

Thus the dissociation energy becomes 2.23 eV.

Solution 23.5

$$E_\phi = \frac{\int \phi^* H \phi\, d\tau}{\int \phi^* \phi\, d\tau}$$

where $H = H_0 + H'$ (see Problem 20.15), $d\tau = d\mathbf{r}_1, d\mathbf{r}_2$ and the integral is over a six-dimensional space. Now,

$$u_0 = u_{100}(\mathbf{r}_1)\, u_{100}(\mathbf{r}_2) = \frac{1}{\pi a_0^3} \exp\left(-\frac{r_1 + r_2}{a_0}\right)$$

and $$H_0 u_0 = E_0 u_0,\; E_0 = -\frac{e^2}{2a_0} - \frac{e^2}{2a_0} = -\frac{e^2}{a_0}$$

Thus, $$\int \phi^* \phi\, d\tau = \int u_0^2\, d\tau + 2A \int u_0\, H' u_0\, d\tau + A^2 \int u_0\, H'^2\, u_0\, d\tau$$

$$= 1 + A^2 \int u_0\, H'^2\, u_0\, d\tau$$

where we have used the relation

$$\int u_0\, H'\, u_0\, d\tau = 0$$

which can be verified by direct integration. Further,

$$\int \phi^* H \phi\, d\tau = \int u_0\, (1 + AH')\,(H_0 + H')\,u_0\,(1 + AH')\,d\tau$$

$$= \int u_0 H_0 u_0 \, d\tau + \int u_0 H_0' u_0 \, d\tau$$

$$+ A\left[\int u_0 H' H_0 u_0 \, d\tau + \int u_0 H_0 (u_0 H') \, d\tau\right.$$

$$\left. + \int u_0 H' u_0 H' d\tau + \int u_0 H'^2 u_0 \, d\tau\right]$$

$$+ A^2\left[\int u_0 H' H_0 (u_0 H') \, d\tau + \int u_0 H'^3 u_0 \, d\tau\right]$$

$$= E_0 + \left[E_0 + A\int u_0 H' u_0 \, d\tau + E_0 \int u_0 u_0 H' \, d\tau + 2\int u_0 H'^2 u_0 \, d\tau\right]$$

$$= E_0 + 2A \int u_0 H'^2 u_0 \, d\tau$$

the other integrals going to zero. Thus,

$$E_\phi = \frac{E_0 + 2A \int u_0 H'^2 u_0 d\tau}{1 + A^2 \int u_0 H'^2 u_0 d\tau}$$

$$\approx E_0 + (2A - A^2 E_0) \int u_0 H'^2 u_0 d\tau$$

Setting $dE/dA = 0$, we get $A = 1/E_0$ thus the minimum value of E_ϕ is given by

$$E_\phi = E_0 \frac{1}{E_0} + \int u_0 H'^2 u_0 d\tau$$

The integral is evaluated in Problem 20.15, on substitution we get

$$E_\phi = E_0 - \frac{6e^2 a_0^5}{R^6}$$

Combining the result of Problem 20.15, we may write

$$E_0 - \frac{8e^2 a_0^5}{R^6} < \text{Ground state Energy} < E_0 - \frac{6e^2 a_0^5}{R^6}.$$

REFERENCES AND SUGGESTED READINGS

1. L. Pauling and E. B. Wilson, *Introduction to Quantum Mechanics*, McGraw-Hill Book Co., New York (1935).
2. L.I. Schiff, *Quantum Mechanics*, McGraw-Hill Book Co., New York (1968).
3. S. Flugge, *Practical Quantum Mechanics*, Springer Verlag, New York (1974).

24 The helium atom and the exclusion principle

24.1 INTRODUCTION

[illegible] Schrödinger's equation [illegible] atom as the Bohr model [illegible] are manifest [illegible] the most important [illegible] fundamental [illegible] shall first [illegible] go on to examine a few [illegible] of atomic structure.

24.2 THE INDEPENDENT PARTICLE MODEL

The potential energy [illegible] nucleus in the [illegible]

[illegible]

Here [illegible] is assumed to be stationary [illegible] Fig. 24.1; for the helium atom, $Z = 2$. Now the term V_{12} is of the same order [illegible] the other two terms in the potential energy; nevertheless, we shall begin by neglecting it. (We shall see later how to incorporate this term.) The resulting approximation [illegible] independent particle model [illegible]

24 The helium atom and the exclusion principle

... According to this point of view the doublet structure of alkali spectra as well as the deviation from Larmor's theorem is due to a particular two-valuedness of the quantum theoretic properties of the electron, which cannot be described from the classical point of view.

— WOLFGANG PAULI (1924).

24.1 INTRODUCTION

Schrödinger's equation, as we have seen in chapter 10, gives the same energies for the hydrogen atom as the Bohr model. It is only when we consider many electron systems that new effects are manifest, the most important being those due to the Pauli exclusion principle which is fundamental to an understanding of the building up of the periodic table. In this chapter, we shall first consider the ground state of a two-electron system, namely the helium atom and then go on to examine a few excited states in order to understand the basic principles of the study of atomic structure.

24.2 THE INDEPENDENT PARTICLE APPROXIMATION

The potential energy of the helium-like atom contains an extra term, apart from the electron nucleus interaction: namely, the repulsive energy of the two electrons. The Hamiltonian is

$$H = \frac{p_1^2}{2m} + \frac{p_2^2}{2m} - \frac{Ze^2}{r_1} - \frac{Ze^2}{r_2} + \frac{e^2}{r_{12}}$$

$$= -\frac{\hbar^2}{2m}\left[\nabla_1^2 + \nabla_2^2\right] - \frac{Ze^2}{r_1} - \frac{Ze^2}{r_2} + \frac{e^2}{r_{12}} \qquad (1)$$

Here $\mathbf{r}_1$ and $\mathbf{r}_2$ represent the coordinates of the two electrons with respect to the nucleus (which is assumed to be stationary) and $r_{12} = |\mathbf{r}_1 - \mathbf{r}_2|$ is the distance between the two electrons—see Fig. 24.1; for the helium atom, $Z = 2$. Now the term e^2/r_{12} is of the same order of magnitude as the other two terms in the potential energy; nevertheless, we shall begin by neglecting it. (We shall see, ater, how to incorporate this term.) The resulting approximation may be called the *independent particle model*. In this approximation, the Schrödinger equation

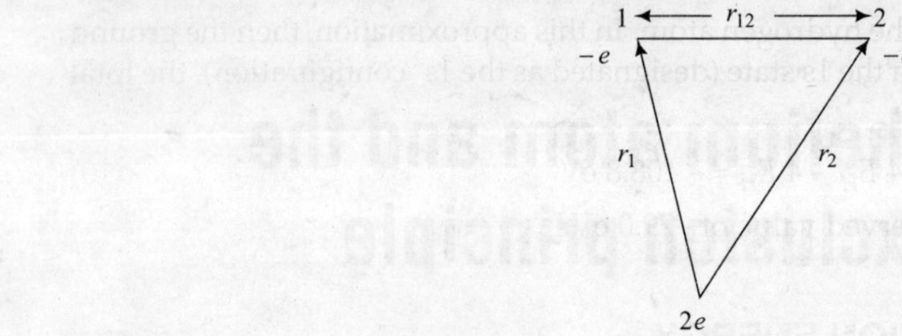

Figure 24.1: The helium atom consists of a nucleus of charge 2*e* and two electrons.

$$\left(-\frac{\hbar^2}{2m}\nabla_1^2-\frac{\hbar^2}{2m}\nabla_2^2-\frac{Ze^2}{r_1}-\frac{Ze^2}{r_2}\right)u\,(\mathbf{r}_1,\mathbf{r}_2)=E\,u\,(\mathbf{r}_1,\mathbf{r}_2) \tag{2}$$

may be separated by writing

$$u\,(\mathbf{r}_1,\mathbf{r}_2)=u_a\,(\mathbf{r}_1)\,u_b\,(\mathbf{r}_2) \tag{3}$$

giving identical equations for u_a and u_b;

$$\left(-\frac{\hbar^2}{2m}\nabla_1^2-\frac{Ze^2}{r_1}\right)u_a\,(\mathbf{r}_1)=E_a\,u_a\,(\mathbf{r}_1) \tag{4}$$

$$\left(-\frac{\hbar^2}{2m}\nabla_2^2-\frac{Ze^2}{r_2}\right)u_b\,(\mathbf{r}_2)=E_b\,u_b\,(\mathbf{r}_2) \tag{5}$$

with
$$E=E_a+E_b \tag{6}$$

Equations (4) and (5) tell us that the two electrons 1 and 2 move independently of each other and the total energy of the system is simply the sum of the energies of the non-interacting electrons.

Since Eqs (4) and (5) are the same as for a hydrogen-like atom, the solutions are immediately written down (see Section 10.2)

$$u_a\,(\mathbf{r}_1)=\mathrm{R}_{n_1 l_1}\,(r_1)Y_{l_1 m_1}(\theta_1,\phi_1) \tag{7}$$

and a similar expression for $u_b(\mathbf{r}_2)$. A particular eigenstate is characterized by specifying *a* and *b*, i.e., by specifying the two sets of quantum numbers n_1, l_1, m_1 and n_2, l_2, m_2.

Similarly, the energy eigenvalues can easily be written down. Both E_a and E_b would be of the form[1]

$$E_n=-\frac{Z^2}{n^2}\frac{me^4}{2\hbar^2}=-\frac{Z^2}{n^2}E_H \quad \text{Energy eigenvalues} \tag{8}$$

where
$$E_H=\frac{me^4}{2\hbar^2}\simeq 13.605\text{ eV} \tag{9}$$

[1] The Hamiltonian given by Eq. (1) assumes infinite mass of the nucleus. The effect of the finite mass of the nucleus can be approximately taken into account by replacing the electron mass *m* by the reduced mass μ (= $mM/(m+M)$) where *M* is the mass of the nucleus—see Problem 24.6 (cf. Section 10.2).

represents the ionization energy of the hydrogen atom. In this approximation, then the ground state is the one with both electrons in the 1s state (designated as the $1s^2$ configuration), the total energy being (for Z = 2)

$$-4\,E_H - 4\,E_H = -\,108.8\text{ eV} \tag{10}$$

This is to be compared with the observed value of –79.0 eV.

24.3 ELECTRON INTERACTION ENERGY

One way to estimate this is to regard e^2/r_{12} as a perturbation. The first order correction to the ground state energy is, then

$$\left\langle \frac{e^2}{r_{12}} \right\rangle = \iint u_{100}^*(\mathbf{r}_1)\, u_{100}^*(\mathbf{r}_2) \frac{e^2}{|\mathbf{r}_1 - \mathbf{r}_2|} u_{100}(\mathbf{r}_1) u_{100}(\mathbf{r}_2)\, d\tau_1\, d\tau_2 \tag{11}$$

which has been evaluated in Problem 24.1, the result is

$$\left\langle \frac{e^2}{r_{12}} \right\rangle = \frac{5Z}{4} E_H \approx 34.02\text{ eV} \tag{12}$$

where we have assumed Z = 2. Thus the total energy is –74.8 eV, only 4.2 eV above the experimental value. Such a large 'correction' makes the perturbation approach suspect and we turn to other methods.

24.4 VARIATIONAL METHOD

The effect of the repulsive term, e^2/r_{12}, is to reduce the nuclear attraction on each electron. One way to acccommodate this is to assume hydrogen-like wavefunctions corresponding to an effective nuclear charge Z′, which will be assumed to be a variable parameter. Thus, we assume the (normalized) wavefunction as [cf. Eq. (65) of chapter 10]:

$$\Phi(\mathbf{r}_1, \mathbf{r}_2) = \left(\frac{Z'^3}{\pi a_0^3} \right) e^{-Z' r_1/a_0}\, e^{-Z' r_2/a_0} \tag{13}$$

The Hamiltonian, however, remains as in Eq. (1) with Z as a fixed number; for He, Z = 2. We have to minimize the integral (see Sections 23.2 and 23.4):

$$E(Z') = \iint \Phi^* \left[-\frac{\hbar^2}{2m} \left(\nabla_1^2 + \nabla_2^2 \right) - Ze^2 \left(\frac{1}{r_1} + \frac{1}{r_2} \right) + \frac{e^2}{r_{12}} \right] \Phi(\mathbf{r}_1, \mathbf{r}_2)\, d\tau_1\, d\tau_2 \tag{14}$$

which has been evaluated in Problem 24.2. The result is

$$\mathrm{E}(Z') = \left[-2Z'^2 + 4Z'(Z' - Z) + \frac{5}{4} Z' \right] E_H \tag{15}$$

The minimum of $E(Z')$ occurs when

$$Z' = Z_{\text{eff}} = Z - \frac{5}{16} \approx 1.69 \tag{16}$$

so that
$$E = -2\,Z_{\text{eff}}^2 E_H$$
$$\approx -76.2 \text{ eV} \tag{17}$$

Other trial functions (having more complicated forms) have also been used, some yielding values within 0.01% of the observed one – see Table 23.1.

24.5 EXCHANGE DEGENERACY AND IDENTICAL PARTICLES

In Section 24.2, we saw that in the independent particle model, the wavefunction

$$u_a(\mathbf{r}_1)\, u_b(\mathbf{r}_2) \tag{18}$$

is a solution of the Schrödinger equation corresponding to the energy $E = E_a + E_b$. Since the Schrödinger equation [Eq. (2)] is completely symmetrical in the two electrons, it is obvious that the wavefunction[2]

$$u_b(\mathbf{r}_1)\, u_a(\mathbf{r}_1) \tag{19}$$

is also a solution of the Schrödinger equation corresponding to the *same* eigenvalue $E_a + E_b$. Thus, we have a degeneracy which is known as the *exchange degeneracy* because it arises due to the possibility of exchanging the two electrons.[3] We could also construct the linear combinations

$$u_+ = \frac{1}{\sqrt{2}}\,[u_a(\mathbf{r}_1)\, u_b(\mathbf{r}_2) + u_a(\mathbf{r}_2)\; u_b(\mathbf{r}_1)] \tag{20}$$

$$u_- = \frac{1}{\sqrt{2}}\,[u_a(\mathbf{r}_1)\, u_b(\mathbf{r}_2) - u_a(\mathbf{r}_2)\; u_b(\mathbf{r}_1)] \tag{21}$$

which are also solutions of Eq. (2) belonging to the same energy $E_a + E_b$; the factor $\frac{1}{\sqrt{2}}$ has been introduced so that the wavefunctions are normalized. The solutions u_+ and u_- are known as the symmetric and antisymmetric solutions because when the indices 1 and 2 are interchanged, u_+ remains the same whereas u_- changes sign. We will shortly see that it is essential that we use either the symmetrical or antisymmetrical solution rather than solutions given by Eq. (18) and Eq. (19) which are neither symmetric nor antisymmetric. This is due to the indistinguishability of identical particles like electrons. It may be mentioned that in classical mechanics, identical particles are always distinguishable in the sense that it is possible to keep track of individual particles. However, in quantum mechanics, since it is not possible to keep track of individual particles (without disturbing the system), identical particles are indistinguishable. Thus considering a two-electron system, if the two electrons are interchanged, we must get the same wavefunction except (possibly) for an unimportant phase factor[4], i.e.,

---

[2] The wavefunction given by Eq. (18) corresponds to the state where the first electron is in the state *a* and the second electron in the state *b* whereas Eq. (19) corresponds to the first electron in state *b* and the second electron in state *a*.

[3] The exchange degeneracy exists as long as we neglect the electron-electron interaction. If the interaction between electrons is taken into account, the levels, in general, will split up (see Section 24.6).

[4] One could also argue that the probability of finding the first electron at $\mathbf{r}_1$ and the second electron at $\mathbf{r}_2$ must be equal to the probability of finding the first electron at $\mathbf{r}_2$ and the second electron at $\mathbf{r}_1$ giving

$$|u(\mathbf{r}_1, \mathbf{r}_2)|^2 = |u(\mathbf{r}_2, \mathbf{r}_1)|^2$$

which would lead to Eq. (22).

$$u(\mathbf{r}_1, \mathbf{r}_2) = e^{i\phi}\, u(\mathbf{r}_2, \mathbf{r}_1) \tag{22}$$

where ϕ is a real constant. Now, if P_{12} represents the particle interchange operator then

$$P_{12}\, u(\mathbf{r}_1, \mathbf{r}_2) = u(\mathbf{r}_2, \mathbf{r}_1) \tag{23}$$

and
$$P_{12}^2\, u(\mathbf{r}_1, \mathbf{r}_2) = P_{12}\, u(\mathbf{r}_2, \mathbf{r}_1) = u(\mathbf{r}_1, \mathbf{r}_2) \tag{24}$$

Further
$$\begin{aligned} P_{12}^2\, u(\mathbf{r}_1, \mathbf{r}_2) &= P_{12}\, u(\mathbf{r}_2, \mathbf{r}_1) = e^{-i\phi}\, P_{12}\, u(\mathbf{r}_1, \mathbf{r}_2) \\ &= e^{-i\phi}\, u(\mathbf{r}_2, \mathbf{r}_1) \\ &= e^{-2i\phi}\, u(\mathbf{r}_1, \mathbf{r}_2) \end{aligned}$$

Thus,
$$e^{-2i\phi} = +1 \text{ or } e^{i\phi} = \pm 1$$

giving
$$u(\mathbf{r}_1, \mathbf{r}_2) = \pm\, u(\mathbf{r}_2, \mathbf{r}_1) \tag{25}$$

The above equation shows that the wavefunction should be either symmetrical or antisymmetrical with respect to the interchange of the indices.

It may be mentioned that if the two particles are in the symmetric (or antisymmetric) state then they will remain in the symmetric (or antisymmetric) state for all times. This follows from the Schrödinger equation

$$\dot{\Psi} = \frac{1}{i\hbar} H \Psi \tag{26}$$

implying that $\dot{\Psi}$ has the same symmetry as Ψ (because H is always symmetrical with respect to the change of indices as the two electrons are identical; see, e.g. Eq. (1)). Hence if $\Psi(\mathbf{r}_1, \mathbf{r}_2, t = 0)$ is symmetric then it will remain symmetric at Δt and so on.

24.6 THE PAULI EXCLUSION PRINCIPLE

Till now we have not considered the spin angular momentum associated with the electron. We denote the spin functions by $|\chi_\pm\rangle$ where $|\chi_+\rangle$ and $|\chi_-\rangle$ represent the spin up and spin down states[5]. For the two electron system, the spin functions are $|\chi_\pm(1)\rangle$ and $|\chi_\pm(2)\rangle$ where the numbers in the parenthesis refer to the first and second electron. Because of the indistinguishability of the electrons, we must construct appropriate linear combinations which are either symmetric or antisymmetric with respect to the interchange of indices; these states are denoted by

$$|\chi_t^+\rangle = |\chi_+^{(1)}\chi_+^{(2)}\rangle = |1\uparrow 2\uparrow\rangle \tag{27}$$

$$|\chi_t^-\rangle = |\chi_-^{(1)}\chi_-^{(2)}\rangle = |1\downarrow 2\downarrow\rangle \tag{28}$$

$$|\chi_t^0\rangle = \frac{1}{\sqrt{2}}\left[|\chi_+(1)\chi_-(2)\rangle + |\chi_-(1)\chi_+(2)\rangle\right]$$

---

[5] Thus, if the electron is in the spin up (or spin down) state then the measurement of the z-component of the spin angular momentum will lead to $+\frac{1}{2}\hbar$ (or $-\frac{1}{2}\hbar$). At some places it will be convenient to designate the states $|\chi_+\rangle$ and $|\chi_-\rangle$ as $|\uparrow\rangle$ and $|\downarrow\rangle$ respectively.

$$= \frac{1}{\sqrt{2}}\left[|1\uparrow 2\downarrow\rangle + |1\downarrow 2\uparrow\rangle\right] \tag{29}$$

$$|\chi_s\rangle = \frac{1}{\sqrt{2}}\left[|\chi_+(1)\chi_-(2)\rangle - |\chi_-(1)\chi_+(2)\rangle\right]$$

$$= \frac{1}{\sqrt{2}}\left[|1\uparrow 2\downarrow\rangle - |1\downarrow 2\uparrow\rangle\right] \tag{30}$$

the first three functions being symmetric and the last one being antisymmetric. The first function, for example, corresponds to both the functions being in the spin up state and so on. Now, in Section 19.3 we had seen that the first three states (known as the triplet state) corresponds to total spin $S = 1$ and the fourth state (known as the singlet state) corresponds to the total spin S = 0; in Eqs (27)–(29), χ_t^+, χ_t^0 and χ_t^- correspond to $S = 1$ (with $m_s = 1, 0, -1$) and χ_s correspond to $S = 0$.

Ground state wavefunction

We consider the ground state ($1s^2$) of the helium atom. The total wavefunction will be

$$u_{100}(\mathbf{r}_1)\, u_{100}(\mathbf{r}_2) \tag{31}$$

multiplied by the appropriate spin function. Now, it is an experimental fact that the ground state of the helium atom does not split up in a magnetic field; consequently it must be a singlet rather than a triplet state, i.e., we must have the following wavefunction representing the ground state

$$u_{100}(\mathbf{r}_1)\, u_{100}(\mathbf{r}_2)\ \frac{1}{\sqrt{2}}\left[|\chi_+(1)\chi_-(2)\rangle - |\chi_-(1)\chi_+(2)\rangle\right] \tag{32}$$

Thus the ground state wavefunction is antisymmetric[6]. Analysis of higher excited states of helium (and also of other atoms) shows that the *wavefunction for a system of electrons must be antisymmetric with respect to the interchange of two electrons*. Consequently, if the space dependent function is symmetric then the spin function should be antisymmetric and *vice versa*. The general wavefunction (for a 2 electron system) should, therefore, be of the form

$$\psi = \frac{1}{\sqrt{2}}\left[\psi_a(1)\,\psi_b(2) - \psi_b(1)\,\psi_a(2)\right] \quad \text{General wavefunction} \tag{33}$$

where the subscripts a and b include the spin state, i.e., they represent n, l, m and m_s. It is obvious that if $a = b$ (i.e., if both the electrons have the same set of values of n, l, m and m_s) then

$$\psi = 0$$

[6] The ground state is denoted by $^1S_0$. The superscript 1 denotes a singlet (total spin = 0), S denotes $L = 0$, the subscript 0 is to indicate $J = 0$. It is unfortunate that an S state here may be confused with total spin; certain coventions have to be adhered to, however!

which is the Pauli exclusion principle[7] according to which in an atom no two electrons can have the same set of values of the four quantum numbers n, l, m and m_s. Thus the antisymmetric nature of the total wavefunction leads automatically to the Pauli exclusion principle.

For a system of N electrons, the antisymmetric wavefunction is given by the Slater determinant

$$\psi = \frac{1}{\sqrt{N!}} \begin{vmatrix} \psi_{a_1}(1)\,\psi_{a_1}(2)\ldots & \psi_{a_1}(N) \\ \psi_{a_2}(1)\,\psi_{a_2}(2)\ldots & \\ \vdots & \\ \psi_{a_N}(1) & \ldots\psi_{a_N}(N) \end{vmatrix} \quad \text{Slater determinant} \tag{34}$$

It is immediately obvious that if $a_1 = a_2$ then the two rows would become the same and the determinant would vanish.

Particles that are described by antisymmetric wavefunctions are known as Fermions and they are said to obey Fermi-Dirac statistics; example's of such particles are electrons, protons, neutrons, etc. On the other hand, particles which are described by symmetric wavefunctions are known as *Bosons* and they are said to obey Bose-Einstein statistics; examples of such particles are photons, pions, etc. Pauli has proved that particles with half integral spin are Fermions and particles with integral spin are Bosons; the proof is, however, much beyond the scope of this book.

24.7 EXCITED STATES OF HELIUM[8]

For the first excited state, we have the possible configurations $1s2s$ and $1s2p$. Once again, if we ignore the electron-electron interaction, we have the independent particle model and the spatial part of the wavefunctions would be of the form

$$u_{100}(\mathbf{r}_1)\, u_{2l0}(\mathbf{r}_2) \tag{35}$$

where $l = 0$ for $2s$ and $l = 1$ for $2p$ states. This wavefunction, however, does not have the symmetry property required and the correct combinations are

$$u_+ = \frac{1}{\sqrt{2}}\left[u_{100}(\mathbf{r}_1)\, u_{2l0}(\mathbf{r}_2) + u_{100}(\mathbf{r}_2)\, u_{2l0}(\mathbf{r}_1)\right] \tag{36}$$

[7] At this stage it may be worthwhile to mention that Pauli had put forward the exclusion principle before Uhlenbeck and Goudsmidt had introduced the idea of the spinning electron. In order to give satisfactory explanations of the spectra of alkali atoms and of the anamolous Zeeman effect, Pauli argued that for a general classification of each of the electron in an atom, four quantum numbers are needed. In his *Nobel Lecture*, Pauli mentions that 'I proposed the assumption of a new quantum theoretic property of the electron, which I called a *two valuedness not describable classically*' He then put forward his exclusion principle according to which there can never be two or more equivalent electrons in an atom which have the same values of all four quantum numbers. Pauli in his *Nobel Lecture* also mentions that 'the physicists found it difficult to understand the exclusion principle, since no meaning in terms of a model was given to the fourth degree of freedom of the electron. The gap was filled by Uhlenbeck and Goudsmit's idea of the electron spin which made it possible to understand the anomalous Zeeman effect simply by assuming that the spin quantum number of one electron is $\frac{1}{2}$ and that the quotient of the magnetic moment to the mechanical angular moment has for the spin a value twice as large as for the ordinary orbit of the electron.' For a nice discussion on the historical development, the reader is referred to the Nobel lecture of Pauli and to Refs 1 and 2; the original article of Pauli has been reprinted in Ter Haar's book (Ref. 1); see also Section 13.7.

[8] The ground state of lithium and the corresponding exchange correction has been discussed Ref. 3.

and $$u_- = \frac{1}{\sqrt{2}}\left[u_{100}(\mathbf{r}_1)\, u_{2l0}(\mathbf{r}_2) - u_{100}(\mathbf{r}_2)\, u_{2l0}(\mathbf{r}_1)\right] \tag{37}$$

(For the ground state $1s^2$ configuration, u_- would vanish identically).

We now incorporate spin wavefunctions. We have already seen that the singlet state $|\chi_s\rangle$ is antisymmetric and the triplet state $|\chi_t\rangle$ is symmetric [see Eqs (27)–(30)]. In accordance with the exclusion principle, only the following antisymmetric combinations are permissible

$$\psi_s = u_+|\chi_s\rangle \quad \text{(singlet)} \tag{38}$$

$$\psi_t = u_-|\chi_t\rangle \quad \text{(triplet)} \tag{39}$$

where the subscripts s and t refer to the singlet and triplet states respectively. Now, if we ignore the interaction of the electrons, these four states are degenerate; its inclusion lifts the degeneracy. Qualitatively, we can see this as follows. For a space symmetric wavefunction, there is a considerable probability for the two electrons to come close together (i.e. u_+ can be large for $\mathbf{r}_1 \approx \mathbf{r}_2$) whereas for the space antisymmetric state, this probability is quite small. Thus, the repulsive energy of the two electrons gives a larger positive contribution for u_+ than for u_-. One, therefore, expects triplet states to lie lower than the singlet ones. We must emphasize that this difference in energy is *not* due to spin dependent forces but is a consequence of the exclusion principle and the electrostatic interaction of the two electrons.

The configuration ls2*s* gives the singlet state $^1S_0$ and the triplet $^3S_1$ since we have $L = 0$ and $S = 0$ or 1, the subscript gives the J value. One can estimate the difference in their energies by a perturbation calculation. The first order correction to their energy is

$$\Delta E_s = \iint u_+^* \frac{e^2}{r_{12}} u_+\, d\tau_1 d\tau_2 \quad \text{(singlet)} \tag{40}$$

$$\Delta E_t = \iint u_-^* \frac{e^2}{r_{12}} u_-\, d\tau_1 d\tau_2 \quad \text{(triplet)} \tag{41}$$

where $$r_{12} = |\mathbf{r}_1 - \mathbf{r}_2| = r_{21}$$

Since[9] $$\langle\chi_s|\chi_s\rangle = 1 = \langle\chi_t|\chi_t\rangle \tag{42}$$

and since e^2/r_{12} does not involve spin, only the space part of the wavefunction appears in the perturbation calculation. Next, we substitute for u_+ and u_- (from Eqs (36) and (37) with $l = 0$) in Eqs (40) and (41) to obtain

$$\Delta E_s = J_0 + K_0 \tag{43}$$

$$\Delta E_t = J_0 - K_0 \tag{44}$$

where $$J_l = e^2 \iint u_{100}^*(\mathbf{r}_1) u_{2l0}^*(\mathbf{r}_2) \frac{1}{r_{12}} u_{100}(\mathbf{r}_1) u_{2l0}(\mathbf{r}_2)\, d\tau_1\, d\tau_2 \tag{45}$$

and $$K_l = e^2 \iint u_{100}^*(\mathbf{r}_1) u_{2l0}^*(\mathbf{r}_2) \frac{1}{r_{12}} u_{100}(\mathbf{r}_2) u_{2l0}(\mathbf{r}_1)\, d\tau_1\, d\tau_2 \quad \text{Exchange integral} \tag{46}$$

and in arriving at Eqs (43) and (44) we have used the fact that in the integrals J_l and K_l, we can interchange $\mathbf{r}_1$ and $\mathbf{r}_2$.

---

[9] It is because of the relation $\langle\chi_s|\chi_s\rangle = 1$ that the perturbation to the ground state energy will be simply Eq. (11) although the ground state wavefunction is given by Eq. (32).

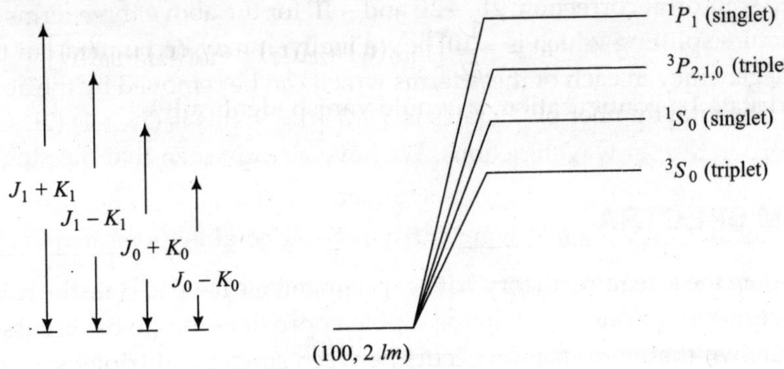

Figure 24.2: The splitting of the first excited state of the helium atom.

The integral J_l (for $l = 0$) is of the same type as we had encountered in the ground state energy shift calculation, the term K_l is known as the *exchange integral*. The evaluation is tedious but straightforward and yields

$$J_0 \approx 9.1 \text{ eV} \qquad (47)$$
$$K_0 \approx 0.4 \text{ eV}$$

A similar treatment can be given for the 1*s* 2*p* configuration. Since $l_1 = 0$ and $l_2 = 1$, $L = 1$ and for the singlet state ($S = 0$) J takes the value 1; on the other hand, for the triplet state ($S = 1$) J takes the values 2, 1, 0. Thus the possible states are

$$^1P_1 \text{ and } ^3P_{2,1,0}$$

(to be read as 'singlet *P* one' and 'triplet *P* two, one and zero'). Again the triplet levels lie lower than the singlet state, in fact

$$J_1 \approx 10.2 \text{ eV} \qquad K_1 \approx 0.1 \text{ eV} \qquad (48)$$

The corresponding energy levels are given in Fig. 24.2 (see also Problem 24.4).

24.8 FINE STRUCTURE

In order to remove the degeneracy within the triplet levels, it is necessary to include the spin-orbit interaction. This is the interaction between the magnetic moment (associated with spin) and the orbital motion of the electron and is represented by a term of the form

$$\sum_i \xi(r_i)\mathbf{L}\cdot\mathbf{S} \qquad (49)$$

in the Hamiltonian. For a state with a particular value of *L*, *S* and *J*, we have

$$\langle \mathbf{L}\cdot\mathbf{S}\rangle = \frac{1}{2}[J(J+1) - L(L+1) - S(S+1)] \qquad (50)$$

which gives 1, –1 and –2 for the $^3P_2$, $^3P_1$ and $^3P_0$ states respectively. Writing

$$2\Gamma = \int d\tau\, \psi^* \left(\sum_i \xi(r_i)\right)\Psi \qquad (51)$$

we have, for the spin-orbit correction, 2Γ, -2Γ and -4Γ for the above three terms giving the so called fine structure splitting which is ~ 10^{-4} eV. Finally, it may be pointed out that there is a further $2J + 1$ degeneracy in each of these terms which can be removed by the application of a magnetic field. The total number of ^3P states is $5 + 3 + 1 = 9$; this is $(2S + 1)(2L + 1)$ as may be expected.

24.9 HELIUM SPECTRA

In order to compare the foregoing theory with experiment, we have to state the rules governing transitions between the various states. In the dipole approximation (to be discussed in Section 26.3), it can be shown that no transition occurs between singlet and triplet states. As a result, the energy level diagram of helium splits into two parts as shown in Fig. 24.3. The figure shows the energy levels when one of the electrons is in the 1*s* state and the other in *nl* state. Note the following features.

1. Singlet levels, belonging to a particular value of *n* and *l* have higher energy than triplet levels of the same *n* and *l*.
2. For $n = 2$, the difference $^1S - {}^3S$ is larger than $^1P - {}^3P$ difference. This is because the exchange integral *K* for the *P* state is smaller; there is a smaller contribution from those regions where $1/r_{12}$ is large.
3. The states 2^3S and 2^1S are *metastable*. The $2^1S \to 1^1S$ transition is forbidden because of the selection rule $\Delta l = \pm 1$ for dipole transitions and the $2^3S \to 1^1S$ for this reason as well as the one mentioned earlier, forbidden triplet-singlet transitions. Thus helium exists as two 'gases' as it were, parahelium[10] atoms with $S = 0$ and orthohelium with $S = 1$.

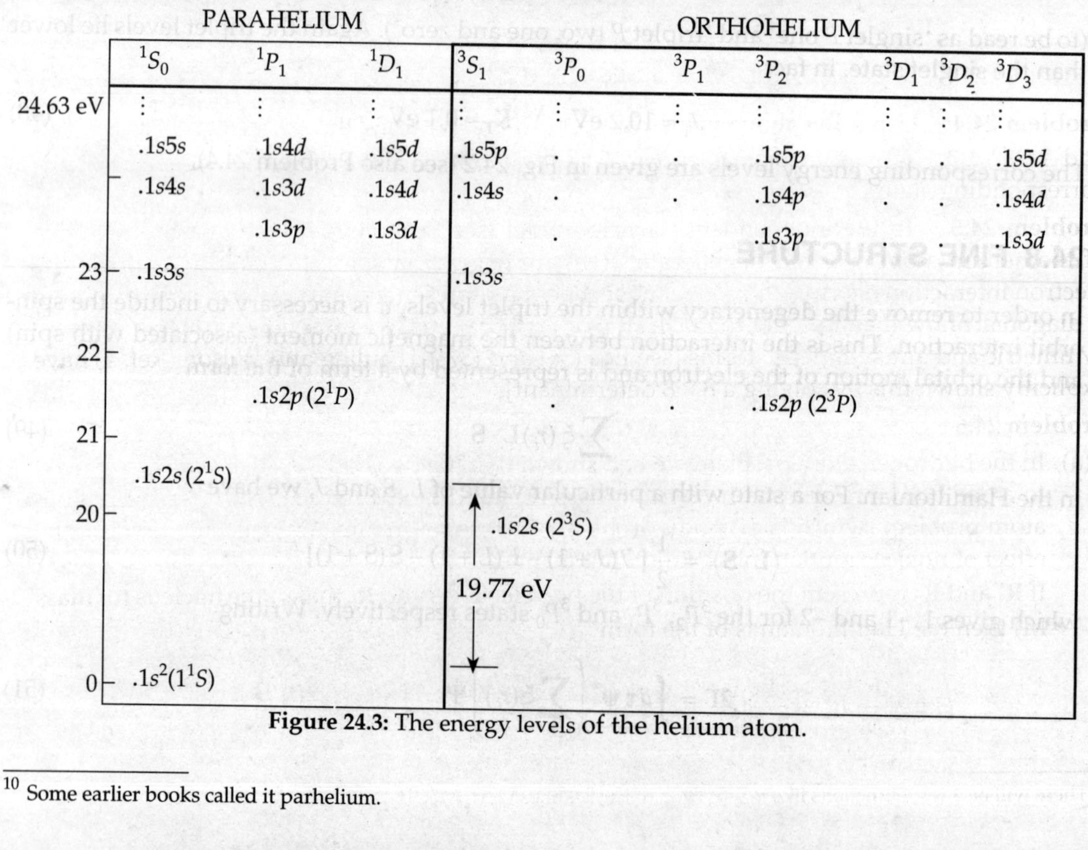

Figure 24.3: The energy levels of the helium atom.

[10] Some earlier books called it parhelium.

24.10 PROBLEMS

Problem 24.1 Evaluate the integral in Eq. (11) and calculate the ground state energy of 2 electron systems like He, Li^+, Be^{++} and B^{+++}. Determine the ionization energy and compare with the corresponding experimental values which are 24.63 eV, 75.64 eV, 153.87 eV and 259.31 eV respectively.

Problem 24.2 Evaluate the integral in Eq. (14) and calculate the ionization energy for 2-electron systems. Compare with the corresponding result obtained by using perturbation theory.

Problem 24.3 Using the expressions

$$u_{100}(\mathbf{r}) = \frac{1}{\sqrt{4\pi}}\left(\frac{Z}{a_0}\right)^{3/2} 2e^{-Zr/a_0} \tag{52}$$

$$u_{200}(\mathbf{r}) = \frac{1}{\sqrt{4\pi}}\left(\frac{Z}{2a_0}\right)^{3/2}\left(2 - \frac{Zr}{a_0}\right)e^{-Zr/2a_0} \tag{53}$$

and

$$u_{210}(\mathbf{r}) = \left(\frac{Z}{a_0}\right)^{3/2}\frac{Zr}{a_0\sqrt{3}}e^{-Zr/2a_0}\sqrt{\frac{3}{4\pi}}\cos\theta \tag{54}$$

evaluate the integrals in Eqs (45) and (46). [You will have to use the relation given by Eq. (63)]. For Z = 2 show that

$$J_0 \approx 9.1 \text{ eV}$$
$$K_0 \approx 0.4 \text{ eV}$$
$$J_1 \approx 10.2 \text{ eV}$$
$$K_1 \approx 0.1 \text{ eV}$$

Problem 24.4 Using the above values of J_0, K_0, J_1 and K_1 draw the energy levels of the 1*s*2*s* and 1*s*2*p* states of helium atom (see Fig. 24.2) and calculate the ionization energies of the corresponding states.

Problem 24.5 In the independent particle model (see Section 24.2), the first excited state of the helium atom is 8-fold degenerate[11] (we are neglecting spin). In order to evaluate the electron interaction energy (e^2/r_{12}) we must choose a representation in which the perturbation is diagonal. Show that, in order to make H' $(= e^2/r_{12})$ diagonal we must choose the appropriate symmetric and antisymmetric forms [see Eqs (36) and (37)]. [Pauling and Wilson (Ref. 4) have explicitly shown this by solving a 8×8 determinant].

Problem 24.6

(a) In the hydrogen atom problem we had shown that the two body problem can be reduced to a one body problem with electron mass replaced by the reduced mass. In the helium atom problem (which is a 3 body problem) one can approximately take into account the effect of nuclear motion in the following manner:

If $\mathbf{R}_1$ and $\mathbf{R}_2$ represent the position of the two electrons and $\mathbf{R}_n$ that of the nucleus (of mass M) then the Hamiltonian is of the form

$$H = \frac{1}{2M}P_n^2 + \frac{1}{2m}\left(P_1^2 + P_2^2\right) - e^2\left(\frac{2}{|\mathbf{R}_1 - \mathbf{R}_n|} + \frac{2}{|\mathbf{R}_2 - \mathbf{R}_n|} - \frac{1}{|\mathbf{R}_1 - \mathbf{R}_2|}\right) \tag{55}$$

[11] There will be 4 wavefunctions like $u_{100}(\mathbf{r}_1)\, u_{2lm}(\mathbf{r}_2)$ and 4 like $u_{2lm}(\mathbf{r}_1)\, u_{100}(\mathbf{r}_2)$.

where $$P_i^2 = -\hbar^2\left[\frac{\partial^2}{\partial X_i^2}+\frac{\partial^2}{\partial Y_i^2}+\frac{\partial^2}{\partial Z_i^2}\right]$$

$$= -\hbar^2\frac{\partial^2}{\partial \mathbf{R}_i^2} \qquad i = n,\ 1 \text{ and } 2 \tag{56}$$

Introduce the coordinates

$$\mathbf{r}_1 = \mathbf{R}_1 - \mathbf{R}_n,\ \mathbf{r}_2 = \mathbf{R}_2 - \mathbf{R}_n \tag{57}$$

and the center of mass coordinate **R**

$$(M + 2m)\mathbf{R} = M\mathbf{R}_n + m\,(\mathbf{R}_1 + \mathbf{R}_2) \tag{58}$$

and show that

$$H = \frac{P^2}{2(M+2m)} + \frac{1}{2\mu}\left(p_1^2 + p_2^2\right) + \frac{1}{M}\,\mathbf{p}_1\cdot\mathbf{p}_2$$

$$+e^2\left(\frac{2}{r_1}+\frac{2}{r_2}-\frac{1}{|\mathbf{r}_1-\mathbf{r}_2|}\right) \tag{59}$$

where $$\mu = \frac{mM}{m+M} \tag{60}$$

is the reduced mass of an electron with the nucleus. Since $M \gg \mu$ the term $\frac{1}{M}\,\mathbf{p}_1\cdot\mathbf{p}_2$ will be small in comparison to the preceeding term and in the first approximation can be neglected. In this approximation, the resulting Hamiltonian is of the same form as Eq. (1) except for m being replaced by μ and for an additional term involving the translational motion of the center of mass.

(b) Use first order perturbation theory to show that using a wavefunction of the form given by Eq. (3), the effect of the term $\frac{1}{M}\,\mathbf{P}_1\cdot\mathbf{P}_2$ is zero.

24.11 SOLUTIONS

Solution 24.1 The normalized hydrogen atom ground state wavefunction is given by Eq. (52). Thus,

$$u_{100}(\mathbf{r}_1)\ u_{100}(\mathbf{r}_2) = \frac{1}{\pi}\left(\frac{Z}{a_0}\right)^3 \exp\left[-Z\frac{(r_1+r_2)}{a_0}\right] \tag{61}$$

Substituting in Eq. (11), we get

$$\left\langle\frac{e^2}{r_{12}}\right\rangle = \frac{Z^6e^2}{\pi^2a_0^6}\int d\tau_1 e^{-2Zr_1/a_0}\int d\tau_2 e^{-2Zr_2/a_0}\frac{1}{|\mathbf{r}_1-\mathbf{r}_2|} \tag{62}$$

Now,
$$\frac{1}{|\mathbf{r}_1-\mathbf{r}_2|} = \begin{cases}\dfrac{1}{r_1}\displaystyle\sum_l\left(\frac{r_2}{r_1}\right)^l P_l(\cos\theta) & \text{for } r_1 \ge r_2\\[2ex] \dfrac{1}{r_2}\displaystyle\sum_l\left(\frac{r_1}{r_2}\right)^l P_l(\cos\theta) & \text{for } r_2 \ge r_1\end{cases} \tag{63}$$

where θ is the angle between $\mathbf{r}_1$ and $\mathbf{r}_2$ (see, e.g. p. 176 of Ref. 5). Further, in evaluating the second integral, if we choose the z_2-axis along $\mathbf{r}_1$, then

$$d\tau_2 = r_2^2\, dr_2 \sin\theta\, d\theta\, d\phi$$

Introducing
$$\rho_1 = \frac{2Zr_1}{a_0} \text{ and } \rho_2 = \frac{2Zr_2}{a_0}$$

we get
$$\left\langle \frac{e^2}{r_{12}} \right\rangle = \frac{Ze^2}{32\pi^2 a_0} \iiint \rho_1^2 d\rho_1 \sin\theta_1\, d\theta_1 d\phi_1\, e^{-\rho_1} \left[\int_0^{\rho_1} \rho_2^2\, d\rho_2 e^{-\rho_2} \right.$$

$$\times \frac{1}{\rho_1} \sum_l \left(\frac{\rho_2}{\rho_1}\right)^l \iint P_l(\cos\theta) \sin\theta\, d\theta\, d\phi + \int_{\rho_1}^{\infty} \rho_2^2\, d\rho_2 e^{-\rho_2}$$

$$\left. \times \frac{1}{\rho_2} \sum_l \left(\frac{\rho_1}{\rho_2}\right)^l \iint P_l(\cos\theta) \sin\theta\, d\theta\, d\phi \right]$$

$$= \frac{Ze^2}{8\pi a_0} \iiint \rho_1^2 d\rho_1 \sin\theta_1\, d\theta_1\, d\phi_1\, e^{-\rho_1} \left[\frac{1}{\rho_1} \int_0^{\rho_1} \rho_2^2 e^{-\rho_2} d\rho_2 + \int_{\rho_1}^{\infty} \rho_2 e^{-\rho_2} d\rho_2 \right]$$

$$= \frac{5}{8} Z \frac{e^2}{a_0} = \frac{5Z}{4} E_H \tag{64}$$

where we have used the relation

$$\int_0^{\pi} P_l(\cos\theta) \sin\theta\, d\theta = 2 \quad \text{for} \quad l = 0$$
$$= 0 \quad \text{for} \quad l \geq 1 \tag{65}$$

Thus, according to the first-order perturbation theory, the ground state energy is given by

$$W^{(1)} = -2Z^2 E_H + \frac{5}{4} Z E_H = \left(-Z^2 + \frac{5}{8} Z\right) 2E_H = \left(-Z^2 + \frac{5}{8} Z\right) \times 27.21 \text{ eV} \tag{66}$$

The calculated values are tabulated in Table 24.1. Since the ground state energy of a one-electron atom is $-Z^2 E_H$, the ionization energy is given by

$$\left[\left(Z^2 - \frac{5}{8} Z\right) 2E_H - Z^2 E_H \right]$$

$$= \left[Z^2 - \frac{5}{4} Z \right] E_H \tag{67}$$

In Table 24.1, we have tabulated the ionization energy as given by the above equation and compared with the corresponding experimental values. It is interesting to note that the absolute error remains almost the same; however, the percentage error decreases with increase in Z. Physically this is due to the fact that for large Z, the term $-Ze^2 \left(\frac{1}{r_1} + \frac{1}{r_2}\right)$ would dominate over the term e^2/r_{12}.

Table 24.1: Ground state and ionization energies of two electron atoms (adapted from Ref. 4)

Atom	Z	*Ground state energy using perturbation theory [Eq. 66]*	*Ionization energy*			*Difference between experimental values and perturbation theory results*
			Experimental value	*Perturbation theory [Eq. (67)]*	*Variational method [Eq. (73)]*	
He	2	–74.83 eV	24.63 eV	20.40 eV	23.06 eV	4.23 eV
Li^+	3	–193.87 eV	75.64 eV	71.43 eV	74.08 eV	4.21 eV
Be^{++}	4	–367.30 eV	153.87 eV	149.66 eV	152.31 eV	4.21 eV
B^{+++}	5	–595.20 eV	259.31 eV	255.09 eV	257.75 eV	4.22 eV

Solution 24.2 Since $\exp[-Z'r/a_0]$ is the ground state wavefunction for a hydrogen-like atom with charge $Z'e$, we must have

$$\left[-\frac{\hbar}{2m}\nabla_1^2 - \frac{Z'e^2}{r_1}\right]\phi(\mathbf{r}_1, \mathbf{r}_2) = -Z'^2 E_H \phi(\mathbf{r}_1, \mathbf{r}_2) \tag{68}$$

Thus,
$$E(Z') = \iint \phi^* \left[-2Z'^2 E_H + (Z' - Z)e^2\left(\frac{1}{r_1} + \frac{1}{r_2}\right) + \frac{e^2}{r_{12}}\right]\phi\, d\tau_1\, d\tau_2 \tag{69}$$

Now,
$$e^2 \iint \phi^* \left(\frac{1}{r_1} + \frac{1}{r_2}\right)\phi\, d\tau\, d\tau_2 = 2e^2 \iint \frac{|\phi|^2}{r_1} d\tau_1\, d\tau_2$$

$$= 2Z'\frac{e^2}{a_0} = 4Z'E_H \tag{70}$$

and
$$\iint \phi^* \frac{e^2}{r_{12}} \phi\, d\tau_1\, d\tau_2 = \frac{5}{4} Z' E_H \tag{71}$$

[see Eq. (64)]. On substitution we get Eq. (15) and the condition $dE/dZ' = 0$ leads to

$$Z' = Z_{eff} = Z - \frac{5}{16}$$

Thus the ground state energy is

$$E = -2\left(Z - \frac{5}{16}\right)^2 E_H \tag{72}$$

and the corresponding ionization energy is

$$\left[2\left(Z - \frac{5}{16}\right)^2 - Z^2\right] E_H \tag{73}$$

which gives a much better estimate than the perturbation theory result (see Table 24.1).

Solution 24.4 According to the first order perturbation theory the energy levels for the 1*s* 2*s* and 1 *s* 2*p* configurations are given by

$$W^{(1)} = \left(-Z^2 E_H - \frac{Z^2}{4} E_H \right) + J_l \pm K_l$$

$$= -5\,E_H + J_l \pm K_l \qquad (74)$$

where we have assumed $Z = 2$ and the upper and lower states correspond to singlet (parahelium) and triplet (orthohelium) states respectively. Substituting the values, we get

$$\begin{aligned} W^1_{1s2s} &\approx -58.9 \pm 0.4 \text{ eV} \\ &\approx -58.5 \text{ eV and } -59.3 \text{ eV} \end{aligned} \qquad (75)$$

$$\begin{aligned} W^1_{1s2s} &\approx -57.8 \pm 0.1 \text{ eV} \\ &\approx -57.7 \text{ eV and } -57.9 \text{ eV} \end{aligned} \qquad (76)$$

The ionization energies are given by

$$\begin{aligned} -Z^2 E_H - W^{(1)} &\approx 4.1 \text{ eV} \quad \text{for the } 1s\,2s \text{ parahelium state } (^1S_0) \\ &\approx 4.9 \text{ eV} \quad \text{for the } 1s\,2s \text{ orthohelium state } (^3S_0) \\ &\approx 3.3 \text{ eV} \quad \text{for the } 1s\,2p \text{ parahelium state } (^1P_1) \\ &\approx 3.5 \text{ eV} \quad \text{for the } 1s\,2p \text{ orthohelium state } (^3P_{2,1,0}) \end{aligned} \qquad (77)$$

The corresponding experimental values are 3.97 eV, 4.76 eV, 3.368 eV and 3.623 eV respectively.

REFERENCES AND SUGGESTED READINGS

1. D. Ter Harr, *The Old Quantum Theory*, Pergamon Press, Oxford (1967).
2. M. Jammer, *The Conceptional Development of Quantum Mechanics*, McGraw-Hill Book Co., New York (1966).
3. S. Flugge, *Practical Quantum Mechanics*, Springer Verlag, Berlin (1971).
4. L. Pauling and E.B. Wilson, *Introduction to Quantum Mechanics*, McGraw-Hill Book Co., New York (1935).
5. J. Irving and N. Mullineux, *Mathematics in Physics and Engineering*, Academic Press, New York (1959).

Some select topics

25.1 INTRODUCTION

25.2 TUNNELLING THROUGH [illegible]

25.2.1 [illegible]

25 Some select topics

I saw a medley of haphazard facts fall into line and order... 'But it's true,' I said to myself. 'It's very beautiful. And it's true.'

— C.P. Snow[1]

25.1 INTRODUCTION

In this chapter, we will discuss in detail some select topics which are of considerable importance in quantum mechanics. In Section 25.2, we will discuss the semi-classical as well as the more rigorous method for calculating the lifetime of quasi-bound states. In Section 25.3, we will present a perturbation analysis of a periodic structure characterized with a sinusoidal variation of potential energy variation. In Section 25.4, we will describe a versatile matrix method for the analysis of a quantum mechanical problems. Finally, in Section 25.5, we will discuss the Thomas-Fermi model of the atom.

25.2 TUNNELLING LIFETIME CALCULATIONS

25.2.1 Semi-classical Approximate Analysis

We first consider the bound states corresponding to the following potential energy variation (see Fig. 25.1)

$$V(x) = \begin{cases} \infty & x < 0 \\ 0 & 0 < x < a \\ V_0 & x > a \end{cases} \tag{1}$$

In Section 17.2, we had discussed the solution of the Schrödinger equation for the above profile and had shown that the energy eigenvalues are determined by solving the following transcendental equation

$$L = \sin ka + \frac{k}{\kappa} \cos ka = 0 \tag{2}$$

where
$$k^2 = \frac{2\mu E}{\hbar^2} \quad \text{and} \quad \kappa^2 = \frac{2\mu(V_0 - E)}{\hbar^2} \tag{3}$$

---

[1] The authors found this quotation in John Gribbin's book, *In Search of Schrödinger Cat*, Black Swan Books (1984).

Equation (2) can be written in the form

$$-\xi \cot \xi = \sqrt{\alpha^2 - \xi^2} \tag{4}$$

where

$$\xi = ka = \left[\frac{2\mu E a^2}{\hbar^2}\right]^{1/2} \quad \text{and} \quad \alpha = \left[\frac{2\mu V_0 a^2}{\hbar^2}\right]^{1/2} \tag{5}$$

Let us assume $\alpha = 7$. The solution of the transcendental equation [Eq. (2)] gives us bound states for which

$$\xi = 2.739488 \quad \text{and} \quad 5.401717$$

We define the normalized energy eigenvalue

$$\varepsilon = \frac{E}{V_0} = \left(\frac{\xi}{\alpha}\right)^2 \tag{6}$$

Thus the corresponding discrete values of ε are given by

$$\varepsilon = 0.153159\,(=\varepsilon_1) \quad \text{and} \quad 0.595480\,(=\varepsilon_2) \tag{7}$$

The corresponding wavefunction is given by

$$\psi_b(x) = \begin{cases} A_b \sin k_b x, & 0 < x < a \\ A_b \sin k_b a e^{-\kappa_b (x-a)}, & x > a \end{cases} \tag{8}$$

where the subscript b refers to the fact that we are considering particular bound states. If we normalize the wavefunction we would readily get

$$A_b = \left[\frac{2\kappa_b}{1 + \kappa_b a}\right]^{1/2} \tag{9}$$

The discrete eigenvalues and corresponding eigenfunctions are shown in Fig. 25.1.

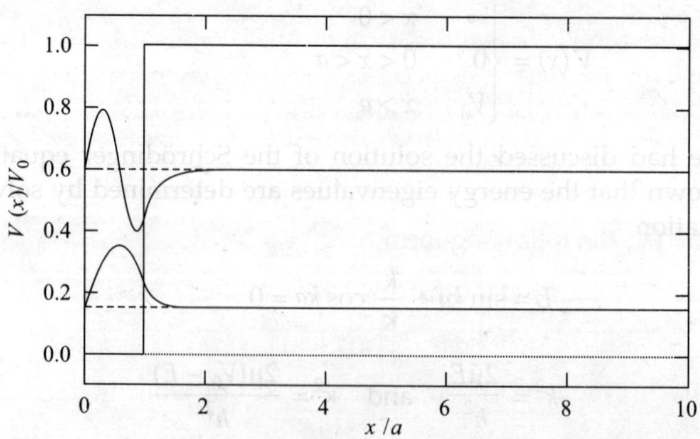

Figure 25.1: The discrete eigenvalues and eigenfunctions corresponding to the potential energy distribution given by Eq. (1) with $\alpha = 7$.

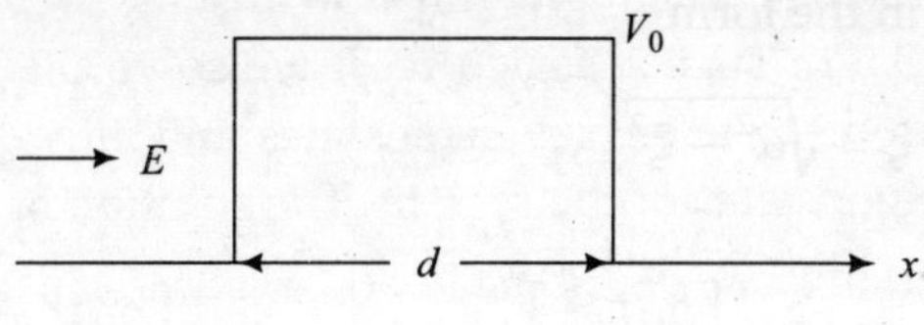

Figure 25.2: Tunnelling through a rectangular potential barrier of height V_0 and width d.

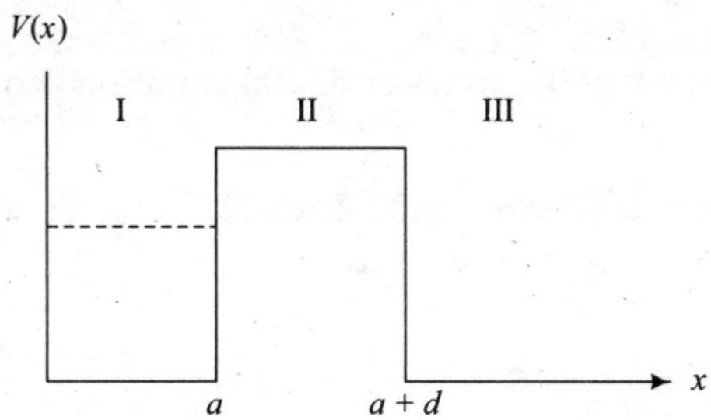

Figure 25.3: A tunnelling potential energy variation as given by Eq. (12).

We next consider tunnelling through a rectangular potential barrier (of width d) as shown in Fig. 25.2. We had discussed this in Section 8.3 and had shown that tunnelling probability is given by

$$T = \frac{4\varepsilon(1-\varepsilon)}{\sinh^2(\kappa d) + 4\varepsilon(1-\varepsilon)} \tag{10}$$

For $\kappa d \gg 1$, the expression for the tunnelling probability takes the form

$$T \approx 16\varepsilon(1-\varepsilon)\, e^{-2\kappa d} \tag{11}$$

We next consider a potential energy variation of the type shown in Fig. 25.3:

$$V(x) = \begin{cases} \infty, & x < 0 \\ 0, & 0 < x < a \\ V_0, & a < x < a + d \\ 0, & x > a + d \end{cases} \tag{12}$$

Now, according to the semi-classical theory, a particle confined inside the well (i.e., within the region $0 < x < a$) will hit the barrier $v/2a$ times a second, where v is the speed of the particle. Thus, the probability per second of escape (which is the inverse of the mean lifetime) is given by

$$\frac{1}{\tau} \approx \frac{v}{2a} T$$

which can be simplified to the following form

$$\tau \approx \frac{1}{16}\left[\frac{2\mu a^2}{\hbar \alpha^2}\right] \frac{\kappa a}{\left[\varepsilon(1-\varepsilon)\right]^{3/2}}\, e^{2\kappa d} \tag{13}$$

(this analysis is similar to the one that we had carried out while studying the alpha decay problem in Section 18.7).

25.2.2 Quasi-bound States

In above we have given a very simple semi-classical analysis for the calculation of the lifetime for a particle to tunnel through a barrier. For a more rigorous analysis, we must understand the concept of quasi-bound states which we will now discuss. We consider again the leaky structure shown in Fig. 25.3. For such a structure, if we solve the Schrödinger equation we would find that there are no discrete energy eigenstates and all values of $E > 0$ are allowed; these form the continuum eigenstates of the system. The solution of the Schrödinger equation in the three regions can be written in the form

$$\psi_E(x) = \begin{cases} A \sin kx & 0 < x < a \\ B e^{\kappa(x-a)} + C e^{-\kappa(x-a)} & a < x < a+d \\ D_+ e^{ik(x-a-d)} + D_- e^{-ik(x-a-d)} & x > a+d \end{cases} \tag{14}$$

where (without any loss of generality) we have assumed $E < V_0$. Continuity of the wavefunction and its derivative at $x = a$ and at $x = a + d$ give us

$$B = \frac{1}{2} A \left[\sin ka + \frac{k}{\kappa} \cos ka \right] \tag{15}$$

$$C = \frac{1}{2} A \left[\sin ka - \frac{k}{\kappa} \cos ka \right] \tag{16}$$

$$D_-^* = D_+ = \frac{1}{2} B \left[1 + \frac{\kappa}{ik} \right] e^{\kappa d} + \frac{1}{2} C \left[1 - \frac{\kappa}{ik} \right] e^{-\kappa d} \tag{17}$$

The orthogonality condition for these continuum eigenstates are given by

$$\int_0^\infty \psi_{E'}^*(x)\, \psi_E(x)\, dx = \delta(E - E') \tag{18}$$

which gives us
$$|D_+| = |D_-| = \frac{1}{\hbar} \left[\frac{\mu}{2\pi k} \right]^{1/2} \tag{19}$$

Now, if we assume $\kappa d. \gg 1$, then the exponentially decaying term in Eq. (17) can be neglected, and in this approximation

$$D_-^* = D_+ \approx \frac{1}{4} LA \left[1 + \frac{\kappa}{ik} \right] e^{\kappa d} \tag{20}$$

where L has already been defined in Eq. (2). It is readily seen that since $\kappa d \gg 1$, unless $L \approx 0$,

$$\left| \frac{D_\pm}{A} \right| \gg 1 \tag{21}$$

implying that the amplitude of the wavefunction is large in the region $x > (a + d)$. However, for values of E such that $L = 0$ [i.e., when the condition for discrete bound states (corresponding to Fig. 25.1) are satisfied], then $B = 0$ and we have only the exponentially decaying solution in the region $a < x < a + d$ [see Eq. (14)] and the amplitude of the oscillatory solution in the region $x > (a + d)$ is very weak. Thus when

$$L = 0 = B \tag{22}$$

the wavefunction has the properties of a discrete eigenfunction in regions I and II having very weak oscillations in the region III. We refer the eigenfunctions corresponding to $B = 0$ as quasi-bound states. In Figs 25.4 (a) and 25.5 (a) we have plotted the eigenfunctions corresponding to $\alpha = 7$ with

$$\varepsilon = 0.153159\ (= \varepsilon_1) \quad \text{and} \quad 0.595480\ (= \varepsilon_2)$$

respectively. As can be seen the wavefunctions are the same as for the bound state except for the weak oscillatory solution in the region $x > (a + d)$. These are the quasi-bound states of the system. In Figs 25.4 (b) and 25.5 (b) we have plotted the eigenfunctions corresponding to

$$\varepsilon = 0.157159 \quad \text{and} \quad 0.675480$$

respectively. Thus, we find that if we go slightly away from the quasi-bound state, there is a drastic change in the behaviour of the eigenfunction. In fact, if we plot

$$\left|\frac{A}{D_\pm}\right|^2$$

as a function of E (see Fig. 25.6), we will get very sharply peaked Lorentzians, the width of which is directly related to the lifetime of the quasi-bound state. All the plots correspond to $d/a = 0.75$.

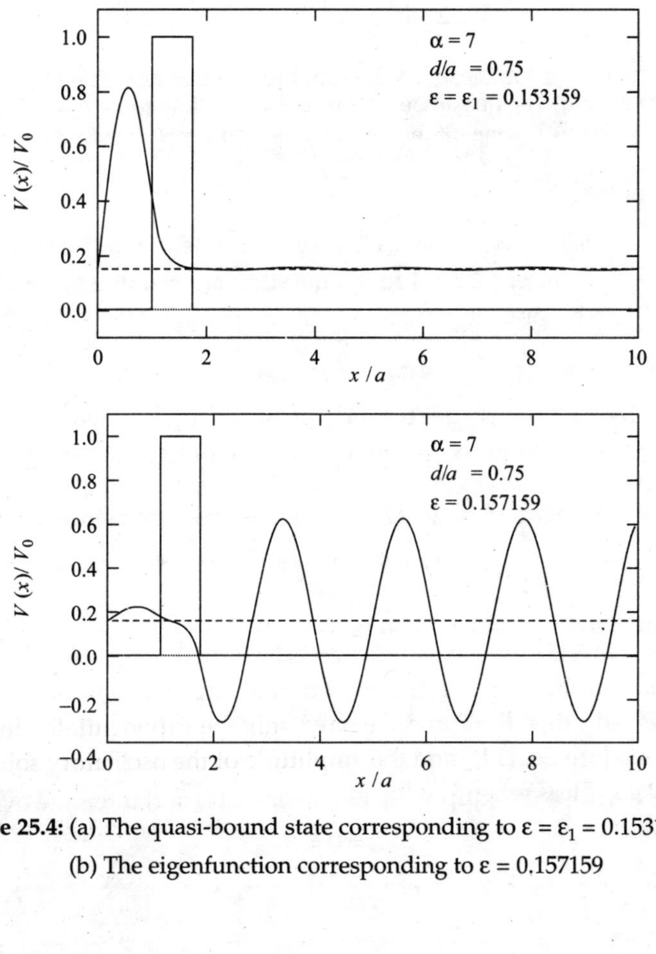

Figure 25.4: (a) The quasi-bound state corresponding to $\varepsilon = \varepsilon_1 = 0.153159$.
(b) The eigenfunction corresponding to $\varepsilon = 0.157159$

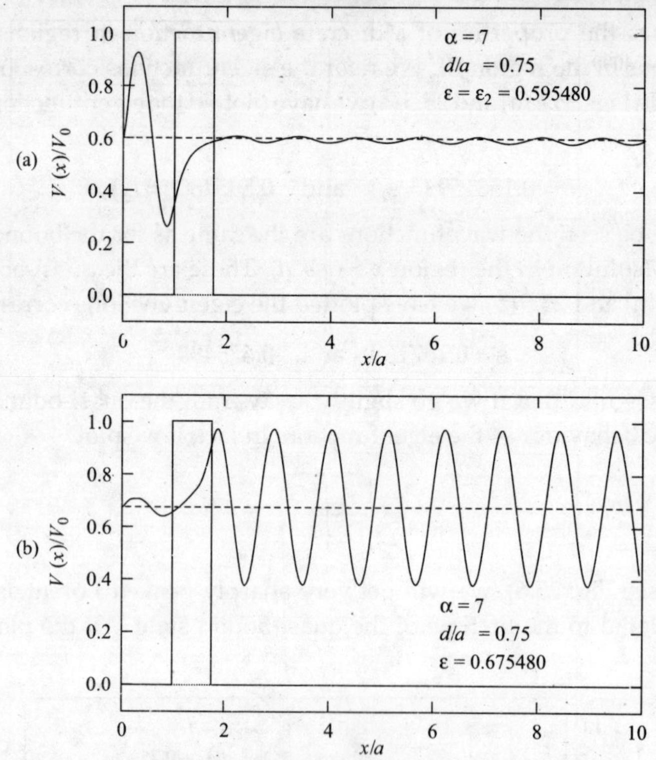

Figure 25.5: (a) The quasi-bound state corresponding to $\varepsilon = \varepsilon_2 = 0.595480$
(b) The eigenfunction corresponding to $\varepsilon = 0.675480$

25.2.3 Lifetime Calculation

In this section, we will discuss the rigorous theory for tunnelling of a particle from a potential well similar to that shown in Fig. 25.3. We assume that at $t = 0$, the particle is described by a wavefunction which corresponds to the bound state corresponding to Fig. 25.1:

$$\psi(x, t = 0) = \psi_b(x) \tag{23}$$

where $\psi_b(x)$ is given by Eq. (8). In order to study its time evolution we must expand it in terms of the (continuum) eigenfunctions corresponding to the potential energy variation shown in Fig. 25.3. Thus, we should write

$$\Psi(x, t = 0) = \psi_b(x) = \int \phi(E)\psi_E(x)\,dE \tag{24}$$

For $t > 0$, the wavefunction would be given by

$$\Psi(x, t) = \int_0^\infty \phi(E)\psi_E(x)e^{-iEt/\hbar}\,dE \tag{25}$$

In order to determine $\phi(E)$, we multiply Eq. (24) by $\psi * E'(x)$ and integrate over x to obtain

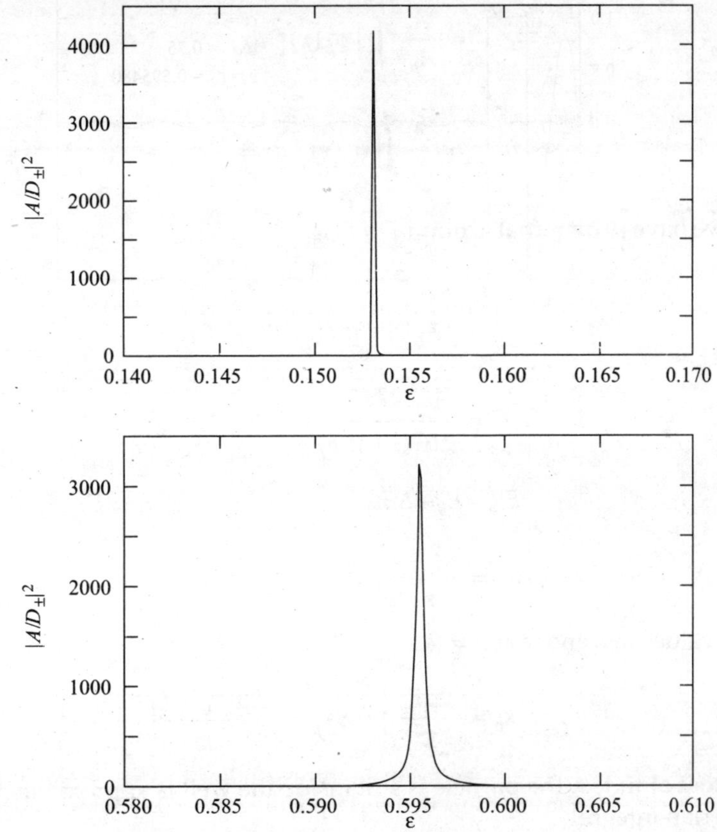

Figure 25.6: A plot of $|A/D_{\pm}|^2$ which shows two sharply peaked Lorentzians the widths of which are directly related to the lifetime of the quasi-bound states.

$$\begin{aligned}\int_0^\infty \psi_{E'}^*(x)\psi_b(x)dx &= \int dE\phi(E)\int dx\psi_{E'}^*(x)\psi_E(x)\\ &= \int dE\phi(E)\,\delta(E-E') = \phi(E')\end{aligned}$$

where use has been made of Eq. (18);. Thus,

$$\phi(E) = \int_0^\infty \psi_E^*(x)\,\psi_b(x)\,dx \tag{26}$$

Now, it is only around $E \approx E_b$ that $\psi_E(x)$ has almost the same spatial dependence as $\psi_b(x)$ in regions I and II (see Figs. 25.4 and 25.5) and, therefore, in the above equation, $\phi(E)$ will be appreciable around $E \approx E_b$; this we will explicitly find later. Thus we may write

$$\psi_E(x) \approx \frac{A}{A_b}\psi_b(x) \qquad \text{Regions I and II} \tag{27}$$

Since in region III $(x > a + d)$, $\psi_b(x)$ has a negligible value, we may neglect the contribution from region III to the integral in Eq. (26) to write

$$\phi(E) \approx \int_0^{a+d} \psi_E^*(x)\,\psi_b(x)\,dx$$

$$\approx \frac{A}{A_b} \int_0^{a+d} |\psi_b(x)|^2 dx$$

$$\approx \frac{A}{A_b} \int_0^{\infty} |\psi_b(x)|^2 dx = \frac{A}{A_b} \tag{28}$$

In Appendix P we have shown that around $E \approx E_b$

$$|\phi(E)|^2 \approx \left|\frac{A}{A_b}\right|^2 \approx \frac{\Gamma}{\pi} \frac{1}{(E - E_b')^2 + \Gamma^2} \tag{29}$$

where
$$\Gamma = \frac{4k_b^3 \kappa_b^3 a^4 \hbar^2}{\mu\alpha^4 (1 + \kappa_b a)} e^{-2\kappa_b d} \tag{30}$$

$$E'_b = E_b - \Delta E_b \tag{31}$$

$$\Delta E_b = \frac{\kappa_b^2 - k_b^2}{2\kappa_b k_b} \Gamma \tag{32}$$

k_b and κ_b are the values of k and κ at $E = E_b$:

$$k_b = \left[\frac{2\mu E_b}{\hbar^2}\right]^{1/2} ; \kappa_b = \left[\frac{2\mu(V_0 - E_b)}{\hbar^2}\right]^{1/2}$$

The probability that at time t, the particle is still inside the well is approximately given by the square of the overlap integral

$$P(t) \approx \left|\int_0^{\infty} \Psi^* (x, 0)\, \Psi(x, t)\, dx\right|^2$$

The above integral has been evaluated in Appendix P and has been shown to be given by

$$P(t) \approx \left|\int dE |\phi(E)|^2 e^{-iEt/\hbar}\, dE\right|^2 \tag{33}$$

where $|\phi(E)|^2$ has been given by Eq. (29). On substituting the expression for $|\phi(E)|^2$ and carrying out the integration (see Appendix P), we get

$$\text{P}(t) \approx e^{-t/\tau} \tag{34}$$

Tunnelling lifetime of quasi-bound states

where
$$\tau = \frac{\hbar}{2\Gamma} = \frac{1}{16}\left[\frac{2\mu a^2}{\hbar\alpha^2}\right] \frac{\kappa_b \left(a + \dfrac{1}{\kappa_b}\right)}{[\varepsilon_b(1 - \varepsilon_b)]^{3/2}} e^{2\kappa_b d} \tag{35}$$

is the mean lifetime of the quasi-bound state and

$$\varepsilon_b = \frac{E_b}{V_0} = \frac{(k_b a)^2}{\alpha^2} \tag{36}$$

Equation (35) gives an analytical expression for the tunnelling lifetime of the quasi-bound states. It should be noted that as d becomes large, the time that the particle spends inside the well becomes large. Equation (35) may be compared with the approximate expression given by Eq. (13). If we replace a by $(a + 1/\kappa_b)$ in Eq. (13) we would get Eq. (35). Since $1/\kappa_b$ represents the perturbation depth of the wavefunction in the 'classically forbidden region' (see Section 8.3) $(a + \frac{1}{\kappa_b})$ can be interpreted as the effective width of the well.

Now, in Appendix P we have also shown that [see Eq. (14) of Appendix P]

$$\left|\frac{A}{D_\pm}\right|^2 \approx 8\left(\frac{\hbar^2}{2\mu}\right)\frac{k_b\,\kappa_b\,\Gamma}{1+\kappa_b a}\frac{1}{\left(E - E_b'\right)^2 + \Gamma^2} \tag{37}$$

Thus if we are able to calculate $\left|\frac{A}{D_\pm}\right|^2$ as a function of E, we would get a series of Lorentzians, each Lorentzian corresponding to a quasi-bound state of the structure. By fitting each peak to a Lorentzian, we would be able to get E_b' and Γ and, therefore, the lifetime. Indeed for an arbitrary potential energy variation one can develop a matrix approach to calculate $\left|\frac{A}{D_\pm}\right|^2$ as a function of E from which the lifetime of a quasi-bound state can be calculated (see, e.g. Refs 2 and 3).

In Appendix P we have also shown that

$$|D_\pm| = \frac{1}{4}A\left|\frac{\mu\alpha(1+\kappa_b a)\,e^{\kappa_b d}}{a\hbar^2\,k_b\,\kappa_b^2}\right|\left|1 \pm \frac{\kappa_b}{ik_b}\right|\left|E - E_b' \mp i\Gamma\right| \tag{38}$$

Thus if we solve the eigenvalue equation

$$D_-(E) = 0 \tag{39}$$

in the complex E plane, then the imaginary part of the root will give the value of Γ and hence the lifetime of the state.

25.3 PERTURBATION THEORY FOR SINUSOIDALLY VARYING POTENTIALS

In Section 17.5, we had discussed the solution of the Schrödinger equation corresponding to a periodic potential described by a series of delta functions: this was the Krönig-Penney model of the lattice. In this section, we present a perturbation approach to solve the Schrödinger equation corresponding to a sinusoidal variation of the potential energy function. The results have been compared with numerical results obtained by solving the Schrödinger equation using the Runge-Kutta method and the perturbation theory results have been shown to be very accurate for rapid oscillations. In fact if the oscillations are very rapid (this will be quantified later) then the number of bound states are determined not by the number of wells but by the average potential energy variation. The perturbation method can easily be extended to potential energy variations which can be expressed as a sum of sine or cosine functions.

We start with the one-dimensional Schrödinger equation

$$\frac{d^2\psi}{dx^2} + \frac{2\mu}{\hbar^2}\left[E - V(x)\right]\psi(x) = 0 \tag{40}$$

We assume a sinusoidal variation of the potential energy variation given by the following equation (see Fig. 25.7)

$$V(x) = \begin{cases} -V_0\left[1 + \Delta\cos(2gx)\right], & |x| < a/2 \\ 0, & |x| > a/2 \end{cases} \tag{41}$$

We present here a perturbation approach to solve the Schrödinger equation. We assume

$$ga = \left(m + \frac{1}{2}\right)\pi\,;\; m = 0, 1, 2, 3,... \tag{42}$$

so that $V(x) = -V_0$ at $x = +a/2$. For the potential energy variation given by Eq. (41), Eq. (40) can be written in the form

$$\left.\begin{aligned} &\frac{d^2\psi}{d\xi^2} + \left[\gamma^2 + 2\varepsilon\cos(2\xi)\right]\psi(\xi) = 0, && |\xi| < \xi_0 \\ &\frac{d^2\psi(\xi)}{d\xi^2} - \sigma^2\psi(\xi) = 0, && |\xi| > \xi_0 \end{aligned}\right\} \tag{43}$$

where we have introduced the following dimensionless variables:

$$\xi = gx;\; \xi_0 = \frac{1}{2}ga = \left(m + \frac{1}{2}\right)\frac{\pi}{2} \tag{44}$$

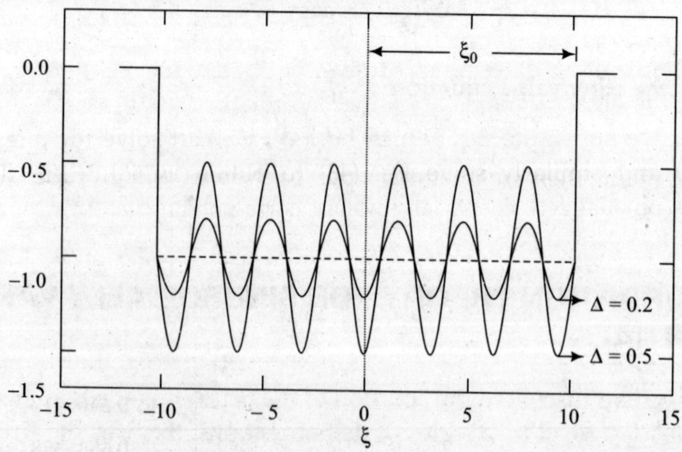

Figure 25.7: The sinusoidal variation of the potential energy function [Eq. (41)]; the average potential energy function [Eq. (51)] has been shown by dotted line.

and $$\varepsilon = \frac{1}{2}\frac{\alpha^2\Delta}{\left[\left(m+\frac{1}{2}\right)\frac{\pi}{2}\right]^2}$$

$$\gamma^2 = 2(1-b)\frac{\varepsilon}{\Delta}$$

$$\sigma^2 = 2b\,\frac{\varepsilon}{\Delta}$$

$$\alpha^2 = \frac{2\mu V_0 a^2}{4\hbar^2}$$

$$b = -\frac{E}{V_0} = \frac{|E|}{V_0}$$

For $\xi < \xi_0$, Eq. (43) is similar to Mathieu's equation (see, e.g. Chapter 11 of Ref. 4). Following Ref. 4, we write the solution of Eq. (43) in the following form.

$$\psi = \psi_0 + \varepsilon\psi_1 + \varepsilon^2\psi_2 + \ldots \tag{45}$$

If we substitute the above expression in Eq. (43) and equate the coefficients of various powers of ε, we would get

$$\frac{d^2\psi_0}{d\xi^2} + \gamma^2\psi_0 = 0 \tag{46}$$

$$\frac{d^2\psi_0}{d\xi^2} + \gamma^2\psi_1 = -2\psi_0\cos(2\xi) \tag{47}$$

$$\frac{d^2\psi_0}{d\xi^2} + \gamma^2\psi_2 = -2\psi_1\cos(2\xi) \tag{48}$$

The solution of Eq. (46) is taken to be

$$\psi_0(\xi) = A\cos(\gamma\xi + \phi) \tag{49}$$

where the constant ϕ is determined from the boundary conditions. However, for a symmetric $V(x)$, the eigenfunctions are either symmetric ($\phi = 0$) or antisymmetric ($\phi = \pi/2$) functions of ξ. If we substitute Eq. (49) in Eq. (47) we can solve for the particular integral to obtain $\psi_1(\xi)$ and similarly solve Eq. (48) to obtain $\psi_2(\xi)$..., etc. The final result is

Table 25.1: The normalized energy eigenvalues corresponding to the average potential energy distribution [given by Eq. (51)] for $\alpha = 7$.

Mode number	*b-value*
0	0.9615098
1	0.8468413
2	0.6588388
3	0.4045254
4	0.1067140

$$\psi(x) = A\cos(\gamma\xi+\phi) + \frac{1}{4}\varepsilon A\left\{\frac{\cos[(2+\gamma)\xi+\phi]}{1+\gamma} + \frac{\cos[(-2-\gamma)\xi+\phi]}{1-\gamma}\right\} + \frac{1}{32}\varepsilon^2 A\left\{-\frac{8\xi\sin[\gamma\xi+\phi]}{\gamma(1-\gamma^2)} + \frac{\cos[(4+\gamma)\xi+\phi]}{(1+\gamma)(2+\gamma)} + \frac{\cos[-(4-\gamma)\xi+\phi]}{(1-\gamma)(2-\gamma)}\right\}+\cdots \quad \text{for } |\xi|<\xi_0$$

$$= C\exp[-\sigma(|\xi|-\xi_0)] \quad \text{for } |\xi|>\xi_0 \tag{50}$$

The continuity of ψ and $d\psi/d\xi$ at $\xi = \xi_0$ would lead to the transcendental equation which would determine the energy eigenvalues.

Obviously, the perturbation theory would be accurate for small values of ε. However, we obtain the general result that for a given value of α, one can always choose a large enough value of m to give a large number of oscillations of the potential energy variation in the well and obtain a sufficiently small value of ε so that the bound states are accurately determined by solving for the average potential energy distribution given by

$$V(x) = \begin{cases} -V_0, & |x| < a/2 \\ 0, & |x| > a/2 \end{cases} \tag{51}$$

We next present some numerical calculations to determine the validity of the perturbation method. We assume

$$\alpha = 7.0 \tag{52}$$

We first consider the average potential energy variation given by Eq. (51). For such a value of α, the potential well will support 3 symmetric and 2 antisymmetric bound states. Table 25.1 gives the corresponding values of b (which represent the normalized energy eigenvalues). Tables 25.2 gives the b values for the ground state and the 3rd excited state for different Δ values for a fixed value of m (=10) and Table 25.3 give the b values for different m-values corresponding to $\Delta = 1$. The tables give a comparison of the results using perturbation theory with the more accurate results obtained using the Runge-Kutta method.

From the tables we make the following observations:

1. The perturbation method, as expected, becomes more accurate for smaller Δ values.

Table 25.2: The normalized energy eigenvalues of the ground state and the 3rd excited state corresponding to the sinusoidally varying potential energy distribution [given by Eq. (41)] for different values of Δ for $m = 10$ with $\alpha = 7$.

Δ	ε	*b-values*			
		Ground state		*3rd excited state*	
		R-K method	*Pert. theory*	*R-K method*	*Pert. theory*
1.0	0.090063	0.986268	0.995867	0.460043	0.462060
0.5	0.045032	0.968165	0.968969	0.425971	0.426323
0.25	0.022158	0.963408	0.963489	0.413646	0.413693
0.1	0.009006	0.961928	0.961930	0.407784	0.407787

Table 25.3: The normalized energy eigenvalues of the ground state and the 3rd excited state corresponding to the sinusoidally varying potential energy distribution [given by Eq. (41)] for different values of m for $\Delta = 1.0$ with $\alpha = 7$.

m	ε	*b-values*			
		Ground state		*3rd excited state*	
		R-K method	*Pert. theory*	*R-K method*	*Pert. theory*
5	0.328247	1.042736	1.046624	0.490174	0.446715
10	0.090063	0.986268	0.995867	0.460043	0.462060
20	0.023628	0.968511	0.969370	0.425882	0.426204
50	0.003893	0.962920	0.962960	0.411655	0.411675

2. For each bound state, the perturbation method becomes more accurate for higher values of m; this is to be expected as a large value of m implies a small value of ε.
3. As the value of m increases, i.e., as the number of oscillations in the well increases, the value of b for each bound state converges to the b value corresponding to the average potential energy variation which are tabulated in Table 25.1.

Figures 25.8–25.10 show the eigenfunctions corresponding to the ground state for various values of m and Δ using the perturbation theory and the (more accurate) Runge-Kutta method. The figures show that the perturbation theory gives an accurate description of the eigenfunction for smaller Δ values. It may be mentioned that even for maximum perturbation; the eigenfield is very accurate for $m \geq 20$. Figures 25.11 shows the eigenfunction of the 3rd excited states for $m = 20$. Comparison of Figs 25.10 and 25.11 show that the perturbation theory works even better for the excited states. A similar behavior can be seen for the other excited states also. Thus, we may conclude that the simple perturbation approach can be effectively used to analyze a sinusoidally varying potential energy distribution.

25.4 THE MATRIX METHOD

In this section, we will describe a versatile matrix method to obtain solutions of the (one-dimensional) Schrödinger equation for a variety of problems. In Ref. 6, we have used this matrix method to develop a software to solve one-dimensional problems in quantum mechanics. The only approximation made would be to replace the actual potential energy variation by a piecewise step variation. Indeed, the method can be used to solve the general scalar wave equation

$$\frac{d^2\psi}{dx^2} + k^2(x)\psi(x) = 0 \tag{53}$$

where, for problems in quantum mechanics

$$k^2(x) = \frac{2\mu}{\hbar^2}\left[E - V(x)\right] \tag{54}$$

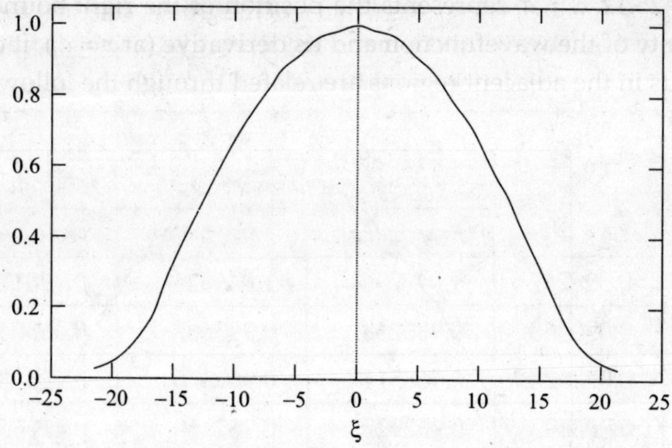

Figure 25.8: Ground state eigenfunction for $m = 10$ (and $\Delta = 0.1$) using Runge-Kutta method (continuous curve) and perturbation theory (dashed curve); the two curves almost completely overlap each other.

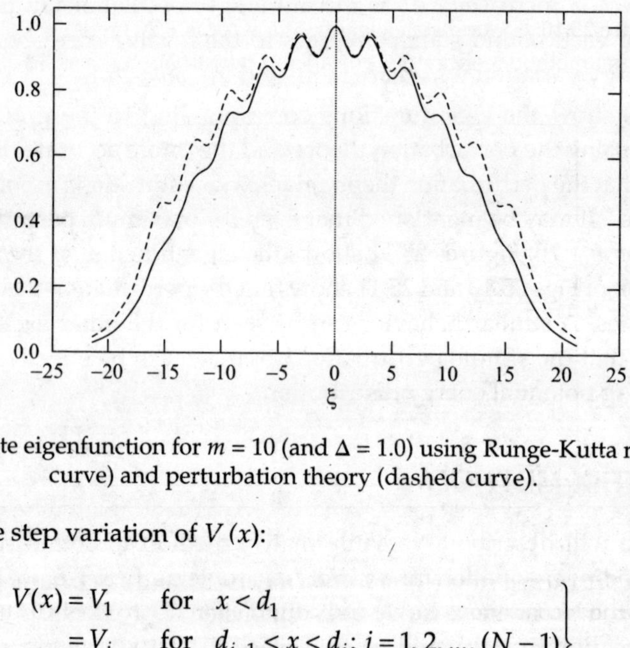

Figure 25.9: Ground state eigenfunction for $m = 10$ (and $\Delta = 1.0$) using Runge-Kutta method (continuous curve) and perturbation theory (dashed curve).

We assume piecewise step variation of $V(x)$:

$$\left.\begin{aligned} V(x) &= V_1 \quad \text{for} \quad x < d_1 \\ &= V_j \quad \text{for} \quad d_{j-1} < x < d_j;\ j = 1, 2, \ldots, (N-1) \\ &= V_N \quad \text{for} \quad x > d_{N-1} \end{aligned}\right\} \tag{55}$$

(see Fig. 25.12). Indeed an arbitrary variation of $V(x)$ can always be approximately described by the above equation. The solution of Eq. (53) in the jth region can be written in the following form

$$\psi_j(x) = A_j e^{i\left[k_j(x-d_{j-1})\right]} + B_j e^{-i\left[k_j(x-d_{j-1})\right]} \tag{56}$$

where $$k_j^2(x) = \frac{2\mu}{\hbar^2}\left[E - V_j\right] \tag{57}$$

As shown in Fig. 25.12, $x = d_j$ represents the position of the right boundary of the jth layer. Using the continuity of the wavefunction and its derivative (at $x = d_j$), it can be readily shown that the coefficients in the adjacent regions are related through the following matrix equation:

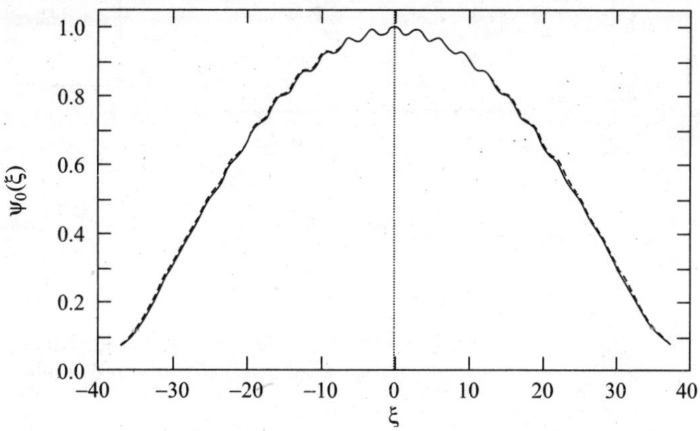

Figure 25.10: Ground state eigenfunction for $m = 20$ (and $\Delta = 1.0$) using Runge-Kutta method (continuous curve) and perturbation theory (dashed curve).

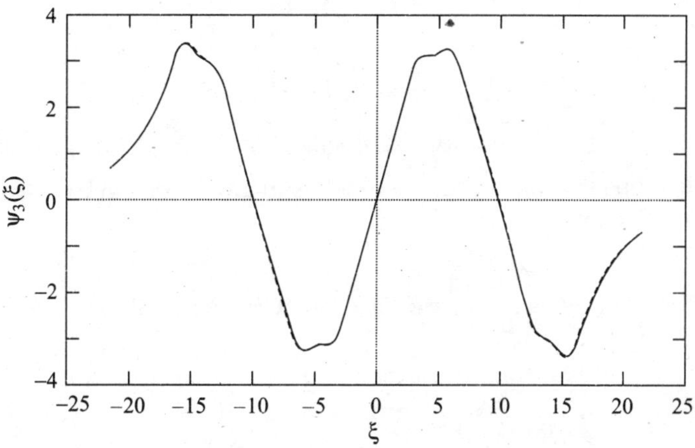

Figure 25.11: 3rd excited state eigenfunction for $m = 20$ (and $\Delta = 1.0$) using Runge-Kutta method (continuous curve) and perturbation theory (dashed curve); the two curves almost completely overlap each other.

$$\begin{pmatrix} A_{j+1} \\ B_{j+1} \end{pmatrix} = S^{(j)} \begin{pmatrix} A_j \\ B_j \end{pmatrix} \tag{58}$$

where

$$S^{(j)} = \begin{pmatrix} S_{11}^{(j)} & S_{12}^{(j)} \\ S_{21}^{(j)} & S_{22}^{(j)} \end{pmatrix} \tag{59}$$

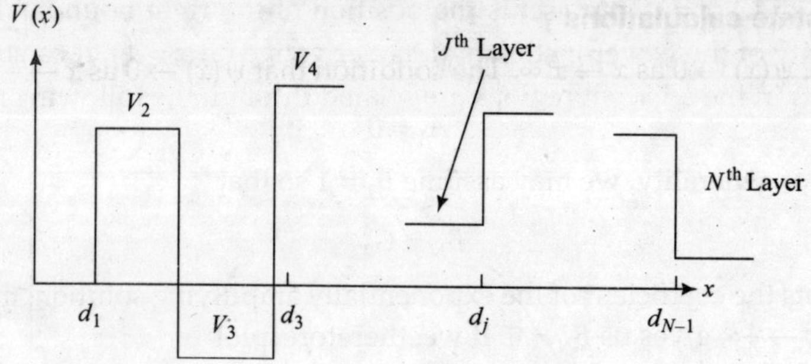

Figure 25.12: Piecewise step variation of the potential energy distribution.

with

$$S_{11}^{(j)} = \frac{1}{2}\left[1 + \frac{k_j}{k_{j+1}}\right] e^{i\alpha_j}$$

$$S_{12}^{(j)} = \frac{1}{2}\left[1 - \frac{k_j}{k_{j+1}}\right] e^{-i\alpha_j}$$

$$S_{21}^{(j)} = \frac{1}{2}\left[1 - \frac{k_j}{k_{j+1}}\right] e^{i\alpha_j}$$

$$S_{22}^{(j)} = \frac{1}{2}\left[1 + \frac{k_j}{k_{j+1}}\right] e^{-i\alpha_j}$$

and
$$\alpha_j = k_j\,(d_j - d_{j-1}) \tag{60}$$

By successive applications of Eq. (58), the coefficients in any layer can be expressed in terms of the coefficients of the first layer:

$$\begin{pmatrix} A_{j+1} \\ B_{j+1} \end{pmatrix} = S^{(j)} S^{(j-1)} \ldots S^{(1)} \begin{pmatrix} A_1 \\ B_1 \end{pmatrix} \tag{61}$$

Finally,
$$\begin{pmatrix} A_N \\ B_N \end{pmatrix} = \begin{pmatrix} G_{11} & G_{12} \\ G_{21} & G_{22} \end{pmatrix} \begin{pmatrix} A_1 \\ B_1 \end{pmatrix} \tag{62}$$

where the G matrix is given by

$$G = S^{(N-1)}\, S^{(N-2)} \ldots S^{(1)} \tag{63}$$

Thus, for given values of A_1 and B_1 and of energy E we can obtain the coefficients A_N and B_N. We should mention that if $E > V_j$, k_j^2 is positive; however, if $E < V_j$, k_j^2 is negative and k_j will be replaced by $(i\kappa_j)$ where

$$\kappa_j^2 = -k_j^2 = \frac{2\mu}{\hbar^2}\left[V_j - E\right] \tag{64}$$

Thus, Eq. (56) takes the form

$$\psi_j(x) = A_j e^{-\kappa_j (x - d_{j-1})} + B_j e^{\kappa_j (x - d_{j-1})} \tag{65}$$

We now consider specific cases:

Case 1: Bound state calculations

For a bound state, $\psi(x) \to 0$ as $x \to \pm\infty$. The condition that $\psi(x) \to 0$ as $x \to -\infty$ gives us

$$A_1 = 0 \tag{66}$$

Without any loss of generality, we may assume $B_1 = 1$ so that

$$B_N = G_{22} \tag{67}$$

Now, B_N represents the coefficient of the exponentially amplifying solution; thus the condition that $\psi(x) \to 0$ as $x \to +\infty$ gives us $B_N = 0$. If we, therefore, plot

$$G_{22}(E)$$

as a function of E, then the values of E where $G_{22}(E)$ becomes zero represent the energy eigenvalues of the system. Thus the roots of the equation

$$G_{22}(E) = 0 \tag{68}$$

would give us the energy eigenvalues (see Problem 25.3).

Case 2: Scattering problem

If k_1^2 and k_N^2 are both positive, we will have a scattering problem similar to what we have discussed in Chapter 8. For a wave incident from the left, we must have

$$B_N = 0$$

(see discussion in Section 8.3). Thus we would have

$$A_N = G_{11}A_1 + G_{12}B_1 \tag{69}$$

and

$$0 = G_{21}A_1 + G_{22}B_1 \tag{70}$$

Therefore, the reflection coefficient would be given by

$$R = \left|\frac{B_1}{A_1}\right|^2 = \left|\frac{G_{21}}{G_{22}}\right|^2 \tag{71}$$

Case 3: Tunnelling structure

For a typical tunnelling structure shown in Fig. 25.3, $A_1 = -B_1$ (so that the wavefunction vanishes at $x = 0$) and a plot of

$$\left|\frac{A_1}{A_N}\right|^2$$

as a function of E would give us approximate Lorentzians, the widths of which would enable us to calculate lifetimes of quasi-bound states.

25.5 THE THOMAS-FERMI MODEL OF THE ATOM

In an atom consisting of a large number of electrons, the electrons can approximately be assumed to be in a degenerate state (a degenerate electron gas is discussed in Section 6.9). The electron gas is in a potential field and in the Thomas-Fermi model, we make two assumptions:

1. The potential function varies very little over a wavelength.
2. Further, the field is assumed to vary slowly enough so that we can consider a volume element $d\tau$ which contains a large number of electrons and at the same time the field can be assumed to be approximately constant in this volume.

We recall that the maximum kinetic energy for a degenerate electron gas is given by [see Eq. (94) of Chapter 6]

$$(\text{K.E.})_{\max} = E_{F_0} = \frac{\hbar^2}{2\mu}\left[2\pi^2 n(r)\right]^{2/3} \tag{72}$$

where $n(r)$ represents the number of electrons per unit volume. Further,

$$\text{Total Energy } E = \frac{p^2}{2\mu} + V(r)$$

$$= \frac{p^2}{2\mu} - q\Phi(r) \tag{73}$$

where $\Phi(r)$ represents the potential and $-q$ represents the charge of the electron. Now, for a bound state, we must have $E < 0$; thus

$$q\Phi(r) = (\text{K.E.})_{\max} = \frac{\hbar^2}{2\mu}\left[3\pi^2 n(r)\right]^{2/3} \tag{74}$$

(This is a rather oversimplified model—for a more rigorous analysis one may look up Refs 7 and 8.) Thus,

$$n(r) = \frac{\left[2\mu q\Phi(r)\right]^{3/2}}{3\pi^2\hbar^3} \tag{75}$$

Now, for the theory to be self-consistent, the potential and the charge density should be related through the Poisson equation

$$\nabla^2\Phi = -\frac{\rho(r)}{\varepsilon_0} \tag{76}$$

where ε_0 is the dielectric permittivity of free space and $\rho(r) = -qn(r)$ represents the electron charge density. Substituting for $n(r)$ from Eq. (75), we get

$$\nabla^2\Phi = \frac{q}{3\pi^2\hbar^3\varepsilon_0}(2\mu q)^{3/2}\,\Phi^{3/2} \tag{77}$$

We assume the nucleus (of charge Zq) to be at the origin of the coordinate system so that the potential is spherically symmetric and we may write

$$\nabla^2\Phi = \frac{1}{r^2}\cdot\frac{d}{dr}\left[r^2\frac{d\Phi}{dr}\right] \tag{78}$$

Further, as $r \to 0$, the potential will be essentially due to the nucleus so that (as $r \to 0$) it will behave as $Zq/(4\pi\varepsilon_0 r)$. This suggests that we introduce the variable $\chi(r)$ defined through the following equation

$$\Phi(r) = \frac{Zq}{4\pi\varepsilon_0 r}\chi(r) \tag{79}$$

so that we have the boundary condition

$$\chi(0) = 1 \tag{80}$$

Using Eqs (78) and (79), Eq. (77) becomes

$$\frac{d^2\chi}{dr^2} = \frac{q}{3\pi^2\hbar^3\varepsilon_0}(2\mu q)^{3/2}\frac{(Zq)^{1/2}}{(4\pi\varepsilon_0)^{1/2}}\frac{1}{r^{1/2}}\chi^{3/2} \tag{81}$$

In order to put the above equation in a more convenient form we introduce the dimensionless variable x defined by the following equation

$$r = \gamma x$$

where the parameter γ is to be defined later. Substituting in Eq. (81), we get

$$\frac{d^2\chi(x)}{dx^2} = \left[\gamma^{3/2}\frac{4}{3\pi\hbar^3}Z^{1/2}\left(\frac{2\mu q^2}{4\pi\varepsilon_0}\right)^{3/2}\right]\frac{1}{x^{1/2}}\chi(x) \tag{82}$$

We choose γ so that the quantity inside the square brackets is unity:

$$\gamma = \frac{1}{2}\left(\frac{3\pi}{4}\right)^{2/3}\frac{1}{Z^{1/3}}\frac{\hbar^2}{(\mu q^2/4\pi\varepsilon_0)} \simeq \frac{0.8853 a_0}{Z^{1/3}} \tag{83}$$

where $$a_0 = \frac{\hbar^2}{(\mu q^2/4\pi\varepsilon_0)} \simeq 0.53 \times 10^{-10}\,\text{m}$$

represents the Bohr radius. Thus Eq. (82) takes the form

$$\frac{d^2\chi}{dx^2} = \frac{1}{x^{1/2}}[\chi(x)]^{3/2} \quad \text{Dimensionless Thomas-Fermi equation} \tag{84}$$

which is known as the dimensionless Thomas-Fermi equation; from now on it will be abbreviated as the TF equation. For an isolated atom, the boundary conditions will be

$$\chi(0) = 1 \quad \text{and} \quad \chi(\infty) = 0 \tag{85}$$

we will denote the solution of the TF equation which satisfies the boundary conditions given by Eq. (85) by $\chi_i(x)$. Now, using Eqs (75) and (79), we may write

$$n(r) = \frac{1}{3\pi^2\hbar^3}\left[\frac{2\mu Zq^2}{4\pi\varepsilon_0 r}\chi_i(r)\right]^{3/2} = \frac{1}{3\pi^2}\left[\frac{2Z}{a_0}\cdot\frac{1}{\gamma x}\chi_i(x)\right]^{3/2} \tag{86}$$

or, using Eq. (83), we get

$$n(r) = \frac{32}{9\pi^3}\frac{Z^2}{a_0^3}\left[\frac{\chi_i(x)}{x}\right]^{3/2} \tag{87}$$

Thus, if $\chi_i(x)$ is known, one can determine the charge distribution in the atom. It may be mentioned that Eq. (86) predicts an infinite electron density at the origin: this is one of the basic errors of the TF model–see, e.g. Ref. 9. Indeed, for an atom the TF model is correct only in the domain

$$\frac{a_0}{Z} \lesssim r \lesssim a_0 \quad . \tag{88}$$

where a_0 is the Bohr radius. The condition $a_0/Z < r$ comes from the fact that at distances smaller than a_0/Z the quasi-classical approximation (which is the basis of the TF model) breaks down. On the other hand, for $r > a_0$ the de-Bröglie wavelength of the electron becomes large. Fortuitously, in a complex atom most of the electrons are in the domain

$$\frac{a_0}{Z} \lesssim r \lesssim a_0$$

so that the TF model can be used to describe the complex atom system.

In Problem 25.4, we have shown that a power series solution of the TF equation is given by

$$\chi(x) = 1 + a_2 x + \frac{4}{3} x^{3/2} + \frac{2}{5} a_2 x^{5/2} + \frac{1}{3} x^3 + \dots \tag{89}$$

where a_2 is determined from the initial slope:

$$a_2 = \chi'(0) \tag{90}$$

The power series is convergent for $x < 1$ and is useful (i.e., rapidly convergent) when $x \ll 1$. For $x \gtrsim 1$, one normally uses a numerical method to solve the TF equation. Figure 25.13 gives the numerical solution of the TF equation for various values of the initial slope. When

$$a_2 = \chi'(0) = -1.58808 \tag{91}$$

$\chi(x)$ approaches zero asymptotically and the solution corresponds to an isolated atom. Thus [see Eq. (89)]

$$\chi_i(x) = 1 - 1.58808x + \frac{4}{3} x^{3/2} - 0.635232\, x^{5/2} + \frac{1}{3} x^3 + \cdots \tag{92}$$

where the subscript i corresponds to the fact that we are considering an isolated atom. When

$$|\chi'(0)| < 1.58808$$

$\chi(x)$ decreases to a minimum and then starts increasing again (see Fig. 25.13). One of the important advantages of the Thomas-Fermi model is the fact that the curve corresponding to an isolated atom is universal, i.e. for a given value of Z one has to calculate γ [see Eq. (83)] and simply scale the x-axis [using the equation $r = \gamma x$] to get $\chi_i(r)$ as a function of r. Once $\chi(r)$ is known, $n(r)$ can be determined from Eq. (86). Because of this, detailed tables for the Thomas-Fermi function can be found in many places (see, Refs 11 and 12). A convenient formula (corresponding to an isolated atom) is given by (Ref. 13)

$$\chi_i(x) \approx 0.35\, e^{-0.3x} + 0.55\, e^{-1.2x} + 0.10\, e^{-6.0x} \tag{93}$$

The above formula gives an accurate description in the region $0 < x \lesssim 7$.

In Fig. 25.14, we have plotted the electron distribution function

$$D(r) = 4\pi r^2 \, n(r) \tag{94}$$

as a function of r for the mercury atom ($Z = 80$); the quantity $D(r)\,dr$ represents the number of electrons in the spherical shell whose inner and outer radii are r and $r + dr$ respectively. The solid curve is obtained by using the TF model [e.g. by using the tables given in References 10 and 11– or, may be, the approximate empirical formula given by Eq. (93)] and the dashed curve represents the actual numerical results—the peaks approximately corresponding to the Böhr orbits.

25.6 NUMERICAL INTEGRATION OF SCHRÖDINGER EQUATION

In most problems in quantum mechanics, analytic solutions in terms of known functions cannot bc found. We then have to resort to numerical methods. For complex atoms, the Hartree or Hartree-Fock methods require such techniques to a high degree of sophistication and there is no substitute for actual experience in handling such problems[2]. Nevertheless, it would be useful to learn some basic principles of such computation. We restrict ourselves to the solutions of Schrödinger equation for a central field. The simple exercise treated here can be carried out with a programmable calculator of the type available in many institutions.

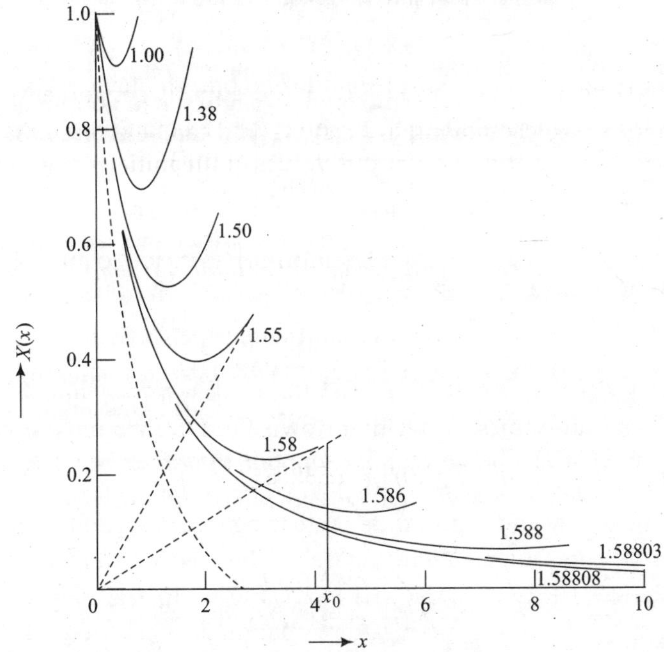

Figure 25.13: The Thomas-Fermi function $\chi(x)$, as a function of x, for various values of the initial slope. The numbers on the curve represent the value of $-a_2$. When $a_2 = -1.58808$, $\chi(x)$ approaches zero asymptotically and the solution corresponds to an isolated atom. The dotted curve represents the qualitative dependence of $\chi(x)$ on x when $|a_2|$ is greater than 1.58808(adapted from Ref. 10).

[2] The reader is referred to the book by Hartree (Ref. 14) for an authoritative account of numerical methods used for atomic structure calculations. For a general account of numerical solutions of ordinary differential equations, we refer to the book by Hartree (Ref. 15)

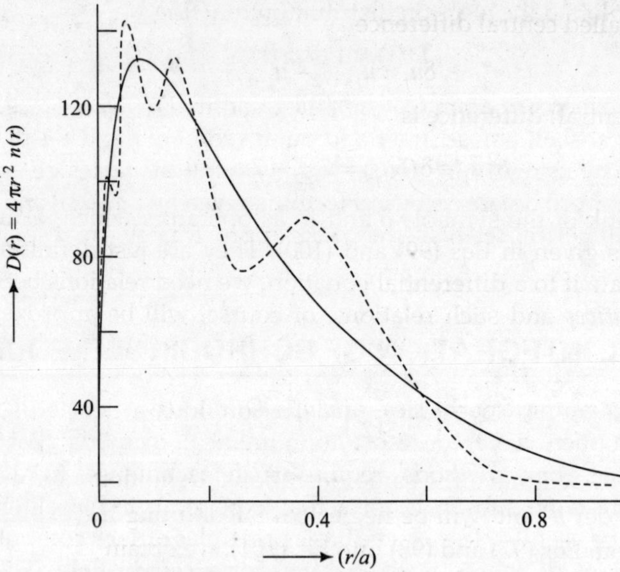

Figure 25.14: The electron distribution function $D(r)[= 4\pi r^2 n(r)]$ as a function of r for the Hg atom (Z = 80). The solid curve is obtained from the TF model and the dashed curve represents the actual numerical results (adapted from Ref. 13).

25.6.1 General principles

The radial equation for a particle moving in a central field can be written as

$$\left\{\frac{d^2}{dr^2} + 2\left[E - V(r)\right] - \frac{l(l+1)}{r^2}\right\}u(r) = 0 \tag{95}$$

the complete wavefunction being

$$\psi(r) = \frac{u(r)}{r} Y_{lm}(\theta, \phi) \tag{96}$$

where Y_{lm} are spherical harmonics. In writing down Eq. (95), we have used natural atomic units ($m = 1$, $|e| = 1$ and $\hbar = 1$). The differential equation may then be written as

$$u'' = F(r)u \tag{97}$$

where

$$F(r) = -2[E - V(r)] + \frac{l(l+1)}{r^2} \tag{98}$$

To solve this numerically, we divide the range of values of r (0 to ∞ in many cases) into fine intervals h, i.e.,

$$r = 0, h, 2h, \ldots$$

a particular value of r being denoted by r_j and the corresponding value of u (to be determined) by u_j. The essence of numerical calculus is to change continuous variations to finite intervals.

We introduce the so-called central difference

$$\delta u_j = u_{j+1/2} - u_{j-1/2} \tag{99}$$

Clearly, the second (central) difference is

$$\delta^2 u_j \equiv \delta(\delta u_j) = u_{j+1} - 2u_j + u_{j-1} \tag{100}$$

It is convenient to think of the symbols δ and δ^2 as operators which, acting on a particular u_j, produce the results given in Eqs (99) and (100), They are just definitions and, hence are exact relations. To relate it to a differential equation, we need relations between these (finite) differences and *derivatives* and such relations, of course, will be approximations. It can be shown[3] that

$$\delta^2 u_j = \left(1 + \frac{\delta^2}{12}\right) h^2 u_j'' + O(h^6) + \cdots \tag{101}$$

where $\delta^2(h^2 u_j'') = h^2 \delta^2 u_j''$ will have the same operational definition as in Eq. (100) and $O(h^6)$ represents terms of order h^6 and will be neglected. Substituting the expression for $\delta^2 u_j$ from Eq. (100) and for u_j'' from Eqs (97) and (98) into Eq. (101), we obtain

$$\left(1 - \frac{h^2}{12} F_{j+1}\right) u_{j+1} = h^2 F_j u_j + 2\left(1 - \frac{h^2}{12} F_j\right) u_j - \left(1 - \frac{h^2}{12} F_{j-1}\right) u_{j-1} \tag{102}$$

where $F_j \equiv F(r_j)$, etc.

Equation (102) gives us a method to iterate values of u for increasing or decreasing values of r. In the former case (called *outward* integration), we start with known values of u_{j-1} and u_j (remember that F_{j-1}, F_j and F_{j+1} are all known values) and calculate u_{j+1}. We the repeat the procedure to calculate u_{j+2} from u_{j+1} and u_j and so on.

25.6.2 Matching the Outward and Inward Solutions

Hartree has given a simple procedure for matching the two solutions unless the value of E is the right one, the two will not match. For a particular value of E, we have

$$\left[\frac{d^2}{dr^2} - 2V(r) + 2E - \frac{l(l+1)}{r^2}\right] u = 0 \tag{103}$$

Let $u + \Delta u$ be the solution for another choice $E + \Delta E$. Then,

$$\left[\frac{d^2}{dr^2} - 2V(r) + 2(E + \Delta E) - \frac{l(l+1)}{r^2}\right](u + \Delta u) = 0 \tag{104}$$

Neglecting the product $\Delta E \Delta u$, these two equations give

[3] This formula can be derived from the Taylor expansions

$$u_{j\pm1} = u_j \pm h u_j' + \frac{h^2}{2} u_j'' \pm \frac{h^3}{6} u_j''' + \ldots$$

so that

$$\delta^2 u_j = h^2 u_j'' + \frac{h^4}{12} u_j^{iv} + \frac{h^6}{360} u_j^{vi} + \ldots$$

$$\left[\frac{d^2}{dr^2} - 2V(r) + 2E - \frac{l(l+1)}{r^2}\right]\Delta u = -2\,u\Delta E \tag{105}$$

Multiplying Eq. (103) by Δu, Eq. (105) by u and subtracting, we obtain

$$\frac{d^2 u}{dr^2}\Delta u - u\frac{d^2(\Delta u)}{dr^2} = 2u^2\,\Delta E$$

The operations d^2/dr^2 and Δ are interchangeable so that we can write

$$\frac{d}{dr}\left[u'\Delta u - u\Delta u'\right] = 2u^2\,\Delta E$$

where $u' = du/dr$. Integrating, we get

$$\left[u'\Delta u - u\Delta u'\right]_{r_1}^{r_2} = 2\Delta E\int_{r_1}^{r_2} u^2\,dr \tag{106}$$

Writing the left-hand side of Eq. (106) as $-u^2\Delta(u'/u)$, we have

$$\left[u^2\Delta\left(\frac{u'}{u}\right)\right]_{r_1}^{r_2} = -2\Delta E\int_{r_1}^{r_2} u^2\,dr \tag{107}$$

For the outward integration, we set $r_1 = 0$, $r_2 = r_0$ and $u(0) = 0$, so that

$$\left[u^2\Delta\left(\frac{u'}{u}\right)\right]_{r_0} = -\,2\Delta E\int_0^{r_0} u^2\,dr \tag{108}$$

Similarly, for the inward integration we set $r_1 = r_0$, $r_2 = \infty$ and $u(\infty) = 0$, so that

$$-\left[u^2\Delta\left(\frac{u'}{u}\right)\right]_{r_0} = -\,2\Delta E\int_{r_0}^{\infty} u^2\,dr \tag{109}$$

Thus (writing $\varepsilon = -E > 0$)

$$\left[\Delta\left(\frac{u'}{u}\right)_{r_0}\right]_{\text{out}} = \frac{2\Delta\varepsilon}{\left[u^2(r_0)\right]_{\text{out}}}\int_0^{r_0} u_{\text{out}}^2\,dr \tag{110}$$

and

$$\left[\Delta\left(\frac{u'}{u}\right)_{r_0}\right]_{\text{in}} = \frac{-\,2\Delta\varepsilon}{\left[u^2(r_0)\right]_{\text{in}}}\int_{r_0}^{\infty} u_{\text{in}}^2\,dr \tag{111}$$

and the values of u'/u should be the same for the correct values of ε. Suppose, for example, $(u'/u)_{\text{out}}$ is larger than $(u'/u)_{\text{in}}$. We clearly have to *reduce* ε, i.e. $\Delta\varepsilon < 0$. Indeed, from Eqs (110) and (111) we obtain

$$\left[\frac{\int_0^{r_0} u_{\text{out}}^2\,dr}{u_{\text{out}}^2(r_0)} + \frac{\int_0^{\infty} u_{\text{in}}^2\,dr}{u_{\text{in}}^2(r_0)}\right] 2\,\Delta\varepsilon = -\left[\left(\frac{u'}{u}\right)_{\text{out}} - \left(\frac{u'}{u}\right)_{\text{in}}\right]_{r_0} \tag{112}$$

a relation easily obtained by setting

$$\left(\frac{u'}{u}\right)_{\text{out}} + \Delta\left(\frac{u'}{u}\right)_{\text{out}} = \left(\frac{u'}{u}\right)_{\text{in}} + \Delta\left(\frac{u'}{u}\right)_{\text{in}}$$

Equation (112) is used to estimate $\Delta\varepsilon$ and the revised estimate ε may be used to repeat the whole procedure, if necessary.

The matched solution may then be normalized to give

$$\int_0^\infty u^2 dr = 1$$

25.6.3 The Hydrogen Atom as an Example

We shall illustrate the method by calculating the ground state energy of the hydrogen atom. Here $V = -1/r$ and $F = -2(E + 1/r)$. Our procedure is to integrate *outwards* from $r = 0$ to $r = 2$ in steps of $h = 0.1$ and to integrate inwards from $r = 10$ to $r = 2$ in steps of $h = 0.5$ and see how to 'match' solutions at $r = 2$. We hope that $u(r = 10)$ is sufficiently small. To start the outward integration, we need a few values of u for small r. We write

$$u(r) = \sum_{n=1}^{\infty} a_n r^n \tag{113}$$

since we know that for small r, the leading term in u behaves like r^{l+1}. Substituting Eq. (113) into Eq. (97) with $F(r) = -2(E + 1/r)$ and comparing coefficients of r, we obtain

$$\begin{aligned} a_2 &= -a_1 \\ 6a_3 &= -2Ea_1 - 2a_2 \\ 12a_4 &= -2Ea_2 - 2a_3 \end{aligned} \tag{114}$$

Choosing $a_1 = 1$, we obtain

$$u = r - r^2 + \frac{(1-E)r^3}{3} + \frac{(4E-1)r^4}{18} + \cdots \tag{115}$$

which can be used to obtain the first few values of u, say up to $r = 0.4$.

To start the inward integration, we note that for large r, $F(r)$ is very nearly constant, so that

$$u(r) \approx A \exp(-\sqrt{F(r)}\, r) \tag{116}$$

Fixing, arbitrarily, the value of $\mu(R)$ at $r = R$, we have

$$A = u(R)\exp(R\sqrt{F})$$

so that

$$u(R-h) = u(R)\exp(h\sqrt{F})$$

We can now proceed to use the iteration formula, Eq. (102) with

$$u_j = u(R-h)$$

and

$$u_{j+i} = u(R)$$

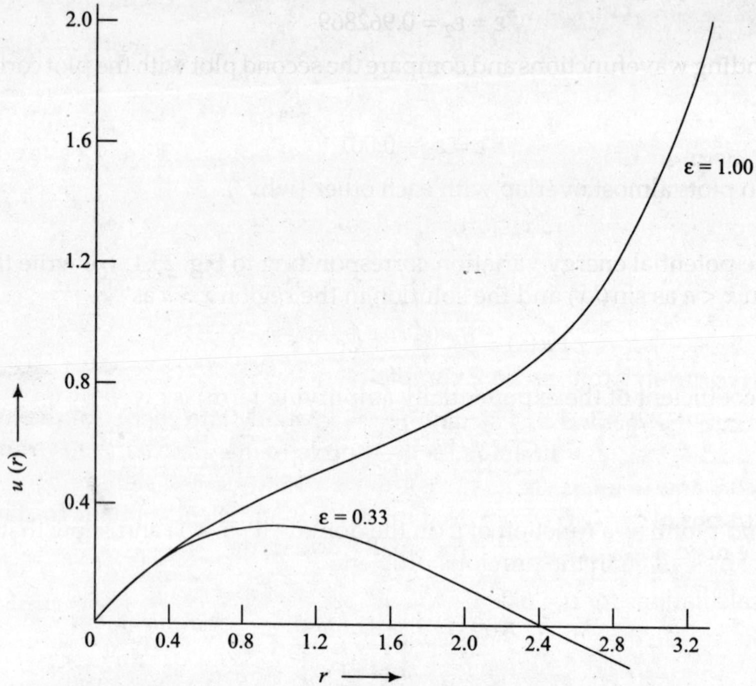

Figure 25.15: The variation $u(r)$ with r for $\varepsilon = 1.00$ and $\varepsilon = 0.33$.

In practice, it is not really necessary to use a large value of R; typically, we have tried $R = 5$, $h = 0.1$ (same interval as for the outward integration) and started with a guess of $\varepsilon = 1.0$. Successive iterations gave $\varepsilon = 0.330, 0.426, 0.484$ and 0.498. (The exact value is $\varepsilon = 0.5$). We have plotted the calculations for $\varepsilon = 1$ and for $\varepsilon = 0.33$ just to show what happens for incorrect values (see Fig. 25.15). We recommend that students try these calculations with a calculator to get a feel for numerical methods.

25.7 PROBLEMS

Note: In Problems 25.1–25.3 all symbols have the same meaning as in Section 25.2.

Problem 25.1 Consider a potential variation corresponding to Fig. 25.3 with $\alpha = 6.0$ and $d/a = 1$. Show that the quasi-modes correspond to

$$\varepsilon = \varepsilon_1 = 0.19932777$$

and

$$\varepsilon = \varepsilon_2 = 0.75863041$$

Plot the corresponding wavefunctions and compare with the plots corresponding to

$$\varepsilon = \varepsilon_1 + 0.0005$$

and

$$\varepsilon = \varepsilon_2 + 0.0005$$

Also plot $|A/D|^2$ around $\varepsilon = \varepsilon_1$ and $\varepsilon = \varepsilon_2$ and find the lifetime of the states.

Problem 25.2 Consider again the potential variation corresponding to Fig. 25.3 with $\alpha = 5.0$ and $d/a = 1$. Show that the quasi-modes correspond to

$$\varepsilon = \varepsilon_1 = 0.269514$$

and $$\varepsilon = \varepsilon_2 = 0.962869$$

Plot the corresponding wavefunctions and compare the second plot with the plot corresponding to

$$\varepsilon = \varepsilon_2 + 0.001$$

Show that the two plots almost overlap with each other (why?).

Problem 25.3

(a) Consider the potential energy variation corresponding to Fig. 25.1; we write the solution in the region $x < a$ as $\sin(kx)$ and the solution in the region $x > a$ as

$$\psi(x) = Be^{\kappa(x-a)} + Ce^{-\kappa(x-a)}$$

Show that B (the coefficient of the exponentially amplifying term) is given by

$$B = \sin\left(\alpha\sqrt{\varepsilon}\right) + \sqrt{\frac{\varepsilon}{1-\varepsilon}}\cos\left(\alpha\sqrt{\varepsilon}\right)$$

Assume $\alpha = 5.0$ and plot B as a function of ε (in the domain $0 < \varepsilon < 1$) and show that the zeroes occur for values of ε as given in the previous problem.

(b) Repeat the calculations for $\alpha = 6.0$.

Problem 25.4 Substitute the power series solution

$$\chi(x) = \sum_{k=0,1,2\ldots} a_k x^{k/2}$$

$$= a_0 + a_1 x^{1/2} + a_2 x + a_3 x^{3/2} + a_4 x^2 + \ldots$$

in the TF equation [Eq. (84)] and assuming $\chi(0) = 1$, derive Eq. (89).

Problem 25.5 Let $\chi_i(x)$ represent the solution of the TF equation corresponding to an isolated atom, i.e.,

$$\chi_i(x) = 1$$

and the slope at $x = 0$ is -1.58808 so that $\chi_i(\infty) = 0$.

(a) Show that the function $D(r)$ [as given by Eq. (94)] is proportional to the function

$$F(x) = \left[x^{1/3}\chi_i(x)\right]^{3/2}$$

(b) Using the empirical formula [Eq. (93)], find the value of x where $F(x)$ attains its maximum value.

(c) Using the above result calculate for the Hg atom (Z = 80) the value of r at which $D\{r)$ attains its maximum value and compare with the value that can be approximately obtained from Fig. 25.14.

[Answer: (b) ≈ 0.39 **(c)** $\approx 0.080a_0$**]**

Problem 25.6 For larger r, $V(r) \to 0$ and the radial equation (for $l = 0$) may be written as

$$u'' = -2|E|u(r)$$

This means that if $u(r) > 0 (< 0)$ the curvature is negative (positive). Perform the outward numerical integration for hydrogen atom ground state for $|E| = 1$ and $|E| = 0.3$ and plot $u(r)$ up to $r \approx 5$. For $|E| = 1$, you will find $u(r) \to -\infty$ and for $|E| = 0.3$, $u(r) \to +\infty$. Study, particularly, the shape of the curves as they diverge from the r axis.

Problem 25.7 Consider the one-dimensional potential well $V = -V_0$ for $|x| < a$ and $V = 0$ for $|x| > a$. How do you expect $\psi(x)$ to behave for large x when

(i) ψ' has too small a value at a and

(ii) ψ' has too large a value at a?

Problem 25.8 Perform the numerical integration to obtain the ground state energy and wavefunction for a deuteron using the square well potential

$$V(r) = \begin{cases} -V_0 & \text{for} \quad r < a \\ 0 & \text{for} \quad r > a \end{cases}$$

with $a = 1.5$ fermi and $V_0 \approx 25$ MeV.

REFERENCES AND SUGGESTED READINGS

1. I.I. Goldman and V.D. Krivchenkov, *Problems in Quantum Mechanics*, Dover Publications, New York (1993).
2. A.K. Ghatak, K. Thyagarajan and M.R. Shenoy, *A novel numerical technique for solving the one-dimensional Schrödinger equation using matrix approach: Application to quantum well structures*, IEEE J. Quantum Electronics, **24**, 1524 (1988).
3. A.K. Ghatak, I.C. Goyal and R.L. Gallawa, *Mean life-time calculations of quantum well structures: A rigorous analysis*, IEEE J. Quantum Electronics, **26**, 305 (1990).
4. A.H. Nayfeh, *Introduction to Perturbation Techniques*, John Wiley, New York (1981).
5. A.K. Ghatak, I.C. Goyal, R. Jindal and C.C. Davis, *Analytical solutions for sinusoidally varying potentials*, J. Phys. Edu. **16**, No. 1, pp 81–86, April 2000.
6. A. Ghatak, *Basic Quantum Mechanics* (with CD), Macmillan India Ltd., (2000).
7. L.D. Landau and E.M. Lifshitz, *Quantum Mechanics: Non-Relativistic Theory*, Pergamon Press, Oxford (1965).
8. R.P. Feynmann, N. Metropolis and E. Teller, *Physical Review*, **75**, 1561 (1949).
9. J. Schwinger, *Physical Review*, **A22**, 1827 (1980).
10. J.C. Slater and H.M. Krutter, *Physical Review*, **47**, 559 (1935).
11. I.M. Torrens, *Interatomic Potentials*, Academic Press, New York (1972).
12. S. Eliezer, A. Ghatak and H. Hora, *An Introduction to Equations of State*, Cambridge University Press, Cambridge (1986).
13. P. Gombas, *Die statistiche theorie des atoms und ihre anwendungen*, Springer Verlag, Wien (1949).
14. D.R. Hartree, *The Calculation of Atomic Structures*, John Wiley, New York (1957).
15. D.R. Hartree, *Numerical Analysis*, Clarendon Press, Oxford (1958).

26 Elementary theory of scattering

Our actual situation in research work in atomic physics is usually this: we wish to understand a certain phenomenon, we wish to recognise how this phenomenon follows from the general laws of nature. Therefore, that part in the phenomenon is the natural 'object' in the theoretical treatment and should be separated in this respect from the tools used to study the phenomenon. This again emphasises a subjective element in the description of atomic events, since the measuring device has been constructed by the observer, and we have to remember that what we observe is not nature in itself but nature exposed to our method of questioning.

— WERNER HEISENBERG in *Physics and Philosophy*, Penguin, London (1989).

26.1 INTRODUCTION

In some of the earlier chapters we had considered solutions of the Schrödinger equation which correspond to bound state problems where the wavefunction vanished at large distances from the origin; the corresponding energy levels were found to form a discrete set. In this chapter, we will consider solutions of the Schrödinger equation where the energy eigenvalues would be continuously distributed; for such a case the wavefunction would not vanish at large distances from the origin. Such solutions correspond to the scattering of a particle by a force field where the energy is specified in advance and the behavior of the wavefunction is found in terms of energy. We should mention here that in Chapter 8 we did consider the scattering by a one-dimensional force field, the corresponding three-dimensional case will be considered here.

Scattering experiments have provided most of the information about interactions among atoms, nuclei and particles. The general technique is to have a collimated beam of projectiles (particles) with well-defined energies and sometimes other properties, such as polarization, as well. The particles are then scattered by a target of atoms and the scattered particles are detected by means of a suitable detector. The measurement could be simply the intensity variation as a function of angle or could involve more detailed information like change in the

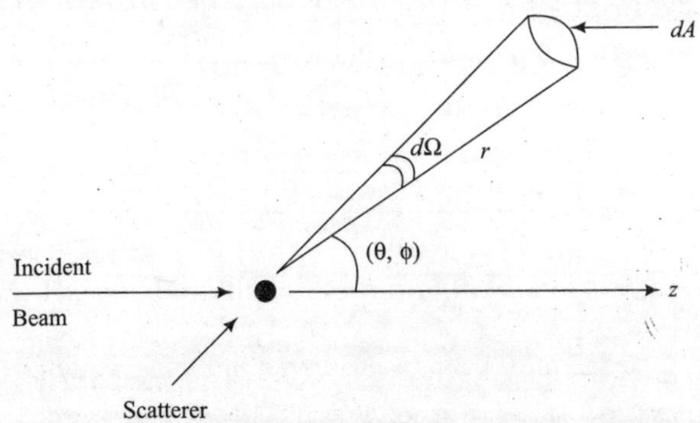

Figure 26.1: The scattering of an incident beam. The particles scattered into the solid angle $d\Omega$ is received by the detector.

energy or the polarization state, etc. The most commonly used parameter which displays the results of scattering experiments is the cross-section. In Section 26.2 we will give the definition of the scattering cross-section according to classical mechanics which will be followed by the corresponding quantum mechanical definition. As mentioned earlier, scattering experiments provide information about interaction among atoms, nuclei and particles. Assuming a certain model for the interaction, one calculates the cross-section which is compared with experimental data. A good agreement between theory and experiment would suggest the validity of the model assumed. However, sometimes a variety of models give good agreement between theory and experiment and it becomes difficult to justify a particular model. In Sections 26.4–26.6, we will develop theoretical methods to calculate the cross-section and wherever possible, comparison will be made with available experimental data.

26.2 THE CLASSICAL DEFINITION OF SCATTERING CROSS-SECTION

We first give the definition of the scattering cross-section according to classical mechanics. We consider a beam of particles moving along z-direction with a flux of I_0 particles per unit area per unit time. The particles undergo scattering by a potential and one measures the angular distribution of scattered particles. The scattering cross-section $\sigma(\theta, \phi)$ is defined by

$$\sigma(\theta, \phi)\, d\Omega = \frac{\text{Number of particles scattered into solid angle } d\Omega \text{ per unit time}}{\text{Incident intensity } I_0} \tag{1}$$

where $d\Omega$ represents an infinitesimal solid angle in the direction (θ, ϕ); the angle θ represents the angle of scattering and ϕ represents the azimuthal angle; the origin has been assumed to be at the center of the force (see Fig. 26.1). For a spherically symmetric potential, $\sigma(\theta, \phi)$ will be independent of ϕ. The total cross-section is given by

$$\sigma = \iint \sigma(\theta, \phi)\, d\Omega = \int_0^{\pi}\int_0^{2\pi} \sigma(\theta, \phi) \sin\theta \, d\theta \, d\phi \tag{2}$$

Thus σI_0 will represent the number of particles scattered out of the beam per unit time. For a spherically symmetric potential

$$\sigma = 2\pi \int_0^{\pi} \sigma(\theta) \sin\theta \, d\theta \tag{3}$$

In classical mechanics, each particle has a well-defined trajectory and if we assume that all particles of the incident beam have the same speed v_0, there is a unique angle of scattering θ corresponding to a particular value of the impact parameter s. For example, for an attractive square well potential defined by

$$\begin{aligned} V(r) &= -V_0, \quad 0 < r < a \\ &= 0, \qquad r > a \end{aligned} \qquad \text{Square well potential} \tag{4}$$

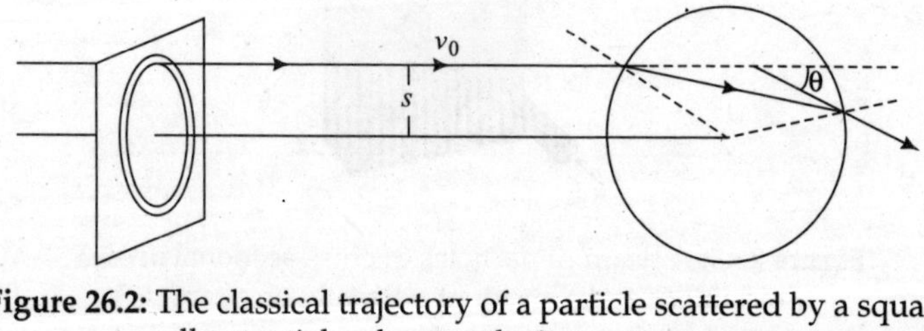

Figure 26.2: The classical trajectory of a particle scattered by a square well potential, s denotes the impact parameter.

the relation between the impact parameter s and the scattering angle θ is given by (see Fig. 26.2 and Problem 26.20):

$$\theta = -2(i-r) = 2\left[\sin^{-1}\left(\frac{s}{na}\right) - \sin^{-1}\left(\frac{s}{a}\right)\right] \quad \text{for } 0 < s < a$$

$$= 0 \quad \text{for } s > a \tag{5}$$

where
$$n = \left[\frac{E+V_0}{E}\right]^{1/2} \quad \text{and} \quad E = \frac{1}{2}mv_0^2 \tag{6}$$

Now, the number of particles which cross the annular area between s and $s + ds$ is $I\ 2\pi s\ ds$ and this must represent the number of particles scattered between θ and $\theta + d\theta$ which from Eq. (1) is simply $\sigma(\theta) I\ 2\pi \sin\theta\ d\theta$. Thus,

$$I\ 2\pi\ sds = -\sigma(\theta) I\ 2\pi \sin\theta d\theta$$

or
$$\sigma(\theta) = -\frac{s}{\sin\theta}\cdot\frac{ds}{d\theta} \tag{7}$$

where the negative sign is due to the fact that θ decreases with increase in the value of the impact parameter s. Equation (7) is quite general; if we now use the result given by Eq. (5), we would obtain

$$\sigma(\theta) = \frac{n^2a^2}{4\cos\frac{\theta}{2}}\cdot\frac{\left(n\cos\frac{\theta}{2}-1\right)\left(n-\cos\frac{\theta}{2}\right)}{\left[1+n^2-2n\cos\frac{\theta}{2}\right]^2} \tag{8}$$

Since no scattering occurs for $s > a$, we must have

$$\theta < \left[\pi - 2\sin^{-1}\left(\frac{1}{n}\right)\right] \tag{9}$$

In a similar manner one can calculate the scattering cross-section for other forms of $V(r)$ (see, e.g. Chapter 3 of Ref. 1).

26.2.1 Physical Interpretation of the Cross-section

Let us consider a beam of particles of cross-sectional area A (and intensity I_0) incident normally on a target as shown in Fig. 26.3. Let there be N scatterers per unit volume inside the target. If $I(x)$ represents the intensity of the beam after penetrating through a distance

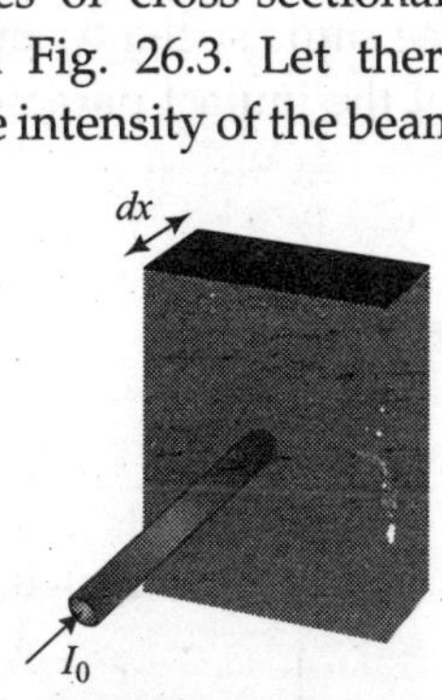

Figure 26.3: A beam of particles of cross-sectional area A (and intensity I_0) incident normally on a target.

x and if $dI(x)$ represents the loss of intensity in traversing an additional distance dx then the total number of interactions (per unit time) in the volume element $A\,dx$ will be given by

$$-A\,dI(x) = \sigma I(x)\,(NA\,dx) \tag{10}$$

where σ represents the total cross-section (see Eq. 3)[1], the quantity $(NA\,dx)$ represents the number of scatterers in the volume element $A\,dx$ and the negative sign is due to the fact that $I(x)$ decreases with x; we have neglected multiple scatterings in the volume element $A\,dx$. Simple integration of Eq. (10) gives

$$I(x) = I_0 e^{-\Sigma x} \tag{11}$$

where $\Sigma = N\sigma$ is known as the macroscopic cross-section and can be interpreted as the inverse of the mean free path. Further, it can be easily seen that the probability that the particle undergoes its first scattering between x and $x + dx$ will simply be given by

$$p(x)\,dx = e^{-\Sigma x}\,\Sigma\,dx$$

where the factor $e^{-\Sigma x}$ represents the probability that it does not undergo scattering in traversing a distance x and $\Sigma\,dx$ represents the probability that it does undergo scattering in traversing the distance dx. Obviously,

$$\int_0^\infty p(x)\,dx = 1$$

as it indeed should be.

26.3 QUANTUM THEORY OF SCATTERING

In quantum mechanics, strictly speaking, the scattering problem needs a temporal description, namely, how a wave packet representing the incident particle as $t \to -\infty$ behaves as it passes through the region of perturbation and preceeds to $t \to \infty$ (see Problem 26.19). However, when there is a steady current of particles in the beam, we often dispense with the temporal description. We merely calculate the current of particles at an infinite distance away from the target (as opposed to behavior at $t \to \infty$). This is done by looking for asymptotic solutions of the time independent Schrödinger equation. Since in this region the particles are free, we are dealing with the continuous spectrum of positive energies.

We start with the time independent Schrödinger equation

$$\nabla^2\psi(\mathbf{r}) + \frac{2\mu}{\hbar^2}\left[E - V(\mathbf{r})\right]\psi(\mathbf{r}) = 0 \tag{12}$$

where, if we are considering a fixed scattering center, μ represents the mass of the particle. On the other hand, if we are considering a two body interaction then μ represents the reduced mass of the system provided the potential energy function depends only on the magnitude of the distance between the two particles, i.e:, $V = V(|\mathbf{r}_1 - \mathbf{r}_2|)$; such potentials are known as spherically symmetric potentials, examples of which are the square well potential [see Eq. (79)], the Coulomb potential (see Section 26.7), etc. Then E (in Eq. (12)) represents the energy in the center of mass system and $\mathbf{r} = \mathbf{r}_1 - \mathbf{r}_2$ represents the relative coordinate (see Sections 10.3 and 26.8 and Appendix H). The cross-section is calculated in the center of mass system and the

[1] We are assuming here that the number of particles scattered is simply the sum of the numbers scattered by each atom. In quantum theory, we will see that if different atoms of the target act coherently, then we cannot simply add scattering cross-sections as above since interference effects are involved. One then has to add the scattering probability *amplitudes* first. An example of this is the scattering of low energy neutrons in crystals (see Problems 26.15 and 27.1).

relation which enables one to calculate the cross-section in the laboratory system is discussed in Section 26.9.

In order to describe the scattering experiment we will obtain the solution of the Schrödinger equation [Eq. (12)] in the form of an incident plane wave and an outgoing scattered wave. Thus, we seek solution of the form:

$$\psi(\mathbf{r}) \underset{r\to\infty}{\longrightarrow} e^{i\mathbf{k}\cdot\mathbf{r}} + f(\theta)\frac{e^{ikr}}{r} \tag{13}$$

where $\mathbf{k} = k\hat{\mathbf{z}}$, $\hat{\mathbf{z}}$ representing a vector along the z-axis. We may mention here that for a short-range force; i.e., when

$$\lim_{r\to\infty} rV(r) = 0$$

the general form of the asymptotic solution of the Schrödinger equation will be of the form (see Appendix O):

$$f(\theta)\frac{e^{ikr}}{r} + g(\theta)\frac{e^{-ikr}}{r} \tag{14}$$

where the second term corresponds to an incoming wave and hence is neglected in Eq. (13). It may be noted that since the potential is assumed to be spherically symmetric, the amplitude of the scattered wave (i.e. the function f) is assumed to be independent of the azimuthal angle ϕ.

If we multiply Eq. (13) by the appropriate time factor $\exp(-iEt/\hbar)$, it will be seen that the first term represents a plane wave propagating along the z-axis. Since $|e^{ikz}|^2 = 1$ the number of particles crossing a unit time is simply

$$v_\infty = \frac{\hbar k}{m} \tag{15}$$

[see Eqs (38) and (40) of Chapter 4]. Corresponding to the second term in Eq. (13) the current density is given by (see Section 4.4)

$$\mathbf{J} = \mathrm{Re}\left[\frac{e^{-ikr}}{r} f^*(\theta)\frac{\hbar}{im}\nabla\left(\frac{e^{ikr}}{r} f(\theta)\right)\right] \tag{16}$$

or

$$\mathbf{J} \simeq v_\infty \frac{|f(\theta)|^2}{r^2}\hat{\mathbf{r}} \tag{17}$$

where we have neglected terms of $O(1/r^3)$ which will make a negligible contribution at large values of r. The $1/r^2$ dependence of $\mathbf{J}$ implies that the intensity of the scattered wave falls off according to the inverse square law (see Problem 4.3). The function $f(\theta)$ describes the angular distribution of the scattered wave and is known as the *scattering amplitude*. It should be pointed out that in our analysis we have calculated the current density due to the two parts of the wavefunction separately. This is justified because in an actual experiment the incident wave is collimated and does not interfere with the scattered wave. Now, in Fig. 26.1 the area dA is given by

$$dA = r^2 d\Omega \tag{18}$$

Therefore, the number of particles scattered in the solid angle $d\Omega$ will be equal to the number that cross the area dA and hence equal to

$$\frac{v_\infty}{r^2}|f(\theta)|^2 r^2 d\Omega \tag{19}$$

Thus[2],

$$\sigma(\theta) = \left[\frac{d\sigma}{d\Omega}\right] = |f(\theta)|^2 \tag{20}$$

Notice that multiplication of the RHS of Eq. (13) by an arbitrary constant would not have changed $\sigma(\theta)$. In order to obtain a solution of Eq. (12) of the form of Eq. (13), we rewrite Eq. (12) in the form

$$(\nabla^2 + k^2)\,\psi(\mathbf{r}) = F(\mathbf{r}) \tag{21}$$

where $k^2 = 2\mu E/\hbar^2$ and $F(\mathbf{r}) = \frac{2\mu}{\hbar^2}(\mathbf{r})\,\psi(\mathbf{r})$

We consider Eq. (21) as an inhomogeneous equation and write the equation in the form

$$\psi(\mathbf{r}) = \phi(\mathbf{r}) + v(\mathbf{r}) \tag{22}$$

where $v(\mathbf{r})$ represents the solution to the inhomogeneous part of Eq. (21) and $\phi(\mathbf{r})$ is a solution of the homogeneous equation

$$(\nabla^2 + k^2)\,\phi(\mathbf{r}) = 0 \tag{23}$$

For the scattering problem in which the incident beam travels along **k**, we choose

$$\phi(\mathbf{r}) = e^{i\mathbf{k}.\mathbf{r}} \tag{24}$$

Now, in order to determine the solution to the inhomogeneous equation, we introduce the Green's function in the next section; an alternative derivation is given in Appendix N.

26.4 THE GREEN'S FUNCTION

The Green's function of an operator L is defined through the following equation:

$$LG(\mathbf{r}, \mathbf{r}') = \delta(\mathbf{r} - \mathbf{r}') \tag{25}$$

Once the Green's function is known, one can show that the solution of the inhomogeneous equation

$$L\psi = F(\mathbf{r}) \tag{26}$$

is given by

$$\psi = \int G(\mathbf{r}, \mathbf{r}')\,F(\mathbf{r}')\,d\mathbf{r}'$$

where the (triple) integral is over the entire three-dimensional space. We will first determine the Green's function of the operator ∇^2 and then of the operator $(\nabla^2 + k^2)$.

26.4.1 The Green's Function for the ∇^2 Operator

We will first show that

[2] In general, the scattering amplitude f depends on θ and ϕ; but for spherically symmetric potentials, the scattering amplitude and therefore σ is independent of the azimuthal angle.

$$\nabla^2\left(\frac{1}{r}\right) = -4\pi\delta(\mathbf{r}) = -4\pi\,\delta(x)\,\delta(y)\,\delta(z) \tag{27}$$

Now, for a function $\psi(r)$ dependent only on the coordinate r,

$$\nabla^2\psi(r) = \frac{1}{r^2}\frac{d}{dr}\left[r^2\frac{d\psi}{dr}\right]$$

Elementary differentiation shows us that

$$\nabla^2\left(\frac{1}{r}\right) = \frac{1}{r^2}\frac{d}{dr}\left[r^2\frac{d}{dr}\left(\frac{1}{r}\right)\right] = 0 \quad \text{for} \quad r \neq 0 \tag{28}$$

At $r = 0$, the above expression is not defined because of the factor $(1/r^2)$. Now,

$$\nabla^2\left(\frac{1}{r}\right) = \text{div } \mathbf{F}$$

where $$\mathbf{F}(r) \equiv \text{grad}\left(\frac{1}{r}\right) = -\frac{1}{r^2}\hat{\mathbf{r}}$$

where $\hat{\mathbf{r}}$ represents a unit vector along $\mathbf{r}$. In order to evaluate div $\mathbf{F}$ at the origin, we consider a sphere of radius R surrounding the origin. Then on the spherical surface, $|\mathbf{F}|$ has a constant value equal to $(1/R^2)$. Thus integrating over this spherical surface, we get

$$\iint \mathbf{F}\cdot d\mathbf{S} = -\frac{1}{R^2}\int_{\theta=0}^{\pi}\int_{\phi=0}^{2\pi} R^2 \sin\theta\, d\theta\, d\phi$$

$$= -4\pi$$

(independent of the value of R). Using Gauss' theorem,

$$\iiint_V \nabla^2\left(\frac{1}{r}\right) d\tau = \iiint_V \text{div } \mathbf{F}\, d\tau = \iint \mathbf{F}\cdot d\mathbf{S}$$

where the volume integration is over a spherical volume of radius R and the surface integral is over the spherical surface of radius R. Thus,

$$\iiint_V \nabla^2\left(\frac{1}{r}\right) d\tau = -4\pi$$

Because of Eq. (28), V can be an *arbitrary* volume surrounding the origin. Thus, the above equation would imply

$$\nabla^2\left(\frac{1}{r}\right) = -4\pi\delta(\mathbf{r}) = -4\pi\delta(x)\,\delta(y)\,\delta(z)$$

If we shift the origin of the coordinate system to $\mathbf{r}'$, we would obtain

$$\nabla^2\left(\frac{1}{|\mathbf{r}-\mathbf{r}'|}\right) = -4\pi\delta(\mathbf{r}-\mathbf{r}')$$

$$= -4\pi\delta(x-x')\,\delta(y-y')\,\delta(z-z')$$

The Green's function $G(\mathbf{r}, \mathbf{r}')$ for the operator ∇^2 is defined through the equation [see Eq. (25)]

$$\nabla^2 G(\mathbf{r}, \mathbf{r}') = \delta(\mathbf{r} - \mathbf{r}') \tag{29}$$

Comparing the above two equations, we set

$$G(\mathbf{r}, \mathbf{r}') = -\frac{1}{4\pi|\mathbf{r} - \mathbf{r}'|} \tag{30}$$

26.4.2 Solution of the Equation $\nabla^2\psi(\mathbf{r}) = F(\mathbf{r})$

We next obtain the solution of the equation

$$\nabla^2\psi(\mathbf{r}) = F(\mathbf{r}) \tag{31}$$

If we multiply Eq. (29) by $\psi(\mathbf{r}')$ and Eq. (31) by $G(\mathbf{r}, \mathbf{r}')$, subtract the resulting equations and then integrate, we obtain

$$\int_V [\psi(\mathbf{r}')\nabla^2 G(\mathbf{r}, \mathbf{r}') - G(\mathbf{r}, \mathbf{r}')\nabla^2\psi(\mathbf{r}')]\, d\tau' = \int_V \psi(\mathbf{r}')\delta(\mathbf{r} - \mathbf{r}')d\tau' - \int_V G(\mathbf{r}, \mathbf{r}')F(\mathbf{r}')d\tau'$$

Now, the integrand on the left-hand side is simply div $(\psi\nabla G - G\nabla\psi)$. Thus, if we use the Gauss' theorem, the above equation would become

$$\int_S (\psi\nabla G - G\nabla\psi)\cdot d\mathbf{S} = \psi(\mathbf{r}) - \int G(\mathbf{r}, \mathbf{r}')\mathbf{F}(\mathbf{r}')\, d\tau'$$

where $\mathbf{S}$ represents the surface bounding the volume V; if this surface is taken at infinity and if the integrand is assumed to fall off faster than $1/r^2$, then the left-hand side of the above equation vanishes and we have

$$\psi(\mathbf{r}) = \int_V G(\mathbf{r}, \mathbf{r}')F(\mathbf{r}')\, d\tau' \tag{32}$$

Using Eq. (30), we get

$$\psi(\mathbf{r}) = -\int \frac{F(\mathbf{r}')}{4\pi|\mathbf{r} - \mathbf{r}'|}\, d\tau' \tag{33}$$

As a simple example, we consider the Poisson equation which is given by

$$\nabla^2\phi = -\frac{\rho(\mathbf{r})}{\varepsilon_0}$$

where ϕ is the electrostatic potential, $\rho(\mathbf{r})$ the charge density and ε_0 the permittivity of free space. Using Eq. (33), we may write the solution of the above equation as

$$\phi(\mathbf{r}) = \int \frac{\rho(\mathbf{r}')}{4\pi\varepsilon_0|\mathbf{r} - \mathbf{r}'|}\, d\tau' \tag{34}$$

The above equation is essentially Coulomb's law in electrostatics because due to a point charge Q at the origin, the electrostatic potential at a distance r is simply

$$\frac{Q}{4\pi\varepsilon_0 r}$$

Thus, since $\rho(\mathbf{r}')d\tau'$ represents the total charge in the infinitesimal volume element $d\tau'$, its contribution to the electric potential at $\mathbf{r}$ would be

$$\frac{\rho(\mathbf{r}')}{4\pi\varepsilon_0|\mathbf{r} - \mathbf{r}'|}\, d\tau'$$

from which Eq. (34) readily follows.

26.4.3 Green's Function for the ($\nabla^2 + k^2$) Operator

Proceeding in a manner similar to that discussed in the previous section, we can show that

$$(\nabla^2 + k^2)\frac{e^{\pm ikr}}{r} = 0 \quad \text{for} \quad r > 0$$

Because of the above equation

$$\int_V \left[(\nabla^2 + k^2)\frac{e^{ikr}}{r}\right] d\tau = \int_R \left[(\nabla^2 + k^2)\frac{e^{ikr}}{r}\right] r^2\, dr\, d\Omega$$

where V is an arbitrary volume surrounding the origin and the second integral is over a sphere of radius R. Using Gauss's theorem, we get

$$\text{RHS} = \iint \nabla\left(\frac{e^{ikr}}{r}\right)\cdot \hat{\mathbf{r}}\, R^2\, d\Omega + k^2 \int_0^R e^{ikr}\, r\, dr \iint d\Omega$$

$$= \left[ikRe^{ikR} - e^{ikR}\right]4\pi + 4\pi k^2\left[r\frac{e^{ikr}}{ik} - \frac{e^{ikr}}{(ik)^2}\right]_0^R$$

$$= -4\pi$$

Thus,
$$\left(\nabla^2 + k^2\right)\frac{e^{\pm ikr}}{r} = -4\pi\delta(\mathbf{r})$$

If we shift the origin, we obtain

$$\left(\nabla^2 + k^2\right)\frac{\exp\left[\pm ik|\mathbf{r} - \mathbf{r}'|\right]}{-4\pi|\mathbf{r} - \mathbf{r}'|} = \delta(\mathbf{r} - \mathbf{r}')$$

Thus the Green's function of the operator ($\nabla^2 + k^2$) is given by

$$G(\mathbf{r}, \mathbf{r}') = -\frac{1}{4\pi}\frac{e^{\pm ik|\mathbf{r} - \mathbf{r}'|}}{|\mathbf{r} - \mathbf{r}'|} \tag{35}$$

and the solution of the equation

$$(\nabla^2 + k^2)\psi(\mathbf{r}) = F(\mathbf{r})$$

is given by

$$\psi(\mathbf{r}) = \phi(\mathbf{r}) - \int \frac{e^{\pm ik|\mathbf{r} - \mathbf{r}'|}}{4\pi|\mathbf{r} - \mathbf{r}'|} F(\mathbf{r}')\, d\tau' \tag{36}$$

where $\phi(\mathbf{r})$ is the solution to the homogeneous part of the equation. If we use Eq. (24) and the fact that

$$F(\mathbf{r}) = \frac{2\mu}{\hbar^2} V(\mathbf{r})\psi(\mathbf{r})$$

we would get $$\psi^{\pm}(\mathbf{r}) = e^{i\mathbf{k}\cdot\mathbf{r}} - \frac{\mu}{4\pi\hbar^2}\int \frac{\exp(\pm ik|\mathbf{r}-\mathbf{r}'|)}{|\mathbf{r}-\mathbf{r}'|} V(\mathbf{r}')\psi^{\pm}(\mathbf{r}')\, d\tau' \tag{37}$$

[An alternative derivation of the above equation is given in Appendix N].

We make two remarks about these solutions. First, any linear combination of ψ^+ and ψ^- will also be a solution. Secondly, we really have not solved anything since Eq. (37) is an integral equation with the unknown $\psi^{\pm}$ appearing on both sides. Actually we need the asymptotic forms, i.e., the form of $\psi(\mathbf{r})$ for $r \to \infty$. If we assume that $V(\mathbf{r}') \neq 0$ only for a finite region, say $|\mathbf{r}'| < a$ then for $r \to \infty$, the main contribution of the integral in Eq. (37) comes from a region where $r' \ll r$; thus we may write

$$|\mathbf{r}-\mathbf{r}'| = r\left[1 - \frac{2\mathbf{r}\cdot\mathbf{r}'}{r^2} + \frac{r'^2}{r^2}\right]^{1/2} \approx r - \hat{\mathbf{r}}\cdot\mathbf{r}'$$

and $$\frac{1}{|\mathbf{r}-\mathbf{r}'|} \approx \frac{1}{r} + \frac{1}{r^2}\hat{\mathbf{r}}\cdot\mathbf{r}' + \cdots$$

so that $$\psi^{\pm}(\mathbf{r}) \xrightarrow[r\to\infty]{} e^{i\mathbf{k}.\mathbf{r}} - \frac{\mu}{4\pi\hbar^2}\frac{1}{r}\int e^{\pm ik(r-\hat{\mathbf{r}}.\mathbf{r}')} V(\mathbf{r}')\psi^{\pm}(\mathbf{r}')\, d\mathbf{r}'$$

or, $$\psi^{\pm}(\mathbf{r}) \xrightarrow[r\to\infty]{} e^{i\mathbf{k}.\mathbf{r}} - \frac{\mu}{4\pi\hbar^2}\frac{e^{\pm ikr}}{r}\int e^{\mp i\mathbf{k}'\cdot\mathbf{r}'} V(\mathbf{r}')\psi^{\pm}(\mathbf{r}')\, d\mathbf{r}'$$

where $\mathbf{k}' \equiv \mathbf{k}\hat{\mathbf{r}}$ is a vector along the direction of scattering. The above equation is of the form

$$\psi^{\pm}(\mathbf{r}) \xrightarrow[r\to\infty]{} e^{i\mathbf{k}.\mathbf{r}} + f_k(\theta, \phi)\frac{e^{\pm ikr}}{r}$$

where θ, ϕ are the polar angles of the scattered particles. We recall that for the complete solutions we must multiply the above equation by the time part, $e^{-i\omega t}$, where k and ω are related by the free particle equation

$$\frac{\hbar^2 k^2}{2\mu} = \hbar\omega$$

We now recognize that $\psi^+(\mathbf{r})$ represents plane wave term plus an outgoing wave while $\psi^-(\mathbf{r})$ represents a plane wave term plus an incoming wave term. We, now, restrict ourselves to the solution corresponding to the outgoing wave case $\psi^+(\mathbf{r})$. The term

$$f_k(\theta, \phi) = -\frac{\mu}{4\pi\hbar^2}\int \exp(-i\mathbf{k}'\cdot\mathbf{r}')V(\mathbf{r}')\psi^+(\mathbf{r}')d\mathbf{r}' \tag{38}$$

is called the *scattering amplitude.* Generally in scattering experiments the detector is placed outside the region of the incident beam and counts only the scattered part. It can be seen that the amplitude of the scattered beam is reduced by a factor $1/r^2$. Once $f(\theta, \phi)$ is known, the scattering cross-section can be calculated using the relation

$$\frac{d\sigma}{d\Omega} = |f(\theta, \phi)|^2$$

26.5 THE BÖRN APPROXIMATION

The scattering amplitude [Eq. (38)] cannot be evaluated unless we know $\psi^+ (\mathbf{r}')$ in the first place. However, if the interaction energy V is treated as a small perturbation, then we can solve for ψ^+ by the method of successive approximations. From Eq. (37) we obtain as the approximation [setting $\psi^+(\mathbf{r}') \approx e^{i\mathbf{k}.\mathbf{r}'}$ in the integrand]

$$\psi_1^+ (\mathbf{r}) = e^{i\mathbf{k}\cdot\mathbf{r}} - \frac{\mu}{4\pi\hbar^2}\int \frac{\exp(ik|\mathbf{r}-\mathbf{r}'|)}{|\mathbf{r}-\mathbf{r}'|} V(\mathbf{r}')\, e^{i\mathbf{k}\cdot\mathbf{r}'}\, d\mathbf{r}' \tag{39}$$

The next approximation is obtained by using this value of ψ^+ in the integral in Eq. (37), i.e.

$$\psi_2^+ (\mathbf{r}) = e^{i\mathbf{k}\cdot\mathbf{r}} - \frac{\mu}{4\pi h^2}\int d\mathbf{r}' \frac{\exp(ik|\mathbf{r}-\mathbf{r}'|)}{|\mathbf{r}-\mathbf{r}'|} V(\mathbf{r}')\, e^{i\mathbf{k}\cdot\mathbf{r}'}\, d\mathbf{r}'$$

$$+\left(\frac{\mu}{4\pi\hbar^2}\right)^2 \iint d\mathbf{r}'\, d\mathbf{r}'' \frac{\exp(ik|\mathbf{r}-\mathbf{r}'|)}{|\mathbf{r}-\mathbf{r}'|} V(\mathbf{r}') \frac{\exp(ik|\mathbf{r}'-\mathbf{r}''|)}{|\mathbf{r}'-\mathbf{r}''|} V(\mathbf{r}'')\, e^{i\mathbf{k}\cdot\mathbf{r}''}$$

The term $\psi_n^+ (\mathbf{r})$ obtained by this procedure is called the nth *Börn approximation*. Frequently, one uses only the first term and it is referred to as the Börn approximation. The first Börn approximation for the scattering amplitude is obtained by setting $\psi^+(\mathbf{r}') = e^{i\mathbf{k}.\mathbf{r}'}$ in Eq. (38) to obtain

$$f(\theta, \phi) = -\frac{\mu}{2\pi\hbar^2}\int V(\mathbf{r}')e^{-i\mathbf{q}.\mathbf{r}'}\, d\mathbf{r}' \tag{40}$$

where $\hbar\mathbf{q} = \hbar\,(\mathbf{k}' - \mathbf{k})$ is the momentum transfer in scattering. We may also write

$$f(\theta, \phi) = -\frac{\mu}{2\pi\hbar^2} V_{\mathbf{q}}$$

where
$$V_{\mathbf{q}} = \int V(\mathbf{r})e^{-i\mathbf{q}\cdot\mathbf{r}}\, d\mathbf{r} \tag{41}$$

is the Fourier transform of the potential. For spherically symmetric potentials

$$V(\mathbf{r}') = V(r')$$

and one obtains

$$f(\theta, \phi) = -\frac{\mu}{2\pi\hbar^2}\iiint r'^2\, dr' \sin\theta'\, d\theta'\, d\phi'\, V(r')e^{-iqr'\cos\theta'}$$

where we have assumed the z'-axis to be along $\mathbf{q}$, and

$$q = |\mathbf{k}' - \mathbf{k}| = 2k \sin\frac{\theta}{2}$$

θ being the angle of scattering (the angle between $\mathbf{k}'$ and $\mathbf{k}$). Carrying out the integrations, we get

$$f(\theta) = -\frac{2\mu}{\hbar^2}\int_0^\infty V(r')\frac{\sin qr'}{qr'} r'^2\, dr' \tag{42}$$

and
$$\sigma(\theta) = \left(\frac{2\mu}{\hbar^2}\right)^2 \left| \int_0^\infty V(r') \frac{\sin qr'}{qr'} r'^2 \, dr' \right|^2 \tag{43}$$

Suppose now $V(\mathbf{r}')$ is a potential that has the same sign everywhere. Then, Eq. (42) shows that for large values of q, there will be large phase differences from different regions of the scatterer (i.e. the exponential term oscillates) and effects from different regions will tend to reduce the scattering amplitude. On the other hand, for $q = 0$ (i.e. $\mathbf{k}'$ parallel to $\mathbf{k}$) the amplitude will be the largest. The scattering amplitude in this case is

$$f(0) = -\frac{\mu}{2\pi\hbar^2} \int V(\mathbf{r}') d\mathbf{r}'$$

The ratio $f(\mathbf{q})/f(0)$ is called the *form factor*. Another way of looking at the problem is to study the difference between a 'point' scatterer and a 'spread out' one. It is easy to see that, in general, the effect of a spread is to reduce the scattering.

The Börn approximation is useful when the scattered wave is negligible compared to the incident wave. Looking at Eq. (39), this implies that

$$\left| \frac{\mu}{2\pi\hbar^2} \int \frac{e^{ik|\mathbf{r}-\mathbf{r}'|}}{|\mathbf{r}-\mathbf{r}'|} V(\mathbf{r}') e^{i\mathbf{k}\cdot\mathbf{r}'} d\mathbf{r}' \right| \ll 1$$

For short-range central potentials, we can put $r \approx 0$ and obtain a crude limit

$$\frac{2\mu}{\hbar^2 k} \left| \int_0^\infty e^{ikr'} \sin kr' \, V(r') dr' \right| \ll 1 \tag{44}$$

a result that is likely to hold for weak potentials and high energies.

26.5.1 Application of Börn Approximation for the Shielded Coulomb Potential

As an example, we consider the shielded Coulomb potential for which the potential energy function is given by (see, e.g. Section 14.1 of Ref. 3)[3]

$$V(r) = \frac{Z_1 Z_2 e^2}{r} \exp\left(-\frac{r}{\lambda}\right) \tag{45}$$

where λ is known as the shielding distance. Substituting in Eq. (43), we get

$$\sigma(\theta) = \frac{4\mu^2}{\hbar^4 q^2} \left(Z_1 Z_2 e^2\right)^2 \left| \int_0^\infty e^{-r/\lambda} \sin qr \, dr \right|^2$$

The integral is the imaginary part of $\int_0^\infty e^{-r/\lambda} e^{iqr} \, dr$ and gives

$$\frac{q}{q^2 + \dfrac{1}{\lambda^2}}$$

[3] We are, for the sake of convenience, using the CGS units. In MKS units e^2 should be replaced by $q^2/4\pi\varepsilon_0$ but then there would be confusion between the charge q and the momentum transfer vector $\hbar\mathbf{q}$.

Thus, $$\sigma(\theta) = \frac{4\mu^2 (Z_1 Z_2 e^2)^2}{\hbar^4 \left(q^2 + \frac{1}{\lambda^2}\right)^2} = \frac{4\mu^2 (Z_1 Z_2 e^2)^2}{\left(4p^2 \sin^2 \frac{\theta}{2} + \frac{\hbar^2}{\lambda^2}\right)^2} \tag{46}$$

If we let $\lambda \to \infty$, we get

$$\sigma = \frac{(Z_1 Z_2 e^2)^2}{16E^2 \sin^4 \frac{\theta}{2}} \tag{47}$$

which is exactly the same as obtained using classical mechanics (see Problem 26.22). Indeed even the exact analysis (see Section 26.7) gives the same result.

26.6 METHOD OF PARTIAL WAVES

In the previous section, we have discussed the Börn approximation and have also given approximate conditions for its validity. When the Börn approximation fails, it is necessary to use other methods; one such method is the method of partial waves which is applicable to spherically symmetric potentials. For such a potential, the angular momentum of the scattered particle is a constant of motion. It is therefore advantageous to develop solutions in terms of angular momentum eigenfunctions. If the potential is of sufficiently short range (or the energy of the incident particles is low), it will turn out that only a few eigenfunctions with small angular momenta (that is only a few partial waves) will really be affected by the potential. Assuming that the incident momentum is $\hbar k$, large angular momenta correspond roughly to large values of the impact parameter[4] and hence the particle is hardly deflected. If the range of the potential is a, it usually suffices to consider angular momenta $l\hbar$ with

$$l \lesssim ka$$

Before we give the details of the partial wave analysis, it is necessary to study the properties of spherical Bessel functions.

26.6.1 Spherical Bessel Functions

For spherically symmetric potential, the radial part of the Schrödinger equation is given by [see Section 10.2]

$$\frac{1}{r^2}\frac{d}{dr}\left[r^2 \frac{dR}{dr}\right] + \frac{2\mu}{\hbar^2}\left[E - V(r) - \frac{l(l+1)\hbar^2}{2\mu r^2}\right] R(r) = 0 \tag{48}$$

For a free particle $V(r) = 0$, and the above equation can be written in the form

$$\frac{1}{\rho^2}\frac{d}{d\rho}\left(\rho^2 \frac{dR}{d\rho}\right) + \left[1 - \frac{l(l+1)}{\rho^2}\right] R(\rho) = 0 \tag{49}$$

[4] Classically the angular momentum of the particle is $mv_0 s$ (see Fig. 26.2) and clearly for $s > a$ (where a is the range of the potential) there is no scattering of the particle.

where $$\rho = kr \quad \text{and} \quad k = \left(\frac{2\mu E}{\hbar^2}\right)^{1/2}$$

Equation (49) is known as the spherical Bessel equation and the solutions are denoted by

$$j_l(\rho) \quad \text{and} \quad n_l(\rho)$$

which are known as the spherical Bessel functions. For $l = 0$, if we make the transformation

$$R(\rho) = \frac{u(\rho)}{\rho}$$

then, $$\frac{1}{\rho^2}\frac{d}{d\rho}\left(\rho^2 \frac{dR}{d\rho}\right) = \frac{1}{\rho^2}\frac{d}{d\rho}\left[\rho \frac{du}{d\rho} - u(\rho)\right] = \frac{1}{\rho}\frac{d^2u}{d\rho^2}$$

Thus Eq. (49) [for $l = 0$] simplifies to

$$\frac{d^2u}{d\rho^2} + u(\rho) = 0$$

the solutions of which are sin ρ and cos ρ. Thus, the two independent solutions of Eq. (49) [for $l = 0$] are

$$j_0(\rho) = \frac{1}{\rho}\sin\rho \quad \text{and} \quad n_0(\rho) = \frac{1}{\rho}\cos\rho$$

Obviously, $n_0(\rho)$ is singular at the origin. The spherical Bessel functions are related to Bessel functions through the following equations (see Appendix J)

$$j_l(\rho) = \sqrt{\frac{\pi}{2\rho}}\, J_{l+\frac{1}{2}}(\rho) \quad \text{and} \quad n_l(\rho) = (-1)^l \sqrt{\frac{\pi}{2\rho}}\, J_{-l-\frac{1}{2}}(\rho)$$

Using the expressions for $j_0(\rho)$ and $n_0(\rho)$, one can calculate higher order functions using the following recurrence relations

$$j_{l+1}(\rho) = \frac{l}{\rho}\, j_l(\rho) - j_l'(\rho); \quad n_{l+1}(\rho) = \frac{l}{\rho}\, n_l(\rho) - n_l'(\rho)$$

Thus, $$j_0(\rho) = \frac{1}{\rho}\sin\rho$$

$$j_1(\rho) = \frac{1}{\rho^2}\sin\rho - \frac{1}{\rho}\cos\rho$$

$$j_2(\rho) = \left(\frac{3}{\rho^3} - \frac{1}{\rho}\right)\sin\rho - \frac{3}{\rho^2}\cos\rho$$

etc. Similarly, $$n_0(\rho) = \frac{1}{\rho}\cos\rho$$

$$n_1(\rho) = \frac{1}{\rho^2}\cos\rho + \frac{1}{\rho}\sin\rho$$

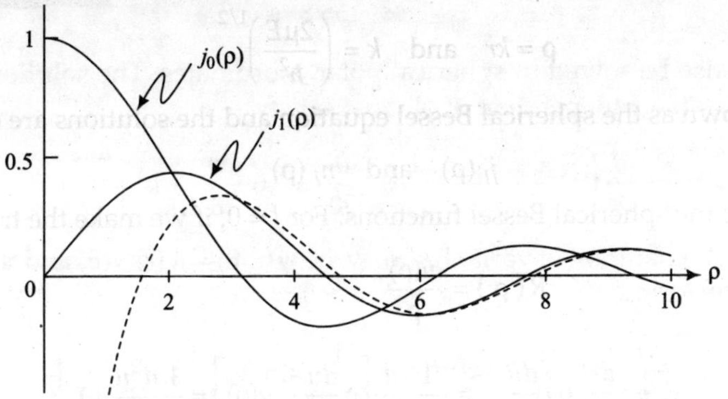

Figure 26.4: The solid curves represent the variations of $j_0(\rho)$ and $j_1(\rho)$ with ρ. The dashed curve represents the asymptotic form of $j_1(\rho)$. The asymptotic form of $j_0(\rho)$ coincides with the exact variation.

$$n_2(\rho) = \left(\frac{3}{\rho^3} - \frac{1}{\rho}\right)\cos\rho - \frac{3}{\rho^2}\sin\rho$$

etc. One can see that the asymptotic forms are

$$\left.\begin{aligned} j_l(\rho) &\underset{\rho\to\infty}{\longrightarrow} \frac{1}{\rho}\sin\left(\rho - \frac{l\pi}{2}\right) \\ n_l(\rho) &\underset{\rho\to\infty}{\longrightarrow} \frac{1}{\rho}\cos\left(\rho - \frac{l\pi}{2}\right) \end{aligned}\right\} \tag{50}$$

The variations of $j_0(\rho)$ and $j_1(\rho)$, along with their asymptotic forms are shown in Fig. 26.4.

26.6.2 Expansion Formula for a Plane Wave

We assume that particles of momenta $\hbar k$ are incident along the z-direction. The plane wave representing the incident particle can be expanded as[5]

$$e^{i\mathbf{k}\cdot\mathbf{r}} = e^{ikz} = e^{ikr\cos\theta} = \sum_{l=0}^{\infty} i^l (2l+1) P_l(\cos\theta)\, j_l(\rho) \tag{51}$$

where $\rho = kr$; notice that there is no azimuthal angular dependence in the expansion due to symmetry around the z-axis. The derivation of Eq. (51) is as follows:

We start with the Schrödinger equation for a free particle

$$\nabla^2\psi + k^2\psi = 0 \tag{52}$$

where $k^2 = 2\mu E/\hbar^2$. The plane wave solution of the above equation is given by

---

[5] A more general formula is given by

$$e^{i\mathbf{k}.\mathbf{r}} = 4\pi \sum_{l=0}^{\infty}\sum_{m=-\infty}^{+\infty} i^l Y_{lm}^*(\theta_k, \phi_k)\, Y_{lm}(\theta_r, \phi_r)\, j_l(kr)$$

where (θ_k, ϕ_k) and (θ_r, ϕ_r) are the polar angles of the vectors **k** and **r**, respectively.

$$\psi = e^{ikz} = e^{ikr\cos\theta}$$

Equation (52) can also be solved in spherical polar coordinates. The solution of Eq. (52) with azimuthal symmetry (i.e. independent of ϕ) must satisfy

$$\frac{1}{r^2}\frac{\partial}{\partial r}\left(r^2\frac{\partial\psi}{\partial r}\right)+\frac{1}{r^2\sin\theta}\frac{\partial}{\partial\theta}\left(\sin\theta\frac{\partial\psi}{\partial\theta}\right)+k^2\psi=0 \tag{53}$$

Using the method of separation of variables we write $\psi(r,\theta)=R(r)\Theta(\theta)$ and the above equation can be written in the form

$$\frac{1}{R}\frac{d}{dr}\left(r^2\frac{dR}{dr}\right)+k^2r^2=-\frac{1}{\Theta(\theta)}\sin\theta\frac{d}{d\theta}\left(\sin\theta\frac{d\Theta}{d\theta}\right)$$

$$= l(l+1) \text{ (say)} \tag{54}$$

If we define $\mu = \cos\theta$, $\Theta(\mu)$ can be shown to satisfy the Legendre's equation

$$(1-\mu^2)\frac{d^2\Theta}{d\mu^2}-2\mu\frac{d\Theta}{d\mu}+l(l+1)\Theta(\mu)=0$$

For the solutions to be bounded at $\mu=\pm 1$ $(\theta = 0, \pi)$, l must be 0, 1, 2, ... [see Section 9.3] and the well behaved solution would be the Legendre's polynomials:

$$\Theta(\mu) = P_l(\mu)$$

Returning to Eq. (54), the function $R(r)$ satisfies the equation

$$\frac{1}{\rho^2}\frac{d}{d\rho}\left(\rho^2\frac{dR}{d\rho}\right)+\left[1-\frac{l(l+1)}{\rho^2}\right]R(\rho)=0$$

where $\rho = kr$. The above equation is the spherical Bessel equation [Eq. (49)] and the solution which is finite at $\rho = 0$ will be $j_l(\rho)$. Thus, any bounded solution of Eq. (53) can be written as a linear combination of $j_l(\rho)P_l(\mu)$. Thus,

$$e^{i\rho\mu}=\sum_{l=0}^{\infty}a_l\, j_l(\rho)P_l(\mu);\ \rho = kr$$

If we multiply both sides by $P_m(\mu)\,d\mu$, integrate and use the orthogonality relation of Legendre polynomials, we get

$$a_m\frac{2}{2m+1}j_m(\rho)=\int_{-1}^{+1}e^{i\rho\mu}P_m(\mu)\,d\mu \tag{55}$$

The above equation must be valid for all values of ρ; we take the limit $\rho\to\infty$, then

$$\text{LHS} \sim a_m\frac{2}{2m+1}\frac{1}{\rho}\sin\left(\rho-\frac{m\pi}{2}\right)$$

$$\text{RHS} \sim \frac{2i^m}{\rho}\sin\left(\rho-\frac{m\pi}{2}\right)-\frac{1}{i\rho}\int_{-1}^{+1}e^{i\rho\mu}\frac{d}{d\mu}P_m(\mu)\,d\mu \qquad \text{[integrating by parts]}$$

If we integrate the second term by parts, we would obtain terms proportional to $1/\rho^2$; thus in the limit of $\rho\to\infty$, we may write

$$\text{RHS} \sim \frac{2i^m}{\rho}\sin\left(\rho - \frac{m\pi}{2}\right)$$

Thus, $$a_m = (2m+1)\, i^m$$

and we obtain

$$e^{i\rho\cos\theta} = \sum_{l=0,1,\ldots}^{\infty} (2l+1)\, i^l\, j_l(\rho) P_l(\mu)$$

From Eq. (55), we also get the integral representation of the spherical Bessel functions

$$j_l(\rho) = \frac{1}{2i^l}\int_{-1}^{+1} e^{i\rho\mu}\, P_l(\mu)\, d\mu \tag{56}$$

For any given value of l, we can substitute the expression for $P_l(\mu)$ and evaluate the integral to obtain the expression for $j_l(\rho)$. Now the asymptotic form of $j_l(\rho)$ can be written as [see Eq. (50)]:

$$j_l(\rho) \underset{\rho\to\infty}{\longrightarrow} \frac{1}{2i\rho}\left[\exp\left[i\left(\rho - \frac{l\pi}{2}\right)\right] - \exp\left[-i\left(\rho - \frac{l\pi}{2}\right)\right]\right]$$

$$\underset{\rho\to\infty}{\longrightarrow} \frac{i^{-l}}{2i\rho}\left[e^{i\rho} - (-1)^l e^{-i\rho}\right]$$

Thus the asymptotic form of Eq. (51) will be

$$e^{ikz} = e^{ikr\cos\theta} \to \sum_l \frac{(2l+1)}{2ikr}\left[e^{ikr} - (-1)^l e^{-ikr}\right] P_l(\cos\theta) \tag{57}$$

Now for spherically symmetric potentials, the Schrödinger equation is given by

$$[\nabla^2 + k^2 - U(r)]\psi(\mathbf{r}) = 0$$

where $U(r) = 2\mu V(r)/\hbar^2$. The solution of the above equation (with azimuthal symmetry) can be written in the form

$$\psi(r) = \sum_l \frac{(2l+1)}{k}\, i^l\, \frac{u_l(r)}{r} P_l(\cos\theta) \tag{58}$$

In writing Eq. (58), we make the following observations:

(a) Since, in the Schrödinger equation [Eq. (21)], the potential is spherically symmetric, the spherical harmonics $Y_{lm}(\theta, \phi)$ are solutions (see Section 10.2); in this problem $m = 0$, there being no ϕ-dependence. We can, therefore, take the general solution to be linear combination of Y_{l0} or $P_l(\cos\theta)$.

(b) The factors $(2l+1)i^l/k$ are introduced for convenience.

Clearly, for Eq. (58) to be correct, $u_l(r)$ must satisfy the Schrödinger equation (21). Substituting Eq. (58) into Eq. (21), we obtain

$$\frac{d^2u_l}{dr^2} + \left[k^2 - U(r) - \frac{l(l+1)}{r^2}\right]u_l(r) = 0 \tag{59}$$

In Appendix O, we have shown that for a short-range force for which

$$\lim_{r\to\infty} rV(r) = 0 \tag{60}$$

the asymptotic form of $u_l(r)$ will be exp ($\pm\, ikr$); thus we may write[6]

$$u_l(r) \xrightarrow[r\to\infty]{} A_l e^{ikr} + B_l e^{-ikr}$$

$$= C_l \sin\left(kr - \frac{l\pi}{2} + \delta_l\right) \tag{61}$$

where the factor $\frac{l\pi}{2}$ has been introduced for the sake of convenience and the quantities δ_l are known as phase shifts. We see that when $V(r) = 0$ everywhere, $u_l(r)$ is simply $kr\, j_l(kr)$ [see Eq. (59)] the asymptotic form of which is $\sin(kr - l\pi/2)$. Thus δ_l is the 'phase shift' introduced by the potential. Equation (61) can be written in the form

$$u_l \sim \frac{1}{2i} C_i (-i)^l \left[e^{i\delta_l}\, e^{ikr} - (-1)^l\, e^{-i\delta_l}\, e^{-ikr} \right] \tag{62}$$

Substituting Eq. (62) into Eq. (58), we obtain

$$\psi(\mathbf{r}) \xrightarrow[r\to\infty]{} \sum_l \frac{(2l+1)}{2ikr} C_l \left[e^{i\delta_l}\, e^{ikr} - (-1)^l\, e^{-i\delta_l}\, e^{-ikr} \right] P_l(\cos\theta) \tag{63}$$

This is to be compared with the form

$$\psi(\mathbf{r}) \longrightarrow e^{ikz} + \frac{e^{ikr}}{r} f(\theta)$$

$$= \sum_l \frac{(2l+1)}{2ikr} P_l(\cos\theta) \left[e^{ikr} - (-1)^l\, e^{-ikr} \right] + \frac{e^{ikr}}{r} f(\theta) \tag{64}$$

where we have used Eq. (57). Equations (63) and (64) represent the same function and hence the coefficients of e^{ikr} and e^{-ikr} should be identical; equating the coefficients gives us

$$C_l = e^{i\delta_l}$$

and

$$f(\theta) = \sum_{l=0}^{\infty} (2l+1) \frac{\exp(2i\delta_l) - 1}{2ik} P_l(\cos\theta) \tag{65}$$

---

[6] This may be very easily seen for a potential which vanishes beyond a certain distance $r = a$; because then for $r > a$, $R_l(r) (= u_l(r)/r)$ satisfies the spherical Bessel equation [see Eq. (49)] and therefore

$$u_l(\rho) \xrightarrow[r\to\infty]{} \rho\left[a_l\, j_l(\rho) + b_l\, n_l(\rho) \right]$$

Using the asymptotic forms of j_l and n_l [see Eq. (50)]:

$$u_l(\rho) \xrightarrow[r\to\infty]{} \frac{1}{\rho}\left(a_l \sin\left(\rho - \frac{l\pi}{2}\right) + b_l \cos\left(\rho - \frac{l\pi}{2}\right)\right)$$

which is identical in form to Eq. (61) with $\delta_l = \tan^{-1}(b_l/a_l)$.

or

$$f(\theta) = \sum_{l=0}^{\infty} \frac{(2l+1)}{k} \sin\delta_l \, e^{i\delta_l} P_l(\cos\theta) \quad \text{Scattering amplitude} \tag{66}$$

The total cross-section can be shown to be (see Problem 26.16)

$$\sigma = \int |f(\theta)|^2 \, d\Omega = \frac{4\pi}{k^2} \sum_{l=0}^{\infty} (2l+1)\sin^2\delta_l \quad \text{Total cross-section} \tag{67}$$

The above expression shows that as far as the total cross-section is concerned, the different partial waves contribute independently, there being no interference. This, however, is not the case for the differential cross-section.

As an application of the method of partial waves, we will consider two examples.

26.6.3 Scattering by a Hard Sphere

We first consider the scattering by a hard sphere for which $V(\mathbf{r})$ will be given by

$$\begin{aligned} V(r) &= 0 \quad r > a \\ &= \infty \quad r < a \end{aligned} \tag{68}$$

Since the region $r < a$ is inpenetrable, the wavefunction must vanish at $r = a$. Now, for $r > a$ the radial part of the Schrödinger equation is [see Eq. (48)]

$$\frac{d^2 u_l}{dr^2} + \left[k^2 - \frac{l(l+1)}{r^2}\right] u_l(r) = 0 \tag{69}$$

For $l = 0$, the solution (which vanishes at $r = a$) is given by

$$u_0(r) = A \sin k(r-a) \tag{70}$$

Comparing with Eq. (61), we get

$$\delta_0 = -ka \tag{71}$$

which is known as the *s*-wave (or $l = 0$) phase shift. In general, (for arbitrary l) the solution is (see Appendix J):

$$u_l(r) = (kr)[A_l j_l(kr) + B_l n_l(kr)] \quad \text{for } r > a \tag{72}$$

Using the boundary condition at $r = a$, we get

$$u_l(r) = (kr) A_l \left[j_l(kr) - \frac{j_l(ka)}{n_l(ka)} n_l(kr) \right]$$

$$\xrightarrow[r\to\infty]{} A_l \left[\sin\left(kr - \frac{l\pi}{2}\right) - \frac{j_l(ka)}{n_l(ka)} \cos\left(kr - \frac{l\pi}{2}\right) \right] \tag{73}$$

Comparing with Eq. (61) we immediately get

$$\tan\delta_l = -\frac{j_l(ka)}{n_l(ka)} \tag{74}$$

The above equation is an exact result; for $l = 0$, one immediately gets Eq. (71). For $ka \ll 1$ (i.e., for low energies), if we use Eq. (13) of Appendix *J*, we would get

$$\tan \delta_l \approx -\frac{(ka)^{2l+1}}{(2l+1)!!\,(2l-1)!!} \tag{75}$$

Since $ka \ll 1$, we get $\delta_1/\delta_0 \ll 1$, $\delta_2/\delta_1 \ll 1$, etc. This result is consistent with our earlier statement that at low energies only few of the angular momenta contribute. Thus, for sufficiently low energies, only δ_0 effectively contributes and using Eqs (67) and (71), we get

$$\sigma \approx \frac{4\pi}{k^2}\sin^2 \delta_0 \approx 4\pi a^2 \tag{76}$$

which is 4 times the result obtained by using classical mechanics (see Problem 26.21). At this point it may be worthwhile to discuss the conditions for the validity of the classical theory in describing the scattering process. The classical theory will be valid when one is justified in describing the scattering through the 'trajectories' of the classical particle. The situation is analogous to the justification of the 'ray-picture' in optics which is valid when the refractive index changes very slowly in distances of the order of wavelength. This is also the condition for the validity of the JWKB approximation (see Chapter 18). When diffraction effects become important it is necessary to have a quantum mechanical description of the scattering process. For the hard sphere case, when $ka \gg 1$

$$\tan \delta_l \approx -\tan\left(ka - \frac{l\pi}{2}\right) \tag{77}$$

so that

$$\sigma \approx \frac{4\pi}{k^2}\sum_{l=0}^{l \approx ka}(2l+1)\sin^2\left(-ka + \frac{l\pi}{2}\right)$$

$$\approx \frac{4\pi}{k^2}\int_0^{ka}(2x+1)\sin^2\left(-ka + \frac{x\pi}{2}\right)dx$$

$$\approx 2\pi a^2 \tag{78}$$

which is still a factor of 2 more than the classical result. This is due to the fact that from the region $\theta \approx 0$ there is a contribution of πa^2; this contribution arises from the famous Poisson-diffraction phenomenon according to which when a plane wave is incident normally on an opaque disc, one always obtains a bright spot on the center of the geometric shadow (for further details one may look up Ref. 4, p. 215).

26.6.4 Scattering by a Square Well and the Ramsauer Effect

We next consider the *s*-wave scattering by a square well

$$\begin{aligned} V(r) &= -V_0 \qquad 0 < r < a \\ &= 0 \qquad\quad r > a \end{aligned} \tag{79}$$

For $l = 0$, the radial part of the Schrödinger equation would be

$$\frac{d^2u_0}{dr^2} + K^2\, u_0(r) = 0 \qquad \text{for} \qquad 0 < r < a \tag{80}$$

$$\frac{d^2u_0}{dr^2} + k^2\, u_0(r) = 0 \qquad \text{for} \qquad r > a \tag{81}$$

where $K^2 = 2m(E + V_0)/\hbar^2$ and $k^2 = 2mE/\hbar^2$. For $r < a$ we must have $u_0 = A \sin kr$ (the cosine solution will make $R_0(r) = u_0(r)/r$ go to infinity at $r = 0$). For $r > a$ we write the solution in the form $u_0(r) = B \sin(kr + \delta_0)$. Continuity of u and its derivative at $r = a$ gives

$$\tan(ka + \delta_0) = \frac{k}{K} \tan Ka$$

or

$$\delta_0 = \tan^{-1}\left(\frac{k}{K} \tan Ka\right) - ka \tag{82}$$

Thus,

$$\sigma \simeq \frac{4\pi}{k^2} \sin^2 \delta_0 = \frac{4\pi}{k^2}\left[1 + \frac{1}{\tan^2 \delta_0}\right]^{-1}$$

$$= 4\pi\left[k^2 + \left(\frac{g + k \tan ka}{1 - g\dfrac{\tan ka}{k}}\right)^2\right]^{-1} \tag{83}$$

where $g = K \cot Ka$. Returning to Eq. (82) we notice that $\tan \delta_0$ (and hence the scattering cross-section) would vanish if

$$\frac{\tan Ka}{K} = \frac{\tan ka}{k};$$

we are assuming $ka \ll 1$ so that δ_1, δ_2, etc., are negligible. Although the above equation has been derived for a square well potential, similar vanishing of the cross-section would occur for a localizable' attractive potential. This effect was first observed by Ramsauer (and hence known as Ramsauer effect) who found that at low energies (~ 0.7 eV) the scattering cross-section of electrons by noble atoms was extremely small. The effect was later explained by quantum theory.

26.6.5 Scattering of Neutrons by Protons

The neutron-proton interaction can be approximately described by the square well interaction [see Eq. (79)]. Thus, we may use the results of Section 26.6.4 to study the scattering of neutrons by protons. For low energies $ka \to 0$ and Eq. (83) becomes

$$\sigma \approx \frac{4\pi}{k^2 + \dfrac{g^2}{(1 - ga)^2}} \tag{84}$$

In order to determine the value of $g (= K \cot Ka)$ we use the data on the binding energy for the ground state of deuteron; this is denoted by B and its value is $\approx$ 2.223 MeV. Now, for the ground state of deuteron $E = -B$ and $l = 0$ so that

$$u_0(r) = A \sin k_1 r \quad \text{for } r < a$$

$$= Ce^{-\kappa r} \quad \text{for } r > a$$

where $\kappa^2 = 2\mu B/\hbar^2$, $k_1^2 = 2\mu(-B + Vo)/\hbar^2$,

$$\mu = \frac{m_n m_p}{m_n + m_p} \approx \frac{1}{2} m_n$$

being the reduced mass of the neutron-proton system. Continuity conditions at $r = a$ gives us

$$k_1 \cot k_1 a = -\kappa \tag{85}$$

If $B \ll V_0$ (which is indeed the case because $V_0 \approx 40$ MeV),

$$k_1 \approx [2\mu V_0/\hbar^2]^{1/2} \approx K$$

Thus,

$$g = K \cot Ka \approx k_1 \cot k_1 a = -\kappa = -\left(\frac{2\mu}{\hbar^2} B\right)^{1/2} \tag{86}$$

Substituting in Eq. (84), we get

$$\sigma \approx \frac{4\pi}{\dfrac{2\mu E}{\hbar^2} + \dfrac{2\mu B}{\hbar^2 (1 - ga)^2}} \approx \frac{4\pi\hbar^2}{m_n (E + B)} \tag{87}$$

The last step is not very accurate because $ga \sim 0.3$, nevertheless the above expression should give a reasonable estimate for σ. However, if we assume $E \ll B = 2.223$ MeV then the above expression gives $\sigma \approx 2.4 \times 10^{-24}$ cm^2 whereas the experimental value is about 21×10^{-24} cm^2. This large discrepancy cannot be accounted for by the approximation made in the analysis. In 1935, Wigner made a suggestion which removed this difficulty. He pointed out that the ground state of the deuteron was a triplet; however, there is no reason why the neutron and proton could not interact with their spins antiparallel (i.e. through the singlet state). Let E_s denote the energy of the neutron-proton system associated with the singlet state then

$$\sigma \approx \frac{4\pi\hbar^2}{m_n}\left[\frac{3}{4}\frac{1}{E+B} + \frac{1}{4}\frac{1}{E+|E_s|}\right] \tag{88}$$

where $\frac{3}{4}$ and $\frac{1}{4}$ are the statistical weights of the triplet and singlet states. If we assume $|E_s| \approx 40$ KeV we get the right order of magnitude for σ. However, it is not possible to decide whether the state is real or virtual[7]. For low energies ($E \lesssim 4-5$ MeV) the experimental dependence of σ on E is in reasonably good agreement with Eq. (88).

26.6.6 Approximate Expression for Phase Shifts in the Born Approximation

In this section, we will obtain approximate expressions for the phase shifts when they have small values. We start out with the equations satisfied by $g_l(r)$ [$= kr\, j_l(kr)$] and $u_l(r)$ [see Eqs (14) and (15) of Appendix J and Eq. (59) of this chapter]:

$$\frac{d^2 g_l}{dr^2} + \left[k^2 - \frac{l(l+1)}{r^2}\right] g_l(r) = 0 \tag{89}$$

[7] Corresponding to the neutron-proton interaction in the triplet state $V_0 \approx 21$ MeV and it is possible to have a bound state (B ≈ 3 MeV); on the other hand, when the spins are antiparallel $V_0 \approx 12$ MeV and it is not possible to have a bound state. However, for $E \approx 40$ KeV it is possible to have a virtual (or a metastable) state with a finite lifetime (see, e.g. Refs 5, 6 and 7).

$$\frac{d^2 u_l}{dr^2} + \left[k^2 - \frac{l(l+1)}{r^2} - U(r)\right] u_l(r) = 0 \tag{90}$$

We multiply Eq. (84) by $u_l(r)$ and Eq. (90) by $g_l(r)$, subtract and integrate to obtain:

$$\left[g_l(r)\frac{du_l(r)}{dr} - u_l(r)\frac{dg_l}{dr}\right]_0^r = \int_0^r U(r) g_l(r) u_l(r)\, dr$$

Both $g_l(r)$ and $u_l(r)$ vanish at $r = 0$, further, we let $r \to \infty$, use the asymptotic forms [see Eq. (12) of Appendix J and Eq. (61)] to obtain for the left-hand side:

$$g_l \frac{du_l}{dr} - u_l(r)\frac{dg_l}{dr} \xrightarrow[r\to\infty]{} k \sin\left(kr - \frac{l\pi}{2}\right)\cos\left(kr - \frac{l\pi}{2} + \delta_l\right)$$

$$- k\sin\left(kr - \frac{l\pi}{2} + \delta_l\right)\cos\left(kr - \frac{l\pi}{2}\right) = -k\sin\delta_l$$

Thus,

$$\sin\delta_l = -\frac{1}{k}\frac{2\mu}{\hbar^2}\int_0^\infty V(r)\, u_l(r) g_l(r)\, dr \tag{91}$$

The above equation is *exact*. Now, if the phase shifts are small (which is likely to be valid for high energies or weak potentials) then in carrying out the above integration we may assume

$$u_l(r) \approx g_l(r) = kr\, j_l(kr)$$

Thus,

$$\sin\delta_l \approx \delta_l \approx -\frac{2\mu}{k\hbar^2}\int_0^\infty V(r)\left[kr\, j_l(kr)\right]^2 dr$$

$$\approx -\frac{\pi\mu}{\hbar^2}\int_0^\infty \left[J_{l+\frac{1}{2}}(kr)\right]^2 r\, dr \tag{92}$$

where use has been made of Eq. (10) of Appendix J. Since Born approximation is also valid under the same conditions, the calculation of δ_l through Eq. (83) is known as the Born approximation. In Table 26.1, we have given the value for δ_0 (for electron scattering by hydrogen–see Problem 26.6) obtained by using the Born approximation [Eq. (92)] and have compared with the exact values obtained by numerical solution of the differential equation. Notice that the higher values of k better is the agreement between the two numbers. The agreement with the experimental data is also satisfactory as discussed in Ref. 7.

Table 26.1: Comparison of exact and approximate values for phases

		δ_0	
ka_0	*Energy (eV)*	*Exact*	*Born Approximation*
1.0	13.5	0.905	0.596
2.0	54	0.695	0.602
3.0	122	0.568	0.534
4.0	215	0.490	0.472
5.0	340	0.432	0.432

* Table adapted from Ref. 7 see also Problem 26.6.

If we use Eq. (92) for δ_l, we should expect $\sigma(\theta)$ to be given by Eq. (43); this is indeed seen from the following:

$$f(\theta) = \frac{1}{2ik} \sum_{l=0,1,2,\ldots}^{\infty} (2l+1) P_l(\cos\theta)\left[e^{2i\delta_l} - 1\right] \quad \text{[see Eq. (66)]}$$

$$\simeq \frac{1}{k} \sum_l (2l+1) P_l(\cos\theta)\,\delta_l$$

$$\simeq -\frac{1}{k} \sum_l (2l+1) P_l(\cos\theta) \frac{\pi\mu}{\hbar^2} \int_0^\infty V(r) r^2 \left[j_l(kr)^2\right] dr \quad \text{[Using Eq. (92)]}$$

$$\simeq -\frac{2\mu}{\hbar^2} \int_0^\infty \frac{\sin qr}{qr} V(r) r^2\, dr \tag{93}$$

where $q = 2k \sin\theta/2$ and use has been made of the identity (see, e.g. Ref. 8),

$$\frac{\sin qr}{qr} = \sum_{l=0,1,2,\ldots} (2l+1) P_l(\cos\theta)\left[j_l(kr)\right]^2 \tag{94}$$

Equation (93) is identical to Eq. (42).

26.7 COULOMB SCATTERING

The Coulomb potential

$$V(r) = \frac{Z_1 Z_2 e^2}{r} \tag{95}$$

is such that

$$\lim_{r\to\infty} rV(r)$$

does not tend to zero and hence does not describe a short-range force (see Appendix O). Thus the results of the previous sections are not applicable. However, the scattering problem corresponding to the Coulomb potential can be solved directly in parabolic coordinates, the direction of the beam being an axis of symmetry. The coordinate system is

$$x = \sqrt{\xi\eta}\, \cos\phi;\; y = \sqrt{\xi\eta}\, \sin\phi;\; z = \frac{1}{2}(\xi - \eta) \quad \text{Parabolic coordinates} \tag{96}$$

We also have

$$r = \frac{1}{2}(\xi + \eta) \tag{97}$$

In terms of the parabolic coordinates, the Schrödinger equation [Eq.(12)] for the Coulomb potential becomes

$$-\frac{\hbar^2}{2\mu}\left[\frac{4}{\xi+\eta}\cdot\frac{\partial}{\partial\xi}\left(\xi\frac{\partial}{\partial\xi}\right)+\frac{4}{\xi+\eta}\cdot\frac{\partial}{\partial\eta}\left(\eta\frac{\partial}{\partial\eta}\right)+\frac{1}{\xi\eta}\frac{\partial^2}{\partial\phi^2}\right]\psi+\frac{2Z_1Z_2e^2}{(\xi+\eta)}\psi=E\psi \quad (98)$$

Since we want solutions independent of ϕ, we write

$$\psi=f_1(\xi)f_2(\eta) \quad (99)$$

Substituting from Eq. (99) into Eq. (98), we obtain

$$\frac{1}{f_1}\frac{d}{d\xi}\left(\xi\frac{df_1}{d\xi}\right)+\frac{1}{f_2}\frac{d}{d\eta}\left(\eta\frac{df_2}{d\eta}\right)-\frac{\mu Z_1Z_2e^2}{\hbar^2}+\frac{k^2}{4}(\xi+\eta)=0$$

so that we can separate the equations

$$\frac{d}{d\xi}\left(\xi\frac{df_1}{d\xi}\right)+\left(\frac{1}{4}k^2\xi-\beta_1\right)f_1=0 \quad (100)$$

and

$$\frac{d}{d\eta}\left(\eta\frac{df_2}{d\eta}\right)+\left(\frac{1}{4}k^2\eta-\beta_2\right)f_2=0 \quad (101)$$

with

$$(\beta_1+\beta_2)=\frac{\mu Z_1Z_2e^2}{\hbar^2} \quad (102)$$

The solution we want must be of the form of a plane wave moving along z-direction plus an outgoing spherical wave. These are proportional to

$$e^{ikz}=e^{ik(\xi-\eta)/2} \quad (103)$$

and

$$e^{ikr}=e^{ik(\xi+\eta)/2} \quad (104)$$

Both these forms suggest that we look for a solution of the form

$$e^{ik\xi/2}f_2(\eta).$$

In other words, we choose

$$f_1(\xi)=e^{ik\xi/2} \quad (105)$$

Substituting Eq. (105) into Eq. (100), we obtain

$$\left(\frac{ik}{2}-\frac{k^2}{4}\xi\right)e^{ik\xi/2}+\left(\frac{1}{4}k^2\xi-\beta_1\right)e^{ik\xi/2}=0$$

so that

$$\beta_1=ik/2$$

Substituting this value of β_1 in Eq. (102) we obtain, from Eq. (101),

$$\frac{d}{d\eta}\left[\eta\frac{df_2}{d\eta}\right]+\left[\frac{1}{4}k^2\eta+\left(\frac{ik}{2}-\frac{\mu Z_1Z_2e^2}{\hbar^2}\right)\right]f_2=0 \quad (106)$$

We try a solution of the form

$$f_2(\eta) = e^{-ik\eta/2} g(\eta)$$

We have

$$\frac{df_2}{d\eta} = \frac{-ik}{2} e^{-ik\eta/2} g + e^{-ik\eta/2} \frac{dg}{d\eta}$$

$$\frac{d^2 f_2}{d\eta^2} = -\frac{k^2}{4} e^{-ik\eta/2} g - ik\, e^{-ik\eta/2} \frac{dg}{d\eta} + e^{-ik\eta/2} \frac{d^2 g}{d\eta^2}$$

so that Eq. (106) becomes

$$\eta \frac{d^2 g}{d\eta^2} + (1 - ik\eta) \frac{dg}{d\eta} - nkg = 0$$

where we have written

$$n = \frac{\mu Z_1 Z_2 e^2}{\hbar^2 k}$$

We want to reduce it to the equation satisfied by the confluent hypergeometric function,

$$\eta' \frac{d^2 g}{d\eta'^2} + (1 - \eta') \frac{dg}{d\eta'} - ag = 0 \tag{107}$$

We identify $ik\eta$ with η', $-in$ with a. Now, the solution of Eq. (107) is (see Appendix *F*)

$$g = {}_1F_1(a; 1; -\eta') = {}_1F_1(-in; 1; ik\eta) \tag{108}$$

For large values of z, the confluent hypergeometric function has the following behavior (see Appendices K and L)

$${}_1F_1(a, c, z) = (c-1)! \left[\frac{(-z)^{-a}}{\Gamma(c-a)} + \frac{z^{a-c} e^z}{\Gamma(a)} \right] \tag{109}$$

Thus, the asymptotic form of g can be written as

$$g \sim g_1 + g_2$$

where

$$\left. \begin{aligned} g_1 &= \frac{(-ik\eta)^{in}}{\Gamma(1+in)} = \frac{[-ik(r-z)]^{in}}{\Gamma(1+in)} \\ g_2 &= \frac{(ik\eta)^{-in-1} e^{ik\eta}}{\Gamma(-in)} = \frac{[ik(r-z)]^{-in-1} e^{ik(r-z)}}{\Gamma(-in)} \end{aligned} \right\} \tag{110}$$

Writing

$$[-ik(r-z)]^{in} = e^{in \ln[-ik(r-z)]}$$

$$= e^{in \ln[k(r-z)]} e^{in \ln(-i)}$$

Since,

$$(-i) = e^{-i\pi/2}$$

we get

$$e^{in \ln(-i)} = e^{in[-i\pi/2]} = e^{n\pi/2}$$

Thus,

$$g_1 = \frac{e^{n\pi/2} e^{in \ln[k(r-z)]}}{\Gamma(1+in)} \tag{111}$$

Similarly,
$$g_2 = \frac{e^{n\pi/2}\, e^{-in\ln[k(r-z)]}\, e^{ik(r-z)}}{ik(r-z)\,\Gamma(-in)} \tag{112}$$

We recall that the solution required is
$$\Psi = e^{ik\xi/2}\, e^{-ik\eta/2}\, g(\eta) \to e^{ikz}(g_1 + g_2) \tag{113}$$

From Eqs (111)–(113) it is clear that $e^{ikz}g_1$ gives a modified plane wave and $e^{ikz}g_2$ gives a spherical wave. We can write

$$\psi \to \frac{e^{n\pi/2}}{\Gamma(1+in)}\Big\{\exp[in\ln[k(r-z)] + ikz]$$
$$+ \frac{\Gamma(1+in)}{\Gamma(-in)}\frac{\exp[ikr - in\ln[k(r-z)]]}{ik(r-z)}\Big\}$$

We have
$$\frac{\Gamma(1+in)}{\Gamma(-in)} = \frac{in\,\Gamma(1+in)}{\Gamma(1-in)}$$

We can write
$$\Gamma(1+in) = |\Gamma(1+in)|\exp(i\sigma_n)$$

Finally,
$$\psi \to \frac{e^{n\pi/2}}{\Gamma(1+in)}\Big\{\exp[in\ln[k(r-z)] + ikz]$$
$$+ \frac{n}{k(r-z)}\exp[ikr - in\ln k(r-z) + i2\sigma_n]\Big\} \tag{114}$$

We note that only the phases of the incident and scattered waves are modified. However, the phase of the scattered wave is of the form $-in\ln[k(r-z)] + 2i\sigma_n$ so that it increases indefinitely for large r, although only logarithmically. We can still identify (putting $z = r\cos\theta$)

$$f(\theta) = \frac{n}{k(1-\cos\theta)}\exp[-in\ln(1-\cos\theta)] \tag{115}$$

as the part containing the angular dependence. This gives the differential cross section as

$$\frac{d\sigma}{d\Omega} = |f(\theta)|^2 = \left(\frac{Z_1 Z_2 e^2}{4\left(\frac{1}{2}\mu v^2\right)}\right)^2 \frac{1}{\sin^4\frac{\theta}{2}} \tag{116}$$

a result that agree with the classical result (see Problem 26.22) as well as with the Born approximation result [see Eq. (47)].

26.8 CONSIDERATIONS FOR IDENTICAL PARTICLES

In this section, we will discuss the scattering of identical particles like scattering of protons by protons or of α-particles by α-particles. However, in order to point out the peculiarities for identical particles we first consider non-identical particles like scattering of protons by α-particles. In the center of mass system the two particles approach each other with the same momentum (see Fig. 26.5 and Section 26.9) and after scattering they are again moving in opposite directions. If the detector is such that it could detect either the α-particle or the proton then the detector will register a count if the particle gets scattered along the direction (θ, ϕ) or along (π – θ, π + ϕ) (see Fig. 26.5). Thus, the cross-section will be given by

$$\sigma(\theta) = |f(\theta)|^2 + |f(\pi - \theta, \pi + \phi)|^2 \tag{117}$$

However, since most scattering potentials are spherically symmetric, one gets

$$\sigma(\theta) = |f(\theta)|^2 + |f(\pi - \theta)|^2 \tag{118}$$

The above equations gives the correct result for non-identical particles like scattering of α-particles by protons; however, it gives wrong results for scattering of protons by protons or of alpha particles by alpha particles. This is because of the indistinguishability of identical particles due to which it is impossible to say whether it is the incident particle or the target particle which has reached the detector. Due to this indistinguishability, we must construct appropriate wavefunctions which are either symmetric or antisymmetric with respect to the interchange of indices (see Section 24.5).

We first consider identical Fermions (like protons) for which the total wavefunction, space and spin, must be antisymmetric. This means that even the asymptotic form of the wavefunction of a two-particle system must obey this rule. However, the space part of the wavefunction alone may be either symmetric or antisymmetric, depending on whether the spin part is antisymmetric or symmetric. Thus for the *p–p* system, the space wavefunction must be symmetric for the singlet and antisymmetric for the triplet case. In the center of mass system the collision may be represented as in Fig. 26.5. The asymptotic form of the wave—function must therefore be a symmetric or an antisymmetric linear combination of the two cases. If

$$\psi(\mathbf{r}) = e^{i\mathbf{k}\cdot\mathbf{r}} + f(\theta, \phi)\frac{e^{ikr}}{r}$$

describing the form for the first, then

$$\psi(-\mathbf{r}) = e^{-i\mathbf{k}\cdot\mathbf{r}} + f(\pi - \theta, \pi + \phi)\frac{e^{ikr}}{r} \tag{119}$$

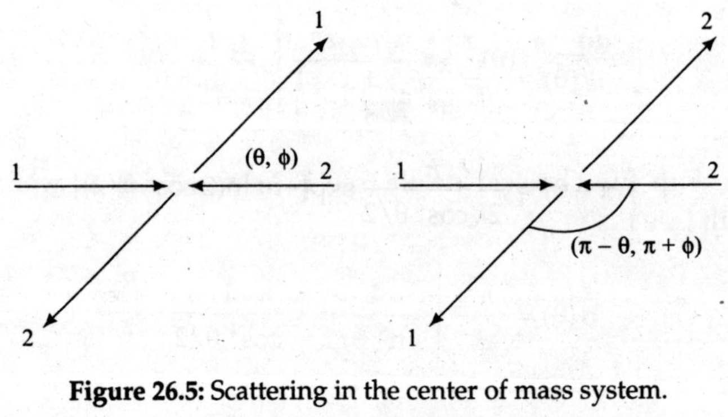

Figure 26.5: Scattering in the center of mass system.

describes the second. The combinations are

$$[\psi(\mathbf{r}) \pm \psi(-\mathbf{r})]$$

The scattering cross-section is now

$$\sigma(\theta, \phi) = |f(\theta, \phi) \pm f(\pi - \theta, \phi + \pi)|^2$$

$$= |f(\theta, \phi)|^2 + |f(\pi - \theta, \phi + \pi)|^2 \pm 2\,\mathrm{Re}[f(\theta, \phi) f^*(\pi - \theta, \phi + \pi)] \quad (120)$$

We note that in the absence of interference we get the classical result

$$\sigma(\theta, \phi) = |f(\theta, \phi)|^2 + |f(\pi - \theta, \phi + \pi)|^2 \quad (121)$$

This result merely says that the detector counting the scattered particles will not be able to say whether it is counting particle 1 or 2 so that the number counted at a particular angle is a sum of both probabilities.

However, the complete quantum mechanical result, viz. Eq. (120), is more subtle than just this. It is actually the probability amplitudes that interfere. (This situation is quite similar to electron diffraction by the double slit see Chapter 3).

At $\theta = \pi/2$, for a central potential, the results are particularly interesting. We obtain,

$$\sigma\left(\theta = \frac{\pi}{2}\right)_{sym} = 4\left|f\left(\frac{\pi}{2}\right)\right|^2$$

which is twice the result predicted by the classical equation (121). Moreover,

$$\sigma\left(\theta = \frac{\pi}{2}\right)_{\text{anti}} = 0$$

Thus the *p-p* triplet scattering at $\pi/2$ should be absent. For an unpolarized beam of protons, the cross-section is

$$\sigma(\theta) = \frac{3}{4}\sigma(\theta)\Big|_{\text{triplet}} + \frac{1}{4}\sigma(\theta)\Big|_{\text{singlet}}$$

because the singlet state has statistical weight one and the triplet state three. Using Eq. (120), we obtain

$$\frac{d\sigma}{d\Omega} = |f(\theta)|^2 + |f(\pi - \theta)|^2 - \mathrm{Re}\,[f(\theta) f^*(\pi - \theta)]$$

Equation (115) gives

$$f(\theta) = \frac{n}{2k\sin^2\theta/2}\exp[-in\ln(2\sin^2\theta/2)]$$

$$f(\pi - \theta) = \frac{n}{2k\cos^2\theta/2}\exp[-in\ln(2\cos^2\theta/2)]$$

Thus,

$$\sigma(\theta) = \frac{n^2}{4k^2}\left\{\frac{1}{\sin^4\theta/2} + \frac{1}{\cos^4\theta/2} - \frac{\mathrm{Re}\exp\left[-in\ln(\tan^2\theta/2)\right]}{\sin^2\theta/2\cos^2\theta/2}\right\}$$

$$= \frac{n^2}{4k^2}\left\{\frac{1}{\sin^4\theta/2} + \frac{1}{\cos^4\theta/2} - \frac{\cos\left[n\ln(\tan^2\theta/2)\right]}{\sin^2\theta/2\cos^2\theta/2}\right\} \tag{122}$$

Here $n = \mu Z_1 Z_2 e^2/\hbar^2 k$. This formula agrees with the experimental data at low energies which are not large enough so as to let the nuclear interaction become effective. For further details see Ref. 9.

It is of interest to mention that for spinless particles (α-particles) the total wavefunction must be symmetric and, therefore, we must have

$$\sigma(\theta) = |f(\theta) + f(\pi - \theta)|^2 \tag{123}$$

The above equation gives

$$\sigma\left(\theta = \frac{\pi}{2}\right) = 4\left|f\left(\frac{\pi}{2}\right)\right|^2 \tag{124}$$

which is to be compared with the classical result [Eq. (118)] which gives $\sigma\ (\theta = \frac{\pi}{2}) = 2\,|f\left(\frac{\pi}{2}\right)|^2$. Thus the quantum theory predicts twice as much scattering as we would expect from classical theory; the experimental data confirms the quantum result.

26.9 LABORATORY AND CENTER OF MASS COORDINATE SYSTEMS

In Section 10.3, we had shown that the two body problem can always be reduced to a one body problem provided the potential has spherical symmetry, i.e. the potential energy depended only on the magnitude of the distance between the two particles. It was shown that the Schrödinger equation could be broken up into two one-particle equations, one describing the free motion of the center of mass and the other describing the relative motion of the two particles (see Section 10.3 and Appendix H). Thus the cross-sections calculated in this chapter correspond to the center of mass system and since all measurements are carried out in the laboratory system we must know how to relate the cross-sections in the two systems[8]. It is the purpose of this section to know the relation between the two systems.

We consider a particle of mass m_1 moving with velocity v_1 towards the target particle of mass m_2 which is initially at rest (see Fig. 26.6 a). The center of mass (shown by C in the figure) moves with velocity

$$v_c = \frac{m_1 v_1}{m_1 + m_2} \tag{125}$$

After scattering, the incident particle moves in a direction given by the angle (θ_0, ϕ_0) the target particle also suffers a recoil and moves in a different direction. In the center of mass system, the center of mass is at rest and the two particles will approach each other with velocities $m_2 v_1/(m_1 + m_2)$ and $-m_1 v_1/(m_1 + m_2)$. It is easy to see that the total momenta of the two particles before scattering is zero and hence they must fly off in opposite directions after scattering as shown in Fig. 26.6 (b). In the center of mass system, if the scattering angle is denoted by (θ, ϕ)

[8] It may be mentioned here that while calculating the energies of the bound states in Chapter 5 we could ignore the motion of the center of mass.

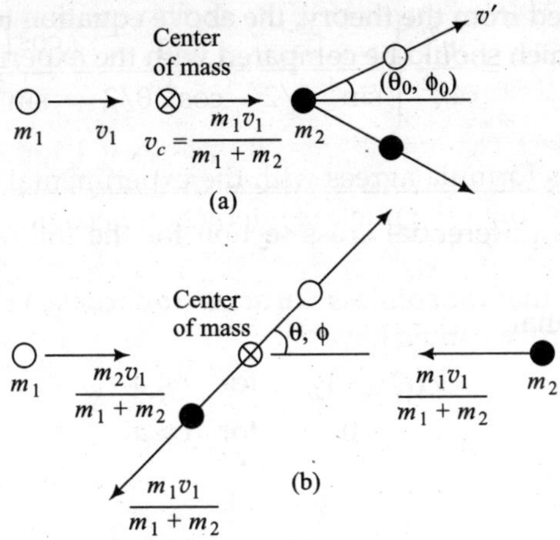

Figure 26.6: (a) The laboratory coordinate system in which the target particle (of mass m_2) is initially at rest; v_c [as given by Eq. (125)] represents the velocity of the center of mass. After scattering, the two particles fly off in different directions. (b) In the center of mass system, the two particles approach each other with the same moments; the center of mass is at rest.

then for elastic scattering (i.e., when kinetic energies are conserved) the particles should move away with their original speeds as shown in Fig. 26.6 (b). In order to get back to the laboratory system we must superpose the velocity of the center of mass which will readily give

$$v' \sin \theta_0 = \frac{m_2 v_1}{m_2 + m_1} \sin \theta$$

$$v' \cos \theta_0 = \frac{m_2 v_1}{m_2 + m_1} \cos \theta + v_c$$

$$\phi = \phi_0 \tag{126}$$

where v' is the speed of the particle of mass m_1 after scattering. Thus,

$$\tan \theta_0 = \frac{\sin \theta}{\dfrac{m_1}{m_2} + \cos \theta} \tag{127}$$

Notice that when $m_2 \gg m_1$, the center of mass is almost at rest, $\theta_0 = \theta$ and the two system of coordinates coincide.

Since the laboratory and center of mass scattering angles are not equal, the angular distribution of the cross-section will be different in the two systems; the relation between the two scattering cross-sections can be obtained by noting that the number of particles scattered in a given element of solid angle must be the same in two systems; thus

$$\sigma(\theta_0, \phi_0) \sin \theta_0 \, d\theta_0 \, d\phi_0 = \sigma(\theta, \phi) \sin \theta \, d\theta \, d\phi \tag{128}$$

or using Eqs (126) and (127)

$$\sigma(\theta_0, \phi_0) = \frac{\left(1 + \gamma^2 + 2\gamma \cos \theta\right)^{3/2}}{|1 + \gamma \cos \theta|} \sigma(\theta, \phi) \tag{129}$$

It is σ (θ, ϕ) which is obtained from the theory, the above equation tells us how to transform it to the laboratory system which should be compared with the experimental data.

26.10 PROBLEMS

Problem 26.1 Calculate the differential cross-section for the following potentials, using the Born approximation:

(a) The square well potential

$$V(r) = -V_0 \quad \text{for} \quad r < a$$
$$= 0 \quad \text{for} \quad r > a$$

(b) The Gaussian potential

$$V(r) = -V_0 \exp\left[-\frac{1}{2}\left(\frac{r}{a}\right)^2\right] \tag{130}$$

You may have to use the relation

$$\int_0^\infty e^{-\alpha t^2} \cos(2\beta t)\, dt = \frac{1}{2}\sqrt{\frac{\pi}{\alpha}}\, e^{-\beta^2/\alpha}$$

Problem 26.2 The Yukuwa potential is given by

$$V(r) = -g^2 \frac{e^{-\alpha r}}{r} \tag{131}$$

which is of the same form as the shielded Coulomb potential [see Eq. (45)] with g^2 playing the role of e^2 so that the dimensionless parameter $g^2/\hbar c$ measures the strength of the interaction. Assuming that the neutron-proton interaction can be described by a Yukuwa potential (which is, however, not strictly true) and that $1/\alpha \sim 1.4 \times 10^{-13}$ cm, estimate $g^2/\hbar c$ from the observed zero energy cross-section of about 3.6×10^{-24} cm^2 (this is the triplet cross-section, see Section 26.8).

Problem 26.3 Assume the charge distribution of a scatterer to be given by

$$Q(\mathbf{r}) = Z_1 e F(\mathbf{r}) \tag{132}$$

where $\int F(\mathbf{r})\, d\mathbf{r} = 1$ implying that the total charge of the scatterer is $Z_1 e$. Consider the scattering of a point charge $Z_2 e$, calculate $V_{\mathbf{q}}$ [see Eq. (41)] by using the Poisson equation and show that

$$f(\theta) = -\frac{\mu}{2\pi\hbar^2} V_{\mathbf{q}} = \left(-\frac{\mu}{2\pi\hbar^2}\right)\frac{4\pi Z_2 e}{q^2}\Lambda(q) \tag{133}$$

where

$$\Lambda(\mathbf{q}) = \int Q(\mathbf{r}) \exp\left[-i\mathbf{q}\cdot\mathbf{r}\right] d\mathbf{r} \tag{134}$$

Problem 26.4 Using the results of the previous problem calculate the cross-section for a point charge for which

$$F(\mathbf{r}) = \delta(\mathbf{r})$$

Compare with the result obtained in Section 26.5.1.

Problem 26.5 Use the results of Problem 26.3 to calculate $\sigma(\theta)$ for the following charge distributions:

(a)
$$Q(\mathbf{r}) = \left(\frac{4\pi}{3}a^3\right)^{-1}(Z_1 e) \quad \text{for} \quad r < a \tag{135}$$
$$= 0 \quad \text{for } r > a \quad \text{(uniformly charged sphere)}$$

(b)
$$Q(\mathbf{r}) = \frac{Z_1 e}{(2\pi a^2)^{3/2}} e^{-r^2/2a^2} \quad \text{(Gaussian)} \tag{136}$$

(c)
$$Q(\mathbf{r}) = \frac{Z_1 e}{4\pi a^2}\frac{1}{r}e^{-r/a} \quad \text{(Shielded Coulomb)} \tag{137}$$

where $Z_1 e$ represents the total charge of the scatterer. Notice that all the cases will go over to the result of Problem 26.4 in the limit of $a \to 0$.

Problem 26.6 The scattering of fast electrons by a complex atom can be, in many cases, represented fairly accurately by assuming the following form for the potential energy distribution:

$$V = -\frac{Ze^2}{r} + Ze^2 \int \frac{\rho(\mathbf{r}')}{|\mathbf{r}-\mathbf{r}'|} d\mathbf{r}' \tag{138}$$

where the first term is the potential due to the nucleus and the second due to the atomic structure. For the hydrogen atom in the ground state, we may write

$$\rho(r) = |\psi(1s)|^2 = \frac{1}{\pi a_0^3}\exp(-2r/a_0)$$

Use the result of Problem 26.3 to calculate $\sigma(\theta)$.

Problem 26.7 Modify the analysis of the previous problem to study the scattering by helium atom, the ground state function of which can be represented by (see Section 24.4)

$$\psi_0 = \frac{Z^3}{\pi a_0^3}\exp\left[-Z(r_1 + r_2)/a_0\right], \; Z = \frac{27}{16}$$

Problem 26.8 Referring to Fig. 26.6 show that the total energy in the laboratory and center of mass systems are related by

$$E_c = \frac{m_2}{m_1 + m_2}E_l \tag{139}$$

Problem 26.9 In the neutron-proton scattering, assuming that the range of the attractiveforce, a, to be approximately 3×10^{-13} cm show that the scattering will be almost isotropic (in the center of mass system) for neutron energy $\lesssim 10$ MeV. (This is indeed borne out by experimental data).

Problem 26.10 Assuming the mass of neutron to be equal to the mass of the proton, prove that there cannot be any back scattering in the scattering of neutron by protons.

Problem 26.11 For neutron-proton scattering, make an angular distribution for $l = 0$ (*s*-wave), $l = 1$ (*p*-wave) and $l = 2$ (*d*-wave) scattering in the center of mass and laboratory coordinate systems.

Problem 26.12 For a short force, use Eq. (92) to prove that

$$\frac{\delta_1}{\delta_0} \sim \frac{(ka)^2}{9}; \quad \frac{\delta_2}{\delta_0} \sim \frac{(ka)^4}{225}, \quad \text{etc.} \tag{140}$$

Problem 26.13 Consider a repulsive potential given by

$$\begin{aligned} V(r) &= V_0 \qquad 0 < r < a \\ &= 0 \qquad r > a \end{aligned} \tag{141}$$

Show that:

$$\delta_0 = \tan^{-1}\left[\frac{k}{K}\tanh Ka\right] - ka \quad \text{for} \quad E < V_0$$

$$= \tan^{-1}\left[\frac{k}{K_1}\tan K_1 a\right] - ka \quad \text{for} \quad E > V_0$$

where $$k^2 = \frac{2mE}{\hbar^2}, \; K^2 = 2m\,(V_0 - E)/\hbar^2$$

and $$K_1^2 = 2m(E - V_0)/\hbar^2$$

Problem 26.14 For the square well potential discussed in Section 26.6.2 show that

$$\tan\delta_l = -\frac{kj_l'(ka) - \alpha_l j_l(ka)}{kn_l'(ka) - \alpha_l n_l(ka)} \tag{142}$$

where $$\alpha_l = \frac{Kj_l'(Ka)}{j_l(Ka)}$$

For $ka \ll 1$, using the forms of $j_l(\rho)$ and $n_l(\rho)$ for $\rho \to 0$, show

$$\delta_0 \approx \tan^{-1}\left[-\frac{\alpha_0 ka^2}{1 + \alpha_0 a}\right]$$

$$\delta_1 \approx \tan^{-1}\left[\frac{(ka)^3}{3}\frac{1 - \alpha_1 a}{2 + \alpha_1 a}\right], \quad \text{etc.}$$

Problem 26.15 Obtain an expression for the coherent scattering cross-section for many scattering centers.

Problem 26.16 Show that the total cross-section, Eq. (67), follows from Eq. (66) on using the orthogonality of Legendre polynomials.

Problem 26.17 In elastic scattering there is no absorption of the incident particle by the scatterer. Thus the outgoing and incoming intensities must be equal. Equation (63) shows that this implies that δ_l must be real. Show that the reaction cross-section, defined as the number of particles removed from the beam per unit flux, is given by

$$\sigma_r = \sum_l \sigma_{r,l} = \sum_l \frac{\pi}{k^2}(2l+1)(1-|\eta_l|^2) \tag{143}$$

where $\eta_l = \exp(2i\delta_l)$ and η_l is called the collision amplitude. Show that while there can be scattering without absorption, there can be no absorption without scattering.

Problem 26.18 Prove the optical theorem

$$\sigma_{\text{total}} = \frac{4\pi}{k}\text{Im}\left[f(0)\right] \tag{144}$$

where Im[f (0)] is the imaginary part of the forward scattering amplitude.

Problem 26.19 Consider the scattering of the wave packet

$$\Psi(\mathbf{r}, t_0) = \frac{1}{(2\pi)^{3/2}}\int d\mathbf{k}\, a(\mathbf{k}) e^{i\mathbf{k}\cdot\mathbf{r}} \tag{145}$$

by a fixed short-range potential. The wave packet described by the above equation is assumed to be localized well out of the range of the potential as shown in Fig. 26.7. Assuming that the particle is reasonably well localized in position space and at the same time has a fairly well defined momentum, calculate Ψ (**r**, t) at a time after the particle has been scattered by the target. **[Hint:** Express Ψ (**r**, t_0) in terms of the exact eigen states of the Schrödinger equation; see Ref. 10.]

Problem 26.20 For the square well potential discussed in Section 26.2, show that the classical trajectory of the particle is the same as the path of a ray of light incident on a sphere of radius a with refractive index given by

$$n = \left(\frac{E + V_0}{E}\right)^{1/2} \tag{146}$$

Problem 26.21 In the classical expression for the scattering cross-section for a square well potential [see Eq. (8)] if we let $V_0 \to \infty$ show that $\sigma(\theta)$ becomes independent of θ, i.e., the scattering becomes isotropic. For such a case show that the total cross-section is πa^2.

Problem 26.22 For a Coulomb potential [see Eq. (95)] show that the relation between the impact parameter and the scattering angle is given by

$$s(\theta) = \frac{Z_1 Z_2 e^2}{2E}\cot\frac{\theta}{2} \tag{147}$$

(see, e.g. Section 3.7 of Ref. 1). Hence using Eq. (7) calculate the scattering cross-section.

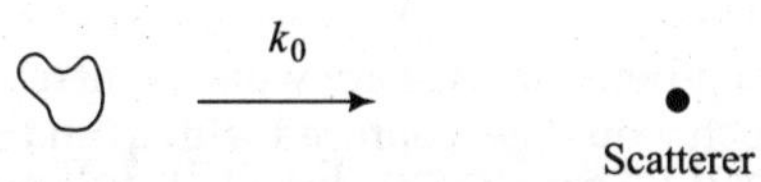

Figure 26.7

26.11 SOLUTIONS

Solution 26.1

(a)

$$\sigma(\theta) = \frac{4\mu^2 V_0^2}{\hbar^4 q^2}\left|\int_0^a \sin qr\, r\, dr\right|^2$$

$$= \frac{4\mu^2 V_0^2}{\hbar^4 q^6}(\sin qa - qa\cos qa)^2$$

where the symbols have the same meaning as in Section 26.4.

(b) Simple integration gives

$$\sigma(\theta) = \left(\frac{\mu}{2\pi\hbar^2}\right)^2 \left|V_{\mathbf{q}}\right|^2 = \frac{2\pi\mu^2 a^6 V_0^2}{\hbar^4} e^{-q^2a^2}$$

Solution 26.2 Using Eq. (46), we get

$$\sigma = 2\pi\int_0^\pi \sigma(\theta)\sin\theta\, d\theta = 2\pi\int_0^{2k} \frac{4\mu^2 g^4}{\hbar^4(q^2+\alpha^2)^2}\cdot\frac{q\,dq}{k^2}$$

where we have used the relation $q = 2k\sin\dfrac{\theta}{2}$. On carrying out the integration, we get

$$\sigma = 16\pi\left(\frac{g^2}{\hbar c}\right)^2\left(\frac{\mu c}{\hbar}\right)^2 \frac{1}{\alpha^4}\frac{1}{(1+4k^2/\alpha^2)}$$

or,
$$\sigma \xrightarrow[k\to 0]{} 16\pi\left(\frac{g^2}{\hbar c}\right)^2\left(\frac{\mu c}{\hbar}\right)^2 \frac{1}{\alpha^4}$$

Using the data given in the problem and the fact that

$$\mu \approx \frac{m_n m_p}{m_n + m_p} \approx \frac{1}{2} m_p \approx 0.836\times 10^{-24}\ \text{gm}$$

we get $g^2/\hbar c \sim 1.3$. This is very much larger than the 'strength' of electromagnetic coupling $e^2/\hbar c \sim 1/137$. However, the Yukawa potential is hardly a sufficiently accurate representation of the nuclear interaction and one should treat this value of $g^2/\hbar c$ as only a rough estimate.

Solution 26.3 The Poisson equation is given by[9]

[9] Two points ought to be mentioned here: first we are working here in the CGS units (because in MKS units e should be replaced by q which would get confused with **q**) and secondly in MKS units the Poisson equation is

$$\nabla^2 U = -[\text{charge density}]/\varepsilon_0$$

where U is the potential (which is the potential energy per unit charge) and ε_0 is the free space permittivity (see, e.g. Ref. 3). The final result obviously remains unchanged.

$$\nabla^2 V(\mathbf{r}) = -4\pi Z_2 e Q(\mathbf{r}) \tag{148}$$

Now, the Fourier transform of Eq. (41) gives

$$V(\mathbf{r}) = \frac{1}{8\pi^3}\int V_{\mathbf{q}} e^{i\mathbf{q}\cdot\mathbf{r}} d\mathbf{q} \tag{149}$$

or,
$$\nabla^2 V = \frac{1}{8\pi^3}\int (-q^2) V_{\mathbf{q}} e^{i\mathbf{q}\cdot\mathbf{r}} d\mathbf{q}$$

Thus,
$$\frac{1}{8\pi^3}\int (-q^2 V_{\mathbf{q}}) e^{i\mathbf{q}\cdot\mathbf{r}} d\mathbf{q} = -4\pi Z_2 e Q(\mathbf{r}) = -4\pi Z_2 e \frac{1}{8\pi^3}\int \Lambda(\mathbf{q}) e^{i\mathbf{q}\cdot\mathbf{r}} d\mathbf{q}$$

where in the last step we have used the Fourier transform of Eq. (134). Thus,

$$V_{\mathbf{q}} = \frac{4\pi Z_2 e}{q^2}\Lambda(\mathbf{q}) \tag{150}$$

or,
$$\sigma(\theta) = |f(\theta)|^2 = \left(\frac{\mu}{2\pi\hbar^2}\right)^2 \left|\frac{4\pi Z_2 e}{q^2}\Lambda(\mathbf{q})\right|^2 \tag{151}$$

Solution 26.5

(a)

$$\Lambda(\mathbf{q}) = \int Q(\mathbf{r})\, e^{-i\mathbf{q}\cdot\mathbf{r}}\, d\mathbf{r}$$

$$= \frac{Z_1 e}{\frac{4\pi}{3}a^3}\int_0^{2\pi}\int_0^{\pi}\int_0^{a} e^{-iqr\cos\theta} r^2\, dr \sin\theta\, d\theta\, d\phi$$

$$= \frac{3(Z_1 e)}{(qa)^3}\left[\sin qa - qa\cos qa\right] \tag{152}$$

Hence,
$$\sigma(\theta) = \left(\frac{2\mu}{\hbar^2}\right)^2 \frac{\left(Z_1 Z_2 e^2\right)^2}{q^4}\cdot\frac{9}{(qa)^6}\left[\sin qa - qa\cos qa\right]^2 \tag{153}$$

(b)
$$\Lambda(\mathbf{q}) = Z_1 e e^{-q^2a^2/2} \tag{154}$$

(c)
$$\Lambda(\mathbf{q}) = \frac{Z_1 e}{(1+q^2a^2)} \tag{155}$$

Solution 26.6 Comparing with the notation of Problem 26.3, we have $Z_2 = -1$ and for the hydrogen atom $Q(\mathbf{r}) = e\delta(\mathbf{r}) - e\rho(\mathbf{r})$; therefore

∴
$$\Lambda(\mathbf{q}) = e[1 - F(q)] \tag{156}$$

where
$$F(q) = \int \rho(\mathbf{r}) e^{-i\mathbf{q}\cdot\mathbf{r}}\, d\mathbf{r}$$

$$= \frac{1}{\pi a_0^3}\iiint e^{-2r/a_0} e^{-iqr\cos\theta} r^2\, dr \sin\theta\, d\theta\, d\phi$$

$$= \frac{1}{(1+q^2 a_0^2/4)^2} \tag{157}$$

is sometimes called the atomic form factor. Substituting for $\Lambda(\mathbf{q})$ in Eq. (133), we get

$$\sigma(\theta) = \left(\frac{2\mu e^2}{\hbar^2 q^2}\right)^2 [1-F(q)]^2 = \frac{4}{a_0^2 q^4}[1-F(q)]^2$$

$$= 4a_0^2 \frac{(8+q^2a_0^2)^2}{(4+q^2a_0^2)^4}; q = 2k\sin\frac{\theta}{2} \tag{158}$$

The total cross-section will be

$$\sigma = \int_0^\pi \sigma(\theta)\, 2\pi \sin\theta\, d\theta = \int_0^{2ka_0} \sigma(q) \frac{2\pi q\, dq}{k^2}$$

$$= \frac{8\pi}{k^2} \int_0^{2ka_0} \frac{(8+y^2)^2}{(4+y^2)^4} y\, dy$$

where $y = qa_0$. Writing

$$(8 + y^2)^2 = [4 + (4 + y^2)]^2 = 16 + 8(4 + y^2) + (4 + y^2)^2$$

we can easily carry out the integration. For $ka_o \gg 1$, we obtain

$$\sigma \simeq \frac{7\pi}{3k^2} \tag{159}$$

Solution 26.7 For the helium atom we have

$$V(r) = -\frac{2e^2}{r} + e^2 \int \sum_{n=1}^{2} \frac{1}{|\mathbf{r} - \mathbf{r}_n|} |\psi_0(\mathbf{r}_1, \mathbf{r}_2)|^2\, d\tau_1\, d\tau_2 \tag{160}$$

or
$$Q(\mathbf{r}) = 2e\delta(\mathbf{r}) - \frac{2Z^3 e}{\pi a_0^3} \exp[-2Zr/a_0] \tag{161}$$

Proceeding as in the previous problem, we get

$$\sigma(\theta) = 16a_0^2 \frac{\left(8\times(1.68)^2 + q^2a_0^2\right)^2}{\left(4\times(1.68)^2 + q^2a_0^2\right)^4} \tag{162}$$

Comparison of the theoretical results with experimental data show good agreement at electron energies above 100 eV (for further details, see Ref. 7).

Solution 26.8 We refer to Fig. 26.6

$$E_{\text{lab}} = \frac{1}{2} m_1 v_1^2$$

and
$$E_{cm} = \frac{1}{2} m_1 \left(\frac{m_2 v_1}{m_1 + m_2} \right)^2 + \frac{1}{2} m_2 \left(\frac{m_1 v_1}{m_1 + m_2} \right)^2$$
$$= \frac{1}{2} \mu v_1^2 = \frac{m_2}{m_1 + m_2} E_{lab} \tag{163}$$

Solution 26.9 Scattering will be predominantly *s*-wave ($l = 0$) for $ka \lesssim 1$

or
$$E_{lab} \approx 2E_{cm} \approx 2\left(\frac{\hbar^2 k^2}{2\mu}\right) \lesssim \frac{\hbar^2}{\mu a^2}$$

Thus,
$$E_{lab} \lesssim \frac{\left(10^{-27}\right)^2}{\left(0.8 \times 10^{-24}\right) \times \left(3 \times 10^{-13}\right)^2} \approx 1.6 \times 10^{-5} \text{ erg} \approx 10 \text{ MeV} \tag{164}$$

Solution 26.10 Equation (127) gives
$$\theta_0 = \theta/2$$
and since θ goes from 0 to π, θ_0 will be less than $\pi/2$.

Solution 26.12 For a short-range force, only small values of r will contribute to the integral; therefore
$$\delta_l \approx -\frac{2\mu}{\hbar^2} \cdot \frac{(ka)^{2l+1}}{\left[1 \cdot 3 \ldots (2l+1)\right]^2} \int_0^a V(r) \left(\frac{r}{a}\right)^{2l+1} r\, dr \tag{165}$$
where a denotes the range of the force, and we have made use of the formula
$$j_l(kr) \approx \frac{(kr)^l}{1 \cdot 3 \ldots (2l+1)}$$
Evaluation of δ_l requires a knowledge of the potential, but an order of magnitude of the ratios is
$$\frac{\delta_1}{\delta_0} \sim \frac{(ka)^2}{9}; \frac{\delta_2}{\delta_0} \sim \frac{(ka)^4}{225}, \text{ etc.}$$

Solution 26.15 For a single scatterer, the asymptotic wave is given by
$$\psi \sim e^{ikz} - \frac{f}{r} e^{ikr}$$
If the scatterer is located at $\mathbf{r}_i$ then the distance to the detector is $|\mathbf{r} - \mathbf{r}_i|$ (see Fig. 26.8). There is, however, an extra phase change suffered by the incident wave by the time it reaches the scatterer so that
$$\psi_{sc}^i = -\frac{f_i}{|\mathbf{r} - \mathbf{r}_i|} e^{ikz_i} e^{ik|\mathbf{r} - \mathbf{r}_i|} \tag{166}$$
The total intensity of the scattered wave is
$$|\psi_{sc}|^2 = \left|\sum \psi_{sc}^i\right|^2 = \sum_{ij} f_i f_j^* \frac{\exp[ik(z_i - z_j)]}{\rho_i \rho_j} \exp[ik(\rho_i - \rho_j)]$$

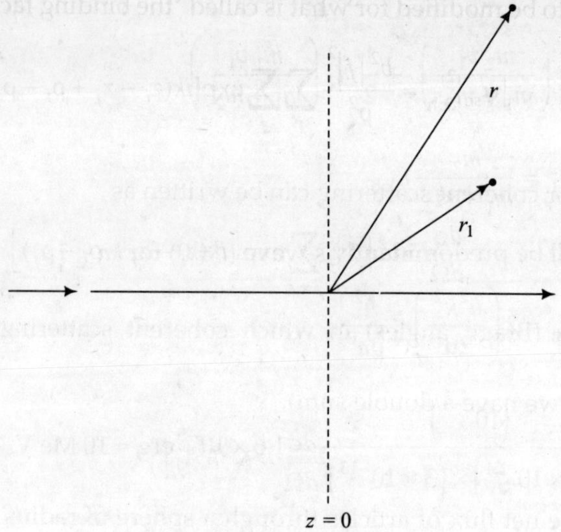

Figure 26.8: If the scatterer is at $\mathbf{r}_1$, then the distance to the detector is $|\mathbf{r}-\mathbf{r}_1|$.

$$= \sum_i \frac{|f_i|^2}{\rho_i^2} + \sum_{i\neq j} f_i f_j^* \frac{\exp\left[ik(z_i - z_j + \rho_i - \rho_j)\right]}{\rho_i \rho_j} \tag{167}$$

where $\rho_i = |\mathbf{r}-\mathbf{r}_i|$. If we assume that all the scatterers are the same and their spread in much smaller than the distance of the detector, we can write

$$|\psi_{sc}|^2 \approx \frac{N|f|^2}{\rho^2} + \frac{|f|^2}{\rho^2} \sum_i \sum_{\substack{j \\ i\neq j}} \exp\left[ik(z_i - z_j + \rho_i - \rho_j)\right] \tag{168}$$

where N is the number of scatterers. The first sum is the part that gives the incoherent scattering and the second the coherent scattering. Notice that the second term in Eq. (167) is a double sum. The coherent part can give large variations with angle. If the scatterers are all randomly spaced, then the sum will average out to zero, if

$$k\Delta\rho \gg 1$$

where $\Delta\rho$ is roughly the spread of the scatterers. For $k\Delta\rho \sim 1$ one can get large values for coherent scattering. In the forward direction,

$$\rho_i = z - z_i \qquad \text{so that} \qquad \rho_i - \rho_j = z_j - z_i$$

Thus the coherent intensity is

$$|\psi_{sc}|^2_{coh} \approx \frac{|f|^2}{\rho^2} N^2 \tag{169}$$

Again, if there is a regular arrangement of scatterers as in a crystal satisfying the condition

$$k\,(z_i - z_j + \rho_i - \rho_j) = 2\pi n$$

one gets maxima.

Equation (168) has to be modified for what is called 'the binding factor' in crystals. One writes

$$|\psi_{sc}|^2_{coh} \approx \frac{b^2|f|^2}{\rho^2}\sum_{\substack{i \\ i\neq j}}\sum_{j} \exp[ik(z_i - z_j + \rho_i - \rho_j)] \tag{170}$$

The cross-section for coherent scattering can be written as

$$\frac{d\sigma}{d\Omega} = b^2|f|^2 \sum_{i\neq j} \exp\left[ik(z_i - z_j + \rho_i - \rho_j)\right]$$

In those directions (Bragg angles) in which coherent scattering is appreciable, $\sum_{i\neq j} \sim N^2$

(remembering that we have a double sum).

Thus, $$\frac{d\sigma}{d\Omega} \sim N^2 b^2 |f|^2$$

Solution 26.17 The net flux of articles through a sphere of radius R is given by [see Eq. (34) of Chapter 4)]

$$J_R = \frac{\hbar}{2\mu i}\int\left[\psi^*\frac{\partial\psi}{\partial r} - \psi\frac{\partial\psi^*}{\partial r}\right]_{r=R} r^2\, d\Omega \tag{171}$$

Now, $$\frac{\partial\psi}{\partial r} \underset{R\to\infty}{\rightarrow} \sum_l \frac{(2l+1)}{2ikR} P_l\left[e^{2i\delta_l} ik\, e^{ikR} + (-1)^l (ik) e^{-ikR}\right]$$

$$= \sum_l \frac{(2l+1)}{2R} P_l\left[\eta_l\, e^{ikR} + (-1)^l e^{-ikR}\right]$$

and $$\psi^* \longrightarrow \sum_l \frac{(2l+1)}{-2ikR} P_l\left[\eta_l^*\, e^{-ikR} - (-1)^l e^{ikR}\right]$$

We now cannot assume that η_l, η_l^* to be the same. Thus,

$$\left[\psi^*\frac{\partial\psi}{\partial r}\right]_R = \sum_l\sum_{l'} \frac{(2l+1)P_l\,(2l'+1)P_{l'}\left[\eta_l\,\eta_{l'}^* - 1\right]}{-4ikR} + \text{terms involving } e^{\pm 2ikR}$$

We can ignore the other two terms which oscillate rapidly and will average to zero. Thus, (remembering that we want the net flux inward)

$$J_R = \frac{\hbar}{4\mu k}\sum_l\sum_{l'}\int (2l+1)(2l'+1)P_l\, P_{l'}\left[1 - \eta_l\,\eta_{l'}^*\right] d\Omega. \tag{172}$$

Due to orthogonality relations

$$\int P_l(\cos\theta)\, P_{l'}(\cos\theta)\, d(\cos\theta) = \frac{2}{2l+1}\delta_{ll'}$$

we obtain

$$J_R = \frac{4\pi\hbar}{4k\mu}\sum_l (2l+1)\left[1 - |\eta_l|^2\right]$$

The current in the lth partial wave is

$$J_R = \frac{\pi\hbar}{\mu k}(2l+1)\left[1-|\eta_l|^2\right] \tag{173}$$

Thus the reaction cross-section

$$\sigma_r^l = \frac{\pi}{k^2}(2l+1)\left[1-|\eta_l|^2\right] \tag{174}$$

In Eq. (174), $|\eta_l| < 1$, for otherwise the outgoing current will exceed the incoming one. Since the scattering cross-section is

$$\sigma_{sc}^l = \frac{\pi}{k^2}(2l+1)|1-\eta_l|^2$$

and sees that σ_{sc}^l is largest for $\eta_l = -1$. In that case σ_r^l vanishes. On the other hand, $\eta_l = 0$ gives the maximum reaction cross-section for which

$$\sigma_{sc}^l = \sigma_r^l = \frac{(2l+1)\pi}{k^2}$$

If there is no scattering ($\eta_l = 1$), there clearly is no reaction either.

Solution 26.18 The total cross-section (from the previous problem) is

$$\sigma = \sigma_{el} + \sigma_r = \sum_l \sigma^l$$

where

$$\sigma^l = \frac{\pi}{k^2}(2l+1)\left[2-\eta_l-\eta_l^*\right]$$

$$= \frac{2\pi}{k^2}(2l+1)\left[1-\text{Re}\,\eta_l\right]$$

and Re η_l is the real part of η_l. From Eq. (65) we have for the forward scattering amplitude,

$$f(0) = \sum_l (2l+1)\frac{\eta_l - 1}{2ik}$$

$$= i\sum_l (2l+1)\frac{(1-\eta_l)}{2ik}$$

This establishes the optical theorem

$$\sigma = \frac{4\pi}{k}\,\text{Im}\, f(0).$$

REFERENCES AND SUGGESTED READINGS

1. H. Goldstein, *Classical Mechanics*, Addison-Wesley Publishing Co., Reading, Massachusetts (1950).
2. L.A. Pipes, *Applied Mathematics for Engineers and Physicists*, McGraw-Hill Book Co., New York (1958).

3. J.R. Reitz and F.J. Milford, *Foundations of Electromagnetic Theory*, Addison-Wesley Publishing Co., Reading, Massachusetts (1960).
4. A. Sommerfeld, *Optics*, Academic Press, New York (1950).
5. D. Bohm, *Quantum Theory*, Prentice-Hall, Englewood Cliffs, New Jersey (1951).
6. H.A. Bethe and P. Morrison, *Elementary Nuclear Theory*, John Wiley, New York (1956).
7. N.F. Mott and H.S.W. Massey, *The Theory of Atomic Collisions*, Oxford University Press, London (1949).
8. G.N. Watson, *A Treatise on the Theory of Bessel Functions*, Cambridge University Press, London (1944).
9. R.R. Roy and B.N. Nigam, *Nuclear Physics*, John Wiley, New York (1967).
10. G. Baym, *Lectures on Quantum Mechanics*, W.A. Benjamin, New York (1969).

3. [illegible]

[illegible]

27 Time dependent perturbation theory

... I think I can safely say that nobody Understands quantum mechanics,

— RICHARD FEYNMAN.

27.1 TIME DEVELOPMENT OF STATES

When a system is isolated and the Hamiltonian is independent of the time, the energy eigenstates are true stationary states. If the system initially is in a particular state Ψ_n which is an eigenstate of the Hamiltonian H_0, it will stay there for all time. The time dependence of Ψ_n is of the form

$$\Psi_n(t) = \Psi_n(0)e^{-iE_nt/\hbar} \tag{1}$$

where E_n is the energy for the state n. If, however, the Hamiltonian is time dependent the energy is no longer a constant of motion. Knowledge of the behaviour of the system is obtained by solving the time dependent Schrödinger equation

$$i\hbar\frac{\partial}{\partial t}\Psi = H\Psi \tag{2}$$

Unfortunately the structure of the Hamiltonian H in most cases does not permit a solution of this equation in closed form.

In many physical problems, the time dependence of the Hamiltonian arises because of what may be thought of an external agency. An atom irradiated by light is an example. The time dependent part in such cases is usually a small perturbation representing the interaction of the system with the external agency. We may then write

$$H = H_0 + V(t) \tag{3}$$

where H_0 may be thought of as the sum of Hamiltonians of two systems in isolation and V, their interaction. Instead of attempting a direct solution of Eq. (2), we pose the problem in a slightly different manner. At a particular instant, say $t = 0$, the system may be regarded as being approximately in an eigenstate of H_0, labelled i. We would like to calculate the probability of finding the system in another state f, also an eigenstate of H_0 at a later time[1] t. Under these circumstances, we shall talk of the system being in quasi-stationary states. In most of the physical problems, we only have quasi-stationary states. For example, an excited state of the hydrogen atom *does* decay and is not strictly a stationary state. At the other extreme, we may

[1] If $V(t)$ acts for a finite duration, then the state i and f may be exact eigenstates of H_0 provided the 'perturbation is zero at the initial and final times of observation.

regard a state that lasts only for 10^{-20} sec as quasi-stationary; the point is that this lifetime may still be much longer than the natural frequencies of the system.

27.2 THE TRANSITION PROBABILITY

Let ψ_n denote the complete set of eigenkets of H_0

$$H_0 \psi_n = E_n \psi_n = \hbar\omega_n \psi_n \tag{4}$$

where $E_n (= \hbar\omega_n)$ represent the corresponding energy eigenvalues. The wavefunction representing the actual state $\psi(t)$ can be expanded terms of ψ_n:

$$\Psi(t) = \sum_n b_n(t)\psi_n(\mathbf{r})$$

where $|b_n(t)|^2$ can be interpreted as the probability of finding the system in the state n at time t. For the sake of convenience we write[2]

$$b_n(t) = C_n(t)\, e^{-i\omega_n t}$$

so that

$$\Psi(t) = \sum_n C_n(t) e^{-i\omega_n t} \psi_n(\mathbf{r}) \tag{5}$$

Obviously, $|C_n(t)|^2$ will also represent the probability of finding the system in the state n at time t. Our objective is to develop a perturbation theory for the calculation of $C_n(t)$ for a given $V(t)$. Now, from Eqs (2), (3) and (5), we have

$$i\hbar \frac{\partial}{\partial t}\Psi(t) = (H_0 + V)\,\Psi(t)$$

$$= (H_0 + V) \sum_n C_n(t) e^{-i\omega_n t} \psi_n$$

$$= \sum_n (E_n + V)\, C_n(t)\, \psi_n(\mathbf{r})\, e^{-i\omega_n t} \tag{6}$$

where in the last step we have used Eq. (4). If we use Eq. (5), the left-hand side of the above equation becomes

$$i\hbar \frac{\partial}{\partial t}\psi(t) = i\hbar \frac{\partial}{\partial t} \sum_n C_n(t) e^{-i\omega_n t} \psi_n$$

$$= i\hbar \sum_n \left[\frac{d\,C_n(t)}{dt} - i\omega_n C_n \right] e^{-i\omega_n t} \psi_n$$

Thus, we have

$$i\hbar \sum_n \left[\frac{dC_n(t)}{dt} - i\omega_n C_n \right] e^{-i\omega_n t} \psi_n = \sum_n (E_n + V) C_n(t) e^{-i\omega_n t} \psi_n \tag{7}$$

where the second term on the left-hand side cancels the first term on the right-hand side

---

[2]The rapidly varying term in $b_n(t)$ is the exponential term. Indeed if $V = 0$, $C_n(t)$ is independent of time.

Multiplying on the left by ψ_s^* and remembering the orthonormality relation

$$\int \psi_s^* \psi_n d\mathbf{r} = \langle s|n\rangle = \delta_{sn} \tag{8}$$

we get

$$i\hbar \frac{dC_s}{dt} = \sum C_n(t) V_{sn}(t) e^{i\omega_{sn}t} \tag{9}$$

where

$$\omega_{sn} = \omega_s - \omega_n = \frac{E_s - E_n}{\hbar} \tag{10}$$

and

$$V_{sn}(t) = \int \psi_s^* V(t) \psi_n \, d\tau = \langle s|V|n\rangle \tag{11}$$

The set of equations (corresponding to various values of $|s\rangle$ represented by Eq. (9) are exact. To solve them, we usually have to resort to perturbation techniques; there are, however, situations where one can obtain exact solutions of the time dependent equations; see, e.g. Section 14.6, see also Problem 28.8.

Let us assume that at time t the amplitudes have not changed very much from their values at $t = 0$; i.e. we replace $C_n(t)$ by $C_n(0)$ on the right-hand side of Eq. (9). We can then integrate Eq. (9) to give

$$i\hbar[C_s(t) - C_s(0)] = \sum_n C_n(0) \int_0^t V_{sn} e^{i\omega_{sn}t} dt \tag{12}$$

We can apply this to the case where we know that the system is in a particular state i at $t = 0$. Thus,

$$\begin{aligned} C_s(0) &= 1 \quad \text{for} \quad s = i \\ &= 0 \quad \text{for} \quad s \neq i \end{aligned} \tag{13}$$

so that

$$C_f(t) = \frac{1}{i\hbar} \int_0^t V_{fi}(t) e^{i\omega_{fi}t} dt, \qquad f \neq i \tag{14}$$

Equation (14) gives the probability amplitudes at later times of states other than the initial one. Suppose now that at $t = 0$ the perturbation is switched on and for $t > 0$, is independent of time[3].

Thus,

$$C_f(t) = \frac{1}{i\hbar} V_{fi} \int_0^t e^{i\omega_{fi}t} dt = -V_{fi} \frac{e^{i\omega_{fi}t} - 1}{\hbar\omega_{fi}} \tag{15}$$

Thus the probability for reaching the state f (at time t) is given by

$$P_f(t) = |C_f(t)|^2 = \left|V_{fi}\right|^2 \frac{1}{\hbar^2} \frac{\sin^2(\omega_{fi}t/2)}{(\omega_{fi}/2)^2} \tag{16}$$

We rewrite the above equation in terms of E_f and E_i:

[3] In the next chapter we will consider harmonic perturbations where $V(t)$ will vary sinusoidally with time and will describe the atom-field interaction.

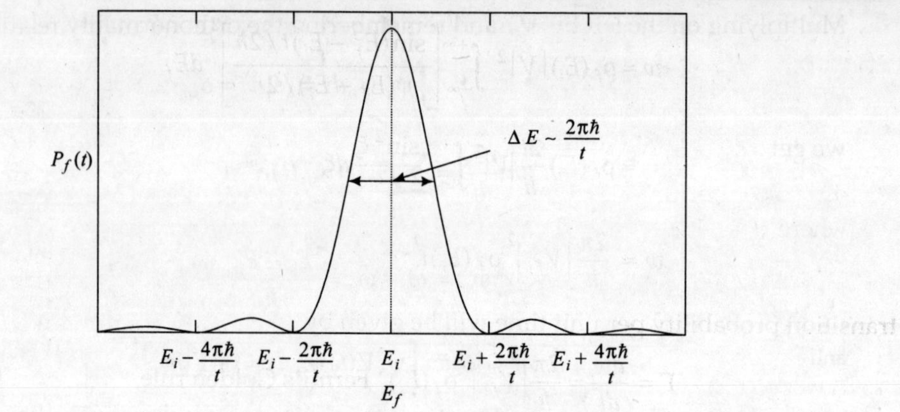

Figure 27.1: The variation of $P_f(t)$ [as given by Eq. (17)] as a function of E_i. Notice that the function is peaked at $E_f = E_i$ with a width $\Delta E \sim \hbar/t$.

$$P_f(t) = |V_{fi}|^2 \left[\frac{\sin(E_f - E_i)t/2\hbar}{(E_f - E_i)/2}\right]^2 \quad \text{Probability of transition to state } \psi_f \tag{17}$$

Figure 27.1 shows the behaviour of $P_f(t)$ as a function of E_f. We note the following:

(i) The perturbation theory result, viz., Eq. (17) is valid only when $P_f(t) \ll 1$.

(ii) As long as t is small enough such that

$$(E_f - E_i)\frac{t}{2\hbar} \ll 1 \tag{18}$$

the quantity $P_f(t)$ is proportional to t^2 for all values of E_f, in fact

$$P_f(t) \approx |V_{fi}|^2 \frac{t^2}{\hbar^2} \tag{19}$$

(iii)
$$\underset{E_f \to E_i}{\text{Lt}} \; P_f(t) = |V_{fi}|^2 \frac{t^2}{\hbar^2} \tag{20}$$

Equation (17) represents the probability of transition to one particular state ψ_f. In the applications that we would discuss, we would be interested in transitions to a group of closely spaced states. Thus, the probability of transition to any of the states is given by

$$w = \int P_f(t)\rho_f(E_f)dE_f \tag{21}$$

where $\rho_f(E_f)\, dE_f$ represents the number of final states in the energy interval dE_f. The density of states is usually a very smoothly varying function and since $P_f(t)$ is very sharply peaked around $E_f \approx E_i$, we may write

$$w \approx \rho_f(E_i)\int_{\Delta E} P_f(t)dE_f \tag{22}$$

where ΔE represents the central peak. We may now replace the limits of integration by $-\infty$ and $+\infty$ to obtain

$$w \approx \rho_f(E_i)|V|^2 \int_{-\infty}^{+\infty}\left[\frac{\sin(E_f - E_i)t/2\hbar}{(E_f - E_i)/2}\right]^2 dE_f$$

$$\approx \rho_f(E_i)\frac{2t}{\hbar}|V|^2 \int_{-\infty}^{+\infty}\frac{\sin^2\xi}{\xi^2}d\xi$$

or
$$w \approx \frac{2\pi}{\hbar}\left|V_{fi}\right|^2 \rho_f(E_i)t \tag{23}$$

The transition probability per unit time will be given by

$$\Gamma = \frac{dw}{dt} = \frac{2\pi}{\hbar}\left|V_{fi}\right|^2 \rho_f(E_i) \quad \text{Fermi's Golden rule} \tag{24}$$

which is usually referred to as *Fermi's Golden Rule.* We illustrate the above procedure with a specific example. Suppose a system A undergoes a transition to two parts $A \to a + b$ due to a perturbation. Our question may be framed thus: If at time $t = 0$ we know that the system is in a specific state A, what is the probability that at a subsequent time $t > 0$, the system has 'decayed' into parts a and b. According to Eq. (17), there is a finite probability that the final state will have an energy $E_f \neq E_i$ (E_i was the energy at $t = 0$) where $E_f - E_i \lesssim \hbar/t$. Thus we must sum probabilities over a set of states f within the interval—in fact, we shall sum over all states whatever the value of E_f. But, Eq. (17) ensures that the contribution from states for which $E_f - E_i \gg \hbar/t$ is negligible.

We are, therefore, inevitably interested in the density of states f, i.e. the more states there are satisfying the relation $E_f - E_i \lesssim \hbar/t$ the more will be the transition probability.

If we assume that the entire system is in a box of volume V, we shall then have to calculate the number of states of a and b within an interval dE_f of a particular value E_f, say $\rho(E_f)\,dE_f$, compute the total probability for all these states and then integrate over all values of E_f. Equation (17) will ensure that we do not have to go far from $E_f = E_i$; i.e., we are essentially interested in the density of states $\rho(E_f = E_i)$.

We may mention here that in carrying out the integration in Eq. (23) we have assumed that t is large enough so that the function $P_f(t)$ is very sharply peaked around $E_f = E_i$; (so that ρ_f can be taken outside the integral); on the other hand, for the validity of perturbation theory, $P_f(t) \ll 1$, i.e., t should be small enough so that

$$\frac{\left|V_{fi}\right|^2}{\hbar^2}t^2 \ll 1 \tag{25}$$

We digress here to mention that since $P_f(t)$ is very sharply peaked around $E_f = E_i$, it should be possible to express it in the form of a delta function. Indeed, since

$$\underset{g\to\infty}{\text{Lt}}\ \frac{1}{\pi g}\cdot\frac{\sin^2 gx}{x^2} = \delta(x) \tag{26}$$

we may write

$$\underset{t\to\infty}{\text{Lt}}\ \frac{\sin^2(E_f - E_i)t/2\hbar}{\left[(E_f - E_i)/2\right]^2} = \frac{\pi t}{\hbar}\delta\left(\frac{E_f - E_i}{2}\right) = \frac{2\pi t}{\hbar}\delta(E_f - E_i) \tag{27}$$

Thus Eq. (17) gives

$$P_f(t) = \frac{2\pi}{\hbar}t\left|V_{fi}\right|^2 \delta(E_f - E_i) \tag{28}$$

a formula which is often used. Equation (28) does appear nonsensical because it is zero for all values of $E_f \neq E_i$ and ∞ for $E_f = E_i$ and we know that $P_f(t)$ should be very small compared to unity. However, Eq. (28) has meaning only when it is multiplied by the smoothly varying function $\rho_f(E_f)$ and integrated over all energies to give Eq. (24).

27.3 APPLICATION TO THE THEORY OF SCATTERING

In Chapter 26, we discussed elastic scattering in which we considered the scattering of a wave packet by a fixed potential $V(\mathbf{r})$; this was done by obtaining a solution of the Schrödinger equation in the form of an incident plane wave and an outgoing scattered wave. Here we will consider another approach to the scattering problem by considering the scattering of a particle as a transition from an initial plane wave state to another plane wave state corresponding to a different direction of propagation. The two approaches will be shown to be equivalent although each approach has its own domain of application.

We consider a particle in a cube of volume L^3 and initially in a plane wave state corresponding to momentum $\mathbf{p}_i$. Thus, the initial state of the particle is described by the wavefunction

$$\psi_i = \frac{1}{L^{3/2}} \exp\left[i\mathbf{k}_i \cdot \mathbf{r}\right] \tag{29}$$

where

$$\mathbf{k}_i = \frac{1}{\hbar}\mathbf{p}_i \tag{30}$$

It may be mentioned that by confining the particle inside a large box and by imposing the periodic boundary conditions (see Appendix M), the states become discrete; in fact the allowed values of k_x, k_y and k_z are

$$k_x = \frac{2n_x\pi}{L},\ k_y = \frac{2n_y\pi}{L},\ k_z = \frac{2n_z\pi}{L} \tag{31}$$

where

$$n_x, n_y, n_z = 0, \pm 1, \pm 2, \pm 3, \ldots \tag{32}$$

We now assume that the perturbation $V(\mathbf{r})$ is switched on at $t = 0$. The probability of transition to the final state

$$\psi_f = \frac{1}{L^{3/2}} \exp(i\mathbf{k}_f \cdot \mathbf{r}) \tag{33}$$

is given by [see Eq. (17)]

$$P_f(t) = \left|C_f(t)\right|^2 = \left|V_{fi}\right|^2 \left[\frac{\sin(E_f - E_i)t/2\hbar}{(E_f - E_i)/2}\right]^2 \tag{34}$$

where

$$V_{fi} = \int \psi_f^* V \psi_i \, d\mathbf{r}$$

$$= \frac{1}{L^3}\int V(\mathbf{r}) \exp(-i\mathbf{q}\cdot\mathbf{r})\, d\mathbf{r} \tag{35}$$

and

$$\mathbf{q} = \mathbf{k}_f - \mathbf{k}_i \tag{36}$$

quation (34) represents the probability of the particle getting scattered to a particular state orresponding to $\mathbf{k}_f$. If we wish to get the total probability of the particle getting scattered into ne solid angle $d\Omega$, which we will denote by $w d\Omega$, then we must multiply $P_f(t)$ by $\rho_f(E_f)\, d\Omega$ nd integrate over all energies; here $\rho_f(E_f)\, dE_f d\Omega$ represents the number of states whose final nergy lies between E_f and $E_f + dE_f$ and the direction of propagation in the solid angle $d\Omega$. Thus, sing the method used in the previous section, we get

$$w d\Omega = \int P_f(t) \rho_f(E_f)\, dE_f\, d\Omega$$

$$\approx \frac{2\pi}{\hbar} \left| V_{fi} \right|^2 \rho_f(E_i)\, t\, d\Omega \tag{37}$$

hus the transition probability per unit time into the solid angle $d\Omega$ is given by

$$\Gamma\, d\Omega \approx \frac{2\pi}{\hbar} \left| V_{fi} \right|^2 \rho_f(E_i) d\Omega \tag{38}$$

ow, the density of final states is given by [see Eq. (10) of Appendix M]:

$$\rho_f(E)\, dE\, d\Omega = \frac{L^3}{16\pi^3 \hbar^3} (2m)^{3/2} E^{1/2}\, dE\, d\Omega \tag{39}$$

urther, since there is only one particle in volume L^3, the number of particles per unit volume ill be $1/L^3$; thus the incident intensity, I, will be $1/L^3$ multiplied by their velocity $\hbar k/m$:

$$I = \frac{1}{L^3} \frac{\hbar k}{m} = \frac{1}{L^3} \frac{(2mE)^{1/2}}{m} \tag{40}$$

om the definition of the cross-section (see Section 26.3) the number of particles scattered into e solid angle $d\Omega$ per unit time $= I\sigma\, d\Omega = \Gamma\, d\Omega$; or,

$$\sigma d\Omega = \frac{\Gamma d\Omega}{I} = \frac{1}{(\hbar k / m L^3)} \left[\frac{2\pi}{\hbar} \left| V_{fi} \right|^2 \rho_f(E_i) d\Omega \right] \tag{41}$$

$$\sigma\, d\Omega = \left(\frac{m L^3}{2\pi \hbar^2} \right)^2 \left| V_{fi} \right|^2 d\Omega \tag{42}$$

bstituting for V_{fi} from Eq. (35), we get

$$\sigma = \left(\frac{m}{2\pi \hbar^2} \right)^2 \left| \int V(\mathbf{r}) e^{-i\mathbf{q}\cdot\mathbf{r}}\, d\mathbf{r} \right|^2 \quad \text{Scattering cross-section} \tag{43}$$

ich is identical to the expression obtained by using the Born approximation [Eqs. (20) and) of Chapter 26]. This is to be expected because both employ first-order perturbation theory.

.4 ADIABATIC APPROXIMATION

his section, we will consider the case of a perturbation varying slowly with time. For such se if we can solve the eigenvalue equation

$$H(t) \psi_n(t) = E_n(t)\, \psi_n(t) \tag{44}$$

then we should expect that if at $t = 0$ the system is in the (non-degenerate) state $\psi_m(0)$ then at a later time the system will be in the state $\psi_m(t)$. We will show this explicitly in this section and also calculate the probability of transitions to other states.

We assume the potential to be turned on from $t = -\infty$ so that the initial condition is [cf. Eq. (13)]

$$C_s(-\infty) = 1 \quad \text{for } s = i$$
$$= 0 \quad \text{for } s \neq i \tag{45}$$

so that [cf. Eq. (14)] $\quad C_f(t) = \dfrac{1}{i\hbar}\displaystyle\int_{-\infty}^{t} V_{fi}(t)^{i\omega_{fi}t}\, dt,\ f \neq i$

Partial integration of the above equation gives

$$C_f(t) = \left[-\frac{V_{fi}e^{i\omega_{fi}t}}{\hbar\omega_{fi}}\right]_{-\infty}^{t} + \int_{-\infty}^{t} \frac{1}{\hbar\omega_{fi}}\frac{\partial V_{fi}}{\partial t}e^{i\omega_{fi}t}\, dt \tag{46}$$

If we assume that $V_{fi}(t) \to 0$ as $t \to -\infty$, we can write

$$C_f(t) = -\frac{V_{fi}(t)}{\hbar\omega_{fi}}e^{i\omega_{fi}t} + \int_{-\infty}^{t} \frac{1}{\hbar\omega_{fi}}\frac{\partial V_{fi}}{\partial t}e^{i\omega_{fi}t}\, dt \tag{47}$$

The state of the system at the time t is given by [see Eq. (5)]

$$\Psi(t) = C_i(t)\psi_i e^{-i\omega_i t} + \sum_{f \neq i} C_f(t)\psi_f e^{-i\omega_f t}$$

Assuming that the initial state has not been depleted very much in time t, we can put $C_i(t) \approx 1$. Substituting the values of $C_f(t)$ from Eq. (47), we obtain

$$\Psi(t) = \left[\psi_i - \sum_{f \neq i} \frac{V_{fi}}{\hbar\omega_{fi}}\psi_f\right]e^{-i\omega_i t} + \sum_{f \neq i}\int_{-\infty}^{t} \frac{1}{\hbar\omega_{fi}} \cdot \frac{\partial V_{fi}}{\partial t}e^{-i\omega_{fi}t}\, dt \tag{48}$$

The terms in the parenthesis in Eq. (48) represents the change in the state due to the application of a constant perturbation V [see Eqs (13), (19) and (20) of Chapter 20] and has nothing to do with transitions. It is, therefore, the second term in Eq. (48) that gives transitions.

The *adiabatic approximation* is the assumption that the perturbation is turned on slowly, i.e., $\partial V_{fi}/\partial t$ is small. Equation (47) suggests the transitions to other states are extremely unlikely in this case. Assuming that $\partial V/\partial t \neq 0$ in an interval t_1 to t_2, we may write

$$C_f(t) \approx \frac{1}{\hbar\omega_{fi}}\left(\overline{\frac{\partial V_{fi}}{\partial t}}\right)\int_{t_1}^{t_2} e^{i\omega_{fi}t}\, dt$$

$$= \frac{1}{i\hbar\omega_{fi}^2}\left(\overline{\frac{\partial V_{fi}}{\partial t}}\right)\left[e^{i\omega_{fi}t_2} - e^{i\omega_{fi}t_1}\right] \tag{49}$$

The condition that $|C_f(t)| \ll 1$ gives

$$\frac{1}{\hbar\omega_{fi}^2}\frac{\partial V}{\partial t} \ll 1 \tag{50}$$

Since $\tau_{fi} = 1/\omega_{fi}$ is the period of the radiation associated with the transition, this statement implies that

$$\frac{1}{\hbar\omega_{fi}}\frac{\partial V}{\partial t} \ll \frac{1}{\tau_{fi}} \tag{51}$$

27.5 THE SUDDEN APPROXIMATION

We next consider the case when the potential is turned on very rapidly, i.e.,

$$\tau \ll \tau_{fi} \sim \frac{1}{\omega_{fi}} \tag{52}$$

which implies that the time in which the perturbation changes appreciably is very small in comparison to the period involved in the transition. As an example, we assume

$$\begin{aligned} H &= H_0 \quad \text{for } t < 0 \\ &= H_1 \quad \text{for } t > 0 \end{aligned} \tag{53}$$

where H_0 and H_1 are independent of time. Let ψ_n and ϕ_n represent the orthonormal set of wavefunctions of H_0 and H_1 respectively. For $t < 0$, let the system be in an eigenstate of H_0 (say ψ_m). Since the Schrödinger equation is a first order equation in time, the wavefunction should be continous at all times (although its time derivative need not be continuous). Now, at $t = 0$ we express ψ_m as a linear combination of ϕ_n:

$$\psi_m = \sum_n a_n \phi_n \tag{54}$$

where
$$a_n = \int \phi_n^* \psi_m \, d\tau \tag{55}$$

represents the probability amplitude for transition to the state ϕ_n.

In general, in the sudden approximation we assume that $\partial V_{fi}/\partial t$ is essentially finite (and large) for a very small period of time say from T to $T + \tau$. During this time the exponent in the integral in Eq. (47) changes very little so that

$$C_f(t) \simeq \frac{1}{\hbar\omega_{fi}} e^{i\omega_{fi}T} \int_T^{T+\tau} \frac{\partial V_{fi}}{\partial t} dt$$

$$= \frac{1}{\hbar\omega_{fi}} e^{i\omega_{fi}T} V_{fi} \tag{56}$$

Thus,
$$\left|C_f(t)\right|^2 = \frac{1}{\hbar^2\omega^2_{fi}}\left|V_{fi}\right|^2 \tag{57}$$

From Eq. (56) we can see that the criterion for validity of sudden approximation is

$$\omega_{fi}\tau \ll 1 \tag{58}$$

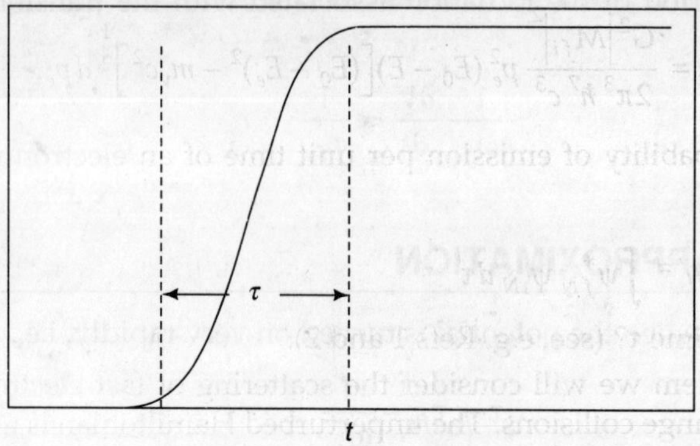

Figure 27.2: A typical variation of the perturbation in the sudden approximation.

The change of the perturbation in the sudden approximation looks somewhat as in Fig. 27.2.

27.6 PROBLEMS

Problem 27.1 The interaction of the neutron with a lattice can be well represented by the Fermi pseudo-potential

$$H' = \frac{2\pi\hbar^2}{m} \sum_n a_n \delta(\mathbf{r} - \mathbf{r}(n)) \tag{59}$$

where m is the mass of the neutron, a_n is known as the bound scattering length of the nth nucleus, $\delta(\mathbf{r})$ is the three-dimensional Dirac delta-function, $\mathbf{r}(n)$ denotes the position of the nucleus and the summation is over all the nuclei in the lattice. Assume all the nuclei to have zero spin and no isotopes so that a_n is independent of n. Calculate the elastic scattering cross-section of neutrons by a simple cubic structure of lattice spacing d (Neglect lattice vibrations). Show that the condition for non-vanishing scattering is that Bragg law be satisfied.
[**Hint:** For a simple cubic structure

$$\mathbf{r}(n) = d\left[n_1 \hat{\mathbf{x}} + n_2 \hat{\mathbf{y}} + n_3 \hat{\mathbf{z}}\right]$$

where n_1, n_2 and n_3 take integral values.]

Problem 27.2 In the elementary theory of beta decay, the form of the interaction operator, H', is simply assumed to be a constant, which we denote by G. Thus the matrix element H'_{fi} of the interaction is defined as

$$H'_{fi} = G \int \psi^*_{fN}\, \psi^*_e\, \psi_v\, \psi_{iN}\, d\tau,$$

where ψ_{iN} and ψ_{fN} are the normalized wavefunctions for the parent and daughter nucleus; ψ_e and ψ_ν are the normalized wavefunctions for the electron (positron) and the antineutrino (neutrino) respectively. Experiments show that G ~ 10^{-48}gm cm^5 sec^{-2}. Assuming the wavelengths of the leptons being considerably longer than nuclear dimensions, show from phase space considerations that

$$P(p_e)\,dp_e = \frac{G^2 \left|M_{fi}\right|^2}{2\pi^3 \hbar^7 c^3}\, p_e^2 (E_0 - E)\left[(E_0 - E_e)^2 - m_\nu c^2\right]^{\frac{1}{2}} dp_e,$$

where $P(p_e)\,dp_e$ is the probability of emission per unit time of an electron with momentum between p_e and $p_e + dp_e$ and

$$M_{fi} = \int \psi_{fN}^* \, \psi_{iN} \, d\tau$$

What will be the mean lifetime τ? (see, e.g. Refs 1 and 2).

Problem 27.3 In this problem we will consider the scattering of fast electrons by hydrogen atom; we will neglect exchange collisions. The unperturbed Hamiltonian is given by

$$H_0 = -\frac{\hbar^2}{2m}\nabla_1^2 - \frac{\hbar^2}{2m}\nabla_2^2 - \frac{e^2}{r_2} \tag{60}$$

where $\mathbf{r}_1$ and $\mathbf{r}_2$ represent the coordinates of the incident electron and the electron bound to the hydrogen atom respectively. The perturbation which causes the scattering will be given by

$$H' = \frac{e^2}{r_{12}} - \frac{e^2}{r_1} \tag{61}$$

where the two terms represent the incident electron-atomic electron interaction and the incident electron-atomic nucleus interaction respectively. Using the theory developed in Section 27.3 show that the scattering cross-section corresponding to the excitation of the hydrogen atom from 1s state to the 2s state is given by

$$\sigma(\theta) = \frac{k_1}{k_0} \frac{128\, a_0^2}{\left[q^2 a_0^2 + \frac{9}{4}\right]^2} \tag{62}$$

where $a_0 = \hbar^2/me^2$, $\qquad q^2 = (\mathbf{k}_0 - \mathbf{k}_1)^2 = k_0^2 + k_1^2 - 2k_0 k_1 \cos\theta \qquad (63)$

θ is the scattering angle, $\hbar\mathbf{k}_0$ and $\hbar\mathbf{k}_1$ represent the momenta of the electron before and after the collision. (We may recall here that in Problem 26.6, we had discussed the elastic scattering of electrons by hydrogen atom; the present exercise is an example of inelastic scattering in which the energy of the scattered electron is different from the energy of the incident electron.)

Problem 27.4 A proton is travelling with an energy of 1 MeV in hydrogen gas. Estimate the minimum impact parameter for which there is appreciable energy transfer.

Problem 27.5 Assuming that a heavy ion (with charge $Z_1 e$) of velocity v passes undeflected as it suffers a collision with an atom, calculate the probability of transition of the atom from the initial state m to the final state n and hence obtain a formula for the rate of energy loss of the heavy charged particle.

Problem 27.6 The deuteron is a very loosely bound structure, the mean distance between the neutron and the proton being of the order of 4.3 fermi, which is considerably larger than the range of the nuclear forces which is about 1.4 fermi (1 fermi = 10^{-13} cm). It is, therefore, possible for a deuteron to undergo a grazing collision with a nucleus in such a way that the proton is stripped off the deuteron and is absorbed by the nucleus while the neutron emerges as a free particle. This process is called ‘stripping’.

(a) Show that for a deuteron energy of 200 MeV, this process is an example of a sudden transition.

(b) Calculate the energy distribution of the neutrons.

Problem 27.7 Show that when perturbation is small (sudden approximation) the probability amplitude given by Eq. (55) reduces to the expression given by Eq. (57).

Problem 27.8 The radioactive H^3 nucleus decays to He^3,

$$H^3 \to He^3 + e^- + \nu$$

The electron energy is of the order of 10 KeV. (It is actually a continuous spectrum but for most of it the above estimate is reasonable.) When a tritium atom decays, the He^3 is a structure with a single atomic electron which has to readjust its motion since the nuclear charge has changed. Show that the sudden approximation holds in this case and calculate the probability that the helium ion is in its ground state.

Problem 27.9 Calculate the probability of excitation to the 2p state of a hydrogen atom, originally in its ground state, due to a homogeneous electric field with time dependence

$$E = \frac{E_0}{\pi} \cdot \frac{\tau}{t^2 + \tau^2}$$

Discuss the limits of small and large values of τ and their significance.

Problem 27.10 Let $|\psi(t)\rangle$ denote the state of a system at time t. Let $|n\rangle$ denote one of a complete set of eigenkets of an observable A, considered time independent. Derive the relation

$$i\hbar \frac{\partial C_n}{\partial t} = \sum_{n'} H_{nn'} C_{n'} \tag{64}$$

where $C_n(t)$ is the probability amplitude that the system is in the state $|n\rangle$ at the time t and $H_{nn'}$ is the matrix element $\langle n|H|n'\rangle$ of the Hamiltonian.

Problem 27.11 The K° is a meson produced in strong interactions such as

$$\pi + p \to \Lambda^\circ + K^\circ$$

the K° state being an eigenstate of the strong interaction Hamiltonian. Similarly, the antiparticle $\bar{K}^\circ$ is also an eigenstate of the strong interaction Hamiltonian. On the other hand, in the absence of strong interactions (i.e., when the K° is all by itself and has no means of participating in strong interactions), the K° can decay via a weak interaction, the Hamiltonian for which may be regarded as very much smaller than the strong interaction Hamiltonian, $H_w \ll H_s$. As a result of H_w the K° and $\bar{K}^\circ$ are not eigenstates of the full Hamiltonian. As in Section 17.4, we choose as the two base states (eigenstates of H_s) $|K^\circ\rangle$ and $|\bar{K}^\circ\rangle$. It turns out that the proper eigenstates are

$$\left|K_1^\circ\right\rangle = \frac{1}{\sqrt{2}}\left[|K^\circ\rangle + |\bar{K}^\circ\rangle\right] \tag{65}$$

and

$$\left|K_2^\circ\right\rangle = \frac{1}{\sqrt{2}}\left[|K^\circ\rangle - |\bar{K}^\circ\rangle\right] \tag{66}$$

This state $|K_1^\circ\rangle$ decays into $\pi^+ + \pi^-$ or $\pi^\circ + \pi^\circ$ with a lifetime $\tau_1 \approx 0.9 \times 10^{-10}$ sec. The lifetime of $|K_2^\circ\rangle$ which we denote by τ_2 is about 5.6×10^{-8} sec. Thus we may write

$$|\Psi(t)\rangle = \left[C_1\left|K_1^\circ\right\rangle e^{-iE_1 t/\hbar} e^{-t/2\tau_1} + C_2\left|K_2^\circ\right\rangle e^{-iE_2 t/\hbar} e^{-t/2\tau_2}\right] \tag{67}$$

Since $\tau_2 \gg \tau_1$, assume $\tau_2 = \infty$ and calculate the probability that starting from a K° particle, we end up with $\bar{K}^\circ$ particle at time t.

Problem 27.12

(a) A quantum mechanical system (say a linear oscillator) is in a state $\psi_i(\mathbf{r})$ at $t = 0$ when it is subjected to a large impulse $\mathbf{F}(t)$ of the form

$$\mathbf{F}(t) = \mathbf{P}\delta(t) \tag{68}$$

Show that immediately after the impulse, the wavefunction of the system is

$$\psi_f(\mathbf{r}) = \psi_i(\mathbf{r}) \exp[i\mathbf{P}\cdot\mathbf{r}/\hbar]$$

(b) Obtain an expression for the probability that, following the impulse, the system is found in the state n.

Problem 27.13 Consider the reaction

$$A + a \rightarrow B + b$$

and the inverse reaction

$$B + b \rightarrow A + a$$

Show that, for the same center of mass energies

$$\frac{\sigma_{A+a\rightarrow B+b}}{\sigma_{B+b\rightarrow A+a}} = \left(\frac{p_a}{p_b}\right)^2 \frac{(2S_B+1)(2S_b+1)}{(2S_A+1)(2S_a+1)} \tag{69}$$

where the momenta p_a and p_b are measured in the center of mass system (cms).

Problem 27.14 Consider the nuclear reactions

$$p + p \rightarrow \pi^+ + d \tag{70}$$

and

$$\pi^+ + d \rightarrow p + p \tag{71}$$

where p, d and π^+ represent the proton, deuteron and π^+ meson respectively. If $\sigma_{pp\rightarrow\pi^+d}$ and $\sigma_{\pi^+\rightarrow pp}$ represent the cross-sections for the processes described by Eqs (70) and (71) at the same center of mass energy, then using the results of the previous problem prove that:

$$\sigma_{pp\rightarrow\pi^+d} = \frac{3}{2}\sigma_{\pi^+d\rightarrow pp}(2s_\pi+1)\left(\frac{p_{\pi^+}}{p_p}\right)^2 \tag{72}$$

where s_π denotes the spin of the π–meson, p_p and p_π^+ are the momenta of the proton and π^+ meson respectively.

Problem 27.15 For an incident proton energy of 340 MeV, the cross-section of the reaction $pp \rightarrow \pi^+ d$ is found to be 0.18 ± 0.06 mb. In another experiment for an incident deuterium energy of 29 MeV, the cross-section of the reaction $\pi^+ d \rightarrow pp$ is found to be[4] 3.1 ± 0.3 mb. Show that

(a) both the reactions correspond approximately to the same center of mass energy, and

(b) that the pion spin is 0.

$$(m_{\pi^+} = 139.6 \text{ MeV}/c^2;\ m_p = 938.3 \text{ MeV}/c^2;\ m_d = 1875.6 \text{ MeV}/c^2)$$

Problem 27.16 The K^+ meson has a decay mode

$$K^+ \rightarrow \pi^+ + \pi^+ + \pi^-$$

[4] Data quoted from Ref. 3. Experimental results are of Cartwright et al. (*Physical Review*, **91**, 667, 1953) and of Durbin et al. (*Physical Review*, **83**, 646,1951; **84**, 581,1951)

Assume that the decay rate can be calculated by the Golden Rule and that the matrix element causing the transition depends on the pion energies as

$$M \sim \frac{1}{\sqrt{\varepsilon_1 \varepsilon_2 \varepsilon_3}}$$

where $\varepsilon_1^2 = \sqrt{p_1^2 + m^2}$ is the total energy (including the rest energy) of the first π meson, etc. Calculate the energy distribution of the pions (say π^-) in the *rest system* of K.

27.7 SOLUTIONS

Solution 27.1 The initial and final states of the neutron are given by

$$\frac{1}{L^{3/2}} \exp\left[i\mathbf{k}_0 \cdot \mathbf{r}\right]$$

and

$$\frac{1}{L^{3/2}} \exp\left[i\mathbf{k}_1 \cdot \mathbf{r}\right]$$

Since the scattering is elastic, the initial and final states of the crystal are the same and $|\mathbf{k}_0| = |\mathbf{k}_1|$. The matrix element of transition is

$$H'_{fi} = \frac{2\pi\hbar^2 a}{mL^3} \int \exp\left[-i\mathbf{k}_1 \cdot \mathbf{r}\right] \sum_n \delta(\mathbf{r} - \mathbf{r}_n) \exp(i\mathbf{k}_0 \cdot \mathbf{r})\, dr$$

$$= \frac{2\pi\hbar^2 a}{mL^3} \sum_n \int \exp\left[i\mathbf{k} \cdot \mathbf{r}\right] \delta(\mathbf{r} - \mathbf{r}_n)\, dr$$

$$= \frac{2\pi\hbar^2 a}{mL^3} \sum_n \exp\left[i\mathbf{k} \cdot \mathbf{r}_n\right] \tag{73}$$

where

$$\mathbf{k} = \mathbf{k}_0 - \mathbf{k}_1 \tag{74}$$

The differential cross-section for scattering is, therefore, given by

$$\frac{d\sigma}{d\Omega} = \frac{1}{N} \left[\frac{1}{\dfrac{\hbar k_0}{m} \dfrac{1}{L^3}} \right] \left[\frac{2\pi}{\hbar} |H'_{fi}|^2 \right] \left[\frac{m_0 k_1}{8\pi^3 \hbar^2} L^3 \right] \tag{75}$$

where the first bracketed term represents the incident flux and the last bracketed term the density of states. A factor $1/N$ is introduced to imply that $d\sigma/d\Omega$ is the differential cross-section per atom (N is the number of atoms in the crystal). For elastic scattering $|\mathbf{k}_0| = |\mathbf{k}_1|$ and one obtains

$$\frac{d\sigma}{d\Omega} = \frac{1}{N} a^2 \left| \sum_n \exp\left[i\mathbf{k} \cdot \mathbf{r}_n\right] \right|^2 \tag{76}$$

For a simple cubic lattice, the lattice points are given by

$$\mathbf{r}_n = d\left[n_1 \hat{\mathbf{x}} + n_2 \hat{\mathbf{y}} + n_3 \hat{\mathbf{z}} \right]$$

where $n_\alpha = 0, 1, 2, ..., N_\alpha - 1;\ \alpha = 1, 2, 3,...$

i.e., the crystal consists of N_1, N_2 and N_3 atoms along the x–, y– and z-axes, respectively and $N = N_1 N_2 N_3$ represents the total number of atoms. The origin is chosen at one of the corners of the crystal. Thus,

$$\Delta = \left|\sum_n \exp[i\mathbf{k}\cdot\mathbf{r}_n]\right|^2$$

$$= \left|\sum_{n_1=0}^{N_1-1}\sum_{n_2=0}^{N_2-1}\sum_{n_3=0}^{N_3-1} \exp\left[id(n_1k_x + n_2k_y + n_3k_z)\right]\right|^2$$

$$= \Delta_1\,\Delta_2\,\Delta_3 \tag{77}$$

where
$$\Delta_1 = \left|\sum_{n_1=0}^{N_1-1} e^{idn_1k_x}\right|^2 = \left|\frac{1-e^{iN_1k_xd}}{1-e^{ik_xd}}\right|^2$$

$$= \frac{\sin^2\dfrac{N_1k_xd}{2}}{\sin^2\dfrac{k_xd}{2}} \tag{78}$$

For large values of N_1 ($\approx 10^8$ in a typical crystal) the value of Δ_1 will be large ($= N_1^2$) if

$$\frac{k_xd}{2} = l\pi$$

l being an integer. If $k_xd/2\pi$ differs from an integer by more than a few times $1/N_1$ then Δ_1 will be negligible. Thus, one obtains maximum scattering when

$$\left.\begin{aligned} k_xd &= 2\pi l\\ k_yd &= 2\pi m\\ k_zd &= 2\pi n\end{aligned}\right\} \tag{79}$$

where l, m and n are integers (these are the Miller indices). Thus,

$$k^2 = \frac{4\pi^2}{d^2}(l^2 + m^2 + n^2) \tag{80}$$

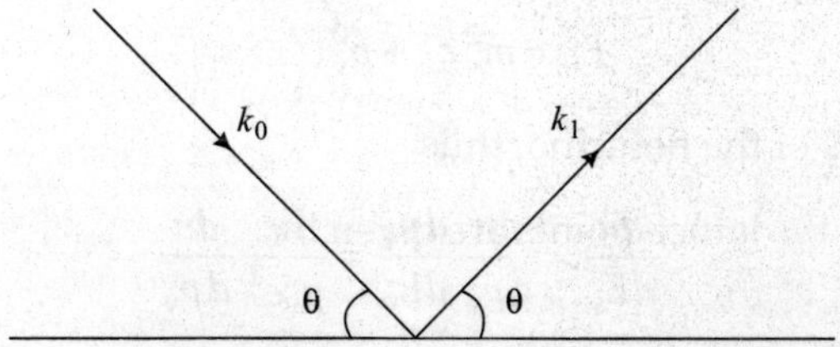

Figure 27.3: Bragg scattering of neutrons by a crystal.

Further, using Eq. (74) we get

$$k^2 = k_0^2 - 2k_0k_1 \cos^2\theta + k_1^2 = 4k_0^2 \sin^2\theta \tag{81}$$

where we have used the fact that $|\mathbf{k}_0| = |\mathbf{k}_1|$ and 2θ represents the scattering angle (see Fig. 27.3). Equating Eqs (80) and (81), we get

$$2d \sin\theta = [l^2 + m^2 + n^2]^{1/2} \lambda \tag{82}$$

where $\lambda = 2\pi/k_0$ The above equation represents the Bragg condition for reflection from a set of planes of spacing $d/(l^2 + m^2 + n^2)^{1/2}$. A more detailed theory for elastic (as well as inelastic) scattering of neutrons from a simple Bravais lattice taking into consideration lattice vibrations has been given in Ref. 4.

Solution 27.2 Since the wavelength of the leptons are much larger than nuclear dimensions ($\approx 10^{-12}$ cm), we can replace the lepton wavefunctions with their value at the center of the nucleus[5]. Now,

$$\psi_e(\mathbf{r}) = \frac{1}{\sqrt{V}} e^{i\mathbf{k}_e\cdot\mathbf{r}}$$

and

$$\psi_\nu(\mathbf{r}) = \frac{1}{\sqrt{V}} e^{i\mathbf{k}_\nu\cdot\mathbf{r}}$$

so that at $r = 0$

$$\psi_e(0) = \psi_\nu(0) = \frac{1}{\sqrt{V}}$$

Thus,

$$|H'_{fi}|^2 = \frac{G^2}{V^2} |M_{fi}|^2$$

where

$$M_{fi} = \int \psi_{fN}^* \psi_{iN}\, d\tau$$

Therefore, the probability of emission per unit time w, is

$$w = \frac{2\pi}{\hbar} \frac{G^2}{V^2} |M_{fi}|^2 \frac{dn}{dE_0}$$

where

$$E_0 = E_e + E_\nu$$

E_e and E_ν representing the energies of electron and neutrino respectively. We first calculate the probability for the electron going into the particular state; then E_e is fixed and

$$\frac{dn}{dE_0} = \frac{dn}{dE_\nu}$$

Now,

$$E_\nu^2 = m_\nu^2 c^4 + p_\nu^2 c^2$$

where m_ν is the rest mass of the neutrino, thus

$$\frac{dn}{dE_\nu} = \frac{dn}{dp_\nu}\frac{dp_\nu}{dE_\nu} = \frac{E_\nu}{p_\nu c^2}\frac{dn}{dp_\nu}$$

---

[5] We are neglecting the electrostatic distortion of the electron wavefunction caused by the charge on the nucleus; see p. 315 of Ref. 1.

In Appendix M we have shown that the number of states for which the magnitude of the momentum has between p_ν and $p_\nu + dp_\nu$ is given by

$$\frac{V}{(2\pi\hbar)^3} 4\pi p_\nu^2 d p_\nu = \frac{V p_\nu^2 \, d p_\nu}{2\pi^2 \hbar^3} \tag{83}$$

Thus,

$$\frac{dn}{dE_0} = \frac{dn}{dE_\nu} = \left(\frac{E_\nu}{p_\nu c^2}\right)\frac{V p_\nu^2}{2\pi^2 \hbar^3}$$

$$= \frac{V}{2\pi^2 \hbar^3 c^3}(E_0 - E_\nu)\left[(E_0 - E_\nu)^2 - (m_\nu c^2)^2\right]^{1/2}$$

and

$$w = \frac{G^2 \left| M_{fi} \right|^2}{\pi \hbar^4 c^3 V}(E_0 - E_\nu)\left[(E_0 - E_\nu)^2 - (m_\nu c^2)^2\right]^{1/2} \tag{84}$$

If $P(p_e)\, dp_e$ represent the probability (per unit time) of the momentum of the emitted electron to lie between p_e and $p_e + d\, p_e$, then in order to obtain $P(p_e)dp_e$ we must multiply the above expression for w by the number of electron states whose momenta lie between p_e and $p_e + dp_e$, this is given by [see Eq. (83)] $Vp_e^2 dp_e / 2\pi^2\hbar^3$. Thus,

$$P(p_e)\, dp_e = \frac{G^2 \left| M_{fi} \right|^2}{2\pi^3 \hbar^7 c^3} p_e^2 (E_0 - E_e)\left[(E_0 - E_e)^2 - (m_\nu c^2)^2\right]^{1/2} d p_e \tag{85}$$

The experimental data on the energy distribution of the emitted electrons show that the plot of $[P(p_e)/p_e^2]^{1/2}$ as a function of E_e is very nearly a straight line, which suggests that m_ν is zero or extremely small. A typical plot of $[P(P_e)/p_e^2]^{1/2}$ as a function of E_e corresponding to the $n \to p + e^- + \bar{\nu}$ process is shown in Fig. 27.4; such a plot is known as the Kurie plot. The departure from the linearity is due to experimental errors and due to the finite velocity of the neutrons which undergo decay. We may mention here that for the β decay from nuclei of large Z values one should also consider the electrostatic distortion of the electron wavefunction (see, e.g. Ref. 1).

The expression for the mean lifetime is obtained by integrating Eq. (85) over all values of p:

$$\frac{1}{\tau} = \frac{G^2 \left| M_{fi} \right|^2}{2\pi^3 \hbar^7 c^3} \int_0^{p_{\max}} p^2 \left[(m^2 c^4 + p_{\max}^2 c^2)^{1/2} - (m^2 c^4 + p^2 c^2)^{1/2}\right]^{1/2} d p$$

where p_{max} represents the maximum kinetic energy of the emitted electrons and in writing the last equation we have assumed $m_\nu = 0$. Carrying out the integration

we obtain

$$\frac{1}{\tau} = \frac{G^2 \left| M_{fi} \right|^2}{2\pi^3 \hbar^7 c^3} F(\eta_0) \tag{86}$$

where

$$F(\eta_0) = -\frac{1}{4}\eta_0 - \frac{1}{12}\eta_0^3 + \frac{1}{30}\eta_0^5 + \frac{1}{4}(1 + \eta_0^2)^{1/2} \sinh^{-1} \eta_0 \tag{87}$$

and $\eta_0 = p_{max}/mc$. Equation (86) tells us that if $|M_{fi}|^2$ does not change, the value of $F\tau$ must be a constant for different β-emitters; this is approximately borne out from experimental data (see, e.g. Refs 1 and 5).

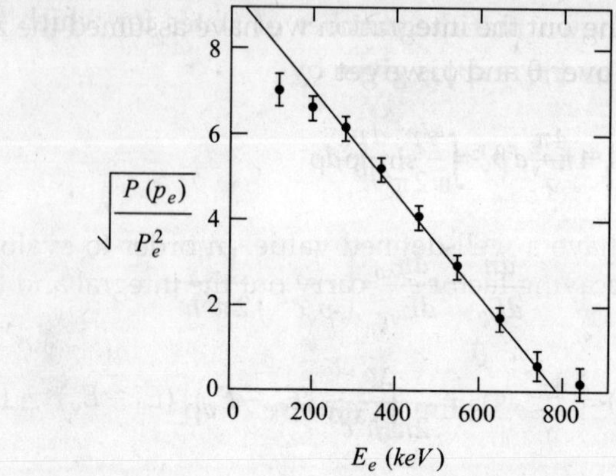

Figure 27.4: The Kurie plot for the $n \to p + e^- + \bar{\nu}$ process. The disagreement between theory and experiment at low energies is due to experimental errors and also due to the fact that neutrons form a beam rather than a solid or liquid [Adapted from Ref. 1; the original data is of I.M. Robson, *Phys. Rev.*, 83, 349 (1951)].

Solution 27.3 The initial and final states of the system, which are eigenfunctions of H_0, are given by[6]

$$\psi_i = \frac{1}{\sqrt{V}} e^{i\mathbf{k}_0\cdot\mathbf{r}_1}\,\psi_{100}(\mathbf{r}_2)$$

and
$$\psi_f = \frac{1}{\sqrt{V}} e^{i\mathbf{k}_1\cdot\mathbf{r}_1}\,\psi_{200}(\mathbf{r}_2)$$

Energy conservation demands that

$$\frac{\hbar^2 k_0^2}{2m} - \frac{me^4}{2\hbar^2} = \frac{\hbar^2 k_1^2}{2m} - \frac{me^4}{8\hbar^2}$$

or
$$k_1^2 = k_0^2 - \frac{3m^2e^4}{8\hbar^4}$$

Now, the matrix element for the transition is given by

$$H'_{fi} = \frac{1}{V}\iint e^{-i\mathbf{k}_1\cdot\mathbf{r}_1}\psi^*_{200}(\mathbf{r}_2)\left[\frac{e^2}{|\mathbf{r}_1-\mathbf{r}_2|} - \frac{e^2}{r_1}\right] e^{i\mathbf{k}_0\cdot\mathbf{r}_1}\psi_{100}(\mathbf{r}_2)\,d\mathbf{r}_1\,d\mathbf{r}_2$$

Since,
$$\int \psi^*_{200}(\mathbf{r}_2)\,\psi_{100}(\mathbf{r}_2)\,d\mathbf{r}_2 = 0$$

the integral involving e^2/r_1 would vanish. Thus,

$$H'_{fi} = \frac{e^2}{V}\int \psi^*_{200}(\mathbf{r}_2)\,\psi_{100}(\mathbf{r}_2)\,I(\mathbf{r}_2)\,d\mathbf{r}_2 \tag{88}$$

where
$$I(\mathbf{r}_2) = \int \frac{e^{i\mathbf{q}\cdot\mathbf{r}_1}}{|\mathbf{r}_1-\mathbf{r}_2|}\,d\mathbf{r}_1 = e^{i\mathbf{q}\cdot\mathbf{r}_2}\iiint \frac{e^{iq\rho\cos\theta}}{\rho}\,\rho^2\,d\rho\sin\theta\,d\theta\,d\phi$$

---
[6] This solution is adapted from Ref. 6 with kind permission of McGraw-Hill Book Company.

where $\rho = \mathbf{r}_1 - \mathbf{r}_2$ and in carrying out the integration we have assumed the z-axis to be along $\mathbf{q}$. Carrying out the integrations over θ and ϕ, we get

$$I(\mathbf{r}_2) = \frac{4\pi}{q} e^{i\mathbf{q}\cdot\mathbf{r}_2} \int_0^\infty \sin q\rho \, d\rho$$

The above integral does not have a well-defined value. In order to evaluate the integral we introduce an exponentially decaying factor $e^{-\lambda\rho}$ carry out the integral and take the limit $\lambda \to 0$ to obtain[7]

$$I(\mathbf{r}_2) = \frac{4\pi}{q} e^{i\mathbf{q}\cdot\mathbf{r}_2} \lim_{\lambda\to 0} \int_0^\infty \sin q\rho \, e^{-\lambda\rho} \, d\rho$$

$$= \frac{4\pi}{q^2} e^{i\mathbf{q}\cdot\mathbf{r}_2} \tag{89}$$

Substituting the above result and the hydrogen atom wavefunctions in Eq. (88), we get

$$H'_{fi} = \frac{e^2}{V} \int d\mathbf{r}_2 \left[\left\{ \left(\frac{1}{32\pi a_0^3} \right)^{\frac{1}{2}} \left(2 - \frac{r_2}{a_0} \right) e^{-r_2/2a_0} \right\} \times \left\{ \left(\frac{1}{\pi a_0^3} \right)^{\frac{1}{2}} e^{-r_2/a_0} \right\} \frac{4\pi}{q^2} e^{i\mathbf{q}\cdot\mathbf{r}_2} \right]$$

where $a_0 = \hbar^2/me^2$. Once again we choose the z_2-axis along $\mathbf{q}$ and carry out the straightforward integrations to obtain

$$H'_{fi} = \frac{1}{V} \frac{16\sqrt{2}\,\pi a_0^2 e^2}{\left(q^2 a_0^2 + \frac{9}{4} \right)^3}$$

The differential cross-section can now be obtained by using the formulae given in Section 27.3; however, we must remember that in calculating I [see Eq. (40)] we should use k_0 and in calculating ρ_f [see Eq. (39)] we should use k_1; consequently

$$\sigma(\theta) = \frac{k_1}{k_0} \left(\frac{m}{2\pi\hbar^2} \right)^2 V^2 \left| H'_{fi} \right|^2 \tag{90}$$

from which one immediately gets Eq. (62). A nice discussion on the total cross-section has been given Section 30 of Ref. 6. One can similarly calculate the scattering cross-section for the $1s \to 2p$ transition.

Solution 27.4 Assume that the impact parameter is d (Fig. 27.5); the collision time, the time for which the potential energy changes appreciably, is of the order of $2d/v$ where v is the velocity of the proton. The change in potential energy is of the order e^2/d. If $\hbar\omega$ is the energy required for excitation to the $n = 2$ state of the atom, Eq. (51) suggests that a transition will occur only if

---

[7] Such a procedure is justified because when $r_1 \gg r_2$

$$\frac{1}{|\mathbf{r}_1 - \mathbf{r}_2|} \to \frac{1}{r_1}\left[1 + \hat{\mathbf{r}}_1 \cdot \hat{\mathbf{r}}_2 \frac{r_2}{r_1} \right] = \frac{1}{r_1} \hat{\mathbf{r}}_1 \cdot \hat{\mathbf{r}}_2 \frac{r_2}{r_1^2}$$

and, therefore, if we first carry out the integration over $\mathbf{r}_2$ then the result will fall off as $1/r_1^2$; the integral containing the term $1/r_1$ will vanish because of orthogonality of ψ_{100} and ψ_{200}. Thus for large r_1 the integrand will fall of as $\sin(qr_1)/r_1$ making the integral convergent.

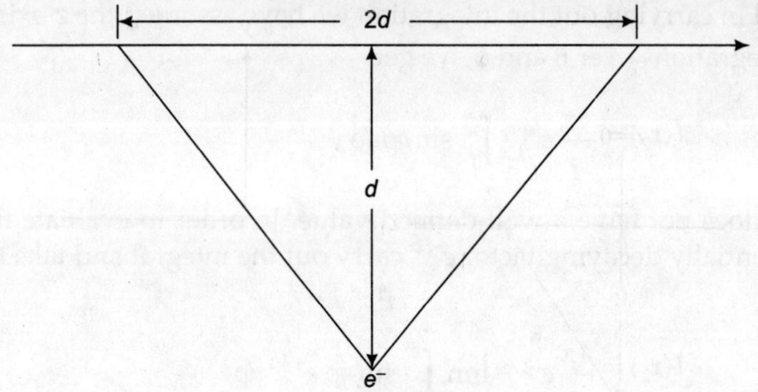

Figure 27.5

$$\frac{\partial V}{\partial t} \gtrsim \hbar\omega^2$$

or

$$\frac{ve^2}{2d^2} \gtrsim \hbar\omega^2$$

Thus,

$$2d^2 \lesssim \frac{ve^2\,\hbar}{(\hbar\omega)^2}$$

Putting $\hbar\omega \sim 10$ eV, we get the estimate

$$d \lesssim 2.5\text{Å}$$

Solution 27.5 We assume that the heavy ion of charge Z_1e, velocity v passes undeflected even as it suffers a collision with an atom (Fig. 27.6). If each atom of the medium (assumed to be of a single element) has Z_2 electrons, the potential energy is

$$V = Z_1e^2 \sum_{i=1}^{Z_2} \frac{1}{|\mathbf{r}_i - \mathbf{R}|} \tag{91}$$

Assuming $|\mathbf{r}_i| \ll |\mathbf{R}|$, we can write

$$V \approx \frac{Z_1e^2}{R} \sum_i \left(1 + \frac{r_i}{R}\hat{\mathbf{r}}_i \cdot \hat{\mathbf{R}}\right) \tag{92}$$

In this expression, we treat r_i as stationary and R (t) alone as a function of time. This is in the spirit of the sudden approximation; it can be shown that if the collision is adiabatic there is no loss of energy. We assume that at $t = 0$, the ion passes closest to the center of the atom, the impact parameter being ρ. We take the coordinates of $\mathbf{R}$ to be $(vt, \rho, 0)$, of $\mathbf{r}_i$ to be (x_i, y_i, z_i) so that

$$V = \frac{Z_1e^2}{R} \sum_i \left[1 + \frac{1}{R^2}(x_i vt + \rho y_i)\right] \tag{93}$$

We have to calculate matrix element $\langle n|V|m\rangle$ where m, n are the initial and final states of the atom. It is clear that the first term in Eq. (93) contributes nothing. We obtain

$$V_{nm} = \frac{Z_1e^2}{R^3}\left[\left(\sum x_i\right)_{nm} vt + \left(\sum y_i\right)_{nm} \rho\right] \tag{94}$$

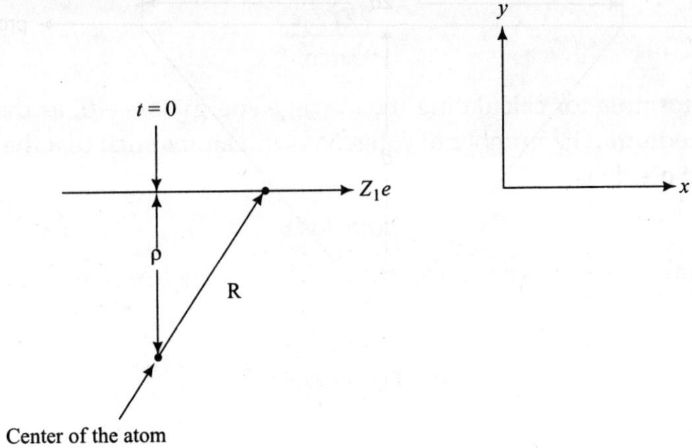

Figure 27.6: Collision of a heavy ion with an atom.

The transition probability is $|c_{nm}|^2$ where

$$c_{nm} = -\frac{i}{\hbar}\int_{-\infty}^{\infty} V_{nm}\exp(i\omega_{nm}t)dt \tag{95}$$

For $\omega_{nm}t \ll 1$, we can replace the exponential by unity. The first term in Eq. (94) being odd in t, vanishes and

$$c_{nm} = -\frac{iZ_1e^2}{\hbar}\int_{-\infty}^{\infty}\frac{\rho\left(\sum y_i\right)_{nm}}{\left(\rho^2+v^2t^2\right)^{3/2}}dt \tag{96}$$

The integral is easily evaluated by setting $vt = \rho\tan\theta$. We get

$$c_{nm} = -\frac{i}{\hbar}Z_1e^2\frac{1}{v\rho}2y_{nm}$$

where $y_{nm} = \left(\sum y_i\right)_{nm}$ is the y-component of the dipole matrix element. The transition rate is

$$|c_{nm}|^2 = \frac{4Z_1^2e^4}{\hbar^2}\cdot\frac{1}{v^2\rho^2}\cdot y_{nm}^2$$

The average energy loss by the ion in such excitations of the atom (assuming that m is always the ground state)

$$\bar{\varepsilon} = \sum_n (E_n - E_m)\left|c_{nm}\right|^2 = \frac{4Z_1^2e^4}{\hbar^2}\frac{1}{v^2\rho^2}\sum_n y_{nm}^2(E_n - E_m) \tag{97}$$

It is shown in Problem 29.9 that the term

$$\sum_n y_{nm}^2(E_n - E_m) = \frac{\hbar^2}{2m}Z_2$$

a relation which is called the Thomas-Reiche-Kuhn sum rule; Z_2 is the charge of the target atom. Hence,

$$\bar{\varepsilon} = \frac{2Z_1^2 Z_2 e^4}{mv^2\rho^2} \tag{98}$$

We now use this formula for calculating the average energy loss $-dE$ as the ion transverses a length dx of the medium. The number of collisions with atoms such that the impact parameter lies between ρ and $\rho + d\rho$ is

$$2\pi n\rho \, d\rho dx$$

where n is the number of atoms per unit volume. The energy loss in dx is

$$-dE = \int_{\rho_{min}}^{\rho_{max}} \frac{2Z_1^2 Z_2 e^4}{mv^2\rho^2} 2\pi n\rho \, d\rho \, dx$$

or

$$-\frac{dE}{dx} = \frac{4\pi Z_1^2 Z_2 e^4 n}{mv^2} \ln\left(\frac{\rho_{max}}{\rho_{min}}\right) \tag{99}$$

In order to decide on the limits ρ_{max} and ρ_{min} we use the following arguments. Referring to Eqs (95) and (96), we can write

$$c_{nm} \sim \int \frac{\exp(i\omega_{nm}t)}{(t^2 + \rho^2/v^2)^{3/2}} dt$$

so that if $\rho/v \gg 1/\omega_{nm}$, we can in fact pull out the denominator from the integral and $c_{nm} \to 0$ since we shall be averaging over several periods even for small t. This, in fact, is the adiabatic limit. We, therefore, set ρ at this limit

$$\rho_{max} = \frac{v}{\omega_{nm}}$$

It can be shown that when relativistic effects are included we should correct this to

$$\rho_{max} = \frac{v}{\omega_{nm}\sqrt{1-\beta^2}}; \quad \beta = \frac{v}{c} \tag{100}$$

For ρ_{min}, we note that the heavy ion 'sees' electrons moving towards it with momenta $\sim mv$. The de Broglie wavelength is $\sim h/mv$. This must be the minimum value of the impact parameter; when the relativistic expression for momentum is introduced,

$$\rho_{min} = \frac{\hbar\sqrt{1-\beta^2}}{mv} \tag{101}$$

Inserting Eqs (100) and (101) in Eq. (99), we get

$$-\frac{dE}{dt} = \frac{4\pi Z_1^2 Z_2 e^4 n}{mv^2} \ln \frac{mv^2}{\hbar\varpi(1-\beta^2)} \tag{102}$$

where $\hbar\varpi$ is an average of the factors $\hbar\varpi_{nm}$. Formula (102) is a simplified version of Böhr's treatment of the loss of energy of heavy ions in matter; it has important application in nuclear physics where it is often necessary to know how much energy is deposited by particles as they go through matter[8].

[8] A more complete treatment was given by H.A. Bethe, *Ann. Phys.*, **5**, 325 (1930).

Solution 27.6 [Reference: R. *Serber*, *Phys*. Rev. **72**, 1008 (1947)] We neglect the effect of Coulomb interaction and assume that the collision time $\tau = r_0/v$ where r_0 is the range of the nuclear forces and v the deuteron velocity. The transition here is from the bound deuteron to the continuum, the relevant period being about $\omega \sim \varepsilon_d/\hbar$ where ε_d is the binding energy of the deuteron. Now $E_d = \frac{1}{2}(2M)v^2$ so that

$$\omega\tau = \frac{r_0}{v}\cdot\frac{\varepsilon_d}{\hbar} = r_0\sqrt{\frac{M}{E_d}}\frac{\varepsilon_d}{\hbar}$$

The deuteron binding energy can be expressed in terms of its 'radius' $r_d = \hbar/\sqrt{M\varepsilon_d}$ so that

$$\omega\tau = \frac{r_0}{r_d}\sqrt{\frac{\varepsilon_d}{E_d}} \ll 1 \quad (\text{for } E_d \sim 200 \text{ MeV} \quad \text{and} \quad \varepsilon_d \sim 2.2 \text{ MeV})$$

which suggests that the collision may be regarded as sudden.

In the center of mass system of the deuteron, the state of the neutron changes from

$$\psi_i = \frac{\alpha}{\sqrt{2\pi}}\cdot\frac{e^{-\alpha r}}{r} \quad \text{to} \quad \psi_f \sim Ae^{i\mathbf{p}\cdot\mathbf{r}/\hbar} \tag{103}$$

where $\alpha = \sqrt{M\varepsilon_d}/\hbar$, M is the mass of the neutron, ψ_i is the wavefunction of the deuteron outside the range of the nuclear force. But, as mentioned in the problem, since the deuteron size is much larger than the range, we can treat ψ_i as valid for the entire space. ψ_f is the wavefunction for the free neutron, A, a constant of normalization. The probability for the transition is

$$P(\mathbf{p})\,d\mathbf{p} = \frac{A^2\alpha^2}{2\pi}\left|\int e^{-i\mathbf{p}\cdot\mathbf{r}/\hbar}\,\frac{e^{-\alpha r}}{r}\,d\mathbf{r}\right|^2 d\mathbf{p}$$

$$= \frac{A^2\alpha^2}{2\pi}\left|\int_0^\infty\int_0^{2\pi} e^{-ipr\cos\theta/\hbar}\,e^{-\alpha r}\,r\,dr\,2\pi\sin\theta\,d\theta\right|^2 d\mathbf{p}$$

$$= \frac{A^2\alpha^2}{2\pi}\left|4\pi\hbar\int_0^\infty dr\,\frac{e^{-\alpha r}}{p}\sin\frac{pr}{\hbar}\right|^2 d\mathbf{p} \tag{104}$$

Now, $$\int_0^\infty \sin kr\,e^{-\alpha r}\,dr = \mathrm{Im}\left[\int_0^\infty e^{ikr}\,e^{-\alpha r}\,dr\right] = \mathrm{Im}\,\frac{1}{ik-\alpha} = \frac{k}{k^2+\alpha^2}$$

Thus, $$P(\mathbf{p})\,d\mathbf{p} = B\,\frac{d\mathbf{p}}{\left(\dfrac{p^2}{\hbar^2}+\alpha^2\right)^2} \tag{105}$$

We now wish to transform the expression to the lab system in which the center of mass of the deuteron moves along the z-direction with a velocity v where

$$\frac{1}{2}(2M)v^2 = E_d$$

If we write
$$p^2 = p_z^2 + p_\perp^2 \tag{106}$$

then, to the z-component we have to add the momentum of the center of mass,

$$p_0 = Mv = \sqrt{ME_d}$$

(we neglect relativistic effects). The distribution in p_x is [from Eq. (105)]

$$P(p_z)\,dp_z = B\,dp_z \int_0^\infty \frac{p_\perp\,dp_\perp}{\left(\frac{p_\perp^2 + p_z^2}{\hbar^2} + \alpha^2\right)^2} = \frac{B'\,dp_z}{(\hbar^2\alpha^2 + p_z^2)} \tag{107}$$

We can relate $\mathbf{p}_z$ to the neutron and deuteron energies in the laboratory system. If $\mathbf{p}_n$ is the neutron momentum in the laboratory system, we have

$$\mathbf{p}_n = \mathbf{p} + \mathbf{p}_0 \quad \text{where} \quad p \ll p_0 \tag{108}$$

(The momenta $\mathbf{p}$ are the Fourier components of the deuteron wavefunction. Since the deuteron is a very loosely bound structure, these momenta are very much smaller then the kinetic energy of the deuteron.) From Eq. (108), we get

$$p_n^2 = p^2 + p_0^2 + 2p_0p_z \approx p_0^2 + 2p_0p_z$$

In terms of the neutron energy $E = p_n^2/2M$ deuteron energy $E_d = p_0^2/M$ this gives

$$E = \frac{E_d}{2} + \sqrt{\frac{E_d}{M}}\,p_z \tag{109}$$

Thus Eq. (107) gives

$$P(E)\,dE \propto \frac{dE}{\hbar^2\alpha^2 + \frac{M}{E_d}\left(E - \frac{1}{2}E_d\right)^2}$$

Writing $\hbar^2\alpha^2 = M\varepsilon_d$, we obtain

$$P(E)\,dE = C\frac{dE}{\left(E - \frac{1}{2}E_d\right)^2 + \varepsilon_d E_d} \tag{110}$$

This is a distribution which is peaked around $E = \frac{1}{2}E_d$ with a width of $\Delta E = 2\sqrt{\varepsilon_d E_d}$. Moreover, it goes to a very small value for $E = 0$ so that we can normalize it by integrating (110) from $E = -\infty$ to $+\infty$. This gives

$$1 = C\frac{\pi}{\sqrt{\varepsilon_d\,E_d}}$$

so that the normalized distribution is

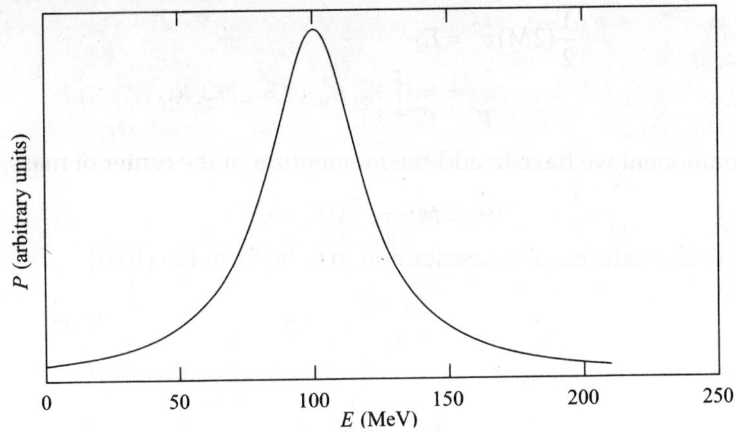

Figure 27.7: The variation of P (E) with E.

$$P(E)\,dE = \frac{\sqrt{\varepsilon_d E_d}}{\pi} \cdot \frac{dE}{\left(E - \frac{1}{2}E_d\right)^2 + \varepsilon_d E_d}$$

The experimental distribution for E_d = 200 MeV agrees very closely with this expression with the peak at 95 MeV and a width of about 45 MeV. The above theory predicts a width of $2\sqrt{\varepsilon_d E_d} \approx 42$ MeV (see Fig. 27.7).

Solution 27.7 The probability amplitude is

$$\langle \phi_f | \psi_i \rangle = \frac{\langle \phi_f | E_f - E_i | \psi_i \rangle}{E_f - E_i} = \frac{\langle \phi_f | H_1 - H_0 | \psi_i \rangle}{E_f - E_i}$$

If $V = H_1 - H_0$ is small, we may replace $\langle \phi_f |$ by $\langle \psi_f |$ so that

$$|\langle \phi_f | \psi_i \rangle|^2 \approx \frac{|V_{fi}|^2}{\hbar^2 \omega_{fi}^2}$$

Solution 27.8 That the sudden approximation holds may be seen on comparing the beta energy with the atomic binding energies. The change of the nuclear charge, hence the potential occurs in approximately the time that the beta particle leaves the atom. This time is of the order of $10^{-8}/(c/5) \sim 10^{-18}$ sec; $\omega \sim 10^{17}$ sec^{-1} so that $\omega\tau \sim 0.1$. The initial wavefunction is

$$Y_{00}R_{10}(Z=1) = \frac{1}{\sqrt{4\pi}}\left(\frac{1}{a_0}\right)^{3/2} 2e^{-r/a_0}$$

The final wavefunction is

$$Y_{00}R_{10}(Z=2) = \frac{1}{\sqrt{4\pi}}\left(\frac{2}{a_0}\right)^{3/2} 2e^{-2r/a_0}$$

The probability is

$$P = \left| \int Y_{00}^* R_{10}^* (Z=2) Y_{00} R_{10} (Z=1) d\tau \right|^2$$

The integral is

$$\frac{2^{3/2}\,4}{a_0^3} \int_0^\infty e^{-3r/a_0} r^2 \, dr = \frac{2^{3/2}\,4}{a_0^2} \frac{2!}{(3/a_0)^3}$$

or

$$P = \left(\frac{8}{9}\right)^3 \approx 0.703$$

Solution 27.9 The perturbation is

$$V = -eEz \qquad \text{(assuming } E \text{ is along } z\text{)}$$

i.e.,

$$V = -eEr\cos\theta \tag{111}$$

The transition probability is

$$\left| C_{n0} \right|^2 = \frac{1}{\hbar^2} \left| \int_{-\infty}^{\infty} \langle n|V|0\rangle e^{i\omega t} \, dt \right|^2 \tag{112}$$

where $|0\rangle$ is the ground state; from Eq. (111) it is obvious that there are no transitions to the states $m = \pm 1$ of $2P$. The matrix element $\langle 210|r\cos\theta|100\rangle$ is easily evaluated as $\left(2^7\sqrt{2}\,a_0/3^5\right)$. This gives

$$|C|^2 = \frac{1}{\hbar^2} \frac{2^{15}}{3^{10}} e^2 a_0^2 E_0^2 \frac{\tau^2}{\pi^2} \left| \int_{-\infty}^{\infty} \frac{e^{i\omega t}}{\tau^2 + t^2} dt \right|^2$$

The integral can be evaluated by going to the complex t plane on which poles occur at $t = \pm i\tau$. Choosing the contour as the real axis (Fig. 27.8) closed by the upper half semicircle, it is easy to check that the contribution over the semicircle vanishes. Thus,

$$\int_{-\infty}^{\infty} \frac{e^{i\omega t}}{\tau^2 + t^2} dt = 2i\pi \left(\frac{(i - i\tau)e^{i\omega t}}{t^2 + \tau^2} \right)_{t=i\tau} = \frac{\pi}{\tau} e^{-\omega\tau}$$

Thus the probability is

$$|C|^2 = \frac{1}{\hbar^2} \frac{2^{15}}{3^{10}} e^2 a_0^2 E_0^2 e^{-2\omega\tau}$$

The limit $\tau \to \infty$ corresponds to a turning on the perturbation slowly, i.e., $\omega\tau \gg 1$, when the transition probability vanishes. The other limit $\omega\tau \to 0$ corresponds to the application of an impulsive perturbation, with

$$\lim_{\tau \to 0} \frac{\tau}{\pi(t^2 + \tau^2)} = \delta(t).$$

Solution 27.10 Since $|n\rangle$ form a complete set, we may write.

$$|\Psi(t)\rangle = \sum_n c_n(t)\,|n\rangle$$

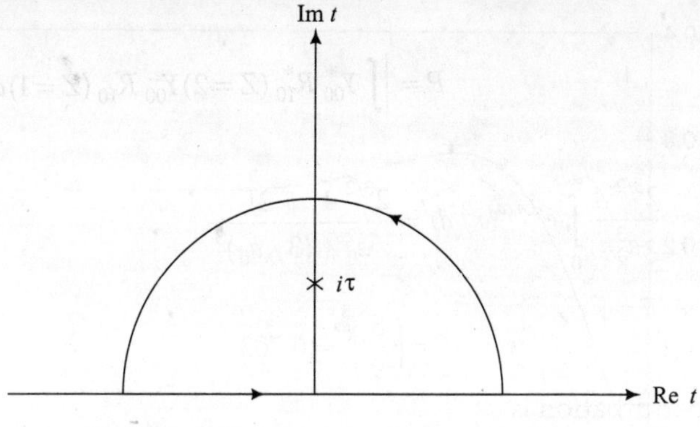

Figure 27.8: The contour of integration.

If we substitute the above expression in the Schrödinger equation

$$i\hbar\frac{\partial}{\partial t}|\Psi(t)\rangle = H|\Psi\rangle$$

and operate by $\langle n'|$, we would get Eq. (64).

Solution 27.11 Since $|\Psi(0)\rangle = |K^\circ\rangle, C_1 = C_2 = 1/\sqrt{2}$, thus,

$$|\Psi(t)\rangle = \frac{1}{\sqrt{2}}\left[\frac{1}{\sqrt{2}}\left(|K^\circ\rangle + |\bar{K}^\circ\rangle\right)e^{-iE_1t/\hbar}\,e^{-t/2\tau_1} + \frac{1}{\sqrt{2}}\left(|K^\circ\rangle - |\bar{K}^\circ\rangle\right)e^{-iE_2t/\hbar}\right] \quad (113)$$

Thus the probability that starting from a K° particle at $t = 0$ we end up with $\bar{K}^\circ$ particle at time t is given by

$$|C|^2 = \left|\langle \bar{K}^\circ|\Psi(t)\rangle\right|^2 = \frac{1}{4}\left[1 + e^{-t/\tau_1} - 2e^{-t/2\tau_1}\cos(E_2 - E_1)t/\hbar\right] \quad (114)$$

If the particles are at rest, then $E_2 - E_1 = c^2\Delta m$ where Δm is the mass difference of K_1° and K_2°. In Eq. (114) if we substitute

$$E_2 - E_1 \approx c^2\Delta m \quad (115)$$

then the mass difference can be determined from the variation of $|C|^2$ with time (see Fig. 27.9).

The behaviour of the neutral K is analogous to that of polarized light. To see this, we first summarize the results of this problem. First, at production the neutral K, produced in strong interactions, is in the state K°. As time passes by, the K° which, according to Eqs (65) and (66), is an equal mixture of K_1° and K_2° decays; but it is only the K_1° part that decays–the K_2° has a long lifetime. The residue, which is practically pure K_2° is, according to Eqs (65) and (66) again a mixture of K° and $\bar{K}^\circ$. We can detect either component by setting up an appropriate detector. In this case, the difference in the behavior of K° and $\bar{K}^\circ$ is in their strong interactions, and we, therefore, study a strong reaction induced by the neutral K.

One can construct an analogy identifying the K° and $\bar{K}_2^\circ$ with the two plane polarized beams of light, one of which is selected by a polarizer. The K_1° and $\bar{K}^\circ$ then correspond to right and left circularly polarized light, with an absorber which selectively absorbs only the right circularly polarized light.

Returning to Eq. (114), if we substitute experimental values one finds

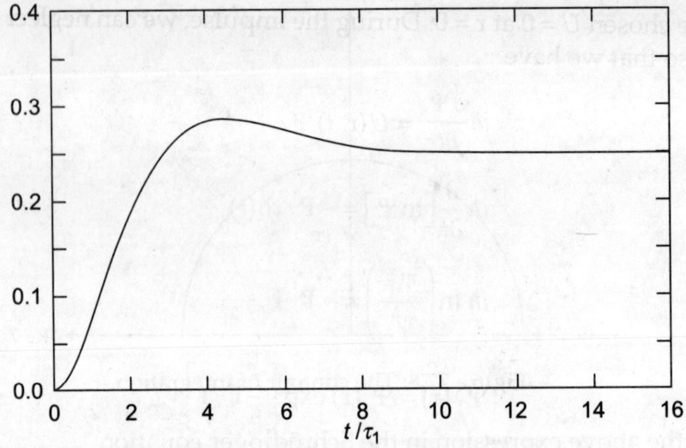

Figure 27.9: The dependence of $|C|^2$ with time.

$$\frac{E_2 - E_1}{\hbar} \approx \frac{1}{2\tau_1}$$

Thus,

$$|C|^2 \approx \frac{1}{4}\left[1 + e^{-t/\tau_1} - 2e^{-t/2\tau_1}\cos(t/2\tau_1)\right] \tag{116}$$

The time dependence of $|C|^2$ is shown in Fig. 27.9 which is in agreement with experimental data. Experimentally, since K° and $\overline{K}^\circ$ are both uncharged particles they leave no tracks; they are distinguished from the fact that K° does not interact with matter (except elastically) whereas $\overline{K}^\circ$ can interact with proton to produce Λ° and π^+

$$\overline{K}^\circ + p \rightarrow \Lambda^\circ + \pi^+ \tag{117}$$

Now in a hydrogen bubble chamber, K° is produced through the following reaction

$$\pi^- + p \rightarrow \Lambda^\circ + K^\circ \tag{118}$$

The trajectory of Λ° can be inferred from its subsequent decay to $\pi^- + p$ because both π^- and proton leave tracks. When K° is produced in the bubble chamber [see Eq. (118)] both K_1° and K_2° are created [see Eqs (65) and (66)]; however, K_1° decays very quickly (either to $\pi^+ + \pi^-$ or $\pi^\circ + \pi^\circ$). K_2° has a large $\overline{K}^\circ$ component and the $\overline{K}^\circ$ interacts with hydrogen through the reaction given by Eq. (117). Indeed analysis of the experiment allows one to determine the difference. We conclude this problem by noting that conversion of K° to $\overline{K}^\circ$ is a beautiful demonstration of quantum interference effects.

Solution 27.12

(a) The Schrödinger equation for the particle is

$$i\hbar\frac{\partial \Psi}{\partial t} = (T + V + U)\Psi \tag{119}$$

where T, V are the kinetic and potential energy operators without the impulse and U represents the energy operator due to the impulse;

$$-\nabla U = \mathbf{F}(t)$$

Integrating, we obtain

$$U = -\mathbf{F} \cdot \mathbf{r} = -\mathbf{P} \cdot \mathbf{r}\delta(t)$$

where we have chosen $U = 0$ at $\mathbf{r} = 0$. During the impulse, we can neglect T, V in comparison to U in Eq. (119) so that we have

$$i\hbar\frac{\partial \Psi}{\partial t} = U(\mathbf{r}, t)\,\Psi$$

or

$$i\hbar\frac{\partial}{\partial t}\left[\ln\Psi\right] = -\,\mathbf{P} \cdot \mathbf{r}\delta(t)$$

Integrating,

$$i\hbar \ln\left(\frac{\Psi_f}{\Psi_i}\right) = -\,\mathbf{P} \cdot \mathbf{r}$$

or

$$\Psi_f(\mathbf{r}) = \Psi_i(\mathbf{r}) \exp\left[\frac{i}{\hbar}\mathbf{P} \cdot \mathbf{r}\right] \tag{120}$$

(b) The final state given by Eq. (120) is not one of the eigenstates of the system. We can expand Ψ_f in terms of these eigenstates $\Psi_n(\mathbf{r})$

$$\Psi_f = \Psi_i \exp\left[\frac{i}{\hbar}\mathbf{P} \cdot \mathbf{r}\right] = \sum_n C_n \Psi_n(\mathbf{r})$$

Since Ψ_i, Ψ_n are eigenstates, we can use orthonormality to obtain the coefficients

$$C_n = \int \Psi_n^*(\mathbf{r})\Psi_i(\mathbf{r})\exp\left[\frac{i}{\hbar}\mathbf{P} \cdot \mathbf{r}\right]d\mathbf{r}$$

$|C_n|^2$ is the probability that the system is in the eigenstate n after the impulse. The above results have a ready application for the so called 'Mossbauer effect'. Here, the nucleus of an atom in a crystal lattice emits a γ-ray, thus delivering an impulse to the crystal. If the crystal is originally in the ground state of its vibrational modes, we can calculate, as above, the probability that it remains in the ground state after the γ-emission. In this situation, it is not the atom, which emits γ, which will vibrate (since this will leave the crystal in a different state) but the crystal as a whole will recoil without changing its internal state. The probability

$$W = |C_i|^2 = \int \Psi_i^* \exp\left[\frac{i}{\hbar}\mathbf{P} \cdot \mathbf{r}\right]\Psi_i\, d\mathbf{r} \tag{121}$$

for such an emission is called the *Debye-Waller factor*. If the system we are considering is, for example, a linear harmonic oscillator, the ground state is

$$\Psi_0(x) = \left(\frac{1}{\sqrt{\pi}\, a}\right)^{1/2} \exp\left(-\frac{x^2}{2a^2}\right)$$

where a is the classical amplitude. For this one-dimensional case we have, in place of Eq. (121)

$$W = \exp\,(-k^2a^2) \tag{122}$$

where $\hbar k = P$. The recoil energy is

$$E_R = \frac{P^2}{2M} = \frac{\hbar^2 k^2}{2M}$$

where M is the mass of the entire recoiling system. For a crystal, M is very large and the energy E_R is negligible. Thus the γ-ray emitted by the excited nucleus appears with the full energy, i.e., the energy difference between the excited and ground states of the nucleus. For a fuller account of Mossbauer effect, the reader is referred to Ref. 7.

Solution 27.13 We have the two reactions

$$A + a \rightarrow B + b \tag{123}$$

and

$$B + b \rightarrow A + a \tag{124}$$

We assume that the matrix element for the processes

$$(\mathbf{p}_A, \mathbf{p}_a, \mathbf{s}_A, \mathbf{s}_a) \rightarrow (\mathbf{p}_B, \mathbf{p}_b, \mathbf{s}_B, \mathbf{s}_b)$$

and

$$(-\mathbf{p}_B, -\mathbf{p}_b, -\mathbf{s}_B, -\mathbf{s}_b) \rightarrow (-\mathbf{p}_A, -\mathbf{p}_a, -\mathbf{s}_A, -\mathbf{s}_a)$$

are the same (this is referred to as the invariance under time reversal). We further assume that the initial beams are unpolarized so that in calculating the cross- section we must sum over the final spin states and average over the initial spin states. Thus, for the reaction given by Eq. (123), instead of $|V|^2$ in Eq. (24) we must have

$$\frac{1}{(2s_A+1)(2s_a+1)} \sum |V|^2$$

where the summation is over all the spin states. Similarly, for the reaction given by Eq. (124) we must have

$$\frac{1}{(2s_B+1)(2s_b+1)} \sum |V|^2$$

We next calculate the density of states. The number of states with magnitude of momentum lying between p and $p + dp$ is given by [see Eq. (8) of Appendix P]

$$d_n = \frac{4\pi p^2 \, dp \, V}{(2\pi\hbar)^3} \tag{125}$$

or

$$\frac{dn}{dp} = \frac{V p^2}{2\pi^2 \hbar^3}$$

For the reaction given by Eq. (123) $p = |\mathbf{p}_B| = |\mathbf{p}_b|$ and if E_B and E_b represent the total particle energies then[9]

9

$$E^2 = c^2 p^2 + m^2 c^4$$

Therefore,

$$E \frac{dE}{dp} = c^2 p$$

or

$$\frac{dE}{dp} = \frac{c^2 p}{E} = \frac{c^2 (m_0 v \gamma)}{(m_0 c^2 \gamma)} = v$$

where

$$\gamma = \left(1 - \frac{v^2}{c^2}\right)^{-1/2}$$

$$\frac{dE}{dp} = \frac{dE_B}{dp} + \frac{dE_b}{dp} = v_{Bb}$$

where v_{Bb} is the relative velocity in the final state. Thus, the density of states is given by

$$\frac{dn}{dE} = \frac{V p^2}{2\pi^2 \hbar^3 v_{Bb}} \quad (126)$$

Substituting in the expression for the cross-section [Eq. (43)] and using Eq. (126), we get

$$\sigma_{A+a\to B+b} \propto \frac{|p_B|^2 V}{v_{aA}\, v_{bB}} \cdot \frac{1}{(2s_A+1)(2s_a+1)} \sum |V|^2 \quad (127)$$

Similarly,

$$\sigma_{B+b\to A+a} \propto \frac{|p_A|^2 V}{v_{bB}\, v_{aA}} \frac{1}{(2s_B+1)(2s_b+1)} \sum |V|^2 \quad (128)$$

(we are assuming the same center of mass energies). Thus,

$$\frac{\sigma_{A+a\to B+b}}{\sigma_{B+b\to A+a}} = \frac{|p_B|^2}{|p_A|^2} \frac{(2s_B+1)(2s_b+1)}{(2s_A+1)(2s_a+1)} \quad (129)$$

where $|\mathbf{p}_A|$ $(= |\mathbf{p}_a|)$ refers to the reaction given by Eq. (129).

Solution 27.14 Using the results of the previous problem we get (at any scattering angle θ)

$$\sigma_{pp\to\pi^+d} = \sigma_{\pi^+d\to pp} \frac{3}{4}(2s_\pi+1)\left(\frac{p_\pi}{p_p}\right)^2 \quad (130)$$

where we have used the facts that $S_d = 1$ and $s_p = \frac{1}{2}$. If we integrate over all angles, we must multiply by a factor of 2 to take into account the indistinguishability of protons in the final state of the $\pi^+ d \to pp$ reaction. Then,

$$\sigma_{pp\to\pi^+d} = \sigma_{\pi^+d\to pp} \frac{3}{2}(2s_\pi+1)\left(\frac{p_\pi}{p_p}\right)^2 \quad (131)$$

Solution 27.15 In the experiment of Cartwright et al. the incident proton energy of 340 MeV corresponds to a meson energy of 22.3 MeV in the cms (see Section 26.8). On the other hand in the experiment of Durbin et al. the incident meson energy of 29 MeV corresponds to meson energy of 25 MeV in the cms. Thus, both the experiments correspond approximately to the same center of mass energies. Further, since the energies involved are small compared to the rest mass energies we may use the non-relativistic expressions giving

$$\frac{p_\pi}{p_p} = \sqrt{\frac{2m_\pi E_\pi}{2m_p E_p}} \approx 0.20$$

Substituting in Eq. (131), we get

$$(0.18 \pm 0.06) = (3.1 \pm 0.3) \times 1.5 \times 0.04 \times (2s_\pi + 1)$$

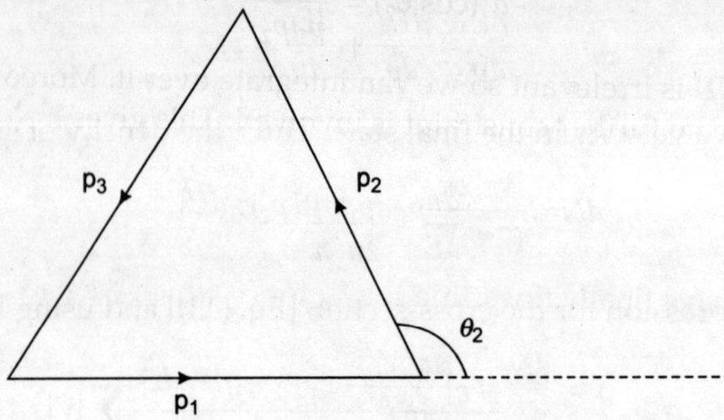

Figure 27.10

giving

$$2s_\pi + 1 \approx 1$$

or the spin of the π-meson is zero. For further details on the spin and parity of elementary particles see Ref. 3.

Solution 27.16 This problem is really an exercise in the calculation of the density of states. Let E be the rest energy of the K^+. We then have the constraints:

$$E = \varepsilon_1 + \varepsilon_2 + \varepsilon_3$$
$$0 = \mathbf{p}_1 + \mathbf{p}_2 + \mathbf{p}_3 \qquad (132)$$

Let us assume that $\mathbf{p}_1$ is held fixed. We now calculate the number of final states with this constraint. (Throughout these calculations we shall use the simple rule that the number of states is $\frac{1}{(2\pi\hbar)^3} \times$ phase volume.)

We are interested in the number of states dN such that the energies of two of the pions are fixed, within infinitesimal intervals, and the remaining parameter varied. If we fix $\mathbf{p}_1$ and $\mathbf{p}_2$, it is clear that energy conservation will automatically fix the magnitude of $\mathbf{p}_3$ (see Fig. 27.10). Looking at the diagram, we see that this also fixes θ_2. Now the number dN is

$$dN = \frac{p_1^2\, dp_1 d\Omega_1}{(2\pi\hbar)^3} \frac{p_2^2\, dp_2 d\Omega_2}{(2\pi\hbar)^3} V^2 \qquad (133)$$

where V is the volume of the box in which the event takes place and $d\Omega_1$ and $d\Omega_2$ are solid angle intervals within which $\mathbf{p}_1$ and $\mathbf{p}_2$ lie. Once we consider $\mathbf{p}_1$, $\mathbf{p}_2$ fixed, we merely have to count dN as the number of states within the energy interval dE, a variation which is related directly to the interval $(d\Omega_2)$. To get this relation we use momentum conservation

$$p_1^2 + p_2^2 + 2p_1 p_2 \cos\theta_2 = p_3^2 \qquad (134)$$

Since $p_i^2 + m_i^2 = \varepsilon_i^2$ (we have used $c = 1$) we can write Eq. (134) as

$$p_1^2 + p_2^2 + 2p_1 p_2 \cos\theta_2 = \varepsilon_3^2 - m_3^2$$

or

$$p_1^2 + p_2^2 + 2p_1 p_2 \cos\theta_2 = (E - \varepsilon_1 - \varepsilon_2)^2 - m_3^2 \qquad (135)$$

Since we keep $p_1, p_2, \varepsilon_1, \varepsilon_2$, fixed, we get

$$2p_1 p_2\, d(\cos\theta_2) = 2(E - \varepsilon_1 - \varepsilon_2)\, dE = 2\varepsilon_3\, dE$$

or
$$d(\cos\theta_2) = \frac{\varepsilon_3\, dE}{p_1 p_2} \tag{136}$$

The orientation ϕ_2 in $d\Omega_2$ is irrelevant so we can integrate over it. Moreover, we can now also integrate over $d\Omega$, since no further constraint is involved. This gives

$$dN = \frac{4\pi.2\pi}{(2\pi\hbar)^6} p_1\, dp_1\, p_2 d\, p_2 \varepsilon_3 dEV^2 \tag{137}$$

Since $p_i\, dp_i = \varepsilon_i\, d\varepsilon_i$, we have finally (inserting the factors c which we had dropped)

$$\rho = \frac{dN}{dE} = \frac{8\pi^2 V^2}{(2\pi\hbar)^6 c^6} \varepsilon_1\, \varepsilon_2\, \varepsilon_3\, d\varepsilon_1\, d\varepsilon_2 \tag{138}$$

The transition probability is, therefore,

$$\Gamma = \frac{2\pi}{\hbar}|M|^2 \rho \sim d\varepsilon_1\, d\varepsilon_2 = dt_1\, dt_2 \tag{139}$$

where t_1 and t_2 are kinetic energies of any two pions. The energy distribution of the π^- is obtained by integrating over the energy of the other. But Eq. (139) tells us that there must be an equal number of π^- emitted per unit energy for the entire energy range allowed by energy momentum constraint.

It can be shown that if the pions are emitted in the $l = 0$ state (low energy) and the interaction of the pions in the final state can be neglected, (which means that they can be represented by plane waves) then the matrix element will have the simple dependence M ~1 $h(\varepsilon_1\, \varepsilon_2\, \varepsilon_3)^{1/2}$ as assumed. This means that any deviation from the prediction Eq. (139), which can be called the 'phase space spectrum', may be attributed to some interaction of the pions in the final state. In the extreme case where two of the pions really are the products of the *decay* of a particle which emerged from the original reaction, there would be a very strong deviation from the phase space spectrum. As an example, we may have

$$\pi^+ + d \to \omega + p + p$$
$$\qquad \llcorner\!\!\to (\pi^+ + \pi^- + \pi^\circ)$$

In a bubble chamber, only the 3 pions are seen but from their energy spectrum one can infer the existence and properties of ω. A very short-lived particle such as the ω is sometimes referred to as a resonance.

REFERENCES AND SUGGESTED READINGS

1. H. Enge, *Introduction to Nuclear Physics*, Addison-Wesley Publishing Co., Reading, Massachusetts (1966).
2. E. Fermi, *Quantum Theory of Radiation*, Reviews of Modern Physics, **4**, 87 (1932).
3. L.S. Hughes, *Elementary Particles*, Penguin, London (1972).
4. A.K. Ghatak and L.S. Kothari, *An Introduction to Lattice Dynamics*, Addison-Wesley Publishing Co., Reading, Massachusetts (1972).
5. M. Börn, *Physics in my Generation*, Springer Verlag, New york (1962).
6. L.I. Schiff, *Quantum Mechanics*, McGraw-Hill Book Co., New York (1955).
7. W.E. Lamb, *Lectures on Mossbauer Effect*, Tata Institute of Fundamental Research, Bombay (1960).

The semi-classical theory of radiation and the Einstein coefficients

28.1 INTRODUCTION

28 The semi-classical theory of radiation and the Einstein coefficients

If we have an atom that is in an excited state and so is going to emit a photon, we cannot say **when** *it will emit the photon. It has a certain amplitude to emit the photon at any time, and we can predict only a probability for emission; we cannot predict the future exactly. This has given rise to all kinds of nonsense and questions on the meaning of freedom of will, and of the idea that the world is uncertain.*

— R.P. FEYNMAN in *Feynman Lectures on Physics*, Vol. Ill, p. 1422.

28.1 INTRODUCTION

In the previous chapter, we developed time dependent perturbation theory which we will now use to study the interaction of an atom with electromagnetic field. The theory presented in this chapter is a curious mixture of classical and quantum pictures. Matter, i.e., atoms, are assumed to exist in discrete (quasi-stationary) states. But the electromagnetic radiation with which it interacts, is described classically. This semi-classical theory of atom-field interaction will be discussed in Section 28.3 which will enable us to calculate probabilities of absorption and induced emission. It may be mentioned that the semi-classical theory is insufficient to describe the spontaneous emission of radiation because the electromagnetic field is described classically; the fully quantum theory of radiation, which we will discuss in the next chapter, describes the spontaneous emission automatically.

An ingenious idea of Einstein enabled him to obtain a relation which would determine the rate at which spontaneous emissions take place. The idea was based on consideration of matter in equilibrium with radiation assuming that the resultant radiation must obey Planck's law. Einstein introduced coefficients, which are now known as Einstein coefficients, which described induced emission, absorption and spontaneous emission. In Section 28.2, we will give the original argument of Einstein (Ref. 1) which gave rise to the relation between different coefficients. In Section 28.3, we will discuss the atom-field interaction using the semi-classical

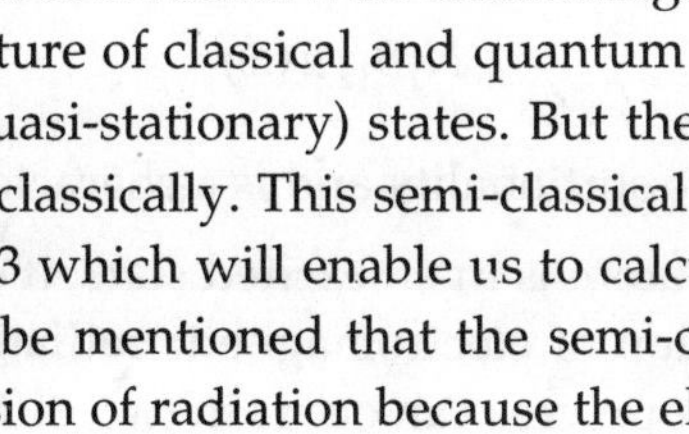

Figure 28.1: The two states of an atomic system. The transition frequency is $\omega_{21} = (E_2 - E_1)/\hbar$.

theory and will derive expressions for stimulated emission rate and stimulated absorption rate. In Section 28.4 we will use the Einstein relations to calculate the spontaneous emission rate and hence the lifetime of atomic states. In Section 28.5, we will discuss the selection rules governing the transitions.

28.2 THE EINSTEIN COEFFICIENTS

Figure 28.1 represents two of the energy levels of an atomic system corresponding to energies E_1 and E_2. Let N_1 and N_2 represent the number of atoms (per unit volume) in levels 1 and 2 respectively. An atom in the lower energy level can absorb radiation and get excited to the level E_2. This excitation process can occur only in the presence of radiation. Such a process is known as *stimulated absorption* or simply as absorption. The rate of absorption would depend on the energy density, $u(\omega)$, associated with the radiation field corresponding to the frequency

$$\omega = \frac{E_2 - E_1}{\hbar} \tag{1}$$

The energy density $u(\omega)$ is defined such that $u(\omega)\, d\omega$ represents the radiation energy per unit volume within the frequency interval ω and $\omega + d\omega$. The rate of absorption would be proportional to N_1 and also to $u(\omega)$. Thus, the number of absorptions per unit time per unit volume can be written as

$$N_1 B_{12}\, u(\omega) \tag{2}$$

where B_{12} is the coefficient of proportionality and is a characteristic of the energy levels.

On the other hand, when the atom is in an excited state, it can make a transition to a lower energy state through the emission of electromagnetic radiation; however, in contrast to the absorption process, the emission process can occur in two different ways:

(i) The first is referred to as *spontaneous emission* in which an atom in the excited state emits radiation even in the absence of any incident radiation. It is thus not stimulated by any incident signal but occurs spontaneously. Further, the rate of spontaneous emissions is proportional to the number of atoms in the excited state; thus if we represent the coefficient of proportionality by A_{21} then

$$N_2 A_{21} \tag{3}$$

would represent the number of spontaneous emissions per unit volume per unit time to the lower energy level.

(ii) The second is referred to as *stimulated emission* in which an incident signal of appropriate frequency triggers an atom in an excited state to emit radiation. The rate of transition to the lower energy level is directly proportional to the energy density of the radiation at the frequency ω. Thus, the number of stimulated emissions per unit time per unit volume would be given by

$$N_2 B_{21}\, u(\omega) \tag{4}$$

The quantities A_{21}, B_{12} and B_{21} are known as Einstein coefficients and are determined by the atomic system.

At thermal equilibrium, the number of upward transitions must be equal to the number of downward transitions. Thus, we may write

$$N_1 B_{12} u(\omega) = N_2 A_{21} + N_2 B_{21} u(\omega)$$

or

$$u(\omega) = \frac{A_{21}}{\dfrac{N_1}{N_2} B_{12} - B_{21}} \tag{5}$$

From Boltzmann's law, we have the following expression for the ratio of the populations of two levels at temperature T:

$$\frac{N_1}{N_2} = \exp\left[\frac{E_2 - E_1}{k_B T}\right] = \exp\left[\frac{\hbar\omega}{k_B T}\right] \tag{6}$$

where k_B represents the Boltzmann constant. Thus, we may write

$$u(\omega) = \frac{A_{21}}{B_{12}\exp(\hbar\omega / k_B T) - B_{21}} \tag{7}$$

Now, according to Planck's law the energy density of radiation (at thermal equilibrium) is given by (see, e.g. Refs 2 and 3):

$$u(\omega) = \frac{\hbar\omega^3}{\pi^2 c^3} \cdot \frac{1}{\exp(\hbar\omega / k_B T) - 1} \tag{8}$$

Comparing Eqs (7) and (8), we obtain[1]

$$B_{12} = B_{21} = B \text{ (say)} \tag{9}$$

and

$$\frac{A_{21}}{B_{21}} = \frac{\hbar\omega^3}{\pi^2 c^3} \tag{10}$$

Thus, the probabilities of stimulated absorption and stimulated emission are the same and the ratio of the A and B coefficients is given by Eq. (10). It is of interest to mention that at thermal equilibrium corresponding to ordinary temperatures ($T \sim 1000°K$) the spontaneous emission rate for optical sources far exceeds the stimulated emission rate (see Problem 28.1).

28.3 THE ATOM-FIELD INTERACTION

In order to calculate the Einstein coefficients we consider an atom in the presence of an oscillating electric field given by[2]

$$\mathcal{E}(t) = \hat{\mathbf{e}}\, \mathcal{E}_0 \cos \omega t \tag{11}$$

[1] If the levels 1 and 2 are g_1-and g_2-fold degenerate, then $N_1/N_2 = (g_1/g_2) \exp(\hbar\omega/k_B T)$, $B_{12} = B_{21}(g_2/g_1)$ and $A_{21}/B_{21} = \hbar\omega^3/\pi^2 c^3$.

[2] The electric field associated with an electromagnetic wave will be of the form $\hat{\mathbf{e}} \cos(\omega t - \mathbf{k.r})$; however, for radiation in the visible region, the wavelength associated ($\sim 5 \times 10^{-5}$ cm) is much larger than atomic dimensions ($\sim 10^{-8}$ cm) and we can use the value of ε at $\mathbf{r} = 0$. This is immediately obvious from Eq. (17) where the atomic wavefunctions are almost zero for $r \gtrsim 10^{-8}$ cm and since $k \simeq 10^5 \text{cm}^{-1}$, the quantity $\mathbf{k.r} \ll 1$ in the domain of integration.

which is switched on at $t = 0$; $\hat{\mathbf{e}}$ represents the unit vector along the direction of the electric field and the oscillation frequency ω is assumed to be close to the resonant frequency ω_{21} [$= (E_2 - E_1)/\hbar$] corresponding to the transition from state 2 to state 1 (see Fig. 28.1). Now, the interaction energy of the electron with the electric field is given by

$$H'(\mathbf{r}, t) = q\,\mathcal{E}(t) \cdot \mathbf{r} = q\,\mathcal{E}_0\,(\hat{\mathbf{e}} \cdot \mathbf{r}) \cos \omega t \tag{12}$$

where $q(> 0)$ represents the magnitude of the electronic charge[3]. Let H_0 represent the Hamiltonian corresponding to the atomic system and let ψ_n denote the eigenfunctions of H_0 belonging to the energy E_n $(= \hbar\omega_n)$:

$$H_0\psi_n = E_n\psi_n = \hbar\omega_n\psi_n \tag{13}$$

Our objective is to solve the equation

$$i\hbar\frac{\partial \Psi(\mathbf{r}, t)}{\partial t} = H\Psi = \left[H_0 + H'(\mathbf{r}, t)\right]\Psi(\mathbf{r}, t) \tag{14}$$

Following the approach developed in Section 27.2, we write

$$\Psi(\mathbf{r}, t) = \sum_n C_n(t)e^{-i\omega_n t}\psi_n(\mathbf{r}) \tag{15}$$

and obtain [see Eq. (9) of the previous chapter]

$$i\hbar\frac{dC_s}{dt} = \sum_n C_n(t)H'_{sn}(t)e^{i\omega_{sn}t} \tag{16}$$

where

$$H'_{sn}(t) = \int \psi_s^*(\mathbf{r})H'(\mathbf{r}, t)\psi_n(\mathbf{r})d\tau = \langle s|H'|n\rangle$$

$$= \frac{1}{2}q\mathcal{E}_0\,\hat{\mathbf{e}}.\langle s|\mathbf{r}|n\rangle\left[e^{i\omega t} + e^{-i\omega t}\right] \tag{17}$$

Substituting for $H'(\mathbf{r}, t)$ from Eq. (12), we get

$$i\hbar\frac{dC_s}{dt} = \frac{1}{2}\mathcal{E}_0\sum_n D_{sn}\,C_n(t)\left[e^{i(\omega_{sn}+\omega)t} + e^{i(\omega_{sn}-\omega)t}\right] \tag{18}$$

where

$$D_{sn} = \hat{\mathbf{e}}\cdot\mathbf{P}_{sn} \tag{19}$$

and

$$\mathbf{P}_{sn} = q\int \psi_s^*(\mathbf{r})\,\mathbf{r}\,\psi_n(\mathbf{r})\,d\tau = q\langle s|\mathbf{r}|n\rangle \tag{20}$$

is known as the dipole matrix element. We assume that at $t = 0$ the atom is in the state ψ_k, *i.e.*,

$$\left.\begin{aligned} C_k(t=0) &= 1 \\ C_n(t=0) &= 0 \text{ for } n \neq k \end{aligned}\right\} \tag{21}$$

---

[3] We are considering here a single electron atom with $\mathbf{r}$ representing the position vector of the electron with respect to the nucleus. Thus the electric dipole moment of the atom is given by $\mathbf{p} = -q\mathbf{r}$ because the direction of the dipole moment is from the negative charge to the positive charge. The interaction energy of a dipole placed in an electric field is $-\mathbf{p}.\mathcal{E}$ which is precisely Eq. (12).

Equation (18) represents an infinite set of coupled equations and as a first approximation we replace $C_n(t)$ by $C_n(0)$ on the right-hand side of Eq. (18) giving

$$i\hbar\frac{dC_s}{dt} \simeq \frac{1}{2}\mathcal{E}_0 D_{sk}\left[e^{i(\omega_{sk}+\omega)t} + e^{i(\omega_{sk}-\omega)t}\right]$$

Integrating, we obtain

$$C_s(t) - C_s(0) = -\frac{i\mathcal{E}_0}{2\hbar}D_{sk}\left[\frac{e^{i(\omega_{sk}+\omega)t}-1}{i(\omega_{sk}+\omega)} + \frac{e^{i(\omega_{sk}-\omega)t}-1}{i(\omega_{sk}-\omega)}\right]$$

or, for $s \neq k$

$$C_s(t) = -\frac{i\mathcal{E}_0}{2\hbar}D_{sk}\left[e^{i(\omega_{sk}+\omega)t/2}\frac{\sin(\omega_{sk}+\omega)t/2}{(\omega_{sk}+\omega)/2} + e^{i(\omega_{sk}-\omega)t/2}\frac{\sin(\omega_{sk}-\omega)t/2}{(\omega_{sk}-\omega)/2}\right] \tag{22}$$

It can be easily seen that for large values of t, the function

$$\frac{\sin(\omega_{sk}-\omega)t/2}{(\omega_{sk}-\omega)/2} \tag{23}$$

is very sharply peaked around $\omega \simeq \omega_{sk}$ and negligible everywhere else (see Fig. 28.2). Thus for states for which ω_{sk} is significantly different from ω, $C_s(t)$ would be negligible, and transitions between suchstates will not be stimulated by the incident field. This implies that in the summation appearing on the right-hand side of Eq. (18) we need only consider those states which correspond closely to the resonance frequency.

In an emission process, $E_k > E_s$ and hence $\omega_{sk}\left(=\frac{E_s - E_k}{\hbar}\right)$ is negative; thus it is the first term on the right-hand side of Eq. (22) which contributes. On the other hand, in an absorption process, $E_k < E_s$ and hence ω_{sk} is positive and consequently it will be the second term on the right-hand side of Eq. (22) which contributes.

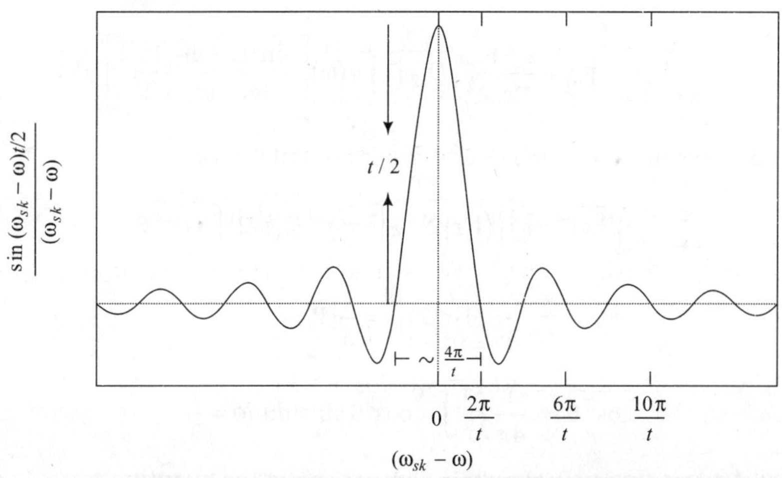

Figure 28.2: For large values of t, the function given by Eq. (23) is a very sharply peaked function of ω about $\omega = \omega_{sk}$.

Let us consider the emission of radiation and assume that at $t = 0$ the atom is in state 2 (see Fig. 28.1). We also assume ω to be close to ω_{21} $(= [E_2 - E_1/\hbar)$. The probability for the transition to state 1 is given by

$$|C_1(t)|^2 = \frac{|D_{12}|^2 \mathcal{E}_0^2}{4\hbar^2}\left[\frac{\sin\left(\frac{-\omega_{21}+\omega}{2}\right)t}{\frac{-\omega_{21}+\omega}{2}}\right]^2 \quad \text{Probability for stimulated emission} \tag{24}$$

Equation (24) represents the probability for stimulated emission of radiation. In deriving this equation, we have assumed that $|C_1(t)|^2 \ll 1$; thus the result will be valid when either

$$\left(\frac{D_{12}\,\mathcal{E}_0\,t}{\hbar}\right)^2 \ll 1 \quad \text{or} \quad \left(\frac{D_{12}\,\mathcal{E}_0}{\hbar}\right)^2 \Big/ (\omega - \omega_{21})^2 \ll 1 \tag{25}$$

A more accurate result for a two state system will be discussed in Problem 28.8.

Now, the radiation energy (per unit volume) associated with a near monochromatic field at frequency ω is given by

$$u_\omega = \frac{1}{2}\varepsilon_0\,\mathcal{E}_0^2 \tag{26}$$

We assume that there is a continuous spectrum of frequency[4] and that there is no correlation between the polarization vectors of different components (as in the case of black body radiation). Thus if $u(\omega)\,d\omega$ represents the radiation energy per unit volume in the frequency interval ω and $\omega + d\omega$, then we may replace [in Eq. (24)]

$$\mathcal{E}_0^2 \to \frac{2}{\varepsilon_0}u(\omega)\,d\omega \tag{27}$$

and integrate the expression over all frequencies to obtain the following expression for the transition probability

$$\Gamma_{21} = \frac{1}{2\varepsilon_0\hbar^2}\overline{|D_{12}|^2}\int u(\omega)\left[\frac{\sin(\omega-\omega_{21})t/2}{(\omega-\omega_{21})/2}\right]^2 d\omega \tag{28}$$

where the bar denotes the averaging over different orientations:

$$\overline{|D_{12}|^2} = q^2\,\overline{\left|\langle 1|\mathbf{r}|2\rangle\cdot\hat{\mathbf{e}}\right|^2} = q^2\left|\langle 1|\mathbf{r}|2\rangle\right|^2\overline{\cos^2\theta}$$

$$= \frac{q^2}{3}\left|\langle 1|\mathbf{r}|2\rangle\right|^2 = \frac{1}{3}|\mathbf{P}_{12}|^2$$

because

$$\overline{\cos^2\theta} = \frac{1}{4\pi}\int_0^\pi\int_0^{2\pi}\cos^2\theta\,\sin\theta\,d\theta\,d\phi = \frac{1}{3} \tag{29}$$

Assuming that $u(\omega)$ varies much slowly in comparison to the quantity

---

[4] One could also have an interaction of a near monochromatic wave (as in a laser) with an atom having a broad frequency spectrum. This is discussed in Problem 28.10.

$$\left[\frac{\sin(\omega-\omega_{21})t/2}{(\omega-\omega_{21})/2}\right]^2 \tag{30}$$

we replace $u(\omega)$ by its value at $\omega = \omega_{21}$ and take it out of the integral (in Eq. 28) to obtain

$$\Gamma_{21} \simeq \frac{1}{6\varepsilon_0\hbar^2}|\mathbf{P}_{12}|^2 u(\omega_{21})G \tag{31}$$

where $$G \simeq \int_{-\infty}^{+\infty}\left[\frac{\sin(\omega_{21}-\omega)t/2}{(\omega_{21}-\omega)/2}\right]^2 d\omega = 2t\int_{-\infty}^{+\infty}\frac{\sin^2\xi}{\xi^2}d\xi = 2\pi t$$

Thus,

$$\Gamma_{21} \simeq \frac{\pi}{3\varepsilon_0}\frac{|\mathbf{P}_{12}|^2}{\hbar^2}u(\omega_{21})t = \beta t \quad \text{(say) Transition probability} \tag{32}$$

The above expression shows that the probability of transition is proportional to time with $\beta\left(=\frac{\pi}{3\varepsilon_0\hbar^2}|\mathbf{P}_{12}|^2 u(\omega_{21})\right)$ representing the proportionality constant.

This immediately implies the radioactive decay law, because if there are $N_2(t)$ atoms (per unit volume) in state 2 at time t and if $-dN_2$ represents the number of transitions (per unit volume) in time dt then according to Eq. (32)

$$-\frac{dN_2}{N_2} = \beta\, dt \tag{33}$$

the negative sign implying that N_2 will decrease with times. Equation (33) immediately leads to

$$N_2(t) = N_2(0)e^{-\beta t} \tag{34}$$

which is the radioactive decay law with $1/\beta$ representing the mean lifetime of the state.

It may be noted that Eq. (32) predicts an indefinite increase in the transition probability with time; however, the first-order perturbation theory itself breaks down when Γ_{21} is not appreciably less than unity. Thus, Eq. (32) is valid for times for which

$$\beta t \ll 1 \tag{35}$$

If the lifetime of the state is ~ 10^{-9}sec, then $\beta \sim 10^9\ \text{sec}^{-1}$ and we must have

$$t \ll 10^{-9}\text{sec} \tag{36}$$

However, the large time behaviour[5] is given by Eq. (34).

It should also be noted that in our analysis we have assumed the quantity given by Eq. (30) to be very sharply peaked around $\omega \approx \omega_{21}$, for this to happen t should be large enough so that[6]

[5]Equation (32) may, therefore, be interpreted as: If the state has not made a transition upto time t then the probability that it will make a transition in dt will be $\beta\, dt$ from which Eq. (34) follows.

[6] This follows from the fact that the maximum of (30) occurs at $\omega = \omega_{21}$ and the first zero at $\omega = \omega_{21} \pm \frac{\pi}{t}$.

$$\frac{1}{t\,\omega_{21}} \ll 1$$

or
$$t \gg \frac{1}{\omega_{21}} \tag{37}$$

Now in the optical region

$$\omega_{21} \sim 10^{15}\ \text{sec}^{-1}$$

then we must have

$$t \gg 10^{-15}\ \text{sec}$$

Thus t in Eq. (32) should satisfy Eq. (37).

Returning to Eq. (32), we get the following expression for the transition probability per unit time (which we denote by w_{21}):

$$w_{21} \simeq \frac{\pi}{3\varepsilon_0 \hbar^2} |\mathbf{P}_{12}|^2\, u(\omega_{21}) \tag{38}$$

If there are N_2 atoms per unit volume in state 2 then the number of stimulated emissions per unit time per unit volume would be given by

$$W_{21} = N_2 w_{21} = N_2 \frac{\pi}{3\varepsilon_0 \hbar^2} |\mathbf{P}_{12}|^2\, u(\omega_{21}) \tag{39}$$

Comparing the above equation with Eq. (4), we obtain

$$B_{21} = \frac{\pi}{3\varepsilon_0} \frac{|\mathbf{P}_{12}|^2}{\hbar^2} = \frac{4\pi^2}{3\hbar^2}\left(\frac{q^2}{4\pi\varepsilon_0}\right)\left|\int \psi_1^* \,\mathbf{r}\, \psi_2\, d\mathbf{r}\right|^2$$

$$= \frac{4\pi^2 \alpha c}{3\hbar} |\langle 1|\mathbf{r}|2\rangle|^2 \quad \text{Einstein's B coefficient} \tag{40}$$

where
$$\alpha = \frac{q^2}{4\pi\varepsilon_0 \hbar c} \simeq \frac{1}{137} \tag{41}$$

represents the fine-structure constant.

The corresponding expression for stimulated absorption is obtained by starting with the first term on the right-hand side of Eq. (22) and proceeding in a similar manner. The final expression is identical to Eq. (40) except for an interchange of indices 1 and 2 and since

$$\left|\int \psi_2^* \,\mathbf{r}\, \psi_1\, d\mathbf{r}\right|^2 = \left|\int \psi_1^* \,\mathbf{r}\, \psi_2\, d\mathbf{r}\right|^2 \tag{42}$$

we get
$$B_{12} = B_{21}$$

consistent with Eq. (9).

28.4 SPONTANEOUS EMISSION RATE

Using Eqs (9), (10) and (40), we get the following expression for the A coefficient

$$A = \frac{4}{3}\alpha\frac{\omega^3}{c^2}\left|\int \psi_2^* \mathbf{r}\,\psi_1\, d\tau\right|^2 \quad \text{Einstein's } A \text{ coefficient} \tag{43}$$

For the $2P \to 1S$ transition in the hydrogen atom, i.e. for the transition from any one of the ($n = 2, l = 1, m = +1, 0, -1$) states to the ($n = 1, l = 0, m = 0$) state, we get (see Problem 28.3)

$$\left|\int \psi_{1S}^* \mathbf{r}\,\psi_{2P}\, d\tau\right|^2 = 2^5\left(\frac{2}{3}\right)^{10} a_0^2 \tag{44}$$

independent of the initial value of m; here $a_0\left(=\frac{\hbar^2}{me^2} \simeq 0.5\times 10^{-10}\,\text{m}\right)$ is the Bohr radius. Further, for $2P \to 1S$ transition

$$\hbar\omega = \left(-\frac{1}{4}+1\right)E_H = \frac{3}{8}\cdot\frac{e^2}{a_0}$$

or

$$\omega = \frac{3c}{8a_0}\alpha \tag{45}$$

On substitution in Eq. (43), we get

$$A = \left(\frac{4}{9}\alpha\right)^4 \frac{c}{a_0} \simeq 6\times 10^8\ \text{sec}^{-1}$$

The coefficient A gives the probability per unit time for spontaneous emission from state 2 to state 1. Thus, if there are N_2 atoms per unit volume in state 2 and if $-dN_2$ represents the number of atoms undergoing spontaneous transitions in time dt then

$$dN_2 = -AN_2\, dt \tag{46}$$

giving once again the radioactive decay law

$$N_2(t) = N_2(0)\, e^{-At}$$

Thus the mean lifetime of the state is given by

$$\tau = \frac{1}{A} \simeq 1.6\times 10^{-9}\ \text{sec} \quad \text{Mean lifetime} \tag{47}$$

In Eq. (46), dt should be small enough so that $Adt \ll 1$. Equation (47) tells us that the mean lifetime of the hydrogen atom in the upper level corresponding to the $2P \to 1S$ transition is about 1.6×10^{-9} sec. Transitions having such small lifetimes are referred to as strongly allowed transitions[7]. One can similarly calculate lifetimes for other atomic states.

---

[7] In contrast, levels used in the laser transition are such that the upper level has a very long lifetime ($\sim 10^{-3}$ to 10^{-5} sec). They are referred to as metastable states. The strength of an atomic transition is usually expressed in terms of the f-value defined by the equation

$$f = \frac{2}{3}\cdot\frac{m\omega_{21}}{\hbar}|D_{21}|^2$$

For strongly allowed transitions $f \sim 1$ and for transitions from metastable states $f \sim 10^{-3} - 10^{-6}$.

The spontaneous emission rate will give rise to finite widths of the spectral lines which can be estimated from the uncertainty principle

$$\Delta E \sim \frac{\hbar}{\tau} \tag{48}$$

In practice, the observed widths are much larger because of other factors such as Doppler broadening, collision broadening, etc.

We should also mention here that the A coefficient for the $2S \rightarrow 1S$ transition is zero. Such transitions are said to be forbidden transitions.

28.5 THE SELECTION RULES

Whenever the quantity

$$\hat{\mathbf{e}} \cdot \int \psi_f^* \, \mathbf{r} \, \psi_i \, d\tau$$

(also written as $\hat{\mathbf{e}} \cdot \langle (f|\mathbf{r}|i) \rangle$ vanishes, the transition probability is zero [see Eq. (24)] and the corresponding transition is said to be forbidden (in the dipole approximation); the subscripts i and f refer to initial and final states respectively. If we assume that the electron moves in a central field, then the angular part of the wavefunction will be of the form $Y_{lm}(\theta, \phi)$; thus, the initial and final states will be given by

$$\left.\begin{aligned} \psi_i &= R_{n_i l_i}(r) Y_{l_i m_i}(\theta, \phi) \\ \psi_f &= R_{n_f l_f}(r) Y_{l_f m_f}(\theta, \phi) \end{aligned}\right\} \tag{49}$$

Now, it can be shown that (see Problem 28.2)

$$\int \psi_f^* z \psi_i \, d\tau = G\left[\delta_{l_f, l_i+1}\left(\frac{(l_i+m_i+1)(l_i-m_i+1)}{(2l_i+1)(2l_i+3)}\right)^{1/2} + \delta_{l_f, l_i-1}\left(\frac{(l_i+m_i)(l_i-m_i)}{(2l_i+1)(2l_i+3)}\right)^{1/2}\right]\delta_{m_f, m_i} \tag{50}$$

and

$$\int \psi_f^* (x \pm iy) \psi_i \, d\tau = G\left[\delta_{l_f, l_i+1}\left(\frac{(l_i+m_i+1)(l_i+m_i+2)}{(2l_i+1)(2l_i+3)}\right)^{1/2} \pm \delta_{l_f, l_i-1}\left(\frac{(l_i-m_i)(l_i-m_i-1)}{(2l_i+1)(2l_i-1)}\right)^{1/2}\right]\delta_{m_f, m_i \pm 1} \tag{51}$$

where

$$G = \int_0^\infty r^3 R_{n_f l_f}(r) R_{n_i l_i}(r) \, dr \tag{52}$$

where R_{nl} represents the radial part of the wavefunctions. Equations (50) and (51) tell us that for a transition to occur, we must have

$$\left.\begin{aligned} \Delta l &= l_f - l_i = \pm 1 \\ \Delta m &= m_f - m_i = 0 \end{aligned}\right\} \text{ Selection rules} \tag{53}$$

and

$$\left.\begin{aligned} \Delta l &= l_f - l_i = \pm 1 \\ \Delta m &= m_f - m_i = \pm 1 \end{aligned}\right\} \text{ Selection rules} \tag{54}$$

which are known as the *selection rules*. Since spin angular momenta are not involved, we have

$$\Delta j = \pm 1 \tag{55}$$

as well.

For the $\Delta m = 0$ transition, the dipole moment vector

$$\mathbf{P} = q \int \psi_f^* \mathbf{r} \psi_i \, d\tau \tag{56}$$

is along the z-direction [because $P_x = P_y = 0$ — see Eq. (51)].

Now, if we consider an oscillatory dipole in the z-direction

$$p = p_0 \, e^{-i\omega t} \hat{\mathbf{z}}$$

then at large distances from such a dipole, the electric and magnetic fields are of the form (see, e.g. Ref. 5).

$$\mathcal{E} = -\left(\frac{k^2 p_0}{4\pi\varepsilon_0}\right) \sin\theta \, \frac{e^{i(kr - \omega t)}}{r} \hat{\theta} \tag{57}$$

$$\mathcal{H} = -\left(\frac{\omega k \, p_0}{4\pi}\right) \sin\theta \, \frac{e^{i(kr - \omega t)}}{r} \hat{\phi} \tag{58}$$

where θ represents the polar angle that $\mathbf{r}$ makes with the z-axis. Further, the Poynting vector (associated with the emitted radiation field) is given by

$$\mathbf{S} = \frac{\omega k^3 p_0^2}{16\pi^2 \varepsilon_0} \sin^2\theta \, \frac{\cos^2(kr - \omega t)}{r^2} \hat{\mathbf{r}} \tag{59}$$

Thus the electric field associated with the emitted radiation will be linearly polarized with its electric vector oscillating in the plane defined by $\hat{\mathbf{z}}$ and the direction of the emitted radiation $\mathbf{k}$ (see Fig. 28.3). Further, the intensity of the radiation will vary as $\sin^2\theta \left(= \left|\hat{\mathbf{e}} \cdot \hat{\mathbf{z}}\right|^2 \right)$. For $\theta = \pi/2$ (i.e., for the emitted radiation propagating along any direction which is perpendicular to the z-axis), the electric field vector will be oscillating in the z-direction.

For the

$$m_f = m_i + 1$$

transition

$$P_z = 0 \quad \text{and} \quad P_x - iP_y = 0$$

[see Eqs (50) and (51)]. Thus,

$$P_y = -iP_x \tag{60}$$

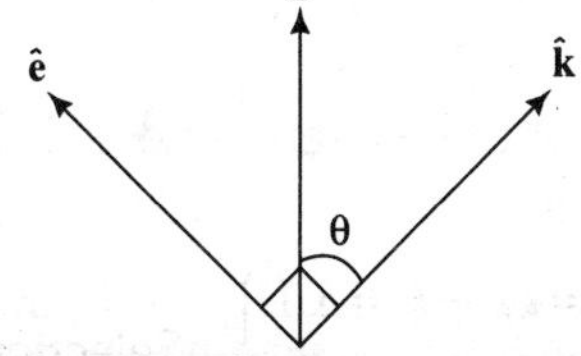

Figure 28.3: The vector $\mathbf{k}$ represents the direction of propagation of the emitted radiation. The plane of the paper is defined by the vectors $\mathbf{k}$ and $\hat{\mathbf{z}}$. For the $\Delta m = 0$ transition, the polarization vector of the emitted radiation $\hat{\mathbf{e}}$ will lie in the $\mathbf{k} - \hat{\mathbf{z}}$ plane with the intensity of radiation maximum for $\theta = \pi/2$ and zero for $\theta = 0$.

The above equation implies that the emitted radiation will be superposition of radiation from two dipoles oscillating (with a phase difference of $\pi/2$) in the x-and y-directions. Using formulae similar to Eqs (57) and (58) one can easily see that if $\mathbf{k}$ is along $\hat{\mathbf{x}}$ and $\hat{\mathbf{y}}$, the emitted radiation is plane polarized along the $\hat{\mathbf{y}}$ and $\hat{\mathbf{x}}$ directions respectively and if $\mathbf{k}$ is along $\hat{\mathbf{z}}$, the radiation is left circularly polarized. For $\mathbf{k}$ in any other direction, the polarization is elliptical.

As an example, we consider the ${}^1D_2 \to {}^1P_1$ transition (as in the case of the Cd 6438.17 Å line). Since we have neglected spin, we consider only singlet states. If we place the atom in a magnetic field in the z-direction, the levels will split up as shown in Fig. 28.4 (this is the Zeeman effect). Using the selection rules, we have

Viewed $\perp$ to $\mathbf{B}$ [$i.e.\ \mathbf{k}$ along $\hat{\mathbf{x}}$ (say)]

- $m_f = m_i$ ($i.e.$ $\Delta m = 0$) transitions plane polarized $\parallel$ to $\mathbf{B}$ (p–components)
- $m_f = m_i \pm 1$ ($i.e.$ $\Delta m = \pm 1$) transitions plane polarized $\perp$ to $\mathbf{B}$ (s–components)

Viewed $\parallel$ to $\mathbf{B}$ [$i.e.\ \mathbf{k}$ along $\hat{\mathbf{z}}$ (say)]

- $m_f = m_i$ ($i.e.$ $\Delta m = 0$) transitions not seen (p-components)
- $m_f = m_i \pm 1$ ($i.e.$ $\Delta m = \pm 1$) transitions circularly polarized (s-components)

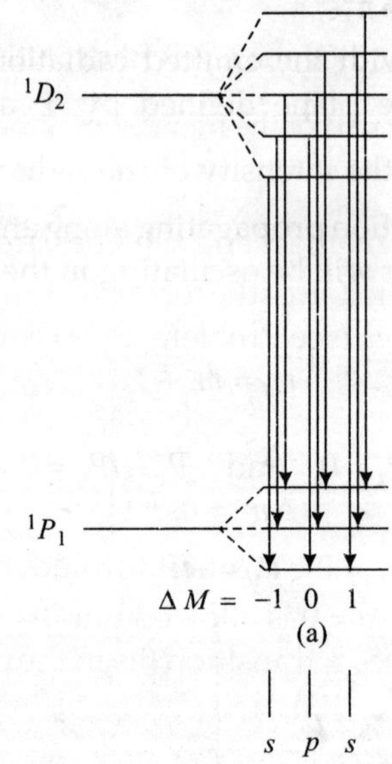

Figure 28.4: (a) The Zeeman pattern for the ${}^1D_2 \to {}^1P_1$ transition. (b) If viewed $\perp$ to $\mathbf{B}$, three lines are seen; if light polarized $\parallel$ to $\mathbf{B}$ is allowed to reach the film then only the p components are seen and if light polarized $\perp$ to $\mathbf{B}$ is allowed to reach the film then only the s components are seen.

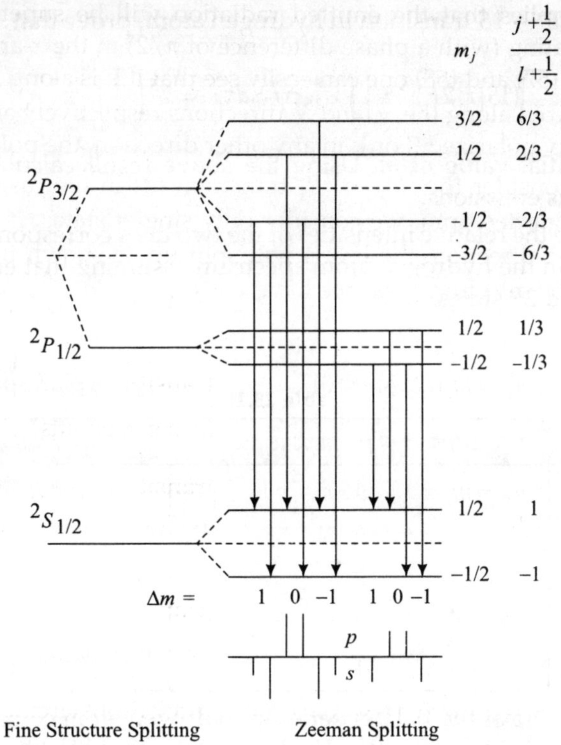

Figure 28.5: The Zeeman pattern for the $^2P_{3/2,\,1/2} \rightarrow {}^2S_{1/2}$ doublet in sodium. Notice that the Zeeman splitting is different for different levels [the splitting is proportional to $\frac{j+\frac{1}{2}}{l+\frac{1}{2}}\, m_j$, see Solution 21.2. The lower part of the figure shows that in the transition there are two p components and four s components.

This is known as the normal Zeeman effect and the lines are known as the Lorentz triplet.

In the presence of spin-orbit interaction of the form $\xi(r)\, \mathbf{L}\cdot\mathbf{s}$, the atomic states are characterized by the quantum numbers l, s j and m_j (see Problem 21.2). The corresponding selection rules are

$$\left.\begin{aligned} \Delta j &= 0, \pm 1 \\ \Delta l &= \pm 1 \\ \Delta m_j &= 0, \pm 1 \end{aligned}\right\} \tag{61}$$

However, $(m_j)_i = 0 \rightarrow (m_j)_f = 0$ transition forbidden if $\Delta j = 0$.
It may be noted that the transition $\Delta j = 0$ is allowed but $j_i = j_f = 0$ transition is not allowed. The Zeeman pattern for the $^2P_{3/2,\,1/2} \rightarrow {}^2S_{1/2}$ transition (as in the case of D lines of Sodium) is shown in Fig. 28.5 (see also Problem 28.7).

28.6 PROBLEMS

Problem 28.1 Show that for optical frequencies, at thermal equilibrium (corresponding to $T \sim 1000°\ K$), the number of spontaneous emissions far exceeds the number of stimulated emissions.

Problem 28.2 Derive Eqs (50) and (51) and hence the selection rules.

Problem 28.3 For the $2P \to 1S$ transition in hydrogen atom, prove that

$$|\langle 1S|\mathbf{r}|2P\rangle|^2 = \left|\int \psi_{1S}^* \mathbf{r} \psi_{2P} d\tau\right|^2 = 2^5 \left(\frac{2}{3}\right)^{10} a_0^2$$

independent of the initial value of m. Using the above result calculate the corresponding lifetime for spontaneous emissions.

Problem 28.4 Calculate the relative intensities of the two lines corresponding to the transitions $3P \to 1S$ and $2P \to 1S$ in the hydrogen atom spectrum assuming that equal number of atoms are excited to the $3P$ and $2P$ states.

Table 28.1:

Transition	*Number of p-components*	*Number of s-components*
${}^2D_{5/2} \to {}^2P_{3/2}$	4	8
${}^2G_{7/2} \to {}^2F_{5/2}$	6	12
${}^3P_2 \to {}^3S_1$	3	6
${}^3P_1 \to {}^3S_1$	2	4
${}^3P_0 \to {}^3S_1$	1	2

Problem 28.5 The first line of the Balmer series in hydrogen atom corresponds to the $n = 3$ to $n = 2$ transition, this line is known as the H_α line. What would be the selection rules for the $n = 3$ to $n = 2$ transitions and the corresponding lifetimes.

Problem 28.6 Obtain the angular distribution of the radiation emitted by the $nP \to n'S$ transition.

Problem 28.7 For $L - S$ coupling the selection rules are

$$\left.\begin{aligned} &\Delta J = 0, \pm 1 \quad (J = 0 \to J = 0 \text{ transition forbidden}) \\ &\Delta L = \pm 1 \\ &\Delta S = 0 \\ &\Delta M = 0, \pm 1 \quad (M = 0 \to M = 0 \text{ transition forbidden for } \Delta J = 0) \end{aligned}\right] \tag{62}$$

Using the above selection rules, show that in the Zeeman splitting of the transitions given in the first column of Table 28.1, the number of p- and s-components would be as given in the second and third columns.

Problem 28.8 Assuming that the atom can exist in two possible states characterised by the wavefunctions ψ_1 (**r**) and ψ_2 (**r**) and considering the atom-field interaction to be of the form given by Eq. (12), obtain a solution of the Schrödinger equation in the rotating wave approximation (i.e., neglect terms whose time dependences are of the form exp $[\pm i (\omega + \omega_{21}) t]$).

Problem 28.9 Using the expression for $\phi\,(n, l, \frac{1}{2}, j, m_j)$ in Solution 21.2, derive the selection rules given by Eq. (61).

Problem 28.10 In Section 28.3, we had considered interaction of an atom with a radiation field having continuous spectrum of frequency. Carry out a similar analysis for a near monochromatic wave interacting with an atom having a broad frequency response.

28.7 SOLUTIONS

Solution 28.1 At thermal equilibrium, the ratio of the number of spontaneous to stimulated emissions is given by

$$\frac{A}{Bu(\omega)} = \exp\left(\frac{\hbar\omega}{k_B T}\right) - 1$$

Now,
$$\frac{k_B T}{\hbar} \simeq \frac{1.38 \times 10^{-23}\,(\text{J/K}) \times 10^3\,(\text{K})}{1.054 \times 10^{-34}\,(\text{J.s})} \simeq 1.3 \times 10^{14}\ \text{sec}^{-1}$$

For the optical region $\lambda \simeq 6000$Å and hence $\omega \sim 3 \times 10^{15}\ \text{s}^{-1}$. Thus $\frac{A}{Bu(\omega)} \gg 1$ and the transitions are mainly due to spontaneous emissions.

Solution 28.2 The selection rule $\Delta m = 0, \pm 1$ can be derived easily from the fact that the ϕ-dependence of Y_{lm} is of the form $\exp\,(im\phi)$ and that

$$z = r\cos\theta, \quad x \pm iy = r\sin\theta\, e^{\pm i\phi}$$

Since,
$$\int_0^{2\pi} e^{i(m'-m)\phi}\, d\phi = 0 \quad \text{for} \quad m' \neq m$$

the ϕ integration in $\int \psi_f^* z \psi_i\, d\tau$ leads to the selection rule $m_f = m_i$ and the ϕ integration in $\int \psi_f^* (x \pm iy)\psi_i\, d\tau$ leads to $m_f = m_i \pm 1$.

In order to obtain Eqs (50) and (51), we note that

$$\left.\begin{aligned} z = r\cos\theta &= r\left(\frac{4\pi}{3}\right)^{1/2} Y_{10} \\ x \pm iy = r\sin\theta\, e^{\pm i\phi} &= \mp r\left(\frac{8\pi}{3}\right)^{1/2} Y_{1,\pm 1} \end{aligned}\right\} \qquad (63)$$

Further, if we use the following recurrence relations

$$\left(\frac{8\pi}{3}\right)^{1/2} Y_{1,1} Y_{l,m-1} = \left[\frac{(l+m)(l+1+m)}{(2l+1)(2l+3)}\right]^{1/2} Y_{l+1,m} - \left[\frac{(l-m)(l+1-m)}{(2l+1)(2l-1)}\right]^{1/2} Y_{l-1,m} \qquad (64)$$

$$\left(\frac{4\pi}{3}\right)^{1/2} Y_{1,0} Y_{l,m} = \left[\frac{(l+1+m)(l+1-m)}{(2l+1)(2l+3)}\right]^{1/2} Y_{l+1,m} - \left[\frac{(l-m)(l+m)}{(2l+1)(2l-1)}\right]^{1/2} Y_{l-1,m} \qquad (65)$$

$$\left(\frac{8\pi}{3}\right)^{1/2} Y_{1,-1} Y_{l,m+1} = \left[\frac{(l-m)(l+1-m)}{(2l+1)(2l+3)}\right]^{1/2} Y_{l+1,m} - \left[\frac{(l+m)(l+1+m)}{(2l+1)(2l-1)}\right]^{1/2} Y_{l-1,m} \qquad (66)$$

and also the orthonormality condition of spherical harmonics, we immediately get Eqs (50) and (51) from which the selection rules [Eqs (53) and (54)] follow.

Solution 28.3 We may directly use Eqs (50) and (51); however, in this case, it is possible to calculate the matrix elements without using the recurrence relations [Eqs (64) to (66)]. The eigenfunctions for the 2*P* and 1*S* states are

$$\psi_i = \Psi_{2P} = R_{21}(r) Y_{1m}(\theta, \phi) \tag{67}$$

and

$$\psi_f = \Psi_{1S} = R_{10}(r) Y_{00}(\theta, \phi) = \frac{1}{\sqrt{4\pi}} R_{10}(r) \tag{68}$$

Substituting for $R_{21}(r)$ and $R_{10}(r)$ (see Section 10.4) in Eq. (52) and carrying out the integration, we get

$$G = 4\sqrt{6}\left(\frac{2}{3}\right)^5 a_0$$

Thus,

$$z_{fi} = \int \psi_f^* z \psi_i \, d\tau = G \iint \frac{1}{\sqrt{4\pi}} \sqrt{\frac{4\pi}{3}} Y_{10} Y_{1m} \sin\theta \, d\theta \, d\phi$$

$$= 4\sqrt{2}\left(\frac{2}{3}\right)^5 a_0 \delta_{m,0} \tag{69}$$

Since,

$$Y_{1,-m} = (-1)^m Y_{1,m}^*(\theta, \phi)$$

we have

$$(x \pm iy)_{fi} = G \iint \frac{1}{\sqrt{4\pi}} \left(\pm \sqrt{\frac{8\pi}{3}} Y_{1,\mp 1}^* \right) Y_{1,m} \sin\theta \, d\theta \, d\phi$$

$$= \pm 8\left(\frac{2}{3}\right)^5 a_0 \delta_{m,\mp 1} \tag{70}$$

Thus,

$$x_{fi} = 4\left(\frac{2}{3}\right)^5 a_0 \left[\delta_{m,-1} - \delta_{m,1}\right] \tag{71}$$

and

$$y_{fi} = -4i\left(\frac{2}{3}\right)^5 a_0 \left[\delta_{m,-1} + \delta_{m,1}\right] \tag{72}$$

It can be easily verified that

$$\left|\mathbf{r}_{fi}\right|^2 = \left|x_{fi}\right|^2 + \left|y_{fi}\right|^2 + \left|z_{fi}\right|^2 = 2^5\left(\frac{2}{3}\right)^{10} a_0^2 \tag{73}$$

for $m = +1, 0$ and -1 which gives the result as given by Eq. (44). The lifetime calculation is given in Section 28.4.

Solution 28.4 Since the intensity of the emitted line is proportional to the energy emitted per second, the required ratio is given by

$$R = \frac{\hbar\omega_1 A_1}{\hbar\omega_2 A_2} \tag{74}$$

where

$$\hbar\omega_1 = E_{2P} - E_{1S} = \left[-\frac{1}{4} + 1\right] E_H = \frac{3}{4} E_H$$

$$\hbar\omega_2 = E_{3P} - E_{1S} = \left[-\frac{1}{9}+1\right]E_H = \frac{8}{9}E_H$$

$$A_1 = A\,(2P\rightarrow 1S) = \frac{4}{3}\alpha\frac{\omega_1^3}{c^2}\left|\int \psi_{1S}^* \,\mathbf{r}\,\psi_{2P}\,d\tau\right|^2$$

and
$$A_2 = A\,(3P\rightarrow 1S) = \frac{4}{3}\alpha\frac{\omega_2^3}{c^2}\left|\int \psi_{1S}^* \,\mathbf{r}\,\psi_{3P}\,d\tau\right|^2$$

Using the expressions for R_{31} (r), R_{21} (r) and R_{20} (r) (given in Section 10.4) in Eq. (52) and carrying out the integrations, we get

$$G_1 = G(2P\rightarrow 1S) = 4\sqrt{6}\left(\frac{2}{3}\right)^5 a_0 = \frac{256}{81\sqrt{6}}a_0$$

$$G_1 = G(3P\rightarrow 1S) = \frac{1}{4\sqrt{6}}\left(\frac{3}{2}\right)^4 a_0 = \frac{81}{64\sqrt{6}}a_0$$

The angular integrals will be the same for both the transitions and the values will be the same as calculated in the previous solution. Thus,

$$R = \left(\frac{\frac{3}{4}E_H}{\frac{8}{9}E_H}\right)^4\left(\frac{256\times 64}{81\times 81}\right)^2 = \frac{256}{81}\simeq 3.16$$

Solution 28.5 For the $n = 3$ to $n = 2$ transition one could have the following possible transitions: $3S \rightarrow 2P$, $3P \rightarrow 2S$, $3D \rightarrow 2P$. In order to determine the total transition probability for a given initial state, one must sum over the possible final states. Thus, for the $|\ 3, 0, 0>$ state, the final state could be either of the following three states: $|\ 2, 1, 1>$, $|\ 2, 1, 0>$ and $|\ 2, 1, -1>$ and the total transition probability (per unit time) would be given by

$$A\,(3S\rightarrow 2P) = \frac{4}{3}\alpha\frac{\omega^3}{c^2}\sum_{m=1,0,-1}\left|\langle 2,1,m|\mathbf{r}|3,0,0\rangle\right|^2 \tag{75}$$

where
$$\omega = \frac{E_3 - E_2}{\hbar}$$

Thus, using Eqs (50) and (51)

$$A\,(3S\rightarrow 2P) = \frac{4}{3}\alpha\frac{\omega^3}{c^2}\Big[\left|\langle 2,1,1|x|3,0,0\rangle\right|^2 + \left|\langle 2,1,1|y|3,0,0\rangle\right|^2$$
$$+\left|\langle 2,1,0|z|3,0,0\rangle\right|^2 + \left|\langle 2,1,-1|x|3,0,0\rangle\right|^2 + \left|\langle 2,1,-1|y|3,0,0\rangle\right|^2\Big] \tag{76}$$

Each of the matrix elements could be calculated easily. For the $3P\rightarrow 2S$ transition one should consider a particular initial state, say $|3, 1, 0>$ and calculate the transition probability, which would be

$$A(|3,1,0>\rightarrow|2,0,0>) = \frac{4}{3}\alpha\frac{\omega^3}{c^2}\left|\langle 2,0,0|\mathbf{r}|3,1,0\rangle\right|^2$$

$$= \frac{4}{3}\alpha\frac{\omega^3}{c^2}\left|\langle 2,0,0|z|3,1,0\rangle\right|^2$$

[because $m_f = m_i$, see Eqs (50) and (51)]

$$= \frac{4}{3}\alpha\frac{\omega^3}{c^2}|G|^2 \tag{77}$$

where we have used the relation

$$\iint Y_{00}^* \cos\theta\, Y_{10} \sin\theta\, d\theta\, d\phi = \frac{1}{\sqrt{3}}$$

etc.

Solution 28.6 Let the direction of propagation of the emitted radiation be denoted by **k** which makes the polar angles Θ, Φ with respect to the *z*-axis. Let $\hat{\mathbf{e}}_1$ represent an unit vector in the $\mathbf{k}-\hat{\mathbf{z}}$ plane and perpendicular to **k** and let $\hat{\mathbf{e}}_2$ be the unit vector normal to the $\mathbf{k}-\hat{\mathbf{z}}$ plane. Thus, the Cartesian components of $\hat{\mathbf{e}}_1$ and $\hat{\mathbf{e}}_2$ are

$$\lambda = 1,\ \hat{\mathbf{e}}_1 : -\cos\Theta\cos\Phi \quad -\cos\Theta\sin\Phi \quad \sin\Theta$$

$$\lambda = 2,\ \hat{\mathbf{e}}_2 : \sin\Phi \qquad -\cos\Phi$$

The angular distribution of the emitted radiation is proportional to

$$\left|\int \psi_f^* \hat{\mathbf{e}}_\lambda \cdot \mathbf{r}\, \psi_i\, d\tau\right|^2 \sin\Theta\, d\Theta\, d\Phi$$

Let us first consider λ = 1 for which

$$\hat{\mathbf{e}}_\lambda \cdot \mathbf{r} = r\left[e_{\lambda x}\sin\theta\cos\phi + e_{\lambda y}\sin\theta\sin\phi + e_{\lambda z}\cos\theta\right]$$

$$= r\left[\left(\frac{8\pi}{3}\right)^{1/2}\frac{1}{2}(-Y_{1,1} + Y_{1,-1})(-\cos\Theta\cos\Phi)\right.$$

$$\left. + \left(\frac{8\pi}{3}\right)^{1/2}\frac{1}{2i}(-Y_{1,1} - Y_{1,-1})(-\cos\Theta\sin\Phi) + \left(\frac{4\pi}{3}\right)^{1/2} Y_{1,0}\sin\Theta\right]$$

Thus for the $n\text{P} \to n'S$ transition

$$\int \psi_f^* \hat{\mathbf{e}}_1 \cdot \mathbf{r}\, \psi_i\, d\tau = G\iint Y_{00}^*\left(\frac{2\pi}{3}\right)^{1/2}\left[\left(Y_{1,1} - Y_{1,-1}\right)\cos\Theta\cos\Phi\right.$$

$$-i\left(Y_{1,1} - Y_{1,-1}\right)\cos\Theta\sin\Phi$$

$$\left. + \sqrt{2}\, Y_{10}\sin\Theta\right] Y_{1,m}\sin\theta\, d\theta\, d\phi$$

where *G* is given by Eq. (52). Simple integrations give

$$\int \psi_f^* \hat{\mathbf{e}}_1 \cdot \mathbf{r}\, \psi_i\, d\tau = \begin{cases} \dfrac{G}{\sqrt{6}}\cos\Theta\, e^{i\Phi} & \text{for } m = 1 \\ \dfrac{G}{\sqrt{3}}\sin\Theta & \text{for } m = 0 \\ -\dfrac{G}{\sqrt{6}}\cos\Theta\, e^{-i\Phi} & \text{for } m = -1 \end{cases} \tag{78}$$

Similarly for $\lambda = 2$

$$\int \psi_f^* \hat{\mathbf{e}}_2 \cdot \mathbf{r}\psi_i \, d\tau = \begin{cases} \frac{i}{\sqrt{6}} e^{i\Phi} & \text{for } m=1 \\ 0 & \text{for } m=0 \\ -\frac{i}{\sqrt{6}} e^{-i\Phi} & \text{for } m=-1 \end{cases} \tag{79}$$

If we write the intensity distribution as

$$I_\lambda d\Omega = I_0 f_\lambda (\Theta, \Phi) \sin \Theta \, d\Theta \, d\Phi \tag{80}$$

m	λ=1	λ=2
1	$\frac{1}{2}\cos^2 \Theta$	$\frac{1}{2}$
0	$\sin^2 \Theta$	0
–1	$\frac{1}{2}\cos^2 \Theta$	$\frac{1}{2}$

the values of f_λ are given in the table above. Notice that $\lambda = 2$ case is isotropic.

Solution 28.8 Assuming that the atom can exist in only two possible states described by the wavefunctions ψ_1 (**r**) and ψ_2 (**r**) we get from Eq. (18)

$$i\hbar \frac{dC_1}{dt} = \frac{1}{2} \mathcal{E}_0 D_{12} C_2(t) \left[e^{-i(\omega' - \omega)t} + e^{-i(\omega' + \omega)t} \right] \tag{81}$$

and

$$i\hbar \frac{dC_2}{dt} = \frac{1}{2} \mathcal{E}_0 D_{21} C_1(t) \left[e^{i(\omega' + \omega)t} + e^{i(\omega' - \omega)t} \right] \tag{82}$$

where

$$\omega' = \frac{E_2 - E_1}{\hbar} = \omega_{21} = -\omega_{12}$$

and use has been made of the fact that

$$\int \psi_1^* \mathbf{r} \psi_1 \, d\tau = \int \psi_2^* \mathbf{r} \psi_2 \, d\tau = 0 \tag{83}$$

and therefore $D_{11} = D_{22} = 0$. In the rotating wave approximation[8], we neglect the terms $\exp [\pm i (\omega' + \omega) t]$ because of their very rapid oscillations to obtain

$$\frac{dC_1}{dt} = -\frac{i}{2\hbar} \mathcal{E}_0 D_{12} C_2(t) e^{i(\omega - \omega')t} \tag{84}$$

and

$$\frac{dC_2}{dt} = -\frac{i}{2\hbar} \mathcal{E}_0 D_{21} C_1(t) e^{-i(\omega - \omega')t} \tag{85}$$

We assume a solution of the type

$$C_1(t) = e^{i\Omega t} \tag{86}$$

then from Eq. (84)

[8] Salzman (see Ref. 5) has obtained exact numerical solution of the Schrödinger equation without using the rotating wave approximation. The final result for $|C_2(t)|^2$ is the same except for the presence of a weak high frequency oscillation superimposed on the solution as shown in Fig. 28.6.

$$C_2(t) = -\frac{2\hbar\Omega}{\mathcal{E}_0 D_{12}} e^{i(\Omega-\omega+\omega')t} \tag{87}$$

Substituting Eq. (86) and Eq. (87) in Eq. (85), we get

$$\Omega^2 + (\omega' - \omega)\Omega - \frac{1}{4}\Omega_0^2 = 0 \tag{88}$$

where
$$\Omega_0^2 = \frac{D_{12} D_{21} \mathcal{E}_0^2}{\hbar^2} = \frac{D^2 \mathcal{E}_0^2}{\hbar^2} \tag{89}$$

and
$$D = D_{12} = D_{21} \tag{90}$$

Equation (88) gives

$$\Omega_{1,2} = \frac{1}{2}\left[-(\omega' - \omega) \pm \left\{(\omega' - \omega)^2 + \Omega_0^2\right\}^{1/2}\right] \tag{91}$$

Thus the general solutions of Eqs (84) and (85), will be

$$C_1(t) = A_1 e^{i\Omega_1 t} + A_2 e^{i\Omega_2 t} \tag{92}$$

and
$$C_2(t) = -\frac{2}{\Omega_0} e^{i(\omega'-\omega)t}\left[A_1 \Omega_1 e^{i\Omega_1 t} + A_2 \Omega_2 e^{i\Omega_2 t}\right] \tag{93}$$

A_1 and A_2 can be determined from initial conditions. For example, if the atom is initially in the ground state then

$$C_1(0) = 1,\ C_2(0) = 0 \tag{94}$$

which gives
$$|C_2(t)|^2 = \left[\frac{\sin(\Omega' t/2)}{\Omega'/2}\right]^2 \left(\frac{\Omega_0}{2}\right)^2 \tag{95}$$

where
$$\Omega' = \left[(\omega' - \omega)^2 + \Omega_0^2\right]^{1/2} \tag{96}$$

Equation (95) gives the transition probability for absorption which has been shown in Fig. 28.6.

Now, at resonance, $\omega = \omega'$ and

$$|C_2(t)|^2 = \sin^2 \frac{\Omega_0 t}{2} \tag{97}$$

which shows that the system flip-flops between states 1 and 2.

Solution 28.10 We consider a nearly monochromatic field (which is indeed true for a laser) interacting with atoms characterized by the line shape function $g(\omega)$. Thus $N_1 g(\omega)\, d\omega$ and $N_2 g(\omega)\, d\omega$ represent the number of atoms per unit volume in levels 1 and 2 respectively which are capable of interacting with radiation of frequency between ω and $\omega + d\omega$ Obviously[9],

$$\int g(\omega)\, d\omega = 1 \tag{98}$$

Now, for a near monochromatic field, the energy (per unit volume) associated with the field (at frequency ω) is given by

---

[9] Different types of line shape functions have been discussed in Ref. 3 – see also Problem 29.3.

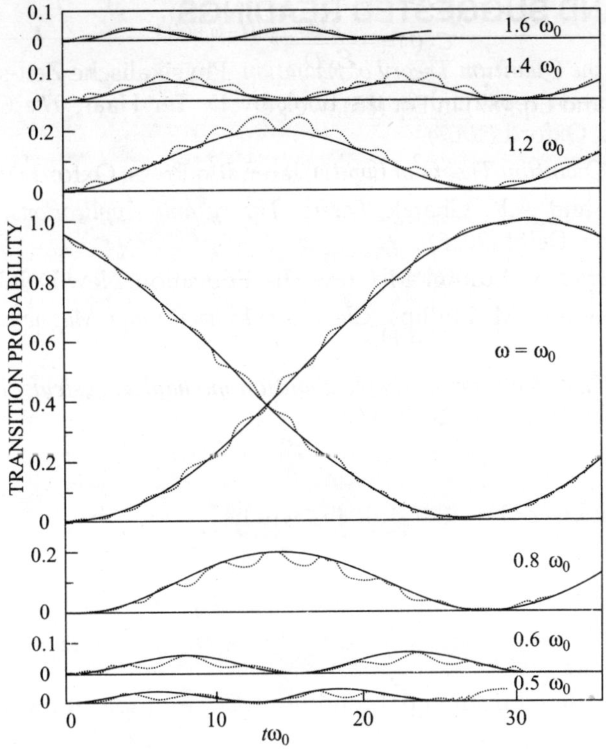

Figure 28.6: Variation of the transition probability for a two level system for different frequencies of the electromagnetic field ω as a function of time. The curves correspond to $D\mathcal{E}_0/\hbar = 0.1\ \omega'$. The solid line corresponds to Eq. (95) and the dotted curve corresponds to an exact numerical computation. Notice the presence of a weak high frequency oscillation present in the exact calculation which is due to the presence of the $\exp[\pm i(\omega + \omega')\ t]$ terms in Eqs (81) and (82). The graph for $\omega = \omega_0$ has been cut at $t\omega_0 = 36$ and shifted to the origin (after Ref. 6).

$$u_\omega = \frac{1}{2}\varepsilon_0\, \mathcal{E}_0^2 \tag{99}$$

Thus, for such a case the probability for stimulated emission is given by

$$\Gamma_{21} = \frac{1}{4}|D_{12}|^2 \frac{\mathcal{E}_0^2}{\hbar^2}\int g(\omega')\left[\frac{\sin\dfrac{\omega'-\omega}{2}t}{\dfrac{\omega'-\omega}{2}}\right]^2 d\omega'$$

$$\simeq \frac{1}{4}\frac{|D_{12}|^2}{\hbar^2}\mathcal{E}_0^2 g(\omega)\, 2\pi t = \frac{\pi P^2}{3\hbar^2\varepsilon_0} g(\omega) u_\omega t \tag{100}$$

where in the last step we have replaced $|D_{12}|^2$ by its average value. Since,

$$B_{21} = \frac{\pi}{3\varepsilon_0}\frac{P^2}{\hbar^2} = \frac{\pi^2 c^3}{\hbar\omega^3}A = \frac{\pi^2 c^3}{\hbar\omega^2\tau} \tag{101}$$

(where τ is the spontaneous emission lifetime), we obtain the following expression for the number of stimulated emissions per unit time per unit volume

$$N_1 \frac{\pi^2 c^3}{\hbar\omega^2\tau} u_\omega g(\omega) \tag{102}$$

REFERENCES AND SUGGESTED READINGS

1. A. Einstein, *On the Quantum Theory of Radiation*, Physikalische Zeit-schrift **18**, 121 (1917); this paper has been reprinted in the book by D. Ter Haar, *The Old Quantum Theory*, Pergamon Press, Oxford (1967).
2. R. Loudon, *The Quantum Theory of Light*, Clarendon Press, Oxford (1973).
3. K. Thyagarajan and A.K. Ghatak, *Lasers: Theory and Applications*. 2nd Edition, Laxmi Publications, New Delhi (2016).
4. A.K. Ghatak, *Optics*, 6th Edition, McGraw Hill Education, New Delhi (2017).
5. W.K.H. Panofsky and M. Phillips, *Classical Electricity and Magnetism*, Addison-Wesley (1962).
6. W.R. Salzman, *Time evolution of simple quantum mechanical systems: II Two state system in intense fields*, Phys. Rev. Lett. **26**, 220 (1971).

29 The quantum theory of radiation and its interaction with matter

Quantum mechanics is very impressive. But an inner voice tells me that it is not yet the real thing. The theory produces a good deal but hardly brings us closer to the secret of the Old One. I am at all events convinced that **He** *does not play dice.*

— ALBERT EINSTEIN in a letter to Max Born, 4 December 1926.

Neils Böhr's response is somewhat less well known[1]: *But still, it cannot be for us to tell God, how he is to run the world.*

29.1 INTRODUCTION

In the previous chapter we developed the semi-classical theory of interaction of radiation with matter in which matter, i.e., atoms, were assumed to exist in discrete (quasi-stationary states) but the electromagnetic field with which it interacts was described classically. In this chapter we will develop the fully quantum theory of radiation and discuss its interaction with matter.

We will first consider the Hamiltonian of an atomic system in a radiation field (Section 29.2). We will next show that the electromagnetic field in a closed cavity can be considered as an infinite set of oscillators, each corresponding to a particular value of the wave vector and particular direction of polarization. By imposing the commutation relations between the canonical variables, it will be shown that the energy of each oscillator can increase or decrease by integral multiples of certain quantum of energy; this quantum of energy is known as the *photon*. In Section 29.4, we will use the quantum mechanical description of the radiation field to study its interaction with an atom and thereby obtain explicit expressions for the Einstein A and B coefficients which will be shown to be identical to the results obtained in the previous chapter. It may be mentioned that the theory developed in Section 29.4 automatically leads to spontaneous emissions which in the semiclassical theory had to be introduced in an ad hoc manner through the Einstein coefficients (see Section 28.4). In Section 29.5, we will discuss the properties of the eigenstates of the Hamiltonian of the radiation field and will show that the state which corresponds to a given number of photons for a particular mode does *not* correspond to the classical plane wave. Indeed, in Section 29.6, we will show that the eigenstates of the annihilation operator (which are known as the *coherent states*) resemble the classical plane wave for large intensities. In fact when a laser is operated much beyond the threshold, it generates a coherent state excitation of a cavity mode. Finally in Section 29.7, we will show that it is difficult to give a quantum mechanical description of the phase of the electromagnetic wave.

---

[1] The authors found this quotation in Jim Baggott's book, *The Meaning of Quantum* Theory, Oxford University Press, Oxford (1992).

29.2 THE HAMILTONIAN

The Hamiltonian of an atomic system in a radiation field can be written as

$$H = H_0 + H'$$
$$= H_a + H_r + H' \qquad (1)$$

where H_a represents the Hamiltonian of the atom, H_r the Hamiltonian corresponding to the pure radiation field, which is given by

$$H_r = \frac{1}{2}\int (\varepsilon_0 \mathcal{E}\cdot\mathcal{E} + \mu_0 \mathcal{H}\cdot\mathcal{H})\, d\tau \quad \text{Radiation field Hamiltonian} \qquad (2)$$

and H' represents the interaction between the atom and the radiation field. In Eq. (2) ε_0 and μ_0 represent the dielectric permittivity and magnetic permeability of free space, $\mathcal{E}$ and $\mathcal{H}$ represent the electric and magnetic fields associated with the radiation field. For example, for a non-relativistic electron (of charge $-q$) in a radiation field, we have

$$H = \frac{1}{2m}(\mathbf{p} + q\mathbf{A})^2 + V + H_r \qquad (3)$$

since the recipe is to replace the electron momentum opertator $\mathbf{p}$ by $\mathbf{p} + q\mathbf{A}$ (see Section 21.2), where $\mathbf{A}$ is the vector potential and V is the potential energy of the electron. We can rewrite Eq. (3) as

$$H = H_0 + H' \qquad (4)$$

where

$$\left.\begin{aligned} H_0 &= H_a + H_r \\ H_a &= \frac{p^2}{2m} + V(r) \\ H' &= \frac{q}{2m}(\mathbf{A}\cdot\mathbf{p} + \mathbf{p}\cdot\mathbf{A}) + \frac{q^2}{2m}A^2 \end{aligned}\right\} \qquad (5)$$

and H_r is given by Eq. (2). In Section 29.4, we will treat H' as a perturbation and study the transitions between eigenstates of H_0. The term $\frac{q^2}{2m}A^2$ appearing in the expression for H' usually represents a very small perturbation and hence will be neglected. Thus the interaction term is given by

$$H' = +\frac{q}{m}\;(\mathbf{A}\cdot\mathbf{p} + \mathbf{p}\cdot\mathbf{A}) \qquad (6)$$

Now,

$$\begin{aligned} \mathbf{p}\cdot\mathbf{A}\psi &= -i\hbar\nabla\cdot(\mathbf{A}\psi) \\ &= -i\hbar[(\nabla\cdot\mathbf{A})\psi + \mathbf{A}\cdot\nabla\psi] \\ &= \mathbf{A}\cdot\mathbf{p}\,\psi \end{aligned}$$

where we have chosen the Coulomb gauge in which $\nabla\cdot\mathbf{A} = 0$ (see chapter 21).

Thus, $$H' = \frac{q}{m}\mathbf{A}\cdot\mathbf{p} = \frac{q}{m}\mathbf{p}\cdot\mathbf{A} \tag{7}$$

There are now two ways of proceeding further. The first is to treat the electromagnetic field classically; this has indeed been done in the previous chapter. The second and more powerful method is to quantize the radiation field and calculate rates of transition between states of matter and quanta; this will be done in Sections 29.3 and 29.4. However, before we do so, we will first show (in Section 29.2.1) that the interaction term given by Eq. (12) of Chapter 28 is consistent with Eq. (7) under certain approximation which is known as the dipole approximation.

29.2.1 The Interaction Term in the Semi-classical Theory of Radiation

For a plane electromagnetic wave, the vector potential can be written in the form

$$\mathbf{A} = \hat{\mathbf{e}} A_0 \cos(\mathbf{k}\cdot\mathbf{r} - \omega t) \tag{8}$$

where $\hat{\mathbf{e}}$ represents the unit vector along $\mathbf{A}$ (denoting the polarization of the wave), and

$$\mathbf{k} = \frac{\omega}{c}\hat{\mathbf{k}} \tag{9}$$

where $\hat{\mathbf{k}}$ represents the unit vector along the direction of propagation. The condition $\nabla\cdot\mathbf{A} = 0$ gives

$$\hat{\mathbf{e}}\cdot\mathbf{k} = 0 \tag{10}$$

implying that $\hat{\mathbf{e}}$ is at right angles to the direction of propagation, i.e., the wave is transverse. The electric field is given by

$$\mathcal{E} = -\frac{\partial \mathbf{A}}{\partial t} = -\hat{\mathbf{e}}\,\mathcal{E}_0 \sin(\mathbf{k}\cdot\mathbf{r} - \omega t) \tag{11}$$

where $$\mathcal{E}_0 = \omega A_0 \tag{12}$$

Now,
$$H' = \frac{q}{m}\mathbf{A}\cdot\mathbf{p}$$
$$= \frac{q\mathcal{E}_0}{2m\omega}\hat{\mathbf{e}}\left[e^{-i(\mathbf{k}\cdot\mathbf{r}-\omega t)} + e^{i(\mathbf{k}\cdot\mathbf{r}-\omega t)}\right]\cdot\mathbf{p}$$

The matrix element of transition is given by [cf. Eq. (17) of Chapter 28]:

$$H'_{sn} = \langle s|H'|n\rangle = \frac{q\mathcal{E}_0}{2m\omega}\hat{\mathbf{e}}\cdot\left[\langle s|e^{i\mathbf{k}\cdot\mathbf{r}}\mathbf{p}|n\rangle e^{-i\omega t} + \langle s|e^{-i\mathbf{k}\cdot\mathbf{r}}\mathbf{p}|n\rangle e^{-i\omega t)}\right] \tag{13}$$

where the kets $|\,n>$ and $|\,s>$ are eigenstates of the atomic Hamiltonian H_a and represent the initial and final states. Now,

$$e^{\pm i\mathbf{k}\cdot\mathbf{r}} = 1 \pm i\mathbf{k}\cdot\mathbf{r} - \frac{1}{2!}(\mathbf{k}\cdot\mathbf{r})^2 + \ldots$$

Since atomic dimensions are $\sim 10^{-8}$ cm and for optical wavelength

$$k\left(=\frac{2\pi}{\lambda}\right) \sim 10^5\ \text{cm}^{-1}$$

we obtain (in the region of integration) $kr \ll 1$. Thus, negligible error will be involved if the exponential is replaced by unity[2] and the transition is said to be an electric dipole transition—a very important case. In this approximation

$$\langle s|e^{\pm i\mathbf{k}\cdot\mathbf{r}}\mathbf{p}|n\rangle \simeq \langle s|\mathbf{p}|n\rangle \tag{14}$$

Now the kets $|n\rangle$ and $|s\rangle$ are eigenkets[3] of the atomic Hamiltonian H_a:

$$H_a|n\rangle = E_n|n\rangle = \hbar\omega_n|n\rangle \tag{15}$$

$$H_a|s\rangle = E_s|s\rangle = \hbar\omega_s|s\rangle \tag{16}$$

Further, $$H_a = \frac{p^2}{2m} + V(\mathbf{r})$$

Thus, $$[x, H_a] = \left[x, \frac{p_x^2 + p_y^2 + p_z^2}{2m} + V(\mathbf{r})\right]$$

$$= \frac{1}{2m}([x, p_x]p_x + p_x[x, p_x])$$

$$= \frac{i\hbar}{m}p_x \tag{17}$$

or, $$\mathbf{p} = \frac{m}{i\hbar}[\mathbf{r}, H_a] \tag{18}$$

Hence, $$\langle s|\mathbf{p}|n\rangle = \frac{m}{i\hbar}\langle s|\mathbf{r}H_a - H_a\mathbf{r}|n\rangle$$

$$= \frac{m}{i\hbar}(E_n - E_s)\langle s|\mathbf{r}|n\rangle$$

$$= \frac{m\omega_{ns}}{i}\langle s|\mathbf{r}|n\rangle \tag{19}$$

where we have used Eq. (15) and conjugate of Eq. (16) and

$$\hbar\omega_{ns} = E_n - E_s$$

On substitution in Eq. (13), we get

$$\langle s|H'|n\rangle = \frac{q\mathcal{E}_0\omega_{ns}}{2\omega}\hat{\mathbf{e}}\cdot\langle s|\mathbf{r}|n\rangle\left[e^{i\omega t} - e^{-i\omega t}\right] \tag{20}$$

The above equation is of a form similar (although not identical) to that of Eq. 17 of Chapter 28. However, for $\omega \approx |\omega_{ns}|$ after one carries out the integration which leads to Eq. (22) of the Chapter 28, one obtains identical expressions for probabilities of absorption and of emission [see Eq. (24) of Chapter 28].

---

[2] However, if the replacement of $e^{\pm i\mathbf{k}\cdot\mathbf{r}}$ by unity leads to a vanishing matrix element, one must use higher order terms like ($\pm i\mathbf{k}\cdot\mathbf{r}$), etc.; these are said to corresond to higher order transitions (see Problem 29.1.)

[3] We may warn the reader of some confusion of notation. The initial and final states of the atomic Hamiltonian are also sometimes denoted by $|i\rangle$ and $|f\rangle$ and sometimes by $|a\rangle$ and $|b\rangle$.

29.3 QUANTIZATION OF THE RADIATION FIELD

We now develop the second method mentioned in Section 29.1. In this method, the electromagnetic field is described in terms of numbers of photons in various states. We consider first a pure radiation field. In Coulomb gauge, such a field can be derived from the vector potential **A** only,with $\Phi = 0$ and

$$\nabla \cdot \mathbf{A} = 0 \tag{21}$$

In this gauge, the magnetic and electric fields are given by (see Section 21.2)

$$\mathcal{B} = \mu_0 \mathcal{H} = \nabla \times \mathbf{A} \tag{22}$$

and

$$\mathcal{E} = -\frac{\partial \mathbf{A}}{\partial t} \tag{23}$$

where we have assumed $\Phi = 0$ since we are considering free space. Substituting for $\mathcal{H}$ and $\mathcal{E}$ in the Maxwell equation

$$\nabla \times \mathcal{H} = \frac{\partial \mathbf{D}}{\partial t} = \varepsilon_0 \frac{\partial \mathcal{E}}{\partial t} \tag{24}$$

we get

$$\nabla \times (\nabla \times \mathbf{A}) = -\varepsilon_0 \mu_0 \frac{\partial^2 \mathbf{A}}{\partial t^2} \tag{25}$$

If we now use the identity[4]

$$\nabla \times (\nabla \times \mathbf{A}) \equiv \nabla (\nabla \cdot \mathbf{A}) - \nabla^2 \mathbf{A} = -\nabla^2 \mathbf{A} \tag{26}$$

(because $\nabla \cdot \mathbf{A} = 0$), we finally obtain

$$\nabla^2 \mathbf{A} = \frac{1}{c^2} \cdot \frac{\partial^2 \mathbf{A}}{\partial t^2} \tag{27}$$

where

$$c = (\varepsilon_0 \mu_0)^{-1/2} \tag{28}$$

represents the speed of light in free space. Equation (27) represents the three-dimensional wave equation. In order to solve the wave equation, we use the method of separation of variables:

$$\mathbf{A}(\mathbf{r}, t) = \mathbf{A}(\mathbf{r})\, q(t) \tag{29}$$

Thus,

$$q(t) \nabla^2 \mathbf{A}(\mathbf{r}) = \mathbf{A}(\mathbf{r}) \frac{1}{c^2} \frac{d^2 q}{dt^2} \tag{30}$$

We next consider a Cartesian component (say the x-component) of **A**(**r**) which we denote by $A_x(\mathbf{r})$; thus

$$\frac{c^2}{A_x(\mathbf{r})} \nabla^2 A_x(\mathbf{r}) = \frac{1}{q(t)} \cdot \frac{d^2 q}{dt^2} = -\omega^2 \quad \text{(say)} \tag{31}$$

Thus,

$$q(t) \sim e^{-i\omega t} \tag{32}$$

---

[4] It may be worthwhile to point out that the operator ∇^2 is *defined* by the following equation

$$\nabla^2 \mathbf{A} \equiv \nabla(\nabla \cdot \mathbf{A}) - \nabla \times (\nabla \times \mathbf{A})$$

However,

$$(\nabla^2 \mathbf{A})_x = \nabla \cdot (\nabla A_x)$$

ie., a Cartesian component of ∇^2 **A** is div grad of the Cartesian component but

$$(\nabla^2 \mathbf{A})_r \neq \nabla \cdot (\nabla \mathbf{A}_r)$$

and $$\nabla^2 A_x(\mathbf{r}) + k^2 A_x(\mathbf{r}) = 0 \tag{33}$$

where $k^2 = \omega^2/c^2$. The solutions of Eq. (33) are plane waves; and similarly if we consider the y- and z-components we obtain

$$\mathbf{A}(\mathbf{r}) = e^{i\mathbf{k}\cdot\mathbf{r}}\,\hat{\mathbf{e}} \tag{34}$$

where $\mathbf{k}\cdot\mathbf{k} = k^2$, $\hat{\mathbf{e}}$ is the unit vector along $\mathbf{A}$. The condition $\nabla\cdot\mathbf{A} = 0$ gives us $\mathbf{k}\cdot\hat{\mathbf{e}} = 0$, implying the transverse character of the wave [see Eq. (10)].

The allowed values of $\mathbf{k}$ (and hence of ω) are determined from the boundary conditions. If we assume the radiation to be confined in a cubical cavity of volume V (= L^3) and use the periodic boundary condition then

$$\mathbf{A}(x = 0, y, z) = \mathbf{A}(x = L, y, z) \qquad \text{etc.} \tag{35}$$

giving $$e^{ik_xL} = 1 = e^{ik_yL} = e^{ik_zL} \tag{36}$$

Thus, $$\left.\begin{aligned} k_x &= \frac{2\pi\nu_x}{L} \\ k_y &= \frac{2\pi\nu_y}{L} \\ k_z &= \frac{2\pi\nu_z}{L} \end{aligned}\right\} \nu_x, \nu_y, \nu_z = 0, \pm 1, \pm 2, \ldots \tag{37}$$

The complete solution of Eq. (27) is, therefore, given by

$$\mathbf{A}(\mathbf{r}, t) = \sum_\lambda \left[q_\lambda(t)\mathbf{A}_\lambda(\mathbf{r}) + q_\lambda^*(t)\mathbf{A}_\lambda^*(\mathbf{r}) \right] \tag{38}$$

where $$\mathbf{A}_\lambda(\mathbf{r}) = e^{i\mathbf{k}_\lambda\cdot\mathbf{r}}\hat{\mathbf{e}}_\lambda \tag{39}$$

$$q_\lambda(t) = |q_\lambda| e^{-i\omega_\lambda t} \tag{40}$$

and the subscript λ signifies the various modes of the field [see Eq. (37)] including the two states of polarization. Thus, a particular value of λ corresponds to a particular set of values of ν_x, ν_y, ν_z and a particular direction of $\hat{\mathbf{e}}$. In Eq. (38) the second term on the RHS is complex conjugate of the first term making $\mathbf{A}$ necessarily real. Because of the allowed values of $\mathbf{k}_\lambda$ [see Eq. (37)], we readily obtain

$$\iiint_V \mathbf{A}_\lambda \cdot \mathbf{A}_\mu^*\, d\tau = \iiint_V \mathbf{A}_\lambda \cdot \mathbf{A}_{-\mu}\, d\tau = V\delta_{\lambda,\mu} \tag{41}$$

where the integration is over the entire volume of the cavity. Using Eq. (38), we obtain the following expressions for the electric and magnetic fields:

$$\mathcal{E} = -\frac{\partial \mathbf{A}}{\partial t} = \sum_\lambda \mathcal{E}_\lambda \tag{42}$$

$$\mathcal{H} = \frac{1}{\mu_0}\nabla \times \mathbf{A} = \sum_\lambda \mathcal{H}_\lambda \tag{43}$$

where $$\mathcal{E}_\lambda = i\omega_\lambda \left[q_\lambda(t)\mathbf{A}_\lambda(\mathbf{r}) - q_\lambda^*(t)\mathbf{A}_\lambda^*(\mathbf{r}) \right] \tag{44}$$

and
$$\mathcal{H}_\lambda = \frac{i}{\mu_0}\mathbf{k}_\lambda \times \left[q_\lambda \mathbf{A}_\lambda - q_\lambda^* \mathbf{A}_\lambda^*\right] \tag{45}$$

The total energy of the radiation field is given by

Now,
$$H_r = \frac{1}{2}\int (\varepsilon_0 \mathcal{E}\cdot\mathcal{E} + \mu_0 \mathcal{H}\cdot\mathcal{H})\, d\tau \tag{46}$$

$$\frac{1}{2}\varepsilon_0\int \mathcal{E}\cdot\mathcal{E}\, d\tau = -\frac{1}{2}\varepsilon_0 \sum_\lambda\sum_\mu \omega_\lambda\,\omega_\mu \Big[q_\lambda q_\mu \int \mathbf{A}_\lambda\cdot\mathbf{A}_\mu\, d\tau$$

$$-q_\lambda q_\mu^* \int \mathbf{A}_\lambda\cdot\mathbf{A}_\mu^*\, d\tau - q_\lambda^* q_\mu \int \mathbf{A}_\lambda^*\cdot\mathbf{A}_\mu\, d\tau + q_\lambda^* q_\mu^* \int \mathbf{A}_\lambda^*\cdot\mathbf{A}_\mu^*\, d\tau\Big]$$

$$= -\frac{1}{2}\varepsilon_0 V \sum_\lambda\sum_\mu \omega_\lambda\,\omega_\mu \left[q_\lambda q_\mu\, \delta_{\lambda,-\mu} - q_\lambda q_\mu^* \delta_{\lambda,\mu} - q_\lambda^* q_\mu \delta_{\lambda,\mu} + q_\lambda^* q_\mu^*\, \delta_{\lambda,-\mu}\right]$$

$$= -\frac{1}{2}\varepsilon_0 V \sum_\lambda \omega_\lambda^2 \left[q_\lambda q_{-\lambda} + q_\lambda^* q_{-\lambda}^* - 2 q_\lambda q_\lambda^*\right] \tag{47}$$

Similarly, one can evaluate $\int \mathcal{H}\cdot\mathcal{H}\, d\tau$. The final result is

$$\frac{1}{2}\mu_0 \int \mathcal{H}\cdot\mathcal{H}\, d\tau = \frac{1}{2}\varepsilon_0 V \sum_\lambda \omega_\lambda^2 \left[q_\lambda q_{-\lambda} + q_\lambda^* q_{-\lambda}^* + 2 q_\lambda q_\lambda^*\right] \tag{48}$$

where use has to be made of the vector identity

$$(\mathbf{a}\times\mathbf{b})\cdot(\mathbf{c}\times\mathbf{d}) = (\mathbf{a}\cdot\mathbf{c})\,(\mathbf{b}\cdot\mathbf{d}) - (\mathbf{b}\cdot\mathbf{c})\,(\mathbf{a}\cdot\mathbf{d}) \tag{49}$$

and the relation

$$k_\lambda^2 = \frac{\omega_\lambda^2}{c^2} = \varepsilon_0\mu_0\omega_\lambda^2 \tag{50}$$

Thus,
$$H_r = 2\varepsilon_0 V \sum_\lambda \omega_\lambda^2\, q_\lambda(t) q_\lambda^*(t) \tag{51}$$

We next introduce the dimensionless variables Q_λ and P_λ defined through the equations

$$Q_\lambda(t) = (\varepsilon_0 V)^{1/2} (q_\lambda(t) + q_\lambda^*(t)) \tag{52}$$

and
$$P_\lambda(t) = \frac{1}{i}\left(\varepsilon_0 V \omega_\lambda^2\right)^{1/2} (q_\lambda(t) - q_\lambda^*(t)) \tag{53}$$

Thus,
$$q_\lambda(t) = (4\varepsilon_0 V \omega_\lambda^2)^{-1/2} (\omega_\lambda Q_\lambda(t) + iP_\lambda(t)) \tag{54}$$

$$q_\lambda^*(t) = (4\varepsilon_0 V \omega_\lambda^2)^{-1/2} (\omega_\lambda Q_\lambda(t) - iP_\lambda(t)) \tag{55}$$

$$H_r = \sum_\lambda H_\lambda \tag{56}$$

where[5]

$$H_\lambda = \frac{1}{2}\left(P_\lambda^2 + \omega_\lambda^2 Q_\lambda^2\right) \tag{57}$$

The Hamiltonian given by Eq. (57) is identical to that of the linear harmonic oscillator (see Section 12.2) which suggests that the electromagnetic field can be regarded as an infinite set of harmonic oscillators, one corresponding to each value of $\mathbf{k}_\lambda$ and to a particular direction of polarization.

Quantization of electromagnetic field

In order to quantize the electromagnetic field we use the same approach as in Section 12.2. We consider Q_λ and P_λ to be real operators satisfying the commutation relations

$$[Q_\lambda(t), P_\lambda(t)] \equiv Q_\lambda(t)\, P_\lambda(t) - P_\lambda(t)\, Q_\lambda(t) = i\hbar \tag{58}$$

$$[Q_\lambda(t), P_{\lambda'}(t)] = 0;\ \lambda \neq \lambda' \tag{59}$$

$$[Q_\lambda(t), Q_\mu(t)] = 0 = [P_\lambda(t), P_\mu(t)] \tag{60}$$

where all the operators are in the Heisenberg representation (see Section 12.9). We next introduce the dimensionless variables

$$a_\lambda(t) = \frac{1}{\sqrt{2\hbar\omega_\lambda}}\left[\omega_\lambda Q_\lambda(t) + iP_\lambda(t)\right] \tag{61}$$

$$\bar{a}_\lambda(t) = \frac{1}{\sqrt{2\hbar\omega_\lambda}}\left[\omega_\lambda Q_\lambda(t) - iP_\lambda(t)\right] \tag{62}$$

Since $[\omega_\lambda Q_\lambda(t) + iP_\lambda(t)]$ is proportional to $q_\lambda(t)$ [see Eq. (54)] which has a time dependence of the form $e^{-i\omega_\lambda t}$, we may write (see also Section 12.8)

$$a_\lambda(t) = a_\lambda e^{-i\omega_\lambda t} \tag{63}$$

Similarly,

$$\bar{a}_\lambda(t) = \bar{a}_\lambda\, e^{+i\omega_\lambda t} \tag{64}$$

where

$$a_\lambda \equiv a_\lambda(0) \text{ and } \bar{a}_\lambda \equiv \bar{a}_\lambda(0) \tag{65}$$

Solving Eqs (61) and (62) for $Q_\lambda(t)$ and $P_\lambda(t)$, we obtain

$$Q_\lambda(t) = \sqrt{\frac{\hbar}{2\omega_\lambda}}\left[\bar{a}_\lambda(t) + a_\lambda(t)\right] \tag{66}$$

---

[5] Notice that

$$\frac{\partial H_\lambda}{\partial Q_\lambda} = \omega_\lambda^2 Q_\lambda = \omega_\lambda^2(\varepsilon_0 V)^{1/2}(q_\lambda + q_\lambda^*) = i(\varepsilon_0 V\omega_\lambda^2)^{1/2}(\dot{q}_\lambda - \dot{q}_\lambda^*) = -\dot{P}_\lambda$$

Similarly,

$$\frac{\partial H_\lambda}{\partial P_\lambda} = +\dot{Q}_\lambda$$

which are nothing but Hamilton's equations of motion [see Ref. 1]. Thus Q_λ and P_λ are the canonical coordinates.

$$P_\lambda(t) = i\sqrt{\frac{\hbar\omega_\lambda}{2}}\left[\bar{a}_\lambda(t) - a_\lambda(t)\right] \tag{67}$$

Substituting the above expressions for $Q_\lambda(t)$ and $P_\lambda(t)$ in Eq. (57), we obtain

$$H_r = \sum_\lambda H_r = \sum_\lambda \frac{1}{2}\hbar\omega_\lambda(\bar{a}_\lambda(t)a_\lambda(t) + a_\lambda(t)\bar{a}_\lambda(t))$$

$$= \sum_\lambda \frac{1}{2}\hbar\omega_\lambda(\bar{a}_\lambda a_\lambda + a_\lambda\bar{a}_\lambda) \tag{68}$$

If we now carry out an analysis similar to that followed in Section 12.2, we obtain

$$\left(n_\lambda + \frac{1}{2}\right)\hbar\omega_\lambda;\ n_\lambda = 0, 1, 2, \ldots \tag{69}$$

as the eigenvalues of H_λ and

$$\sum_\lambda \left(n_\lambda + \frac{1}{2}\right)\hbar\omega_\lambda \tag{70}$$

as the eigenvalues of the total Hamiltonian H_r (= ΣH_λ). Thus, quantum mechanically, we can visualize the radiation field as consisting of an infinite number of simple harmonic oscillators; the energy of each oscillator can increase or decrease by integral multiples of $\hbar\omega_\lambda$. If we consider $\hbar\omega_\lambda$ as the energy of a photon, each oscillator can have energy corresponding to n_λ photons.

The eigenkets of the total Hamiltonian would be

$$|n_1\rangle|n_2\rangle|n_3\rangle\ldots = |n_1, n_2, n_3, \ldots, n_\lambda, \ldots\rangle \tag{71}$$

where n_λ represents the number of photons in the mode characterized by λ. Thus,

$$H_r|n_1, n_2, \ldots, n_\lambda, \ldots\rangle = \left(\sum_\lambda \left(n_\lambda + \frac{1}{2}\right)\hbar\omega_\lambda\right)|n_1, n_2, \ldots, n_\lambda, \ldots\rangle \tag{72}$$

Further [see Eqs (24) and (25) of Chapter 12]:

$$a_\lambda|n_1, n_2, \ldots, n_\lambda, \ldots\rangle = \sqrt{n_\lambda}|n_1, n_2, \ldots, n_\lambda - 1, \ldots\rangle \tag{73}$$

$$\bar{a}_\lambda|n_1, n_2, \ldots, n_\lambda, \ldots\rangle = \sqrt{n_\lambda + 1}|n_1, n_2, \ldots, n_\lambda + 1, \ldots\rangle \tag{74}$$

and

$$\langle n_1', n_2', \ldots, n_\lambda', \ldots|n_1, n_2, \ldots, n_\lambda \ldots\rangle = \delta_{n_1, n_1'}\delta_{n_2, n_2'}\ldots\delta_{n_\lambda, n_\lambda'}\ldots \tag{75}$$

Finally, the state of the radiation field need not be an eigenstate of H_r, it could be a superposition of the eigenstates like that given by the following equation.

$$|\psi\rangle = \sum_{n_1, \ldots} C_{n_1, n_2, \ldots, n_\lambda \ldots}|n_1, n_2, \ldots, n_\lambda, \ldots\rangle \tag{76}$$

Physically $\left|C_{n_1, n_2, \ldots, n_\lambda \ldots}\right|^2$ would represent the probability of finding n_1 photons in the first mode, n_2 in the second mode, etc.

29.4 SPONTANEOUS AND STIMULATED EMISSIONS

We next express the vector potential **A** (and hence the interaction term H') in terms of the operators a_λ and $\bar{a}_\lambda$. Now using Eqs (54) and (61), we get

$$q_\lambda(t) = \left(\frac{\hbar}{2\varepsilon_0 V\omega_\lambda}\right)^{1/2} a_\lambda(t) \tag{77}$$

which is now to be considered as an operator. Using Eq. (38), we get

$$\mathbf{A} = \sum_\lambda \left(\frac{\hbar}{2\varepsilon_0 V\omega_\lambda}\right)^{1/2} \left[a_\lambda(t)\mathbf{A}_\lambda(\mathbf{r}) + \bar{a}_\lambda(t)\mathbf{A}_\lambda^*(\mathbf{r})\right]$$

$$= \sum_\lambda \left(\frac{\hbar}{2\varepsilon_0 V\omega_\lambda}\right)^{1/2} \left[a_\lambda(t)e^{i\mathbf{k}_\lambda\cdot\mathbf{r}} + \bar{a}_\lambda(t)e^{-i\mathbf{k}_\lambda\cdot\mathbf{r}}\right]\hat{\mathbf{e}}_\lambda \tag{78}$$

where all the operators are in the Heisenberg representation (see Section 12.9). In the Schrödinger representation, we will have

$$\mathbf{A} = \sum_\lambda \left(\frac{\hbar}{2\varepsilon_0 V\omega_\lambda}\right)^{1/2} \left[a_\lambda e^{i\mathbf{k}_\lambda\cdot\mathbf{r}} + \bar{a}_\lambda e^{-i\mathbf{k}_\lambda\cdot\mathbf{r}}\right]\hat{\mathbf{e}}_\lambda \tag{79}$$

which will be independent of time. The interaction energy is

$$H' = \frac{q}{m}\mathbf{A}\cdot\mathbf{p} \tag{80}$$

or

$$H' = \frac{q}{m}\sum_\lambda \left(\frac{\hbar}{2\varepsilon_0 V\omega_\lambda}\right)^{1/2} \left[a_\lambda e^{i\mathbf{k}_\lambda\cdot\mathbf{r}} + \bar{a}_\lambda e^{-i\mathbf{k}_\lambda\cdot\mathbf{r}}\right]\hat{\mathbf{e}}_\lambda\cdot\mathbf{p} \quad \text{Interaction energy} \tag{81}$$

Now, the eigenvalue equations for H_a and H_r are

$$H_a|\Psi_i\rangle = E_i|\Psi_i\rangle \tag{82}$$

and

$$H_r|n_1, n_2 \ldots\rangle = \left[\sum_\lambda \left(n_\lambda + \frac{1}{2}\right)\hbar\omega_\lambda\right]|n_1, n_2, \ldots, n_\lambda \ldots\rangle \tag{83}$$

where $|\Psi_i\rangle$ and E_i represent respectively the eigenkets and energy eigenvalues of the isolated atom and $|n_1, n_2, \ldots\rangle$ represent the eigenket of the pure radiation field with $\sum_\lambda \left(n_\lambda + \frac{1}{2}\right)\hbar\omega_\lambda$ representing the corresponding eigenvalue [see Eq. (72)]. Thus the eigenvalue equation for H_0 will be

$$H_0|u_n\rangle = W_n|u_n\rangle \tag{84}$$

where

$$W_n = E_n + \sum\left(n_\lambda + \frac{1}{2}\right)\hbar\omega_\lambda \tag{85}$$

and

$$|u_n\rangle = |i\rangle|n_1, n_2, \ldots, n_\lambda, \ldots\rangle = |i; n_1, n_2, \ldots, n_\lambda, \ldots\rangle \tag{86}$$

represents the ket corresponding to the atom being in state $|i\rangle$ and the radiation being in the state $|n_1, n_2, ..., n_\lambda, ...\rangle$. Now the Schrödinger equation for the system consisting of the atom and the radiation field is

$$i\hbar \frac{\partial |\Psi\rangle}{\partial t} = (H_0 + H')|\Psi\rangle \tag{87}$$

The solution of the above equation can be written as a linear combination of the eigenkets of H_0 [cf. Eq. (15) of Chapter 28]:

$$|\Psi\rangle = \sum_n C_n(t)\, e^{-iW_n t/\hbar} |u_n\rangle \tag{88}$$

Substituting in Eq. (87), we obtain

$$i\hbar \sum_n \left[\frac{dC_n}{dt} - i\frac{W_n}{\hbar} C_n\right] e^{-iW_n t/\hbar} |u_n\rangle = \sum_n C_n(t)\, W_n e^{-iW_n t/\hbar} |u_n\rangle + H' \sum_n C_n(t)\, e^{-iW_n t/\hbar} |u_n\rangle$$

where we have used Eq. (84). Premultiplying by $\langle u_s |$ we get [cf. Eq. (16) of Chapter 28]

$$i\hbar \frac{dC_s}{dt} = \sum_n \langle u_s | H' | u_n \rangle e^{i(W_m - W_n)t/\hbar} C_n(t) \tag{89}$$

Now using Eq. (81)

$$\langle u_s | H' | u_n \rangle = \frac{q}{m} \sum_\lambda \left(\frac{\hbar}{2\varepsilon_0 V \omega_\lambda}\right)^{1/2} \hat{\mathbf{e}}_\lambda \cdot \langle u_s | \left(a_\lambda e^{i\mathbf{k}_\lambda \cdot \mathbf{r}} + \bar{a}_\lambda e^{-i\mathbf{k}_\lambda \cdot \mathbf{r}}\right) \mathbf{p} | u_n \rangle \tag{90}$$

Because of the appearance of a_λ and $\bar{a}_\lambda$ in the expression for H', the various terms in $\langle u_s | H' | u_n \rangle$ will be non-zero only if the number of photons in $| u_s \rangle$ differs by unity from the number of photons in $| u_n \rangle$. If we write out completely the right-hand side of Eq. (89) it will lead to a coupled set of an infinite number of equations which would be impossible to solve. We employ the perturbation theory and consider the absorption of one photon (of energy $\hbar\omega_i$) from the ith mode. Further, if we assume the frequency ω_i to be very close to the resonant frequency corresponding to the transition from the atomic state $| a \rangle$ to $| b \rangle$, then Eq. (89) reduces to the following two coupled equations (see also Solution 29.3):

$$i\hbar \frac{dC_1}{dt} = H'_{12}\, e^{i(W_1 - W_2)t/\hbar}\, C_2(t) \tag{91}$$

$$i\hbar \frac{dC_2}{dt} = H'_{21}\, e^{-i(W_1 - W_2)t/\hbar}\, C_1(t) \tag{92}$$

where
$$|1\rangle = |a; n_1, n_2, \ldots, n_\lambda, \ldots, n_i, \ldots\rangle \tag{93}$$

and
$$|2\rangle = |b; n_1, n_2, \ldots, n_\lambda, \ldots, n_i - 1, \ldots\rangle \tag{94}$$

represent the initial and final states of the system. Obviously, because of relations like Eq. (115), $H'_{11} = H'_{22}$. Further,

$$\left.\begin{aligned} W_1 &= E_a + \sum_{\substack{\lambda \\ \lambda \neq i}} \left(n_\lambda + \frac{1}{2}\right)\hbar\omega_\lambda + \left(n_i + \frac{1}{2}\right)\hbar\omega_i \\ W_2 &= E_b + \sum_{\substack{\lambda \\ \lambda \neq i}} \left(n_\lambda + \frac{1}{2}\right)\hbar\omega_\lambda + \left(n_i - 1 + \frac{1}{2}\right)\hbar\omega_i \end{aligned}\right\} \tag{95}$$

Thus,
$$W_1 - W_2 = (E_a - E_b) + \hbar\omega_i \tag{96}$$

Now,
$$H_{21}'^{*} = H_{12}' = \frac{q}{m}\sum_\lambda \left(\frac{\hbar}{2\varepsilon_0 V\omega_\lambda}\right)^{1/2} \hat{\mathbf{e}}_\lambda \cdot \langle a; n_1, n_2, \ldots, n_i, \ldots|$$

$$(a_\lambda e^{i\mathbf{k}_\lambda\cdot\mathbf{r}} + \bar{a}_\lambda e^{-i\mathbf{k}_\lambda\cdot\mathbf{r}})\mathbf{p}|\, b; n_1, n_2, \ldots, n_i - 1, \ldots\rangle$$

$$= \frac{q}{m}\left(\frac{\hbar}{2\varepsilon_0 V\omega_i}\right)^{1/2} \sqrt{n_i}\, \hat{\mathbf{e}}_\lambda \cdot \langle a\,|e^{-i\mathbf{k}_\lambda\cdot\mathbf{r}}\mathbf{p}|\, b\rangle$$

$$\simeq \frac{q}{m}\left(\frac{\hbar}{2\varepsilon_0 V\omega_i}\right)^{1/2} \sqrt{n_i}\, \hat{\mathbf{e}}_\lambda \cdot \langle a|\mathbf{p}|b\rangle \tag{97}$$

where in the last step we have used the dipole approximation in which we have replaced $e^{-i\mathbf{k}_\lambda\cdot\mathbf{r}}$ by unity [see Eq. (14)]. If we now use Eq. (19), we get

$$\begin{aligned} \langle a\,|\mathbf{p}|\,b\rangle &= \frac{m}{i\hbar}(E_b - E_a)\langle a|\mathbf{r}|b\rangle \\ &= -im\omega_{ba}\langle a\,|\,\mathbf{r}\,|\,b\rangle \end{aligned} \tag{98}$$

Thus,
$$H_{21}'^{*} = H_{12}' = -i\left(\frac{\hbar}{2\varepsilon_0 V\omega_i}\right)^{1/2} \omega_{ba}\sqrt{n_i}\, D_{ab}$$

where
$$D_{ab} = q\,\hat{\mathbf{e}}_\lambda \cdot \langle a|\mathbf{r}|b\rangle = \frac{iq}{m\omega_{ba}}\hat{\mathbf{e}}_\lambda \cdot \langle a|\mathbf{p}|b\rangle \tag{99}$$

[see Eq. (20) of Chapter 28]. We next try to solve Eqs (91) and (92) by using a method similar to that employed in Problem 28.8. We assume that at $t = 0$ the system is in the state represented by $|\,1\rangle$, i.e.,

$$C_1(0) = 1,\ C_2(0) = 0 \tag{100}$$

On working out the solution one obtains

$$|C_2(t)|^2 \simeq \left(\frac{\Omega_0}{2}\right)^2 \left[\frac{\sin \Omega' t/2}{\Omega'/2}\right]^2 \tag{101}$$

where
$$\Omega_0 = \left[\frac{2n_i |D_{ab}|^2}{\hbar\varepsilon_0 V}\cdot\frac{\omega_{ab}^2}{\omega_i}\right]^{1/2} \tag{102}$$

and
$$\Omega' = \left[(\omega_{ba} - \omega_i)^2 + \Omega_0^2\right]^{1/2} \tag{103}$$

For $\Omega_0 t/\hbar \ll 1$, we obtain

$$|C_2(t)|^2 \simeq \frac{n\omega |D_{ab}|^2}{2\hbar\varepsilon_0 V}\left[\frac{\sin(\omega_{ba}-\omega)t/2}{(\omega_{ba}-\omega)/2}\right]^2 \tag{104}$$

where we have dropped the subscript i and have taken into account the fact that $|C_2(t)|^2$, is negligible except when $\omega \approx \omega_{ba}$ and have thus replaced ω_{ba} by ω. Equation (104) is the same as Eq. (24) of Chapter 28 provided we replace[6] ε_0^2 by $2n\hbar\omega/V\varepsilon_0$. Using Eq. (104) and proceeding as in Section 28.4, we would get an expression for the Einstein B coefficient identical to Eq. (40) of Chapter 28.

In a similar manner, if we consider the emission process, we would obtain

$$|C_2(t)|^2 \simeq \frac{\omega(n+1)}{2\hbar\varepsilon_0 V}|D_{ab}|^2\left[\frac{\sin(\omega_{ba}-\omega)t/2}{(\omega_{ba}-\omega)/2}\right]^2 \tag{105}$$

where the initial and final states $|1\rangle$ and $|2\rangle$ are now given by[7]

$$\left.\begin{aligned} |\text{initial}\rangle &= |1\rangle = |b; n_1, n_2, \ldots, n_i, \ldots\rangle \\ |\text{final}\rangle &= |2\rangle = |a; n_1, n_2, \ldots, n_i+1, \ldots\rangle \end{aligned}\right\} \tag{106}$$

Notice the presence of the term $(n + 1)$ in Eq. (105). The appearance of the term $(n + 1)$ is because of the relation $\left|\langle n+1|\bar{a}|n\rangle\right|^2 = \left|\sqrt{n+1}\langle n+1|n+1\rangle\right|^2 = (n+1)$. This implies that even if the number of photons were zero originally, the emission probability is finite. The term proportional to n in Eq. (105) gives the probability for induced or stimulated emission since the rate at which it occurs is proportional to the intensity of the applied radiation. On the other hand, the second term which is independent of n gives the spontaneous emission rate into the mode (see also Problem 29.3). It may be noted that the spontaneous emission probability into a particular mode is exactly the same as the stimulated emission probability caused by a single photon into the same mode.

We next calculate the probability per unit time for spontaneous emission of radiation. If we consider the emission to be in the solid angle $d\Omega$ then the number of modes for which the photon frequency lies between ω and $\omega + d\omega$ is (see Appendix M)

$$N(\omega)\,d\omega\,d\Omega = \frac{V\omega^2 d\omega}{8\pi^3 c^3}\,d\Omega \tag{107}$$

Thus, the total probability of emission in the solid angle $d\Omega$ would be given by

---

[6] This is justified because the energy density associated with an electromagnetic field is $\frac{1}{2}\varepsilon_0 E_0^2$ and the energy density is also equal to $n\hbar\omega/V$.

[7] We would have

$$i\hbar\frac{dC_2}{dt} \simeq H'_{21}\,e^{-i(W_1-W_2)t/\hbar}\,C_1(t) \simeq H'_{21}\,e^{-i(\omega_{ba}-\omega_i)t}$$

and integration of which would lead to Eq. (105). However, it may be mentioned that the equation for $C_1(t)$ would be a sum over states as discussed in Solution 29.3.

$$\Gamma = \frac{|D_{ab}|^2}{2\hbar\varepsilon_0 V}\int\left[\frac{\sin(\omega_{ba}-\omega)t/2}{(\omega_{ba}-\omega)/2}\right]^2 \omega\,\frac{V}{8\pi^3c^3}\,\omega^2 d\omega\, d\Omega$$

$$\simeq \frac{|D_{ab}|^2}{2\hbar\varepsilon_0 8\pi^3 c^3}\, d\Omega\,\omega_{ba}^3\int\left[\frac{\sin(\omega_{ba}-\omega)t/2}{(\omega_{ba}-\omega)/2}\right]^2 d\omega \tag{108}$$

where use has been made of the fact that the quantity inside the square brackets is a sharply peaked function around $\omega = \omega_{ba}$. Carrying out the integration, using the fact that

$$\int_{-\infty}^{+\infty}\frac{\sin^2 x}{x^2}\,dx = \pi$$

we obtain

$$\Gamma \cong \frac{1}{2\pi}\left[\frac{q^2}{4\pi\varepsilon_0\hbar c}\right]\frac{\omega^3}{c^2}\left|\langle a|\mathbf{r}|b\rangle\cdot\hat{\mathbf{e}}\right|^2 d\Omega\, t \tag{109}$$

Thus the transition rate is given by

$$w_{sp} = \frac{1}{2\pi}\left[\frac{q^2}{4\pi\varepsilon_0\hbar c}\right]\frac{\omega^3}{c^2}\left|\langle a|\mathbf{r}|b\rangle\cdot\hat{\mathbf{e}}\right|^2 d\Omega \quad \text{Transition rate} \tag{110}$$

In order to calculate the total probability per unit time for the spontaneous emission to occur (the inverse of which will give the spontaneous lifetime of the state), we must sum over the two independent states of polarization and integrate over the solid angle. Assuming the direction of **k** to be along the z-axis, we may choose $\hat{\mathbf{e}}$ to be along the x-or y-axes. Thus, if we sum

$$\left|\langle a|\mathbf{r}|b\rangle\cdot\hat{\mathbf{e}}\right|^2$$

over the two independent states of polarization, we obtain

$$\left|\langle a|\mathbf{r}|b\rangle\cdot\hat{\mathbf{x}}\right|^2 + \left|\langle a|\mathbf{r}|b\rangle\cdot\hat{y}\right|^2 = P_x^2 + P_y^2 = P^2\sin^2\theta$$

where $\mathbf{P} \equiv \langle a\,|\mathbf{r}|\,b\rangle$ and θ is the angle that **P** makes with the z-axis. Thus in order to obtain the Einstein A coefficient (which represents the total probability per unit time for the spontaneous emission to occur), in Eq. (110), we replace $\left|\langle a|\mathbf{r}|b\rangle\cdot\hat{\mathbf{e}}\right|^2$ by $|\langle a\,|\mathbf{r}|\,b\rangle|^2\sin^2\theta$ and integrate over the solid angle $d\Omega$ to obtain

$$A = \frac{1}{2\pi}\left[\frac{q^2}{4\pi\varepsilon_0\hbar c}\right]\frac{\omega^3}{c^2}|\langle a|\mathbf{r}|b\rangle|^2\iint\sin^2\theta\sin\theta\, d\theta\, d\phi$$

$$= \frac{4}{3}\left[\frac{q^2}{4\pi\varepsilon_0\hbar c}\right]\frac{\omega^3}{c^2}|\langle a|\mathbf{r}|b\rangle|^2 \quad \text{Einstein's } A \text{ coefficient} \tag{111}$$

which is identical to Eq. (43) of Chapter 28.

29.5 PROPERTIES OF THE EIGENSTATES OF THE HAMILTONIAN OF THE RADIATION FIELD

If we substitute for $q_\lambda(t)$ from Eq. (77) in Eq. (44), we would obtain

$$\mathcal{E} = \sum_\lambda \mathcal{E}_\lambda \tag{112}$$

where
$$\mathcal{E}_\lambda = i\left(\frac{\hbar\omega_\lambda}{2\varepsilon_0 V}\right)^{1/2}\left[a_\lambda e^{i\mathbf{k}_\lambda\cdot\mathbf{r}} - \bar{a}_\lambda e^{-i\mathbf{k}_\lambda\cdot\mathbf{r}}\right]\hat{\mathbf{e}}_\lambda \tag{113}$$

and all operators are in the Schrödinger representation. We consider the state of the radiation field for which there are n_λ photons in the λth state. The expectation value of $\mathcal{E}_\lambda$ in this state would be given by

$$\langle n_1, n_2, \ldots, n_\lambda, \ldots | \mathcal{E}_\lambda | n_1, n_2, \ldots, n_\lambda, \ldots\rangle = \langle n_1 | n_1\rangle\langle n_2|n_2\rangle \ldots \langle n_\lambda | \mathcal{E}_\lambda | n_\lambda\rangle = 0 \tag{114}$$

Expectation value of $\mathcal{E}_\lambda$

because
$$\langle n_\lambda | a_\lambda | n_\lambda\rangle = 0 = \langle n_\lambda | \bar{a}_\lambda | n_\lambda\rangle \tag{115}$$

Similarly,

$$\langle n_1, n_2, \ldots, n_\lambda, \ldots | \mathcal{E}_\lambda^2 | n_1, n_2, \ldots, n_\lambda, \ldots\rangle = \langle n_1|n_1\rangle\langle n_2|n_2\rangle\ldots\langle n_\lambda|\mathcal{E}_\lambda^2|n_\lambda\rangle \ldots$$

$$= -\left(\frac{\hbar\omega_\lambda}{2\varepsilon_0 V}\right)\langle n_\lambda|(a_\lambda e^{i\mathbf{k}_\lambda\cdot\mathbf{r}} - \bar{a}_\lambda e^{-i\mathbf{k}_\lambda\cdot\mathbf{r}})(a_\lambda e^{i\mathbf{k}_\lambda\cdot\mathbf{r}} - \bar{a}_\lambda e^{-i\mathbf{k}_\lambda\cdot\mathbf{r}})|n_\lambda\rangle$$

$$= \left(\frac{\hbar\omega_\lambda}{\varepsilon_0 V}\right)\left(n_\lambda + \frac{1}{2}\right) \tag{116}$$

where use has been made of relations like (see Section 12.2):

$$\langle n_\lambda | a_\lambda \bar{a}_\lambda | n_\lambda\rangle = \sqrt{n_\lambda + 1}\,\langle n_\lambda | a_\lambda | n_\lambda + 1\rangle = n_\lambda + 1 \tag{117}$$

$$\langle n_\lambda | \bar{a}_\lambda a_\lambda | n_\lambda\rangle = \sqrt{n_\lambda}\,\langle n_\lambda | \bar{a}_\lambda | n_\lambda - 1\rangle = n_\lambda \tag{118}$$

$$\langle n_\lambda | a_\lambda a_\lambda | n_\lambda\rangle = 0 \tag{119}$$

$$\langle n_\lambda | \bar{a}_\lambda \bar{a}_\lambda | n_\lambda\rangle = 0 \tag{120}$$

The uncertainty in $\mathcal{E}_\lambda$, $\Delta\mathcal{E}_\lambda$ can be defined through the relation

$$(\Delta\mathcal{E}_\lambda)^2 = \langle\mathcal{E}_\lambda^2\rangle - \langle\mathcal{E}_\lambda\rangle^2 = \left(\frac{\hbar\omega_\lambda}{\varepsilon_0 V}\right)\left(n_\lambda + \frac{1}{2}\right) \tag{121}$$

Equation (114) tells us that the expectation value of the electric field in the state $|n\rangle$ (= $|n_1, n_2, \ldots, n_\lambda, \ldots\rangle$) is zero. Since the average of sine waves with random phases is zero, we may loosely say that the state $|n\rangle$ does not specify the phase[8]. Some authors tend to explain this by resorting to the uncertainty principle

[8] We say it loosely because it is not possible to define a phase operator which is real (see Section 29.7).

$$\Delta E\, \Delta t \gtrsim \hbar \tag{122}$$

where ΔE is the uncertainty in the energy of the radiation field and Δt is related to the uncertainty in the phase angle through the relation

$$\Delta\phi = \omega\Delta t \tag{123}$$

Since $E = (n + 1/2)\hbar\omega$, $\Delta E = \hbar\omega\Delta n$ and we obtain

$$\Delta n\, \Delta\phi \gtrsim 1 \tag{124}$$

If the number of photons is exactly known, then $\Delta n = 0$ and consequently there is no knowledge of the phase. However, such arguments are not rigorously correct because it is not possible to give a precise definition of $\Delta\phi$ (see Section 29.7). Nevertheless, we can say that the states described by $| n \rangle$ do not, correspond to the classical electromagnetic wave with a certain phase.

Returning to Eq. (116), we notice that the states $|n\rangle$ have a definite amplitude $(2\hbar\omega_\lambda/\varepsilon V)^{1/2}(n_\lambda + 1/2)^{1/2}$ for the mode λ which is directly related to the number of photons.

29.6 THE COHERENT STATES

We next consider the radiation field to be in one of the coherent states which are, the eigenkets of the operator a_λ (see Section 12.5). We will show that when the radiation field is in a coherent state, the field has properties very similar to that of a classical electromagnetic wave with a certain phase and amplitude. However, before we do so, we would like to discuss some of the properties of the coherent state.

The coherent states satisfy the equation

$$a_\lambda |\alpha_\lambda\rangle = \alpha_\lambda |\alpha_\lambda\rangle \tag{125}$$

where α_λ, which represents the eigenvalues of a_λ, can be an arbitrary complex number. In Section 12.5 we had shown that

$$|\alpha_\lambda\rangle = \exp\left[-\frac{1}{2}|\alpha_\lambda|^2\right] \sum_{n_\lambda = 0,1,2,\ldots} \frac{\alpha_\lambda^{n_\lambda}}{\sqrt{n_\lambda!}}|n_\lambda\rangle$$

For convenience, we drop the subscript λ and write the above equation as

$$|\alpha\rangle = \exp\left[-N/2\right]\sum_n \frac{\alpha^n}{\sqrt{n!}}|n\rangle \tag{126}$$

where $N(= |\alpha|^2)$ represents the expectation value of the number operator N_{op} (see Section 12.7). Since $| H\rangle$ are eigenkets of H, if the field is in the coherent state at $t = 0$, then at a later time t, the state will be given by (see Section 12.6)

$$|\Psi(t)\rangle = e^{-N/2}\sum_n \frac{\alpha^n}{\sqrt{n!}}\exp\left[-i\left(n+\frac{1}{2}\right)\omega t\right]|n\rangle \tag{127}$$

It is easy to see that

$$|\Psi(0)\rangle = |\alpha\rangle$$

Further[9]

$$\langle \Psi(t)|\bar{a}|\Psi(t)\rangle = e^{-N}\sum_m\sum_n \frac{\alpha^{*m}\alpha^n}{\sqrt{m!\,n!}} e^{i(m-n)\omega t}\sqrt{n+1}\langle m\,|\,n+1\rangle$$

$$= e^{i\omega t}\, e^{-N}\sum \frac{\alpha^*|\alpha|^{2n}}{n!} = \alpha^* e^{i\omega t} \tag{128}$$

Similarly, or taking the complex conjugate of the above equation

$$\langle \Psi(t)|a|\Psi(t)\rangle = \alpha e^{-i\omega t} \tag{129}$$

We now consider the radiation field to be in the coherent state and calculate the expectation value of $\mathcal{E}$ and $\mathcal{E}^2$:

$$\langle \Psi(t)|\mathcal{E}|\Psi(t)\rangle = i\left(\frac{\hbar\omega}{2\varepsilon_0 V}\right)^{1/2}\left[\langle \Psi(t)|a|\Psi(t)\rangle\, e^{i\mathbf{k}\cdot\mathbf{r}} - \langle \Psi(t)|\bar{a}|\Psi(t)\rangle\, e^{-i\mathbf{k}\cdot\mathbf{r}}\right]\hat{\mathbf{e}}$$

$$= i\left(\frac{\hbar\omega}{2\varepsilon_0 V}\right)^{1/2}\left[\alpha e^{i(\mathbf{k}.\mathbf{r}-\omega t)} - \alpha^* e^{-i(\mathbf{k}.\mathbf{r}-\omega t)}\right]\hat{\mathbf{e}}$$

$$= -2\left(\frac{\hbar\omega}{2\varepsilon_0 V}\right)^{1/2}|\alpha|\sin(\mathbf{k}.\mathbf{r}-\omega t+\phi)\,\hat{\mathbf{e}} \tag{130}$$

where $$\alpha = |\alpha| e^{i\phi} \tag{131}$$

Thus the coherent state can be interpreted to represent a harmonic wave with phase ϕ. In a similar manner, we can calculate the expectation value ε^2. The result is

$$\langle \Psi(t)|\mathcal{E}\cdot\mathcal{E}|\Psi(t)\rangle = \left(\frac{\hbar\omega}{2\varepsilon_0 V}\right)\left[4|\alpha|^2\sin^2(\mathbf{k}\cdot\mathbf{r}-\omega t+\phi)+1\right] \tag{132}$$

Finally, the uncertainty in $\mathcal{E}$ would be given by [cf. Eq. (121)]:

$$(\Delta\mathcal{E})^2 = \langle \Psi(t)|\mathcal{E}^2|\Psi(t)\rangle - \langle \Psi(t)|\mathcal{E}|\Psi(t)\rangle^2$$

$$= \left(\frac{\hbar\omega}{2\varepsilon_0 V}\right) \tag{133}$$

Notice that the uncertainty $\Delta\mathcal{E}$ is independent of the amplitude $|\alpha|$; thus, greater the intensity of the beam greater will be the proximity of the radiation field (corresponding to the coherent state) to the classical plane wave. Indeed when a laser is operated much beyond the threshold, it generates a coherent state excitation of a cavity mode.

---

[9] In the Heisenberg representation, the expectation value of $\bar{a}$ would have been

$$\langle \Psi(0)|\bar{a}(t)|\Psi(0)\rangle = \langle \alpha|\bar{a}e^{i\omega t}|\alpha\rangle = \alpha^* e^{i\omega t}$$

which is the same as expressed by Eq. (128).

29.7 THE PHASE OPERATOR

In classical mechanics the displacement and momentum of a linear harmonic oscillator are given by

$$x = Ae^{i\phi} + Ae^{-i\phi}, \ \phi = \omega t \tag{134}$$

and

$$p = m\dot{x} = im\omega A[e^{i\phi} - e^{-i\phi}] \tag{135}$$

where A has been assumed to be real. In quantum mechanics, we have [see Eqs (30) and (31) of Chapter 12]:

$$x = \left(\frac{\hbar}{2m\omega}\right)^{1/2}\left[\bar{a}+a\right] \tag{136}$$

$$p = im\omega\left(\frac{\hbar}{2m\omega}\right)^{1/2}\left[\bar{a}-a\right] \tag{137}$$

The above four equations suggest the phase operator ϕ to be defined through the following equations.

$$\bar{a} = Re^{i\phi} \tag{138}$$

and

$$a = e^{-i\phi} R \tag{139}$$

where R and ϕ are assumed to be Hermitian operators (see, e.g. Refs 2 and 3). However, such a definition will lead to inconsistent results as discussed below. Now, the number operator is given by

$$N_{op} = \bar{a}\,a = Re^{i\phi} e^{-i\phi} R = R^2$$

or

$$R = N_{op}^{1/2} \tag{140}$$

where the square root of an operator is defined through the relation

$$N_{op} = N_{op}^{1/2}\ N_{op}^{1/2}$$

Since,

$$\left[a, \bar{a}\right] = a\bar{a} - \bar{a}a = 1$$

so,

$$e^{-i\phi} RRe^{i\phi} - Re^{i\phi} e^{-i\phi} R = 1$$

or, premultiplying by $e^{i\phi}$, we get

$$N_{op}\, e^{i\phi} - e^{i\phi} N_{op} = e^{i\phi} \tag{141}$$

The above equation is satisfied if ϕ and N_{op} satisfy the commutation relation[10]

$$[N_{op}, \phi] = N_{op,}\, \phi - \phi\, N_{op} = -i \tag{142}$$

Thus,

$$\langle m|N_{op}\,\phi - \phi N_{op}\,|n\rangle = -i\langle m|n\rangle = -\,i\delta_{mn}$$

or

$$(m-n)\langle m|\phi|n\rangle = -\,i\delta_{mn} \tag{143}$$

---

[10] This follows from the repeated application of Eq. (142), which gives

$$N_{op}\,\phi^{\,m} - \phi^{\,m} N_{op} = -\,im\phi^{m-1}$$

Now,

$$\left[N_{op}, e^{i\phi}\right] = \sum_m \frac{i^m}{m!}\left[N_{op}, \phi^m\right] = \sum_m \frac{i^{m-1}\phi^{m-1}}{(m-1)!} = e^{i\phi}$$

which is certainly an impossibility! Thus the definition of a Hermitian ϕ through Eq. (138) leads to inconsistent results.

29.8 PROBLEMS

Problem 29.1 Consider the matrix element $\langle s|e^{-i\mathbf{k}\cdot\mathbf{r}}\hat{\mathbf{e}}\cdot\mathbf{p}|n\rangle$ [see Eq. (13)]. Since in many cases of interest $kR \ll 1$ (where R represents the dimension of the system), we may write

$$e^{-i\mathbf{k}\cdot\mathbf{r}} = 1 - i\mathbf{k}\cdot\mathbf{r} - \frac{(\mathbf{k}\cdot\mathbf{r})^2}{2!} + \cdots \tag{144}$$

It can be shown that the term $(\mathbf{k}\cdot\mathbf{r})^n$ gives the electric 2^{n+1} pole and the magnetic 2^n pole transitions. Verify this statement for the case $n = 1$.

Problem 29.2 Obtain the selection rules for allowed transitions of a linear harmonic oscillator. Calculate the lifetime of an oscillator in the first excited state.

Problem 29.3 Show that the frequency distribution of the emitted radiation from an isolated atom is given by the natural line shape function

$$f(\omega)\,d\omega = \frac{K\,d\omega}{(\omega_0 - \omega)^2 + A^2/4} \tag{145}$$

where A represents the Einstein coefficient.

Problem 29.4 Prove that the average number of photons in a coherent state is $|\alpha|^2$, *i.e.*;

$$N \equiv \langle\alpha|N_{op}|\alpha\rangle = |\alpha|^2 \tag{146}$$

Problem 29.5 Show that the probability of finding n photons in a coherent state is a Poisson distribution[11] about the mean $|\alpha|^2$:

$$|\langle n|\alpha\rangle|^2 = \frac{N^n e^{-N}}{n!} \tag{147}$$

where $N\,(= |\alpha|^2)$ is defined in the previous problem.

Problem 29.6 Prove that the uncertainty in the number of photons in a coherent state is $\sqrt{N}$

$$\Delta N \equiv \left[\langle\alpha|N_{op}^2|\alpha\rangle - \langle\alpha|N_{op}|\alpha\rangle^2\right]^{1/2} = N^{1/2} \tag{148}$$

Thus the fractional uncertainty in the average number of photons $\Delta N/N\ (= 1/\sqrt{N})$ goes to zero with increase in intensity. It may be worthwhile pointing out that even in nuclear counting, the uncertainty in the actual count is $N^{1/2}$ (see, e.g. Ref. 5).

Problem 29.7 Show that for $|\psi(t)\rangle$ given by Eq. (127)

$$\langle\psi(t)|N_{op}|\psi(t)\rangle = N \qquad \text{(independent of time)}$$

[11] The photon counting experiment is of considerable importance in determining statistical properties of the light source. In a typical experiment, the light beam is allowed to fall on a phototube for a specific time interval T by having a shutter open in front of the detector for time T and one registers the number of photoelectrons so liberated. Then the shutter is again opened for an equal time interval T after a certain time delay and the experiment is repeated a large number of times. From the experiment one obtains the probability distribution of counting n photons in time T. From this one can derive the statistical properties of the sources (see, *e.g*. Ref. 4),

Problem 27.8 Obtain an expression for the photodisintegration of the deuteron in the electric dipole approximation. Assume that the wavefunction of the deuteron may be approximated by

$$\psi = \sqrt{\frac{K}{2\pi}} \cdot \frac{e^{-Kr}}{r} \tag{149}$$

where $\hbar K = (2\mu B)^{1/2}$, $\mu = \frac{1}{2}M$ is the reduced mass of the system, $B \simeq 2.2$ MeV is the binding energy. Neglect the interaction between neutron and proton in the final state [**Hint:** Use the Golden rule.]

Problem 29.9

(a) Derive the Thomas-Reiche-Kuhn sum rule

$$\sum_s (E_s - E_n)\left|\langle s|x|n\rangle\right|^2 = \frac{\hbar^2}{2m} \tag{150}$$

(b) Using the above sum rule show that for dipole transitions

$$\int \sigma\, dE = 2\pi^2 \alpha \hbar^2 / m \tag{151}$$

Problem 29.10 If the quantum energy $\hbar\omega$ of light falling on an atom is greater than the ionization energy, I, of the atom the electron is raised into a state of the continuous spectrum. In this case light of all frequencies can be absorbed, and the absorption spectrum is continuous. The kinetic energy T of the electron after leaving the atom is determined by Einstein's equation

$$T = \hbar\omega - I \tag{152}$$

Show that for the incident radiation polarized with its electric vector along $\hat{\mathbf{e}}$, the photoelectric absorption cross-section of an atom, for the ejection of an electron from the K-shell, is given by

$$\frac{d\sigma}{d\Omega} = \frac{32\hbar k}{m\omega}\left(\frac{q^2}{4\pi\varepsilon_0 \hbar c}\right)(\hat{\mathbf{e}}\cdot\mathbf{k})^2 \left(\frac{Z}{a_0}\right)^5 \left[\left(\frac{Z}{a_0}\right)^2 + q^2\right]^{-4}, \tag{153}$$

where $\mathbf{k}$ is the wave vector of the ejected electron, $\mathbf{q}$ is the momentum transferred to the atom in units of $\hbar$:

$$\mathbf{q} = \mathbf{k} - \frac{\omega}{c}\hat{\mathbf{n}} \tag{154}$$

$\hat{\mathbf{n}}$ is the unit vector along the direction of propagation. Assume $\hbar\omega \gg I$. [Use the Golden rule.]

29.9 SOLUTIONS

Solution 29.1 The relevant term in the matrix element is

$$M = \langle s|(i\mathbf{k}\cdot\mathbf{r})(\hat{\mathbf{e}}\cdot\mathbf{p})|n\rangle \tag{155}$$

Let us assume that the photon moves along the x-direction, and the polarization $\hat{\mathbf{e}}$ is along the z-direction. Thus,

$$M = \langle s|ikxp_z|n\rangle = ik\langle s|xp_z|n\rangle$$

We write xp_z as

$$xp_z = \frac{1}{2}\left[xp_z + zp_x\right] + \frac{1}{2}\left[xp_z - zp_x\right]$$

Now, according to Eq. (17)

$$p_z = \frac{m}{i\hbar}\left[zH_a - H_a z\right]$$

Thus,

$$xp_z = \frac{m}{i\hbar}\left[xzH_a - xH_a z\right]$$

Similarly,

$$zp_x = p_x z = \frac{m}{i\hbar}\left[xH_a - H_a x\right]z$$

so that

$$\frac{1}{2}\left(xp_z + zp_x\right) = \frac{m}{2i\hbar}\left[xzH_a - H_a xz\right] \tag{156}$$

and

$$\frac{1}{2}\left(xp_z - zp_x\right) = -\frac{1}{2}L_y \tag{157}$$

where L_y is the y-component of the angular momentum. Thus,

$$M = -ik\langle s|\frac{m}{2i\hbar}(H_a xz - xzH_a) + \frac{1}{2}L_y|n\rangle$$

$$= -\frac{m}{2\hbar}(E_s - E_n)\langle s|xz|n\rangle - \frac{1}{2}ik\langle s|L_y|n\rangle \tag{158}$$

The first term is the electric quadrupole term while the second is the magnetic dipole term (the magnetic moment is related to the angular momentum).

Solution 29.2 The relevant matrix element for allowed transition is

$$M = \langle m|x|n\rangle$$

The only non-vanishing matrix elements correspond to

$$m = n \pm 1$$

(see Problem 12.3) and the values are given by

$$\left.\begin{aligned}\langle n+1|x|n\rangle &= \sqrt{\frac{\hbar}{2m\omega}}\sqrt{n+1}\\ \langle n-1|x|n\rangle &= \sqrt{\frac{\hbar}{2m\omega}}\sqrt{n}\end{aligned}\right\} \tag{159}$$

Thus the selection rule is $\Delta n = \pm 1$. The transition rate is given by [see Eq. (109)]

$$w_{sp} = \frac{1}{2\pi}\left[\frac{q^2}{4\pi\varepsilon_0 \hbar c}\right]\frac{\omega}{m^2c^2}\left|\langle 0|p_x|1\rangle\right|^2 d\Omega \tag{160}$$

Now,

$$\left|\langle 0|p_x|1\rangle\right| = -i\sqrt{\frac{m\hbar\omega}{2}} \tag{161}$$

(see Problem 12.3). Thus the integrated rate over the whole solid angle is

$$\frac{1}{\tau} = \frac{\alpha \hbar \omega^2}{mc^2} \tag{162}$$

which gives the mean life τ; here $\alpha \left(\approx \frac{1}{137} \right)$ is the fine structure constant.

Solution 29.3 A spontaneous transition from a state b to a lower energy state a does not give radiation at a single frequency ω_{ba}. A finite lifetime τ of the excited state gives it an energy width of the order $\hbar/\tau$ so that the emitted radiation has a frequency distribution. This argument may be made more precise by considering the simple case of an atom with only 2 states a and b $(E_b > E_a)$ undergoing spontaneous transition from b to a. We will calculate here the frequency distribution of the emitted radiation from such a spontaneous transition; the analysis is based on the treatment given in Chapter 18 of Ref. 3 and is known as Weisskopf-Wigner theory of the natural line width.

It is assumed that at $t = 0$, the atom is in the excited state b and the radiation field has no photons. We denote this state by

$$|1\rangle = |b; 0, 0, 0, \ldots\rangle$$

[see Eq. (93)]. The atom makes a transition to the state a emitting a photon of frequency ω_λ in the mode characterized by λ [see Eq. (71)]. We denote this state by

$$|a\lambda\rangle = |a; 0, 0, \ldots, 1_\lambda, 0, \ldots\rangle$$

If we denote the corresponding probability amplitudes by

$$C_1(t) \quad \text{and} \quad C_{a\lambda}(t)$$

then

$$\left.\begin{aligned} C_1(0) &= 1 \\ C_{a\lambda}(0) &= 0 \end{aligned}\right\} \tag{163}$$

Further [see Eqs (89)–(92)]

$$\begin{aligned} i\hbar \frac{dC_1}{dt} &= \sum_\lambda \langle 1|H'|a\lambda\rangle e^{i(W_1 - W_{a\lambda})t/\hbar} C_{a\lambda}(t) \\ &= \sum_\lambda \langle 1|H'|a\lambda\rangle e^{i(\omega_0 - \omega_\lambda)t} C_{a\lambda}(t) \end{aligned} \tag{164}$$

and

$$i\hbar \frac{dC_{a\lambda}}{dt} = \langle a\lambda|H'|1\rangle e^{-i(\omega_0 - \omega_\lambda)t} C_1(t) \tag{165}$$

where

$$\hbar\omega_0 = E_b - E_a \tag{166}$$

and

$$W_1 - W_{a\lambda} = E_b - (E_a + \hbar\omega_\lambda) = \hbar(\omega_0 - \omega_\lambda) \tag{167}$$

We try to solve Eqs (164) and (165) by assuming[12]

$$C_1(t) = e^{-\gamma t/2} \tag{168}$$

---

[12] It should be mentioned that the present treatment is consistent with the treatment given in Section 29.4. Indeed in Section 29.4 we have shown that A [as given by Eq. (111)] represents the probability for unit time for the spontaneous emission to occur from which the decay given by Eq. (108) follows with $\gamma = A$; this is explicitly shown later [see Eqs (178) and (179)].

which satisfies the condition $C_1(0) = 1$. It may noted that since $|C_1(t)|^2$ represents the probability of finding the atom in the upper state, the quantity $1/\gamma$ represents the mean lifetime of the state. If we substitute for $C_1(t)$ in Eq. (165) and carry out the integration, we get

$$C_{a\lambda}(t) = \frac{\langle a\lambda|H'|1\rangle}{\hbar} \frac{e^{-[i(\omega_0 - \omega_\lambda) + \gamma/2]t} - 1}{(\omega_0 - \omega_\lambda) - i\gamma/2} \tag{169}$$

We substitute Eqs (168) and (169) in Eq. (164) to obtain

$$-i\hbar\frac{\gamma}{2} = \sum_\lambda \frac{|\langle H'\rangle|^2}{\hbar} \frac{1 - \exp[i(\omega_0 - \omega_\lambda)t + \gamma t/2]}{(\omega_0 - \omega_\lambda) - i\gamma/2} \tag{170}$$

where $\langle H'\rangle \equiv \langle a\lambda|H'|1\rangle$. The summation on the right-hand side of the above equation is over a number of states λ with very nearly the same frequency. Under these circumstances, the summation can be replaced by an integral

$$\sum_\lambda \to \iint N(\omega)\, d\omega\, d\Omega \tag{171}$$

where $N(\omega)$ represents the density of states [see Eq. (107)]. Equation (170) then becomes

$$\gamma = \frac{2i}{\hbar}\left[\frac{1}{\hbar}\iint |\langle H'\rangle|^2 N(\omega) J(\omega)\, d\omega\, d\Omega\right] \tag{172}$$

where

$$J(\omega) = \frac{1 - \exp[i(\omega_0 - \omega)t + \gamma t/2]}{(\omega_0 - \omega) - i\gamma/2} \tag{173}$$

We will assume that $\gamma \ll \omega_0$; i.e. the inverse of the lifetime ($\leq 10^9 s^{-1}$) is much smaller than the characteristic frequencies ($\sim 10^{15}\, s^{-1}$), this will indeed follow from the final result. Under this assumption γ can be neglected in Eq. (173) to obtain

$$J \simeq \frac{1 - e^{i(\omega_0 - \omega)t}}{\omega_0 - \omega}$$

$$\simeq \frac{1 - \cos[(\omega_0 - \omega)t]}{\omega_0 - \omega} - i\frac{\sin[(\omega_0 - \omega)t]}{\omega_0 - \omega} \tag{174}$$

The first term has very rapid variation around $\omega \simeq \omega_0$ and gives negligible contribution to any integral over ω except around $\omega \simeq \omega_0$ where the function itself vanishes. Thus this term will lead to a small imaginary value of γ. We are in any case interested in the real part of γ (which would give the lifetime, etc.) which will be given by

$$\gamma \approx \frac{2}{\hbar}\left[\frac{1}{\hbar}\int |\langle H'\rangle|^2 N(\omega_0)\, d\Omega \int_{-\infty}^{+\infty} \frac{\sin(\omega_0 - \omega)t}{\omega_0 - \omega}\, d\omega\right] \tag{175}$$

We have made two approximations here. First, we have pulled out factors which are essentially constant in the neighbourhood of $\omega \simeq \omega_0$. Secondly, the limits of the integral have been taken from $-\infty$ to $+\infty$ since in any case the contribution vanishes except around ω_0. Thus,

$$\gamma \simeq \frac{2\pi}{\hbar}\left[\frac{1}{\hbar}\int |\langle H'\rangle|^2 N(\omega_0)\, d\Omega\right]$$

$$= \frac{2\pi}{\hbar}\frac{1}{\hbar}\frac{V\omega_0^2}{8\pi^3 c^3}\int \left|\langle H'\rangle\right|^2 d\Omega \tag{176}$$

Now,
$$\langle H'\rangle = \langle a\lambda|H'|1\rangle$$

$$= \langle a; 0, 0, \ldots, 1_\lambda, 0, \ldots| \frac{q}{m}\sum_\lambda \left(\frac{\hbar}{2\varepsilon_0 V\omega_\lambda}\right)^{1/2}$$

$$\times \left[a_\lambda e^{i\mathbf{k}_\lambda\cdot\mathbf{r}} + \bar{a}_\lambda e^{-i\mathbf{k}_\lambda\cdot\mathbf{r}}\right]\hat{\mathbf{e}}_\lambda\cdot\mathbf{p}|b, 0, 0, \ldots, 0_\lambda, \ldots\rangle$$

$$= iq\left(\frac{\hbar\omega_\lambda}{2\varepsilon_0 V}\right)^{1/2}\langle a|\mathbf{r}|b\rangle\cdot\hat{\mathbf{e}}_\lambda \tag{177}$$

where in the last step we have used the dipole approximation [see the discussion after Eq. (97)]. On substitution in Eq. (176), we get

$$\gamma \simeq \frac{1}{2\pi}\left[\frac{q^2}{4\pi\varepsilon_0\hbar c}\right]\frac{\omega_0^3}{c^2}\int \left|\hat{\mathbf{e}}_\lambda\cdot\langle a|\mathbf{r}|b\rangle\right|^2 d\Omega \tag{178}$$

[cf. Eq. (110)]. On carrying out the integration and summing over the two states of polarization we get [see Eq. (111)],

$$\gamma = A \tag{179}$$

i.e. γ is simply the Einstein A coefficient corresponding to spontaneous emissions as it indeed should be!

Now the probability of a photon being emitted in the mode λ is given by [see Eq. (169)]:

$$\left|C_{a\lambda}(t=\infty)\right|^2 = \frac{\left|\langle H'\rangle\right|^2}{\hbar^2}\cdot\frac{1}{(\omega_0-\omega_\lambda)^2+\gamma^2/4} \tag{180}$$

If we multiply the above equation by $N(\omega)\,d\omega d\Omega$ which gives the number of modes in the frequency interval ω to $\omega + d\omega$ and in the solid angle $d\Omega$ and carry out the integration over $d\Omega$ and sum over the two states of polarization, we would get

$$g(\omega)d\omega = \frac{\gamma}{2\pi}\frac{d\omega}{(\omega_0-\omega)^2+\gamma^2/4} \tag{181}$$

where we have used Eq. (178). The above equation indeed gives us the probability that the spontaneously emitted photon has its frequency between ω and $\omega + d\omega$, which is nothing but the Lorentzian line shape. Notice that the above expression is normalized which implies that the probability that the atom makes a spontaneous transition from the upper state to the lower state is unity (as $t \to \infty$) as it indeed should be.

Solution 29.6

$$\langle\alpha|N_{op}|\alpha\rangle = e^{-|\alpha|^2}\sum_m \frac{\alpha^{*m}}{\sqrt{m!}}\left\langle m\left|\sum_n \frac{\alpha^n}{\sqrt{n!}}n\right|n\right\rangle$$

(because $N_{op}|n\rangle = n|n\rangle$; see Section 12.7)

$$= e^{-|\alpha|^2} \sum_n \frac{|\alpha|^{2n}\, n}{n!}$$

$$= N e^{-N} \sum_{n=1} \frac{N^{n-1}}{(n-1)!} = N \tag{182}$$

Similarly, $$\langle \alpha | N_{op}^2 | \alpha \rangle = e^{-N} \sum_n \frac{N^n n^2}{n!} = N e^{-N} \sum \frac{N^{n-1}(n-1+1)}{(n-1)!}$$

$$= N^2 + N \tag{183}$$

from which Eq. (148) follows.

Solution 29.8 The differential cross-section $d\sigma$ is given by (see Section 27.3)

$$cn\, d\sigma = w\, d\Omega \tag{184}$$

where $w\, d\Omega$ represents the probability (per unit time) of scattering into the solid angle $d\Omega$ and n is the number of photons per unit volume corresponding to one photon in a box of volume, V, i.e;.

$$n = \frac{1}{V} \tag{185}$$

Further, $w\, d\Omega$ is given by the Golden rule [see Eq. (38) of Chapter 27]:

$$w\, d\Omega = \frac{2\pi}{\hbar} N(E)\, d\Omega \left| \langle f | H' | i \rangle \right|^2 \tag{186}$$

where $N(E)$ represents the density of states and for the neutron-proton system in a final state of relative momentum **p**, it is given by (see Appendix M).

$$N(E)\, d\Omega = \frac{\mu V}{(2\pi\hbar)^3} p\, d\Omega \tag{187}$$

This $N(E)\, d\Omega$ represents the number of states (per unit energy interval) such that the direction of **p** lies within a solid angle $d\Omega$ and $\mu = M/2$ is the reduced mass of the n-p system. The kets $|i\rangle$ and $|f\rangle$ represent the initial and final states and since the initial state contains one photon and the final state none, we have [cf. Eq. (97)]

$$\langle f | H' | i \rangle \simeq \frac{q}{M} \left(\frac{\hbar}{2\varepsilon_0 V \omega} \right)^{1/2} \hat{\mathbf{e}} \cdot \langle a | \mathbf{p} | b \rangle \tag{188}$$

where M is the proton mass. Thus,

$$d\sigma = \frac{V\alpha}{4\pi\hbar^2 \omega M} p \left| \hat{\mathbf{e}} \cdot \langle a | \mathbf{p} | b \rangle \right|^2 d\Omega \tag{189}$$

where $$\alpha = \frac{q^2}{4\pi\varepsilon_0 \hbar c} \simeq \frac{1}{137}$$

is the fine structure constant and we have used the fact that $\mu = M/2$. Integration over all directions gives a factor $4\pi/3$, thus

$$\sigma = \frac{1}{3}\alpha \frac{pV}{\hbar^2 \omega M} \left| \langle a|\mathbf{p}|b\rangle \right|^2 \tag{190}$$

The two states are given by

$$|b\rangle = \sqrt{\frac{K}{2\pi}} \cdot \frac{e^{-Kr}}{r} \tag{191}$$

$$|a\rangle = \frac{1}{\sqrt{V}} e^{i\mathbf{p}\cdot\mathbf{r}/\hbar} \tag{192}$$

The final state corresponds to a free neutron-proton system. We must remember that in Eq. (190), **p** is the operator $-i\hbar\nabla$. The integration, however, poses no problem because[13]

$$\left|\langle a|\mathbf{p}|b\rangle\right|^2 = \left|\langle b|\mathbf{p}|a\rangle\right|^2$$

$$= p^2 \left|\langle b|a\rangle\right|^2$$

$$= p^2 \frac{K}{2\pi V} \left| \int \frac{e^{-Kr}}{r} e^{i\mathbf{p}\cdot\mathbf{r}/\hbar} r^2 dr \sin\theta \, d\theta \, d\phi \right|^2$$

$$= \frac{Kp^2}{2\pi V} (2\pi)^2 \left| \int_0^\infty \int_0^\pi e^{-Kr} e^{ipr\cos\theta/\hbar} r \, dr \sin\theta \, d\theta \right|^2$$

where we have chosen the polar axis along **p**. Carrying out the integrations, we get

$$\sigma = \frac{8\pi}{3} \frac{\alpha}{M\hbar^2\omega} \frac{p^3 K}{(p^2/\hbar^2 + K^2)^2} \tag{193}$$

If B is the binding energy of the deuteron, then

$$\hbar\omega - B = p^2/M \tag{194}$$

(This relation is slightly incorrect since the incident quantum carries a momentum $\hbar\omega/c$ which will give the center of mass some energy as well.) Also it can be shown that the binding energy is related to the parameter K though the equation $\hbar K = (MB)^{1/2}$. We have finally,

$$\sigma \simeq \frac{8\pi}{3} \alpha \frac{\hbar^2}{M} \frac{B^{1/2}(\hbar\omega - B)^{3/2}}{(\hbar\omega)^3} \tag{195}$$

The above equation gives the cross-section for photo disintegration of deuteron corresponding to the electric dipole moment interaction and is in good agreement with the experimental data at high γ-ray energies (see, e.g. Ref. 6 for a good discussion on comparison with experimental data).

---

[13] Notice that the final state is an eigenstate of **p**, i.e.

$$\mathbf{p} \exp\left(\frac{i\mathbf{p}\cdot\mathbf{r}}{\hbar}\right) = -i\hbar\nabla \exp\left(\frac{i\mathbf{p}\cdot\mathbf{r}}{\hbar}\right) = \mathbf{p} \exp\left[i\mathbf{p}\cdot\mathbf{r}/\hbar\right]$$

where on the right-hand side, **p** is simply a number.

Solution 29.9

$$\sum_s (E_s - E_n)|\langle s|x|n\rangle|^2 = \sum_s (E_s - E_n)\langle n|x|s\rangle\langle s|x|n\rangle$$

$$= \sum_s \langle n|xH_a - H_a x|s\rangle\langle s|x|n\rangle$$

[see Eqs (15) and (16)]

$$= \frac{i\hbar}{m}\sum_s \langle n|p_x|s\rangle\langle s|x|n\rangle \quad \text{[Using Eq. (17)]}$$

$$= \frac{i\hbar}{m}\langle n|p_x x|n\rangle \tag{196}$$

(because $\sum_s |s\rangle\langle s| = 1$)

We could also write

$$\sum_s (E_s - E_n)|\langle s|x|n\rangle|^2 = \sum_s \langle n|x|s\rangle\langle s|H_a x - xH_a|n\rangle$$

$$= -\frac{i\hbar}{m}\langle n|x p_x|n\rangle \tag{197}$$

Thus,

$$\sum_s (E_s - E_n)|\langle s|x|n\rangle|^2 = -\frac{i\hbar}{2m}\langle n|x p_x - p_x x|n\rangle$$

$$= \hbar^2/2m$$

This can be generalized to more than one electron in which case, we get

$$\sum_i\sum_s (E_s - E_n)\,|\langle s|x_i|n\rangle|^2 = \frac{\hbar^2}{2m}Z \tag{198}$$

where Z is the atomic number. Equation (198) is referred as the dipole sum rule or sometime also as the Thomas-Reiche-Kuhn sum rule. The factors

$$\frac{2m}{\hbar^2}(E_s - E_n)\,|\langle s|x|n\rangle|^2 = f_{sn} \tag{199}$$

are called oscillator strengths. The reason for this terminology is that classically, f_{sn} gives the number of electrons moving with oscillator frequency ω_{sn}. The quantum mechanical result (199) shows that they need not be integers.

(b) We start with the formula for the cross-section in the dipole approximation [see Eq. (184)–(186)]:

$$\frac{d\sigma}{d\Omega} = \frac{2\pi}{\hbar}\frac{N(E)}{c/V}|\langle f|H'|i\rangle|^2 \tag{200}$$

or,

$$\frac{d\sigma}{d\Omega} = 4\pi^2\alpha\,(E_s - E_n)|\hat{\mathbf{e}}.\langle s|\mathbf{r}|n\rangle|^2\,N(E) \tag{201}$$

If we integrate this over dE, the right-hand side of Eq. (201) may be regarded as sum over final states to give

$$\int \sigma\,dE = 4\pi^2\alpha\sum_s (E_s - E_n)\,|\langle s|\hat{\mathbf{e}}\cdot\mathbf{r}|n\rangle|^2 \tag{202}$$

We consider a beam of photons with a fixed linear polarization, say $\hat{\mathbf{e}}$ along the x-direction. We then have

$$\int \sigma\, dE = 4\pi^2\alpha \sum_s (E_s - E_n)|\langle s|x|n\rangle|^2 = \frac{2\pi^2\alpha\hbar^2}{m} \tag{203}$$

Solution 29.10 For the electron initially in the K shell (ejection from which is comparatively probable), the initial wavefunction is given by

$$\psi_i = \left(\frac{1}{\pi a^3}\right)^{1/2} e^{-r/a} \tag{204}$$

where
$$a = \frac{a_0}{Z} = \frac{\hbar^2}{m}\left(\frac{4\pi\varepsilon_0}{q^2}\right)\frac{1}{Z} \tag{205}$$

For the evaluation of the matrix element the assumption will be made that $\hbar\omega$ is much larger than the ionization potential of the atom so that to a good approximation, the final state can be represented by

$$\psi_f = \frac{1}{\sqrt{V}} e^{i\mathbf{k}_1\cdot\mathbf{r}} \tag{206}$$

where $\mathbf{k}_1$ represents the electron wave vector; we have put a subscript to $\mathbf{k}$ so that there is no confusion with the wave vector associated with the electromagnetic wave [see Eq. (208)]. Now the perturbing potential may be taken as [see Eq. (7)]

$$H' = \frac{q}{m}\mathbf{A}\cdot\mathbf{p} \tag{207}$$

where
$$\mathbf{A} = \hat{\mathbf{e}} A_0 e^{i(\mathbf{k}\cdot\mathbf{r}-\omega t)} + \text{cc} \tag{208}$$

or,
$$\mathbf{A} = 2\hat{\mathbf{e}}|A_0|\cos(\mathbf{k}\cdot\mathbf{r} - \omega t + \phi) \tag{209}$$

In Eq. (208) 'cc' denotes the complex conjugate of the first term so that $\mathbf{A}$ is necessarily real. Now, according to first order perturbation theory, the amplitude of the final state at time t is given by [see Eq. (14) of Chapter 27]

$$C_f(t) = \frac{1}{i\hbar}\int_0^t H'_{fi}(t) e^{i\omega_{fi}t}\, dt$$
$$= \frac{M_1}{\hbar}\frac{e^{i(\omega_{fi}-\omega)t} - 1}{i(\omega_{fi}-\omega)} + \frac{M_2}{\hbar}\frac{e^{i(\omega_{fi}-\omega)t} - 1}{i(\omega_{fi}+\omega)} \tag{210}$$

where
$$M_1 = -\frac{iqA_0}{m}\hat{\mathbf{e}}\cdot\int \psi_f^* e^{i\mathbf{k}\cdot\mathbf{r}}\,\mathbf{p}\,\psi_i\, d\tau$$

and
$$M_2 = -\frac{iqA_0^*}{m}\hat{\mathbf{e}}\cdot\int \psi_f^* e^{-i\mathbf{k}\cdot\mathbf{r}}\,\mathbf{p}\,\psi_i\, d\tau$$

As discussed in Section 27.3, the first term on the right-hand side of Eq. (210) contributes in an absorption process and the second term contributes in an emission process. Since we are interested in an absorption of a photon, we have [cf. Eq. (16) of Chapter 27]

$$P_f(t) = |C_f(t)|^2 = |M_1|^2 \frac{1}{\hbar^2}\left[\frac{\sin\left[(\omega_{fi}-\omega)t/2\right]}{(\omega_{fi}-\omega)/2}\right]^2 \tag{211}$$

Following a procedure similar to that discussed in Section 24.2, we get the following expression for the transition probability to solid angle $d\Omega$:

$$W\,d\Omega = \frac{2\pi}{\hbar}|M_1|^2 N(E)\,d\Omega \tag{212}$$

where
$$N(E)\,d\Omega = \frac{mk_1V}{8\pi^3\hbar^2}\,d\Omega \tag{213}$$

represents the density of final states (see Appendix M). Thus,

$$W\,d\Omega = \frac{2\pi}{\hbar}\cdot\frac{mk_1V}{8\pi^3\hbar^2}\cdot|M_1|^2\,d\Omega \tag{214}$$

Now, the electric field associated with the electromagnetic wave is given by

$$\mathcal{E} = -\frac{\partial \mathbf{A}}{\partial t} = -2\omega|A_0|\,\hat{\mathbf{e}}\sin(\mathbf{k}\cdot\mathbf{r}-\omega t+\phi)$$

The time averaged energy density is given by [see Eq. (46)]

$$\varepsilon_0 4\omega^2|A_0|^2\cdot\frac{1}{2}$$

Thus,
$$\text{Incident Intensity} = 2\varepsilon_0\omega^2 c|A_0|^2$$

and
$$\text{Incident Flux} = \frac{\text{Incident Intensity}}{\hbar\omega} = \frac{2\varepsilon_0\omega c}{\hbar}|A_0|^2 \tag{215}$$

Hence,
$$d\sigma = \frac{W\,d\Omega}{\text{Incident Flux}}$$

or
$$\frac{d\sigma}{d\Omega} = \frac{mk_1V}{8\pi^2\hbar^2\varepsilon_0\omega c|A_0|^2}|M_1|^2 \tag{216}$$

Now,
$$M_1 = -\frac{iq}{m}A_0\frac{1}{V^{1/2}}\hat{\mathbf{e}}\cdot\int e^{-i\mathbf{k}_1\cdot\mathbf{r}}e^{i\mathbf{k}\cdot\mathbf{r}}\,\mathbf{p}\psi_i\,d\tau$$

But
$$\int \phi p_x\psi\,d\tau = -i\hbar\int\phi\frac{\partial\psi}{\partial x}d\tau = i\hbar\int\psi\frac{\partial\phi}{\partial x}d\tau$$
$$= -\int\psi p_x\phi\,d\tau$$

Thus,
$$\int\phi\mathbf{p}\psi\,d\tau = -\int\psi\,\mathbf{p}\phi\,d\tau = i\hbar\int\psi\nabla\phi\,d\tau \tag{217}$$

This also follows from the Hermitian property of $\mathbf{p}$ (see Problem 8.14). Using Eq. (217), we get

$$M_1 = \frac{q\hbar}{m}A_0\frac{1}{V^{1/2}}\hat{\mathbf{e}}\cdot\int\psi_i\nabla e^{i(\mathbf{k}_1-\mathbf{k})\cdot\mathbf{r}}\,d\tau$$

$$= -\frac{iq\hbar A_0}{m}(\hat{\mathbf{e}}\cdot\mathbf{k}_1)\frac{1}{V^{1/2}}\int\left[\frac{1}{\pi a^3}\right]^{1/2} \times \exp\left[-i\mathbf{K}\cdot\mathbf{r} - \frac{r}{a}\right]d\tau \qquad (218)$$

where we have used the transverse character of light waves (i.e. $\hat{\mathrm{e}}\cdot\mathrm{k} = 0$) and

$$\mathbf{K} = \mathbf{k}_1 - \mathbf{k} \qquad (219)$$

is the momentum transfer in units of $\hbar$. In carrying out the integration, we choose the polar axis along **K** to obtain an expression for $|M_1|^2$ which on substitution in Eq. (216) gives

$$\frac{d\sigma}{d\Omega} = 32\,\alpha\,\frac{\hbar k_1}{m\omega}(\hat{\mathbf{e}}\cdot\mathbf{k}_1)^2\left(\frac{Z}{a_0}\right)^5\left[\left(\frac{Z}{a_0}\right)^2 + K^2\right]^{-4} \qquad (220)$$

where α is the fine structure constant. The above expression is in agreement with the experimental data. For a more thorough treatment of the photoelectric effect, see Ref. 3, p. 21.

REFERENCES AND SUGGESTED READINGS

1. H. Goldstein, *Classical Mechanics*, Addison-Wesley Publishing Co., Reading, Massachusetts (1950).
2. P.A.M. Dirac, *The Principles of Quantum Mechanics*, Oxford University Press, Oxford (1958).
3. W. Heitler, *The Quantum Theory of Radiation*, Oxford University Press, Oxford (1954).
4. F.T. Arecchi, Quantum Optics and Photon Statistics, in *Lasers and Their Applications*, (A. Sona, ed.), p. 497, Gordon and Breach, New York (1976).
5. E. Bleuler and G.J. Goldsmith, *Experimental Nucleonics*, Holt, Rinehart and Winston, New York (1952).
6. H.A. Bethe and P. Morrison, *Elementary Nuclear Theory*, John Wiley, New York (1956).

30 Relativistic theory

The new theories, if one looks apart from their mathematical setting, are built up from physical concepts which cannot be explained in terms of things previously known to the student, which cannot even be explained adequately in words at all.

— P.A.M. Dirac in the Preface to the first edition of *The Principles of Quantum Mechanics.*

30.1 INTRODUCTION

The quantum theory developed so far is non-relativistic. This invariance under Lorentz transformations is obvious if one notes that the time-dependent Schrödinger equation contains the first derivative in time and second derivatives in space coordinates. A simple procedure for building a relativistic theory is to follow an approach analogous to the one followed for the non-relativistic theory, which is to start with the Hamiltonian formulation, with a relativistically correct Hamiltonian, and to replace the classical observables by operators. This is essentially the procedure we adopt in this chapter. From the point of view of specific predictions, the non-relativistic theory fails in some major features. It fails to explain the spin of an elementary system, such as the electron, and the consequences thereof. Moreover, the theory cannot give a satisfactory account of high-energy phenomena such as pair production, bremmstrahlung, etc.

30.2 THE KLEIN-GORDON EQUATION

We start with the relativistic Hamiltonian for a free particle

$$H = \sqrt{c^2p^2 + m^2c^4} \tag{1}$$

where $\mathbf{p}$ is the momentum, m the rest mass. Introducing the operators

$$E \to i\hbar\frac{\partial}{\partial t} \quad \text{and} \quad \mathbf{p} \to -i\hbar\nabla$$

and writing $H\psi = E\psi$ we obtain

$$\left[-c^2\hbar^2\nabla^2 + m^2c^4\right]^{1/2}\psi = i\hbar\frac{\partial\psi}{\partial t}$$

an equation beyond which theory can proceed only if we have a consistent scheme for interpreting the square root of the above operator. This can be done but leads to difficulties, (see, e.g., Ref. 1). We avoid them by squaring Eq. (1) and replacing by the operators to obtain

$$\left(-c^2\hbar^2\nabla^2 + m^2c^4\right)\psi = -\hbar^2\frac{\partial^2\psi}{\partial t^2}$$

Dividing by c^2h^2 and rearranging the terms, we obtain the Klein-Gordon equation

$$\left(\nabla^2 - \frac{1}{c^2}\frac{\partial^2}{\partial t^2} - \frac{m^2c^2}{\hbar^2}\right)\psi = 0 \tag{2}$$

The Klein-Gordon equation, it turns out, is the correct one for spinless particles and not for electrons. Although this equation is an important one for further developments of relativistic field theories, we shall not discuss it further here. The reason is that the equation is valid for certain classes of particles, such as the π-and *K*-mesons, which are strongly interacting. The approximation methods of elementary quantum mechanics, such as the perturbation theory, are usually inadequate for these anyway.

30.3 THE DIRAC EQUATION

In this section, we will discuss the Dirac equation and in subsequent sections of this chapter we will give some simple consequences of the Dirac equation; for more thorough treatments, we refer the reader to References 1–8. The original argument of Dirac, in search of a relativistic equation was that it must be a first-order differential equation in time because knowledge of the wavefunction alone at an initial time t_0 should suffice to predict its value at a later time. Since in relativity, time and space coordinates are treated symmetrically, the equation (he argued) must be a first-order differential equation in space coordinates as well. Dirac started with a Hamiltonian of the form

$$H = c\,(\alpha_1 p_x + \alpha_2 p_y + \alpha_3 p_z) + \beta mc^2 \tag{3}$$

where the coefficients α_1, α_2, α_3 and β are not mere numbers but (square) matrices. For brevity, we shall often write Eq. (3) as

$$H = c\alpha\cdot\mathbf{p} + \beta mc^2$$

It must be clear that α is not to be thought of as a vector; α.**p** is defined to be equal to $\alpha_1 p_x + \alpha_2 p_y + \alpha_3 p_z$. The wave equation corresponding to this Hamiltonian is

$$i\hbar\frac{\partial\psi}{\partial t} = -i\hbar c\left(\alpha\cdot\nabla\psi\right) + \beta mc^2\psi \tag{4}$$

Since α and β are allowed to be matrices of numbers, it follows that the wave-function ψ must have more than one component; i.e. at each point the state of the particle will be described by more than one wavefunction. It follows that Eq. (4) will be capable of describing a particle with additional intrinsic degrees of freedom. The Hamiltonian given by Eq. (3) must still satisfy the energy-momentum relation

$$E^2 = H^2 = c^2p^2 + m^2c^4 \tag{5}$$

We introduce the notation

$$p_1 = p_x,\ p_2 = p_y,\ p_3 = p_z,\ p_4 = iE/c$$

$$x_1 = x,\ x_2 = y,\ x_3 = z,\quad \text{and}\quad x_4 = +ict$$

Thus Eq. (3) becomes

$$H = c\sum_{j=1}^{3} \alpha_j p_j + \beta mc^2$$

and $$H^2 = c^2 \left\{ \sum_j \sum_k (\alpha_j\alpha_k + \alpha_k\alpha_j) p_j p_k + \sum_{j=1}^{3} \alpha_j^2 p_j^2 + mc^3 \sum_{j=1}^{3} (\alpha_j\beta + \beta\alpha_j) p_j + m^2c^4\beta^2 \right\} \tag{6}$$

Equations (5) and (6) are consistent if

$$\alpha_j\alpha_k + \alpha_k\alpha_j = 0 \quad (\text{for } j \neq k) \tag{7}$$

$$\alpha_j\beta + \beta\alpha_j = 0 \tag{8}$$

$$\alpha_j^2 = 1 \tag{9}$$

and $$\beta^2 = 1 \tag{10}$$

Since Eqs (7)–(10) are to be matrix relations, the symbols 0 and 1 on the right ought to be regarded as the null matrix and the unit matrix, respectively.

We first show that these matrices must be of even dimension. We have for example,

$$\alpha_1\beta = -\beta\alpha_1$$

Multiplying by α_1 on the left, setting $\alpha_1^2 = 1$ and taking the trace, we obtain

$$\text{Trace } \beta = -\text{Trace } (\alpha_1\, \beta\alpha_1) = -\text{Trace } (\alpha_1^2\beta) = -\text{Trace } \beta$$

where we have used the result Trace (ABC) = Trace (CAB). We, therefore, have Trace $\beta = 0$, a result which applies to each of the α matrices as well. Equations (9) and (10) also tell us that the eigenvalues of the matrices must be ±1. Since the trace is just the sum of all eigenvalues, it follows that the +1 and –1 eigenvalues must occur the same number of times. Hence each of the matrices must be of even dimension N. Now N cannot be 2, since we can only have three 2×2 independent non-commuting matrices. We shall see that 4×4 matrices satisfy all the conditions.

We may mention here that the Pauli spin matrices are given by (see Section 13.4)

$$\sigma_x = \begin{pmatrix} 0 & 1 \\ 1 & 0 \end{pmatrix} \tag{11}$$

$$\sigma_y = \begin{pmatrix} 0 & -i \\ i & 0 \end{pmatrix} \tag{12}$$

and $$\sigma_z = \begin{pmatrix} 1 & 0 \\ 0 & -1 \end{pmatrix} \tag{13}$$

we can readily see that

$$\sigma_x^2 = \sigma_y^2 = \sigma_z^2 = 1 \tag{14}$$

where the RHS respresents the unit matrix. Further, the spin matrices satisfy the anti-commutative realtion [cf. Eqs (8) and (9)]

$$\begin{aligned}\sigma_x\sigma_y + \sigma_y\sigma_x &= 0\\ \sigma_y\sigma_z + \sigma_z\sigma_y &= 0\\ \sigma_z\sigma_x + \sigma_x\sigma_z &= 0\end{aligned} \tag{15}$$

Comparing Eqs. (14) and (15) with Eqs (7)–(10), we may choose the following representation for the matrices α and β:

$$\alpha_1 = \begin{pmatrix} 0 & \sigma_x \\ \sigma_x & 0 \end{pmatrix} = \begin{pmatrix} 0 & 0 & 0 & 1 \\ 0 & 0 & 1 & 0 \\ 0 & 1 & 0 & 0 \\ 1 & 0 & 0 & 0 \end{pmatrix} \tag{16}$$

$$\alpha_2 = \begin{pmatrix} 0 & \sigma_y \\ \sigma_y & 0 \end{pmatrix} = \begin{pmatrix} 0 & 0 & 0 & -i \\ 0 & 0 & i & 0 \\ 0 & -i & 0 & 0 \\ i & 0 & 0 & 0 \end{pmatrix} \tag{17}$$

$$\alpha_3 = \begin{pmatrix} 0 & \sigma_z \\ \sigma_z & 0 \end{pmatrix} = \begin{pmatrix} 0 & 0 & 1 & 0 \\ 0 & 0 & 0 & -1 \\ 1 & 0 & 0 & 0 \\ 0 & -1 & 0 & 0 \end{pmatrix} \tag{18}$$

and

$$\beta = \begin{pmatrix} 1 & 0 \\ 0 & -1 \end{pmatrix} = \begin{pmatrix} 1 & 0 & 0 & 0 \\ 0 & 1 & 0 & 0 \\ 0 & 0 & -1 & 0 \\ 0 & 0 & 0 & -1 \end{pmatrix} \tag{19}$$

Obviously, each element of the 2 × 2 matrices shown above is itself a 2 × 2 matrix. One can immediately see that the above matrices satisfy Eqs (7)–(10). For example,

$$\alpha_1\,\alpha_3 + \alpha_3\,\alpha_1 = \begin{pmatrix} 0 & \sigma_x\sigma_z + \sigma_z\sigma_x \\ \sigma_x\sigma_z + \sigma_z\sigma_x & 0 \end{pmatrix}$$

where the RHS is indeed a 4 × 4 null matrix. Equations (16)–(18) can be written as

$$\alpha = \begin{pmatrix} 0 & \sigma \\ \sigma & 0 \end{pmatrix} \tag{20}$$

we note that all these matrices are Hermitian, i.e.,

$$\alpha_j^\dagger = \alpha_j, \quad \beta^\dagger = \beta$$

Now, Eq. (4) can be written as

$$-\frac{\partial\psi}{\partial x_4} = -\,i\alpha\cdot\nabla\psi + \beta\frac{mc}{\hbar}\psi$$

Multiplying by β on the left side, we have

$$-i\beta\alpha\cdot\nabla\psi+\beta\frac{\partial\psi}{\partial x_4}+\frac{mc}{\hbar}\psi=0$$

We define the following four γ matrices

$$\gamma_j=-i\beta\alpha_j;\qquad j=1,2,3$$

and

$$\gamma_4=\beta$$

Note that these γ matrices are also Hermitian, i.e. $\gamma_\mu^\dagger=\gamma_\mu$. The Dirac equation now takes the form

$$\left(\sum_{\mu=1}^{4}\gamma_\mu\frac{\partial}{\partial x_\mu}+\kappa\right)\psi=0 \tag{21}$$

where $\kappa=mc/\hbar$. It is usual to drop the summation over the repeated index μ and write simply

$$\gamma_\mu\frac{\partial\psi}{\partial x_\mu}+\kappa\psi=0$$

(A convention we shall adopt is to use Greek indices λ, μ, ν, etc., when their values range 1 to 4 and Roman ones *i*, *j*, *k*, etc. for values ranging from 1 to 3). In terms of the momentum operator, we can write

$$(\gamma_\mu p_\mu-i\,m\,c)\psi=0 \tag{22}$$

Equations (21) and (22) are the covariant forms of Dirac equation; they do not appear to single out the space and time parts. However, writing them in those forms does not automatically guarantee that the equation is covariant under Lorentz transformations.

We may mention here that the above equations are meaningful only if we interpret ψ as a 4-row, 1-column matrix (i.e., a spinor with 4 components). Equation (22) is a shorthand for the four equations (λ = 1, 2, 3, 4)

$$\sum_{\mu=1}^{4}\sum_{\sigma=1}^{4}\left\{(\gamma_\mu)_{\lambda\sigma}\,p_\mu-i\,m\,c\,\delta_{\lambda\sigma}\right\}\psi_\sigma=0 \tag{23}$$

where $(\gamma_\mu)_{\lambda\sigma}$ is the (λσ)th component of the matrix γ_μ and $\delta_{\lambda\sigma}$ is the (λσ)th component of the unit matrix. The summation over μ runs over 4-terms because these are 4 space-time components. On the other hand, λ and σ takes values 1 to 4 because the Dirac matrices happen to be four dimensional. Thus, if one is not alert, there can be some confusion in the algebraic work. It is never really necessary to write the equation in this elaborate form, but if ever in doubt, the reader is advised to write it out in full.

30.4 PROBABILITY AND CURRENT DENSITY

We introduce the Hermitian conjugate (adjoint) matrix $\psi^\dagger=(\psi_1^*\;\psi_2^*\;\psi_3^*\;\psi_4^*)$ which is a 1-row 4-column matrix. Multiplying Eq. (4) on the left by $\psi^\dagger$, we have

$$i\hbar\psi^\dagger\frac{\partial}{\partial t}=-i\,\hbar\,c\,\psi^\dagger\,\alpha\cdot\nabla\psi+mc^2\psi^\dagger\,\beta\,\psi \tag{24}$$

Similarly, starting with the adjoint of Eq. (4),

$$-i\hbar \frac{\partial \psi^\dagger}{\partial t} = i\hbar c\, \nabla\psi^\dagger .\alpha + m c^2 \psi^\dagger \beta \tag{25}$$

(recall that α, β are Hermitian) and after multiplying by ψ, we obtain

$$-i\hbar \frac{\partial \psi^\dagger}{\partial t} \psi = i\hbar c\, \nabla\psi^\dagger .\alpha\, \psi + m c^2 \psi^\dagger \beta\, \psi \tag{26}$$

Subtracting Eq. (26) from Eq. (24), we can obtain the following:

$$\frac{\partial}{\partial t}(\psi^\dagger \psi) + c\,\nabla.(\psi^\dagger \alpha\, \psi) = 0 \tag{27}$$

Equation (27) can be written in the form of the continuity equation

$$\frac{\partial \rho}{\partial t} + \nabla \cdot \mathbf{J} = 0 \tag{28}$$

provided we make the identification

$$\rho = \psi^\dagger\ \psi = \sum_{\lambda=1}^{4} \psi_\lambda^* \psi_\lambda \tag{29}$$

and

$$\mathbf{J} = c\psi^\dagger\ \alpha\psi = c \sum_{\lambda,\sigma=1}^{4} \psi_\lambda^* (\alpha)_{\lambda\sigma}\, \psi_\sigma \tag{30}$$

30.5 AN IDENTITY

In this section, we will prove the identity

$$(\sigma\cdot\mathbf{A})(\sigma\cdot\mathbf{B}) = \mathbf{A}\cdot\mathbf{B} + i\sigma\cdot(\mathbf{A}\times\mathbf{B}) \tag{31}$$

which will be extensively used in this chapter. In the above equation σ_x, σ_y and σ_z represent the Pauli spin matrices and components of **A** and **B** are assumed to commute with components of σ; however, components of **A** need not commute with components of **B**. Now,

$$\begin{aligned}(\sigma\cdot\mathbf{A})(\sigma\cdot\mathbf{B}) &= (\sigma_x A_x + \sigma_y A_y + \sigma_z A_z)(\sigma_x B_x + \sigma_y B_y + \sigma_z B_z)\\ &= (A_x B_x + A_y B_y + A_z B_z) + i\sigma_x (A_y B_z - A_z B_y) + \cdots + \cdots\end{aligned}$$

where we have used the relations

$$\sigma_x^2 = \sigma_y^2 = \sigma_z^2 = 1$$

and $$\sigma_x\sigma_y = -\sigma_y\sigma_x = i\sigma_z \quad \text{etc.}$$

Thus, $$\begin{aligned}(\sigma\cdot\mathbf{A})(\sigma\cdot\mathbf{B}) &= \mathbf{A}\cdot\mathbf{B} + i[\sigma_x(\mathbf{A}\times\mathbf{B})_x + \sigma_y(\mathbf{A}\times\mathbf{B})_y + \ldots]\\ &= \mathbf{A}\cdot\mathbf{B} + i\sigma\cdot(\mathbf{A}\times\mathbf{B})\end{aligned}$$

proving the identity.

30.6 PLANE WAVE SOLUTIONS

We start with the free particle Dirac equation:

$$(c\,\alpha\cdot\hat{\mathbf{P}} + \beta mc^2)\,\psi(\mathbf{r}, t) = \hat{\mathbf{E}}\,\psi(\mathbf{r}, t) \tag{32}$$

where we have explicitly indicated that $\hat{\mathbf{P}}$ and $\hat{\mathbf{E}}$ are operators. We try plane wave solutions

$$\psi(\mathbf{r}, t) = \chi \exp\left[\frac{i}{\hbar}(\mathbf{p}.\mathbf{r} - Et)\right] \tag{33}$$

where $\mathbf{p}$ and E are now numbers, not operators and χ are spinors. Clearly, Eq. (32) is satisfied if χ obeys the equation

$$(c\,\alpha\cdot\mathbf{p} + \beta mc^2)\chi = E\chi \tag{34}$$

Again in Eq. (34), $\mathbf{p}$ and E are now numbers. Since α and β can be conveniently expressed in terms of 2×2 matrices, we start by writing

$$\chi = \begin{pmatrix} u \\ v \end{pmatrix}$$

where u, v are themselves 2-component spinors. Using Eqs (19) and (20), we obtain

$$\left(c\begin{pmatrix} 0 & \sigma.\mathbf{p} \\ \sigma.\mathbf{p} & 0 \end{pmatrix} + \begin{pmatrix} 1 & 0 \\ 0 & -1 \end{pmatrix} mc^2\right)\begin{pmatrix} u \\ v \end{pmatrix} = E\begin{pmatrix} u \\ v \end{pmatrix} \tag{35}$$

We can recover the four equations, if we want, by writting

$$u = \begin{pmatrix} u_1 \\ u_2 \end{pmatrix}, \quad v = \begin{pmatrix} u_3 \\ u_4 \end{pmatrix}$$

after solving for u and v. Equation (35) gives

$$c\,\sigma\cdot\mathbf{p}\,v = (E - mc^2)\,u \tag{36}$$

$$c\,\sigma\cdot\mathbf{p}\,u = (E + mc^2)\,v \tag{37}$$

We can decouple u, v by solving (say) for v from Eq. (37) and substituting in Eq. (36). This gives us

$$v = \frac{c\sigma.\mathbf{p}}{E + mc^2}u \tag{38}$$

and
$$\frac{c^2(\sigma.\mathbf{p})(\sigma.\mathbf{p})}{E + mc^2}u = (E - mc^2)\,u$$

Since,
$$\sigma\cdot\mathbf{p}\,\sigma\cdot\mathbf{p} = p^2 \quad \text{(a number)},$$

[see Eq. (31)], we have

$$c^2 p^2 u = (E^2 - m^2c^4)u \tag{39}$$

Apart from the trivial solution $u = 0$ (and hence $v = 0$) we have the result

$$E = \pm\sqrt{c^2 p^2 + m^2 c^4} = E_{\pm} \tag{40}$$

We note that for small momenta [see Eqs (36) and (37)]

$$|v| \sim \frac{cp}{2mc^2}|u| \sim \frac{V}{2c}|u| \qquad \text{(for } E = E_+\text{)}$$

and

$$|u| \sim \frac{V}{2c}|v| \qquad \text{(for } E = E_-\text{)}$$

where V is the velocity of the particle. Thus, for positive energies, u are large components and for negative energies, v are large components. For $E = E_+$, we express v in terms of u

$$v = \frac{c\boldsymbol{\sigma}.\mathbf{p}}{E_+ + mc^2}u \tag{41}$$

Similarly for $E = E_-$, we express u in terms of v

$$u = \frac{c\boldsymbol{\sigma}.\mathbf{p}}{E_- - mc^2}v \tag{42}$$

We can obtain four linearly independent solutions from Eqs. (41) and (42) as follows. Writing

$$u = \begin{pmatrix}1\\0\end{pmatrix} \quad \text{or} \quad u = \begin{pmatrix}0\\1\end{pmatrix}$$

and substituting in Eq. (41), in turn, we obtain two solutions. For example, $u = \begin{pmatrix}1\\0\end{pmatrix}$ gives

$$v = \begin{pmatrix}u_3\\u_4\end{pmatrix} = \left(\frac{c}{E_+ + mc^2}\right)\begin{pmatrix}p_z & p_x - ip_y\\ p_x + ip_y & -p_x\end{pmatrix}\begin{pmatrix}1\\0\end{pmatrix}$$

$$= \left(\frac{c}{E_+ + mc^2}\right)\begin{pmatrix}p_z\\ p_x + ip_y\end{pmatrix}$$

This solution is, therefore,

$$\chi_{\uparrow}^{+} = N\begin{pmatrix}u\\v\end{pmatrix} = N\begin{pmatrix}1\\0\\cp_z/(E_+ + mc^2)\\cp_+/(E_+ + mc^2)\end{pmatrix} \tag{43}$$

The other solution for $E = E_+$ can be constructed in the same way.

$$\chi_{\downarrow}^{+} = N\begin{pmatrix}0\\1\\cp_-/(E_+ + mc^2)\\-cp_z/(E_+ + mc^2)\end{pmatrix} \tag{44}$$

Similarly, for $E = E_-$, we choose $v = \begin{pmatrix}1\\0\end{pmatrix}, \begin{pmatrix}0\\1\end{pmatrix}$ in turn and substitute in Eq. (42)

to obtain
$$\chi_{\uparrow}^{-} = N\begin{pmatrix} cp_z/(E_- - mc^2) \\ cp_+/(E_- - mc^2) \\ 1 \\ 0 \end{pmatrix}, \quad \chi_{\downarrow}^{-} = N\begin{pmatrix} cp_-/(E_- - mc^2) \\ -cp_z/(E_- - mc^2) \\ 0 \\ 1 \end{pmatrix} \tag{45}$$

In these equations, we have used the symbols

$$p_\pm = p_x \pm ip_y$$

and the normalization N is chosen to satisfy

$$\chi^+ \chi = 1$$

for each of the solutions. It is easily shown that

$$N = \left[1 + \frac{c^2 p^2}{\left(|E| + mc^2\right)^2}\right]^{-\frac{1}{2}} \tag{46}$$

and the different solutions Eq. (34) are orthogonal,

$$\chi_a^+ \chi_b = 0$$

The complete plane wave solutions take the form

$$\psi_{p\sigma}(r, t) = \chi_{p\sigma} \exp\left[\frac{i}{\hbar}(\mathbf{p}\cdot\mathbf{r} - Et)\right] \tag{47}$$

The suffix p is to tell us that we are dealing with plane waves of momentum p and the suffix σ is to identify one of the 4 solutions.

30.7 THE ELECTRON IN AN ELECTROMAGNETIC FIELD

We modify the Dirac equation simply by replacing the operators

$$\mathbf{p} \text{ by } \mathbf{p} - QA$$

and

$$E \text{ by } E - Q\phi$$

where $\mathbf{A}$ and ϕ are the vector and scalar potentials (see Chapter 21). The solutions will no longer be of the plane wave form. However, we still write

$$\psi(\mathbf{r}, t) = \begin{pmatrix} u(\mathbf{r},t) \\ v(\mathbf{r},t) \end{pmatrix}$$

and remember that the spinors u and v now incorporate the space time dependence. Thus the Dirac equation can be written in the form

$$(c\alpha\cdot(\mathbf{p} - Q\mathbf{A}) + \beta mc^2)\,\psi(\mathbf{r}, t) = (E - Q\phi)\,\psi(\mathbf{r}, t) \tag{48}$$

Using Eqs (19) and (20), the above equation can be written in the form

$$c\begin{pmatrix} 0 & \sigma.\pi \\ \sigma.\pi & 0 \end{pmatrix}\begin{pmatrix} u \\ v \end{pmatrix} = \begin{pmatrix} E - Q\phi - mc^2 & 0 \\ 0 & E - Q\phi + mc^2 \end{pmatrix}\begin{pmatrix} u \\ v \end{pmatrix}$$

where $$\boldsymbol{\pi} \equiv \mathbf{p} - Q\mathbf{A} \tag{49}$$

Thus, $$c\boldsymbol{\sigma} \cdot \boldsymbol{\pi} v = (E - Q\phi - mc^2)\, u \tag{50}$$

and $$c\boldsymbol{\sigma} \cdot \boldsymbol{\pi} u = (E - Q\phi + mc^2)\, v \tag{51}$$

We are interested in positive energy solutions and write

$$E = mc^2 + E' \tag{52}$$

where E' is the energy that appears in the (non-relativistic) Schrödinger equation. For small values of E' and $Q\phi$ ($\ll mc^2$), we may write

$$E - Q\phi + mc^2 \approx 2mc^2 \tag{53}$$

Thus Eq. (51) would give

$$v \approx \frac{1}{2mc}\boldsymbol{\sigma} \cdot \boldsymbol{\pi} u \tag{54}$$

Substituting in Eq. (50) we obtain for the large component

$$\frac{1}{2m}\,(\boldsymbol{\sigma} \cdot \boldsymbol{\pi})\,(\boldsymbol{\sigma} \cdot \boldsymbol{\pi})\, u = (E' - Q\phi)\, u \tag{55}$$

If we now use the identity given by Eq. (31), we would get

$$(\boldsymbol{\sigma} \cdot \boldsymbol{\pi})(\boldsymbol{\sigma} \cdot \boldsymbol{\pi}) = \boldsymbol{\pi} \cdot \boldsymbol{\pi} + i\boldsymbol{\sigma} \cdot (\boldsymbol{\pi} \times \boldsymbol{\pi}) \tag{56}$$

Now,
$$(\boldsymbol{\pi} \times \boldsymbol{\pi}) = (\mathbf{p} - Q\mathbf{A}) \times (\mathbf{p} - Q\mathbf{A})$$
$$= -Q\,(\mathbf{p} \times \mathbf{A} + \mathbf{A} \times \mathbf{p})$$

where we have used the fact that the components p_x, p_y and p_z commute with each other; however, since $\mathbf{p}$ is an operator, $\mathbf{p} \times \mathbf{A} \neq -\mathbf{A} \times \mathbf{p}$. Indeed

$$(\mathbf{p} \times \mathbf{A} + \mathbf{A} \times \mathbf{p})_x \psi = (p_y A_z - p_z A_y + A_y p_z - A_z p_y)\, \psi$$
$$= -i\hbar\left[\frac{\partial}{\partial y}(A_z \psi) - \frac{\partial}{\partial z}(A_y \psi) + A_y \frac{\partial \psi}{\partial z} - A_z \frac{\partial \psi}{\partial y}\right]$$
$$= -i\hbar\left[\frac{\partial A_z}{\partial y} - \frac{\partial A_y}{\partial z}\right]\psi$$
$$= -i\hbar\,(\nabla \times \mathbf{A})_x\, \psi$$
$$= -i\hbar\, B_x\, \psi$$

Thus, $$\boldsymbol{\pi} \times \boldsymbol{\pi} = i\hbar\, Q\mathbf{B} \tag{57}$$

Substituting in Eq. (56), we get

$$(\boldsymbol{\sigma} \cdot \boldsymbol{\pi})\,(\boldsymbol{\sigma} \cdot \boldsymbol{\pi}) = (\mathbf{p} - Q\mathbf{A})^2 - Q\hbar\, \boldsymbol{\sigma} \cdot \mathbf{B}$$

Equation (55), therefore, becomes

$$\left[\frac{1}{2m}(\mathbf{p}-Q\mathbf{A})^2 - \frac{Q\hbar}{2m}\sigma\cdot\mathbf{B} + Q\phi\right]u = E'u \tag{58}$$

If we compare the above equation with the corresponding non relativistic Schrödinger equation, we find that we now have an important extra term. This term suggests that the electron in a magnetic field, acquires an extra energy

$$-\mu\cdot\mathbf{B} = -\frac{Q\hbar}{2m}\sigma\cdot\mathbf{B}$$

so that it behaves as if it has a moment μ associated with its spin, namely

$$\mu = \frac{Q\hbar}{2m}\sigma = -\frac{q\hbar}{2m}\sigma \tag{59}$$

where $Q = -q$ represents the actual charge of the electron; $q \approx + 1.6 \times 10^{-19}$C represents the magnitude of the electronic charge. Thus, the Dirac equation predicts that for a spin 1/2 particle, the magnetic moment is always given by the relation [Eq. (59)]. While this prediction is in very good agreement with experiment for the electron, it fails badly for the proton and the neutron. It is possible to modify the equation keeping it relativistically invariantly introducing, *ad hoc*, an extra term that accounts for the anomalous magnetic moment of these particles (see, e.g. Ref. 6).

30.8 THE SPIN-ORBIT INTERACTION

Another new feature that may be expected is the interaction between the spin and orbital motions. We examine the higher order corrections to the Pauli equation in an electrostatic field.[1] Writing $\mathbf{A} = 0$ we obtain from Eqs (50) and (51)

$$c\,(\sigma\cdot\mathbf{p})\,v = (E' - Q\phi)\,u \tag{60}$$

$$c\,(\sigma.\,\mathbf{p})\,u = (2mc^2 + E' - Q\phi)\,v \tag{61}$$

The charge density is

$$\rho = \psi^\dagger\psi = u^\dagger u + v^\dagger v$$

From Eq. (61), we have

$$v \approx \frac{1}{2mc}\sigma\cdot\mathbf{p}u \tag{62}$$

so that,

$$v^\dagger v \approx u^\dagger \frac{p^2}{4m^2c^2}u$$

This gives $\rho = u^\dagger(1 + p^2/4m^2c^2)u$ (to order v^2/c^2) which suggests that to this order we *cannot* regard u as the wavefunction in the non-relativistic approximation. Instead, if we take

[1] We follow the analysis given in Ref. 7.

$$u' = \left(1 + \frac{p^2}{8m^2c^2}\right)u$$

and
$$u = \left(1 - \frac{p^2}{8m^2c^2}\right)u' \tag{63}$$

then $\rho = (u')^\dagger u'$ to order v^2/c^2 so that it is u' that we want for the wavefunction. Our aim is to obtain an equation for u', taking care to retain terms up to order $1/c^2$. From Eq. (61), we have

$$c\boldsymbol{\sigma}\cdot\mathbf{p}u = (2mc^2)\,v + (E' - Q\phi)\frac{\boldsymbol{\sigma}\cdot\mathbf{p}}{2mc}u$$

where in the term multiplying $(E' - Q\phi)$, which is small in comparison to $2mc^2$, we have used the lower approximation (62) for v. Thus,

$$v = \frac{1}{2mc}\left[1 - \frac{(E' - Q\phi)}{2mc^2}\right]\boldsymbol{\sigma}\cdot\mathbf{p}u \tag{64}$$

Substituting this into Eq. (60), we obtain

$$(E' - Q\phi)u = \frac{1}{2m}(\boldsymbol{\sigma}\cdot\mathbf{p})\left(1 - \frac{(E' - Q\phi)}{2mc^2}\right)\boldsymbol{\sigma}\cdot\mathbf{p}u$$

$$= \frac{1}{2m}\left[p^2 - \frac{E'}{2mc}p^2 + \frac{Q}{2mc^2}(\boldsymbol{\sigma}\,.\,\mathbf{p})\phi(\boldsymbol{\sigma}\,.\,\mathbf{p})\right]u$$

Rearranging terms, we get

$$E'\left(1 + \frac{p^2}{4m^2c^2}\right)u = \left[Q\phi + \frac{1}{2m}\left\{p^2 + \frac{Q}{2mc^2}(\boldsymbol{\sigma}\,.\,\mathbf{p})\phi(\boldsymbol{\sigma}\,.\,\mathbf{p})\right\}u\right]$$

We multiply this on the left by $1 - p^2/8m^2c^2$, retain the terms up to $1/c^2$ and use Eq. (63)

$$E'u' = \left(1 - \frac{p^2}{8m^2c^2}\right)\left[Q\phi + \frac{p^2}{2m} + \frac{Q}{4m^2c^2}(\boldsymbol{\sigma}.\mathbf{p})\phi(\boldsymbol{\sigma}.\mathbf{p})\right]\left(1 - \frac{p^2}{8m^2c^2}\right)u' \tag{65}$$

The above equation gives us the required eigenvalue relation

$$E'u' = Hu'$$

Retaining terms up to the order $1/c^2$, we have

$$H = Q\phi + \frac{p^2}{2m} - \frac{p^4}{8m^3c^2} - \frac{Q}{8m^2c^2}(p^2\phi + \phi p^2) + \frac{Q}{4m^2c^2}(\boldsymbol{\sigma}\,.\,\mathbf{p})\phi(\boldsymbol{\sigma}\,.\,\mathbf{p})$$

Now,
$$(\boldsymbol{\sigma}\cdot\mathbf{p})\,\phi\,(\boldsymbol{\sigma}\cdot\mathbf{p}) = \phi(\boldsymbol{\sigma}\cdot\mathbf{p})\,(\boldsymbol{\sigma}\cdot\mathbf{p}) + (\boldsymbol{\sigma}\cdot\mathbf{p}\phi)\,(\boldsymbol{\sigma}\cdot\mathbf{p})$$

$$= \phi\,p^2 + i\hbar(\boldsymbol{\sigma}\cdot\mathbf{E})\,(\boldsymbol{\sigma}\cdot\mathbf{p})$$

$$= \phi\,p^2 + i\hbar\,[\mathbf{E}\cdot\mathbf{p} + i\boldsymbol{\sigma}\cdot\mathbf{E}\times\mathbf{p}] \tag{66}$$

where $\mathbf{E} = -\nabla\phi$ is the electric field.

We also have

$$(p^2\phi - \phi p^2)u' = -\hbar^2 [\nabla^2(\phi u') - \phi\nabla^2 u']$$
$$= -\hbar^2 [\nabla \cdot (u'\nabla\phi + \phi\nabla u') - \phi\nabla^2 u']$$
$$= -\hbar^2 [2\nabla\phi \cdot \nabla u' + (\nabla^2\phi)u']$$

Writing this in terms of **E** and **p**, we obtain

$$p^2\,\phi - \phi\,p^2 = 2\,i\,\hbar\,\mathbf{E}.\mathbf{p} + \hbar^2\nabla\cdot\mathbf{E} \tag{67}$$

Substituting (66) and (67) into expression of H, we obtain

$$H = V + \frac{p^2}{2m} - \frac{p^4}{8m^3c^2} - \frac{Q\hbar}{4m^2c^2}\boldsymbol{\sigma}\cdot\mathbf{E}\times\mathbf{p} - \frac{Q\hbar^2}{8m^2c^2}\nabla\cdot\mathbf{E} \tag{68}$$

where
$$V = Q\phi \tag{69}$$

represents the potential energy function. Now, the first two terms on the RHS of Eq. (68) represent the usual non relativistic terms. The third term is the kinetic energy correction up to order $1/c^2$ as may be seen by expanding

$$mc^2\left[1 + \frac{p^2}{m^2c^2}\right]^{\frac{1}{2}} - mc^2 \approx \frac{p^2}{2m} - \frac{p^4}{8m^3c^2}$$

(see Section 20.5.1). The fourth term represents spin-orbit interaction; this can be seen by noting that for a spherically symmetric potential $V(r)$

$$Q\mathbf{E} = -\nabla V = -\frac{1}{r}\frac{dV}{dr}\mathbf{r} \tag{70}$$

Thus, the spin-orbit interaction energy is given by

$$H_{s-o} = -\frac{Q\hbar}{4m^2c^2}\boldsymbol{\sigma}\cdot\mathbf{E}\times\mathbf{p}$$
$$= \frac{1}{2m^2c^2r}\frac{dV}{dr}\mathbf{s}\cdot\mathbf{r}\times\mathbf{p}$$

where
$$\mathbf{s} = \frac{1}{2}\hbar\boldsymbol{\sigma}$$

represents the spin angular momentum (Chapter 14). Thus,

$$H_{s-o} = \frac{1}{2m^2c^2}\frac{1}{r}\frac{dV}{dr}\mathbf{s}.\mathbf{L} \tag{71}$$

For a hydrogen-like atom problem, $V(r)$ is given by the Coulomb potential

$$V(r) = -\frac{Ze^2}{r} \tag{72}$$

where, to be consistent with our notations in Chapter 20, we have used cgs units.

Thus,
$$H_{s-o} = \frac{Z\alpha\hbar}{2m^2 c}\frac{1}{r^3}\mathbf{s}\cdot\mathbf{L} \tag{73}$$

where
$$\alpha = \frac{e^2}{\hbar c} \approx \frac{1}{137} \tag{74}$$

is the fine structure constant. Equation (73) is the perturbation due to spin-orbit interaction that we had used in Section 20.5.2.

The last term in Eq. (68) is a correction to the potential energy V, and is known as the Darwin term (see, e.g., Ref. 1 for more details). We may mention here that for the Coulomb potential [see Eq. (72)] the Darwin term is given by

$$H_D = \frac{\hbar^2}{8m^2c^2}\nabla^2 V$$
$$= \frac{\pi\hbar^3 Z\alpha}{2m^2 c}\delta(\mathbf{r}) \tag{75}$$

where we have used the relation [see Section 26.3.1]:

$$\nabla^2\left(\frac{1}{r}\right) = -4\pi\delta(\mathbf{r}) \tag{76}$$

Since $R_{nl}(r = 0) = 0$ for $l = 1, 2, 3, \ldots$(see Section 10.4.2); the Darwin term contributes only for s ($l = 0$) states. Indeed, for the ground state of the hydrogen-like atom, the perturbation due to the Darwin term is given by

$$(\Delta E)_D = \frac{\pi\hbar^3 Z\alpha}{2m^2 c}|\psi_{n00}(r=0)|^2$$

$$= \frac{\pi\hbar^3 Z\alpha}{2m^2 c}\cdot\frac{Z^3}{n^3\pi a_0^3} = \frac{1}{2}\cdot\frac{mc^2 Z^4\alpha^4}{n^3} \tag{77}$$

30.9 CENTRAL POTENTIAL

Let us now consider the solution of the Dirac equation for a particle in a central potential $V(r)$; we follow the general method described in Ref. 8. The Hamiltonian is

$$H = c\boldsymbol{\alpha}\cdot\mathbf{p} + \beta mc^2 + V(r) = H_0 + V(r) \tag{78}$$

It can be shown that the free particle Hamiltonian H_0 does not commute with the orbital angular momentum $\mathbf{L}$ (see Problem 30.2). However, it does commute with $\mathbf{J} = \mathbf{L} + \mathbf{S}$. Here,

$$\hbar\,\mathbf{S} = \frac{1}{2}\hbar\Sigma = \frac{1}{2}\hbar\begin{pmatrix}\sigma & 0\\ 0 & \sigma\end{pmatrix} \tag{79}$$

For the central potential the Hamiltonian still commutes with J^2 and J_z so that eigenfunctions can be conveniently chosen so that they belong to definite states of J^2 and J_z. We label these states by j and m, the eigenvalue of J^2 is $j(j+1)\hbar^2$ and of J_z is $m\hbar$ for the particular state.

We shall proceed by separating the variables in spherical polar coordinates, the appropriate coordinates in view of the spherical symmetry of $V(r)$. To do this, it is convenient to introduce the operators (see Problem 30.5)

$$p_r = \frac{1}{r}\left[xp_x + yp_y + zp_z - i\hbar\right]$$

$$= \frac{1}{r}\left[\mathbf{r}\cdot\mathbf{p} - i\hbar\right] = -i\hbar\left[\frac{\partial}{\partial r} + \frac{1}{r}\right] \tag{80}$$

and
$$\alpha_r = \frac{1}{r}(\alpha.\mathbf{r}) \tag{81}$$

Note that

$$\alpha_r^2 = 1$$

Using the relation

$$(\alpha\cdot\mathbf{B})(\alpha\cdot\mathbf{C}) = \mathbf{B}\cdot\mathbf{C} + i\,\Sigma\cdot(\mathbf{B}\times\mathbf{C}) \tag{82}$$

and putting $\mathbf{B} = \mathbf{r}$, $\mathbf{C} = \mathbf{p}$, we obtain

$$(\alpha\cdot\mathbf{r})\,(\alpha\cdot\mathbf{p}) = rp_r + i\,\hbar + i\,\Sigma\cdot(\mathbf{r}\times\mathbf{p})$$

$$= rp_r + i\,\hbar\,(\Sigma\,\mathbf{L} + 1) \tag{83}$$

where we have used the definition $\mathbf{L} = (\mathbf{r}\times\mathbf{p})/\hbar$. Thus,

$$r\alpha_r\,\alpha\cdot\mathbf{p} = rp_r + i\,\hbar\,(\Sigma\cdot\mathbf{L} + 1)$$

Multiplying by α_r/r on the left, we obtain

$$\alpha\cdot\mathbf{p} = \alpha_r p_r + i\hbar\frac{\alpha_r}{r}\left(\Sigma\cdot\mathrm{L} + 1\right) \tag{84}$$

It is convenient to introduce the operator

$$K = \beta(\Sigma\cdot\mathbf{L} + 1) \tag{85}$$

It is easily shown that this commutes with the Hamiltonian. (This is the reason for the β in the definition of K)

$$K^2 = [\beta(\Sigma\cdot\mathbf{L} + 1)]^2 = (\Sigma\cdot\mathbf{L} + 1)^2 \quad \text{(since } \beta \text{ commutes with } \Sigma)$$

$$= (\Sigma\cdot\mathbf{L})^2 + 2\,\Sigma\cdot\mathbf{L} + 1$$

Now,
$$(\Sigma\cdot\mathbf{L})^2 = L^2 - \Sigma\cdot\mathbf{L}$$

using Eq. (82) and

$$\Sigma\times\Sigma = i\,\Sigma$$

Thus,
$$K^2 = L^2 + \Sigma\cdot\mathbf{L} + 1$$

$$= \left(L + \frac{\Sigma}{2}\right)^2 + \frac{1}{4} = J^2 + \frac{1}{4} \tag{86}$$

since the eigenvalue of Σ^2 is 3. The eigenvalues of K^2 are, therefore,

$$j(j+1)+\frac{1}{4}=\left(j+\frac{1}{2}\right)^2$$

Since K commutes with the Hamiltonian, we can set up a representation in which K and H are diagonal. Now,

$$H = c\alpha_r\, p_r + \frac{i\hbar c}{r}\alpha_r\, \beta K + \beta mc^2 + V(r) \tag{87}$$

We can now choose (matrix) representations of α_r and β which satisfy the conditions

$$\alpha_r^2 = \beta^2 = 1 \text{ and } [\alpha_r, \beta] = 0 \tag{88}$$

A simple one is $\beta = \begin{pmatrix} 1 & 0 \\ 0 & -1 \end{pmatrix}$ and $\alpha_r = \begin{pmatrix} 0 & -i \\ i & 0 \end{pmatrix}$. Writing $\psi = \begin{pmatrix} \psi_a \\ \psi_b \end{pmatrix}$, the relation $H\psi = E\psi$ gives the two coupled equations

$$(E - mc^2 - V)\,\psi_a + \left(icp_r + \frac{\hbar c}{r}K\right)\psi_b = 0 \tag{89}$$

and

$$(E + mc^2 - V)\,\psi_b - \left(icp_r - \frac{\hbar c}{r}K\right)\psi_a = 0 \tag{90}$$

The angle and spin dependent parts are included in K. If we choose ψ_a, ψ_b to be eigen functions of K (with eigenvalue denoted by the same symbol K), then in Eqs. (89) and (90), we need only concern ourselves with the radial parts of ψ_a and ψ_b. We set

$$\psi_a = \frac{F}{r}, \qquad \psi_b = \frac{G}{r} \tag{91}$$

Noting that

$$p_r\left(\frac{F}{r}\right) = -\frac{i\hbar}{r}\left(r\frac{\partial}{\partial r}+1\right)\left(\frac{F}{r}\right) = -\frac{i\hbar}{r}\frac{dF}{dr} \quad \text{etc.}$$

we obtain

$$(E - mc^2 - V)F + \hbar c\left(\frac{dG}{dr} + \frac{KG}{r}\right) = 0 \tag{92}$$

and

$$(E + mc^2 - V)G + \hbar c\left(\frac{dF}{dr} - \frac{KF}{r}\right) = 0 \tag{93}$$

30.10 ENERGY LEVELS OF THE HYDROGEN ATOM

For the hydrogen-like atom, we set $V(r) = -Ze^2/r$. Before trying a series solutions of Eqs (92) and (93) let us look at the asymptotic behavior of F and G. We may write (as $r \to \infty$),

$$(E - mc^2)\,F + \hbar c\,\frac{dG}{dr} = 0$$

$$(E + mc^2)\,G - \hbar c\,\frac{dF}{dr} = 0$$

whence
$$\frac{d^2F}{dr^2} - \frac{m^2c^4 - E^2}{(\hbar c)^2}F = 0 \tag{94}$$

Since we expect, for the bound state, $F \to 0$ as $r \to \infty$, we must have $mc^2 > |E|$ – otherwise, the solution to Eq. (94) will be oscillatory. For positive energies, this merely states that we expect the total energy of the atom to be somewhat less than the rest energy of the electron; one must not forget that we have not included the rest energy of the proton in the Hamiltonian in the first place. Thus, the asymptotic behavior is

$$F, G \sim e^{-\rho} \tag{95}$$

where
$$\rho = \frac{\sqrt{m^2c^4 - E^2}}{\hbar c} r \tag{96}$$

is an obviously convenient dimensionless variable. Equations (92) and (93) become

$$\left(-v + Z\frac{\alpha}{\rho}\right)F + \frac{dG}{d\rho} + \frac{KG}{\rho} = 0 \tag{97}$$

$$\left(\frac{1}{v} + Z\frac{\alpha}{\rho}\right)G - \frac{dF}{d\rho} + \frac{KG}{\rho} = 0 \tag{98}$$

where we have put

$$\sqrt{\frac{mc^2 - E}{mc^2 + E}} = v, \qquad \frac{e^2}{\hbar c} = \alpha \tag{99}$$

The asymptotic forms (95) suggest a solution of the following form

$$F(\rho) = f(\rho)\, e^{-\rho}, \qquad G(\rho) = g(\rho)\, e^{-\rho} \tag{100}$$

so that we have

$$\left(-v + Z\frac{\alpha}{\rho}\right)f + \left(\frac{d}{d\rho} + \frac{K}{\rho} - 1\right)g = 0 \tag{101}$$

$$\left(\frac{1}{v} + Z\frac{\alpha}{\rho}\right)g - \left(\frac{d}{d\rho} - \frac{K}{\rho} - 1\right)f = 0 \tag{102}$$

Substituting
$$f(\rho) = \rho^s \sum_{j=0}^{\infty} a_j p^j, \qquad g(\rho) = \rho^s \sum_{j=0}^{\infty} b_j p^j \tag{103}$$

and comparing the coefficients of ρ^{s+j-1}, we shall get the following indicial equations:

$$-va_{j-1} + Z\alpha a_j + (j+s)\, b_j + Kb_j - b_{j-1} = 0 \tag{104}$$

$$\frac{1}{v}b_{j-1} + Z\alpha b_j - (j+s)a_j + Ka_j + a_{j-1} = 0 \tag{105}$$

From the coefficients of the lowest term ρ^{s-1}, we obtain

$$Z\alpha a_0 + (s+K)\,b_0 = 0 \text{ and } Z\alpha b_0 - (s-K)\,a_0 = 0 \tag{106}$$

Since a_0 and b_0 are not zero, we must have

$$s^2 = K^2 - Z^2\alpha^2$$

or

$$s = \sqrt{K^2 - Z^2\alpha^2} \tag{107}$$

where the positive root is chosen to keep the functions F and G regular at the origin. Multiplying Eq. (105) by v and adding it to Eq. (104), we obtain

$$a_j\,[(j+s-K) - Z\alpha] = b_j\,[(j+s+K) + Z\alpha)] \tag{108}$$

It can be shown, as was done for the Schrödinger equation solution of the hydrogen atom, that if we let the series go to an infinite number of terms, the functions $f(\rho)$ and $g(\rho)$ will behave like $e^{2\rho}$ so that the solutions $F(\rho)$ and $G(\rho)$ will diverge. We must, therefore, assume that $j = n'$, where n' is a fixed integer, the series terminates and $a_{n'+1} = 0$ [and hence $b_{n'+1} = 0$, according to Eq. (108)]. Setting $j = n' + 1$ in Eqs (104) and (105), we obtain

$$-va_{n'} - b_{n'} = 0 \quad \text{or} \quad b_{n'} = -\,va_{n'} \tag{109}$$

Using Eq. (108), with $j = n'$, we obtain

$$(n' + s - K) - Z\alpha = -\,v\,[(n' + s + K) + Z\alpha] \tag{110}$$

or

$$2v\,(n' + s) = Z\alpha\,(1 - v^2)$$

$$= \frac{Z\alpha 2E}{mc^2 + E}$$

remembering that $v^2 = (mc^2 - E)/(mc^2 + E)$. On squaring, we have $(m^2c^4 - E^2)(n' + s)^2 = Z^2\alpha^2E^2$. This gives

$$E = mc^2\left[1 + \frac{Z^2\alpha^2}{(n'+s)^2}\right]^{-1/2} \tag{111}$$

it may be recalled that α is the fine structure constant,

$$\alpha = \frac{e^2}{\hbar c} \approx \frac{1}{137}$$

Now, the eigenvalues of K^2 are $(j + 1/2)^2$ so that $s = \sqrt{(j+1/2)^2 - Z^2\alpha^2}$. To appreciate the change from the non-relativistic theory, let us introduce the small number

$$\varepsilon_j = \left(j + \frac{1}{2}\right) - \sqrt{\left(j + \frac{1}{2}\right)^2 - Z^2\alpha^2}$$

so that $s = j + 1/2 - \varepsilon$. We also introduce the principal quantum number

$$n = j + \frac{1}{2} + n' \begin{pmatrix} n = 1, 2, 3, \ldots \\ j = \frac{1}{2}, \frac{3}{2}, \ldots, n - \frac{1}{2} \end{pmatrix}$$

Thus $n' + s = n - \varepsilon_j$, so that

$$E = mc^2 \left[1 + \frac{Z^2\alpha^2}{(n - \varepsilon_j)^2} \right]^{-1/2} \tag{112}$$

Remembering that ε_j is also a function of $Z^2\alpha^2$, one can expand E as a power series in $Z^2\alpha^2$ to obtain

$$E_{n,j} = mc^2 \left[1 - \frac{Z^2\alpha^2}{2n^2} - \frac{Z^4\alpha^4}{2n^4} \left(\frac{n}{j + \frac{1}{2}} - \frac{3}{4} \right) \right] \tag{113}$$

Apart from the constant term mass, the second term in Eq. (113) is identical to the non-relativistic expression while the third gives the relativistic corrections which is the same as we had obtained in Section 20.5.2. We label the energy levels as

$$2S_{1/2} (\text{for } n = 2, j = 1/2 \text{ and } l = 0)$$
$$2P_{1/2} (\text{for } n = 2, j = 1/2 \text{ and } l = 1)$$
$$2P_{3/2} (\text{for } n = 2, j = 3/2 \text{ and } l = 1)$$

We note that, while non-relativistic theory predicts that these three states are all degenerate, Eq. (113) predicts that only $2S_{1/2}$ and $2P_{1/2}$ are degenerate while $2P_{3/2}$ lies lower by about 10^{-4} eV; for detailed calculations see Section 20.5. In 1947, a crucial experiment by Lamb and Rutherford established that $2S_{1/2}$ lies about 1058 MHz below $2P_{1/2}$ (the frequency corresponds to the energy difference $\sim 4.4 \times 10^{-9}$ eV). Although this might appear as a very small discrepancy, the experiment initiated very important developments which led to a consistent theory of quantum electrodynamics[2].

30.11 COVARIANCE OF DIRAC EQUATION

Let us first see what covariance demands. Assume that there are two observers in Lorentz frames Σ and Σ'. We require that scalar quantities remain the same in the two systems while four vectors must transform like the space-time coordinates. Finally, although the wavefunction for the two observers need not be the same, the recipe for writing down Dirac equation must be the same for both observers. The equations in Σ and Σ' frames are therefore

$$\left(\gamma_\mu \frac{\partial}{\partial x_\mu} + \kappa \right) \psi(x_\mu) = 0 \tag{114}$$

[2] For a simple picture of the origin of the Lamb Shift see Chapter 4 of Ref.4.

$$\left(\gamma_\mu \frac{\partial}{\partial x'_\mu} + \kappa\right)\psi'(x'_\mu) = 0 \tag{115}$$

Notice that γ_μ are the same in both frames; they define the recipe and ought not to be changed. The coordinates x and x' are related through

$$x'_\nu = a_{\nu\mu} x_\mu \tag{116}$$

and the inverse transformation

$$x_\mu = a_{\nu\mu} x'_\nu \tag{117}$$

with the restriction

$$a_{\lambda\mu} a_{\mu\nu} = \delta_{\lambda\nu} \tag{118}$$

We have

$$\frac{\partial\psi}{\partial x_\mu} = \frac{\partial\psi}{\partial x'_\nu}\frac{\partial x'_\nu}{\partial x_\mu} = a_{\nu\mu}\frac{\partial\psi}{\partial x'_\nu} \tag{119}$$

Writing

$$\psi'(x'_\mu) = S\psi(x_\mu) \tag{120}$$

and the inverse relation

$$\psi(x_\nu) = S^{-1}\psi'(x'_\nu) \tag{121}$$

we can rewrite Eq. (114) as

$$\left(\gamma_\mu a_{\nu\mu} \frac{\partial}{\partial x'_\nu} + \kappa\right) S^{-1}\psi'(x'_\mu) = 0$$

Multiplying on the left by S, we have

$$\left(a_{\nu\mu} S\gamma_\mu S^{-1} \frac{\partial}{\partial x'_\nu} + \kappa\right)\psi'(x'_\mu) = 0 \tag{122}$$

This is identical with Eq. (115) provided we can find an S which satisfies

$$a_{\nu\mu} S\gamma_\mu S^{-1} = \gamma_\nu$$

Multiplying by $a_{\lambda v}$, summing over v and using relation (118), we obtain

$$S\gamma_\lambda S^{-1} = a_{\lambda\nu}\gamma_\nu \tag{123}$$

For the complete proof of covariance, we have to show that such an S exists. Here we merely assert that this is true and refer the interested reader to other works for explicit construction of S for every type of Lorentz transformation (see, e.g. Refs 4 and 5). Eq. (123) suggests that a set of matrices $\gamma'_\lambda = a_{\lambda\nu}\gamma_\nu$ will obey all the commutation properties of γ_ν and would serve just as well. One could, therefore, have written down the equation in Σ' frame as

$$\left(\gamma'_\lambda \frac{\partial}{\partial x'_\lambda} + \kappa\right)\psi'(x'_\lambda) = 0 \tag{124}$$

and γ'_λ and γ_ν are related exactly as four vectors are in the two frames Neverthless, one must remember that the four γ_λ are matrices and cannot be regarded as a physical vector.

30.12 THE HOLE THEORY AND POSITRONS

We have seen that the Dirac equation admits both positive and negative energy solutions. In the presence of a field one would expect in fact, the electron to make radiative transitions to a negative energy state, that is, $E = -\infty$. There are other difficulties with negative energies. An electron at rest in a negative energy state has a negative rest mass. Clearly, such electrons would move opposite to an applied force. The theory presented so far, is thus an incomplete *physical* theory.

Dirac made a remarkable suggestion to avoid these difficulties. He assumed that all the negative energy states are normally occupied by electrons; i.e. vacuum is the state of fully occupied negative energy states and empty positive ones. An electron with positive energy now introduced into this system is prevented from making a transition to a negative energy state because of the exclusion principle. In vacuum, one therefore has this negative energy 'sea' of electrons giving an infinite charge density. Dirac, therefore, made a further assumption, namely, that electrons filling up negative energy states do not contribute to the external field, the total energy and momentum of the system. Only *departures* from the distribution in a vacuum will contribute to observable quantities.

To see how this works, suppose a negative energy electron with an energy $-|E_i|$ and momentum **p** is removed from the negative energy sea, under the action of an external field, to a positive energy state (see Fig. 30.1).

The system (the sea) is now *deficient in energy* $-|E_i|$ and momentum **p**. This means that the residual sea behaves as though it has a *hole* with energy $|E_i|$ and momentum $-$**p**. (Notice that the hole has a *positive* energy $|E_i|$.) Also, with the removal of charge $-e$, the net effect of the hole is that of a positive charge $+e$. This hole, therefore, behaves like a particle with the electronic mass m and charge e and is interpreted as a positron.[3, 4]

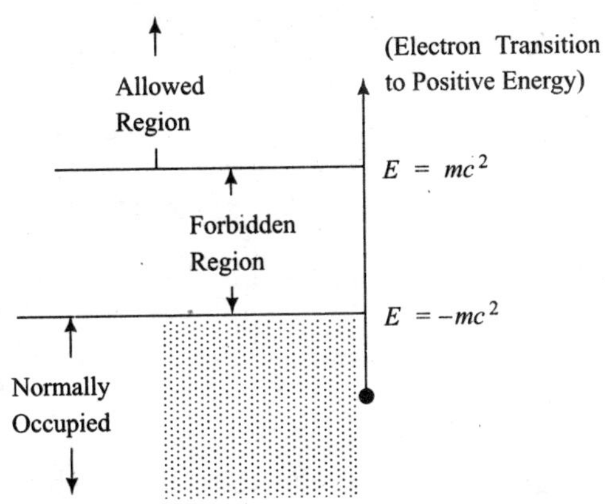

Figure 30.1: The positive and negative energy states of the electron

If the external field removes an electron of momentum $\mathbf{p}_i$ from a negative energy state $-|E_i|$ to a state of momentum $\mathbf{p}_f$ and positive energy $+|E_f|$, then the hole will have an energy $+|E_i|$ and momentum $-\mathbf{p}_i$. In other words, the final system now consists of an electron of energy $|E_f|$ and

[3] All observed states are states of positive energy. When making calculations for physical quantities it is necesssary to select out final states of positive energy.

[4] The hole theory treats the positron on a different footing (at least in the physical interpretation) from the electron. Modern theories of quantum electrodynamics have avoided this distinction.

momentum $\mathbf{p}_f$ and a positron of energy $|E_i|$ and momentum $-\mathbf{p}_i$. Clearly, the total energy $|E_i| + |E_f|$ and the total momentum $\mathbf{p}_f - \mathbf{p}_i$ of the two particles in the final states must have been supplied by the external field. Moreover, since

$$|E_i| + |E_f| > 2mc^2$$

this is the minimum required energy for this process, called pair production. Thus, the hole theory provides a simple explanation for the positron and phenomena such as the creation or annihilation of pairs.

However, what has been gained in interpretation has made the theory far more complicated. Electron-electron scattering, for example is no longer to be thought of a simple two body process. The vacuum[5], with its infinite number of particles can participate in the phenomenon making the theory a many-body one. While it is true that for low energies generally the situation is not as drastic, nevertheless one should expect new phenomena to manifest themselves.

30.13 PROBLEMS

Problem 30.1 Show that the four spinors described by Eqs (43)–(45) are orthonormal, i.e

$$\chi_a^\dagger \chi_b = \delta_{ab}$$

where $\chi_a^\dagger$, χ_b are any two of them.

Problem 30.2 Show that $\mathbf{L} = \mathbf{r} \times \mathbf{p}$ does not commute with the free particle Dirac Hamiltonian, while $\mathbf{J} = \mathbf{L} + \mathbf{S}$ does and hence $\mathbf{J}$ is a constant of the motion. Here $\mathbf{S} = \frac{1}{2}\hbar\Sigma$, where

$$\Sigma = \begin{pmatrix} \sigma & 0 \\ 0 & \sigma \end{pmatrix}$$

Problem 30.3 Prove

$$(\alpha\cdot\mathbf{B})\,(\alpha\cdot\mathbf{C}) = \mathbf{B}\cdot\mathbf{C} + i\Sigma\,(.\mathbf{B}\times\mathbf{C}) \tag{125}$$

Problem 30.4 Verify that the rate of change of position operator (in the Heisenberg picture) is given by

$$\frac{d\mathbf{r}}{dt} = c\alpha \tag{126}$$

Discuss this result.

Problem 30.5

(a) Use the operator representations of p_x, p_y and p_z (see Section 9.2) to show that

$$\frac{1}{r}\left[xp_x + yp_y + zp_z - i\hbar\right]\psi = -i\hbar\left[\frac{\partial\psi}{\partial r} + \frac{\psi}{r}\right]$$

(b) Show that

$$[r, p_r] = i\hbar$$

[5] Diarc's original idea of the particle as an excitation for the strange 'vacuum' has been very fruitful and the idea is exploited in the theory of condensed states. (See Kaempffer, Concepts in Quantum Mechanics, Academic Press, New York, 1965.)

Problem 30.6 Assume that the frame Σ' introduced in Section 30.11 is obtained by a simple rotation of Σ about the *z*-axis by an angle φ. Show that the following matrix *S* satisfies Eq. (123).

$$S = \exp\left[\frac{i\varphi}{2}\sigma_{12}\right] \tag{127}$$

where
$$\sigma_{12} = \frac{i}{2}\left[\gamma_1, \gamma_2\right] = i\gamma_1\gamma_2 \tag{128}$$

Problem 30.7 Suppose Σ' moves along the *x*-axis with respect to Σ with a velocity βc. Show that the matrix *S* is given by

$$S = \exp\left[\frac{i\varphi}{2}\sigma_{14}\right], \qquad \sigma_{14} = i\gamma_1\gamma_4 \tag{129}$$

Writing $\varphi = i\theta$, show that $\tanh\theta = \beta$.

30.14 SOLUTIONS

Solution 30.2 We consider first $[L_x, H]$. This is

$$c\alpha\cdot[(yp_z - zp_y)\,\mathbf{p} - \mathbf{p}(yp_z - zp_y)] = c\alpha_2\,(yp_y - p_y y)\,p_z + c\alpha_3\,(-zp_z + p_z z)\,p_y$$
$$= ic\hbar\,(\alpha_2 p_z - \alpha_3 p_y)$$

To calculate $[\Sigma_x, H]$, we note that

$$\Sigma_x = \begin{pmatrix}\sigma_x & 0\\ 0 & \sigma_x\end{pmatrix} \quad\text{and}\quad \alpha = \begin{pmatrix}0 & \sigma\\ \sigma & 0\end{pmatrix}$$

so that
$$\Sigma_x\alpha_2 - \alpha_2\Sigma_x = 2i\alpha_3$$

and
$$\Sigma_x\alpha_3 - \alpha_3\Sigma_x = -2i\alpha_2$$

This gives
$$[\Sigma_x, H] = ic(\alpha.\mathbf{p}\Sigma_x - \Sigma_x\,\alpha.\mathbf{p})$$
$$= 2ic(\alpha_3 p_y - \alpha_2 p_z)$$

Thus, $J_x = L_x + \frac{1}{2}\hbar\Sigma_x$ (and similarly other components) commutes with *H*.

Solution 30.3 From Heisenberg's equation, we have

$$\frac{d\mathbf{r}}{dt} = \frac{i}{\hbar}\left[\mathbf{H}, \mathbf{r}\right]$$

We have
$$\frac{dx}{dt} = \frac{i}{\hbar}c\left[\alpha\,.\,\mathbf{p}x - x\alpha\,.\,\mathbf{p}\right]$$

We need only consider $\alpha_x p_x$ in $\alpha\cdot\mathbf{p}$, the rest commute. Thus,

$$\frac{dx}{dt} = \frac{ic}{\hbar}\alpha_x\left[p_x x - xp_x\right] = c\alpha_x$$

Since the eigenvalues of α_x are ± 1, $\pm c$ are the only possible eigenvalues of each of the operators dr/dt.

A precise velocity measurement requires the precise measurement of position at two different times which means the uncertainties in momenta approach infinity. In relativity, these correspond to the speed of light.

Solution 30.6 We have

$$S = 1 + \left(\frac{i\phi}{2}\right)\sigma_{12} + \frac{1}{2!}\left(\frac{i\phi}{2}\right)^2 \sigma_{12}^2 + \ldots$$

Now,
$$\sigma_{12}^2 = (i)^2 \gamma_1 \gamma_2 \gamma_1 \gamma_2 = -(-\gamma_1 \gamma_2 \gamma_2 \gamma_1) = +1$$

$$\sigma_{12}^3 = \sigma_{12}\sigma_{12}^2 = \sigma_{12} \text{ etc.}$$

It is easily shown that

$$S = \cos\frac{\phi}{2} - \gamma_1\gamma_2 \sin\frac{\phi}{2} \tag{130}$$

We, now, have to show that

$$S\gamma_\lambda S^{-1} = a_{\lambda\nu}\gamma_\nu \tag{131}$$

or
$$S\gamma_\lambda = a_{\lambda\nu}\gamma_\nu S \tag{132}$$

Here,
$$a_{\lambda\nu} = \begin{pmatrix} \cos\phi & \sin\phi & 0 & 0 \\ -\sin\phi & \cos\phi & 0 & 0 \\ 0 & 0 & 1 & 0 \\ 0 & 0 & 0 & 1 \end{pmatrix}$$

as may be seen from working out the coordinate transformations under the desired rotation

$$x' = ax$$

For $\lambda = 3$ and 4, Eq. (132) merely says that S and γ_l commute. For example, the second term in S [see Eq. (130)] gives a term proportional to $\gamma_1\gamma_2\gamma_3 = -\gamma_1\gamma_3\gamma_2 = \gamma_3\gamma_1\gamma_2$. We, therefore, merely have to prove the result for $\lambda = 1, 2$. We have (for $\lambda = 1$)

$$S\gamma_1 = \gamma_1 \cos\frac{\phi}{2} + \gamma_2 \sin\frac{\phi}{2}$$

and
$$a_{1\nu}\gamma_{\nu S} = (\gamma_1 \cos\phi + \gamma_2 \sin\phi)_S$$

$$= \gamma_1 \cos\phi \cos\frac{\phi}{2} - \gamma_2 \cos\phi \sin\frac{\phi}{2} + \gamma_2 \sin\phi \cos\frac{\phi}{2} + \gamma_1 \sin\phi \sin\frac{\phi}{2}$$

$$= \gamma_1 \cos\frac{\phi}{2} + \gamma_2 \sin\frac{\phi}{2}$$

Similarly, we can establish the result for $\lambda = 2$.

Solution 30.7 The Lorentz transformation is defined by

$$a_{\lambda\nu} = \begin{pmatrix} \gamma & 0 & 0 & i\beta\gamma \\ 0 & 1 & 0 & 0 \\ 0 & 0 & 1 & 0 \\ -i\beta\gamma & 0 & 0 & \gamma \end{pmatrix} \tag{133}$$

where $\gamma = (1 - \beta^2)^{-1/2}$, not to be confused with the γ matrices. Proceeding as in the previous problem, we have

$$S\gamma_1 = \gamma_1 \cos\frac{\phi}{2} + \gamma_4 \sin\frac{\phi}{2}$$

and

$$a_{1\nu}\gamma_\nu S = (\gamma\gamma_1 + i\beta\gamma\,\gamma_4)S$$

$$= \gamma_1\left(\gamma\cos\frac{\phi}{2} + i\beta\gamma\sin\frac{\phi}{2}\right) + \gamma_4\left(i\beta\gamma\cos\frac{\phi}{2} - \gamma\sin\frac{\phi}{2}\right)$$

from which we obtain (equating the two expressions)

$$\cos\phi = \cos\theta = \gamma$$

$$\sin\phi = i\sinh\theta = i\beta\gamma \quad (\text{where } \phi = i\theta)$$

Thus,

$$\tanh\theta = \beta$$

and

$$S = \cosh\frac{\theta}{2} - i\gamma_1\gamma_4\sin\frac{\theta}{2}. \tag{134}$$

REFERENCES AND SUGGESTED READINGS

1. G. Baym, *Lectures on Quantum Mechanics*, W. A. Benjamin, New York (1969).
2. P.A.M. Dirac, *The Principles of Quantum Mechanics*, Fourth Edition, Oxford University Press, London (1958).
3. T.P. Das, *Relativistic Quantum Mechanics of Electrons*, Harper & Row, New York (1973).
4. J.D. Bjorken and S. D. Drell, *Relativistic Quantum Mechanics*, McGraw-Hill Book Co., New York (1964).
5. J.A. Eisele, *Modern Quantum Mechanics with Applications to Elementary Particle Physics*, Wiley Interscience, New York (1969).
6. A.S. Davydov, *Quantum Mechanics*, Pergamon Press, Oxford (1965).
7. V.B. Berestetskii, E. M. Lifshitz and L. P. Pitaevskii, *Relativistic Quantum Theory*, Pergamon Press, Oxford (1971).
8. L.I. Schiff, *Quantum Mechanics*, Third Edition, McGraw-Hill Book Co., New York (1968).

ppendix A Gamma functions and integrals involving Gaussian functions

We will first show that

$$\int_{-\infty}^{+\infty} e^{-\alpha x^2 + \beta x}\, dx = \sqrt{\frac{\pi}{\alpha}} \exp\left[\frac{\beta^2}{4\alpha}\right] \quad \text{Re}\, \alpha > 0 \tag{1}$$

We consider the integral

$$I = \int_{-\infty}^{+\infty} e^{-x^2}\, dx \tag{2}$$

Thus,

$$I^2 = \int_{-\infty}^{+\infty} e^{-x^2}\, dx \int_{-\infty}^{+\infty} e^{-y^2}\, dy$$

$$= \int_{-\infty}^{+\infty}\int_{-\infty}^{+\infty} e^{-(x^2+y^2)}\, dx\, dy$$

Transforming to polar coordinates, we get

$$I^2 = \int_0^{\infty} e^{-r^2}\, r\, dr \int_0^{\pi} d\theta$$

$$= \left[-\frac{1}{2} e^{-r^2}\right]_0^{\infty} 2\pi$$

$$= \pi$$

Thus,

$$I = \int_{-\infty}^{+\infty} e^{-x^2}\, dx = \sqrt{\pi} \tag{3}$$

Now,

$$\int_{-\infty}^{+\infty} e^{-\alpha x^2 + \beta x}\, dx = \exp\left[\frac{\beta^2}{4\alpha}\right] \int_{-\infty}^{+\infty} \exp\left[-\alpha\left(x - \frac{\beta}{2\alpha}\right)^2\right] dx$$

$$= \exp\left[\frac{\beta^2}{4\alpha}\right] \int_{-\infty}^{+\infty} e^{-\alpha z^2}\, dz$$

where $z = x - \dfrac{\beta}{2\alpha}$. Using Eq. (3), we get

$$\int_{-\infty}^{+\infty} e^{-\alpha z^2}\, dz = \sqrt{\frac{\pi}{\alpha}} \tag{4}$$

using which we obtain Eq. (1). We also get

$$\sqrt{\pi} = 2\int_0^{\infty} e^{-x^2}\, dx$$

$$= \int_0^{\infty} y^{-1/2} e^{-y}\, dy$$

Thus,

$$\Gamma\left(\frac{1}{2}\right) = \sqrt{\pi} \tag{5}$$

where $\Gamma(z)$ is defined through the equation

$$\Gamma(z) = \int_0^{\infty} x^{z-1} e^{-x}\, dx \qquad \text{Re } z > 0 \tag{6}$$

For Re $z > 1$, if we integrate by parts we would obtain

$$\Gamma(z) = (z-1)\,\Gamma(z-1) \tag{7}$$

Since,

$$\Gamma(1) = \int_0^{\infty} e^{-x}\, dx = 1$$

we obtain,

$$\Gamma(n+1) = n!\,;\; n = 0, 1, 2, \ldots \tag{8}$$

Further since $\Gamma\left(\frac{1}{2}\right) = \sqrt{\pi}$, we obtain

$$\left.\begin{aligned} \Gamma\left(\frac{3}{2}\right) &= \frac{1}{2}\Gamma\left(\frac{1}{2}\right) = \frac{1}{2}\sqrt{\pi} \\ \Gamma\left(\frac{5}{2}\right) &= \frac{3}{2}\Gamma\left(\frac{3}{2}\right) = \frac{3}{2}\cdot\frac{1}{2}\sqrt{\pi} \\ \Gamma\left(\frac{7}{2}\right) &= \frac{5}{2}\cdot\frac{3}{2}\cdot\frac{1}{2}\sqrt{\pi} \end{aligned}\right\} \tag{9}$$

etc. Finally for $n = 0, 1, 2, \ldots$

$$\int_{-\infty}^{+\infty} x^{2n} e^{-x^2}\, dx = \Gamma\left(n + \frac{1}{2}\right) = \frac{1\cdot 3\cdot 5\ldots(2n-1)\sqrt{\pi}}{2^n} \tag{10}$$

and

$$\int_{-\infty}^{+\infty} x^{2n+1} e^{-x^2}\, dx = 0 \tag{11}$$

Appendix B Equation of continuity

According to Gauss's theorem,

$$\int_{\Omega} \operatorname{div} \mathbf{J}\, d\tau = \int_{S} \mathbf{J}\cdot\hat{\mathbf{n}}\, dS \tag{1}$$

where Ω represents an arbitrary volume bounded by the surface S with $\hat{\mathbf{n}}$ representing the outward unit normal on the surface. Now, if $\mathbf{J}$ is assumed to represent the current density then

$$\mathbf{J}\cdot\hat{\mathbf{n}}\, dS$$

would represent the number of particles going *out* of the area dS (per unit time). Thus the right-hand side of Eq. (1) would represent the total number of particles going out of the volume per unit time. This is nothing but

$$-\frac{\partial}{\partial t}\int_{\Omega} \rho\, d\tau$$

where ρ represents the number of particles per unit volume. Thus,

$$-\frac{\partial}{\partial t}\int_{\Omega} \rho\, d\tau = \int_{\Omega} \operatorname{div}\ \mathbf{J}\, d\tau \tag{2}$$

Since the above equation holds for an arbitrary volume, we get

$$\operatorname{div}\ \mathbf{J} + \frac{\partial \rho}{\partial t} = 0 \tag{3}$$

which represents the equation of continuity.

Appendix C The Fresnel Integrals

If we substitute Eq. (90) [of Chapter 5] in Eq. (76) [of Chapter 5] we would obtain

$$\Psi(x, y, t) = \Psi(x, t)\,\Psi(y, t) \tag{1}$$

where $\Psi(x, t)$ is given by Eq. (35) [of Chapter 5] and

$$\Psi(y, t) = \sqrt{\frac{m}{2\pi i\hbar t}}\frac{1}{\sqrt{b}}\int_{-b/2}^{+b/2}\exp\left[i\frac{(y'-y)^2 m}{2\hbar t}\right]dy' \tag{2}$$

We would like to express the right-hand side of the above equation in terms of the Fresnel integrals

$$C(v) = \int_0^{v}\cos\left(\frac{\pi x^2}{2}\right)dx \tag{3}$$

and

$$S(v) = \int_0^{v}\sin\left(\frac{\pi x^2}{2}\right)dx \tag{4}$$

Thus we define the variable

$$u = (y'-y)\sqrt{\frac{m}{\pi\hbar t}} \tag{5}$$

to obtain

$$\Psi(y, t) = \frac{1}{\sqrt{2ib}}\int_{-\mu-\nu}^{-\mu+\nu}\exp\left[i\frac{\pi u^2}{2}\right]du \tag{6}$$

where

$$\mu = y\sqrt{\frac{m}{\pi\hbar t}} = \frac{1}{\tau\sqrt{\pi}}\left(\frac{y}{\sigma_{x_0}}\right) \tag{7}$$

and

$$\nu = \frac{b}{2}\sqrt{\frac{m}{\pi\hbar t}} = \frac{1}{\tau\sqrt{\pi}}\left(\frac{b}{2\sigma_{x_0}}\right) \tag{8}$$

where τ is given by Eq. (36) [of Chapter 5] with σ_0 replaced by σ_{x0}. Straightforward algebra gives

$$|\Psi(y,t)|^2 = \frac{1}{2b}\left\{\left[C(-\mu+\nu)-C(-\mu-\nu)\right]^2+\left[S(-\mu+\nu)-S(-\mu-\nu)\right]^2\right\} \tag{9}$$

Obviously, the probability distribution will be symmetric about the y-axis and therefore we may replace μ by $-\mu$ to obtain

$$|\Psi(y,t)|^2 = \frac{1}{2b}\left\{\left[C(\mu+\nu)-C(\mu-\nu)\right]^2+\left[S(\mu+\nu)-S(\mu-\nu)\right]^2\right\} \tag{10}$$

For a given value of ν, the probability distribution (as a function of μ) is a universal curve. The total probability distribution is therefore given by

$$|\Psi(x,y,t)|^2 = \left[\frac{1}{\sqrt{\pi}\,\sigma(t)}\exp\left[-\frac{\left(x-\frac{p_0}{m}t\right)^2}{\sigma^2(t)}\right]\right]$$

$$\times\left[\frac{1}{2b}\left\{\left[C(\mu+\nu)-C(\mu-\nu)\right]^2+\left[S(\mu+\nu)-S(\mu-\nu)^2\right\}\right] \tag{11}$$

where
$$\sigma(t) = \sigma_{x_0}\left[1+\frac{\hbar^2}{m^2\sigma_{x_0}^4}t^2\right]^{1/2} \tag{12}$$

The analysis (and the corresponding intensity plots) are very similar to the calculation of the Fresnel diffraction pattern for a long narrow slit in wave optics (see, e.g. Section 5.10 of Ref. 1).

REFERENCE

1. A.K. Ghatak and K. Thyagarajan, *Optical Electronics*, Cambridge University Press, Cambridge (1989); Reprinted by Foundation Books, New Delhi (1991).

Appendix D Airy functions

Airy functions are the two independent solutions of the differential equation

$$\frac{d^2y}{dx^2} - xy(x) = 0 \tag{1}$$

The above equation is usually referred to as the Airy equation (after the well-known British astronomer G.B. Airy). In order to solve Eq. (1) we try out a power series solution

$$y(x) = a_0 + a_1 x + a_2 x^2 + \cdots \tag{2}$$

Substituting in Eq. (1), we get

$$2a_2 + 3 \cdot 2a_3 x + 4 \cdot 3a_4 x^2 + \cdots + n(n-1)\, a_n x^{n-2} + \cdots$$

$$= a_0 x + a_1 x^2 + \cdots + a_{n-3} x^{n-2} + \cdots$$

Thus, $$a_2 = 0,\ a_3 = \frac{a_0}{2 \cdot 3},\ a_4 = \frac{a_1}{3 \cdot 4},\ \ldots,\ a_n = \frac{a_{n-3}}{n(n-1)},\ldots$$

and $$y(x) = a_0\left[1 + \frac{x^3}{3 \cdot 2} + \frac{x^6}{2 \cdot 3 \cdot 5 \cdot 6} + \cdots\right] + a_1\left[x + \frac{x^4}{3 \cdot 4} + \frac{x^7}{3 \cdot 4 \cdot 6 \cdot 7} + \cdots\right] \tag{3}$$

where a_0 and a_1 are arbitrary constants. Thus we may write

$$y(x) = a_0 f(x) + a_1 g(x) \tag{4}$$

where $$f(x) = 1 + \frac{1}{3!}x^3 + \frac{1 \cdot 4}{6!}x^6 + \frac{1 \cdot 4 \cdot 7}{9!}x^9 + \cdots \tag{5}$$

$$g(x) = x + \frac{2}{4!}x^4 + \frac{2 \cdot 5}{7!}x^7 + \frac{2 \cdot 5 \cdot 8}{10!}x^{10} + \cdots \tag{6}$$

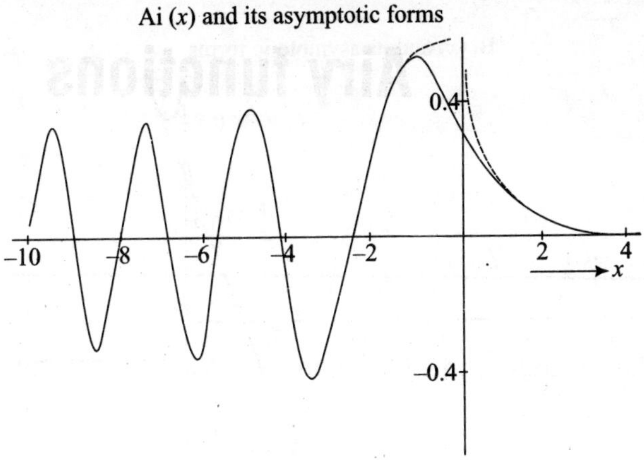

Figure D.1: The solid curve represents the Airy function Ai (x). The dashed curves on the left and on the right side of the origin represent the asymptotic forms as given by Eqs (21) and (20) respectively.

The Airy functions Ai (x) and Bi (x) are defined as

$$\text{Ai}(x) = c_1 f(x) - c_2 g(x) \tag{7}$$

$$\text{Bi}(x) = \sqrt{3}\,[c_1 f(x) + c_2 g(x)] \tag{8}$$

where

$$c_1 = \text{Ai}(0) = \left[3^{2/3}\Gamma\left(\frac{2}{3}\right)\right]^{-1} = 0.35503 \tag{9}$$

$$c_2 = -\text{Ai}'(0) = \left[3^{1/3}\Gamma\left(\frac{1}{3}\right)\right]^{-1} = 0.25882 \tag{10}$$

Further,

$$\text{Bi}(0) = \sqrt{3}\,c_1 = 0.61493 \tag{11}$$

and

$$\text{Bi}'(0) = \sqrt{3}\,c_2 = 0.44829 \tag{12}$$

The two functions Ai (x) and Bi (x) represent the two independent solutions of Eq. (1). Actually, originally, Ai (x) and Bi (x) were defined through the following integrals:

$$\begin{aligned}
\text{Ai}(x) &= \frac{1}{2\pi}\int_{-\infty}^{+\infty} \exp\left[i\left(\eta x + \frac{1}{3}\eta^3\right)\right] d\eta = \frac{1}{\pi}\int_0^{\infty} \cos\left[\eta x + \frac{1}{3}\eta^3\right] d\eta \\
&= \left[3^{2/3}\Gamma\left(\frac{2}{3}\right)\right]^{-1}\left[1 + \frac{1}{3!}x^3 + \frac{1\cdot 4}{6!}x^6 + \frac{1\cdot 4\cdot 7}{9!}x^9 + \cdots\right] \\
&\quad - \left[3^{1/3}\Gamma\left(\frac{1}{3}\right)\right]^{-1}\left[x + \frac{2}{4!}x^4 + \frac{2\cdot 5}{7!}x^7 + \frac{2\cdot 5\cdot 8}{10!}x^{10} + \cdots\right]
\end{aligned} \tag{13}$$

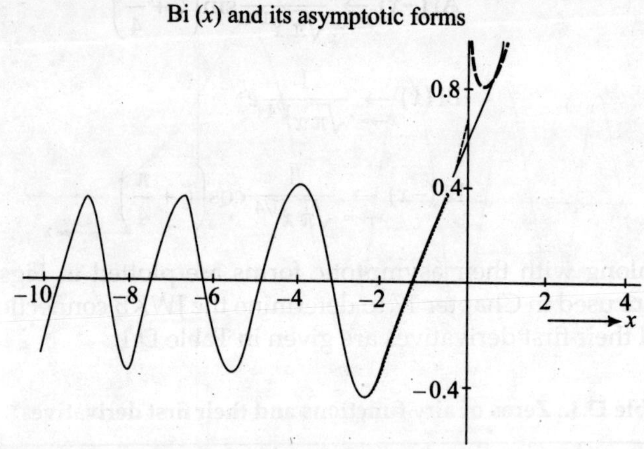

Figure D.2: The solid curve represents the Airy function Bi (x). The dashed curves on the left and on the right side of the origin represent the asymptotic forms as given by Eqs (23) and (22) respectively.

$$\text{Bi}\,(x) = \frac{1}{\pi}\int_0^\infty \left[\exp\left(-\frac{1}{3}\eta^3 + x\eta\right) + \sin\left(\frac{1}{3}\eta^3 + x\eta\right)\right] d\eta$$

$$= \sqrt{3}\left[3^{2/3}\,\Gamma\left(\frac{2}{3}\right)\right]^{-1}\left[1 + \frac{1}{3!}x^3 + \frac{1\cdot 4}{6!}x^6 + \cdots\right]$$

$$+\sqrt{3}\left[3^{1/3}\,\Gamma\left(\frac{1}{3}\right)\right]^{-1}\left[x + \frac{2}{4!}x^4 + \frac{2\cdot 5}{7!}x^7 + \cdots\right] \tag{14}$$

The Airy functions (and their derivatives) are related to Bessel functions through the following equations

$$\zeta = \frac{2}{3}x^{3/2}\ , \quad x > 0 \tag{15}$$

$$\text{Ai}\,(-x) = \frac{1}{3}\sqrt{x}\left[J_{-1/3}(\zeta) + J_{+1/3}(\zeta)\right] \tag{16}$$

$$\text{Bi}\,(-x) = \sqrt{\frac{x}{3}}\left[J_{-1/3}(\zeta) - J_{+1/3}(\zeta)\right] \tag{17}$$

$$\text{Ai}\,(x) = \frac{1}{3}\sqrt{x}\left[I_{-1/3}(\zeta) - I_{+1/3}(\zeta)\right] = \frac{1}{\pi}\left(\frac{x}{3}\right)^{1/2} K_{1/3}(\zeta) \tag{18}$$

$$\text{Bi}\,(x) = \sqrt{\frac{x}{3}}\left[I_{-1/3}(\zeta) + I_{+1/3}(\zeta)\right] \tag{19}$$

Using the asymptotic forms of Bessel functions, we have the following asymptotic (i.e. large $|x|$) behaviour of the Airy functions:

$$\text{Ai}\,(x) \underset{x\to\infty}{\longrightarrow} \frac{1}{2\sqrt{\pi}\,x^{1/4}}\,e^{-\zeta} \tag{20}$$

$$\mathrm{Ai}(-x) \underset{x\to\infty}{\to} \frac{1}{\sqrt{\pi}\, x^{1/4}} \sin\left(\zeta + \frac{\pi}{4}\right) \tag{21}$$

$$\mathrm{Bi}(x) \underset{x\to\infty}{\to} \frac{1}{\sqrt{\pi}\, x^{1/4}} e^{\zeta} \tag{22}$$

$$\mathrm{Bi}(-x) \underset{x\to\infty}{\to} \frac{1}{\sqrt{\pi}\, x^{1/4}} \cos\left(\zeta + \frac{\pi}{4}\right) \tag{23}$$

The Airy functions along with their asymptotic forms are plotted in Figs D.l and D.2. The asymptotic forms were used in Chapter 17 to determine the JWKB connection formulae. Zeros of Airy functions and their first derivatives are given in Table D.1.

Table D.1.: Zeros of airy functions and their first derivatives*.

	Ai $(a_s) = 0$	Bi $(b_s) = 0$	Ai$'(a'_s) = 0$	Bi$'(b'_s) = 0$
S	a_s	b_s	a'_s	b'_s
1.	− 2.33811	− 1.17371	− 1.01879	− 2.29444
2.	− 4.08795	− 3.27109	− 3.24820	− 4.07316
3.	− 5.52056	− 4.83074	− 4.82010	− 5.51240
4.	− 6.78671	− 6.16985	− 6.16331	− 6.78129
5.	− 7.94413	− 7.37676	− 7.37218	− 7.94018
6.	− 9.02265	− 8.49195	− 8.48849	− 9.01958
7.	− 10.04017	− 9.53819	− 9.53545	− 10.03770
8.	− 11.00852	− 10.52991	− 10.52766	− 11.00646
9.	− 11.93602	− 11.47695	− 11.47506	− 11.93426
10.	− 12.82878	− 12.38642	− 12.38479	− 12.82726

* Adapted from Reference 2.

REFERENCES

1. H. Jeffreys and B. Jeffreys, *Methods of Mathematical Physics*, Cambridge University Press, Cambridge (1972).
2. M. Abramowtiz and I.A. Stegun, *Handbook of Mathematical Functions*, National Bureau of Standards, United States Department of Commerce, Washington, D.C (1964).
3. A.K. Ghatak, I.C. Goyal and S.J. Chua, *Mathematical Physics*, Laxmi Publications, New Delhi (2017).

Appendix E Integral and summation involving Hermite-Gauss functions

We first write the generating function associated with Hermite polynomials

$$e^{-s^2+2s\xi} = \sum_{m=0,1,2,\dots}^{\infty} \frac{s^m}{m!} H_m(\xi) \qquad (1)$$

Thus,

$$e^{-s^2+2s\xi}\, e^{-t^2+2t\xi} = \sum_n \sum_m \frac{s^m t^n}{m!\,n!} H_m(\xi) H_n(\xi)$$

If we multiply both sides by $e^{-\xi^2}$ and integrate we would obtain

$$\text{RHS} = \sum_n \sum_m \frac{s^m t^n}{m!\,n!} \int_{-\infty}^{+\infty} H_m(\xi) H_n(\xi) e^{-\xi^2}\, d\xi \qquad (2)$$

and

$$\text{LHS} = e^{-s^2-t^2} \int_{-\infty}^{+\infty} e^{-\xi^2-2(s+t)\xi}\, d\xi$$

$$= e^{-s^2-t^2} \sqrt{\pi}\, e^{(s+t)^2}$$

where we have used Eq. (1) of Appendix A. Thus,

$$\text{LHS} = \sqrt{\pi}\, e^{2st}$$

$$= \sqrt{\pi} \sum_{n=0,1,\dots}^{\infty} \frac{2^n s^n t^n}{n!} \qquad (3)$$

If we use Eqs (2) and (3) we would obtain

$$\frac{1}{n!\,m!} \int_{-\infty}^{+\infty} H_m(\xi) H_n(\xi) e^{-\xi^2}\, d\xi = \sqrt{\pi} \frac{2^n}{n!} \delta_{mn}$$

or

$$\int_{-\infty}^{+\infty} u_m(\xi) u_n(\xi)\, d\xi = \delta_{mn} \qquad (4)$$

where $$u_n(\xi) = \left[\frac{1}{2^n\, n!\sqrt{\pi}}\right]^{1/2} H_n(\xi) e^{-\frac{1}{2}\xi^2} \tag{5}$$

Equation (4) represents the orthonormality condition of Hermite-Gauss functions. We next evaluate the integral

$$I_n = \int_{-\infty}^{+\infty} H_n(\xi) e^{-\left[\xi^2 - \xi\xi_0 + \frac{1}{2}\xi_0^2\right]} d\xi \tag{6}$$

We multiply both sides of Eq. (1) by the exponential factor in the integrand of Eq. (1) and carrying out the integration, we get

$$\sum_{n=0,1,2,\ldots}^{\infty} \frac{s^n}{n!} I_n = \int_{-\infty}^{+\infty} e^{-s^2+2s\xi} e^{-\left[\xi^2 - \xi\xi_0 + \frac{1}{2}\xi_0^2\right]} d\xi$$

$$= e^{-s^2 - \frac{1}{2}\xi_0^2} \int_{-\infty}^{+\infty} e^{-\xi^2} e^{(2s+\xi_0)\xi} d\xi = e^{-s^2 - \frac{1}{2}\xi_0^2} \sqrt{\pi} \exp\left[\frac{(2s+\xi_0)^2}{4}\right]$$

where we have used Eq. (1) of Appendix A. Thus,

$$\sum_{n=0,1,2,\ldots}^{\infty} \frac{s^n}{n!} I_n = \sqrt{\pi} e^{-\frac{1}{4}\xi_0^2} e^{s\xi_0}$$

$$= \sqrt{\pi} e^{-\frac{1}{4}\xi_0^2} \sum_{n=0,1,2,\ldots}^{\infty} \frac{s^n}{n!} \xi_0^n$$

Comparing coefficients of s^n on both sides, we get

$$I_n = \sqrt{\pi}\, \xi_0^n e^{-\frac{1}{4}\xi_0^2}$$

using which we would readily get Eq. (57) of Chapter 7. We next consider the summation [see Eq. (58) of Chapter 7]

$$\Psi(x,t) = \sum_{n=0,1,2,\ldots}^{\infty} \left[\frac{1}{\sqrt{n!}} \left(\frac{1}{2}\xi_0^2\right)^{n/2} \exp\left(-\frac{1}{4}\xi_0^2\right)\right] \times \left[\frac{\gamma}{2^n n! \sqrt{\pi}}\right]^{1/2} H_n(\xi) e^{-\frac{1}{2}\xi^2} e^{-i\left(n+\frac{1}{2}\right)\omega t}$$

$$= \left(\frac{\gamma}{\sqrt{\pi}}\right)^{1/2} \exp\left[-\frac{1}{4}\xi_0^2 - \frac{1}{2}\xi^2 - \frac{i}{2}\omega t\right] \times \sum_{n=0,1,\ldots}^{\infty} \frac{1}{n!}\left[\frac{1}{2}\xi_0 e^{-i\omega t}\right]^n H_n(\xi)$$

$$= \left(\frac{\gamma}{\sqrt{\pi}}\right)^{1/2} \exp\left[-\frac{1}{4}\xi_0^2 - \frac{1}{2}\xi^2 - \frac{i}{2}\omega t\right] \times \exp\left[-\frac{1}{4}\xi_0^2 e^{-2i\omega t} + \xi\xi_0 e^{-i\omega t}\right]$$

$$= \left(\frac{\gamma}{\sqrt{\pi}}\right)^{1/2} \exp\left[-\frac{1}{2}(\xi - \xi_0\cos\omega t)^2 - i\left(\frac{1}{2}\omega t + \xi\xi_0 \sin\omega t - \frac{1}{4}\xi_0^2 \sin 2\omega t\right)\right]$$

Appendix F The confluent hypergeometric equation

The confluent hypergeometric equation is given by

$$x\frac{d^2y}{dx^2} + (c-x)\frac{dy}{dx} - ay(x) = 0 \tag{1}$$

where a and c are constants. If we substitute a power series solution

$$y(x) = \sum_{r=0}^{\infty} d_r\, x^{p+r}, \quad d_0 \neq 0 \tag{2}$$

in Eq. (1) we would obtain (see, e.g. Ref. 1)

$$p = 0 \quad \text{or} \quad p = 1-c \qquad \text{Roots of indicial equation.} \tag{3}$$

$$d_r = \frac{p+r+a-1}{(p+r+c-1)\,(p+r)}\,d_{r-1} \quad \text{for} \quad r \geq 1 \text{ Recurrence relation.} \tag{4}$$

Case 1: $c \neq 0, \pm1, \pm2, \pm3, \ldots$

For $c \neq 0, \pm1, \pm2, \pm3, \ldots$, the two roots of the indicial equation will yield linearly independent solutions of Eq. (1); these are

$$y_1(x) = {}_1F_1(a, c, x) = 1 + \frac{a}{c}\cdot\frac{x}{1!} + \frac{a(a+1)}{c(c+1)}\cdot\frac{x^2}{2!} + \cdots \tag{5}$$

and

$$y_2(x) = x^{1-c}\,{}_1F_1(a-c+1, 2-c, x) \tag{6}$$

The function ${}_1F_1(a, c, x)$ as defined by Eq. (5) is known as the confluent hypergeometric function. Thus for $c \neq 0, \pm1, \pm2, \pm3, \ldots$ the general solution of Eq. (1) is given by

$$y(x) = C_1\,{}_1F_1(a, c, x) + C_2\,x^{1-c}\,{}_1F_1(a-c+1, 2-c, x) \tag{7}$$

Case 2: $c = 1$

This is the case of repeated roots of the indicial equation. In this case, we write the solution of Eq. (1) as

$$y(x, p) = x^p\left[1 + \frac{p+a}{(p+1)^2}x + \frac{(p+a)(p+a+1)}{(p+1)^2(p+2)^2}x^2 + \cdots\right] \tag{8}$$

The two independent solutions are given by [see, e.g. Chapter 8 of Ref. 1]

$$y_1(x) = [y(x,p)]_{p=0} = 1 + \frac{a}{(1!)^2}x + \frac{a(a+1)}{(2!)^2}x^2 + \ldots$$

$$= {}_1F_1(a, 1, x) \tag{9}$$

and
$$y_2(x) = \left(\frac{\partial y}{\partial p}\right)_{p=0} = y_1(x)\ln x + \Big[(1-2a)x + \left(\frac{1}{a} + \frac{1}{a+1} - \frac{2}{1} - \frac{2}{2}\right)$$

$$\times \frac{a(a+1)}{(2!)^2}x^2 + \left(\frac{1}{a} + \frac{1}{a+1} - \frac{2}{1} - \frac{2}{2} - \frac{2}{3}\right)\frac{a(a+1)(a+2)}{(3!)^2}x^3 + \cdots\Big] \tag{10}$$

Case 3: c = 0, –1, ±2, ±3, …

It can be seen that in this case the coefficients of one of the series [Eq. (5) or (6)] become infinite. Let us assume $c = 0$ which gives $p = 0$ and 1. Equation (4) can now be written as

$$d_r = \frac{(p+r+a-1)}{(p+r-1)(p+r)}d_{r-1}; \quad r \geq 1 \tag{11}$$

Thus,
$$y(x,p) = d_0 x^p \left(1 + \frac{p+a}{p(p+1)}x + \frac{(p+a)(p+a+1)}{p(p+1)^2(p+2)}x^2 + \cdots\right) \tag{12}$$

which we rewrite as

$$y(x,p) = b_0 x^p \left(p + \frac{p+a}{(p+1)}x + \frac{(p+a)(p+a+1)}{(p+1)^2(p+2)}x^2 + \cdots\right) \tag{13}$$

The two linearly independent solutions are:

$$y_1(x) = y(x,p)\Big|_{p=0} = a\left(x + \frac{a+1}{1^2 \cdot 2}x^2 + \frac{(a+1)(a+2)}{1^2 \cdot 2^2 \cdot 3}x^3 + \cdots\right)$$

$$= ax\,{}_1F_1(a+1, 2, x) \tag{14}$$

and
$$y_2(x) = \frac{\partial y}{\partial p}\Big|_{p=0}$$

$$= y_1(x)\ln x + \left[1 + (1-a)x + \left(\frac{1}{a} + \frac{1}{a+1} - \frac{2}{1} - \frac{1}{2}\right) \times \frac{a(a+1)}{1^2 \cdot 2}x^2 + \cdots\right] \tag{15}$$

Similarly, for $c = -2$, the two roots of the indicial equation are

$$p = 0 \quad \text{and} \quad 3$$

with

$$y(x,p) = b_0 p\, x^p \left(1 + \frac{(p+a)}{(p-2)(p+1)}x + \frac{(p+a)(p+a+1)}{(p-2)(p-1)(p+1)(p+2)}x^2 + \ldots\right) \tag{16}$$

Since $p = 0$ represents the lower root, the two independent solutions would be

$$y(x,p)\Big|_{p=0} \quad \text{and} \quad \frac{\partial y(x,p)}{\partial p}\Big|_{p=0}$$

Obviously,
$$y(x,p)\Big|_{p=0} = b_0 \frac{a(a+1)(a+2)}{2\cdot 1\cdot (3!)} x^3 \left[1 + \frac{(a+3)}{1\cdot 4} x + \frac{(a+3)(a+4)}{1\cdot 2\cdot 4\cdot 5} x^2 + \cdots\right] \tag{17}$$

which is proportional to the second solution on the right-hand side of Eq. (7). On the other hand, for $c = +2$, the two roots of the indicial equation are

$$p = 0 \text{ and } -1$$

and the two independent solutions would be

$$y(x, p)\Big|_{p=-1} \quad \text{and} \quad \frac{\partial y(x,p)}{\partial p}\Big|_{p=-1}$$

where
$$y(x, p) = b_0 (p+1)\, x^p \left[1 + \frac{(p+a)}{(p+2)(p+1)} x + \frac{(p+a)(p+a+1)}{(p+2)(p+1)(p+1)(p+2)} x^2 + \cdots\right] \tag{18}$$

We may mention that the associated Laguerre polynomials and the Hermite polynomials can be expressed in terms of the confluent hypergeometric functions and are given by

$$L_{n+1}^{2l+1}(x) = -\frac{\left[(n+l)!\right]^2}{(n-l-1)!\,(2l+1)!}\ {}_1F_1(l+1-n, 2l+2, x) \tag{19}$$

$$H_n(x) = \begin{cases} (-1)^{n/2} \dfrac{n!}{\frac{n}{2}!}\ {}_1F_1\left(-\dfrac{n}{2}, \dfrac{1}{2}, x^2\right) & \text{for } n = 0, 2, 4\ldots \\[2ex] (-1)^{n-1/2} \dfrac{n!}{\frac{n-1}{2}!} 2x\ {}_1F_1\left(-\dfrac{n-1}{2}, \dfrac{3}{2}, x^2\right) & \text{for } n = 1, 3, 5\ldots \end{cases} \tag{20}$$

As mentioned in Section 10.6, many eigenvalue equations can be transformed to an equation of the following form

$$\frac{d^2u}{dx^2} + \left[-\frac{1}{4} + \frac{\lambda}{x} + \frac{\frac{1}{4} - \mu^2}{x^2}\right] u(x) = 0 \tag{21}$$

Making the transformation

$$u(x) = x^{\mu + \frac{1}{2}} e^{-x/2} y(x) \tag{22}$$

we readily obtain

$$x \frac{d^2y}{dx^2} + (2\mu + 1 - x)\frac{dy}{dx} - \left(\mu + \frac{1}{2} - \lambda\right) y(x) = 0 \tag{23}$$

which is the confluent hypergeometric equation with

$$a = \mu + \frac{1}{2} - \lambda \quad \text{and} \quad c = 2\mu + 1 \tag{24}$$

Thus, for $c \neq 0, \pm 1, \pm 2, \ldots$ the two independent solutions are

$$u_1(x) = x^{\mu+\frac{1}{2}} e^{-x/2} {}_1F_1\left(\mu + \frac{1}{2} - \lambda, 2\mu + 1, x\right) \tag{25}$$

and

$$u_2(x) = x^{-\mu+\frac{1}{2}} e^{-x/2} {}_1F_1\left(\frac{1}{2} - \mu - \lambda, -2\mu + 1, x\right) \tag{26}$$

For $c = 0, \pm 1, \pm 2, \ldots$ one has to proceed as detailed earlier.

REFERENCES

1. A.K. Ghatak, I.C. Goyal and S.J. Chua, *Mathematical Physics,* Laxmi Publications, New Delhi (2017).
2. J. Irving and N. Mullineux, *Mathematical Physics and Engineering*, Academic Press, New York (1959).

Appendix G Associated Legendre polynomials

We first consider positive values of m. Now in order to obtain the solution of Eq. (35) [of Chapter 9] for $m > 0$, we differentiate Eq. (36) [of Chapter 9] m times to obtain

$$(1-\mu^2)\frac{d^2P^{(m)}}{d\mu^2} - 2(m+1)\mu\frac{dP^{(m)}}{d\mu} + \left[\lambda - m(m+1)\right]P^{(m)}(\mu) = 0 \tag{1}$$

where

$$P^{(m)}(\mu) = \frac{d^m P}{d\mu^m} \tag{2}$$

Further, in Eq. (35) [of Chapter 9], if we make the substitution

$$F(\mu) = (1-\mu^2)^{m/2} G(\mu) \tag{3}$$

then $G(\mu)$ will be found to satisfy the same equation as satisfied by $P^{(m)}$ [i.e. Eq. (1)]. Thus, for $m = 1, 2, 3, \ldots$ the solution of Eq. (35) [of Chapter 9] is the associated Legendre polynomial

$$P_l^{(m)}(\mu) = (1-\mu^2)^{m/2}\frac{d^m P_l(\mu)}{d\mu^m} \tag{4}$$

defined only for non-negative integers. Since $P_l(\mu)$ is a polynomial of degree l, we must have $m \le l$. Obviously,

$$P_l^0(\mu) = P_l(\mu) \tag{5}$$

Thus, the well-behaved solutions of Eq. (28) [of Chapter 9]

$$Y_{lm}(\theta,\phi) = (-1)^m\left[\frac{2l+1}{4\pi}\cdot\frac{(l-m)!}{(l+m)!}\right]^{1/2} P_l^m(\cos\theta)e^{im\phi}; \quad m \ge 0 \tag{6}$$

For negative values of m, Eq. (35) [of Chapter 9] and therefore the part dependent on θ remains the same, only the ϕ dependent part changes. Thus, for $m < 0$, we may define the solution as

$$Y_{lm}(\theta,\phi) \equiv (-1)^m Y^*_{l,-m}(\theta,\phi); \quad m < 0. \tag{7}$$

Further,

$$l = 0, 1, 2, \ldots;\ m = -l, -l+1, \ldots, l-1, l \tag{8}$$

ppendix H Schrödinger equation in center of mass and relative coordinates

The x-components of Eqs (26) and (27) of Chapter 10 are

$$X = \frac{m_1 x_1 + m_2 x_2}{m_1 + m_2} \tag{1}$$

and
$$x = x_1 - x_2 \tag{2}$$

Thus,
$$\frac{\partial \Psi}{\partial x_1} = \frac{\partial \Psi}{\partial x} \cdot \frac{\partial x}{\partial x_1} + \frac{\partial \Psi}{\partial X} \cdot \frac{\partial X}{\partial x_1}$$
$$= \frac{\partial \Psi}{\partial x} + \frac{m_1}{m_1 + m_2} \cdot \frac{\partial \Psi}{\partial X} \tag{3}$$

and
$$\frac{\partial^2 \Psi}{\partial x_1^2} = \frac{\partial}{\partial x}\left[\frac{\partial \Psi}{\partial x}\right]\frac{\partial x}{\partial x_1} + \frac{\partial}{\partial X}\left[\frac{\partial \Psi}{\partial x}\right]\frac{\partial X}{\partial x_1}$$
$$+ \frac{m_1}{m_1 + m_2}\left[\frac{\partial}{\partial x}\left(\frac{\partial \Psi}{\partial x}\right)\frac{\partial x}{\partial x_1} + \frac{\partial}{\partial X}\left(\frac{\partial \Psi}{\partial x}\right)\frac{\partial X}{\partial x_1}\right]$$

or
$$\frac{\partial^2 \Psi}{\partial x_1^2} = \frac{\partial^2 \Psi}{dx^2} + \frac{2m_1}{(m_1 + m_2)}\frac{\partial^2 \Psi}{\partial X \partial x} + \left(\frac{m_1}{m_1 + m_2}\right)^2 \frac{\partial^2 \Psi}{\partial X^2}$$

Similarly,
$$\frac{\partial^2 \Psi}{\partial x_2^2} = \frac{\partial^2 \Psi}{\partial x^2} - \frac{2m_2}{(m_1 + m_2)} \cdot \frac{\partial^2 \Psi}{\partial X \partial x} + \left(\frac{m_2}{m_1 + m_2}\right)^2 \frac{\partial^2 \Psi}{\partial X^2}$$

Thus,
$$-\frac{\hbar^2}{2m_1}\frac{\partial^2 \Psi}{\partial x_1^2} - \frac{\hbar^2}{2m_2}\frac{\partial^2 \Psi}{\partial x_2^2} = -\frac{\hbar^2}{2\mu}\frac{\partial^2 \Psi}{\partial x^2} - \frac{\hbar^2}{2M}\frac{\partial^2 \Psi}{\partial X^2} \tag{4}$$

where
$$\mu = \frac{m_1 m_2}{m_1 + m_2} \tag{5}$$

represents the reduced mass and

$$M = m_1 + m_2 \tag{6}$$

We therefore, finally obtain

$$-\frac{\hbar^2}{2m_1}\nabla_1^2 \Psi - \frac{\hbar^2}{2m_2}\nabla_2^2 \Psi = -\frac{\hbar^2}{2\mu}\left(\frac{\partial^2}{\partial x^2} + \frac{\partial^2}{\partial y^2} + \frac{\partial^2}{\partial z^2}\right)\Psi - \frac{\hbar^2}{2M}\left(\frac{\partial^2}{\partial X^2} + \frac{\partial^2}{\partial Y^2} + \frac{\partial^2}{\partial Z^2}\right)\Psi \tag{7}$$

Appendix I Hydrogen atom wavefunctions

For the hydrogen-like atom $V = -Ze^2/r$ (see Section 10.4) and the eigenfunctions are

$$\psi_{nlm}(r, \theta, \phi) = R_{nl}(r)\, Y_{lm}(\theta, \phi) \tag{1}$$

where

$$\begin{aligned} n &= 1, 2, 3, \ldots \\ l &= 0, 1, 2, \ldots, n-1 \\ m &= -l, -l+1, \ldots, +l \end{aligned} \tag{2}$$

and $Y_{lm}(\theta, \phi)$ are the spherical harmonics which have been discussed in Sections 9.3 and 15.3 and in Appendix G. The normalized radial functions are given by

$$R_{nl}(r) = -\left\{\left(\frac{2Z}{na_0}\right)^3 \frac{(n-l-1)!}{2n[(n+l)!]^3}\right\}^{1/2} e^{-(\rho/2)}\, \rho^l\, L_{n+l}^{2l+1}(\rho)$$

$$= \frac{1}{(2l+1)!}\left\{\left(\frac{2Z}{na_0}\right)^3 \frac{(n+l)!}{2n(n-l-1)!}\right\}^{1/2} e^{-\rho/2}\, \rho^l\, {}_1F_1(l+1-n, 2l+2, \rho) \tag{3}$$

where

$$a_0 = \frac{\hbar^2}{\mu e^2}, \quad \rho = \frac{2Z}{na_0}\, r \tag{4}$$

and

$$L_{n+l}^{2l+1}(\rho) = \sum_{k=0}^{n-l-1} (-1)^{k+2l+1} \frac{[(n+l)!]^2 \rho^k}{(n-l-1-k)!(2l+1+k)!k!} \tag{5}$$

are known as the associated Laguerre polynomials. It may be noted that ρ depends on n. The orthonormality condition is

$$\int_0^\infty R_{n'l'}(r) R_{nl}(r) r^2 dr = \delta_{nn'}\, \delta_{ll'} \tag{6}$$

and first few $R_{nl}(r)$ are given by

$$R_{10} = 2\left(\frac{Z}{a_0}\right)^{3/2} e^{-\xi} \tag{7}$$

$$\left.\begin{aligned} R_{20} &= \frac{1}{\sqrt{2}}\left(\frac{Z}{a_0}\right)^{3/2} e^{-\xi/2}\left(1-\frac{\xi}{2}\right) \\ R_{21} &= \frac{1}{2\sqrt{6}}\left(\frac{Z}{a_0}\right)^{3/2} e^{-\xi/2}\, \xi \end{aligned}\right\} \tag{8}$$

$$
\left.\begin{aligned}
R_{30} &= \frac{2}{3\sqrt{3}}\left(\frac{Z}{a_0}\right)^{3/2} e^{-\xi/3}\left(1-\frac{2\xi}{3}+\frac{2\xi^2}{27}\right)\\
R_{31} &= \frac{8}{27\sqrt{6}}\left(\frac{Z}{a_0}\right)^{3/2} e^{-\xi/3}\left(\xi-\frac{\xi^2}{6}\right)\\
R_{32} &= \frac{4}{81\sqrt{30}}\left(\frac{Z}{a_0}\right)^{3/2} e^{-\xi/3}\,\xi^2, \text{ etc.}
\end{aligned}\right\} \tag{9}
$$

where $\xi = Zr/a_0$. Further

$$
\begin{aligned}
\left\langle \frac{1}{r} \right\rangle &= \frac{1}{n^2}\left(\frac{Z}{a_0}\right)\\
\left\langle \frac{1}{r^2} \right\rangle &= \frac{1}{n^3\left(l+\frac{1}{2}\right)}\left(\frac{Z}{a_0}\right)^2\\
\left\langle \frac{1}{r^3} \right\rangle &= \frac{1}{n^3 l\left(l+\frac{1}{2}\right)(l+1)}\left(\frac{Z}{a_0}\right)^3\\
\langle r \rangle &= n^2\left[1+\frac{1}{2}\left\{1-\frac{l(l+1)}{n^2}\right\}\right]\frac{a_0}{Z}
\end{aligned} \tag{10}
$$

where
$$
\langle r^m \rangle = \int_0^\infty r^m \left[R_{nl}(r)\right]^2 r^2\, dr \tag{11}
$$

In general, for $p > -2l - 3$

$$
\frac{p+1}{n^2}\langle r^p \rangle - (2p+1)\frac{a_0}{Z}\langle r^{p-1} \rangle + \frac{p}{4}\left[(2l+1)^2 - p^2\right]\left(\frac{a_0}{Z}\right)^2 \langle r^{p-2} \rangle = 0 \tag{12}
$$

The above relation has been derived in Chapter 3 of Ref. 1.

REFERENCE

1. F. Constantinescu and E. Magyari, *Problems in Quantum Mechanics*, Pergamon Press, Oxford (1971).

Appendix J Bessel functions and spherical Bessel functions

Bessel functions

The differential equation

$$x^2 \frac{d^2y}{dx^2} + x\frac{dy}{dx} + (x^2 - n^2)y(x) = 0 \tag{1}$$

is known as the Bessel equation. For $n \neq 0, 1, 2, \ldots$ the two linearly independent solutions are

$$J_n(x) \quad \text{and} \quad J_{-n}(x)$$

where

$$J_n(x) = \sum_{r=0}^{\infty} (-1)^r \frac{1}{r!\,\Gamma(n+r+1)} \left(\frac{x}{2}\right)^{n+2r} \tag{2}$$

For $n = 0, 1, 2, \ldots, J_{-n}(x)$ is a multiple of $J_n(x)$; indeed

$$J_{-n}(x) = (-1)^n J_n(x);\ n = 0, 1, 2, \ldots \tag{3}$$

For $n = 0, 1, 2, \ldots$ the other independent solution is

$$Y_n(x) = \lim_{\nu \to n} \frac{\cos \nu\pi\, J_\nu(x) - J_{-\nu}(x)}{\sin \nu\pi} \tag{4}$$

In fact $Y_n(x)$ always represents an independent solution[1].

---

[1] The series representing $Y_n(x)$ is never really needed but is given below for the sake of completeness

$$Y_n(x) = \frac{2}{\pi}\left[\left(\ln\frac{x}{2} + \gamma\right) J_n(x) - \frac{1}{2}\sum_{r=0}^{\infty} \frac{(n-r-1)!}{r!}\left(\frac{x}{2}\right)^{-n+2r} - \frac{1}{2}\sum_{r=0}^{\infty} (-1)^r \frac{(x/2)^{n+2r}}{r!(r+n)!}\{\phi(r) + \phi(r+n)\}\right]$$

where $\gamma = \lim_{n\to\infty}\left[1 + \frac{1}{2} + \frac{1}{3} + \cdots + \frac{1}{n} - \ln n\right] \approx 0.5772$, $\phi(r) = 1 + \frac{1}{2} + \cdots + \frac{1}{r}$ with $\phi(0) = 0$.

For $n = \pm\frac{1}{2}, \pm\frac{3}{2}$, ... the infinite series given by Eq. (2) can be summed[2] and one has

$$\left.\begin{aligned} J_{\frac{1}{2}}(x) &= \left(\frac{2}{\pi x}\right)^{1/2} \sin x,\ J_{-\frac{1}{2}}(x) = \left(\frac{2}{\pi x}\right)^{1/2} \cos x \\ J_{\frac{3}{2}}(x) &= \left(\frac{2}{\pi x}\right)^{1/2} \left(\frac{\sin x}{x} - \cos x\right), \\ J_{-\frac{3}{2}}(x) &= \left(\frac{2}{\pi x}\right)^{1/2} \left(\frac{-\cos x}{x} - \sin x\right) \end{aligned}\right\} \tag{5}$$

$J_{\pm 5/2}(x)$, $J_{\pm 7/2}(x)$, etc., can be found from the recurrence relation

$$J_{n+1}(x) = \frac{2n}{x} J_n(x) - J_{n-1}(x) \tag{6}$$

The asymptotic forms are

$$\left.\begin{aligned} J_n(x) &\underset{x\to\infty}{\to} \left(\frac{2}{\pi x}\right)^{1/2} \cos\left(x - \frac{n\pi}{2} - \frac{\pi}{4}\right) \\ Y_n(x) &\underset{x\to\infty}{\to} \left(\frac{2}{\pi x}\right)^{1/2} \sin\left(x - \frac{n\pi}{2} - \frac{\pi}{4}\right) \end{aligned}\right\} \tag{7}$$

We would like to introduce the Hankel functions which are often used [as independent solutions of Eq. (1)] for their convenient asymptotic forms:

$$\left.\begin{aligned} H_n^{(1)}(x) &= J_n(x) + iY_n(x) \underset{x\to\infty}{\to} \left(\frac{2}{\pi x}\right)^{1/2} e^{i\left(x - \frac{n\pi}{2} - \frac{\pi}{4}\right)} \\ H_n^{(2)}(x) &= J_n(x) - iY_n(x) \underset{x\to\infty}{\to} \left(\frac{2}{\pi x}\right)^{1/2} e^{-i\left(x - \frac{n\pi}{2} - \frac{\pi}{4}\right)} \end{aligned}\right\} \tag{8}$$

Spherical Bessel functions

In the Bessel equation [Eq. (1)] corresponding to $n = l + \frac{1}{2}$ (l = 0, 1, 2,...) if we make the transformation $f(x) = \frac{1}{\sqrt{x}}\, y(x)$, we would obtain

$$\frac{d^2 f}{dx^2} + \frac{2}{x}\frac{df}{dx} + \left[1 - \frac{l(l+1)}{x^2}\right] f(x) = 0 \tag{9}$$

which is known as the spherical Bessel equation. Equation (9) is of the same form as the radial part of the Schrödinger equation [Eq. (15) of Chapter 10] with $V(r) = 0$ and $x = \left(\frac{2mE}{\hbar^2}\right)^{1/2} r$. For l = 0, 1, 2,... the two independent solutions are known as the spherical Bessel functions and

[2] This can also be seen by using the transformation $w = x^{1/2}\, y(x)$ in Eq. (1), the resulting equation for $n = \frac{1}{2}$ is $w'' + w(x) = 0$, the solutions of which are $\sin x$ and $\cos x$.

are defined through the following equations.

$$\left.\begin{aligned} j_l(x) &= \left(\frac{\pi}{2x}\right)^{1/2} J_{l+1/2}(x) \\ n_l(x) &= (-1)^l \left(\frac{\pi}{2x}\right)^{1/2} J_{-l-1/2}(x) \end{aligned}\right\} \tag{10}$$

Using the results given in the previous section it is easy to show that

$$\left.\begin{aligned} j_0(x) &= \frac{\sin x}{x},\ j_1(x) = \frac{\sin x}{x^2} - \frac{\cos x}{x} \\ n_0(x) &= \frac{\cos x}{x},\ n_l(x) = \frac{\cos x}{x^2} - \frac{\sin x}{x},\ \text{etc.} \end{aligned}\right\} \tag{11}$$

$j_2(x), n_2(x)$, etc., can be obtained from the recurrence relation [see Eq. (6)]. Further, the asymptotic forms are:

$$\left.\begin{aligned} j_l(x) &\underset{x\to\infty}{\longrightarrow} \frac{1}{x}\sin\left(x - \frac{l\pi}{2}\right) \\ n_l(x) &\underset{x\to\infty}{\longrightarrow} \frac{1}{x}\cos\left(x - \frac{l\pi}{2}\right) \end{aligned}\right\} \tag{12}$$

and

$$\begin{aligned} j_l(x) &\underset{x\to 0}{\rightarrow} \frac{x^l}{(2l+1)!!} \\ n_l(x) &\underset{x\to 0}{\rightarrow} \frac{(2l+1)!!}{(2l+1)} \frac{1}{x^{l+1}} \end{aligned} \tag{13}$$

where $(2l + 1)!! = 1.3.5.... (2l +1)$. Finally, the function $u(x) = x f(x)$ satisfies the equation

$$\frac{d^2u}{dx^2} + \left[1 - \frac{l(l+1)}{x^2}\right] u(x) = 0 \tag{14}$$

which is of the same form as Eq. (17) of Chapter 10 and Eq. (59) of Chapter 26 with $V(r) = 0$ and $x = \left(\frac{2mE}{\hbar^2}\right)^{1/2} r$. Obviously, the solution of Eq. (14) is given by

$$u(x) = A[xj_l(x)] + B[xn_l(x)] \tag{15}$$

Modified Bessel functions

In the Bessel equation [Eq. (1)] if we make the transformation $\xi = -ix$, we would obtain

$$\xi^2 \frac{d^2y}{d\xi^2} + \xi\frac{dy}{d\xi} - (\xi^2 + n^2)y(\xi) = 0 \tag{16}$$

The solutions are $J_n(i\xi)$ and $Y_n(i\xi)$. For $n \neq 0, 1, 2,...$ the two independent solutions are written as $I_{\pm n}(\xi)$ where

$$I_n(\xi) = i^{-n} J_n(i\xi) = \sum_{r=0}^{\infty} \frac{(\xi/2)^{n+2r}}{r!\Gamma(n+r+1)} \tag{17}$$

The functions $I_n(\xi)$ are known as modified Bessel functions with the asymptotic form

$$I_n(\xi) \underset{x\to\infty}{\rightarrow} \frac{1}{(2\pi\xi)^{1/2}} e^{\xi} \tag{18}$$

These functions are not oscillatory in character [cf. Eq. (7)]. Often it is more convenient to choose

$$K_n(x) = \frac{\pi}{2} \frac{I_{-n}(x) - I_n(x)}{\sin n\pi} \underset{x\to\infty}{\rightarrow} \left(\frac{\pi}{2x}\right)^{1/2} e^{-x} \tag{19}$$

as an independent solution. The functions $K_n(x)$ are known as modified Bessel function of the second kind.

Appendix K Asymptotic series

In order to study the behavior of various functions for large values of x, it is often convenient to expand them in inverse powers of x:

$$f(x) = \phi(x)\left[a_0 + \frac{a_1}{x} + \frac{a_2}{x^2} + \ldots\right] \tag{1}$$

where $\phi(x)$ is a known function whose behaviour for large x is known. The series, which is often divergent, is usually of great practical importance not only for a qualitative understanding of a function, but even in its computation for large values of x.

Definition

The series $a_0 + (a_1/x) + (a_2/x^2) + \ldots$ is said to represent $f(x)/\phi(x)$ asymptotically, if

$$\lim_{x\to\pm\infty}\left\{x^N\left[\frac{f(x)}{\phi(x)} - \sum_{n=0}^{N}\frac{a_n}{x^n}\right]\right\}\to 0 \tag{2}$$

i.e. for a given value of N, the first N terms of the series can be made as close as may be desired to the ratio $f(x)/\phi(x)$ by making x large enough. When the series diverges, there will be an optimal number of terms of the series to be used to represent $f(x)/\phi(x)$ for a given x; and, therefore, there will be an *unavoidable error*. This *unavoidable error* would decrease as $|x|$ increases. We will now consider a few examples:

Example 1 Consider the exponential integral

$$f(x) = \int_x^\infty \frac{e^{-y}}{y}\,dy$$

Integrating by parts

$$f(x) = \frac{e^{-x}}{x} - \int_x^\infty \frac{e^{-y}}{y^2}\,dy$$

$$= \frac{e^{-x}}{x} - \frac{e^{-x}}{x^2} + 2\int_x^\infty \frac{e^{-y}}{y^3}\,dy$$

Continuing this procedure, we get

$$f(x) = \frac{e^{-x}}{x}\left[1 - \frac{1}{x} + \frac{2!}{x^2} - \frac{3!}{x^2} + \ldots + \frac{(-1)^n n!}{x^n}\right] + (-1)^{n+1}(n+1)!\int_x^\infty \frac{e^{-y}}{y^{n+2}}\,dy$$

Thus, in this problem $\phi(x) = e^{-x}/x$; $a_0 = 1$, $a_1 = -1$, ..., $a_n = (-1)^n n!$. The limit of the ratio of two successive terms of the series is given by

$$\lim_{n\to\infty}\left|\frac{T_{n+1}}{T_n}\right| = \lim_{n\to\infty}\frac{n}{x} \to \infty$$

Thus, the series diverges. Further, the two successive terms become equal in magnitude for $n = x$, indicating that the optimum number of terms for a given x is roughly the integer closest to x. Typically, for $x = 4$, the exact value of $xe^{+x} f(x)\,|_{x=4} = 0.82533$.

However, the sum of the series including the 1st, 2nd, 3rd, 4th, 5th, 6th, 7th, 8th and 9th terms in the series are 1.00000, 0.75000, 0.87500, 0.78125, 0.87500, 0.75781, 0.93360, 0.62347 and 1.24373 respectively. Thus the optimal value is 0.78125 which occurs at $N = 3$ and 0.82533 – 0.78125 is the unavoidable error. Further, by taking a large number of terms the accuracy becomes poorer.

Example 2 In a similar manner, one may discuss the asymptotic series of the error function,

$$\text{erfc}\,(x) = 1 - \text{erf}\,(x) = \frac{2}{\sqrt{\pi}}\int_x^\infty e^{-y^2}\,dy$$

$$= \frac{1}{\sqrt{\pi}}e^{-x^2}\left(\frac{1}{x} - \frac{1}{2x^3} + \frac{1\cdot 3}{2^2 x^5} - \frac{1\cdot 3\cdot 5}{2^3 x^7} + \dots\right)$$

Example 3 Finally, we will consider the asymptotic form of the solutions of the confluent hypergeometric equation: (see also Appendix L)

$$x\frac{d^2y}{dx^2} + (c - x)\frac{dy}{dx} - ay = 0$$

We substitute the series

$$y = e^{mx}\sum_{r=0}^{\infty} a_r\, x^{-s-r}$$

On substitution and equating the highest power of x, we obtain

$$m = 0 \quad \text{or} \quad m = 1$$

(i) For $m = 0$, $s = a$, $a_1 = -a\,(a - c + 1)\,a_0$, etc., giving

$$y = a_0\, x^{-a}\left[1 - \frac{a(a-c+1)}{1!x} + \frac{a(a+1)\,(a-c+1)\,(a-c+2)}{2!x^2}\dots\right]$$

(ii) For $m = 1$, $s = c - a$, $a_1 = (c - a)\,(1 - a)\,a_0$ etc., giving

$$y = a_0\, e^x\, x^{a-c}\left[1 + \frac{(c-a)(1-a)}{1!x} + \frac{(c-a)(c-a+1)(1-a)(2-a)}{2!x^2}\dots\right]$$

The asymptotic form of ${}_1F_1\,(a, c, x)$ is a linear combination of the above solutions.

Appendix L Complex variable

Complex integration

In the theory of functions of a complex variable there is a famous theorem due to Cauchy which states that, if $f(z)$ is an analytic function of z in a certain region around a point $z = a$, then

$$\int_C f(z)\,dz = 0 \tag{1}$$

and

$$\frac{1}{2\pi i}\int_C \frac{f(z)}{z-a}\,dz = f(a) \tag{2}$$

where C is a closed contour taken around $z = a$ in a contour-clockwise direction (Fig. L. 1). If $f(z)$ can be expanded around $z = z_0$ in the form

$$f(z) = \frac{a_{-m}}{(z-z_0)^m} + \frac{a_{-m+1}}{(z-z_0)^{m-1}} + \cdots + \frac{a_{-1}}{z-z_0} + a_0 + a_1(z-z_0) + \cdots \tag{3}$$

where m is a *finite* integer, then it can be shown that

$$\frac{1}{2\pi i}\int_C f(z)\,dz = a_{-1} \tag{4}$$

where C is a contour around $z = z_0$ the counter-clockwise direction. Note that in Eqs (3) and (4), $f(z)$ is *not* analytic at $z = z_0$; $f(z)$ has a pole of order m. If $m = 1$, the pole is said to be a simple pole.

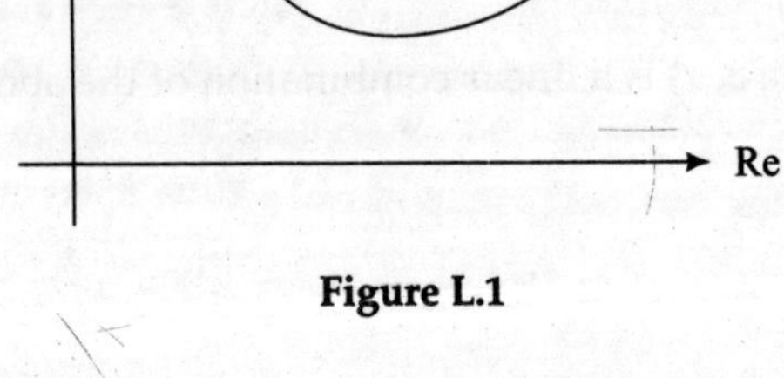

Figure L.1

The coefficient a_{-1} in the expansion (3) is called the residue of the function $f(z)$. If the pole at $z = z_0$ is a simple pole, then

$$[(z - z_0) f(z)]_{z = z_0} = a_{-1} \tag{5}$$

Equation (5) is the recipe for evaluating at simple poles. When there are n poles within a contour C, we can generalize (4) to

$$\frac{1}{2\pi i}\int_C f(z)\, dz = \sum_{r=1}^{n} \alpha_r \tag{6}$$

where α_r is the residue at the rth pole.

Equation (6) remains unaltered, no matter how we distort the counter C as *long as we do not cross any new poles*. This technique is exploited frequently in mathematical manipulations.

A remarkable property of an analytic function $f(z)$ is that all its derivatives are also analytic in the same region. It can be shown that Eq. (2) can be generalized to

$$f^n(a) = \frac{n!}{2\pi i}\int_C \frac{f(z)}{(z-a)^{n+1}} dz \tag{7}$$

Singularities

A function $f(z)$ which can be expanded with $m \le -1$ *but finite* is said to have a singularity at $z = z_0$.

It may turn out that an expansion of the form (3) is not possible for *finite m*. For example

$$f(z) = e^{1/z} = \sum_{m=0}^{\infty} \frac{z^{-m}}{m!}$$

In this case the function $f(z)$ is said to have an *essential singularity* at $z = 0$.

One should not get the idea that a singularity is merely a point at which the function goes to infinity. For example, $f(z) = z^{1/2}$ cannot be expanded near $z = 0$ in the form (3). Such singularities arise when we have multivalued functions and are called branch points.

Multivalued functions

We discuss the example

$$f(z) = z^{1/2}$$

Writing

$$z = re^{i\phi} \tag{8}$$

and

$$f(z) = Re^{i\theta}$$

we see that $R = r^{1/2}$ and $\theta = \phi/2$. As ϕ goes from $-\pi + \varepsilon$ to $\pi - \varepsilon$ where ε is a small positive quantity, θ goes from $-\pi/2 + \varepsilon/2$ to $\pi/2 - \varepsilon/2$. As $\varepsilon \to 0$, ϕ goes from $-\pi$ to $+\pi$ and z returns to its original value. However f does not. To describe this situation, one draws a *double line* such as *I* (Fig. L.2), around the branch point ($z = 0$ in this case). Whenever, one crosses this line (called *branch line*) on the z-plane, the function $f(z)$ will not return to its original value.

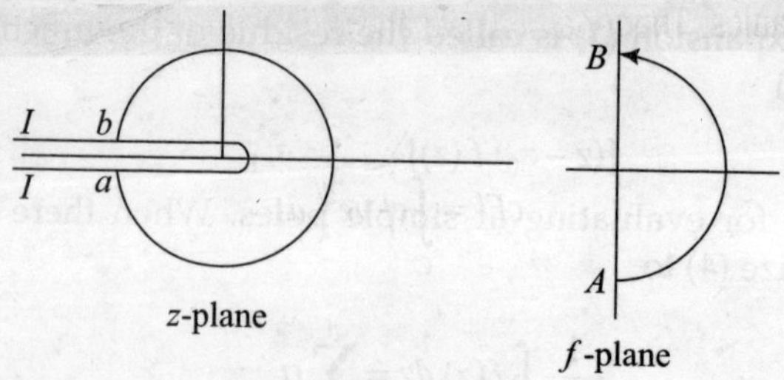

Figure L.2

It can be shown that branch points, occur in pairs and a branch line joins a pair of branch points. In the above example, $z = \infty$ is the other branch point.

Application to solution of confluent hypergeometric equation

We apply the methods of contour integrals to solve the confluent hypergeometric equation

$$zF'' + (c - z)\,F' - aF = 0 \tag{9}$$

This equation is of great importance in quantum mechanical problems, since many equations such as Hermite's equation, Legendre and associated Legendre equation and Bessel's equation can be converted to this form (see Appendix F). We write tentatively, its solution as $F(a, c, z)$. The method employed is to take the Laplace transform

$$F = \int_C e^{zt}\,\nu(t)\,dt \tag{10}$$

where C is a contour in the complex t plane and $\nu(t)$ a function to be chosen so as to give a solution of Eq. (9) with the desired behaviour. We note that

$$F'(z) = \int_C t e^{zt}\,\nu(t)\,dt$$

$$= \left[\frac{e^{zt}}{z}\,t\nu(t)\right]_C - \int_C \frac{e^{zt}}{z}\cdot\frac{d}{dt}\big[t\nu(t)\big]\,dt$$

and

$$F''(z) = \int_C t^2\,e^{zt}\,\nu(t)\,dt$$

$$= \left[\frac{e^{zt}}{z}\,t^2\,\nu(t)\right]_C - \int_C \frac{e^{zt}}{z}\,\frac{d}{dt}\big[t^2\,\nu(t)\big]\,dt$$

This permits us to write

$$zF'' = \big[e^{zt}\,t^2\,\nu\big]_C - \int_C e^{zt}\,\frac{d}{dt}\big[t^2\,\nu\big]\,dt$$

$$zF' = \big[e^{zt}\,t\nu\big]_C - \int_C e^{zt}\,\frac{d}{dt}\big[t\nu\big]\,dt$$

and
$$cF' = \int_C ct\, \nu e^{zt}\, dt$$

so that Eq. (9) becomes

$$\left[e^{zt} t(t-1)\nu(t)\right]_C - \int_C e^{zt}\left[t^2 \frac{d\nu}{dt} + 2t\nu - t\frac{d\nu}{dt} - \nu - ct\nu + a\nu\right] dt = 0 \tag{11}$$

We now make the restriction that the contour C will be so chosen that

$$[e^{zt} t\,(t-1)\,\nu\,(t)]_C = 0 \tag{12}$$

and the function $\nu(t)$ will be chosen so that the integrand in Eq. (11) vanishes; i.e.

$$\frac{d\nu}{dt} t(t-1) = \nu\left[1 - a + (c-2)t\right] \tag{13}$$

Equation (13) is readily integrated[1] to give

$$\nu \sim t^{a-1}\,(t-1)^{c-a-1}$$

and the condition for the contour, Eq. (12) becomes

$$\left[\phi(t)\right]_C = \left[e^{zt}\, t^a\,(t-1)^{c-a}\right]_C = 0 \tag{14}$$

The solution is [from Eq. (10)]

$$F = \int_C e^{zt}\, t^{a-1}(t-1)^{c-a-1}\, dt \tag{15}$$

Before we proceed further, we note that if one substitutes

$$F = z^{1-c} F_1 \tag{16}$$

into Eq. (9), we obtain

$$zF_1'' + (2-c-z)F_1' - (a-c+1)F_1 = 0 \tag{17}$$

which is of the same form as Eq. (9). Thus, if we write the solution of Eq. (9) in the form

$$F = F\,(a, c, z)$$

then $z^{1-c} F(a-c+1, 2-c, z)$ is another independent solution of the confluent hypergeometric equation. Thus this solution is

$$z^{1-c} \int e^{zt} t^{a-c}(t-1)^c\, dt \tag{18}$$

Now the contour C need not be a closed one; all that is necessary is that Eq. (14) is obeyed so that the value of the function $\phi(t)[= e^{zt}\, t^a(t-1)^{a-c}]$ must be the same at the two ends of the curve C. For Re $z > 0$, any contour which starts and ends at $t = -\infty$ clearly satisfies this. However, we shall consider the case Re $c >$ Re $a > 0$ so that $\phi(t)$ vanishes at $t = 0$ and $t = 1$. Hence we choose, for our contour, the portion of the real axis extending from 0 to 1. Thus,

[1] This is the point of the whole exercise; the equation in ν is a simpler than the original one in F.

$$F = \int_0^1 e^{zt}\, t^{\alpha-1}(t-1)^{c-a-1}\, dt = (-1)^{c-a-1}\sum_{n=0}^{\infty}\frac{z_n}{n!}\int_0^1 t^{n+a-1}(1-t)^{c-a-1}\, dt \tag{19}$$

This series can be expressed in terms of Γ functions by using the relation

$$\int_0^1 t^{\alpha-1}(1-t)^{\beta-1}\, dt = \frac{\Gamma(\alpha)\Gamma(\beta)}{\Gamma(\alpha+\beta)} \tag{20}$$

Thus, $$F = (-1)^{c-a-1}\sum_n \frac{z^n}{n!}\frac{\Gamma(n+a)\Gamma(c-a)}{\Gamma(c+n)}$$

The series[2] $${}_1F_1(a, c, z) = \frac{\Gamma(c)}{\Gamma(a)}\sum_{n=0}^{\infty}\frac{\Gamma(a+n)\; z^n}{\Gamma(c+n)\; n!} \tag{21}$$

is called the **confluent hypergeometric** series. Further,

$$F = (-1)^{a-c-1}\frac{\Gamma(c-a)\Gamma(a)}{\Gamma(c)}\,{}_1F_1(a, c, z) \tag{22}$$

A general solution of the hypergeometric equation can, therefore, be expessed as a linear combination of

$${}_1F_1(a, c, z) \quad \text{and} \quad z^{1-c}\,{}_1F_1(a-c+1, 2-c, z)$$

Asymptotic behavior

The function (21) written out in a series as

$${}_1F_1(a, c, z) = 1 + \frac{a}{c}z + \frac{a(a+1)z^2}{c(c+1)2!} + \dots \tag{23}$$

is convergent for all finite values of $|z|$; tne only restriction is that c is not a negative integer. We now study the behaviour of this series for large values of $|z|$ when a and c are integers. In this series, the term containing z^n is

$$\frac{(a+n)!\;(c-1)!}{(a-1)!\;(c+n)!}\cdot\frac{z^n}{n!}$$

For large $n \gg a, c$ we have

$$\frac{(a+n)!}{(a-1)!} \sim n^{a-c} \sim n(n-1)\cdots(n-a+c+1)$$

This gives, for the term with z^n,

$$\frac{(c-1)!\; n^{a-c}}{(a-1)!\;\; n!}z^n \approx \frac{(c-1)!}{(a-1)!}\frac{z_n}{(n-a+c)!}$$

Thus, for $z \to \infty$, we have

[2] The peculiar notation ${}_1F_1$ is because this is a special form of generalized hypereometric function ${}_nF_m(a, b, z)$.

$$ {}_1F_1 \sim \frac{(c-1)!}{(a-1)!}\sum_n \frac{z^n}{(n-a+c)!} = \frac{(c-1)!}{(a-1)!} z^{a-c} e^z $$

Thus,
$$ {}_1F_1\,(a, c, z) = \rightarrow \frac{\Gamma(c)}{\Gamma(a)} z^{a-c}\, e^z \tag{24} $$

We have put the results in terms of the Γ functions because the result it turns out, is valid for non-integral values of a and c as well. While Function (24) gives the right asymptotic behaviour for Re $z \to \infty$ it can be seen that for Re $z \to -\infty$, the limit goes to zero exponentially. In this case the limit given by Function (24) is not valid; however, the following result may be established

$$ {}_1F_1\,(a, c, z) = e^z\,{}_1F_1\,(c-a, c, -z) \tag{25} $$

so that, for $z \to -\infty$, we substitute the asymptotic form (24) on the RHS of Eq. (25), to obtain

$$ {}_1F_1(a,c,z) \rightarrow \frac{\Gamma(c)}{\Gamma(c-a)}(-z)^{-a} \tag{26} $$

Connection with other functions

It can be easily shown that the Hermite polynomials and the associated Laguerre polynomials are related to the confluent hypergeometric function (see Appendix F). Further, the Bessel functions are given by

$$ J_\nu(z) = \frac{z^\nu e^{-iz}}{2^\nu \Gamma(1+\nu)}\,{}_1F_1\left(\nu+\frac{1}{2}, 2\nu+1, 2iz\right) \tag{27} $$

For Re $\left(\nu+\frac{1}{2}\right) > 0$, the following integral representation can be established by the methods discussed:

$$ J_\nu(z) = \frac{1}{\sqrt{\pi}\,\Gamma\left(\nu+\frac{1}{2}\right)}\left(\frac{z}{2}\right)^\nu \int_{-1}^{1} e^{izt}\,(1-t^2)^{\nu-1/2}\,dt \tag{27} $$

Appendix M The periodic boundary conditions and the density of states

In developing scattering theory, we consider the particle inside a cube of volume L^3 and initially in a plane wave state corresponding to momentum $\mathbf{p}$ (see Section 25.3). Thus the particle is described by the wavefunction

$$\psi = \frac{1}{L^{3/2}} e^{i\mathbf{k}\cdot\mathbf{r}} \tag{1}$$

where

$$\mathbf{k} = \frac{1}{\hbar}\mathbf{p} \tag{2}$$

and the factor $L^{-3/2}$ normalizes the wavefunction

$$\int_0^L \int_0^L \int_0^L |\psi|^2 \, dx\, dy\, dz = 1$$

The allowed values of $\mathbf{k}$ are determinated from the boundary conditions. It is convenient to use the periodic boundary conditions (see also Section 29.3):

$$\begin{aligned}\psi(x=0, y, z) &= \psi(x=L, y, z)\\ \psi(x, y=0, z) &= \psi(x, y=L, z)\end{aligned} \tag{3}$$

and

$$\psi(x, y, z=0) = \psi(x, y, z=L)$$

giving

$$\exp[ik_xL] = 1 = \exp[ik_yL] = \exp[ik_zL] \tag{4}$$

Thus the allowed values of k_x, k_y and k_z are given by [cf. Eg. (37) of Chapter 27]:

$$k_x = \frac{2\pi n_x}{L};\ k_y = \frac{2\pi n_y}{L};\ k_z = \frac{2\pi n_z}{L} \tag{5}$$

where

$$n_x, n_y, n_z = 0, \pm 1, \pm 2, \cdots \tag{6}$$

Now, the number of states whose x-component of $\mathbf{k}$ lies between k_x and $k_x + dk_x$ would simply be the number of integers lying between $Lk_x/2\pi$ and $L(k_x + dk_x)\ /2\pi$; this number would be approximately equal to $Ldk_x/2\pi$. Similarly, the number of states whose y- and z-components of $\mathbf{k}$ lie between k_y and $k_y + dk_y$ and k_z and $k_z + dk_z$ would, respectively, be

$$\frac{L}{2\pi} dk_y \quad \text{and} \quad \frac{L}{2\pi} dk_z$$

Thus there will be

$$\frac{L^3}{8\pi^3}\, dk_x\, dk_y\, dk_z \tag{7}$$

states in the range $dk_x\, dk_y\, dk_z$ of **k**. Obviously, the number of states in the range $dp_x\, dp_y\, dp_z$ of **p** will be

$$\frac{L^3}{8\pi^3\hbar^3}\, dp_x\, dp_y\, dp_z \tag{8}$$

Now if $\rho(k)\, dk\, d\Omega$ represents the number of states whose $|\mathbf{k}|$ lies between k and $k + dk$ and the direction of **k** lies in the solid angle $d\Omega$ (= $\sin\theta\, d\theta\, d\phi$) then

$$\rho(k)\, dk\, d\Omega = \frac{L^3}{8\pi^3}\, k^2\, dk\, d\Omega \tag{9}$$

For a non-relativistic particle of mass m we have $E = \hbar^2 k^2/2m$ giving

$$\rho(E)\, dE\, d\Omega = \frac{L^3}{16\pi^3\hbar^3}(2m)^{3/2}\, E^{1/2}\, dE\, d\Omega \tag{10}$$

If we integrate over $d\Omega$ we would get the total number of states whose energies lie between E and $E + dE$

$$g(E)\, dE = \frac{L^3}{4\pi^2\hbar^3}(2m)^{3/2}\, E^{1/2}\, dE \tag{11}$$

which is identical to the expression for $g(E)$ derived in Section 6.8.1 (except for the additional factor of 2 introduced there) where we had assumed end point boundary conditions, i.e. ψ was assumed to vanish on the surface of the cube, thus the expression for density of states is independent of the boundary conditions used.

Returning to Eq. (9) we note that for an electromagnetic field inside a cube (see Section 27.3) $\omega = ck$ and one has

$$\rho(\omega)\, d\omega\, d\Omega = \frac{L^3}{8\pi^3 c^3}\, \omega^2\, d\omega\, d\Omega \tag{12}$$

If we sum over two independent modes of polarization and integrate over the solid angle we would get

$$\rho(\omega)\, d\omega = \frac{L^3}{\pi^2 c^3}\, \omega^2\, d\omega \tag{13}$$

Appendix N Alternative derivation of the Green's function for the $(\nabla^2 + k^2)$ operator

The free particle Green's function is defined through the equation

$$(\nabla^2 + k^2)\, G(\mathbf{r},\mathbf{r}') = \delta(\mathbf{r}-\mathbf{r}') \tag{1}$$

We Fourier analyse G and δ:

$$G(\mathbf{r}-\mathbf{r}') = \frac{1}{(2\pi)^3}\int g(\mathbf{k}')\exp\left[i\mathbf{k}'\cdot(\mathbf{r}-\mathbf{r}')\right]d\mathbf{k}' \tag{2}$$

$$\delta(\mathbf{r}-\mathbf{r}') = \frac{1}{(2\pi)^3}\int \exp\left[i\mathbf{k}'\cdot(\mathbf{r}-\mathbf{r}')\right]d\mathbf{k}' \tag{3}$$

where we have used Eq. (46) of Chapter 1. Substituting these in Eq. (1), we get

$$(-k'^2 + k^2)\, g(\mathbf{k}') = 1$$

or

$$g(\mathbf{k}') = \frac{1}{k'^2 - k^2} \tag{4}$$

Thus,

$$G(\mathbf{r}-\mathbf{r}') = -\frac{1}{8\pi^3}\int \frac{1}{k'^2-k^2}\exp\left[i\mathbf{k}'\cdot(\mathbf{r}-\mathbf{r}')\right]d\mathbf{k}' \tag{5}$$

which is the Green's function for the operator $(\nabla^2 + k^2)$. We may mention that the integrals in Eqs (2), (3) and (5) are 3 dimensional integrals. We define a new variable $\rho = \mathbf{r} - \mathbf{r}'$ and choose the direction of the polar axis (i.e., the direction of k_z') along ρ to get

$$\mathbf{k}'\cdot(\mathbf{r}-\mathbf{r}') = \mathbf{k}'\cdot\rho = k'\rho\cos\theta \tag{6}$$

so that

$$G(\rho) = -\frac{1}{8\pi^3}\int_0^\infty \frac{k'^2\,dk}{k'^2-k^2}\int_0^\pi e^{ik'\rho\cos\theta}\sin\theta\, d\theta\int_0^{2\pi} d\phi$$

$$= -\frac{1}{8\pi^3}\cdot 2\pi\int_0^\infty \frac{k'^2\,dk'}{k'^2-k^2}\left[\frac{e^{ik'\rho}-e^{-ik'\rho}}{ik'\rho}\right]$$

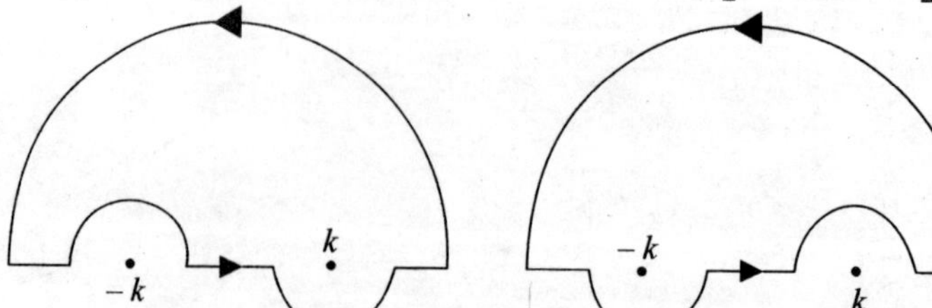

Figure N.1: The two contours of integration chosen.

or $$G(\rho) = -\frac{1}{4\pi^2 i\rho}\int_{-\infty}^{+\infty} dk'\left[\frac{k'e^{ik'\rho}}{(k'+k)(k'-k)}\right] \tag{7}$$

We integrate this by going over to the complex k'-plane. The integrand has poles on the real axis at $k' = \pm k$. For the contour we shall choose the real axis closed by the infinitely far upper half semicircle (see Fig. N.l) since the factor $e^{ik'\rho}$ ensures that the integral vanishes on the upper half semicircle; this follows from Jordan's lemma (see, for example, Ref. 1). There, however, remains the important question of how to go around the two poles. We shall choose the two contours shown in Figs N.l (a) and (b). In case (a), the result is

$$G(\rho) = -\frac{1}{4\pi^2 \rho i} 2\pi i\left[\frac{k'e^{ik'\rho}}{(k'+k)}\right]_{k'=k}$$

This gives $$G_+(\rho) = -\frac{e^{ik\rho}}{4\pi\rho} \tag{8}$$

Similarly case (b) gives $$G_-(\rho) = -\frac{e^{-ik\rho}}{4\pi\rho} \tag{9}$$

Another way of formulating the rules for evaluating $G_\pm$ is that for G_+ we replace k in the denominator in the integral (30) by $k + i\varepsilon$ and for G_- we replace k by $k - i\varepsilon$ always remembering that ε is positive and letting $\varepsilon \to 0$.

REFERENCE

1. J. Irving and N. Mullineux, *Mathematics in Physics and Engineering*, Academic Press, New York (1959).

Appendix O Asymptotic form of the radial part of the Schrödinger equation for a short-range force

For a spherically symmetric potential, the radial part of the Schrödinger equation can be put in the form [see Eq. (59) of Chapter 26]:

$$\frac{d^2 u_l(r)}{dr^2} + \left[k^2 - U(r) - \frac{l(l+1)}{r^2}\right] u_l(r) = 0 \tag{1}$$

where $u_l(r) = rR_l(r)$, $k^2 = 2mE/\hbar^2$, $U(r) = 2mV(r)/\hbar^2$. In this appendix we will show that for a short-range force for which

$$\lim_{r\to\infty} rV(r) = 0 \tag{2}$$

the asymptotic form of $u_l(r)$ is $\exp(\pm ikr)$.

We first assume that at large values of r, the terms $U(r)$ and $l(l+1)/r^2$ in Eq. (1) can be neglected. Then the solution will be $u_l \sim \exp(\pm ikr)$, this suggests that we try a solution of the form

$$u_l(r) = A\exp\left[\int_a^r f(r')\,dr'\right]\exp[\pm ikr] \tag{3}$$

If we substitute in Eq. (1), we would obtain

$$\frac{df}{dr} + f^2 \pm 2ik\, f(r) = U(r) + \frac{l(l+1)}{r^2} \equiv W(r) \tag{4}$$

If $W(r)$ behaves as $e^{-\alpha r}$ for $r\to\infty$ (then l is necessarily zero), $f(r)$ will also behaves as $e^{-\alpha r}$ as can be seen by substitution (the term f^2 will be negligible). Consequently the integral $\int_a^r f(r')dr'$ will tend to a constant value and $u_l(r)$ will behave as $\exp(\pm ikr)$. On the other hand if $W(r)(\sim A'/r^\alpha$ $(\alpha > 0)$ then $f(r) \sim \mp (A'/2ik)(1/r^\alpha)$ (the terms involving df/dr and f^2 would be negligible). However, only for $\alpha > 1$, the integral $\int_a^r f(r')dr'$ would tend to a constant value.

Indeed for $\alpha = 1$ (i.e. for the Coulomb potential) we will have

$$u_l(r) \underset{r\to\infty}{\to} \exp\left(\pm ikr \pm \frac{A'}{2ik}\ln r\right) \tag{5}$$

Thus the form of $u_l(r)$ for $r\to\infty$ does not approach a definite phase; however, we can still obtain an expression for the scattering cross-section (see Section 26.7). Thus, if $\alpha > 1$, the asymptotic form of $u_l(r)$ is indeed of the form of $\exp(\pm ikr)$.

Appendix P Tunnelling calculations

The probability that at time t, the particle is still inside the well is approximately given by the overlap integral

$$P(t) \approx \left| \int_0^\infty \psi^*(x, 0)\psi(x, t)\,dx \right|^2 \tag{1}$$

Now,

$$\psi(x, t) = \int_0^\infty dE\,\phi(E)\,\psi_E(x)\,e^{-iEt/\hbar} \tag{2}$$

[see Eq. (25) of Chapter 25]. Thus,

$$P(t) \approx \left| \int dx \left[\int dE'\,\phi^*(E')\,\psi_{E'}^*(x) \right] \left[\int dE\,\phi(E)\,\psi_E(x)\,e^{-iEt/\hbar} \right] \right|^2$$

$$\approx \left| \int dE'\,\phi^*(E') \int dE\,\phi(E)\,e^{-iEt/\hbar} \int dx\,\psi_{E'}^*(x)\,\psi_E(x) \right|^2$$

Since the last integral is $\delta(E - E')$, we readily get

$$P(t) \approx \left| \int dE \left|\phi(E)\right|^2 e^{-iEt/\hbar}\,dE \right|^2 \tag{3}$$

In order to evaluate the above integral, we must evaluate $|\phi(E)|^2$ which is given by [see Eq. (28) of Chapter 25]:

$$|\phi(E)|^2 \approx \left| \frac{A}{A_b} \right|^2 \tag{4}$$

Now, A_b is given by [see Eq. (9) of Chapter 25]

$$A_b = \left[\frac{2\kappa_b}{1 + \kappa_b a} \right]^{1/2} \tag{5}$$

In order to evaluate A, we express it in terms of D and then use Eq. (19) of Chapter 25. Since the wave packet is a superposition of states around the *quasibound state*, $|\phi(E)|^2$ is very sharply peaked around $E \approx E_b$ and, therefore, all calculations will be carried out around $E \approx E_b$. We begin with calculation of C around $E = E_b$; we first note that at resonance

$$-k_b a \cot k_b a = \kappa_b a \tag{6}$$

[see Eq. (2) of Chapter 25]. Thus,

$$\sin k_b a = \pm\frac{k_b a}{\alpha}, \quad \cos k_b a = \mp\frac{\kappa_b a}{\alpha} \tag{7}$$

where
$$\alpha^2 = \frac{2\mu V_0 a^2}{\hbar^2} \tag{8}$$

[see Eq. (5) of Chapter 25]. Now,

$$\begin{aligned} C &= \frac{1}{2}A\left[\sin ka - \frac{k}{\kappa}\cos ka\right]_{\substack{k \approx k_b \\ \kappa \approx \kappa_b}} \\ &\approx \frac{1}{2}A\left[\sin k_b a - \frac{k_b}{\kappa_b}\cos k_b a\right] \\ &\approx \pm\frac{1}{2}A\left[\frac{k_b a}{\alpha} + \frac{k_b}{\kappa_b}\frac{\kappa_b a}{\alpha}\right] \\ &\approx \pm A\frac{k_b a}{\alpha} \end{aligned} \tag{9}$$

Since,
$$B\,|_{E = E_b} = 0$$
we must make a Taylor series expansion of B around $E = E_b$:

$$\begin{aligned} B &\approx \left.\frac{dB}{dE}\right|_{E = E_b}(E - E_b) \\ &\approx \frac{1}{2}A\left[\left(a\cos ka + \frac{1}{\kappa}\cos ka - \frac{ka}{\kappa}\sin ka\right)\frac{dk}{dE} - \frac{k}{\kappa^2}\cos ka\frac{d\kappa}{dE}\right]_{E = E_b}(E - E_b) \end{aligned}$$

Since,
$$k^2 = \frac{2\mu E}{\hbar^2} \quad \text{and} \quad \kappa^2 = \frac{2\mu}{\hbar^2}(V_0 - E) \tag{10}$$

we get
$$\frac{dk}{dE} = \frac{\mu}{\hbar^2 k} \quad \text{and} \quad \frac{d\kappa}{dE} = -\frac{\mu}{\hbar^2\kappa} \tag{11}$$

Using the above equations [and Eq. (7)] we readily obtain

$$B \approx -\frac{1}{2}A\frac{\mu\alpha}{a\hbar^2 k_b\kappa_b^2}(1 + \kappa_b a)(E - E_b) \tag{12}$$

where we have used the fact that

$$k^2 + \kappa^2 = \frac{2\mu V_0}{\hbar^2} = \frac{\alpha^2}{a^2} \tag{13}$$

Substituting for B and C from Eqs (12) and (9) in the expression for $D_\pm$ [see Eq. (17) of Chapter 25], we get

$$|D_\pm| = \frac{1}{4}A\frac{\mu\alpha(1 + \kappa_b a)e^{\kappa_b d}}{a\hbar^2 k_b\kappa_b^2}\left|1 \pm \frac{\kappa_b}{ik_b}\right|\left|E - E_b' \mp i\Gamma\right| \tag{14}$$

where
$$\Gamma = \frac{4k_b^3 \kappa_b^3 a^4 \hbar^2}{\mu\alpha^4(1+\kappa_b a)} e^{-2\kappa_b d} \tag{15}$$

$$E_b' = E_b - \frac{\kappa_b^2 - k_b^2}{2\kappa_b k_b}\Gamma \tag{16}$$

Since,
$$|D_+| = \frac{1}{\hbar}\left[\frac{\mu}{2\pi k}\right]^{1/2} \quad \text{and} \quad A_b = \left[\frac{2\kappa_b}{1+\kappa_b a}\right]^{1/2} \tag{17}$$

[see Eqs (19) and (9) of chapter 25], we may write

$$|\phi(E)|^2 \approx \left|\frac{A}{A_b}\right|^2 \approx \frac{\Gamma}{\pi}\frac{1}{(E-E_b')^2+\Gamma^2} \tag{18}$$

Thus Eq. (3) becomes

$$P(t) \approx \left|\int \frac{\Gamma}{\pi}\frac{e^{-iEt/\hbar}}{(E-E_b')^2+\Gamma^2}dE\right|^2 \tag{19}$$

We may evaluate the integral from $-\infty$ to $+\infty$ since most of the contribution will cone from the region near resonance ($E \approx E_b'$). We introduce the variable

$$\eta = \frac{(E-E_b')}{\hbar} \tag{20}$$

to write
$$P(t) \approx \left|\frac{\Gamma}{\pi\hbar}\int_{-\infty}^{+\infty} \frac{e^{-i\eta t}}{\left(\eta + i\frac{\Gamma}{\hbar}\right)\left(\eta - i\frac{\Gamma}{\hbar}\right)}d\eta\right|^2 \tag{21}$$

For $t > 0$, the integral may be evaluated by using complex variable techniques and Jordan's lemma. In the complex η-plane we choose a contour which consists of the real axis and a semicircle in the lower-half plane where the integral vanishes. There is a simple pole within the contour at

$$\eta = -\frac{i\Gamma}{\hbar}$$

so that
$$P(t) \approx \left|\frac{\Gamma}{\pi\hbar}\cdot 2\pi i \frac{e^{-\Gamma t/\hbar}}{-\frac{2i\Gamma}{\hbar}}\right|^2$$

$$\approx e^{-t/\tau} \tag{22}$$

where
$$\tau = \frac{\hbar}{2\Gamma} \tag{23}$$

represents the mean lifetime of the particle.

Index

Index

D

E

F

G

H

I

J

K

L

M

N

O